Proceedings of the 27TH International Conference on Biomechanics in Sports

August 17 – 21 2009
University of Limerick – Ireland

Edited By

Andrew J. Harrison, Ross Anderson & Ian Kenny

Biomechanics Research Unit & Dept of PE and Sport Sciences
Faculty of Education and Health Sciences
University of Limerick

Proceedings of the 27th International Conference on Biomechanics in Sports

Edited By
Andrew J. Harrison, Ross Anderson, & Ian Kenny

All rights reserved. No part of this publication may be reproduced in any form or by any means—graphic, electronic or mechanical, including photocopying, recording, taping or information storage and retrieval systems—without the prior written permission of the author.

Distributed & Published By
Original Writing Ltd, 2010
in conjunction with
Biomechanics Research Unit
Department of PE & Sport Sciences
Faculty of Education & Health Sciences
University of Limerick
Castletroy, Limerick
Ireland
Tel +353 (0) 61 213632
email – info@isbs2009.com or ross.anderson@ul.ie

Copyright © 2009 International Society of Biomechanics in Sport

ISBN - 978-1-907179-65-5

A cip catalogue for this book is available from the National Library.

Printed in Great Britain by MPG Books Group, Bodmin and Kings Lynn.

WITHDRAWN

PREFACE

The 27th International Conference on Biomechanics in Sports is the official annual meeting of the International Society of Biomechanics in Sports (ISBS) whose primary purposes are:

- To provide a forum for the exchange of ideas for sports biomechanics researchers, coaches and teachers
- To bridge the gap between researchers and practitioners
- To gather and disseminate information and materials on biomechanics in sports

These proceedings, presented at the University of Limerick, Ireland, from 17th to 21st August 2009 include seven keynote papers, 110 oral papers, 180 poster papers and 5 thematic applied sessions. The conference proceedings present cutting edge research on sports biomechanics and its applications from many internationally renowned researchers and academics from more than 35 countries and 5 continents.

344 papers were submitted online for blind review by two expert members of the ISBS 2009 Scientific Committee and were required to satisfy strict requirements for scientific merit and relevance to sports biomechanics. For inclusion in the proceedings, authors had to respond appropriately to the reviewers' comments and make requested changes to the satisfaction of the ISBS 2009 Editorial Board. 46 papers were rejected or withdrawn by authors during the review process. Accepted papers were edited and formatted to ensure they complied fully with ISBS submission guidelines. Papers are published as either four-page short communications or as one-page work in progress reports.

The proceedings also include over 30 applied session papers under the themes of rowing, strength & conditioning, swimming, coaching biomechanics and novel data analysis techniques. These sessions provide opportunities for academics and researchers to communicate important findings in sports biomechanics to practitioners in coaching and teaching.

We are grateful to the ISBS 2009 Keynote Speakers, the ISBS 2009 Scientific Committee and the ISBS 2009 Applied Session Chairs for their important work that assured the quality of the conference proceedings.

The ISBS 2009 Editorial Board is confident that this process has ensured that these proceedings retain the high standards of scientific merit and relevance typical of the International Society of Biomechanics in Sports and will be an important reference for academics, researchers and practitioners in sports biomechanics.

Drew Harrison, Ross Anderson & Ian Kenny

University of Limerick
August 2009

ISBS 2009 Conference Chairs
Ross Anderson
Drew Harrison

ISBS 2009 Conference Executive Committee
Ian Kenny
Rhoda Sohun
Caroline Teulier
Joseph O'Halloran

ISBS 2009 Conference Planning Committee

DJ Collins	Maeve Gleeson
Joseph Hamill	Dolores Hanley
Randall Jensen	Duane Knudson
Susan McDonagh	Louise Mulcahy
Ross Sanders	Richard Smith

ISBS 2009 Conference Organisation Team

Sarah Breen	Joe Costello
Deirdre Donohoe	Derek Doran
Kieran Dowd	Laura-Anne Furlong
Darragh Graham	Caroline Hansberry
Deirdre Harrington	Diarmuid Horgan
Loran Hughes	Siobhan Leahy
Will McCormack	Elaine McDermott
Brian McGowan	Ciaran McInerney
Niamh NiCheilleachair	Kim Siekerman
Paul Talty	Catherine Tucker

ISBS 2009 Editorial Board and Scientific Committee

Drew Harrison	(Editorial Board)	Ireland
Ross Anderson	(Editorial Board)	Ireland
Ian Kenny	(Editorial Board)	Ireland

ISBS 2009 Scientific Committee

Adamantios Arampatzis	Germany	Glenn Fleisig	USA
JingXian Li	Canada	Pamela Russell	USA
Kevin Ball	Australia	Paul Grimshaw	Australia
Young-Tae Lim	Korea	Aki Salo	UK
Athanassios Bissas	UK	Joseph Hamill	USA
Chris Low	UK	Hermann Schwameder	Austria
Elizabeth Bradshaw	Australia	Michael Hanlon	Ireland
Laurie Malone	USA	Amir Shafat	Ireland
Nick Caplan	UK	Gareth Irwin	UK
Wayne Marino	Canada	Richard Smith	Australia
Amanda Clifford	Ireland	Randall Jensen	USA
HansJoachim Menzel	Brazil	Lothar Thorwesten	Germany
Tom Comyns	Ireland	Norelee Kennedy	Ireland
Claire Milner	USA	Antonio Veloso	Portugal
Susan Coote	Ireland	Justin Keogh	New Zealand
Kieran Moran	Ireland	Cassie Wilson	UK
Orna Donoghue	UK	Duane Knudson	USA
Kieran O'Sullivan	Ireland	Maurice Yeadon	UK
Bruce Elliott	Australia	YoungHoo Kwon	USA
Karen Roemer	USA		

ISBS Executive Committee

Youlian Hong	President
Manfred Vieten	President Elect
Richard Smith	VP Awards
Mario Lamontagne	VP Conferences
Duane Knudson	VP Publications
Hermann Schwameder	VP Public Relations
Ross Sanders	VP Projects & Research
John Ostarello	Secretary General
Manfred Vieten	Treasurer

ISBS Directors 2007 – 2009

Gareth Irwin	UK
Uwe Kersting	New Zealand
Elizabeth Bradshaw	Australia
Justin Keogh	New Zealand
Young-Tae Lim	Korea
Hans-Joachim Menzel	Brazil
Spiros Prassas	USA
Pamela Russell	USA
Lothar Thorwesten	Germany
Qing Wang	China

ISBS Directors 2008 – 2010

Randall Jensen	USA
Cassie Wilson	UK
Karen Roemer	USA
Kevin Ball	Australia
Young-Hoo Kwon	USA
Jing Xian Li	Canada
Wolfgang Potthast	Germany
Antonio Veloso	Portugal
Mark Walsh	USA
Chenfu (Peter) Huang	Tiawan

Sports Medicine & Performance

Sports Medicine and performance analyses go hand in hand with analyses aimed at preventing injury. While the protocols are often very similar, the analysis of sports related movements often entails analyzing a variety of highly dynamic movements. The magnitudes for angular velocities and accelerations associated with sports activities are often large and occur while translating through large fields of view.

Motion Analysis provides the tools for the Sports Medicine & Performance professional to perform accurate functional evaluations/analysis for clinical and research oriented purposes such as: Performance enhancement, injury prevention, return to activity, and the effects of orthopaedic/athletic devices.

Choosing the correct camera and number of cameras for your motion capture system is dependent on various factors including: Size of capture area, size of physical room, complexity of movement, speed of movement and current and future needs.

Your Motion Analysis account representative will work with you to determine the best system configuration. Our systems integrate fully with EMG and forceplate data, as well as many other hardwares. Customers have full access to an SDK in order to develop their own software interfaces.

Motion Analysis
European Sales and Service Office (UK)
Lucy Keighley
lucy.keighley@motionanalysis.com
+44 (0) 1932 213 152
Limerick - Ireland
Visit us online at www.motionanalysis.com

Recommended Cameras

EAGLE DIGITAL CAMERA

Resolution of 1.3 million pixels
1-500 Hz selectable frame rates at full resolution
Up to 2,000 frames per second at reduced resolutions
High quality 55mm lens for low optical distortion
Built-in zoom provides for more visual options
237 LED's for brighter and better light uniformity
Four body mount points for variable positioning
Software controlled electronic shutter
Software controlled adjustable light output
Separate zoom, iris and focus settings independent of ringlight

HAWK DIGITAL CAMERA

Resolution of 0.3 million pixels
1-200 Hz selectable frame rates
C-Mount or zoom lenses
Built-in zoom provides for more visual options
237 LED's for brighter and better light uniformity
Four body mount points for variable positioning
Software controlled electronic shutter
Software controlled adjustable light output
Separate zoom, iris and focus settings independent of ringlight

Recommended Software

CORTEX - Our core motion capture software comprises tracking, editing, scripting and modeling functions in a *single integrated package*.

SKELETON BUILDER - Creates skeletons that are relatively simple, direct and fast calculations of segments (bones) that are defined and calculated from one marker center to another.

KINTOOLS RT - A full-body, three-dimensional kinetics and kinematics analysis package.

ORTHOTRAK - A fully automated, three dimensional, clinical gait measurement, evaluation and database management system.

UE TRAK - Calculates three-dimensional upper extremity kinematics for upper body movement measurement.

MOTION COMPOSER - Tools for collating, integrating and presenting interactive motion capture data.

SIMM - Create computer models of musculoskeletal structures.

BTS Technology working with SPORT:

The technological innovations by BTS have allowed developing products to be used for Sport Applications thanks to their high versatility, compactness and easy to use interface. BTS are the only company able to offer Motion Analysis, EMG, Video and forceplate with seamless integration and with management and control of the whole lab by using one single unit.

Movement analysis technology contributes so much to the Biomechanics Science and is a fundamental part in supporting the definition of optimized sport training programs.
With BTS technology you retain the flexibility to work in and out of the lab.

BTS have remained the leaders in cutting edge technology presenting the first wire-less EMG system named the BTS FREEEMG, also BTS developed a Motion Analysis System named BTS SMART-D that allows you to work in full light including outdoors in natural sun light.

BTS FREEEMG

BTS Bioengineering lead the way in cutting edge technologies, the latest system designed by BTS represents a generation leap in the design of diagnostic devices for surface electromyography analysis.
BTS FREEEMG the first entirely wireless electromyography unit in the world using miniature probes weighing less than 10 grams and measuring only a few centimetres across.
- Lighter probes this means minimal motion artefact
- Active electrodes and not fixed geometry
- Unique on-board memory for loss of network
- Personal area network protocol, no outside interference
- Only system with Patient/ Receiving Unit for outdoor use
- First time the subject has full range of movement

BTS SMART-D

The new high frequency digital optoelectronic system that allows all types of movements to be analysed with an unprecedented level of precision and accuracy.
- Up to 16 digital infrared cameras per SMART-D unit
- CCD Sensors for better study of movement
- Operates in all lighting conditions, including daylight
- Acquisition Frequency 70 – 500Hz
- Quick and Easy calibration
- Accurate Synchronisation of data

BTS SMART-Performance

Is the software tools for Biomechanical Analysis in of all types of sport movements.
- Standard protocols for cycling, tennis, running, golf etc.
- Optimisation, Performance, Motion Strategy and Rehabilitation
- Tre Process Module runs and performs on all motion analysis systems
- Import/export and archiving of files in different formats (c3d)

  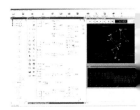

For further information on the BTS SMART-D or BTS FREEEMG or any other of the BTS product range please visit us on **Stand 7** in the main exhibition hall or send an e-mail at *sales.uk@bts.it*.

Thank you and best wishes for a good conference.

BTS Bioengineering

www.btsbioengineering.com

We would like to acknowledge the following companies and organisations

ISBS 09 Platinum Sponsor

ISBS 09 Silver Sponsor

BTS Bioengineering

ISBS 09 Sponsors

ISBS 09 Exhibitors

ISBS 09 Associated Companies

Biometrics Ltd

ISBS Sponsors

CONTENTS

21 KEYNOTE LECTURE #1 – DAVID KERWIN
BIOMECHANICS, TECHNOLOGY AND COACHING

22 KEYNOTE LECTURE #2 – CLARE MILNER
GAIT BIOMECHANICS AND TIBIAL STRESS FRACTURE IN RUNNERS

26 GEOFFREY DYSON KEYNOTE LECTURE #3 – ALBERT GOLLHOFER
FUNCTIONAL ROLE OF PROPRIOCEPTIVE FEEDBACK IN BALANCE AND IN REACTIVE MOVEMENT

31 KEYNOTE LECTURE #4 – YOUNG-HOO KWON
KINEMATIC ANALYSIS OF SPORTS MOVEMENTS: GOLF SWING PLANE ANALYSIS

35 DAVID WADDELL MEMORIAL KEYNOTE LECTURE #5 – BRUCE ELLIOTT
THEORY INTO PRACTICE: THE KEY TO ACCEPTANCE OF SPORT BIOMECHANICS

39 KEYNOTE LECTURE #6 – ADAMANTIOS ARAMPATZIS
PLASTICITY OF HUMAN TENDON'S MECHANICAL PROPERTIES: EFFECTS ON SPORT PERFORMANCE

43 KEYNOTE LECTURE #7 – CASSIE WILSON
APPROACHES FOR OPTIMISING JUMPING PERFORMANCE

ORAL SESSION S01 – ROWING

49 • THE EFFECT OF VARIABLE NUMBERS OF REPETITIONS ON PEAK TORQUE IN ROWER AND NON-ATHLETE FEMALES WHEN USING ISOKINETIC TESTING - SARAH MOODY, DEANNA MALIKIE AND BARBARA WARREN

53 • MECHANICAL LOADING OF THE LUMBAR SPINE OF ELITE ROWERS WHILE ROWING FIXED AND SLIDING ERGOMETERS - RICHARD SMITH, MICHAEL DICKSON AND FLOREN COLLOUD

57 • THE EFFECTS OF HORIZONTAL AND VERTICAL FORCES ON SINGLE SCULL BOAT ORIENTATION WHILE ROWING - PETER SINCLAIR, ANDREW GREENE AND RICHARD SMITH

61 • OAR BLADE FORCE COEFFICIENTS AND A MATHEMATICAL MODEL OF ROWING - ANNA COPPEL, TREVOR GARDNER, NICHOLAS CAPLAN AND DAVID HARGREAVES

65 • VALIDITY OF THE POWERLINE BOAT INSTRUMENTATION SYSTEM - JENNIE COKER, PATRIA HUME AND VOLKER NOLTE

ORAL SESSION S02 – GOLF

70 • A COMPARISON OF ACCURACY AND STROKE CHARACTERISTICS BETWEEN TWO PUTTING GRIP TECHNIQUES - YONGHYUN PARK, DAVID O'SULLIVAN, INSIK SHIN, CHULSOO CHUNG, GYESAN LEE AND HAKSOO SHIN

74 • ANALYSIS OF THE JOINT KINEMATICS OF THE 5 IRON GOLF SWING - AOIFE HEALY, KIERAN MORAN, JANE DICKSON, CILLIAN HURLEY, MADS HAAHR AND NOEL E. O CONNOR

78 • ANALYSIS OF ELITE GOLFERS' KINEMATIC SEQUENCE IN FULL-SWING AND PARTIAL SWING SHOTS - FREDRIK TINMARK, JOHN HELLSTRÖM, KJARTAN HALVORSEN AND ALF THORSTENSSON

79 • GOLF DRIVE LAUNCH ANGLES AND VELOCITY: 3D ANALYSIS VERSUS A COMMERCIAL LAUNCH MONITOR - MATTHEW SWEENEY, JACQUELINE ALDERSON, PETER MILLS AND BRUCE ELLIOTT

83 • A STUDY ON THE RELATIONSHIP BETWEEN EYE MOVEMENT AND GRIP FORCE DURING GOLF PUTTING - HYUNG-SIK KIM, JIN-SEUNG CHOI, GYE-RAE TACK, YOUNG-TAE LIM AND JEONG-HAN YI

ORAL SESSION S03 – SWIMMING

85 • MARKERLESS ANALYSIS OF SWIMMERS' MOTION: A PILOT STUDY - ELENA CESERACCIU, ZIMI SAWACHA, SILVIA FANTOZZI, GIULIA DONÀ, STEFANO CORAZZA, GIORGIO GATTA AND CLAUDIO COBELLI

86 • A COMPARISON BETWEEN THE VALUES OBTAINED FROM ACTIVE DRAG ANALYSIS COMPARED TO FORCES PRODUCED IN TETHERED SWIMMING - BRUCE MASON, DANIELLE FORMOSA AND SHANNON ROLLASON

90 • MOTOR CONTROL PATTERNS IN ELITE SWIMMERS' FREESTYLE STROKE DURING DRYLAND SWIMMING - TRACY SPIGELMAN, TOM CUNNINGHAM, SCOTT MAIR, ROBERT SHAPIRO, TIM UHL AND DAVID MULLINEAUX

94 • STROKE PARAMETERS AND ARM COORDINATION IN COMPETITIVE UNILATERAL ARM AMPUTEE FRONT CRAWL SWIMMERS - CONOR OSBOROUGH, CARL PAYTON AND DANIEL DALY

98 • MEASURING PROPULSIVE FORCE WITHIN THE DIFFERENT PHASES OF BACKSTROKE SWIMMING - DANIELLE FORMOSA, BRUCE MASON AND BRENDAN BURKETT

102 • EFFECTS OF HANDLE AND BLOCK CONFIGURATION ON SWIM START PERFORMANCE - PETER VINT, SCOTT MCLEAN, RICHARD HINRICHS, SCOTT RIEWALD AND ROBYN MASON

ORAL SESSION S04 – MODELLING

107 • AN ALGORITHM TO COMPUTE ABSOLUTE 3D KINEMATICS FROM A MOVING MOTION ANALYSIS SYSTEM - FLOREN COLLOUD, MICKAËL BEGON, VINCENT FOHANNO AND TONY MONNET

108 • DEVELOPMENT, EVALUATION AND APPLICATION OF A SIMULATION MODEL OF A SPRINTER DURING THE FIRST STANCE PHASE - NEIL BEZODIS, AKI SALO AND GRANT TREWARTHA

112 • RECOGNITION AND ESTIMATION OF HUMAN LOCOMOTION WUTH HIDDEN MARKOV MODELS – ANDREAS FISCHER, DIRK GEHRIG, THORSTEN STEIN, HERMANN SCHWAMMEDER AND TANJA SCHULTZ

113 • TOWARDS UNDERSTANDING HUMAN BALANCE - SIMULATING STICK BALANCING - MANFRED VIETEN

117 • APPLICATION OF SIMULATION TO FREESTYLE AERIAL SKIING - MAURICE YEADON

121 • COMPUTER SIMULATION OF THE OPTIMAL VAULTING MOTION DURING THE HORSE (TABLE) CONTACT PHASE - HUI-CHIEH CHENG, CHIH-YING YU AND KUANGYOU B. CHENG

ORAL SESSION S05 – JUMPING I

127 • THE JOINT KINETICS OF THE LOWER EXTREMITY ASSOCIATED WITH DROP JUMP HEIGHT INCREMENTS - HSIEN-TE PENG, PO-HAN CHU, CHI LIANG, NING LEE, PEI-YU LEE, TING-CHIEH TSAI AND TAI-FEN SONG

131 • DOES A CONCENTRIC ONLY SQUAT ILLICIT SIMILAR POTENTIATION OF SQUAT JUMP PERFORMANCE AS A BACK SQUAT? - DARRAGH GRAHAM, ANDREW HARRISON AND EAMONN FLANAGAN

132 • ANKLE AND KNEE COORDINATION FOR SINGLE-LEGGED VERTICAL JUMPING COMPARED TO RUNNING - TORSTEN BRAUNER, THORSTEN STERZING AND THOMAS L. MILANI

136 • COORDINATION IN TRACK & FIELD SPRINTERS WHILE PERFORMING THE COUNTERMOVEMENT JUMP - EZIO PREATONI, ELENA DEVODIER, GIUSEPPE ANDREONI AND RENATO RODANO

140 • MECHANICS OF AMPUTEE JUMPING – LOADING - MARLENE SCHOEMAN, CERI DISS AND SIOBHAN STRIKE

ORAL SESSION S06 – INDOOR SPORTS

145 • A BIOMECHANICAL COMPARISON OF JUMPING TECHNIQUES IN THE VOLLEYBALL BLOCK AND SPIKE - ROBERTO LOBIETTI, SILVIA FANTOZZI, RITA STAGNI AND SIMON G.S. COLEMAN

149 • ELITE OUTSIDE HITTERS IN VOLLEYBALL DO NOT MEET THEIR INDIVIDUAL POSSIBLE MAXIMUM IMPACT HEIGHT IN HIGH SPIKE JUMPS - CLAAS KUHLMANN, KAREN ROEMER AND THOMAS MILANI

153 • BIOMECHANICAL ASPECTS OF BADMINTON SHOE DURING A LUNGE - SEUNGBUM PARK, DARREN STEFANYSHYN, JAY WOROBETS, SANG-KYOON PARK, JUNGHO LEE, KYUNGDEUK LEE AND JONGJIN PARK

154 • A COMPARISON OF COMPENSATION FOR RELEASE TIMING AND MAXIMUM HAND SPEED IN RECREATIONAL AND COMPETITIVE DARTS PLAYERS - DAVE BURKE AND FRED YEADON

158 • INVERSE KINEMATICS IN THE SUCCESS OF THE THROW IN BASKETBALL GAME - EMIL BUDESCU, EUGEN MERTICARU AND RADU MIHAI IACOB

ORAL SESSION S07 – TRACK & FIELD I

163 • ANGULAR KINEMATICS IN ELITE RACE WALKING PERFORMANCE - BRIAN HANLEY, ATHANASSIOS BISSAS AND ANDI DRAKE

167 • 3D KINEMATICAL ANALYSIS OF BRAZILIAN FEMALE POLE VAULTERS - TIAGO RUSSOMANNO, NATHALIA ROSSETTO AND RICARDO BARROS

168 • HIGH SPEED ULTRASONIC DETECTION SCHEME FOR SPORTS PERFORMANCE MONITORING – MUHANNAD ZAKIR HOSSAIN

172 • ELECTROMYOGRAPHIC ANALYSIS ON LOWER EXTREMITY MUSCLES DURING OVERGROUND AND TREADMILL RUNNING - MANDY MAN LING CHUNG, KAM MING MOK, JUSTIN WAI YUK LEE AND YOULIAN HONG

ORAL SESSION S08 – EQUIPMENT

177 • WEDGED FOOTWEAR PERTURBATIONS AFFECT LOWER EXTREMITY COORDINATION DYNAMICS - ELIZABETH RUSSELL, ALLISON GRUBER, KRISTIAN O'CONNOR, RICHARD VAN EMMERIK AND JOSEPH HAMILL

181 • IMPACT OF WII-FIT TRAINING ON NEURO-MUSCULAR CONTROL - DANIEL BASSETT, ERIKA DE CANONICO, ANTONIO ARCHILLETTI AND GIULIANO CERULLI

182 • THE EFFECTS OF IMPACT AVOIDANCE TECHNIQUES ON HEAD INJURY RISK - PHILIPPE ROUSSEAU AND BLAINE HOSHIZAKI

186 • NOVEL ENERGY ABSORBING MATERIALS WITH APPLICATIONS IN HELMETED HEAD PROTECTION - LIANG CUI, STEPHEN KIERNAN AND MICHAEL GILCHRIST

ORAL SESSION S09 – METHODS

192 • TRUNK POSTURE AND STATO-DYNAMICAL SPINE ANALYSIS - COMPARING ULTRASOUND BASED VS. OPTICAL BASED MEASUREMENT SYSTEM - LOTHAR THORWESTEN, DANIEL SCHNIEDERS, MARKUS SCHILGEN AND KLAUS VÖLKER

196 • SPATIO-TEMPORAL PARAMETERS AND INSTANTANEOUS VELOCITY OF SPRINT RUNNING USING A WEARABLE INERTIAL SENSING UNIT - ELENA BERGAMINI, PIETRO PICERNO, MICHELE GRASSI, VALENTINA CAMOMILLA AND AURELIO CAPPOZZO

197 • VALIDATION OF MODEL-BASED IMAGE-MATCHING TECHNIQUE WITH BONE-PIN MARKER BASED MOTION ANALYSIS ON ANKLE KINEMATICS: A CADAVER STUDY - KAM-MING MOK, DANIEL TIK-PUI FONG, TRON KROSSHAUG AND KAI-MING CHAN

201 • CUSTOM-BUILT WIRELESS PRESSURE SENSING INSOLES FOR DETERMINING CONTACT-TIMES IN 60M MAXIMAL SPRINT RUNNING - GREGOR KUNTZE, MARCELO PIAS, IAN BEZODIS, DAVID KERWIN, GEORGE COULOURIS AND GARETH IRWIN

205 • EFFECT OF SPORT ACTIVITY ON COUNTER MOVEMENT JUMP PARAMETERS IN JUVENILE STUDENTS - ANNE RICHTER, DIANA LANG, GERDA STRUTZENBERGER AND HERMANN SCHWAMEDER

ORAL SESSION S10 – GYMNASTICS

207 • A SIMULATION STUDY OF THE INTERNAL TWISTING TORQUE IN THE FOUETTÉ TURN - AKIKO IMURA AND M.R. YEADON

211 • BIOMECHANICAL ANALYSIS OF GYMNASTIC BACK HANDSPRING - CHENFU HUANG AND GIN-SHU HSU

215 • CALCULATION FORCES FROM BAR MOVEMENT ON PARALLEL BARS IN GYMNASTICS - FALK NAUNDORF, KLAUS KNOLL AND STEFAN BREHMER

219 • OPTIMISATION OF THE FELGE TO HANDSTAND ON PARALLEL BARS - MICHAEL HILEY, ROGER WANGLER AND GHEORGHE PREDESCU

223 • RELIABILITY OF A GYMNASTICS VAULTING FEEDBACK SYSTEM - ELIZABETH BRADSHAW, PATRIA HUME, MARK CALTON AND BRAD AISBETT

ORAL SESSION S11 – INJURY I

228 • EFFECT OF LUMBAR LORDOTIC ANGLE ON LUMBOSACRAL JOINT DURING ISOKINETIC EXCERCISE - TAE SOO BAE, JONG KWON KIM, SHIN KI KIM, JEI CHEONG RYU AND MU SEONG MUN

232 • ANTICIPATION EFFECT ON KNEE JOINT STABILITY DURING PLANNED AND UN-PLANNED MOVEMENT TESTS IN LABORATORY - MAK-HAM LAM, PIK-KWAN LAI, DANIEL TIK-PUI FONG AND KAI-MING CHAN

236 • AN ANKLE SPRAIN RECOGNITION SYSTEM FOR IDENTIFYING ANKLE SPRAIN MOTION FROM OTHER NORMAL MOTIOIN USING MOTION SENSOR - YUE-YAN CHAN, DANIEL TIK-PUI FONG AND KAI-MING CHAN

240 • THE EFFECTS OF SPORTS INJURY PREVENTION TRAINING ON THE RISK FACTORS OF ANTERIOR CRUCIATE LIGAMENT INJURY IN FEMALE ATHLETES - BEEOH LIM, YONGSEUK LEE AND YOUNGHOO KWON

241 • THE EFFECT OF WARM-UP, STATIC STRETCHING AND DYNAMIC STRETCHING ON HAMSTRING FLEXIBILITY IN PREVIOUSLY INJURED SUBJECTS - KIERAN O'SULLIVAN, ELAINE MURRAY AND DAVID SAINSBURY

ORAL SESSION S12 – FIELD SPORTS

243 • VELOCITY AND ACCURACY AS PERFORMANCE CRITERIA FOR THREE DIFFERENT SOCCER KICKING TECHNIQUES - THORSTEN STERZING, JUSTIN S. LANGE, TINO WÄCHTLER, CLEMENS MÜLLER AND THOMAS L. MILANI

247 • FIELD TESTING TO PREDICT PERFORMANCE IN RUGBY UNION PLAYERS - BRIAN GREEN, CATHERINE BLAKE AND BRIAN CAULFIELD

251 • THE PLANARITY OF THE STICK AND ARM MOTION IN THE FIELD HOCKEY HIT - ALEXANDER WILLMOTT AND JESUS DAPENA

255 • THE INFLUENCE OF STANCE WIDTH ON MOVEMENT TIME IN FIELD HOCKEY GOALKEEPING - KEVIN BALL AND GEORGIA GIBLIN

259 • THE EFFECTS OF LOCALISED QUADRICEPS FATIGUE ON BOWLING PERFORMANCE IN FEMALE COUNTY CRICKETERS - ALAN DE ASHA, MARK ROBINSON, AMITY CAMPBELL AND MARK LAKE

260 • THE EFFECT OF INDIVIDUALISED COACHING INTERVENTIONS ON ELITE YOUNG FAST BOWLERS' TECHNIQUE - PETER WORTHINGTON, CRAIG RANSON, MARK KING, ANGUS BURNETT AND KEVIN SHINE

ORAL SESSION S13 – STRENGTH & CONDITIONING

266 • BARBELL ACCELERATION ANALYSIS ON VARIOUS INTENSITIES OF WEIGHTLIFTING - KIMITAKE SATO, PAUL FLESCHLER AND WILLIAM SANDS

270 • COMPARISSON OF ANGLES AND THE CORRESPONDING MOMENTS IN KNEE AND HIP DURING RESTRICED AND UNRESTRICTED SQUATS - SILVIO LORENZETTI, MIRJAM STOOP, THOMAS UKELO, HANS GERBER, ALEX STACOFF AND EDGAR STÜSSI

274 • THE EFFECTS OF DYNAMIC AND STATIC STRETCHING METHODS ON SPEED, AGILITY AND POWER - DANIEL BISHOP

275 • RELATIONSHIP BETWEEN MUSCULAR OUTPUTS AND THE HORIZONTAL PERTURBATION IN THE EARLY PHASE OF BENCH PRESS MOVEMENT UNDER STABLE AND UNSTABLE CONDITIONS - SENTARO KOSHIDA, KOJI MIYASHITA, KAZUNORI IWAI, AYA KAGIMORI AND YUKIO URABE

279 • ABDOMINAL PRESSURIZATION AND MUSCLE ACTIVATION DURING SUPINE TRUNK CURLS - ALF THORSTENSSON, ANNA BJERKEFORS, MARIA EKBLOM AND FREDRIK TINMARK

ORAL SESSION S14 – OTHER SPORTS I

281 • ACUTE EFFECTS OF DENTAL APPLIANCES ON UPPER AND LOWER ISOKINETIC MUSCLE FUNCTION - NATHAN DAU, DONALD SHERMAN, RICHARD BOLANDER, CYNTHIA BIR AND HERMANN ENGELS

285 • ACCUMULATED FATIGUE AFFECTS RUNNING BIOMECHANICS DEPENDING ON THE PERFORMANCE LEVEL DURING A TRIATHLON COMPETITION - ANTONIO CALA AND ENRIQUE NAVARRO

289 • CAN ELITE TENNIS PLAYERS JUDGE THEIR SERVICE SPEED? - KIERAN MORAN, COLM MURPHY, CAHILL GARRY AND BRENDAN MARSHALL

293 • OPTIMISATION OF ENERGY ABSORBING LINER FOR EQUESTRIAN HELMETS – LIANG CUI, MANUEL FORERO RUEDA AND

MICHAEL GILCHRIST

297 • THE EFFECT OF SELECTED KINEMATICS ON BALL SPEED AND GROUND REACTION FORCES IN FAST BOWLING - PETER WORTHINGTON, MARK KING AND CRAIG RANSON

ORAL SESSION S15 – WINTER SPORTS

302 • PERCEIVED DIFFERENCES IN SKATING CHARACTERISTICS RESULTING FROM THREE CROSS SECTIONAL SKATE BLADE PROFILES - MARSHALL KENDALL, SCOTT FOREMAN, PHILIPPE ROUSSEAU AND BLAINE HOSHIZAKI

306 • ANALYSIS OF COMBINED EMG AND JOINT ANGULAR VELOCITY FOR THE EVALUATION OF ECCENTRIC/CONCENTRIC CONTRACTION IN SKIING - NICOLA PETRONE AND GIUSEPPE MARCOLIN

310 • EFFECT OF SKI BOOT TIGHTNESS ON SHOCK ATTENUATION TIME AND JOINT ANGLE WITH ANTERIOR-POSTERIOR FOOT POSITIONING IN DROP LANDINGS - SHINYA ABE AND RANDALL L. JENSEN

314 • THE EFFECT OF WALKING SPEED AND SKILL LEVELS ON ELBOW FLEXION AND UPPER LIMB EMG SIGNALS IN NORDIC WALKING: A PILOT STUDY - NICOLA PETRONE, MANUEL ORSETTI AND GIUSEPPE MARCOLIN

ORAL SESSION S16 – TRAINING METHODS

319 • GAIT DOES NOT RETURN TO NORMAL FOLLOWING TOTAL HIP ARTHROPLASTY: IMPLICATIONS FOR A RETURN TO ATHLETIC ACTIVITIES - MARIO LAMONTAGNE, MELANIE BEAULIEU AND PAUL BEAULE

323 • THE EFFECTS ON STRENGTH, POWER, AND GENU VALGUM FOLLOWING A FIVE WEEK TRAINING PROGRAM WITH WHOLE-BODY VIBRATION - SARAH HILGERS AND BRYAN CHRISTENSEN

324 • THE EFFECTS OF RECOVERY MODALITIES INCLUDING ICE BATH IMMERSION ON RECOVERY FROM EXERCISE AND SUBSEQUENT PERFORMANCE - PAULA FITZPATRICK AND GILES WARRINGTON

328 • SPECIFIC TRAINING CAN IMPROVE SENSORIMOTOR CONTROL IN TYPE 2 DIABETIC PATIENTS - LOTHAR THORWESTEN, ARMIN EICHLER, CHRISTOPH SPERLBAUM, ERIC EILS, DIETER ROSENBAUM AND KLAUS VÖLKER

ORAL SESSION S17 – VARIABILITY

331 • A COMPARISON OF VARIABILITY IN GROUND REACTION FORCE AND KNEE ANGLE PATTERNS BETWEEN MALE AND FEMALE ATHLETES - SARAH BREEN AND MICHAEL HANLON

335 • INTERSEGMENTAL COORDINATION DIFFERENCES BETWEEN BEGINNING PERFORMERS EXECUTING A BADMINTON SMASH FOR ACCURACY OR VELOCITY - SCOTT STROHMEYER, COLE ARMSTRONG, YURIY LITVINSKY, ROBERT NOONEY, JABAN MOORE AND KEVIN SMITH

336 • SKILL LEVEL AND PARTICIPATION OF UNIVERSITY STUDENTS IN RECREATIONAL SPORT - PHILIP KEARNEY, PJ SMYTH AND ROSS ANDERSON

340 • AUGMENTED KNOWLEDGE OF RESULTS FEEDBACK IN TENNIS SERVE TRAINING - KIERAN MORAN, COLM MURPHY, GARRY CAHILL AND BRENDAN MARSHALL

343 • THE SPANNING SET AS A MEASURE OF MOVEMENT VARIABILITY - MICHAEL HANLON, JOAN CONDELL AND PHILIP KEARNEY

347 • ACTUE EFFECTS OF WHOLE-BODY VIBRATION ON KNEE JOINT DROP LANDING KINEMATICS AND DYNAMIC POSTURAL STABILITY - EAMONN DELAHUNT, DOMENICO CROGNALE, BRIAN GREEN, CHRIS LONSDALE AND DENISE MCGRATH

ORAL SESSION S18 – OTHER SPORTS II

349 • THE EFFECT OF MAXIMAL FATIGUE ON THE MECHANICAL PROPERTIES OF SKELETAL MUSCLE IN STRENGTH & ENDURANCE ATHLETES - NIAMH NI CHEILLEACHAIR AND ANDREW HARRISON

353 • A STUDY OF SCULLING SWIMMING PROPULSIVE PHASES AND THEIR RELATIONSHIP WITH HIP VELOCITY – RAUL ARELLANO, BLANCA DE LA FUENTE AND ROCÍO DOMINGUEZ

356 • PROPOSAL OF AN ADDITIONAL-INERTIA TUNING METHOD TO VISCOUS LOAD FOR HIGH SPEED MOTION TRAINING - KENJI SHIGETOSHI, SADAO KAWAMURA AND TADAO ISAKA

360 • FACTORS IN UPPER EXTREMITY LOADING IN THE POWER DROP EXERCISE - DUANE KNUDSON AND CHENTU HSIEH

364 • MECHANICAL EFFICIENCY IN BASEBALL PITCHING - DAVE FORTENBAUGH AND GLENN S. FLEISIG

368 • INTRA-LIMB JOINT COUPLING PATTERNS DURING THE USE OF THREE LOWER EXTREMITY EXERCISE MACHINES - THOMAS J. CUNNINGHAM, DAVID R. MULLINEAUX AND SHAUN K. STINTON

ORAL SESSION S19 – TRACK & FIELD II

373 • THE EFFECT OF EXHAUSTIVE RUNNING ON POSTURAL DYNAMICS - MARK WALSH, ADAM STRANG, MATHIAS HIERONYMUS AND JOSHUA HAWORTH

374 • GRAVITY'S ROLE IN ACCELERATED RUNNING - A COMPARISON OF AN EXPERIENCED POSE® AND HEEL-TOE RUNNER - GRAHAM FLETCHER, MARCUS DUNN AND NICHOLAS ROMANOV

378 • ATHLETE-SPECIFIC ANALYSES OF LEG JOINT KINETICS DURING MAXIMUM VELOCITY SPRINT RUNNING - IAN N. BEZODIS, AKI, I.T. SALO AND DAVID G. KERWIN

382 • EFFECTS OF FATIGUE ON TECHNIQUE DURING 5 KM ROAD RUNNING - BRIAN HANLEY AND LAURA SMITH

386 • OPTIMIZATION OF DISCUS FLIGHT - AXEL SCHUELER, FALK HILDEBRAND AND JOHANNES WALDMANN

390 • STABILISING THE HIP AND PELVIS DURING RUNNING: IS THERE AN EXPLOSIVE SOLUTION FOR UNINJURED ATHLETES? - DANIEL MEEHAN, MORGAN WILLIAMS, CAMERON WILSON AND ELIZABETH BRADSHAW

ORAL SESSION S20 – MARTIAL ARTS & DANCE

396 • ANALYSIS OF PERFORMANCE OF THE KARATE PUNCH (GYAKU-ZUKI) - EDIN SUWARGANDA, RUHIL RAZALI AND BARRY WILSON

400 • ROUNDHOUSE KICK WITH AND WITHOUT IMPACT IN KARATEKA OF DIFFERENT TECHNICAL LEVEL - VALENTINA CAMOMILLA, PAOLA SBRICCOLI, FEDERICO QUINZI, ELENA BERGAMINI, ALBERTO DI MARIO AND FRANCESCO FELICI

404 • A KINEMATIC ANALYSIS OF THE NAERYO-CHAGI TECHNIQUE IN TAEKWONDO - MICHAEL KLOIBER, ARNOLD BACA, EMANUEL PREUSCHL AND BRIAN HORSAK

408 • A KINEMATIC ANALYSIS OF FINGER MOTION IN ARCHERY - BRIAN HORSAK, MARIO HELLER AND ARNOLD BACA

412 • BALLET DANCER INJURIES DURING PERFORMANCE AND REHEARSAL ON VARIED DANCE SURFACES - LUKE HOPPER, NICK ALLEN, MATTHEW WYON, JACQUELINE ALDERSON, BRUCE ELLIOTT AND TIM ACKLAND

ORAL SESSION S21 – JUMPING II

417 • QUANTIFYING THE ONSET OF THE CONCENTRIC PHASE OF THE FORCE-TIME RECORD DURING JUMPING – RANDALL L. JENSEN, SARAH K. LEISSRING, LUKE R. GARCEAU, ERICH J. PETUSHEK AND WILLIAM P. EBBEN

421 • KNEE SEPARATION DISTANCE AND QUADRICEPS AND HAMSTRINGS STRENGTH DURING DROP VERTICAL JUMP LANDINGS - ORNA DONOGHUE, LORRAINE STEEL, RACHEL COLLINS, HARRIET YOUNG AND SIMON COLEMAN

422 • LATERALITY IN VERTICAL JUMPS - DIANA LANG, GERDA STRUTZENBERGER, ANNE RICHTER, GUNTHER KURZ AND HERMANN SCHWAMEDER

423 • MECHANICS OF AMPUTEE JUMPING - JOINT WORK - MARLENE SCHOEMAN, CERI DISS AND SIOBHAN STRIKE

427 • THE INFLUENCE OF SAND SURFACE ON KINEMATICS DURING VOLLEYBALL SPIKE JUMPS – MARKUS TILP, HERBERT WAGNER AND ERICH MUELLER

ORAL SESSION S22 – INJURY II

429 • EFFICACY OF FUNCTIONAL KNEE BRACES IN SPORT: A REVIEW – MARIO LAMONTAGNE AND NICHOLAS BRISSON

433 • ESTIMATION OF THE POTENTIAL RISK OF ACL RUPTURE AND THE EFFECTIVENESS OF A PREVENTION TRAINING PROGRAM - THOMAS JÖLLENBECK, BRITTA GREBE AND DOROTHEE NEUHAUS

437 • LEG PRESS EXERCISE IN PATELLOFEMORAL PAIN- A ONE-YEAR FOLLOW-UP STUDY - CHEN-YI SONG AND MEI-HWA JAN

441 • THE EFFECT OF INJURY AND DOMINANCE ON SAGGITAL PLANE KINEMATICS DURING A DROP LAND - AMANDA CLIFFORD, KIERAN O'SULLIVAN AND MARIE TIERNEY

442 • LOW BACK PAIN IN GOLF: DOES THE CRUNCH FACTOR CONTRIBUTE TO LOW BACK INJURIES IN GOLFERS? - MICHAEL COLE AND PAUL GRIMSHAW

ORAL SESSION S23 – OTHER SPORTS III

444 • KINEMATIC ANALYSIS OF THE SLIDING STOP IN WESTERN RIDING AT THE MALLORCA WESTERN REINING TROPHY 2006 - MAREN FROEGER AND CHRISTIAN PEHAM

448 • TOWARDS UNDERSTANDING HUMAN BALANCE - ANALYZING STICK BALANCING - MANFRED VIETEN

452 • THE USE OF PASSIVE DRAG TO INTERPRET VARIATION IN ACTIVE DRAG MEASUREMENTS - BRUCE MASON, DANIELLE FORMOSA AND VINCE RALEIGH

456 • SURFACE ELECTROMYOGRAPHY OF ABDOMINAL AND SPINAL MUSCLES IN ADULT HORSERIDERS DURING RISING TROT - ANNETTE PANTALL, SHARRON BARTON AND PETER COLLINS

460 • ACCURACY OF A PORTABLE (PTZ DIGITAL) CAMERA SYSTEM DESIGNED FOR AQUATIC THREE-DIMENSIONAL ANALYSIS - GEORGIOS MACHTSIRAS AND ROSS H. SANDERS

POSTER SESSION PS1

462 • THE ROLE OF ANXIETY IN GOLF PUTTING PERFORMANCE - IAN KENNY, AINE MACNAMARA, AMIR SHAFAT, ORLA DUNPHY, SINEAD MURPHY, KENNETH O'CONNOR, TARA RYAN AND GERRY WALDRON

463 • FATIGABILITY OF TRUNK MUSCLES WHEN SIMULATING PUSHING MOVEMENT DURING TREADMILL WALKING - YI-LING PENG, YANG-HUA LIN AND HEN-YU LIEN

464 • A COMPARATIVE STUDY BETWEEN BLADES AND STUDS IN FOOTBALL BOOTS - JAMES NUTT, RAMI ABBOUD, GRAHAM ARNOLD AND WEIJIE WANG

465 • STEP HEIGHT EFFECTS ON LOWER LIMB BIOMECHANICS AND BODY CENTRE OF MASS MOTION DURING ELLIPTICAL EXERCISE - YEN-PAI CHEN, CHU-FEN CHANG, HUILIEN CHIEN, YI-CHENG CHEN AND TUNG-WU LU

466 • EFFECTS OF TAI-CHI CHUAN ON THE CONTROL OF BODY'S CENTER OF MASS MOTION DURING OBSTACLE-CROSSING IN THE ELDERLY - TSUNG-JUNG HO, SHENGCHANG CHEN, CHU-FEN CHANG AND TUNG-WU LU

467 • KINEMATICS OF TACTICS IN THE FINAL MEN'S 1500 M FREESTYLE SWIMMING OF THE BEIJING 2008 OLYMPIC GAMES - PATRYCJA LIPINSKA AND WLODZIMIERZ S ERDMANN

471 • EFFECT OF FATIGUE ON THE COORDINATION VARIABILITY IN ROWERS - PAUL TALTY AND ROSS ANDERSON

- 472 • BIOMECHANICS AND POTENTIAL INJURY MECHANISMS OF WRESTLING - TSONG-RONG JANG, SHENG-CHANG CHEN, CHU-FEN CHANG, YANG-CHIEH FU AND TUNG-WU LU
- 473 • COMPARISON OF BALL-AND- RACKET IMPACT FORCE IN TWO-HANDED BACKHAND BETWEEN DIFFERENT DIRECTIONS OF STROKE - CHIN-FU HSU, KUO-CHENG LO, YI-CHIEN PENG AND LIN-HWA WANG
- 474 • INFLUENCES OF THE MASS OF BOXING GLOVES ON THE IMPACT FORCE OF A REAR HAND STRAIGHT PUNCH - GENKI NAKANO, YOICHI IINO AND TAKEJI KOJIMA
- 475 • THE EFFECT OF VARYING CLUB HEAD MASS ON VELOCITY AND KINETIC ENERGY - CATHERINE TUCKER, IAN KENNY, DEREK BYRNE AND ROSS ANDERSON
- 476 • RELIABILITY OF FORCES DURING VARIATIONS OF PLYOMETRIC EXERCISES - SARAH K LEISSRING, WILLIAM P EBBEN, LUKE GARCEAU, ERICH JR PETUSHEK AND RANDALL L JENSEN
- 480 • AN INVESTIGATION INTO THE EFFECTS OF A SIMULATED EFFUSION IN HEALTHY SUBJECTS ON KNEE KINEMATICS AND LOWER LIMB MUSCLE ACTIVITY DURING A SINGLE LEG DROP LANDING - GARRETT COUGHLAN, ULRIK MC CARTHY PERSSON, ROD MC LOUGHLIN AND BRIAN CAULFIELD
- 484 • VARIABILITY OF STRIDE FREQUENCY AND PRONATION VELOCITY DURING A 16 DAY RELAY-RUN AROUND GERMANY – A CASE STUDY - NINA GRAS, THORSTEN STERZING, TORSTEN BRAUNER, DORIS ORIWOL, JENS HEIDENFELDER, THOMAS L MILANI
- 488 • EFFECT OF BILATERAL OR SINGLE LEG LANDING ON KNEE KINEMATICS - SHINYA ABE, MATTHEW K D LEWIS, KRISHNAKUMAR MALLIAH, PARIS L MALIN AND RANDALL L JENSEN
- 489 • LANDING STRATEGY MODULATION IN BACKWARD ROTATING PIKED AND TUCKED SOMERSAULT DISMOUNTS FROM BEAM - MARIANNE GITTOES, GARETH IRWIN, DAVID MULLINEAUX AND DAVID KERWIN
- 493 • THE EFFECT OF IMPACT CONDITION ON THE RELATIONSHIP BETWEEN LINEAR AND ANGULAR ACCELERATION - PHILIPPE ROUSSEAU, EVAN WALSH, SCOTT FOREMAN AND BLAINE HOSHIZAKI
- 494 • JOINT KINEMATIC VARIABILITY IN THE AERIAL AND LANDING PHASES OF BACKWARD ROTATING DISMOUNTS FROM BEAM - MARIANNE GITTOES, GARETH IRWIN, DAVID MULLINEAUX AND DAVID KERWIN
- 498 • A COMPARISON OF RUNNING KINEMATICS BETWEEN TOP 6 AND HONG KONG ELITE TRIATHLETES IN 2008 ASIAN CHAMPIONSHIPS - ANGUS TAK KAI LAM, DANNY PAK KEUNG CHU AND PAK MING CHEUNG
- 499 • COMPARISON OF MALE AND FEMALE PEAK TORQUE USING A VARIABLE NUMBER OF REPETITIONS - BARBARA WARREN, SARAH MOODY AND DEANNA MALIKIE
- 503 • KINEMATIC ANALYSIS OF THE UPPER LIMB AT THE HEIGHT OF HITTING POINT IN A BASEBALL BATTING - TAKAHITO TAGO, MICHIYOSHI AE, DAISUKE TSUCHIOKA, NOBUKO ISHII AND TADASHI WADA
- 507 • TREND ANALYSIS OF COMPLEX RELEASE AND RE-GRASP SKILLS ON THE HIGH BAR - MARK SAMUELS, GARETH IRWIN, DAVID KERWIN AND MARIANNE GITTOES
- 511 • EFFECT OF JUMP PERFORMANCE ON DIFFERENT COMPRESSION GARMENTS - SHAO-YI LIU, WAN-CHIN CHEN AND TZYY-YUANG SHIANG
- 512 • VARIABILITY IN COMPETITIVE PERFORMANCE OF ELITE TRACK CYCLISTS - NICK FLYGER
- 516 • DYNAMIC BALANCE IN ALPINE SKIERS - TOM CRESSWELL AND ANDREW MITCHELL
- 517 • COMPARISON OF TURN TECHNIQUES IN PERFORMING THE BASKET WITH HALF TURN TO HANDSTAND ON PARALLEL BARS - TETSU YAMADA, DAISUKE NISHIKAWA, YUSUKE SATO AND MAIKO SATO
- 518 • THE EFFECT OF DIFFERENT EXTERNAL ELASTIC COMPRESSION ON MUSCLE STRENGTH, FATIGUE, EMG AND MMG ACTIVITY - YU LIU, WEIJIE FU AND XIAOJIE XIONG
- 522 • COMPARISON OF SINGLE- AND MULTILAYER MATERIALS USED AS DAMPENING ELEMENTS IN KNEE-PROTECTORS - ALEXANDER KOTSCHWAR AND CHRISTIAN PEHAM
- 526 • LOWER LIMB JOINT STIFFNESS IN THE SPRINT START PUSH-OFF - LAURA CHARALAMBOUS, GARETH IRWIN, IAN BEZODIS, DAVID KERWIN AND ROBERT HARLE
- 530 • A KINETIC COMPARISON OF RUNNING ON TREADMILL AND OVERGROUND SURFACES: AN ANALYSIS OF PLANTAR PRESSURE - JUSTIN WAI YUK LEE, MANDY MAN LING CHUNG, KAM MING MOK AND YOULIAN HONG
- 534 • THE EFFECTS OF ADIDAS POWERWEB COMPRESSION SHORTS ON MUSCLE OSCILLATION AND DROP JUMP PERFORMANCE - RUSSELL PETERS, NEAL SMITH AND MIKE LAUDER
- 538 • A KINEMATIC DESCRIPTION OF THE POST PUBESCENT WINDMILL SOFTBALL PITCHING MOTION - DAVID KEELEY, GRETCHEN OLIVER, PRISCILLA DWELLY AND HIEDI HOFFMAN
- 542 • ALAN SWANTON AND ROSS ANDERSON - DEVELOPMENT OF A RECORDING SYSTEM TO EMPIRICALLY ANALYSE THE SHOOTING CHARACTERISTICS OF A CLAY PIGEON SHOOTER
- 543 • COMPLEX TRAINING: THE OPTIMUM REST INTERVAL FOR POTENTIATION BETWEEN A 3RM BACK SQUAT AND A SQUAT JUMP - DARRAGH GRAHAM AND ANDREW HARRISON
- 544 • RELIABILITY OF DROP JUMP VARIATIONS IN PERFORMANCE DIAGNOSTICS - GUNTHER KURZ, DIANA LANG, ANNE RICHTER AND HERMANN SCHWAMEDER
- 545 • TENNISSENSE: A MULTI-SENSORY APPROACH TO PERFORMANCE ANALYSIS IN TENNIS - LUKE CONROY, CIARÁN Ó CONAIRE, PHILIP KELLY, SHIRLEY COYLE, PADDY NIXON, GRAHAM HEALY, DAMIAN CONNAGHAN, NOEL O'CONNOR, BRIAN CAULFIELD AND ALAN SMEATON
- 546 • RELIABILITY OF TIME TO STABILIZATION IN SINGLE LEG STANDING - MAHENDRAN KALIYAMOORTHY AND RANDALL L JENSEN
- 550 • EFFECT OF FATIGUE ON DYNAMIC BALANCE AFTER MAXIMUM INTENSITY CROSS-COUNTRY SKIING - MAHENDRAN KALIYAMOORTHY, PHILLIP B WATTS, RANDALL L JENSEN AND JAMES A LACHAPELLE

POSTER SESSION PS2

- 57 • IS PASSIVE PLANTAR FLEXION TORQUE DETERMINANT OF LOWER LIMB STIFFNESS? - KEISUKE WATANABE AND TOSHIO YANAGIYA
- 553 • THE EFFECT OF FATIGUE ON REACTIVE STRENGTH IN ANTERIOR CRUCIATE LIGAMENT RECONSTRUCTED INDIVIDUALS - EAMONN FLANAGAN, RANDALL L JENSEN, DREW HARRISON AND DAN RICKABY
- 557 • A WARM-UP INCLUDING A 5RM SQUAT PROTOCOL INCREASED BLOOD LACTATE, WITHOUT ALTERING THE SUBSEQUENT JUMP PERFORMANCE - ALEX DINSDALE, ATHANASSIOS BISSAS AND SOPHIE REYNOLDS
- 558 • INFLUENCE OF ANKLE TAPING ON JUMP PERFORMANCE - IAN KENNY AND SELVA PRAKASH JEYARAM
- 559 • DOUBLE KNEE BEND IN THE POWER CLEAN - LAURA-ANNE FURLONG, GARETH IRWIN, CASSIE WILSON, HUW WILTSHIRE AND DAVID KERWIN
- 560 • STUDY ON DEVELOPMENTS OF BODY COMPOSITION AND PHYSICAL FITNESS FOR A YEAR IN JAPANESE ADOLESCENT TRACK AND FIELD ATHLETES - AYA MIYAMOTO, JOJI UMEZAWA AND TOSHIO YANAGIYA
- 561 • A COMPARISON OF THE AEROBIC ENERGY DEMANDS OF TWO COMMERCIALLY AVAILABLE CYCLE ERGOMETERS IN TRAINED CYCLISTS - GREGORY MAY AND GILES WARRINGTON
- 565 • EFFECTS OF STEP LENGTH ON THE BIOMECHANICS OF LOWER LIMBS DURING ELLIPTICAL EXERCISE - TUNG-WU LU, CHIH-HUNG HUANG, YEN-PAI CHEN, CHU-FEN CHANG AND HUI-LIEN CHIEN
- 566 • PREPARATORY LONGSWING TECHNIQUES FOR DISMOUNTS ON UNEVEN BARS - DAWN TIGHE, GARETH IRWIN, DAVID KERWIN AND MARIANNE GITTOES
- 570 • PRACTICE AND TALENT EFFECTS IN SWING HIGH BAR INTER-JOINT COORDINATION OF NOVICE ADULTS - ALBERT BUSQUETS, MICHEL MARINA, ALFREDO IRURTIA AND ROSA M ANGULA-BARROSO
- 574 • THE RELATIONSHIP BETWEEN GLUTEAL ACTIVITY AND PELVIC KINEMATICS DURING THE WINDMILL SOFTBALL PITCH - HIEDI HOFFMAN, GRETCHEN OLIVER, DAVID KEELEY, KASEY BARBER AND PRISCILLA DWELLY
- 575 • HIP MOMENT PROFILES DURING CIRCLES IN SIDE SUPPORT AND IN CROSS SUPPORT ON THE POMMEL HORSE - TOSHIYUKI FUJIHARA AND PIERRE GERVAIS
- 579 • THE EFFECTS OF SPORTS TAPING ON IMPACT FORCES AND MECHANICAL BEHAVIORS OF SOFT TISSUE DURING DROP LANDING - NYEON-JU KANG, WOEN-SIK CHAE, CHANG-SOO YANG AND GYE-SAN LEE
- 580 • FOOT FUNCTION IN SPRINTING: BAREFOOT AND SPRINT SPIKE CONDITIONS - GRACE SMITH AND MARK LAKE
- 581 • THE VELOCITY DEPENDENCE OF TECHNNIQUES COMMONLY LINKED WITH LOWER BACK INJURY IN CRICKET FAST BOWLING - KANE MIDDLETON, POONAM CHAUHAN, BRUCE ELLIOTT AND JACQUELINE ALDERSON
- 585 • RELATIONSHIPS BETWEEN HIP AND SHOULDER ROTATION DURING BASEBALL PITCHING - DAVID KEELEY, GRETCHEN OLIVER, PRISCILLA DWELLY, HIEDI HOFFMAN AND CHRISTOPHER DOUGHERTY
- 586 • ACUTE EFFECTS OF STATIC STRETCHING ON FORCE OUTPUT OF DORSI FLEXORS IN DANCE STUDENTS - MAYUMI KUNO-MIZUMURA AND MAKIKO NAGAKURA
- 589 • THE DIFFERENCE OF THE BALANCE ABILITIY BETWEEN THE FUNCTIONAL ANKLE INSTABILITY AND HEALTHY SUBJECTS - YUKIO URABE, YUKI NODA, SHINJI NOMURA, TAKESHI AKIMOTO, HIROE SHIDAHARA AND YUKI YAMANAKA
- 590 • A BIOMECHANICAL COMPARISON OF THE LOWER EXTREMITY DURING FRONT AND BACK SQUATS IN HEALTHY TRAINED INDIVIDUALS - ERIN LEAROYD AND KATHRYN LUDWIG
- 591 • QUALITY OF DINGHY HIKING: EFFECTS ON SPEED AND HEADING - ROBERT MARSHALL
- 592 • MECHANICAL COMPARISON BETWEEN ROUNDHOUSE KICK TO THE CHEST AND TO THE HEAD IN FUNCTION OF EXECUTION DISTANCE IN TAEKWONDO - ISAAC ESTEVAN, CORAL FALCO, OCTAVIO ALVAREZ, FERNANDO MUGARRA AND ANTONIO IRADI
- 596 • KINETIC AND KINEMATICAL ANALYSIS BETWEEN DOMINANT AND NONDOMINANT KICKING LEG IN A ROUNDHOUSE KICK IN TAEKWONDO - CORAL FALCO, OCTAVIO ALVAREZ, ISAAC ESTEVAN, JAVIER MOLINA-GARCIA, FERNANDO MUGARRA AND ANTONIO IRADI
- 600 • RECONSTRUCTION ACCURACY FOR VISUAL CALIBRATION METHOD - MARC ELIPOT, NICOLAS HOUEL, PHILIPPE HELLARD AND GILLES DIETRICH
- 604 • MOTOR UNIT FIRINGS DURING VOLUNTARY ISOMETRIC RAMP AND BALLISTIC CONTRACTIONS IN HUMAN VASTUS MEDIALIS MUSCLE - SHINJI MIZUMURA, YOSHIHISA MASAKADO, KATSUHIKO MAEZAWA AND KELLY F MCGRATH
- 608 • EXAMINATION OF GLUTEAL MUSCLE FIRING AND KINETICS OF THE LOWER EXTREMITY DURING THE WINDMILL SOFTBALL PITCH - GRETCHEN OLIVER, PRISCILLA DWELLY, DAVID KEELEY AND HIEDI HOFFMAN
- 612 • BIOMECHANICAL SETTINGS OF TWO WAYS TO EXECUTE THE GOLF SWING - FERDINAND TUSKER AND FLORIAN KREUZPOINTNER
- 616 • DETERMINATION OF ARMS AND LEGS CONTRIBUTION TO PROPULSION AND PERCENTAGE OF COORDINATION IN BUTTERFLY SWIMMING - MORTEZA SHAHBAZI
- 620 • THE DETERMINATION OF DRAG IN THE GLIDING PHASE IN SWIMMING - DANIEL MARINHO, FILIPE CARVALHO, TIAGO BARBOSA, VICTOR REIS, FRANCISCO ALVES, ABEL ROUBOA AND ANTÓNIO SILVA
- 621 • ASYMMETRY IN FRONTCRAWL SWIMMING WITH AND WITHOUT HAND PADDLES - MIKE LAUDER AND REBECCA NEWELL
- 625 • DIFFERENCES IN SEGMENTAL MOMENTUM TRANSFERS BETWEEN TWO STROKE POSTURES FOR TENNIS TWO-HANDED BACKHAND STROKE - LIN-HWA WANG, KUO-CHENG LO, YUNG-CHUN HSIEH AND FONG-CHIN SU
- 626 • BIOMECHANICAL ANALYSIS OF ROUNDED OUTSOLE DESIGN SHOE DURING WALKING - PARK SEUNGBUM, PARK SANGKYOON, LEE JUNGHO, LEE KYUNGDEUK, KIM DAEWOONG, SEO KOOKEUN AND SHIN HAKSOO
- 627 • PREPARATORY LONGSWINGS PRECEDING TKACHEVS ON UNEVEN BARS - MICHELLE MANNING, GARETH IRWIN, DAVID KERWIN AND MARIANNE GITTOES

628	• RELIABILITY OF INVERSE DYNAMICS OF THE WHOLE BODY IN THE TENNIS FOREHAND - YOICHI IINO AND TAKEJI KOJIMA
629	• THE INFLUENCE OF EXTRA LOAD ON TIME AND FORCE STRUCTURE OF VERTICAL JUMP - FRANTISEK VAVERKA, ZLATAVA JAKUBSOVA AND DANIEL JANDACKA
630	• ESTIMATION OF HORSE LEGS MUSCLES FORCES DURING JUMPING - MORTEZA SHAHBAZI AND NARGES KHOSRAVI
634	• THE ANALYSIS OF PEDALING FORCE AND LOWER EXTREMITY EMG USING DIFFERENT PEDALING RATES AND LOADS - CHENG-SHUAN CHANG AND TZYY-YUANG SHIANG

POSTER SESSION PS3

636	• ON-BOARD AND PRE-FLIGHT MECHANICAL MODEL OF YURCHENKO 1/1 TWIST ON VAULT: IMPLICATIONS FOR PERFORMANCE - GABRIELLA PENITENTE, FRANCO MERNI AND SILVIA FANTOZZI
640	• GROUND REACTION FORCE OF BASEBALL FLAT GROUND PITCHING - CHUNLUNG LIN AND CHEN-FU HUANG
644	• NON-LINEAR CAMERA CALIBRATION FOR 3D RECONSTRUCTION USING STRAIGHT LINE PLANE OBJECT - AMANDA P SILVATTI, MARCEL M ROSSI, FÁBIO A S DIAS, NEUCIMAR J LEITE AND RICARDO M L BARROS
645	• THE CONTRIBUTION OF LOWER TORSO, UPPER TORSO AND UPPER LIMBS SEGMENTAL MOTION TO HAMMER HEAD VELOCITY DURING ACCELERATION PHASE - HIROAKI FUJII, KEIGO OHYAMA BYUN, MITSUGI OGATA, NORIHISA FUJII
646	• ASYMMETRIC LOADING DURING THE HANG POWER CLEAN - THE EFFECT THAT SIDE DOMINANCE HAS ON BARBELL POWER SYMMETRY - JASON LAKE, MIKE LAUDER AND NEAL SMITH
650	• FOOT-TO-BALL INTERACTION IN PREFERRED AND NON-PREFERRED LEG AUSTRALIAN RULES KICKING - JASON SMITH, KEVIN BALL AND CLARE MACMAHON
654	• THE STUDY AND APPLICATION OF THE INERTIA GAIT - WASSF ISAAC AND A A KAMAL
655	• LOWER LIMB BIOMECHANICAL ADAPTATIONS TO TOTAL HIP ARTHROPLASTY EXIST DURING SITTING AND STANDING TASKS - MELANIE BEAULIEU, MARIO LAMONTAGNE, DANIEL VARIN AND PAUL BEAULE
659	• FRONT AND PIPE SPIKES IN FEMALE ELITE VOLLEYBALL PLAYERS: IMPLICATIONS FOR THE IMPROVEMENT OF PIPE SPIKE TECHNIQUES - MASANAO MASUMURA, MARQUES WALTER QUISPE AND MICHIYOSHI AE
660	• SECURITY OF RUNNING OF COMPETETIVE COURSE IN ALPINE SKIING ACCORDING TO ITS GEOMETRY OF SETTING - PIOTR ASCHENBRENNER AND WLODZIMIERZ ERDMANN
664	• LOWER LIMB LANDING BIOMECHANICS ON NATURAL AND FOOTBALL TURF - PHILIPPA JONES, DAVID KERWIN, GARETH IRWIN, LEN NOKES AND RAJIV KAILA
668	• FALL AND INJURY INCIDENCE RATES OF JOCKEYS WHILE RACING IN FRANCE, GREAT BRITAIN AND IRELAND - MANUEL FORERO RUEDA, WALTER HALLEY AND MICHAEL GILCHRIST
669	• ACUTE EFFECTS OF WHOLE-BODY VIBRATION TRAINING ON ELASTIC CHARGE TIME IN TRAINED MALE ATHLETES - LAURA-ANNE FURLONG AND ANDREW HARRISON
672	• AN INVESTIGATION INTO THE IMPACT FORCE EXPERIENCED BY DIFFERENT TYPES OF FOOTBALLS - YO CHEN, JIA-HAO CHANG, TAI-YEN HSU AND YU-PEI KAO
673	• THE SYMMETRY IN GAIT KINEMATICS OF ADOLESCENTS' CASE STUDIES - MATILDE ESPINOSA-SANCHEZ
677	• ANKLE JOINT LOADING DURING THE DELIVERY STRIDE IN CRICKET MEDIUMFAST BOWLING - KATHLEEN SHORTER, NEAL SMITH AND MIKE LAUDER
681	• EMG ANALYSIS OF THE LOWER EXTREMITY BETWEEN VARYING STANCE SQUAT WIDTHS IN BASEBALL CATCHER THROWING - YI-CHIEN PENG, KUO-CHENG LO, HWAITING LIN AND LIN-HWA WANG
682	• RATE FORCE DEVELOPMENT DURING BENCH PRESS IS ONLY RELATED TO THROWING VELOCITY WHEN USING LIGHT LOADS - DANIEL MARINHO, RICARDO FERRAZ, ROLLAND TILLAAR, ALDO COSTA, VICTOR REIS, ANTÓNIO SILVA, JUAN GONZÁLEZ-BADILLO AND MÁRIO MARQUES
683	• WHAT ARE VALUES OF SHOELACES IN RUNNING? - JING XIAN LI, LIN WANG AND YOULIAN HONG
684	• CHANGES IN SCAPULAR MOTION DURING A SIMULATED BASEBALL GAME - DAISAKU HIRAYAMA, NORIHISA FUJII AND MICHIYOSHI AE
685	• THE MEASUREMENT OF KINETIC VARIABLES IN RACE WALKING - BRIAN HANLEY, ANDI DRAKE AND ATHANASSIOS BISSAS
689	• IN-SHOE PLANTAR PRESSURE MEASURES DURING GOLF SWING PERFORMANCE WEARING METAL AND ALTERNATIVE SPIKED GOLF SHOES - PAUL WORSFOLD, NEAL SMITH AND ROSEMARY DYSON
693	• STATIC AND DYNAMIC ANALYSIS OF THE FOOT IN SOCCER PLAYERS SUSTAINING PROXIMAL 5TH METATARSAL STRESS FRACTURE - MOSHE AYALON, IFTACH HETSRONI AND DAVID BEN SIRA
694	• BIOMECHANICAL CHARACTERISTICS OF GRINDING IN AMERICA'S CUP SAILING - SIMON PEARSON, PATRIA HUME, JOHN CRONIN AND DAVID SLYFIELD
695	• ANALYSIS OF GROUND REACTION FORCE DURING FASTBALL AND CHANGE-UP SOFTBALL PITCHES - JIA-HAO CHANG AND WEI-MING TSENG
696	• THE INFLUENCE OF EXPERIENCE ON FUNCTIONAL PHASE KINEMATICS OF THE LONGSWING - GENEVIEVE WILLIAMS, GARETH IRWIN AND DAVID KERWIN
697	• CONTRIBUTION OF THE SUPPORT LEG TO THE VELOCITY CHANGE IN THE CENTRE OF GRAVITY DURING CUTTING MOTION - YUTA SUZUKI, MICHIYOSHI AE AND YASUSHI ENOMOTO
701	• TIBIAL ROTATIONS DURING STEP UP EXERCISE DO NOT CHANGE KNEE EXTENSOR ACTIVITY IN THE PATELLOFEMORAL PAIN SYNDROME - JULIANA MORENO CARMONA, CRISTINA MARIA NUNES CABRAL AND AMELIA PASQUAL MARQUES
705	• CONDITIONS FOR THE FORWARD MOTION OF A SNAKEBOARD - ALEXANDER KULESHOV

709 • LUMBAR SPINE POSTURE IN SENIOR AND ELITE LEVEL ROWERS – A COMPARISON WITH THE LOW BACK PAIN POPULATION - CAROLINE MACMANUS AND KIERAN O'SULLIVAN

710 • EFFECT OF A SPECIFIC STRENGTH TRAINING ON THE DEPTH SQUAT WITH DIFFERENT LOAD: A CASE STUDY - SIMONE CIACCI AND FABRIZIO PECORAIOLI

714 • EVALUATION OF SADDLE HEIGHT IN ELITE CYCLISTS - MARLENE MAUCH AND ANDREAS GOESELE

715 • LOWER LIMB JOINT KINETICS IN THE SPRINT START PUSH-OFF - LAURA CHARALAMBOUS, GARETH IRWIN, IAN BEZODIS, DAVID KERWIN AND ROBERT HARLE

719 • TEMPORAL CHARACTERISTICS OF THOMAS FLAIRS ON THE POMMEL AND FLOOR - SPIROS PRASSAS, GIDEON ARIEL AND ELEFTHERIOS TSAROUHAS

Poster Session PS4

721 • REAL WORLD HEAD IMPACT DATA MEASUREMENTS ON JOCKEYS - MANUEL FORERO RUEDA AND MICHAEL GILCHRIST

722 • RELATIONSHIP BETWEEN A REDUCTION OF THE FOOT ARCH HEIGHT AND THE KNEE VALGUS - TAKESHI AKIMOTO, YUKIO URABE, YUKI YAMANAKA, NATSUMI KAMIYA AND SHINJI NOMURA

723 • A KINEMATIC COMPARISON OF THE YOUTH PITCHING MOTION ACROSS PREPUBESCENT AND PUBESCENT PITCHERS - DAVID KEELEY, GRETCHEN OLIVER, PRISCILLA DWELLY AND HIEDI HOFFMAN

727 • PREVENTION & REHABILITATION OF DIVERS' CERVICAL VERTEBRA INJURIES FROM PERSPECTIVE OF BIOMECHANICAL BALANCE - RENBO QIAO

731 • DETERMINANT OF LEG SPRING STIFFNESS DURING MAXIMAL HOPPING - HIROAKI HOBARA, KOUKI GOMI AND KAZUYUKI KANOSUE

732 • SURFACE MARKERS VERSUS CLUSTERS FOR DETERMINING LOWER LIMB JOINT KINEMATICS IN SPRINT RUNNING - TIMOTHY EXELL, DAVID KERWIN, GARETH IRWIN AND MARIANNE GITTOES

736 • MEASUREMENT OF THIGH MUSCLE SIZE USING TAPE OR ULTRASOUND IS A POOR INDICATOR OF THIGH MUSCLE STRENGTH - KIERAN O'SULLIVAN, DAVID SAINSBURY AND RICHARD O'CONNOR

737 • PEAK VELOCITY OF NORDIC SKI DOUBLE POLE TECHNIQUE: STAND-UP VS - JODI TERVO AND RANDALL L JENSEN

738 • HIGH- AND LOW-ARCHED ATHLETES EXHIBIT SIMILAR STIFFNESS VALUES WITHIN THE LOWER EXTREMITY - DOUGLAS POWELL, CAROLYN ALBRIGHT AND BENJAMIN LONG

739 • ARCHERY TRAINING IMPROVE POSTURAL CONTROL IN YOUNG CHILDREN - ALEX J Y LEE, YU-CHI CHIU, YING-FANG LIU AND WEI-HSIU LIN

740 • KNEE AND ANKLE JOINT KINEMATICS IN KENDO MOVEMENT AND THE REPEATABILITY - SENTARO KOSHIDA, TADAMITSU MATSUDA AND KYOHEI KAWADA

741 • EFFECTS OF FATIGUE ON THE GROUND REACTION FORCES AND LEG KINEMATICS IN ALL-OUT 600 METES RUNNING - HIROSUKE KADONO, MICHIYOSHI AE, YUTA SUZUKI AND KAZUHITO SHIBAYAMA

745 • IMPROVED MUSCLE ACTIVATION IN PERFORMING A FUNCTIONAL LUNGE COMPARED TO THE TRADITIONAL SQUAT - PRISCILLA DWELLY, GRETCHEN OLIVER, HEATHER ADAMS-BLAIR, DAVID KEELEY AND HIEDI HOFFMAN

749 • THE EFFECT OF STRENGTH TRAINING ON THE KINEMATICS OF THE GOLF SWING - AMY SCARFE, FRANCOIS-XAVIER LI AND MATTHEW BRIDGE

750 • TRUNK KINEMATICS DURING THE TEE-SHOT OF MALE AND FEMALE GOLFERS - MICHAEL COLE AND PAUL GRIMSHAW

751 • ANALYSIS OF BILATERAL ASYMMETRIES BY FLIGHT TIME OF ONE LEG COUNTERMOVEMENT JUMP - HANS-JOACHIM MENZEL, SILVIA RIBEIRO ARAÚJO AND MAURO HELENO CHAGAS

752 • ACUTE STATIC STRETCH EFFECTS ON MULTIPLE BOUTS OF VERTICAL JUMP - CHELSEA WALTER AND MICHAEL BIRD

756 • DYNAMIC STABILIZATION IN COLLEGIATE FEMALE VOLLEYBALL PLAYERS: EFFECTS OF LEG DOMINANCE AND OFF-SEASON - KIMITAKE SATO, GARY D HEISE AND KAHY LIU

757 • KINEMATIC AND KINETIC ANALYSIS IN TEAM-HANDBALL JUMP THROW - HERBERT WAGNER, MICHAEL BUCHECKER AND ERICH MÜLLER

758 • INFLUENCE OF BODY WEIGHT ON JOINT LOADING IN STAIR CLIMBING - GERDA STRUTZENBERGER, ANNE RICHTER, DIANA LANG AND HERMANN SCHWAMEDER

759 • ANALYSIS OF HUMAN MOTION WITH METHODS FROM MACHINE LEARNING - WOLFGANG SEIBERL, MICHELLE KARG, KOLJA KÜHNLENZ, MARTIN BUSS AND ANSGAR SCHWIRTZ

760 • POSTURAL CONTROL IN ELITE ARCHERS DURING SHOOTING - WEI-HSIU LIN, GUO-TANG HUANG, PING-KUN CHIU AND ALEX J Y LEE

761 • EVALUATION OF MECHANICAL POWER OUTPUT MEASUREMENT AT BENCH PRESS EXERCISE UNDER VARIABLE LOAD - DANIEL JANDACKA AND FRANTISEK VAVERKA

762 • RESPONSE TIME AND JAB FORCE PUNCH OF THAI FEMALE AMATEUR BOXERS: A PRELIMINARY STUDY - RAT TONGAIM, WEERAWAT LIMROONGREUNGRAT, SIRIRAT HIRUNRAT AND SUMETHEE THANANGKUL

763 • A PLANE-BASED CALIBRATION PROCEDURE FOR THE 3D ANALYSIS OF VIDEO RECORDINGS IN DISCUS THROWING DURING COMPETITION - VOLKER DRENK

767 • UPPER EXTREMITY AND CORE MUSCLE FIRING PATTERNS DURING THE WINDMILL SOFTBALL PITCH - GRETCHEN OLIVER, PRISCILLA DWELLY, DAVID KEELEY AND HIEDI HOFFMAN

771 • A CASE STUDY OF THE EFFECTS OF INSTRUCTION USING MOBILE PHONE'S ANIMATION FEEDBACK ON THROWING KINEMATICS - KENGO SASAKI, SOUICHI SHIMIZU, AMI USHIZU, KENJI KAWABATA, TAKAHIKO SATO, KAZUHIRO MATSUI, YU NAKASHIMA AND HIROH YAMAMOTO

775 • AN ANALYSIS OF THE IMPACT FORCES OF DIFFERENT MODES OF EXERCISE AS A CONTRIBUTING FACTOR TO THE LOW BONE MINERAL DENSITY IN JOCKEYS - SARAHJANE CULLEN, GILES WARRINGTON, EIMEAR DOLAN AND KIERAN MORAN

779 • DAILY KNEE JOINT LAXITY IN FEMALES ACROSS A MENSTRUAL CYCLE - DANIEL MEDRANO, JR, DARLA SMITH, DAYANAND KIRAN AND MARY CARLSON

783 • DEVELOPMENT OF A CRITERION METHOD TO DETERMINE PEAK MECHANICAL POWER OUTPUT IN A COUNTERMOVEMENT JUMP - NICHOLAS OWEN, JIM WATKINS, LIAM KILDUFF, DAN CUNNINGHAM, MARK BENNETT AND HUW BEVAN

784 • ACUTE EFFECTS OF HOPPING WITH WEIGHTED VEST ON VERTICAL STIFFNESS - LAWSON STEELE AND ORNA DONOGHUE

785 • GENDER DIFFERENCES IN INSTRUMENTED TREKKING POLE USE DURING DOWNHILL WALKING - JULIANNE ABENDROTH, GREG DIXON AND MICHAEL BOHNE

789 • DURABILITY OF RUNNING SHOES WITH EVA AND PU MIDSOLE - YOULIAN HONG, YAU CHOI NGAI, LIN WANG AND JING XIAN LI

POSTER SESSION PS5

791 • EFFECT OF KINETIC MECHANISMS OF LOWER LIMBS ON TORSO MOTION IN BASEBALL BATTING FOR DIFFERENT BALL SPEEDS - TOKIO TAKAGI, NORIHISA FUJII, SEKIYA KOIKE AND MICHIYOSHI AE

792 • MODERATION OF LOWER LIMB MUSCULAR ACTIVITY DURING JUMP LANDING BY THE APPLICATION OF ANKLE TAPING - ASHLEIGH DODD, ROSEMARY DYSON AND RUSSELL PETERS

796 • KINEMATIC ANALYSIS OF THE TRADITIONAL BACK SQUAT AND SMITH MACHINE SQUAT EXERCISES - ANTHONY GUTIERREZ AND RAFAEL BAHAMONDE

800 • DO PEOPLE WITH UNILATERAL CAM FAI FAVOUR THEIR SYMPTOMATIC LEG DURING MAXIMAL DEPTH SQUATS? - MATTHEW J KENNEDY, MARIO LAMONTAGNE AND PAUL E BEAULÉ

804 • EFFECTS OF SHORT-TERM SLED TOWING AND TRADITIONAL SPRINT TRAINING ON LEG POWER AND STIFFNESS - PEDRO E ALCARAZ, JOSE L L ELVIRA, JOSÉ M PALAO AND VICENTE ÁVILA PALAO AND VICENTE ÁVILA

805 • EFFECTS OF UPHILL RUNNING ON SPRINTING TECHNIQUE IN FOOTBALL PLAYERS - JOSE L L ELVIRA, VICENTE ÁVILA, JOSÉ M PALAO AND PEDRO E ALZARAZ

806 • THE DEVELOPMENT OF LOW COST SENSOR TECHNOLOGY TO PROVIDE AUGMENTED FEEDBACK FOR ON-WATER ROWING - DJ COLLINS, ROSS ANDERSON AND DEREK O'KEEFFE

807 • QUANTITATIVE ANALYSIS ON THE MUSCULAR ACTIVITY OF LOWER EXTREMITY DURING WATER WALKING - KOICHI KANEDA, DAISUKE SATO, HITOSHI WAKABAYASHI, YUJI OHGI AND TAKEO NOMURA

811 • STUD LENGTH AND STUD GEOMETRY OF SOCCER BOOTS INFLUENCE RUNNING PERFORMANCE ON THIRD GENERATION ARTIFICIAL TURF - CLEMENS MÜLLER, THORSTEN STERZING AND THOMAS MILANI

815 • EFFECTS OF CONCENTRIC VERSUS ECCENTRIC TRAINING ON MUSCLE STRENGTH AND NEUROMUSCULAR ACTIVATION - CHUN-HAN TSENG, YU-RU KUO, CHIHUANG HUANG AND HENG-JU LEE

816 • ASSISTED AND RESISTED SPRINT TRAINING MAY REDUCE ACTIVE DRAG IN SWIMMERS IN AN AEROBIC TRAINING PHASE - PER-LUDVIK KJENDLIE AND TOMMY PEDERSEN

817 • DIFFERENCES BETWEEN CONCENTRIC AND ECCENTRIC CONTRACTION INDUCED MUSCLE FATIGUE - CHUN-HAN TSENG, YU-RU KUO, CHI-HUANG HUANG AND HENG-JU LEE

818 • AN ORIGINAL INVERSE KINEMATICS ALGORITHM FOR KAYAKING - VINCENT FOHANNO, MICKAËL BEGON, FLOREN COLLOUD AND PATRICK LACOUTURE

819 • ESTIMATION OF KNEE EXTENSION MOMENT CONSIDERING VELOCITY EFFECT AND MUSCLE ACTIVATION USING TENDON SLACK LENGTH OPTIMIZATION - HYUN WOO UHM, WOO EUN LEE AND YOONSU NAM

823 • EVALUATION OF HOCKEY HELMET PERFORMANCE BY FINITE ELEMENT MODELING - ANDREW POST, BLAINE HOSHIZAKI AND MICHAEL GILCHRIST

824 • A KINEMATIC COMPARISON OF RUNNING ON TREADMILL AND OVERGROUND SURFACES - KAM MING MOK, JUSTIN WAI YUK LEE, MANDY MAN LING CHUNG AND YOULIAN HONG

828 • EFFECTS OF STATIC STRETCHING, PNF STRETCHING, AND DYNAMIC WARM-UP ON MAXIMUM POWER OUTPUT AND FATIGUE - JOSHUA AMAN AND BRYAN CHRISTENSEN

832 • KINEMATICS ANALYSIS OF TWO STYLES OF BOW FOR MARTIAL ARTISTS AND AVARAGE STUDENTS - KENJI KAWABATA, YUSUKE MIYAZAWA, SOICHI SHIMIZU, TAKAHIKO SATO, YOSHINORI TAKEUCHI, RYOICHI NISHITANI, NAOTOSHI MINAMITANI AND HIROH YAMAMOTO

836 • KINEMATICAL ANALYSIS OF 110M HURDLES - FOCUSING ON THE STEP LENGTH - KAZUHITO SHIBAYAMA, NORIHISA FUJII AND MICHIYOSHI AE

837 • LANDING DEVELOPMENT: A FIRST LOOK AT YOUNG CHILDREN - PAMELA RUSSELL, JEAN ECKRICH AND MADISON HAWKINS

841 • A COMPARISON OF THE PERFECT PUSH-UP™ TO TRADITIONAL PUSH-UP - MICHAEL BOHNE, JASON SLACK, TIM CLAYBAUGH AND JEFF COWLEY

842 • EFFECT OF RELAY CHANGEOVER POSITION ON SKATING SPEED FOR ELITE SHORT TRACK SPEED SKATERS - CONOR OSBOROUGH AND SARAH HENDERSON

843 • MATUATION EFFECTS ON LOWER EXTREMITY KINEMATICS IN A DROP VERTICAL JUMP - CHANGSOO YANG, INSIK SHIN, GYESAN LEE AND BEEOH LIM

844 • THE EFFECTS OF AGING ON THE HIP AND SPINAL MOTIONS IN THE GOLF SWING - CLIVE LATHEY, SIOBHAN STRIKE AND RAYMOND LEE

845 • DEVELOPMENT AND VALIDATION OF A SYSTEM FOR POLING FORCE MEASUREMENT IN CROSS-COUNTRY SKIING AND NORDIC WALKING - LORENZO BORTOLAN, BARBARA PELLEGRINI AND FEDERICO SCHENA

849 • EFFECTS OF DIFFERENT JUMP-LANDING DIRECTIONS ON LOWER EXTREMITY MUSCLE ACTIVATIONS - YU-MING LI AND HENG-JU LEE
850 • BIOMECHANICAL ANALYSIS OF STANDING LONG JUMP: A 3D STUDY - CHENFU HUANG AND RAY-HSIEN TANG
851 • WRIST POSITION, GRIP SIZE AND MAXIMAL FORCE IN TENNIS PLAYERS - DAVIDE SUSTA AND DAVID O' CONNELL
854 • BIOMECHANICAL ANALYSIS OF THE HANDBALL IN AUSTRALIAN FOOTBALL - LUCY PARRINGTON, KEVIN BALL, CLARE MACMAHON AND SIMON TAYLOR
858 • EFFECTS OF INDEPENDENT CRANK ARMS AND SLOPE ON PEDALING MECHANICS - SAORI HANAKI-MARTIN, DAVID MULLINEAUX AND STACY UNDERWOOD
862 • THE MOTION ANALYSIS OF WALK ON RACE WALKING PLAYERS - YOSHINORI TAKEUCHI, YOSHINARI OKA, TATSUYA NISHIMURA, KENJI KAWABATA, AMI USHIZU, KENGO SASAKI, YU NAKASHIMA AND HIROH YAMAMOTO
865 • PRIOR STUDY FOR THE DYNAMICAL ANALYSIS OF A PRECISE STARTING SENSOR FOR SHORT DISTANCE ATHLETIC SPORTS - JEONG-TAE LEE, HAN-WOOK SONG, CHEONGHWAN OH, CHAN-HO PARK, JIN LEE AND EUN-HYE HUH
869 • A PROPOSED METHODOLOGY FOR TESTING ICE HOCKEY HELMETS - BLAINE HOSHIZAKI, EVAN WALSH, PHILIPPE ROUSSEAU AND SCOTT FOREMAN
870 • EFFECTS OF BALANCING HAMSTRING AND QUADRICEPS MUSCLE TORQUE ON RUNNING TECHNIQUE - DOUG ROSEMOND, PETER BLANCH, ROSS SMITH AND TUDOR BIDDER
871 • COMPARISON OF VERTICAL FORCES BETWEEN A PRESSURE MEASUREMENT SYSTEM AND A FORCE PLATE - NICHOLAS BRISSON AND MARSHALL KENDALL
872 • THE DETERMINATION OF NOVEL IMPACT CONDITIONS FOR THE ASSESMENT OF LINEAR AND ANGULAR HEADFORM ACCELERATIONS - EVAN WALSH, PHILIPPE ROUSSEAU, SCOTT FOREMAN AND BLAINE HOSHIZAKI
873 • A CROSS-SECTIONAL STUDY OF GENDER DIFFERENCES IN PULLING STRENGTH OF TOW - TAKAHIKO SATO, SAKI SODEYAMA, RYUJI NAGAHAMA, SOUICHI SIMIZU, KENGO SASAKI, RYOICHI NISHITANI, CAO YULIN AND HIROH YAMAMOTO
877 • POSTURAL EFFECTS ON COMPARTMENTAL VOLUME CHANGES OF BREATHING BY OPTOELECTRONIC PLETHYSMOGRAPHY IN HEALTHY SUBJECTS - RONG-JIUAN LIING, KWAN-HWA LIN, TUNG-WU LU AND SHENG-CHANG CHEN
878 • AN INVESTIGATION OF THE ACTIVATION OF THE SUBDIVISIONS OF GLUTEUS MEDIUS DURING ISOMETRIC HIP CONTRACTIONS - CATRIONA O'DWYER, DAVID SAINSBURY AND KIERAN O'SULLIVAN
879 • APPLICABILITY OF OPERATIONS RESEARCH AND ARTIFICIAL INTELLIGENCE APPROACHES TO NON- CONTACT ANTERIOR CRUCIATE LIGAMENT INJURY STUDIES - NICHOLAS ALI, GORDON ROBERTSON AND GHOLAMREZA ROUHI

APPLIED SESSION AS1 - COACHING BIOMECHANICS

885 • QUALITATIVE BIOMECHANICS FOR COACHING - DUANE KNUDSON, JACQUE ALDERSON, RAFAEL BAHAMONDE AND MICHAEL BIRD

APPLIED SESSION AS2 - ROWING BIOMECHANICS

887 • ROWING APPLIED SESSION @ ISBS 2009 – RICHARD SMITH

APPLIED SESSION AS3 - STRENGTH & CONDITIONING

890 • APPLICATION OF COMPLEX TRAINING WITHIN STRENGTH AND CONDITIONING PROGRAMMES - THOMAS COMYNS
894 • UNDERSTANDING AND OPTIMISING PLYOMETRIC TRAINING - EAMONN FLANAGAN
898 • MEASUREMENT TECHNIQUES IN ASSESSING ATHLETIC POWER TRAINING - RANDALL JENSEN
902 • BASIC PERFORMANCE CUES FOR TEACHING THE POWER SNATCH AND POWER CLEANS TO NON-OLYMPIC WEIGHTLIFTING ATHLETES - ANDREW TYSZ

APPLIED SESSION AS4 - DATA ANALYSIS TECHNIQUES

905 • SYMPOSIUM ON RECENT DEVELOPMENTS IN DATA ANALYSIS - J. HAMILL, R. VAN EMMERIK, R. MILLER, K. O'CONNOR, N. COFFEY AND D. HARRISON

APPLIED SESSION AS5 - SWIMMING

911 • PREVENTING INJURIES IN SWIMMING - KEVIN BOYD
912 • BRITISH SWIMMING - CREATING A PLATFORM FOR ELITE SWIM PERFORMANCES - ANDREW LOGAN
913 • RECOGNISING AND AVOIDING OVERTRAINING - FRANCISCO ALVES
914 • GREEN SWIMMING: GETTING INTO THE ENERGY SAVING RHYTHM - ROSS SANDERS
915 • BODY ROLL: WHAT WE NOW KNOW - CARL PAYTON
916 • HOW TO START IN BACKSTROKE CONSIDERING THE NEW RULES? - J. PAULO VILAS-BOAS, KARLA DE JESUS, KELLY DE JESUS, PEDRO FIGUEIREDO, SUZANA PEREIRA, PEDRO GONÇALVES, LEANDRO MACHADO, RICARDO FERNANDES
917 • A GUIDE FOR THE CO-ORDINATION IN THE FRONT CRAWL VARIANTS - ULRIK PERSYN AND FILIP ROELANDT
918 • PHYSIOLOGICAL MONITORING OF SWIMMING - JOHN BRADLEY
919 • UNDERWATER DOLPHIN KICKING IN STARTS AND TURNS - RAÚL ARELLANO
920 • CFD AND SWIMMING: PRACTICAL APPLICATIONS - ANTÓNIO J. SILVA AND DANIEL A. MARINHO

921 - "I ALWAYS SWIM BADLY IN THE FINAL" - BRIAN DANIEL MARSHALL
922 - CORRECTING PHYSICAL AND TECHNICAL ASYMMETRIES OF SWIMMERS - ALISON FANTOM
923 - BREATHING AND STROKE FREQUENCY STRATEGIES FOR TOP PERFORMANCE - PER-LUDVIK KJENDLIE
924 - SPRINT AND DISTANCE SWIMMERS: THE SAME ANIMAL? - CARLA B. MCCABE
925 - SUPPORTING THE COACH WITH SCIENCE: THE IRISH EXPERIENCE - CONOR OSBOROUGH
926 - LAND TRAINING FOR SWIMMING - NEIL DONALD
927 - USING CRITICAL VELOCITIES TO SET TRAINING INTENSITIES - JEANNE DEKERLE
928 - COORDINATION: HOW ELITE SWIMMERS DIFFER FROM SUB-ELITE - DIDIER CHOLLET AND LUDOVIC SEIFERT
929 - SUCCESSFUL INTERVENTIONS FOLLOWING ANALYSIS - BRUCE R. MASON

ISBS 2009

Scientific Sessions

ISBS 2009

Keynote Lectures

BIOMECHANICS, TECHNOLOGY AND COACHING

David Kerwin

Cardiff School of Sport, University of Wales Institute, Cardiff, UK

KEY WORDS: research, wireless technology, athlete-focused research.

INTRODUCTION: The interactions between biomechanics, technology and coaching are presented via a research example. The purpose of this paper is to review new technologies and applied biomechanics in a coaching setting. Sprinting is used to highlight technological demands in providing meaningful feedback. This is a snapshot of work in progress, which builds on years of combined expertise within a large multi-partner research team. The overall objective is to apply novel wireless technologies in an elite sport setting with the aim of improving performance through the use of enhanced coach and athlete feedback.

APPROACH: Wide ranging methods have been employed, including structured coach interviews, field data collections requiring novel developments (e.g. Kerwin *et al*, ISBS2007, 497-500) and extensive technological development, prototyping and from our point of view evaluation (e.g. Kuntze *et al*, ISBS2009). Sprinting was selected for its clear objective performance measure (time). Sprinting also presents a set of technological challenges including the need to adopt high sampling rates whilst ensuring that athlete worn sensors are small enough not to alter an athlete's running action. Speed is the ultimate goal in sprinting and so an obvious device to deploy would be a 'speedometer'. Remote and accurate measurement of velocity turns out to be particularly difficult to achieve. Laser 'guns' (e.g. Opti-Logic™, LaserTech™) have been used to track athlete's speed in training and occasionally in competition – e.g. the 1997 World Athletics Championships, (Müller & Hommel, 1997). This technology requires a dedicated operator per athlete. The vision within the current project is to produce a turnkey system where a coach flicks a switch on entering the training arena and the technology is 'live' and ready for use. This vision is still someway off, but current examples from the project will illustrate progress.

EXAMPLES: Four examples will highlight different challenges and outcomes. Wireless technology was central to this research project, and so sensors were key components. The problem with any athlete-worn sensor is that it has volume and mass, requires a battery and is generally not popular amongst athletes. Two such systems will be used to illustrate that progress has been made in addressing these issues. By extension, the athlete needs to become part of a ubiquitous computing network, within a training environment – in this case the National Indoor Athletics Centre (NIAC) in Cardiff. Two approaches, where wireless technology has been employed but without athlete-worn sensors, will be highlighted. The four examples comprise insole pressure measurements; a multi-sensor integrated system; automatic video tracking and a multi-lane light gate solution.

CONCLUSION: This paper outlines how wireless technology is addressing problems associated with collecting biomechanical data in a training environment to aid understanding, and enhance feedback with the ultimate goal of improving sporting performance.

REFERENCES:

Müller, H. and H. Hommel (1997). "Biomechanical Research Project at the VIth World Championships in Athletics, Athens, 1997: Preliminary Report." *New Studies in Athletics* **12**(2-3): 43-73.

Acknowledgement

[1] The author acknowledges the support of the Engineering and Physical Sciences Research Council, UK for financial support (Grant No. EP/D076943) and all members of the SESAME project team, The SEnsing for Sports And Managed Exercise, http://www.sesame.ucl.ac.uk

GAIT BIOMECHANICS AND TIBIAL STRESS FRACTURE IN RUNNERS

Clare E Milner

Department of Exercise, Sport, & Leisure Studies, University of Tennessee, Knoxville TN, USA

Many risk factors for tibial stress fracture in runners have been proposed. This presentation will focus on biomechanical differences which remain when factors such as age, height, weight, sex, footstrike pattern and running mileage were matched in runners with previous tibial stress fracture and healthy control runners without a history of bony injuries. Differences found in running biomechanics included both loading-related and kinematic variables.

KEY WORDS: tibial shock, females, injury, kinematics, ground reaction forces,.

INTRODUCTION: Stress fractures are a common injury in athletes in general, and runners in particular. Stress fractures are a chronic, or overuse, injury resulting from fatigue damage to the bone. They occur when the damage accumulated due to the repeated application of physiological loads exceeds the capacity of the bony tissue to repair itself. The incidence of stress fractures in athletic populations is up to 50% (Brukner et al., 1996). Tibial stress fractures in particular are very common in recreational and competitive runners and military recruits. The tibia is the most common site of stress fracture in distance runners, accounting over 40% of all stress fractures (Brukner et al., 1996). The typical recovery time from a stress fracture is between 6 and 12 weeks. This includes periods of rest and reduced activity, to allow the natural reparative process of bone to take place at a rate exceeding that of damage accumulation. Reduced training capacity for a period of two to three months is a significant amount of time for both runners and military recruits. Therefore, identification of injury mechanisms and the prevention of stress fractures in runners is an important area of study.

RISK FACTORS: Many risk factors for stress fracture have been proposed. Some of these factors are intrinsic to the individual athlete; some are extrinsic and related to environmental factors. Many risk factors can be modified, whereas others can only be accommodated. Proposed extrinsic factors include those related to training, including the volume per session and per week, intensity level, running surface, and recent changes in the program (Bennell & Brukner, 2005). These factors can be modified, although other considerations such as competition scheduling or basic training requirements may influence this. Proposed intrinsic factors include anatomical structure of the lower extremity, muscle strength, flexibility, menstrual status, bone density, diet and nutrition, and running biomechanics (Bennell & Brukner, 2005). In order to investigate contributions to stress fracture risk from biomechanical factors, other risk factors should be standardized as much as possible between comparison groups.

GAIT BIOMECHANICS: A series of investigations into the biomechanics of running in relation to tibial stress fracture have been conducted. The aim of these studies was to determine objectively whether running biomechanics are different in those who have sustained a tibial stress fracture compared to those with no previous lower extremity bony injuries. This may enable identification of those at increased biomechanical risk of tibial stress fracture and of potential strategies to reduce their risk by modification of these factors. In cross-sectional studies the influence of potential confounding factors can be reduced by matching characteristics of the participants across comparison groups. Factors including age, height, weight, sex, footstrike pattern, and monthly running mileage can be matched to reduce the variability between groups due to these potentially confounding factors.

Given that tibial stress fracture is an overuse injury due to fatigue of bone tissue, early work in this area focused on loading characteristics of gait, primarily ground reaction forces.

Studies focusing on peak ground reaction forces were inconclusive, with some indicating greater peak values after tibial stress fracture compared to controls (Grimston et al., 1991) and others finding no difference (Crossley et al., 1999; Bennell et al., 2004). However, greater loading rates were found in runners with a previous tibial stress fracture compared to matched control runners (average vertical loading rate 79.0BW/s vs. 66.3BW/s, $P = 0.041$; instantaneous loading rate 92.6BW/s vs. 79.6BW/s, $P = 0.036$; Milner et al., 2006a). Additionally, a more direct measurement of tibial loading, tibial shock (peak acceleration) measured using an accelerometer, also supported the hypothesis of increased loading of the tibia in those susceptible to tibial stress fracture compared to controls (7.7g vs. 5.8g, $P = 0.014$). However, further analysis indicated that tibial shock explained only 17% of the variance between tibial stress fracture and control groups. This finding supported the hypothesis that the risk of tibial stress fracture is multifactoral, and that other biomechanical factors are likely involved.

FREE MOMENT: During running, the tibia is exposed simultaneously to a combination of shearing, bending and torsional loads, in addition to compression (Ekenman et al., 1998). While vertical ground reaction forces and tibial shock may provide an indication of the compressive load applied to the tibia, they do not indicate torque about the vertical axis. The free moment of ground reaction force indicates this torque at the point of contact of the runner with the ground (Milner et al., 2006b). Thus, it may provide an indirect measure of the torque acting on the tibia. Comparison of the magnitude of peak free moment during the stance phase of running between those with previous tibial stress fracture and matched controls indicated greater values in the stress fracture group (9.3×10^{-3} vs. 5.9×10^{-3} Nm/BW*ht, $P < 0.001$). Further analysis suggested that this variable explained 27% of the variance between the two groups. It's important to note that the majority of runners exhibit an adduction bias in the direction of free moment during the stance phase of running (i.e. resisting toe out torque of the foot on the ground). This torque has been associated with pronation (eversion) in the literature (Holden and Cavanagh, 1991). These loading-related variables that exhibit differences during running between runners with a previous tibial stress fracture and matched controls occur in the earlier part of stance phase. During early stance, body weight is shifted rapidly onto the stance limb. Therefore, studying lower extremity biomechanics during this period of rapidly increasing loading may be critical in understanding differences between runners susceptible to tibial stress fracture and healthy controls.

INITIAL LOADING: Higher vertical ground reaction force loading rates and higher tibial shock have been found in runners with previous tibial stress fracture compared to matched controls. The effect of external loading on the body can be modulated by body's the response to it. A good example is jumping off a wall onto the ground: landing in a stiff posture with knees maintained in extension results in higher loads transmitted through the body than landing with a large range of flexion motion at the lower extremity joints. In this respect, the knee is often considered to be of primary importance as a damper during landing. During running, each stance phase can be considered a single-legged landing, since the runner moves from a flight phase to single limb support. Thus, the initial loading part of the stance phase, from foot contact to the impact peak of vertical ground reaction force, may be important in terms of factors relating to tibial stress fracture. Vertical loading rate, which is calculated over the initial loading period, is greater in runners with a previous tibial stress fracture compared to controls (Milner et al., 2006a). Therefore, the body's response, in particular at the knee, to the high rate of loading during early stance may be an important consideration in injury risk. Knee joint stiffness was indeed found to be greater in runners with previous tibial stress fracture compared to controls (0.044Nm/(mass*ht) vs. 0.030 Nm/(mass*ht), $P = 0.015$; Milner et al., 2007). It was also positively correlated with tibial shock. However, there was only a moderate relationship in the stress fracture group between knee stiffness and tibial shock ($r = 0.406$). In the control group, the relationship was weak ($r = 0.161$). Thus, the body's response to loading during the early part of stance phase may also be important in understanding the complex relationship between loading and the occurrence of tibial stress fracture.

PROXIMAL AND DISTAL FACTORS: While several studies have focused on ground reaction forces and tibial shock measures in relation to tibial stress fracture, it should be remembered that the lower extremity is a linked chain with several joints and segments. The position of each lower extremity joint is important in determining the position of each segment. While static alignment factors in general have not shown strong association with tibial stress fracture, there is some evidence that extremes of foot type (very high or very low arches) may increase the risk of tibial stress injuries (Barnes et al., 2008). It has been suggested that dynamic alignment during the stance phase of running may be important in relation to stress fracture (Bennell & Brukner, 2005). Abnormal joint kinematics within the lower extremity chain may contribute to abnormal distribution of musculoskeletal loads, including within the tibia. Both proximal and distal joint kinematics may contribute to the combination of factors predisposing some runners to tibial stress fracture, even in the presence of normal loads. Altered frontal and transverse plane joint positions may change the axial, bending and torsional loads in the tibia. Several important differences were found in a comparison of frontal and transverse plane kinematics at the hip, knee, and ankle in female runners (Milner et al., in review). In particular, peak rearfoot eversion (11.7° vs. 9.0°, $P = 0.015$) and peak hip adduction (11.6° vs. 8.1°, $P = 0.004$) were both several degrees greater in runners with previous tibial stress fracture compared to controls with no previous bony injuries. This ties in with other work that found peak hip adduction, peak rearfoot eversion and the absolute free moment were the most important predictors of previous tibial stress fracture in distance runners (Pohl et al., 2008). Currently, it cannot be determined whether the differences in the frontal plane at the hip are a proximal compensation for the distal differences in the frontal plane at the rearfoot or vice versa.

DISCUSSION: While many factors, both internal and external to the runner, likely play a role in the development of tibial stress fractures, several biomechanical variables have been associated with this injury. It should be noted that the studies reported were retrospective and cross-sectional in design. Therefore, it cannot be determined whether the biomechanics of runners with a previous stress fracture measured after recovery from the injury are the same as prior to the stress fracture. While this is a limitation in relation to predisposing factors for tibial stress fracture, a large proportion of runners suffer multiple stress fractures after the initial occurrence. Thus, the information obtained in these studies is directly applicable to the case of recurring stress fractures. Future prospective studies may be able to confirm whether the high risk biomechanical features of running were also present in runners with tibial stress fracture prior to their injury. Primarily, several loading-related variables and joint angles have been identified and found to be greater in runners with a previous tibial stress fracture compared to runners with no previous bony injury. In terms of loading, the literature is somewhat inconclusive with regard to ground reaction force variables. While it seems intuitive that bony injury and fatigue fracture (i.e. stress fracture) are associated with damaging loads that exceed the body's ability to repair itself, tibial loading is only indirectly linked to ground reaction force variables. However, it appears that vertical loading rates (Milner et al., 2006a) and the absolute free moment (Milner et al., 2006b) may be increased in runners with previous tibial stress fracture. These variables may be providing an indication of the magnitude of compression and torsional loads that the tibia is subjected to during the stance phase of running. Given the lower magnitudes of these variables in healthy runners compared to those with a previous stress fracture, these aspects of running biomechanics may be modifiable. Decreasing their magnitude may decrease the risk of injury in susceptible runners. Similarly, the kinematic differences observed and abnormal peak angles reported at the hip and rearfoot in runners with previous tibial stress fracture (Milner et al., in review) may also be modifiable. Although a detailed consideration of interventions to modify running biomechanics is beyond the scope of this paper, several options may be considered. These may be mechanical or functional interventions. Potential mechanical interventions include orthotics or specialized footwear, or changing the running surface. Possible functional interventions include various types of instruction to retrain gait.

CONCLUSION: While acknowledging the limitations of retrospective cross-sectional studies, several biomechanical variables have been identified as having greater magnitude during running in those with previous tibial stress fracture compared to controls. Loading-related variables include vertical ground reaction force loading rates, the magnitude of peak free moment, and peak tibial shock. Kinematic variables include peak knee flexion stiffness during initial loading, peak rearfoot eversion and peak hip adduction.

REFERENCES:

Barnes, A., Wheat, J. & Milner, C. (2008). Association between foot type and tibial stress injuries: a systematic review. *British Journal of Sports Medicine*, 42, 93-98.

Bennell, K., Crossley, K., Jayarajan, J., Walton, E., Warden S., Kiss, Z.S. & Wrigley, T. (2004). Ground reaction forces and bone parameters in females with tibial stress fracture. *Medicine and Science in Sports and Exercise*, 36, 397-404.

Bennell, K. & Brukner. P. (2005). Preventing and managing stress fractures in athletes. *Physical Therapy in Sport*, 6, 171-80.

Brukner, P., Bradshaw C., Khan, K., White, S., Crossley, K. (1996). Stress fractures: A review of 180 cases. *Clinical Journal of Sports Medicine*, 6, 85-9.

Crossley, K., Bennell, K.L., Wrigely, T., & Oakes, B.W. (1999). Ground reaction forces, bone charactceristics, and tibial stress fracture in male runners. *Medicine and Science in Sports and Exercise*, 31, 1088-93.

Ekenman, I., Halvorsen, K., Westblad, P., Fellander-Tsai, L. & Rolf, C. (1998). Local bone deformation at two predominant sites for stress fractures of the tibia: An in vivo study. *Foot and Ankle International*, 19, 479-84.

Grimston, S.K., Engsberg, J.R., Kloiber, R. & Hanley, D.A. (1991). Bone mass, external loads and stress fractures in female runners. *International Journal of Sports Biomechanics*, 7, 292-302.

Holden, J.P. & Cavanagh, P.R. (1991). The free moment of ground reaction in distance running and its changes with pronation. *Journal of Biomechanics*, 24, 887-97.

Milner, C.E., Ferber, R., Pollard, C.D., Hamill, J. & Davis, I.S. (2006a). Biomechanical factors associated with tibial stress fracture in female runners. *Medicine and Science in Sports and Exercise*, 38, 323-328.

Milner, C.E., Davis, I.S. & Hamill, J. (2006b). Free moment as a predictor of tibial stress fracture in distance runners. *Journal of Biomechanics*, 39, 2819-25.

Milner, C.E., Hamill, J. & Davis, I. (2007). Are knee mechanics during early stance related to tibial stress fractures in runners? *Clinical Biomechanics*, 22, 697-703.

Milner C.E., Hamill J. & Davis I.S. (in review) Abnormal hip and rearfoot kinematics in runners with tibial stress fracture *Journal of Orthopaedic and Sports Physical Therapy*.

Pohl, M.B., Mullineaux, D., Milner, C.E., Hamill, J. & Davis, I.S. (2008). Biomechanics predictors of retrospective tibial stress fractures in runners. *Journal of Biomechanics*, 41, 1160-65.

Acknowledgement:

This work was supported by US Department of Defense Grant #DAMD17-00-1-0515 (PI: Dr Irene Davis).

FUNCTIONAL ROLE OF PROPRIOCEPTIVE FEEDBACK IN BALANCE AND IN REACTIVE MOVEMENT

Albert Gollhofer

Department of Sport Science; University of Freiburg; Schwarzwaldstr. 175; D-79117 Freiburg; Germany

Human movement is generated either by internal muscle forces, or by external forces that are attached to the body system. Most of our muscles act via lever arms on the bone system, thus generating rotational forces that produce consequently torques and joint moments. Therefore, studies dealing with control strategies of joint moments to achieve a desired movement or an intended task are addressing one of the most interesting topics.

In several papers the constraints and relative importance of sensory feedback are investigated. It seems that in a given task a complex interaction of feed-forward- and feed-back-mechanisms adjust the actual joint stiffness. By means of the H-reflex methodology, the spinal excitability for muscles can be determined. Under selective conditions, the inhibitory or facilitatory behavior of spinal reflex contribution can be investigated. Quite recently, the transcranial magnetic stimulation (TMS) has been developed to assess corticospinal excitability during human movement. Selective stimulation of the neurons in the motor cortex allows the determination of the relative contribution of corticospinal activation during movement. Application of both techniques, H-reflex and TMS, allows differentiation of spinally and centrally organized muscle activation.

The present paper highlights recent findings about neuromuscular control in balance and stretch-shortening cycle movements and reflects adaptations induced by balance training.

For both type of movements, the stiffness properties of the involved joint complexes are modulated by spinal and central modulation of the neuromuscular activation. Training adjusts/adapts this motor control specifically for balance tasks and for reactive movements.

Longitudinal training studies in which postural control (balance training) was exercised showed that the spinal and the cortical contributions were reduced after the training. Thus it was assumed that motor control was shifted towards supraspinal centers (Taube et al. 2007).

From stretch shortening cycle (SSC) it is known that high muscular stiffness is a prerequisite to enable proper performances and that feed-forward activation of the extensor muscles prior to ground contact is modulated by effective stretch reflex contributions (feed-back activation). This modulation, however, is largely dependent on the individual stretch load tolerance of the neuromuscular system.

Recent results indicate that the "stereotyped" reflexes are much more modulated than expected previously. It has been shown that modulation of spinal circuitry is achieved by presynaptic inhibition.

BALANCE CONTROL: Human posture or balance control and motion are controlled by a complex interaction of centrally and peripherally organized neuronal networks. Task specific voluntary movement is permanently under the influence of information feedback from various sources of proprioceptive receptors. This control system is highly effective if the feedback is organized "in real-time" and even more effective if feed-forward mechanisms are anticipating the motor requirements. From mechanoreceptors in the fingertips it is known, that for a precision grip the actual forces are slightly higher than necessary to hold the object

(Johansson and Westling 1984, 1987). Further, it is well known (Eliasson et al 1995) that disturbances of load will result in compensatory forces occurring with a latency of 40 ms. From a biomechanical point of view, balance for example is characterized by changes of the centre of pressure (COP) with respect to the actual projection of the centre of mass (COM) to the supporting area during a distinct motor task (Winter 1996). The ability of the nervous system to detect joint positions, movement directions, and force applications is mainly processed by proprioceptive afferents. In addition, precise information is necessary to balance gravitational forces. From postural control and balance it is known, that reflexes largely contribute to keep the COP within the constraints of the supporting area. It has been shown that balancing on a narrow beam reduces the size of the spinal excitability compared to normal walking (Llewellyn et al. 1990). Similarly, spinal excitability is reduced during stance with eyes closed as compared to balancing with open eyes (Earles et al. 2000; Hoffmann and Koceja 1995).

On the basis of altered reflex excitabilities in different motor conditions the challenging question arises: To what extend is the proprioceptive input qualifying the balance capability and what other sources of the neurophysiological system are controlling them?

ADAPTABILITY OF SPINAL REFLEXES: In recent balance training studies, exercising on unstable ground (wobbling boards, discs, soft mats etc.) led to a suppression of the H-reflex (Fig. 1) (Gruber et al. 2007; Taube et al. 2007a, 2007b).

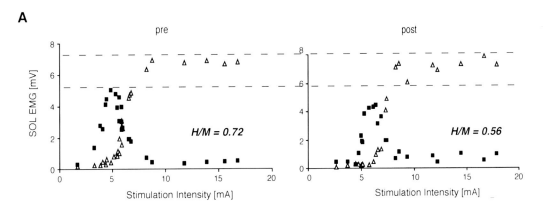

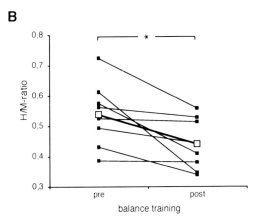

Figure 1: Adaptation of the H-reflex with balance training. H-reflex recruitment curves and resulting H_{max}/M_{max}-ratios recorded during perturbation on the treadmill. In A, H-reflex recruitment curves (■ H-reflexes, △ M-waves) from one subject before (pre) and after balance training (post) illustrate the reduction in H_{max}/M_{max}-ratio. B, H_{max}/M_{max}-ratios from each single subject (■) as well as the mean of all subjects (□) are displayed ($p < 0.05$ *). (Modified from Taube et al., 2007b).

This may imply that balance training modifies spinal reflex circuits and may lead in general to reduced H-reflexes in subjects who face high postural demands over a long time. The reduced H-reflex excitability of ballet dancers might then be explained in this way (Nielsen and Kagamihara 1993). It has been speculated that these reductions are mediated by presynaptic inhibition of the alpha-motoneurons. From postural control studies (Katz et al. 1988; Koceja and Mynark 2000; Mynark et al. 1997) it has been consistently shown that presynaptic modulation of the Ia-afferents by centrally pathways is a major mechanism to modulate spinal excitability.

Recently the transcranial magnetic stimulation technique has been developed for application in human motor experiments. By means of rapid magnetic stimulation the neurons in the motor cortex can be artificially activated and, based on elaborated electrophysiological techniques, the relative contribution of the fastest corticospinal projections from the motor cortex down to the spinal motoneurons can be assessed (Taube et al. 2007a). These authors investigated these pathways before and after balance training, besides a reduced spinal excitability they also revealed reduced cortical influence after balance training.

In a functional interpretation they stated that reduced spinal excitability plus reduced corticospinal influence me be explained by a gradual shift from spinal and central control before balance training to an increased control by subcortical centres. Functionally such shifts would reduce reflex generated oscillations and, concomitantly, deliberate the motor cortex for other tasks.

STRETCH SHORTENING CYCLE (SSC): From studies investigating the reactive movement capacity it is known that proprioceptive reflexes are basically responsible to explain supramaximal muscular forces during stretch-shortening cycle (SSC) contractions. Especially the immediate transfer from the preactivated and eccentrically stretched muscle-tendon complex to the concentric push-off determines the efficacy of motor output in the SSC. Analysis of EMG-profiles of the leg extensor muscles during hopping, jumping or running revealed that the reflex contribution appears interindividually rather consistently with a latency of 30 – 40 ms after landing on force platforms (Figure 2).

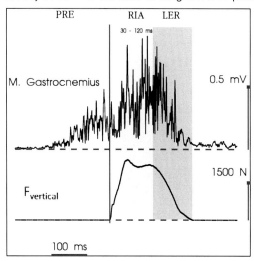

Figure 2: Schematic representation of an averaged (n=10) activation pattern of M. Gastrocnemius and vertical ground reaction force record. Drop Jumps were performed from 32 cm falling height. Preactivation, phases for reflex induced (RIA) as well as late emg responses (LER) are inserted.

On the basis of the concept that active stretch reflexes are essential for an effective reactive movement capability the challenging question arises: How is the interaction of feedforward organized activation of the muscles to stiffen the joints prior to landing related to the feedback organized reflex contributions and what other sources of the neurophysiological system are controlling them?

In order to sample closer details about the neuronal control during SSC we investigated spinal excitability by peripheral nerve stimulation and cortical excitability by application of TMS (Taube et al. 2008). In these experiments the H-reflexes were applied at distinct phases before and after the ground contact in order to assess the relative distribution of the spinal excitability for the reflex induced activation period. It could be demonstrated that spinal excitability is high at the early phases of ground contact coinciding with the latencies of the short latency, basically monosynaptic component of the stretch reflex contributions. However, as time of ground contact progresses, this excitability was systematically reduced. Conversely, the MEPs elicited by the TMS to study the central excitability was low at the early contact phases and increased towards later contact times (Figure 3).

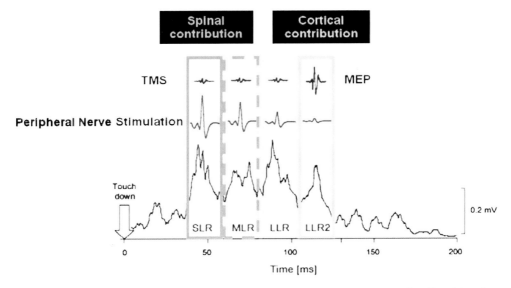

Figure 3: Reflex pattern of M. Soleus represents various components of reflex latencies following touchdown (Short- (SLR); Medium- (MLR) and Long-latency component (LLR; LLR2). Upon each component the relative size of H-Reflex, determined by peripheral nerve stiumaltion and of MEP, determined by transcranial stimulation (TMS) is inserted. Whereas the H-reflex size is decreasing with longer latencies, the amplitudes of the MEPs are increasing. Consequently, this has been interpreted as a reduced spinal excitability in favor of increased cortical control with longer latencies.

Functionally these observations led to the conclusion that the early phases after ground contact, characterized by the short latency reflex activity, are under spinal control. However, as ground contact time proceeds, the later components in the EMG of the leg extensors, characterized by the medium and long latency reflex activity, a more and more subcortical and finally cortical control must be assumed.

In summary, neuronal control of reactive movements in the SSC represent an activation profile that is controlled by different neuronal circuits. On the basis of recent electrophysiological methodological tools the classical distribution in preactivation, reflex and voluntary induced activation is preserved, but a fine-structured picture can be drawn: Especially during the period of reflex induced activity (30 – 120 ms after ground contact) the cortical control is progressively increased. This may be interpreted as an integration of feedback-mechanisms (stretch reflex activation) into the feed-forward-control by centrally organized the motor programs.

CONCLUSION: Proprioceptive feedback is essential not only for transmitting force, position and movement senses under human movement but also to assist neural activation. In balance, these contributions are suppressed as training proceeds. Concomitant to the reduced cortical control a gradual shift towards subcortical control during balance may be assumed. Moreover, in situations when strong activation is desired to achieve high muscle stiffness (SSC) the feed-forward control, already initiated by preactivated extensors, is substantially supported by segmental stretch reflexes. However, these reflex contributions are controlled for the short- and medium-latency component by spinal mechanisms. The later responses have been shown to be under subcortical or even cortical control.

REFERENCES:

Earles DR, Koceja DM and Shively CW. Environmental changes in soleus H-reflex excitability in young and elderly subjects. *Int J Neurosci* 105: 1-13, 2000.

Eliasson AC, Forssberg H, Ikuta K, Apel K, Westling G and Johannsson R. Development of human precision grip. V. Anticipatory and triggered grip actions during sudden loading. Exp. Brain Res. 106, 425- 433, 1995.

Gruber M, Taube W, Gollhofer A, Beck S, Amtage F and Schubert M. Training-specific adaptations of H- and stretch reflexes in human soleus muscle. *J Mot Behav* 39: 68-78, 2007.

Hoffman MA and Koceja DM. The effects of vision and task complexity on Hoffmann reflex gain. *Brain Res* 700: 303-307, 1995.

Johansson RS and Westling G. Roles of glabrous skin receptors and sensimotor memory in automatic control of precision grip when lifting rougher or more slippery objects. Exp. Brain Res. 56, 550-564, 1984.

Johansson RS and Westling G. Signals in tactile afferents from the fingers eliciting adaptive motor responses during precision grip. Exp. Brain Res. 66, 141-154, 1987.

Llewellyn M, Yang JF and Prochazka A. Human H-reflexes are smaller in difficult beam walking than in normal treadmill walking. *Exp Brain Res* 83: 22-28, 1990.

Koceja DM, Markus CA and Trimble MH. Postural modulation of the soleus H reflex in young and old subjects. *Electroencephalogr Clin Neurophysiol* 97: 387-393, 1995.

Koceja DM and Mynark RG. Comparison of heteronymous monosynaptic Ia facilitation in young and elderly subjects in supine and standing positions. *Int J Neurosci* 103: 1-17, 2000.

Mynark RG, Koceja DM and Lewis CA. Heteronymous monosynaptic Ia facilitation from supine to standing and its relationship to the soleus H-reflex. *Int J Neurosci* 92: 171-186, 1997.

Nielsen J, Crone C and Hultborn H. H-reflexes are smaller in dancers from The Royal Danish Ballet than in well-trained athletes. *Eur J Appl Physiol Occup Physiol* 66: 116-121, 1993.

Schubert M, Beck S, Taube W, Amtage F, Faist M and Gruber M. Balance training and ballistic strength training are associated with task-specific cortico-spinal adaptations. *Eur J Neurosci* 27, 2007 – 2018, 2008.

Taube W, Gruber M, Beck S, Faist M, Gollhofer A and Schubert M. Cortical and spinal adaptations induced by balance training: correlation between stance stability and corticospinal activation. *Acta Physiol (Oxf)* 189: 347-358, 2007a.

Taube W, Kullmann N, Leukel C, Kurz O, Amtage F and Gollhofer A. Differential Reflex Adaptations Following Sensorimotor and Strength Training in Young Elite Athletes. *Int J Sports Med* 2007b.

Taube W, Leukel C, Schubert M, Gruber M, Rantalainen T and Gollhofer A. Differential modulation of spinal and corticospinal excitability during drop jumps. *J Neurophysiol* 99: 1243-1252, 2008.

Trimble MH and Koceja DM. Effect of a reduced base of support in standing and balance training on the soleus H-reflex. *Int J Neurosci* 106: 1-20, 2001.

Winter DA. Anatomy, Biomechanics and Control of balance during standing and walking, Waterloo Biomechanics. 1996.

KINEMATIC ANALYSIS OF SPORTS MOVEMENTS: GOLF SWING PLANE ANALYSIS

Young-Hoo Kwon and Jeff Casebolt

Biomechanics Laboratory, Texas Woman's University, Denton, Texas, USA

The swing plane is one of the most controversial and misleading concepts in modern golf vocabulary. Several popular swing theories in regards to the swing plane have emerged in the popular literature by golf professionals, while the majority of the biomechanical studies have been conducted based on the planar double- and triple-pendulum swing models. The purpose of this paper is to provide a comprehensive review of both the scientific and popular golf literature on golf swing mechanics in regards to the concept of swing plane.

KEY WORDS: golf swing mechanics, on-plane swing, one-plane swing, double-pendulum

INTRODUCTION: Kinematics is an area of biomechanics dealing with measurement and description of the human body motion. Quantitative measurement of the motion of interest and subsequent analysis based on the computed kinematic quantities allow investigators an in-depth understanding of the motion itself and the common motion patterns. The human body is a mechanical system with a large number of degrees of freedom and isolating a set of key performance characteristics/components is of crucial importance for effective performance enhancement in complex 3-dimensional (3-D) body motions such as the golf swing.

Golf is one of the most popular sports in the modern world with 35 million participants worldwide (Geisler, 2001; Theriault & Lachance, 1998). The sole objective in a golf competition is to minimize the total number of shots taken to finish an 18-hole course using a variety of clubs and shots. The two most important elements of the performance in golf are accuracy (direction and distance) and consistency and one must develop a consistent fundamental swing pattern to secure these qualities. The direction of a shot and the ball carry distance are essentially determined by the clubhead trajectory around the impact position and the impact conditions such as the clubhead velocity, clubface orientation, impact location on the clubface, coefficient of restitution, and the effective mass involved in the impact.

The 'swing plane', which affects the impact conditions directly, is one of the most frequently used terms in golf coaching lately and is also one of the most controversial and misleading concepts. Since Hogan and Wind (1957) used this term in their book titled "Ben Hogan's five lessons: the modern fundamentals of golf", different swing theories have emerged in the popular literature (e.g. Haney & Huggan, 1999; Hardy & Andrisani, 2005). None of these, however, has truly grasped the essence of the swing plane due to the lack of understanding of the complex nature of the actual 3-D swing motion. Moreover, for last four decades, the majority of biomechanical studies on swing mechanics have been conducted based on the planar double-pendulum model (e.g. Budney & Bellow, 1979; Milburn, 1982; Milne & Davis, 1992; Pickering & Vickers, 1999; Sanders & Owens, 1992), originally proposed by Cochran and Stobbs (1968), or the triple-pendulum model (Sprigings & Mackenzie, 2002; Sprigings & Neal, 2000), a variation of the double-pendulum model. Although Vaughan (1981) and Neal and Wilson (1985) pointed out that the swing plane was not planar, it is only recently that scientists have critically investigated the swing plane (Coleman & Anderson, 2007; Coleman & Rankin, 2005; Nesbit, 2005; Shin, Casebolt, Lambert, Kim, & Kwon, 2008). The purpose of this paper is to provide a comprehensive review of both the scientific and the popular golf literature on golf swing mechanics in regards to the concept of swing plane.

MULTIPLE PENDULUM MODELS: Since Cochran and Stobbs (1968) proposed the planar double-pendulum model, it has been used as the fundamental swing model in numerous biomechanical studies (e.g. Budney & Bellow, 1979; Milburn, 1982; Milne & Davis, 1992; Pickering & Vickers, 1999; Sanders & Owens, 1992; Sprigings & Mackenzie, 2002; Sprigings

& Neal, 2000). The key concepts of the model include: (a) the golfer is considered as a system of two levers hinged in the middle; (b) the upper lever corresponds to the golfer's shoulders and the arms and the club forms the lower lever; (c) the levers are hinged at the wrists/hands; (d) the system is swung around a fixed point (hub) in a single inclined plane. The model has been particularly popular in the modelling and simulation studies due to its simplicity in deriving the equations of motion.

Figure 1: Trajectories (thin black lines) of the clubhead, left hand, left shoulder, right elbow and right shoulder of a scratch golfer during the downswing (driver shot). The body shows the impact position.

Although some modifications have been introduced, such as flexible clubshaft (Milne & Davis, 1992), triple pendulum (Campbell & Reid, 1985; Sprigings & Mackenzie, 2002; Sprigings & Neal, 2000), and moving hub (Jorgensen, 1994), the essence of the model, planar swing, has been remained intact. The assumption of planar swing, however, has never been substantiated in these studies other than subjective observations. In fact, a 3-D analysis of a golf swing at any skill level, regardless of the club used, readily reveals that the downswing is not planar (Figure 1). Trajectories of the major moving parts exhibit complex movement patterns and, in particular, the shoulders show motions on completely different planes and a significant amount of off-plane shoulder girdle motion is observed as a result. A single planar swing plane simply does not exist and the assumption of planar swing is not valid.

SWING PLANE STUDIES: A detailed, systematic analysis of the swing plane is possible only in a 3-D study. Vaughan (1981) was the first investigator who reported that the motion plane of the club was not planar. Using the unit vector normal to the clubshaft plane, he visualized the continuous motion of the clubshaft plane during the downswing. Neal and Wilson (1985) also concluded that the motion of the club was not planar for any substantial period of time during the downswing but no quantitative analysis result of the club motion was provided.

Coleman and Rankin (2005) systematically evaluated the assumption of planar downswing by quantifying the orientation (inclination and direction) of the instantaneous left arm plane, formed by the left arm and the shoulder, and the deviation of the clubhead from the arm plane. It was shown that the left arm plane continuously changed its orientation during the downswing and the deviation of the clubhead from the left arm plane was inconsistent. Coleman and Anderson (2007) further investigated the validity of the assumption of planar swing by finding a single swing plane that best fitted to the clubshaft motion and the instantaneous clubshaft planes. The reported mean goodness of fit values (RMS residuals) of the single swing plane were fairly large (> 8 cm) for all clubs used (driver, 5-iron, and pitching wedge), suggesting the shaft motion was in fact not planar. The driver showed flatter swing plane orientation than the 5-iron and the pitching wedge.

Although it has been clearly shown in these 3-D studies that the downswing plane is not planar, further studies were necessary to understand the swing mechanics in depth. In a study utilizing a full-body multi-link 3-D model of golfer's body, Nesbit (2005) showed that the downswing did not take place in a fixed plane and there was significant pitch motion of the club during the downswing. He also quantified the angle between the planes traced out by the clubhead and the hands (9 to 12°). Further more, in a study using several different phases of the downswing/follow-through (top of the backswing to mid follow-through, vertical shaft to mid follow-through, and horizontal shaft to mid follow-through), Shin et al. (2008) demonstrated that a well-defined "clubhead" swing plane could be obtained from the impact portion of the downswing (horizontal shaft to mid follow-through; mean RMS residual < 1.0 cm) and the swing plane changed continuously during the early phase of the downswing (top of the backswing to horizontal shaft). The slope and the direction angles of the driver, 5-iron, and pitching wedge shots were reported.

POPULAR SWING PLANE THEORIES: Hogan and Wind (1957) defined the backswing plane as "an angle of inclination running from the ball to the shoulders determined by the height of the golfer and the distance he stands from the ball at address" and visualized it as a large pane of glass that rests on the shoulders and inclines upward from the ball with the golfer's head sticking out through a hole. It was also noted that the downswing plane is less steeply inclined than the backswing plane and the swing plane points slightly to the right of the golfer's target as the body moves toward the target. In spite of the superb, detailed description of the downswing mechanics, Hogan's visualization of the swing plane often has caused misconceptions due to the image of the glass pane connecting the ball and the shoulder line and has been subject to criticism.

More recent swing theories that have gained popularity include Hank Haney's 'on-plane' swing and Jim Hardy's 'one-plane' swing (Haney & Huggan, 1999; Hardy & Andrisani, 2005). Maintaining the clubshaft parallel to the original shaft plane, formed by the shaft at the address position, throughout the entire backswing and downswing is the key concept of the on-plane swing. The one-plane swing theory, on the other hand, views the shoulder plane as the swing plane and emphasizes the importance of aligning the shoulder plane perpendicular to the spine while keeping the ball within the shoulder plane. Both theories are flawed mechanically and anatomically in several aspects: (a) the club motion during the downswing occurs in less than 250 ms and bringing the club back to the address position for impact by rotating the club on the shaft plane while translating it in the normal direction simultaneously is a poor conceptualization at best (on-plane swing); (b) the trunk shows a lateral flexion toward the target during the early phase of the downswing (Nesbit, 2005) and maintaining the shaft plane parallel to the original shaft plane regardless of the trunk motion means that the club and the arms move independently from the trunk motion during this phase (on-plane swing); (c) the shoulders exhibit complex 3-D motions during the downswing and a postural plane like the shoulder plane at the top of the backswing is meaningless in regards to the actual motion of the club and the arms during the downswing (one-plane swing); (d) the shoulder girdles provide additional mobility in the trunk and a pure rotation of the trunk and the shoulder about the spine axis without shoulder girdle motion substantially limits the mobility of the trunk (one-plane swing). These recent popular swing theories are in fact a setback from Hogan's original swing model and neither one has truly grasped the essence of the swing plane and mechanics during the downswing.

SUMMARY AND CONCLUSION: From the findings of the recent 3-D studies (Coleman & Anderson, 2007; Coleman & Rankin, 2005; Nesbit, 2005; Shin et al., 2008), it is evident that the downswing does not occur in a single plane and the major moving parts (the hands, elbows, shoulders, and the clubhead) exhibit complex movement patterns (Figure 1). In particular, the findings of Shin et al. (2008) provide several important implications: (a) since the impact portion (horizontal shaft to mid follow-through) is the most important component of the downswing as the motion of the clubhead during this phase directly affects the outcome (direction and distance), it is the swing plane obtained from this phase that truly characterizes a golfer's downswing; (b) as long as the impact portion of the downswing forms a well-defined single plane, it may not be so critical for the entire downswing to be planar; (c) it is of crucial importance to understand how the golfer's body and club move during the early phase of the downswing to secure a well-defined swing plane around the impact position.

In a recent review paper, Farrally et al. (2003) stated "although biomechanical analysis of the swing has attracted considerable research, it has yet to produce a convincing explanation of the physics involved that makes a significant advance on the landmark work of Cochran and Stobbs (1968)". It could be the simple nature of the planar double-pendulum model proposed by Cochran and Stobbs (1968) which actually hindered scientists to produce a "convincing explanation of the swing mechanics". Future studies on golf swing mechanics must be based on a 3-D swing model incorporating full-body, multi-link 3-D representation of golfer's body (Nesbit, 2005). The cocking/uncocking motion, the key wrist motion in the double- and triple-pendulum swing models, must be replaced with the anatomically correct wrist deviation (ulnar/radial) and forearm pronation/supination. The in-plane motion of the club during the impact portion of the downswing and the off-plane motion during the early downswing phase must also be incorporated into the 3-D swing model for an in-depth understanding of the swing mechanics.

REFERENCES:
Budney, D. R., & Bellow, D. G. (1979). Kinetic analysis of a golf swing. *Research Quarterly, 50*, 171-179.
Campbell, K. R., & Reid, R. E. (1985). The application of optimal control theory to simplified models of complex human motions: the golf swing. In Winter, D. A., Norman, R. W., Wells, R. P., Hayes, K. C. & Patla, A. E. (Eds.), *Biomechanics IX-B* (pp. 527-532). Champaign, IL: Human Kinetics Publishers.
Cochran, A., & Stobbs, J. (1968). *The search for the perfect swing.* Philadelphia, PA: J. B. Lippincott.
Coleman, S., & Anderson, D. (2007). An examination of the planar nature of golf club motion in the swings of experienced players. *Journal of Sports Sciences, 25*, 739-748.
Coleman, S. G. S., & Rankin, A. J. (2005). A three-dimensional examination of the planar nature of the golf swing. *Journal of Sports Sciences, 23*, 227-234.
Farrally, M. R., Cochran, A. J., Crews, D. J., Hurdzan, M. J., Price, R. J., Snow, J. T., et al. (2003). Golf science research at the beginning of the twenty-first century. *Journal of Sports Sciences, 21*, 753-765.
Geisler, P. R. (Ed.). (2001). *Golf.* New York: McGraw-Hill.
Haney, H., & Huggan, J. (1999). *The only golf lesson you'll ever need: easy solutions to problem golf swings.* New York, NY: HarperCollins.
Hardy, J., & Andrisani, J. (2005). *The plane truth fro golfers: breaking down teh one-plane swing and the two-plane swing anf finding the one that's right for you.* New York, NY: McGraw-Hill.
Hogan, B., & Wind, H. W. (1957). *Ben Hogan's five lessons: the modern fundamentals of golf.* New York: Simon & Schuster.
Jorgensen, T. (1994). *The physics of golf.* New York: American Institute of Physics Press.
Milburn, P. D. (1982). Summation of segmental velocities in the golf swing. *Med Sci Sports Exerc, 14*, 60-64.
Milne, R. D., & Davis, J. P. (1992). The role of the shaft in the golf swing. *Journal of Biomechanics, 25*, 975-983.
Neal, J. R., & Wilson, B. D. (1985). 3D kinematics and kinetics of the golf swing. *International Journal of Sport Biomechanics, 1*, 221-232.
Nesbit, S. M. (2005). A three dimensional kinematic and kinetic study of the golf swing. *Journal of Sports Science and Medicine, 4*, 499-519.
Pickering, W. M., & Vickers, G. T. (1999). On the double pendulum model of the golf swing. *Sports Engineering, 2*, 161-172.
Sanders, R. H., & Owens, P. C. (1992). Hub movement during the swing of elite and novice golfers. *International Journal of Sport Biomechanics, 8*, 320-330.
Shin, S., Casebolt, J. B., Lambert, C., Kim, J., & Kwon, Y.-H. (2008). A 3-D determination and analysis of the swing plane in golf. In Kwon, Y.-H., Shim, J., Shim, J. K. & Shin, I.-S. (Eds.), *Scientific Proceedings of the XXVIth International Conference on Biomechanics in Sports* (pp. 550). Seoul, Korea: Seoul National University.
Sprigings, E. J., & Mackenzie, S. J. (2002). Examining the delayed release in the golf swing using computer simulation. *Sports Engineering, 5*, 23-32.
Sprigings, E. J., & Neal, R. J. (2000). An insight into the importance of wrist torque in driving the golfball: a simulation study. *Journal of Applied Biomechanics, 16*, 356-366.
Theriault, G., & Lachance, P. (1998). Golf injuries: an overview. *Sports Medicine, 26*, 43-57.
Vaughan, C. L. (1981). A three-dimensional analysis of the forces and torques applied by a golfer during the downswing. In Morecki, A., Fidelus, K., Kedzior, K. & Witt, A. (Eds.), *Biomechanics VII-B.* Baltomore, MD: University Park Press.

THEORY INTO PRACTICE: THE KEY TO ACCEPTANCE OF SPORT BIOMECHANICS
Bruce Elliott
School of Sport Science, Exercise and Health
The University of Western Australia, Perth, Western Australia, Australia

'A legacy of a scientific approach to coaching excellence, with a particular love and talent for badminton coaching'

Acceptance by administrators, coaches and players alike requires sport science research to be valid, repeatable and to meet the perceived needs of the specific sport. David Waddell certainly worked towards these goals in badminton. In this paper I will provide a three stage approach (flow from basic to applied research followed then by research dissemination) that has been shown to be successful in tennis and cricket. Basic research is essential to establish testing and data analysis protocols that are valid, repeatable and answer questions of interest. Applied research then needs to be of sufficient depth to make a difference to coaching philosophy. Applied findings must then be interpreted for coaches in a meaningful manner. In this paper I will focus on the tennis serve to show the depth of understanding required to 'make a difference'. Finally a brief discussion will be included on areas of current research, within the framework of this topic.

KEY WORDS: Tennis, racket sports, applied biomechanics

Figure 1: High performance tennis service action

INTRODUCTION: In thinking about how to present a tribute to David Waddell, it may have seemed logical to review biomechanics research related to badminton. However, to be accepted by coaches, players and administrators requires a greater 'body of knowledge' than is currently available in badminton. For that reason I will focus this paper on the biomechanics of the tennis serve (Figure 1), so that young scientists can see, at least in my opinion, the research process that needs to be followed to 'make a difference'. This needs to happen in much the same way as clinical biomechanics has influenced so many aspects of medical and para-medical practice.

THE START POINT - BASIC RESEARCH
The following three basic research examples, while not exhaustive with reference to the racket sport literature, have in different ways enabled applied studies to be designed and conducted with far greater surety than would have been possible without their input.

- *Impact smoothing*: While racket velocities, pre- and post-impact, have been a problem for many decades, Knudson and Bahamonde (2001) brought this issue to prominence in tennis research. The question that needs to be addressed is; 'do you need to use a different 'data filtering' approach for situations where the oncoming ball has a large amount of momentum (forehand as studied by Knudson and Bahamonde, 2001), compared with the situation where the ball has minimal momentum (tennis serve).

The question is even more important in kicking studies where you have much longer contact periods between the ball and the swinging limb (~5 ms in tennis compared with ~15 ms in kicking). Currently linear and polynomial extrapolation conditions have been shown to produce accurate angular position and velocity estimates for tennis impacts.

- *Segment contributions to velocity generation*: A number of 2D papers have reported percentage contributions to pre-impact racket velocity (in the sagittal plane). I know I always wondered when one followed this procedure, why the wrist flexion was so dominant with reference to percentage contribution to service velocity. Sprigings et al. (1994), published an algorithm that addressed the percentage contribution of various segment movements to final racket velocity (including long axis rotation, such as shoulder internal rotation) This enabled Elliott et al. (1995) to apply this procedure to the service action of high performance players, and it all became clear - shoulder internal rotation occurred late in the swing to impact, at a similar time to wrist flexion (it did not follow the kinetic or kinematic chain). This movement rotation was also identified as a key factor in the development of racket speed.

- *Role of muscle pre-contraction in enhancing performance*: While papers by Walshe et al. (1998) go a long way to explaining the contributions of muscle-pre-stretch, athletes are typically only concerned with 'end results'. For that reason many sport scientists have combined the influences of muscle pre-stretch and elastic energy storage under the title 'contributions to performance following a pre-stretch'. While this is not theoretically correct it certainly permits athletes to focus on the influence of technique on performance outcome. This was the approach we took when investigating the role of 'pause' between back swing (external rotation at the shoulder) and forwardswing (internal rotation at the shoulder) phases of the tennis serve (Elliott et al., 1999)

THE NEXT STEP - APPLIED RESEARCH
The applied research studies reviewed represent a spectrum of research into different aspects of service technique.

- *Lower limb drive and trunk rotations:* Initial work on the forces with the court produced by different serving techniques (Elliott and Wood, 1983), when linked with a paper on trunk angular momentum during the high performance serve (Bahamonde, 2000) provide a clear theoretical construct for understanding and therefore teaching the 'undercarriage' for an effective serve. More recent work (Sweeney et al., in review) has taken this research further by demonstrating the importance of 'back-hip' elevation in this process.

- Velocity generation
    a. Upper limb movements: Research has clearly shown the importance of shoulder internal rotation and wrist flexion to the generation of racket speed in the service action (Elliott et al., 1995). Suddenly coaching and physical preparation of athletes could be structured to create optimal conditions to produce high levels of shoulder internal rotation.

b. Impact locations: Very practical research by Chow and colleagues (2003) at the Atlanta Olympics provided ball impact locations for professional players under tournament conditions. Impact for both male and female players was marginally to the left of the front toe for the power serve as shown in Figure 2. This information then linked the research on leg drive, hip and trunk rotations with impact location. This position also produced an almost ideal situation for shoulder internal rotation to play its role in the service action. The miss-alignment of the racket and the forearm (observe this angle in Figure 2) further demonstrates the ability to internally rotate at the shoulder, such that the racket is assisted in its forward movement through impact.

Figure 2: High performance tennis service action at impact

THE FINAL STEP - DISSEMINATION AND APPLICATION OF RESEARCH FINDINGS

This can obviously take a number of different approaches. In the early days (less academic pressure) I tried to write practical articles to complement papers published in the research literature. A typical flow of these papers is listed below.

- A 3-D kinematic method for determining the effectiveness of arm segment rotations in producing racquet-head speed. Journal of Biomechanics, 27(3): 245-254, 1994.

⇩

- Contributions of upper limb segment rotations during the power serve in tennis, Journal of Applied Biomechanics. 11 (4): 433-442, 1995.

⇩

- The super servers: Pete Sampras and Goran Ivanisevic. Australian Tennis, June: 46-47, 1996

While this is a far more difficult task in today's academic climate that demands research grants and high quality publications, it is important that we do not lose sight of the need to impart knowledge. Books, applied articles and presentations at Sport Specific Conferences all go a long way to enhancing your profile in the world of sport biomechanics. Where possible complement your research papers with review articles such as: *'Biomechanics of the serve in tennis. A biomechanical perspective. Sports Medicine, 6: 285-294, 1998'*, and books *'Biomechanics of Advanced Tennis. International Tennis Federation Press, London, 2003'* that permit you to disseminate your research findings to an even larger population (this book was also published into Spanish and French).

CURRENT RESEARCH

Our current work is similarly based around both basic and applied approaches.

- Basic Science - *3D modelling of the shoulder joint*: More recent work on the modelling process has attempted to better define the shoulder joint - the centre of rotation - to improve reporting upper limb angles during fast rotations, as in the tennis serve. An in-vivo approach using an MRI to establish the glenohumeral joint centre compared a number of commonly used protocols along with a new regression approach (Campbell et al. a, in press). This regression, combining: 3D distance between the *Clavicle, C7* an *Acromion lateral ridge* markers and *the virtual location* between C7 and the sternal notch, subject *height and mass*, was shown to better calculate the 'centre of rotation' during dynamic movements than previously established methods (Meskers et al.,

1998). This work further elucidated that a single centre of rotation can be assumed to occur throughout large ranging, high velocity upper limb movements, such as the tennis serve (Cambell et al., b, in press).

- Applied science - Development sequences in stroke production with a particular interest in differences between male and female players are also a current interest. When does service power become influenced by shoulder internal rotation and do contributions to racket velocity follow a linear increase with chronological age?

CONCLUDING REMARKS: When talking with young clinical biomechanists my advice is always to develop a research theme and build a research reputation in that area. You may take a neuromuscular approach; to working with various populations (e.g. cerebral palsy or children), or with specific areas of the body (e.g. the shoulder). In sport biomechanics the temptation has always been, at least for me, to allow my focus to become a little blurred. However, I have always tried to retain a focus on tennis and cricket research, either from a technique development or from an injury reduction perspective. You need a team to achieve the goals I have promoted in this paper, so do not be afraid to collaborate with your colleagues - I would encourage you to do so. I openly acknowledge the tremendous role that fellow colleagues and graduate students have played in my research endeavours.

REFERENCES:

Bahamonde, R. (2000). Changes in angular momentum during the tennis serve. Journal of Sports Sciences, 18: 579-592.
Campbell, A., Lloyd, D., Alderson, J., Elliott, B. (in press - a) MRI development and validation of a two new predictive methods of glenohumeral joint centre location identification and comparison with established techniques. Journal of Biomechanics.
Campbell, A., Lloyd, D., Alderson, J. & Elliott, B. (in press - b). Effects of different technical coordinate system definitions on the three dimensional representation of the glenohumeral joint centre. Medical and Biological Engineering and Computing.
Chow, J., Carlton, L., Lim, Y., Chae, W., Shim, J., Kuenster, A & Kokubun, K. (2003). Comparing the pre- and post-impact ball and racquet kinematics of elite tennis players' first and second serves: A preliminary study. Journal of Sports Sciences, 21: 529-537.
Elliott, B. (1996). The super severs: Pete Sampras and Goran Ivanisevic. Australian Tennis, June: 46-47.
Elliott, B. (1998). Biomechanics of the tennis serve: A biomechanical perspective. Sports Medicine, 6: 285-294.
Elliott, B. & Wood, G. (1983). The biomechanics of the foot-up and foot-back tennis service techniques. Australian Journal of Sports Sciences, 3:3-6.
Elliott, B., Marshall, R. & Noffal, G. (1995). Contributions of upper limb segment rotations during the power serve in tennis. Journal of Applied Biomechanics, 11: 433-442.
Elliott, B., Baxter, K. & Besier, T. (1999). Internal rotation of upper-arm segment during a stretch-shorten cycle movement. Journal of Applied Biomechanics, 15: 381-395.
Elliott, B., Reid, M. & Crespo, M. (Eds) Biomechanics of Advanced Tennis. International Tennis Federation Press, London, 2003. Knudson, D. & Bahamonde, R. (2001). Effect of endpoint conditions on position and velocity near impact in tennis. Journal of Sports Sciences, 19: 839-844.
Meskers, C. G., van der Helm, F. C., Rozendaal, L. A. and Rozing, P. M., 1998. In vivo estimation of the glenohumeral joint rotation center from scapular bony landmarks by linear regression. Journal of Biomechanics 31(1), 93-96.
Sprigings, E., Marshall, R., Elliott, B. & Jennings, L. (1994). A 3-D kinematic method for determining the effectiveness of arm segment rotations in producing racquet-head speed. Journal of Biomechanics, 27: 245-254, 1994.
Sweeney, M., Reid, M., Alderson, & Elliott, B. (submitted). Lower limb and trunk function in the high performance tennis serve. Journal of Sports Sciences.
Walshe, A. Wison, G. & Ettema, G. (1998). Stretch-shorten cycle compared with isometric preload: Contributors to enhanced muscular performance. Journal of Applied Physiology, 18: 97-106.

PLASTICITY OF HUMAN TENDON'S MECHANICAL PROPERTIES: EFFECTS ON SPORT PERFORMANCE

Arampatzis Adamantios

Department of Training and Movement Sciences, Humboldt-University of Berlin, Germany

INTRODUCTION: In the literature it is often mentioned, that the tendon is very relevant for the work producing capability of the muscle fibers and for the motion and the performance of the human body. During a given movement, strain energy can be stored in the tendon and this way the whole energy delivery of the muscle can be enhanced. Further, the higher elongation capability of the tendon with respect to the muscle fiber, allows a bigger change in length of the muscle-tendon unit. Therefore, the muscle fibers may work on a lower shortening velocity and as a consequence of the force-velocity relationship their force producing potential will be higher. Generally, the main functions of the tendon during locomotion are: (a) to transfer muscle forces to the skeleton (b) to store mechanical energy coming from the human body or/and from muscular work as strain energy and (c) to create favorable conditions for the muscle fibers to produce force as a result of the force-length-velocity relationship. A higher force potential of the muscle fibers due to the force-length-velocity relationship during submaximal contractions would decrease the volume of active muscle at a given force or a given rate of force generation and consequently would decrease the cost of force production. In the same manner during maximal muscle contractions (maximal activation level) the higher force potential of the muscle fibers will allow the muscles to exert higher forces. The reports about the influence of the non rigidity of the tendon on the effectivity of muscle force production reveal the expectation that sport performance during submaximal as well as maximal running intensities may be affected by the mechanical and morphological properties of the tendon.

In a series of experiments we examined the mechanical properties of the lower extremities muscle-tendon units (MTU) from athletes displaying different running economy and sprint performance. The most economical runners showed a higher contractile strength and a higher tendon stiffness in the triceps surae MTU and a higher compliance of the quadriceps tendon and aponeurosis at low level tendon forces (Arampatzis et al., 2006). The faster sprinters exhibited a higher elongation of the vastus lateralis (VL) tendon and aponeurosis at a given tendon force and a higher maximal elongation of the VL tendon and aponeurosis during the MVC (Stafilidis and Arampatzis, 2007). Furthermore, the maximal elongation of the VL tendon and aponeurosis showed a significant correlation with the 100 m sprint times ($r = -0.567$, $P = 0.003$). It has been supposed that, the more compliant quadriceps tendon and aponeurosis will increase the energy storage and return as well as the force potential of the muscle due to the force-velocity relationship. These studies provide evidence that the mechanical properties of the tendons at the lower extremity at least partially explain the performance of the human musculoskeletal system during running activities. However, until now no study exist in reference to the potential for improving running performance by manipulating the tendon mechanical properties.

Mechanical load induced as cyclic strain on connective soft tissues such as tendons is an important regulator of fibroblast metabolic activity as well as for the maintenance of tendon matrix (Chiquet et al., 2003). An increased loading typically stimulates cells for remodelling and, therefore, for increasing the mechanical properties of the tissue (Arnoczky et al., 2002). Whereas, a decreased loading leads to tissue destruction and weak mechanical properties of the tissue (Arnoczky et al., 2004). These reports demonstrate the highly plastic nature of tendons within the muscle-tendon unit of mammals and give evidence that tendon strain is an important mechanical factor regulating tendon properties. Generally, from a mechanobiological point of view strain magnitude, strain frequency, strain rate and strain duration of cells influence the cellular biochemical responses and the mechanical properties of collagen fascicles. Although it is known that mechanical loading induced as cyclic strain affects the mechanical properties of human tendons in vivo, the effect of a controlled modulation in cyclic strain magnitude, frequency, rate or duration applied to the tendon on the plasticity of human tendons in vivo is not well established. Understanding the details of tendon plasticity in response to mechanical loading applied to the tendon *in vivo* may help to improve tendon adaptation, reduce tendon injury risks and increases the performance potential of the human system.

This paper aimed (a) to present the effects of a controlled modulation of strain magnitude and strain frequency applied to the Achilles tendon on the plasticity of tendon mechanical and morphological properties and (b) to investigate whether an exercise induced increase in tendon-aponeurosis stiffness and contractile strength at the triceps surae muscle-tendon unit affect running economy.

TENDON PLASTICITY TO CYCLING LOADING: In order to get relevant knowledge about the dependence of tendon mechanical loading induced as cyclic strain and the mechanical properties of tendon and aponeurosis we examined the effect of two different exercise interventions (14 weeks). In these two exercise interventions we modified the strain magnitude (low: 2.5-3.0 % strain vs. high: 4.5-5.0 % strain) and strain frequency (low: 0.17 Hz vs. high: 0.5 Hz) applied to the Achilles tendon. Thirty two adults (two experimental groups each with n=11 and a control group n=10) participated in the study after giving informed consent to the experimental procedure accomplishing with the rules of the local scientific board.

The participants of group 1 performed 5 sets, 4 times per week of repetitive (0.17 Hz, 3 s loading, 3 s relaxation), isometric plantar flexion contractions. Repetitive isometric plantar flexion contractions were used to induce cyclic strains on the triceps surae tendon and aponeurosis. One leg (randomly chosen) has been exercised at low magnitude tendon-aponeurosis strain (low strain magnitude exercise), and the other leg at high tendon-aponeurosis strain magnitude (high strain magnitude exercise, figure 1). Mechanical loading of the triceps surae tendon and aponeurosis at 55 % and 90 % of the MVC are supposed to cause a tendon aponeurosis strain of 2.5-3.0 % and 4.5-5.0 % respectively (Arampatzis et al. 2005). The participants of group 2 performed again 5 sets, 4 times per weeks of repetive isometric plantar flexion contraction but in the high strain frequency (0.5 Hz, 1 s loading, 1 s relaxation, figure 1). Similar to the first intervention one leg has been exercised at low magnitude tendon-aponeurosis strain (55 % MVC) and the other leg at high tendon-aponeurosis strain magnitude (90 % MVC). In both exercise interventions both legs were trained at the same exercise volume (integral of the plantar flexion moment over time).

After 14 weeks of loading in the low as well as in the high frequency interventions by equal exercise volume we found an increase in tendon-aponeurosis stiffness and tendon elastic modulus only in the leg exercised at high strain magnitude (~4.7 %). These findings suggest that tendon cells may have a threshold, or set point, regarding their deformation for triggering a homeostatic perturbation leading to anabolical responses. Low strain values (2.5 to 3.0%) may lead to a habitual loading of the Achilles tendon (i.e. similar loading as during daily activities). This habitual loading would result in a lack of cell stimulation for further extracellular matrix synthesis and of remodelling leading to constant mechanical and morphological properties over time. Furthermore in the low frequency and high strain magnitude training we found a region specific hypertrophy of the Achilles tendon. However, a threefold increase in the strain frequency (from 0.17 to 0.5 Hz) caused lower adaptational effects on the Achilles tendon mechanical and morphological properties (i.e. lower increase in tendon-aponeurosis stiffness and no effect on tendon CSA, figure 2). The maximum voluntary plantar flexion moment has been increased in both, the exercise protocol at 55% and 90% of the MVC indicating an improvement in muscular capacity of the triceps surae muscles. The consequence of the increased muscle strength was an increase in tendon stress during the MVC indicating a higher mechanical tendon loading after the intervention in both legs. On the other hand, the maximum tendon-aponeurosis strain during the MVCs has been increased only in the leg exercised at low strain magnitude (i.e. exercise at 55% MVC).

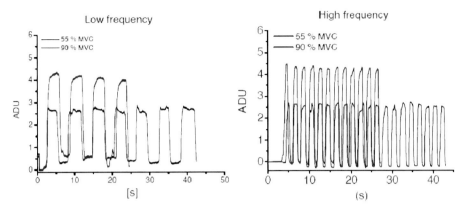

Figure 1: Each training day of the intervention protocol consisted of 5 sets of repetitive (3 s loading, 3 s relaxation, left and 1 s loading, 1 s relaxation, rigth) isometric plantar flexion contractions to induce cyclic strain on the triceps surae tendon and aponeurosis. One leg exercised at low magnitude tendon-aponeurosis strain (55 % of the MVC), whereas the other one exercised at high tendon-aponeurosis magnitude (90 % MVC). The total exercise volume (integral of the plantar flexion moment over time) was identical for both legs.

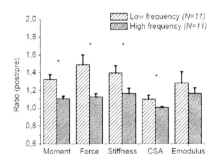

Figure 2: Ratio (post- to pre-exersice values) for the low frequency (0.17 Hz; 3 s loading, 3s relaxation) and high strain-frequency (0.5 Hz; 1 s loading, 1 s relaxation) exercise protocols. The ratios has been tested only for the high strain magnitude exercise.
*: Statistically significant differences between low- and high-frequency exercise protocols ($P<0.05$).

Given that the ultimate failure strain of the tendons cannot be altered significantly (Abrahams, 1967) and the best predictor for tendon damage accumulation for both sustained and cyclic loading is the tendon strain (Wren et al., 2003) higher strain values during the MVC after the intervention for the low strain magnitude exercised leg would decrease the safety factor (ratio of tendon ultimate strain to functional tendon strain) and may increase the risk factors for tendon injury. The maximum strain during the MVC at the leg exercised with the high strain magnitude did not change despite an increase in muscle strength. Therefore, our results show that after the exercise intervention the safety factor against tendon loading has been updated only in the high strain exercised leg.

In conclusion our findings show that (a) there is not necessarily a coordinated adaptation between muscle and tendon by a given exercise loading, (b) the strain magnitude applied to the Achilles tendon should exceed the habitual value which occurs during daily activities to trigger adaptational effects on the tendon mechanical properties and finally (c) a higher tendon strain duration per contraction leads to superior adaptational responses on the mechanical and morphological properties of the tendon.

EFFECTS OF TENDON STIFFNESS AND CONTRACTILE STRENGTH ON RUNNING ECONOMY: In an additional experiment we examined the effects of an increased tendon stiffness and contractile strength at the triceps surae muscle-tendon unit on running economy. Twenty five recreational long-distance runners (experimental group, n = 12, control group, n = 13) participated in the study. The participants of the experimental group performed the 14 weeks exercise program similar to our high magnitude and low frequency protocol (3 s loading, 3 s relaxation, 95 % MVC) in both legs. Running economy (rate of oxygen consumption at a given running velocity), tendon stiffness, maximal voluntary ankle plantarflexion joint moment, and fascicle behavior during running were analyzed before and after the intervention. Running economy was determined by measuring the rate of oxygen consumption at steady state at two running velocities (3.0 m/s and 3.5 m/s) on a treadmill. After the 14 weeks intervention, the maximum voluntary ankle plantarflexion joint moment showed a statistically significant ($p<0.05$) increase of about 6%. The triceps surae tendon-aponeurosis stiffness showed a significant increase of ~15% (figure 3). During submaximal running the subjects showed a 5% reduction ($p<0.01$) and 3% reduction ($p<0.05$) in oxygen consumption for the low and high running velocity, respectively, while the control group showed no changes after 14 weeks (figure 4).

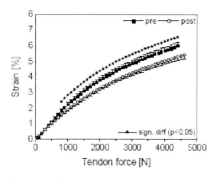

Figure 3: Strain values at every 100 N calculated tendon force (means ± sem)

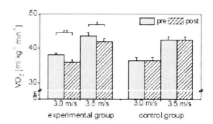

Figure 4: Rate of oxygen consumption (means ± sem)

elocity of the gastrocnemius medialis during the stance phase of running revealed no significant differences before and after the training intervention. Although, we could experimentally not confirm that fascicle behaviour of the gastrocnemius medialis muscle has been changed, the enhanced running economy found after increasing triceps surae tendon stiffness and contractile strength indicate that the force generation within the lower extremities has become more efficient while running.

REFERENCES:

Abrahams, M. (1967). Mechanical behaviour of tendon in vitro. *Medical and Biological Engineering* 5, 433-43.

Arampatzis, A., De Monte, G., Karamanidis, K., Morey-Klapsing, G., Stafilidis, S., Brüggemann, G.-P. (2006). Influence of the muscle–tendon unit's mechanical and morphological properties on running economy. *Journal of Experimental Biology* 209: 3345-57.

Arampatzis, A., Stafilidis, S., De Monte, G., Karamanidis, K., Morey-Klapsing, G., Brüggemann, G.-P. (2005). Strain and elongation of the gastrocnemius tendon and aponeurosis during maximal plantarflexion effort. *Journal of Biomechanics* 38: 833-41.

Arnoczky, S.P., Tian, T., Lavagnino, M., Gardner, K., Schuler, P. and Morse, P. (2002). Activation of stress-activated protein kinases (SAPK) in tendon cells following cyclic strain: the effects of strain frequency, strain magnitude & cytosolic calcium. *Journal of Orthopaedic Research* 20: 947–52.

Arnoczky, S.P., Tian, T., Lavagnino, M. and Gardner, K. (2004). Ex vivo static tensile loading inhibits MMP-1 expression in rat tail tendon cells through a cytoskeletally based mechanotransduction mechanism. *Journal of Orthopaedic Research* 22, 328–333.

Chiquet, M., Renedo, A.S., Huber, F. and Fluck, M. (2003). How do fibroblasts translate mechanical signals into changes in extracellular matrix production? *Matrix Biology* 22: 73–80.

Stafilidis, S., Arampatzis, A. (2007). Muscle-tendon unit mechanical and morphological properties and sprint performance. *Journal of Sports Sciences* 25: 1035-46.

Wren, T. A. Lindsey, D. P. Beaupre, G. S. Carter, D. R., 2003. Effects of creep and cyclic loading on the mechanical properties and failure of human Achilles tendons. *Annals of Biomedical Engineering* 31, 710-17.

APPROACHES FOR OPTIMISING JUMPING PERFORMANCE

Cassie Wilson

School for Health, University of Bath, Bath, UK

The ability to maximise jumping performance can be a critical factor for success in sport. This paper presents a number of studies which have looked at optimising or enhancing jumping performance. The first of these is a computer simulation study which addresses the need for model constraints when optimising high jumping performance. The second study investigates the role of coordination variability in elite triple jumping performance and the final study investigates the effectiveness of training drills in maximising performance in the triple jump.

KEYWORDS: jumping, performance, optimization

INTRODUCTION: Jumping activities are required in many sporting situations and therefore the ability to optimise jumping performance is a key element in performance enhancement. This paper presents different approaches to studying jumping activities in an attempt to optimise or enhance performance. The areas of simulation modelling, motor control and learning and the application of biomechanical principles to training theory are addressed.

SIMULATION MODELLING: When using simulation models to optimise performance care must be taken to avoid obtaining unrealistic solutions. Specifically, in the optimisation of high jumping performance, simply maximising the peak height reached by the centre of mass may result in a theoretical simulation that is inaccurate since various factors will have been neglected. Wilson et al. (2007) investigated the effects of imposing various constraints on optimisations of high jumping. An eight segment simulation model of the contact phase in running jumps for height was developed (Figure 1). The model was torque driven and contained wobbling masses to represent the soft tissue movement within the human body. Following evaluation, the model was used to maximise the height reached by the centre of mass in a series of optimisations with various constraints imposed. The constraints took into account the technical requirements of the skill, the anatomical range of motion and the robustness or consistency of the performance.

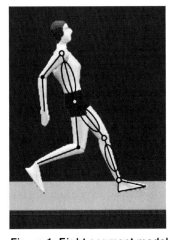

Figure 1. Eight segment model

With no constraints imposed the jump height was unrealistically high when compared to the personal best of the subject to whom the model was specific. By introducing the constraints sequentially, the height reached by the centre of mass was reduced incrementally. The height reached by the centre of mass when all constraints were imposed was very similar to the height achieved in the actual performance against which the model was evaluated. These results highlighted the need for the consideration of (i) technical requirements of the skill, such as the angular momentum at take off, so the performance is representative of an actual performance (ii) the anatomical ranges of movement so the performance is not likely to result in injury and (iii) consistency of performance which is crucial in elite sport. Future work will focus on use the model with constraints to investigate the effect of changes in different parameters or variables on jumping performance.

COORDINATION VARIABILITY AND SKILL DEVELOPMENT: The study of Wilson et al. (2007) highlighted the need for consistency of performance and therefore the need for the system (body) to be able to adapt to small changes or perturbations which may occur during a performance. Coordination is the relationship between the movements of limb segments and the variability of this coordination has been considered to be an essential element to normal healthy function offering flexibility in adapting to perturbations (Hamill et al.,1999). In contrast to this and from a traditional motor learning perspective, variability has been considered to be noise leading to an inconsistent performance. Wilson et al. (2008) investigated how coordination variability changes as a function of skill in the triple jump. Specifically we studied how lower extremity intra-segmental coordination variability in the hop-step transition of the triple jump changes as a function of the skill level in expert performers and how skill level influences the nature of the coordination variability present in the system.

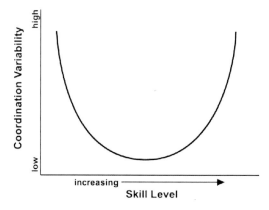

Figure 2. Proposed U-shape profile of coordination variability

The results from this study are consistent with a U-shaped change in the coordination variability, present in a system, as skill increases in expert performers (Figure 2). From a dynamical systems perspective, it has been suggested that the coordination variability in a system allows the flexibility to adapt to perturbations and this can be used to explain the higher coordination variability in the participants with the highest skill. The higher coordination variability found in the participant with the lowest skill compared to the participants with intermediate skill can be explained from a traditional motor learning perspective. This higher level of coordination variability found in the participant with the lowest skill may be surprising considering all the participants are classed as expert and likely to be in the third stage of learning according to Newell's (1985) hierarchy of the stages of learning. It may, however, be the case that even apparently high level performers go through a U-shaped change in coordination variability as skill develops. One such explanation for this is the complexity of the movement being developed. The ability of participants to integrate the three phases of the triple jump into a coordinated process may be similar to that when learning a new movement. The ability to access functional variability that allows the athlete to cope with perturbations, which could be present due to environmental or task constraints, may be indicative of a highly skilled jumper. The reduced level of variability displayed by the intermediate performers may be an indication of the ability to produce a consistent performance without necessarily being able to adapt to perturbations. Differences in coordination variability could be due to individual coordination strategies.

SPECIFICITY OF TRAINING PRINCIPLE: As intimated by Wilson et al. (2008), the study of the individual coordination strategies, in addition to the coordination variability, may provide a more holistic analysis of jumping performance. The use of training drills have previously been used in the development of complex movements, whereby coaches use the concept of specificity to encourage performance-related adaptations (Irwin, Hanton & Kerwin, 2004). As well as developing a complex movement, training drills may also be used to develop and improve movements when the full skill places very high loads on the body and where repetitions should be limited. Quantifying the similarity between a skill such as the triple jump and training practices or drills in terms of coordination patterns rather than single joint kinematics may provide a better overall assessment of their effectiveness as a training drill (Irwin & Kerwin, 2007). The ground contacts preceding the hop, step and jump phases in the triple jump largely determine the flight distance within each phase and it has been suggested that the transition, or contact, between the hop and step phase is the most critical element in successful triple jump performance (Jurgens, 1998). The demands placed on the body during the triple jump are very high with vertical forces of around 18 body weights experienced during the contact between the hop and step phases (Perttunen, 2000). Activities with such high demands might therefore have implications for injury and coaches should ensure that the number of repetitions are limited. The purpose of the study by Wilson et al. (2009) was to examine the differences between the triple jump and four plyometric drills (2 static and 2 dynamic), that are employed in training, in terms of the coordination strategies adopted by the lower extremities during the hop-step transition phase. The similarity between the drills the triple jump was assed using coupling angles quantified through the use of vector coding. Three coupling angles were investigated; ankle flexion/ext – knee flex/ext (stance), knee flex/ext – hip flex/ext (stance) and knee flex/ext – hip flex/ext (swing). For each coupling, a two-way repeated measures analysis of variance (ANOVA) was employed (trial main effect; phase main effect; trial – phase interaction effect) to investigate any differences in movement coordination patterns between jump and drill trials.

Table 1. Interaction effects from an ANOVA for differences in coupling angles (C1-C3) between the phases of the triple jump and drills (D1-D4)

	Coupling 1 ankle flex/ext – knee flex/ext (stance) [°]	Coupling 2 knee flex/ext – hip flex/ext (stance) [°]	Coupling 3 knee flex/ext – hip flex/ext (swing) [°]
Drill 1 static hop-step	p = 0.282	**p = 0.046***	**p = 0.032***
Drill 2 dynamic hop-step	p = 0.465	p = 0.499	P = 0.055
Drill 3 static raised hop-step	p = 0.871	p = 0.159	**p = 0.032***
Drill 4 dynamic raised hop-step	p = 0.996	p = 0.289	p = 0.200

*Significant interaction effects are displayed in bold.

The results of this study show that the dynamic drills are more similar to the triple jump than static drills and that replication of the coordination strategies adopted in the stance leg are better than in the swing leg. Therefore, if the primary purpose of the training drills, as suggested by coaches, is to replicate the movement patterns used in the triple jump then coaches should encourage the use of the dynamic drills. In addition, more attention should be given to the swing leg given previous studies have highlighted the importance of free limbs in jumping activities (Yu & Andrews, 1998).

DISCUSSION: The studies presented have all sought to optimize or enhance jumping performance. They have highlighted key components of jumping activities which may contribute to performance enhancement as well as important considerations for future studies.

REFERENCES:

Hamill, J., van Emmerik, R. E. A, Heiderscheit, B. C. and Li, L. (1999). A dynamical systems approach to lower extremity running injuries. *Clinical Biomechanics, 14,* 297-308.

Irwin, G., Hanton, S. and Kerwin, D. G. (2004). Elite coaching I: The origins of elite coaching knowledge. *Reflective Practice, 5,* 419–436.

Irwin, G. and Kerwin, D.G. (2007). Inter-segmental coordination in progressions for the longswing on high bar. *Sports Biomechanics, 6,* 131-144.

Jurgens, A. (1998). Biomechanical investigation of the transition between the hop and the step. *New Studies in Athletics,* 13(4), 29-39.

Newell, K. M. (1985). Coordination, control and skill. In D. Goodman., R. B. Wilberg, and I. M. Franks (Eds.), *Differing perspectives in motor learning, memory and control.* North-Holland: Elsevier Science Publishers B.V.

Perttunen, J., Kyrolainen, H., Komi, P.V. and Heinonen, A.(2000). Biomechanical loading in the triple jump. *Journal of Sports Sciences, 18,* 363-370.

Wilson, C., Yeadon, M.A. and King, M.A. (2007). Considerations that affect optimised simulation in a running jump for height. *Journal of Biomechanics, 40,* 3155-3161.

Wilson, C., Simpson, S.E.,-van Emmerik, R.E.A. and Hamill, J. (2008). Coordination variability and skill development in expert performers. *Sports Biomechanics, 7,* 2-9.

Wilson, C., Simpson, S.E. and Hamill, J. (2009). Movement coordination patterns in triple jump training drills. *Journal of Sports Sciences, 26,* 845 – 854.

Yu, B. & Andrews, J.G. (1998). The relationship between free limb motions and performance in the triple jump. *Journal of Applied Biomechanics, 14,* 223-237.

ISBS 2009

Scientific Sessions

Tuesday AM

ISBS 2009

Oral Session S01

Rowing

THE EFFECT OF NUMBERS OF REPETITIONS ON PEAK TORQUE IN ROWER ANS NON-ATHLETE FEMALES WHEN USING ISOKINETIS TESTING

Sarah Moody, Barbara Warren, and Deanna Malikie
Department of Exercise Science, University of Puget Sound, Tacoma, United States

Isokinetic training has been used as a successful means for testing and increasing muscle strength. The purpose of this study was to investigate the effect that different numbers of repetitions have on fatigue and force generation in females and specifically between athletes and non-athletes. Thirty college-aged females (15 rowers, 15 non-athletes) were tested using an isokinetic machine to measure peak torque. Each subject was tested 5-6 times on the isokinetic machine. This included 1-2 familiarization tests and four experimental testing sessions during which subjects performed randomly assigned maximal knee extensions of either four, six, eight, or ten repetitions, through a 90 degree range of motion, at 60, 120, 180, 240, and 300deg/s. Rest periods between velocities were kept constant at 60 sec. Using SPSS 14.0 data were analyzed using a 2 X 4 X 5 repeated measures ANOVA with alpha < .05. Group, repetitions, and velocity served as the independent variables and peak torque as the dependent variable. Peak torque in rowers was also compared to the time taken to complete a 2000m distance on a rowing ergometer. No significant difference was found between the various repetitions at different velocities. A significant difference was found between peak torques at the different velocities ($F=1221.37$, $p<.05$). A significant difference was also found between athletes and non-athletes at the different velocities ($F=24.272$, $p<.05$). A correlation of -0.82 was found between peak torque and 2000m time for rowers. The number of repetitions does not appear to effect peak torque production. Athletes appear to produce more torque at all velocities compared to non-athletes. There is a linear relationship between knee extensor peak torque and performance on a rowing ergometer.

KEY WORDS: isokinetics, CYBEX, knee extension, rowing.

INTRODUCTION: Isokinetic training has been used as a successful means for measuring and tracking the progress of peak torque development in muscle strength (Dauty & Rochcongar, 2001) and to identify asymmetry in leg strength (Dauty et al., 2007). Because isokinetic training controls velocity during movement it can be used for safe and effective rehabilitation, its original purpose. Hamstring/quadriceps ratios have been studied using isokinetic machines in regards to ACL injuries and rehabilitation (Hewett et al., 2008). Research has been conducted testing varying velocities (Arnold & Perrin, 1995) and strength gain has been observed across these velocities (Akima et al., 1999). Isokinetic testing, however, does not have a set protocol and standardization needs to be established.

While there are other muscles involved in a rowing stroke, the quadriceps muscles of the leg have a direct impact on the performance power of the leg drive (Tachibana et al., 2007). Rowers have shown greater strength in knee extensors than non-rowers (Parkin et al., 2001); however it is uncertain as to whether peak knee torque has a relationship to rowing performance. Little research has compared the effect on peak torque when using a variable number of repetitions and velocities, nor have many comparisons been made between athletic and non-athletic populations. The purpose of this study was to investigate the effect that different numbers of repetitions have on peak torque generation of the knee extensors in females, specifically between athletes and non-athletes. A secondary purpose was to compare knee extensor strength in rowers to rowing performance.

METHODS: Data Collection: Thirty apparently healthy college-aged females were recruited from the student population at the University of Puget Sound in Tacoma, WA. The 15 athletes were members of the women's varsity crew team and the other 15 subjects were university athletes. The mean age, height, and mass of the subjects respectively were 20±1 years, 1.70±.006m, and 70.63±12.90kg. Subjects with previous knee injuries were excluded from this study. The study was approved by the Internal Review Board and all subjects signed a form of informed consent. Testing was conducted on a Cybex NORM Isokinetic Machine. Subjects completed one-two familiarization sessions and four experimental sessions. Each test included a five minute warm-up on a cycle ergometer at a self-selected resistance and pace and a warm-up on the Cybex machine. This involved four submaximal knee extensions at 60, 120, 180, 240, and 300°/s with 60s rest periods between each velocity. All isokinetic testing occurred through a 90deg range of motion and held knee flexion constant at 300deg/s. Familiarization sessions involved either four or six maximal knee extensions at 60, 180, and 300°/s with a three minute rest period between velocity sets. The four experimental tests incorporated randomly assigned maximal knee extensions of either four, six, eight, or ten repetitions with ascending ordered velocities of 60, 120, 180, 240, and 300°/s. Flexion speed was held constant at 300deg/s and rest periods between velocity sets were 60s. Both visual and verbal feedback was provided throughout the tests. Subjects were asked to abstain from maximal exercise bouts within 24 hours prior to testing and all subjects had a minimum 24-hour rest period between data collection sessions.

Data Analysis: Using SPSS 14.0 data were analyzed using a 2X4X5 repeated measures ANOVA with alpha < .05. Differences in peak torque across velocities between athletes and non-athletes were determined. A Pearson Correlation and linear regression were also computed to compare peak torque at 60°/sec to times recorded to complete a 2000m piece on a rowing ergometer for the 15 athletes.

RESULTS: There were no significant differences in peak torque when varying the number of repetitions. A significant difference in peak torque was found between the different velocities ($F=1221.37$, $p<.05$) (Fig.1).

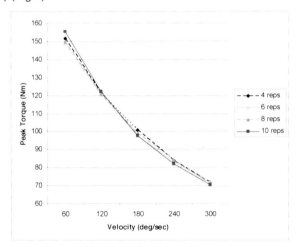

Figure 1: Peak torque versus velocities at various repetitions.

There was also a significant difference in peak torque found between athletes and non-athletes at the different velocities ($F=24.27$, $p<.05$) (Fig. 2).

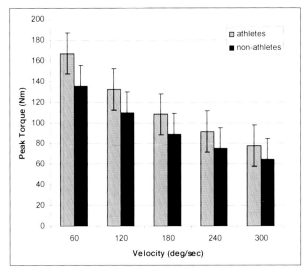

Figure 2: Differences in peak torques between athletes and non-athletes across five different velocities. All velocities were significant.

Pearson's Correlation revealed a -0.82 correlation (α=0.01) between peak torque at 60°/sec and the time taken to complete a 2000m distance on a rowing ergometer for the athletes (Fig. 3).

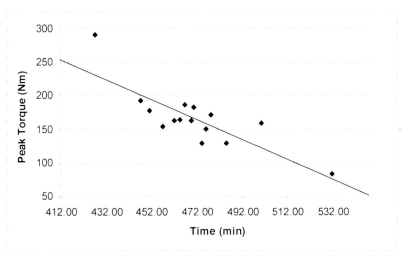

Figure 3: Peak torque at 60°/sec (4 reps) for the 15 rowers compared to the time taken to complete 2000m on a rowing ergometer.

DISCUSSION: The results confirm an inverse relationship between peak torque and velocity, which supports findings in previous studies (Gerodimos et al., 2003; Koutedakis et al., 1998). Similar to past research, athletes produced more torque at all velocities compared to non-athletes (Koutedakis et al., 1998). In addition, the number of repetitions does not appear to effect peak torque production; therefore four repetitions may be adequate when measuring peak torque. There was also an inverse relationship between the peak torque produced at 60°/sec and the time it took the rowers to complete a 2000m distance on a rowing ergometer. This suggests that there is a relationship between quadriceps strength and rowing power performance as reported by Tachibana et al. (2007). As rowers begin a rowing strength they push primarily with the quadriceps muscles. Moreover, this study supports the findings of Parkin et al. (2001), who reported that male rowers exhibited greater knee extensor strength than non-rowers. Although it was not tested, verbal and visual encouragement seemed to impact the effort of the subjects. O'Sullivan & O'Sullivan (2008) found there was a significant difference between conditions with and without verbal and visual feedback. Maximal effort also seemed to be impacted by the competitive nature of the subjects. As rowing is an endurance sport, it is unclear whether the trends found in this study would hold true for non-endurance athletes.

CONCLUSION: The number of repetitions does not seem to effect peak torque production in athletic and non-athletic female populations and therefore four repetitions may be adequate when measuring peak torque. Rowing performance power does appear to be influenced by the strength of the knee extensors.

REFERENCES:

Akima, H., Takahashi, H., Kuno, S., Masuda, K., Shimojo, H., Anno, I., Itai, Y., & Katsuta, S. (1999). Early phase adaptions of muscle use and strength to isokinetic training. *Medicine and Science in Sports and Exercise, 31*(4), 588-94.
Arnold, B. & Perrin, D. (1995). Effect of repeated isokinetic concentric and eccentric contractions on quadriceps femoris muscle fatigue. *Isokinetics and Exercise Science, 5*, 81-5.
Colye, E., Feiring, D., Rotkis, T., Cote, R., Roby, F., Lee, W., & Wilmore, J. (1981). Specificity of power improvements through slow and fast isokinetic training. *Journal of Applied Physiology, 51*(6), 1437-42.
Dauty, M., Dupré, M., Potiron-Josse, M., & Dubois, C. (2007). Identification of mechanical consequences of jumper's knee by isokinetic concentric torque measurement in elite basketball players. *Isokinetics and Exercise, 15*, 37-41.
Dauty, M. & Rochcongar, P. (2001). Reproducibility of concentric and eccentric isokinetic strength of the knee flexors in elite volleyball players. *Isokinetics and Exercise Science, 9*, 129-32.
Gerodimos, V., Mandoa, V., Zafeiridis, A., Ioakimidis, P, Stavropoulos, N., & Kellis, S. (2003). Isokinetic peak torque and hamstring/quadriceps ratios in young basketball players. Effects of age, velocity, and contraction mode. *The Journal of Sports Medicine and Physical Fitness, 43*(4), 444-52.
Hewett, T., Myer, G., & Zazulak, B. (2008) Hamstrings to quadriceps peak torque ratios diverge between sexes with increasing isokinetic angular velocity. *Journal of Science and Medicine in Sport, 11*(5), 452-9.
Koutedakis, Y., Agrawal, A., & Sharp, C. (1998). Isokinetic characteristics of knee flexors and extensors in male dancers, Olympic oarsmen, Olympic bobsleighers, and non-athletes. *Journal of Dance Medicine & Science, 2*(2), 63-6\7.
O'Sullivan, A. & O'Sullivan, K. (2008). The effect of combined visual feedback and verbal encouragement on isokinetic concentric performance in healthy females. *Isokinetics and Exercise, 16*, 47-53.
Parkin, S., Nowicky, A., Rutherford, O., & McGregor, A. (2001). Do oarsmen have asymmetries in the strength of their back and leg muscles? *Journal of Sports Science, 19*(7), 521-6.
Tachibana, K., Yashiro, K., Miyazaki, J., Ikegami, Y, & Higuchi, M. (2007). Muscle cross-sectional areas and performance power of limbs and trunk in the rowing motion. *Sports Biomechanics, 6*(1), 44-58.

Acknowledgement
This research was funded by a research grant from the University of Puget Sound.

MECHANICAL LOADING OF THE LUMBAR SPINE OF ELITE ROWERS WHILE ROWING FIXED AND SLIDING ERGOMETERS

Richard Smith[1], Michael Dickson[1] &Floren Colloud[2]

[1] University of Sydney, Sydney, Australia
[2] University of Poitier, Poitiers, France

Low back injury is common in rowers. This study compared compressive forces of the lumbar spine, while rowing on fixed and sliding ergometers. Fifteen elite male rowers with no history of serious low back injury rowed the Concept2 Fixed (C2F), Concept2 Sliding (C2S) and RowPerfect (RP) ergometers at 32 strokes/min while 3D motion and external force data were recorded. Inverse dynamics analysis was used to find net lumbar moment and a lumbar model used to model compressive forces acting at L4/L5. Compressive force was significantly larger on C2F, at the catch and for 45 % of the stroke. Rowing on the C2F ergometer places greater compressive stress on the lumbar spine.

KEY WORDS: biomechanics, ergometry, rowing, L4/L5 compressive forces.

INTRODUCTION: Rowing is a physically and technically demanding skill that requires the back to act as a transfer for large forces between the upper and lower extremities (Hagerman, 1984). It is characterised by long distance aerobic training sessions which comprise 90 % of training volume and it is considered that the majority of rowing injuries are related to overuse. The most common type of injury in elite male rowers is the lower back (Hickey et al., 1997). About half of specific events causing rowing injuries occur off water.

On a fixed ergometer, the rower is positioned on a sliding seat and during the drive phase the rower is required to accelerate the entire body mass away from a stationary foot stretcher-flywheel complex. In addition to a sliding seat, the sliding ergometer has a foot stretcher-flywheel complex that is mounted on a slide. This allows a transfer of momentum between the rower and the sliding complex.

It is the aim of this study to estimate the spinal compressive forces in elite rowers while using Concept2 Fixed (C2F), Concept2 Sliding (C2S) and Rowperfect (RP) rowing ergometers. It is hypothesised that due to increased acceleration requirements, lumbar compressive forces will be greater when rowing on the C2F, compared to C2S and RP during the drive phase.

METHODS: Subjects: Fifteen injury-free elite male rowers volunteered to participate in this study. Their mean (± SD) age was 25.2 ± 4.4 years, height 1.915 ± 0.072 m and body mass 91.0 ± 7.4 kg. The Human Ethics Review Committee of the University of Sydney approved this study.

Experimental Design: The experiment was a multivariate repeated measures design with two within subject factors: ergometer (three levels: C2F, C2S, RP) and stroke (10 levels: 10 strokes); and one primary (catch lumbar compressive force) and three secondary (trunk acceleration, trunk-pelvis angle and stroke rate) dependent variables. The order of ergometer presentation was balanced to reduce carry over effects.

Rowers were asked to warm up for 5 min then perform 1 min rowing at at 80 % maximal propulsive power at 32 strokes/min, with a 1 min rest period between stroke rate trials. A rest period of 5 min was given between ergometer conditions.

Force data collection: Two new foot stretchers were constructed, each fitted with two 3D force transducers (Model 9067, Kistler Instrument Corp., AG Winterthur, Switzerland). A strain gauge (Model TLL-500, Transducer Techniques Inc., CA, USA) was connected in series at the chain-handle attachment.

Kinematic data collection: Fifty two markers were placed for an initial static trial with 12 of these being removed for the following rowing trials. The 3D trajectories of the joint centers were then calculated for each rowing trial. Nine video cameras and the force transducers provided input for the motion analysis system (EvaRT, Motion Analysis Corporation, USA) at 60 Hz.

Inverse dynamics modelling: The kinematics of the anatomical markers were recorded in 3D to provide more accurate joint center data for the sagittal plane model of the rower. Using a two-dimensional nine-segment whole-body model, the net joint forces and moments were calculated in a custom program. The spectra of position and force data were analysed to determine optimum cutoff frequencies (5 Hz for position and 10 Hz for force data) according to the method of Giakas and Baltzopoulos (1997).

Spinal modelling: The current study used a sagittal model of the lumbar spine to partition trunk forces and moments into extensor muscle force and resultant bone-on-bone force at L4/L5.

Analysis of results: Ten full strokes were analysed from each rowing trial. Each stroke was normalised to 100 % stroke. Ensemble force-time stroke profiles represent the mean of all subjects for one condition and 95 % confidence intervals were included to indicate variability across subjects.

Statistical analysis: Multivariate analysis of variance with repeated measures (SPSS for Windows, SPSS Inc., USA) was used to test the significance of any observed differences in the means. A Bonferroni adjustment was made for pairwise comparisons and multiple dependent variables. Differences were considered significant for continuous data if the mean was outside the 95% confidence interval for more than five consecutive data points.

RESULTS: The ten strokes analyzed within a trial were consistent. There was no effect of stroke on any of the tested variables or interaction with ergometer or stroke rate. The strokes rate across subjects did not differ by more than 1% of the required stroke rate.

Lumbar compressive forces: Compressive force was high at the catch and during early drive phase, increasing to reach a peak at mid drive phase (Figure 1.). The C2F curve shows compressive force on the fixed ergometer to be significantly greater than the two sliding ergometers during early drive phase (0 - 13 % stroke) and again in recovery (from 66 - 100 % stroke) ($p > 0.05$).

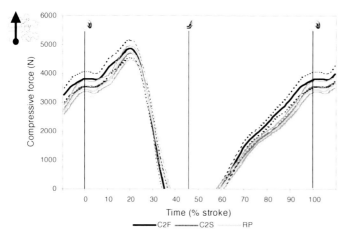

Lumbar compressive forces at the catch: Lumbar compressive force at the catch was significantly different between all ergometer conditions ($p < 0.001$) (Table 1). C2F produced the largest compressive force at the catch, followed by C2S and RP produced the least. Effect size for interaction due to ergometer was large and differences in

Figure 1. Stroke profile of mean compressive force.

means ranged from 150 N (C2F vs C2S) to 435 N (C2F vs RP). Compressive force at the catch on C2F was 77 % of compressive force at maximum and slightly less (~ 70 %) for both the sliding ergometers.

Table 1. Mean and SD for Catch Lumbar Compressive Force and Stroke Rate, comparisons between ergometer conditions, power and effect size.

C2F	C2S	RP	Pair	Sig.	For the effect of ergometer	
					Power	Effect size
Catch Lumbar Compressive Force						
3670±114	3380±88	3220±78	C2F vs RP	0.000	1.000	0.818
Stroke Rate						
31.6±0.22	32.2±0.20	32.0±0.23	C2F vs RP	0.003		

Trunk acceleration: The two sliding ergometers have very similar trunk acceleration profiles (Figure 2.). The C2F has much greater trunk acceleration when compared to the RP and C2S ergometers. C2F trunk acceleration ranges from a negative peak in early drive phase of -7.9 m·s^{-2}, to a positive peak in late drive phase of 7.3 m·s^{-2}, a range of 15.2 m·s^{-2}. Both the sliding ergometers require much less trunk acceleration, a range of 9.2 m·s^{-2} ($-5.6 - 3.6$ m·s^{-2}, which represents about 60 % of C2F range).

Trunk-pelvis angle: There was substantial variability between subjects as indicated by large 95 % confidence intervals for the ensemble mean trunk-pelvis stroke profile. When trunk-pelvis angle profiles were considered for individual subjects the most consistent feature was an increased flexion from ~ 20 - 30 % of the stroke. Slight observable differences in trunk-pelvis angle between ergometers occurred during late recovery, at the catch and early drive phase. At this time the C2F ergometer produced a smaller trunk-pelvis angle compared to both the sliding ergometers (Figure 3).

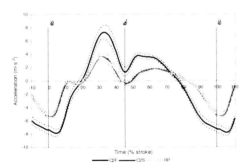

Figure 2. Stroke profile of horizontal acceleration of trunk segment COG.

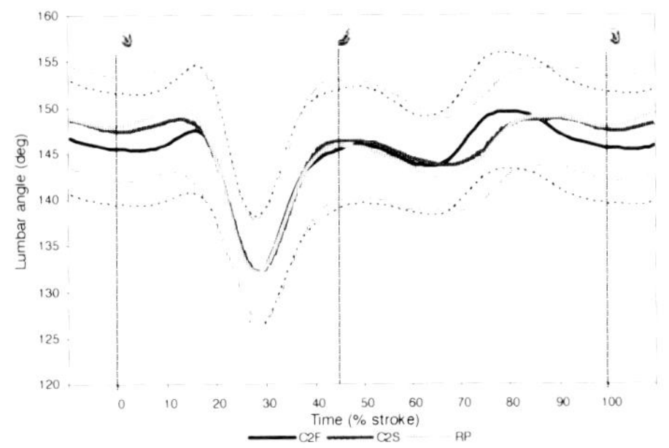

Figure 3. Ensemble mean trunk-pelvis angle at race pace

Mean trunk-pelvis angle at the catch was ~ 146° on C2F and ~ 148° on both C2S and RP. Trunk-pelvis angles at the catch were generally greater than at the time of maximum spinal force production (corresponds to 22 % of stroke).

DISCUSSION: Compressive forces: Compressive force profiles, for the fixed and both the sliding ergometer conditions, indicate that the lumbar spine is under considerable stress during ergometer rowing. The C2F ergometer produced consistently larger compressive forces at the catch, when compared to the C2S and RP ergometers.

Differences in lumbar stress between the fixed and sliding ergometers at the catch cannot be attributed to differences in handle force as the handle force is zero at the catch. When no force is applied to the handle, the force produced by the rower is used only to accelerate the stretcher complex and the body mass (in opposite directions). As the stretcher is stationary on the fixed ergometer, all force produced by the rower at the catch is used to accelerate the rower COG. In comparison to the sliding ergometers, the C2F exhibits far greater horizontal trunk acceleration at the catch. As the handle force is zero, the net lumbar moment at the catch is determined largely by the required acceleration of the trunk, head and arms. Therefore, at the catch, the significantly larger compressive forces produced are the result of the larger body mass acceleration requirements of the fixed ergometer.

Risks of injuries: Mean maximum compressive forces exceeded the NIOSH manual handling recommendations for safe lifting (3400 N) during all ergometer conditions. The vertebral body is the most likely to be injured by a purely compressive load and this recommended maximum compressive force is based on cadaver experiments of vertebral breaking limits. However, this limit considers only static lifting situations. For the trunk, rowing can be considered as a repetitive, dynamic, flexion-extension movement. Repetitive dynamic movements, with large compressive force components, may place structures other than just the vertebral body at risk of injury.

Trunk-pelvis angle: The absolute angles of the trunk and pelvis segments followed similar trends of extension during the drive phase. It seems that despite individual anatomical and technical variability, there are several observations regarding technique. A posture at the catch of increased anterior rotation of the pelvis may help reduce stress on the lumbar spine. During the drive phase, a moderately flexed lumbar posture may be suited to optimal compressive force resistance and technical effectiveness.

CONCLUSION: At the catch, rowing on the C2F ergometer produced consistently larger compressive loading of the lumbar spine, compared to the C2S and RP ergometers. This was due to the large body acceleration requirements of the stroke while rowing on the fixed ergometer. It is possible that these elite male rowers have found a safe posture for the lumbar spine at the time of maximum compressive loading.

At the catch, differences in compressive forces can be attributed to the effects of upper body acceleration which, in turn, depend on the ratio of body mass to ergometer/fan assembly mass. Even greater differences may be obtained if the mass of the instrumentation for the stretcher could be reduced.

While these results emphasize the significant mechanical stress on the lower back during ergometer rowing, similar research is needed during on-water rowing.

REFERENCES:

Giakis, G. and V. Baltzopoulos. Optimal digital filtering requires a different cut-off frequency strategy for the determination of the higher derivatives. *J. Biomech.* 30:851-855, 1997.
Hagerman, F. C. Applied physiology of rowing. *Sports Med.* 1:303-326, 1984.
Hickey, G. J., P. A. Fricker, and W. A. McDonald. Injuries to elite rowers over a 10-yr period. *Med. Sci. Sports Exerc.* 29:1567-1572, 1997.

Acknowledgements

Floren Colloud was granted by a *Bourse Lavoisier* from the French Foreign Office.
We wish to thank the subjects who gave their time and effort to participate as well as coaches of NSWIS, Sydney Uni Boat Club and UTS Haberfield for helping with recruiting subjects. Thanks to Ray Patton for his technical assistance throughout this project.

THE EFFECTS OF HORIZONTAL AND VERTICAL FORCES ON SINGLE SCULL BOAT ORIENTATION WHILE ROWING

Peter J Sinclair, Andrew J Greene and Richard Smith
Faculty of Health Sciences, The University of Sydney, Australia

Forces produced at the oarlock pins and the foot stretcher, and the orientations of a single scull rowing boat, were investigated in eleven male rowers. Rowers were tested at 32 strokes·minute^{-1} on an instrumented single scull boat. The pitch of the boat was shown to be increased by the vertical forces applied to the pin and the foot stretcher, and roll was increased by an imbalance of vertical forces delivered to the bow and stroke side pins. Pitch and roll were largest in the late recovery and early drive phase when forces applied to the boat were greatest. It is suggested that the provision of feedback on boat orientation and force production may enable rowers to be trained to reduce changes in orientation and therefore to reduce energy loss through hydrodynamic drag.

Keywords: biomechanics, kinetics, on-water rowing.

INTRODUCTION: Boat orientation is a vital component to maximising the overall velocity of the boat by reducing the drag forces that act on the hull of the boat during the stroke. Drag forces acting on the boat depend upon the frontal area of the shell relative to the water and are also increased by wave motion caused by changes in orientation (Baudoin and Hawkins, 2002). Therefore, by increasing oscillations in boat orientation in the water, the drag forces present throughout the stroke will be increased. Increases in drag forces that act on the boat throughout the stroke have to be overcome by the rowers before they are able to accelerate the boat (Nolte, 1991). Forces acting on the foot stretcher and oarlock pin have specific effects upon the pitch, roll and yaw of the boat. Changes in boat orientation act to increase the cross sectional area of the boat in the water and thus increasing the drag forces acting on the hull of the boat (reference). The purpose of this study was to quantify the changes in boat orientation during single scull rowing and to investigate the forces responsible for these changes.

METHODS: Eleven male rowers (age 18.5 ± 1.9 years, height 1.87 ± 0.04 m and body mass 82.3 ± 8.2 kg), consisting of national, state and university competitors, participated in the study. Training frequencies ranged from 6 - 10 sessions per week, with all rowers training for both sweep rowing and sculling. Participants were requested to row at 32 strokes·minute^{-1} and were provided with visual information feedback of stroke rate. They were instructed to perform their usual rowing technique, especially in terms of stroke length, over a 250m distance.

A single scull was instrumented to measure the external forces generated by the rower at the pins (the force transferred from the oar through the oarlock, Smith and Loschner, 2002) and at the foot stretcher. A new foot stretcher was constructed using two carbon fibre composite core 'sandwich' sheets (Divinycell H, DIAB International AB) each fitted with three force transducers (Model 9602A, Kistler Instrument Corp., AG Winterthur, Switzerland) to record 3D reaction forces under the feet. Angular velocity was collected using a three dimensional gyroscope (Analog Devices ADXRS300 Gyroscope) and, after integration, was used to measure the angular pitch (about the horizontal axis), roll (about the longitudinal axis) and yaw (about the vertical axis) of the boat.

Data were collected using a high quality telemetry system (ROWBOT, Digital Effects, Australia) at a sampling rate of 100Hz and transmitted to a laptop computer (4700CT, Toshiba). Data were analysed using a custom designed integrated software program written in Visual Basic and a 10Hz zero phase-shift Butterworth filter used to smooth the data. Twelve full strokes were analysed from each 250m rowing trial. Strokes were defined by the minimum oar angle (catch angle) signalling both the start and end of the rowing stroke, and the maximum oar angle (release angle) determining the end of the drive phase and the start of the recovery phase. Each stroke was time normalised to 100% stroke. All 12 strokes were used to form an average stroke profile for each rower, and then ensemble averaged to indicate variability across subjects.

The pitch of the boat was measured as changes in the vertical displacement of the bow of the boat, in cm, from a horizontal axis through the middle of the boat. Roll was measured as vertical displacement of the oarlock pin about the longitudinal axis of the boat. To be consistent with common rowing terminology, stroke side roll is reported as elevation of the right hand side of the boat when facing forward. Bow side roll is defined as elevation of the left hand side of the boat. Yaw of the boat was measured as the change in direction of the stern of the boat about the vertical axis of the boat, again reported in cm from the midline.

RESULTS AND DISCUSSION: The pitch of the boat alters by up to 3.5cm throughout the duration of the stroke. 'Bow up' pitch, where the bow of the boat moves up and the stern of the boat moves down, increases in the late recovery phase as vertical stretcher forces increase rapidly while the rower slides back towards the catch position (Figure 1). As the oar enters the water at the catch position, the rower applies propulsive forces at the handle and foot stretcher to accelerate the boat. At this point, the rower also pushes down on the foot stretcher, increasing the pitch of the boat.

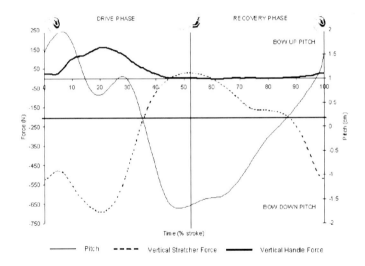

Figure 1: Vertical pin and stretcher forces and their effect upon the pitch of the boat.

Positive vertical forces on the pins also increase during the early part of the drive phase to ensure that the blade remains fully submerged in the water throughout the stroke. The coordinated increase of vertical pin forces and decrease of vertical stretcher force causes the 'bow up' pitch of the boat to reduce throughout the drive phase. Towards the end of the drive, the pitch of the boat changes to a 'bow down' orientation as the centre of mass of the rower moves further towards the bow of the boat. Extension of the trunk at the finish position increases the displacement of the centre of mass of the rower towards the bow of the boat, further increasing the 'bow down' pitch. At the finish position and through the early recovery phase, the vertical forces acting through the stretcher and pins are minimal, so it is the weight force of the rower acting through the seat that maintains the 'bow down' pitch of the boat. The 'bow down' pitch of the boat reduces as the rower initiates the slide back towards the catch. Vertical forces are increased rapidly at the stretcher as the rower moves forwards, so the 'bow up' pitch increases again approaching the catch. The investigation of these forces delivered through the seat has not been conducted in the present study. However, a better understanding of the relationship between the transfer of vertical forces between the stretcher, pin and seat would be of benefit to rowers and coaches due to the high ratio of the mass of the rower to that of the boat.

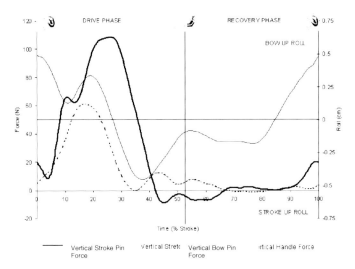

Figure 2: Vertical pin forces and their effect upon the roll of the boat.

While the total vertical forces acting upwards on the pin are coordinated with the forces at the stretcher to reduce the pitch of the boat, the forces applied at the individual bow and stroke side pins are uneven (Figure 2). Vertical forces generated at the bow side pin are of greater magnitude than those of the stroke side pin. This occurs because the gate on the bow side pin is positioned higher during the rigging set-up than that of the stroke side gate to enable the right hand to pass over the left when the oars draw through in the drive phase. Increased bow gate height introduces an asymmetrical aspect to the rowing stroke, suggested by Wagner et al (1993) to be a reason for the repetitive rolling and yawing of the boat. Caution should be used when viewing average roll curves because variance in the data results in the mean is not necessarily representing any given rower.

At the catch position, the vertical bow pin forces increase as the oar enters and is submerged in the water, causing the roll of the boat to increase on the bow side. 'Bow side' roll peaks at the catch position with an average value of 0.5 cm. As the drive phase begins, vertical force is generated at the stroke side pin which counteracts those produced at the bow side pin, reducing the bow side roll of the boat. Bow side roll is again elevated as the hands cross over each other in the mid drive phase, causing an increase in the vertical bow side pin forces and an increase in asymmetry within the stroke.

After the second peak of bow side roll, the boat 'rolls' to the stroke side, despite the vertical bow side pin force continuing to rise, and the stroke side pin force reducing. This may be due to a postural control mechanism employed by the rowers, who adjust the distribution of their weight force through the seat to maintain the balance of the boat and reduce the roll to one side. Once again, instrumentation of the seat to measure forces would enable analysis of the distribution of vertical forces through the seat to see whether coping mechanisms were being implemented by the rowers to maintain boat balance.

The stroke side roll reduces at the finish position, when the 'legs down' position of the stroke enables the rower to provide a stable base by which they can minimise fluctuations of boat movement and maximise the 'run of the boat' in the water (Mazzone, 1988). The roll remains relatively stable until the rower returns to the catch position.

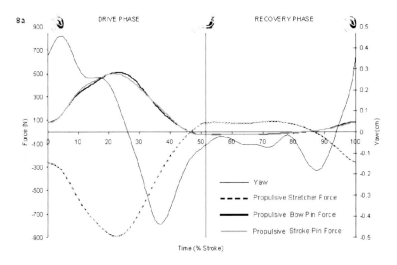

Figure 3: Propulsive pin and stretcher forces and their effect on the yaw of the boat.

Oscillations occur in the yaw of the boat during the late drive and early recovery phase (Figure 3). Changes in yaw of the boat were greatest during the drive phase when propulsive pin and stretcher forces were at their greatest. These oscillations may therefore occur as a result of unequal development of pin and stretcher forces. While the average peak displacement in figure 3 was 0.46 cm, four out of the eleven rowers yawed initially in the opposite direction, depending on the magnitude and timing of horizontal pin forces. The average amount of yaw (ignoring direction) was 1.2 cm. The synchronicity of foot stretcher force application would also have an effect upon the yaw of the boat if a rower delivered a larger proportion of the stretcher forces through one of the legs. Further instrumentation is required to measure this effect as the current study measured only a single force for both feet.

CONCLUSION: On average, boat orientation deviated from a neutral position by about 2 cm for pitch, 0.5 cm for roll and 0.3 cm for yaw. These orientations were greatest just after the catch of the stroke when large forces were applied to the foot stretcher and oarlock pins. These values were averages for the group, with some individuals producing much larger motions than this, and there being variation within each individual's set of strokes. Technology may allow feedback to be provided to rowers and coaches may be an effective way to train rowers to reduce fluctuations in boat orientation and therefore to reduce energy loss through hydrodynamic drag.

REFERENCES

Baudouin, A & Hawkins, D. (2002). A biomechanical review of factors affecting rowing performance. *British Journal of Sports Medicine*, 36, 396-403.

Smith R.M. & Loschner C. (2002). Biomechanics feedback for rowing. *Journal of Sports Sciences*, 20, 783-791.

Mazzone, T. (1988). Kinesiology of the rowing stroke. NSCA Journal, 10, 4-11.

Nolte, V. (1991). Introduction to the biomechanics of rowing. *FISA Coach*, 2, 1-6.

Wagner, J., Bartmus, U. & de Mareees, H. (1993). Three-axis gyro system quantifying the specific balance of rowing. *International Journal of Sports Medicine*, 14, S35-S38.

OAR BLADE FORCE COEFFICIENTS AND A MATHEMATICAL MODEL OF ROWING

Coppel, A.[1], Gardner, T.[1], Caplan, N.[2] & Hargreaves, D.[3]

[1]School of Sport and Exercise Sciences, University of Birmingham, UK.
[2]School of Psychology and Sport Sciences, Northumbria University, UK.
[3]School of Civil Engineering, University of Nottingham, UK.

The aim of this study was to validate the use of computational fluid dynamics (CFD) to determine oar blade force coefficients for use in a mathematical model of rowing mechanics to predict the performance of a boat. Experimental and CFD derived lift and drag force coefficients for a Macon oar blade were taken from previously published research. Each set of coefficients was used to drive a mathematical model of rowing, and predicted instantaneous and mean steady state boat velocity compared. Instantaneous boat velocity was similar throughout the stroke and mean boat velocity varied by only 1.33%. In conclusion, this investigation has demonstrated that lift and drag coefficients obtained by computational methods may be used successfully to predict boat behaviour in a mathematical model of rowing. The use of computational data closely matches model outputs derived from experimental data.

KEY WORDS: rowing, CFD, mathematical modelling, oar blade.

INTRODUCTION: In rowing, the boat is propelled through the water by musculoskeletal forces exerted by each rower on the boat. These are transferred to the water by the oars and oar blades. Due to these physical interactions, this system of rower-boat-oars-water lends itself to being modelled mathematically. Several complete mathematical models of rowing have been presented in the literature (Pope, 1973; Sanderson & Martindale, 1986; Millward, 1987; Brearley & de Mestre, 1996; Lazauskas, 1997; Brearley et al., 1998;Cabrera et al., 2006; Caplan & Gardner, 2007a). The oar blade motion is a key component of these models and how it is represented across the literature sees much variation.

Millward (1987), Brearley and de Mestre (1996) and Brearley et al. (1998) present models where the blade is assumed to remain fixed in the water during the drive phase of the stroke. Assuming that the blade remains fixed neglects the complex interactions between blade and water which produce lift and drag forces on the blade, affecting boat propulsion. Cabrera (2006) extended the work of Pope (1973), improving the model of the oar blade–water interaction, by defining the motion of the blade as moving backward through the water, with the water flow being perpendicular to the blade chord line. The force coefficients were taken from experimental data for a flat plate obtained from Hoerner (1965).

It has been shown, however, that the oar blade does not simply move backward during the stroke, generating only a drag force (Nolte, 1984). Instead the oar blade moves forward in the water during the first third of the drive phase, followed by a period where it slips backwards, and finally it moves forwards again as it approaches the boat (Figure 1). The angle of attack, α, between the line of the blade and its direction of motion varies transiently throughout the stroke and the path which it moves through. Additionally, the blade is known to act as a hydrofoil (Nolte, 1984; Dal Monte & Komor, 1989) and lift forces on the blade are produced as well as drag forces.

The importance of the fluid dynamic interaction between oar blade and water was considered by Caplan & Gardner (2007a) in their model. They used experimentally obtained force coefficients of lift and drag for a number of oar blade designs (Caplan & Gardner, 2007b) as variables in their model. This allowed for a more accurate simulation of the fluid dynamic behaviour of oar blades and therefore of propulsion to the boat, thus developing a more complete model of rowing.

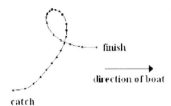

Figure 1: Path of oar blade during the drive phase of a stroke.

The fluid dynamic properties of rowing oar blades have recently been investigated using numerical methods, in particular through the use of computational fluid dynamics (CFD). Some preliminarily investigations have been carried out on flat plates by Leroyer et al. (2008) and Kinoshita et al. (2007). These studies have been able to provide additional details of the flow properties occurring around rowing oar blades, and Coppel et al. (2008) presented a CFD analysis of the Macon oar blade.

The aim of this investigation was to use Caplan & Gardner's (2007a) mathematical model of rowing with both the experimentally derived values of lift and drag coefficients from Caplan & Gardner (2007b) and the CFD predicted results from Coppel et al. (2008). A series of simulations were carried out using the experimental and predicted values, and comparison was made between the two simulations.

METHODS: The mathematical model of rowing used in this investigation (Caplan & Gardner, 2007a) defines the rowing system using Newton's second law ($\sum F = ma$) where the rowing propulsive forces must be applied to the boat to overcome the various sources of resistance. The water resistance is determined experimentally from Wellicome (1967) for an eight, and the air drag is defined from experiments on a seated man by Hoerner (1965) multiplied by the number of rowers. The crew were modelled as a single mass moving back and forth with half of a simple harmonic motion within the boat. The equation of motion for the rowing model can be written as:

$$P - D = m \frac{dv_{shell}}{dt} + M \left(\frac{dv_{crew}}{dt} + \frac{dv_{shell}}{dt} \right) \qquad (1)$$

where, m is the mass of the shell, oars and coxswain, M is the combined mass of the crew, v_{shell} is the absolute velocity of the shell, v_{crew} is the velocity of the crew relative to the boat, P is the propulsive force provided by the rowers and D is the drag force, which is proportional to the velocity of the boat. The rowing model is driven by the change in the angular velocity of the oar shaft about the oarlock caused by the rower pulling the oar handle, and also by the motion of the rower.

The model was built in Simulink (Matlab, Mathworks, USA) and solved using the in-built Runge-Kutta variable rate solver with a maximum time-step of 0.005 s. The model produced a continuous output of all variables with time. Boat velocity was the integral of the modelled boat acceleration. This allowed the performance of the rowing boat to be monitored in terms of attained boat velocity as a function of a given stroke.

A heavyweight men's eight with a combined mass of 740 kg and with a stroke rate of 30 strokes per minute was used. Their drive time was assumed to be 0.92 s and recovery time 1.08 s, giving a total stroke time of 2 s. The Macon oar blade, which is typically used for beginners, and has a projected surface area 0.108 m^2 was used in the simulations and is shown in Figure 2.

In order to determine the lift and drag forces generated through the stroke the lift and drag coefficients of the Macon oar blade as a function of angle of attack were predefined. Here both the values obtained from the experiments of Caplan & Gardner (2007b) and those obtained from the CFD investigation of Coppel et al. (2008) were used, and the outputs compared. Although in rowing the blade moves continuously through a path from catch to finish (as shown in Figure 1), in the experimental and computational work it was only

possible to sample a discrete number of angles of attack. The investigated angles were 0°, 20°, 45°, 90°, 115°, 135°, 160°, 180° and this represented the complete range of angles of attack that the blade goes through during the drive phase of the stroke. A cubic-spline interpolation was used to interpolate the data between each angle of attack. This gave a continuous reading of the values of lift and drag coefficient during the model simulation.
The simulation was started from the catch position and ran for a duration of 20 s to provide data for 10 strokes. A steady state velocity had been achieved by 5 strokes. The last stroke was then isolated and used as comparison between the experimental and CFD results.

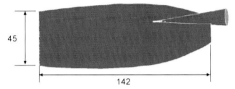

Figure 2: Geometry of Macon blade (dimensions defined in mm).

RESULTS AND DISCUSSION: Table 1 presents the absolute values of lift and drag coefficient for the Macon oar blade at the 9 angles of attack sampled. The values for the CFD predicted and the measured experimental values are given. These coefficients were used as predefined variables in the rowing model (Caplan & Gardner, 2007a).

Table 1: Comparison of measured (Caplan & Gardner, 2007b) and predicted (CFD) lift (CL) and drag (CD) force coefficient values.

Angle (deg)	Experimental		CFD model	
	CL	CD	CL	CD
0	0.02	-0.03	0.00	0.01
20	0.70	0.22	0.64	0.40
45	1.20	1.16	0.95	1.31
70	0.66	1.72	0.53	1.80
90	0.04	1.85	0.04	1.97
115	-0.77	1.74	-0.74	1.76
135	-1.19	1.25	-1.17	1.43
160	-0.80	0.03	-0.38	0.49
180	-0.14	0.04	0.01	0.01

The experimental and CFD model values presented show reasonable correlation with one another, except for some significant variations, for example at 160°. However, it is their influence on boat performance as determined by the output from the rowing model that was considered in this investigation. The boat velocity variation over the same stroke using the experimental values and then the CFD predicted values is shown in Figure 3. There is very little difference in the two boat velocity predictions, with the CFD and experimental coefficients resulting in good agreement in the corresponding boat velocity profiles. In the early part of the stroke, during the initial part of the drive phase, 0<t<0.7 the curves of boat velocity coalesce, although deviate away from each other as the minimum boat velocity is approached. Towards the end of the drive phase and during recovery 0.8<t<2.0 there is a very slight difference between the boat velocity curves.
It was also possible to predict the mean boat velocity over this stroke. The mean boat velocity using the experimental values of force coefficients was 5.25 ms^{-1} and using the CFD predicted values it was 5.32 ms^{-1}. This produced a percentage error of 1.33% between the measured and predicted values. The error could be reduced further by improvements in the CFD predictions of the lift and drag coefficients. This could be achieved by addressing some of the assumptions made in the CFD model, such as employing a free surface condition, rather than an open symmetry boundary, at the top of the flume.

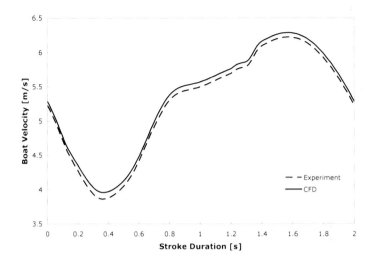

Figure 3: Variation in boat velocity during a rowing stroke. Output when experimental values and CFD predicted values of force coefficient are used.

CONCLUSION: It has been shown that a CFD simulation of the flow around the oar blades can be used in conjunction with a mathematical model of rowing to provide predictions of the motion of a boat in rowing. The results show very good agreement (within 1.33%) with previously published experimental data of boat performance (Caplan & Gardner, 2007b).

REFERENCES:
Brearley, M. N. & de Mestre, N. J. (1996). Modelling the rowing stroke and increasing its efficiency. *3rd conference on mathematics and computers in sport, Bond University, Australia,* 35-46.
Brearley, M. N., de Mestre, N. J., & Watson, D. R. (1998). Modelling the rowing stroke in racing shells. *The Mathematical Gazette, 82,* 389-495.
Cabrera, D., Ruina, A., & Kleshnev, V. (2006). A simple 1+ dimensional model of rowing mimics observed forces and motions. *Human Movement Science, 25,* 192-220.
Caplan, N. & Gardner, T. (2007a). A mathematical model of the oar blade - water interaction in rowing. *Journal of Sports Sciences, 25,* 1025-1034.
Caplan, N. & Gardner, T. (2007b). A fluid dynamic investigation of the Big Blade and Macon oar blade designs in rowing propulsion. *Journal of Sports Sciences, 25,* 643-650.
Coppel, A. L., Gardner, T., Caplan, N., & Hargreaves, D. M. (2008). Numerical modelling of the flow around rowing oar blades. In M. Estivalet & P. Brisson (Eds.), (pp. 353-361). Paris: Springer, Verlag.
Dal Monte, A. & Komor, A. (1989). Rowing and sculling mechanics. In C.L.Vaughan (Ed.), *Biomechanics of sport* (pp. 53-120). Boca Raton, Florida: CRC Press, Inc.
Hoerner, S. F. (1965). *Fluid-dynamic drag.* Albuquerque, New Mexico: Hoerner Fluid Dynamics.
Lazauskas, L. (1997). A performance prediction model for rowing races. *University of Adelaide Department of Applied Mathematics Technical Report, L9702.*
Millward, A. (1987). A study of the forces exerted by an oarsman and the effect on boat speed. *Journal of Sports Sciences, 5,* 93-103.
Nolte, V. (1984). *Die effectivitat des ruderschlages.* Berlin.
Pope, D. L. (1973). On the dynamics of men and boats and oars. In J.L.Bleustein (Ed.), *Mechanics and sport* (pp. 113-130). New York: American Society of Mechanical Engineers.
Sanderson, B. & Martindale, W. (1986). Towards optimizing rowing technique. *Med.Sci.Sports Exerc., 18,* 454-468.
Wellicome, J. F. (1967). Report on resistance experiments carried out on three racing shells. *National Physical Laboratory Ship Division Technical Memorandum, 184.*

VALIDITY OF THE POWERLINE BOAT INSTRUMENTATION SYSTEM

Jennie Coker[1], Patria Hume[1] and Volker Nolte[2]
ISRRNZ, AUT University, Auckland, New Zealand[1]
University of Western Ontario, London, Canada[2]

The PowerLine Boat Instrumentation System[3] is comprised of instrumented oarlocks capable of measuring pin forces in the direction of boat travel and oarlock angles. The aim of this study was to determine the reliability and validity of the force and angle data from the PowerLine Boat Instrumentation System in a laboratory setting. Data were collected with the sculling oarlocks affixed to a horizontally aligned, stabilised wing rigger. For force analysis, signals were collected at 50 Hz from both the PowerLine system and a 1 kN load cell[4] during 10 repetitions at a rate of approximately 30 repetitions per minute. For angular analysis, whilst recording with PowerLine, oarlocks were repositioned for a minimum of two seconds at known angles in a random order using an inclinometer accurate to one tenth of a degree over a range of -80° to +60°, in 20° increments. Linear regression analysis through the origin was used to compare the PowerLine values with known values from the load cell and the inclinometer. Laboratory testing proved the force and angle sensors to be valid throughout the testing range (0 N to 554.8 $\pm$ 20.4 N, and -80° to +60° respectively) when fully functioning. The PowerLine Boat Instrumentation System appears to be appropriate for measuring biomechanical variables in an elite sculling programme. On-water reliability testing is still required to fully evaluate their application in quantifying the effect of interventions made to technique or boat set-up.

KEY WORDS: rowing, biomechanics

INTRODUCTION: The role of an applied rowing biomechanist is to supply coaches with the information they need to analyse rowing technique and boat speed (McBride, 2005). At the elite level, coaches and athletes strive to cut tenths of a second from performance times, thus a high degree of accuracy and reliability is required from any instrumentation used to supply such measures (Baudouin and Hawkins, 2004). Although athlete testing in a laboratory setting will provide a more controlled environment, it will not represent the task as it would be performed in competition (Baca, 2006; Williams and Kendall, 2007). Comparative studies between on-water and ergometer force profiles have highlighted that on-water analysis is the only option for data that truly signifies the rowing performance situation (Dawson et al., 1998; Elliott et al., 2002; Kleshnev, 2008; Lamb, 1989; Li et al., 2007). In providing highly applicable measures, it is also vital that the instrumentation does not interfere with the normal operation of the shell and the sculler (Müller et al., 2000; Smith and Spinks, 1989). The PowerLine Boat Instrumentation System represents a means of providing relevant, on-water data without noticeable change to the athlete set-up. The manufacturers claim accuracy in the force measures of up to two percent of its full scale (an error of up to 40 N) and 0.5° in the angle measures, but independently tested validity of its measures have not previously been documented (Peach Innovations). The aim of this study was therefore to provide independent validity measures for the instrumented sculling oarlocks.

METHODS: To avoid damage to Rowing New Zealand equipment and to control for movement of the pin, all validation was carried out in a laboratory setting. For all procedures, the oarlocks were fixed to a pin, horizontally oriented in a wing rigger as shown in Figure 1. Eight sculling oarlocks were tested in total. Only dynamic force validation could be performed due to an auto-zeroing function built into the oarlocks - the system assumes any force application that remains static is zero and automatic calibration occurs. This was not considered limiting as static forces are not seen in the normal rowing situation.

---

[3] Peach Innovations Ltd, 27 Grantchester Road, Cambridge CB3 9ED, U.K.
[4] Applied Measurement Australia, P.O. Box 159, Oakleigh, Victoria 3166, Australia

Figure 1. Laboratory rigger set-up for all validation procedures.

Using an inclinometer (SmartTool™ Level[5]) the PowerLine logger zeroing function was used to set 0° as the position where the oarlock would be parallel to the midline of the shell – horizontal orientation of the working face of the oarlock in the validation set-up. Dynamic linearity and validity of the force measure was determined by recording sample data whilst a dynamic linear force was manually applied by pulling downwards on a bar hanging from the oarlock with a 1.0 kN load cell suspended in series. At a rate of 30 ± 2 repetitions per minute, 10 repetitions were recorded at 50 Hz from both the PowerLine system and the load cell in a range of 0.0 N to 554.8 ± 20.4 N. Outputs from the load cell were recorded using Labview[6] and output from the logger was downloaded later. The two entire data sets from the 10 repetitions were collated in Excel[7] and analysed, synchronising the data using the first local maximum force reading.

Although all effort was made to apply the force in the vertical direction, some deviation occurred and, because the load value presented by PowerLine is the actual force, resolved in the vertical direction, equation 1 was used to calculate the actual load.

$$L = \frac{L_{vert}}{\cos \alpha} \quad (1)$$

where L is the actual force applied, α is the oarlock angle, and L_{vert} is the resolved vertical force presented by the PowerLine software. Linear regression analyses through the origin were computed in SPSS[8] where the "dependent variable" was the oarlock reading and the "independent variable" was the load cell reading.

For angle validation, the oarlocks were repositioned for a minimum of two seconds at known angles in a random order using an inclinometer (SmartTool™ Level) accurate to 0.1° (MD Building Products, 2007). A range -80° to +60° was used in 20° increments. PowerLine data were downloaded and linear regression analyses through the origin computed in SPSS using the average value from each two second increment in comparison with the known angles.

RESULTS: Table 1 shows the results of the linear regression through the origin of the oarlock reading with the load cell for each of the eight oarlocks. The standard error of the estimate (SEE) was at most 8.9 N for all oarlocks except oarlock #1408 which displayed an SEE of 11.7 N. The R^2 values were all 1.00, except oarlock #1408 that displayed an R^2 value of 0.99. Oarlocks 1401 to 1407 showed a range of 15.5 N to 45.6 N in the maximal error of the estimate for each oarlock.

---

[5] M-D Building Products, OKC Plant, 4041 N. Santa Fe, Oklahoma City, OK 73118, USA
[6] National Instruments, 11500 N Mopac Expwy, Austin, Texas 78759-3504, USA
[7] Microsoft, Washington, WA 98052, USA
[8] SPSS Inc. Headquarters, 233 S. Wacker Drive, 11th floor, Chicago, Illinois 60606

Table 1. Dynamic linearity statistics for the force measures from the sculling oarlocks in the range of 0 N to 554.8 ± 20.4 N compared with force measures from the load cell. Linear regression was computed through the origin.

Oarlock number	Slope	R^2	Standard error of the estimate (N)	Max error (N)
1401	1.01	1.00	8.9	44.5
1402	1.04	1.00	5.1	30.6
1403	0.99	1.00	8.0	39.3
1404	1.04	1.00	7.4	21.2
1405	1.00	1.00	4.3	15.5
1406	1.05	1.00	7.7	45.6
1407	1.01	1.00	4.2	20.3
1408	0.93	0.99	11.7	81.5

Table 2. Static linearity statistics for the angle measures from the sculling oarlocks in the range - 80° to 60° compared with the angle measure from the inclinometer. Linear regression was computed through the origin.

Oarlock number	Slope	R^2	Standard error of the estimate (°)	Max error (°)
1401	1.00	1.00	0.9	1.1
1402	1.01	1.00	0.2	0.7
1403	1.00	1.00	0.4	0.6
1404	1.00	1.00	0.7	1.0
1405	1.00	1.00	0.9	1.4
1406	1.00	1.00	0.3	0.5
1407	1.00	1.00	0.7	1.1
1408	0.98	1.00	3.1	8.2

Results of the regression analyses for oarlock angle versus inclinometer angle are presented in Table 2. The SEE was 0.9° or less for all oarlocks except for oarlock #1408 which had an error of 3.1°. R^2 was 1.00 for all oarlocks. Oarlocks 1401 to 1407 showed a range of 0.5° to 1.4° in the maximal error of the estimate for each oarlock.

DISCUSSION: Apart from sculling oarlock #1408 (which has since been replaced), the force and angle measures proved to have an acceptable level of validity in the range tested in a laboratory setting. In previous repeated short on-water bursts, elite scullers showed typical expected variation between trials of 1.2% in stroke length and 4.9% in peak propulsive force (Soper et al., 2003). This would equate to 1.1° in a sculler with a total arch of 95°, and 29.4 N in a sculler with a peak propulsive force of 600 N. For oarlocks #1401 to #1407, SEE in force was at most 8.9 N, and 0.9° for the angle measure therefore a greater percentage of variation in the overall values will come from the scullers themselves rather than the instrumentation system. For oarlocks #1401 to #1407, the SEE for the force measures falls below the manufacturers' claimed accuracy level of 40 N but the SEE of the angle measure exceeded 0.5° (claimed angle measure error) in four of these oarlocks. The maximal errors are also higher than the manufacturers' error values in some oarlocks. The non-automatic synchronisation method used in this study may account for this higher than anticipated, and potentially over-estimated, maximal error and SEE. Subjective feedback from the scullers who have used the testing system over the past 16 months has shown that there is no alteration to the feel of the boat set-up as long as the pitch of the scullers' usual oarlocks is the same as the instrumented oarlocks. Further investigation is required to determine the on-water reliability of the output variables from the PowerLine Boat Instrumentation System when used by elite scullers.

CONCLUSION: The force and angle measures from the laboratory testing of the PowerLine Boat Instrumentation System for sculling proved to have an acceptable level of validity represented by a standard error of the estimate of 8.9 N or less for force, 0.9° or less for angle, and an R^2 of 1.00 for both variables in all functioning oarlocks over the testing range. Malfunction in one sculling oarlock highlighted the need for regular validity testing.

REFERENCES

Baca, A. (2006). Innovative diagnostic methods in elite sport. *International Journal of Performance Analysis in Sport*, **6**, 148-156.

Baudouin, A., and Hawkins, D. (2004). Investigation of biomechanical factors affecting rowing performance. *Journal of Biomechanics*, **37**(7), 969-976.

Dawson, R.G., Lockwook, R.J., Wilson, J.D., and Freeman, G. (1998). The rowing cycle: Sources of variance and invariance in ergometer and on-the-water performance. *Journal of Motor Behaviour*, **33**(1), 33-43.

Elliott, B.C., Lyttle, A.D., and Birkett, O. (2002). The RowPerfect ergometer: A training aid for on-water single scull rowing. *Journal of Sports Biomechanics*, **1**(2), 123-134.

Kleshnev, V. (2008). How accurately can Concept2 monitor represent the real force curve? *Rowing Biomechanics Newsletter*, **8**(85).

Lamb, D.H. (1989). A kinematic comparison of ergometer and on-water rowing. *American Journal of Sports Medicine*, **17**(3), 367-373.

Li, C.-F., Ho, W.-H., and Lin, H.-M. (2007). Strength curve characteristics of rowing performance from the water and the land. *Journal of Biomechanics*, **40** (Supplement 2), S770

McBride, M.E. (2005). Rowing Biomechanics. In V. Nolte (Ed.), *Rowing Faster* (pp. 111-123). Champaign, IL: Human Kinetics.

MDBuildingProducts. (2007). *SmartTool level specification*. Retrieved 10 December, 2007, from http://www.mdteam.com/products.php?category=1343

Müller, E., Benko, U., Raschner, C., and Schwameder, H. (2000). Specific fitness training and testing in competitive sports. *Medicine and Science in Sport and Exercise*, **32**(1), 216-220.

Peach Innovations. *Oarlock features*. Retrieved 5 December, 2007, from http://www.peachinnovations.com/touroarlocks.htm#

Smith, R., and Spinks, W.L. (1989). *Matching technology to coaching needs: On-water rowing analysis*. Communication to International Symposium on Biomechanics in sports (ISBS)

Soper, C., Hume, P., and Tonks, R. (2003). High reliability of repeated short burst on-water rowing trials. *Journal of Science and Medicine in Sport*, **6**(4, S1), 26-S184.

Williams, S.J., and Kendall, L.R. (2007). A profile of sports science research (1983-2003). *Journal of Science and Medicine in Sport*, **10**(4), 193-200.

ISBS 2009

Oral Session S02

Golf

A COMPARISON OF ACCURACY AND STROKE CHARACTERISTICS BETWEEN TWO PUTTING GRIP TECHNIQUES

Yonghyun Park, Insik Shin, Chulsoo Chung, Gyesan Lee[1] Haksoo Shin[2] and David O'Sullivan

[1]Kwangdon University, Kangwondo, Rep. of Korea
Daegu University, Rep. of Korea,
[2]Sports Biomechanics Lab, Department of Physical Education, Seoul National University, Seoul, Rep. of Korea

Nowadays PGA golfers are experimenting with various golf putting grips. The purpose of this study was to investigate the traits of using two putting grips; reverse overlapping grip and finger bone grip at three different putting distances. 20 subjects with no previous golf experience participated in this study. The kinematic data of the subject and the putter's shaft and head was recorded by 8 Qualisys cameras at 100Hz. There was no significant difference between the success rate of getting the ball in the hole at all distances. The finger bone grip produced statistically smaller radial error values than the reverse overlapping grip at the distances for 7 and 11 metres. The finger bone grip provided straighter putter head trajectories and less change in the movement of the COG, which implies more stability of the player and that the ball will travel in the desired path. In conclusion, the finger bone putting technique gave radial errors less than the reverse overlapping grip technique which seems to be due to the added stability and straighter putter head trajectories.

KEY WORDS: golf putting, grip types, finger bone putting grip, reverse overlap grip.

INTRODUCTION: It is well known that putting plays a large role in the game of golf. According to American PGA statistics from 197 players with an average round of 70.92±0.70 strokes, each player averages 29.3±0.53 putting strokes, which is 41.8% of all strokes in one round (PGA tour homepage). According to Cochran and Stobbs (1968) stated that the most important factors to affect the direction and magnitude of putting is the putter head movement. In previous research, many researchers have focused on the putter's head movement rather than motion or skill of the person controlling the putter. Various studies have demonstrated and discussed the importance of the grip during golf driving and especially putting (Pelz, D. & Frank, J. A., 2000, Paradisis, G. & Rees, J. 2000). With the importance of grip during putting many PGA golfers are continuously researching, developing and testing various golf putting grip techniques, i.e. cross handed, claw, long shaft putter, and crocket grip. The most common of all these techniques is the reverse overlapping grip (Pelz, 1994).

Figure 1: Finger Bone Grip

Lately in Korea a new grip technique has been developed by Il-Ju Na, who called it as 'the finger bone putting grip' shown in Fig. 1. This grip is similar to claw grip but it uses right hand fingers to connect with putter. For right hander players the right hand is to be place at the lower part and likewise for left hander players the opposite.

The artificial lawn made by 100% poly propylene with densely packed 9mm pile length was installed in gymnasium. The lawn's length was 15mX2.8m and by the Stipmeter the rolling length was 19.42. The putter is 33.86 inches in length and weights 500 grams. 20 male university students without any golf experience were recruited for this study. After the explanation of the experimental procedure and signature of the consent form the experiment began. Any left handed subjects were excluded from this study. The experiment started with 5 minutes dedicated to the teaching of putting grip methods, reverse overlapping grip(ROG) and finger bone grip(FBG). Then the subject practiced 15 times for each grip at each distance. The subjects were then partnered up and the putting at 3 different lengths 3m, 7m and 11m for a total time of 60 strokes was randomly assigned. 30 strokes were to be performed with the reverse overlapping grip and the other 30 with the finger bone grip. 10 strokes were performed at each distance with both grips. The full body and the kinematic motion of the putter head and shaft was recorded at 100Hz by 8 Qualisys© OQUS 500 cameras (Sweden). Data was recorded with Qualisys Track Manager (QTM) and then the data was exported as a c3d file and analyzed by Visual 3D (version 4). The events were defined as follows; set up(SU) event was the position at the address, beginning of back swing (BB) was the point at the start of the back swing, transition(TR) was the point at the end of the back swing and the beginning of the forward swing, impact(IM) with the ball and finish(FN) at the end of the front swing.

Radial error is defined a distance between the hole and the ball which putted without regard for direction. The change of centre of gravity shown in Figure 3 was calculated in the axis in the direction of the putting. After averaging the coordinates of the 3 points, 15 cm before the IM, at IM and 15 cm after IM, the radius of curvature was calculated and shown below in Table 3. The angle between the right shoulder, left shoulder and left wrist is used as an index of how independent the movement of the arms from the trunk.

RESULTS:

1. Putting success rate

Table 1 Frequency success according to distance

	3m	7m	11m
Reverse Overlapping Grip(ROG)	29.50%	8.50%	1%
Finger Bone Grip(FBG)	29.50%	12.50%	2.50%

2. Radial Error

Table 2 Radial Error according to distance

		3m	7m	11m
ROG	Mean	72.12	105.07	138.3
	St. deviation	42.09	47.24	31.84
FBG	Mean	74.54	83.03**	118.64**
	St. deviation	30.75	28.74	20.31
p-value		0.390	0.019*	0.009**

(all values are in cm, *p<0.05, **p<0.01)

3. Putter Head Trajectory

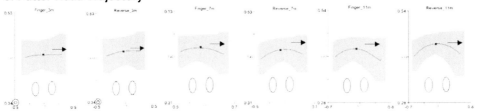

Figure 2: Putter Heads Trajectory for the distance of 3, 7 and 11 m. Finger bone technique on the left side of each(TR-FN, horizontal plane)

Table 3 Radius of curvature (horizontal plane)

	3m	7m	11m
ROG	369.76cm	94.55cm	54.21cm
FBG	1434.27cm	97.60cm	61.00cm

※ Radius of curvature was calculated with 3 points. These 3points which are averaged by all trials are collected at IM-15(cm), IM, IM+15(cm).(Y-axis)

4. Motion Data Analysis

Changes of COG are measured for stability while subjects are putting.

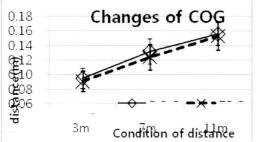

Figure 3 Changes of COG in putting direction

Table 5 Change in angle between the Right Shoulder, left shoulder and left wrist(TR-IM)

		3m	7m	11m
ROG	Mean	1.69	2.17	2.63
	St. deviation	2.11	2.77	3.15
FBG	Mean	4.10	5.05	4.10
	St, deviation	4.43	3.60	2.43
p-value		<0.00**	<0.00**	<0.00**

(All values are in degrees, *p<0.05, **<0.01)

DISCUSSION: There is a not a significant in between the success rate between the distances for the two grip types. As for the radial error there was no difference between the two grip types for the 3 metre putting. However there was a statistical difference between the radial error between the finger bone putting technique and the reverse overlapping technique for both longer distances of 7 m and 11 m. The bone finger putting radial error was smaller for both of these distances. The mean radial error at the 7 m distance was 83.03

cm which was approximately 22.04 cm closer for the finger bone grip. Likewise at the distance of 11m the mean radial error was 118.64 for the finger bone putting as opposed to 138.30 cm for the reverse overlapping grip. At the 11 m distance, the subject using the finger bone grip technique putted the ball 19.66 cm closer to the hole.

It is assumed that the reason that the finger bone grip produces putting closer to the hole is that because the putter head trajectory is straighter and there is less COG movement, the player is more in control of the putter.

By observation of the putter's head trajectory it can clearly be seen that the finger bone grip technique displays a straighter trajectory(shown in table 3). For the change of COG in putting direction, finger bone putting grip is more stable on all condition than reverse overlapping grip but there is no significant difference.(shown by fig. 3) and For the change in angle between the right shoulder, left shoulder and left wrist there was a significant difference for between the two grip techniques. The finger bone technique had a statistically larger change in angle (shown by table 5). The change in angle can be used as a index of how independent the movement of the arms from the trunk.

CONCLUSION: This study identified the different biomechanical characteristics of using the two grip methods finger bone grip and the reverse overlapping grip for putting at different distances. The main advantage of using the finger bone grip technique is that for the further distances it tends to get the ball closer to the hole.

This difference of accuracy is assumed to be caused by the finger bone putting grip to be more stable than overlapping grip, as the change of COG on finger bone grip is smaller than reverse overlapping grip's (shown in Fig.3). This suggests that the finger bone grip has more unlocking feature between trunk and upper arms than reverse overlap grip i.e. the trunk and the arms can move more independently.

In future studies, the evaluation of the effect that the small COG movement has on the performance of putting must be investigated, as it is stated by instructors that the locking of the trunk and arms is supposed to be ideal for putting. However in this research, the Finger Bone grip produced less locking with more accurate putting. It is also recommended that for future research, the balls trajectory path and rotation should be examined to explain what happens the ball after impact and to explain why the Finger Bone grip method putts the ball closer to the hole.

REFERENCES:

Cochran, A. & Stobbs, J. (1968) The Search for the Perfect Swing. Golf Society of Great Britain, Heinemann, London.

Delay, D., Nougier, V., Orliaguet, J.P. & Coello, Y. (1997). Movement control in golf putting. Human Movement Science 16(5): 597-619.

Pelz, D. & Frank, J. A.(2000). Dave Pelz's Putting Bible: The Complete Guide to Mastering the green. NY: Doubleday Publishers.

McCarty, J. D. (2002). A descriptive analysis of golf putting: what variables affect accuracy? Master of Science thesis, Purdue University, USA.

Paradisis, G. & Rees, J. (2000). Kinematic analysis of golf putting for expert and novice golfers. The 18th International Symposium on Biomechanics in Sports: proceedings, Hong Kong, China.

Acknowledgement

This work was supported by the Korea Science and Engineering Foundation(KOSEF) grant funded by the Korea government(MOST) (No. R11-2007-028-02001-0).

ANALYSIS OF THE JOINT KINEMATICS OF THE 5 IRON GOLF SWING

Aoife Healy[1], Kieran Moran[1], Jane Dickson[1], Cillian Hurley[1], Mads Haahr[2], & Noel E. O Connor[3]

[1]School of Health and Human Performance, [3]CLARITY: The Centre for Sensor Web Technologies, Dublin City University, Dublin, Ireland; [2]Department of Computer Science, Trinity College, Dublin, Ireland

> The purpose of this study was to identify the performance determining factors of the 5-iron golf swing. Joint kinematics were obtained from thirty male golfers using a twelve camera motion analysis system. Participants were divided into two groups, based on their ball launch speed (high vs. low). Those in the high ball speed group were deemed to be the more skillful group. Statistical analysis was used to identify the variables which differed significantly between the two groups, and could therefore be classified as the performance determining factors. The following factors were important to performance success: (i) the ability of the golfer to maintain a large X Factor angle and generate large X Factor angular velocity throughout the downswing, (ii) maintain the left arm as straight as possible throughout the swing, (iii) utilise greater movement of the hips in the direction of the target and a greater extension of the right hip during the downswing and (iv) greater flexion of both shoulders and less left shoulder internal rotation during the backswing.
>
> **KEY WORDS:** golf, joint kinematics, 5 iron, ball launch speed.

INTRODUCTION: Golf is played by 10-20% of the adult population in most countries (Thériault and Lachance 1998). The full golf swing using the iron clubs to strike the ball is a key element of success in golf. Therefore in order to help enhance golfing performance it is important to identify the "performance determining factors" of the full golf swing. Comparison of skilled and lesser skilled golfers' joint kinematics allows for the identification of these performance determining factors. Unfortunately, previous research has focused on the driver club despite the fact that either an equal or even a higher proportion of shots for maximum distance in the game of golf are taken with iron clubs. Only two studies (Budney and Bellow 1982, Cheetham et al. 2001) appear to have compared the kinematics of the golf swing of skilled and unskilled golfers using an iron club. There is a need therefore for research that focuses on golfing performance using the iron clubs. The aim of the present study is to identify the performance determining factors of the 5 iron golf swing through analysis of joint kinematics of skilled and lesser skilled golfers.

METHODS: Participants: 30 male right handed golfers (34 ± 16 yrs.) were recruited for the study. Ethical approval was received from the Ethics Committee at Dublin City University. All participants were injury free at the time of the test. Participants underwent a familiarisation session in the laboratory prior to initiation of testing.

Data Collection: Forty one reflective spherical markers were placed on anatomical landmarks on the body. A 12 camera (250 Hz) VICON motion analysis system (VICON 512 M, Oxford Metrics Ltd, UK) was used to record the motion of the participant throughout the golf swing. The testing session consisted of a prescribed warm up, recording of fifteen golf swings and a participant selected cool down period. The prescribed warm up consisted of five minutes walking on a treadmill (2.5 $km.h^{-1}$) followed by 3mins of practice swings. The participants were instructed to 'hit the ball as hard as possible towards the target-line, with the aim to maximize both distance and accuracy, as if in a competitive situation'. The ball was hit from a tee on a Pro V swing analyser (Golftek Inc., USA) into a net located three metres from the swing analyser using their own 5 iron golf club.

Data Analysis: X factor (relative rotation of the shoulders with respect to the hips), shoulder, elbow, wrist, hip and knee angles and angular velocities were calculated using the 'golf' model (Vicon, Oxford Metrics Ltd, UK). The angle and angular velocity of each variable was obtained at each of the eight key events during the swing (Figure 1). Results for each participant's top three trials with regard to ball speed were averaged to create a representative trial of their best swing.

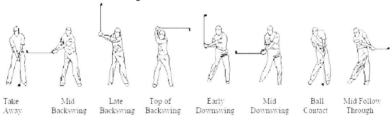

Figure 1: Key swing events. Adapted from Ball and Best (2007).

Statistical Analysis: Participants were divided into two groups, fifteen in each group, based on their ball launch speed [high ball speed (HBS) (52.9 ± 2.1 m.s^{-1}; handicap 0.1 ± 6.0) and low ball speed (LBS) (39.9 ± 5.2 m.s^{-1}; handicap 9.7 ± 7.5)]. Independent t-tests were used to assess differences between the groups (α = 0.05).

RESULTS: Due to the large number of variables analysed and subsequent significant differences found only those variables that showed differences across a minimum of three consecutive key events are presented. Results are presented for the X Factor angle (Table 1), left elbow flexion/extension angle (Table 2), right hip abduction/adduction angle (Table 3), right hip flexion/extension angle (Table 4), left and right shoulder flexion/extension angles (Table 5) and left shoulder internal/external rotation angles (Table 6). A relevant selection of angular velocity results are also included.

Table 1 Significant differences between groups (means ± standard deviations) for X Factor angle (°) and angular velocity (°.s^{-1})

X Factor	Group	Early Downswing	Mid Downswing	Ball Contact	Mid Follow Through
Angle (°)	HBS	-42.6 ± 6.3	-38.5 ± 6.1	-36.3 ± 5.3	-15.5 ± 11.2
	LBS	-35.1 ± 6.5	-30.4 ± 6.3	-26.8 ± 7.4	1.8 ± 13.3

Table 2 Significant differences between groups (means ± standard deviations) for left elbow flexion/extension angle (°) [flexion (+)/extension (-)]

Group	Mid Backswing	Late Backswing	Top of Backswing	Early Downswing
HBS	21.2 ± 5.8	26.5 ± 6.4	42.1 ± 10.6	32.2 ± 8.6
LBS	29.5 ± 11.3	36.2 ± 11.9	52.3 ± 9.8	43.6 ± 8.7

Table 3 Significant differences between groups (means ± standard deviations) for right hip abduction/adduction angle (°) [adduction (+)/abduction (-)]

Group	Early Downswing	Mid Downswing	Ball Contact
HBS	-16.97 ± 6.65	-25.35 ± 5.8	-27.14 ± 5.26
LBS	-3.97 ± 7.79	-14.19 ± 7.52	-18.53 ± 6.02

Table 4 Significant differences between groups (means ± standard deviations) for right hip flexion/extension angle (°) and angular velocity (°.s^{-1}) [flexion (+)/extension (-)]

Variable	Group	Early Downswing	Mid Downswing	Ball Contact	Mid Followthrough
angle (°)	HBS		18.9 ± 9.2	2.3 ± 9.4	-10.5 ± 9.5
	LBS		30.2 ± 13.9	14.5 ± 13.9	-0.2 ± 11.7
angular velocity (°.s^{-1})	HBS	233.5 ± 87.3	443.2 ± 115.2		
	LBS	77.1 ± 115.9	290.4 ± 106.7		

Table 5 Significant differences between groups (means ± standard deviations) for left and right shoulder flexion/extension angle (°) and angular velocity (°.s^{-1}) [flexion (+)/extension (-)]

Shoulder	Variable	Group	Mid Backswing	Late Backswing	Top of Backswing	Early Downswing
Left	angle (°)	HBS	46.9 ± 14.1	78.4 ± 12.3	114.1 ± 17	
		LBS	32.9 ± 15.6	55.8 ± 17.5	90.6 ± 19.8	
	angular velocity (°.s^{-1})	HBS	138.2 ± 62.5	258.2 ± 131.9		494.5 ± 200.3
		LBS	91.3 ± 56.5	150.6 ± 85.4		36.3 ± 144.4
Right	angle (°)	HBS	40.6 ± 10.1	47.1 ± 9.8	57.3 ± 10.6	
		LBS	29.4 ± 8.9	33.9 ± 12.7	44.2 ± 15.9	
	angular velocity (°.s^{-1})	HBS				206.0 ± 69.3
		LBS				114.9 ± 71.7

Table 6 Significant differences between groups (means ± standard deviations) for left shoulder internal/external rotation angle (°) and angular velocity (°.s^{-1}) [internal (-)/external (+)]

Variable	Group	Mid Backswing	Late Backswing	Top of Backswing	Early Downswing
Angle (°)	HBS	-49.5 ± 17.6	-42.5 ± 15.1	-28.3 ± 20.9	
	LBS	-66.9 ± 15.2	-62.9 ± 14.6	-44.4 ± 17.7	
angular velocity (°.s^{-1})	HBS				198.0 ± 238.2
	LBS				36.3 ± 144.4

DISCUSSION: Cheetham et al. (2001) was the only previous study found to examine the X Factor angle when using an iron club. Their study measured the X Factor at our fourth swing event, the top of the backswing. Similar to the present study they found no significant difference between their highly skilled and less skilled golfers. This finding for the 5 iron differs to recent research on the driver which has found significant differences for the X Factor at the top of the backswing between golfers of varying skill level (Myers et al. 2008, Zheng et al. 2008). In the present study the HBS group had a significantly greater X Factor angle at the events of early downswing (ED), mid downswing (MD) and ball contact (BC) (Table 1). They also exhibited a significantly greater X Factor angular velocity at mid downswing and ball contact (Table 1). These results suggest that the X Factor angle at the top of the backswing may not be the most important phase for the X Factor angle. It may be the ability of the golfer to maintain a larger X Factor angle and generate greater X Factor angular velocity throughout the downswing that contributes to producing higher ball speeds. No previous studies using an iron club have examined this. The HBS group kept their left elbow more extended than the LBS group at four consecutive events (from mid backswing to early downswing) (Table 2). The postulated benefit of this is the more extended a golfer keeps their arms the greater the velocity the club head is capable of generating since the club head travels through a longer arc in a given time and therefore moves faster (Broer 1973). The right hip was significantly more abducted for the HBS group (Table 3) resulting in them moving their hips more in the direction of the target during the downswing than the LBS group. The right hip for the HBS group was also more extended from mid downswing to mid follow through (Table 4); this is likely to aid in the faster transfer of weight to the front foot. Greater right hip flexion/extension angular velocity was evident for the HBS group at early and mid downswing (Table 4). This ability to generate higher velocities early in the concentric phase (downswing) by the HBS group possibly contributed to their greater club head speed by employing a more enhanced utilisation of the stretch shortening cycle. The increased velocity of the hips early in the downswing also possibly indicates the HBS group's superior use of proximal to distal sequencing as they reached higher velocity of the proximal segment (hips) early in the concentric movement which possibly led to their higher velocity at the distal segment (club head). Results for left and right shoulder flexion/extension angles (Table 5) showed that the HBS group flexed their shoulders more during the backswing (at events mid backswing, late backswing and top of backswing), thereby utilising a greater range of motion in the backswing. This appears to have allowed the HBS group to produce greater extension angular velocity in both shoulders at early downswing. This is possibly due to effective utilisation of both the impulse-momentum relationship and utilisation of the stretch shortening cycle; with a greater angular velocity of the left shoulder by the HBS group evident at mid and late backswing. Higher velocities during the backswing increase eccentric loading, which increases the potential for enhancement in the concentric phase (downswing) through the stretch shortening cycle (Cavagna et al. 1968, Bosco et al. 1981), as evident by the significantly greater angular velocity during early downswing for the HBS group. The HBS group used less left shoulder internal rotation than the LBS group during the backswing (Table 6), despite producing a significantly greater angular velocity during early downswing. The smaller rotation may make use of an enhanced utilisation of the stretch shortening cycle (Moran and Wallace 2007).

CONCLUSION: Clear performance determining factors are evident for swinging a 5 iron golf club to hit a ball as far and as accurately as possible. It is likely that maintaining a large X Factor angle and increasing X Factor angular velocity during the downswing could benefit golfers in increasing their ball launch velocity with the 5 iron club. In relation to the arms, greater ball velocity may be generated by maintaining the left arm as straight as possible throughout the swing, as this would increase the arc the club head travels through and therefore increase its velocity. A greater transfer of weight from the back to the front foot from early downswing through to ball contact brought about by greater movement of the hips in the direction of the target and a greater extension of the right hip allows greater force generation in the direction of the target, which can be transferred to the club to produce greater ball velocity. A greater flexion of both shoulders and less left shoulder internal rotation during the backswing in the direction of the target could also contribute to greater ball velocity generation.

REFERENCES:
Ball, K. A. and Best, R. J. (2007). Different centre of pressure patterns within the golf stroke II: group-based analysis. *Journal of Sports Sciences*, 25 (7), 771-779.
Bosco, C., Komi, P. V. and Ito, A. (1981). Prestretch potentiation of human skeletal muscle during ballistic movement. Acta Physiologica Scandinavica, 111, 135-140.
Broer, M.R. (1973). *Efficiency of Human Movement*. 3rd. Philadelphia: W.B. Saunders
Budney, D. R. and Bellow, D. G. (1982). On the Swing Mechanics of a Matched Set of Golf Clubs. *Research Quarterly for Exercise and Sport*, 53 (3), 185-192.
Cavagna, G. A., Dusman, B. and Margaria, R. (1968). Positive work done by a previously stretched muscle. *Journal of Applied Physiology*, 24 (1), 21-32.
Cheetham, P.J., Martin, P.E., Mottram, R.E. and St. Laurent, B.F. (2001). The Importance of Stretching the "X factor" in the Downswing of Golf: The "X-Factor Stretch". *IN:* Thomas, P.R. *Optimising Performance in Golf.* Australia: Brisbane Australian Academic Press. 192-199.
Moran, K. A. and Wallace, E. S. (2007). Eccentric loading and range of knee joint motion effects on performance enhancement in vertical jumping. *Human Movement Science*. 26 (6), 824-840.
Myers, J., Lephart, S., Tsai, Y. S. (2008). The role of upper torso and pelvis rotation in driving performance during the golf swing. *Journal of Sports Sciences*, 26 (2), 181-188.
Thériault, G. and Lachance, P. (1998). Golf Injuries: An Overview. *Sports Medicine*, 26 (1), 43-57.
Zheng, N., Barrentine, S. W., Fleisig, G. S. and Andrews, J. R. (2008). Kinematic analysis of swing in pro and amateur golfers. *International Journal of Sports Medicine*, 29 (6), 487-493.

ANALYSIS OF ELITE GOLFERS' KINEMATIC SEQUENCE IN FULL-SWING AND PARTIAL SWING SHOTS

Fredrik Tinmark, John Hellström, Kjartan Halvorsen, Alf Thorstensson

The Swedish School of Sport and Health Sciences (GIH), Stockholm, Sweden

KEY WORDS: Motor control; Multi-joint movements; Skilled golf players

INTRODUCTION: Proximal-to-distal sequencing (PDS) has been observed in full-swing golf shots as in most throwing and striking skills, where the main goal is to maximize speed in the most distal segment of an open-link system (Zheng et al., 2007). Although PDS primarily is associated with mechanical advantage when the speed requirement is high, this temporal order has also been found and ascribed various merits in relatively slow multi-joint movements (Furuya & Kinoshita, 2007). However, no research to date has examined the sequencing pattern in partial golf shots to submaximal distances. The purpose here was to investigate whether PDS is a common characteristic also in partial swing shots of skilled golf players.

METHODS: A total of 47 golfers were investigated, 11 male tournament professionals, 23 male amateurs (HCP 0 ± 2 strokes), and 13 female amateurs (HCP -2 ± 2 strokes) performed partial shots with a wedge to targets at three discrete distances (40, 55 and 70 m), and full-swing shots with a five iron as well as a driver in the same direction for maximal distance. Pelvis, upper torso, and hand movement were recorded in 3D with an electromagnetic tracking system (Polhemus) at 240 Hz. The magnitude of the resultant angular velocity vector of each segment was used to examine the sequencing pattern and the angular speed of segment motions. Movement onset, peak amplitude and time for peak amplitude were analyzed in separate repeated-measure ANOVAs with pre-planned Bonferroni corrected pairwise comparisons. Significance level was set at $P < 0.05$.

RESULTS: This study showed a significant proximal-to-distal temporal relationship of movement onset and maximum angular speed at the pelvis, upper torso and hand segments in the golf swing. The same temporal structure was evident in all test conditions, as well as among different genders and levels of expertise. However, the increment in angular speed from the upper torso to hand were significantly larger for male professionals than for female amateurs at all shot conditions and significantly larger for male amateurs than for female amateurs at full-swing shots.

DISCUSSION AND CONCLUSION: While there exists a body of evidence in support for PDS providing mechanical advantages when the highest possible ball speed is to be achieved, merits of PDS in partial golf shots are less evident. However, it has been proposed that a given torque or force can be more accurately generated by a stronger muscle than a weaker muscle (Hamilton et al., 2004) and a potential role of the observed sequencing pattern in partial shots of skilled golf players could be to improve accuracy and minimize the speed-accuracy tradeoff.

REFERENCES:
Zheng N., Barrentine S.W., Fleizig G.S., Andrews J.R. (2008). Kinematic analysis of swing in pro and amateur golfers, *International Journal of Sports Medicine*, 6, 487-493.
Furuya S., Kinoshita H. (2007). Roles of proximal-to-distal sequential organization of the upper limb segments in striking the keys by expert pianists, *Neuroscience Letters*, 421, 264-269.
Hamilton A., Jones K.E., Wolpert D.M. (2004). The scaling of motor noise with muscle strength and motor unit number in humans, *Experimental Brain Research*, 157, 417-430.

GOLF DRIVE LAUNCH ANGLES AND VELOCITY: 3D ANALYSIS VERSUS A COMMERCIAL LAUNCH MONITOR

Matthew Sweeney[1], Jacqueline Alderson[1], Peter Mills[2], and Bruce Elliott[1]

[1] School of Sport Science, Exercise and Health. University of Western Australia, Perth, Australia; [2] School of Physiotherapy and Exercise Science. Griffith University, Gold Coast, Australia

The purpose of this study was to compare initial ball direction and velocity for a golf drive collected using a commercially available launch monitor and 3D data, collected using a retro-reflective motion analysis system. Six golfers (handicap: 2-20) completed 10 drives each, with data simultaneously recorded by both a 12 camera Vicon MX system (400Hz) and a Vector Pro launch monitor. Both systems produced outputs for launch angle, side angle and ball velocity. The launch monitor data were compared against the 'benchmark' 3D results and showed a high correlation (0.93 – 0.96). Mean errors (launch angle 0.5°, side angle 1.1°, ball velocity 1.1m/s) were also relatively small. The results of the study suggest that if a high speed 3D system is not available or practical, a launch monitor such as the one tested, should provide accurate measurements of golf ball launch data.

KEY WORDS: golf, ball velocity, kinematics

INTRODUCTION: The effectiveness of each golf shot is dictated at all times by the inter-relationship of distance and accuracy. Although external factors including; wind, air density and friction of the landing surface play a role in this outcome, the components controlled by the player are the initial ball velocity and direction, as well as the spin imparted on the ball. The ability to accurately measure these component variables is important in attempting to better explain the outcome of a particular shot. Being able to predict the outcome of a shot in situations where the ball is not able to reach its potential endpoint, such as in a laboratory environment may also be valuable.

While launch monitors are most commonly used by golf coaches and club fitters to produce quantitative measurements of specific components of ball flight, they have also been used to obtain quantitative data that form the basis of scientific studies (Myers et al., 2008). Most launch monitors purportedly measure ball direction, velocity and spin, however their accuracy has yet to be validated. Conversely, while the accuracy of 3D retro-reflective measurement systems in tracking objects over space is established (CMMAS 2002, Richards, 1999), they have yet to be employed in the measurement of ball spin rate or spin axis of rotation in golf. This study will aim to compare the results of a Vicon MX system with a Vector Pro launch monitor in measuring initial golf ball direction and velocity.

METHODS: Six male golfers with handicaps ranging from 2 to 20 were recruited for the study. After a warm up, each golfer was required to hit 10 drives with their own driver, using their natural swing to achieve both distance and accuracy. The drives were performed in a laboratory setting, whilst standing on an artificial golf surface, with a net situated approximately 5 m in front of the participant. Each drive was recorded using a 12 camera Vicon MX (Oxford Metrics, Oxford, UK) system operating at 400 Hz, as well as a Vector Pro 2 launch monitor (Accusport, Winston-Salem, USA). The golf balls were new Titleist NXT's, with retro reflective tape attached to one side and two marks drawn on the other in accordance with the calibration procedures recommended by the manufacturer of the launch monitor (Figure 1).

The Vicon system is a retro-reflective system that uses multiple cameras to track the position of reflected light in a 3D space over time. As there was reflective tape on the golf ball it was treated as a single marker with its position tracked in the laboratory space over 5 m of flight before reaching the net.

At each time point, the velocity was calculated using the change in displacement from the previous frame divided by the change in time between frames. The launch and side angles reported using the 3D system were both taken at the time point of maximum velocity. Both the launch and side angle of the golf ball at every frame was calculated using the following equations.

$$Launch\ angle\ \theta = \tan\theta = (Zc - Zi) \div (Xc - Xi)$$
$$Side\ angle\ \theta = \tan\theta = (Xc - Xi) \div (Yc - Yi),$$

Where Xc is current position of ball in forward direction, Xi is initial position of ball in the forward direction, Yc is current position of ball in lateral direction, Yi is initial position of ball in the lateral direction, Zc is current position of ball in vertical direction and Zi is initial position of ball in the vertical direction

In contrast to the 3D system, the launch monitor uses two photographs of the golf ball taken immediately following impact from two cameras located in the unit to calculate ball direction, velocity and spin. Velocity is differentiated from the forward displacement (horizontal) of the ball between the photos, launch angle is calculated from the vertical angle created of the ball between the two photos and side angle calculated by the difference in size of the ball between the two shots (Figure 1).

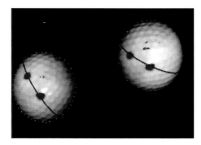

Figure 1: Composite picture of two shots taken by launch monitor with two marks and perimeter of the ball identified by the software.

The common variables examined across both systems were launch angle, side angle and ball velocity. Launch angle was defined as the angle created vertically between the ground and the golf ball direction, whilst the side angle is defined as the angle created horizontally between a line going directly toward the target and the actual ball direction

Due to the accepted accuracy and reliability (CMMAS 2002, Richards, 1999) the 3D system was used as the benchmark for comparing the two approaches.

RESULTS: The reported outputs from each trial (n=60), for launch angle, side angle and velocity are plotted between the two methods in Figure 2.

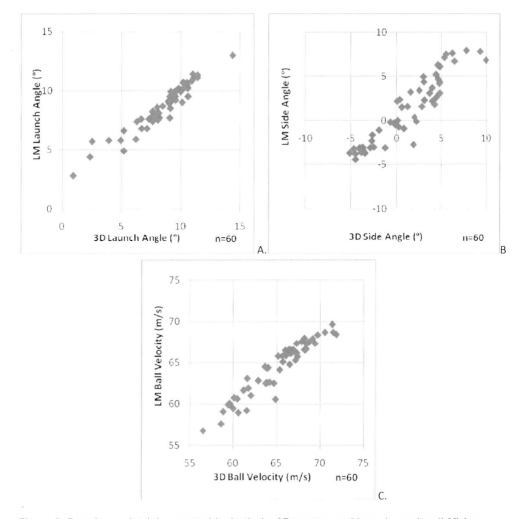

Figure 2: Data for each trial outputted by both the 3D system and launch monitor (LM) for launch angle (A), side angle (B), and ball velocity (C).

Correlations were run between methods, for each of the common measurements to give an indication of the relationship between the outputs (Table 1). Also, the mean error (absolute) and maximum error of the launch monitor outputs, compared with the 3D system, were calculated and reported alongside the standard deviation of the 3D data in Table 1.

Table 1: Mean correlation, mean error, maximum error and 3D standard deviations for the two data collection procedures.

Variable	Correlation	Mean Error	Maximum Error	Std Dev of 3D data
Launch Angle	.96	0.5° (±.6)	3.2°	2.4°
Side Angle	.93	1.1°(±.9)	4.6°	3.9°
Ball Velocity	.95	1.1m/s(±1.0)	4.8m/s	14.3m/s

DISCUSSION: Launch monitor data compared favourably with the benchmark 3D system for all variables analysed. Further, the mean error of each of variable was small in comparison to the standard deviation of the 3D data. This would indicate a good level of agreement between the 3D system and the launch monitor, for the variables analysed.

Amongst the two common variables concerned with direction (launch and side angle), outputs for launch angle produced the highest correlation (.96), as well as the smallest mean error (0.5°). In comparison the outputs for side angle, although also having a very strong correlation (.93) produced a higher mean difference (1.1°). This may be explained by difference in accuracy between predicting an angle based on the distance of a spherical object away from a 2D camera position, rather than one calculated based on the absolute position of the ball in two shots moving in a plane perpendicular to the camera.

CONCLUSION: There was a high level of agreement between the two methods of producing launch angle, side angle and velocity of the golf ball. Caution, however should be taken when collecting the above variables with a launch monitor as the maximum errors reported could produce misrepresentative data.

Although a 3D system with a high frame rate may still be the optimal for collecting launch data, it is not always feasible or practical and therefore a launch monitor, such as the one used in this study could provide suitable measurements for coaching, club fitting and selected research designs.

REFERENCES

Myers, J., Lephart, S., Tsai, Y., Sell, T., Smoliga, J., and Jolly, J. (2008). The role of upper torso and pelvis rotation in driving performance during the golf swing. *Journal of Sports Sciences*, 26(2), 181-8

Results: Basic measurement accuracy, and processing time. Paper presented at Comparison Meeting of Motion Analysis Systems 2002; Japan Technology College, Tokyo, Japan. http://www.ne.jp/asahi/gait/analysis/comparison2002/Result/basic/basic_eng.html

Richards, J. (1999). The measurement of human motion: A comparison of commercially available systems. *Human Movement Science*, 18 (5), 589-602

Acknowledgment
The authors of this paper would like to greatly thank Titleist for the use of the launch monitor and assistance in technical difficulties through the research.

A STUDY ON THE RELATIONSHIP BETWEEN EYE MOVEMENT AND GRIP FORCE DURING GOLF PUTTING

Hyung-Sik Kim[1], Jin-Seung Choi[1], Gye-Rae Tack[1], Young-Tae Lim[2], and Jeong-Han Yi[1]

Biomedical Engineering, Konkuk University, Chungju, Korea[1]
Sports Science, Konkuk University, Chungju, Korea[2]

KEY WORDS: Eye Movement, Grip Force, Head Movement, Putting.

INTRODUCTION: Golf is becoming a very popular sports game. In golf games, there are four kinds of strokes: driving, Iron, chip shots and putting stroke. Among the strokes, putting stroke accounts between 42% and 45% in a golf rounding(Pelz, 2000) and most golfers have difficulties on doing it. There are various parameters affecting the results of a putting stroke but effects of a few parameters were studied(Delay, 1997). However, there is a lack of studies for the simultaneous measurement and analysis of eye movement, grip force, and kinematic parameters with putting results. A purpose of this study was to quantify the parameters and identify the relationship between eye movement, grip force, and putting results.

METHODS: We conducted sets of experiments in our indoor lab facility. We designed the experimental set-up as close to a real putting green as possible. A hole (108mm in diameter) was placed on a 2m x 10m artificial green putting surface. Three groups of participants were recruited: professional (handicap<3avg.), experienced (>1year), and novice (<6month) groups. Before the experiment, all the participants had enough time to adapt to the experiment set-up. The measurement system consists of GF (grip force), EOG (electrooculogram) and 3-axis accelerometers. We developed the system using low power ATMEGA128L micro-processor and it had bluetooth wireless link to transfer the measured data sets to a computer. During the experiments, 3D motion analysis system (Motion Analysis Corp., USA) with 6 infrared cameras was synchronized with the measurement system. All the participants performed identical putting trials of 2.1m distance. The data were analyzed using MATLAB (Mathworks Co., USA).

RESULTS AND DISCUSSION: Brief results of eye movement are shown in Table 1.

Table 1 The result of eye movement with each group

Novice player (n=8)		Professional player (n=8)	
Mean	STD	Mean	STD
7.28%	5.89%	3.19%	2.33%

In this study, relationship between the putting results and three different measured parameters such as eye movement, grip force and kinematic parameters will be more investigated.

CONCLUSION: In this study, we are finding relationships between putting results and measured values of eye movement, grip force during the putting strokes. We expect strong relationship between putting result and eye movement and also different tendencies of grip forces among three participant groups.

REFERENCES:

Pelz, D. (2000). *Dave Pelz's putting bible*. New York: Doubleday.
Delay, D., Nougier, V., Orliaguet, J.P., and Coello, Y. (1997) *Moment Control in Golf Putting*, Human Movement Science, 16(5), 597-619

Joan N. Vickers (2007). *Perception, Cognition, and Decision Training: The Quiet Eye in Action*. Human Kinetics.
Acknowledgement

This research was supported by the Sports Promotion Fund of Seoul Olympic Sports Promotion Foundation grant S07-2008-15 from Ministry of Culture, Sports and Tourism, Korea.

ISBS 2009

Oral Session S03

Swimming

MARKERLESS ANALYSIS OF SWIMMERS' MOTION: A PILOT STUDY

Elena Ceseracciu[1], Zimi Sawacha[1], Silvia Fantozzi[2,3], Giulia Donà[1], Stefano Corazza[4], Giorgio Gatta[3], Claudio Cobelli[1]

Department of Information Engineering, University of Padova, Padova, Italy[1]
DEIS, University of Bologna, Bologna, Italy[2]; Faculty of Exercise and Sport Sciences, University of Bologna, Bologna, Italy[3] Animotion Inc. San Francisco California USA[4]

KEY WORDS: markerless, swimming

INTRODUCTION: Regular laboratory-based motion analysis with skin surface markers is not always feasible. In particular, when studying swimmers kinematics, traditional motion capture techniques cannot be adopted. Although video recordings from swimmers often exist, current methods for biomechanical analysis of these are inadequate. They usually rely on manual digitization of joints' position on a single sagittal view of the subject. Therefore, in this study a method for three dimensional (3D) markerless motion capture of swimmers is presented. The method adopts the markerless motion capture system developed at Stanford University.

METHODS: An elite swimmer performing free style was acquired employing 5 synchronized subaqueous CCTV colour cameras and Canopus ADVC-55 A/D converters (PAL interlaced video, 25frames/sec). Cameras calibration was performed with Bouguet method; intrinsic parameters were obtained with a dry calibration, then corrected for underwater application. Silhouette extraction (Fig.1) was performed employing a Gaussian mixture algorithm implemented in the Intel OpenCV library, which creates an adaptive model of the background; a priori information, in terms of an extra "white" Gaussian component of the model, was included in order to deal with the presence of the foam. From the intersection of these silhouettes' back-projections in space, a visual hull of the subject was obtained at each frame. The joints' position was reconstructed by means of matching the visual hull with a subject-specific mesh model (obtained from a dry and static visual hull of the subject), based on rigid-segments, employing the articulated-ICP algorithm. Only a manual initialization step is required, in which the initial positions of the joints are determined by digitizing their positions on each view, and triangulating them. Finally a comparison was performed with joints' positions obtained by means of manual digitalization.

Fig.1: Silhouette extraction

Fig.2: Shoulder z axis

RESULTS AND DISCUSSION: Trajectories of right shoulder, elbow and wrist joints were estimated automatically for a free-style stroke sequence (Fig.2). The constraints in the space available for placing the cameras is one of the main drawbacks of this methodology: using images obtained from only lateral views in the visual hull reconstruction makes this process very hard. Furthermore trying to overcome this problem by placing cameras on the bottom of the swimming pool makes calibration procedures so difficult, and reduces the calibration volume. Finally the presence of reflexes and foam, compromises the foreground extraction process and a rigid body model affects the accuracy of the joint's position.

CONCLUSION: The developed technique for swimmers 3D markerless kinematic analysis allows athletes quantitative and objective evaluation that can help improve their performance.

REFERENCES:
Corazza S. et al. (2006). A markerless motion capture system to study musculoskeletal biomechanics: visual hull and simulated annealing approach. Ann Biomed Eng. 34(6):1019-29.
Bouguet J.Y. Camera Calibration Toolbox *http://www.vision.caltech.edu/bouguetj/calib_doc/*

A COMPARISON BETWEEN THE VALUES OBTAINED FROM ACTIVE DRAG ANALYSIS COMPARED TO FORCES PRODUCED IN TETHERED SWIMMING

Bruce Mason, Danielle Formosa and Shannon Rollason

Australian Institute of Sport Aquatics programme, Canberra, Australia.

The purpose of this study was to identify in a maximum swim effort by elite freestyle swimmers, if the mean force produced in tethered swimming over a set number of whole strokes could reliably be utilised as an alternative measure for mean propulsive force over the same number of whole strokes. Tethered force can be measured relatively easily. Although mean propulsive force at a maximum swim velocity may be derived, the process of doing so is not direct, is time consuming and requires an extensive setup. Stepwise regression analysis indicated that mean tethered force was not an acceptable alternative for mean propulsive force. Therefore the use of mean propulsive power to monitor training would require the measurement of mean propulsive force rather than simply measuring the mean tethered force in a maximum swim effort.

KEY WORDS: biomechanics, swimming, tethered, active drag, propulsive force

INTRODUCTION: Regular assessment of effective power output may be utilised as an indicator of how an elite competitive swimmer is progressing in the free swimming aspect of performance during the preparation phase for a major swim meet. Mean propelling power may be computed by averaging the product of the swimmer's instantaneous propelling force by the swimmer's velocity at the same instant in time throughout a full stroke cycle. To obtain a reliable measure of propelling force during a swim trial, the swimmer must produce a maximal effort for the measure to be accurate. While swimming at a constant velocity, the measure of the active drag force and that of the swimmer's propelling force are identical in magnitude but opposite in direction. While instantaneous propulsive force or drag force assessment may be obtained throughout the stroke cycle, a measure of instantaneous velocity is very difficult to ascertain during that assessment. However, mean swimming velocity while swimming with maximum effort is easily obtained in an unaided trial by dividing the distance swam over a set interval, by the time taken to do so.

The mean propelling force of the swimmer is not as easily measured as mean velocity. However there are several methods available that can achieve such a measure. The MAD System as described by Toussaint et al. (1988) used in the Netherlands, the velocity perturbation method (VPM) as developed in Russia by Kolmogorov et al. (2000) and the method as used in China by Xin-Feng et al. (2007) are different methods that have been utilised to obtain a measure of mean active drag or mean propelling force in swimming. In a maximal effort, the mean active drag force may be considered as identical in magnitude to the mean propulsive force, as the swimming velocity may be considered as constant. The Australian Institute of Sport has developed a variation on the VPM as reported by Alcock et al. (2007) by towing the swimmer at a constant velocity that is greater than the swimmer's top velocity by a factor of five percent. The powerful dynamometer that pulls the swimmer determines the constant velocity at which the swimmer travels through the water. The force platform on which the dynamometer is mounted measures the force required to pull the swimmer at this velocity. The AIS method assumes that a maximal propulsive effort is applied by the swimmer during the active towing trials as well as the unaided trials used to assess the top swimming velocity. This implies an equal power input by the swimmer in both situations. The active drag and hence propelling force can be computed from the force required to pull the swimmer at the increased velocity. The propelling force so obtained will represent the force required by the swimmer to travel through the water at the swimmer's top unaided swimming velocity. This and previous methods of active drag assessment rely on computing active drag by an indirect method rather than measuring the force independently. The assessment of propelling force and following on from that, power output, requires the use of sophisticated scientific apparatus and a complex testing session. This project was performed to identify whether propelling force and hence mean propelling power may be assessed for a swimmer by utilising tethered swimming using simply a force link to measure propelling force during a maximal effort tethered swim. This would thus make the process of monitoring average power for free swimming in a maximum effort, a far less complicated task. Intuitively, it is feasible proposition that the magnitude of the propelling force in free swimming would be very similar to the force derived from utilising tethered swimming.

METHODS: Data Collection: Thirteen Australian swimmers (6 males and 7 females) were tested in the A.I.S. aquatics laboratory. The calibre of the swimmer was such that each had the ability to reach the final in at least one of the three freestyle events (50m, 100m and 200m) at the Australian National Open Swimming Championships. All testing included only the Australian crawl swimming stroke. The testing included three separate tests.

The first test involved obtaining the peak swim velocity of each swimmer. Here the subject was instructed to swim at their maximum speed through a 10m interval. A swim-in of 10m was used as a lead into the testing interval to enable the swimmer to attain maximum swimming velocity before reaching the start of the timing interval. The timing of each trial was performed utilising a video system (50 hertz) which included two cameras with one focused on the start and the other at the end of the timing interval. Each camera's view included vision of an elapsed electronic timing clock, synchronised in each camera view and accurate to a hundredth of a second, to assess the time taken to swim the 10m interval and hence enable calculation of the subject's maximum swimming velocity. Three such trials were conducted on each swimmer and the trial with the quickest time was chosen to represent the swimmer's maximum swim velocity.

The second test involved obtaining a measurement of active drag for each swimmer at the swimmer's top swimming velocity. Here the swimmer was familiarised with the active drag tow protocols used to obtain the swimmer's active drag at the subject's top swimming velocity. The Kevlar tow attachment to the swimmer was connected to the belt worn around the waist and attached at the anterior side of the body. Five trials were conducted with each swimmer at a tow velocity that was 5% faster than the subject's top swimming pace. The first press button trigger for the collection of kinetic data occurred at the beginning of a stroke (right hand entry) and the data was captured for four complete strokes (each stroke being right hand entry to right hand entry). The termination of data capture was denoted by a second press button trigger at the forth right hand entry after the initial button press. The force data collected in these trials represented the additional force required to tow the swimmer at the subject's 5% higher velocity beyond that required at the swimmer's top pace. It was sampled as the Y component of force from the Kistler force platform. The active drag measurement for the swimmer's maximum velocity was then able to be computed under the assumption that an equal effort was produced by the swimmer in both the maximum swim trials and the active drag trials. The mean value for active drag force over the four complete strokes was computed for each of the five trials. Each of the mean scores representing the five separate trials, was utilised to obtain the mean propulsive force for the swimmer.

The third test involved a measurement of tethered force during which the swimmer performing a maximal swim effort. The swimmer was familiarised with the testing procedure prior to testing. The Kevlar tether attachment to the swimmer was connecting to the belt worn around the waist and was attached at the posterior side of the body. The other end of the Kevlar non stretch cable was attached to the force platform. Three trials were conducted for each swimmer performing at maximum effort. The first trigger for the collection of data occurred at the beginning of a stroke and the kinetic data was captured for four complete strokes as denoted by a second press button trigger. The force data (total Y force from the force platform) collected in these trials represented the swimmer's pulling force during swimming, on the tether cable and against a rigid stationary object that measured the force. The tethered force measurement for the swimmer's maximum effort was computed under the assumption that an equal effort was produced by the swimmer in the free swim maximum swim trials, the active drag trials and the tethered swimming trials. The mean force value for the tethered swimming over the four complete strokes was recorded for each of the three trials. The three mean scores, each representing different trials, were utilised in the computation of the mean tethered force for the swimmer.

Data Analysis: The data representing the swim velocity, the mean propulsive force and the mean tethered force for each of the 13 swimmers was tabulated. A stepwise regression analysis was performed on the data with the mean propulsive force used as the dependent variable and swim velocity, swim velocity squared and mean tethered force used as the independent variables. A significance level of 0.05 was chosen as acceptance into the regression equation and 0.10 for rejection.

RESULTS

Subject Number	Gender	Velocity (m/s)	Mean Propulsive Force (N)	Mean Tethered Force (N)	Mean Power (watts)
1	M	1.9	161.7	179.5	307.3
2	M	1.92	226.4	183.9	434.7
3	M	1.92	151	175.5	289.9
4	M	1.85	235.7	128.2	436
5	M	1.91	256.5	156.4	489.9
6	M	1.89	302.2	181.2	571.1
7	F	1.76	127.4	125.5	224.3
8	F	1.71	77.5	137.7	132.4
9	F	1.74	164.6	136.8	286.4
10	F	1.69	171.3	113	289.5
11	F	1.61	95.3	119.5	153.4
12	F	1.64	89.3	105.4	146.5
13	F	1.64	100.1	106.6	164.2

The correlation coefficient for each of the independent variables with the dependent variable was for velocity 0.751, for velocity squared 0.749 and with mean tethered force 0.604. Velocity squared was utilised in this analyses, as force is a function of velocity squared. The analyses indicated that velocity explained 56% the variance in propulsive force where mean tethered force explained only 36%. The significance level for the correlation between the dependent variable and both velocity and velocity squared was 0.002, making both statistically significantly related to mean propulsive force at the 0.01 level. The significance level for the correlation between the dependent variable and mean tethered force was 0.014, making it not significant statistically at the 0.01 level. In the regression equation only swim velocity was accepted into the equation and mean tethered force was rejected. This was partly due to the fact that the correlation coefficient between velocity and mean propulsive force was very high at 0.890 indicating that mean tethered force contributed very little extra to the equation. The conclusion drawn from the statistical analysis was that mean tethered force could not be used as a reliable alternative for propulsive force in swimming.

DISCUSSION: A regular monitoring of propelling power in free swimming provides valuable information about the state of progress in an elite swimmer's free swim performance. The ideal method to compute mean propelling power would be to compute the mean of the product of instantaneous velocity with instantaneous propelling force produced by the swimmer. It is here that the solution to this problem becomes quite difficult. Instantaneous propelling force is measured by an indirect method. Because the method by which the force values are obtained requires towing of the swimmer through the water at a constant velocity it becomes impossible to obtain a measure of the swimmer's instantaneous velocity in an unaided condition. The mean velocity of the swimmer is used as a substitute for instantaneous velocity and therefore the measure of propulsive force is most readily provided as mean propulsive force. Due to the fear of being inaccurate by using mean velocity and mean force in the computation of mean power, an analysis of the possible inaccuracies was assessed. In this assessment the product of the instantaneous propulsive force with the velocity as represented by a sine wave in and slightly out of phase with the force curve was performed. The sine wave representing velocity was such that the mean value was equal to the mean velocity obtained in the free swim with an amplitude of the curve equal to approximately 7% of the mean velocity. When the sine wave was more out of phase with the force curve, the difference between the original computed value of power derived from using mean values and that derived from the sine wave simulation was more divergent. However, even when the phase shift was as much as 10 deg out of phase with the force curve, the difference between the mean computed value for power and that derived from the sine curve simulation was less than one percentage point. This result indicated that mean values for propelling force and velocity could be used reliably in the computation of mean propelling power. The measurement of mean propulsive force required an extensive setup. The use of mean tethered force as an alternative would have made the task far less difficult. However, this project found that mean tethered force was not a reliable alternative indicator of mean propulsive force to be used in the derivation of mean propulsive power.

CONCLUSION: This study identified that both velocity and mean tethered force were related to some degree with mean active drag and hence mean propulsive force. However, the relationship between mean velocity and mean tethered force was also highly related. When the relationship between mean velocity and mean tethered force was removed from the tethered force variable, mean tethered force was identified not to be closely related to mean active drag and hence mean propulsive force. Therefore, mean tethered force was found not to be a suitable alternative to mean propulsive force in the computation of mean propelling power.

REFERENCES:

Alcock.A., & Mason.B. (2007). Biomechanical Analysis of active drag in swimming, In Menzel.H & Chagas.M (Eds.), *Proceedings of the XXVth International Symposium of Biomechanics in Sports*, Brazil (pp212-215)

Kolmogorov,S., Lyapin,S., Rumyantseva,O., & Vilas-Boas,J. (2000). Technology for decreasing actrive drag at the maximal swimming velocity. In Sanders,R & Hong,Y. (Eds.), *Proceedings of XVIII International Symposium on Biomechanics in Sport.Applied Program:Application of Biomechanical Studies in Swimming, Hong Kong* (pp 39-47).

Toussaint,H., De Groot,G., Savelberg,H., Vervoorn,K., Hollander,A.,& Van Ingen Schenau,G. (1988). Active Drag related to velocity in male and female swimmers. *Journal of Biomechanics*, 21(5), 435-438.

Xin-Feng,W., Lian-Ze,W.,Wei-Xing,Y.,De-Jian,L., & Xiong,S. (2007). A new device for estimating active drag in swimming at maximal velocity. *Journal of Sport Science*, 25(4), 375-379.

MOTOR CONTROL PATTERNS IN ELITE SWIMMERS' FREESTYLE STROKE DURING DRYLAND SWIMMING

Tracy Spigelman[1], Thomas J Cunningham[2], Scott Mair[2], Robert Shapiro[2], Timothy L Uhl[2] and David R Mullineaux[2]

Eastern Kentucky University, Richmond, KY, USA[1]
University of Kentucky, Lexington, KY, USA[2]

The purpose of this study was to compare motor control patterns of elite freestyle swimmers when asked to swim at 100m freestyle pace using a dryland swimbench. Collegiate and masters level swimmers (n=15) whose 100m freestyle time were faster than 75% of the FINA cutoff time, performed four 10 second trials of freestyle swimming on a dryland swimbench. 3-D kinematic analysis was used to calculate displacement in the hand in the cranial-caudal, vertical, and medial-lateral directions. A 2-way repeated measures ANOVA was used to compare hand path between swimmers and within trials (n=58). Data was not statically significant, but three distinct combinations of hand paths were used to perform the 100m freestyle task on the swimbench. These hand paths differed from historical in-water data. Findings imply individual swimmers adjusted kinematics on the swimbench to accommodate for environmental constraints.

KEY WORDS: biomechanics, dynamical systems.

INTRODUCTION: Traditional dynamical systems model explains the bilateral arm motion of freestyle swimming as coordinated structures emerging from the central nervous systems' (CNS) ability to control the redundant degrees of freedom of the shoulder, elbow and hand (Latash, 1998). Coaching for performance enhancement, using this model, means finding a generalized, optimal movement pattern and instructing all athletes to move in the same way. Schleihauf (1979) distinguished three variations of hand path used during the propulsive phase of freestyle swimming in elite swimmers: a medial-lateral path; a deep catch with a straight line pull, and a primarily medial hand path. This implies individual swimmers vary motor control patterns to accomplish the task of swimming freestyle.

The contemporary dynamical systems model (Davids et al., 2003) suggests that environmental, task, and anthropometric constraints contribute to the evolution of movement patterns. Spontaneous pattern reorganization occurs from continuous sensory feedback from the CNS allowing athletes to develop individual movement strategies based on the surrounding constraints. This version of dynamical systems provides an explanation for the variation in elite swimmers' freestyle stroke reported by Schleihauf (1979). Based on this, it would be expected that freestyle stroke technique in elite swimmers when swimming freestyle on a dryland swimbench will also vary among swimmers. Swimbenches are land based training devises that allow swimmers to mimic swimming motions lying prone and using a pulley system. Evidence of stroke technique differences would provide valuable information for coaches trying to enhance swimming performance through correction of stroke technique, possibly coaching each swimmer from a more individualized standpoint. The purpose of this research was to determine if different motor control patterns can be seen among elite swimmers performing simulated 100m freestyle on a dryland swimbench.

METHODS: Data Collection: A sample of 15 elite healthy collegiate and masters' swimmers volunteered to participate in this study (8 males, age=24.7±8.0yrs, height=183.8±4.0cm, mass=78.4±7.0 kg; 7 females, age= 26.0±11.0 yrs, height 170.2± 6.0 cm, mass 64.8± 5.6 kg). Written informed consent was obtained from each participant prior to implementing the study, which was approved by the university's human participant institutional review board. All participants' 100m freestyle times were equal to or greater than 75% FINA national cutoff time.

Freestyle strokes were performed on a swimbench (Swimworks, Inc, Santa Rosa, CA) which allowed kicking and rotation of the trunk about the cranial-caudal axis. Following a 30-minute familiarization session, 30 reflective markers were attached on the upper extremity and torso of the swimmer. Each swimmer performed four ten second trials of freestyle swimming on the swimbench at a 100 m sprint race pace, with each trial separated by 2 seconds with the participant adopting a streamlined position to simulate the aspects of the flip-turn. Kinematic data were collected at 200 Hz using EVaRT 5.0 (Motion Analysis Corporation, Santa Rosa, CA).

Data were analyzed in Matlab 7.1 (The Mathworks, Inc, Natick, MA). Using an interactive algorithm, three phases of the stroke were defined: pull; push, and recovery. In absence of the water surface, the origin was set to the height of the seventh cervical vertebrae (C7) marker for the first five frames of the streamlined position before trial 1. Data were smoothed using a low pass Butterworth filter with cutoff frequencies, determined via residual analysis (Winter, 2005), of 4 Hz for the shoulder, elbow, and hand, and 2 Hz for the trunk. Dependent variables were range of hand displacement in the cranial-caudal (H_d^{cc}), medial-lateral (H_d^{ml}) and vertical (H_d^v) directions.

Data Analysis: Hand displacements(m) for all swimmers (n=15) were analyzed using a 2 way repeated measures ANOVA (Version 12.0; SPSS, Inc, Chicago. IL). Hand displacements(m) were also graphed for the complete stroke cycle, with all trials overlaid (n=58) and visually inspected for variability in stroke patterns.

RESULTS: Means and standard deviations of hand displacement(m) for four trials on the swimbench in the cranial-caudal, medial-lateral and vertical directions are shown in Table 1. No significant differences were found among swimmers and trials (p> 0.05).

Table 1. Average Left and Right Hand Displacements (m) in the Cranial-Caudal (H_d^{cc}), Medial-Lateral (H_d^{ml}) and Vertical (H_d^v) Directions for Elite Swimmers (n=15) Performing Four Trials of Freestyle Swimming on a Dryland Swimbench

		Trial 1	Trial 2	Trial 3	Trial 4
H_d^{cc} (m)	Right	1.0±0.15	0.94±0.23	1.01±0.19	0.92±0.20
	Left	1.0±0.18	0.98±0.19	0.99±0.19	0.97±0.17
H_d^{ml} (m)	Right	0.30±0.17	0.30±0.10	0.32±0.07	0.33±0.10
	Left	0.32±0.13	0.32±0.11	0.31±0.09	0.33±0.13
H_d^v (m)	Right	0.44±0.16	0.41±0.19	0.45±0.23	0.44±0.17
	Left	0.36±0.12	0.39±0.17	0.39±0.19	0.39±0.17

Elite swimmers had similar displacement values; however three slightly different combinations of hand path in the cranial-caudal, medial-lateral and vertical directions were adopted by swimmers within trials and among swimmers. Figure 1 (left side) is a representation of three distinct hand paths used by elite swimmers on the swimbench in the cranial-caudal versus vertical direction. The solid line shows a longer stroke where the hand extended cranially from the starting position, reaching maximum vertical displacement (-0.6m) at 0.4m cranial to C7 then steadily moved upwards toward C7. The dash-dot line is a shorter, shallower hand path. The hand reached maximal vertical displacement (-0.4m) at 0.5m cranial to C7 and remained at that depth from approximately 0.5m cranial to 0.2m caudal of C7, when upward movement toward C7 began. The dotted line shows an overall shallower, flatter hand path that reached maximum vertical hand displacement at 0.2m cranial to C7, then gradually moved upward till 0.6m caudal of C7. Figure 1 (right side) is a representation of hand motion in the cranial-caudal versus medial-lateral directions. The solid line shows the hand moved laterally for the first 0.1m of pull and then followed a medial path for the remainder of the stroke. The dashed line shows the hand extended to maximum range of cranial hand displacement (0.6m) and followed a slightly curved path between -0.2m and -0.4m lateral of C7. The dotted line shows a wider lateral arm path, 0.4m lateral to C7 from maximum cranial to caudal displacement (0.6m to -0.6 m).

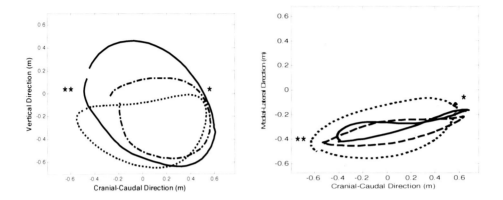

Figure 1: Three primary hand paths used by elite swimmers (n=15) for freestyle swimming in cranial-caudal vs. vertical directions (left), and in the cranial-caudal vs. medial-lateral directions (right) on a dryland swimbench. The hand paths are not exclusive as different combinations of hand paths in the cranial-caudal vs. vertical directions where seen with different combinations of hand paths in the medial-lateral direction between swimmers.
Note: * beginning of the propulsive phase; ** end of the propulsive phase.

DISCUSSION: Elite swimmers had similar ranges of hand displacement during the propulsive phase on a dryland swimbench with standard deviations between 0.1 and 0.2 m. Elite swimmers also displayed three distinct combinations of hand paths during freestyle swimming on a dryland swimbench. The data agrees with previous research reported for in-water swimming by Schleihauf (1979), describing variation in propulsive phase stroke technique among elite swimmers. The present data indicates each individual elite swimmer made different kinematic adjustments to their freestyle swimming when the environmental and task constraints were varied. Hand paths on the swimbench were different to published in-water hand paths indicating swimmers changed hand patterns to swim freestyle.

Dynamical systems theory suggests that the CNS spontaneously reorganizes movements as sensory feedback is received from surrounding constraints (Davids et al., 2003). Researchers agree freestyle swimming technique is integral to peek performance, but that elite swimmers demonstrate differences in stroke styles (Schleihauf, 1979, Seifert et al., 2007). On the swimbench, variations of hand path could be seen during the propulsive phase. Stroke technique variations within trials suggest spontaneous reorganization of the CNS from the task constraints (Davids et al., 2003). Swimbenches use a pulley system where swimmers place their hands in paddles to simulate water resistance. Paddles constrain the movement of the hands to the length of the pulleys. In water, the hand, elbow and shoulder are free to move through all planes of motion which allows swimmers to optimize the forces in the aquatic environment. On the swimbench, swimmers had fewer degrees of freedom at the distal segment (hand attached to paddle), but still adjusted the length and depth of the hand path to compensate for this constraint and to achieve the task. Changing hand paths required the swimmer to readjust shoulder flexion and abduction angles, and elbow flexion angles, while controlling the hand paddle in the pulley system. As each trial lasted 10 seconds, and swimmers performed between seven to eleven strokes in each trial, it is plausible that the variability seen among swimmers is an example of dynamical reorganization of the CNS to accomplish a goal-oriented task.

Data were compared with the three stroke techniques of in-water freestyle swimming described by Schleihauf (1979). In general, on the swimbench, all swimmers used a flatter, more lateral to medial hand path most likely due to differences in the environmental constraints (Schleihauf, 1979). Swimmers were instructed to swim using their 100m freestyle sprint pace. In the water, swimmers receive sensory feedback from the interaction of the hand and forearm with the water during a pulling motion. During a sprint race, swimmers make modifications to their stroke based on how much resistance they feel as they press through the water, pressing harder to increase velocity (Seifert et al., 2007). On the swimbench, sensory feedback was primarily received from the hand paddles, so swimmers adapted their freestyle technique to exploit this feedback. First, swimmers used a shallow hand path as they moved in the cranial-caudal direction. To generate momentum on the swimbench, the swimmer pulls with the hand flat against the paddle. Flexion at both the elbow and the wrist would enable the swimmer to maximize contact of the hand on the paddle, but this would also decrease the downward depth of the pull creating a shallower path of motion. Second, instead of a sculling motion, swimmers moved the hand from a medial to lateral direction until the end of the propulsive phase. Lateral movement of the hand may have been the swimmers' strategies to engage the latissimus dorsi muscle for power as the arm moved cranial to caudal, as this is the primary function of the muscle.

The ability of the elite swimmers to adjust freestyle stroke technique to the task and environmental constraints supports the idea that the CNS can spontaneously reorganize based on feedback during a goal driven task (Davids et al., 2003). The variations in freestyle motion in this sample of elite swimmers imply the use of a swimbench allows for subtle individual variation of motion.

CONCLUSION: On a dryland swimbench, elite swimmers appeared to utilize slightly different movement patterns. This suggests elite level swimmers are able to spontaneously reorganize freestyle propulsive phase movement patterns based on environmental and task constraints. Thus coaches should be aware of this inherent variation in swimmers stroke technique and provide individually based feedback.

REFERENCES

Davids, K., Glazier, P., Araújo, D. & Bartlett, R. (2003). Movement systems as dynamical systems: the functional role of variability and its implications for sports medicine. *Sports Medicine*, 33(4), 245-260.

Latash M.L. (Ed.) (1998). *Progress on Motor Control. Bernstein's Traditions in Movement Studies.* Champaign: Human Kinetics.

Schleihauf, R.E. (1979). Hydrodynamic analysis of swimming propulsion. In: J. Terauds & E.W. Bedingfield (Eds.), *Third International Symposium of Biomechanics in Swimming* (pp 70-109). Edmonton, Canada: University Park Press.

Seifert, L., Chollet, D. & Rouard, A. (2007). Swimming constraints and arm coordination. *Human Movement Science*, 26, 68-86.

Winter, DA. (2005). *Biomechanics and Motor Control of Human Movement.* (3rd ed). Hoboken, NJ: John Wiley and Sons, Inc.

Acknowledgement:

Swimworks, Inc. for the use of the swimbench, and USA Swimming for funding this research.

STROKE PARAMETERS AND ARM COORDINATION IN COMPETITIVE UNILATERAL ARM AMPUTEE FRONT CRAWL SWIMMERS

Conor Osborough [1,2], Carl Payton [2] and Daniel Daly [3]

School of Science and Technology, Nottingham Trent University, Nottingham, UK[1]; Department of Exercise and Sport Science, Manchester Metropolitan University, Alsager, UK[2]; Faculty of Kinesiology and Rehabilitation Sciences, Katholieke Universiteit Leuven, Belgium[3]

> The aims of this study were to: (1) determine the changes in stroke parameters and arm coordination as a function of swimming speed; and (2) examine the relationships between stroke parameters and arm coordination, for competitive unilateral arm amputee front crawl swimmers. Thirteen highly-trained swimmers (3 male, 10 female) were filmed underwater from lateral views during six increasingly faster 25 m front crawl trials. Increases in swimming speed were achieved by an increase in stroke frequency which coincided with a decrease in stroke length. All swimmers showed asymmetric coordination between their affected and unaffected arm pulls, which was not affected by an increase in swimming speed up to maximum. The fastest amputee swimmers used higher stroke frequencies and less catch-up coordination before their affected arm pull, when compared to the slower swimmers. Reducing the time delay before initiating the affected arm pull appears to be beneficial for successful swimming performance.
>
> **KEY WORDS:** aquatics, biomechanics, motor control, disability sport.

INTRODUCTION: Swimming speed (SS) is the product of stroke frequency (SF) and stroke length (SL). Stroke frequency is determined by the rate at which a swimmer's limb segments move, while stroke length is determined by the propulsion generated by a swimmer and the resistance experienced. Success in competitive swimming depends on a swimmer's ability to maximise propulsion and minimise resistance. The timing of propulsion relative to resistance becomes crucial when considering the effectiveness of a swimmer's stroking technique.

The Index of Coordination (IdC) is often used to quantify the coordination of arm movements during front crawl swimming. The IdC measures the time lag (expressed at a percentage of total stroke time) between the beginning of propulsion in one arm stroke and the end of propulsion in the other. As described by Chollet et al. (2000), arm coordination conforms to one of three major models: (1) The model of catch-up describes a time delay between the propulsive phases of the two arms (i.e. IdC < 0%); (2) The model of opposition describes a continuous series of propulsive actions: one arm begins the pull phase when the other is finishing the push phase (i.e. IdC = 0%); (3) The superposition model, describes an overlap, to a greater or lesser extent, in the propulsive phases (i.e. IdC > 0%).

Able-bodied swimmers have been shown to modify their arm coordination with increases in swimming speed (Chollet et al., 2000; Potdevin et al., 2006). Swimmers switched from using catch-up at slow swimming speeds, to opposition or superposition at fast swimming speeds. This change coincided with an increase in stroke frequency and a decrease in stroke length. For swimmers with various loco-motor disabilities, Satkunskiene et al. (2005) reported that "more-skilled" swimmers were characterised by greater amounts of superposition and higher stroke frequencies, when compared to "less-skilled" swimmers.

No such examination of arm coordination has been undertaken with a single homogenous group of highly-trained swimmers with the same physical impairment. Faster unilateral arm amputee front crawl swimmers are able to attain higher stroke frequencies, when compared to their slower counterparts (Osborough et al., 2009). However, large variations in the timing of the underwater arm stroke movements have been observed within this group of swimmers. It is unclear what influence these variations in inter-arm coordination might have on performance. The aims of this study were to: (1) determine the changes in stroke parameters and arm coordination as a function of swimming speed; and (2) examine the relationships between stroke parameters and arm coordination, for competitive unilateral arm amputee front crawl swimmers.

METHODS: Data Collection: Thirteen (3 male and 10 female) competitive swimmers (age 16.9 ± 3.1 yrs) consented to participate in this study. All the participants were single-arm amputees, at the level of the elbow. The mean 50 m front crawl personal best time was 32.7 ± 3.1 s. Twelve of the swimmers competed in the International Paralympic Committee S9 classification for front crawl; one male swimmer competed in the S8 classification. The procedure for the data collection was approved by the Institutional Ethics Committee. All participants provided written informed consent before taking part in the study.

After being randomly allocated into one of two test groups, participants completed six 25 m front crawl trials, counterbalanced from slow to maximum swimming speed (SS_{max}). To control for the effects of the breathing action on the swimming stroke, participants were instructed not to take a breath through a 10 m test section of the pool. Two digital video camcorders (Panasonic NVDS33), sampling at 50 Hz with a shutter speed of 1/350 s were used to film the participants. Each of the camcorders was enclosed in a waterproof housing suspended underwater from one of two trolleys that ran along the side of the pool, parallel to the participants' swimming direction. This set-up enabled the participants to be filmed under the water, from opposite sides, over the 10 m test section.

Data Analysis: The digital video footage was transferred to a laptop computer and analysed using SIMI Motion 7.2 software. Three consecutive, non-breathing stroke cycles for each participant, were then selected for analysis. The estimated locations of the gleno-humeral joint centre and the elbow joint centre of both the affected and unaffected arms were digitised at 50 Hz to obtain the angular position of the limbs, as a function of time.

The following variables were then determined from the digitised data or video recordings at 80%, 85%, 90%, 95% and 100% of each participant's SS_{max}: IdC_{aff} (%) - lag time between the beginning of the pull phase with the affected arm and the end of the push phase with the unaffected arm; IdC_{un} (%) - lag time between the beginning of the pull phase of the unaffected arm and the end of the push phase with the affected arm; IdC_{adapt} (%) - mean of IdC_{aff} and IdC_{un}; SL (m) - distance travelled down the pool with one stroke cycle; SF (Hz) - number of stroke cycles performed in one second; SS (m·s^{-1}) - mean forward speed of the participant over three stroke cycles.

Repeated measures general linear modelling tests were used to compare the changes in the dependent variables between the percentage swimming speeds. Correlations were calculated among the dependent variables at 100% of SS_{max}. In all comparisons, the level of statistical significance was set at $p < 0.05$.

RESULTS:

Table 1: Mean ± S.D. stroke parameters at 80%, 85%, 90%, 95% and 100% of SS_{max}.

	Percentage of maximum swimming speed ($M \pm SD$)				
	80	85	90	95	100
SS (m·s^{-1})	1.10 ± 0.10 [a]	1.16 ± 0.12 [a]	1.22 ± 0.12 [a]	1.29 ± 0.13 [a]	1.36 ± 0.14 [a]
SL (m)	1.78 ± 0.15 [b]	1.77 ± 0.14 [b]	1.73 ± 0.14	1.71 ± 0.15 [c]	1.66 ± 0.16
SF (Hz)	0.57 ± 0.18	0.66 ± 0.08 [d]	0.71 ± 0.08	0.76 ± 0.09 [a]	0.82 ± 0.11 [a]

[a] Significantly different with all percentage SS_{max} values ($p < 0.01$). [b] Significantly different with 100% of SS_{max} value ($p < 0.05$). [c] Significantly different with 85% of SS_{max} value ($p < 0.05$). [d] Significantly different with 90% of SS_{max} value ($p < 0.01$).

The mean and standard deviations for the stroke parameters are presented in Table 1. Between the first swim at 80% of SS_{max} and the last swim at 100% of SS_{max}, mean stroke frequency significantly increased (from 0.57 ± 0.18 Hz to 0.82 ± 0.11 Hz; $p < 0.01$) and mean stroke length significantly decreased (from 1.78 ± 0.15 m to 1.66 ± 0.16 m; $p < 0.05$) in conjunction with a significant increase in mean swimming speed (from 1.10 ± 1.10 m·s^{-1} to 1.36 ± 0.14 m·s^{-1}; $p < 0.01$).

Table 2: Mean ± S.D. arm coordination variables at 80%, 85%, 90%, 95% and 100% of SS_{max}.

	Percentage of maximum swimming speed (M ± SD)				
	80	85	90	95	100
IdC_{adpt} (%)	-16.5 ± 4.5	-16.6 ± 5.9	-17.3 ± 5.6	-17.5 ± 5.3	-17.3 ± 5.2
IdC_{aff} (%)	-24.0 ± 8.5	-24.1 ± 8.8	-23.8 ± 8.5	-24.1 ± 7.7	-24.3 ± 9.1
IdC_{un} (%)	-9.0 ± 9.8 [a]	-9.1 ± 10.4 [a]	-10.8 ± 9.5 [a]	-10.8 ± 8.8 [a]	-10.2 ± 8.7 [a]

[a] Differences between IdC_{aff} and IdC_{un} are statistically significant ($p < 0.01$).

Results dealing with the arm coordination variables are reported in Table 2. There were no significant differences in mean IdC_{adapt} values across the five percentage speed increments. The mean IdC_{aff} values (- 24.1 ± 8.3 %) were significantly lower ($p < 0.01$) than that of the mean IdC_{un} values (- 10.0 ± 9.2 %) at all percentage swimming speeds. The values for both the IdC_{aff} and IdC_{un} were not seen to change as the participants increased their swimming speed (- 24.0 ± 8.5 % vs. - 24.3 ± 9.1 % and - 9.0 ± 9.8 % vs. -10.2 ± 8.7 % for IdC_{aff} and IdC_{un} respectively). There was no significant interaction effect on inter-arm coordination.

Table 3: Inter-correlations among stroke parameters and arm coordination variables at SS_{max}.

	Swimming speed (m·s^{-1})	Stroke frequency (Hz)	IdC_{aff} (%)
Stroke frequency (Hz)	0.72 [b]		
Stroke length (m)	0.01	- 0.68 [a]	
IdC_{adpt} (%)	0.26	0.54	
IdC_{aff} (%)	0.59 [a]	0.66 [a]	
IdC_{un} (%)	- 0.30	- 0.50	- 0.31

[a] Correlations are statistically significant ($p < 0.05$). [b] Correlations are statistically significant ($p < 0.01$).

Inter-correlation coefficients among stroke parameters and arm coordination variables at 100% of SS_{max} are shown in Table 3. At 100% of SS_{max}, stroke frequency was significantly related to swimming speed ($r = 0.72$; $p < 0.01$) whereas stroke length was not ($r = 0.01$). Both swimming speed ($r = 0.59$) and stroke frequency ($r = 0.66$) were significantly related ($p < 0.05$) to IdC_{aff}. There were moderate but non-significant correlations between swimming speed and IdC_{un} ($r = - 0.30$), and stroke frequency and IdC_{un} ($r = - 0.50$).

DISCUSSION: The aims of this study were to: (1) determine the changes in stroke parameters and arm coordination as a function of swimming speed; and (2) examine the relationships between stroke parameters and arm coordination, for competitive unilateral arm amputee front crawl swimmers.

The arm amputee swimmers in this study achieved progressive increases in swimming speed by increasing stroke frequency. The increase in stroke frequency coincided with a decrease in stroke length, this being similar for able-bodied swimmers (Chollet et al., 2000; Potdevin et al., 2006). In comparison, the mean SS_{max} of the amputees was substantially slower than that of able-bodied swimmers (1.81 ± 0.1 m·s^{-1} for 14 high-performing male and females, Chollet et al., 2000; 1.63 ± 0.12 m·s^{-1} for 13 expert males, Potdevin et al., 2006). The amputees had lower stroke frequencies when compared to these able-bodied swimmers (0.90 ± 0.1 Hz, Chollet et al., 2000; 0.92 Hz, Potdevin et al., 2006). The amputees had appreciably shorter stroke lengths, when again compared to these able-bodied swimmers (2.01 ± 0.1 m, Chollet et al., 2000; 1.83 ± 0.14 m Potdevin et al., 2006). These differences can be attributed to the physical impairment of the amputees but might also be influenced by the predominate gender and the relatively small stature of the amputee swimmers.

The IdC_{adapt}, IdC_{aff} and IdC_{un} values of the amputees did not change with an increase in swimming speed. At all swimming speeds arm coordination conformed to the front crawl catch-up model (i.e. IdC < 0%). There was significantly more catch-up before the amputees' affected arm pull than before their unaffected arm pull, at all swimming speeds. Furthermore, this asymmetrical catch-up did not appear to be affected by an increase in swimming speed, suggesting that swimmers maintained stable inter-arm coordination even though they swam faster. This finding contrasts with that of able-bodied front crawl swimmers. Both Chollet et al. (2000) and Potdevin et al. (2006) reported that able-bodied swimmers switched from using catch-up at slow swimming speeds, to opposition or superposition at fast swimming speeds. Being deprived of an important propelling limb and the inability to attain the higher swimming speeds of those tested by Chollet et al. (2000) and Potdevin et al. (2006) might account for the observed differences in the arm coordination values between the amputee and able-bodied front crawl swimmers.

There were significant inter-swimmer correlations between SS_{max} and the stroke frequency and the IdC_{aff} used at SS_{max}. This indicates that the fastest amputee swimmers used higher stroke frequencies and less catch-up before their affected arm pull, when compared to the slower swimmers. This finding has similarities to those reported for able-bodied swimmers and swimmers with loco-motor disabilities. Satkunskiene et al. (2005) reported that "more-skilled" swimmers were characterised by greater amounts of superposition and higher stroke frequencies, when compared to "less-skilled" swimmers. Chollet et al. (2000) showed that for able-bodied swimmers, stroke frequency was significantly correlated ($r = 0.67$) with arm coordination. For the fastest unilateral arm amputee front crawl swimmers in this study, reducing the time delay before initiating the affected arm pull appears to be a motor control strategy for the attainment of the highest stroke frequencies and swimming speeds.

CONCLUSION: In the current study the findings imply that to improve their maximum swimming speed, unilateral arm amputees should focus on increasing their stroke frequency, rather than swimming with the longest possible stroke length. All swimmers showed asymmetric coordination between their affected and unaffected arm pulls. This asymmetry did not appear to be affected by an increase in swimming speed up to maximum. The quickest swimmers exhibited less front crawl catch-up coordination before their affected arm pull and higher stroke frequencies, when compared to their slower counterparts. For successful swimming performance, reducing the time delay before initiating the affected arm pull appears to be beneficial for competitive front crawl swimmers with a single-arm amputation.

REFERENCES

Chollet, D., Chalies, S. & Chatard, J.C. (2000). A new index of coordination for the crawl: Description and usefulness. *International Journal of Sports Medicine, 21*, 54-59.

Osborough, C.D., Payton, C.J. & Daly, D. (2009). Relationships between the front crawl stroke parameters of competitive unilateral arm amputee swimmers, with selected anthropometric characteristics. *Journal of Applied Biomechanics*, in press.

Potdevin, F., Bril, B., Sidney, M. & Pelayo, P. (2006). Stroke frequency and arm coordination in front crawl swimming. *International Journal of Sports Medicine, 27*, 193-198.

Satkunskiene, D., Schega,L., Kunze, K., Birzinyte, K. & Daly, D. (2005). Coordination in arm movements during crawl stroke in elite swimmers with a loco-motor disability. *Human Movement Science, 24*, 54-60.

Acknowledgement

The authors would like to acknowledge the following: British Disability Swimming for their support in this project; Professor Ross Sanders for the use of his facilities at the Centre for Aquatics Research and Education, The University of Edinburgh; Miss Casey Lee for her assistance during data collection.

MEASURING PROPULSIVE FORCE WITHIN THE DIFFERENT PHASES OF BACKSTROKE SWIMMING

Danielle Formosa[1,2], Bruce Mason[1] & Brendan Burkett[2]

Australian Institute of Sport, Canberra[1]
University of Sunshine Coast, Australia[2]

The purpose of this study was to identify the propulsive force profile associated within the different phases of backstroke to provide individual feedback to elite swimmers and coaches. Elite backstrokers (n=4) performed three maximal velocity time trials to determine the swimmers maximum velocity. This was followed by three passive drag trials and three active drag trials using a flux vector drive dynamometer mounted on a force platform to tow them at set velocities (derived from the swimmer's maximum swim pace) while measuring the force to do so. The computed active drag and the propulsive propelling force profile were represented as a dynamic parameter, allowing identification of intra cyclic force fluctuations with respect to time. The force profiles were synchronised to video footage which provided unique quantitative and individual stroke kinematic feedback to the elite swimmers and coaches.

KEY WORDS: biomechanics, swimming, backstroke, propulsion, technique

INTRODUCTION: The propulsive and resistive forces in free swimming are continuously changing, within each stroke. To objectively measure these forces is a complex task. One of the first methods used to measure these forces was the Measuring Active Drag, (MAD) system. The swimmer's hands pull directly on a series of fixed pads 1.35 m apart, at a depth of 0.8 m, while the legs are restricted with a pull buoy (Toussaint, 2002). Another method developed was the Velocity Perturbation Method (VPM) where the swimmer performed a maximal effort swim with and without the added resistance of a towed hydrodynamic body (Kolmogorov & Duplishcheva, 1992). The difference in the two velocities was used to compute the active drag force. In both methods the common assumption was that at a constant velocity, the propulsive force was equal to the opposing, active drag (Hollander et al, 1987; Kolmogorov & Duplishcheva, 1992; Xin-Feng et al, 2007). These methods presented the net force as a single or mean value representing drag force, across each individual stroke cycle, or a number of stroke cycles, thereby neglecting the intra-stroke force fluctuations. Providing a single mean value which represents active drag to elite swimmers and coaches does not highlight specific aspects within the stroke that can guide intervention to enhance performance. To address these limitations, this research investigated the propulsive force profile generated whilst freely swimming backstroke, at maximal velocity. The aim of this study was to identify the propulsive force profile within the different phases of backstroke, and to synchronise this information with above and underwater video to provide unique and valuable feedback to elite athletes and coaches.

METHODS: Four Australian National swimmers (male age; 21 ± 3.6; female age; 18) were tested using a flux vector drive dynamometer positioned directly on a calibrated Kistler™ force platform (Kistler Instruments in Winterthur Switzerland Dimensions: 900 x 600 m Type Z12697). The validity and reliability of the system was determined prior to data collection (Fulton et al, 2008). The dynamometer enabled towing velocity to be accurately set and the force plate allowed the net force to tow the swimmer through the water to be measured. The swimmers completed a typical 20 min individual race preparation warm up, followed by three individual maximal swimming velocity trials. The highest average velocity achieved over a set 10 m interval defined the passive and active testing velocity. The swimmers were towed at their top swim velocity during the passive drag test. For the three passive drag trials the swimmers were instructed to hold the end of the tow line around the middle finger of their dominant hand, with the non-dominant hand interlocking to minimise any additional movement. The criteria for a successful passive trial was that the swimmer maintained a streamline position just below the water surface, with no arm strokes nor kicking nor breathing, and there was visible water flow passing over the head, back and feet. Three

active drag tows were performed at a velocity five percent greater than the swimmer's maximal swimming velocity. The active trials consisted of the swimmers actively swimming and using their typical stroke characteristics with an Eyeline ® tow belt attached to the lumbar region and the dynamometer. The five percent increase in towing velocity was considered to not have any major effect on the swimmer's stroke pattern while still allowing continuous force measurement. Data capture was collected for a total of seven seconds, one second prior to and six seconds after the synchronisation trigger was depressed. The sensitivity of the amplifier was set at 5000 pC for both conditions. Data was processed using a 12 bit A to D card, sampled at 500 Hz, and a 5 Hz Butterworth low pass digital filter was applied to the force data collected. (Alcock & Mason, 2007). Each trial was filmed at 50 Hz using three genlocked cameras; a side-on underwater, side-on above and head-on camera.
The following formulas were used to determine active drag:

$$F_1 = 0.5C \cdot \rho \cdot A \cdot V_1^2 \quad \text{(equation one)}$$
$$F_2 = 0.5C \cdot \rho \cdot A \cdot V_2^2 - F_b$$

Where C is dimensionless coefficient of drag, ρ is the density of water, A is the frontal cross sectional area of the swimmer, F_b is the force needed to pull the swimmer. F_1= the force applied by the swimmer during free swimming (unaided) and F_2 = the force applied by the swimmer during the assisted condition.
If we assume an equal power output in both the free swimming and assisted conditions:
$P_1 = P_2$ and therefore $F_1 \cdot V_1 = F_2 \cdot V_2$
then substitution of F_1 and F_2 gives:

$$0.5C \cdot \rho \cdot A \cdot V_1^3 = 0.5C \cdot \rho \cdot A \cdot V_2^3 - F_b \cdot V_2$$

Rearranging the formula to find C:

$$C = \frac{F_b \cdot V_2}{0.5\rho \cdot A \cdot (V_2^3 - V_1^3)}$$

then substitution of C into equation one gives the following formula for the active drag:

$$F_1 = \frac{F_b \cdot V_2 \cdot V_1^2}{V_2^3 - V_1^3}$$

(Alcock & Mason, 2007)

RESULTS:

Table 1: Mean passive and active forces (mean ±SD) at maximal velocity (participants 1-3 male, 4 female)

Participant	Maximal Velocity (m.s⁻¹)	Mean Passive Force (N) Force (N) (mean ± N)	Mean Propulsive Force (N) Force (N) (mean ± N)
1	1.79	63.72 ± 0.73	235.89 ± 25.48
2	1.75	69.15 ± 1.59	184.84 ± 2.98
3	1.78	74.92 ± 1.60	199.92 ± 15.89
4	1.63	46.19 ± 1.99	128.17 ± 11.11

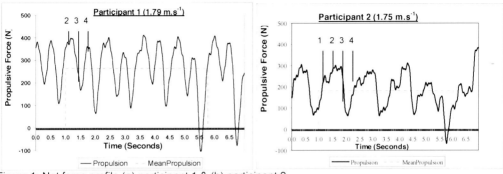

Figure 1: Net force profile (a) participant 1 & (b) participant 2

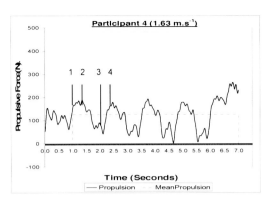

1. Right hand entry and left hand 2nd downsweep
2. Right hand 1st phase
3. Left and entry and right hand 2nd downsweep
4. Left hand 1st phase

First downsweep: begins immediately when the swimmer stops pushing back against the water.
Second downsweep: begins during the transition from the previous sweep and continues until the arm is completely extended and below the body (Maglischo, 1993).

Figure 2: Net force profile

DISCUSSION: The aim of this research was to quantify the propulsive forces within the different phases of backstroke swimming, to provide unique feedback to the swimmer and coach. As shown in figures 1 and 2 the individual swimmer's intra-cyclic propulsive force profiles are presented. These graphs demonstrate the importance of expressing propulsive force as a dynamic parameter, as opposed to single mean value representing the parameter. Each swimmer presented by their own unique profile of generated forces within their stroke, which highlighted the individual's profile. The synchronised head-on and side-on video footage (figure 3) allowed the swimmer and coach to identify specific strengths and weaknesses. This unique information provided objective and quantitative feedback highlighting technical errors within the stroke, as opposed to making judgements based on speculation or opinion. The mean active drag values in this study did not concur with the values established by Kolmogorov et al (1992) using the VPM method. The participants in the current study achieved a higher maximal velocity. Kolmogorov et al (1992) suggested that a female swimming at 1.43 $m.s^{-1}$ produced an average force of 49.78 N, while males, at 1.72 $m.s^{-1}$, produced an average force of 119.92 N. Had the participants in Kolmogorov et al (1992) study achieved faster maximal velocities the force may have been similar to the values obtained in this study due primarily to the fact that force is a function of velocity squared.

Participant one presented a symmetrical stroke pattern compared to participants two, three and four. Based on participant one analysis it was evident that minimal force occurred at the beginning of the left arm down sweep and during the extension kick (Figure 3a). Maximal force occurred during the second downsweep phase. Participant two produced two propulsive force peaks on the right side. The first peak during the right side stroke was during the first downsweep phase and the second occurred during the second downsweep phase. The multiple peaks was supported by Schleihauf et al (1988) research which illustrated three large resultant force phases occurred when examining the propulsive forces associated with the arm stroke during backstroke swimming. A limitation of Schleihauf et al (1988) research was the data only represented the forces associated with the hand not the whole body. In contrast, participant four presented with maximal force during the first down sweep of the right arm phase. However, during this same phase on the left arm minimal force was produced (figure 2). Similarly, participant two presented a weakness on the left side of the body which may have been due to muscular weakness or technique error (figure 1b). Weaknesses in musculature or technique can be highlighted in the force profile and identified from the video analysis prompting changes to technique based on quantitative feedback.

It was evident by examining the force profiles of four swimmers in that each individual produced distinctive propulsive force profiles, therefore strengthening the importance of providing propulsive force as a dynamic parameter synchronised to video footage. This allowed swimmers and coaches to correct technique based upon individual feedback. This effective biomechanics servicing tool provides unique quantitative, stroke kinematics feedback to elite swimmers and coaches. The kinematic data was displayed in an easy to interpret format, providing analysis to each swimmer independent of their biomechanics knowledge. Future research could be directed to identify the actual intra cyclic velocity of breaststroke and how it is related to the propulsive force profile.

Figure 3: Participant One (a) minimal propulsion (b) maximal propulsion

CONCLUSION: Understanding and identifying the dynamic forces generated within backstroke swimming provides further guidance to optimising swimming performance. Previous researchers have tended to neglect investigating propulsive force during backstroke swimming due to limitations in the various methods of data collection. The present study has illustrated the importance of representing the propulsive force as a dynamic parameter synchronised to video footage, thereby highlighting the intra-stroke force variability within and between individuals. This provided a beneficial feedback tool to coaches and swimmers allowing specific technique changes based upon objective and quantitative analysis.

REFERENCES:

Alcock, A., & Mason, B. (2007). Biomechanical analysis of active drag in swimming. In *c2007, p.212-215*. Brazil.

Hollander, A. P., de Groot, G., & van Ingen Schenau, G. J. (1987). Active drag in female swimmers. In *In, Jonsson, B. (ed.), Biomechanics X-B, Champaign, Ill., Human Kinetics Publishers, c1987, p. 717-720*. United States.

Fulton, S. K., Pyne, D. P., & Burkett,B (2008). Influence of kicking velocity and amplitude on net force and kick rate in elite swimmers. *Journal of Applied Biomechanics (In review)*

Kolmogorov, S. V., & Duplishcheva, O. A. (1992). Active drag, useful mechanical power output and hydrodynamic force coefficient in different swimming strokes at maximal velocity. *Journal of Biomechanics, 25*(3), 311-318.

Maglischo, E. (2003). Swimming Fastest. USA: Human Kinetics

Schleihauf, R.E., Higgins, J.R., Hinrichs, R., Luedtke, D., Malglischo, C., Maglischo, E.W., Thayer, A.,(1988) Propulsive techniques: Front crawl stroke, butterfly, backstroke, and breaststroke In, Ungerechts, B.E. et al. (eds.), Swimming science V, Champaign, Ill., Human Kinetics Publishers, p. 53-59.

Toussaint, H. M. (2002). Biomechanics of front crawl swimming: propulsion and drag. (Abstract). In *In 12th Commonwealth International Sport conference, 19-23 July 2002, Manchester, United Kingdom: abstract book, London, Association of Commonwealth Universities, 2002, p.310*. United Kingdom.

Xin-Feng, W., Lian-Ze, W., Wei-Xing, Y., De-Jian, L., & Xiong, S. (2007). A new device for estimating active drag in swimming at maximal velocity. *Journal of Sports Sciences, 25*(4).

Acknowledgements

The researcher would like to thank the Australian Institute of Sport, Aquatics Testing, Training and research team for their contribution to the testing procedure and the Australian Institute of Sport swimming team for their participation.

EFFECTS OF HANDLE AND BLOCK CONFIGURATION ON SWIM START PERFORMANCE

Peter F. Vint[1], Richard N. Hinrichs[2], Scott K. Riewald[1], Robyn A. Mason[1], and Scott P. McLean[3]

[1]United States Olympic Committee, Colorado Springs, CO, USA
[2]Department of Kinesiology, Arizona State University, Tempe, AZ, USA
[3]Department of Kinesiology, Southwestern University, Georgetown, TX, USA

The purpose of this study was to quantify differences in swimming track start performances using side handle and front handle grip techniques and using an inclined platform at the rear of the starting platform. An instrumented starting block system was designed to allow front grip and side grip starting techniques and inclusion of a rear incline. Thirty male and 20 female junior elite swimmers completed three starts in each of four start block configurations: (1) Flat (traditional) block, front handle grip; (2) Flat (traditional) block, side handle grip; (3) Incline (new) block, front handle grip; (4) Incline (new) block, side handle grip. Force and video data were used to quantify parameters related to starting performance. Results indicated that use of side handles had a substantial impact on start performance while the effects of the rear incline were less pronounced. Compared to using a front grip technique, use of the side handles increased horizontal velocity at takeoff up to 18%, resulted in a more horizontal takeoff angle by up to 2°, increased contribution to horizontal impulse from arms up to 12%, increased peak horizontal power up to 28%, decreased time to 6 m by 4% and increased velocity at 6 m by 2.5%. These advantages were achieved at a cost of an 8% increase in propulsion time. Based on this study, we recommend swimmers develop familiarity with the use of side handles when performing a track start and to use this technique if handles are available on a starting block.

KEY WORDS: swimming, track start, incline, side handle, FINA

INTRODUCTION: LaRue (1985) suggested that using a track start in swimming would allow a swimmer to generate a greater horizontal impulse than the grab start thus increasing horizontal takeoff velocity. However, numerous comparisons between the grab and track start techniques used in competitive swimming have yielded equivocal kinematic and kinetic results suggesting that neither of these techniques is superior (Welcher et al., 2008; Breed et al., 2000; Blanksby et al., 2002; Holthe and McLean, 2001; Miller et al., 2002; Kruger et al., 2002; Vilas-Boras et al., 2002). Despite these findings, a preference among elite swimmers for the track start has emerged. During the 2008 Beijing Olympic Games, 77% of the swimmers in 15 of 22 finals heats chose to use a track start over the grab start including all eight of the men's 50m freestyle finalists.

Start technique preference and limited practice with unfamiliar techniques may compromise start performance (Blanksby et al., 2002) such that it may be difficult to identify a particular start as superior. Additionally, LaRue (1985) concluded that one weakness in the track start was the lack of a vertical support for the rear foot against which the swimmer could push to maximize the horizontal impulse. Recent changes in rules governing the design of starting blocks (FINA rule FR 2.7, 2005) now permit the addition of an inclined platform to the rear of a starting block and the addition of handles to the block which may lead to improved start performance. The purpose of this study was to evaluate the effect of using an inclined platform and side handles on the performance of a competitive track start. It was hypothesized that these changes will (1) increase horizontal velocity of the CM at takeoff; (2) improve the trajectory of the CM during the flight phase; (3) increase the contribution of the arms; (4) decrease time to 6 m; 5) increase horizontal velocity at 6 m.

METHODS: Thirty male and 20 female swimmers participating in the 2007 USA Swimming junior elite and national select camps completed a minimum of three maximum effort starts in each of the following four start block configurations: (1) Flat block, front handle grip; (2) Flat block, side handle grip; (3) Incline block, front handle grip; (4) Incline block, side handle grip. A custom-built, instrumented start block system was designed using two AMTI force plates (model OR6-5-1) and two Kistler 3D force transducers (model 9347C) attached to handles which were configurable to allow front grip and side grip start techniques. A rear incline platform of 36° could be bolted to the top of the block such that the distance from the front of plate to the back of the rear incline was approximately 55.5 cm.

Athletes were allowed adequate practice to become familiar with each start block configuration, but received no specific training on any technique or configuration. None of the athletes had any prior experience with the rear incline and only two had prior experience using the side handle grips. Start order was counterbalanced across testing groups. Athletes were instructed to perform maximum effort starts upon hearing an audible "start" tone and, once in the water, to glide as far as possible without stroking or kicking.

During each start performance, force data were sampled at 600 Hz for 3 seconds while sagittal plane underwater video data were collected on the underwater glide phase. Video data were digitized at 60 Hz using Peak Motus and calibrated using a 70-point control object and a multistage, two-dimensional DLT reconstruction algorithm. Video and force data were synchronized by a common signal which actuated the data collection program (via digital trigger), below water LED (visible to the underwater camcorder), and the audible "start" tone heard by the athletes.

Since the start blocks were oriented at a 10° incline, force data from each plate were rotated back to a true horizontal-vertical reference frame. Body mass (kg) was calculated using force data collected during a quiet standing period prior to each start performance. Propulsion time (s) was defined from the first appreciable change in the net horizontal force until the instant of takeoff. Horizontal and vertical impulses (N·s) acting on the body during the start were calculated by numerical integration of the respective net forces over the propulsion time. Horizontal and vertical velocities (m/s) of the CM at takeoff were calculated by dividing the respective impulses by body mass. Takeoff angle (deg) was defined as the orientation of the resultant velocity vector at the instant of takeoff. Peak horizontal power (W/kg) was defined as the maximum value of the instantaneous product of net horizontal force and horizontal CM velocity during the propulsion phase and was normalized by body mass.

Based on procedures described by Welcher et al. (2008), total time and horizontal velocity at 6 meters were extracted from the underwater video data. Time to 6 m was defined as the time from the synchronization pulse until the athlete's greater trochanter passed the 6-m mark. To find this time, the digitized and 2D DLT scaled horizontal position data for this landmark were fit with a 2nd order polynomial and solved for the roots when y (position) = 6.0. This time was subsequently used to find the horizontal velocity at 6 m by differentiation of the same polynomial.

Two-way repeated measures ANOVAs were used to detect statistical differences ($p<.05$) in a set of key swim start performance variables as a function of grip type (front or side handle) and block configuration ("flat" or incline).

RESULTS: Compared to the more traditional track start configuration (flat block, front grip technique), use of the side handles had a significant and substantial effect on several important start performance variables (Table 1). Specifically, use of the side handles increased horizontal velocity at takeoff 16-18%, resulted in a more horizontal takeoff angle, increased contribution to horizontal impulse from arms 10-12%, increased peak horizontal power 18-28%, decreased time to 6 m by 4% and increased velocity at 6 m by 2.5%. These advantages were achieved at a cost of an 8% increase in propulsion time.

Use of the rear incline also affected start performance but not as dramatically. Using the rear incline increased horizontal velocity at takeoff <2% with a 3% reduction in propulsion time. Takeoff angle was more downward by ~2°. Contribution to horizontal impulse from the arms was decreased 2.5-5% but peak horizontal power increased 5-11%. Time to 6 m was essentially unchanged when using the incline with side handles but decreased by 2.6% when the inclines were combined with the front grip. Use of the incline resulted in an insignificant increase (~0.5%) in velocity at 6 m.

Table 1. Mean (± SD) kinematic and kinetic measures of swim start performances as a function of hand orientation and block design (n=50).

	Sig	Flat (traditional) Block		New (rear incline) Block	
		Side handles	Front grip	Side handles	Front grip
Propulsion Time (s)	*†	0.67 ± 0.09	0.62 ± 0.06	0.65 ± 0.08	0.62 ± 0.08
Hor. Takeoff Velocity (m/s)	*	4.72 ± 0.52	4.00 ± 0.31	4.73 ± 0.52	4.07 ± 0.28
Takeoff Angle (deg)	*†‡	−0.36 ± 6.35	−4.73 ± 6.73	−2.51 ± 6.30	−5.68 ± 7.31
Hor. Peak Power (W/kg)	*†	38.99 ± 8.11	30.44 ± 5.55	41.18 ± 9.40	33.66 ± 6.14
Hor. Impulse from Arms (%)	* ‡	24.87 ± 11.81	12.04 ± 8.66	19.49 ± 9.73	9.54 ± 7.55
Time to 6 m (s)		2.97 ± 0.36	3.08 ± 0.40	2.99 ± 0.40	3.00 ± 0.33
Hor. Velocity at 6 m (m/s)	* ‡	2.05 ± 0.32	2.00 ± 0.29	2.06 ± 0.33	2.01 ± 0.29

*Side handle significantly different than front handle ($p<.05$). †Incline block significantly different than flat block ($p<.05$). ‡Significant two-way interaction ($p<.05$). All tests: $F(1,48)$.

DISCUSSION: Prior to the 2008 Beijing Olympics, Omega^{TM} debuted a new block design that included a rear incline feature intended to allow swimmers to perform a track start similar to those used in track and field. However, due to concerns raised within the international swimming community (Nicole, retrieved, 23-03-2009) FINA decided to delay the introduction of this block. In addition to the rear foot incline, FINA had previously approved the use of side handles on the block (FINA rule FR 2.7, 2005). The present study sought to document the effect of each of these design features on a track start.

Use of side handles had a more substantial effect than use of the rear incline on takeoff characteristics in a track start. Horizontal takeoff velocity was vastly improved using the side handles likely due to an increased contribution from the arms during the start. This improvement was accompanied by a slightly longer propulsion time but despite this, time to 6 m was still 0.1 s faster ($p=.161$) with a significantly faster horizontal velocity ($p=.018$) at this point. The minimally improved time to 6 m was unexpected given the improved power and takeoff velocity. However, these changes may be attributable to what occurs underwater. The altered flight dynamics frequently had an observable, although unmeasured, impact on the swimmer's entry and the subsequent body position during the glide phase. It is reasonable to assume that with additional practice swimmers would be able to refine that transition to achieve better times and velocities at 6 m. Furthermore, as Welcher et al. (2008) argued, the fact that when using side handles swimmers reached the 6-m mark in the same time but with higher horizontal velocity suggests that the starts incorporating the use of side handles were best.

To better present the effect of these design features, it is useful to consider the trajectory of the center of mass during the flight phase of the start. Given that the flight of the center of mass is governed by the laws of projectile motion, the flight distance may be estimated if the relative height of projection, projection angle and projection velocity are known. Although not measured in the present study, data do exist to suggest that the relative height of projection is approximately 0.7 m for adult male swimmers competing on a national level (McLean et al., 2000). Using the measured values for projection velocity and angle and McLean et al.'s estimate of relative height, it was possible to compare the trajectories of the various start conditions. Based on this analysis, the use of side handles produced 30-40 cm of additional flight distance. Given that this distance was covered with negligible resistance due to travelling in the air, this portion of the race was covered faster than if it had been travelled underwater. Considering Cossor and Mason's (2001) observation that better starters spent less time from entry to 9.5 m, using the side handles with or without an incline would result in the swimmers needing to cover less distance from entry to a pre-determined distance.

CONCLUSION: In light of the minimal instruction and practice provided in the study, the substantial improvement offered by the use of side handles suggests that side handles should be used with a track start whenever they are available. However, the more modest improvements afforded by the use of the rear foot incline suggests that additional work is required to understand the true potential of this design feature in the swimming track start.

REFERENCES:
2005-2009 FINA Rules and Regulation (2005). http://www.fina.org/rules/rules_index.htm
Blanksby, B., Nicholson, L., and Elliott, B. (2002). Biomechanical analysis of the grab, track and handle swimming starts: An intervention study. *Sports Biomechanics*, *1*, 11–24.
Breed, R. V. P., and McElroy, G. K. (2000). A biomechanical comparison of the grab, swing and track starts in swimming. *Journal of Human Movement Studies*, *39*, 277–293.
Cossor, J., and Mason, B. (2001). Swim start performances at the Sydney 2000 Olympic Games. In J. Blackwell (Ed.) *Proceedings of XIX Symposium on Biomechanics in Sports*. San Francisco. University of Cailfornia at San Francisco, June 19-29, 2001.
Guimaeres and Hay (1985). A mechanical analysis of the grab starting technique in swimming. *International Journal of Sport Biomechanics*, *1*, 25-35.
Holthe, M.J and McLean, S.P. (2001). Kinematic comparison of grab and track starts in swimming. In J. Blackwell (Ed.) *Proceedings of XIX Symposium on Biomechanics in Sports*. San Francisco. University of Cailfornia at San Francisco, June 19-29, 2001.
Kruger, T., Wick, D., Hohmann, A., El-Bahrawi, M., Koth, A. (2002). Biomechanics of the grab and track start technique. In J. Chatard (Ed.) *Biomechanics and Medicine in Swimming IX* (pp. 219-224). Publications de l'Universite de Saint-Etienne.
LaRue, R.J. (1985). Future start: If a track start proves faster, will blocks be modified to accommodate it? *Swimming Technique*, *February-May*, 30-32.
McLean, S.P., Holthe, M., Vint, P.F., Beckett, K.D., and Hinrichs, R.N. (2000). Addition of an approach to a swimming relay start. *Journal of Applied Biomechanics*, *16*, 343-356.
Miller, M., Allen, D., and Pein, R. (2002). A kinetic and kinematic comparison of the grab and track starts in swimming. In J. Chatard (Ed.) *Biomechanics and Medicine in Swimming IX* (pp. 231-236). Publications de l'Universite de Saint-Etienne.
Nicole, J. (retrieved, 23-03-2009). Omega starting block in mix for Beijing Olympics. http://www.news.com.au
Sanders and Bonnar (retrieved, 23-03-2009). Start technique – recent findings. http://www.coachesinfo.com/
Vilas-Boas, J.P., Cruz, J., Sousa, F., Conceicao, F., Fernandes, R., Carvalho, J. (2002). Biomechanical analysis of ventral swimming starts: comparison of the grab with two track start techniques. In J. Chatard (Ed.) *Biomechanics and Medicine in Swimming IX* (pp. 249-254). Publications de l'Universite de Saint-Etienne.
Welcher, R.L., Hinrichs, R.N., and T.R. George (2008). Front- or rear-weighted tack of grab start: Which is the best for female swimmers? *Sports Biomechanics*, *7*, 100-113.

ACKNOWLEDGEMENT:
This work was supported by grants from USA Swimming's Science & Technology Research Grant program and from the United States Olympic Committee Performance Services Division.

ISBS 2009

Oral Session S04

Modelling

AN ALGORITHM TO COMPUTE ABSOLUTE 3D KINEMATICS FROM A MOVING MOTION ANALYSIS SYSTEM

Floren Colloud[1], Mickaël Begon[2], Vincent Fohanno[1] and Tony Monnet[1]

Laboratoire de Mécanique des Solides, Université de Poitiers, Poitiers, France[1]
Département de Kinésiologie, Université de Montréal, Montréal, Canada[2]

KEY WORDS: 3d Kinematics, motion analysis system, gait analysis.

INTRODUCTION: Recently, Colloud *et al.* showed the feasibility of using a *moving* motion analysis system to acquire three dimensional (3d) kinematics over a large volume. They placed a motion analysis system on a rigid rolling frame that followed the displacement of a known object. In this pilot study, Colloud *et al.* obtained accuracy similar to those report for motion analysis systems (Richards, 1999). As a result, the rolling system is accurate enough for capturing the local 3d kinematics. However, the expression of the kinematics in a global frame – *i.e.* the absolute kinematics – has not been assessed. Thus it is impossible to calculate spatial-temporal parameters (*e.g.* step length, step width, walking speed in gait analysis). The purpose of this study is to propose an algorithm for calculating the 3d global kinematic of a subject walking on a 40 m-long pathway.

METHODS: One male participant (age: 21 yr, height: 170 cm, mass: 62 kg) equipped with 22 reflective markers performed five trials on 40 meters. He was followed by a rolling frame (4.4 × 4.0 × 2.5 m) with a 8-camera motion analysis system (T40 series, Vicon, Oxford, UK) sampled at 100 Hz. Forty-one reflective markers were placed every meter on the ground on an horizontal line using a tape measurer and a self levelling laser. The algorithm consists in three steps: (*i*) estimation of the kinematics from the camera frame (A_L) to a local frame (A_i) using two markers (g_i and g_j) seen on the ground, (*ii*) expressed this local kinematics in a global frame (A_G) and (*iii*) calculation of the roto-translation (^iR_j) from this current local frame (A_i) to the next local frame (A_j) before g_i disappears. This last step requires three visible ground markers (g_i, g_j and g_k). An elimination procedure that minimizes the norm of Frobenius is used until 50% of the image remained. The accuracy and precision of the reconstruction were evaluated as the deviation of reconstructed marker position relative to its reference and as the radius of the spheres of 95% confidence for the ground markers express in the global frame, respectively.

RESULTS and DISCUSSION: The accuracy was up to 16 mm in antero-posterior direction but could reach 138 and 163 mm in lateral and vertical directions over the 40 m translation. The deviations differed in direction and magnitude between the trials. The precision was lower than the precision estimated with a rigid object (1.3 mm). Although their position was fixed in the global frame, the markers were shaking, in the worst case, in a sphere of 20 mm. The errors in marker position could be reduced with a reconstruction using at least three cameras.

CONCLUSION: This algorithm is efficient for the analysis of human movement on horizontal ground. It allows the calculation of spatio-temporal parameters related to the performance in ecological environments over many cycles for walking and many sports (e.g. running)

REFERENCES:

Colloud, F., Chèze, L., Andrè, N., Bahuaud, P. (2008). An innovative solution for 3D kinematics measurement for large volume. *Journal of Biomechanics*, 41(S1), S57.

Richards, J., 1999. The measurement of human motion: A comparison of commercially available systems. *Human Movement Science*, 18, 589–602.

Acknowledgement:
The financial support of Région Rhône-Alpes (Projet Emergence) and Région Poitou-Charentes--European Union (CPER 2007-2013) is gratefully acknowledged.

DEVELOPMENT, EVALUATION AND APPLICATION OF A SIMULATION MODEL OF A SPRINTER DURING THE FIRST STANCE PHASE

Neil E. Bezodis, Aki I. T. Salo and Grant Trewartha

Sport and Exercise Science, University of Bath, Bath, United Kingdom

This study aimed to investigate how alterations in kinematics at touchdown could improve the performance of an international-level sprinter during the first stance phase of a sprint. A seven-segment angle-driven simulation model was developed, and evaluation against empirical data revealed the model matched reality to within a mean value of 5.2%. A series of simulations altering the horizontal distance between the stance foot and the CM at touchdown were undertaken. By positioning the foot slightly further behind the CM, performance (external power) was improved due to favourable increases in horizontal force production and only small increases in stance duration. However, continuing to increase this distance between the foot and the CM led to decreased performance due to an inability to generate sufficient force despite continued increases in stance duration.

KEY WORDS: performance, simulation modelling, sprint start, technique.

INTRODUCTION: The start is an important part of an athletics sprint as sprinters strive to rapidly accelerate from a stationary position. Any small improvements in start performance can be critical to overall success – in the men's 100 m final at the 2004 Olympics there were only 0.04 s between the gold-medallist and the fourth placed sprinter – and therefore elite coaches strive to obtain any performance improvements. Several studies (e.g. Guissard et al., 1992) have directly altered the technique of a sprinter during the start, and observed the consequent effects on performance. However, due to the understandable opposition towards such experimental procedures from coaches, these studies have focussed on less-well-trained sprinters, and their application may thus be limited for internationally-competitive sprinters.

In recent years, there has been an increased use of theoretical models to investigate the performances of highly-trained athletes (e.g. Hiley and Yeadon, 2003). Computer simulation models have been applied in an attempt to determine how changes to specific inputs (such as technique variables) can affect certain performance-related outputs, thus allowing the identification of adjustments which could lead to performance improvement. However, simulation models should not be applied before evaluation of how well they represent the true system of interest (Yeadon and King, 2002). The first aim of this study was therefore to develop a simulation model of a sprinter, and evaluate its accuracy against empirical data collected from an international-level sprinter during the first stance phase of a sprint.

Based on empirical data collected from international-level sprinters, it has previously been suggested that positioning the stance foot further behind the whole body centre of mass (CM) at the onset of the first stance phase (i.e. an increasingly *negative* touchdown distance) could be beneficial for performance (Bezodis et al., 2008). In order to further investigate this, the second aim of this study was to use the evaluated simulation model to determine how changes in touchdown distance affected performance.

METHODS: Three international-level sprinters completed a series of maximal 30 m sprints commencing from blocks. Ground reaction forces were collected from the first stance phase using a force platform (Kistler, 9287BA; 1000 Hz). Two digital video cameras (Redlake, MotionPro HS-1; 200 Hz) were positioned in series to obtain high-resolution images from the set position until the end of the first stance phase. Twenty anatomical points were digitised, and the resulting displacement time-histories were scaled, digitally filtered, and combined with individual-specific segmental inertia data (Yeadon, 1990) to calculate CM motion. These kinetic, kinematic and anthropometric data provided sufficient data with which to customise a simulation model and to facilitate the undertaking of a quantitative evaluation of the model.

A seven-segment 2D simulation model (Figure 1a) was developed in Simulink® (v. 7.1), with its structure determined through quantitative and qualitative analyses of the empirical data from all three sprinters. The foot-ground interface was modelled using a two segment foot with horizontal and vertical spring-damper systems at the hallux and the MTP joint (Figure 1b). The model was customised using data from the sprinter exhibiting the highest levels of performance (age = 20 yr, mass = 86.9 kg, height = 1.78 m, 100 m PB = 10.28 s). Input data (joint positions and velocities at touchdown) were obtained from one empirical trial, and the model was driven using angular acceleration time-histories at each of the six joints. Five terms of a Fourier series were combined with each of the empirical angular accelerations, thus allowing slight adjustments to account for any small digitising error. The Fourier series co-efficients and the remaining input data (visco-elastic co-efficients for the foot-ground interface) were obtained through matching optimisations using a pattern search algorithm to determine the values which facilitated the closest match between the model and reality.

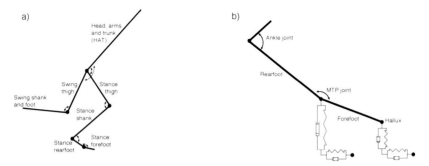

Figure 1. Illustration of the structure of a) the seven segment model, including b) the system used to model ground contact.

The accuracy of the simulation model was evaluated against the empirical data based on five specific criteria: *configuration* (mean RMS difference in angular displacements at the six joints throughout stance), *orientation* (RMS difference in absolute angle of the trunk throughout stance), *force accuracy* (mean RMS difference in horizontal and vertical ground reaction forces throughout stance, expressed as a percentage of force excursion), *impulse* (mean percentage difference in horizontal and vertical impulse production during stance) and *performance* (percentage difference in average horizontal external power production during stance – power was identified as an objective measure of sprint start performance by Bezodis et al., 2007). Assuming errors in percentages were equal to those in degrees (Yeadon and King, 2002), an overall evaluation score was calculated. To address the second aim of this study, knee joint angle at touchdown was manipulated by ±10° at 1° intervals from the empirically-recorded value (88.8°), whilst all other inputs and initial conditions remained constant. This simulated 20 touchdown distances deviating from a median empirical value (-0.073 m; i.e. CM 7.3 cm ahead of the foot), with the extreme touchdown distances being -0.009 m and -0.141 m. Model output data from all 20 simulations were expressed relative to the output associated with the empirical touchdown distance (-0.073 m). In addition to outputting power data to determine how performance levels changed, further variables were also output in an attempt to understand and explain why changes in performance occurred.

RESULTS: The model was able to replicate reality to within a mean value of 5.2% (*configuration* = 5.7°, *orientation* = 8.6°, *force accuracy* = 8.3%, *impulse* = 1.4%, *performance* = 2.0%). Graphical representations of the *force accuracy* (and thus *impulse*) criteria are presented in Figure 2.

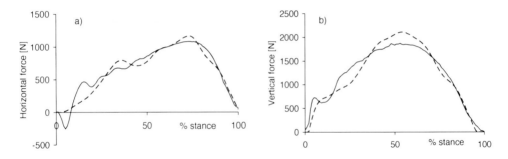

Figure 2. Empirical (solid line) and modelled (dashed line) a) horizontal and b) vertical ground reaction force time-histories throughout the stance phase.

In the simulations using the evaluated model, a curvilinear relationship was observed between touchdown distance and performance (Figure 3a). As the foot was positioned further behind the CM at touchdown, performance levels (i.e. power production) increased towards a peak value at a touchdown distance of -0.093 m. However, performance levels began to decrease beyond this, falling below original levels as touchdown distances exceeded -0.107 m. All simulations starting with the foot less far behind the CM at touchdown than in the empirical trial were associated with reductions in performance. The changes in velocity achieved during stance followed a curvilinear trend similar to that observed for power, whereas stance duration followed a more linear pattern, with stance duration increasing as the distance between the foot and the CM at touchdown also increased (Figure 3a).

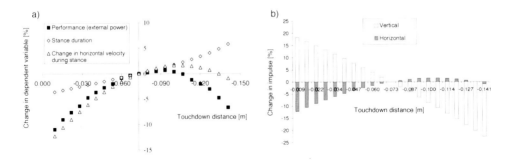

Figure 3. The effect of altering touchdown distance on a) performance (i.e. average horizontal external power), change in horizontal velocity during stance and total stance duration, and b) horizontal and vertical impulse production during stance.

A linear relationship existed between touchdown distance and vertical impulse production. Vertical impulse increased as the foot was placed less far behind the CM at touchdown compared to the empirical trial, and decreased when the foot was further behind the CM (Figure 3b). In contrast to vertical impulse, horizontal impulse production decreased when the foot was placed less far behind the CM than in the empirical trial. However, the overall relationship between touchdown distance and horizontal impulse was curvilinear - the largest horizontal impulses were associated with a touchdown distance of -0.107 m, with magnitudes reducing as the distance between the foot and the CM increased further (Figure 3b).

DISCUSSION: The evaluation results revealed that the chosen model structure was sufficient for simulating the motion of a sprinter during the first stance phase. When compared with previous detailed evaluations of angle-driven models (e.g. Wilson et al., 2006), the overall difference (5.2%) indicated a close match between the model and reality. The five individual criteria indicated that no single aspect of the model was matched considerably worse than the others, and thus both kinematic and kinetic variables were accurately represented. Most importantly, the evaluation yielded confidence in the accuracy of the model, and thus in the results obtained when applying the model to investigate how changes to technique could affect performance.

Model-based simulations suggested that it would be possible for this international-level sprinter to further improve his performance during the first stance phase by commencing stance with his stance foot positioned slightly further behind the CM. This facilitated a slight increase in horizontal impulse production which, although associated with small increases in stance duration, was largely due to greater force production (as evident by the greater power production). However, as the distance between the stance foot and the CM increased further, insufficient impulse could be generated, and performance levels consequently decreased. The model outputs suggested that these performance reductions were due to an inability to generate sufficient force (horizontally and vertically), despite increases in stance duration.

CONCLUSION: The evaluation results revealed that a simulation model could be confidently used to theorise how an international-level sprinter could improve performance during the first stance phase of a sprint. Application of the model confirmed previous suggestions (Bezodis et al., 2008) that landing with the stance foot further behind the CM could lead to performance improvements. By analysing additional outputs, it was identified that the improvements in performance were due to a favourable increase in horizontal impulse production. However, as this distance increased further, performance levels decreased, which appeared to be due to the sprinter being in a less favourable position for generating force against the track. The results of this study can be used to inform the training programme of the international-level sprinter, by encouraging a slight repositioning of the foot relative to the CM at touchdown. The existing model provides a useful framework which, with appropriate input data, can be customised to any individual sprinter. Additionally, the model could be applied to ascertain further potentially beneficial technique adjustments, and algorithms could be used to vary several inputs in an attempt to identify a technique associated with optimal performance.

REFERENCES:

Bezodis, N.E., Trewartha, G. & Salo, A.I.T. (2007). Choice of performance measure affects the evaluation of sprint start performance. *Journal of Sports Sciences*, 25, S72-S73.

Bezodis, N.E., Trewartha, G. & Salo, A.I.T. (2008). Understanding elite sprint start performance through an analysis of joint kinematics. In *Proceedings of XXVI International Symposium on Biomechanics in Sports* (edited by Y-H. Kwon, J. Shim, J.K. Shim & I-S. Shin). pp. 498-501. Seoul, S.Korea, Seoul National University Press.

Guissard, N., Duchateau, J. & Hainaut, K. (1992). EMG and mechanical changes during sprint starts at different front block obliquities. *Medicine and Science in Sports and Exercise*, 24, 1257-1263.

Hiley, M.J. & Yeadon, M.R. (2003). Optimum technique for generating angular momentum in accelerated backward giant circles prior to a dismount. *Journal of Applied Biomechanics*, 19, 119-130.

Wilson, C., King, M.A. & Yeadon, M.R. (2006). Determination of subject-specific model parameters for visco-elastic elements. *Journal of Biomechanics*, 39, 1883-1890.

Yeadon, M.R. (1990). The simulation of aerial movement-II. A mathematical inertia model of the human-body. *Journal of Biomechanics*, 23, 67-74.

Yeadon, M.R. and King, M.A. (2002). Evaluation of a torque-driven simulation model of tumbling. *Journal of Applied Biomechanics*, 18, 195-206.

RECOGNITION AND ESTIMATION OF HUMAN LOCOMOTION WITH HIDDEN MARKOV MODELS

A. Fischer[1], D. Gehrig[2], T. Stein[1], T. Schultz[2], H. Schwameder[1]

Institute for Sports and Sport Science, University of Karlsruhe, Germany[1]
Cognitive Systems Lab, Dept. of Informatics, University of Karlsruhe, Germany[2]

KEY WORDS: Hidden Markov Model, Gait, Pattern Recognition.

INTRODUCTION: The Collaborative Research Centre "Humanoid Robots" situated at the University of Karlsruhe is aimed to construct a learning and cooperating service robot. To cope with its tasks it is necessary that the robot is able to identify diverse objects as well as different persons. Looking at stochastic models for pattern recognition Hidden Markov Models (HMMs) are described to be most suitable to classify time arranged data (Bilmes 2002). The objective of this study is to screen if the HMMs supply satisfying rates of recognition of human trajectory and angle data.

METHOD: Kinematic data of eight men and three women was captured at different walking and running speed (1.2 m/s, 3 m/s, 4 m/s, 5 m/s) on a treadmill. Data acquisition was realised with an infrared camera system with a frequency of 250Hz. For each walking/running speed there were 120 gait cycles of every test person available. The construction and training of the stochastic model was based on the gait data. Due to the fixed sequence of gait phases a HMM with a simple linear topology was chosen. Each state of the HMM represented a phase of the gait cycle. The different states were equipped with Gaussian distributions and transition probabilities to model the run of the angles observed. The HMM modelling human gait best was selected and trained with data of 17 double gait cycles for each data sequence of every test person.

RESULTS: The trained HMMs showed recognition rates from 63% to 100% for the observed data sequences for five male test persons. Highest rates could be obtained with Centre of Mass and head angles. For some test person recognition rates decreased with data of gait cycles that were captured towards the end of one run.

DISCUSSION: The high recognition rates based on kinematic data of Centre of Mass were expected due to the different mean values of the test persons according to their body height. The decrease of recognition rates that could be observed at some of the test person on late data of one run seems to be caused by acclimatisation to treadmill running. The achieved recognition rates exceed rates typical for speech recognition (Rabiner 1989). A combination of different angle data seems to promise increasing recognition rates.

CONCLUSION: The study showed that HMMs seem to be suitable to identify humans based on their kinematic gait data satisfyingly stable. According to dislocation of the Gaussian distributions it could be possible to suggest on systematic changes on patterns over changes in walking-/running speed.

REFERENCES:
Bilmes, J. (2002). What HMMs Can Do. *UWEE Technical Report*, No UWEETR-2002-2003, University of Washington, Dept. of EE.
Rabiner, L. R. (1989). A Tutorial on Hidden Markov Models and Selected Applications in Speech Recognition. *Proceedings of the IEEE*, 77 (2), 257-286

Acknowledgement
V. Wank, Institute of Sport Science, University of Tübingen
German Research Foundation – CRC 588 Humanoid Robots

TOWARDS UNDERSTANDING HUMAN BALANCE – SIMULATING STICK BALANCING

Manfred Vieten

University of Konstanz, Konstanz, Germany

The purpose of this study was to develop a simulation model for stick balancing. Experimental result served as a guide for the developing progress. The progress started with a deterministic approach. We solved the Euler-Langrange equation and received the equation of motion. The controlling variable within this equation is the acceleration of the lower end of the stick. This parameter depends on the balancing strategy and ability of the human subject. We chose the van der Pol equation as an ansatz for describing it. A second attempt included the incorporation of a time delayed parameter. The third form included additional stochastic noise. We found close similarity between the measured and the calculated parameters tilt angle at reversal points, frequency expectation value, and others.

Keywords: balancing, movement pattern, simulation, stability, chaos theory, van der Pol equation

INTRODUCTION: Balance is an ability to maintain the center of gravity of a body within the base of support with minimal postural sway (Shumway-Cook, Anson et al. 1988). It is an essential feature for achievement in most human movement tasks. For the analysis of stick balancing we define two tilt angles. Those tilt angles and the respective angular velocity revealed to be connected with the acceleration of the lower end of the stick. These interrelations together with a characteristic sway frequency seemed to be the main ingredients for successfully balancing a stick. In this study we developed a simulation model based on this context. We draw a comparison between the result from the experiment and the simulation to check the degree of concordance.

METHODS: The movement situation is depicted in Figure 1. From the experiment we know there exist a moderate connection between the two tilt angles. However, for this study we decided to restrict the analysis to one plane of motion, the x-y-plane with x as the horizontal axis and y the vertical axis. The associated angle is β which is defined in equation (1.1).

$$\beta = \arctan\left(\frac{x_{up} - x}{y_{up} - y}\right) \text{ with } x = x_{low} \text{ and } y = y_{low} \quad (1.1)$$

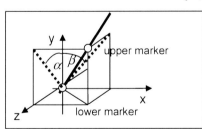

Figure 1: Stick with marker arrangement

We used the Hamilton principle respectively the Euler-Lagrange equation to derive the equation of motion. The stick was treated as a one-dimensional object of length 1 m with its mass evenly distributed along a strait line with the center of gravity symmetrically between the upper and lower ends. The Lagrange function is given in equation (1.2) with T the kinetic energy containing $T_v = \tfrac{1}{2} m \cdot v_{CoG}^2$ being the translation and $T_\omega = \tfrac{1}{2} \dot{\beta}^T I \dot{\beta}$ the rotation part. V stands for the potential energy. Here l is the stick length, m the stick mass, I the inertia tensor, $\vec{v}_{CoG}$ the velocity of the stick's center of

$$L = T - V = T_v + T_\omega - V$$
$$= \frac{1}{2} m \left(\dot{x}^2 + \dot{y}^2 + \dot{x} l \cos(\beta) \dot{\beta} - \dot{y} l \sin(\beta) \dot{\beta} + \frac{l^2}{4} \dot{\beta}^2 \right) + \frac{1}{24} m l^2 \dot{\beta}^2 - m g \frac{l}{2} \cos(\beta) \quad (1.2)$$

gravity, $g = 9.81 \, m/s^2$ the gravitational acceleration, and β the tilt angle as defined above. The Euler-Lagrange equation (1.3)

$$\frac{d}{dt}\left(\frac{\partial L}{\partial \dot{\beta}}\right) - \frac{\partial L}{\partial \beta} = 0 \qquad (1.3)$$

leads to the sought equation of motion (1.4). Here the mass of the stick canceled out. The remaining attribute of the stick is the length l. For practical reasons in an experiment with human subjects mass does play a role. If there is too much mass the subject cannot accelerate the finger holding the stick fast enough, if the stick is too light a subject will have problems to "feel" the movement. However, within a transition range mass does not greatly influence the outcome.

$$\ddot{\beta} = \frac{3}{2l}\left((g + \ddot{y})\sin(\beta) - \ddot{x}\cos(\beta)\right) \qquad (1.4)$$

As a simplification, from here onwards $\ddot{y}$ is suppressed and we are left with the angular acceleration in the form of equation (1.5).

$$\ddot{\beta} = \frac{3}{2l}\left(g\sin(\beta) - \ddot{x}\cos(\beta)\right) \qquad (1.5)$$

We are simulating balancing with the vertical coordinate of finger being constant. The angular acceleration $\ddot{\beta}$ is depending on g, β, and eminently on $\ddot{x}$. g is constant, β we are going to calculate, but the most important parameter is the acceleration $\ddot{x}$. This is the acceleration of the finger and as such exclusively depending on a subject's performance. We are developing an iterative solution (1.6) by writing down the first three terms of a Taylor series.

$$\beta(t + \delta t) = \beta(t) + \dot{\beta}(t)\delta t + \frac{1}{2}\ddot{\beta}(t)\delta t^2 = \beta(t) + \left(\int_0^t \ddot{\beta}(t')dt' + \dot{\beta}_0\right) \cdot \delta t + \frac{1}{2}\ddot{\beta}(t)\delta t^2 \qquad (1.6)$$

Here $\dot{\beta}_0$ is the angular velocity at $t = 0$ and δt the step length of the iteration. $\ddot{x}$ necessarily must allow for a stable result. It must also take into account the empirical relationship between β, $\dot{\beta}$, and $\ddot{x}$. Such an ansatz we found in the van der Pol equation (Goldobin, Rosenblum et al. 2008), which we write here in the form of equation (1.7).

$$\ddot{x} = \mu_1 \cdot (\beta_b - \beta) \cdot \dot{\beta} - \mu_2 \cdot \beta \qquad (1.7)$$

Here μ_1, μ_2 and β_b are constants. Putting (1.7) into (1.5) leads to a deterministic equation. In a next attempt (1.8) time delay is included through the substitution $\dot{\beta}(t) \rightarrow \dot{\beta}(t - \tau)$.

$$\ddot{x} = \mu \cdot (\beta_b - \beta) \cdot \dot{\beta}(t - \tau) - \nu \cdot \beta \qquad (1.8)$$

The solution number three (1.9) introduces stochastic components.

$$\ddot{x} = \xi(\mu_1, \sigma_1) \cdot (\beta_b - \beta) \cdot \dot{\beta}(t - \tau) - \xi(\mu_2, \sigma_2) \cdot \beta \qquad (1.9)$$

$\xi(\mu_1, \sigma_1)$ respectively $\xi(\mu_2, \sigma_2)$ represent normal distributed stochastic noise with the first parameter being the mean and the second the square root of the variance. One of our test parameters is the Frequency expectation value. It is defined as

$$\langle v \rangle = \int_0^{v_s/2} v \cdot |F(v)| \cdot dv \bigg/ \int_0^{v_s/2} |F(v)| \cdot dv \qquad (1.10)$$

Here v is the frequency, $F(v)$ the Fourier transform of β, and v_s the sampling frequency.

RESULTS: All iterations were done with $\delta t = 0.005$ seconds. For the deterministic ansatz (1.5) the iteration returns, as expected, a stable solution Figure 2. There is a transition time, when the movement amplitude changes from the initial value of 1° at an angular velocity of zero as well as an initial acceleration of zero towards a stable situation. Afterwards, the solution is stable at a period two with a frequency of 0.61 Hz and absolute amplitude of 1.99 °.

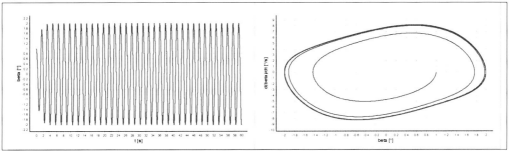

Figure 2: Space-time (left) and phase-space (right) diagrams of a deterministic configuration with $\mu_1 = 1$, $\mu_2 = 10$, **and** $\beta_b = 1$

We delayed the angular velocity term of (1.7) to arrive at (1.8). $\dot{\beta}(t-\tau)$ is delayed with $\tau = 0.15s$. Again, after a transition time, we got a stable two period (Figure 3) with a slightly higher amplitude of 2.09 ° at a frequency of 0.84 Hz.

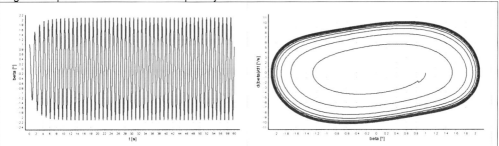

Figure 3: Space-time (left) and phase-space (right) diagrams of a deterministic configuration with time delay $\tau = 0.15$ sec

In the next step we altered (1.8) by adding normal distributed stochastic noise and arrived at equation (1.9). $\xi(\mu,\sigma)$ represents the normal distributed noise. The first parameter in the bracket stands for the mean (as above $\mu_1 = 1$ and $\mu_2 = 10$), the second for the standard deviation. Figure 4 shows a simulation with $\sigma_1 = 1$ and $\sigma_2 = 10$.

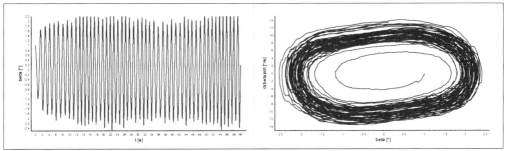

Figure 4: Space-time (left) and phase-space (right) diagrams of a deterministic configuration with time delay $\tau = 0.15$ sec **and normal distributed noise**

We still receive a fairly stable result. The maximal and minimal value of the amplitude within a cycle is not constant anymore, but still with just moderate variations. The mean frequency again slightly increases to about 0.88 Hz with a variation in the order of ± 0.1 Hz. The variables for a comparison between measurement and simulation are given in Table 1.

Table 1: Simulation-measurement comparison of four different parameters

Parameter / Simulation	Deterministic	Delayed	Stochastic	Measurement
$|\bar{\beta}|$ [°] for the reversal points	1.99	2.09	2.16±0.13	2.38±1.41
Correlation coefficient $\beta \square \ \ddot{x}$	0.94	0.95	0.75	0.53±0.09
Time shift to r being maximal [s]	0.41	0.30	0.28	0.13±0.03
Frequency expectation value [Hz]	9.78	6.06	3.81	7.82±3.32

DISCUSSION: The mean tilt angle $|\bar{\beta}|$ for the reversal points is in the right range around 2 °. However, the variation in the simulation results is much smaller than those of the measurements. This hints toward an additional source of uncertainty that is not represented by the equations used in this simulation. The other parameters derived in the simulation, correlation $\beta \square \ \ddot{x}$, time shift, and frequency expectation value are close to those of the measurement. Still, deviations do occur. We varied the time delay to react towards an increased tilt angle β respectively tilt angle's angular velocity $\dot{\beta}$. The solution is fairly stable against time delay with regard to the angular velocity $\dot{\beta}$. Even a time delay of $\tau = 0.2 \text{ sec}$ still results in a stable solution. On the other hand, a small time delay of $\tau = 0.025 \text{ sec}$ with regard to the tilt angle β results in the stick falling.

CONCLUSION: The presented work is a first step towards a proper simulation. The equations allow quantifying the effects within the arrangement. We are able to figure out the influence of the deterministic part compared with the stochastic components. All these findings give confidence for an enhanced model and possibility to apply analogous simulations to other human balancing tasks such as walking.

REFERENCES:
Goldobin, D., M. Rosenblum, et al. (2008). Controlling Coherence of Noisy and Chaotic Oscillators by Delayed Feedback. Handbook of Chaos Control. H. G. Schuster. Weinheim, Viley-VCH: 275-290.
Shumway-Cook, A., D. Anson, et al. (1988). "Postural sway biofeedback: its effect on reestablishing stance stability in hemiplegic patients." Arch. Phys. Med. Rehabil. **69**: 395-400.
Vieten, M. (2006). Statfree: a software tool for biomechanics. In International Society of Biomechanics in Sports, Proceedings of XXIV International Symposium on Biomechanics in Sports 2006, Salzburg, Austria, University of Salzburg, 2006, p.589-593. Austria.

Acknowledgement
All numerical analyses within this study and also the generation of diagrams were done using the software StatFree (Vieten 2006). It is freely available on the internet at www.uni-konstanz.de/FuF/SportWiss/vieten/Software/.

APPLICATION OF SIMULATION TO FREESTYLE AERIAL SKIING

M.R.Yeadon

School of Sport and Exercise Sciences, Loughborough University, UK

The aim of this study was to use a computer simulation model of aerial movement to investigate the ability of asymmetrical arm movements to initiate twist in the flight phase of a triple layout somersault in the aerials event of freestyle skiing. Three arm movements were analysed resulting in triple somersaults with four and five twists together with lead up movements with one twist less. It is concluded that four twists can be initiated during flight using two phases of asymmetrical arm movement and that the production of five twists requires three phases of asymmetrical arm movement.

KEY WORDS: simulation, twisting somersault, freestyle skiing

INTRODUCTION: In the aerials event of freestyle skiing competitors gain speed while skiing down an incline and then enter a curved takeoff "kicker" which provides vertical velocity and angular momentum for the flight phase which lasts for around 2.6 seconds. During this time the aerialist completes three somersaults in an extended configuration with up to five twists. Previous research on the aerials event has shown that competitors can initiate the twist after takeoff using asymmetrical arm movements (Yeadon, 1989). The following question was posed by a national coach: "How can five twists be produced in a triple somersault using asymmetrical arm movements after takeoff?" This study investigated this question using a computer simulation model of aerial movement (Yeadon et al., 1990).

METHODS: Model: An 11-segment computer simulation model of aerial movement was customised to a freestyle skiing competitor using anthropometric measurements to obtain segmental inertia parameters. Measurements of skis, boots and helmet were used to adjust the inertia parameters.

Simulations: Three types of asymmetrical sequential arm movements were investigated for the ability to produce twist in a triple somersault. An empirical process was used in the search for optimum techniques employing manual adjustments to the timing of arm movements with a constraint of a minimum time of 0.25 s for a single arm movement. In keeping with the expectations of competitive skills the arms were extended as each somersault was completed to "show" the completion of each twisting phase. Additionally the twist in the third somersault was constrained to be no more than one revolution in order to allow room for adjustments prior to landing. The head positions in the graphical output were adjusted so that the head looked down when the head was above the feet and the head looked back when upside down. This head orientation enables a continuous view of the landing area.

RESULTS: The first arm sequence comprised a lowering of the left arm to the side followed by a lowering of the right arm to the front (Figure 1). The lowering of the left arm produced a 4° tilt of the body away from the vertical somersault plane and initiated the twist in the triple somersault. The movement of the right arm was timed to occur at around the quarter twist position so that the tilt increased further to 8°. This allowed the completion of one twist in the first somersault, two twists in the second somersault, and one twist in the third somersault. Such phasing of the twist gives rise to the term "full – double full – full" identifying this skill (Figure 2).

Figure 1: Left arm lowered to the side followed by right arm lowered to the front.

Figure 2: A "full – double full – full" produced using the arm sequence depicted in Figure 1.

The second arm sequence comprised a lowering of the left arm to the side followed by a lowering of the right arm to the side (Figure 3). The lowering of the left arm produced a 4° tilt of the body away from the vertical somersault plane and initiated the twist in the triple somersault. The movement of the right arm was timed to occur at around the half twist position so that the tilt increased further to 9°. This allowed the completion of a "full – double full – full" generally similar to that shown in Figure 2.

Figure 3: Left arm lowered to the side followed by right arm lowered to the side.

The third arm sequence comprised a lowering of the right arm across the body, followed by a double arm switch, and finally a lowering of the right arm down the front (Figure 4). The initial movement of the right arm resulted in little tilt which increased to 9° with the double arm switch and increased further to 12° with the final lowering of the right arm to the front around the quarter twist position. This was sufficient to allow the completion of three twists in the second somersault resulting in a "full – triple full – full" (Figure 5).

Figure 4: Arm action in the third simulation: right arm lowered across the body to the side followed by double arm reversal and right arm lowered to the front.

Figure 5: A "full – triple full – full" produced using the arm sequence depicted in Figure 4.

A lead up skill for this third simulation was produced using the same sequence of arm movements to initiate the tilt and twist but using a more abducted arm position in the second somersault so as to obtain two twists rather than three. This resulted in a "full – double full – full" with wide arms (Figure 6). The advantage of practising such a lead up skill before attempting the five twists is that it allows a gradual approach in the learning process in which shortcomings can be adjusted by changing the arm position. The amount that the arms can be abducted whilst twisting also gives a measure of the twist potential.

Figure 6: A "full – double full – full" lead up skill using the same arm sequence as in Figure 5.

DISCUSSION: Three arm sequences were investigated for their ability to produce twist in the airborne phase of triple somersaults in freestyle skiing. Sequences comprising the lowering of one arm followed by the lowering of the other were capable of producing four twists but not five. In order to produce five twists a more complex arm sequence was required involving three phases of arm movement. Although the results are based on a single set of inertia parameters there is sufficient flexibility in the arm positions used in the simulations to accommodate different inertia sets and produce similar outcomes.

It might be thought that the production of aerial twist while wearing skis and boots is made more difficult because the moment of inertia about the twist axis is larger due to the skis and less twist will occur in a somersault. While the moment of inertia about the twist axis is larger, the moment of inertia about the tilt axis is larger still. As a consequence the number of twists per somersault for a tilt angle of 8° is 2.2 compared to 2.0 without skiing equipment and so for a given tilt angle there is an advantage to twisting with ski boots and skis. On the other hand the larger moment of inertia about the tilt axis results in a tilt angle of only 4° when an arm is moved through 180° compared to 8° when there is no ski equipment as in gymnastics dismounts. Thus the difficulty in producing twist in freestyle skiing is really a difficulty in producing tilt. The use of skis and boots with less mass would provide an advantage for creating aerial twist as would gloves with increased mass.

CONCLUSION: It is theoretically possible to initiate five twists in the aerial phase of a triple somersault in freestyle skiing using a sequence of three asymmetrical arm movements.

REFERENCES:

Yeadon, M.R. (1989). Twisting techniques used in freestyle aerial skiing. *Journal of Sport Biomechanics*, 5, 275-281.

Yeadon, M.R., Atha, J. and Hales, F.D. 1990. The simulation of aerial movement - IV: A computer simulation model. *Journal of Biomechanics*, 23, 85-89.

Acknowledgement

The author would like to thank Michel Roth of Swiss Freestyle Skiing for posing the question.

COMPUTER SIMULATION OF THE OPTIMAL VAULTING MOTION DURING THE HORSE (TABLE) CONTACT PHASE

Hui-Chieh Chen[1], Chih-Ying Yu[2], and Kuangyou B. Cheng[1]*

[1]Institute of Physical Education, Health, and Leisure Studies, National Cheng Kung University, Tainan, Taiwan
[2]Department of Athletic Performance, National Taiwan Normal University, Taipei, Taiwan

The purpose of this research is to investigate how the kinematic factors during the horse (table) contact phase influence the post-flight performance in handspring vaulting. A six-segment planar simulation model comprising the lower arm, upper arm, head-trunk, thigh, shank, and foot was customized to an elite gymnast. The body segment parameters, maximum joint torques, and initial kinematic parameters from video analysis of the subject are required for the optimal matching computer simulation. The model was able to match a handspring vault after adjusting the visco-elastic characteristics of the arm-horse interface and joint activation time histories. The model was then used to determine the key factors which influence performance by varying the initial conditions. The objective function was the vertical velocity of the body center of mass at takeoff. The results suggest that smaller wrist angle, greater wrist angular velocity, straighter elbow, greater shoulder angular velocity, greater maximum shoulder torque, and smaller hip angle at horse contact were crucial in achieving the optimal performance. Compared with the five-segment model with a visco-elastic shoulder of a previous study, the six-segment model without a visco-elastic shoulder could still closely match the real performance, and better mimic the actual pushing movement of the arms.

KEY WORDS: gymnastics, modelling, optimization, muscular activation

INTRODUCTION: Most studies about gymnastic vaulting can be characterized into two approaches. One is to record the motions of vaults by cameras and identify the relationship between performance and kinematic parameters. The other is to predict the results by computer simulation. The strength of simulation is not only to examine the sensitivity of initial kinematic variables to give athletes and coaches advices, but also to reduce unnecessary trials/errors to avoid injury. Compared with video analysis, only a few studies employed computer simulation to investigate the vaulting skills. Dainis (1981) used a three-segment human model to describe the motion of handspring vault. The results indicate that the decrease of take-off velocity reduces the after-flight distance. Two-segment models without shoulder torque had also been developed for studying handspring vaults and the Hecht vault. It was found that when the model was limited to one segment by fixing the shoulder, the vault cannot be finished (Sprigings & Yeadon, 1997; King et al., 1999). Koh et al. (2003) used a five-segment model comprising the hand, upper limb, upper trunk, lower trunk, and lower limb to find out the key variables in the Yurchenko vault. The optimal vault displayed greater post-flight amplitude and angular momentum when compared with the gymnast's best trial, and the optimal parameter is within the capacity of the gymnast. King and Yeadon (2005) also used a five-segment model but consider the visco-elastic property of shoulder joint and arm–horse interface. The results show that factors such as shoulder elasticity and the hands which have previously been ignored also have a substantial influence on performance.

Although models of vaulting have been developed from two- to five-segment types with visco-elastic properties, the sensitivity of initial kinematic parameters at horse contact to post-flight performance is still not clear. The difficulty level of a vault is determined by extra spins/somersaults in addition to its basic form. Both greater take-off vertical velocity and post-flight amplitude are necessary for optimal performance. The purpose of this study is to develop a six-segment model for investigating how the initial kinematic factors during the horse contact phase influence the performance during post-flight in handspring vaulting.

METHODS: A six-segment (6S) planar human body model comprising the lower arm, upper arm, head - trunk, thighs, shanks, and feet was developed to simulate the vaulting motion during horse (table) contact. Movement was driven by torque actuators at the ankle, knee, hip, shoulder and elbow. The model was customized to an elite gymnast through subject specific length and strength parameters. Detailed inertia parameters were determined using the data of Taiwanese gymnasts by an MRI method (Chen & Ho, 2006). A high-speed camera operating at 200Hz was used to record the vaulting motion. The trail with the greatest CM velocity at take-off from horse was chosen for kinematical analysis and as the input values of the model. Equations of motion were generated by the software AUTOLEV (www.autolev.com). Each joint torque T was assumed as the product of 3 factors:

$$T = T_{max}(\theta)h(\omega)A(t) \quad (1)$$

$T_{max}(\theta)$ depends on joint angle is the maximum isometric torque (effective torque for both extremities). The dependence on joint angular velocity is modeled by $h(\omega)$.

$$\begin{cases} h(\omega) = (\omega_0 - \omega)/(\omega_0 + \Gamma\omega), & \omega/\omega_0 < 1 \\ h(\omega) = 0, & \omega/\omega_0 \geq 1 \end{cases} \quad (2)$$

Here ω is the instantaneous joint angular velocity, $\omega_0 = \pm 20$ rad/s is maximum angular velocity (positive in extension), and constant $\Gamma = 2.5$ is a shape factor. Joint activation level $A(t)$ characterizing the coordination strategy corresponds to the effective activation of muscles across the joint. The activation level $-1 \leq A(t) \leq 1$ represents maximum effort for flexion and extension respectively. Linear interpolation was used to get the value at every time instant. The visco-elastic properties of the interface between the model and the vaulting horse (arm-horse interface) were modeled by a non-linear spring force F (King & Yeadon, 2005).

$$F = S \times D^2 + K \times D \times V \quad (3)$$

Here S is the stiffness, D is joint displacement, K is the damping coefficient, and V is the joint velocity. This force F is applied in both horizontal and vertical directions. The objective was to maximize the vertical velocity of the center of mass (CM) at takeoff from the horse. The optimization algorithm adopted was the downhill simplex method. Varying initial guesses and re-starting the optimization from a newly found optimum are employed. The model was validated when the averaged angular difference between the model and actual performance was < 5%. Next, the elbow joint was fixed to develop a five-segment (5S) model. By varying the initial kinematic parameters and repeating the optimum calculation, the likelihood of finding the global rather than a local optimum was increased.

RESULTS AND DISCUSSION: The averaged difference between actual performance and model was 4.10%, which met the criterion of model validity in this study. The elbow angle changed from 2.98 to 3.14 rad in the horse contact phase (Fig.1, 2).

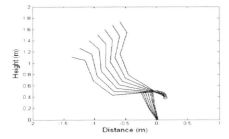

Figure 1: Motion in the horse contact phase of the 6S model

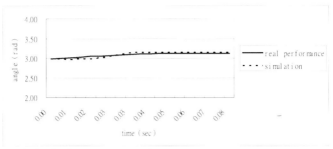

Figure 2: Elbow angle in the horse contact phase

Table1 : Parameters of visco-elastic property at the arm-horse interface. Letter h and v represents the horizontal and vertical direction, respectively.

	6S	5S
Sh	130 Nm^{-2}	130 Nm^{-2}
Sv	25284000 Nm^{-2}	1264200 Nm^{-2}
Kh	146000 Nsm^{-2}	146000 Nsm^{-2}
Kv	900000 Nsm^{-2}	900000 Nsm^{-2}

Table 2 : Influence of initial parameters on the vertical velocity at take-off

	parameter	Initial value	adjustment	Vertical velocity of the CM (Vv-cm) at take-off	Comparison with the nominal 6S results
wrist	angle (rad)	0.9	-10%	5.70	9.08%
		1.1	+10%	4.86	-7.13%
	angular velocity (rad/s)	6.42	-10%	4.88	-6.60%
		7.84	+10%	5.73	9.47%
elbow	angle (rad)	2.68	-10%	3.14	-39.94%
		3.14	+5%*	5.66	8.23%
	angular velocity (rad/s)	0.9	-10%	5.32	1.68%
		1.1	+10%	5.25	0.41%
shoulder	angle (rad)	1.7	-10%	5.39	2.99%
		2.08	+10%	5.03	-3.76%
	angular velocity (rad/s)	5.34	-10%	5.06	-3.24%
		6.52	+10%	5.53	5.70%
hip	angle (rad)	3.53	-10%	5.73	9.62%
		4.31	+10%	4.50	-13.99%
	angular velocity (rad/s)	1.62	-10%	5.28	0.96%
		1.98	+10%	5.14	-1.73%
maximum shoulder joint torque (Nm)		108	-10%	-10%	-2.35%
		132	+10%	+10%	2.00%

* Here a 5% increase is used because a 10% increase will exceed the range of motion.

In order to match with the duration of horse contact, the vertical stiffness (Sv) of the 5S model had to be about half of the 6S model. It is probably because the 5S model lacks the cushioning effect provided by the elbow, and Sv should be smaller to lengthen the contact time. In addition, elbow angle changed slightly in the horse contact phase. The results demonstrate that model 6S can reproduce the actual motion more precisely.

At the wrist joint (contact point), Vv-cm increased by 9.08% if the joint angle was reduced by 10% to be 0.9 rad, implying that the wrist angle of the individual is somewhat too big. This result agrees with the optimal contact angle of 0.87 rad in the previous study (King et al.,1999). Besides, if the angular velocity increased by 10%, Vv-cm also increased by 9.47%. When the elbow angle decreased to 2.68 rad, Vv-cm decreased considerably by 39.94%, resulting in the motion far different from the actual performance. But if the elbow became straight, Vv-cm increased by 8.23%. This result proves the general strategy of keeping the whole arm straight to get more reaction force from the horse. When the shoulder angle increased to 2.08 rad, Vv-cm decreased by 3.76%. This trend disagrees with the optimal simulation shoulder angle of 3.13 rad (King et al., 1999) and 2.27 rad of two elite gymnasts (Xu et al., 2004) in other studies. The wrist angle of the individual was already much greater than the optimal value, so the shoulder angle should be smaller to avoid over-shortened contact time. Actually, the contact time of this simulation is less than that of the real performance. When the shoulder angular velocity increased by 10%, Vv-cm increased by 5.70%. As for maximum shoulder joint torque, 10% increase caused about 2% increase in Vv-cm. Although King and Yeadon (2005) indicated minor difference between simulations with and without shoulder torque, this torque should have certain influence on the performance of handspring vaults. When the hip angle was reduced to 3.53 rad, Vv-cm increased by 9.62%. Compared with the hip angle of 2.69 rad. and 3.40 rad. of the two elite gymnasts (Xu et al., 2004), this initial angle used by our subject was not large enough.

CONCLUSION: The six-segment model can describe the real motion of vault more exactly than models with less segments, and more closely mimics the actual pushing movement of the arms. Although the visco-elastic property was not contained in shoulder joint, the model closely matched the real performance. From the optimal simulations, the suggestion to the individual is to have a smaller wrist angle, greater wrist angular velocity, straighter elbow, greater shoulder angular velocity, greater maximum shoulder torque, and smaller hip angle at horse contact.

REFERENCES:
Chen, K.-H. & Ho, W.-H. (2006). An analysis of kehr with 1/1 turn on 1 pommel: The case of Tsai-Chih Hsiao. *Journal of Physical Education in Higher Education*. 8-1, 201-212.

Dainis, A. (1981). A model for gymnastic vaulting. *Medicine and Science in Sports and Exercise*, 13, 34-43.

King, M. A. & Yeadon, M.R. (2005). Factors influencing performance in the Hecht vault and implications for modeling. *Journal of Biomechanics*, 38, 145-151

King, M.A., Yeadon, M.R., & Kerwin, D.G. (1999). A two segment simulation model of long horse vaulting. *Journal of Sports Sciences*, 17, 313-324.

Koh, M., Jennings, L., Elliott, B., & Lloyd, D. (2003) A predicted optimal performance of the Yurchenko layout vault in women's artistic gymnastics, *Journal of applied biomechanics*, 19, 187-204.

Sprigings, E.J. & Yeadon, M.R. (1997). An insight into the reversal of rotation in the Hecht vault. *Human Movement Science*, 16, 517-532.

Xu, Yuan-yu, Yao, Xia-wen, Ji, Zhong-qiu, Huang, Yu-bin, & Chen, Xiong (2004). *China Sport Science and Technology*. 40, 73-77.

Acknowledgement
This study was supported by National Science Council, Taiwan (NSC 96-2413-H-006-014-).

ISBS 2009

Scientific Sessions

Tuesday PM

ISBS 2009

Oral Session S05

Jumping I

THE JOINT KINETICS OF THE LOWER EXTREMITY ASSOCIATED WITH DROP JUMP HEIGHT INCREMENTS

Hsien-Te Peng[1] Po-Han Chu[1] Chi Liang[1] Ning Lee[1] Pei-Yu Lee[1] Ting-Chieh Tsai[1] and Tai-Fen Song[2]

Department of Physical education of Chinese Culture University, Taipei, Taiwan[1]; Institute of Sport Performance, National Taiwan Sport University, Taichung, Taiwan[2]

The purpose of this study was to identify the joint kinetics of the lower extremity associated with drop jump height increments. Sixteen subjects performed the drop jumps from the 20, 30, 40, 50 and 60-cm heights. Eleven Eagle cameras (200 Hz) and two AMTI force platforms (2000 Hz) were synchronized to collect the data. The study showed the impact was relatively larger in the higher height of drop jump. The greater joint absorption in drop jump from 40, 50 and 60-cm heights was found in the study. Moreover, subjects had greater knee and ankle absorption in drop jump during the eccentric movement. Drop jumps from 20 and 30-cm heights were advisable.

KEY WORDS: plyometric, biomechanics, power, work.

INTRODUCTION: Many athletes practice the drop jump as a kind of plyometric exercise for training their lower extremities. Plyometric training is widely used to enhance the neuromuscular ability since the stretch-shortening supplies the elastic energy and elicit the stretch reflex for greater force output (Bosco, Viitasalo, Komi, & Luhtanen, 1982). Athletes who need explosive jumping performances are often trained with drop jumps from different heights of platforms. Commonly, athletes perform the drop jumps at increased heights for a greater training stimulus. However, the intensity of drop jump is based on anecdotal evidence rather than scientific evidence. In these circumstances, they may be exposed to a higher incidence of joint injuries. Previous research has quantified various plyometric exercises (Jensen & Ebben, 2007). Few studies examined the incremental height of drop jump. The purpose of this study was to investigate the joint kinetics of the lower extremity associated with drop jump height increments.

METHODS: Data Collection: Sixteen college students of the department of physical education – eleven males (age: 21.8 ± 1.8 years; height: 172.8 ± 8.1 cm; mass: 73.6 ± 15.5 kg) and five females (age: 21.2 ± 1.1 years; height: 162.4 ± 3.8 cm; mass: 57.2 ± 7.2 kg) voluntarily participated in this study. All volunteers had no prior knee pain or any history of trauma on other joints of the lower extremities. Subjects changed into specific footwear (Model s.y.m. B9025, Lurng Furng, Inc., Taipei, Taiwan) to control for the different shoe-sole absorption properties before testing. A standardized dynamic warm-up of five-minute cycling on a stationary bicycle at a self-selected pace was performed prior to the testing protocol. Following the warm-up, subjects rested for five minutes. Each subject performed three bounce drop jumps from each of the 20, 30, 40, 50 and 60-cm heights (DJ20, DJ30, DJ40, DJ50 and DJ60). The testing sequence was randomly determined. They were asked to immediately and maximally jump off the ground after landing (Bobbert, Huijing, & van Ingen Schenau, 1987a, b). Their hands were put on their waist during the drop jumps. A sixty-second rest was practiced between jumps.

The movement data were collected with eleven Eagle cameras (Motion Analysis Corporation, Santa Rosa, CA, USA) at 200 Hz sampling rate which were positioned around the performance area. Cameras were synchronized to two force platforms (AMTI Inc., Watertown, MA, USA) which sampling rate was 2000 Hz. One platform collected the right leg data, and another collected the left leg data. Both kinematic and kinetic data were recorded in EVaRT software (Version 4.6, Motion Analysis Corporation, Santa Rosa, CA, USA).

Data Analysis: The data were analyzed in Orthotrak software (Version 6.2, Motion Analysis Corporation, Santa Rosa, CA, USA). The dominant leg, determined in relation to the foot normally used to kick a ball, was analyzed for all subjects. The landing and jumping off during the impact phase was determined where the vertical ground reaction force (vGRF) exceeded a 10 N threshold. The impact phase was then divided into the eccentric and concentric phases where the eccentric phase was from the landing to maximal knee flexion, and the concentric phase was from the maximal knee flexion to jumping off the ground.

The power of joints were calculated from the inverse dynamics. The work of joints was calculated from the integration of power-time curve. The average power of joints was the work divided by the impact time. These variables were normalized by each subject's body weight (BW). The impulse was calculated from the integration of force-time curve. The vertical ground reaction force was also normalized to subjects' body weight.

One-way repeated measures ANOVAs were used to compare the differences between drop jump heights in peak vGRF, time to peak vGRF, peak knee flexion angle, impulse, and duration variables during the phases of the jump. Two-way (3 joints × 5 heights) repeated measures ANOVAs were used to compare the differences between joints and drop jump heights during eccentric and concentric phases in power and work. The significance level was set at α=0.05. The *post-hoc* analysis was performed with the Bonferroni test.

RESULTS: Mean peak vGRF, time to peak vGRF, peak knee flexion angle, impulse, and duration variables are shown in Table 1. The peak vGRF in DJ50 and DJ60 was significantly greater than that in DJ20, DJ30 and DJ40 ($P=.000$). The impulse in DJ40, DJ50 and DJ60 during eccentric phase was significantly greater than that in DJ20 and DJ30 during eccentric phase ($P=.000$). No difference was found in impulse during the concentric phase. The time to peak vGRF in DJ40, DJ50 and DJ60 was significantly smaller than that in DJ20 ($P=.000$). No difference was found in the duration of both eccentric and concentric phases and in the peak knee flexion angle.

Table 1 Peak vGRF, time to peak vGRF, peak knee flexion angle, impulse, and duration variables; MEAN (SD). (N=16)

Variables	DJ20	DJ30	DJ40	DJ50	DJ60
Peak vGRF (BW)	2.00 (0.22)	2.27 (0.47)[d]	2.47 (0.50)[d]	3.07 (0.75)[b]	3.78 (0.85)[a]
Eccentric Impulse (BW · s)	0.185 (0.035)	0.197 (0.039)	0.221 (0.028)[c]	0.232 (0.025)[c]	0.265 (0.037)[a]
Concentric Impulse (BW · s)	0.189 (0.021)	0.191 (0.026)	0.189 (0.026)	0.198 (0.032)	0.194 (0.027)
Total Impulse (BW · s)	3.682 (0.486)	3.878 (0.552)	4.075 (0.488)[d]	4.271 (0.494)[c]	4.552 (0.543)[a]
Eccentric Time (ms)	143 (42)	142 (42)	146 (36)	148 (37)	156 (41)
Concentric Time (ms)	165 (30)	168 (37)	168 (35)	178 (46)	184 (46)
Contact Time (ms)	309 (68)	309 (77)	315 (70)	325 (81)	340 (86)
Time to peak vGRF (ms)	119 (51)	84 (44)	70 (28)[f]	60 (24)[f]	49 (8)[e]
Peak knee flexion (degree)	79.0 (12.0)	77.1 (13.6)	78.3 (13.0)	80.9 (16.9)	83.4 (16.7)

[a] Significantly greater than 20, 30, 40, 50 cm; [b] Significantly greater than 20, 30, 40 cm; [c] Significantly greater than 20, 30 cm; [d] Significantly greater than 20 cm; [e] Significantly smaller than 20, 30, 40 cm; [f] Significantly smaller than 20 cm. ($P<.05$)

The results of comparing the differences between joints and drop jump heights during eccentric and concentric phases are shown in Figure 1. Negative power and work during eccentric phase are expressed in absolute value in the figure. The negative power indicates absorption power while the positive power indicates generation power (Winter, 1990). The peak absorption power and average absorption power of the hip, knee and ankle joint in DJ50 and DJ60 during eccentric phase were significantly greater than those in DJ20 and DJ30 ($P=.000$). A significant interaction was found in work between joints and heights ($P=.025$), then the simple main effect was tested in joints and heights, respectively. The work at the ankle joint in DJ40 and DJ60 during eccentric phase was significantly greater than that in DJ20 and DJ30. The work at the knee joint in DJ40, DJ50 and DJ60 during eccentric phase was significantly greater than that in DJ20 and DJ30. The work at the hip joint in DJ60 during eccentric phase was significantly greater than that in DJ20 and DJ30. The peak

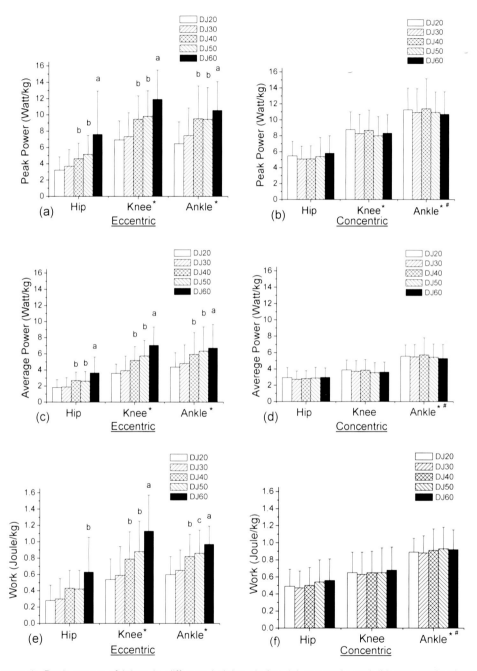

Figure 1: Peak power of joints in different heights during (a) eccentric and (b) concentric phase; Average power of joints in different heights during (c) eccentric and (d) concentric phase; Work of joints in different heights during (e) eccentric and (f) concentric phase. [a] Significantly greater than 20, 30, 40 cm; [b] Significantly greater than 20, 30 cm; [c] Significantly greater than 20 cm; * Significantly greater than Hip; # Significantly greater than Knee. ($P<.05$)

absorption power, average absorption power and work of the knee and ankle joint during eccentric phase were significantly greater than those of the hip joint. The peak generation power, average generation power and work at the ankle joint were significantly greater than those at the knee and hip joint during the concentric phase. The peak generation power at the knee joint was significantly greater than that at the hip joint during concentric phase.

DISCUSSION: The study showed the impact was relatively greater in the drop jump of higher heights. If a subject stands on a higher platform, theoretically his body will have higher potential energy. The potential energy transfers to the kinetic energy which is proportional to the square of velocity during the fall (Hall, 2002). After landing, the downward velocity of his body has to be reduced to zero before push-off can start (Bobbert, Huijing, & van Ingen Schenau, 1987b). The eccentric impulse could reflect the profile that as a subject tried to decelerate while the concentric impulse would be the effort of push-off. Subjects had greater impact peak force and eccentric impulse when jumping from 40, 50 and 60-cm heights. The higher the drop jump height is; the greater joint muscle effort may be needed to decelerate during the landing.

Subjects utilized hip, knee and ankle joint muscles to absorb the impact during the eccentric movement. A higher rate of joint power absorption was performed with the raised platform to reduce the direct impact to the body. Greater joint power absorption in drop jumps from 40, 50 and 60-cm heights were found in the study which maybe with regard to high injury risk. Moreover, the knee and ankle had greater power absorption in drop jumps during the eccentric movement which was in agreement with studies by Bobbert, et. al. (1987b) and Walsh, Arampatzis, Schade, & Bruggemann (2004).

Following the absorption, the joint muscles generated power to push off. In this study, subjects could not generate more power during the push-off due to the contact time limitation. The joint power generation which mainly contributed to the jump performance during the concentric movement showed less variation with respect to drop height increments. Bobbert, et. al. (1987b) indicated that there was no advantage of performing drop jumps from a height of 60 cm. The contact time of drop jump was the factor for joint power generation rather than the height (Bobbert, et. al., 1987a, b). In addition, the greater ankle joint power generation output was found to contribute to the push-off.

CONCLUSION: Athletes should be cautious when practicing drop jumps from 40, 50 and 60-cm heights because of the greater impact at the knee and ankle. In terms of injury prevention, drop jumps from 20 and 30-cm heights were advisable in this study. The further study could look into the muscles activation.

REFERENCES:
Bobbert, M. F., Huijing, P., & van Ingen Schenau, G. (1987a). Drop jump I. The influence of dropping technique on the biomechanics of jumping. *Med. Sci. Sports Exerc, 19*, 332-338.
Bobbert, M. F., Huijing, P., & van Ingen Schenau, G. (1987b). Drop jump I. The influence of dropping height on the biomechanics of jumping. *Med. Sci. Sports Exerc, 19*, 339-346.
Bosco,C. Viitasalo, J.T., Komi, P.V., & Luhtanen, P. (1982). Combines effect of elastic energy and myoelectrical potentiation during stretch-shortening cycle exercise. *Acta. Physiol Scand, 114*, 557-565.
Jensen, R. L., & Ebben, W. P. (2007). Quantifying plyometric intensity via rate of force development, knee joint, and ground reaction forces. *Journal of Strength and Conditioning Research, 21*(3), 763-767.
Walsh, M., Arampatzis, A., Schade, F., & Bruggemann, G. P. (2004). The effect of drop jump starting height and contact time on power, work performed, and moment of force. *Journal of Strength and Conditioning Research, 18*(3), 561-566.
Hall, S. J. (2002). *Basic biomechanics (4th edition).* Columbus: McGraw-Hill.
Winter, D. A. (1990). *Biomechanics and motor control of human Movement (2nd edition).* New York: John Wiley & Sons.

Acknowledgement
Authors would like thank the National Science Council (NSC) in Taiwan for funding this study.

DOES A CONCENTRIC ONLY SQUAT ILLICIT SIMILAR POTENTIATION OF SQUAT JUMP PERFORMANCE AS A BACK SQUAT?

Darragh Graham, Andrew Harrison and Eamonn Flanagan

Biomechanics Research Unit, University of Limerick, Limerick, Ireland

KEY WORDS: complex training, starting strength, stretch shortening cycle, rugby, sledge.

INTRODUCTION: The stretch shortening cycle (SSC) involves the stretching of musculature immediately prior to being rapidly contracted. This eccentric/concentric coupling produces a more powerful contraction than concentric action alone (Flanagan, 2007). Complex training (CT) hypothesises that a near maximal muscle contraction will enhance the explosive capabilities of the muscle (Docherty et al., 2004). In order for a complex pair (CP) to be effective the exercises must be biomechanically similar (Ebben, 2002). A traditional example of a CP is a back squat (BS) and a depth jump. If the goal of the CP is to improve starting strength (SS) a suitable exercise is the squat jump (SJ). Theoretically a concentric squat (CS) is more biomechanically similar to a SJ than a BS, although not as commonly practiced. A CS does not invoke the SSC, hence motor unit recruitment may be attenuated. The aim of this paper was to examine if the biomechanically similar CP of a CS and SJ will illicit similar potentiation as a BS and SJ CP.

METHODS: Twenty male rugby players, proficient with the technique of the BS and SJ participated in this study. All subjects were part of a professional (n=13) or semi-professional (n=7) rugby academy. Physical characteristics of subjects were age 20.3 years $\pm$ 1.1, height 1.84 $\pm$ 0.07m and mass 96.6 $\pm$ 10.7kg. Day one of testing was a SJ familiarisation session and a 1RM BS and CS test. Day 2 was a pre test, a 3RM BS and a post test. Day 3 was a pre test, a 3RM CS and a post test. All pre and post tests consisted of one SJ every minute for 10 minutes. The SJ starting position was 90° flexion of the knee. All SJ's were performed on a sledge apparatus inclined at 30° as described by Harrison et al. (2004). Each jump was recorded on an AMTI OR6-5 force platform mounted at right angles to the sledge apparatus sampling at 1000 Hz. The first post test SJ was performed one minute after respective 3RM squats. Depth of BS's was 90° flexion of the knee. All CS's were performed on weightlifting jerk boxes. The starting position was 90° flexion of the knee. Days 2 and 3 3RM scores were calculated as 93% of Day 1 1RM scores. Dependent variables for each jump were height jumped, peak ground reaction force, rate of force development and SS. Max and min dependent variables were compared between pre and post tests. A GLM ANOVA examined for differences between the BS and CS conditions from the pre to post test.

RESULTS:
Results from this study will elucidate whether biomechanically similar exercises will potentiate to the same extent as a higher loaded activity.

REFERENCES:
Docherty, D., Robbins, D., Hodgson, M. (2004). Complex Training Revisited: A Review of its Current Status as a Viable Training Approach. National Strength and Conditioning Association. 26(6), 52-57.
Ebben, W.P., (2002). Complex Training: A Brief Review. Journal of Sports Science and Medicine. 1, 42-46
Flanagan, E.P. (2007). An Examination of the Slow and Fast Stretch Shortening cycle in Cross Country Skiers and Runners. Proceedings of the XXV International Symposium of Biomechanics in Sports Ouro Preto, Brazil. 51-54.
Harrison, A.J., Keane, S.P., and Coglan, J. (2004). Force-velocity relationship and stretch-shortening cycle function in sprint and endurance athletes. Journal of Strength and Conditioning Research. 18(3), 473-479.

Acknowledgements: Joseph O'Halloran and Dr. Tom Comyns.

ANKLE AND KNEE COORDINATION FOR SINGLE-LEGGED VERTICAL JUMPING COMPARED TO RUNNING

Torsten Brauner, Thorsten Sterzing, Thomas L. Milani

Department of Human Locomotion, Chemnitz University of Technology, Chemnitz, Germany

Similar basic movement patterns in ground reaction forces and ankle frontal plane kinematics of single-legged vertical jumping (JUMP) compared to running (RUN) have already been identified in earlier investigations. To broaden these findings, lower extremity kinematics of 25 subjects were recorded executing RUN and JUMP. Special focus was laid on the knee and ankle coordination of tibial endorotation and ankle eversion as well as on knee and ankle flexion/extension by applying a modified vector coding technique. RUN and JUMP demonstrated similar knee and ankle joint coordination patterns. However, differences in coupling angles unveiled phases, where joint coordination of ankle eversion/tibial endorotation was adjusted in JUMP. By comparing knee and ankle coordination of JUMP in healthy athletes with athletes suffering from anterior knee pain, common in sports with high jumping occurrences, key differences in execution leading to this overuse injury might be unveiled.

KEY WORDS: running, jumping, single-legged jump, joint coupling, vector coding

INTRODUCTION: Jumping is an athletic skill that is used in various sports (e.g. basketball, volleyball) and is carried out in a variety of techniques (e.g. single-/two-legged, standing/running). When jumping, high loads act on the human body during take-off and landing. Thus, jumps need to be analyzed with respect to performance and injury prevention criteria, since Cumps et al. (2007) found Jumper's knee diagnosis to have the highest overuse injury rate in high level basketball athletes.

Numerous biomechanical investigations have focused on lower extremity kinematics and kinetics in (McClay, 2000). Especially walking and running have been studied intensively, also because their movement patterns serve as a basis to more complex movements, e.g. jumping. We were able to demonstrate, that athletes use similar rearfoot striking patterns in single-legged vertical jumps (JUMP) and heel-to-toe running (RUN) (Brauner et al., 2009). McClay et al. (1994) analyzed the kinematics of professional basketball players in sport basketball specific tasks. They found no significant differences in sagittal and frontal knee and ankle kinematics in lay-up takeoffs compared to running. Other than that, little attention has been spent on the kinematical analysis of JUMP, so far.

Whereas these previous studies looked only at single joint kinematics, more recently, kinesiologists have concentrated on the combined analysis of motion of different joints by investigating interactive mechanisms, potentially being the cause of injuries or different levels of performance (DeLeo et al., 2004). Heiderscheidt et al. (2002) proposed the vector coding technique (VC) for the analysis of joint coupling and investigated the coordination of internal tibia rotation to rearfoot eversion since asynchrony between these two motions is discussed to increase the occurrences of overuse injuries (Tiberio, 1987; DeLeo et al., 2004).

However, to our knowledge, VC has not been used to analyze sagittal plane kinematics of knee and ankle joint in jumping so far. This is surprising, since coordination of multiple joint motions is of special interest in the research of jumping (Jacobs et al., 1996) Therefore, as new insights into the lower extremity coordination of jumping can be obtained by application of VC, an optimization of athletes jumping performance might be possible.

Based on these thoughts, the purpose of this study was to analyze the basic patterns in ankle and knee kinematics of RUN and JUMP with specific consideration of the movement coupling of these joints. The technique of VC was chosen for this goal since it allows a complex consideration of joint coordination.

METHODS: 25 subjects (9♀, 16♂; 25.1±4.2yrs; 1.78±0.11m; 72.7±11.1kg) participated in this laboratory study. Written consent was obtained of each subject prior to data collection. All subjects wore the same running shoe model (*PUMA Complete Eutopia*) for running and jumping.

Subjects familiarized themselves with the two movement task during warm-up. Thereby, their dominant leg was determined (20 left, 5 right). For data collection, the two movement tasks, heel-toe running (RUN) and single-legged running jumps (JUMP), were performed in randomized order. Six valid trials were recorded for each condition:

- RUN: heel-toe running on a 13m runway at a given speed of 3.5 m/s (±0.1) controlled by two light barriers, dominant foot on the force plate
- JUMP: single-legged jumps using a three-step running approach, take-off from the force plate with the dominant leg, goal was maximum vertical jumping height

Kinetic data were recorded at 960Hz by a force plate (*Kistler 9287BA*). An 11-camera (MX-3) (*Vicon, Oxford Metrics, UK*) system was used to collect kinematic data at 240Hz. For kinematic analysis, a three segment approach was applied (thigh, shank, foot) using calibration markers on anatomical landmarks and tracking markers on the segments (Heiderscheidt et al., 2002). Calibration markers were fixed to left/right great trochanter, lateral/medial epicondyllus, lateral/medial malleoli, and first /fifth metatarsal head. 4-Marker cluster were used as tracking markers on the lateral side of thigh and shank, a 3-marker cluster was attached to the heel cap to represent foot movement.

Kinematic and kinetic time series data was analyzed using Visual 3D™ (*Version 3.99, C-Motion Inc, Rockville, MD, USA*). Joints were assigned six degrees of freedom and knee and ankle angles were calculated using Cardan sequence of rotations in X-Y-Z order (Cole et al., 1993) with X-axis being medio-lateral oriented, Y-axis anterior-posterior, and Z-axis proximal-distal.

Kinetic and kinematic data of ground contact were normalized to 101 data points. Angle-time curves of all trials were averaged and used for VC analysis. Angle-angle diagrams of the knee (KAX) and ankle (AAX) in the sagittal plane as well as for knee in transverse (KAZ) and ankle in frontal plane (AAY) were created. Coupling angles (CA) were calculated as the orientation of a vector adjoining to consecutive data points relative to the right horizontal (Heiderscheidt et al., 2002) by using the below given formula:

$$CA_i = \tan^{-1}\left(\frac{y_{i+1} - y_i}{x_{i+1} - x_i}\right), \quad \text{where} \quad i = 0, 1, 2, \ldots, 99$$

CA were analyzed by applying four coordination patterns (in-phase, anti-phase, knee phase and ankle phase) following a modified approach of the one introduced by Chang et al. (2008). Whereas knee and ankle phases are characterized by single joint movements of knee respectively ankle only, movements in both joints can be observed in Anti-phase and In-phase. In Anti-phase, the joints move in anatomically "opposite" direction (e.g. knee extension and ankle plantar flexion). Thus, in-phase CA show a movement in the "same" direction to a similar extent (e.g. knee extension and ankle dorsal extension).

Table 1: Coordination patterns categorized by a scheme adapted from Chang et al. (2008)

Coordination pattern	Coupling angle definitions
Anti-phase	112.5° ≤ CA < 157.5°, 292.5° ≤ CA < 337.5°
In-phase	22.5° ≤ CA < 67.5°, 202.5° ≤ CA < 247.5°
Knee phase	0° ≤ CA < 22.5°, 157.5° ≤ CA < 202.5°, 337.5° ≤ CA ≤ 360°
Ankle phase	67.5° ≤ CA < 112.5°, 247.5° ≤ CA < 292.5°

RESULTS AND DISCUSSION: Ankle and knee joint kinematics show similar general movement patterns in RUN and JUMP. In JUMP, however, midstance phase seems to be prolonged and push-off phase shortened, whereas impact phase is similar in length between conditions.

In contradiction to McClay et al. (1994), we found significant higher maximum ankle eversion, ankle eversion excursions and ankle eversion velocity values in JUMP compared to RUN (Figure 1).

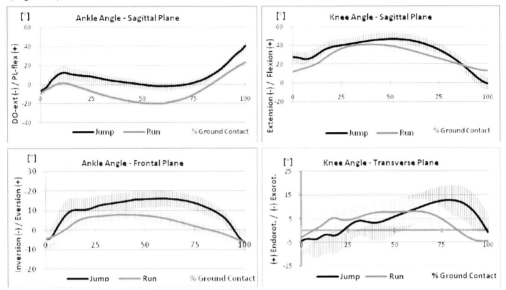

Figure 1: Angle-time diagrams of AAX (top-left), KAX (top-right), AAY (bottom-left), and KAZ (bottom right).

	AAY-vel	AAY-max	AAY-rom
JUMP	1062.3 °/s	13.5°	17.8°
RUN	544.6 °/s	7.1°	11.7°
t-test	p<0.001	p<0.001	p<0.001

Table 2: Comparison of discrete parameters ankle frontal plane kinematics between JUMP and RUN

Knee and ankle joint coordination show similar patterns in RUN and JUMP (Figure 2) Interestingly, sagittal plane coupling shows almost no In-phase movement, but long periods of Anti-phase movement during midstance and push-off (Figure 2-top).

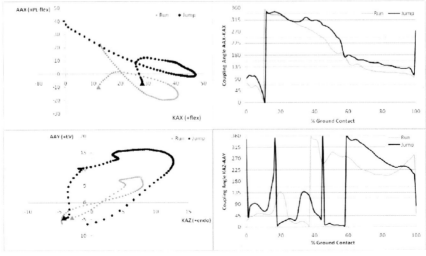

Figure 2: Angle-angle diagram and coupling angles AAX & KAX (top) and AAY & KAZ (bottom). (EV: Eversion, endo: endorotation, PL-flex: plantarflexion)

During the first 10% of ground contact, CA of AAX & KAX in RUN show In-phase motion with ankle and knee flexing simultaneously. In JUMP, however, the ankle is flexed, while the knee joint is stiffened, resulting in Ankle-phase coordination pattern.

Transverse plane knee and frontal plane ankle kinematics show basically a similar coordination pattern in RUN & JUMP, too. However, the angle-angle diagram illustrates the highly increased joint excursion in JUMP in both joints (Figure 2-bottom). In contrast to the sagittal CA, AAY & KAZ show longer periods of In-phase motion throughout stance phase. During the initial 10% of stance, however, CA of AAY&KAZ show Ankle- phase pattern in JUMP, whereas the initial 10% in RUN are executed with In-phase motion.

CONCLUSION: General similarities were found in the coordination patterns of knee and ankle joints in run single-legged vertical jumping compared to running. These similarities confirm the idea of running as a basic human movement pattern that can be adjusted for specific purposes.

Joint coupling in the sagittal plane seems to be more robust across different movement tasks than coupling of ankle frontal and knee transverse plane motion. However, especially in single-legged jumping, internal tibia rotation and ankle eversion show periods of Anti-phase motion, possibly causing high torque in the ankle and knee joint. For this reason and due to the more dynamic nature, single-legged jumping should be used to study coordination of internal tibia rotation and ankle eversion in future investigations.

Vector coding was successfully utilized to compare different sports techniques. By comparing group average curves, we chose a very general approach. Further investigations should focus on individual coordination patterns of athletes as it was proposed originally by Heiderscheidt et al. (2002). Especially, knee and ankle coordination patterns of athletes with anterior knee pain should be compared to those without symptoms.

REFERENCES:

Brauner, T., Sterzing, T., Milani, T.L. (2009). Der einbeinige Vertikalsprung zeigt laufähnliche Pronationsmuster mit stark erhöhten Belastungsmaxima. dvs Konferenz – Sektion Biomechanik, Tübingen, Germany.
Chang, R., Emmerik, R. E. van, & Hamill, J. (2008). Quantifying rearfoot-forefoot coordination in human walking. Journal of Biomechanics, 41(14), 3101–3105.
Cole, G. K., Nigg, B. M., Ronsky, J. L., & Yeadon, M. R. (1993). Application of the joint coordinate system to three-dimensional joint attitude and movement representation: a standardization proposal. Journal of biomechanical engineering, 115(4A), 344–349.
Cumps, E., Verhagen, E., & Meeusen, R. (2007). Prospective epidemiological study of basketball injuries during one competitive season: Ankle sprains and overuse knee injuries. Journal of Sports Science and Medicine, (6), 204–211.
DeLeo, A., Dierks, T., Ferber, R., & Davis, I. (2004). Lower extremity joint coupling during running: a current update. Clinical Biomechanics, 19(10), 983–991.
Heiderscheit, B. C., Hamill, J., & Emmerik, R. E. van (2002). Variability of Stride Characteristics and Joint Coordination Among Individuals With Unilateral Patellofemoral Pain. Journal of Applied Biomechanics, 18, 110–121.
Jacobs, R., Bobbert, M. F., & van Ingen Schenau, G. J. (1996). Mechanical output from individual muscles during explosive leg extensions: The role of biarticular muscles. Journal of Biomechanics, 29(4), 513–523.
McClay, I. S., John R. Robinson, Thomas P. Andriacchi, Edward C. Frederick, Ted Gross, Philip E. Martin, et al. (1994). A Kinematic Profile of Skills in Professional Basketball Players. Journal of Applied Biomechanics, 10(3), 205–221.
McClay, I. (2000). The evolution of the study of the mechanics of running. Relationship to injury. Journal of the American Podiatric Medical Association, 90(3), 133–148.
Tiberio, D. (1987). The effect of excessive subtalar joint pronation on patellofemoral mechanics: a theoretical model. The Journal of orthopaedic and sports physical therapy, 9(4), 160–165.

ACKNOWLEDGEMENT:

This study was supported by Puma Inc., Germany.

COORDINATION IN TRACK & FIELD SPRINTERS WHILE PERFORMING THE COUNTERMOVEMENT JUMP

Ezio Preatoni[1,2], Elena Devodier[1], Giuseppe Andreoni[2] and Renato Rodano[1]

Dipartimento di Bioingegneria, Politecnico di Milano, Milan, Italy [1]
Dipartimento INDACO, Politecnico di Milano, Milan, Italy [2]

The aim of this study was to assess coordination, coordination variability and their evolution with time, during the countermovement jump. For this purpose a population of track & field sprinters was analysed through a Dynamic Systems approach. Five testing sessions over the year were considered. The kinematics of lower limbs was recorded by an optoelectronic system, and the continuous relative phase of the hip-knee and knee-ankle joints was considered. Results showed different behaviours for the two couplings across the functional phases of the movement, with an increased variability and a less in-phase relationship during transitions between phases. No relevant changes were reported over the subsequent testing sessions.

KEY WORDS: joint coupling, countermovement jump, performance monitoring.

INTRODUCTION: Coordination in sports activities plays a fundamental role for the achievement of successful performances. Sports movements usually involve a large number of body segments, which are subjected to high intensity biomechanical demands and have to act synergically in order to produce the desired outcome. A poor organisation of the elements that concur to the realisation of an harmonious action may cause a bad result and may increase the risk of injury (Hamill et al., 1999; Kurz & Stergiou, 2004). Traditional biomechanical analyses make use of kinematic and kinetic variables to describe the characteristics of the movement and to understand the underlying factors that generated it. Although very useful, these approaches are not very effective in addressing motor coordination, because they describe measures from single joints or segments rather than investigating the interaction between multiple elements of the system. Dynamic Systems Theory (DST) has given new means for inspecting the organisation of the locomotor system and for gaining more insight into the multifactorial nature of human motion (Hamill et al., 1999; Kurz & Stergiou, 2004). According to DST human limbs are seen as a system of coupled pendulums that oscillate about joints. Quantitative information regarding how joint coordination evolves may be drawn by observing the continuous phasing relationships (CRP) between the different elements that participate to the movement (Hamill et al., 1999; Kurz & Stergiou, 2004). Changes in the mutual relations between body segments or adjacent joints may give important indications about the inherent coordinative factors of the neuro-musculo-skeletal system. In particular, the amount of variability in the relative phase relationships over many repetition of the same task has been used by some authors to understand how external perturbations, developmental stages, pathologies or detrimental behaviours may influence the choice of a particular motor solution (Hamill et al., 1999; Kurz & Stergiou, 2004). DST may be exploited for the analysis of sports movements, too. Similar performances in sporting events are often the result of different motor strategies, both within and between individuals. These subtle discrepancies are typically less detectable than the ones that emerge in clinical studies, and are often concealed by the presence of variability. Hence, the observation of discrete variables and time varying measures are not always effective, while the exploration of motor coordination might unveil either hidden changes or anomalous functionalities. The athlete's phase portraits and CRP measures derived thereof are very likely to be influenced by training programs and motor learning. Furthermore, they may manifest the presence of detrimental behaviour. DST tools may represent a valuable tool either for gauging the progresses that are achieved over time or for injury prevention purposes.

To our knowledge, there is no published research about a DST analysis of the countermovement jump (CMJ), which is a common field test for explosive force. Furthermore there is a lack of information concerning longitudinal monitoring thorugh DST. Therefore the aim of this work was to assess coordinative patterns in a population of track and field athletes that performed CMJ and that were followed over a whole competitive year.

METHODS: Participants: Four male and 2 female track & field sprinters (age: 21.2±4.7 years; height: 1.74±0.06 m, weight: 65.8±6.4 kg) of national and regional level were the subjects of this study. Their season best in the 100 m event ranged between 10.71 s and 11.19 s for males, between 13.14 s and 13.15 s for females. All the participants had the same coach and were used to perform from 5 to 8 sessions a week.

Instrumentation: An 8-TVCs optoelectronic system (Elite2002, BTS, Italy), working at 100 Hz, and a force platform (AMTI OR6-7-1000, USA), whose sampling rate was 500 Hz, were used to capture the 3D kinematics of body segments and ground reaction forces (GRF) during vertical jumping.

Data Collection: The SAFLo (Frigo et al., 1998) marker set was chosen. It is a total body protocol that matches experimental needs for practicality and freedom of movement, to reliability of measures. After a standard 20 min warm up, each subject was asked to perform 24 (4 sets of 6 reps) double-leg maximal countermovement jumps, keeping their arms akimbo. Trials were executed alternating the right and left limb on the force platform. Two and 9 minutes recovery was respected between subsequent repetition and sets to avoid fatigue. 5 testing sessions (TS1-TS5) were collected over the year, corresponding to different phases of the training programme: TS1 in April, TS2 in June, TS3 in September, TS4 in November and TS5 in March of the next year.

Data Analysis: Anthropometric measures and specially designed algorithms were used to estimate and filter (D'Amico & Ferrigno, 1990; Frigo et al., 1998) 3D coordinates of internal joint centres and joint angles. Lower limb joint angles (hip, knee and ankle) and angular velocities in the sagittal plane were considered for this study. They were selected because they may be considered the most reliable and representative measures of lower limb kinematics during vertical jumping. The analysed movement was defined as the interval (Δt) between the beginning of the countermovement (t_i) and the instant the toes lose contact with the ground (t_f). Kinematic time series were time normalised to 100 points, so that t_i corresponded to 0% and t_f to 100% of the movement. Phase portraits (angular velocity – angle) and phase angles of the hip (ϕ_h), knee (ϕ_k) and ankle (ϕ_a) were estimated and normalised according to Hamill et al. (1999). Continuous relative phase (CRP) between adjacent joints ($\theta_{pd}(t) = \phi_p(t) - \phi_d(t)$, where p=proximal and d=distal) was studied to assess coordination patterns: hip-knee and knee-ankle intra-limb couplings were considered. These couplings were chosen in accordance with the proximal to distal activation sequence suggested by some authors' studies on vertical jumping performances (Bobbert & van Ingen Schenau, 1988). The mean±standard deviation curves of CRP ($\Theta_{pd}(t) \pm \sigma_{pd}(t)$) were created for each subject, intra-limb coupling and session. Coordination and coordination variability between intra-limb joints was evaluated by measuring, respectively: the Mean Absolute Relative Phase (MARP), which is the mean $\Theta_{pd}(t)$ across the whole of the movement; the Deviation Phase (DP), which is the average standard deviation between trials, over Δt. Furthermore, the movement was divided into 3 functional phases (Δt_1, Δt_2, Δt_3) and the same analysis was carried out on each of them. Δt_1 was between the beginning of the countermovement and the minimum of GRF; it represented the passive joint flexion under gravity force. Δt_2 was between the minimum of GRF and maximal knee flexion; it represented the braking phase and the eccentric muscular action. Δt_3 was between maximal knee flexion and take off; it represented the propulsive phase with concentric muscular efforts. Nonparametric statistics (median and IQR) were used to describe individual and group measures. Nonparametric tests (Kruskal-Wallis and Friedman, $P<0.05$) and Bonferroni post-hoc comparisons ($P<0.05$) were used to assess significance of changes between functional phases or testing sessions.

RESULTS: The overall average (median and IQR) transitions from Δt_1 to Δt_2 and from Δt_2 to Δt_3 occurred, respectively, at 38.5% (7%) and at 67.5% (5%) of the movement. Hip-knee and knee-ankle couplings presented different CRP, concerning both the shape and the magnitude of patterns (Figure 1a and 1b). $\Theta_{hk}(t)$ ran about the baseline during Δt_1 and showed 2 peaks (10 and 22 deg respectively) at 45% and 82% of the movement. $\Theta_{ka}(t)$ was close to antiphase (160 deg) at the beginning, it decreased rapidly between Δt_1 and Δt_2 (50 deg at the minimum of GRF), and stayed around the zero line till the end of the movement. MARP$_{hk}$ and MARP$_{ka}$ were (Figure 1c): 11.4 (2.2) and 132.1 (10.6) in Δt_1; 7.9 (1.3) and 19.3 (2.5) in Δt_2; 13.4 (2.4) and 8.6 (2.0) in Δt_3. CRP variability manifested more similar behaviours between the two couplings (Figure 1d and 1e). Both $\sigma_{hk}(t)$ and $\sigma_{ka}(t)$: increased during the transition between Δt_1 and Δt_2; decreased during Δt_2, with a "valley" between Δt_2 and Δt_3; raised again in late Δt_3. The magnitude (36 deg for $\sigma_{hk}(t)$ and 28 deg for $\sigma_{ka}(t)$) and occurrence (next to minimum GRF) of peaks was comparable. CRP variability was higher at the beginning of the movement for the knee-ankle coupling, while $\sigma_{hk}(t)$ showed increased values at the end of Δt_3. DP$_{hk}$ and DP$_{ka}$ were (Figure 1f): 7.6 (2.0) and 9.5 (5.0) in Δt_1; 6.3 (2.0) and 6.4 (0.6) in Δt_2; 6.1 (1.8) and 4.3 (0.5) in Δt_3.

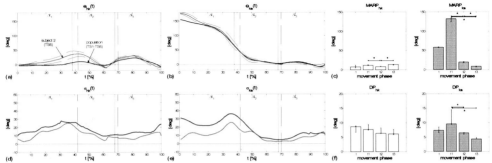

Figure 1 CRP ($\Theta(t)$) and CRP variability ($\sigma(t)$) concerning the hip-knee (a and d) and knee-ankle (b and e) couplings. Solid black curves represent the overall mean (of the whole set of subjects and sessions). Dashed red lines represent an example of individual result during a single testing session (i.e. subject 2 during TS5). Individual CRP data are presented as mean±STD. Vertical lines subdivide the three functional phases of the movement. Figures c and f report bar histograms of MARP and DP for the two articular couplings. (*) indicates P<0.05.

The longitudinal monitoring did not manifest remarkable trends. CRP variability (of the population) tended to slightly decrease over testing sessions in both couplings, but the only significant change emerged in DP$_{hk}$ of the whole movement, between TS2 and TS5. All the other parameters concerning both Δt and the single functional phases did not changed significantly.

DISCUSSION: Lower limbs coordination in the execution of countermovement jumps was studied through a DST approach. A population of track & field sprinters of the same team was analysed five times over a competitive year. Results concerning the average hip-knee and knee-ankle couplings manifested peculiar coordinative behaviour over the whole movement, and in correspondence of the functional phases into which it had been subdivided.

The phasing relation between intra-limb joints was measured through the mean absolute relative phase: the greater the MARP, the more out of phase joints are (Kurz & Stergiou, 2004). The overall MARP was 7.3 (1.9) deg for the hip-knee coupling and 58.2 (1.6) deg for the knee-ankle one, thus evidencing a more in-phase relationship between proximal joints. This findings were in contrast with reports from other authors (Kurz & Stergiou, 2004) who registered increased tuning from proximal to distal segments, both in walking and in running. The higher values of MARP$_{ka}$ are mainly caused by the different dynamics of the knee and the ankle during the first phase of the movement. In Δt_1, in fact, the ankle was more "static" than the knee. This caused a shift of the CRP toward an out-of-phase relation. The tuning between the knee and the ankle was progressively recovered during Δt_2 and Δt_3. In contrast, MARP$_{hk}$ manifested an in-phase relationship throughout the movement, with two peaks just after each transition between the functional phases. The transition between Δt_1 and Δt_2 was interesting even for CRP variability. Variability of relative phase is a measure of the stability in the organisation of the neuromuscular system. Increased variability has been assumed to correspond to transitions during which the neuromotor system is in search for the most appropriate strategy among the possible coordinative patterns. Despite it may appear as uncertainty, higher variability in CRP has been interpreted as a form of flexibility to overcome local and global perturbations or to redistribute detrimental loads (Hamill et al., 1999; Kurz & Stergiou, 2004). $\sigma_{hk}(t)$ and $\sigma_{ka}(t)$ evidenced a rise of variability about the instant when lower limbs started to contrast the descent of the centre of mass due gravity force. Furthermore, DP decreased across the passage from Δt_1 to Δt_3 in both couplings (significantly for the knee-ankle one). This may be explained by the lack of muscular control during the initial, passive, joint flexion, and by the quick transition to an eccentric action in order to invert the downward movement. In contrast, the transition between the eccentric and the concentric phase did not involve an increase of variability, but only a less in-phase relation between adjacent joints that may derive by the proximal to distal activation described by (Bobbert & van Ingen Schenau (1988) and confirmed by the measured angular time series.

The coordination patterns and the corresponding parameters did not change over the five testing sessions that were collected, both concerning the whole movement and the single phases. This may be interpreted in many ways: (i) the athletes were very familiar with the movement; (ii) the training program they underwent did not change their coordinative characteristics; (iii) individual changes were masked by the analysis of the population.

CONCLUSION: The DST analysis of sports movements may represent an effective mean for investigating the coordinative proprieties of the neuro-musculo-skeletal system. In this study intra-limb coordination during vertical jump exercises was addressed and described by measuring the phasing relationship (MARP) and its variability (DP) between adjacent lower-limb joint. Relations between functional phases of the movement and joint-coupling patterns were evidenced. Furthermore, the possible evolution of coordinative features over a year of training was investigated. However, further efforts must be spent for interpreting data, creating reference databases and provide useful information to athletes and coaches. In particular, further potentialities may come by monitoring coordinative peculiarities and time evolutions of the single individual. This was carried out but was not presented in this work.

REFERENCES:
Bobbert, M.F., & van Ingen Schenau, G.J. (1988). Coordination in vertical jumping. *Journal of Biomechanics*, 21(3), 249-262.

D'Amico, M., & Ferrigno, G. (1990). Technique for the evaluation of derivatives from noisy biomechanical displacement data using a model-based bandwidth-selection procedure. *Med Biol Eng Comput*, 28(5), 407-415.

Frigo, C., Rabuffetti, M., Kerrigan, D.C., Deming, L.C., & Pedotti, A. (1998). Functionally oriented and clinically feasible quantitative gait analysis method. *Med Biol Eng Comput*, 36(2), 179-185.

Hamill, J., van Emmerik, R.E.A., Heiderscheit, B.C., Li, L. (1999) A dynamical systems approach to lower extremity running injuries. *Clinical Biomechanics*, 14(5), 297-308.

Kurz, M.J., & Stergiou, N. (2004). Applied Dynamic Systems Theory for the analysis of movement. In N.Stergiou (Ed.), *Innovative analysis of human movement* (pp. 93-119). Champaign: Human Kinetics.

Acknowledgement
This research has been supported by "Easy Speed 2000 Associazione Sportiva Dilettantistica", Italy.

MECHANICS OF AMPUTEE JUMPING – CONSIDERATION FOR LOADING

Marlene Schoeman, Ceri Diss and Siobhan Strike
School of Human and Life Sciences, Roehampton University, London, UK

Amputees must develop compensatory mechanisms to overcome the constraints imposed by a mechanical prosthesis. In completing a bilateral countermovement jump, amputees must accommodate the limited ankle dorsiflexion angle and adapt to the limited plantar-flexor moment that occurs at the prosthetic joint. The aim of this research was to determine the loading on the limbs and the joint kinetics adopted by transtibial amputees in order to achieve a jump. Six amputee (AMP) and 10 able-bodied (AB) participants performed maximal vertical jumps on two force plates while kinematic data was recorded using a 9-camera VICON infrared system. The amputees did not jump as high as the AB participants. The AMPs raised the prosthetic heel from the floor to compensate for the restricted motion at the ankle. Consequently, kinematic symmetry was maintained at the knee and the hip. The knee flexion places the prosthetic shank in a more horizontal position. This is a vulnerable position due to the reduced strength in the knee extensors as a consequence of the amputation. In order to reduce the instability and loading at the knee, the maximum propulsive vGRF on the prosthetic side was reduced and the intact limb assumed a dominant role. Until amputees can take the loading on the prosthetic side, it is not recommended that they participate in jumping.

KEY WORDS: loading, jump, amputee.

INTRODUCTION: A below-knee amputation causes a major disruption to the musculoskeletal system. Prosthetic design has attempted to reduce the consequences of the amputation by developing dynamic elastic response (DER) prostheses. The aim of these devices is to absorb strain energy when a load is applied and then to return this energy as elastic energy when the prosthesis is unloaded. However, because the prostheses are passive, they are not capable of replacing the adaptable function of the intact limb. As a result compensatory mechanisms develop in the residual limb and in the intact limb. Asymmetry in amputee biomechanics have been reported in various walking and running studies usually to maintain the prosthetic limb in an upright position and to ensure stability at the knee (Nolan et al., 2003; Sanderson and Martin, 1997). Vertical ground reaction force (vGRF) analyses indicate that the intact limb generally adopts a dominant role in gait with increased vGRF loading when compared to the prosthetic limb and other able bodied participants. It is suggested that Amputees load the intact side to protect the residual limb (Nolan et al., 2003). Possibly as a consequence of this asymmetrical loading, amputees evidence greater joint degeneration in their intact limb compared to both the residual limb and able bodied (AB) subjects. Melzer et al., (2001), reported that both highly active amputees and those who are not active had a 65.5% greater incidence of osteoarthritis (OA) in their intact limb compared to AB equivalents. Walking and running gait are unilateral movements which have different mechanics to the bilateral contermovement jump (CMJ) which requires simultaneous contribution from both limbs to generate the required forces to achieve flight. The aim of this study was to determine the levels of loading and joint kinetics required by amputees to produce a bilateral CMJ.

METHODS: Participants: Six unilateral transtibial AMPs (5 males and 1 female) who were between 18 and 50 years, more than 12 months post-operative, with no secondary pathology and had an amputation of a traumatic nature were recruited. All the AMPs wore patellar tendon-bearing sockets with rigid pylons and their own prosthesis. Ten AB participants (9 male and 1 female) of the same age range with no pathology were used to facilitate the comparison of results. All participants (AMPs and AB) were recreationally active with similar proficiency in jumping and wore their own footwear (athletic trainers). All participants signed an informed consent form approved by the University and the National Health Services' Ethics Committee.

Data Collection: Data were collected in a single session. Following a 5 minute warm-up on a treadmill at a self-selected fast walking velocity, participants were given the opportunity to practise the movement and become familiar with the laboratory setting. Ten maximum effort bilateral countermovement jumps (CMJ) were performed with hands on hips with 1 minute rest between each trial. The only instruction given was to jump as high as possible. The jumps were performed with each foot on a separate force plate. Trials were excluded if the participants used their arms or if they missed the force plate during landing. On average 13 jumps were required to collect 10 successful trials. Data were collected using two Kistler (model 9581B, sampling at 1080Hz) force plates synchronized with a 9-camera Vicon (model 612, sampling at 120Hz) infra red system. Thirty four 25mm diameter reflective markers were attached to specific anatomical landmarks according to Vicon's Plug-in-Gait full body gait model (Oxford Metrics). Measurements were taken for each individual according to the Vicon requirements for full body modelling.

Data Analysis: The jump with the highest flight height determined by the CoM displacement (maximum height less height at take-off) was chosen for analysis. Jump kinetics and kinematics were calculated with Vicon Workstation software. Kinematic data were filtered using a Woltering quintic spline (MSE = 15mm) filter. Inverse dynamics using standard procedures determined the net joint reaction components and the net joint moments (normalised to body mass) at the ankle, knee and hip from the ground reaction force data associated with each foot. All other variables are presented for the AMPs as intact and prosthetic limb separately. For the AB, the results are for the preferred and non-preferred jumping limb. The symmetry index (SI) was calculated $SI = \frac{(X_r - X_l)}{0.5(X_r + X_l)} * 100\%$

Where X_r is the variable for the right limb, and X_l is the corresponding variable for the left limb. The magnitude of the SI indicates the degree of symmetry and the sign indicates the direction (positive is greater on the intact side). A value of zero indicated perfect symmetry.

RESULTS: The flight height and maximum vertical ground reaction force for each limb is presented in Table 1. SI results outside one SD of the AB participants SI are indicated in bold.

Table 1 Flight height and maximum vGRF experienced by the AMPs and AB participants

	Flight Height (m)	Max vGRF during propulsion (N.kg^{-1})		
		Intact	Prosthetic	SI
AMP1	0.24	1.141	1.010	12
AMP2	0.19	1.182	0.971	**19**
AMP3	0.17	1.292	0.591	**75**
AMP4	0.13	1.228	0.861	**35**
AMP5	0.10	1.177	0.731	**47**
AMP6	0.09	1.178	0.844	**33**
x̄ (AMP) (± sd)	0.15 ±0.06	0.20 ±0.05	0.83 ±0.16	37 ±22
x̄ (AB) (± sd)	0.31±0.04	1.134±0.15	1.093±0.12	3±6

AMPs jumped lower than the AB participants. Each AMP loaded the intact limb more than the prosthetic limb. The magnitude of the loading on the intact limb was similar to the loading for the AB group, preferred limb. The two AMPs who jumped the highest experienced loading on the prosthetic limb that was similar to the AB group non-preferred limb loading.

Table 2 Heel rise and maximum flexion experienced by the AMPs and AB participants

	Heel Rise (m)	Ankle Angle (°)			Knee Angle (°)			Hip Angle (°)		
		Intact	Prosthetic	SI	Intact	Prosthetic	SI	Intact	Prosthetic	SI
AMP$_1$	0.07	44	16	34	91	100	-9	105	116	-10
AMP$_2$	0.06	38	19	22	94	89	5	87	95	-9
AMP$_3$	0.07	33	6	21	89	93	-4	85	96	-12
AMP$_4$	0.05	41	17	28	84	90	-7	91	104	-13
AMP$_5$	0.09	43	13	34	93	107	-14	92	101	-9
AMP$_6$	0.06	26	12	11	74	65	13	92	89	3
$\bar{x}$ (AMP) (sd)	0.07	38±7	14±5	25±9	88±8	91±14	-3±9	92±7	100±9	-8±6
$\bar{x}$ (AB) (sd)	0.03±0.02	39±5	37±8	2±6	111±12	111±12	0±12	94±10	95±13	1±11

The mechanical restraints of the prosthesis prevented adequate ankle dorsiflexion in the eccentric phase. To maintain symmetry at the knee and hip, the prosthetic heel rose from the floor. Negative SI indicates more flexion on the prosthetic side.

Table 3 Maximum joint moments experienced by the AMPs and AB participants

	Ankle Moment (Nm.kg^{-1})			Knee Moment (Nm.kg^{-1})			Hip Moment (Nm.kg^{-1})		
	Intact	Prosthetic	SI	Intact	Prosthetic	SI	Intact	Prosthetic	SI
AMP$_1$	1.86	1.28	37	1.08	1.18	-9	2.45	2.55	-4
AMP$_2$	1.57	1.08	37	1.37	2.16	-45	1.67	2.16	-26
AMP$_3$	1.96	0.78	86	1.08	0.39	94	2.06	1.08	62
AMP$_4$	1.77	0.88	67	1.57	1.18	28	1.47	1.47	0
AMP$_5$	1.57	1.08	37	1.28	0.98	27	2.55	1.47	54
AMP$_6$	1.86	1.08	53	1.08	0.69	44	1.28	1.28	0
$\bar{x}$ (AMP) (sd)	1.8±0.2	1±0.2	52.8	1.2±0.2	1.1±0.6	23.2	1.9±0.5	1.7±0.6	14.3
$\bar{x}$ (AB) (sd)	1.57±0.2	1.47±0.2	7±0.2	1.57±0.3	1.57±0.3	0±0.3	1.77±0.5	1.86±0.06	-5±0

The ankle moments were asymmetrical and always smaller on the prosthetic side. At the intact knee the moments for the AMPs who jumped the highest were less on the intact side, while for the rest, the intact knee moments were greater. Overall, there was no clear trend for each joint related to height. The AMP joint moments were generally similar or lower at the ankles and knees but were similar at the hip compared to the AB participants.

DISCUSSION: Vertical jumping is a fundamental skill common to numerous recreational activities and training strategies. As jumping is a multi-joint action that requires substantial muscular effort about the ankle, knee and hip joints, it was expected that biomechanical compensations would result from the amputation. The results can be used to develop prosthetic design, to enhance rehabilitation and to develop exercise programmes for amputees for movements which require different loading and alternative movement patterns to walking.

The prosthetic ankle and the compensatory motor pattern adopted by these AMPs did not overcome the muscular and sensory loss of the anatomical limb. The AMPs did not achieve flight heights equivalent to the AB participants. In bilateral jumping kinematic asymmetry is restricted compared to walking and running due to the side-by-side positioning of the feet and the constraint of the pelvis. In order to produce a countermovement, dorsiflexion of the ankle and flexion of the knees and hips are required. As the AMPs were restricted in the magnitude of dorsiflexion, the heel rose to allow flexion at the knee and hip to occur. The resulting more horizontal positioning of the shank is usually avoided by AMPs who prefer a straight knee to maintain an upright prosthesis in walking (Barr et al., 1992), running (Sanderson and Martin, 1996), stair descent (Jones et al., 2006). In a unilateral jump, where AMPs cannot compensate at the contralateral limb, relatively straight knees were maintained, indicating this reluctance to flex the knee joint under loading (Strike and Diss, 2005). In the bilateral jump the intact limb can dominate. Although the prosthetic side knee flexed, the loading on the limb was reduced to prevent it collapsing (due to its reduced strength as a consequence of the amputation) and to stop excessive loading on it in this position. A consequence of this reduced loading was a smaller plantarflexor moment at the prosthetic ankle in the concentric phase, since, as a passive device, the moment is related to the loading on the prosthesis. This prosthetic plantarflexor moment was similar to the results often seen in gait, and indicates the limitation of the prosthesis to adapt to different demands, as would be expected from a passive, but responsive device. On the intact side, the reduced knee extensor moment could be as a result of a modulation to match the prosthetic limb to reduce asymmetry (Sanderson and Martin, 1996) or it could be to avoid loading the knee. Royer and Koenig (2005) illustrated the link between the increased frontal plane joint moments and bone mineral density in transtibial amputees when walking and suggested that the potential exists for premature intact-side knee joint degeneration in transtibial amputees.

CONCLUSION: AMPs did not jump as high as AB participants. The loading, as represented by the vGRF was lower on the prosthetic side and indicates that the intact side assumes a dominant role in producing the jump. These results imply that the AMPs were both promoting residual knee stability and protecting the residual joints by not loading the affected side. If amputees are to actively use the prosthetic limb when jumping, they need to learn to maintain knee stability in order to load the prosthesis. Until amputees can achieve this, it is not recommended that they participate in activities which excessively load the affected side with the knee in a flexed position. The contralateral loading may have consequences for the intact limb health given that amputees hare known to have increased levels of OA on this side.

REFERENCES:

Barr, A., Siegel, K., Danoff, J., McGarvey, C., 3rd, Tomasko, A., Sable, I., et al. (1992). Biomechanical comparison of the energy-storing capabilities of SACH and Carbon Copy II prosthetic feet during the stance phase of gait in a person with below-knee amputation. *PHYS THER, 72,* 344-354.
Jones, S. F., Twigg, P. C., Scally, A. J., and Buckley, J. G. (2006). The mechanics of landing when stepping down in unilateral lower-limb amputees. *Clinical Biomechanics,* 21, 184-193.
Melzer, I., Yekutiel, M., and Sukenik, S. (2001). Comparative study of osteoarthritis of the contralateral knee joint of male amputees who do and do not play volleyball. *The Journal of Rheumatology,* 28, 169-172.
Nolan, L., Wit, A., Dudziński, K., Lees, A., Lake, M., and Wychowañski, M. (2003). Adjustments in gait symmetry with walking speed in trans-femoral and trans-tibial amputees. *Gait and Posture,* 17, 142-151.
Royer, T., and Koenig, M. (2005). Joint loading and bone mineral density in persons with unilateral, trans-tibial amputation. *Clinical Biomechanics,* 20, 1119-1125.
Sanderson, D. J., and Martin, P. E. (1996). Joint kinetics in unilateral below-knee amputee patients during running,. *Archives of Physical Medicine and Rehabilitation,* 77, 1279-1285.
Sanderson, D. J., and Martin, P. E. (1997). Lower extremity kinematic and kinetic adaptations in unilateral below-knee amputees during walking. *Gait and Posture,* 6, 126-136.
Strike, S., and Diss, C. (2005). The biomechanics of one-footed vertical jump performance in unilateral transtibial amputees. *Prosthetics and Orthotics International,* 29, 39-51.

ISBS 2009

Oral Session S06

Indoor Sports

A BIOMECHANICAL COMPARISON OF JUMPING TECHNIQUES IN THE VOLLEYBALL BLOCK AND SPIKE

Roberto Lobietti[1], Silvia Fantozzi[2], Rita Stagni[2], Simon G.S. Coleman[3]

Department of Histology, Embryology and Applied Biology- Faculty of Exercise and Sport Science, University of Bologna, Italy[1]
Department of Electronics, Computer Science and Systems, University of Bologna, Italy[2]
Department of Physical Education and Sport Leisure Studies, Edinburgh University, Scotland (UK)[3]

The present case study kinematically analysed the spike and the block movements of a single volleyball player. The aim was to verify the hypothesis that for a right handed player the spike approach and the cross-over step in blocking after a move to the left are similar in coordination, whereas moving to the right before blocking requires a different movement pattern. The spatial and temporal variables of the jumps and the joint angles of the lower limbs during the countermovement were analysed. The results showed a high repeatability of the collected data. The similarity between the spike and the block when moving to the left confirmed the hypothesis. These results from a single subject should be extended by further studies of more athletes of varying skill levels.

KEY WORDS: stereophotogrammetry, variability, laterality, coordination.

INTRODUCTION: The spike, block and serve are the three most important skills to score points in volleyball (Lobietti et al., 2006). Jumping techniques in these three skills has become increasingly important to offence in advanced volleyball. The biomechanics of the spiking technique have been investigated by Coleman et al. (1993) and more recently by Kuhlmann et al. (2007) and Shabbazi et al. (2007). As described by these authors, the spike consists of four phases: a three-step approach (left-right-left), the jump, the hitting action and the landing. Recently, biomechanical analyses of the block (Lobietti et al., 2005; Donà et al., 2006) have focussed on the footwork techniques used by blockers. One of the major limitations is that none of these previous studies have analysed the same subjects performing spiking and blocking actions. To our knowledge, only Quade (1993) and Lawson et al. (2006) have previously presented data relative to the kinematics of both the block and the spike when executed by the same players. However, the movements acquired in these latter studies (a spike executed with only one step of approach before the jump, and a vertical block with no prior lateral movement) are used only very infrequently in game situations. During the game, it is much more likely that the spikers will have at least three steps of approach and that the blockers will have to move laterally before jumping. Lobietti et al. (2005) looked at blocking, and found in case of an outside set, the majority of the previous investigations showed an advantage in terms of time taken and height reached when using cross-over steps (XS) as opposed to slide-steps (SS) when moving prior to block jumping at the left side of the net. These findings were explained as result of the similarity in coordination and rhythm between this movement (XS to the left) and the spike approach but no data were presented to support this argument.

Therefore, the aim of the present is to test that explanation by comparing the biomechanical parameters of the jumps performed by a player when blocking after a XS to left and also to the right to those obtained when spiking with a three step approach.

METHODS: A Qualisys Capture System using 6 infrared cameras (Proreflex MCU 500 Hz) was used to acquire the execution of spikes and blocks of a single female hitter (25 years old, 174 cm height, one-handed reach 219 cm, mass 65 kg, 10 years of experience) who plays regularly in Scottish National League Division 1.

The lower limbs were considered as a kinematic chain of seven rigid body segments: right and left feet, right and left shanks, right and left thighs and pelvis. Anatomical landmarks and bone-embedded anatomical reference frames were defined according to the CAST protocol (Cappozzo et al, 1995; Benedetti et al., 1998)). For each body segment, four spherical reflective markers (19 mm) were attached to the skin. The trajectories of all 32 skin markers were collected by the motion capture system in each trial. In figure 1 an example of the CAST volleyball marker set (left) and a reconstruction of the spiking approach performed by the system (right) are shown.

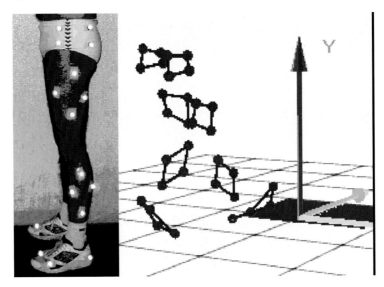

Figure 1: The marker set CAST volleyball and the reconstruction of a spiking trial

The markers were fixed to the player's skin and shoes and a series of anatomical calibrations were performed. After a 10 minute warm-up, the players executed the following trials in order:
1. 20 spikes: players had to touch a Vertec jump monitor (Sports Imports, USA) as high as they possibly could, and the measurement of the height reached in each trial was registered.
2. 10 blocks performed moving laterally to the left to a target (ball hung from the ceiling over the net at a height of 2.5 m) and 10 blocks moving to the right with the same target.

Movements were repeated twice for a total of 60 jumps: 40 blocks and 20 spikes. The total number of jumps, plus warm-up simulated a normal match or training session of volleyball. To prevent any effects of fatigue, a rest of 30 seconds between trials and a pause of 3 minutes between series was performed.

After tracking, because there was marker occlusion or trajectory gaps longer than 20 samples only 14 spikes, 9 trials blocking to the left and 13 moving to the right were analysed.

The jump height was calculated as the difference between the height of the COM of the lower limbs at the peak of the jump and at the take-off. The Range of Motion (ROM) of the COM was defined as the difference between maximum and minimum values of the COM.
Joint angles were calculated in the sagittal plane according to Grood and Suntay convention, (1983).

Mean, Standard Deviation and Coefficient of Variation of the dependent variables (jump height, flight time, ankle knee and hip joint angles at the moment of the countermovement) were calculated.

RESULTS: Table 1 shows the mean, standard deviation and coefficient of variance of the temporal and spatial variables of the 3 movements.

Table 1 Temporal and spatial variables

	Flight time (msec)	Upward motion time (msec)	COM max (mm)	COM min (mm)	ROM (mm)	Jump Height (mm)
Block to the left	618±41	316±31	667±15	315± 4	353±13	283±38
Block to the right	589±28	292±25	661±9	304± 4	357±11	252±34
Spike	634±12	293±13	732±10	323± 11	410±11	291±18
CV for block to the left	7	10	2	1	4	14
CV for block to the right	5	9	1	1	3	14
CV spike	2	4	1	4	3	6

In Table 2 displays the flexion-extension angles (according to Grood and Suntay convention, 1983) of the ankles, the knees and the hips at the moment of the minimum value (lowest vertical point) reached by the COM during the countermovement executed before the jump.

Table 2 Ankle, knee and hip angles (degrees) during the countermovement

Joint Angles (degree)	Right ankle	Left ankle	Right knee	Left knee	Right hip	Left hip
Block to the left	30 ± 2	34 ± 3	61 ± 2	66 ± 3	28 ± 4	39 ± 3
Block to the right	33 ± 2	41 ± 3	42 ± 3	92 ± 4	22 ± 2	52 ± 8
Spike	21 ± 5	20 ± 4	63 ± 7	69± 5	30 ± 4	43 ± 7
CV for block to the left	7	9	2	3	13	7
CV for block to the right	8	8	2	4	8	16
CV spike	23	21	6	5	15	17

DISCUSSION: The results of the jump height confirmed the previous findings of Lobietti et al (2005): this hitter performed the highest jump when spiking (291 ± 18 mm), then blocking to the left (283 ± 38 mm) and the lowest elevation when blocking to the right (252 ± 34 mm). She showed a higher repeatability in jumping to spike (CV=6%) whereas when blocking, the variability was higher than spiking but similar in the two block directions (CV=14%). Furthermore, the ROM of the COM showed a higher repeatability than jump height in all movements acquired. Thirdly the reliability of the countermovement was similar between trials (spiking CV=2.6%; block in 2 CV=2.9%; block in 4 CV= 3.7%).

Despite the limitations of a case study, it is still interesting to observe that the two legs show similar values for the joint angles when spiking and blocking to the left, whereas going to the right, the left leg (leading the move in a XS) shows larger flexion by all three joints. The knee joint angles showed higher flexion going to the left (92° ± 4) suggesting this might be one of the causes of the player's poorer performance when moving in this direction. These results are in agreement with the previous biomechanical description of volleyball jumps by Gollhofer & Brun (2003) showing knee flexion data in the countermovement phase of around 70°. The elastic muscular strength is much more effective when the knees are flexed around 65° (as in the spike and moving to the left), whereas going to the right the higher countermovement requires a greater concentric muscular activation during the push-off phase of the jump due to the greater flexion.

CONCLUSION: This study compared the three typical jump movements performed by a volleyball player: the spike, the block moving to the left and the block going to the right. In the present study, only one subject was analysed but the findings confirmed the hypothesis: the XS in blocking to the left is similar in coordination with the spiking approach but different to that when moving to the left. More subjects will be required to reinforce the findings of this study.

REFERENCES:
Benedetti, M.G., Catani, F., Leardini, A., Pignotti, E., Giannini, S. (1998), Data management in gait analysis for clinical application. *Clinical Biomechanics. 13(3)*, 204-215.

Cappozzo, A., Catani F., Della Croce U., and Leardini A., (1995), Position and orientation in space of bones during movement: anatomical frame definition and determination. *Clinical Biomechanics. 10(4)*, 171-178.

Coleman S.G.S., Benham A.S.,and Northcott S.R. (1993). A 3D cinematographical analysis of the volleyball spike. *Journal of Sports Sciences, 11*(4), 295-302.

Donà, G., Zorzi, E., Petrone, N., Sawacha Z.,and Cobelli, C., (2006). Biomechanical Analysis of three different blocking footwork techniques in volleyball: a pilot study. *Proceedings of the XXIV International Symposium on Biomechanics in Sport.* (H. Schwameder, G. Strutzenberger, V. Fastenbauer, S. Lindinger, E. Müller, Editors) pp. 327-330.Salzburg: Department of Sport Science and Kinesiology University of Salzburg, Salzburg.

Gollhofer A. and Bruhn S.(2003). The biomechanics of jumping. *In Handbook of Sports Medicine and Science: Volleyball* (pp. 18-28) Blackwell Science Pub Publisher.

Grood, E.S., Suntay, W.J. (1983). A joint coordinate system for the clinical description of three-dimensional motions: application to the knee. *J Biomech Eng.* 135(2), 136-144.

Huang K.C., Hu L.-H., Huang C., Sheu T.-Y. and Tsue C.M.(2002). Kinetic and Kinematic differences of two volleyball-spiking jump. *Proceedings of the XX International Symposium on Biomechanics in Sport.* (pp. 148-151)Caceres-Extremadura-Spain.

Kuhlmann C., Roemer K. and Milani T.L., (2007). Aspect of the three dimensional motion analysis of the volleyball spike in high level competition. In H.J. Menzel & M.H. Chagas (Eds.) *Proceedings of the XXV International Symposium on Biomechanics in Sport.* (pp. 47-50) Ouro Preto-Brazil.

Lawson B.R., Stephens T.M., DeVoe D.E. and Reiser R.F. (2006) Lower-Extremity differences during step-close and no-step countermovement jumps with concern for gender. *Journal of Strength and Conditioning Research*, 20(3), 608-619.

Lobietti R., Di Michele R., Merni F. "Relationships between performance parameters and final ranking in professional volleyball" Proceedings of WCPAS 2006 SZOMBATHELY 24-28 august 2006 World Congress of the Society of Performance Analysis in Sport

Lobietti R., Merni F. and Ciacci S., (2005). A 3D Biomechanical Analysis of Volleyball Block. In W.Starosta & S.Squatrito (Eds.) *Scientific Fundaments of Human Movement and Sport Practice 21*(2), (pp. 413-415). Bologna: Edizioni Centro Universitario Sportivo Bolognese.

Quade, K., (1993). *Zur Funktion ud belastung der unteren Extremitäten bei volleyballspezifischen Sprüngen*. SFT-Verlag Erlensee.

Shahbazi M.M., Mirabedi A. and Gaeini A.A.(2007). The volleyball approach: an exploration of run-up last stride length with jump height and deviation landing. *Proceedings of the XXV International Symposium on Biomechanics in Sport.* (pp. 574-577) Ouro Preto Brazil.

Tokuyama M., Ohashi H., Iwamoto H., Takaoka K. and Okubo M., (2005). Individuality and reproducibility in high speed motion of volleyball spike jumps by phase-matching and averaging. *Journal of Biomechanics*, 38 (10), 2050-2057.

ELITE OUTSIDE HITTERS IN VOLLEYBALL DO NOT MEET THEIR INDIVIDUAL POSSIBLE MAXIMUM IMPACT HEIGHT IN HIGH SPIKE JUMPS

Claas H. Kuhlmann[1], Karen Roemer[2], Thomas L. Milani[1]

[1]Chemnitz University of Technology, Department of Human Locomotion, Chemnitz, Germany
[2]Michigan Technological University, Department of Exercise Science, Health and Physical Education

It is assumed that a high impact height is a relevant factor for success in volleyball spikes. The purpose of the study was to investigate whether outside hitters hit the ball at the highest possible impact height. Spikes from position IV were analysed at a tournament of the European League. The posture of the athletes was less extended in the trunk and upper limb with increasing jump height. Regarding the body posture at the moment of impact, there was no effect on the post impact ball speed. It is concluded that there could be enhancement with respect to the impact height as jump height increases even in elite athletes without reducing ball speed. This should be addressed within the training process.

Keywords: Volleyball, Spike, Motion analysis, Coordination

INTRODUCTION: Based on the complexity of the volleyball spike movement, an in depth analysis is difficult. It is easier to investigate such movements in a laboratory setting than during competition. This may be the reason why most of the studies were performed in laboratory conditions or during training when analysing volleyball spikes (e.g.: Tokuyama et al., 2005). Nevertheless, spike jumps in a competition are likely to differ from spike jumps in a laboratory setting. In the laboratory different environmental conditions can be found to influence the movement as they can be detected during competition. Therefore, findings from Forthomme et al. (2005) under laboratory conditions are reasonable showing a higher impact height and a higher post impact ball velocity of Belgian first league players than findings from Coleman et al.(1993) or Kuhlmann et al. (2008) both with international elite outside hitters during competition.
Because the optional target area increases with rising impact height, it is assumed that a higher impact height is a relevant factor for success in volleyball spikes (Neef & Heuchert, 1978). Hence, the aim of this study was to investigate, if outside hitters of different national teams reached their individual potential maximum impact height during spike jumps from position IV. This study was performed in order to close the gap of analysing volleyball spikes in a competitive setting. The data acquisition for this study took place during an international volleyball tournament.

METHODS: The database included spike movements performed by male outside hitters of the national teams of Croatia, Estonia, Germany and the Netherlands. The data were recorded during a European League tournament. Since the most important spike-position used by international top level teams is position IV (Kuhlmann et al., 2008), only spikes from this position were considered. Position IV is the position of the outside hitter at the left side of the court in front of the net. The pass was always played half high and without combination. Fast or unscheduled actions due to a bad pass were excluded. Therefore, only actions were accepted when the outside hitters always had enough time to prepare themselves for an optimal spike jump including the approach and the takeoff. All subjects performed a step-close technique, identified by Coutts (1982). The flight angle of the ball after impact had to be 110° to 145° to the net to improve the standardisation of the boundary conditions.

10 elite outside hitters of different national teams of the highest order were analysed. This is equivalent to approximately 10% of the outside hitters playing in national teams on highest international level. The sample had a mean body height of 198.8 ± 4.4cm and a mean body mass of 92.0 ± 5.3kg. From each player one spike jump was analysed. The study concurred in the exigencies of the ethics committee for human research and in current local laws and regulations. Due to the camera positions, no effect on the players or on the results occurred.

Four digital cameras were positioned around the volleyball court capturing the spikes with a frame rate of 100 Hz. The cameras were activated and triggered externally. The frame rate was also externally controlled. The accuracy of the frequency controlling was previously tested. No relevant irregularity in the frequency was reported.

The calibration of the measurement setup and the methodology of the data-processing was described by Kuhlmann et al. (2007) in detail. A verification of this method by calculating points with known 3D coordinates showed an accuracy of 9.9 ± 7.7 mm (x-direction), 4.7 ± 1.4 mm (y-direction), 8.3 ± 4.4 mm (z-direction).

All digitisers had to digitise the same standardised test videos before they were allowed to start the original digitising process for this study. Inter-digitiser reliability was investigated by calculating differences of angles and changes in segment length of selected angles and segments of these test-videos. Inter-digitiser reliability was calculated as 0.07 ± 0.06° for angles and 1.3 ± 0.9mm for changes in segment length in the mentioned test videos.

The centre of mass (COM) was calculated using the HANAVAN-Model. The values of impact height, COM-height and the difference in height between the COM and the impact height at the moment of impact (ΔCOM-height / impact height) were calculated from the 3D-coordinates, provided by the software SIMI-Motion. Jump height was calculated as the difference of COM-height at the last frame with ground contact and the highest COM in the flight phase. The extension of body angles or the elongation of the upper body position was calculated as the difference between the COM-height and the impact height.

To calculate post impact ball speed (PIBS) a vector was calculated out of the 3D-coordinates of the ball two frames and 8 frames after the ball contact. The length of this vector represented the mileage of the ball. Due to the camera frequency the used time period is known and velocity was calculated out of mileage and the dedicated time.

Pearson's correlation coefficient was calculated to detect coherence between those parameters. Statistical evaluation of the data was conducted using SPSS 16.0.

RESULTS: The values of impact height, COM-height and the difference in height between the COM and the impact height at the moment of impact are presented in Table 1. The mean jump height of the sample was 63.17 ± 6.2 cm.

Table 1: Parameters representing the efficiency of impact height with respect to jump height

Subject	Impact height in cm	COM-height at impact in cm	ΔCOM-height / impact height in cm
1	323.6	208.0	115.6
2	313.1	202.2	110.9
3	305.5	196.7	108.8
4	317.7	206.7	111.0
5	310.0	191.1	118.9
6	307.6	188.5	119.1
7	317.4	176.2	141.3
8	317.5	212.4	105.1
9	312.7	201.3	111.5
10	314.9	207.3	107.5
$\bar{x}$	314.0	199.0	115.0
S.D.	5.4	11.1	10.3

Pearson's correlation coefficients were calculated between those different parameters. The parameters and the appendant coefficients are shown in Table 2.

Table 2: Pearson's correlation coefficients of the analysed parameters

Parameters	Pearson's correlation coefficient
COM height / Impact height	0.37
COM height / Jump height	-0.19
Jump height / Impact height	0.45
COM-height at impact / Impact height	0.38
COM-height at impact / ΔCOM-height - Impact height	**-0.87**
COM height / ΔCOM-height - Impact height	0.20
ΔCOM-height - Impact height / Post impact ball speed	0.43

As shown in Table 2 a high correlation coefficient can be calculated only between COM-height at impact and ΔCOM-height / impact height. The higher the COM-height at impact the lower is ΔCOM-height / impact height.

DISCUSSION: Impact height is an important factor for success in volleyball spikes. Increasing impact height automatically increases the target area on the opponent's court (Neef & Heuchert, 1978). Therefore, the chance of a successful spike increases. It was the aim of this study to analyse the impact height and whether the individual maximum possible impact height of the athletes could be reached. A detector for the individual maximum impact height is the difference between the COM-height at the moment of impact and the impact height as a value for body angle extension. The larger this difference, the more extended are the body angles at the moment of impact. The athlete is able to hit the ball in a higher position when body angles are more extended. As a result, the athlete gets closer to the individual maximum impact height.

According to the results of the correlation analysis, it cannot be stated that taller attackers will automatically hit the ball higher or jump higher than smaller outside hitters under standardised conditions. Also the COM-height at the moment of impact showed just a small correlation coefficient to impact height (Table 2).

The difference in height between the COM height at the moment of impact and the impact height showed a high negative correlation ($r = -0.87$) to the COM height at the moment of impact. As a consequence, outside hitters with less jump height hit the ball with more extended body angles than those with a higher jump height. This leads to the assumption, that high jumping attackers may increase their impact height even more by extending their upper body.

The parameter ΔCOM-height / Impact height showed no correlation to body height and post impact ball speed. This parameter seems to be independent from body height. Therefore, it can be assumed, that more elongation of the trunk and upper limbs will not reduce the post impact ball speed, but will increase the target area.

Outside hitters might use different elongated postures and changes in movement techniques for tactical reasons. Regarding to the discussed importance of a high impact height (Neef & Heuchert, 1978), it could be useful to modulate the coordination pattern in higher jumps for reaching higher impact heights. Therefore, it is essential that the outside hitters are able to "read" the set with high precision and coordinate their jumping movement according to it.

CONCLUSION: Elite outside hitters of different volleyball national teams did not meet their individual maximum impact height in volleyball spikes from position IV for higher jump heights. The posture of the upper part of the body was more extended within spike jumps showing lower COM-height than in jumps with a higher COM-height. Because of the small coherency between elongated posture and post impact ball speed it is assumed that a different posture will not reduce the ball speed. Hence, no negative effect from an elongated posture in higher jumps is anticipated. The individual impact height could be increased easily in higher spike jumps. This study provides important insight about the hitting technique from elite outside-hitters, which may be used to improve their performance during competition.

REFERENCES
Coleman, SG., Benham, AS., Northcott, SR. (1993). A three-dimensional cinematographical analysis of the volleyball spike. *Journal of Sport Sciences*, 11, 4, 295-302.
Coutts, KD. (1982). Kinetic differences of two volleyball jumping techniques. *Medicine and Science in Sports and Exercise*, 14, 1, 57-59.
Forthomme, B., Croisier, J., Ciccarone, G., Crielaard, J., Cloes, M. (2005). Factors correlated with volleyball spike velocity. *The American Journal of Sports Medicine*, 33, 10.
Kuhlmann, CH., Roemer, K., Milani, TL. (2007). Aspects of a three dimensional motion analysis of the volleyball spike in high level competition. In: HJ. Menzel, MH. Chagas. *Proceedings*. XXV. International Symposium on Biomechanics in Sports. Belo Horizonte: Segrac.
Kuhlmann, CH., Roemer, K., Zimmermann, B,. Milani, TL., Fröhner, B. (2008). Vergleichende Analyse von Technikparametern beim Angriff in definierten Spielsituationen im Volleyball. *Leistungssport*, 38, 5, 29-34.
Neef, W., Heuchert, R. (1978). Kennzeichnung der Abhängigkeiten und Beziehungen zwischen den Handlungshöhen und Handlungspositionen von Angreifern und Block einerseits und der Trefffläche im gegnerischen Feld andererseits im Volleyball. *Wiss. Zeitschrift der Deutschen Hochschule für Körperkultur Leipzig*, 19, 2, 127–136.
Tokuyama, M., Ohashi, H., Iwamoto, H., Takaoka, K., Okubo, M. (2005). Individuality and reproducibility in high-speed motion of volleyball spike jumps by phase-matching and averaging. *Journal of Biomechanics*, 38, 10, 2050–2057.

Acknowledgment:
Funded by: Federal Institute of Sports Science, Bonn, Germany.
German Volleyball Association (DVV), European Volleyball Association (CEV).
Institute of Applied Training Sciences (IAT), Leipzig, Germany.
Simi Reality Motion GmbH, Munich, Germany.

BIOMECHANICAL ASPECTS OF BADMINTON SHOE DURING A LUNGE

Seungbum Park[1], Darren Stefanyshyn[2], Jay Worobets[2], Sang-Kyoon Park[1], Jungho Lee[1], Kyungdeuk Lee[1] and Jongjin Park[3]

Footwear Industrial Promotion Centre, Busan, S. Korea[1]; University of Calgary, Canada[2], Kyungsung University, Busan, S. Korea[3]

KEY WORDS: foot slippage, cushion, impact force, peak pressure, traction, heel height

INTRODUCTION: Athletic footwear can influence performance, injury and comfort. Badminton is a dynamic sport characterized by quick movements such as lunges, side cuts and jumps. As a result, many aspects are important for badminton footwear. Therefore, the purpose of this project was to determine biomechanical differences between newly developed badminton shoes and its competitive badminton shoes.

METHODS: A variety of analyses were performed on the two different shoe models (Figure 1). Type B was selected as a high quality shoe for benchmarking of Type A. Measurements of shoe shape and dimensions (both internal and external), foot movement within the shoe (using high speed camera at a sampling rate of 500 Hz), cushioning of ground reaction forces (using Kistler force plate at a sampling rate of 1000 Hz), in-shoe pressure (using a Pedar pressure sys. at a sampling rate of 100 Hz) and outsole traction (using Automated Footwear Testing sys.) were performed. In addition, subjective feedback (VAS: Mündermann et al., 2002) of the fit and function of the shoes was quantified for 17 recreational badminton players (age: 28.12 ± 13.44 yrs, height: 174.53 ± 10.36 cm, mass: 67.53 ± 9.49 kg).

Figure 1: the increased sliding of the foot within Type A shoe compared to Type B

Type A: Newly designed badminton shoe   **Type B:** Competitive badminton shoe

RESULTS: Type A shoe had a much higher heel (103.4 mm vs 101.9 mm) and shallower heel cup, so the heel was not secured well in the shoe and the ankle joint was higher off the ground. Figure 1 shows relative translation of the foot with respect to the shoe from the markers on the shoe and the lateral malleolus: left (initial foot plate), right (maximum foot movement. Foot slippage was up to 40% greater in Type A shoe than Type B shoe ($p<0.01$). The flexion axis of Type B shoe occurred in the midfoot, not at the ball of the foot like Type B, where they would want the shoe flexion to occur. Impact forces and peak pressures under the foot were higher with Type A shoe compared to Type B shoe ($p<0.05$).

DISCUSSION & CONCLUSION: Initial biomechanical tests performed on newly designed shoe compared to its competitive shoe led to some recommended design changes. These suggestions include: 1) lowering the heel height and improving the influence of the heel counter to provide better stability, 2) improving the rearfoot cushioning, 3) a narrower toe box and/or modified lacing system to decrease foot slippage in the shoe and 4) a stiffer arch to provide the appropriate flexion axis at the toes rather than under the arch of the foot. It is recommended that in order to improve the shoe in the future, focus should be given to making it a lower profile (closer to the ground).

REFERENCES: Mündermann, A., Nigg, B.M., Stefanyshyn, D.J. & Humble, R.N. (2002). *Gait & Posture,* 16(1), 38-45.

A COMPARISON OF COMPENSATION FOR RELEASE TIMING AND MAXIMUM HAND SPEED IN RECREATIONAL AND COMPETITIVE DARTS PLAYERS

D.J. Burke and M.R. Yeadon
Loughborough University, Loughborough, UK.

The level of accuracy achieved by darts players is dependent on their timing capabilities, judgement of velocity and their use of technique. This paper examines the techniques used by a recreational and a competitive darts player in a throwing task. The inaccuracy of the throws by both players were attributed to variations in release timing and maximum hand speed, with variations in release timing found to be the primary factor in both subjects (94.8% and 99.2% of total variance). The degree of compensation for variations in release timing and maximum hand speed were calculated for both competitors. The greater accuracy achieved by the competitive player was found to be the result of better compensation from technique rather than less variation in release timing and maximum hand speed.

KEY WORDS: throwing, technique.

INTRODUCTION: In targeted throwing tasks, such as darts, variations in landing height arise from variations in release timing and maximum hand speed; such variations should be controlled or interact with each other in order to achieve accurate throws. Variation in the release conditions can be controlled to some extent but can never be eliminated and the throwing action used in darts is too brief (<200 ms) for feedback control mechanisms based on proprioception to be implemented to compensate for such variations (Müller and Loosch, 1999). Throwing technique can be optimised to minimise the effects of variations in release timing and hand speed on landing height using feedforward control. We hypothesised that the difference in throwing accuracy between recreational and competitive darts players can be attributed to the level of compensation between release speed and release angle achieved by the technique employed.

METHODS: Three-dimensional position data were collected on two subjects using 12 Vicon MX13 cameras operating at 800 Hz. The subjects were a recreational darts player with over 40 years experience and a current county player with over 20 years experience. The subjects threw a small ball (mass 16.5 g) 18 times from the official darts distance aiming at the centre of a dartboard placed at the official height. This task was similar to throwing a dart accurately and had the advantage that the projectile could be tracked by the Vicon motion capture system. The landmarks of the subject's throwing arm and hand were identified by three pairs of reflective markers placed to track the positions of shoulder, elbow and wrist joint centres and three markers attached to the first three digits (Figure 1).

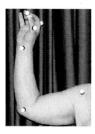

Figure 1: Reflective marker set used to track the motion of the arm and hand (posterior shoulder, lateral elbow and posterior wrist markers not shown).

When the ball was in the hand its position relative to the hand was determined and this was used to reconstruct the position of a hypothetical ball in the hand after the time of release. Quintic splines were fitted to the time histories of the horizontal and vertical positions of the actual ball and hypothetical ball with closeness of fit based upon a noise estimate determined from the data (0.4 mm). Release time was estimated as the average of the latest time that the ball was in the hand and the earliest time the ball was in flight as determined by comparison of spline value of actual vertical ball height with heights from spline to hypothetical ball and parabolic fit to flight. Pronounced movements of the fingers at the time of release could possibly lead to a discrepancy between the release velocities calculated from the spline to the hypothetical ball and the velocity from the parabolic fit. In this case velocity offsets were calculated and these offsets were used in all subsequent hypothetical releases. The accuracy of the release time estimates were assessed by calculating the landing height from the release time estimate and comparing this with the measured landing height.

To determine the vertical variation at the target arising from variation in maximum hand speed it was assumed that the percentage variations in hand speed at release and in maximum hand speed were equal. Horizontal and vertical release position and velocity were regressed against maximum hand speed and the resulting regression equations were used to calculate release conditions. The landing height was calculated using equations of constant acceleration under gravity:

Horizontally: $x = x_0 + u_0 t$ giving t

Vertically: $z = z_0 + v_0 t - \tfrac{1}{2}gt^2$ giving landing height z

The variance arising from hand speed variation was subtracted from the total variance in landing height to determine the variance in landing height arising from timing variation and the corresponding release timing window was determined for each trial.

The expected landing height variation arising from the above variation in maximum hand speed when timing of release was unchanged was calculated for each trial. The horizontal and vertical release coordinates and velocities were regressed against maximum hand speed. The regression equations were used to calculate hypothetical release conditions and landing heights for each recorded throw with the hand speed varied by the observed percentage variation in maximum hand speed. The percentage reduction in variance arising from the actual technique was calculated to give the level of velocity compensation. The landing height variation corresponding to release timing variation was calculated for a hypothetical circular hand trajectory at constant angular velocity. The percentage reduction in variance from this value to that of the actual technique was calculated to give the level of release timing compensation.

RESULTS: The vertical accuracy of the 18 throws at the target centre by the recreational and competitive subjects was -29.8 ± 37.1 mm and -8.6 ± 25.3 mm respectively (see Figure 2). The standard deviations of the vertical accuracy are comparable with those achieved by the subjects in the studies of Smeets *et al.* (2002) [29mm and 61mm] and Müller and Loosch (1999) [31mm and 44mm achieved by the most experienced subjects]. (The difference between calculated and actual landing height was -13.8 ± 17.9 mm for the recreational player and -3.8 ± 3.8 mm for the competitive player. The variations in maximum hand speeds were 1.6% (6.18 ± 0.10 ms^{-1}) for the recreational player and 1.5% (6.45 ± 0.10 ms^{-1}) for the competitive player, which resulted in vertical variations at the target of 8.5 mm and 2.2 mm respectively. These variations correspond to 5.2% and 0.8% of the total variance for the recreational player and competitive player respectively. The remainder of the vertical variation, 36.1 mm for the recreational player and 25.2 mm for the competitive player, equivalent to 94.8% and 99.2% of the total variances, corresponded to release timing windows of 3.4 ± 0.5 ms and 5.9 ± 5.8 ms (i.e. a change in release time of ±1.7 ms would result in a change in landing height of 36.1 mm).

Variation in landing height corresponding to uncompensated maximum hand speed variations of 1.6% and 1.5% were found to be 22.8 mm and 18.7 mm for the recreational and competitive players respectively; showing that the techniques used by the two subjects reduced the variance due to velocity variation by 86.1% and 98.6% respectively. Variation in landing height when timing variation is without compensation was 107.5 mm for the recreational player and 250.1 mm for the competitive player so that the techniques used reduced the variance due to release timing variation by 88.7% and 99.0% respectively.

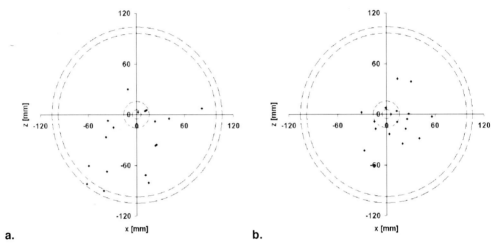

Figure 2. Illustration of the accuracy of the throws of the recreational (a.) and competitive (b.) subjects, including scale representations of the inner and outer bulls eye and the treble ring.

DISCUSSION: The contributions of release timing variations (>95%) and maximum hand speed variations (<5%) to the variance of the vertical accuracy of the throws indicate that variations in timing of release is the primary factor in darts performance in these two subjects. This is in agreement with Müller and Loosch (1999) who found that subjects' technique advanced with practice towards an 'equifinal path' that has reduced timing sensitivity. However this result conflicts with Smeets et al. (2002) who found that hand speed variation was the major contributor. The conflicting conclusion of Smeets et al. (2002) probably arose from incorrect estimation of release time since estimated release time for the best participant apparently released his darts close to the horizontal which would not have lead to a landing close to the centre of the board. The release time estimates in this study were calculated to be accurate to within 1.2 ± 2.2 ms for the recreational player and 0.0 ± 1.0 ms for the competitive player.

In this study the technique, timing capabilities and kinaesthetic awareness of each subject combined to result in the overall accuracy of their throws (shown in Figure 2). The competitive player was found to have similar ability when comparing the variation in maximum hand speed (1.5%) to that of the recreational player (1.6%). However his timing capabilities (5.9 ± 5.8 ms) were found to be inferior to those of the recreational player (3.4 ± 0.5 ms). Despite the large contribution (>95%) of timing to overall throwing accuracy the competitive player was able to achieve a more accurate throwing performance than the recreational player.

The competitive player was able to compensate for variations in maximum hand speed (98.6%) and timing of release (99.0%) using a better technique than the recreational player (86.1% and 88.7%). The effect of the superior technique used by the competitive player is that the inferior timing abilities of the player were counteracted by his better technique resulting in superior performance. Both techniques compensated for variations in timing of release more than the variations of maximum hand speed, illustrating that the importance of timing variations relative to that of hand speed variation has been recognised during the learning process and optimisation of technique. Possible compensatory techniques include coordinated movements of the shoulder and elbow affecting the radius of curvature of the hand path as well as changing the horizontal and vertical accelerations of the projectile. Further investigation into the mechanisms used for compensatory techniques should have coaching applications to the improvement of performance by identifying how compensation can be improved.

CONCLUSIONS: The variation in landing height was primarily due to variation in release timing rather than variation in maximum hand speed. Compensatory mechanisms have been employed by both subjects within their technique in order to minimise the expressions of their innate variations in release timing and maximum hand speed in the accuracy of their throws. The greater accuracy of the competitive player is a result of better use of compensatory mechanisms within his technique rather than less variation in release timing and maximum hand speed.

REFERENCES

Müller, H. and Loosch, E. (1999). Functional variability and an equifinal path of movement during targeted throwing. *Journal of Human Movement Studies, 36, 103-126.*

Smeets, J.B.J., Frens, M.A. and Brenner, E. (2002). Throwing darts: timing is not the limiting factor. *Experimental Brain Research, 144, 268-274.*

INVERSE KINEMATICS IN THE SUCCESS OF THE THROW IN BASKETBALL GAME

Emil Budescu[1], Eugen Merticaru[1], Radu Mihai Iacob[2]

[1]**Biomechanics Laboratory, Technical University "Gh. Asachi" Iasi, Romania,**
[2]**Physical Education and Sports Faculty, "Al. I. Cuza" University, Iasi, Romania**

The paper analyses the problematic of the human arm's movement in order to get a successful direct throw in basketball. For the human arm we analyzed the mechanic model of the double pendulum, formed by the arm and the forearm, with the hand being immobilized by the forearm. The classic equations of throwing the ball (dependent on the initial speed and throwing angle vector) have been correlated with the corresponding equations of the arms' movement. After imposing the mathematical conditions necessary for the ball to get into the basket, meaning certain inequality-type restrictions, we determined the kinematic dependents between the human arms' movement and the trajectory of the ball. The numeric analysis for several initial values of the throw allowed the assessment, through validation or non-validation, of the kinematic conditions for the success of the throw, and also several discussions regarding the influences of various kinematic parameters upon the throw.

KEY WORDS: geometrical conditions, mathematic model, trajectory, upper limb

INTRODUCTION: The ball trajectory during a throw depends upon the initial throwing conditions (the throwing angle, the initial speed vector, the geometrical coordinates of the throwing point) and on the type of movement (uniform, diversified etc.), representing a classic problem of the mechanics. Nevertheless, the success of the throw in basketball depends not only on the initial throwing conditions and on the type of movement, but also on the final coordinates of the ball and the incidence angle towards the basket. Schroder and Bauer (1996) treat, among others, the aspects related to finalizing the throw in basketball, putting into evidence the geometrical parameters for the success of the throw. Budescu and Iacob (2005) present, in the chapter called "Kinematics", the classic equations of throwing a body, with applications in sports, and with an emphasis on the kinematic dependents of the body trajectory. Budescu et al. (2005) talk about certain issues regarding the dependence between the arm's movement and the kinematic conditions for the success of the throw in basketball. We mention the fact the arm is considered here as a simple pendulum kinematic model.
In this paper, our aim was to analyze the mathematical dependence between the flexion-extension movement of the human arm, seen as a double pendulum, when throwing the basket ball and the success of the direct free throw towards the basket. The main objectives of the study were: emphasizing the geometrical parameters of the flexion-extension movement of the human arm, which have a great influence upon the success of the direct free throw in basketball; determining the kinematic restrictions which may ensure the success of a throw, as well as correlating these restrictions with the arm's movement parameters; mathematically emphasizing the relation between the movement of the arm (double pendulum type) and that of the ball during the direct free throw in basketball.
The purpose of the study was to offer the theoretic support for specialists in biomechanics for the individualized calculus of the optimal angular amplitudes for the flexion-extension of the arm and of the forearm during the direct free throw at the basket, which may ensure the success of the throw in basketball.

METHODS: The study has used the analytical method in order to study the throw in the basketball game. There is a schematic presentation of the basketball player and of the ball trajectory in Figure 1.

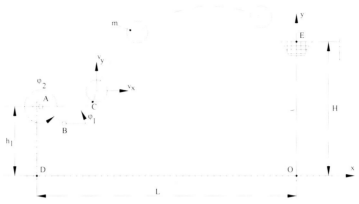

Fig. 1 Schematic of basketball player and ball trajectory

In Figure 1 letters represent as follows: h_1 is the height of the basketball player, ABC is the player's arm, modeled as a double pendulum, L is the distance from which the ball is thrown, m is the ball, v_y and v_x are the initial vertical and horizontal speeds of throwing the ball, H is the height of the basket.

The position and velocity equations for point C are the following:

$$x_C = -L + AB \cdot \cos(\varphi_2) + BC \cdot \cos(\varphi_1) \tag{1}$$

$$y_C = h_1 + AB \cdot \sin(\varphi_2) + BC \cdot \sin(\varphi_1) \tag{2}$$

$$\dot{x}_C = -\omega_2 \cdot AB \cdot \sin(\varphi_2) - \omega_1 \cdot BC \cdot \sin(\varphi_1) \tag{3}$$

$$\dot{y}_C = \omega_2 \cdot AB \cdot \cos(\varphi_2) + \omega_1 \cdot BC \cdot \cos(\varphi_1) \tag{4}$$

The equations (1) to (4) represent the initial conditions for the ball in its fly toward the basket.

Thus:
$$v_x = \dot{x}_C = -\omega_2 \cdot AB \cdot \sin(\varphi_2) - \omega_1 \cdot BC \cdot \sin(\varphi_1) \tag{5}$$

$$v_y = \dot{y}_C = \omega_2 \cdot AB \cdot \cos(\varphi_2) + \omega_1 \cdot BC \cdot \cos(\varphi_1) \tag{6}$$

The throwing angle is: $\alpha = \tan^{-1}\left(\dfrac{v_y}{v_x}\right)$.

It is well known that the trajectory of the ball is a parabola. In the reference system xOy shown in Figure 1, the equation of the ball trajectory is:

$$y = A \cdot x^2 + B \cdot x + C \tag{7}$$

where:
$$A = -\frac{g}{2 \cdot v_x^2} \;;\; B = \frac{g \cdot C_3}{v_x^2} + \frac{C_1}{v_x} \;;\; C = C_2 - \frac{g \cdot C_3^2}{2 \cdot v_x^2} - \frac{C_1 \cdot C_3}{v_x} \tag{8}$$

In relations (8), C_1, C_2 and C_3 are the following:

$$C_1 = v_y = \omega_2 \cdot AB \cdot \cos(\varphi_2) + \omega_1 \cdot BC \cdot \cos(\varphi_1) \tag{9}$$

$$C_2 = y_C = h_1 + AB \cdot \sin(\varphi_2) + BC \cdot \sin(\varphi_1) \tag{10}$$

$$C_3 = x_C = -L + AB \cdot \cos(\varphi_2) + BC \cdot \cos(\varphi_1) \tag{11}$$

The top of the ball trajectory (the parabola) is, in our case, the point U of the coordinates x_u and y_u.

$$x_u = -\frac{B}{2 \cdot A} \text{ and } y_u = -\frac{B^2 - 4 \cdot A \cdot C}{4 \cdot A} \tag{12}$$

In Figure 2 there is shown the ball when it reaches the basket. The segment D_p represents the D_m diameter of the ball, projected along the D_c diameter of the basket:

$$D_p = \frac{D_m}{\sin \alpha_E} \qquad (13)$$

where α_E is the falling angle of the ball through the basket, at point E (see Figure 1).

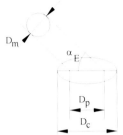

Fig. 2 Schematic showing ball entry to basket

Several conditions have to be met in order for the ball to pass through the basket. These conditions are as follows:

$$D_p < D_c \qquad (14)$$

$$-\frac{D_c - D_p}{2} < x_E < \frac{D_c - D_p}{2} \qquad (15)$$

$$x_u < x_E \text{ and } y_u > y_E \qquad (16)$$

where: $\qquad x_E = 0 \text{ and } y_E = H \qquad (17)$

From relations (14), (15) and (17) we can see that conditions (14) and (15) are equivalent.

From condition (16), $y_u > y_E = H$, and relations (8), (9), (10), (11) and (12) we may conclude that:

$$v_y^2 + 2 \cdot g \cdot y_c > 2 \cdot H \cdot g \Rightarrow v_y > \sqrt{2 \cdot g \cdot (H - y_c)} \qquad (18)$$

From condition (16), $x_u < x_E$, it results that $x_u = -\frac{B}{2 \cdot A} < x_E = 0$. But from relation (8) it can be observed that $A < 0$, thus B has to be $B < 0$. From relation (8), (11) and Figure 1, we can conclude that:

$$B = \frac{g \cdot x_c + v_x \cdot v_y}{v_x^2} < 0 \text{ , hence: } g \cdot x_c + v_x \cdot v_y < 0 \Rightarrow v_x < -\frac{g \cdot x_c}{v_y} \qquad (19)$$

By analyzing relations (18), (19) and Figure 1, it can be observed that x_c and y_c are the horizontal and vertical distances from which the ball is thrown toward the basket. In comparison to the dimensions of the basketball ground, the lengths of the player's arm and forearm may be neglected. Thus, the possibility of the ball to fall into the basket is more likely influenced by the initial speeds v_x and v_y, than by the initial position s x_c and y_c. Therefore, y_c could be approximated with the height h_1 of the player, and x_c could be approximated with the distance L (see Figure 1).

The falling angle α_E of the ball through the basket, at point E (see Figures 1 and 2), must then be $0 < \alpha_E < \frac{\pi}{2}$ and may be calculated as follows:

$$\tan(\alpha_E) = |2 \cdot A \cdot x_E + B| \qquad (20)$$

but $\qquad x_E = 0 \text{ and } \tan(\alpha_E) = |B| \qquad (21)$

By taking into account relations (8), (9), (11), (13), (14), and (21), it yields:

$$\left| \frac{g \cdot x_c + v_x \cdot v_y}{v_x^2} \right| > \tan\left[\sin^{-1}(\frac{D_m}{D_c}) \right] \qquad (22)$$

Thus, the velocities v_x and v_y have to meet the conditions (18), (19) and (22), in order for the ball to pass through the basket.
From relations (5) and (6), the angular velocities of the arm and forearm of the player are the following:

$$\omega_1 = \frac{v_y \cdot \sin\varphi_2 + v_x \cdot \cos\varphi_2}{BC \cdot \sin(\varphi_2 - \varphi_1)} \qquad (23)$$

$$\omega_2 = -\frac{v_x}{AB \cdot \sin\varphi_2} - \frac{\omega_1 \cdot BC \cdot \sin\varphi_1}{AB \cdot \sin\varphi_2} \qquad (24)$$

RESULTS: In Table 1 we present several cases of throws obtained for certain initial throwing values, with the following input data: H=3.05 (m), L=6.25 (m), h1=1.8 (m), D_c=0.45 (m), D_m=0.24 (m), g=9.81 (m/s^2) and the initial ball velocities given in Table 1.

Table 1. The assessment of the mathematical conditions for the throw

Crt. no.	v_y (ms^{-1})	Cond. (18)	v_x (m.s^{-1})	Cond. (19)	α_E (°)	α (°)	Cond. (14) $D_p < D_c$
1	6.00	true	3	true	78.261	63.43	0.24<0.45 true
2	6.50	true	4	true	65.624	58.39	0.26<0.45 true
3	7.00	true	5	true	46.465	54.46	0.33<0.45 true
4	7.50	true	6	true	24.376	51.34	0.58<0.45 false
5	8.00	true	7	true	6.1877	48.81	2.22<0.45 false
6	8.50	true	8	false	5.9653	46.73	2.30<0.45 false

The initial throwing speeds of the ball proposed in Table 1 are realistic, having similar magnitude with the speeds experimentally determined in paper (Budescu et al., 2005).

DISCUSSION: From Table 1 we can see that condition (19) is not fulfilled for the last throw, and that condition (14) is not fulfilled for the three last throws. Also, as the throwing angle decreases, the condition (14) is not fulfilled. The theoretical calculus concluded that:
- if the ball is thrown at an angle smaller than 40° or bigger than 50°, the conditions (14)-(17) are no longer fulfilled and the ball does not pass through the basket;
- condition (15) is the most restrictive one, so that, during the basketball practices, we should teach the player how to throw towards a fixed point;
- an acceptable throwing distance, taking into account the results of the study, is equal or smaller than 6.25 [m];
- the chance for a successful throw increases if the player jumps before throwing with a speed as big as possible.

CONCLUSIONS: The paper may be useful to sportsmen, trainers and researchers in biomechanics in sports, to determine the optimum angles of arms and throwing speed of the ball. The angles may be personalized for the sportsman's anthropometrical dimensions, to ensure the success of throwing at the basket. After getting to know these angles, we may adjust the corresponding mobile orthoses for the arm, used during trainings, especially for practicing the throw at the basket.

References:
Budescu, E., Merticaru, E., and Iacob, R. (2005). Biomechanical study regarding the success of the throw in basketball game, *Exercise & Society Journal of Sport Science, 13*, 100-101, Greece.

Budescu, E., Iacob, I. (2005). *Fundamentals of biomechanics in sports* (pp.153-159), SedcomLibris Press, Iasi, Romania.

Schroder, J., Bauer, C. (1996). *Basketball trainiren und spielen*, Rowohlt Taschenbuch Verlag GmbH, Germany.

ISBS 2009

Oral Session S06

Track & Field I

ANGULAR KINEMATICS IN ELITE RACE WALKING PERFORMANCE

Brian Hanley, Athanassios Bissas, and Andi Drake
Carnegie Research Institute, Leeds Metropolitan University, Leeds, UK

The purpose of this study was to measure and analyse the important angular kinematic variables in elite race walking. Research has shown that these variables include knee angle at contact and midstance, rotation of the hips and shoulders, and hip extension velocity. Eighty elite race walkers were videoed during competition and analysed using 3D-DLT with SIMI Motion. The knee angle was found to be almost straight at contact in most athletes and hyperextended by the vertical upright position. Athletes varied in the amount of rotation at the hips and shoulders, with 50 km men having greater hip rotation and 20 km women having greater shoulder rotation. There was much more variation in the values found for elbow and shoulder angles. Very few angular measurements correlated with key race walking variables such as speed, step length and cadence.

KEY WORDS: race walking, angular kinematics, gait, athletics.

INTRODUCTION: Angular kinematics are of particular importance in race walking. First, the hip angle will determine how far the foot is in front or behind the body. Second, the knee is the most important joint to analyse during race walking as it is the only joint which has specific technical rules applied to it. The rules state that it must be straightened from the moment of first contact with the ground until the vertical upright position (IAAF rule 230.1), although the definition of 'straightness' is unclear. For example, Knicker and Loch (1990) defined angles in a range between 175 and 185° as representing a straight knee. Although an extended knee is abnormal when used in normal walking or running, a straight knee at landing may be of benefit to the race walker (Murray *et al.*, 1983). Effective control of the pelvic girdle is important in increasing race walking speed (Hoga *et al.* 2000). This is partly because increased walking speed is achieved through decreasing support time by increasing hip extension velocity (Lafortune, Cochrane & Wright, 1989). Murray *et al.* (1983) also found that this increased hip extension velocity began during late swing so that momentum was increased prior to contact. The purpose of this study was to measure and analyse the important angular kinematic variables in elite race walking.

METHODS: Data collection: The 7th European Cup Race Walking was held at Leamington Spa (GBR) in May 2007. Video data of the men's 50 km race and the men's and women's 20 km races were collected. Two stationary cameras (Canon, Tokyo) were placed on one side of the course. The cameras were mounted on rigid tripods and placed at approximately 45° and 135° respectively to the plane of motion. The reference volume was 5 m long, 2 m wide, and 2.16 m high; this ensured data collection of at least three successive steps and provided a calibration reference for 3D-DLT. The sampling rate was 50 Hz and the shutter speed 1/500 s. In total, twenty-nine men were analysed in the 20 km race, twenty-one in the 50 km race, and thirty women in their race. For the 20 km men, the mean age was 27 yrs (± 5) and stature 1.80 m (± .06); for the 20 km women, mean age was 26 yrs (± 5) and stature 1.64 m (± .05); and for the 50 km men, mean age was 31 yrs (± 7) and stature 1.78 m (± .08). For the men's 20 km race, analysis occurred at 8.5 km. Because the women's 20 km field was not well spread out at this point, 13.5 km was chosen instead. In the 50 km race, the men were analysed at 28.5 km.

Data analysis: The video data were downloaded and digitised to obtain kinematic data using motion analysis software (SIMI, Munich). The recordings were filtered using a Butterworth 2nd order low-pass filter and De Leva's (1996) body segment parameter models for males and females was used. Pearson's product moment correlation coefficient was used to find associations within each group of athletes. One-way ANOVA was conducted to compare values between the 20 km men, 20 km women, and 50 km men, with post hoc pairwise comparisons using Bonferroni adjustments. An alpha level of 0.05 was selected with Greenhouse-Geisser correction if Mauchly's test for sphericity was violated.

RESULTS: The average values found for speed, step length, and cadence are shown in Table 1. Step length is also expressed as a percentage of stature. The 20 km men were the fastest group, followed by the 50 km men, and then the women. Step length was significantly longer for both groups of men compared to women ($p < .01$), while cadence was significantly higher for 20 km athletes compared to the 50 km athletes ($p < .01$). Step length was correlated with speed in both groups of 20 km athletes ($p < .01$) but not in the 50 km men ($p = .14$). However, speed was correlated with step length in all groups of walkers when it was expressed as a percentage of the participants' statures ($p < .01$).

Table 1 Speed, step length, and cadence (mean ± SD)

	Speed (km/hr)	Step length (m)	Step length (%)	Cadence (Hz)
20 km Women	13.29 (± .78)	1.08 (± .05)	66.1 (± 3.2)	3.41 (± .12)
20 km Men	14.80 (± .52)	1.23 (± .05)	68.4 (± 2.4)	3.35 (± .13)
50 km Men	14.14 (± .55)	1.22 (± .06)	68.4 (± 3.4)	3.23 (± .17)

The knee angle was calculated as the angle between the thigh and leg segments. In Table 2, average absolute straightness (180°) has not been achieved in any group at initial contact. At midstance, all groups had hyperextended knees, with women having significantly greater angles than 50 km men ($p < .01$) but not 20 km men. Three athletes showed slightly flexed knees at this point (the minimum was 177°). Toe-off angles for the knee were significantly lower for 50 km men than for the other two groups ($p < .01$). Swing phase knee angle was the amount of knee flexion during swing. The women had significantly more flexion than the male groups ($p < .01$). With regard to the knee angle, the main concern of the race walker is to have it straight from contact to midstance. However, it was also found that the knee contact angle was negatively correlated with contact time in both sets of men ($p < .01$) and hence also positively with cadence ($p < .05$). There was no such correlation in women ($p = .21$). The maximum knee flexion angles found during the swing phase were correlated with speed in the 20 km women's group only ($p < .05$). The knee angle at toe-off was positively correlated with step length in both the women's and 50 km men's groups ($p < .01$).

Table 2 Knee joint angles (mean ± SD)

	Initial contact (°)	Midstance (°)	Toe-off (°)	Swing phase (°)
20 km Women	178 (± 3)	189 (± 4)	158 (± 4)	99 (± 5)
20 km Men	178 (± 3)	188 (± 3)	156 (± 3)	103 (± 5)
50 km Men	180 (± 3)	185 (± 5)	149 (± 4)	102 (± 4)

In Table 3, the figures show an average of between 9 and 13° of hip flexion at contact, and between 7 and 13° of hyperextension at toe-off. There was a significant difference in hip contact angles between 20 km men and women, and between 20 km men and 50 km men ($p < .01$). At toe-off, the hip angle was significantly larger in women than in both groups of men, and larger in 20 km men than 50 km men ($p < .01$). The hip angle at toe-off was negatively correlated with cadence in 50 km men ($p < .05$). Similarly to hip and knee contact angles, the ankle contact angle was negatively correlated with contact time and positively correlated with cadence in the 20 km men ($p < .01$). There were correlations between the ankle contact angle and that of the knee (20 km women: $p < .01$; 20 km men: $p < .01$; 50 km men: $p = .08$). In effect, those athletes with the greatest amount of knee extension had larger ankle angles.

Table 3 Hip and ankle joint angles (mean ± SD)

	Hip		Ankle	
	Initial contact (°)	Toe-off (°)	Initial contact (°)	Toe-off (°)
20 km Women	169 (± 3)	193 (± 2)	109 (± 3)	127 (± 4)
20 km Men	167 (± 4)	190 (± 2)	107 (± 3)	123 (± 4)
50 km Men	171 (± 2)	187 (± 3)	103 (± 2)	124 (± 4)

The pelvic and shoulder rotation values are shown in Table 4. The hip-shoulder distortion angle is also shown. This is the maximum amount of torsion in the trunk caused by these rotations at a given instant. The value for pelvic rotation was significantly higher in both groups of men than in women ($p < .01$) and higher in 50 km men than in 20 km men ($p < .01$). Women's shoulder rotation was higher than in both groups of men ($p < .01$) but the men's groups were not significantly different. The distortion angle was significantly lower in women than in both groups, and in 20 km men compared to 50 km men ($p < .01$). The distortion angle was correlated with step length in the women ($p < .05$), but not in either the 20 km men's group ($p = .08$) or the 50 km men ($p = .21$). Although it might be expected that the angle of the hip joint was associated with the distortion angle, there were only correlations between hip contact angle ($p < .05$) and between hip toe-off angle and distortion angle ($p < .05$) in 20 km men. The hip contact angle was positively correlated and the toe-off angle negatively correlated.

Table 4 Rotation values for the hip and shoulders (mean ± SD)

	Pelvic rotation (°)	Shoulder rotation (°)	Distortion angle (°)
20 km Women	16 (± 4)	21 (± 4)	37 (± 5)
20 km Men	24 (± 3)	17 (± 3)	40 (± 4)
50 km Men	32 (± 4)	18 (± 3)	49 (± 4)

The values for the angles of the shoulder and elbow are shown in Table 5. The shoulder was considered to be 0° in the anatomical standing position. The figures show between 68 and 77° of ipsilateral shoulder hyperextension at contact, and between 37 and 41° of flexion at toe-off. Whereas the standard deviations for the lower limb and rotation angles ranged between 2 and 5°, there was a much greater degree of variation in the values for these angles. 50 km men had a significantly greater angle at the shoulder at contact (and the elbow at toe-off) compared to both sets of 20 km walkers ($p < .01$). At contact, the elbow angle was significantly lower in women than in either group of men ($p < .01$), although the male groups were not different from each other. There were no correlations found between elbow angles and any other variables in any group. However, the shoulder contact angle was found to be positively associated with step length in the women's group ($p < .01$) and there was a tendency towards significance in the 50 km men ($p = .09$). The shoulder toe-off angle was also associated with step length in both these groups (women $p < .01$; 50 km men $p < .05$).

Table 5 Shoulder and elbow joint angles (mean ± SD)

	Shoulder at contact (°)	Shoulder at toe-off (°)	Elbow at contact (°)	Elbow at toe-off (°)
20 km Women	-69 (± 7)	41 (± 7)	71 (± 9)	65 (± 7)
20 km Men	-68 (± 7)	37 (± 5)	77 (± 8)	67 (± 5)
50 km Men	-77 (± 7)	38 (± 5)	82 (± 10)	73 (± 9)

DISCUSSION: The aim of this study was to measure and analyse the important angular kinematic variables in elite race walking. Differences were found between men and women, and between men competing over 20 km and 50 km. There were very few correlations between any angular values and the main performance variables within any single group and it is therefore difficult to ascertain which angles were particularly important. It may be that as all athletes analysed were elite, the differences that allow particular athletes to perform better are more subtle. Analysis of the angular velocities and accelerations may provide more useful information.

There were no differences in the angle of the knee at initial contact between groups, which is not surprising as athletes attempt to straighten the knee at this point. While on average the knee was not entirely extended in 20 km athletes, it may be that the knees (measured slightly below 180°) were considered straight when judged. The knee was hyperextended in all groups by midstance, as the knee was kept straightened. That only three of the eighty athletes had slight flexion at this point shows that this rule is predominantly adhered to. The correlation between knee extension angle at contact and cadence in 20 km men shows that a straight knee is not necessarily disadvantageous. However, it may have been simply the case that the walkers with the highest cadences were also the best athletes technically.

The amount of hip rotation (to the left and right) measured in women was significantly lower than in either group of men. This may be due to a lack of muscular strength in the abdominal, pelvic, and hip muscles responsible for this movement. In contrast, the women had the highest amount of shoulder rotation in counterbalancing pelvic rotation. As women had the greatest amount of hip joint hyperextension at toe-off, it is possible that this compensated for their lower levels of pelvic rotation in generating step length. The distortion angle created by the opposing movements of the pelvis and shoulders is partially what gives race walkers their distinctive gait. 50 km men had the largest amount of distortion. It may be that it is through this movement that 50km men achieve their great step lengths, as opposed to achieving them by way of the longer flight times associated with 20 km walkers.

A much greater range of values was found for the elbow and shoulder angles compared to the lower limbs. The standard deviations found for the elbow were quite large. Coaches have recommended elbow angles of approximately 90° during race walking (e.g. Markham, 1989). No group of athletes reached this value on average at either contact or toe-off, although some individuals did. Both sets of 20 km athletes had particularly low elbow angles at toe-off. It may be that these angles are of little importance or are neglected by athletes.

CONCLUSION: Most athletes in this study did conform to the straight knee rule as defined by IAAF rule 230.1. There was a great deal of variation between groups of athletes (and the individual athletes themselves) that suggests that no particular technique is optimal. Basing technical models of walking using particular walkers is unwise in developing young athletes. Recommendations for future studies include analysis of junior athletes and the changes in angular kinematics with the onset of fatigue.

REFERENCES:

Hoga, K., Ae, M., Enomoto, Y., and Fujii, N. (2003). Mechanical energy flow in the recovery leg of elite race walkers. *Sports Biomechanics*, 2(1), 1-13

Knicker, A. and Loch, M. (1990). Race walking technique and judging – the final report to the International Athletic Foundation research project. *New Studies in Athletics*, 5(3), 25-38

Lafortune, M., Cochrane, A., and Wright, A. (1989). Selected biomechanical parameters of race walking. *Excel*, 5(3), 15-17

Markham, P. (1989). *Race Walking.* Birmingham: British Amateur Athletics Board

Murray, M. P., Guten, G. N., Mollinger, L. A., and Gardner, G. M. (1983). Kinematic and electromyographic patterns of Olympic race walkers. *The American Journal of Sports Medicine*, 11(2), 68-74

3D KINEMATICAL ANALYSIS OF BRAZILIAN FEMALE POLE VAULTERS

Tiago G. Russomanno, and Nathalia P. Rossetto, Ricardo M. L. Barros
Faculty of Physical Education, University of Campinas, Campinas, Brazil

KEY WORDS: pole vault, athletics, kinematics

INTRODUCTION: Pole Vault is one of the most technical events in track and field. Schade et al. (2005) evaluated kinematical and dynamic variables during the 2005 World Championship in Helsinki providing data of world class vaulters for comparison. Recently, Brazilian pole vaulters have obtained world class results but, despite this, no biomechanical analysis has been conducted with such athletes. Therefore, the aim of this work is the 3D kinematical analysis of the best female athletes during the "XXVII Brazilian Trophy" in 2008.

METHODS: Five female vaulters with jumps ranging from 3.90 to 4.80 meters were analysed, totalling 13 jumps. The Dvideo kinematic analysis system (Figueroa et. al., 2003) was used for 3D kinematical analysis. Four Basler cameras (A602fc,100Hz) were located in the field beside the runaway corridor, 2 cameras to record the approach and maximum pole bend position and two others to record the bar clearance. The follow variables were analyzed: Grip height at the takeoff (GHTO), takeoff distance (DTO), Grip Height (GH) defined as the distance between the middle of the upper grip hand at the pole and the deepest point of the planting box at the moment of the pole straight position, Pole chord length at maximum pole bend (MPB) defined as the distance between the middle of the upper grip hand at the pole and the deepest point of the planting box at MPB, % MPB that is (1-(MPB/GH)) * 100. The results were compared to those presented by Schade et al. (2005) on the final report of the 2005 World Championship. A statistical test (Mann-Whitney $p < 0.05$) was done to compare the average height jumped in both events.

RESULTS: Table 1 shows the kinematical data of the three best pole vaults of Brazilian Trophy in 2008 and the data of the winner of the 2005 World Championship in Helsinki.

Table 1 Kinematical variables of the vaulters

Vaulter	Height (m)	GHTO(m)	DTO(m)	MPB(m)	GH (m)	%MPB
1st Fabiana Murer 2008	4.80	2.13	3.84	3.21	4.30	25.4
2nd Joana Costa 2008	4.35	2.11	3.26	3.28	4.20	21.9
3rd Carolina Torres 2008	4.15	2.11	3.16	3.50	4.12	15
1st Yelena Isibayeva 2005	5.01	#	3.41	2.98	4.37	31.7

The overall mean value of pole vault in the World Championship in 2005 (N=10) was 4.47m ± 0.22m and in Brazilian Trophy in 2008 (N=13) was 4.31m ± 0.24m with no statistical difference (P<0.05). Comparing the kinematical data of the winner of the World Championship (5.01m) with data of the winner of the Brazilian Trophy (4.80m), in both cases, the best results occurred when they achieved the best use of the pole %MPB.

DISCUSSION: According to Schade, the more effective the interaction with the pole the better is the performance. Considering our overall results, this conclusion was confirmed with just one exception. However, this relationship should be carefully interpreted since the pole stiffness could interfere with the results and this factor was not controlled in both studies.

CONCLUSION: The present study show original data obtained of Brazilian female pole vaulters and these kinematical variables can be useful for academic and applied purposes.

REFERENCES: Schade, F., Brüggemann, G.-P, Isoletho, J., Komi, P., Arampatzis, A., Pole vault at the world championship in athletics Helsinski 2005. Final Report

Acknowledgement: Research supported by FAPESP (00/01293-1), CNPq (451878/2005-1; 309245/2006-0; 473729/2008-3) and PRODOC-CAPES (0131/05-9), CBAt and COB.

HIGH SPEED ULTRASONIC DETECTION SCHEME FOR SPORTS PERFORMANCE MONITORING

M. Zakir Hossain and Wolfgang Grill

Institute of Experimental Physics II, University of Leipzig
Linnéstr. 5, D-04103, Leipzig, Germany

To observe muscle performance of athletes with high resolution a novel ultrasonic detection scheme has been developed. It is based on bulk waves passing the monitored muscle. The detection is obtained along a line between two acoustic transducers with similar size and shape as stick-on electrodes, mounted on the skin. The time-of-flight from which all the data is derived is observed with the aid of a computer controlled arbitrary function generator and a synchronized transient recorder. An available separate channel can be used for synchronous monitoring of the force or pressure or the EMG-signals. The demonstrated movement and time resolution is ± 0.02 mm and 0.01 ms respectively. The equipment of lap-top size is battery operated and suitable for on-field monitoring.

KEY WORDS: Time resolved detection, ultrasonic detection scheme, quantification of sports performance, non-invasive monitoring of muscle performance.

INTRODUCTION: The high-level athletic performance achievement is increasingly associated with careful monitoring of all key dynamic and metabolic functions, both during the effort and during the recuperation phases. Novel ultrasonic detection schemes for motion monitoring have been reported lately. These include non contact monitoring by airborne ultrasonics (Barany, L. P. 1993) operated in reflection to detect the vertical component of any motion of the body surface (skin). Furthermore ultrasonic waves travelling at or near the surface have been used to monitor the dilatation of the skin (Juergen, 1993). Detection in the volume of the body has been realized for respiration related movements (Friedrichs, A.,Voegeli, F. 2006) with a demonstrated temporal resolution of about 80 ms.

Lately we have contributed to these developments with a detection scheme suitable for monitoring of muscle dynamics with a demonstrated temporal resolution down to 0.01 ms (Hossain, 2008). The system has also been used to monitor the sonic velocity under voluntarily activated muscle (Hossain, 2009). The observed variations of the velocity below 0.1% justify the here employed scheme to detect the muscle extension by variations of the observed time of flight (TOF). Monitoring is based on chirped ultrasonic wave-trains passing through the observed muscle. The TOF from which all the data is derived, is observed with the aid of a computer controlled arbitrary function generator (AFG) and a synchronized transient recorder (TR). The implemented software allows rapid and objective quantitative determination of relevant training parameters for evaluation and support of the optimization of the training procedures. The aim of this paper is to demonstrate the developed schemes including applications where the high temporal resolution is essential. Empirical fitting has been implemented to derive parameters demonstrating the capability to deliver quantitative and objective information to assess the monitored athlete's reaction time, muscle performance and energy expenditure.

METHODS: For the demonstrated measurements chirped pulses were employed to allow signal compression following the detection with the TR. The Lab-view based software facilitates selection of appropriate chirp signals, repetition rates and averaging during monitoring to enhance signal to noise ratio prior to processing.The athlete was instructed to isometrically contract the monitored gastrocnemius muscle with maximum possible effort and to hold the maximal contraction as long as possible and then suddenly cease the effort. The contraction was initiated by visual sensing of a thrown ball (Figure 1, left; enclosed in the small circle) at unpredictable time to determine the reaction time. To ensure the maximal isometric contraction, the knee joint was flexed at about 90° while sitting on a chair. To keep the upper body in fixed position, the athlete was instructed to grasp the chair from below with both hands (figure 1, left). Monitoring and evaluation was performed by custom developed LabVIEW software. Empirical curve fitting has been employed for data analysis.

 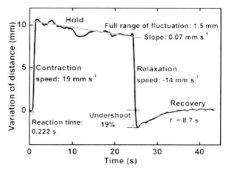

Figure 1. The image (left) shows the arrangement and data acquisition procedure of the continuous ultrasonic monitoring of the gastrocnemius muscle of an athlete and the graph (right) represents the results of the monitoring and data analysis.

RESULTS: From respective readings and fits as illustrated in Figure 1 (right) the reaction time was found to be 222 ms, the hold time 22.62 s, holding variations are within 1.5 mm and the slope of the observed drop is 0.07 mm s^{-1}. For the monitored muscle a contraction and relaxation speed of 19 mm s^{-1} and 14 mm s^{-1} respectively are determined. A fit to the recovery phase allows the determination of a time constant (τ) of 8.7 s for recovery from the initiated action which is a quantitative measure for the ability to recover from post isometric tetanus effect. The observed post isometric 19% undershoot represents also an example for a scientifically relevant result related to post isometric stretch (Brenner, 1990; Alter, 1996).

Contraction and relaxation data has been evaluated for a range of 2.25 s (Figure 2) together with the second order derivative to characterise the dynamics of the monitored muscle. For contraction three different stages, namely acceleration, constant speed and deceleration, are identified whereas for the muscle relaxation phase only two different stages, acceleration and deceleration, are observed and quantified.

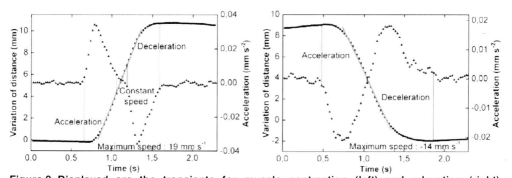

Figure 2. Displayed are the transients for muscle contraction (left) and relaxation (right) together with a second order derivative to evaluate the dynamics of the gastrocnemius muscle. The interpretation of the different phases is indicated.

Comparative study of pre and post physical loading session:

Studies have been conducted on maximal isometric contraction of the gastrocnemius muscle performance of five hockey players with an average age (18.4 ± 0.55) years, height (165.32 ± 1.55) cm, weight (60.38 ± 4.67) kg, and body mass index (BMI) 22.08 ± 1.5. Data have been recorded to observe the effect of pre and post exercise loading on the athletes. The session was organised as follows: gradual increase of speed from walking to running up to a speed of 12 km h^{-1} for 3 min after that 15 min of continuous running on a treadmill at zero grade Just after finishing the physical loading the monitoring was repeated. The athlete's total reaction time and the muscle dynamics were evaluated. The reaction was initiated with a thrown ball (figure 1, left; enclosed in the small circle).The time dependent lateral extension of the monitored gastrocnemius muscle is displayed for one of the athletes in figure 3.

The lateral extension of the muscle with elapsed time displays a sharp raise at the initial inflection point. The time from the visual stimulation to the onset of muscle movement of the monitored athlete, that is total reaction time for pre and post physical loading, is found to be 140 ms and 222 ms respectively. The following time span up to the second inflection point of the sharp rising curve represents the muscle contraction phase. A contraction speed of 5.3 m s^{-1} and 9.8 m s^{-1} is derived from the respective slope of the pre and post loading curve respectively. This is followed up to the third inflection point by a comparatively stable slope with a value of 0.001 mm s^{-1} and -0.12 mm s^{-1}, which quantify the respective holding phases.

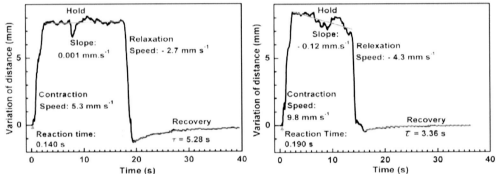

Figure 3. Monitoring for pre (left) and post (right) physical loading session. Displayed is the time dependence of the lateral movement of the gastrocnemius muscle.

The holding time was found to be 18 s and 16 s respectively.The falling slope between the third and fourth inflection points represents the relaxation phase. The muscle relaxation speed is -2.7 mm s^{-1} and -4.3 mm s^{-1} obtained from the respective slopes. Undershoots of 20% and 12% and recovery time constants of 5.28 s and 3.36 s are observed.

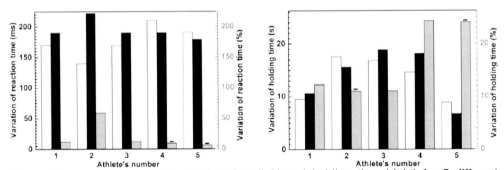

Figure 4. Pre and post exercise reaction time (left) and holding time (right) for 5 different athletes prior (light grey) and post (black) loading with respective variations (in %, right scales) indicated in dark grey. A minus sign above the bar indicates negative deviations.

The reaction times for the 5 monitored athletes are displayed in figure 4 (left). Only athlete number 2 shows a significant variation of the reaction time. The quantification of the holding time (Figure 4, right) for maximal isometric contraction depicts that athlete 1, 3 and 4 hold longer and 2 and 5 hold shorter after standard loading. Variations reach from ±10% to ±20%.

DISCUSSION: The system has been used for temporally resolved monitoring of the muscle performance during isometric movement of the muscle. The values of the reaction time obtained from the monitoring (140 to 222 ms) comply with reported reaction times to visual stimulus (Fieandt et al., 1956; Brebner and Welford, 1980). The developed ultrasonic monitoring scheme is suitable for isometric contraction which cannot easily be achieved by high speed camera monitoring. Different to EMG monitoring the additional delay from the nerve signal to the actual movement is included and can even be observed by synchronized detection of EMG signals. The result of the comparative study indicates with respect to the reaction time that the athlete number 2 is not well trained, since the exhaustion delayed the reaction after exercise significantly. Even though the holding times show a rather large variation for the different athletes, it is clearly indicated that athlete number 5 is not conditioned for endurance since the initially observed already comparatively short holding time dropped significantly by 20% after exercise.

CONCLUSION: The developed monitoring scheme has been applied for monitoring of the gastrocnemius muscle performance of athlete during isometric and isotonic contraction. It has been demonstrated that a quantitative analysis of the different stages of muscle contraction can be achieved. The comparative study results illustrate the possibility to obtain unbiased quantitative parameters to evaluate the performance of athletes. Empirical fitting procedures based on mathematical schemes have been employed for data analysis. Novel accessible parameters like muscle contraction speed, relaxation speed, slope and steadiness of the holding phase, undershoot due to the tetanus effect followed by a recovery. The respective results demonstrate that sports specific quantitative results can be derived from the data. The developed scheme provides a so far not reported temporal resolution.

REFERENCES:

Fieandt, K., Huhtala, A., Kullber, P., and Saarl, K.(1956). Personal tempo and phenomenal time at different age levels. Reports from the Psychological Institute, No. 2, University of Helsinki.

Welford, A. T. (1980). Choice reaction time: Basic concepts. In: A. T. Welford (Ed.), Reaction Times (pp. 73-128). Academic Press, New York.

Brebner, J. T. (1980). Reaction time in personality theory. In: A. T. Welford (Ed.), Reaction Times (pp. 309-320). Academic Press, New York.

Brenner, B. (1990). Muscle mechanics and biochemical kinetics. In: Squire JM (ed) Molecular mechanisms of muscular contraction (pp 77-149). MacMillan Press, London.

Barani, L. P. (1993). Ultrasonic non-contact motion monitoring system. Patent: US 5,220,922.

Juergen, M. (1993). Method for measurement of posture and body movements. Patent: DE 4214523.

Alter, M. J. (1996). Science of Flexibility, Third edition (pp 19-87). Human Kinetics.

Friedrichs, A., Voegeli, F. (2006). Device and method for producing respiration-related data. Patent: US 7,041,062 B2.

Zakir Hossain, M., Twerdowski, E., and Grill, W. (2008). High speed ultrasound monitoring in the field of sports biomechanics. Health monitoring of structural and biological systems, SPIE 6935-70.

Zakir Hossain, M., Voigt H., and Grill, W. (2009). Monitoring of variations in the speed of sound in contrac-ting and relaxing muscle. Health monitoring of structural and biological systems, SPIE 7295-16.

Acknowledgement
The development of the hard- and software employed here has been supported by ASI Analog Speed Instruments GmbH.

ELECTROMYOGRAPHIC ANALYSIS ON LOWER EXTREMITY MUSCLES DURING OVERGROUND AND TREADMILL RUNNING

Mandy Man-Ling Chung, Kam-Ming Mok, Justin Wai-Yuk Lee, Youlian Hong

Human Movement Laboratory, Department of Sports Science and Physical Education, Faculty of Education, The Chinese University of Hong Kong, Hong Kong SAR, China

The goal of this study was to compare treadmill running with overground running, so as to investigate if the shoe testing and training on treadmill can reflect the overground performance. Thirteen Chinese male subjects were instructed to run on the four conditions (1.Treadmill, 2.Tartan, 3.Grass, 4.Concrete). Comparisons between running conditions were made for muscle activity. The electromyography (EMG) signals from rectus femoris, tibialis anterior, biceps fermoris and medial gastrocnemius in one stride were evaluated. Results of the root mean square of EMG signal in rectus femoris, biceps femoris and gastrocnemius were found to be significantly different between overground and treadmill running during stance phase and the time point of toe off.

KEY WORDS: EMG, electromyography, muscle activity, shoe testing.

INTRODUCTION: Treadmill has been a common instrument in the research field in which different investigations were evaluated under a speed-controlled running condition. Many running scientific investigations, including running gait pattern at different speeds and sports shoes research, have been performed on a treadmill (Frishberg 1983, Nigg 1995, Wank 1998, Gazendam and Hof 2007, van Ingen Schenau 1980). However, from the biomechanical point of view, whether the data collected on a treadmill can be applied to overground surface is still unknown. Previous studies on different surfaces running have focused primarily on the kinematics analysis and limited information about muscle activity pattern had been presented. With muscle activity parameters taken into account, it can provide more insight into the validity of treadmill running to reflect the situation in overground running.

METHODS: Thirteen male subjects (age = 22.4 ± 3.9 years, body mass = 63.6 ± 9.2 kg, body height = 170.6 ± 6.2 cm) with no known running gait disorders were recruited in the study. They were all heel-toe runners and familiar with treadmill running with shoe size 40 (EURO Standard). A pair of running shoes (TN 600, ASICS, Japan) were provided for them during testing. Each subject was requested to perform running on each of the four situations with a controlled speed of 3.8 ms^{-1} on treadmill (6300HR, SportsArt, US) and the acceptable range was 3.6 ~ 4.0 ms^{-1} on overground surfaces with the use of light gate monitor. Figure 1 showed the experimental set up of the overground on field.

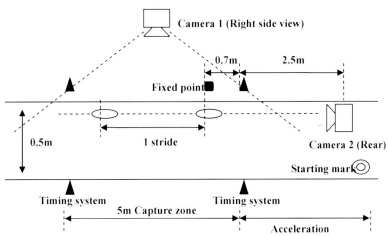

Figure 1: Experimental set up of overground surface on field

Four lower extremity muscles including rectus femoris, tibialis anterior, biceps femoris and gastrocnemius were tested. Each EMG signal was magnitude normalized to maximum magnitude recorded in reference activity. Four controlled reference postures were selected for normalizing each muscle group respectively (Fong, Hong and Li 2008). They were then trimmed into four phases and time normalized to 0-100% stance or swing phase. Band pass filter of 20-500Hz and full wave rectification were applied. An One-way repeated measure ANOVA was used to compare the value of four different running conditions, and followed by Post-hoc Tukey tests.

RESULTS: Significant differences were found on the stance phase and toe off of the stride.

Table 1 Mean and SD of muscle activity parameters (magnitude normalized to reference contraction)

Mean and SD of muscle activity parameters	Treadmill	Tartan	Grass	Concrete	p value	Tukey
Rectus Femoris TO	1.839 (0.540)	14.576 (8.706)	7.003 (5.584)	8.477 (2.990)	< 0.05	(T-Ta)**
Gastrocnemius TO	8.534 (5.950)	39.011 (16.098)	19.091 (5.013)	21.200 (0.278)	< 0.05	(T-Ta)**
Rectus Femoris Phase 1	3.673 (2.311)	21.279 (7.611)	15.432 (4.541)	24.729 (13.011)	< 0.01	(T-Ta)*, (T-G)*, (T-C)*, (G-C)**
Biceps Femoris Phase 1	4.776 (2.791)	13.310 (7.165)	9.854 (6.210)	12.783 (12.594)	< 0.05	(T-Ta)**, (T-G)**, (T-C)**, (G-C)**

T:Treadmill, Ta:Tartan, G:Grass, C:Concrete

Phase 1:Stance phase, Phase 2:Early swing, Phase 3:Middle swing, Phase 4:Late swing, TO:Toe off

*$p < 0.01$, **$p < 0.05$

Throughout stance, EMG in rectus femoris (P < 0.01) and biceps femoris (P < 0.05) were greater on overground compared with the treadmill. Also, EMG in rectus femoris (P < 0.05) and biceps femoris (P < 0.05) were greater running on concrete compared with grass. At toe off (TO), the EMG in rectus femoris (P < 0.05) and gastrocnemius (P < 0.05) were greater running on Tartan compared with treadmill. Significant differences found in mean and SD of muscle activity was showed in Table 1.

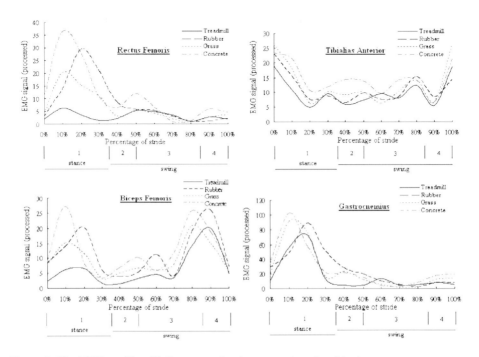

Figure 2: The EMG profile with the normalized average for all subjects

DISCUSSION: At toe off, the muscle activity of rectus femoris and gastrocnemius were lower when running on treadmill compared with tartan surface. In treadmill running, the belt is moving backward continuously, this can be treated as a slightly slippery surface. When we are running on wet or slippery surface, it is easier to slide and more attention is needed. We try to run with smaller steps with higher frequency. The kinematics finding in previous studies (Schache 2001, Wank 1998) reported that the stride length and stride frequency were found to be shorter and higher respectively in treadmill when compared to overground. Therefore, there is not enough time for runner to extend the knee and plantar-flex the ankle. As a result, lower muscle activity of rectus femoris and gastrocnemius were recorded on treadmill. (Fong et al. 2008) showed that when the leg was in fully extended situation, it would be difficult for a recovery action in case of slip. Putting the knee in a slightly flexed position could be a strategy to prepare for correction on slippery surfaces. As the muscle activity rate on treadmill running is lower than that on overground at toe off, it is proposed that the drag force applied by runner at toe off is smaller on treadmill than overground. It is promising to investigate the drag force when running on treadmill and overground comparing with overground.

During stance phase, the muscle activity of rectus femoris and biceps femoris showed lower magnitude on treadmill running as compared with overground running. In treadmill running, there is no need to move the body forward continuously, not much energy is needed to provide forward centre of gravity of the body (CG) movement comparing with overground running during the heel touch down to toe off period. With lower muscle activity rates on treadmill than overground running during stance phase, it is proposed that the foot sole pressure should be different on treadmill and overground running. It is promising to evaluate the sole pressure distribution on treadmill and overground running.

CONCLUSION: The results obtained showed that the muscle activity reacted significant differently between treadmill and overground running. The treadmill running results for any research design should be checked carefully when applying to other overground surfaces. With different muscle activity rates on different running surfaces, it is important to evaluate the training effects as well. Moreover, we proposed that drag force provided by the running shoe sole of treadmill testing shoe may differ from the overground one. The sole pressure distribution may also be different between treadmill and overground surfaces. The drag force and pressure cushioning effects were not reflected truly when treadmill running was used for running shoe testing, sole material selection and sole design.

REFERENCES:
Fong, D.T.P., Hong, Y. and Li, J.X. (2008). Lower extremity preventive measures to slips – joint moments and myoeletric analysis. *Ergonomics*, 51, 1830-1846.
Frishberg, B.A. (1983). An analysis of overground and treadmill sprinting. *Medicine & Science in Sports & Exercise*, 15, 478-485.
Gazendam, M.G.J. and Hof, A.L. (2007). Averaged EMG profiles on jogging and running at different speeds, *Gait and Posture*, 25, 604-614.
Nigg, B.M., De Boer, R.W. and Fisher, V. (1995). A kinematic comparison of overground and treadmill running. *Medicine and Science in Sports and Exercise*, 27(1), 98-105.
Schache, A.G., Blanch, P.D., Rath, D.A., Wrigley, T.V., Starr, R. and Bennell, K.L. (2001). A comparison of overground and treadmill running for measuring the three-dimensional kinematics of the lumbo-pelvic-hip complex. *Clinical Biomechanics*, 16, 667-680.
Van Ingen Schenau, G.J. (1980). Some fundamental aspects of the biomeachanics of overground versus treadmill locomotion. *Medicine and Science in Sports and Exercise*, 12, 257-261.
Wank, V., Frick, U. and Schmidtbleicherm, D. (1998). Kinematics and Electromyography of Lower Limb Muscles in Overground and Treadmill Running, *International Journal of Sports Medicine*, 19, 455-461.

ISBS 2009

Oral Session S08

Equipment

WEDGED FOOTWEAR PERTURBATIONS AFFECT LOWER EXTREMITY COORDINATION DYNAMICS

Elizabeth Russell[1], Allison Gruber[1], Kristian O'Connor[2], Richard Van Emmerik[1], Joseph Hamill[1]

Biomechanics Laboratory, University of Massachusetts, Amherst, MA U.S.A.[1]
Human Performance Laboratory, University of Wisconsin, Milwaukee, WI, USA[2]

The purpose of this study was to investigate the coordinative changes that occur with a footwear perturbation consisting of a neutral shoe and varus and valgus wedged shoes. This type of footwear is often prescribed for clinical use. Lower extremity kinematics were collected as six male subjects ran overground at 3.6 m·s^{-1}±5%. A modified vector coding technique assessed coordination between rearfoot motion and leg rotation. It was determined that there were clinically relevant differences between the footwear during the middle and late stance period. The differences were most evident between the varus and valgus conditions. However, the varus condition was closer in coordination structure to the neutral condition. The difference in coordination during the wedged conditions indicated that the valgus wedge perturbation may have implications in producing soft tissue injury.

KEY WORDS: coordination, vector coding, wedged shoes, running

INTRODUCTION: Wedged shoes, orthotics and insoles are commonly prescribed to runners and clinical populations to treat injury or pathology in the lower extremity. Altering rearfoot motion can adjust the kinematics farther up the chain in the limb (Bates et al., 1978). Examining the coordinated motion of the lower extremity segments in space and time gives insights that traditional time series plots of segment motion cannot. The purpose of this study was to examine how a wedged shoe intervention influenced coordination patterns between the rearfoot and leg. We hypothesized that the varus and valgus wedged footwear would result in coordinative patterns that were significantly different from each other and both would be different from the neutral shoe.

METHODS: Six healthy male subjects participated in this study. The mean age, height and body mass of the subjects was 31.2±4.9 years years, 177.3±5.99 cm and 77.2±10.3 kg respectively. All subjects gave written, informed consent to participate in the study, were injury-free and had normal values for pronation during walking and running. Subjects wore custom shoes with sole wedges made of EVA. Neutral, 8° varus and 8° valgus shoe wedges ran the entire length of the sole (see Figure 1).

Figure 1: Illustration of shoe sole wedges for the right shoe from a posterior view.

Lower extremity kinematics were captured (240 Hz) as subjects ran overground across a force platform (1200 Hz) at 3.6 m·s^{-1}±5%. Five trials were collected in each shoe condition. The vertical ground reaction force component was used to identify heel contact and toe-off. Kinematic data were low-pass filtered (8 Hz) and interpolated to 101 data points in Visual 3D. Rearfoot eversion/inversion and leg internal/external rotation segment angles were calculated in the global coordinate system and scaled relative to the standing posture.

A modified vector coding approach assessed the coordination between the segments. Angle-angle plots of the leg (x) relative to the rearfoot (y) were constructed. Inter-segment coordination was inferred from the vector angle (θ_i) between adjacent points relative to the right horizontal (Sparrow et al., 1987; Heiderscheit et al., 2002).

$$\theta_i = \left| \tan^{-1} \frac{(y_{i+1} - y_i)}{(x_{i+1} - x_i)} \right|, \text{ where } \theta_i \text{ is the coupling angle}$$

Due to the directional nature of the vector angles, each trial's coupling angles were calculated from the mean horizontal and vertical components during each 1% of stance and averaged using circular statistics (Batschelet, 1981). Angles were grouped into bins to identify the range of specific coordinative patterns the segments could undergo (Table 1).

Table 1: Coordination patterns defined by coupling angle ranges

Coordination pattern	Coupling angle definitions	
Proximal	$157.5° < \gamma \leq 202.5°$	$337.5° < \gamma \leq 22.5°$
Distal	$67.5° < \gamma \leq 112.5°$	$247.5° < \gamma \leq 292.5°$
In-phase	$22.5° < \gamma \leq 67.5°$	$202.5° < \gamma \leq 247.5°$
Anti-phase	$112.5° < \gamma \leq 157.5°$	$292.5° < \gamma \leq 337.5°$

Vector angles classified as in-phase indicated the same degree of rearfoot eversion rotation as leg internal rotation. Anti-phase coordination indicated the same amount of opposing rotations of rearfoot eversion/inversion with leg external/internal rotation. Distal coordination indicated rearfoot motion relative to a stationary leg segment and proximal indicates leg rotation relative to a stationary rearfoot segment (Figure 2).

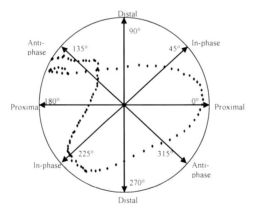

Figure 2: An exemplar angle-angle plot of hip and knee joint motion during the stance phase of a single trial. A polar plot is overlaid to illustrate the four different coordinative patterns.

Circular statistics were used to calculate mean coupling angles during each third of stance (Batschelet, 1981) and effect sizes assessed differences between coordination patterns in the three shoe conditions (Cohen, 1991).

RESULTS and DISCUSSION: The purpose of this study was to examine coordinative patterns between the rearfoot and leg as a result of a wedged shoe perturbation. Rearfoot segment angles (Figure 3) resulted in greater differences than leg angle among the conditions. For example, the valgus wedge had lesser peak eversion, when referenced to a standing position, than the varus wedge. The valgus wedge also had greater range of rearfoot motion than the neutral or varus wedge.

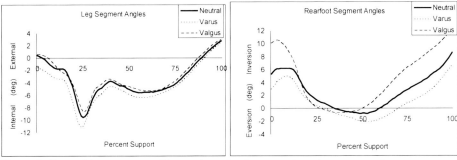

Figure 3: Ensemble averages of leg and rearfoot segment angles across stance.

A vector coding approach assessed four different coordination patterns during the stance phase of running. Coordination patterns differed between the three shoe conditions with the valgus wedge having a greater impact on the coordination dynamics then the varus wedge. Figure 4 shows the mean coupling angles from heel stride to toe-off.

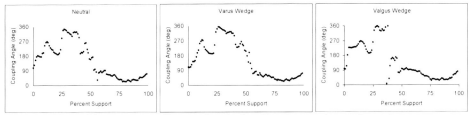

Figure 4: Ensemble mean coupling angles during the stance phase of running.

On heel-strike, a brief period of distal coordination predominated when the rearfoot everted relative to a stationary leg segment. This lasted until the foot was flat on the running surface. Subsequently, the segments shifted towards more in-phase pattern as the rearfoot began to evert while the leg internally rotated. For a brief period during early-mid stance, the leg was externally rotated while the rearfoot continued to evert; thus an anti-phase coordination is shown at approximately 25% of stance. At toe-off, the rearfoot inverted closely followed by leg external rotation; in-phase coordination is shown during the last 50% of stance. The differences between the two wedged insole conditions was particularly pronounced during mid and late stance. Coupling angles during early, mid and late stance are presented in Table 2.

The coordination was in-phase during early stance. During mid-stance, proximal segment coordination predominated in the neutral and varus wedged shoes, but a mean anti-phase coordination was present in the valgus wedge, suggesting opposing rotations of the two segments. During late stance, the segments were in-phase as the leg externally rotated with rearfoot inversion. The valgus wedge imposed a greater perturbation to the lower extremity rearfoot-leg dynamics than the varus wedge, particularly in mid and late stance.

Table 2: Mean vector coding angles across tertiles of stance. Numbers indicate differences from other shoe conditions. Neutral = 1; Varus wedge = 2; Valgus wedge = 3. Plain text numbers indicate an effect size of 0.5 or higher, suggesting a moderate effect of shoe condition. Bold numbers indicate an effect size of 0.7 or higher, suggesting a large effect of shoe condition.

	Early Stance	Mid Stance	Late Stance
Neutral (1)	231.2°	180.6°	42.7° [3]
Varus (2)	232.1°	190.9° [3]	42.4° [3]
Valgus (3)	242.3°	136.9° [2]	45.7° [1,2]

During mid-stance, there was a difference in the timing of the initiation of rearfoot inversion between the varus and valgus wedged shoes. There was, however, no difference in timing of initiation of leg external rotation.

CONCLUSIONS: Despite the same degree of perturbation through wedging, a valgus wedge had a greater impact on the rearfoot-leg coordination than a varus wedge. These differences were greatest in the latter portions of stance. The differences in rearfoot range of motion in late stance and lack of differences in leg range of motion among the three shoe conditions accounts for the differences in coordination just prior to toe-off. The opposing rotations shown during mid-stance in the valgus wedge may have important implications for soft-tissue or ligament injury at the ankle joint or in other joints of the lower extremity.

REFERENCES:
Bates B.T., James, S.L., and Osternig, L.R. (1978). Foot function during the support phase of running. *Running*, Fall, 24-31.
Batschelet, E. (1981) *Circular Statistics in Biology*. New York: Academic Press.
Cohen, J. (1991). *Statistical Power Analysis for the Behavioral Sciences* (2nd edition). Erlbaum, New Jersey.
Heiderscheit, B.C., Hamill, J., and Van Emmerik R.E.A. (2002). Variability of stride characteristics and joint coordination among individuals with unilateral patellofemoral pain. *Journal of Applied Biomechanics*, 18, 110-121.
Sparrow, W.A., Donovan, E., van Emmerik, R.E.A., and Barry, E.B. (1987). Using relative motion plots to measure changes in intra-limb and inter-limb coordination. *Journal of Motor Behavior*, 19, 115-129.

IMPACT OF WII-FIT TRAINING ON NEURO-MUSCULAR CONTROL

Daniel N. Bassett[1,2,3,4], Erika De Canonico[1], Antonio Archilletti[1], Giuliano Cerulli[1,2,4]

Let People Move Biomechanical Laboratory, Perugia, Italy[1], Nicola's Foundation Onlus, Arezzo, Italy[2], Center for Biomedical Engineering Research, University of Delaware, Newark, USA[3] and Ortopedia e Traumatologia, Università degli Studi di Perugia, Perugia, Italy[4]

KEY WORDS: rehabilitation, muscle training, co-contraction, synergy.

INTRODUCTION: In the past year, the interactive exercise video game Wii Fit (Nintendo, Tokyo, Japan) has achieved worldwide popularity. This system could be a potential asset for both training and physical therapy purposes; however, there is a lack of scientific validation to justify such applications. As a first step in ascertaining the advantages of the Wii Fit system, the present study is focused on the neuromuscular control changes that occur after 8 weeks of daily training.

METHOD: Two healthy subjects (25.5±2.1 years, 177.8±14.37 cm, 71.5±16.26 kg) trained for 30 minutes a day for 8 consecutive weeks using standard Wii Fit strength training, aerobic, and yoga exercises. Before and after the training period, a series of tests were performed (gait, hop, isometric, and one leg stability) while collecting EMG data from the quadriceps (rectus femoris, vastus lateralis, vastus medialis), the hamstrings (biceps femoris and semitendinosus), and the grastrocnemii (lateralis and medialis). The EMG data was linear-enveloped and normalized by a maximum isometric voluntary contraction (MVIC). Similarly to Lloyd et al. (2005), the electromyographic activations were then summed by muscle group to calculate the co-contraction ratio (CCR), which is a value between 0 and 1 that indicates equalizing activation as it increases.

RESULTS: Only the right leg data is being reported in this paper. Table 1 displays the CCR for the antagonist coactivations of the hamstrings and quadriceps and also the synergistic activations of the knee flexor muscle groups. It is worth noting that during gait and hopping motions, the ratios are decreased after training, while during the stability tests they increased. Finally, no trend emerged for the isometric data.

Table 1 Co-contraction Ratio Maximums (Ext 60 and Flex 60 refer to isometric extesion and flexion at 60°)– *values in italics are reciprocals*

		Ext 60	Flex 60	Gait	Hop	Stability
Ham/Quad Coactivation	Pre	0.21	0.41	0.17	0.20	0.06
	Post	0.46	0.47	0.02	0.06	0.88
Ham/Gast Synergy	Pre	0.34	0.47	0.02	0.08	0.04
	Post	0.47	0.28	0.01	0.08	*0.35*

DISCUSSION: After training, the CCR data for dynamic activity indicated more focused muscle control. During the stability tests, much higher CCR values were reported, indicating the muscles were doing a better job achieving a intra-articular equilibrium.

CONCLUSION: These preliminary results indicate a promising use of the Wii Fit system for training and physical therapy as on a small population they demonstrated neuromuscular control improvement during dynamic and static trials.

REFERENCES:
Lloyd, D. G., Buchanan, T. S., and Besier, T. F. (2005). Neuromuscular Biomechanical Modeling to Understand Knee Ligament Loading. *Medicine & Science in Sports & Exercise*, 37, 1939-1947.

THE EFFECTS OF IMPACT AVOIDANCE TECHNIQUES ON HEAD INJURY RISK

Philippe Rousseau and T. Blaine Hoshizaki

Neurotrauma Impact Science Laboratory, University of Ottawa, Ottawa, Canada

The purpose of this study was to determine the influence of impact deflection and neck compliance on peak linear and peak angular accelerations during a front impact to a Hybrid III head using a pneumatic linear impactor. Impact deflection was done by translating the headform laterally and showed to be effective at reducing linear and angular accelerations as well as GSI. Neck compliance was altered using one Hybrid III 50th percentile neck and two modified Hybrid III necks. A less compliant neck increased linear acceleration but decreased angular acceleration. When compared to estimated injury thresholds, the results demonstrated that an increase in lateral translation or a decrease in neck compliance resulted in a significant decrease in the risk of head injury as reflected by peak linear and angular accelerations.

KEY WORDS: sports, biomechanics, mTBI, linear acceleration, angular acceleration.

INTRODUCTION: Unanticipated impacts to the head occurring in a sporting event can have particularly devastating consequences. In fact, it is common for athletes to suffer from mild traumatic brain injuries (mTBI) following such collisions (Pellman et al., 2004). Athletes unaware of an imminent collision are unable to properly execute protective techniques placing them in a vulnerable position. Conversely, athletes aware of an impending collision prepare themselves to avoid, deflect or resist forces generated by an impact.

Impact Deflection: Athletes have the capacity to diminish head injury risk by deflecting a portion of the force resulting from an impact. When two bodies collide, the force transmitted is in part related to the angle of transmitted force vector during contact. As a result, the impacts closer to the center of gravity of the head are more severe. Athletes anticipating the hit are able to either avoid the hit altogether or deflect the blow by adjusting their head position; thus, reduce the force transmitted to their head.

Neck Compliance: When forces are transferred from one body to another, the resulting accelerations can be reduced by increasing the effective mass of the receiving body (Aubry et al., 2002). Athletes unaware of an impending collision are not prepared to resist the shock; thus, the force transmitted is resisted mostly by the mass of the head. In contrast, athletes anticipating the hit have sufficient time to contract their neck muscles, allowing them to resist the impact with the mass of their head, neck and upper torso. Assuming that the impact forces sustained by both players are equal, the lower impact mass of the unaware player will result in higher head peak acceleration of the head post-impact (Viano & Pellman, 2005). Linear and rotational acceleration of the head are common predictors of mTBI; hence, when they increase, there is an associated increase in risk of injury. Thus, it was proposed in this study that a noncompliant neck will mitigate head injuries.

METHODS: Data Collection: The effect of impact deflection was evaluated by impacting a 50th percentile Hybrid III head with a linear impactor. The headform was hit nine times per impact location at the following two velocities: 5 and 7 m/s, which are comparable to impact velocities seen in football (Pellman et al., 2003). Impact locations were chosen using the width of the headform as a reference (15.5 cm) and were the following: through the center of gravity; 3.875 cm lateral displacement; 7.75 cm lateral displacement; and 11.625 cm lateral displacement. All impacts were located 30 ± 1 mm above the reference plane of the headform.

The effect of neck compliance was evaluated by impacting a 50th percentile Hybrid III head attached to three different Hybrid III necks with a linear impactor. The impacts were located 30 ± 1 mm above the intersection of the longitudinal plane and the reference plane of the headform (centre of gravity). Three sets of three impacts were performed for each neck at the following velocities: 5 m/s, 7 m/s and 9 m/s.

Data Analysis: To determine the effects of impact location, an analysis of covariance (ANCOVA) was performed on each dependant variable, using velocity as a covariate. Further analysis was performed at each velocity (5 m/s and 7 m/s) using simple one-way analysis of variance (ANOVA), followed by a post hoc test using Tukey's method. The same process was performed to determine the effects of neck compliance on each dependant variable. All statistical analysis will be performed using SPSS software (SPSS Inc., Chicago IL, USA).

RESULTS: Impact Deflection: The results showed that an increase in lateral displacement caused a decrease in peak linear acceleration and peak angular acceleration as demonstrated on Figure 1. This signifies that impact deflection can successfully reduce injury risk.

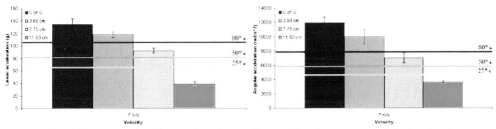

Figure 1: Peak linear acceleration and peak angular acceleration measured during front impacts through the centre of gravity plus three displacements at 7 m/s. The lines represent a 25%, 50% and 80% probability of sustaining an mTBI.

Neck Compliance: The results showed that under the conditions studied in this research an increase in neck compliance caused a decrease in peak linear acceleration but an increase in peak angular acceleration as demonstrated on Figure 2. This signifies that neck compliance may influence injury risk.

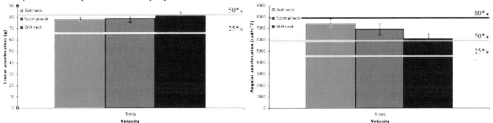

Figure 2: Peak linear acceleration and peak angular acceleration measured during a direct impact to the front of a Hybrid III head attached to necks with three different compliances at 5 m/s. The lines represent a 50% and 80% probability of sustaining an mTBI.

DISCUSSION: Impact Deflection: As expected, peak linear acceleration and peak angular acceleration decreased when the impacts got further from the centre of gravity. This was expected because these types of collisions do not engage the headform fully, meaning that the total momentum from the impactor was not fully transferred to the head.

Figure 1 shows a comparison of peak linear and angular acceleration using estimated brain injury thresholds reported by Zhang and his colleagues (Zhang, Yang & King, 2004). At 7

m/s, impacts through the centre of gravity and with a 3.875 cm displacement were above the 80% risk of injury.
Impacts with a 7.75 cm and an 11.625 cm lateral displacement were slightly above 50% and below 25%, respectively. This means that even a small head movement away from the centre of impact can effectively reduce the risk of sustaining a head injury.

Neck Compliance: Contrary to what has been previously reported in the literature (Pellman et al., 2008; Johnston et al., 2001), a stronger, less compliant neck did not reduce peak linear acceleration of the head. It was expected that the higher effective mass generated by a less compliant neck would help absorb the impact, yet these results presented a negative relationship between neck compliance and linear acceleration. However, a positive relationship was found between neck compliance and angular acceleration. In other words, more compliance will lead to higher angular acceleration, while less compliance will produce more linear acceleration. This implies that forces acting on the head upon contact need to somehow be dispersed as linear and angular acceleration, with the neck compliance determining its distribution.

Figure 2 shows a comparison of peak linear and angular acceleration as well as thresholds published by Zhang and colleagues (2004). It shows that peak linear accelerations measured when using the stiffer neck were associated with a probability of 50% of sustaining an mTBI at 5 m/s. This would suggest that players with tense neck muscles prior to an impact would be at a higher risk of sustaining a head injury then those who have not contracted their muscles. Conversely, peak angular accelerations measured when using the softer neck were near an 80% probability of sustaining an mTBI. Even though a compliant neck produces lower linear accelerations, it can produce high angular accelerations that may put a player at a higher risk of sustaining a head injury.

Injury Prevention: The data presented in this study provides a better understanding of the relationship between risk of brain injury and strategies for injury prevention. Training methods (strength and training), administrative intervention (rules and regulations) and protective equipment design (helmet) all benefit from a better understanding of the mechanism of injury. These results support training neck muscles to improve strength in order to reduce the risk of mTBI (Bailes & Cantu, 2001; Barthe et al., 2001). Impact deflection or avoidance should be considered as a strategy to prevent brain injuries resulting from head impacts. Skill training contributes to better ball tracking, helps avoid contact and increases the chance of impact deflection, thus reduces the risk of mTBI (McIntosh & McCrory, 2005).

Awareness of an impending collision is crucial because of the time required by the neck muscles to achieve full contraction, between 130 and 263 ms (Stemper et al., 2005). This indicates that athletes initiating neck muscle contraction as a result of an impact cannot protect themselves appropriately. Thus, efforts should be made to protect them through the establishment of rules and regulations, which have already shown to be effective at reducing injuries (Cantu & Mueller, 2003).

CONCLUSION: The objective of this paper was to determine the influence of impact deflection and neck compliance on head injury risk. As expected, an increase in lateral displacement caused a reduction in peak linear and angular accelerations, meaning that any head motion away from the impacting mass will reduce risk of head injury. Furthermore, a reduction in neck compliance caused a decrease in peak angular acceleration; however, it also resulted in an increase in peak linear acceleration. Considering that the peak angular accelerations were much higher, it is fair to say that a decrease in neck compliance will reduce risk of head injury. Hence, players should be encouraged to use both strategies as they showed effective at reducing the risk of suffering from a concussive injury.

REFERENCES:

Aubry, M., Cantu, R., Dvorak, J., Graf-Baumann, T., Johnston, K. M., Kelly, J., et al. (2002). Summary and agreement statement of the 1st International Symposium on Concussion in Sport, Vienna 2001. *Clinical Journal of Sport Medicine*, 12(1), 6-11.

Bailes, J. E., and Cantu, R. C. (2001). Head injury in athletes. *Neurosurgery*, 48(1), 26-45.

Barthe, J. T., Freeman, J. R., Broshek, D. K., and Varney, R. N. (2001). Acceleration-deceleration sport related concussion: The gravity of it all. *Journal of Athletic Training*, 36(3), 253-256.

Cantu, R. C., and Mueller, F. O. (2003). Brain injury-related fatalities in American football, 1945-1999. *Neurosurgery*, 52(4), 846-852.

Gerberich, S. G., Priest, J. D., Boen, J. R., Straub, C. P., and Maxwell, R. E. (1983). Concussion incidences and severity in secondary school varsity football players. *American Journal of Public Health*, 73(12), 1370-1375.

McIntosh, A. S., and McCrory, P. (2005). Preventing head and neck injury. *British Journal of Sports Medicine*, 39(6), 314-318.

Johnston, K. M., McCrory, P., Mohtadi, N. G., and Meuwisse, W. (2001). Evidence-based review of sport-related concussion: clinical science. *Clinical Journal of Sport Medicine*, 11(3), 150-159.

Pellman, E. J., Powell, J. W., Viano, D. C., Casson, I. R., Tucker, A. M., Feuer, H., et al (2004). Concussion in professional football: epidemiological features of game injuries and review of the literature - part 3. *Neurosurgery*, 54(1), 81-94.

Pellman, E. J., Viano, D. C., Tucker, A. M., Casson, I. R., and Waeckerle, J. F. (2003). Concussion in professional football: reconstruction of game impacts and injuries. *Neurosurgery*, 53(4), 799-812.

Powell, J. W., and Barber-Foss, K. D. (1999). Traumatic brain injury in high school athletes. *Journal of the American Medical Association*, 282(10), 958-963.

Stemper, B. D., Yoganandan, N., Rao, R. D., and Pintar, F. A. (2005). Reflex muscle contraction in the unaware occupant in whiplash injury. *Spine*, 30(24), 2794-2798.

Toth, C., McNeil, S., and Feasby, T. (2005). Central nervous system injuries in sport and recreation: a systematic review. *Sports Medicine*, 35(8), 685-715.

Viano, D. C., Casson, I.R., and Pellman, E. J. (2008) Concussion in professional football: biomechanics of the struck player – part 14. *Neurosurgery*, 61(2), 313-327.

Viano, D. C., and Pellman, E. J. (2005). Concussion in professional football: biomechanics of the striking player - part 8. *Neurosurgery*, 56(2), 266-280.

Zhang, L., Yang, K. H., and King, A. I. (2004). A proposed injury threshold for mild traumatic brain injury. *Journal of Biomechanical Engineering*, 126(2), 226-236.

Acknowledgement

The Authors would like to thank Xenith for supporting the laboratory.

NOVEL ENERGY ABSORBING MATERIALS WITH APPLICATIONS IN HELMETED HEAD PROTECTION.

Stephen Kiernan, Liang Cui and Michael D. Gilchrist

School of Electrical, Electronic and Mechanical Engineering, University College Dublin, Belfield, Dublin 4, Ireland

A finite element, functionally graded foam model (FGFM) is proposed, which is shown to provide more effective energy absorption management, compared to homogenous foams, under low energy impact conditions. The FGFM is modelled by discretising a virtual foam into a large number of element layers through the foam thickness. Each layer is described by a unique constitutive cellular response, which is derived from the initial foam density, ρ, unique to that layer. Large strain unixial compressive tests at a strain rate of 0.001 s^{-1} are performed on expanded polystyrene (EPS), and their σ -ε response is used as input to a modified constitutive model from the literature. It is found that under low energy impacts an FGFM can outperform a uniform foam of equivalent density terms of reducing peak accelerations, while performing almost as effectively as uniform foams under high energy conditions. These novel materials, properly manufactured, could find use as next generation helmet liners in answer to recent, more rigorous equestrian helmet standards, e.g. BS EN 14572:2005.

KEY WORDS: constitutive model, functionally graded foam, energy absorption.

INTRODUCTION: Cellular foams are widely used in energy absorbing applications where it is important to minimise the peak acceleration of the impacting body (Hilyard & Djiauw, 1971), e.g packaging of fragile goods, helmets and head protection systems (Doorly & Gilchrist, 2006) and body garments. This is due to their low solid volume fraction and complex microstructure, which allows large degrees of plastic crushing to occur at a fairly constant plateau stress value.

The ability of a uniform foam in reducing the peak acceleration of an impacting body is dependant on both volume and density. It is shown in this study that by varying the density spatially throughout the volume of a foam, it is possible to dramatically reduce peak accelerations without the need to increase its volume. This technique could have reaching implications in helmet manufacture, whereby such foams could provide additional head protection without sacrificing helmet aesthetics in the form of increasing the liner thickness.

METHODS: Data Collection: The ABAQUS crushable foam model (ABAQUS, 2007), in conjunction with an existing model from the literature (Schraad & Harlow, 2006) was used to describe the σ-ε behaviour of each element layer through a virtual foam's thickness. In order to calibrate the Schraad & Harlow model for any given ρ, large strain uniaxial compression tests on expanded polystyrene (EPS) specimens of density 15, 20, 25, 50 and 64 kg.m^{-3} were performed. Using data from these results, the model could be calibrated to generate a complete σ - ε curve from an arbitrary ρ, value as an input argument. Table 1 shows the material gradients and density ranges used in the simulations for a density difference of $\Delta\rho$ = 40 kg.m^{-3}. For all simulations the material gradients decreased monotonically from the striker to the anvil face as preliminary results showed this to be the more favourable gradient orientation for reducing peak accelerations. A single striker impact velocity of 5.4 m.s^{-1}, for striker masses of 1, 2, 4, 6, 8, 10, 12 and 14 kg, was used in all simulations and rate independent plasticity was assumed in order to quantify the influence of the material gradients alone. Figure 1 shows the model geometry used during the simulations. The foam relative density is the controlling parameter in describing the shape of each $\sigma - \varepsilon$ curve. By varying this parameter in an incremental manner, it is possible to generate multiple $\sigma - \varepsilon$ curves and calibrate the ABAQUS crushable foam model for a range of foam densities. Each calibrated crushable foam model for a given density may then be assigned to a given element layer through the specimen thickness, creating a quasi-graded cellular constitutive response. This methodology has previously been used by Cui et al, 2009a and Kiernan et al, 2009b for investigating the dynamic behaviour of FGFMs.

Table 1. Material gradients with density ranges used in striker impact simulations.

Gradients	Density Range (kg.m^{-3}) $\Delta\rho$ = 40 kg.m^{-3}				
Uniform	44	54	64	84	104
Logarithmic	74.4 – 34.4	84.4 – 44.4	94.4 – 54.4	114.4 – 74.4	134.4 – 94.4
Square Root	70.6 – 30.6	80.6 – 40.6	90.6 – 50.6	110.6 – 70.6	130.6 – 90.6
Linear	64.0 – 24.0	74.0 – 34.0	84.0 – 44.0	104.0 – 64.0	124.0 – 84.0
Quadratic	57.5 – 17.5	67.5 – 27.5	77.5 – 37.5	97.5 – 57.5	117.5 – 77.5
Cubic	54.2 – 14.2	64.2 – 24.2	74.2 – 34.2	94.2 – 54.2	114.2 – 74.2

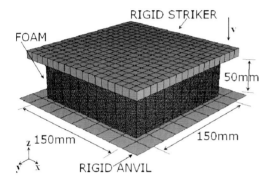

Figure 1. Modelling of the FGFM is achieved by approximating a specimen with a continuous variation in material properties as fifty discrete, finely meshed element layers though the thickness, with a unique $\sigma - \varepsilon$ curve associated with each layer along the z-direction.

RESULTS: By plotting the peak acceleration (normalised against the peak acceleration of the equivalent uniform foam) of each simulation against input parameters such as material gradient and kinetic impact energy, it can be easily seen in Figure 2 under what conditions a FGFM is most advantageous in reducing the peak acceleration of an impact. The average density of all gradients shown here is 54 kg.m^{-3}, with a density difference of 20 kg.m^{-3} and 40 kg.m^{-3} in Figure 2(a) and 2(b) respectively. It is clear that increasing this density range has a considerable influence on peak acceleration values in the low energy region while only a slight influence at higher energies.

DISCUSSION: The surface plots of Figure 2 are indicative of an FGFMs impact response for the different average densities examined. For low kinetic energy impacts, a graded foam performs more effectively than an equivalent uniform foam and the convex gradients (e.g. quadratic) perform better than the concave gradients (e.g. square root). However, as the impacting mass (and therefore KE) is increased to 14 kg, an opposite trend is observed. The marked improvement of the FGFM over the uniform foam in reducing the peak acceleration of the lower energy impacts can be explained as follows.

A homogenous foam is most efficient at absorbing impact energy when it works within the plateau strain region, up to densification, as it is here where it absorbs most energy under large plastic strains with little corresponding increase in stress.

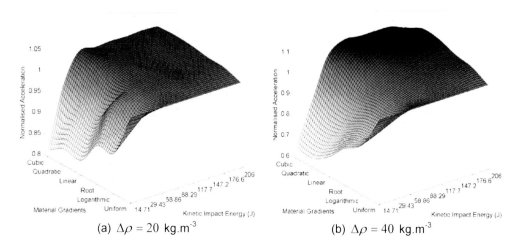

Figure 2. Surface plots of normalised (against uniform foams) peak accelerations of five material gradients across a range of striker impact energies. Significant reduction of peak acceleration is seen in the low energy regions of the plot.

From simulation it was found that for a uniform foam of 44 kg.m^{-3}, the stress imparted at the time of peak acceleration was 198 kPa for a striker energy of 206J and was 581 kPa for a striker energy of 14.71J. From the experimental $\sigma - \varepsilon$ compression tests it can be deduced that 44 kg.m^{-3} EPS foam will yield at about 310 kPa and thus will not yield when struck with a striker of 1 kg at 5.4 m.s^{-1} (14.71J), but rather will behave elastically with very little deformation, resulting in high peak accelerations. However, when struck with a striker of 14 kg at 5.4 m.s^{-1} it will absorb the corresponding kinetic energy within the plateau stress region up to 0.6 strain. The FGFMs perform distinctly better than the uniform foam when absorbing the lower energies due to their spatially varying yield surface, a direct result of the density gradient. From Table 1, for example, the density of a cubically varying foam with an average density of 44 kg.m^{-3} will vary from 54.2 kg.m^{-3} to 14.2 kg.m^{-3}. At 14.2 kg.m^{-3}, local plastic deformation was found from simulations to initiate at about 100 kPa, deforming to almost 0.7 strain, and approximately 20% by volume (14.2 - 28 kg.m^{-3}) of the graded foam will yield plastically at a stress of 198 kPa. This is in stark contrast to the equivalent uniform foam, which exhibits no yielding at this stress level. As the kinetic energy of the striker is increased the advantage gained by a varying yield surface diminishes rapidly. Low yielding regions of the FGFM are no longer effective and local deformation beyond their densification strains occurs while mitigating only a small fraction of the total energy.

Results show that a uniform 44 kg.m^{-3} foam experiences 0.54 strain at the incident surface and 0.52 strain at the distal surface when impacted by a 14 kg striker at 5.4 m.s^{-1}. In contrast, the cubically varying FGFM deforms locally to only 0.2 strain at the incident surface and yet there is 0.98 strain at the distal face. Intuitively, and from previous work (Avalle et al. 2001), it is more advantageous for a foam's entire volume to deform up to, but not beyond, its densification strain if it is to act most effectively as a cushioning structure.

CONCLUSION. A functionally graded polymeric foam model was proposed and its energy absorbing ability has been analysed using the finite element method. The influence of material distribution, controlled by various explicit gradient functions, was studied. The main findings can be summarised as:

It is shown that a functionally graded foam can exhibit superior energy absorption over equivalent uniform foams under low energy impacts, and that convex gradients perform better than concave gradients. This advantage is negated when the impact energy becomes significantly high such that low-density regions of the graded foam become ineffective at bearing the higher load and they densify after absorbing only a small fraction of the total energy. What constitutes a 'high energy impact' is somewhat difficult to define but will depend on the average density of the foam, matrix composition, and the density gradient.

For a specified density range the energy absorption performance of a functionally graded foam under low energy impacts can be improved if the density range is increased. For higher energy impacts, increasing the density range can reduce the performance of the graded foams due to a higher volume fraction deforming beyond the densification strain. Functionally graded foams are capable of reducing the duration of the high acceleration during an impact event. This property could have wide implications in the head protection industry as many head injury criteria rely on acceleration durations as indicators of the likelihood for a person suffering significant head trauma. In this respect, protective headgear, e.g., safety helmets, employing functionally graded foams as the liner constituent may be advantageous to the wearer in reducing the risk of brain injury after a fall.

Traditionally, many helmet certification standards require a helmet to keep the acceleration of a headform dropped from a single drop height below some certain target level – achieving this is quite simple. However, recent helmet standards demand that helmets be effective at multiple drop heights, thus simulating both high and low energy impacts. This can be more difficult to achieve with current helmet liner technologies. Functionally graded foams have been shown to exhibit significant advantages under low energy impact conditions while still performing nearly as well as their uniform counterpart under high energy conditions. These foams, carefully manufactured, may be one possible answer to the more stringent requirements of emerging helmet standards such as BS EN 14572:2005.

REFERENCES:

Hilyard, N.C., and Dijauw, L.K. (1971). Observations on the Impact Behaviour of Polyurethane Foams: I. The Polymer Matrix. Journal of Cellular Plastics. 7:33 - 42

Doorly, M.C., and Gilchrist, M.D. (2006) The use of accident reconstruction for the analysis of traumatic brain injury due to head impacts arising from falls. *Computer Methods in Biomechanics and Biomechanical Engineering* 9 371–377.

ABAQUS. (2007). ABAQUS Analysis User's Manual, v 6.7. ©ABAQUS, inc.

Schraad, M.W., and Harlow, F. H., (2006) A stochastic constitutive model for disordered cellular materials: Finite-strain uni-axial compression. *International Journal of Solids and Structures.* 43, 3542 – 3568

Avalle, M., Belingardi, G., and Montanini, R. (2001) Characterization of polymeric structural foams under compressive impact loading by means of energy-absorption diagram. *International Journal of Impact Engineering,* 25, 455 – 472

Cui, L., Kiernan, S., and Gilchrist, M. D., (2009a) Designing the energy absorption capacity of functionally graded foam materials, *Materials Science and Engineering A* 507,215–225

Kiernan, S., Cui, L., and Gilchrist, M. D., (2009b), Propagation of a stress wave through a virtual functionally graded foam, *International Journal of Non-Linear Mechanics,* 44: 456 – 468

Acknowledgement
This study has been funded by Enterprise Ireland (PC/2005/071), Science Foundation Ireland (08/RFP/ENM/1169), and the Turf Club of Ireland.

ISBS 2009

Scientific Sessions

Wednesday AM

ISBS 2009

Oral Session S09

Methods

TRUNK POSTURE AND STATICO-DYNAMICAL SPINE ANALYSIS– COMPARING ULTRASOUND BASED VS. OPTICALLY BASED MEASUREMENT SYSTEM

Thorwesten, L.[1], Schnieders, D.[1], Schilgen, M.[2], Völker K.[1]

[1]Institute of Sports Medicine, University Hospital Münster
[2]Academy of Manual Medicine, Westfälische Wilhelms-University Münster

The purpose of this study was to compare two different methods measuring trunk posture and statico-dynamical spine analysis. 32 patients participated in this cross sectional study. Comparing measured values a wide congruence could be demonstrated with marginal underestimating in kyphosis and lordosis data for the ultrasound based system. The largest deviation could be shown for pelvic obliquity measured in mm. Trunk inclination, vertical deflection and pelvic obliquity measured in degree showed proper analogy for both measuring systems. Validity, reliability based on particular technical principles could be verified.

KEY WORDS: trunk, posture, spine analysis

INTRODUCTION: Posture relevant parameters can be measured by numerous measuring systems using different technologies and allow the registration of asymmetries in skeleton-axis. There is a lack of comparison of different technologies in the literature. The aim of the study was to evaluate different methods (ultrasound based versus optically based) in statico-dynamical spine analysis and posture (zebris® CMS HS versus Formetric®) regarding accuracy and elaborate benefits as well as disadvantages in application.

METHODS: 32 patients (17 male; 15 female) within the age of 27.7(± 6.2) years participated in this cross sectional study. Anamnesis' questionnaires were used to assess information on sport injury as well as pain. Patients were measured thrice with both systems. Statistic evaluation was done according to Bland/Altman (1986) as well as Spearmans correlation calculation using SPSSv11.5 and Excel 2003.

Table 1. Anthropometric data

	Age	Height	Mass
Male	27.4 y	1.83 m	79.1 kg
(n=17)	(±6.8)	(±0.1)	(±8.8)
Female	28.1 y	1.72 m	63.9 kg
(n=15)	(±5.6)	(±0.1)	(±8.1)
Total	27.7 y	1.78 m	72.0 kg
(n=32)	(±6.2)	(±0.1)	(±11.3)

There is no comparison possible to radiologic data for the ZEBRIS® System. Objectivity (r=0.93-0.98) as well as retest-reliability (r=0.97-0.99) has been evaluated (Asamoah 2000; Himmelreich et al. 1998). The Formetric® system with its triangular principle is a fully developed method with a close match to radiograph. It is free of retroactivity but poor on distinctive muscle relief and obesity. At the moment the „gold standard" in non invasive spine analysis. (Drerup et al 2001; Hackenberg and Hierholzer 2002; Lilienquist et al. 1998).

RESULTS: Comparing measured values a wide congruence could be demonstrated with marginal underestimating in kyphosis and lordosis data for the ultrasound based system. The largest deviation could be shown for pelvic obliquity measured in mm. Trunk inclination, vertical deflection and pelvic obliquity measured in degree showed proper analogy for both measuring systems. Validity, reliability based on particular technical principles could be verified. The Bland & Altman plot (Bland & Altman, 1986 and 1999) is a statistical method to compare two measurements techniques. In this graphical method the differences (or alternatively the ratios) between the two techniques are plotted against the averages of the two techniques.

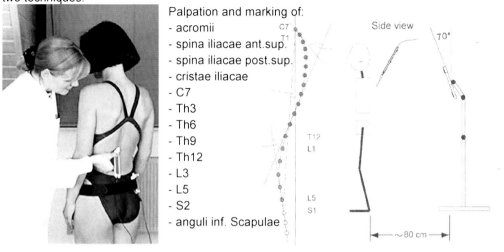

Figure 1. Static spinal column and back measurements using Zebris CMS HS

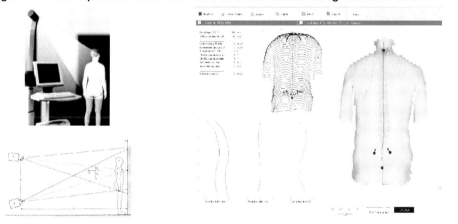

Figure 2. Formetric® optical based measuring system. Patient positioning and calculating of parameters.(Screenshot)

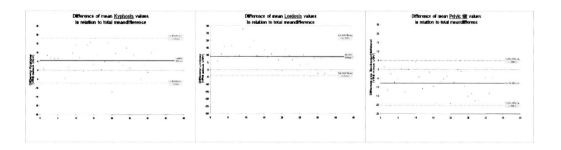

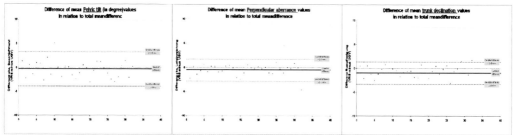

Figure 3. Bland & Altman plots of kyphosis, lordosis, pelvic tilt, perpendicular aberrance and trunk declination values of both measurement techniques.

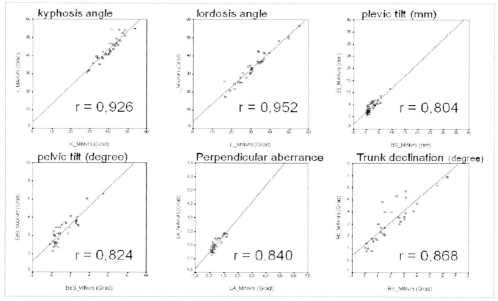

Figure 4. Reliability of the Formetric ® System

DISCUSSION: In summary both systems revealed usable quality in specific applications. The manual performance in using the ultrasound based system bears the risk of cumulated errors during measurement. The analysis of system-quality produced an error in measurement of 0.65% (0.58 +/- 1.29 mm). Himmelreich et al. described the ultrasound based system as a screening method which is not a substitute for case-history or classical anthropometry, but it offers useful parameters which facilitate decision making for further diagnostic procedures. (Himmelreich et al. 1998)

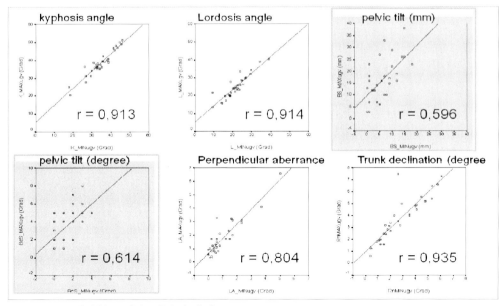

Figure 5. Reliability of the Zebris ® System

Working with the ultrasound based system, pelvic oblique is an accident-sensitive parameter. Versatile applications such as static or dynamic measurements could be done with the ultrasound based system.

Differences in axial balance in the range of physiologic motion can be calculated under dynamic conditions, and compared to normative data. The contact free and quick done rasterstereography allows with the help of surface back analysis to supplement radiological and clinical examinations of the spine in orthopedic and biomechanics questions.

References
Asamoah V, Mellerowicz H, Venus J, and Klöckner C. Measuring the surface of the back. Value in diagnosis of spinal diseases. *Orthopade*. 2000 Jun;29(6):480-9
Drerup B, Ellger B, Meyer zu Bentrup FM, and Hierholzer E.: Functional rasterstereographic images. A new method for biomechanical analysis of skeletal geometry. *Orthopade*. 2001 Apr;30(4):242-50
Hackenberg L, and Hierholzer E.: 3-D back surface analysis of severe idiopathic scoliosis by rasterstereography: comparison of rasterstereographic and digitized radiometric data. *Stud Health Technol Inform*. 2002; 88:86-9.
Himmelreich H, Stefanicki E, and Banzer W.: Ultrasound-controlled anthropometry--on the development of a new method in asymmetry diagnosis. *Sportverletz Sportschaden*. 1998 Jun;12(2):60-5
Liljenqvist U, Halm H, Hierholzer E, Drerup B, and Weiland M: 3-dimensional surface measurement of spinal deformities with video rasterstereography. *Z Orthop Ihre Grenzgeb*. 1998 Jan-Feb;136(1):57-64

SPATIO-TEMPORAL PARAMETERS AND INSTANTANEOUS VELOCITY OF SPRINT RUNNING USING A WEARABLE INERTIAL MEASUREMENT UNIT

E. Bergamini, P. Picerno, M. Grassi, V. Camomilla and A. Cappozzo

Department of Human Movement and Sport Sciences, University of Rome "Foro Italico", Rome, Italy

KEY WORDS: sprint running, biomechanics, inertial sensors

INTRODUCTION: Wearable inertial measurement units (IMU) provide movement-related data without any space limitation or cumbersome setup. They can be proficiently used to perform an in-field biomechanical analysis of sprint running providing information useful for performance optimisation and injury prevention. Mechanical key quantities characterizing sprint running performance are instantaneous velocity and displacement of the athlete (Cavagna et al., 1971). However, the process of determining velocity and position by numerical integration of acceleration is jeopardized by the noise characterizing the signal of micro-machined accelerometers (Thong et al., 2002). The aim of this study was to compensate these errors by reducing the integration interval, taking advantage of *a priori* known laws of motion, and by cyclically determining the initial conditions of the integration process, in order to yield reliable spatio-temporal parameters during sprint running.

METHODS: A male subject (26 yrs, 73 kg, 1.73 m) performed 7 in-lab sprints, starting from a standing position. Due to limited lab volume (12*9*4 m) only the first 3 steps were considered. 3D linear acceleration and orientation of a wearable IMU positioned on the upper back trunk (MTx, Xsens; m=30g) were collected and the following parameters were estimated over each cycle: 1) stance time (ST); 2) centre of mass progression displacement (d); 3) variation of vertical and progression velocity (Δv_v, Δv_p). Reference data were obtained as follows: ST from a contact-sensitive mat (stance 1) and two force platforms (Bertec) (stance 2-3); Δv and d from stereophotogrammetry (Vicon MX, Plug-in-Gait protocol). The average of the absolute percentage difference ($e_{abs\%}=|(reference-inertial)*100/reference|$), referred to as error ($e_\%$), was calculated for each parameter.

RESULTS: Reference and sensor estimates and percentage error are reported in Table 1.

Table 1. Parameter and percentage error values (mean ± standard deviation) for three stance

	stance 1			stance 2			stance 3		
	reference	inertial	$e_\%$	reference	inertial	$e_\%$	reference	inertial	$e_\%$
ST [s]	0.23±0.01	0.21±0.01	7±4	0.18±0.01	0.17±0.01	6±2	0.17±0.01	0.19±0.01	8±3
d [m]	1.31±0.06	1.22±0.07	7±2	1.37±0.18	1.30±0.18	6±2	0.89±0.28	0.89±0.27	3±2
Δv_p [ms^{-1}]	0.95±0.06	0.99±0.11	7±4	0.67±0.02	0.78±0.04	17±7	0.42±0.11	0.34±0.07	21±11
Δv_v [ms^{-1}]	1.41±0.11	1.27±0.11	11±7	1.68±0.27	1.38±0.16	17±8	1.65±0.37	1.30±0.25	27±6

DISCUSSION AND CONCLUSION: The obtained Δv percentage errors are consistent with respect to the literature (Vetter et al., 2008). Even though these errors still increase at each stance phase, the methodology is sensitive to the variations of velocity determined by the reference measurement system. As regards ST and d, no similar previous study has been reported. However since the methodology relies on the identification of foot contact timings for reducing the integration interval, small errors in the determination of these parameters, are encouraging. Future developments concern in-field sprint running experimental sessions.

REFERENCES:
Cavagna, G. et al. (1971). The mechanics of sprint running. *J. Physiol*, 217, 709-721.
Thong, Y.K. et al. (2002). Dependence of inertial measurements of distance on accelerometer noise. *Meas. Sci. Technol*, 13, 1163-1172.
Vetter R. et al. (2008). Estimation of a runner's speed based on chest-belt integrated inertial sensors. *The Engineering of Sport 7*, Pub. Springer Paris

VALIDATION OF MODEL-BASED IMAGE-MATCHING TECHNIQUE WITH BONE-PIN MARKER BASED MOTION ANALYSIS ON ANKLE KINEMATICS: A CADAVER STUDY

Kam-Ming MOK[1], Daniel Tik-Pui FONG[1,2], Tron KROSSHAUG[3], Kai-Ming CHAN[1,2]

Department of Orthopaedics and Traumatology, Prince of Wales Hospital, Faculty of Medicine, The Chinese University of Hong Kong, Hong Kong, China [1]
The Hong Kong Jockey Club Sports Medicine and Health Sciences Centre, Faculty of Medicine, The Chinese University of Hong Kong, Hong Kong, China [2]
Oslo Sports Trauma Research Center, Norwegian School of Sport Sciences, Ullevaal Stadion, Oslo, Norway [3]

Krosshaug (2005) introduced a model-based image-matching (MBIM) technique for 3D reconstruction of human motion from uncalibrated video sequences. The aim of this study is to validate the MBIM technique on ankle joint movement with the reference to bone-pin marker based motion analysis on a cadaver. One cadaveric below-hip specimen was prepared for performing full-range plantarflexion/dorsiflexion, inversion/eversion and relative circular motion between two segments. The videos were recorded and analyzed by the MBIM technique and bone-pin marker based motion analysis. The results are presented as the qualitative visual evaluation and the root mean square (RMS) error. In general, the validation results showed good agreement between the MBIM estimation and bone-pin marker based motion analysis results. This technique will contribute to the motion measurement of ankle joint kinematics in the future, for instance, the motion analysis in real game situations and understanding the injury mechanisms of real injury cases.

KEY WORDS: video analysis, ankle joint movement

INTRODUCTION: Skin marker based motion analysis is the most common method to calculate joint kinematics. However, skin marker based motion analysis is not always available in all situations. Sportsmen will not wear skin markers in real game situations and the target motion happens unexpectedly (Fong, 2009). In order to develop a novel biomechanical analysis to produce continuous estimates of joint kinematics from video recordings, Krosshaug and Bahr (2005) introduced a model-based image-matching (MBIM) technique for investigating human motion from uncalibrated video sequences. For the MBIM technique, only the validations on hip and knee joint movements were done, thus the validation on ankle joint movement is needed. Skin marker based motion analysis was regarded as the golden standard in previous validations. However, the results from skin marker based motion analysis were influenced by the skin artefact (Reinschmidt, 1997). The kinematics data deduced from skin markers is not a prefect standard for validating the MBIM technique. Thus, bone-pin marker based motion analysis was utilized to calculate the ankle joint kinematics in current study. The aim of this study was to validate the MBIM technique on ankle motion measurement with reference to bone-pin marker based motion analysis on cadaver.

METHODS: One cadaveric below-hip specimen was prepared for testing. Achilles tendon was cut to increase the joint flexibility. Hofmann II external fixation 5.0mm bone-pins (Stryker, US) were inserted into the posterolateral side of the calcaneus and into the lateral tibial condyle in the cadaver (Reinschmidt, 1997). Triads of reflective markers (14.0mm diameter spheres) were attached to the bone-pins (Figure 1). Skin makers were attached to lateral femoral epicondyle, medial femoral epicondyle, lateral malleolus and medial malleolus for defining knee and ankle joint center (Wu, 2002). After that, the specimen in upright position was mounted on a jig. Four high speed cameras (Casio EX-F1, Japan) were utilized to record the ankle motion in 30Hz with 640x480 resolutions from different views. A static calibration trial in the anatomical position served as the offset position to determine the segment embedded axes of the shank and foot segment. The foot segment was embedded with the Laboratory Coordinate System (LCS). The line connecting knee joint center and ankle joint center was the longitudinal axis of the shank segment (X1). The anterior-posterior axis of the shank segment (X2) was the cross product of X1 and the line joining the lateral femoral epicondyle and medial femoral epicondyle. The medial-lateral axis of the shank segment was the cross product of X1 and X2. Full-range pure plantarflexion/dorsiflexion, pure inversion/eversion and shank circular motion were performed on the ankle joint manually. A motion analysis system (Ariel Performance Analysis System, USA) was used to calculate the reflective marker's 3D coordinates by direct linear transformation. A singular value decomposition method was employed to calculate the transformation from triad reference frame to anatomical shank and foot reference frame (Soderkvist, 1993). Joint kinematics was deduced by the Joint Coordinate System (JCS) method (Grood, 1983). On the other hand, the videos were analyzed by the MBIM technique (Krosshaug, 2005). Using a commercialized animation software Poser (Poser4, Curious Lab, US), a virtual environment was built and matched with the video images in every camera view by adjusting the camera calibration parameters. A skeleton model (Zygote Media Group Inc, USA) was customized to match the anthropometry of the specimen. The skeleton matching started with the shank segment and then distally matching the foot, and toe segments

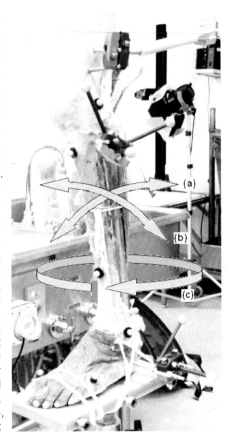

Figure 1. Three testing motions
(a) plantarflexion/dorsiflexion
(b) inversion/eversion
(c) relative circular motion between two segments

frame by frame. The joint angle time histories were read into Matlab (MathWorks, USA) with a customized script for data processing. The kinematics results from both MBIM technique and bone-pin marker based motion analysis were filtered by Butterworth low pass filter with 5Hz cut-off frequency, in order to filter out the high frequency white noise.

RESULTS: Figure 2 presents the curve fitting of the MBIM technique to bone marker based motion analysis method in measuring ankle kinematics. Good agreement was found for the plantarflexion/dorsiflexion and inversion/eversion. And internal/external rotation was less precise.

Table 1. Root mean square errors on ankle joint for the whole testing motion and the percentage differences to maximum range. Results from a similar study of comparing bony marker and skin marker (Reinschmidt,1997) are shown

	Plantarflexion/Dorsiflexion		Inversion/Eversion		Internal/External Rotation	
	MBIM	Skin marker	MBIM	Skin marker	MBIM	Skin marker
R.M.S. error (°)	4.6	4.7	3.1	4.6	4.5	3.6
Percentage difference to max. range (%)	10.7	14.1	11.7	34.7	30.5	51.2

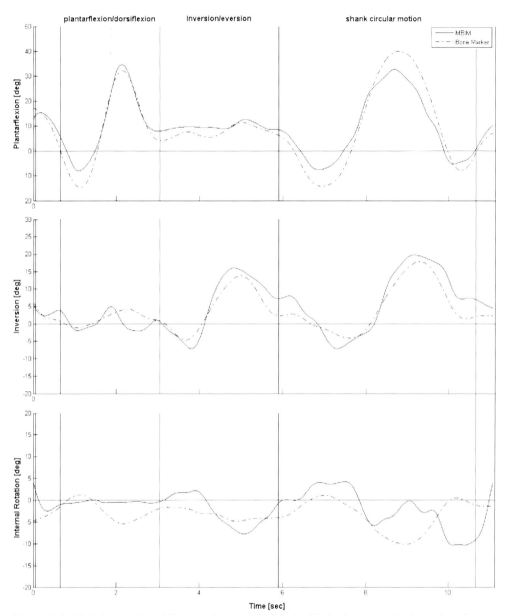

Figure 2. Ankle joint angles of the specimen, calculated with the bone marker based motion analysis (dotted lines) and the model-based image-matching (MBIM) technique (solid lines).

DISCUSSION: The aim of the study was to evaluate the MBIM technique for the estimation of ankle movements from uncalibrated video sequences. From the qualitative visual evaluation, good agreements were found in plantarflexion/dorsiflexion and inversion/eversion while internal/external rotation was less precise. The RMS errors of the three kinematic parameters were less than 5 degrees for the whole testing motion. The percentage differences to range are about 10% for plantarflexion/dorsiflexion and inversion/eversion results. The small RMS error coupled with good curve agreements, the MBIM technique was adequate to produce the ankle joint kinematics based on uncalibrated video recordings. Compared to a similar study of investigating the accuracy of skin-marker based motion analysis, MBIM technique performed relatively better than skin-maker based motion analysis (Table 1). However, it is not sufficient to conclude MBIM technique is more accurate because of the inconsistence of testing protocol.

Good agreement in plantarflexion/dorsiflexion was expected. The flexion axis of ankle joint was the cross product of the shank longitudinal axis and the foot longitudinal axis (Wu, 2002). As the longitudinal axis orientations of the shank and the foot could be accurately defined from the video images, plantarflexion/dorsiflexion result was expected to have a good fitting. Regarding the inversion/eversion result, it was highly depended on the orientation of the foot segment. Foot segment could be regarded as a flat rectangular board. And, the orientation of the plantar foot would be key information to match the foot skeleton on the video images. Using top view camera and front view camera in Poser, the detail orientation of the foot segment could be seen and further fine tuning was possible. For the shank, it was comparably difficult to be perfectly matched because it was in a cylindrical shape. The internal/external rotation result was highly depended on the internal rotation orientation of the shank segment. While only the patella position was a decisive landmark to define the internal rotation orientation of the shank. Therefore, it was understandable to have a less precise agreement on internal/external rotation.

The validation of the MBIM technique on ankle joint kinematics was considered achieved. The ankle kinematics information will contribute to different research areas in the future, for instance, the motion analysis in real game situations and understanding the injury mechanisms of real injury cases.

CONCLUSION: MBIM technique on ankle joint movements has been validated. This technique can produce ankle joint angle histories from uncalibrated video sequences. In the future, the MBIM technique would be regarded as a promising motion analysis approach for cases without skin marker based motion analysis in real game situations. Future works will be the investigation on the repeatability and reliability of MBIM technique.

REFERENCES:

Fong, D.T.P., Hong, Y., Shima, Y., Krosshaug, T., Yung, P.S.H., and Chan, K.M. (2009). Biomechanics of Supination Ankle Sprain: A Case Report of an Accidental Injury Event in the Laboratory. *The American Journal of Sports Medicine*, 37, 822-827.

Grood, E.S. and Suntay, W.J. (1983). A joint coordinate system for the clinical description of three-dimensional motions: application to the knee. *Journal of Biomechanical Engineering*, 105, 136-144.

Krosshaug, T. and Bahr, R. (2005). A model-based image-matching technique for three-dimensional reconstruction of human motion from uncalibrated video sequences. *Journal of Biomechanics*, 38, 919-929

Reinschmidt, C., van den Bogert, A.J., Murphy, N., Lundberg, A., and Nigg, B.M. (1997). Tibiocalcaneal motion during running, measured with external and bone markers. *Clinical Biomechanics*, 12, 8-16.

Soderkvist, I. and Wedin, P-A. (1993). Determining the movements of the skeleton using well-configured markers. *Journal of Biomechanics*, 26, 1473-1477.

Wu, G., Siegler, S., Allard, P., Kirtley, C., Leardini, A., Rosenbaum, D., Whittle, M., D'lima, D.D., Cristofolini, L., Witte, H., Schmid, O., and Stokes, I., (2002). ISB recommendation on definitions of joint coordinate system of various joints for the reporting of human joint motion - part I: ankle, hip, and spine. *Journal of Biomechanics*, 35, 543-548.

Acknowledgement
This research project was made possible by equipment and resources donated by The Hong Kong Jockey Club Charities Trust

CUSTOM-BUILT WIRELESS PRESSURE SENSING INSOLES FOR DETERMINING CONTACT-TIMES IN 60M MAXIMAL SPRINT RUNNING

Gregor Kuntze[1], Marcelo Pias[2], Ian Bezodis[1], David Kerwin[1], George Coulouris[2] and Gareth Irwin[1]

Cardiff School of Sport, University of Wales Institute, Cardiff, Wales[1]
Computer Laboratory, University of Cambridge, Cambridge, England[2]

The purpose of this study was to evaluate a custom-built wireless pressure sensing insole system for recording ground contact times in sprinting. Despite interest in the foot contact time/running velocity relationship, no study has examined the contact times in a maximal 60 m sprint. Insole data were collected on three athletes during maximal indoor sprint runs. Simultaneous kinematic data for start and maximum velocity phases were recorded with a CODA system to validate insole contact times and determine velocity. Insole derived contact times were accurate to ±4 ms. Preliminary data indicate a usable contact time/velocity relationship. It is anticipated that these data will provide support for the use of wireless technology in sprint performance monitoring, and facilitate novel insights into the contact time/sprint running velocity relationship.

KEY WORDS: athletics, remote sensing, performance.

INTRODUCTION: The recording of foot contact times during sprint running presents an opportunity to gain insight into sprint-specific performance parameters. Previous research has shown that contact times decrease with increasing velocity as an athlete changes from a walk or jog to a fast run or sprint (Luhtanen & Komi, 1978; Nilsson & Thorstensson, 1987). Separate studies have investigated the velocity-time curve of a sprint run (e.g. Henry & Trafton, 1951; Arsac & Locatelli, 2002) and the relationship between ground contact time and sprint running velocity has been investigated in the past for specific sections of the 200m sprint (Mann & Herman, 1985) and treadmill sprinting (Weyand et al., 2000). However, to the knowledge of the authors, no previous attempts have been made to link ground contact times and velocity throughout a maximal sprint run. To achieve this, a reliable, lightweight wireless pressure sensing insole system was developed to ensure minimal interference with the athlete and task. It is anticipated that these data will enhance our understanding of the relationship of foot contact times and sprint velocity and form a vital component in providing performance related feedback. The first aim of this study, therefore, is to validate the insole system and the second is to investigate the relationship between contact times and velocity.

METHODS: Participants: Three competitive male sprint athletes volunteered for participation in the study (age = 21.7 yr, height = 182.8 ± 2.5 cm, body mass = 76.8 ± 6.9 kg). These athletes facilitated appropriate maximum sprint velocities (~10 m/s). All procedures were approved by the University's Research Ethics Committee and written informed consent was obtained from the participants before data collection.

Data Collection: Motion data were captured unilaterally from each athlete's left side, using four cx1 CODA scanners (Charnwood Dynamics Ltd, UK), sampling at 800 Hz, with six active markers located according to Bezodis et al. (2007). Custom built pressure sensing insoles were used for the sampling foot contacts at ~2.5 kHz. Each insole was equipped with Force Sensing Resistor (FSR) – a polymer thick film, which exhibits decreasing resistance with increasing applied force. Two FSR sensor pads (6 cm square) were fitted per insole (*middle* sensor = metatarsophalangeal for touch-down detection & *toe* sensor = great toe area for toe-off (Figure 1a). The FSRs were connected with fine wires to a lightweight (28 gm) custom-built WiFi wireless data logger[1] (IMote2 with CSK WiFi board, Figure 1b) in an athlete-worn pouch at rear of the pelvis.

---

[1] Cambridge Sensor Kit (CSK): http://imote2-linux.wiki.sourceforge.net/UCAM-WSB100

 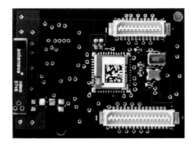

Figure 1a: FSR insoles; 1b: Cambridge Sensor Kit (CSK)

Participants were given time to perform their normal warm-up. The insoles and motion analysis markers were then attached. Following familiarisation, athletes performed six maximal 60 m sprint runs along a 120 m straight track. Motion capture data were recorded for the start (0 to 20 m) and the maximum speed phase (30 to 50 m) for three trials respectively, while the wireless insoles recorded throughout the duration of all sprints. Between runs the athletes were given self-selected recovery periods to allow for consistent performance and minimise the likelihood of fatigue effects.

Data Analysis: Contact durations were computed from the CODA data using marker acceleration maxima, (Bezodis et al., 2007). Contact durations for the wireless insoles were computed using touchdown to toe-off events detected with an automated algorithm implemented in Matlab. Low-pass filtered data were differentiated and local maximum and minimum of first derivatives computed respectively. Individual step velocities were determined using the mean velocity of a greater trochanter marker throughout the duration of a step (touch-down of one foot to touch-down of the opposite foot).

RESULTS: Contact times computed from FSRs were compared to those from CODA. The absolute RMS error was 4 ms for a total of 96 steps (53 in starts and 43 in maximal velocity). The distribution of the error (FSR ct – CODA ct) follows a standard normal distribution with mean around zero, (Figure 2). Contact time and step velocity data for a representative participant are displayed in Figure 2. Regression analysis statistics of contact time against step velocity for individual and group data is presented in Table 1.

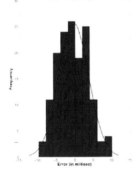

Figure 2: Distribution of error

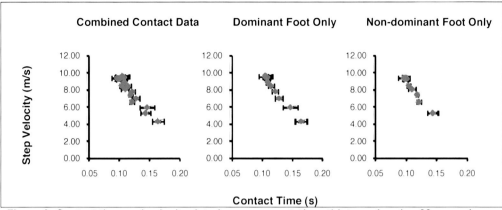

Figure 3: Contact time and velocity data for a representative athlete performing 60 m maximum speed sprints.

Table 1: Coefficient of determination values for regression analysis of individual and group contact time against velocity data

Participant	Combined Contact Data			
	Regression analysis (R^2)			
	Exponential	Linear	Power	Exponential Equation
1	0.920	0.898	0.907	$y = 32.218e^{-12.09x}$
2	0.977	0.971	0.968	$y = 26.491e^{-10.16x}$
3	0.930	0.900	0.927	$y = 28.402e^{-10.51x}$
All	0.927	0.908	0.919	$y = 27.69e^{-10.53x}$
	Dominant Limb Contact Data			
1	0.991	0.984	0.985	$y = 37.758e^{-13.02x}$
2	0.982	0.978	0.974	$y = 28.432e^{-10.73x}$
3	0.909	0.887	0.898	$y = 25.943e^{-9.88x}$
All	0.950	0.936	0.945	$y = 29.675e^{-11x}$
	Non-dominant Limb Contact Data			
1	0.979	0.964	0.979	$y = 33.358e^{-12.95x}$
2	0.979	0.962	0.976	$y = 23.907e^{-9.307x}$
3	0.978	0.941	0.989	$y = 32.806e^{-11.56x}$
All	0.902	0.877	0.895	$y = 25.705e^{-10.02x}$

DISCUSSION: The aim of this study was to validate a lightweight, wireless insole pressure sensing system for measuring foot contact times in sprinting and investigate the relationship between ground contact times and velocity.

Contact times for the start and maximum sprint phases can be determined with a high level of accuracy using the custom built insole system. For the mean data of three runs, there appears to be a usable relationship of contact time and velocity (Figure 3). Regression analysis showed that an exponential relationship resulted in the best data fit (Table 1). Continuous bilateral monitoring of foot contact times provides novel insight into the fundamentals of sprint running. In addition to the relationship with sprint velocity the tracking of contact times over the entire length of a sprint run may eventually allow for the evaluation of additional performance parameters such as transitional phases of the sprint run. This relationship was strongest for the data from individual athletes compared to group results. Furthermore, the dominant limb displayed a slightly closer relationship than the non-dominant limb.

These findings suggest that athlete-specific contact and velocity data recorded for multiple sprint runs may allow for the creation of individual contact time/velocity relationships and the prediction of velocity based on the regression equation (Table 1) and contact times only. The prediction of velocity from contact time would provide an essential tool for the sprint coach to track athlete development and enhance coach-athlete communication. Moreover, these findings show that the inclusion of lightweight, wireless technology in the sporting setting allows for continuous and un-intrusive recording of performance measures

CONCLUSION: Moreover, the use of contact time data and its inclusion as a performance measure in athlete feedback may provide vital information to enhance task comprehension and performance. Using the simple measurement of individual ground contact times may therefore allow for enhanced performance related feedback and act as an essential tool in sprint coaching through athlete monitoring in competition and training.

REFERENCES:
Arsac, L.M. and Locatelli, E. (2002). Modelling the energetics of 100 m running by using speed curves of world champions. *Journal of Applied Physiology,* **92** (5), 1781-1788.

Bezodis, I.N., Thomson, A., Gittoes, M.J.R. and Kerwin, D.G. (2007). Identification of Instants of Touchdown & Take-Off in Sprint Running Using an Automatic Motion Analysis System. In H.-J. Menzel and M.H. Chargas (Eds). *Proceedings of the XXVth Symposium of the International Society of Biomechanics in Sports* (pp 501-504). Ouro Preto, Brazil. ISBS.

Henry, F.M. and Trafton, I.R. (1951). The Velocity Curve of Sprint Running. *Research Quarterly,* **22**, (4), 409-422.

Luhtanen, P. and Komi, P.V. (1978). Mechanical factors influencing running speed. In E. Asmussen and K. Jørgensen (Eds). *Sixth International Congress of Biomechanics* (pp 23-29). Baltimore, University Park Press.

Mann, R.V. and Herman, J. (1985). Kinematic Analysis of Olympic Sprint Performance: Men's 200 Meters. *International Journal of Sport Biomechanics,* **1**, 151-162.

Nilsson, J. and Thorstensson, A. (1987). Adaptability in Frequency and Amplitude of Leg Movements During Human Locomotion at Different Speeds. *Acta Physiologica Scandinavica,* **129**, *1*, 107-114.

Weyand, P.G., Sternlight, D.B., Bellizzi, M.J. and Wright, S. (2000). Faster top running speeds are achieved with greater ground forces not more rapid leg movements. *Journal of Applied Physiology,* **89** (5), 1991-1999.

Acknowledgement
The authors would like to thank the athletes who kindly agreed to participate in the study. This work was funded by EPSRC grant number EP/D076943.

EFFECT OF SPORT ACTIVITY ON COUNTER MOVEMENT JUMP PARAMETERS IN JUVENILE STUDENTS

Anne Richter, Diana Lang, Gerda Strutzenberger and Hermann Schwameder

BioMotion Center, Department of Sport and Sport Science, Karlsruhe Institute of Technology (KIT), Germany

FoSS – Research Center for Physical Education and Sports of Children and Adolescents, Karlsruhe, Germany

KEY WORDS: counter movement jump, arm-swing, variability.

INTRODUCTION: The counter movement jump (CMJ) is a commonly used method in performance diagnostics to measure leg power (Frick et al., 1991). The use of arm movements during jumping can increase the release velocity and thereby the jump height (e.g. Harman et al., 1990, Gerodimos et al., 2008). Despite the consensus, that arm-swing enhances the performance in the CMJ, still disagreement exists which jumping technique (with or without arm-swing) should be used in performance diagnostics. Marcovic et al. (2004) showed a good reliability for squat jump (SJ) and CMJ without arm-swing for physical education students, who had sufficient experience in explosive activities such as jumping. In contrast, for untrained individuals both, the use and the avoidance of the arm-swing can lead to differences in the jumping performance. The fixation of the arms at the hips might be unfamiliar and might cause variability in jump height. Unskilled use of the arm-swing, however, can lead to differences in jump height (Marcovic et al., 2004). To our knowledge, limited research has been done regarding the difference in variability between the two jumping techniques (with or without arm-swing). Therefore, one aim of this study was to describe differences between these two jumping techniques under consideration of the athletic experience. Furthermore, a specific focus should be set on the variability of the different jumps for experienced (more than 6 hours sports activity a week) compared with less experienced individuals, which may give some more information about the application of these two methods to the performance diagnostics.

METHODS: 380 students (12.7 ± 2.0 yrs, 47.9 ± 12.7 kg) of the secondary school category participated in this study. They were divided in two groups. The first group included athletic experienced students participating in special sports programs (more than 6 hours a week), and the second group consisted of students with less or nearly no special sports experience. All subjects were asked to perform three maximum vertical jumps while using an arm-swing (CMJA) and afterwards while holding their arms at the hip (CMJ). The instruction in both conditions was to jump as high as possible. Vertical ground reaction forces were measured with a divided force plate ("Leonardo Force Platform", 800 Hz). Jump height (h), maximum force and coefficient of variance (cv) were calculated. t-tests and ANOVA were used for statistical analysis.

RESULTS AND DISCUSSION: CMJAs show significantly greater jump heights compared to CMJ (Δh=3.3 cm ± 2.1). This corresponds with previously reported results (e.g. Harman et al., 1990, Gerodimos et al., 2008). Jump height of the experienced group was higher than the jump height of the less experienced group (CMJA: Δh=5.0 cm, CMJ: Δh=3.8 cm). The variability of the jump height is higher for CMJA compared with CMJ (Δcv=1.4% ± 5.8). The experienced group shows lower variability than the less experienced group (CMJA: Δcv=-1.8%, CMJ: Δcv= -1.0%).

REFERENCES:
Frick, U., Schmidtbleicher, D. and Wörn, C. (1991). Vergleich biomechanischer Meßverfahren zur Bestimmung der Sprunghöhe bei Vertikalsprüngen. *Leistungssport*, 2/91, 48-53.
Harman, E.A., Rosenstein, M.T., Frykman, P.N. and Rosenstein, R.M. (1990). The effects of arms and countermovement on vertical jumping. *Medicine and Science in Sports and Exercise*, 22(6), 825-833.
Gerodimos, V., Zafeiridis, A., Perkos, S., Dipla, K., Manou, V. and Kellis, S. (2008). The contribution of stretch-shortening cycle and arm-swing to vertical jumping performance in children, adolescents, and adult basketball players, Pediatric Exercise Science, 20, 379-389.
Markovic, G., Dizdar, D., Jukic, I. and Cardinale, M. (2004). Reliability and factorial validity of squat and countermovement jump tests. *Journal of Strength and Conditioning Research*, 18(3), 551-555.

ISBS 2009

Oral Session S10

Gymnastics

A SIMULATION STUDY OF THE INTERNAL TWISTING TORQUE IN THE FOUETTÉ TURN

Akiko Imura[1] and M.R. Yeadon[2]

Graduate School of Arts and Sciences, University of Tokyo, Tokyo, Japan[1]
School of Sport and Exercise Sciences, Loughborough University, Loughborough, UK[2]

The purpose of this study was to investigate the effects of the magnitude of the twisting torque for one revolution of a Fouetté turn. Simulations were performed using a simple model comprising the supporting leg and the remainder of the body. It is shown that when the dancer turns more than one revolution with a small twisting torque, the turn will be decelerated and will finally stop. A large twisting torque is required at the start of each turn in order to increase the angular momentum which will subsequently decrease during the turn due to friction.

KEY WORDS: turn, simulation, angular momentum.

INTRODUCTION: Skilled ballet dancers can continuously perform repeated Fouetté turns (Figure 1) for more than 30 revolutions, starting from one or two revolutions of the pirouetté which is started with both feet in contact with the floor. Friction during slipping reduces the initial angular momentum at the beginning of each turn. The dancers have to regain the angular momentum for the next revolution from the swing of the free leg and the arms enabled by a large frictional torque (T_F) exerted on the supporting foot (Laws, 1984, Imura et al., 2008). This is achieved using a torque T to produce the twisting motion of the free leg, upper body and arms relative to the supporting leg. The net external rotational impulse of T_F should be zero or positive after one revolution in order to maintain or increase the angular momentum. The behaviour of T_F depends on the magnitude of T relative to the limiting frictional torque which is the product of the friction coefficient, the normal ground reaction force and the radius of the foot contact. Thus, the magnitude of T regulates the continuity and speed of the turn under a certain friction coefficient between shoes and floor.

The purpose of this study was to investigate the effects of the magnitude of T on one revolution of Fouetté turn using a simple model comprising the supporting leg and the remainder of the body.

Figure 1: Sequential view of one revolution of Fouetté turn. Each picture is shown every 10% time of one revolution. The left most picture shows the configuration of the body at the middle of full foot contact (pictures from Imura et. al, 2008).

METHODS: Model: One typical dancer's body (mass 50 kg) was used for a model comprising two cylinders (Figure 2): the supporting leg (L) and the remainder of the body (B). The twisting torque T which rotates B relative to L, the radius (r) of the foot contact area, the moments of inertia (I_B and I_L) of the bodies B and L and the normal ground reaction force (N) were specified using monotonic quintic functions based on experimental data (Imura et al., 2008) as shown in Figure 3. The averaged N through the rotation was one body weight. The coefficient of friction (μ) was estimated to be 0.2 from experimental data (Imura et al., 2008).

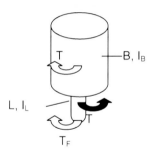

Figure 2: The model comprised the supporting leg L and the remainder of the body B. Initial torque directions are shown.

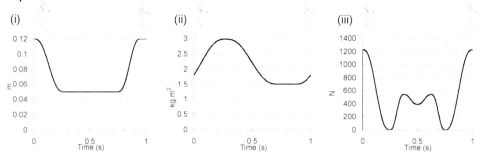

Figure 3: Time histories of: (i) Radius of the foot contact area (ii) moment inertia of body B (iii) normal ground reaction force.

Simulations: $T_F = T$ when $T \leq \mu \cdot N \cdot r$ and $T_F = \mu N r$ once L slipped. The angular accelerations of the body B and leg L ($\ddot{\phi}_B$ and $\ddot{\phi}_L$, respectively) were determined from the equations: $\ddot{\phi}_B = T / I_B$ and $\ddot{\phi}_L = (T_F - T) / I_L$, from which the angles and angular velocities of the bodies ($\phi_B, \dot{\phi}_B$ and $\phi_L, \dot{\phi}_L$) were calculated using stepwise integration. One revolution was simulated from midstance when the supporting foot fully contacted the floor. The initial value of T was varied from 6 to 30 Nm which was the limiting frictional torque and the initial angular velocity of B was varied from 6.0 to 8.0 radians/s. A search was made for turns which satisfied the following conditions: (1) The leg L turns approximately 6.28 radians. (2) The peak angle between B and L is less than 1.57 radians. (3) The relative angle between B and L at the end of the turn is within 0.1 rad of the value at the beginning of the turn. An empirical process was used in which simulations were run individually by hand and input was varied.

RESULTS: Two simulations were found satisfying the required conditions:(a) one with large initial torque and (b) one with small initial torque. Figures 4 and 5 show representative data for (a) and (b). In (a) the initial angular velocity of B and the magnitude of T were 6.2 radians/s and 28 Nm. In (b) corresponding values were 6.7 radians/s and 15 Nm. In both cases, T had to become negative before limiting friction became zero in order to accelerate the leg L in the direction of the turn. Twisting torque T during slipping was not different between (a) and (b) in order to rotate one revolution. In (b) the rotational impulse of T_F was negative at the end of the turn while in (a) the impulse was positive and close to zero(Figure 6). The whole body rotated one revolution in each turn but the angular velocity of body B decreased in (b) while it remained the same in (a).

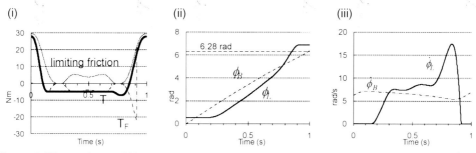

Figure 4: Kinematics and kinetics of the larger torque turn (a): (i) T, T_F and limiting T_F (ii) $\dot\phi_B$ and $\dot\phi_L$ (iii) $\dot\phi_B$ and $\dot\phi_L \cdot \dot\phi_B$ is the same at the start and end of the simulation.

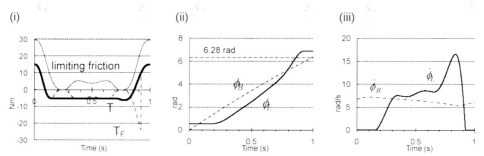

Figure 5: Kinematics and kinetics of the small torque turn (b): (i) T, T_F and limiting T_F (ii) $\dot\phi_B$ and $\dot\phi_L$ (iii) $\dot\phi_B$ and $\dot\phi_L \cdot \dot\phi_B$ is smaller at t=1 compared to t=0.

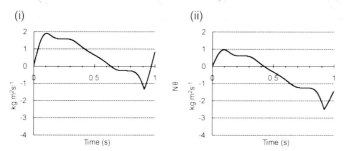

Figure 6: The rotational impulse of the frictional torque: (i) larger torque turn (a), (ii) smaller torque turn (b).

DISCUSSION: To start to slip in the direction of the turn, T has to be negative and have magnitude greater than the limiting friction ($\mu N r$). This explains the mechanism to generate the angular momentum for the next revolution: T_F is exerted on the supporting foot and gives the rotational impulse in the direction of the turn when the dancer swings the free leg before the limiting friction becomes small. If T is still positive when it exceeds limiting friction the supporting leg of the dancer will slip in the opposite direction to the turn.

The difference between the large initial T and small initial T caused the difference in the angular momentum of the remainder of the body at the end of the simulation. T was changed at the same timings in (a) and (b), the rotational impulse of T_F was positive in (a) and negative in (b) though B and L rotated one revolution in each case while the angular velocity of B decreased in (b). Before slipping T should be less than limiting friction, and T_F is equal to

T. Thus, the rotational impulse in (b) did not increase as much as that in (a) before slipping. Because the decrease in rotational impulse of T_F during slipping was the same in (a) and (b), the rotational impulse over one turn depends on the value of T at the start and end of the simulation. Thus, the Fouetté turn will continue but will be reduced in speed because the rotational impulse cannot be recovered with little torque as in (b). In such a case, the dancer has to swing the leg quicker and with more force from the muscles in the next turn to regain the angular momentum, recognizing that there is insufficient angular momentum to complete the turn.

To start the Fouetté turn, the dancer begins to turn around the supporting leg by exerting force on the floor with the free leg. This will produce more angular momentum than in the following turn because the moment arm between both feet is larger than the radius of the supporting foot (Laws, 1978). Thus the initial angular velocity of B reflects the angular momentum the dancer already has before the following Fouetté turn. For one revolution, the initial velocity was larger in (b) than in (a). This means the dancer adjusts the magnitude of T for one revolution according to the initial angular momentum. However, the magnitude of T for the swing is practically limited because the dancer has to face the front after one revolution. The dancer has to sense the angular momentum gained by the initial movement of the turn and exert the appropriate T for the following turns.

The time course of the angular velocities will be changed according to the input pattern of T. The restriction of the angle difference between B and L restricts the magnitude of T both initially and during slipping. More intricate input of T would provide another way of producing one revolution of the Fouetté turn, reflecting more complicated movement of the free leg. However, the magnitude of T for the swing regulates the speed and the continuity of the Fouetté turn.

More T will be required when the friction coefficient is larger for continuing the turn because the decrease in the angular momentum during slipping will be greater. However, the dancer will slip in the opposite direction if T exdeeds limiting friction. There is a limit as to how much friction can be accommodated since there is a limit to the amount of relative rotation that can occur between B and L.

CONCLUSION: To perform a number of consecutive Fouetté turns, it is important to maintain the angular momentum and swing the free leg with appropriate torque just sufficient to recover the angular momentum lost by friction. Hence the Fouetté turn starting with some pirouetté turns is difficult because the dancer has to regulate the magnitude of T for the swing just after the pirouetté and generate additional angular momentum. In a practical situation, the dancer pays attention to the friction of the floor, sensing the appropriate magnitude of the required twisting torque when performing turns. The dancer might be able to estimate the magnitude from the amount of rotation of the upper body, regulating the swing of the free leg and arms.

REFERENCES:

Imura A., Iino Y. and Kojima T. (2008). Biomechanics of the continuity and speed change during one revolution of the Fouetté turn. *Human Movement Science*. 27, 903-913.

Laws K (1978-79). An Analysis of Turns in Dance. *Dance Research Journal*.11,12-19.

Laws K (1984). *The physics of dance* (pp78-81.). New York: Schermer Books.

BIOMECHANICAL ANALYSIS OF GYMNASTIC BACK HANDSPRING

Chenfu Huang and Gin-Shu Hsu
National Taiwan Normal University

The purpose of this study is to compare the kinematics and kinetics of back handspring of skilled and non-skilled performers. Eight gymnasts and eight cheerleaders participted in the study. A JVC 9800 DV camera (60 Hz) was synchronized by using a LED light with a Kistler force platform (600 Hz) to collect the data. The results indicated the peak vertical GRF before take-off for gymnasts and cheerleaders are 2.3 BW and 2.19 BW; and the peak horizontal GRF before take-off are 0.67 BW and 0.53 BW respectively; the gymnasts have greater jump height, take-off center of mass (CM) velocity, horizontal CM velocity at hand push off, hip angle at take-off and longer first phase flight time than the cheerleaders. It suggests that the greater jump height and longer flight time are required for good handspring performance.

KEY WORDS: performance, take-off, kinematics, kinetics.

INTRODUCTION: The gymnastics is an official event in the first modern Olympics Games in 1896. The floor exercise is the foundation for learning the basic gymnastics movement including jumping, turning, rolling, and flipping. The athletes have to master the floor exercise before advance to other instrument events. The back handspring is one of the important skills in floor exercise and is a basic building block for many gymnastic routines (Payne and Barker, 1976). The back handspring also frequently performed in cheerleading. Until now, no detailed biomechanical analysis of back handspring has been reported. There is a need to determine the difference in the biomechanics of performance of back handspring in skilled and unskilled performers. This purpose of this study is to compare the biomechanics of back handspring between skilled gymnasts with 12 years experience and unskilled cheerleaders with one year experience.

METHODS: Eight gymnasts (height 167.9 ± 4.2 cm, age 21.9 ± 1.9 yrs, mass 65 ± 3.6 kg) and eight cheerleaders (height 174.4 ± 4.2 cm, age 19.8 ± 3.7 yrs, mass 75.3 ± 11.7 kg) participated in this study. All subjects were informed of the experimental procedures and gave their consent before participating. The subjects performed a 10-mimute warm-up session consisting of stretching upper and lower limb muscles before data collection. A JVC 9800 DV camera (60 Hz) was synchronized by using a LED light with a Kistler force platform (600 Hz) to collect the data of subject performing the back handspring. For each trial, the subjects were instructed to initially stand on a force platform and did a back handspring once given a verbal signal. The subject performed three successful trials and one trial was selected for analysis based on best judging score. Nine body landmarks (ear, shoulder, elbow, wrist, hip, knee, ankle, toe and heel) were digitized by the Kwon3d motion system (www.kwon3d.com). Based on a frequency content analysis of the digitized coordinate data, marker trajectories were filtered at 6 Hz using a Butterworth fourth order zero-lag filter. The second central different differentiation method was used to determine velocities. The segment COM, and body COM were calculated by using the Dempster data provided by Winter (1990). The ground reaction forces and impulses of back handspring were analyzed by Kwon GRF software. An independent t-test was used to test the variables between the gymnasts and cheerleaders on back handspring. The variables are time, body CM velocity, joint angles and velocities, force and impulse.

RESULTS AND DISCUSSION:

There are significant differences on first flight phase and push-off phase and height of CM between gymnasts and cheerleaders (Table 1). The gymnasts have a longer first flight time and a shorter push-off phase and body CM height than the cheerleaders. The gymnasts have a longer first flight phase indicated the gymnasts' using both feet forceful push-off ground which results a longer CM height than the cheerleaders. The greater CM height after takeoff helped the gymnasts perform the turn and execute the back handspring. The gymnasts also have a shorter push-off phase suggest the hands quickly push-off floor which helps body swing and landing. Two unskilled cheerleaders performed the movement without the second flight phase. Their hands push-off the ground after their feet contact the floor. Only six cheerleaders' second flight phase was used to run the t-test. No difference was found on other time variables between two groups.

Table 1 Time of back handspring   unit: s

	Gymnast (N=8)	Cheerleader (N=8)	t	p
Total time	1.98 ± 0.18	1.98 ± 0.22	.04	.968
Eccentric phase	1.05 ± 0.18	1.00 ± 0.29	.44	.669
Concentric phase	0.26 ± 0.05	0.25 ± 0.06	.38	.707
I flight phase	0.22 ± 0.02	0.17 ± 0.03	3.8	.002*
Push-off phase	0.32 ± 0.04	0.49 ± 0.08	-5.3	.000*
II flight phase	0.14 ± 0.02	0.13 ± 0.05(N=6)	.32	.759
Height of CM(cm)	89.2 ± 3.3	81.7 ± 8.8	2.25	.041*

*p<.05

The gymnasts have the greater horizontal and vertical body CM velocities than the cheerleaders at feet takeoff which also indicated the gymnasts have greater resultant body CM velocity at take-off. The greater body CM velocity at takeoff for gymnasts result the greater CM jumping height which gives a longer time for performing back handspring. The gymnasts also have greater horizontal body CM velocity at hand pushoff which help the body for the landing. The smaller values of body CM velocities at takeoff and hands push-off results the poor back handspring performance of the cheerleaders.

Table 2   Horizontal and vertical CM velocities at take-off and pushoff   unit : ms^{-1}

	Gymnast (N=8)	Cheerleader (N=8)	t	p
Horizontal CM velocity take-off	1.76 ± 0.32	1.22 ± 0.47	2.65	.019*
Vertical CM velocity take-off	0.88 ± 0.23	0.39 ± 0.25	4.02	.001*
Horizontal CM velocity hand pushoff	1.58 ± 0.24	1.15 ± 0.34	2.96	.010*
Vertical CM velocity hand pushoff	-0.06 ± 0.25	-0.04 ± 0.68	-.07	.943

*p<.05

Joint angles and velocities of shoulder, hip, knee, and ankle of both groups at takeoff are listed in Table 3 and 4. The gymnasts have greater knee angle and hip and knee angular velocities than the cheerleaders at feet take-off. The greater knee extension and faster hip and knee angular velocities at take-off are the important variables for indentify good and average back handspring performance.

Table 3 Joint angle of back handspring at take-off   unit : deg

	Gymnast (N=8)	Cheerleader (N=8)	t	p
Shoulder	162.9 ± 7.9	165.2 ± 4.3	-.74	.474
Hip	215.9 ± 46.2	217.7 ± 8.5	-.33	.749
Knee	135.5 ± 13.2	126.2 ± 5.4	2.87	.012*
Ankle	142.1 ± 6.3	140.5 ± 2.8	.65	.528

*p<.05

Table 4 Joint angular velocity of back handspring at take-off   unit : deg/s

	Gymnast (N=8)	Cheerleader (N=8)	t	p
Shoulder	333.8 ± 86.1	366.1 ± 127.7	-.59	.562
Hip	767.1 ± 47.4	672.0 ± 61.3	3.47	.004*
Knee	456.5 ± 106.4	364.0 ± 59.3	2.15	.050*
Ankle	466.3 ± 103.9	379.9 ± 90.8	1.77	.098

*$p<.05$ *

There are significant differences on shoulder and knee angles between groups at hands touchdown. The overextension of shoulder observed on cheerleaders indicated lack of balance control during the handstand position. The cheerleaders less knee extension at hands touchdown may due to smaller knee angle at takeoff (Table 5). The control of shoulder angle and more knee extension indicated good handstand position during back handspring. The faster knee angular velocity than the cheerleaders suggest that the gymnast continue fast knee extension at hands touchdown (Table 6).

Table 5 Joint angles of back handspring at hands touchdown   unit：deg

	Gymnast (N=8)	Cheerleader (N=8)	t	p
Shoulder	174.8 ± 7.2	186.23 ± 3.3	-4.08	.001*
Hip	243.8 ± 6.6	247.28 ± 11.6	-.73	.476
Knee	138.0 ± 9.6	110.67 ± 17.9	3.81	.002*
Ankle	146.9 ± 7.1	147.56 ± 5.4	-.20	.846

*$p<.05$

Table 6 Angular velocities of back handspring at hands touchdown   unit：deg/s

	Gymnast (N=8)	Cheerleader (N=8)	t	p
Shoulder	-195.6 ± 121.2	-208.1 ± 72.5	.25	.807
Hip	-214.2 ± 88.0	-153.8 ± 92.9	-1.34	.202
Knee	193.3 ± 82.9	-42.8 ± 138.3	4.14	.001*
Ankle	-76.7 ± 75.0	-53.3 ± 85.3	-.58	.569

*$p<.05$

The gymnasts have greater hip angle and velocity than the cheerleaders at hands push off.(Table 7,8) The greater hip angle and velocity show the gymnasts fast extend the hip forceful hands push off to increase the rotation of trunk for the control landing. The less hip extension angle and slow hip angular velocity for the cheerleaders show lack of fast hip extension at hands push-off.

Table 7 Joint angles of back handspring at hands push-off   unit：deg

	Gymnast (N=8)	Cheerleader (N=8)	T	p
Shoulder	140.9 ± 7.5	134.1 ± 20.8	.87	.399
Hip	137.3 ± 12.1	107.0 ± 32.5	2.47	.027*
Knee	187.4 ± 9.9	164.8 ± 37.4	1.65	.121
Ankle	126.9 ± 12.4	112.6 ± 19.1	1.78	.097

*$p<.05$

Table 8 Joint angular velocities of back handspring at hands push-off   unit：deg/s

	Gymnast (N=8)	Cheerleader (N=8)	t	p
Shoulder	-17.9 ± 98.8	-58.7 ± 106.6	.79	.441
Hip	-383.1 ± 110.7	-194.6 ± 133.5	-3.08	.008*
Knee	67.6 ± 37.4	58.0 ± 95.72	.27	.794
Ankle	-39.7 ± 63.0	29.0 ± 77.8	-1.95	.072

*$p<.05$

The peak vertical and horizontal forces before back handspring take-off for gymnasts and cheerleaders are 2.3, 2.2 BW and 0.7, 0.5 BW respectively (Table 9). No sifnificant difference was found on peak force between two groups. Only one gymnast show all positive horizontal force before take-off which indicated most subjects produce braking horizontal force during eccentric phase to prevent backward fall before take-off.

Table 9 Peak vertical and horizontal force at back handspring take-off   unit: BW

	Gymnast (N=8)	Cheerleader (N=8)	t	p
Peak Vertical force	2.3 ± 0.3	2.2 ± 0.1	.92	.371
Peak Horizontal force	0.7 ± 0.1	0.5 ± 0.2	1.88	.082

*p<.05

No significant was found on total vertical and horizontal impulse and vertical and horizontal impulse during eccentric and concentric phase before back handspring take-off for gymnasts and cheerleaders. (Table 10)

Table 10 Eccentric, concentric and Total impulse at back handspring take-off   unit: N*s

	Gymnast (N=8)	Cheerleader (N=8)	t	p
Total vertical impulse	933.2 ± 122.8	973.7 ± 168.1	-.55	.590
Total horizontal impulse	-108.4 ± 15.7	-99.6 ± 28.8	-.76	.459
Vertical impulse eccentric	665.4 ± 124.1	683.3 ± 167.0	-.24	.811
Horizontal impulse eccentric	-44.7 ± 9.4	-45.1 ± 17.3	.07	.946
Vertical impulse concentric	268.5 ± 44.4	291.1 ± 75.2	-.73	.476
Horizontal impulse concentric	-64.1 ± 12.9	-54.7 ± 13.7	-1.41	.178

*p<.05

CONCLUSION: The purpose of this study is to compare the kinematics and kinetics of back handspring of skilled and non-skilled performers. The results indicated the peak vertical GRF before take-off for gymnasts and cheerleaders are 2.3 BW and 2.19 BW; and the peak horizontal GRF before take-off are 0.67 BW and 0.53 BW respectively; The gymnasts have greater jump height, take-off center of mass (CM) velocity, horizontal CM velocity at hand push off, hip angle at take-off and longer first air time than the cheerleaders. It suggests that the greater jump height and longer flight time are required for good handspring performance.

REFERENCES:

Payne, A. H. and Barker, P. (1976). Comparison of the take-off forces in the flic flac and the back somersault in gymnastics. *Biomechanics V : proceedings of the Fifth International Congress of Biomechanics* (pp 314-321). Jyvaskyla, Finland: 1975 International Congress of Biomechanics.

Winter, D. A. (1990). *Biomechanics and motor control of human movement (2nd Ed.)*. New York: John Wiley & Sons.

Calculation forces from bar movement on parallel bars in gymnastics

Falk Naundorf, Klaus, Knoll and Stefan Brehmer

Institute for Applied Training Science, Leipzig, Germany

Modern artistic gymnastics apparatus have elastic properties, which the gymnast should use. It is important to know how a gymnast can give energy to the apparatus, especially to the bar(s) and how the stored energy can be used by the gymnast. The parallel bars were not included in such questions in the research yet. A static calibration at different positions of one bar was utilized as a precondition for the calculation of the forces during gymnastics exercises. Using synchronized 2D-video-analysis of the bar movement and the gymnasts performance (2 cameras) we calculate the forces based on our calibration. Examples of force-time-curves from parallel bars dismounts from German national gymnastics team will be shown. Using force-time-characteristics for supporting motor learning is a difficult task for the future.

KEY WORDS: artistic gymnastics, parallel bars, bar movement, forces, dismounts

INTRODUCTION: Optimal technique in artistic gymnastics depends not only on the movement of the gymnasts body parts. It is also important to use the elastics properties of the gymnastics apparatus. Differences of hard and elastic surfaces (Krug, Minow & Jassmann, 2001), use of elastic springboards (Sano, Ikegami, Nunome, Apriantono & Sakurai, 2004, 2007), coordination of gymnasts and bar movement on high bar (Krug, Knoll, Wagner & Bronst, 1998) and the energy exchange between the body of the gymnast and the high bar (Arampatzis & Brüggemann, 1999) are major fields of research in gymnastics. The properties of modern parallel bar are fundamental for the elements performed by the gymnasts at international competition. Elements on parallel bar like giant swing (Prassas, Ostarello & Inouye 2004) or basket or rather felge to handstand (Veličković, Kolar & Petković, 2006; Hiley, Wangler & Predescu, 2009) have been investigated. However, there is not enough information about the elastic properties and the coordination of the gymnasts' movements on parallel bars. Bringing the gymnast's movement and bar movement into resonance will lead to better performance.

The aim of our preliminary study is to take the first step to build a measuring system which could be used for supporting gymnast's technique training. A dynamometric measuring system on parallel bars was first invented by Schmidt (1973). The instrumented parallel bars were used to measure vertical and horizontal forces. Current parallel bars have more elastic bars than in the 1970s.

METHODS: To calculate forces from the bars movement a calibration of one bar was necessary. For this investigation we used a static calibration with defined weights. The parallel bars were located at the national training camp for the German national gymnastics team. The apparatus was from the official supplier of the Olympic Games in Athens 2004 and Beijing 2008, Janssen-Fritsen Gymnastics from Helmond, Netherlands. We fixed weights on three positions on one bar (Figure 1 and 2). The first position was the centre of the bar, second 40 cm out of the centre and third end of

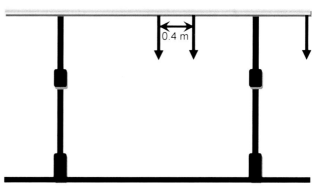

Figure 1: Location of the Calibration points on the bar.

the bar. Often gymnasts start their routines at the end of the bar, that's why this position is also important. We used weights of 50 kg, 100 kg and 150 kg. We took pictures from every weight level and measured the bar movement in vertical direction. Based on this data a formula was created to calculate the vertical forces from the bar position. An example formula for the centre position is shown (Formula 1).

Force [N] = $-0{,}2847b^2 + 35{,}784b$ (where b is bar position) (1)

We also carried out the calibration for a bar from another European manufacturer. The bar was made by Spieth, Esslingen, Germany (supplier of the World Championships in Stuttgart 2007) and it has slightly different properties.

Figure 2: Fixation of the weights on the bar.   Figure 3: Gymnast near the marked bar.

Applying this formula to bar movements during exercises of the gymnasts (Figure 3) the estimated forces on the bar were calculated. We used two synchronized cameras (Basler) with 640 x 480 pixel resolution and 100 Frames per second. Video recording was done by the software Templo (Contemplas). The cameras were set up for a 2D-Analysis. The first camera recorded the movement of the gymnast and the second camera was focused on the bar movement using only a small part of the first camera view. For data analysis we digitized manually the movement of the bar and the gymnast with the Software MESS2DDV from Drenk (IAT Leipzig).
Different elements with high value in the gymnastics rules, the Code of points (Fédération Internationale de Gymnastique, 2009) were analyzed (Giant swing (Value C), basket or rather felge to handstand (Value D), roll backward with somersault backward tucked to upper arm hang (Dimitrenko, Value E), dismounts like double somersaults backward pike (Value D)). Force-time-curves of absolute and relative vertical forces were calculated. For overall forces it is necessary to double the forces. This is only valid under the precondition of symmetric element (symmetric forces on both bars) as named before. Other elements with single support on only one bar or no simultaneous release of the hands (elements: Diamidov, swing forward with ½ turn to handstand) were not included into our analysis.

RESULTS: The example in figure 4 shows the force-time-velocity-curves for the dismount of two different gymnasts. The starting position for both gymnasts was the handstand. Both relieve the bars, but gymnast 1 (blue) pull the bar up. Gymnast 2 (red) has a higher and earlier force maximum.

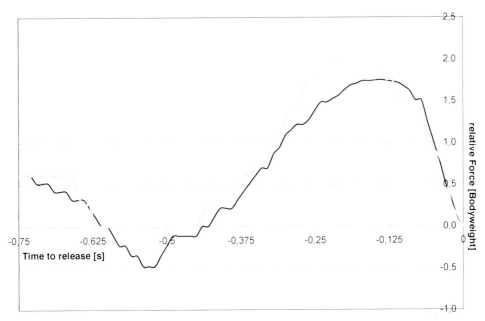

Figure 4: Force-time-curve from two gymnasts doing a double somersault backwards dismount from parallel bars (vertical force on one bar).

DISCUSSION: The study shows the possibility of calculation forces from bar movement. This first step is only usable for vertical forces. Differences between two gymnasts for the dismount are visible. These may be characteristics for a better technique. Gymnast 2 (grey) is a more successful gymnast. For a better understanding of the force-time-characteristics detailed analyses of the gymnasts' movement are necessary.

CONCLUSION: Manual digitizing of the bar movement for calculation forces can only be the first step. Further developments are necessary to get quick information about the bar movement and the forces. Automatically detecting of a marker on the bar and direct measuring of bar flexion e.g. with strain gauges are challenges for the future. However there is one problem: Forces of the gymnast result not only in flexion of the bar, also the other parts of a modern parallel bar take forces in. Using force-time-characteristics for supporting motor learning of gymnasts is a difficult task for the future. Typical force-time-curves from gymnasts of different performance level will be used for detecting movement errors. More data about the elastic properties of the bars are necessary for a better understanding of energy exchange between the gymnast and the parallel bars.

REFERENCES:

Arampatzis, A., and Brüggemann, G.-P. (1999). Mechanical energetic processes during the giant swing exercise before dismounts and flight elements on the high bar and the uneven parallel bars. *Journal of Biomechanics*, 32 (8), 811–820.

Fédération Internationale de Gymnastique. (2009). Code de Pointage – Gymnastique Artistique Masculine - Code of points Men's artistic gymnastics. Lausanne.

Hiley, M.J., Wangler, R. and Predescu, G. (2009). Optimization of the felge on parallel bars. *Sports Biomechanics*, 8 (1), 39–51.

Krug, J., Knoll, K., Wagner, R., and Bronst, A. (1998). Differences in basic elements between juniors and seniors and between women and men in gymnastics. In H. J. Riehle & M. M. Vieten (Eds.), *Proceedings II of the XVI ISBS Symposium* (pp. 221-224). Konstanz: UVK - Universitätsverlag Konstanz.

Krug, J., Minow, H.-J., and Jaßmann, P. (2001). Differences between jumps on hard and elastic surfaces. In J. R. Blackwell (Ed.), *Proceedings of Oral Sessions XIX International Symposium on Biomechanics in Sports June 20-26, 2001* (pp. 139-142). San Francisco: Exercise and Sport Science Department, University of San Francisco.

Prassas, S., Ostarello, J. and Inouye, C. (2004). Giant swings on the parallel bars: a case study. In M. Lamontagne, D. G. E. Robertson & H. Sveistrup (Eds.), *Proceedings XXIInd International Symposium on Biomechanics in Sports 2004* (pp. 133). Ottawa: Faculty of Health Sciences University of Ottawa.

Sano, S., Ikegami, Y., Nunome, H., Apriantono, T. and Sakurai, S. (2004). An accurate estimation of the springboard reaction force in vaulting table of gymnastics. In M. Lamontagne, D. G. E. Robertson & H. Sveistrup (Eds.), *Proceedings XXIInd International Symposium on Biomechanics in Sports 2004* (pp. 61-63). Ottawa: Faculty of Health Sciences University of Ottawa.

Sano, S., Ikegami, Y., Nunome, H., Apriantono, T., and Sakurai, S. (2007). The continuous measurement of the springboard reaction force in gymnastic vaulting. *Journal of Sports Science*, 25 (4), 381–391.

Schmidt, D. (1973). *Zur Bestimmung dynamischer und kinematischer Kennwerte am Barren in Verbindung mit der Entwicklung und Erprobung eines dynamographischen Meßverfahrens (dargestellt am Beispiel ausgewählter Vorschwungelemente der Pflicht- und Kürübungen aus dem Bereich der Meisterklasse)*. Unveröffentlichte Dissertation, Deutsche Hochschule für Körperkultur, Leipzig.

Veličković, S., Kolar, E. and Petković, D. (2006). The kinematic of the basket with ½ turn to handstand on the parallel bars. *Physical Education and Sport*, 4, (2), 137–152.

Acknowledgement

The Institute for Applied Training Science, Leipzig, Germany is supported by the Federal Ministry of the Interior of the Federal Republic of Germany.

OPTIMISATION OF THE FELGE TO HANDSTAND ON PARALLEL BARS

Michael Hiley, Roger Wangler and Gheorghe Predescu

School of Sport and Exercise Sciences, Loughborough University, UK

The felge, or undersomersault, to handstand on parallel bars has become an important skill in Men's Artistic Gymnastics as it forms the basis of many complex variations. To receive no deductions from the judges, the felge must be performed without demonstrating the use of strength to achieve the final handstand position. Two male gymnasts each performed nine trials of the felge from handstand to handstand while data were recorded using an automatic motion capture system. The highest and lowest scoring trials of each gymnast, as determined by four international judges, were chosen for further analysis. The technique used by each gymnast was optimised using a computer simulation model so that the final handstand position could be achieved with straight arms. Two separate optimisations found different techniques identified in the coaching literature that are used by gymnasts. Although the stoop stalder technique used by the two gymnasts was found to be more demanding than the clear circle technique in terms of the strength required, it offered the potential for more consistent performance and future developments in skill complexity.

KEY WORDS: gymnastics, simulation, technique, undersomersault

INTRODUCTION: In the new Code of Points (Fédération Internationale de Gymnastique (FIG), 2006) the felge, or undersomersault, on parallel bars has become an important skill in Men's Artistic Gymnastics as it forms the basis for many variations of the skill. Although the basic skill is performed to support, it is the felge from handstand to handstand (Figure 1a) that provides the basis for the more complex variations. From the handstand position the gymnast lowers the body by closing the shoulder angle and allowing the shoulders to move forwards relative to the hands (Figure 1b). The gymnast then rotates backwards about the point of contact with the bars and circles below the bar. Release occurs shortly after the gymnast's mass centre has passed above the level of the bars (Figure 1b). The gymnast re-grasps the bars before reaching the handstand position. In order to receive no deductions from the judges, the gymnast must perform the felge without demonstrating the use of strength to achieve the final handstand position.

Figure 1. The felge from hand stand to handstand using (a) the "clear circle" (adapted from the FIG Code of Points, 2006) and (b) the "stoop stalder" technique.

The technique depicted in the Code of Points (FIG, 2006), Figure 1a, closely resembles a backward clear circle to handstand as performed on the high bar. During this technique the gymnast maintains quite an extended hip angle throughout the majority of the circle, in particular whilst the gymnast is below the bars (Figure 1a). It has been recommended that this technique is used during the initial stages of learning the felge (Davis, 2005). However, the technique used by many senior gymnasts more closely resembles a "stoop stalder" (Davis, 2005). As the gymnast passes beneath the bars a deep pike position is adopted from which the gymnast rapidly extends passing through release and into the final handstand position (Figure 1b).

The aim of the present study was to optimise the existing technique of gymnasts performing the felge from handstand to handstand so that the final position could be achieved with straight arms. The optimisations would be used to gain an insight into which of the two techniques described above is the most appropriate.

METHODS: Data collection: Two senior male gymnasts competing at national level (gymnast 1: mass 61.2 kg, height 1.65 m; gymnast 2: mass 63.5 kg, height 1.75 m) each performed 9 trials of the felge from handstand to handstand. All trials were captured using 13 Vicon M2 cameras operating at 100 Hz. In addition all trials were recorded with a standard 50 Hz digital video camcorder (Panasonic NV-GS200EB). Three-dimensional marker coordinates were reconstructed from which arm orientation and joint configuration angles were calculated. A set of 95 anthropometric measurements were taken on each gymnast and inertia parameters were calculated using the model of Yeadon (1990b). Four judges with international accreditation (FIG) scored each felge from the video recordings. The highest and lowest scoring trials of each gymnast were chosen for further analysis. None of the chosen trials achieved the final handstand position with straight arms.

Matching Process: A four segment model including damped linear springs at the shoulder and hands for the elastic structures of the gymnast and high bar was used (Hiley and Yeadon, 2003). The simulation model was angle driven using joint angle time histiories in the form of Fourier series, which were matched to the recorded angle data during a matching procedure. During the matching optimisation the bar and gymnast spring parameters were allowed to vary together with the initial orientation and angular momentum of the model. The optimisation was required to produce a close match between the recorded and simulated rotation angles, bar displacements, joint angle time histories and absolute linear and angular momentum at release. Each simulation started once the angular velocity of the arm segment was in the positive direction (anti-clockwise).

Optimisation: The cost function was based on minimising the peak joint torques at the hip and shoulder joints whilst seeking an acceptable felge through the use of appropriate penalties. Joint torque limits were obtained from the matching simulations. The simulations started from the same point as in the matching process and finished once the torso segment had rotated 40° past the vertical. The cost function was calculated from when the torso segment reached the vertical through to 40° past the vertical. The value returned to the optimisation was the lowest value of the cost function during this period. The optimum technique was required to produce sufficient vertical velocity at release to achieve a mass centre height in flight of at least 90% of the final handstand position measured above the bars. The simulation incurred penalties if the horizontal velocity and normalised angular momentum at release exceeded the range obtained from the analysis of the 18 trials and values reported for high scoring performances (Takei and Dunn,1996). A further penalty was imposed for excessive hip flexion angles at release from the bars as this was likely to result in poor body configurations on re-grasping the bars (Takei and Dunn, 1996). The optimisations were run twice: the first set of four optimisations with no limits placed on the joint angle time histories (other than those described above) and the second set where they were constrained to produce a stoop stalder technique – this was achieved by creating a penalty for flexing too early in the felge and not producing the characteristic deep pike position.

RESULTS: Over the approximate 270° rotation of the four matching simulations the model was able to reproduce the whole body rotation angle to within 2° root mean squared (rms) difference and the displacements of the bar to within 0.005 m rms difference (Figure 2). The matches between the measured joint angle time histories and those determined using Fourier series were close with an average rms difference of 4°. The simulation model matched the mass centre velocity at release to within 1%.

In the first set of optimisations where the joint angle time histories were not constrained the model was able to achieve the appropriate vertical velocity at release whilst satisfying the criteria for a successful performance. At release the model had a higher mass centre position and vertical velocity than in the actual performances. The peak hip and shoulder joint torques from the optimisations did not exceed either of the limits. The technique in the first set of optimised simulations differed from the gymnasts' technique (Figure 3b). In the optimisations where the joint angle time histories were encouraged to produce a technique similar to the gymnasts' own technique the model was still able to achieve the appropriate vertical velocity at release whilst satisfying the criteria for a successful performance (Figure 3c). However, the peak joint torques were higher than those obtained from the first set of optimisations.

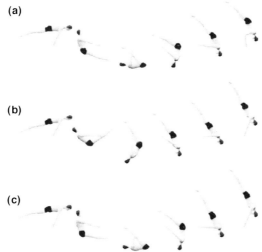

Figure 2. Graphical sequences of the (a) gymnast's technique, (b) optimised technique and (c) constrained optimised technique.

DISCUSSION: In all four cases the first set of optimisations was able to improve the gymnasts' performances with the mass centre able to reach over 90% of handstand height in the flight phase. The improvement was achieved through a combination of increased vertical velocity and an increase in mass centre height at release. The higher mass centre location and more extended body configuration at release have been shown to be desirable from a judging perspective (Takei and Dunn, 1996). However, the optimised technique differed from the stoop stalder technique and more closely represented the clear circle technique. When the optimisation used a constraint to encourage a stoop stalder technique the second set of optimal solutions was still able to achieve the required increase in vertical velocity at release. The increased vertical velocity at release was produced predominantly by a more rapid extension of the hip angle.

Both sets of optimisations were able to achieve the improved performance whilst staying within the joint torque limits defined by the gymnasts' actual performances. When choosing which technique to use, it was found that in terms of peak joint torque the stoop stalder was more demanding of the gymnast. This explains why in the early stages of learning, the clear circle technique is adopted, as recommended by Davis (2005). Why then is the stoop stalder technique adopted by the majority of senior gymnasts?

The path of the mass centre during the optimal felges is shown in Figure 3. In the optimisations encouraged to produce the stoop stalder the path of the mass centre is flatter and more vertical as the gymnast approaches release. This has two advantages: firstly the direction of the mass centre velocity changes less near to release, when compared to the clear circle technique, and this should lead to a more consistent performance, when compared to the clear circle technique. Secondly, the felge to handstand forms the basis of more complex skills: typically the felge to handstand with either a half or full twist. In these skills there is not a flight phase as such, rather the gymnast makes hand changes whilst the force on the bars is low. Having a vertical mass centre velocity while the body is twisting reduces the task complexity of the hand changes (i.e. less correction for non-vertical alignment).

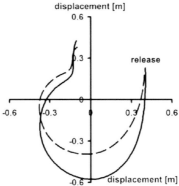

Figure 3. Path of the mass centre during the optimised (dashed line) and constrained optimised (solid line) felge to handstand.

CONCLUSION: It was found that both good and poor performances of the felge from handstand to handstand could be improved. The technique used by the gymnasts could be improved by extending the hip angle more rapidly and over a larger range. The minimisation of joint torque resulted in a global optimum similar to the clear circle technique. Since the strength requirements of the clear circle are lower than those of the stoop stalder the clear circle is more appropriate for the early stages of development as suggested by Davis (2005). Although the optimum technique that closely resembled the technique used by the gymnasts was found to be more demanding in terms of the strength required, it does offer the potential for more consistent performance and future developments in skill complexity.

REFERENCES:
Davis, J. (2005). Undersomersaults on parallel bars. Gym Craft, 14, 6-7.
Fédération Internationale de Gymnastique, (2006). Code of Points. Moutier, Switzerland: F.I.G.
Hiley, M.J. and Yeadon, M.R. (2003). Optimum technique for generating angular momentum in accelerated backward giant circles prior to a dismount. *Journal of Applied Biomechanics*, 19, 119-130.
Takei, Y. and Dunn, J.H. (1996). A comparison of techniques used by elite gymnasts in performing the basket-to-handstand mount. *Journal of Sports Sciences*, 14, 269-279.
Yeadon, M.R. (1990). The simulation of aerial movement - II. A mathematical inertia model of the human body. *Journal of Biomechanics*, 23, 67-74.

Acknowledgement
The authors wish to acknowledge the support of the British Gymnastics World Class Programme.

RELIABILITY OF A GYMNASTICS VAULTING FEEDBACK SYSTEM

Elizabeth Bradshaw[1,2], Patria Hume[2], Mark Calton[3] and Brad Aisbett[4]

[1]Centre of Physical Activity Across the Lifespan, School of Exercise Science, Australian Catholic University, Melbourne, Australia
[2]Institute of Sport and Recreation Research New Zealand, School of Sport and Recreation, Auckland University of Technology, Auckland, New Zealand
[3]Queensland Academy of Sport Gymnastics Centre, Sleeman Centre, Brisbane
[4]School of Exercise and Nutrition Sciences, Deakin University, Melbourne, Australia.

The current study assessed the intra- and inter-day reliability of a custom-built gymnastics vaulting feedback system. The system is a coach-friendly customized infra-red timing gate and contact timing mat system operated by the coach to augment the feedback provided to gymnasts on their vaulting performance during regular training practice. Thirteen Australian high performance gymnasts (eight males and five females) aged 11-23 years were assessed during two training sessions (Day 1 and Day 2) at their regular training centre. The approach velocity and board contact time measures were found to be reliable measures during vault training, with measures of pre-flight and table contact time less consistent. Future research should examine the validity of these measures as a tool for monitoring vault training.

KEY WORDS: gymnastics, vault, velocity, biomechanics, performance.

INTRODUCTION: The men's and women's vault are a feature of any artistic gymnastics competition. Competitors sprint towards a take-off board, where they launch themselves onto a vaulting table and then into the air completing various acrobatic manoeuvres before landing on their feet. Successful performance requires the optimization of each aspect of the action; the run up (or approach), pre-take-off hurdle, take-off, pre-flight, table contact, post-flight, and landing. Recent research has attempted to quantify the relationship between each or multiple aspect(s) of the vault and overall performance (i.e., judges' score). For instance, Krug, Knoll and Zocher (1998) reported a moderate correlation ($r = 0.68$, $P < 0.01$) between average approach velocity and judges' score at the 1997 World Gymnastics Championships. Similarly, Bradshaw and Sparrow (2001) indicated that vaulting score ($r = 0.59$, $P<0.05$) was significantly correlated with a short board contact time, which in turn, was significantly correlated with post-flight time ($r = -0.41$, $P < 0.05$). A brief contact time on the take-off board and / or vaulting table is likely to translate the gymnasts approach velocity into a longer post flight time (Bradshaw, 2004) or distance, allowing the gymnast more time to complete more complex acrobatic manoeuvres in the air. This increases the degree of difficulty and the potential for high scores. The significant relationships between vaulting score and specific aspects of the gymnast's vault should compel coaches to monitor these variables as a part of training or routine testing.

Evaluating changes in performance predictive variables could highlight the gymnast's training progress between competitions. Instantaneous measurements of variables such as approach and take-off velocity, board and table contact time have not however been part of routine gymnastics aptitude testing. The reliance on descriptive or qualitative measures of vault ability has arisen, at least in part, from the absence of a valid, reliable system for measuring velocity and power in a training environment. The variability associated with predictors of vaulting aptitude using a novel timing system during training is unclear. Highly sensitive sports science measurements are characterized by little variation in consecutive measures of performance (Hopkins, 2000). The advantage of small test-retest variability between testing sessions is that any change in the athlete's performance can be confidently attributed to their recent training history, and not random fluctuations (Hopkins, 2000). Minimizing variability is also an advantage when monitoring fatigue or trialling interventions to improve performance. The change in performance due to the intervention has to be greater than the normal day-to-day training variation before coaches can conclude that the intervention has had a meaningful impact on the athlete's performance (Soper and Hume, 2004). The primary aim of the current study was to assess inter- and intra-session reliability of gymnast's approach velocity, board and vaulting table contact times, and pre-flight time using a novel timing system.

METHODS: Thirteen Australian high performance gymnasts (eight males and five females) aged 11-23 years who performed vaulting routinely participated in this study. The females were aged 14.6±2.5 years, 1.48±0.12 m tall, 42.2±12.1 kg in mass, and had a leg length of 0.78±0.08 m. The males were aged 15.8±3.9 years, 1.58±0.13 m tall, 50.2±12.2 kg in mass, and had a leg length of 0.79±0.05 m. All gymnasts were injury-free at the time of testing and capable of performing 2.4 or higher graded vaults according to the International Gymnastics Federation Code of Points 2005-2008. A Yurchenko layout (stretched) salto, for example, is graded as a 4.4 difficulty rating in women's gymnastics. The criteria for injury was when an athlete had not participated in training for more than seven days and/or had not participated in two sequential competitions at the time of testing (Noyes, Lindenfield, and Marshall, 1988). All procedures used in this study complied with the guidelines of the Australian Catholic University Ethics Committee and Auckland University of Technology Ethics Committee.

All participating gymnasts were measured for standing height, body mass, and leg length using the International Society for the Advancement of Kinanthropometry (ISAK) protocols (Marfell-Jones, 2006). The gymnasts were assessed during two training sessions (Day 1 and Day 2) at their regular training centre after a preceding familiarization training session (familiarization with the timing system). The gymnasts completed their general and vaulting warm-up under the supervision of their coach. Each gymnast completed a number of vault repetitions as per their normal vaulting training session. Vault category groups performed included: Handspring entry; Tsukahara entry; and Yurchenko entry. All vaults were maximal effort and separated by a 2-3 minute rest period, with a verbal "go" signal the only in-trial feedback. Each training session was completed within 60 minutes.

The experimental setup consisted of a set of seven Fusion Sport infra-red timing lights, a beat board and vaulting table with contact mats included. The Fusion Sport system (Fusion Sport, Brisbane, Australia) included single beam, timing gates with error correction, and contact timing mats. These devices provide timing information at a rate of 1.8 MHz. The beat board was an American Athletic Incorporated men's Stratum beat board (480 mm wide and 800 mm long). The Jansen Fritsen vaulting table had a customised Acromat vaulting table cover made of 35 mm aqualite foam (50 mm thick). The upper surface was of synthetic suede with the no-touch zone indicated with a red marking as usual for an Acromat vaulting table. Vinyl foot switch contact mats were inserted underneath the upper surfaces of both the beat board (Fusion Sport, Brisbane, Australia, Jump Mat 226554, 45.72 cm wide, 60.96 cm long, 1.3 cm deep, 6.35 kg) and the vaulting table (Fusion Sport, Brisbane, Australia, Jump Mat 210005, 69.85 cm wide, 82.55 cm long, 1.3 cm deep, 12.70 kg).

The variables of interest to this study were run-up velocities (at -18 to -12 m, -12 to -6 m, -6 to -2 m, -2 to 0 m from the beat board) and board contact time, pre-flight time (board take-off to table contact) and table contact time which could be used to quantify performance. Descriptive statistics for all variables are represented as mean and standard deviations. Data were log transformed to provide measures of reliability (e.g. difference in mean, typical error of measurement as a coefficient of variation percentage, Pearson correlation coefficients) (Hopkins, 2000). Measures of reliability were determined using a repeated measures analysis of variance. Mixed modelling statistical procedures were performed using SAS (Statistical Analysis System).

RESULTS AND DISCUSSION: This study quantified the variability in vaulting performance between two daily training sessions with the main measures of interest, in terms of reliability, being the typical error of measurement as a coefficient of variation (CV%) and 90% limits of agreement (LOA) for the approach velocity, pre-flight time, and board and table contact time variables of vault performance.

Table 1. Mean, standard deviation and variability statistics for the handspring vault for males and females combined performed across two consecutive days of training.

Male and Female	Day 1 (n=22)		Day 2 (n=16)			Typical error as a CV (%)	Limits of agreement (%)	Pearson r	ICC r
Handspring vault	Mean	SD	Mean	SD	Mdiff (%)				
-18 to -12 m	4.33	0.83	4.28	0.82	1.8	16.0	56.4	0.52	0.45
-12 to -6 m	6.32	0.25	6.23	0.17	-1.7	1.9	5.8	0.72	1.00
-6 to -2 m	6.51	0.26	6.46	0.28	-2.4	3.6	11.1	0.18	0.29
-2 to 0 m	6.07	0.22	6.02	0.28	-2.5	3.4	10.5	0.22	0.32
Board contact time	0.12	0.01	0.12	0.01	0.4	4.3	13.4	0.54	0.46
Pre-Flight time	0.28	0.04	0.29	0.05	1.7	15.6	54.9	0.25	0.32
Table contact time	0.14	0.05	0.13	0.03	4.7	22.7	85.2	0.30	0.48

Table 2. Mean, standard deviation and variability statistics for the Yurchenko layout full twist for females across two consecutive days of training.

Female	Day 1 (n=7)		Day 2 (n=7)			Typical error as a CV (%)	Limits of agreement (%)	Pearson r	ICC r
Yurchenko layout vault	Mean	SD	Mean	SD	Mdiff (%)				
-18 to -12 m	5.30	0.79	4.98	0.64	-5.74	16.7	70.6	-0.10	-0.10
-12 to -6 m	6.64	0.10	6.80	0.09	2.33	1.0	3.6	0.50	0.49
-6 to -2 m	6.86	0.07	6.83	0.58	-0.82	5.6	20.8	0.71	0.17
-2 to 0 m	5.17	0.45	5.88	0.51	12.42	8.0	32.2	0.25	0.24
Board contact time	0.14	0.00	0.14	0.01	1.13	6.9	30.2	-0.82	-0.69
Pre-Flight time	0.23	0.08	0.18	0.03	-20.26	28.6	139.0	0.08	0.07
Table contact time	0.10	0.01	0.11	0.02	27.0	5.5	160.6	0.00	0.07

Table 3. Mean, standard deviation and variability statistics for the Tsukahara layout for males across two consecutive days of training.

Male	Day 1 (n=5)		Day 2 (n=10)			Typical error as a CV (%)	Limits of agreement (%)	Pearson r	ICC r
Tsukahara layout vault	Mean	SD	Mean	SD	Mdiff (%)				
-18 to -12 m	5.75	0.05	6.05	0.21	3.2	1.4	5.6	-0.82	0.76
-12 to -6 m	7.30	0.19	7.29	0.38	-4.9	2.1	8.4	-0.02	0.80
-6 to -2 m	7.65	0.31	7.97	0.44	0.4	3.2	13.1	-0.09	0.63
-2 to 0 m	7.49	0.18	6.92	0.26	-5.9	2.6	10.5	-0.14	0.44
Board contact time	0.12	0.00	0.12	0.01	-4.1	4.0	16.6	0.51	0.50
Pre-Flight time	0.11	0.03	0.10	0.02	4.9	33.6	211.5	-0.27	-0.55
Table contact time	0.22	0.03	0.23	0.03	-4.6	16.4	81.7	-0.57	-0.37

The variation in gymnastics performance for the handspring (males and females combined), Yurchenko layout (females only), and Tsukahara layout (males only) are summarised in Tables 1-3 respectively. When testing during a training session, a much higher level of technical (biological) variation could be expected due to the gymnast and coach focusing upon different aspects of their vault performance. The vault tests are therefore not pure repeats, but working trials to improve aspects of technical execution. A standard error of measurement of 10% or less is considered small in pure test-repeats of three or more trials (Bennell et al., 1999).

The performance variation was generally small across days for the velocity measures of the vault approach (Average CV= 5.45%). The least variation (CV=1.4%, LOA=5.6%) was observed during the first 6 m of the run-up (-18-12 m segment) of the males Tsukahara tuck vault and the least reliable (CV=16.7%, LOA=70.6%) being the same segment during the females Yurchenko layout vault. The difference in reliability for velocity during the initial phase of the vault could have been due to the younger age of some of the female gymnasts performing the Yurchenko vault and different acceleration rates. Measures of board contact time from the contact mat built into the beat board were considered adequately reliable for all vaults tested (CV=4.0-6.9%, LOA=16.6-30.2%). Whereas pre-flight and table contact time revealed less favourable results with a CV as high as 33.6% found for pre-flight during the Tsukahara layout vaulting. High pre-flight variation could be due to variations in angular rotation between gymnasts and also vault techniques. Large variation in hand placement technique on the vault table was observed. The youngest females (aged 11 years) often brushed the table with their fingers ('fingered'), as opposed to contacting the table with the full hands. Differing table contact techniques was also observed between the male and female gymnasts, with great variation within the male gymnasts. Table contact time was not adequately reliable for the handspring vaults performed by the males and females (CV=22.7%) but was reliable for the Yurchenko layout vaults performed by the females (CV=5.5%).

CONCLUSION: This study has revealed that velocity measures from timing gates when combined with a contact mat embedded into the beat board can be reliably used to assess vaulting performance during training. Improvements in testing precision are required for measurements of pre-flight and table contact time. Test of these two factors may require competition simulation- style testing where the gymnast performs pure repeats, as opposed to working (training) trials. Further research should examine the validity of the vault timing system for individual gymnasts' performance by assessing changes in these measures with training.

REFERENCES:
Bennell, K., Crossley, K., Wrigley, T. and Nitschke, J. (1999), Test-retest reliability of selected ground reaction force parameters and their symmetry during running. *Journal of Applied Biomechanics, 15*, 330-336.

Bradshaw, E.J. (2004). Target-directed running in gymnastics: A preliminary exploration of vaulting. *Sports Biomechanics. 3*,125-144.

Bradshaw, E.J. and W.A. Sparrow (2001). The approach, vaulting performance, and judge's score in women's artistic gymnastics. In; *International Society of Biomechanics in Sport 2001: Scientific Proceedings.* University of San Francisco.

Hopkins, W.G., (2000). Reliability from consecutive pairs of trials. In; *A new view of statistics.* sportsci.org: International Society for Sport Science.

Krug, J., K. Knoll, and H.-D. Zocher. (1998). Running approach velocity and energy transformation in difficult vaults in gymnastics. In; *XVI Symposium on Biomechanics in Sport Proceedings*. Konstanz, Germany.

Marfell-Jones, M.J., Olds, T., Stewart, A.D. and Carter, L. (2006). *International Standards for Anthropometric Assessment.* Potchefstroom, South Africa: International Society for the Advancement of Kinanthropometry (ISAK).

Noyes, F.R., T.N. Lindenfeld, and M.T. Marshall (1988). What determines an athletic injury (definition)? Who determines an injury (occurrence)? *American Journal of Sports Medicine, 8,* S65-68.

Soper, C. and P.A. Hume (2004). Reliability of power output during rowing changes with ergometer type and race distance. *Sport Biomechanics. 3*, 237-247.

ISBS 2009

Oral Session S11

Injury I

EFFECT OF LUMBAR LORDOTIC ANGLE ON LUMBOSACRAL JOINT DURING ISOKINETIC EXCERCISE

Tae Soo Bae, Jong Kwon Kim, Shin Ki Kim, Jei Cheong Ryu and Mu Seong Mun

Biomechanics Lab., Korea Orthopedics & Rehabilitation Engineering Center, Incheon, South Korea

The purpose of this study was to to analyze the biomechanical impact of the level of lumbar lordosis angle during isokinetic exercise through dynamic analysis using a 3-dimensional musculoskeletal model. We made each models for normal lordosis, excessive lordosis, lumbar kyphosis, and hypo-lordosis according to lordotic angle and inputted experimental data as initial values to perform inverse dynamic analysis. Comparing the joint torques, the largest torque of EL was 16.6% larger than that of NL, and LK was 11.7% less than NL. There existed no significant difference in the compressive intervertebral forces of each lumbar joint ($p>0.05$), but statistically significant difference in the anterioposterior shear force (LK>HL>EL>NL, $p<0.05$). For system energy, LK required the least and most energy during flexion and extension respectively. Therefore during the rehabilitation process, more efficient training will be possible by taking into consideration not simply weight and height but biomechanical effects on the skeletal muscle system according to lumbar lordosis angles.

KEY WORDS: lumbar lordotic angle, musculoskeletal model, isokinetic exercise

INTRODUCTION: Lumbar lordosis is especially known to be an important and decisive factor in flexibility during lifting, lowering and other movements. There are various factors affecting lumbar lordosis. Pre-existing studies discovered that the level of lumbar lordosis is affected by age and sex, and it is reportedly largely affected by movement in the center of mass (COM) such as pregnancy or obesity. Muscle imbalance and expected back pain caused by disorder in the lumbar lordosis has been disputable. In the case of rehabilitation exercises for lumbar lordosis disorder, muscle-building exercises are usually carried out, and these exercises are subcategorized into static and dynamic muscle-building exercises, actualized by isometric and isokinetic exercises. There have been reports made by previous researchers claiming that lumbar lordosis is only related to static muscle power in upright posture and statistically non-related to dynamic muscle power, but as reports on statistical irrelevance of static muscle power have been made, exercises strengthening dynamic muscle power for rehabilitation exercise for lumbar lordosis disorder have been developed. However, many exercise guides on muscle-building exercise devices take only body weight and height into consideration, not age and sex. Therefore if one carries out inappropriate muscle building through a training protocol that does not take lumbar lordosis into consideration, unexpected posture deviation and back pain could be caused due to using incorrect positions. Therefore this study intends to analyze the biomechanical impact of the level of lumbar lordosis curvature during isokinetic exercise through dynamic analysis using a 3-dimensional musculoskeletal model according to the curvature.

METHODS: Four healthy males (171.8 ± 4.0 cm, 70.5 ± 9.0 kg) were selected as subjects who have normal lordosis by radiography. These males were ordinary people with no pathological diagnosis of the nervous and skeletal muscle systems, and who do not perform any periodic muscle building exercises. In order to materialize the musculoskeletal model of each subject's entire body, 35 Helen Hayes marker sets (16 on the upper body, 19 on the lower body) were attached to each subject's body, and their static poses were shot by camera; then the full body musculoskeletal models were acquired through a conversion program (Motion module & SIMM ver. 4.2, Motion Analysis Inc., CA, USA). Due to the purpose of the study, the neck spine and lumbar spine were made to operate separately, especially when modeling the skeletal muscle. The muscles around the lumbar spine are composed of six pairs – one pair of flexors and five pairs of extensors. Rectus abdominis was defined as a flexor, and iliocostalis lumborum, longissimus thoracis, spinalis thoracis, quadratus lumborum (L1), quadratus lumborum (L2) were selected as extensors. A governing equation composed of Hill-type functions was applied to each muscle, and existing research data were used for each muscle variable. In addition, to apply the activation mechanism of the muscles, a 1st degree differential equation was used in modeling, and muscle activation levels were set at maximum values of each, in accordance to isokinetic exercise.

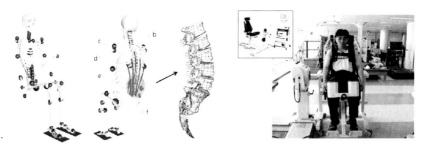

Fig. 1 Fullbody model with back muscles (left) and Isokinetic testing with Multi-Joint System 3:

In order to do a dynamics analysis using a whole-body musculoskeletal model, the lengths and weights of the subjects were measured, then the mass, center of mass, and moment of inertia needed for dynamics analysis were calculated by referencing existing research results. Measurement of the isokinetic muscular strength of the lumbar spine joint was carried out using the Multi-Joint System 3 pro (BIODEX, U.S.A.). (Fig.1) The dynamic variables on each part of the subject were put in, and after appropriately modifying the operating range on each joint, an equation of motion was induced using the Dynamic Pipeline Module (Ver3.2.1, Motion Analysis, USA) and SD/FAST (B2.8, PTC Inc, USA), a dynamics analysis program for mechanical systems. The initial values for all joints were set at the angular value of each joint of the subject during isokinetic exercise, and since this is an analysis on isokinetic exercise, all joint motions except spinal joints at sagittal plane were restricted. Also, the simulations were performed after putting in the pre-measured joint angles and angular velocity measured over time during isokinetic exercise as prescribed motions. For verification on the whole-body musculoskeletal model, an inverse dynamics analysis using SIMM was used to calculate lumbar joint torque, to be compared with torque value measured through isokinetic experiment on actual subjects. The results showed a small difference in the size of the lumbar torque in extension and flexion, but the overall torque patterns were similar to each other. (Fig. 2) Lumbar lordosis is defined by the angle between the upper plane of the L1 lumbar vertebrae and the upper plane of the S1 sacral vertebrae.

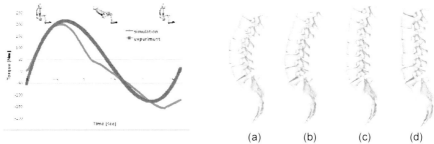

(a)   (b)   (c)   (d)

Fig. 2 Result of inverse dynamic analysis for validation(left) and Lumbosacral Joint model with respect to lordosis(right)   (a) EL, (b) NL, (c) HL (d) LK

Normal lordosis (NL) is in the 31°~50° range, excessive lordosis (EL) is over 70°, lumbar kyphosis (LK) is less than 10°, and hypo-lordosis (HL) is set at 11°~30° to materialize the musculoskeletal model of lumbar lordosis for each case. Their whole-body musculoskeletal model for various lumbar lordosis angles can be arbitrarily modified using bone editing tool in SIMM. In order to take into consideration the impact of pelvic movement due to lumbar lordosis, 3 degrees of freedom was applied to the pelvic movement of the L5 lumbar vertebrae in redefining the skeletal parts. (Fig.2)

RESULTS: Comparing the joint torques when the trunk is flexed and extended, the largest torque was created in the case of EL, then in NL, HL and LK in that order. In the case of EL, the torque amount was 16.6% larger than that of NL, and LK was 11.7% less than NL. The results from calculating compressive force in the normal direction and shear force in the anterioposterior direction showed that there existed no significant difference in the compressive forces of each joint in the vertical direction at each lordosis angle. ($p > 0.05$) Also, in shear force in the anterioposterior direction, LK showed the largest force in all lumbar joints, followed by HL, EI and NL; there existed a statistically significant difference ($p < 0.05$). However, in the pelvic and L5 lumbar there was no statistically significant difference in both compressive and shear forces. During trunk flexion, LK required the least energy, but had no statistically significant difference ($p > 0.05$), and during trunk extension, NL and HL required the least energy, while LK required the most energy, with statistically significant differences ($p < 0.05$).

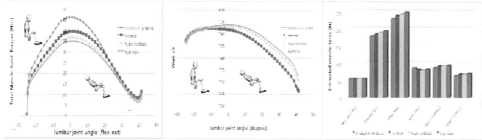

Fig. 3 Result of Total muscle joint torque (right), system energy (mid), and muscle forcess as lordosis curvature

The required muscle forces during isokinetic exercise for muscles surrounding the lumbar area - one pair of flexors and five pairs of extensors – were separately calculated. In the case of rectus abdominus, there were few differences in each case of lumbar lordosis angle. For spinalis thoracis, quadrates lumborum (L1,L2), EL showed a smaller value compared to other angles, but with no statistically significant difference.($p < 0.05$) However, in cases of iliocostalis lumborum and longissimus thoracis, the muscle force dropped as the angle widened (EL), and led to lumbar kyphosis (LK), which increased muscle force.

Table 1. Intervertebral compressive and anterioposterior shear force from pelvic to thorax joint

	Pelvic-L5		L5-L4		L4-L3		L3-L2		L2-L1		L1-thorax	
	comp	shear	comp	shear	compr	shear	comp	shear	comp	shear	comp	shear
EL	791.8	240.1	1600.5	66.2*	1574.9	165.4*	1538.5	251.0*	1494.0	322.1*	1471.2	285.4
NL	790.3	241.4	1581.2	31.4	1553.1	112.4	1527.6	208.9	1513.8	291.0	1512.0	287.0
HL	783.7	246.9	1576.8	110.7*	1562.0	155.9*	1547.4	252.4*	1533.2	333.6*	1533.2	351.0*
LK	785.7	248.6*	1582.0	141.3*	1567.3	165.3*	1552.5	265.0*	1537.8	350.5*	1540.7	398.6*

* P-values < 0.05 revealed significant differences between NL and other cases through statistical analysis.   (unit:newton)

DISCUSSION: Most cases of back pain are said to be caused by incorrect posture. This posture deviation is known to be caused by changes in muscle length and subsequent strength of muscles surrounding the skeleton over time. Change in muscle length directly affects posture alignment, and is known to cause excessive posture deviation such as lumbar lordosis, kyphosis, and scoliosis.[25] These changes in lumbar lordosis angle lead to back pain over time, but most physiotherapists overlook trunk muscle strength test when assessing spine-related data. This is due to the fact that good trunk muscle strength does not always mean the patient has a well-functioning spine. However, considering the fact that the results from this study showed that EL, HL and LK showed statistically significant differences in shear forces of front and rear lumbar, system energy and in some parts of extensors, there is a possibility these quantitative differences could ultimately lead to skeletal muscle disorders caused by posture imbalance. Most muscle building training devices do not take lumbar lordosis angle into consideration when setting up programs, which could lead to the possibility of additional problems due to trunk exercise over time.

The best model is to test and verify actual subjects according to various lumbar lordosis angle, but considering the range of lumbar lordosis angles, it is very difficult to find subjects who match the description. Also, in the case of subjects who fall away from normal lordosis angles, there is a possibility that various types of pain could be caused during exercise; therefore in this study is a whole-body musculoskeletal model on which the various lumbar lordosis angles can be arbitrarily modified. Due to the fact that not all muscles around lumbar area were modelled, however, while the overall lumbar flexing and extending torques showed similar tendencies during verification comparisons, quantitative difference existed between the experiment and calculated results. Further studies should include a more precise modeling of lumbar area muscle to heighten reality, and a more systemized verification process is considered necessary by comparing various clinical results on lumbar lordosis disorder.

CONCLUSION: During lumbar muscle strengthening training, in order to examine its impact according to lumbar lordosis angle, a 3-dimensional whole-body musculoskeletal model on EL, NL, HL and LK was used in a simulated experiment to examine biomechanical impact according to lordosis angles. In joint torques, EL showed the highest value, followed by NL, HL and LK, but in anterioposterior shear force, LK was the largest, followed by HL, EL then NL. The compressive intervertebral force inside the joint showed statistically to not be affected by lordosis angles. In terms of system energy, HL and LK required more energy than EL. Therefore, during the rehabilitation process, more efficient training will be possible by taking into consideration not simply weight and height but biomechanical effect on the skeletal muscle system according to lumbar lordosis angles.

REFERENCES:

Farfan H.F. (1978). The biomechanical advantage of lordosis and hip extension for upright activity. *Spine*, 3, pp336-342.

White, A. A. and Punjabi, M. M. (1990). *Clinical Biomechanics of the Spine*, Lippincott, Philadelphia, Winters, D. A. (1990). Anthropometry In: David A. Winter, *Biomechanics and Motor Control of Human Movement* (pp 51-74). John Wiley & Sons.

De Leva, P.(1996). Adjustments to Zatsiorsky-Seluyanov's Segment Inertia Parameters. *Journal of Biomechanics*, Vol. 29, No. 9, pp1223-1230.

Acknowledgement

This study was supported by Ministry of Culture, Sports, and Tourism. Grant # S07-2008-03

ANTICIPATION EFFECT ON KNEE JOINT STABILITY DURING PLANNED AND UN-PLANNED MOVEMENT TESTS IN LABORATORY

Mak-Ham LAM[1], Pik-Kwan LAI[1], Daniel Tik-Pui FONG[1,2], Kai-Ming CHAN[1,2]

Department of Orthopaedics and Traumatology, Prince of Wales Hospital, Faculty of Medicine, The Chinese University of Hong Kong, Hong Kong, China [1]
The Hong Kong Jockey Club Sports Medicine and Health Sciences Centre, Faculty of Medicine, The Chinese University of Hong Kong, Hong Kong, China [2]

The purpose of this study was to investigate the anticipation effect on knee stability during functional test in laboratory. Ten healthy male subjects were recruited and instructed to perform a series of planned and un-planned stop-jumping tasks. Knee joint kinematics was measured by a motion analysis system. The subjects demonstrated different abduction and rotation angles for reactive tasks. This suggested that if knee abduction or rotational stability is considered as a primary measurement in documenting knee stability, such as in the investigation of rehabilitation progress after anterior cruciate ligament reconstruction, both planned and un-planned tasks should be considered as to take the anticipation effect into account.

KEY WORDS: stop-jumping, knee kinematics, unanticipated

INTRODUCTION: Anterior cruciate ligament (ACL) rupture is a common sport-related trauma which causes knee instability (Hurd and Snyder-Mackler, 2007). It is often treated operatively to reconstruct the ligament in order to restore the joint stability (Zaffagnini et al, 2007). Ten years ago, orthopaedics surgeons managed to reconstruct the ACL in a single-bundle fashion, with either patellar tendon or hamstring tendon as the graft (Eriksson, 2007). However, anatomical studies showed that the intact ACL contains two bundles which provide translational and rotational stability respectively (Petersen and Zantop, 2007). Therefore, in the recent decade, surgeons began to consider reconstructing the ACL in a double-bundle fashion (Zella et al, 2006). The effect has been well demonstrated in cadaveric study to be successful in restoring mechanical stability in both translational and rotational directions (Woo et al, 2002).

However, for athletes, the functional stability is much more important than the mechanical stability. Therefore, functional test has to be included during follow-up consultation to evaluate if a patient is adequately rehabilitated (Risberg and Ekeland, 1994). Such functional test often includes jogging, running, jump-landing, and most importantly, pivoting and cutting tasks. These functional tests allow subjects to preplan movement patterns and it may not reflect the movement patterns performed in competition during which athletes must react to unanticipated events (Besier et al, 2001). Numerous studies showed the anticipation effect on knee joint loading and muscle activation patterns during these tests (Ford et al, 2005). However, such effect on knee rotational stability is not yet demonstrated, and is awaited by orthopaedics surgeons to evaluate the functional outcome after ACL reconstruction with the recent double-bundle technique. The current study aims to investigate the anticipation effect on knee rotational stability during functional test in laboratory. Kinematics at foot strike (Cham and Redfern 2002) was considered in our study as ACL injury was reported to occur 17-50 milliseconds after initial foot strike during landing (Krosshaug et al, 2007). We then hypothesis that there is no difference for the landing maneuver between planned and un-planned stop-jumping tasks. Such information is important for biomechanists to decide whether to conduct both planned and un-planned tasks to document the rehabilitation progress of ACL patients.

METHODS: Data Collection: Ten healthy male subjects without any previous injury history on lower limbs were recruited. Informed consent in written format was obtained from each subject. The study was conducted in a motion analysis laboratory. Fifteen reflective markers were attached to the lower extremity of the subject at the major anatomical positions, including sacrum, left and right greater trochanter, anterior superior iliac spine, lateral femoral condyle, tibial tuberosity, lateral malleolus, calcaneus, and fifth metatarsal heads. The lower extremity motion was captured by an 8-camera high-speed motion analysis system (Vicon, UK) at 120Hz.

A series of functional tasks were performed planned and un-planned randomly for each subject. For each task, the subject was instructed to run straight on a 5-meter walkway approaching a ground-mounted force plate (AMTI, USA), with a running speed of 3.1 to 3.5 m/s (De Cock et al, 2005) as monitored by the forward speed of the sacrum marker by the motion analysis system. Trials with running speed out of the range were discarded.

In the planned tasks, the subjects were instructed to stop with both feet on the force plate and jump immediately either to four directions, including anterior, vertical, left and right as far as they could. The order of the four directions is selected randomly. In un-planned tasks, a light gate was set at a distance of 0.5 meters in front of the force plate. When the subject passed through the gate, a signal was delivered to a computer to trigger an instruction to change the movement direction as shown on a 17-inch monitor in front of the walkway. Upon the availability of the instruction signal, the subject stepped on the force plate and jumped to the instructed direction in the shortest time he could. Four directions were delivered to the subject in a random sequence. Five successful trials for each direction for planned and un-planned tasks were collected.

Data Analysis: The lower extremity biomechanics were calculated with a standard procedure (Soderkvist and Wedin, 1993). The primary measurement was the knee rotational displacement at the time of foot strike, as this suggests if the anticipatory effect is significant to the preparatory stage of the functional task. Knee adduction/abduction angle and flexion/extension angle were calculated as well. Stratified paired t-tests were conducted for each of the four tasks to determine the anticipatory effect on knee rotational stability during planned and un-planned movement tests. A 0.05 significant level was chosen a priori to denote statistical significance for the comparisons.

RESULTS: Ten male subjects with mean age 26.4 (1.78) were recruited. Their mean height and body mass were 1.73m (0.72 m) and 70.9 kg (15.62 kg). The result of the three knee angles at the time of foot strike was summarized in the Table 1. Significant differences were found in adduction/abduction angle and internal/external rotation angle for anterior jump.

Table 1 knee kinematics for planned and un-planned tasks at the time of foot strike

Knee joint angle at landing	Jump direction	Planned	Un-planned	Statistical analysis p-value[a]
	Anterior	25.8° (7.4)	23.8° (7.4)	No significant differences (p = 0.367)
Flexion (+ve) or	Vertical	27.0° (12.0)	23.5° (6.5)	No significant differences (p = 0.252)
extension (-ve)	Right	28.9° (3.6)	23.5° (7.2)	No significant differences (p = 0.068)
	Left	21.4° (4.4)	22.4° (4.5)	No significant differences (p = 0.491)
	Anterior	9.4° (8.9)	6.0° (10.9)	Significant differences (p = 0.049)*
Abduction (+ve) or	Vertical	8.9° (10.4)	6.5° (8.1)	No significant differences (p = 0.112)
adduction (-ve)	Right	9.3° (10.2)	6.9° (10.1)	No significant differences (p = 0.125)
	Left	5.7° (10.3)	7.9° (8.9)	No significant differences (p = 0.193)
	Anterior	20.2° (8.8)	13.9° (6.4)	Significant differences (p = 0.030)*
External rotation (+ve) or	Vertical	17.4° (8.8)	14.1° (5.4)	No significant differences (p = 0.231)
internal rotation (-ve)	Right	18.3° (9.7)	14.5° (4.9)	No significant differences (p = 0.173)
	Left	12.1° (10.6)	15.9° (6.8)	No significant differences (p = 0.111)

* Significant difference observed between planned and un-planned tasks (p < .05)

DISCUSSION: Currently, research investigating anticipated effect was limited. Pollard et al (2007) and Landry et al (2007) reported that both male and female performed similarly in randomly cued cutting maneuver. However, neither research group focused on stop-jumping maneuver, in which most of the ACL injuries occurs during such landing movement with a change in direction. Sell et al did comparison between planned and un-planned stop-jumping tasks and demonstrated increased knee joint loading characteristics such as greater knee valgus and flexion moments. They suggested that directional and reactive jumps should be included in research methodology.

In the current study, we included anterior jumping addition to vertical and horizontal side jumping. Kinematics of the knee joint was measured at the time of foot strike as investigators believe that ACL injuries typically occur at this moment (Olsen et al, 2004). Subjects in this study performed similarly in flexion angle and differently in abduction angle between planned and un-planned tasks. There was a significant difference on abduction during anterior jumping task, which suggests anterior jumping should be considered as one of the jumping directions when performing stop-jumping tasks.

Most importantly, we measured knee rotational displacement, which is the major laxity the ACL patients suffer. The healthy subjects in our study demonstrated decreased knee internal rotation for unanticipated tasks and significant difference was found in anterior stop-jumping task. Since the current study incorporated a few high risk movements of ACL injury including a sharp deceleration, a change in direction and a landing maneuver, the result of this study may indicate that both planned and un-planned stop-jumping tasks should be considered when evaluating knee rotational stability after ACL reconstruction and before safe return-to-sports.

CONCLUSION: Different kinematics results were demonstrated in healthy male subjects between planned and un-planned stop-jumping tasks. This suggests that both planned and un-planned tasks should be conducted to document the rehabilitation progress of ACL patients, especially for evaluating ACL reconstruction with double-bundle fashion which claimed to better restore rotational stability.

REFERENCES:
Besier TF, Lloyd DG, Ackland TR, and Cochrane JL (2001). Anticipatory effects on knee joint loading during running and cutting maneuvers. Medicine and Science in Sports and Exercise, 33(7), 1176-1181.

Cham R and Redfern MS (2002) Changes in gail when anticipating slippery floors. Gait and Posture, 15(2), 159-171.

De Cock A, De Clercq D, Willems T, and Witvrouw E (2005). Temporal characteristics of foot roll-over during barefoot jogging: reference data for young adults. Gait and Posture, 21(4), 432-439.

Eriksson E (2007). Hamstring tendon or patellar tendon as graft for ACL reconstruction? Knee Surgery Sports Traumatology Arthroscopy, 15(2), 113-114.

Ford KR, Myer GD, Toms HE, and Hewett TE. Gender differences in the kinematics of unanticipated cutting in yound athletes. Medicine and Science in Sports and Exercise, 37(1), 124-129.

Hurd WJ, and Snyder-Mackler L (2007). Knee instability after acute ACL rupture affects movement patterns during the mid-stance phase of gait. Journal of Orthopaedic Research, 25(10), 1368-1377.

Krosshaug T, Nakamae A, Boden BP, Engebretsen L, Smith G, Slauterbeck JR, Hewett TE, and Bahr R (2007) Mechanisms of anterior cruciate ligament injury in basketball, video analysis of 39 cases. American Journal of Sports Medicine, 35(3), 359-367.

Landry SC, McKean KA, Hubley-Kozey CL, et al. Neuromuscular and lower limb biomechanical differences exist between male and female elite adolescent soccer players during an unanticipated run and crosscut maneuver. Am J Sports Med. 2007. 35(11):1901-11.

Olsen OE, and Myklebust G, Engebretsen L, et al. Injury mechanisms for anterior cruciate ligament injuries in team handball: a systematic video analysis. American Journal of Sports Medicine. 2004. 32(4):1002-12.

Petersen W, and Zantop T (2007). Anatomy of the anterior cruciate ligament with regard to its two bundles. Clinical Orthopaedics and Related Research, (454), 35-47.

Pollard CD, Sigward SM, and Powers CM. Gender differences in hip joint kinematics and kinetics during side-step cutting maneuver. Clinical Journal of Sport Medicine. 2007. 17(1):38-42.

Risberg MA, and Ekeland A (1994). Assessment of functional tests after anterior cruciate ligament surgery. Journal of Orthopaedic and Sports Physical Therapy, 19(4), 212-217.

Soderkvist I, and Wedin PA (1993). Determining the movements of the skeleton using well-configured markers. Journal of Biomechanics, 26(12), 1473-1477.

Woo SL, Kanamori A, Zeminski J, Yagi M, Papageorgiou C, and Fu FH (2002). The effectiveness of reconstruction of the anterior cruciate ligament with hamstrings and patellar tendon – a cadaveric study comparing anterior tibial and rotational loads. Journal of Bone and Joint Surgery – American Volume, 84A(6), 907-914.

Zaffagnini S, Bignozzi S, Martelli S, Lopomo N, and Marcacci M (2007). Does ACL reconstruction restore knee stability in combined lesions? An in vivo study. Clinical Orthopaedics and Related Research, (454), 95-99.

Zelle BA, Brucker PU, Feng MT, and Fu FH (2006). Anatomic double-bundle anterior cruciate ligament reconstruction. Sports Medicine, 36(2), 99-108.

Acknowledgement
This research project was made possible by equipment/resources donated by The Hong Kong Jockey Club Charities Trust.

AN ANKLE SPRAIN RECOGNITION SYSTEM FOR IDENTIFYING ANKLE SPRAIN MOTION FROM OTHER NORMAL MOTION USING MOTION SENSOR

Yue-Yan Chan[1,2], Daniel Tik-Pui Fong[1,2] and Kai-Ming Chan[1,2]

The Hong Kong Jockey Club Sports Medicine and Health Sciences Centre, Faculty of Medicine, The Chinese University of Hong Kong, Hong Kong, China[1]
Department of Orthopaedics and Traumatology, Faculty of Medicine, The Chinese University of Hong Kong, Hong Kong, China[2]

The purpose of this study was to develop an ankle sprain recognition system which identifies ankle sprain motions from other normal motions. Six healthy male subjects performed a total of 600 simulated ankle sprain motions and normal sports motions. Eight motion sensors were attached to cover the whole foot segment to monitor the linear velocity and angular accelerations of the segment. The data obtained from the motion sensor at the medial calcaneus selected to train up the Support Vector Machine (SVM). The trained SVM model was then verified by another 600 trials from other six healthy male subjects. Among the 300 sprain trials, 291 (97.0%) of them were identified correctly. However, there was still a 14.3% false alarm which normal trials being identified as sprain trails. In general, a good accuracy of 91.3% was achieved.

KEY WORDS: accelerometer, gyrometer, signal identification.

INTRODUCTION: Ankle sprain is one of the most common ankle injuries in sports (Fong et al., 2007), accounting about 12% of total sport-related cases admitted to the accident and emergency department (Fong et al., 2008). In order to protect the ankle from sprain injury, our research team is developing an intelligent sprain-free shoe. The idea of the sprain-free shoes is to protect from ankle sprain injury while allowing freedom of motion during normal activities. The sprain-free shoes consist of a recognition system which identifies ankle sprain motion to activate the protection mechanism while the ankle is at risk of sprain injury. In this paper, the ankle sprain recognition system using motion sensor and Support Vector Machine (SVM) is introduced.

Fig 1: A subject performing simulated sprain motion on a mechanical ankle sprain simulator

METHODS: Data Collection: Six male subjects (age = 21.2 ± 1.7 yr, height = 1.72 ± 0.05m, body mass = 61.5 ± 3.1kg, foot length = 255.3 ± 10.6mm) with healthy ankles were recruited. The university ethics committee approved the study. Each subjects contributed to 100 trials, including 50 trials of simulated ankle sprain motion and 50 trails of non-sprain normal motions. The simulated sprain motions were conducted on a mechanical ankle sprain simulator (Fig 1, Chan et al., 2008). Different combination of inversion and plantarflexion (total inversion, 23 degrees supination, 45 degrees supination, 67 degrees supination and total plantarflexion) were performed. The sprain simulator performed 30 degrees angular perturbation along the rotation axis when the shutter released. Non-sprain motion included walking, running, jump-landing, cutting and stepping-down were performed in a random sequence in a motion laboratory. Ten trials from each motion were done.

Eight wired motion sensors (Sengital Ltd., Hong Kong, China) were attached as shown in Figure 2. These attachment positions allowed a full coverage of right foot and ankle segment. Each motion sensors consisted of a tri-axial gyrometer which measured angular velocities (G_x, G_y, G_z), and a tri-axial accelerometer that measured linear accelerations (A_x, A_y, A_z). Therefore a total of 48 signals from all sensors were collected at a frequency of 500Hz.

Data Analysis: The data collected were used to train up the Support Vector Machine for the development of the identification system. The learning theory of SVM can be expressed as a function:

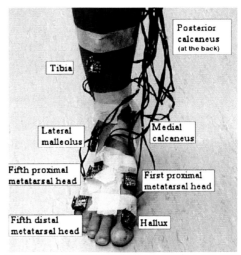

Fig 2: The attachment of the 8 motion sensors

$$f : \Re^n \rightarrow \pm 1$$

Where $y = f(x)$. This function maps patterns x to the classification y. the function $f(x)$ can be expressed as:

$$f(x) = \sum_{i=1}^{N} y_i \alpha_i k(x, x_i) + b \qquad - (1)$$

where N is the number of training patterns, (x_i, y_i) is training pattern i with its classification as y_i, α_i and b are learned weights, and k is a kernel function (Cristianini & Shawe-Taylor, 2000):

$$K(\bar{x}_i, \bar{x}) = \Phi(\bar{x}_i) \cdot \Phi(\bar{x}_j)$$

k can be any symmetric kernel function that satisfy the Mercer's condition corresponds to a dot product in some feature space (Bernhard et al., 1998). (x_i, y_i) with $\alpha_i > 0$ are denoted as support vectors. The surface where $f(x) = 0$ is a hyperplane through the feature space as defined by the kernel function. Optimal parameters α_i and b are selected to minimized the number of incorrect classifications by maximizing the distance of the support vectors to the hyperplane $f(x) = 0$. $y_i > 0$ indicate a simulated supination sprain trail, where $y_i < 0$ indicate a non-sprain trial.

Maximize:

$$L_D \equiv \sum_{i=1}^{N} \alpha_i - \frac{1}{2} \sum_{i=1}^{N} \sum_{j=1}^{N} y_i y_j \alpha_i \alpha_j k(x_i, x_j) \text{-(2)}$$

subject to:

$$0 \leq \alpha_i \leq C, \quad \sum_{i=1}^{N} y_i \alpha_i = 0 \qquad -(3)$$

The constant C denotes the penalty to errors, therefore it affects the tolerance to incorrect classifications. After solving the equation (2) and find α_i, we can use any other support vector (x_i, y_i) to find b.

A value of signal strength (unitless) was calculated for each of the eight sensors to quantify its ability to identify the spraining and non-spraining motions. SVM training was then done with the data from the motion sensor with highest signal strength. One second (500 frames) of data from all the six channels of that sensor were processed by Discrete Fourier Transform (DFT) and was converted to frequency domain. The converted data was then used to train up the SVM (Joachims, 1999).

Model Validation: Another six subjects (age = 22.0 ± 1.7 yr, height = 1.75 ± 0.04 m, body mass = 69.7 ± 2.8 kg, foot length = 262.0 ± 9.9 mm) took part in the validation test and contributed to a total of 600 trials. The same protocol was done for data collection as in the previous part. The accuracy of the recognition system was calculated by the percentage of trials being correctly identified. The SVM training was considered successful when the accuracy reached 90%. If the SVM training was not successful, the training process would be performed again with the sensor with the second highest signal strength and so on. If SVM training with a single sensor was not successful, combinations of two or more sensors will be performed.

RESULTS: The sensor located at medial calcaneus was found to possess the highest signal strength among all the sensor location. Therefore, data from the sensor located at medial calcaneus were chosen to train up the SVM. After training the SVM with 600 simulated sprain and non-sprain trials, 521 support vectors and the threshold b = 0.46397071 were selected to built up the SVM model in equation (1).

The SVM model was then went through the validation test. 600 trails simulated sprain and non-sprain trials from another six subjects were feed into the SVM model. 548 out of 600 trials were identified correctly. Therefore the accuracy was 91.3%. Details of results of validation test were shown in table 1. The SVM model built was considered successful, and further training using other data was abandon.

Table 1 Results of validation test

	Correctly identified trials	Incorrectly identified trials	Total
Simulated sprain trials	291 (97.0%)	9 (3.0%)	300 (100%)
Non-sprain trials	257 (85.7%)	43 (14.3%)	300 (100%)
Total	548 (91.3%)	52 (8.7%)	600 (100%)

DISCUSSION:

There was 91.3% accuracy for the trained SVM model which was considered to be very good for biomechanics studies. For the 300 simulated sprain cases, 291 trials (97.0%) were identified correctly. As this device is going to act as the activation signal of the intelligent sprain-free shoes, the alarm can be activated at 97% of the cases while the ankle is at risk of sprain motion. However, for the 300 non-sprain cases, there were 43 trials incorrectly identified trials, which made the false alarm rate be 14.3%. This indicates a gap and needs for improvement.

In order to improve the accuracy, data from two or more sensors can be adopted to train up the SVM model. The trade off of using more than one sensor is the increased amount of data to be processed. Hence, the process time and system requirement will be increased.

Therefore we have to make a balance between accuracy and process time. On the other hand, the SVM model developed was for young male subjects. In order to fit the model to individual of different homogenous groups, the whole procedure can be repeated. Therefore, we can come up with different SVM models which suits different homogenous groups.

The current recognition system only allows data processing after data collection, but not real time recognition. Data trimming and discrete fourier transform were need to be done after data collection. In order to achieve real time identification, further investigation has to be done on immediate analysis. Application of sliding window would be possible approach. The SVM model real time data analysis can be built on a printed circuit board. The wireless prototype can be made at a cost of around US$100. In order to lower the production cost for mass production in the future, further investigation have to be done, for example, reducing the sampling rate of the sensor and sliding window, which can lower the cost of the sensor, as well as the processing unit of the recognition module.

All the simulated sprain motion was performed on a mechanical sprain simulator. The sprain simulator can only perform sub-injury motions instead of real sprain cases due to ethical reasons. No ligamentous injury was introduced. Therefore we could only rely on the simulated sprain motion which is less vigorous to train up the SVM model.

CONCLUSION: This study developed an ankle sprain recognition system which identifies ankle sprain motions from other normal motions. The system consists of one motion sensor of 500 Hz sampling frequency and a recognition model. An accuracy of 91.3% was achieved. The system can be further developed for the real time identification of ankle sprain injury in the intelligent ankle sprain free shoes.

REFERENCES:
Bernhard, S., Burges, C.J.C., & Smola, A. (1998). Advanced in kernel methods support vector learning. Cambridge, MA: MIT Press.
Chan, Y.Y., Fong, D.T.P., Yung, P.S.H., Fung, K.Y. & Chan. K.M. (2008). A mechanical supination sprain simulator for studying ankle supination sprain kinematics. *Journal of Biomechanics*. 41(11), 2571-2574.
Christianini, N., & Shawe-Taylor, K. (2000). An introduction to support vector machines. Cambridge, UK: Cambridge University Press.
Fong, D.T.P, Hong, Y., Chan, L.K., Yung, P.S.H., Chan, K.M. (2007). A systematic review on ankle injury and ankle sprain in sports. *Sports Medicine*, 37(1), 73-94.
Fong, D.T.P., Man, C.Y., Yung, P.S.H., Cheung, S.Y., Chan, K.M. (2008). Sport-related ankle injuries attending an accident and emergency department. *Injury*, 39(10), 1222-1227.

Acknowledgement
This research project was made possible by equipment/ resources donated by The Hong Kong Jockey Club Charities Trust.
We would like to acknowledge Dr Alan Lam and Mr Joe Wong from the Sengital Limited, Hong Kong, China for the development of the motion sensors.

THE EFFECTS OF SPORTS INJURY PREVENTION TRAINING ON THE BIOMECHANICAL RISK FACTORS OF ANTERIOR CRUCIATE LIGAMENT INJURY IN FEMALE ATHLETES

BeeOh Lim[1], YongSeuk Lee[2] and YoungHoo Kwon[3]

Sports Science Institute, Seoul National University, Seoul, Korea[1]
Korea University Ansan and Guro Hospital, Korea University, Ansan, Korea[2]
Department of Kinesiology, Texas Woman's University, Denton, USA[3]

KEY WORDS: anterior cruciate ligament, sports injury prevention, female athletes.

INTRODUCTION: The purpose of this study were to investigate the effects of sports injury prevention training (SIPT) on the biomechanical risk factors of ACL Injury in high school female basketball players.

METHOD: A total of 22 high school female basketball players were recruited and randomly divided into 2 groups (the experimental group and the control group, 11 participants each). The experimental group was instructed in the 6 parts (warm-up, stretching, strengthening, plyometrics, agilities, and alternative exercise-warm down) of the sports injury prevention training program and performed it during the first 20 minutes of team practice for the next 8 weeks, while the control group performed their regular training program. Both groups were tested with a rebound-jump task before and after the 8-week period. A total of 21 reflective markers were placed in pre assigned positions. In this controlled laboratory study, a 2-way analysis of variance (2 × 2) experimental design was used for the statistical analysis ($P < .05$) using the experimental group and a testing session as within and between factors, respectively. Post hoc tests with Sidak correction were used when significant factor effects and/or interactions were observed.

RESULTS: A comparison of experimental group's pre- and post-training results identified training effects on all strength parameters ($p=0.004 \sim 0.043$) and on knee flexion, which reflects increased flexibility ($p=0.022$). Concerning biomechanical risk factors, the experimental group showed higher knee flexion angles ($p=0.024$), greater inter-knee distances ($p=0.004$), lower H-Q ratios ($p=0.023$), and lower maximum knee extension torques ($p=0.043$) after training. In the control group, no statistical differences were observed between pre- and post-training findings ($p=0.084 \sim 0.873$). At pre-training, no significant differences were observed between the two groups for any parameter ($0.067 \sim 0.784$). However, a comparison of the two groups after training revealed that the experimental group had significantly higher knee flexion angles ($p=0.023$), greater knee distances ($p=0.005$), lower H-Q ratios ($p=0.021$), lower maximum knee extension torques ($p=0.124$) and higher maximum knee abduction torques than the control group ($p=0.043$).

DISCUSSION: We believe that these effects can be attributed to SIPT because all of the training sessions were observed and subjects were instructed by three well trained 3 coaches throughout the SIPT. It would appear that all successful programs contain one or several of the following components; traditional stretching and strengthening activities, aerobic conditioning, agilities, plyometrics, and risk awareness training. The SIPT has the advantage that it requires little additional time on the athlete's part because it only requires 20 minutes per training program (from warm-up to the warm-down exercises).

CONCLUSION: The sports injury prevention training program improved the strength and flexibility of the competitive female basketball players tested and biomechanical properties associated with anterior cruciate ligament injury as compared with pre-training parameters and with post-training parameters in the control group.

THE EFFECT OF WARM-UP, STATIC STRETCHING AND DYNAMIC STRETCHING ON HAMSTRING FLEXIBILITY

Kieran O'Sullivan[1], Elaine Murray[1] and David Sainsbury[1]

[1]Physiotherapy Department, University of Limerick, Ireland

KEY WORDS: stretching, hamstrings, flexibility, warm-up

INTRODUCTION: Warm-up and stretching may increase flexibility and reduce injury risk, yet there is disagreement on the benefits of stretching and which technique is best (Thacker et al 2004). This study examined the short-term effects of warm-up, static and dynamic stretching on hamstring flexibility in individuals with previous hamstring injury and uninjured controls.

METHOD: A randomised crossover study was performed over 2 days. Hamstring flexibility was assessed in supine using passive knee extension range of motion (PKE ROM). The reliability of assessing PKE ROM using a 'Myrin' goniometer, with a crossbar maintaining the hip at 90° flexion, was examined in a preliminary pilot study (n= 25) and was excellent (ICC = 0.945, SEM 1.84°). 36 individuals (18 injured, 18 controls) participated. Only previously injured subjects who were now painfree, but with a residual reduction of 5° of PKE ROM compared to the other leg were included. On both days, four measurements of PKE ROM were recorded: (1) at baseline; (2) after a 5-minute aerobic warm-up; (3) after stretch (static or dynamic) and (4) after a 15-minute rest. Both stretches were performed for 30 seconds and repeated 3 times for each leg. Participants carried out both static and dynamic stretches, but on different days. Data were analysed using a 1-way repeated measures anova.

RESULTS: <u>Across both groups</u>, there was no interaction effect (p=0.344). There was a significant main effect for time (p<0.001). PKE ROM significantly increased with warm-up (p<0.001). From warm-up, PKE ROM further increased with static stretching (p=0.04) but significantly decreased after dynamic stretching (p=0.013). The increased flexibility after warm-up and static stretching reduced significantly (p<0.001) after 15 minutes of rest, but remained significantly greater than at baseline (p<0.001). <u>Between groups</u>, there was no main effect for group (p=0.462), with no difference in mean PKE ROM values at any individual stage of the protocol (p>0.05). Using ANCOVA to adjust for the non-significant (p=0.141) baseline differences, the previously injured group demonstrated a greater response to warm-up and static stretching, which was not statistically significant (p=0.05)

DISCUSSION: Warm-up significantly increased hamstring flexibility. Static stretching also increased hamstring flexibility, whereas dynamic did not, in agreement with previous findings on uninjured controls. The greater effect of warm-up and static stretching on flexibility in those with reduced flexibility post-injury is worthy of further study.

CONCLUSION: Further prospective research is required to validate the hypothesis that increased flexibility improves injury or performance outcomes, as this study only examined short-term flexibility. The results contrast with research demonstrating the greater benefits of dynamic stretching on performance measures. There may be a need to consider both forms of stretching during training and rehabilitation, but for different purposes.

REFERENCES:

Thacker, SB, Gilchrist, J, Stroup, DF and Dexter, C (2004) The Impact of Stretching on Sports Injury Risk: A Systematic Review of the Literature. Medicine and Science in Sports and Exercise 36, 371-378.

ISBS 2009

Oral Session S12

Field Sports

VELOCITY AND ACCURACY AS PERFORMANCE CRITERIA FOR THREE DIFFERENT SOCCER KICKING TECHNIQUES

Thorsten Sterzing, Justin S. Lange, Tino Wächtler, Clemens Müller, Thomas L. Milani

Department of Human Locomotion, Chemnitz University of Technology, Chemnitz, Germany

Kicking velocity (KV) and kicking accuracy (KA) of 19 experienced male soccer players were examined for the full instep, the inner instep, and the side foot kick. Measurements were performed simultaneously by a radar gun (KV) and a newly introduced high-speed-video camera set-up (KA). Subjects had two different tasks: to kick as fast as possible (Max KV) and to kick as accurate as possible (Max KA) with each kicking technique. Six repetitive kicks were performed for each required condition. The full instep and the inner instep kick were faster compared to the side foot kick for both performance tasks. In contrast, the side foot kick was the more accurate technique compared to the inner instep and the full instep kick, also for both performance tasks. Kicking variability between and within subjects was generally low for KV and generally high for KA for all kicking. It is concluded that velocity control is easier to achieve than accuracy control for soccer kicks.

KEY WORDS: soccer, kicking performance, kicking velocity, kicking accuracy, variability

INTRODUCTION: The different kicking techniques in soccer are the most characteristic technical skills of the game. Especially the full instep kick has been biomechanically studied in detail defining its typical components including the foot/ball contact phase (Barfield, 1998). This phase is characterized as a mixture of an impact-like and a throwing-like movement (Tsaousidis & Zatsiorsky, 1996). Due to the relatively short contact time of about 9 ms for instep kicking (Shinkai et al., 2009), the success of the kick, ball velocity and ball accuracy, is already determined at ball impact. Generally, successful kicks need to be fast and accurate, especially when kicking on goal. This allows the goal keeper less time to react and also makes it difficult to reach the ball. Thus, KV and KA are important performance criteria for soccer kicking. Recent research showed the influence of soccer footwear on KV and KA. Astonishingly, decreased KV was found when kicking shod compared to barefoot (Sterzing & Hennig, 2008). However, for KA shod kicking increases performance compared to barefoot kicking (Hennig, Althoff & Hömme, 2009). Only few studies have investigated the general influence of different kicking techniques on these performance criteria. Side foot kicks were shown to be slower compared to full instep kicks (Levanon & Dapena, 1998; Nunome et al., 2002). Among the different instep kicking techniques the full instep kick is the fastest, followed by the inner and outer instep kicks (Neilson & Jones, 2005). Here, reduced ball velocities for the inner and outer instep kicking techniques are traded to ball spin achieved by off-centered foot/ball contact. Kicking performance studies have mainly focused on KV. One study compared KV and KA of the instep compared to the less frequently used toe kick stating the toe kick to be less precise than the instep kick at 90 % of maximum KV (Kristensen, Andersen & Sorensen, 2005). The lack of KA studies may be due to the lack of suitable and easy to use measurement procedures. One protocol was proposed by Finnoff, Newcomber and Laskowski (2002). Their approach was based on the use of carbon paper sheets attached to a wooden target board, thereby providing imprints of the ball at impact. Hennig, Althoff and Hömme (2009) determined KA by usage of a circular electronic target fixed to a wooden board. Here, the ball creates electrostatic charges that allow identifying the ball impact location relative to the board center.

The purpose of our research was to quantify KV and KA for three frequently used soccer kicking techniques. Between and within subject variability was examined for different kicking performance tasks and kicking techniques. In order to carry out this research, an innovative KA measurement procedure was introduced.

METHODS: Research was conducted on a 5 x 20 m artificial turf outdoor testing area. 19 experienced soccer players (4th - 6th German league, 23.7 ± 3.4 yrs, 1.80 ± 0.05 m, 74.8 ± 5.6 kg) participated in this study. After warming-up and familiarization trials, subjects performed twelve repetitive kicks with each of three kicking techniques, full instep, inner instep and side foot. For each kicking technique, subjects had to execute two different performance tasks. Six kicks were meant to maximize ball velocity (Max KV) and the remaining six kicks were meant to maximize ball accuracy (Max KA). This resulted in altogether 36 single kicks per subject. Different performance tasks and kicking techniques were randomized between subjects. For Max KV, KV was the main variable whereas KA was regarded as dependent variable and vice versa for Max KA. All kicks were performed with a stationary ball from a distance of 6 m to the target goal construction in subjects' own soccer shoes.

Kicking was directed towards a bull's eye target (1 m height) in the middle of a 5 x 2 m goal construction. The goal construction was covered with a spanned, slightly transparent sheet that was hanging down from the cross-bar. KV and KA were recorded simultaneously for identical kicks. KV was measured by usage of a *Stalker Pro* radar gun (*Applied Concepts Inc., TX, USA*) positioned behind the goal, according to Sterzing and Hennig (2008). KA measurements were performed by recording the ball impact on the sheet by usage of a high-speed-video camera at 200 Hz (*CMOS Camera, HCC-1000, VDS Vosskühler, Germany*) also positioned behind the goal construction. KA was measured as the distance from bull's eye to ball impact location of the ball center. Absolute distance was determined with *MaxTRAQ 2.06* software (*Innovisions Systems, MI, USA*).

  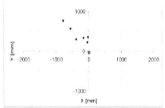

Figure 1: Experimental set-up (left), ball impact relative to bull's eye captured by high-speed-video (middle), visualization of KA for six kicks relative to bull's eye (right)

Means, standard deviations (SD) and coefficients of variability (CoV) for KV and KA for the two performance tasks applied to the three kicking techniques were calculated across all subjects. Repeated measures ANOVA and Bonferroni post-hoc tests were utilized to compare KV and KA between kicking techniques for both tasks. Additionally, respective effect sizes based on partial eta squared (eta²) were calculated. RMSE of KV and KA were performed to assess the relation of intraindividual and interindividual variability of kicks for performance tasks and kicking techniques. Also, the mean of within subjects $CoV_{6\ kicks}$ was calculated across all subjects. Thereby, $CoV_{6\ kicks}$ refers to the variability of the six repetitive kicks for each performance task/kicking technique.

RESULTS AND DISCUSSION: Independent of the required performance task, KV was significantly influenced by the different kicking techniques (Max KV: p<0.01, eta^2: 0.88; Max KA: p<0.01, eta^2: 0.43) (Figure 2, Table 1). Bonferroni post-hoc tests revealed significant differences (p<0.01) between side foot and both instep kicks but not between the two instep kicking techniques. In contrast but also independent of the required performance task, KA was generally higher for the side foot kick compared to the inner instep and the full instep kick (Max KV: p<0.01, eta^2: 0.37; Max KA: p<0.01, eta^2: 0.56) (Figure 2, Table 2). However, Bonferroni post-hoc tests revealed significant differences for Max KV only between side foot and full instep kicks and for Max KA between side foot and both instep kicks. As expected, subjects kicked with sub maximal velocity in all different kicking techniques (full instep: 85 %, inner instep: 82 %, side foot: 86 %) when KA was required. Thereby, the presented percentages have to be regarded as specific for the given study design requiring kicks over a distance of only 6 m. These KV percentages might be increased when kicking over a longer distance is required.

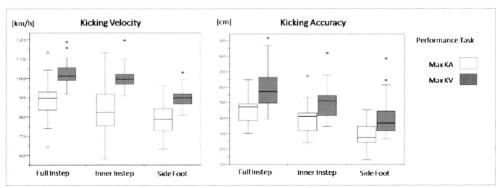

Figure 2: Kicking velocity [km/h] and kicking accuracy [cm]

CoV between subjects revealed that KV was the more homogenous variable compared to KA, regardless of the required performance task. The lowest CoV were present when subjects had to kick as fast as possible (Table 1 & 2). Mean within subject $CoV_{6\ kicks}$ showed that kicks were performed with strikingly homogenous KV (Table 1) for all required performance tasks. This is not astonishing when subjects were required to kick as fast as possible. However, when required to kick as accurate as possible individual subject's kicking velocity was similarly homogenous. This indicates that soccer players use a stable KV strategy when required to perform precision kicks. In contrast, individual KA variability was considerably high for all required performance tasks (Table 2).

Table 1: Kicking Velocity [km/h]

Technique	Task	n	Range	Min	Max	Mean	SD	COV	RMSE	$CoV_{6\ kicks}$
Full Instep	Max KV	19	27,00	91,83	118,83	103,16	6,51	0,063	1,49	0,034
Full Instep	Max KA	19	48,90	64,43	113,33	87,76	11,17	0,127	2,56	0,044
Inner Instep	Max KV	19	28,77	90,83	119,60	100,75	6,90	0,068	1,58	0,032
Inner Instep	Max KA	19	55,03	58,17	113,20	82,99	14,45	0,174	3,32	0,052
Side Foot	Max KV	19	22,33	80,67	103,00	89,79	5,65	0,063	1,30	0,029
Side Foot	Max KA	19	45,83	50,17	96,00	77,38	11,27	0,146	2,59	0,047

Table 2: Kicking Accuracy [cm]

Technique	Task	n	Range	Min	Max	Mean	SD	COV	RMSE	$CoV_{6\ kicks}$
Full Instep	Max KV	19	52,36	39,33	91,69	59,51	14,04	0,236	3,22	0,633
Full Instep	Max KA	19	34,64	30,16	64,80	45,93	9,90	0,216	2,27	0,550
Inner Instep	Max KV	19	47,54	34,48	82,02	50,85	12,29	0,242	2,82	0,477
Inner Instep	Max KA	19	43,29	23,98	67,27	39,65	11,02	0,278	2,53	0,613
Side Foot	Max KV	19	51,81	26,58	78,39	40,53	14,38	0,355	3,30	0,573
Side Foot	Max KA	19	32,34	13,05	45,38	28,83	7,90	0,274	1,81	0,572

CONCLUSION: A new protocol for simultaneous KV and KA measurements of soccer kicks was introduced and shown to be suited and practicable for the determination of kicking performance. This allowed carrying out a first soccer kicking performance study that comprehensively investigated the relationship of KV and KA for three different kicking techniques. Generally, our KV results of the examined techniques are in line with the literature knowledge which reported the full instep kick to be the fastest kicking technique followed by inner instep and side foot kicks (Levanon & Dapena, 1998; Nunome et al., 2002; Neilson & Jones, 2005). Previous knowledge was enhanced by quantification of KA of the different soccer techniques, stating the side foot kick to be most accurate followed by the inner instep and full instep kicking techniques. However, in soccer, kicking requirements differ with respect to the game situation, giving priority to concentrate either on KV or KA. In this study, soccer players, on average, reduced their KV to 82 - 86 % of their maximum KV for a given kicking technique when KA was the ultimate priority. It seems that KV, in contrast to KA, can be more easily tuned by soccer players. The fact that subjects had considerable low individual KV variability during all different performance tasks of this study is interesting. It shows that experienced soccer players have developed a solid motor performance pattern, which features a stable KV, when executing a given kicking task. In contrast, KA compared to KV must be regarded the much more variable aspect, also when referring to within subjects analysis. Thus, although when highly standardized kicking tasks are required to be performed by the players, a solid motor performance pattern does not guarantee KA success.

The results of this research call for a follow-up study which should aim towards identification of the biomechanical mechanisms that are responsible for the observed findings. Therefore, further research should investigate full 3D kinematics of the kicking tasks analyzed in the present study in order to link performance criteria to variation in skill execution. This might help to improve soccer kicking skills and consequently playing performance of players. Also, with a viable protocol to assess performance criteria of soccer kicking now available, one should aim to examine specific player groups, e.g. of different playing level, age and gender. Furthermore, the general relationship of ball velocity and ball accuracy is of inherent interest in numerous types of ball/team/ sports. Thus, sports, which feature different ball propulsion characteristics, e.g. handball, volleyball, tennis or field hockey should be investigated too.

ACKNOWLEDGEMENT:

This research was supported by Puma Inc. Germany.

REFERENCES:

Barfield, W.R. (1998). The biomechanics of kicking in soccer. *Clinics in Sport Medicine*, 17(4), 711-728.

Finnoff, J.T., Newcomber, K. and Laskowski, E.R. (2002). A valid and reliable method for measuring the kicking accuracy of soccer players, *Journal of Science and Medicine in Sport*, 5(4), 348-353.

Hennig, E.M., Althoff, K. and Hömme, A.-K. (2009). Soccer footwear and ball kicking accuracy, In *Proceedings 9. Footwear Biomechanics Symposium*, Stellenbosch, South Africa.

Kristensen, L.B., Bull Andersen, T. and Sorensen, H. (2005). Comparison of precision in the toe and instep kick in soccer at high kicking velocities, In T. Reilly, J. Cabri & D. Araújo (Eds.) *Science and Football V*, 70-72.

Levanon, J., and Dapena, J. (1998). Comparison of the kinematics of the full-instep and pass kicks in soccer, *Medicine & Science in Sports & Exercise*, 30(6), 917-927.

Neilson, P.J. and Jones, R. (2005). Dynamic Soccer Ball Performance Measurement, In T. Reilly, J. Cabri & D. Araújo (Eds.) *Science and Football V*, 21-27.

Nunome, H., Asai, T., Ikegami, Y. and Sakurai, S. (2002). Three-dimensional kinetic analysis of side-foot and instep soccer kicks, *Medicine & Science in Sports & Exercise*, 34(12), 2028-2036.

Shinkai, H., Nunome, H., Isokawa, M., and Ikegami, Y. (2009). Ball impact dynamics of instep soccer kicking. *Medicine & Science in Sports & Exercise*, 41(4), 889-897.

Sterzing, T., and Hennig, E.M. (2008). The influence of soccer shoes on kicking velocity in full instep soccer kicks, *Exercise and Sport Sciences Reviews*, 36(2), 91-97.

Tsaousidis, N. and Zatsiorsky, V. (1996). Two types of ball-effector interaction and their relative contribution to soccer kicking, *Human Movement Science*, 15, 861-876.

FIELD TESTING TO PREDICT PERFORMANCE IN RUGBY UNION PLAYERS

Brian Green[1], Catherine Blake[1] and Brian Caulfield[1]

School of Physiotherapy and Performance Science, Health Sciences Centre, University College Dublin, Dublin, Republic of Ireland

The purpose of this study was to identify the relationships between various field tests for leg power, sprinting speed and agility in a group of rugby union players. A group of 26 semi professional rugby union players each completed a test protocol consisting of a unilateral horizontal triple jump, 10 and 30M linear speed sprint tests, a change of direction test and a reactive agility speed test. Simple correlation analysis revealed multiple strong relationships between measured variables, most notably between linear speed and reactive agility speed (r=0.72-0.83). Multiple regression analysis indicated that linear speed and change of direction speed could predict 81% of the variance in reactive agility speed.

KEY WORDS: performance, testing, agility, rugby, jump.

INTRODUCTION: Professionalism in Rugby union has placed emphasis on the strength and conditioning field as a key component for success. Strength and conditioning staff formulate physical development programs based around the specificity of rugby union, thus focusing on the critical components of performance in order to optimize outcomes.

Data from American football demonstrate a link between performance testing outcome and achievement on the field (Garstecki et al., 2004). It has also been observed that performance testing that replicate sport specific tasks are more effective in distinguishing players of varying skill and playing level (Gabbett et al, 2008). Rugby union is a complex game that requires frequent short duration sprints with changes in multiple directions in reaction to other player movements during play (Deutch et al., 2007). Therefore, it seems logical to incorporate anticipated and unanticipated change of direction components to test protocols in order to account for the cognitive and perceptual factors that contribute to agility performance (Sheppard & Young, 2006).

Indeed, in recent studies (Green et al Unpublished Data) we have demonstrated that reactive agility testing can be used in rugby union to distinguish between players at different levels of ability. However, rugby specific reactive agility speed testing is a complex procedure that requires the use of expensive timing gates with random triggering/signaling functionality. Therefore, identification of factors that could effectively predict reactive agility in rugby players would be advantageous to coaches and trainers. Several investigators have examined relationships between various field tests yet there is a need to establish relationships between simple field tests and reactive agility. The purpose of this study is to investigate potential relationships between measures of leg power, linear speed, and agility. Furthermore, we wish to determine whether triple jump (TJ) performance, linear speed (LS), change of direction speed (CODS) can be used to effectively predict reactive agility speed (RAS) in rugby union players.

METHODS: Study design. This was cross-sectional observational study based on comparison of TJ, LS, CODS, RAS performance in a group of semi professional rugby players.

Participants. Data was collected from 26 rugby union players who were members of All-Ireland League squads. The average (±SD) age, height and body mass of the subjects were 19.5 ± 1.5 years, 1.84 ± 0.07 m, and 96.2 ± 14.9 Kg, respectively. This is a professional development league immediately below full-time professional competitions. The study was approved by the institutional ethical review committee and written informed consent was obtained from each subject. Subjects completed a physical screening questionnaire, anthropometric measurements, a functional movement screen, and a structured warm-up. Inclusion criteria were (1) within the ages of 18-23, (2) healthy and physically active, (3) currently playing rugby union for a club participating in the All-Ireland League. Exclusion criteria included (1) a lower extremity injury within the past 3 months resulting in loss of participation for 3 consecutive practice sessions or games, (2) current spine, hip, knee or ankle pain, (3) history of lower extremity neurovascular symptoms.

Procedures. Each subject performed testing in a multi-station non-randomized format in the following order: a structured warm up, TJ, 10/30m LS, CODS, and RAS trials. Subjects were given a demonstration prior to 3 sub maximal practice attempts. All testing was performed in one session 4 weeks into the pre-season.

Unilateral Horizontal Triple Jump. Jumping distance was measured for 3 trials of left and right foot TJ. Subjects stood on the test leg and jumped forward on the same leg, landing on both feet after the third jump. Arm movement was allowed. A measurement from the back of the heel closest to the starting line was taken using a T-square aligned perpendicular to the tape measure. Subjects had 2 minutes rest between trials.

10/30 Meter Linear Speed. 3 trials of LS over 0m-10m-30m were measured using electronic timing gates (Fusion Sport Smart Speed, Brisbane, Australia). The starting line was placed 0.7m behind the starting. Subjects performed a standing start and sprinted to cones that were placed 5m beyond the 30m timing gate. No hopping or backward movement was allowed prior to the start. Subjects were given 3 minutes rest between trials.

Change of Direction Speed Test. CODS was measured using electronic timing gates (Fusion Sport Smart Speed, Brisbane, Australia). White tape and cones were placed on the floor to guide and facilitate a sharp 45^0 degree cut. Subjects sprinted forward 5m then changed direction in order to pass through either the left or right finish gate placed 5m away. Each subject performed 3 cutting trials to the left and right, respectively with 2 minutes rest between trials.

Reactive Agility Speed Test. RAS was measured using electronic timing gates arranged in the same fashion as the CODS test. A timing gate was placed 0.5m from the start gate triggering either the left or right finish gates to flash. The subject would react to the flashing gate and change direction 45^0 to run through the finish gate. Each subject performed 6 trials (3 left, 3 right) in a random order with 2 minutes rest between trials.

Data analysis and Statistical Testing. Means for all variables were first calculated for each subject. In the case of LS measures the average of 3 trials was calculated whereas in the case of all other variables an average was calculated from 6 (3 left and 3 right) trials. To establish relationships between individual pairs a simple correlation analysis was performed by calculating Pearson's correlation coefficient (R). The level of significance was calculated using a t-test. Relationships of R > 0.6 were considered 'strong' (Swinscow, 1997). Following this, we performed a multiple regression analysis in order to explain the relationships between separate sets of independent variables and RAS (the dependent variable). Significance was established in each case using an ANOVA F-test.

RESULTS: The results of the simple correlation analysis are outlined in Table 1. Strong relationships were observed between most of the measured variables. The strongest relationships existed between measures of LS, CODS and RAS. The relationships between TJ and 10m LS and RAS were only moderate in nature.

Multiple regression statistics are outlined in Table 2. LS measures taken alone were good predictors of RAS ($R^2=0.72$, $P<0.0001$). Further analysis revealed a stronger relationship when LS and CODS were used to predict RAS ($R^2=0.81$, $P<0.0001$). Addition of TJ to the multiple regression process did not improve the strength of the prediction, when taken alone with measures of LS or with measures of LS and CODS.

Table 1. Simple Correlation Analysis

	TJ	10M LS	30M LS	CODS	RAS
TJ AVG	1.00				
10M LS AVG	-0.47	1.00			
30M LSAVG	-0.66*	0.95*	1.00		
CODS	-0.54	0.81*	0.79*	1.00	
RAS	-0.69*	0.72*	0.83*	0.78*	1.00

Value denotes Pearson's correlation coefficient indicating strength of relationship between opposing variables. Correlation coefficients greater than 0.6 have been shaded in grey in order to highlight areas where relationship between variables is strong. * indicates statistically significant Pearson's correlation coefficient as determined using a t-test at $P<0.05$) level.

Table 2. Multiple Regression Analysis between TJ, LS, CODS (independent variables) and RAS

Variables	R^2	Level of Significance
10m LS + 30m LS	0.72	P< 0.0001
10m LS, 30m LS +TJ	0.72	P< 0.0001
10m LS, 30m LS, CODS	0.81	P< 0.0001
10m LS, 30m LS, TJ, Agility	0.81	P< 0.0001

(dependent Variable)

DISCUSSION: In this study we have identified strong relationships between measures of LS, TJ, CODS, and RAS. In particular, we have established that relatively simple field tests of LS and CODS can be used to effectively predict RAS in rugby union players, accounting for 81% of the variance in RAS. This is an important finding in light of our previous work that identified that RAS is a strong indicator of playing level in rugby union players.

The only previous attempt to establish relationships between measures of LS, CODS and RAS was performed by Gabbett et al. (2008) who performed an analysis of relationships between LS, CODS and RAS in 42 rugby league players. They demonstrated statistically significant relationships between 10 and 20m sprint times and reactive agility speed. However, the strength of relationship was not strong with Pearson's correlation coefficients of $r = 0.41$ and 0.51 respectively. In our study, we have observed much stronger relationships between LS and RAS, with Pearson's correlation coefficients of 0.72 and 0.83 describing the relationship between RAS and 10 and 30M LS respectively. Differences in results may be related to differences in the test protocol for RAS testing. The Gabbett protocol involved reaction to the movement of another player whereas our protocol was based on reaction to a light that was timed to illuminate at a fixed delay from the starting point and could be considered to be less complex in nature.

Several investigators have examined horizontal jumping tests and their relationship to LS. Maulder and Cronin (2005) performed various horizontal and vertical jumping tests and investigated their relationship to 20m linear speed. Unilateral horizontal triple jump demonstrated the highest correlation to linear speed ($r = -0.86$ $P<0.001$). Our results are not as strong as this with Pearson's correlation coefficients of $r = -0.47$ and -0.66 explaining the relationship between TJ and 10m and 30n LS respectively. It is possible that differences in outcome could be related to differences in athlete training and body type profiles. Nesser et al. (1996) reported that the 5 step alternating bound for distance test highly correlates ($r = -0.810$) to 40m LS. Taken alone, TJ in our study was strongly related to 30M LS and RAS performance and can be considered to account for 43 and 47% its variance.

CONCLUSION: The objective for field-testing is to place a player in a sports specific environment in order to identify skill level and measure training effects. Identifying relationships between performance variables allows strength and conditioning staff to choose appropriate tests in order to maximize time and equipment. These results suggest that LS, TJ, and CODS field tests can identify potential for performance on more complex sport specific tasks (RAS).

References:

Duetsch, M.U., Kearney, G.A., and Rehrer, N.J. (2007). Time-motion analysis of professional rugby union players during match-play. *Journal of Sport Sciences*, 25, 461-472.

Gabbett, T.J., Kelly, J.N., and Sheppard, J.M. (2008). Speed, change of direction speed, and reactive agility of rugby league players. *Journal of Strength and Conditioning Research*, 22, 174-181.

Garstecki, M.A., Latin, R.W., and Cuppett, M.M. (2004) Comparisons of selected physical fitness and performance variables between NCAA division I and II football players. *Journal of Strength and Conditioning Research*, 18, 292-297.

Green, B., Blake, C., and Caulfield, B. Comparisons of linear speed, change of direction speed and reactive agility speed in rugby union players of varying skill level. Submitted for publication.

Maulder, P., and Cronin, J. (2005). Horizontal and vertical jump assessment: reliability, symmetry, discriminative and predictive ability. *Physical Therapy in Sport*, 6, 74-82.

Nesser, T.W., Latin, R.W., Berg, K., and Prentice, E. (1996). Physiological determinants of 40-meter sprint performance in young male athletes. *Journal of Strength and Conditioning Research*, 10, 263-267.

Sheppard, J.M., and Young, W.B. (2006). Agility literature review: classifications, training and testing. *Journal of Sport Sciences*, 24, 919-932.

Swinscow, T.D. (1997). *Statistics at square one* (9th edn). University of Southhampton, UK: BMJ Publishing Company.

THE PLANARITY OF THE STICK AND ARM MOTION IN THE FIELD HOCKEY HIT

Alexander Willmott[1,2] and Jesús Dapena[1]

Department of Kinesiology, Indiana University, Bloomington, IN, USA[1]
Department of Sport, Coaching & Exercise Science, University of Lincoln, Lincoln, UK[2]

The development of relevant simulation models is one way in which our knowledge of the field hockey hit may be improved. The aim of this study was to test the appropriateness of a planar pendulum model for the motion of the stick and arms during the downswing. The hits of 13 experienced female players were filmed, and swing planes were fitted to the motion of the stickface during the downswing. Low variability in the length of a segment's projection onto the swing plane was taken as evidence for the validity of a planar model. Coefficients of variation of less than 5% for the stick and forearm lengths supported the use of such a model for these segments, but its validity for the upper arms is less certain.

KEY WORDS: field hockey, hit, modelling, swing plane.

INTRODUCTION: Despite its importance for long-range passing and shooting at goal (Anders and Myers, 2008), the field hockey hit is surprisingly poorly understood. Three-dimensional kinematics of the stick and body have been reported only by Chivers and Elliott (1987) and Elliott and Chivers (1988). Improvements in our knowledge of the stroke, which would provide a basis for better-informed recommendations on technique, may possibly come from a combination of further empirical investigations and computer simulation. The most obvious candidate in the latter area might be a planar pendulum model of the kind first popularised for golf by Cochran and Stobbs (1968). For field hockey, Elliott and Chivers (1988) suggested that the left arm and stick function as a double pendulum featuring a single arm segment, but that the right arm would need to be treated as two separate segments.

The validity of planar models for the golf swing has recently been challenged, with a single plane not always being appropriate for the club only, let alone the club *and* the arms (see Coleman and Anderson, 2007, for a summary). For the field hockey hit, and following Cochran and Stobbs' (1968) original definition of the swing plane, Willmott and Dapena (2008) found that the motion of the stickface during the downswing was remarkably planar. This plane might form the basis of a pendulum model for the arms and stick either if these segments move directly within the same plane, or if they share a common axis of rotation that it is perpendicular to the plane. In the latter case, each segment's projection onto the swing plane would be of constant length and would appear to move in the plane (Figure 1).

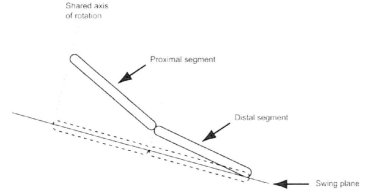

Figure 1. A 'planar' double pendulum model in which the two segments rotate about a shared axis, with their projections (dashed lines) moving in the swing plane.

The purpose of this study was to investigate the validity of a planar model for the motion of the stick and arms during a field hockey hit. This was done by determining the angles at which the stick and arm segments are held relative to the stickface swing plane during the downswing of the field hockey hit. The cosines of these angles are a measure of the length of the projected segments, and the consistency of these projected lengths can be used to test the appropriateness of a planar representation of the hit.

METHODS: Data Collection: Thirteen experienced female field hockey players (height = 1.67 ± 0.06 m; mass = 64 ± 6 kg; mean ± sd) were asked to hit a stationary ball after a single approach step. Six of these players used a straight backswing in which the stickface path was similar to that of the subsequent downswing; the remaining seven players adopted a looped backswing in which the stick was taken back in a pronounced curve above the plane of the downswing. The hits were filmed with two Locam motion-picture cameras at 200 fps, and the DLT method (Abdel-Aziz and Karara, 1971) was used to reconstruct the three-dimensional positions of three stick markers as well as the left and right wrists, elbows and shoulders. The elbows and shoulders were tracked as part of a wider study in which the whole body was digitised in every other frame during the downswing; due to their higher velocities, the wrists and stick markers were digitised in every frame.

The data for all landmarks were smoothed from the start of the downswing to impact using quintic splines. Cutoff frequencies were selected on a subject-by-subject basis, and ranged from 30 to 42 Hz for the stick markers, 22 to 30 Hz for the wrists, and 16 to 18 Hz for the elbows and shoulders. Finally, the three stick markers were used to reconstruct the position of the centre of the stickface and the two ends of the stick shaft at each instant.

Data Analysis: Stroke planes were fitted to the stickface motion using Total Least Squares Regression, as described in Willmott and Dapena (2008). The stickface motion was resampled at 0.10 m intervals to give equal weight to all parts of the downswing, and the motion was considered to be planar where the mean absolute residual between the stickface coordinates and the fitted plane was less than 0.5% of the path length travelled by the stickface. Working backwards from impact, the longest portion of the downswing that met this criterion was selected.

Quintic spline interpolation was used to determine the positions of the shoulders, elbows, wrists and stick shaft endpoints at the instant of each of the resampled stickface positions. The angles between the stroke plane and the following segments were calculated: the stick shaft, the left and right upper arms, and the left and right forearms. Positive angles indicated that the proximal end of the segment was higher than the distal end, relative to the swing plane. The cosine of a segment's angle gave the length of its projection onto the swing plane as a proportion of the true segment length. For each segment, the mean cosine at every stickface position was determined for each backswing group. The consistency of these mean cosines across the downswing was quantified using the coefficient of variation (CV, the standard deviation expressed as a percentage of the mean value).

RESULTS & DISCUSSION: The planar portion of the stickface motion had a length of 2.45 ± 0.28 m for the straight backswing group, covering the entire downswing for five players and 95% of it for the sixth player. The length of the planar portion for the looped backswing group was 2.87 ± 0.29 m, which was 86 ± 8% of the downswing length.

Figure 2 shows how the projected length of each segment varied over the portion of the downswing for which the stickface motion for all members of a particular backswing group was planar: the last 2.1 m for the straight backswing group, and the last 2.3 m for the looped backswing group. The variation in projected lengths was small for the stick (CV <1% for both backswing groups) and the forearms (CV <5%), supporting the use of a planar model for these segments.

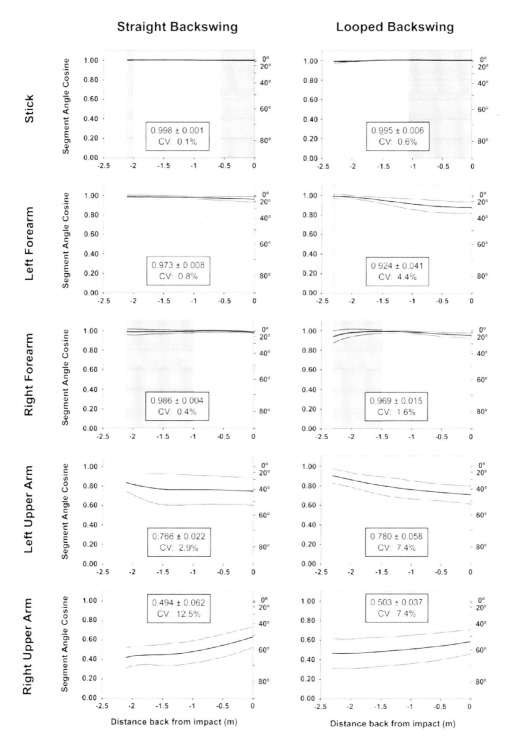

Figure 2. Variation in the cosines of the segment angles relative to the swing plane with increasing distance of the stickface back from impact. The lines shown are the mean ± sd for each group of subjects. The shaded areas represent periods when the angles were negative. The inset boxes show the distribution of each mean cosine and its coefficient of variation.

The variability was greater, in general, for the projected lengths of the upper arms, with the CV reaching 12.5% for the right upper arm in the straight backswing group.

The angles on the right side of each graph in Figure 2 are a guide to the segment angles that correspond to particular cosine values. Given the non-linear relationship between angles and their cosines, however, the variability in the mean segment angles cannot be read directly from Figure 2. The distribution of the values of each mean segment angle across the shared planar section of the downswing is therefore listed in Table 1.

Table 1. The distributions of the mean segment angles, calculated for each backswing group. (mean ± sd, all values in degrees.)

	Straight Backswing	Looped Backswing
Stick	0.0 ± 0.5	1.7 ± 5.0
Left Forearm	12.3 ± 2.2	17.0 ± 9.4
Right Forearm	0.5 ± 5.9	3.6 ± 11.1
Left Upper Arm	38.0 ± 1.9	37.5 ± 5.9
Right Upper Arm	60.0 ± 4.2	59.1 ± 2.5

The stick and the right forearm moved close to the swing plane; the left forearm was maintained at a small positive angle. Large ranges of angles for the forearms in the looped backswing group did not lead to large variation in the projected lengths because the absolute angles were comparatively small, and thus the cosines did not change much. The upper arms were held at considerable angles to the stroke plane: approximately 40° and 60° for the left and right upper arms, respectively.

CONCLUSION: Investigation of the consistency of segments' projected lengths in the swing plane has demonstrated that the motion of the stick shaft and forearms could be approximated by a planar pendulum model. For the stick and right forearm, this is because the segments are moving close to the plane itself; the left forearm is held at a positive angle to the plane but its projected length is very consistent. The validity of a planar model for the upper arms is less certain: these segments are at much larger angles to the swing plane, where small changes in these angles result in larger changes in the projected lengths.

REFERENCES:

Abdel-Aziz, Y.I. and Karara, H.M. (1971). Direct linear transformation from comparator coordinates into object coordinates in close-range photogrammetry. In *Proceedings of the ASP Symposium on Close-Range Photogrammetry* (pp. 1-18). Falls Church: American Society of Photogrammetry.

Anders, E. and Myers, S. (2008). *Field hockey: steps to success* (2nd edn). Champaign: Human Kinetics.

Chivers, L. and Elliott, B. (1987). The penalty corner in field hockey. *Excel*, 4, 5-8.

Cochran, A. and Stobbs, J. (1968). *The search for the perfect swing*. Philadelphia: J.B. Lippincott.

Coleman, S. and Anderson, D. (2007). An examination of the planar nature of golf club motion in the swings of experienced players. *Journal of Sports Sciences*, 25, 739-748.

Elliott, B.C. and Chivers, L. (1988). A three dimensional cinematographic analysis of the penalty corner hit in field hockey. In G. de Groot, A. P. Hollander, P. A. Huijing and G. J. van Ingen Schenau (Eds.) *Biomechanics XI-B* (pp. 791-797). Amsterdam: Free University Press.

Willmott, A.P. and Dapena. J. (2008). Quantifying the planarity of the field hockey hit [online]. *North American Congress of Biomechanics*, Ann Arbor, August 5-9. Available from: http://www.x-cdtech.com/nacob/abstracts/401.pdf [Accessed 20 March 2009].

THE INFLUENCE OF STANCE WIDTH ON MOVEMENT TIME IN FIELD HOCKEY GOALKEEPING

Kevin Ball and Georgia Giblin

School of Sport and Exercise Science, Victoria University, Melbourne

Minimising movement time is essential for a field hockey goalkeeper and stance width is considered important to agility. The aim of this study was to examine if an optimal stance width exists for field hockey goalkeepers and if so, does it vary for different movement directions and for different individuals. Ten state and national level goalkeepers made simulated saves from ten different stance widths ranging from 0.4 m to 1.2 m. AMTI force plate data was used to identify start of movement time and timing gates in the corner of the goals recorded the end of movement time. On a group basis, a stance width of 1.1 m was optimal for minimising movement time for high and low saves and for right and left saves. On an individual basis, 1.1 m was the optimal stance for eight of ten subjects. Only two subjects performed optimally at their preferred stance width. Where shots to the corner of the goals are likely, goalkeepers should adopt a wide stance.

KEYWORDS: Agility, preparatory stance

INTRODUCTION: A game of field hockey is won by outscoring the opposition. The goalkeeper forms the last line of defence for a team and their task is to intercept shots that are made from within a 14.6m radius from the goal. Given the ball can travel at up to speeds of 33 m/s (Ball, 1994), the goalkeeper can have less than 1 s to react to a shot from the edge of this area and move to stop it. Further, the distance from the shooter to the goalkeeper is typically shorter than the maximum distance. As such, the ability of a goalkeeper to minimise movement time is paramount.

There are few studies examining goalkeeping technique and none exist in field hockey. In soccer, technique differences have been identified between goalkeepers of different skill levels (Suzuki et al., 1987) and between saves to the dominant and non-dominant sides (Spratford et al., 2007). These studies reported that a more direct path towards the save point was evident in more elite performers (Suzuki, et al., 1987) and on the dominant compared to the non-dominant side (Spratford et al., 2007). In ice hockey, Wiliberg (1979) identified three classifications of stance among 300 goalkeepers from junior to senor level; an open or "V" stance (feet parallel, legs wide apart), a closed or parallel stance (feet parallel, legs together) or a broken "V" or butterfly stance (knees angled in toward each other, legs wide apart). However, none of these studies used performance measures or focussed specifically on stance width.

Links between stance and performance have been examined in the scientific literature in a number of forms although none have specifically looked at stance width and movement time. Preparatory stance with bent knees compared to straight knees (e.g. Yamamoto, 1996) and with weight evenly balanced over flat feet compared with on the toes (Stater-Hammel, 1953) have been linked to faster reaction-movement times. A closed stance has also been reported as being better for reaction-movement time in tennis (Lockerman, 1973), although Hopkins (1984) reported that the open stance allowed players to recover from a wide forehand shot to a backhand shot significantly faster than the closed stance.

In spite of the importance of preparatory stance in numerous sports, it is perhaps surprising that there are only a few studies focussing on this feature of sport skills. The aims of this study were to examine if an optimal stance width existed for field hockey goalkeepers and if this optimal stance width differed for saves in different directions.

METHODS: Eight male and two female hockey goalkeepers competing at state or national level at the time of testing participated in this study. Table 1 reports subject details including two measures used to normalise stance width (leg and arm length). Arm length was measured from the acromion process to the tip of the middle digit on each arm while fully extended. Leg length was measured from the anterior superior iliac spine to the lateral malleolus while standing.

Table 1: Subject characteristics (N = 10)

Age (years)	Height (m)	Leg Length (m)	Arm Length (m)	Mass (kg)
21.6 ± 2.5	1.80 ± 0.06	0.94 ± 0.08	0.77 ± 0.03	86.2 ± 18.7

Each subject wore their full goalkeeping gear as used in games (pads, kickers, helmet, gloves protective equipment and stick) and performed simulated saves, one to each corner of the goal at each of 10 stance widths (total of 44 saves). Goalkeepers stood on two AMTI force plates (Advanced Mechanical Technologies, Inc, Massachusetts, USA), one under each foot. From this position, each goalkeeper's stance width was adjusted using a specialised metre ruler with arms extending perpendicularly. These arms were set apart to the required width and goalkeepers increased or decreased their stance width until the two arms were touching the lateral malleolus of each foot. Stance widths ranged from narrow (0.4m) through increments of 0.1 m up to wide (1.2 m) and included the goalkeepers preferred stance width. A flat wooden signal board (1 m x 0.76 m, figure 3.2) with an LED in each of the four corners was positioned on a camera tripod 0.87 m off the ground 4 m directly in front of the goalkeeper. For each trial, one LED was lit to indicate the diving direction (e.g. top right, bottom left). Subjects were instructed to react as fast as possible to the signal and to use their normal save movement when completing each trial. The order of stance widths and the direction of the required response were randomised. One trial was performed for each stance width-direction condition.

To provide a 'target' for the goalkeepers to save, Four tennis balls were suspended 1.67 m either side of the centre of the force plates, and 0.3 m and 2.0 m above the ground approximately corresponding to shots to the corner of the hockey goal. Four sets of custom built timing gates were mounted on steel rods and were aligned just to the goalkeeper's side of each tennis ball such that a beam of light was broken once the goalkeeper reached the target tennis ball.

All data including data from the force plates, the timing gates, the directional LED indicator and the 1s flashing LED was passed through 16 analog channels into the Optotrak Certus Motion Capture System (Northern Digital Inc, Canada). All data was sampled at a rate of 1200Hz. Movement time was defined as the difference between the onset of forces (Fz) associated with the save to the point at which the timing gates positioned near the tennis balls were broken. A 50 Hz camera recorded all trials and was used to correct any trials where the timing gates did not function correctly and to evaluate any inconsistencies with breaking the timing gates such as when the hockey stick rather than the glove broke the timing gate beans.

RESULTS: Figure 1 shows movement time for each of the set stance widths and figure 2 compares movement time for high and low saves (figure 2a) and for saves to the right compared to the left (figure 2b). The fastest mean movement time (0.6 s) occurred at a stance width of 1.1 m and the slowest movement time (0.774 s) occurred at the narrowest stance of 0.4 m. Movement time progressively decreased from the 0.4 m to the 1.1 m stance width after which it increased again at 1.2m. For all save directions, 1.1 m produced the fastest movement times. For three of the four individual corners, the lowest movement time was produced at a stance width of 1.1 m, with producing the lowest movement time for the remaining corner.

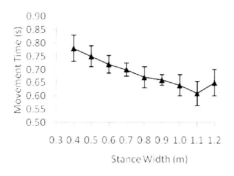

Figure 1: Mean Group Movement Time (s) at each stance interval.

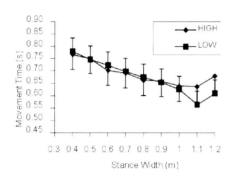 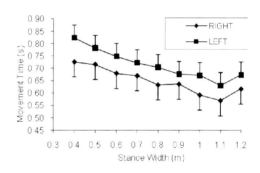

(a) High compared to low saves   (b) Right compared to left saves

Figure 1: Mean Group Movement Time (s) comparing high and low saves and comparing right and left saves

On an individual basis, the 1.1 m stance width produced the lowest movement times for eight subjects, while for the remaining two subjects, stance widths of 0.9 m and 1.0 m produced the lowest times. Seven subjects produced a smaller minimum movement time at a stance width other than their preferred stance width (figure 3). There was no association between anthropometric measures and optimal stance width.

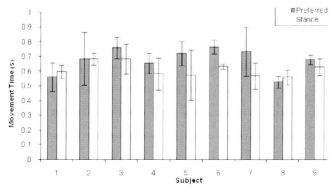

Figure 3: Movement time (s) for preferred and test-optimal stance

DISCUSSION: An optimal stance width exists for hockey goalkeeping to minimise movement times. A stance width of 1.1 m was optimal for saves to the corner of the goals. This stance width was optimal on an overall basis, for saves to the right and left side, for high and low saves and for eight of the ten subjects. Further, this stance width was optimal for three of the four corners of the goal and while differences in total movement times occurred between males and females, optimal stance widths were the same. Based on this very strong support, it would seem that this stance width should be employed by hockey goalkeepers where wide shots are likely.

Optimal stance width was not related to height, mass, leg length or arm length. This was an interesting finding as it might be expected, for example, that a longer legged goalkeeper might have an optimal stance width that was wider than a shorter legged subject. However, the data in this study indicated that no relationship existed between the anthropometric measures used in this study and optimal stance width. Rather, very strong support existed for an absolute (1.1 m) rather than a relative (normalised to body dimensions) relationship between stance width and movement time.

Only two subjects produced faster movement times at their preferred stance compared to their optimal stance as identified in the specified stance widths (test-optimal stance widths). The remaining seven subjects adopted a stance, which was sub optimal for shots directed to each of the four corners. Based on this finding, a sound recommendation for hockey goalkeepers is to evaluate their stance width to determine if they are optimising their preparatory position for wide saves.

This research has identified a wide stance of 1.1m as optimal for saving shots directed at the four corners of the goal. Future work needs to look at the kinematic and kinematic factors associated with this movement and examining saves to different areas of the goal, such as between the legs or nearer to the goalkeeper. To assist with this analysis, a profile of common shot placements evaluated from games would provide information on the most common shot targets in the goal. Another important direction is to examine stance width in specific game situations such as where shots nearer the goalkeeper or in situations where the point of shot is not imminent (i.e. if a player passes rather than shooting). Finally, the target of the movement was a stationary ball and while this was appropriate for this study as it was concerned with movement time only, it did not allow for evaluation of save success. The addition of 'real' saves would allow for evaluation of movement time along with success of the save and perceptual and reaction timing aspects of the skill.

CONCLUSION: An optimal stance width exists for hockey goalkeeping to minimise movement times for shots to the corner of the goal. A stance width of 1.1 m was optimal on a group basis for all save directions and on an individual basis for eight of ten subjects. Only two subjects produced faster movement times at their preferred stance width. Where wide shots are likely, goalkeepers should adopt a wide stance width.

REFERENCES:
Hopkins, P. W. (1984). A comparison of movement times between the open and the closed stance for the tennis forehand groundstroke. United States, Microform Publications, University of Oregon.
Lockerman, W. (1973). A Comparison of the open and closed foot stance for reaction and movement times. *Journal of Motor Behavior 5*: 57-63.
Spratford, W., Burkett, B. and Mellifont, R. (2007). Biomechanical Symmetry Differences in the Goalkeeping Diving Save. *Journal of Sports Science and Medicine 6, Supplement 10*: 175-180.
Suzuki, S., Togari, H., Isokawa, M., Ohashi, J. and Ohgushi, T. (1987). Analysis of the goalkeeper's diving motion. *In, Reilly, et al. (eds.), Science and football: proceedings of the First World Congress of Science and Football. pp.468-475*. United States.
Wilberg, R.B. (1979). Basic stance of age-group ice hockey goalkeepers. *Canadian Journal of Applied Sport Sciences*, 4, 66-70
Yamamoto, Y. (1996). The relation between preparatory stance and trunk rotation movements. *Human Movement Science 15*: 899-908.

THE EFFECTS OF LOCALISED QUADRICEPS FATIGUE ON BOWLING PERFORMANCE IN FEMALE COUNTY CRICKETERS

Alan De Asha [1], Mark Robinson [1], Amity Campbell [2], Mark Lake [1]

Research Institute for Sport & Exercise Sciences, L.J.M.U., Liverpool, UK [1]
School of Physiotherapy, Curtin University of Technology, Bentley, Australia [2]

KEY WORDS: Cricket, Bowling, Knee Angle, Fatigue, Technique.

INTRODUCTION: Several studies in cricket bowling indicate that a more extended front knee at ball release contributes to increased ball release speed but reported correlation co-efficients vary from 0.34 to 0.52. Portus et al. (2000) found an 8-over spell had no significant effect on knee angle or ball release speed. This study manipulated knee angle by fatiguing the quadriceps to explore its relationship with ball release speed and technique.

METHOD: Five female medium-fast bowlers (aged 16 ± 1.3 yrs) bowled 24 deliveries whilst their movement was tracked opto-electronically using 44 retro-reflective markers (Qualisys, Sweden). The ball was tracked with retro-reflective tape on one hemisphere and ball release was determined from ball velocity. Subsequently, a fatigue protocol was administered to the quadriceps of the front leg using an isokinetic dynamometer (LIDO, Loredan Biomedical Inc., USA). Once quadriceps torque decreased to <30% MVC, 4 more deliveries were bowled. A whole-body biomechanical model was constructed following a static calibration in Visual 3D (C-Motion, USA) and non-parametric correlation analysis was undertaken in SPSS.

RESULTS & DISCUSSION: When grouped according to knee angle at ball release; two bowlers had a straight front leg at ball release whereas the three others had a more flexed knee (Fig. 1). This technique influenced their response to fatigue and subsequent ball release speeds. Pre-fatigue there was a significant negative correlation between knee flexion angle and ball release speed ($r = -0.66$, $P < 0.01$). This supported the trend shown in Portus et al. (2000) that bowlers with straighter legs have higher ball release speeds. Post-fatigue however there was no correlation ($r = -0.21$, $P = 0.40$) as bowlers with straighter legs were more greatly affected by quadriceps fatigue. Their ball release speed decreased by 1.95 ms^{-1}, which was 0.55 ms^{-1} more than the flexed knee group. In comparison to the fatigue induced in Portus et al. (2000), quadriceps fatigue led to reductions in ball speed in both techniques.

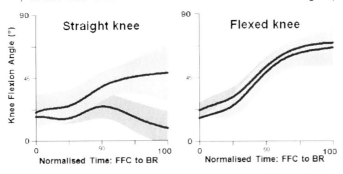

Figure 1. Knee flexion angles from front foot contact (FFC) to ball release (BR). Quadriceps fatigue caused bowlers with straighter legs at ball release (blue) to increase knee flexion post-fatigue (red).

CONCLUSION: This study found that minimal knee flexion during the delivery stride was beneficial to achieving maximal ball release speed. The effect of quadriceps fatigue on knee flexion and ball release speed was dependent on the bowler's knee angle at ball release. Female fast / medium bowlers should endeavour to reduce knee flexion in the delivery stride and ensure sufficient quadriceps endurance to maximise and maintain ball release speed.

REFERENCE:

Portus,M.R., Sinclair,P.J., Burke, S.T. et al. (2000). Cricket fast bowling performance and the influence of selected physical factors during an 8-over spell. *Journal of Sports Sciences*, 18, 999-1011.

THE EFFECT OF INDIVIDUALISED COACHING INTERVENTIONS ON ELITE YOUNG FAST BOWLERS' TECHNIQUE

Peter Worthington[1], Craig Ranson[2], Mark King[1], Angus Burnett[3] & Kevin Shine[4]

School of Sport and Exercise Science, Loughborough University, UK[1]
UK Athletics Chief Physiotherapist Loughborough, UK[2]
School of Exercise, Biomedical & Health Sciences, Edith Cowan University, Australia[3]
ECB National Lead Fast Bowling Coach, Loughborough, UK[4]

Fast bowling in cricket is an activity well recognised as having a high injury prevalence. Previous research has associated lower back injury with aspects of fast bowling technique. Coaching interventions that may decrease the likelihood of injury, whilst maintaining or increasing ball speed, remain a priority within the sport. Selected kinematics of the bowling action of 14 elite young fast bowlers were measured using an 18 camera Vicon Motion Analysis System. Subjects were tested before and after a two year coaching intervention period, during which subject-specific coaching interventions were provided. Mann-Whitney tests were used to identify significant differences in the change in the selected kinematics between those bowlers who were coached or un-coached on each specific aspect. Coached athletes demonstrated a significant change in shoulder alignment at back foot contact (more side-on, P = 0.002) and shoulder counter-rotation (decreased, P = 0.001) relative to un-coached athletes. There was no difference in the amount of change in flexion angles of the front or back knee or lower trunk side-flexion between those who received coaching intervention and those that did not. This study shows that specific aspects of fast bowling technique in elite players can change over a two year period and may be attributed to coaching intervention.

KEY WORDS: fast bowling, cricket, injury, coaching intervention.

INTRODUCTION: Fast bowlers in cricket have injury rates comparable to contact sports such as Australian Rules football and Rugby football. Lower back stress fracture has consistently accounted for the most lost playing time in elite cricket (Newman, 2003; Orchard, James & Portus, 2006). Previous research has associated lower back injury and the appearance of abnormal radiological features with: a large shoulder counter-rotation (SCR) angle between back foot contact (BFC) and front foot contact (FFC) in the delivery stride (Foster, et al., 1989; Portus, et al., 2004); large amounts of trunk side-flexion during FFC in female bowlers (Stuelcken, et al., 2008); non-significantly larger lower trunk extension during FFC (Ranson, et al., 2008b); and the use of a straight knee during FFC (Portus, et al., 2004). However, maintaining a straight knee during FFC has also been linked to increased ball speeds (Foster, et al., 1989; Portus, et al., 2004).

Although interventions targeted at reducing injury risk and increasing ball speed are undertaken by fast bowling coaches, only one study (Elliott & Khangure, 2002) (aimed to reduce SCR during the delivery stride) has investigated the efficacy of attempted technique modification. Knowledge of the changeability of fast bowling technique in elite bowlers, a highly complex motor skill, may give some insight into the ability of coaching to alter technique characteristics.

The purpose of this study was to determine whether a two-year coaching intervention resulted in significant changes to certain key aspects of fast bowling technique. This study was conducted on a group of elite young fast bowlers and aimed to improve the kinematics of the bowling action, focussing on risk factors related to low back injury and increased ball speed.

METHODS: Data Collection: Subjects consisted of 14 elite young fast bowlers, with age, height and mass of 18.5 (± 2.3) years, 1.90 (± 0.06) m and 82 (± 5) kg, respectively, at the time of initial testing. Baseline testing was conducted late in the summer season and follow up testing was performed two years later. An 18 camera Vicon Motion Analysis System (OMG Plc, Oxford UK) operating at 300 Hz was used to capture: a static trial; a range of motion (ROM) trial (as described by Ranson, et al., 2008a); and six maximum velocity fast bowling deliveries for each bowler. All testing was conducted in an indoor practice facility, allowing subjects to bowl using their normal length run-up on a standard size artificial cricket pitch. Fourteen millimetre diameter, spherical retroflective markers were positioned on anatomical landmarks to define: shoulder; lower trunk; pelvic; thigh; shank; upper arm; and forearm motion. A square of reflective tape (1.5 cm × 1.5 cm) was fixed to one side of the ball, enabling the instant of ball release and ball velocity to be determined.

Trials were manually labelled and the best 3 selected for each bowler (maximum velocity with minimal marker loss). Hip joint centres were located according to the methodology of Davis, et al. (1991) and the lower trunk and pelvis joint centres were calculated as described by Ranson, et al. (2008a). All other joint centres were defined as the mid-point of a pair of strategically positioned markers. Local reference frames were defined and joint angles calculated using the methodology of Ranson, et al. (2008a). Orientation angles of the lower back relative to the pelvis were normalised using a neutral position from the ROM trial. Shoulder and pelvis angles were defined by projecting the respective joint centres onto a horizontal plane (180° = side-on, 270° = aligned with the bowling crease), corresponding to previous fast bowling research (Portus, et al., 2004).

Data Analysis & Coaching Intervention: Parameters determined for each trial consisted of: shoulder alignment at BFC; amount of SCR; back knee angle at BFC and the amount of flexion occurring during the BFC phase; front knee angle at FFC and the amount of knee flexion from the instant of FFC to ball release; and ball velocity. The time histories of each kinematic descriptor were fitted using quintic splines (Wood & Jennings, 1979). The closeness of fit at each point was based on the difference between the descriptor value and the average value from the two adjacent times (Yeadon & King, 2002).

Aspects of technique requiring coaching intervention were identified for each bowler, based on these kinematic parameters. Remediation, utilising coaching techniques such as verbal feedback, video feedback and part drills (exercises focused on specific part components of the delivery stride) were instituted during both Club and ECB coaching sessions over the ensuing two years. Coaching interventions encouraged: a reduced shoulder angle at BFC (causing a consequent reduction in SCR); decreased knee flexion during FFC and/or BFC; a more upright trunk alignment during FFC (i.e. less side-flexion). Interventions were only provided to those bowlers identified as requiring them, the remainder of the bowlers formed the un-coached group for that specific aspect of technique.

Figure 1 – Illustration of a bowler with a large shoulder counter-rotation (SCR).

Statistical Analysis: The inter-trial reliability of the kinematic data was assessed using the Intra-Class Correlation Coefficient (ICC) and the Standard Error of Measurement (SEM). Good reliability was found for all kinematic variables (Initial Testing: ICC = 0.27-0.98, SEM = 1.2-7.9; Follow-up Testing: ICC = 0.54-0.98, SEM = 0.9-5.8). Consequently, the three trials selected for each bowler were averaged to provide representative data. Mann-Whitney tests were performed (Statistical Package for Social Sciences V15) to identify significant differences in the amount of change in the kinematic variables over the two year period, between bowlers who received specific coaching on that variable and those who did not (from the group of 14).

RESULTS:

Table 1 Kinematic parameters for the coached and un-coached subjects, pre and post intervention period. P-values were calculated using a Mann-Whitney non-parametric test.

Kinematic Variable	Coached? (group size)	Initial testing Mean (°)	SD (°)	Follow-up testing Mean (°)	SD (°)	Difference (°)	P Value
Shoulder angle at BFC	Yes (n = 8)	254	16	233	12	21	0.002
	No (n = 6)	228	13	228	11	0	
SCR	Yes (n = 8)	52	11	32	12	20	0.001
	No (n = 6)	37	16	37	14	0	
Back knee angle at BFC	Yes (n = 9)	41	8	36	10	5	0.226
	No (n = 5)	33	5	31	7	2	
Back knee flexion	Yes (n = 9)	22	10	25	12	-3	0.332
	No (n = 5)	32	7	33	5	-1	
Front knee angle at FFC	Yes (n = 9)	14	8	18	3	-4	0.358
	No (n = 5)	5	7	11	2	-6	
Front knee flexion	Yes (n = 9)	20	13	17	15	3	0.072
	No (n = 5)	5	3	6	6	-1	
Lower trunk peak side-flexion	Yes (n = 8)	39	5	37	6	2	0.247
	No (n = 6)	29	6	29	4	0	

DISCUSSION: The use of a mixed type action, characterised by a large amount of SCR, has previously been associated with low back injury in fast bowlers (Foster, et al., 1989; Portus, et al., 2004). Increased SCR has been linked to large shoulder angles at BFC (a more front-on alignment) (Portus, et al., 2004; Ranson, et al., 2008a). Bowlers receiving coaching interventions targeted at reducing their SCR and shoulder angle at BFC evidenced a significant change in these parameters over the two year period (P = 0.002 and 0.001, respectively), relative to the un-coached bowlers.

The specific pathomechanics of the highly prevalent non-bowling arm side lumbar stress injuries are thought to be related to the repetition of end range lower trunk side-flexion, rotation and extension typically adopted during the FFC phase of the delivery stride (Elliott, 2000; Ranson, et al., 2008a). Although the amount of SCR and the number of mixed action bowlers decreased over the two year period, there was no significant change in the lower trunk peak side-flexion of the coached subjects. Ranson, et al. (2008a) suggested a greater coaching emphasis on obtaining less stressful (less extreme side-flexion, extension and rotation) lower trunk postures during the FFC phase may have a greater influence on back injury pathomechanics than a focus on shoulder alignment and SCR during BFC.

Actions in which the front knee remains relatively straight during FFC have been associated with an ability to produce faster ball speeds (Portus, et al., 2004; Loram, et al., 2005). Bowlers in this study tended to display a more flexed front knee at the instant of FFC in the follow-up testing than in the baseline testing. However, there was no significant difference in the change in either front knee angle at FFC, or the amount of knee flexion occurring up to ball release, between the the coached and un-coached groups. Ball velocity increased by an average of 3 mph over the course of the study, this is likely to be due to the physical maturation of the subjects, rather than any coaching interventions. Bowlers receiving coaching intervention regarding back knee technique did not exhibit a significant change in back knee kinematics in comparison with the un-coached subjects.

Although small group sizes are a perennial problem when investigating elite sporting subgroups such as fast bowlers in cricket, the cohort investigated in this study was homogenous and the effect size for significant statistics were large. Apart from coaches and participants agreeing to follow general principles, it was not possible to standardise the amount and quality of coaching intervention delivered. As previously mentioned, aside from coaching, extraneous variables such as age, physical development and injury might have had a significant impact on participants' technique and ball velocity. Although there was no specific "control" group in this investigation, bowlers who did not receive a particular intervention essentially acted as a control group. Neither outcome data (e.g. change in injury status), nor other deliveries (e.g. Yorkers or bouncers) were examined in this study.

CONCLUSION: This study showed that specific aspects of fast bowling technique in elite young players can change over a two year period. Shoulder alignment at BFC and the amount of SCR demonstrated a significant change over the two year intervention period in the coached bowlers relative to those who were un-coached on these aspects of technique. This may be due to coaching intervention and warrants future investigation.

REFERENCES:

Davis, R. B., Ounpuu, S., Tyburski, D., and Gage, J. R. (1991). A gait analysis data collection and reduction technique. *Human Movement Science, 10*, 575-587.

Elliott, B. C. (2000). Back injuries and the fast bowler in cricket: A review. *Journal of Sports Sciences, 18*, 983-991.

Elliott, B. C., and Khangure, M. (2002). Disk degeneration and fast bowling in cricket: An intervention study. *Medicine and Science in Sports and Exercise, 34*, 1714-1718.

Foster, D., John, D., Elliott, B. C., Ackland, T. and Fitch, K. (1989). Back injuries to fast bowlers in cricket: A prospective study. *British Journal of Sports Medicine, 23*(3), 150-154.

Loram, L. C., McKinon, W., Wormgoor, S., Rogers, G. G., Nowak, I., and Harden, L. M. (2005). Determinants of ball release speed in schoolboy fast-medium bowlers in cricket. *Journal of Sports Medicine and Physical Fitness, 45*(4), 483-490.

Newman, D. (2003). *A prospective survey of injuries at first class counties in England and Wales 2001 and 2002 seasons.* Paper presented at the Science and Medicine in Cricket: A collection of papers from the Second World Congress of Science and Medicine in Cricket, Cape Town, South Africa.

Orchard, J. W., James, T. and Portus, M. R. (2006). Injuries to elite male cricketers in Australia over a 10-year period. *Journal of Science and Medicine in Sport, 9*(6), 459-467.

Portus, M. R., Mason, B. R., Elliott, B. C., Pfitzner, M. C., and Done, R. P. (2004). Technique factors related to ball release speed and trunk injuries in high performance cricket fast bowlers. *Sports Biomechanics, 3*(2), 263-283.

Ranson, C., Burnett, A., King, M., Patel, N., and O'Sullivan, P. (2008a). The relationship between bowling action classification and three-dimensional lower trunk motion in fast bowlers in cricket. *Journal of Sports Sciences, 26*(3), 267-276.

Ranson, C., Burnett, A., King, M., O'Sullivan, P., Cornish, R., and Batt, M. (2008b). *Acute lumbar stress injury, trunk kinematics, lumbar MRI and paraspinal muscle morphology in fast bowlers in cricket.* Paper presented at the *XXVI International Conference on Biomechanics in Sports*, Seoul National University.

Stuelcken, M. C., Ginn, K., and Sinclair, P. J. (2008). Musculoskeletal profile of the lumbar spine and hip regions in cricket fast bowlers. *Physical Therapy in Sport, 9*(2), 82.

Wood, G. A., and Jennings, L. S. (1979). On the use of spline functions for data smoothing. *Journal of Biomechanics 12*, 477-479.

Yeadon, M. R., and King, M. A. (2002). Evaluation of a torque driven simulation model of tumbling. *Journal of Applied Biomechanics, 18*, 195-206.

ISBS 2009 Scientific Sessions Wednesday PM

ISBS 2009

Oral Session S13

Strength & Conditioning

BARBELL ACCELERATION ANALYSIS ON VARIOUS INTENSITIES OF WEIGHTLIFTING

Kimitake Sato[1], Paul Fleschler[2], and William A. Sands[3]

[1]University of Northern Colorado, Greeley, CO USA
[2]USA Weightlifting, Colorado Springs, CO USA
[3]Mesa State College, Grand Junction, CO USA

The purpose of this study was to examine how various intensity levels influence the peak barbell acceleration in weightlifting. USA weightlifting resident team members ($n=9$, men:5 & women:4) participated in this study. They performed two repetitions at intensities of 80, 85, and 90% of 1 repetition maximum (total six repetitions). The peak barbell acceleration was measured at the 2nd pull phase of the snatch/clean. A one-way repeated measure ANOVA was used to analyze the effects of the intensity levels ($p = .05$). The results showed that intensity has a significant effect on the peak barbell acceleration ($F(2,16) = 11.49$, $p < .001$). The peak barbell acceleration decreased as the intensity level increased (80%: 19.63±3.04, 85%: 16.78±3.56, 90%: 13.65±3.50). Comparison between elite and beginners or other power-oriented athletes can be considered in future studies.

KEY WORDS: barbell acceleration, weightlifting kinematics, weightlifters.

INTRODUCTION: Biomechanical characteristics of weightlifting (in both snatch and clean & jerk) have been studied for a decade. Studies have specifically focused on barbell path in relation to body position, barbell velocity, and mechanical work and power output (Barton, 1997; Gourgoulis, Aggeloussis, Mavromatis, & Garas, 2000; Gourgoulis, Aggeloussis, Kalivas, Antoniou, & Mavromatis, 2004; Haff, et al., 2003; Isaka, Okada, & Funato, 1996; Schilling, et al., 2002; Stone, O'Bryant, Williams, Pierce, & Johnson, 1998). The primary intention of analyzing both kinematic and kinetic variables was to distinguish the difference between a good and bad lift, and analyze the typical lifting techniques of elite weightlifters. However, Stone et al. (2006) concluded that based on these biomechanical variables, it is difficult to predict a perfect lifting technique that is accepted by all weightlifters. For example, it is commonly thought that the barbell path should be close to the body and relatively S-shaped throughout the lift. But some studies reported different barbell paths among some elite weightlifters (Hiskia, 1997). Based on the reviews, the barbell path seems to vary by individual depending on different anthropometric measurement and lifting preference. Another example is that a fast barbell velocity is thought to be a characteristic of good and strong lifters (Gourgoulis et al., 2004; Haff et al., 2003; Stone et al., 1998). However, the fast barbell velocity requires lifters to squat down quickly to be in the catch position, which may lead to an unsuccessful lift. It is questionable to conclude that the faster barbell velocity is a good indication of a successful lift. Rather, the fast barbell velocity may be just one's lifting style. The barbell velocity should be reviewed more carefully in future studies.

Even though the biomechanics of weightlifting have been a well-studied subject, a report of barbell acceleration is limited to only three studies, and no discussion was made regarding the interpretation of the barbell acceleration graph and table in their studies (Gourgoulis, et al., 2000; Haff, et al., 2003; Isaka, et al., 1996). Specifically, the present study focused on the peak barbell acceleration during the 2nd pull phase. The 2nd pull phase is a critical part of the lift to determine whether the barbell is being pulled up to the desired height to catch (Stone, et al., 2006). Furthermore, acceleration is directly proportional to force production while mass is a constant value. Thus, when a barbell increases its rate of velocity, the body is producing the force to accelerate the barbell to an upward direction. This acceleration measurement can be a valuable assessment for weightlifters and other athletes, and it is necessary to examine how various intensities change the peak barbell acceleration. Therefore, the purpose of the study was to examine how various intensity levels influence the peak barbell acceleration in weightlifting. This study hypothesized that the peak barbell acceleration at the 2nd pull phase decreases as the intensity level increases.

METHODS: Participants: Men's and women's weightlifting resident team members at Colorado Springs Olympic Training Center participated in this study (see Table 1). They were free of injuries at the time of data collection. They were also in the middle of their strength development phase leading up to the competition. Data were collected in compliance with policies of the United States Olympic Committee on the testing of athletic subjects.

Table 1 Demographic data of participants ($N = 9$)

	Men ($n = 5$)	Women ($n = 4$)
Age (yr)	22.2 ± 3.6	20.3 ± 1.5
Height (m)	1.77 ± 0.12	1.62 ± 0.10
Mass (kg)	100.1 ± 30.2	73.1 ± 19.1

Procedures: All participants reported to the training facility of USA weightlifting for data collection, and were provided the procedure of the testing protocol. They had an adequate amount of stretching and warm-up in a similar fashion as they normally do before the training session. A 3-axis accelerometer (PS-2119, Pasco Scientific, Roseville, CA) was used to measure the barbell acceleration, and was attached to a Bluetooth™ wireless device (Pasco Pasport Airlink SI (PS-2005)). The total weight of the unit is 170.1 grams, which is equivalent to a plastic barbell collar. Thus, the weight of the accelerometer should not interrupt a lifter's ability to sense asymmetry of weight between the left and right sides of the barbell. Recently published data reported that this device accurately measured acceleration as well as a high-speed camera at the same sampling rate (Sato, et al., 2009). Data were collected at sampling rate of 100Hz. In order to minimize the external shock when the lifter drops the barbell, the foam unit was designed to secure the accelerometer (see Figure 1). The accelerometer unit was attached to the end of the barbell. It is important to note that the orientation of the sensor must remain in the constant position throughout the lift to avoid misrepresentation of the resultant acceleration. Therefore, the unit was securely attached directly below the barbell (see Figure 2).

Figure 1 Accelerometer in the foam pad   **Figure 2** Accelerometer placement

The barbell acceleration data was collected at the intensities of 80, 85, and 90% of one repetition maximum (1RM) (Baechle, & Earle, 2008). Four participants performed snatch and the other five participants performed clean with two repetitions of each intensity level.

Data Analysis: Data Studio™ software (Pasco Scientific, Roseville, CA) was used to acquire, display, and analyze the data. The peak barbell acceleration at the 2nd pull phase was captured from each participant who performed snatch or clean at three different intensity levels. The previous study validated that this peak barbell acceleration is occurring at the 2nd pull phase (Sato, et al., 2009). Each intensity level of the data were then averaged and analyzed with one-way repeated measure ANOVA ($p = .05$) to indentify if there are any effects on various intensity levels on the peak barbell acceleration. Follow-up T test was performed with p-value of .017 (.05/3). The Statistical Package for Social Sciences (SPSS) was used for the analyses (SPSS, Inc., Chicago, IL).

RESULTS: A one-way repeated measure ANOVA was calculated comparing the intensity levels of 80, 85, and 90% of 1RM. A significant effect was found ($F(2,16) = 11.49$, $p < .001$). Paired-sample T tests were used as a protected follow-up T test. It revealed that the peak barbell acceleration decreased significantly from 80-85% and 80-90%, but not from 85-90%. Table 1 shows the averaged peak barbell acceleration at each intensity level.

Table 1 Mean peak barbell acceleration at the 2nd pull (m/s^2)

	80%	85%	90%
Average	19.63 ± 3.04	16.78 ± 3.56	13.65 ± 3.50

DISCUSSION: The purpose of the study was to examine how various intensity levels influence the peak barbell acceleration in weightlifting. The results of this study supported the hypothesis that the peak barbell acceleration decreased as the intensity level increased. It is understandable that the increase of the mass of the barbell has an effect on decreasing the barbell acceleration at the 2nd pull phase. The investigators were interested in identifying what intensity level the peak barbell acceleration significantly decreases. During the pilot study, the peak barbell acceleration showed no change from 50 to 80% of 1RM among elite and experienced weightlifters, indicating that the force production becomes greater while the mass of the barbell increased and the peak barbell acceleration remains relatively constant. In this study, the peak barbell acceleration significantly decreased as the intensity level increased from 80 to 85% of 1RM. The results demonstrated that the force affecting barbell acceleration at the 2nd pull phase reaches near maximal level around 85% of 1RM. In other words, the force production remains relatively the same while the peak acceleration decreases and the mass of the barbell increases. The main training intensity during the strength development phase is between 80 to 90% of 1RM (Baechle, & Earle, 2008). These results showed that roughly 80% of 1RM is the threshold for the elite level weightlifters to be able to maintain the peak barbell acceleration. A resultant acceleration was calculated in this study even though the other studies reported linear vertical acceleration (Gourgoulis, et al., 2000; Haff, et al., 2003). Measuring the resultant acceleration was believed to be appropriate since the 2nd pull phase is not typically a linear fashion. Rather, the 2nd pull phase of the barbell path is displayed in curvilinear in many studies (Gourgoulis et al., 2000; Gourgoulis et al., 2004; Haff et al., 2003; Isaka, et al., 1996). The typical resultant acceleration sequence from this study seems consistent with the acceleration figure from Isaka et al. (1996) that the 2nd pull phase exerted the highest barbell acceleration value.

In this study, all participants were experienced and elite level in this sport. Participants were a mix of female and male weightlifters. The difference in the peak barbell acceleration between the genders was not observed. In future studies, it would be appropriate to compare this data with beginner level weightlifters (mainly in youth) and other athletes who require power components in their sports to identify how intensity level influences the barbell acceleration.

Since Olympic weightlifting and its modified versions (power snatch/power clean) are well-utilized in the strength and conditioning field, the investigators discussed possible benefits that some coaches may gain from this analysis. First, attempting the maximal weight is one way to measure how athletes are improving the strength over the long-term training, but tracking the peak barbell acceleration can be another useful assessment to observe progression of the peak acceleration values which equal to the progression of force production capability. Another benefit is that when tracking the peak barbell acceleration throughout a single training session, significant decrease in the acceleration value in later stage of the training session can be an indicator of fatigue (less force is being produced to accelerate the barbell). If the lifter continues to lift after the fatigue sets in, it may lead to over-training/over-use injuries. Describing and identifying fatigue is sensitive and difficult, but the barbell acceleration test may be a suitable assessment to create a better communication environment to re-evaluate the training program between coaches and athletes.

CONCLUSION: Overall, this study tested the peak value of the barbell acceleration at the 2nd pull phase of weightlifting with three different intensity levels. The peak barbell acceleration decreased as the intensity level increased. Since this study was conducted with elite level weightlifters, comparing the data with beginners or other power-typed athletes would be interesting to examine how various intensity levels influence the barbell acceleration.

REFERENCES:

Baechle, T. R., & Earle, R. W. (2008). *Essentials of Strength Training and Conditioning* (3rd Ed.). Champaign, IL: Human Kinetics, *p*. 250.
Barton, J. (1997). Are there general rules in snatch kinematics?. Proceedings of the Weightlifting Symposium. Ancient Olympia, Greece.
Gourgoulis, V., Aggeloussis, N., Mavromatis, G., & Garas, A. (2000). Three-dimentional kinematic analysis of the snatch of elite Greek weightlifters. *Journal of sports sciences. 18*, 643 – 652.
Gourgoulis, V., Aggeloussis, N., Kalivas, V., Antoniou, P. & Mavromatis, G. (2004). Snatch lift kinematics and bar energetics in male adolescent and adult weightlifters. *The Journal of Sports Medicine and Physical Fitness. 44*, 126 – 131.
Haff, G. G., Whiteley, A., McCoy, L. B., O'Bryant, H. S., Kilgore, J. L., Haff, E. E., Pierce, K., & Stone, M. H. (2003). Effects of different set configurations on barbell velocity and displacement during a clean pull. *Journal of Strength and Conditioning Research, 17*(1), 95 – 103.
Hiskia, G. Biomechanical analysis on performance of world and Olympic champion weight lifters. (1997). Proceedings of IWF Coaching, Referring & Medical Symposium. Ancient Olympia, Greece.
Isaka, T., Okada, J. & Funato, K. (1996). Kinematics analysis on the barbell during the snatch movement of elite Asian weight lifters. *Journal of Applied Biomechanics, 12*, 508 – 516.
Sato, K., Smith, S. L., & Sands, W. A. (2009). Validation of accelerometer for measuring sport performance. *Journal of Strength and Conditioning Research. 23*(1), 341 – 347.
Schilling, B. K., Stone, M. H., O'Bryant, H. S., Fry, A. C., Coglianese, R. H. & Pierce, K. (2002). Snatch technique of collegiate national level weightlifters. *Journal of Strength and Conditioning Research. 16*(4), 551 – 554.
Stone, M. H., O'Bryant, H. S., Williams, F. E., Pierce, K., & Johnson, R. L. (1998). Analysis of bar paths during the snatch in elite male weightlifters. *Strength and Conditioning Journal.* August, 30 – 38.
Stone, M. H., Pierce, K., Sands, W. A., & Stone, M.E. (2006). Weightlifting: A brief overview. *Strength and Conditioning Journal. 28*(1), 50 – 66.

COMPARISON OF ANGLES AND THE CORRESPONDING MOMENTS IN KNEE AND HIP DURING RESTRICTED AND UNRESTRICTED SQUATS

Silvio Lorenzetti[1], Mirjam Stoop[1], Thomas Ukelo[1], Hans Gerber[1], Alex Stacoff [†1] and Edgar Stüssi[1]

Institute for Biomechanics, ETH Zürich, Switzerland[1]

The aim of this study is the comparison of angles and the corresponding moments in knee and hip during squatting. The five subjects performed restricted and unrestricted squats. The experimental set-up consisted of a motion capture system and two force plates. The loading conditions were 0, ¼ and ½ BW. The moments and the force were calculated using inverse dynamics. Overall, the maximal moments were observed in the knee during unrestricted squats and in the hip during restricted squats. Comparing the moments at a knee angle of 60°, the loading conditions have a larger influence than the type of execution. The moment in the knee is 10.4%, respectively 11.2% lower with ¼ and ½ body weight during restricted squats. In the hip, the moment is 15.5 %, respectively 14 % higher for the same conditions. The angle of the hip remains rather constant. This most likely implies a higher load to the lower back. Hence, the exercise instruction should be adapted to the aims and the training condition of the athlete.

KEY WORDS: squats, movement analysis, force, moments, exercise.

INTRODUCTION: The squat is one of the most common exercises in sport and rehabilitation. In Switzerland, squatting has one of the highest risks during training because of overload or wrong execution (Müller, 1999). Fifty percent of all injuries occuring during training concern the lower extremities or the back. Therefore, a correct execution of the squat exercise is important if not to compromise the positive effects of the training (Dunn et al., 1984, Cappozzo et al., 1985, Chandler et al., 1989, Fry et al., 2003). A widespread guideline for the barbell squat is the need to keep the knees from moving forward past the toes. To our knowledge, the instruction for squats regarding the position of the knee are based on studies of McLaughlin et al. (1977 and 1978) and Ariel (1974) and are established in Europe and in the NSCA (Fry et al., 2003).

From a biomechanical point of view it has been discussed that during the movement of the knee beyond the toes, shear forces accrue that might harm the knee (Ariel, 1974). This shear force is below the ultimate load of healthy cruciate ligaments (Andrews et al., 1983 and Woo et al., 1991). As a second argument, the pressure between the patella and the femur rises with the flexion of the knee (Escamilla, 1998, Nisell & Eckholm, 1984 and 1986). In general, this pressure seems to be within the limit of the tolerated load (Woo et al., 1991).

This study was designed to compare the angles of the knee and the hip and the corresponding moment during unrestricted (UR) and restricted (R) squats.

METHODS: Data Collection: Kinematics and kinetics of squat exercise was evaluated using a 12 camera 3D Vicon (Oxford, UK) system. The five subjects were all students of movement science with experiences in weight lifting. The average weight was 71.0 ± 12.5 kg, the age 25.4 ± 5.3 and the height 1.75 ± 0.05 m. They performed restricted and unrestricted squats with zero, ¼ bodyweight (BW) and ½ BW loading using a barbell. The barbell position was on the trapecius muscle. When performing a restricted squat, the knee is not allowed to go beyond the toes. The side view of the knee and the toes was projected onto a screen in front of the subject. The subject was able to control the position of the knee in respect to a vertical line in front of the toes. No external force was applied to the knee. For the unrestricted squat, there was no restriction on the movement of the knee. To determine the force for each foot, two Kistler force plates (Winterthur, CH) were used. The marker set consisted of 53 skin markers including 20 for the spine (Bachmann et al., 2008).

Data Analysis: Joint centers were functionally defined and the estimation of the joint rotations was based on a least-square fit of two point clouds and orthogonal anatomically defined joint coordinate systems. For the calculation of the moments, an inverse dynamics based on the position of the body and the ground reaction force was performed. In respect to the body weight of the subjects, the moments were normalized to BW.

RESULTS: The moments of the knee and hip at a knee angle of 60° and the maximal moments are given in Table 1 and 2. As expected, the maximal moments during restricted squats are lower in the knee and higher in the hip (Table 3). At a knee flexion of 60°, the moment in the knee is the same with no extra load, 10.4 % smaller with ¼ BW load, and 11.2 % smaller with ½ BW for the restricted squat (Figure 1). In the hip, the maximal moment is 6.3 % and 6 % higher for restricted squats (Figure 2). Looking at a knee angle of 60°, corresponding to a similar condition for the length of the muscle, the moment is just 10% smaller with the restricted squats, whereas the moment in the hip is 15.5 %, respectively 14.0 % higher for ¼ and ½ BW extra load. When comparing the results of this work, the values are in good agreement with Fry et al. (2003) except for the maximal moment of the hip.

Table 1 Average moments of the knee and hip at a knee angle of 60 ° normalized to the body weight [Nm/Kg]

Load BW	M_{KNEE} UR(60°)	M_{KNEE} R(60°)	M_{HIP} UR(60°)	M_{HIP} R(60°)
0	0.60±0.09	0.61±0.07	0.30±0.06	0.31±0.04
¼	0.69±0.11	0.63±0.07	0.49±0.09	0.58±0.06
½	0.79±0.15	0.71±0.13	0.74±0.11	0.86±0.12

Table 2. Average maximum moments of the knee and hip [Nm]

Load BW	$MaxM_{KNEE}$ UR	$MaxM_{KNEE}$ R	$MaxM_{HIP}$ UR	$MaxM_{HIP}$ R
0	61.5±9.3	57.9±7.9	47.2±5.7	47.2±5.7
¼	78.7±10	66.5±5.7	73.6±7.0	78.7±4.3
½	93.7±12	77.9±10.7	100±13	106.5±8.6
1*	150.1±50.8	117.3±34.2	28.2±65.0	302.7±71.2

* Fry et al. (2003)

Table 3. Comparison between unrestricted and restricted squats (unrestricted is equal to 100 %).

Load BW	M_{KNEE} 60°	M_{HIP} 60°	Max Knee angle	Max Hip angle	$MaxM_{KNEE}$	$MaxM_{HIP}$
0	0%	3.2%	-11.1%	1.4%	-5.8%	0
¼	-8.7%	15.5%	-14.0%	1.8%	-15.5%	6.3%
½	-10.1%	14.0%	-12.0%	2.7%	-16.8%	6%
1*			-6.9%	5.1%	-28.0%	Factor 10 higher

* Fry et al. (2003)

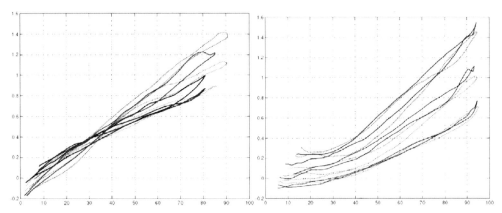

Figure 1: left: Knee flexions [°] vs. moments [Nm/Kg], red 0 BW load, black ¼ BW load, blue ½ BW load, line for restricted (R) and dots for unrestricted (UR) squats. Right: Hip flexions [°] vs. moments [Nm/Kg].

DISCUSSION: The squat is mainly an exercise for the quadriceps. Hence a certain moment in the knee is needed to set a stimulus. The restriction of the position of the knee implies either a shift of the center of pressure (COP) toward the heel or a compensation mechanism of the upper body. Shifting the COP towards the heel reduces the stability of the stands. Especially with high weights a stable stand is required.

The compensation mechanism leads to higher moments in the hip. One of this compensation mechanism is the bending forward of the trunk during restricted squatting. Assuming a simple kinematic chain model, the higher moments in the hip are leading to a higher load to the lower spine.

Performing unrestricted squats results in higher maximal moments in the knee and lower moments in the hip. Whereas the angle of the knee is higher for unrestricted squatting, in the hip, the flexion angle is similar in both conditions. Therefore compensation mechanisms such as flexion of the spine are expected to counteract these differences.

The higher maximal moment in the knee during unrestricted squatting can be explained by the higher angle of the knee. This shows the importance of choosing the proper depth of the squat.

A rise in the torque for the hip by a factor of 10 between un- and restricted squats as given by Fry et al. (2003) was not observed and seems rather unlikely given the changes of the angles in the knee and the hip.

Of course great care must be taken on the depth of the squat not only the moment of the knee rise, but the stress on the hip and the lower back too.

CONCLUSION: In this study the angles and the corresponding moments of the knee and the hip were determined. Not surprisingly, the moment in the knee rises with the angle and the load. Even though the maximum moment is higher in the knee, the unrestricted squat has comparable moments at the same angle of the knee. The stress on the hip and most likely on the lower back, is lower during an unrestricted squat. For these reasons, the unrestricted squat may be the right choice for most athletes.

REFERENCES:

Andrews J.G., Hays J.G. & Vaughan C.L. (1983). Knee shear forces during a squat exercise using barbell and a weight machine. In Matsui H. & Kobayashi K. (Eds), *Biomechanics* (pp 923-927) Vol. 8B Human Kinetics, Champaign, IL.

Ariel B.G. (1974). Biomechanical analysis of the knee joint during deep knee bends with heavy load. In R. C. Nelson & C. Morehouse (Eds.), *Biomechanics* IV (pp44-52). Baltimore: Universitiy Park Press.

Bachmann C., Gerber H. & Stacoff A. (2008). Messsysteme, Messmethoden und Beispiele zur instrumentierten Ganganalyse. *Schw. Z. Sportmed. Sporttraum.* 2, p.29-34.

Cappozzo A., Felici F., Figura F. & Gazzani F. (1985). Lumbar spine loading during half-squat exercises. *Med Sci Sports Exerc.*17(5), 613-20.

Chandler T.J., Wilson G.D. & Stone M.H., 1989. The effect of the squat exercise on knee stability. *Med Sci Sports Exerc.* 21(3), 299-303.

Dunn B., Klein K., Kroll B., McLaughun T., O'Shea P. & Wathen D. (1984). Coaches round table: The squat and its application to athletics performance. *Strength Cond. J.* 6, 68, 10-22.

Escamilla R.F., Fleisig G.S., Zheng N., Barrentine S.W., Andrews J.R., Bergmann B.W. & Moorman C.T. (1998). Effects of technique variations of knee biomechanics during the squat and leg press. *Med Sci Sport Exerc,* 30, 556–569.

Fry, A., Smith, J. & Schilling, B. (2003). Effect of knee position on hip and knee torques during the barbell squat. *Journal of Strength and Conditioning Research*, 17:629-633.

McLaughlin T.M., Dillman C.J. & Lardner T.J. (1977). A kinematic model of performance in the parallel squat by champion powerlifers. *Med Sci Sports,* 9(2),128-33.

McLaughlin T.M., Lardner T.J., & Dillman C.J. (1978). Kinetics of the parallel squat. Res Q. 49(2),175-89.

Müller R. (1999) Fitness-Center – Verletzungen und Beschwerden beim Training. *BFU Report.* Switzerland

Nisell R. & Ekholm J. (1984). Patellar forces during knee extension. Scand. J. Rehabil Med. 17(2), 63-74.

Nisell R. & Ekholm J. (1986). Joint load during the patellar squat in powerlifting and force analysis of in vivo bilateral quadriceps tendon rupture. Scand. J. Sports Sci., 8, 63-70.

Woo S.L., Hollis J.M., Adams D.J., Lyon R.M. & Takai S. (1991). Tensile properties of the human femur-anterior cruciate ligament-tibia complex. The effects of specimen age and orientation. *Am J Sports Med.* 19(3),217-25.

Acknowledgement

This work is supported by the Eidgenössische Sport Kommission. Alex Stacoff [†] was an investigator in this project. We miss him.

THE EFFECTS OF DYNAMIC AND STATIC STRETCHING METHODS ON SPEED, AGILITY AND POWER.

Daniel C. Bishop

Department of Sport, Coaching and Exercise Science, University of Lincoln, UK

KEY WORDS: Warm-up, Illinois, Vertical Jump,

INTRODUCTION: Warm-ups are integral to coaches' and athletes' preparations, yet current research and practice provides conflicting advice on the most effective warm-up procedure for aiding performance. The value of a warm-up is not in question but the role of static stretching within the warm-up is contentious. A number of studies have suggested that static stretching is detrimental to performance (Shrier, 2004), though these have not always employed stretching protocols that reflect those actually used by performers. Many athletes continue to include static stretching as part of their routine, often following a dynamic warm-up. The purpose of the study was to investigate, using realistic protocols, the effects on speed, agility and power of additional static stretching following a dynamic warm-up.

METHODS: To date 16 University sports students (11 male, 5 female) have been recruited to the study, which was approved by the University ethics commitee. Following familiarisation with the warm-up methods and sports performance measures, all participants performed two warm-up protocols one week apart. The protocols concentrated on the quadriceps, hamstrings, gastrocnemius, soleus, gluteals, adductors and hip flexors. The Dynamic Warm-up (DW) protocol used a series of specific progresive exercises lasting 10 minutes over a distance of 20m with a jog recovery. The Dynamic Warm-up plus Static Stretching (DWS) protocol used the same DW protocol followed by a 5 minute period during which 7 stretches were held at a point of moderate discomfort for 20 seconds. After an intial rest period of 2 minutes the subjects performed a countermovement vertical jump (CMJ), 20m sprint and Illinios agility test. A one minute rest period was used between performance measures to mimic competitive performance. The order the performance measures were conducted were randomised for each subject; for an individual subject the order was maintained for both protocols. Paired t-tests were used to identify differences between the 3 performance measures over the two warm-up protocols, with a significance level of $p \leq 0.05$.

RESULTS & DISCUSSION: 20m sprint performance was significantly reduced ($p=0.03$) by the use of the DWS protocol (3.4 ± 0.32s, mean $\pm$ SD) when compared to DW only (3.29 ± 0.29s). The CMJ (DW =0.51 ± 0.09m vs DWS =0.48 ± 0.09m) and Illinois agility (DW =17.31 ± 1.0s vs DWS =17.37 ± 1.04s) performance showed no significant difference with the addition of static stretching. Previous studies have used static stretching for periods in excess of levels used in conventional warm-ups but the present study indicates that static stretching for as little as 20 seconds after a dynamic warm-up can reduce 20m sprinting performance. Possible mechanisms for this include a decrease in the stiffness of the musculotendinous unit and an acute neural inhibition leading to a decrease in muscle activation levels (Young, 2007).

CONCLUSION: The use of static stretching following a dynamic warm-up decreased sprinting performance compared with the dynamic warm-up alone. Practitioners and athletes should use static stretching prior to performance with caution.

REFERENCES:

Shrier, I. (2004). Does stretching improve performance? a systematic and critical review of the literature. *Clinical Journal of Sport Medicine*, 14(5), 267-273.

Young, W.B. (2007). The use of static stretching in warm-up for training and competition. *International Journal of Sports Physiology and Performance*, 2, 212-216.

RELATIONSHIP BETWEEN MUSCULAR OUTPUTS AND THE HORIZONTAL PERTURBATION IN THE EARLY PHASE OF BENCH PRESS MOVEMENT UNDER STABLE AND UNSTABLE CONDITIONS

Sentaro Koshida[1], Koji Miyashita[2], Kazunori Iwai[3], Aya Kagimori[4] and Yukio Urabe[5]

Faculty of Health Sciences, Ryotokuji University, Urayasu, Japan[1]
Research Institute of Life and Sciences, Chubu University, Kasugai, Japan[2]
Hiroshima National College of Maritime Technology, Hiroshima, Japan[3]
Ryukoku University Training Centre, Fushimi, Japan[4]
Graduate School of Health Sciences, Hiroshima University, Hiroshima, Japan[5]

We demonstrated the relationship between the change rates of muscular outputs and horizontal perturbation under stable and unstable conditions in dynamic bench press movement. Twenty-seven male collegiate athletes attended the study. We used a tri-axis accelerometer attached to the barbell shaft to obtain the acceleration data in the bench press and computed peak force output, rate of force development (RFD), and horizontal acceleration trajectory length for 0.2 seconds after the initiation. Significant reduction was found in the peak force output and RFD under stable and unstable conditions, but not in the horizontal acceleration trajectory length. Significant correlation was found between the change rate of RFD and the horizontal acceleration trajectory length under stable and unstable conditions ($r=0.55$, $P<0.01$).

KEY WORDS: instability resistance training, coordination, accelerometer

INTRODUCTION: It has been suggested that resistance training exercise under unstable conditions (also known as instability resistance training) may improve trunk stability. However, its benefit on athletic performance improvement remains a matter of debate because of significant loss of muscular outputs under unstable conditions. (Behm et al, 2002; Anderson et al, 2004; Koshida et al, 2008).

Previous literature indicates that the loss of muscular outputs under unstable conditions is attributed to altered neuromuscular coordination during the movement. Anderson et al (2004) reported that the force output in chest press movement was significantly decreased, but EMG activity of the prime movers was maintained. The authors explained that the unchanged EMG activity may be due to greater postural and stabilization roles in the prime movers. Konecki et al (2001) also reported that the motor contribution of the wrist muscles was decreased when the external object became unstable in the push movement. The result suggested that greater joint stabilization requirement may affect the muscular outputs in dynamic movements. Based on those previous findings, we anticipated that the degree of perturbation in the other planes which are not engaged in a desired movement may be related to the loss of muscular outputs during instability resistance training exercise. However, to our best knowledge, how the perturbation is related to loss of muscular outputs under unstable conditions is little quantified. Therefore, the purpose of the present study was to demonstrate how the change of muscular outputs correlates with the change of horizontal perturbation in the dynamic movement by using a tri-axis accelerometer.

METHODS: Data Collection: Twenty-seven male collegiate athletes (age, 21.5 ± 1.9 years; height, 171.2 ± 4.6 cm; weight, 79.0 ± 12.8 kg) volunteered to participate in the study. All the participants had experienced moderate to extensive resistance training for at least 12 months. The participants executed 3 sets of single bench press movements as quickly as possible under stable and unstable conditions. Previous studies have reported that the maximal power output in bench press movement was observed when using the weight of 45% to 55% of 1 repetition maximum (1RM)(Mayhew et al, 1997). Therefore, we used a weight of 50%1RM for the test. We used a tri-axis accelerometer (Microstone Co., Saku, Japan) attached to the center of the barbell shaft to obtain the acceleration data in the bench press movement (200Hz). A flat bench provided the stable condition. The participant's feet were placed on the flat bench to minimize the influence of lower extremity on postural control. The experimental setting of the unstable condition is shown in Figure 1. Two inflatable rubber discs were placed underneath the upper thoracic and buttock areas provided the unstable condition.

Data Analysis:
We calculated peak force output (N) based on Newton's second law. Rate of force development (RFD; N/s) was computed by the equation below:

RFD (N/s)
 = {Peak force output(N) − [g × Barbell weight(N)]} / TTP(s)
 Where: *g:* gravitational acceleration, TTP: time to peak force from the onset of movement

Moreover, the acceleration trajectory length on the horizontal plane assuming the representative variables of horizontal perturbation was calculated for 0.2 seconds after initiating the movement (Figure 2). The greater the horizontal acceleration trajectory length becomes, the greater perturbation is assumed.
The change rates of these variables (%) under the two conditions were also computed by the equation below:

Change rate (%)
 = [(The value in the unstable condition− the value in the stable condition)
  / the value in the stable condition] × 100

We used a paired t test with Bonferroni correction to compare the muscular outputs between the two conditions. In addition, we had a Pearson coefficient analysis performed to demonstrate the relationships of the differences in the outputs and horizontal acceleration between the two conditions. Statistical significance was set as $P<0.05$ in this study.

RESULTS: The mean value (±SD) of 1 RM was 85.0±13.4kg; therefore, the participants used the average weight of 42.5 ±6.7kg in the experiment. The mean values of maximum force outputs in the bench press under stable and unstable conditions were 565.1±104.6N and 559.9±148.0N respectively (Figure 3A). The mean values of RFD were 1323.5±605.1N/s under the stable condition and 1038.0±434.2N/s under the unstable condition. Maximum force output and RFD under the stable condition were significantly greater than those under the unstable condition. However, the horizontal acceleration trajectory lengths were not significantly different between the stable and the unstable conditions.

There is significant correlation between the change rate of RFD and that of horizontal acceleration ($r=0.55$, $P<0.01$)(Figure 4) ; however, no significant correlation was found between the change rate of peak force output and that of the horizontal acceleration trajectory length.

Figure1. The experimental setting of the unstable condition: A red circle represents the position of the accelerometer. Three arrows represent the xyz coordinate in the study

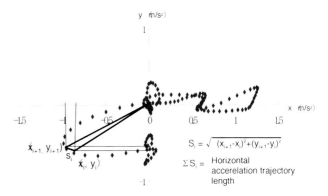

Figure2. Calculation method of horizontal acceleration trajectory length

$$S_i = \sqrt{(x_{i+1}-x_i)^2+(y_{i+1}-y_i)^2}$$

ΣS_i = Horizontal accelelation trajectory length

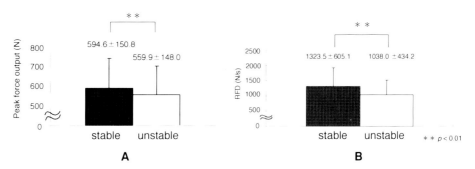

Figure 3. The comparisons of (A) peak force output and (B) RFD between the two conditions

** $p < 0.01$

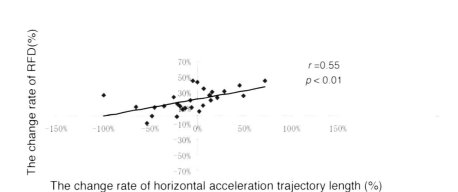

$r = 0.55$
$p < 0.01$

The change rate of horizontal acceleration trajectory length (%)

Figure 4. The scatter diagram of the change rates of RFD and horizontal acceleration trajectory length

DISCUSSION: The results showed that the peak force output in the unstable condition was lower than that in the stable condition. Koshida et al (2008) reported that peak force output in dynamic bench press with the weight of 50%1RM on the Swiss ball was significantly lower than that on the flat bench. The current result supports the previous finding. In bench press movement, regardless of static or dynamic movements, significant reduction of peak force output is observed under unstable conditions. RFD was also decreased under the unstable condition when compared to that under the stable condition. In previous studies, the efferent neuromuscular drive during muscle contraction is reported to be a factor associated with RFD value (Aagaard et al, 2002). The neuromuscular coordination altered by the given instability may have led to the RFD reduction as well as the peak force output reduction.

The difference in the horizontal acceleration trajectory length between the stable and the unstable conditions was not statistically significant in this study. It has been accepted that postural control was prioritized to force production. Our result supports the findings.

A moderate correlation was found only between the reduced rates of RFD and the horizontal acceleration trajectory length in this study. The study of Anderson et al (2004) suggested that prime movers may function as stabilizers under unstable conditions. In addition, previous studies have shown that postural and joint stabilities under unstable conditions were reinforced by muscle co-contraction of antagonistic muscle. We speculated that those changes of neuromuscular patterns might be associated with the RFD reduction under the unstable condition. Interestingly, there was no significant correlation between the reduction rates of peak force output and the perturbation. Approximately 2.5% of the small change rate of the peak force outputs might affect the result.

The result suggests that excessive stabilization activity might decrease RFD and that the instability resistance training could aim at improving muscular outputs by better coordinating force production with postural and joint stabilization. Some of resistance training exercises such as bench press on a flat bench and most of machine exercises do not require much of stabilizing activity during the movements. Because actual sports movements require the whole body to be stabilized during dynamic movement and a large amount of force to be generated in a very short period of time simultaneously, instability resistance training could be a more functional method for improving muscular outputs in sports activities. However, the significant loss of muscular outputs under unstable conditions was constantly reported; therefore, instability resistance training may not be suitable to induce maximal or submaximal force on limbs.

CONCLUSION: The change of horizontal perturbation caused by unstable conditions may be associated with the reduced rate of force development in the early phase of bench press movement.

REFERENCES:
Behm, D.G., Anderson, K. and Curnew, R.S. Muscle force and activation under stable and unstable conditions. *J. Strength Cond. Res.* 16:416-422, 2002.
Anderson, K., et al. Maintenance of EMG activity and loss of force output with instability. *J. Strength Cond. Res.* 18:637-640, 2004.
Koshida, S., et al. Muscular outputs during dynamic bench press under stable versus unstable conditions. J. Strength Cond. 22:1584-1588, 2008.
Kornecki, S., Kabel, A. and Siemienski, A. Muscular co-operation during joint stabilization, as reflected by EMG. *Eur. J. Appl. Physiol.* 83:453-461, 2001.
Mayhew, J.L.,et al. Changes in upper body power following heavy-resistance strength training in college men. *Int. J. Sports Med.* 18:516-520.1997.
Aagaard, P., Simonsen, E.B., Anderson, J.L., Magnusson, P. and Dyhre-Poulsen, P. Increased rate of force development and neural drive of human skeletal muscle following resistance training. *J. Appl Physiol.* 93:1318-1326, 2002.

ABDOMINAL PRESSURIZATION AND MUSCLE ACTIVATION DURING SUPINE TRUNK CURLS

Alf Thorstensson, Anna Bjerkefors, Maria Ekblom and Fredrik Tinmark

The Swedish School of Sport and Health Sciences and Department of Neuroscience, Karolinska Institutet, Stockholm, Sweden

KEY WORDS: sit-ups, intra-muscular EMG, training exercises

INTRODUCTION: Trunk curls are common exercises in sports training. A multitude of practical recommendations exist on how to best perform this type of exercise to reach a specific effect. Since abdominal muscles can contribute to breathing, it was thought of interest to study the effects of systematic variations in breathing on the pattern of abdominal muscle activation during trunk curls. Of particular interest was the influence of fluctuations in the intra-abdominal pressure (IAP), since changes in IAP have been shown to be coupled to the activation of the innermost abdominal muscle, the Transversus abdominis (TrA), and also to be able to contribute to an extensor moment of the trunk.

METHODS: Ten healthy young adult females performed straight trunk curls from a supine position with bent knees. The movement speed was set by a metronome and the upward and downward phases each lasted approximately 2 s. Six different variants were carried out: 1) spontaneous breathing, 2) inhalation during the upward and downward phases, 3) exhalation during the upward and downward phases, 4) breath-holding on a sub-maximal level, 5) breath-holding on a maximal inhalation level, and 6) breath-holding on a maximal exhalation level. Kinematics was obtained with a position transducer and air flow with a respiratory flow-meter. Intramuscular electromyography (EMG) was recorded via bipolar fine-wire electrodes placed under the guidance of ultrasound in the TrA, Obliquus Internus (OI), Obliquus Externus (OE), and Rectus Abdominis (RA) muscles on the right side. IAP was obtained with a pressure-sensitive tip-transducer introduced into the gastric ventricle via the nose. Root mean square EMG and mean IAP were calculated for a 1-s-interval in the middle of the upward and downward phases, respectively, and normalized to the EMG obtained in an isometric maximal voluntary sagittal trunk curl in a supine position and of the IAP in a maximal voluntary pressurization of the abdominal cavity (Valsalva) in the same position.

RESULTS: The overall range of relative EMG-levels was 6-72%. The activation of TrA was lower than that of the other muscles (6-29% versus 19-72%). All muscles showed higher activation during the upward than the downward phase (ranges: 13-72% versus 6-56%). Breath-holding in maximal exhalation showed higher EMG-values than the spontaneous breathing for OE and RA both in the upward and downward phases. There was no difference in IAP between spontaneous breathing and steered breathing on a sub-maximal level, whereas breath-holding on a maximal level caused higher IAPs, both for inhalation and exhalation.

DISCUSSION AND CONCLUSION: There was no or only minor effects on abdominal muscle activation levels of breathing in or out or holding breath on a sub-maximal level during the execution of a straight trunk curl in a supine position. Holding breath in maximal exhalation caused an increase in trunk flexor muscle activity, possibly to overcome a combined effect of an increased intra-abdominal pressure and decreased leverage due to the compressed abdomen. In the maximal inhalation situation, a possible opposing effect of increased abdominal pressure might have been counteracted by an improved leverage caused by an expanded abdominal cavity, resulting in an unchanged need for trunk flexor activation.

ISBS 2009

Oral Session S14

Other Sports I

ACUTE EFFECTS OF DENTAL APPLIANCES ON UPPER AND LOWER ISOKINETIC MUSCLE FUNCTION

Nathan Dau[1], Don Sherman[1], Richard Bolander[1], Cynthia Bir[1], Hermann Engels[2]

Biomedical Engineering Department, Wayne State University, Detroit, USA
Kinesiology, Health and Sport Studies, Wayne State University, Detroit, USA

The possibility that athletic performance can be affected by a person's jaw posture during the activity has been of interest to sports practitioners for many years. Using established elbow and knee flexion/extension testing protocols on a calibrated isokinetic dynamometer (Biodex System 2, Shirley, NY), this study examined selected muscle function characteristics in male NCAA II college football players (n=18) under test conditions in which they wore a professionally-fitted dental appliance (PowerPlus) designed for optimal maxilla-mandibular spacing, a common "boil-and-bite"-type mouth guard (Shock Dr.), and conditions in which they were instructed to have their teeth touch while keeping the jaws relaxed (Relax) or clenched (Clench) without wearing any oral appliances. Results indicated a significant improvement in total work (+9.8%), peak torque/body weight (+10.5%), and average power (+11.25%) for elbow flexion in the PowerPlus relative to the Relax condition. Similarly, knee flexion total work for the PowerPlus was significantly higher compared to both Relax and Clench test conditions.

KEY WORDS: strength, isokinetic, mouth guard.

INTRODUCTION: Strength and strength training are critical to performance in competitive sports. Many studies have been conducted that evaluate the effect jaw position may have on acute muscle strength. These studies have shown that subjects who suffer from temporomandibular pain dysfunction (TMD) or have a severe vertical dimension of occlusion (VDO) can benefit from wearing a Bite Elevating Appliance (BEA) (Abdallah, et al., 2004). One such appliance that has been studied in depth is the MORA. This device has shown to improve isometric strength when worn in athlete and non-athlete populations compared to a habitual occlusion and placebo appliance (Gelb, et al., 1995). The data has shown there is an optimal maxilla-mandibular spacing that can create maximum isometric strength (Chakfa, et al., 2002). Making the spacing larger or smaller can decrease a subject's strength. One study compared a BEA created for optimal spacing to a common mouth guard for American football players with TMD. The results showed that isometric strength was improved by using a common mouth guard and even more using the BEA (Abduljabbar, et al., 1997).

Unfortunately, BEA's have not proven effective in all studies. In particular, isokinetic strength has not proved to increase with the use of BEA's (Forgione, et al., 1991). These tests have been conducted at various speeds (60,120, and 240 °/s) without significant results. If BEA's can only improve isometric strength they will offer minimal benefit to athletes who rarely perform isometric tasks in competition. The goal of this study was to test the effect of selected oral appliances on isokinetic strength in athletes. The hypothesis is that wearing a BEA designed for optimal maxilla-mandibular spacing can increase isokinetic strength in an athlete population without screening for VDO. The BEA was compared to a common mouth guard in addition to no appliance conditions to determine if similar results could be obtained with a simple, off the shelf alternative.

METHODS: Bite Appliance Fitting: The effect of wearing the dental appliances on strength was tested in eighteen NCAA Division II American football athletes. The mean age of the subjects was 20 +/- 1.4 years. The subject's average height and weight was 1.87 +/- 0.07 m and 101.2 +/- 18.4 kg. The athletes were fitted for two mouth guards typically worn during sport and practice. One mouth guard was the commercially available Shock Doctor Pro (Plymouth, MN). These mouth guards were fitted according to the manufacturer's instructions, by heating in water and biting the appliance to fit each subject. The other mouth guard was a PowerPlus (Traverse City, MI) mouth guard. This appliance can be customized in many ways. For the purpose of this study, lower dental impressions were taken, and the appliance was fabricated on a stone model for each test subject. The mouth guard was then custom-fitted by a dentist. The PowerPlus mouth guard is a BEA that is designed to optimize the non restrictive jaw position. Each subject's vertical dimension of occlusion was measured and the mouth guard thickness adjusted to create vertical dimension of occlusion of 19 mm.

Data Collection: The subjects participated in two different strength tests under four different conditions. The conditions included both mouth guards and two conditions without any dental appliance. These conditions required subjects to clench their teeth for one condition (clench) and to have their teeth touching while relaxing their jaw for the other condition (relax). Upper and lower body isokinetic tests were conducted on a Biodex System 2 (Shirley, NY). The subjects performed four sets for both tests, one for each condition. The order of the sets was randomized for each participant.

The tests were concentric-concentric tests at a velocity of 60°/s on the subjects' dominant side. The upper body test was an elbow flexion/extension, and the lower body exercise was knee extension/flexion. Prior to isokinetic exercise testing, each subject completed a standard warm-up protocol that consisted of a 5-min sub maximal upper or lower body cycle ergometry exercise at a moderate intensity. The Biodex device was calibrated before each day of testing and the device was warmed-up according to the manufacturer's instructions. The Biodex positioning was adjusted for each test subject to ensure proper alignment with joint position. Once the device was positioned properly, the subject was secured in place and allowed to familiarize themselves with the device and the exercise. They were instructed to perform the exercise motion without resistance to familiarize them with the range of motion and the device. The subjects performed a set of five repetitions for each of the conditions listed above. The subjects were given a three minute rest period between each of the sets. To diminish effects of position changes on results, the subjects remained in the Biodex chair with the restraints in place during the rest periods. here was no verbal encouragement provided during any of the testing, and the monitor of the Biodex was adjusted so that the test subjects could not observe their performance during testing. The results were not made available to the test subjects until all testing had been completed.

Data Analysis: Four variables were evaluated for the isokinetic tests: Peak Torque/Body Weight (PT/BW), Max Repetition Work (MRW), Total Work (TW), and Average Power (AP). Since the subjects varied considerably in stature and strength, the data was normalized using the relax condition before statistical analysis was conducted. The relaxed condition was considered the baseline, and was subtracted from all conditions. The other three conditions were evaluated as improvements or degradations from the relax condition.

The results were analyzed using SPSS 17.0 (Chicago, IL). MANOVA's were conducted for the upper and lower body isokinetic tests in flexion and extension. Bonferroni post hoc tests were conducted for any conditions that demonstrated a statistical significance ($p \leq 0.05$).

RESULTS: In extension, the isokinetic tests did not show statistical significance for any variables in either test. However, both tests had statistical significance in flexion (Table 1 and Table 2). The upper body tests showed that using the PowerPlus yielded significantly higher results than the relax condition for PT/BW, TW, and AP (p = 0.007, 0.02, and 0.011 respectively). The lower isokinetic tests showed that using the PowerPlus yielded significantly higher than the relax and the clench conditions for TW (p = 0.021 and 0.031 respectively).

Table 1 Upper Body Isokinetic Data Means

	Upper Body			
Condition	PT/BW (%)	RW (J)	TW (J)	AP (W)
Clench	1.69	3.99	25.58	3.59
PowerPlus	3.19†	8.30†	41.95†	8.30†
Shock Dr.	2.21	5.50	22.15	5.72
Relax	0.00†	0.00†	0.00†	0.00†

Results marked with † are statistically significant ($p \leq 0.05$)

Percent increases were calculated for the data that proved significantly different. Using the PowerPlus BEA showed mean improvements of 10.5%, 9.8%, and 11.2% from the relax condition in the upper body tests for PT/BW, TW, and AP respectively. The total work for the PowerPlus lower body tests was significantly improved from both the relaxed and the clenched conditions. It provided an increase of 16.8% and 16.0% from the relaxed and clenched conditions, respectively.

Table 2 Lower Body Isokinetic Data Means

	Lower Body			
Condition	PT/BW (%)	RW (J)	TW (J)	AP (W)
Clench	0.21	-0.62	5.02†	3.59
PowerPlus	6.54	20.03	113.40†	21.70
Shock Dr.	2.68	4.03	35.06	8.34
Relax	0.00	0.00	0.00†	0.00

Results marked with † are statistically significant ($p \leq 0.05$)

DISCUSSION: These results indicate using BEA devices can increase isokinetic strength in athletes without screening for severe VDO. This implies that an average athlete population could benefit from the device. Some subjects may observe a larger benefit due to their predisposition, but it appears there is a benefit to the general population. During the fitting of the BEA, the VDO was measured for each subject, and these measurements could be utilized to compare groups based on their VDO. The improvements in strength could be compared between large and small VDO groups to determine if there is a significant difference between the groups.

Future studies could examine the long term effects of BEA's on strength and performance. Athletes could be evaluated over the length of an off-season strength conditioning program. Strength improvements could be compared between subjects who are provided a BEA to wear during strength training to a control group participating in the same conditioning program. Other studies have required subjects to wear their BEA for a week before testing (Allen, et al., 1984). This study did not find any statistical significance, but week to week performance variability for test subjects could have diminished results.

The ultimate goal of improving an athlete's strength is to improve their sport-specific performance. One study examined a test more applicable to sports, the in-flight velocity of golf balls (Egret, et al., 2002). This study found that velocity increased and was more consistent with the BEA. Future studies should examine other sport-specific tests to determine if BEA's can improve performance. American football athletes could perform a battery of tests included in the NFL combine (40 yard dash, standing long jump, bench press, etc.).

CONCLUSION: These data indicate wearing dental appliances designed for optimal maxilla-mandibular spacing can improve isokinetic strength in flexion. Using the PowerPlus BEA provided improvements in both upper and lower body isokinetic flexion. These results differ from previous studies. Using the PowerPlus device provided statistically significant increases in three of the four variables evaluated (PT/BW, TW, and AP) compared to the relax condition in elbow flexion. It provided increases of 10.5%, 9.8%, and 11.2% for these variables respectively. In isokinetic knee flexion, the PowerPlus increased TW by 16.8% and 16.0% compared to the relaxed and clenched conditions.

These improvements were not present with a common mouth guard. These data show that an average athlete population can benefit from a specifically designed BEA, and that a common mouth guard does not provide the same benefit.

REFERENCES:
Abdallah EF, Mehta NR, Forgione AG, et al. (2004). Affecting upper extremity strength by changing maxillo-mandibular vertical dimension in deep bite subjects. *Journal of Craniomandibular Practice*, 22, 268-275.

Abduljabbar T, Mehta NR, Forgione AG, et al. (1997). Effect of increased maxillo-mandibular relationship on isometric strength in TMD patients with loss of vertical dimension of occlusion. *Journal of Craniomandibular Practice*, 15, 57-67.

Allen ME, Walter P, McKay C, Elmajian A. (1984). Occlusal splints (MORA) vs. Placebos show no difference in strength in symptomatic subjects: double blind/cross-over study. *Canadian Journal of Applied Sport Science*, 9, 148-152.

Chakfa AM, Mehta NR, Forgione AG, et al. (2002). The effect of stepwise increases in vertical dimension of occlusion on isometric strength of cervical flexors and deltoid muscles in nonsymptomatic females. *Journal of Craniomandibular Practice*, 20, 264-273.

Egret C, Leroy D, Loret A, Chollet D, Weber J. (2002). Effect of mandibular orthopedic repositioning appliance on kinematic pattern in golf swing. *International Journal Sports Medicine*, 23, 148-152.

Forgione AG, Mehta NR, Westcott WL. (1991). Strength and bite, Part 1: An analytical review. *Journal of Craniomandibular Practice*, 9, 305-315.

Gelb H, Mehta NR, Forgione AG. (1995). Relationship of muscular strength to jaw posture in sports dentistry. *The New York State Dental Journal*, 61, 58-66.

Gelb H, Mehta NR, Forgione AG. (1996). The relationship between jaw posture and muscular strength in sports dentistry: a reappraisal. *Journal of Craniomandibular Practice*, 14, 320-325.

Acknowledgement
The authors would like to thank Bulent Ozkan for his assistance with statistical analysis, Paul Harker and Paul Winters for allowing their players to participate in the study, all the athletes who took the time to participate in the study, and Dr. Thomas Birk for the use of his lab equipment and time.

ACCUMULATED FATIGUE AFFECTS RUNNING BIOMECHANICS DEPENDING ON THE PERFORMANCE LEVEL DURING A TRIATHLON COMPETITION

Antonio Cala[1,2] and Enrique Navarro[1]

Sports Biomechanics Laboratory, Polytechnic University of Madrid, Spain[1]
Institute of Sport & Recreation Research, AUT University, New Zealand.[2]

The purposes of the present study were 1) to examine the different responses to the accumulated fatigue between international and national level triathletes in competition, and 2) to compare the profile of the running part presented by the two different performance levels. 32 participants at Madrid 2008 Triathlon World Cup and 32 participants at Spanish National Championships 2008, made up the sample. We found higher values ($p<0.05$) of stride frequency, stride length and flight time in international level triathletes. Also, lower values ($p<0.05$) in contact time and knee angles at toe-off were obtained. International level triathletes showed a consistent tendency in some of the analyzed variables ($p = 1.00$ among the laps) whereas the national level participants presented significant differences ($p<0.05$).

KEY WORDS: Triathlon, Cycle-run transition, competition, performance level.

INTRODUCTION: One of the most difficult parts (strategically and physically) of a triathlon is the transition from cycling to running (Rowlands & Downey, 2000). In fact, the performance of a triathlete may decrease due to a discomfort during the first part of the running provoked by the previous cycling (Quigley & Richards, 1996; Hue et al., 1998, 2000).
The cycle-run transition has been widely studied in laboratory-conditions (Quigley & Richards, 1996; Hausswirth et al., 1996; Millet et al., 2001; Millet & Bentley, 2004; Palazzetti et al., 2005) but only one study has been carried out in competition (Cala et al., 2009). A Triathlon World Cup event was analyzed and they found no effect of the previous cycling on the subsequent running kinematics in elite triathletes.
However, the influence of the performance level is not clear. Millet et al. (2000) analyzed the running off the bike in elite and middle-level triathletes and they found differences in the vertical displacement of the centre of gravity between the two groups. But the study was performed in laboratory conditions and the situation in competition remains unclear.
Therefore, the aims of the present study are: (1) to examine the different responses to the previous cycling between international and national level in competition, and 2) to compare the profile of the running part presented by the two different performance levels.

METHODS: Data Collection: Two different competitions were analyzed: Madrid 2008 Triathlon World Cup and Spanish National Championships 2008. Both events took place in the same circuit at the same time in two consecutive days. The sample size was 64 triathletes: 16 men and 16 women, ranked among the first sixteen competitors at the end of the cycling part in each event, were selected for this study.
A video camera (JVC GY-DV500E) was positioned perpendicular to the longitudinal direction of the track, recording at 50 Hz of sampling frequency according to other studies (Amico et al., 1989; Palazzetti et al., 2005; Cala et al., 2009). The 10km run was broken into four 2,5km laps, i.e. the triathletes were recorded four times. The kinematic analysis was performed through a photogrammetric technique in the saggital plane (2D). As the triathletes did not follow the same line, five different planes of movement were calibrated in order to choose the nearest to the trajectory of each one of athletes. The calibration system covered a surface of 7 meters width and 2 meters high. A Clausser-based kinematic model (Clausser et al., 1969) was used to analyse the running biomechanics. Eight anatomical landmarks (markerless digitisation method) were selected: hip, knee, ankle and toe-cap (both right and left side). An algorithm with 2D direct linear transformation (2D-DLT) was used (Abdel-Aziz & Karara, 1971) and the coordinates obtained were smoothed using quintic spline functions with the Cross Generalized Validation procedure as a method for evaluating the adjusting factor

(Woltring, 1985). The root mean error (RMS) (Allard et al., 1995) in the reconstruction of the coordinates in the x and y axis was 0.02 and 0.03 m, respectively. The RMS error when reconstructing the distance between two points was 1.23%. Once the coordinates of the anatomical landmarks were obtained, the following gait variables were calculated: "Stride frequency" (cycles/minute), "Stride length" (meters), "Contact time" (in seconds and in percentage), "Flight time" (in seconds and in percentage), "Knee angles at toe-off" (degrees) and "Ankle angles at toe-off" (degrees).

Intra-rater reliability of measurement was evaluated by asking the same investigator to repeat the digitizing of the same sequence 30 times. The coefficient of variation (CV) was under 2% in all the variables measured. Inter-rater reliability of measurements was assessed by three investigators who digitized the same video sequence (each video include a series of 200 frames). There was no significant difference among the operators in terms of digitizing (x-, y- coordinates recording). Validity of the measurement was evaluated analyzing the same athlete with a 3D protocol (2 cameras and a 3D-DLT algorithm) and with the 2D-DLT protocol (the same one used in the present study). There was no significant difference between the two protocols used.

Data Analysis: A one-way repeated measures analysis of variance (ANOVA) was performed. A Bonferroni adjustment was performed for multiple comparisons. All statistical measures were conducted at α < 0.05.

RESULTS: Significant differences (p<0.05) were found in many of the analyzed variables between the two levels of performance. The variables that showed those differences were stride frequency, stride length, contact time (seconds and percentage), flight time (seconds and percentage), knee angle of the support and non-support leg and the ankle angle of the support leg. Higher values were found in stride frequency, stride length and flight time (percentage) for international level triathletes. However, contact times (seconds and percentage), knee angle of the support and non-support leg and ankle angle of the support leg values were higher for national level triatheles.

Table 1 Results of the variables obtained for national level triathletes in the different laps.

NATIONAL LEVEL	LAP 1	LAP 2	LAP 3	LAP 4
STRIDE FREQUENCY	86.92 ± 3.35	85.30 ± 4.17*	85.49 ± 3.71*	85.99 ± 3.30
STRIDE LENGTH	3.08 ± 0.53	3.12 ± 0.41	3.06 ± 0.45	3.06 ± 0.47
CONTACT TIME (S)	0.46 ± 0.03	0.48 ± 0.03*	0.48 ± 0.03*	0.48 ± 0.03*
CONTACT TIME (%)	67.32 ± 5.55	68.67 ± 4.89*	68.92 ± 5.67*	69.84 ± 5.60*
FLIGHT TIME (S)	0.23 ± 0.04	0.22 ± 0.03*	0.22 ± 0.04*	0.22 ± 0.04*
FLIGHT TIME (%)	32.68 ± 5.55	31.34 ± 4.89*	31.08 ± 5.67*	29.86 ± 5.60*
KNEE ANGLE NON-SUPPORT LEG	111.90 ± 12.41	112.28 ± 13.01	112.00 ± 11.13	113.27 ± 11.51
KNEE ANGLE SUPPORT LEG	166.17 ± 4.41**	166.82 ± 4.32**	166.56 ± 4.58**	166.41 ± 4.95**
ANKLE ANGLE NON-SUPPORT LEG	124.24 ± 9.33	125.72 ± 9.59*	126.04 ± 7.29*	125.84 ± 9.23*
ANKLE ANGLE SUPPORT LEG	137.33 ± 6.91**	140.77 ± 6.00**	139.33 ± 6.02**	136.91 ± 6.76**

* Significant differences (p<0.05) with the lap 1
** P value of 1.00

Two different tendencies were found during the 10km running according to the performance level of the triathletes. Significant differences (p<0.05) between the lap 1 and the other laps were found in stride frequency, contact time (seconds and percentage), flight time (seconds and percentage) and in the ankle angle of the non-support leg for national level triathletes (table 1). On the other hand, international level triathletes showed a p value of 1.00 among the 4 laps in contact time (percentage), flight time (percentage) and ankle angle of non-support leg (table 2).

Table 2 Results of the variables obtained for international level triathletes in the different laps.

INTERNATIONAL LEVEL	LAP 1	LAP 2	LAP 3	LAP 4
STRIDE FREQUENCY (cycles/min)	91.13 ± 3.56	90.15 ± 2.94*	89.84 ± 3.30*	90.00 ± 3.20*
STRIDE LENGTH (m)	3.47 ± 0.32	3.44 ± 0.38	3.35 ± 0.38	3.38 ± 0.37
CONTACT TIME (s)	0.43 ± 0.02	0.43 ± 0.02	0.44 ± 0.02	0.44 ± 0.02
CONTACT TIME (%)	65.22 ± 3.53**	64.85 ± 3.41**	65.52 ± 3.10**	65.54 ± 4.32**
FLIGHT TIME (s)	0.22 ± 0.02	0.23 ± 0.02	0.23 ± 0.02	0.23 ± 0.03
FLIGHT TIME (%)	34.77 ± 3.53**	35.14 ± 3.41**	34.48 ± 3.10**	34.45 ± 4.32**
KNEE ANGLE NON-SUPPORT LEG	101.63 ± 10.19	99.38 ± 10.66	101.44 ± 11.39	102.89 ± 12.87
KNEE ANGLE SUPPORT LEG	162.60 ± 3.24**	163.69 ± 4.63**	163.94 ± 4.05**	163.27 ± 3.74**
ANKLE ANGLE NON-SUPPORT LEG	126.52 ± 5.76**	126.18 ± 6.52**	127.00 ± 5.79**	128.53 ± 6.82**
ANKLE ANGLE SUPPORT LEG	135.76 ± 7.77**	137.05 ± 5.62**	136.67 ± 4.59**	135.38 ± 5.27**

* Significant differences (p<0.05) with the lap 1
** P value of 1.00

DISCUSSION: The most important result found in the present study was the existence of two different tendencies of the biomechanics during the 10km running depending on the performance level of the triathlete. International level triathletes presented similar values (p = 1.00) in the four laps in several variables: contact time (percentage), flight time (percentage) and ankle angle of non-support leg. On the other hand, national level triathletes showed significant differences (p<0.05) between the lap 1 and the other laps in stride frequency, contact time (seconds and percentage), flight time (seconds and percentage) and in the ankle angle of the non-support leg. These two different tendencies can be explained by the differences in the performance level of the triathletes. Millet et al. (2000) found a significant effect of the triathlon performance level on the change of the running energy cost after cycling, but the study was carried out in laboratory conditions. To the best of our knowledge, only one study in the literature has performed a biomechanical analysis of the running part during a triathlon competition (Cala et al., 2009) and they only focused in elite triathletes. The results of the present study show how top-level group could maintain the same values during the 10km running, whereas national-level group could not do so. It seems the accumulated fatigue would be the reason of these two different trends.

Other important result found in this study was the existence of significant differences (p<0.05) in absolute values found in many of the analyzed variables between the two levels of performance. Cala et al. (2009) found that differences in some variables could be related to the triathlete's gender, as stride frequency, stride length and ankle angles values. In the present study, we found stride frequency, stride length, ankle angles may be related to performance level as well. Furthermore, contact time (seconds and percentage), flight time (percentage) and knee angles may be related only to the performance level. These differences can be explained by the different velocities values reached by each group of performance. International-level triathletes showed an average velocity of 3min 15sec per kilometre, while national-level triathletes presented an average velocity of 3min 50sec per kilometre. Probably, the biomechanical requirements will be different to achieve each velocity and that would be the reason that could explain the differences found.

CONCLUSION: Two different tendencies on the biomechanics during the 10km-running were found depending on the performance level of the triathlete. International-level triathletes presented similar values (p = 1.00) in the four laps in several variables while national-level triathletes showed significant differences (p<0.05) between the lap 1 and the other laps. The accumulated fatigue could be the reason to explain this phenomenon. Significant differences (p<0.05) in absolute values were found in many of the analyzed variables between the two levels of performance. The main reason that could explain this situation was the different velocities achieved by the two level of performance.

REFERENCES:

Abdel-Aziz, Y.I., Karara, H.M. (1971). Direct linear transformation from comparator coordinates into space coordinates in close range photogrammetry. In the American Society of Photogrammetry (Ed.), *Proceedings of the Symposium on close range photogrammetry* (pp.1-18).

Amico, A.D., Ferrigno, G., Rodano, R. (1989). Frequency content of different track and field basis movements. In W. Morrison (Ed), *Proceedings of the VII International Symposium of Biomechanics in sports* (pp.177-193). Melbourne, Australia.

Allard, P., Blanchi, J.P., Aïssaqui, R. (1995). Bases of three-dimensional reconstruction. In P. Allard, I.A.F. Stokes & J.P. Bianchi (Eds.), *Three Dimensional Analysis of Human Movement*, (pp.19-40). Champaign, IL: Human Kinetics.

Cala, A., Veiga, S., Garcia, A., Navarro, E. (2009). Previous cycling does not affect running efficiency during a triathlon World Cup competition, *Journal of Sport Medicine and Physical Fitness*, In press

Clauser, C. E., McConville, J. T., Young, J. W. (1969). Weight, volume and centre of mass of segments of the human body. *Wright-Patterson Air Force Base* (pp. 69-70).

Hausswirth, C., Bigard, A.X., Berthelot, M., Thomaidis, M., Guezennec, C.Y. (1996). Variability in energy cost of running at the end of a triathlon and a marathon. *International Journal of Sport Medicine*, 17, 572-579.

Hue, O., Le Gallais, D., Boussana, A., Galy, O., Chamari, K., Mercier, B. (2000). Catecholamine, blood lactate and ventilatory responses to multi-cycle run blocks. *Medicine & Science in Sport & Exercise*, 32, 1582-1586.

Hue, O., Le Gallais, D., Chollet, D., Boussana, A., Préfaut, C. (1998). The influence of prior cycling on biomechanical and cardiorespiratory response profiles during running in triathletes. *European Journal of Applied Physiology*, 77, 98-105.

Millet, G. P., Bentley, D. J. (2004). The physiological responses to running after cycling in elite junior and senior triathletes. *International Journal of Sports Medicine*, 25(3), 191-197

Millet G.P., Millet, G.Y., Candau, R.B. (2001). Duration and seriousness of running mechanics alterations after maximal cycling in triathletes. Influence of the performance level. *Journal of Sport Medicine and Physical Fitness*, 41, 147-153.

Millet G.P., Millet, G.Y., Hofmann, M.D., Candau, R.B. (2000). Alterations in running economy and mechanics after maximal cycling in triathletes: Influence of performance level. *International Journal of Sport Medicine*, 21, 127-132.

Palazzetti, S., Margaritis, I., Guezennec, C.Y. (2005). Swimming and cycling overloaded training in triathlon has no effect on running kinematics and economy. *International Journal of Sport Medicine*, 26(3), 193-199.

Quigley, E. J., & Richards, J. G. (1996). The effects of cycling on running mechanics. *Journal of Applied Biomechanics*, 12(4), 470-479.

Rowlands D.S., Downey B. (2000). Physiology of Triathlon. In K. D. Garret WEJ (Ed.). *Exercise and Sport Science*. (pp. 919-39).Philadelphia: Lippincott Williams & Wilkins,

Woltring, H.J. (1985). An optimal smoothing and derivate estimation from noisy displacement data in biomechanics. *Human Movement Sciences*, 4, 229-45.

CAN ELITE TENNIS PLAYERS JUDGE THEIR SERVICE SPEED?

Kieran Moran[1], Colm Murphy[1], Garry Cahill[2], Brendan Marshall[1]

[1]School of Health and Human Performance, Faculty of Science and Health, Dublin City University, Ireland; [2] Tennis Ireland, Dublin, Ireland

The purpose of this study was to examine if elite tennis players could accurately determine whether successive serves were faster or slower than the preceding serve. Eleven national standard junior tennis players completed 10 acceptable maximum effort serves, aiming to land the ball with-in a 1m square area adjacent to the service box T. A Wilcoxen signed rank non-parametric test was employed ($\alpha = 0.05$). Results indicated players were no more likely to correctly differentiate serves (4.9 ± 1.5) than that which would be expected by chance (5 out of 10) [$p = 0.92$]. The average speed of serve was 46.9 ± 4.5 $m.s^{-1}$ and the variation in each player's service was 1.1 ± 0.5 $m.s^{-1}$ (approximately 2.3%). The implications of these findings is that it is not possible for elite junior tennis players to use service speed (knowledge of results) as a means of guiding and fine-tuning their technique when they rely on gaining this information from purely internal physiological systems (e.g. vision).

KEY WORDS: augmented feedback, tennis, service speed

INTRODUCTION: The serve in tennis is commonly considered the most important stroke in the game (Roetert and Groppel, 2001), with a high service speed being of paramount importance. In many sports, including tennis, players attempt to use knowledge of results (outcome measures) to guide positive changes in technique (Adams, 1987; Newell, 1991). In essence, if a serve is faster than previous attempts then the player will aim to more perminantly adopt the technique that produced it. This requires tennis players to be able to judge whether one serve attempt results in a faster or slower serve than a previous attempt. At the very least they should be able to judge from consecutive attempts which is faster or slower. However, as no previous studies could be found that examined this; the present study aims to do so. It was hypothesised that they would be able to coorectly identify whether consecutive attempts were faster or slower.

METHODS: Participants: Eleven national standard junior tennis players, 7 male and 4 female, between the age of thirteen and eighteen (15.7 ± 1.6 years) volunteered for the study. Players were free from any injury that would have prevented them from using maximum effort. All participants were training between 20 and 26 hours per week as part of the Tennis Ireland national squad. Ethical approval was received by Dublin City University.

Data Collection: Following a warm-up, players served fifteen acceptable serves to the T of the deuce service box. Attempts were deemed acceptable if they were (i) within a 1m x 1m area of the T in the service box (Figure 1), and (ii) judged by the player to be maximum effort. Players were shown by means of a large digital display the speed of the first five serves and were asked to state whether each of the subsequent 10 serves were faster or slower than the immediately preceding one. Players were allowed up to 20 seconds between serves in accordance with ITF guidelines (ITF, 2009).
The warm-up consisted of: 3 minutes jogging at a self-selected 'slow' pace and 2 minutes at a 'fast' pace; 8 minutes of whole body dynamic stretching; 10 minutes of rallying increasing in intensity and 3 minutes service practice.
Service speed was measured using a StalkerPro speed gun (Stalker, USA) placed 4m behind the end base-line and in line with the intended direction of serve. Reported absolute errors for the speed gun are very small (± 0.04 $m.s^{-1}$). It is acknowledged however that serves to either side of the sensor of the speed gun would contain inaccuracies; these were estimated to be up to 0.25%.

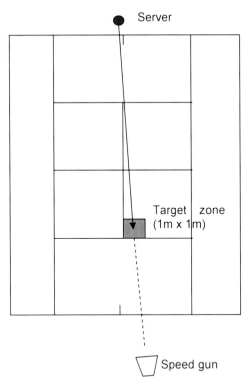

Figure1: Experimental set-up

Data Analysis: A Wilcoxen signed rank non-parametric test was employed ($\alpha = 0.05$). The two variables used were (i) the number of correct differentiation (faster/slower) out of 10 and (ii) the number of correct differentiations expected due to chance (5).

RESULTS: No significant difference was evident between the number of correctly differentiated serves (4.9 ± 1.5) and the number of serves expected to be differentiated correctly due to chance (5) [$p = 0.92$]. This indicates that tennis players could not correctly determine if consecutive serves were faster or slower than each other. The average speed of serve was 46.9 ± 4.5 m.s^{-1}, and the average difference between serves was 1.1 ± 0.5 m.s^{-1} (Table 1).

DISCUSSION: Newell (1991) suggests that knowledge of the service speed in tennis is extremley important in technique enhancement, as it provides information by which an individual can judge whether a particular movement action was more or less appropriate than previous attempts. The present study indicates that players can not accurately differentiate service speed between consecutive serves. If they can not do this between consecutive serves, it is very unlikely that they can differentiate between non-consecutive serves. The consequence of this is that tennis players can not use knowledge of results, based on their own perception of service speed, as a means of guiding and fine-tuning their technique. From the available data it is not possible explain why two subjects (participant 1 and 2, Table 1] had a higher number of correct identifications, but this could clearly be by chance only.

Table 1. Participant ability to identify if their service speed was faster or slower than their preceding serve

Participant	Correct out of 10	Ball Speed ($m.s^{-1}$)	Difference in speed ($m.s^{-1}$)
1	7	41.5	1.3 (0.7)
2	8	43.6	2.4 (1.1)
3	4	53.3	1.5 (1.0)
4	5	51.6	0.8 (0.6)
5	4	44.3	1.1 (0.9)
6	3	46.3	1.0 (0.8)
7	5	53.7	0.8 (0.7)
8	5	45.4	0.7 (0.6)
9	5	47.3	0.8 (0.9)
10	5	48.2	1.2 (0.8)
11	3	40.6	0.8 (0.7)
Mean ± SD	4.9 ± 1.5	46.9 ± 4.5	1.1 ± 0.5

Unfortunately, no previous studies appear to have investigated the ability of elite standard tennis players (or any standard) to differentiate between the speeds of consecutive serves hit as fast as possible. Similarly, no studies could be found that investigated this ability in other high speed ball-projection based sports where the full flight path of the ball is impeded (e.g. by a wall, net etc).

For novel (non-sporting) tasks Magill (1998) suggests that in performing a movement as quickly as possible a person can initially differentiate whether the movement was faster or slower than the previous one and subsequently utilise enahnced techniques; however, such enhancements seem to stop at a certain level of performance due to a person's inexperience and therefore decreased capability to discriminate small movement-speed differences (Magill, 1998). It is unclear from the present study if the inability to differentiate between serves in terms of speed is related to either, or both, the relatively high speed of serve (46.9 ± 4.5 $m.s^{-1}$) or the small variation in each player's service speed (1.1 ± 0.5 $m.s^{-1}$, which is approximately 2.3% of the ball speed); however, given the level of expertise and prior training of the participants, it is unlikely due to inexperience.

The findings from the present study would suggest that if speed of serve is to be used to optimise service technique, it may be necessary to use an appropriate measuring device (such as a speed gun), but this needs investigating.

Finally, it is worth noting that a number of sports/sports actions make it diifficult to identify ball projection speed (e.g. pitching in baseball, penalty kicks in soccer, forehand drive in tennis) bacause a playes can not see how far the ball would travel without interference from an another player or the environment. It would be worthwhile to determine if athletes from these sports could accurately differentiate ball projection speeds and whether this is dependent on the skill level or experience of the player.

CONCLUSION: Elite junior tennis players are unable to accuractely judge whether their tennis serves are faster or slower than their preceding serves. It is therefore unlikely that without the help of external augmented feedback (e.g. from the coach or a measuring device) they would be able to use this information to effectively guide the development of a more optimal technique.

REFERENCES:

Adams, J. A. (1987). Historical review and appraisal of research on the learning, retention and transfer of human motor skill. *Psychological Bulletin* 101, 41-47
ITF (2009) Rules of Tennis. ITF Limited, London. 14
Magill, R. A. (1998). Motor Learning, Concepts and Applications. McGraw-Hill Companies. New York. 171-224
Newell, K. M. (1991). Motor skill acquisition. *Annual Review of Psychology*. 42, 213-237
Roetert, P. & Groppel, J. (2001). World Class Tennis Technique. Human Kinetics Publishers. Leeds. 207

OPTIMISATION OF ENERGY ABSORBING LINER FOR EQUESTRIAN HELMETS

L. Cui, M. A Forero Rueda and M. D. Gilchrist*

UCD School of Electrical, Electronic and Mechanical Engineering, University College Dublin, Belfield, Dublin 4, Ireland

The density of foam used as energy absorbing liner material in safety helmets was optimised in this paper using Finite Element Modelling (FEM). FEM simulations of impact tests from certification standards were carried out to obtain the best performing configurations of helmet liner. For each test condition, two best liner configurations were identified as minimising peak impact accelerations: one was composed of layers of uniform foam and the other of functionally graded foam (FGF). It was found that the observed decreases in the peak accelerations for the best performing helmets in various test conditions are directly related to the contact area, the distribution of internal stresses, and the dissipated plastic energy density (DPED). Application of the methods described in this study could help increase energy absorption for current and future equestrian helmet designs.

KEY WORDS: Functionally graded foam material, helmets, impact.

INTRODUCTION: Epidemiological statistical studies across the world have shown that horse racing is a particularly risky sport (Forero Rueda et al. 2009), particularly when head injury is concerned. EN 1384:1996 is the current European standard to certify equestrian helmets. The new high performing helmet standard EN 14572:2005 is intended for helmets for "high-risk" activities, but it does not supersede EN 1384:1996. No helmet currently available in the market complies with EN 14572:2005. The EN1384:1996 standard specifies an impact speed of 5.4 $m.s^{-1}$, while the new standard EN 14572:2005 specifies a "high energy" impact velocity (7.7 $m.s^{-1}$), as well as a "low energy" impact velocity (4.4 $m.s^{-1}$). This is with the intention of stimulating the construction of helmets that reduce head injury risk for both high and low impact energies. This study aims to suggest a possible solution to manufacturing helmets conforming to standard EN14572:2005 by optimising the liner density.

METHODS: Model description: The current study developed a FE model of an equestrian helmet based on the geometry of commercially available helmets using ABAQUS (ABAQUS 2009). The helmet model consists of an outer shell, foam liner, foam block and ring. The outer shell is modelled as linear elastic material and the ring is modelled as an incompressible rubber elastomer. The foam block between the shell and foam liner is modelled as a hyperelastic elastomeric compressible foam with material constants specified by experimental test data. The expanded polystyrene (EPS) foam liner material is modelled using the crushable foam model with a volumetric hardening rule in conjunction with the linear elastic model (ABAQUS 2007). The stress-strain curve for the polymeric foam is a function of foam density. Constants for the constitutive model used in the current study have been tested and determined in a previous study (Cui et al. 2009). The curve is tri-linear in form, corresponding to elastic, plateau, and densification stages (Figure 1). It is more efficient that the foam liner absorbs energy within the plateau stage as the stress remains nearly constant over a large strain.

The headform is simulated as a rigid body. The helmeted head is impacted against a flat rigid anvil. The impact positions, crown impact (Fig 2(a)) and 45° side impact (Fig 2(b)), are as recommended in both standards. Impact velocities of 5.4 (EN1384:1996), 4.4 and 7.7 $m.s^{-1}$ (EN 14572:2005) are used. ABAQUS/Explicit was used for the finite element helmet dynamic impact tests. The headform is modelled using three dimensional four node elements (R3D4) with a rigid body constraint at the centre of mass where the linear headform accelerations were read. The liner and foam block is modelled as three dimensional eight node linear brick elements with reduced integration and hourglass control (R3D8R). The shell is modelled with four node doubly curved thin shell, reduced integration, hourglass control, finite membrane strain model elements (S4R) with a section thickness of 2mm.

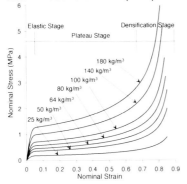

Figure 1: Stress-strain curves for representative densities of EPS foam

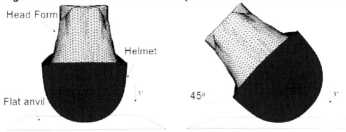

Figure 2: Representative impact configurations (a) 45° side impact; (b) normal crown impact

Simulation parameters: The EPS foam liner material is typically of density 64 kg/m^3. To optimise the liner density, a three equally layered liner and a functionally graded foam (FGF) liner are introduced. The FGF is a type of material, the characteristics of which (e.g. density, strength) vary through the thickness according to various gradient functions. A FGF liner can eliminate issues regarding crack initiation and propagation that discrete interfaces of different foam densities could generate. It is possible to make a liner with different density layers with current manufacturing techniques, while FGF manufacturing methods are still under development. Therefore, both types of liner are considered in this study. Density of each layer is selected from the values of 25, 50, 60, 80, and 100 kg/m^3. The FGF used in the current simulations has its density varied through the thickness according to a power-law gradient function as

$$\rho(y) = \rho_1 + (\rho_2 - \rho_1)\left(\frac{y}{d}\right)^n$$

where ρ_1 and ρ_2 are the densities at the inner and outer surfaces of the liner and d is the liner thickness. The FGF liners are set to have the same average density as the corresponding uniform foam liner (64 kg/m^3) to give parallel comparisons. Power index, n of 1, 0.25, and 4, and $\Delta\rho$ ($\rho_2-\rho_1$) of 20, 40, 60, 80, 100, 120, 140, and 160 kg/m^3 are selected for simulations.

RESULTS AND DISCUSSION: The peak accelerations of the best performing helmets with layered foam liners and FGF liners are listed in Table 1 and 2. As there was negligible improvement for three impact positions for the high energy impact, these improvements are still insufficient to make the helmet pass standard EN 14572:2005. However, the best performing helmets of each type substantially improved the energy absorbing performance in the low energy impact and the 1384 impact.

The contact area between the inner surface of the liner and the headform, and the contact area between the outer surface of the liner and the shell are analysed for the 45° side impact. Representative comparisons of contact area are shown in Figure 3. Larger contact areas are consistently related to the lower peak accelerations.

Table 1 Peak accelerations of best performing helmets with layered foam liner.

Impact position	Energy	Layered density configuration (kg/m^3)	Acceleration (g = 9.81 m/s^2)	Reduction in Acceleration
45° side	1384	Uniform 64	199.0g	--
		Uniform 50	167.6g	15.8%
	Low	Uniform 64	165.0g	--
		Inner 50-25-25 outer	108.5g	34.2%
	High	Uniform 64	317.5g	--
		Inner 80-64-64 outer	327.4g	-3.1%
Crown	1384	Uniform 64	211.8g	--
		Uniform 50	192.6g	9.1%
	Low	Uniform 64	161.9g	--
		Inner 50-25-25 outer	124.8g	22.9%
	High	Uniform 64	428.2g	--
		Inner 80-64-64 outer	403.6g	5.7%

Table 2 Peak accelerations of best performing helmets with FGF liner (* higher density outside and lower density inside)

Impact position	Energy	FGF density configuration (kg/m^3)	Acceleration (g = 9.81 m/s^2)	Reduction in Acceleration
45° side	1384	Uniform 64	199.0g	--
		n=4 [40.63, 140.63] $\Delta\rho$=100	186.4g	6.5%
	Low	Uniform 64	165.0g	--
		n=4 [26.61, 186.61] $\Delta\rho$=160	136.5g	17.3%
	High	Uniform 64	317.5g	--
		n=1 [54, 74] $\Delta\rho$=20	315.9g	0.5%
Crown	1384	Uniform 64	211.8g	--
		n=4 [59.33, 79.33] $\Delta\rho$=20*	208.0g	1.8%
	Low	Uniform 64	161.9g	--
		n=4 [26.61, 186.61] $\Delta\rho$=160	151.8g	6.2%
	High	Uniform 64	428.2g	--
		n=4 [59.33, 79.33] $\Delta\rho$=20	426.7g	0.4%

The distribution of stress and plastic energy absorption through the thickness of different types of foam liner for the 45° side impact are shown in Figure 4 to explore how the non-uniform foam liners improve the energy absorption. By comparing Figure 4 and the plateau stresses in Figure 1, a relationship between the energy absorption, the stress level and the peak acceleration is found. For the uniform liner in the low energy impact, the majority of the form absorbs the energy at the early plateau stage; the energy absorbed is lower than the layered liner and is proportional to the volume of material plastically deformed. For the layered foam liner, the outer layer and the middle layer of the liner reaches the middle and late plateau stage so the plastic energy absorbed by them reaches high values; the inner layer reaches the initial plateau stage so the energy absorbed only reaches a lower value. Therefore, the layered foam liner in the low energy impact substantially improves the energy absorption efficiency and reduces the peak acceleration imparted to the head. The comparison for the high energy impact shows that both the uniform and layered liners absorb energy at initial plateau stage. The layered liner neither improved the energy absorption efficiency nor reduced the peak acceleration. Similar findings are obtained for the FGF liner.

CONCLUSION: The observed decreases in the peak accelerations for the best performing helmets in various test conditions are found to be related to the increase of contact area between the liner and either the inner headform or the outer shell. The peak acceleration is reduced if the foam liner absorbs the energy in the late plateau stage or if a larger part of the liner contributes to energy absorption; the peak acceleration is reduced when the DPED in the foam liner is increased. This study suggests a possible approach to manufacturing helmets that would conform to EN14572:2005 while keeping overall size and weight. Future helmets that comply with EN14572:2005 could help attenuate injury risk for a wider range of impact energies.

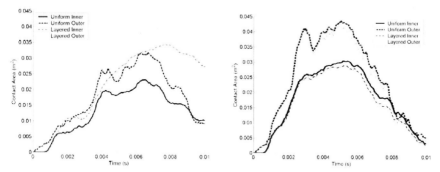

Figure 3: Evolution of contact areas at the inner and outer surfaces of helmet liner using either a uniform liner or a layered liner: (a) Low energy impact; (b) High energy impact

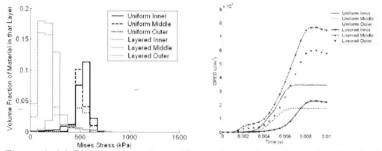

Figure 4: (a) Distributions of von Mises stress at peak acceleration in three layers for helmet liner of uniform density and of layered density at low energy impact; (b) Evolution of average DPED in three layers for helmets of uniform density foam and of layered density foam at low energy impact

REFERENCES:

ABAQUS. (2009). ABAQUS, Inc., Rising Sun Mills, 166 Valley Street, Providence, RI 02909, USA.

ABAQUS. (2007). ABAQUS Analysis User's Manual, v 6.7. ©ABAQUS, Inc.

Cui, L., Kiernan, S., and Gilchrist, M. D. (2009). Designing the energy absorption capacity of functionally graded foam materials. *Materials Science and Engineering A, 507(1-2)*, 215-225.

EN 1384:1996: Specification for helmets for equestrian activities.

EN 14572:2005: High performance helmets for equestrian activities.

Forero Rueda, M. A., Halley, W. L., and Gilchrist, M. D. (2009). Fall and injury incidence rates of jockeys while racing in Ireland, France and Britain. *Injury*, doi:10.1016/j.injury.2009.05.009

Acknowledgement

This study has been funded by Enterprise Ireland (PC/2005/071), Science Foundation Ireland (08/RFP/ENM/1169), and the Turf Club of Ireland.

THE EFFECT OF SELECTED KINEMATICS ON BALL SPEED AND GROUND REACTION FORCES IN FAST BOWLING

Peter Worthington[1], Mark King[1] and Craig Ranson[2]

School of Sport and Exercise Science, Loughborough University, UK[1]
UK Athletics Chief Physiotherapist, Loughborough, UK[2]

Lumbar stress fractures and lumbar injury account for the greatest amount of lost playing time in international cricket. Previous research has associated lower back injury with large peak ground reaction forces occurring during the front foot contact phase of the fast bowling action. Selected kinematics of the bowling action of 16 elite male fast bowlers were measured using an 18 camera Vicon Motion Analysis System. Ground reaction forces during front foot contact and ball release speed were recorded; correlations with kinematic factors were identified using Pearson's correlation coefficient. Ball release speeds were correlated with run-up speed, plant angle and the motion of the front knee during the period of front foot contact. Knee flexion during the first 15 frames of the front foot contact phase was correlated with increased peak vertical force and decreased peak horizontal loading rate. The use of a heel strike technique at the instant of front foot contact was correlated with decreased peak vertical force and loading rates. All correlations observed were moderate in strength, representing the multifactorial nature of the generation of ball speed and ground reaction forces. This study motivates future investigation of the effects of these selected kinematic factors on forces occurring above the knee, and the effect of kinematic factors on the performance of an individual bowler.

KEY WORDS: fast bowling technique, cricket, ball release speed, ground reaction forces.

INTRODUCTION: Fast bowlers have the highest injury prevalence in professional cricket (Newman, 2003; Orchard, et al., 2006), the most common cause being lumbar stress fractures and lumbar injury (Newman, 2003). These injuries occur predominantly on the opposite side to the bowling arm (contralateral side) (Gregory, et al., 2004; Ranson, et al., 2005). The high incidence of lower back injuries in fast bowlers is thought to be the result of a combination of factors, including: incorrect technique; poor preparation; overuse; age; and clinical features (Bell, 1992). Research, however, has focussed on bowling technique due to reported relationships between aspects of technique and the appearance of radiological abnormalities (Elliott, et al., 1992, 1993; Burnett, et al., 1996).

Previous researchers (Bartlett, et al., 1996; Ranson, et al., 2008) suggested large peak ground reaction forces during the front foot contact (FFC) phase, together with lateral flexion, hyperextension and rotation of the lower back, could be the major cause of lower back injuries. The purpose of this investigation was to further the understanding of the effect of selected kinematic variables on ball speed as well as the magnitude and rate of build up of ground reaction forces during the period of FFC.

METHODS: Data Collection: Sixteen elite male fast bowlers (mean ± standard deviation: age 22.3 ± 2.6 years; height 1.88 ± 0.08 m; body mass 81.5 ± 6.1 kg) performed 6 maximum pace deliveries in an indoor practice facility, using their full length run-up. Kinematic data were collected using an 18 camera Vicon Motion Analysis System (OMG Plc, Oxford, UK) operating at 300 Hz. Forty-three 14 mm spherical retroflective markers were attached to each subject, positioned over bony landmarks, in accordance with an 18 segment full-body model developed for the analysis of fast bowlers. An additional marker, in the form of a 1.5 cm x 15 cm patch of 3M Scotch-Lite reflective tape was attached to the ball, enabling ball speed and the instant of ball release to be identified. A Kistler force plate (900 x 600 mm), with a thin layer of artificial grass fixed to its surface, was used to measure ground reaction forces during the FFC phase (1008 Hz).

Data Analysis: Trials were manually labelled and the best three selected for each bowler (maximum velocity with minimal marker loss). The instants of back foot contact (BFC), FFC and ball release (BR) were identified in the kinematic data and used to synchronise the force data. Kinematic data were filtered using a fourth-order low pass Butterworth filter (double pass) with a cutoff frequency of 30 Hz.

Kinematic parameters were determined for each bowler, describing: run-up velocity (at BFC); plant angle (Figure 1) and motion of the front knee during FFC; the manner of footplant at the instant of FFC (heel strike or otherwise); and the ball release speed. Subject-specific segmental properties were determined using Yeadon's (1990) geometric model.

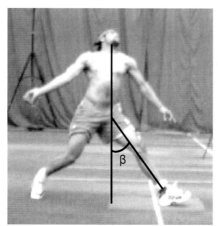

Figure 1: Illustration of the plant angle (β) at the instant of front foot contact.

Descriptive parameters calculated for the ground reaction forces (Figure 2) consisted of: horizontal (braking) and vertical peak forces; peak vertical loading rate (PVLR) and peak initial loading rate (PILR) (as defined by Hurrion, et al., 2000); peak horizontal loading rate (PHLR); and horizontal and vertical impulse up to BR. Correlations between parameters were assessed using Pearson's correlation coefficient.

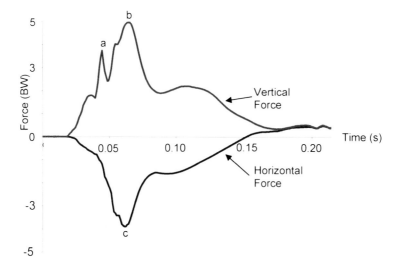

Figure 2: A typical force trace during the front foot contact phase.
[a – initial peak vertical force; b – peak vertical force; c – peak horizontal force]

RESULTS:

Table 1 Details of significant correlations (P<0.02)

Variable	Correlated to	R	P-Value
Ball Speed	Run-up Speed	0.449	0.001
	Plant Angle	0.450	0.001
	Knee Flexion up to BR	-0.434	0.002
	Knee Flexion in Initial 15 Frames of FFC Phase	-0.497	0.000
	Horizontal Impulse to BR	0.561	0.000
	Vertical Impulse to BR	0.339	0.018
Peak Vertical Force	Knee Flexion in Initial 15 Frames of FFC Phase	0.364	0.011
	Heel Strike? (1 = Yes; 0 = No)	-0.486	0.000
Peak Horizontal Force	Run-up Speed	0.378	0.008
PVLR	Heel Strike? (1 = Yes; 0 = No)	-0.618	0.000
PILR	Heel Strike? (1 = Yes; 0 = No)	-0.481	0.001
PHLR	Knee Flexion in Initial 15 Frames of FFC Phase	-0.525	0.000
	Heel Strike? (1 = Yes; 0 = No)	-0.623	0.000

DISCUSSION: Previous research into ground reaction forces during the FFC phase of the fast bowling action have reported mean peak vertical and braking forces of 3.8 – 9.0 BW and 1.8 – 3.54 Body Weights (BW), respectively (Hurrion, et al., 2000). The mean peak forces of 6.66 (± 1.48) BW vertically and 4.57 (± 0.90) BW horizontally in this group of bowlers, although higher than the mean values reported in previous studies, are similar to those of individuals reported in the literature. These relatively high peak forces may be attributed to the elite nature of the subjects (mean ball speed 35.10 (± 1.77) $m.s^{-1}$ and mean approach speed 5.83 (± 0.60) $m.s^{-1}$).

The goal of a fast bowler is to release the ball as quickly as possible, whilst retaining accuracy, ideally using a technique which does not predispose them to injury. Within this group of fast bowlers, correlations were observed between ball release speed and: run-up speed; plant angle; and the motion of the front knee during the FFC phase.

Ball speed was positively correlated with run-up speed; bowlers with a faster approach speed have more kinetic energy which can be potentially transmitted to the ball. However, run-up speed was also related to the peak horizontal force produced during FFC. A positive correlation was also observed between ball speed and plant angle. By adopting a wide stance (characterised by a large plant angle), bowlers are well positioned to generate large forces with their front leg, enabling run-up speed to be converted into rotational energy about the plant foot.

Bowlers exhibiting greater knee flexion between the instant of FFC and BR tended to release the ball at lower speeds. This appears to support previous relationships reported between release height (as a percentage of standing height) and ball speeds (Bartlett, et al., 1996). However, no direct relationship between the knee angle at BR and ball release speeds was observed.

Knee flexion during the first 15 frames of the FFC phase was correlated with increased peak vertical force and decreased PHLR, as well as ball speed. This appears to contradict previous suggestions that knee flexion during the FFC phase dissipates ground reaction forces (Foster, et al., 1989; Portus, et al., 2004). However, the correlations observed in this investigation should remain in context – they represent relationships betweeen variables in a group of bowlers, rather than the effect of changing the kinematics of a particular bowler. Further investigation into the inter-relationships between these variables is warranted and should investigate the effect of the front knee kinematics on the forces above the knee, as well as at the foot-ground interface.

These relationships observed between kinematic factors and ball release speed are perhaps explained by the correlations observed between the horizontal and vertical impulses (between the instant of FFC and BR) and ball speed. Smaller peak loading rates (PVLR, PILR and PHLR) and peak vertical force were associated with the use of a heel strike technique at the instant of FFC. The moderate correlations observed in this investigation are indicative of the multifactorial nature of the generation of ball speed and ground reaction forces.

CONCLUSION: This study shows that ball release speeds are related to run-up speed, plant angle and the motion of the front knee during the period of FFC. The use of a heel strike technique at the instant of FFC was found to decrease peak vertical force and loading rates. Further investigation is required to establish the effect of changing particular kinematic variables on the performance of an individual subject, rather than relationships within a population. Future work should investigate peak forces and loading rates above the knee as well as at the foot-ground interface.

REFERENCES:

Bartlett, R.M., Stockill, N.P., Elliott, B.C. and Burnett, A.F. (1996). The biomechanics of fast bowling in men's cricket: A review. *Journal of Sports Sciences*, 14, 403-424.

Bell, P. (1992). Spondylolysis in fast bowlers: principles of prevention and a survey of awareness among cricket coaches. *British Journal of Sports Medicine*, 26(4), 273-275.

Burnett, A.F., Khangure, M.S., Elliott, B.C., Foster, D.H., Marshall, R.N. and Hardcastle, P.H. (1996). Thoracolumbar disc degeneration in young fast bowlers in cricket: a follow-up study. *Clinical Biomechanics*, 11(6), 305-310.

Elliott, B. C., Hardcastle, P., Burnett, A., and Foster, D. (1992). The influence of fast bowling and physical factors on radiological features in high performance fast bowlers. *Sports Medicine Training and Rehabilitation*, 3(2), 113-130.

Elliott, B.C., Davis, J.W., Khangure, M.S., Hardcastle, P. and Foster, D. (1993). Disc degeneration and the young fast bowler in cricket. *Clinical Biomechanics*, 8(5), 227-234.

Gregory, P.L., Batt, M.E. and Kerslake, R.W. (2004). Comparing spondylolysis in cricketers and soccer players. *British Journal of Sports Medicine*, 38, 737-742.

Hurrion, P.D., Dyson, R. and Hale, T. (2000). Simultaneous measurement of back and front foot ground reaction forces during the same delivery stride of the fast-medium bowler. *Journal of Sports Sciences*, 18, 993-997.

Newman, D. (2003). *A prospective survey of injuries at first class counties in England and Wales 2001 and 2002 seasons.* Paper presented at the Science and Medicine in Cricket: A collection of papers from the Second World Congress of Science and Medicine in Cricket, Cape Town, South Africa.

Orchard, J. W., James, T. & Portus, M. R. (2006). Injuries to elite male cricketers in Australia over a 10-year period. *Journal of Science and Medicine in Sport*, 9(6), 459-467.

Portus, M. R., Mason, B. R., Elliott, B. C., Pfitzner, M. C., and Done, R. P. (2004). Technique factors related to ball release speed and trunk injuries in high performance cricket fast bowlers. *Sports Biomechanics*, 3(2), 263-283.

Ranson, C.A., Kerslake, R.W., Burnett, A.F., Batt, M.E. and Abdi, S. (2005). Magnetic resonance imaging of the lumbar pine in asymptomatic professional fast bowlers in cricket. *The Journal of Bone and Joint Surgery (Br)*, 87-B(8), 1111-1116.

Ranson, C., Burnett, A., King, M., Patel, N., and O'Sullivan, P. (2008). The relationship between bowling action classification and three-dimensional lower trunk motion in fast bowlers in cricket. *Journal of Sports Sciences*, 26(3), 267-276.

Yeadon, M.R. (1990). The simulation of aerial movement – II: A mathematical inertia model of the human body. *Journal of Biomechanics*, 23, 67-74.

ISBS 2009

Oral Session S15

Winter Sports

PERCEIVED DIFFERENCES IN SKATING CHARACTERISTICS RESULTING FROM THREE CROSS SECTIONAL SKATE BLADE PROFILES

Marshall Kendall, Scott Foreman, Philippe Rousseau and Blaine Hoshizaki
International Ice Hockey Research Academy, University of Ottawa, Ottawa, Canada

The purpose of this study was to document differences in perceived skating characteristics resulting from three unique cross sectional skate blade profiles. Sixteen (n=16) University level hockey players were used in this double blind study looking at the perceived performance differences of four different skate blade profiles. No significant differences were found between skate blade profiles, preferred skate blade profile and time to complete given drills. Future research should look at different blade profiles and their interaction at ice level.

KEY WORDS: skate, blade, hockey

INTRODUCTION: The ability to skate faster, start and stop more proficiently, and be more agile than your opponent is a big advantage in ice hockey (Hoshizaki et al., 1989). The skate blade's ability to bite or hold on the ice during high intensity skills is crucial to allow the skater to perform required agility skills during a hockey game (Federolf et al., 2008). Very little research looking at cross sectional profiles has been done with the goal of optimizing the performance of skate blade/ice interface. Blackstone Sports has developed a new and innovative method for resurfacing skate blade profiles using a flat-bottom V shape rather than the more popular circular-shaped cut. The objective of this study was to document differences in perceived skating characteristics resulting from three different unique cross sectional skate blade profiles.

METHODS: Data Collection: Eighteen university level male hockey players (87.32 ± 6.04 kg) were recruited for this study. Four profiles were investigated: the subject's original profile, *105 x .05*, *90 x .75* and *80 x 2 (where the first number represents the width of the blade(in 100th of an inch) and the second number representing the depth of the cut (in 1000th of an inch))* .

Skating drills were also developed to isolate the important skating performance characteristic involved in ice hockey (Appendix B). These drills included starting, stopping, agility, short radius and long radius cornering both forward and backwards as well as acceleration and high velocity. These drills were also timed to identify performance differences between specific skills.

Drill description

#1 – Agility: Used Progressive high speed cornering, starts and stops with changes in direction around obstacles.
#2 – Power: Involved players taking tight corners. Quick acceleration-deceleration profiles which forced players to push skate blade cornering capabilities.
#3⁻ – Start/Stop: Demanded the player accelerate and decelerate with Blade edge manipulation to help evaluate blade "bite."
#4 – Control: evaluated agility in backwards, lateral and forward skating following an arc in both directions.

Each participant repeated the drill three times under each of the different blade profiles including their own pattern. It should be noted that the players profiles were determined by asking each subject what profile they were using. Eight of the skaters identified a radius of ½ inch as their preferred sharpening profile. The others ranged from *3/8* inch radius to *1 ½* inch radius.

A five point Likert scale (performance perception) questionnaire consisting of eleven questions was developed to measure the athletes perceived performance differences during the various drills using the four different blade profiles. Upon completion of each drill, participants were asked to fill out the performance questionaire. Players were also encouraged to add their own comments regarding the skate blade performance during the skating drills.

Timing data was also collected for each participant during each drill for all four cross-sectional blade profiles.

A completely crossed repeated measures design with randomly assigned conditions was used. This study used a double blind protocol, meaning that neither the subjects nor the testers knew which profile was being used at any time in the study. This protocol was chosen to ensure that no artificial bias by either the subjects or the experimenters would influence the results.

Questionnaire Analysis:

The answers from the questionnaire were broken down in to 3 categories; negative, neutral and positive answers. Negative answers were defined by an answer to a question being either 'strongly disagree' or 'disagree', neutral answers were defined by 'neither agree nor disagree' and positive answers defined by answering either 'agree' or 'strongly agree'. These categories were then summed together for each of the four blade profiles. A Performance Index with this data was determined by dividing the number of positive responses by the number of negative responses.

Statistical Analysis:

A spearman correlation was performed in order to see if there was any correlation between, the blade profiles, the preferred blade profile as chosen by the skater and the combined timing data.

RESULTS:

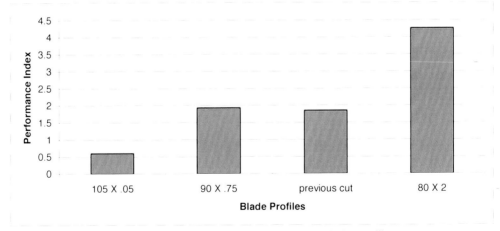

Figure 1: Shows the performance index scores for each blade profile.

The perception questionnaire identified the 80 x 2 profile as the one that player felt performed the best during the four drills (figure 1). It resulted in a performance index of 4.26. The next profile was the 90 x .75 at 1.93 followed by the prior profile at 1.85 and the 105 x .05 only scoring a score of .60. These results are telling in that the players prior cut was third in perceived performance, which is unusual as players become very attached to their sharpening profile and often choose their existing profile over novel profiles.

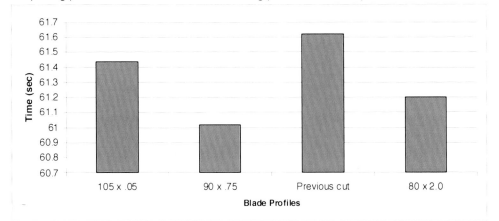

Figure 2: Shows the total average time to complete all four drills for each blade profile.

Timing data was collected on all subjects during all drills for each of the four blade profiles. It was thought that the players would perform faster on their previous profile due to the fact that they are familiar with this profile. The sum of the averaged times (figure 2) shows that all of the new blade profiles outperformed the previous blade profiles of the skaters. The 90 x .75 profile had the best combined total of 61.02 seconds followed by the 80 x 2 profile at 61.20 seconds, 105 x .05 at 61.44 and last was the subject's previous profile at 61.62.

At the end of the test session each player was asked to rate each profile in order of perceived performance and then to choose a profile as their final sharpening. The highest preference was their previous cut (38%) and this was expected as it is the profile they are most familiar. However the 90 x .75 rated second with 34% ahead of the 80 x 2 at 21%. When the subjects chose a profile to continue playing with after they left the study we found an interesting result. Tied for first choice was the 80 x 2 profile and their previous profile (29%). The third choice after their original profile was 90 x .75 (24%) as expected with the lowest choice being the 105 x .05 (18%).

After running spearman correlation, there were no significant differences found between the skate blade profile, the preference chosen by the skater and the timing data.

DISCUSSION: The results shown with the performance index demonstrated some very interesting findings. Surprisingly, the players previous cut was ranked third in perceived performance, which is unusual as players tend to become very attached to their sharpening profile and are often more resistant to equipment change. This being said, when given a choice of blade profiles that they could continue playing with, the player's previous cut ranked tied for first, proving that there may be a comfort zone among hockey players when it comes to the skate blade profile that they are used to playing with.

Though there were no significant differences in the timing data between the four blade profiles, it was quite surprising to see that the players' previous cut actually performed the worst. It was believed prior to the study that the 105 X .05 profile might perform the best in terms of time (speed) since this cut offered the least amount of resistance (according the Blackstone Inc.), however this was not the case. A cause for the higher than expected timing data could be related to the drills used in the study. The drills did not really include pure gliding, which would likely be this profile's forte. It is believed that this profile did not give the players the balance they required to complete the drills efficiently, as it was described by many of the skaters as "too slippery".

The players expressed that there was an easily perceived difference in performance in the profiles tested. When we look at the timing data however, it would seem that the preferred profile chosen by the players didn't necessarily match their fastest time for a given drill. This is probably because in ice hockey overall skating speed is not necessarily the highest predictor of skating performance. Speed is important,however, it is unlikely the deciding factor if it meets a certain performance threshold. In this case, all profiles may have met the speed performance threshold. Once the speed threshold was met, other performance characteristics like purchase and control became more important in determining skating performance. This may also be a reflection of the drills chosen to test the performance of the different profiles. The drills were chosen to reflect the performance nature of the game of ice hockey and were not specifically designed to distinguish the contribution of the different sharpening profiles to velocity. This was important in order for the new innovation to be recognized as a real innovation that will make a difference in skating performance. The results definitely supported this with both the questionnaire and post test interviews. A number of subjects remarked that even when they chose their previous blade profile the Blackstone sharpening was superior to their existing sharpening.

CONCLUSION: As in most well crafted research this study created more questions then it answered. Even though we were able to establish that the Black Stone profile system provided real performance benefits the interaction between the blade profile characteristics and skating performance benefits remains vague.

It seems clear that there is an opportunity to tie the runner profile characteristics to specific skating characteristics. Research needs to be undertaken to identify the relationship between skate blade profile characteristics and specific improvements in skating performance. Presently the vast majority of ice hockey player use a ½ inch radius because it is what is available.

Future research

The first part would entail the development of precise measures to characterize the physical interaction between the various blade profiles (width of the flat bottom, the height of the edge and the angle of the edge) and the ice. There is also a need to document the expected performance characteristics of the profiles being developed. The second part would look at the development of a more refined test protocol to document the perceived performance benefits as well as the actual skating improvements. Once these relationships are understood it would allow Black Stone the opportunity to develop profiles for player using different styles of skating, different weight and needs.

REFERENCES:

Federolf, P.A., Mills, R. and Nigg, B., (2008), Ice Friction of Flared ice hockey skate blades, Journal of Sports Sciences, September; 26(11); 1201-1208.
Hoshizaki T.B., Kirchner R.K. & Hall K. (1989). Kinematic analysis of the talocrural and subtalar joints during the hockey skating stride. In C.R. Castaldi and E.F. Hoerner (Eds.), *Safety in Ice Hockey, ASTM STP 1050*, pp. 141-149.

Acknowledgement: Special thanks to Black Stone Sports for their support in this study.

ANALYSIS OF COMBINED EMG AND JOINT ANGULAR VELOCITY FOR THE EVALUATION OF ECCENTRIC/CONCENTRIC CONTRACTION IN SKIING

Nicola Petrone[1], Giuseppe Marcolin[2]

Department of Mechanical Engineering University of Padova, Italy [1]
Department of Anatomy and Physiology University of Padova, Italy [2]

The purpose of this study was the introduction of a method combining the joint angular velocity and the EMG signals for a biomechanical evaluation of the eccentric/concentric contraction patterns of major muscles used in alpine skiing. The adopted approach was applied to three types of slalom courses: racing, training and recreational slaloms. The introduction of a pseudo-Muscular-Work allowed to define a Ratio between Eccentric and Concentric muscular activation states that, compared to simple on-off timing criteria, resulted to reveal more consistently the amount of eccentric muscle mechanical work.

KEY WORDS: joint angular velocity, EMG, eccentric, concentric, index.

INTRODUCTION: In alpine skiing, the knowledge of the type and intensity of muscle contraction during racing or recreational skiing events is able to give important information about the skiing technique, the effect of equipment tuning and the proper training program and exercise definition. Several authors have studied muscle contraction patterns in alpine skiing: Tesch (1995) described the physiological demands in competitive alpine skiing and the physiological profile of elite skiers, Berg et al. (1999) studied concentric and eccentric muscle use in relation to positions and angular movement velocities of the hip and knee joints, Hintermeister et al. (1995) investigated muscle activity in slalom and giant slalom finding a similarity in muscle activation between the two disciplines with ample evidence of co-contraction suggesting a quasistatic component to skiing. Hoppeler et al. (2009) defined a method for applying an eccentric ergometer suitable for evaluating and training the eccentric activation of anti-gravitational muscles. Considering these previous investigations and the variables analyzed we decided to develop a method suitable for quantifying eccentric and concentric muscle contractions actions. This analysis was applied in two skiing sessions to three types of courses, namely racing slalom, training slalom and free field recreational slalom.

METHODS: Data Collection: Two test sessions were performed to study three different skiing conditions with the same tester. In Session1, a training slalom on a medium steep slope and a free-field recreational slalom were recorded; in Session2, a racing slalom on a steep, icy snow was recorded. Data of each course were recorded by means of a 16 channels PDA PocketEMG (BTS-Italy) placed on the chest of the skier (145x95x20mm, mass 0.3kg). In Session1, eight channels were used to collect EMG data of the right leg and the back of the subject (inter-electrode distance 25mm). Muscles analyzed were: Erector Spinae, Gluteus Maximum, Vastus Lateralis, Vastus Medialis, Biceps Femoris, Semimembranosus, Tibialis Anterior and Gastrocnemius Medialis. Two Biometrics goniometers were laterally placed at the right knee and right hip. Four more channels were employed for the acquisition of two strain gauged accelerometers stuck on the shin guards. In Session2, the eight EMG channels were symmetrically placed at the two sides of the tester: Vastus Medialis, Biceps Femoris, Tibialis Anterior and Gastrocnemius Medialis were recorded. The two Biometrics goniometers were placed on both knees and all data were synchronously recorded at 1Khz.

A professional ski tester was involved in the study. Before the tests he was asked to read and sign an informed consent. In Session1, ten short poles were placed in the snow (span 4 m, pace 12 m, named S0-S9 with odd numbers external turns for right leg) along a medium steep slope. Two additional poles at a span of 60 m were employed to mark the free slalom area. In Session2, 26 short poles were placed in the snow (*span 4 m, pace 12 m*, named S0-

S25 with odd numbers external turns for right leg) along a steep slope.

Data Analysis: Analysis of courses between poles was based on the identification of impact instants of the leg with the pole shown by the shin guard accelerometers, in such a way that internal/external turns of each leg could be easily identified. EMG raw signals of the recorded muscles were rectified, integrated with a mobile window of 200ms and filtered with a 5Hz low pass 3rd order Butterworth filter.

Aim of the work was the definition of a method for muscle eccentric/concentric contraction analysis including not only the sign of the angular velocity, but also the instantaneous level of muscle activation that is experimentally expressed by an integrated iEMG signal. To quantify knee flexor and extensor muscle activity and to relate them to the knee joint angular velocity, data were therefore analyzed as follows.

The relative angle between thigh and shank axes (θ) was set at 180° with knee fully extended and its time derivative, the angular velocity (ω) of the knee joint, was coherently assumed positive during extension and negative during flexion. Assuming that the muscle Activation signal (iEMG) is correlated to the Force exerted at the muscle insertion, and that the force multiplied by the tendon lever arm is in turn producing the muscle Torque applied to the joint, then the angular velocity (after low pass filtering at 5Hz) was multiplied by the integrated EMG signal (iEMG) of flexors muscles (Biceps Femoris and Semimembranosus) and extensors muscles (Vastus Medialis and Vastus Lateralis) to obtain a new set of curves: the product between Activation (iEMG) and joint angular velocity (ω) was named *pseudo Muscular Power* (pMP). The "pseudo" prefix is due to the complex relationship between Activation and Force, the nonlinear variation of the tendon Lever arm with the Joint relative angle (θ) and the motion at other proximal/distal joints (like the hip or the ankle) in the case of bi-articular muscles. Nevertheless, the *pseudo Muscular Power* (pMP) was supposed to be possibly adopted as a meaningful parameter to discriminate eccentric and concentric muscle actions from a mechanical point of view.

Coherently with the angular convention, the muscle knee Torque contribution was set positive if acting in extension, both for extensor and flexor muscles: therefore the pMP resulting curves showed positive parts corresponding to concentric contractions and negative parts for eccentric contractions for all the muscles. Subsequently, the area enclosed between the positive part of the pMP curve and the time axis, calculated after integration versus time on a pre-defined time period, was associated with a *concentric pseudo Muscular Work* (pMW_C); conversely the area enclosed between the negative part of the pMP curve and the time axis was associated with an *eccentric pseudo Muscular Work* (pMW_E). Finally the ratio R_{MWEC} between pMW_E and pMW_C was introduced as an immediate index useful to estimate in a certain interval of time the relative amount of eccentric and concentric muscular work.

A different method for evaluating the eccentric/concentric contraction amount, based on the simple analysis of the sign of angular velocity, was applied to find how much time the muscle spent in a concentric or eccentric phase during a given time period. Looking at the angular velocity sign (positive for concentric, negative for eccentric contraction), the ratio R_{TEC} of eccentric and concentric contraction intervals was calculated. Both the two indexes R_{MWEC} and R_{TEC} were calculated for Session 1 over the interval between S2 and S7 poles as well as over the whole 60 m free slalom duration for each course; for Session2 the indexes were calculated from the 5th to the 15th pole.

RESULTS: Knee relative angle (θ), angular velocity (ω), pseudo muscular concentric and eccentric power (pMP_C and pMP_E) of each muscle were calculated for the whole 60 m free slalom, then for the training slalom in the interval between S2 and S7 poles, and finally between S5 and S15 poles for the racing slalom as reported in Figure 1 for vasti muscles as an example.

In Table 1 the values of R_{MWEC} and R_{TEC} are presented for the three types of slalom for direct comparison.

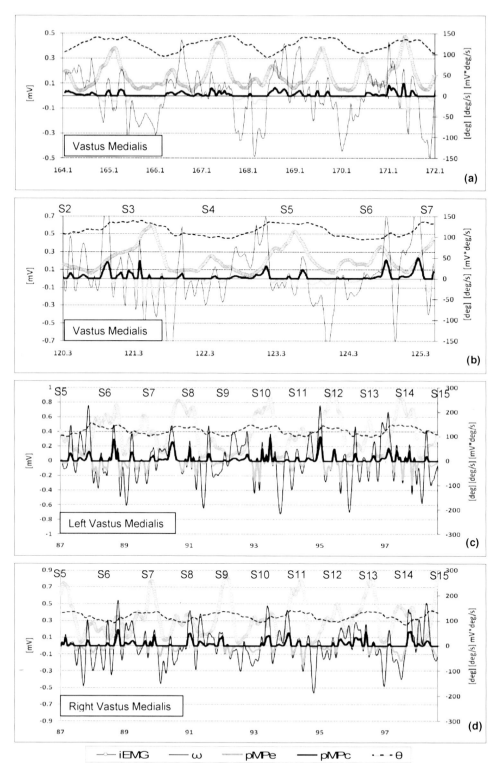

Figure 1. Results of the analysis of Vastus Medialis in the three different skiing conditions: (a) free slalom, (b) training slalom, (c & d) racing slalom.

Table 1 R_{TEC} versus R_{MWEC} for eccentric and concentric muscle evaluation in the three skiing conditions.

60 m FREE SLALOM	TIME = 7.52 sec	
	R_{MWEC}	R_{TEC}
Vastus Medialis	1.07	1.08
Vastus Lateralis	1.06	
Biceps Femoris	0.93	0.93
Semimembranosus	0.57	
TRAINING SLALOM	TIME = 5.12 sec	
	R_{MWEC}	R_{TEC}
Vastus Medialis	0.66	0.78
Vastus Lateralis	0.57	
Biceps Femoris	1.85	1.28
Semimembranosus	1.03	
RACING SLALOM	TIME = 11.50 sec	
	R_{MWEC}	R_{TEC}
Left Vastus Medialis	1.07	0.96
Right Vastus Medialis	0.95	1.06
Left Biceps Femoris	1.25	1.04
Right Biceps Femoris	1.23	0.95

DISCUSSION: The ratio between eccentric and concentric knee muscles action elaborated with the two methods showed appreciable differences. The ratio between eccentric and concentric contraction calculated in terms of time duration using angular velocity (R_{TEC}) showed similar values for both flexors and extensors in the free slalom with a minimal predominance of the eccentric action. In training slalom there is a great eccentric contraction of flexors muscles and concentric contraction of extensor muscles. Finally in racing slalom the eccentric and concentric knee muscle actions were very close. R_{MWEC} allowed to appreciate an increase of the eccentric action of the biceps femoris in training and racing slalom and this result may have relevance for proper training and injury prevention.

The introduction of concentric and eccentric pseudo Muscular Work (pMW_C and pMW_E) was useful to better understand the intensity of the eccentric and concentric activation. In fact, eccentric and concentric contraction calculated in terms of time duration using only the sign of angular velocity missed important functional information: how hard muscle fibres were working in that condition. The consequence is that a muscle which spent short times in eccentric action with a high activation shouldn't be trained as a muscle which spent for example short times eccentrically but with a low level of activation.

CONCLUSION: A marked predominance of eccentric over concentric muscle actions was not observed in the quadriceps muscles. The evaluation of concentric and eccentric muscle action during three types of slalom can give important information for athletes and trainers in the development of specific training programs in term of intensity of the stimulus, time duration and knee angular velocity. Further investigations with a higher number of professional skiers and with a higher number of courses could help to better understand knee flexors and extensors eccentric and concentric actions with respect to time and intensity of activation.

REFERENCES:

Tesch PA. (1995), Aspects on muscle properties and use in competitive Alpine skiing, *Med Sci Sports Exerc*. 27: 310-314, 1995.

Berg HE, Eiken O. (1999), Muscle control in elite alpine skiing. *Med Sci Sports Exerc*. 31: 1065-1067.

Hintermeister RA, O'Connor DD, Dillman CJ, Suplizio CL, Lange GW, Steadman JR. (1995), Muscle activity in slalom and giant slalom skiing. *Med Sci Sports Exerc*. 27: 315-322.

Hoppeler H., Vogt M., (2009), Eccentric exercise in Alpine skiing, *Science & Skiing IV*, pp. 33-42, Meyer & Meyer Sport.

EFFECT OF SKI BOOT TIGHTNESS ON SHOCK ATTENUATION TIME AND JOINT ANGLE WITH ANTERIOR-POSTERIOR FOOT POSITIONING IN DROP LANDINGS

Shinya Abe and Randall L. Jensen

Dept. HPER, Northern Michigan University, Marquette, MI, USA

Eight recreational skiers performed drop landings barefoot and wearing ski boots. The purpose of this study was to examine if different ski boot tightness affects the shock attenuation time, and minimal joint angles during landings. Shock attenuation time and joint angles were obtained via an AMTI force plate and video analysis of subjects' drop landings. Repeated Measures Analysis of Variance (ANOVA) was used to analyze differences between means for the dependent variables. Our results indicate that differences in ski boot tightness do not affect shock attenuation time during landing. Results also indicate that minimal ankle joint angle can be affected by different ski boot tightness. This implies that a tighter ski boot condition can restrict dorsiflexion of the ankle during landing.

KEYWORDS: anterior cruciate ligament, ski boot, knee, ankle, alpine skiing

INTRODUCTION: Alpine skiing is one of the most popular winter sports in the world. Despite the popularity of the sport, a large number of injuries occur each year. In past decades, injuries to the lower extremities excluding anterior cruciate ligament (ACL) have been reduced as the quality of ski bindings and ski boots have been improved (McConkey, 1986). In contrast, the incidence of ACL injuries have increased (McConkey, 1987; Pujol & Blanchi, 2007).

Several mechanisms have been suggested as causes of an ACL injury during skiing. According to McConkey (1987), a small balance disturbance during the landing phase following a jump in alpine skiing can easily cause the skier to fall backwards. Once skiers fall backward, quadriceps muscles contract forcefully to regain balance. This large quadriceps contracture pulls the tibia anteriorly via the patellar tendon (Geyer & Wirth, 1991; McConkey, 1986). This anterior movement of tibia can create strong enough tension of the ACL to rupture it (McConkey, 1987).

Previous researchers have also noted that off-balance conditions such as backward falling can create an impact force that drives the ski boots anteriorly with respect to the femur, resulting in rupturing of the ACL (Hame et al., 2002; McConkey, 1986; 1987). This is termed "boot induced" mechanism of injury.

Despite the fact that many studies were conducted to find correlations between ACL injuries and movements involved in alpine skiing, the effect of ski boot tightness (by changing the number of hooks on the buckles) on ACL injury has not yet been examined. Therefore, the main aim of this study was to examine if different ski boot tightness can affect shock attenuation time in drop landings. The secondary purpose of this study was to observe if differences in ski boot tightness can alter minimum knee and ankle joint angles during recovery time.

METHODS: Eight recreational skiers volunteered as subjects (three female and five male; mean ± SD: 23.3±4.0 years; height = 172.6±11.3 cm; body mass = 69.1±8.9 kg). Subjects completed a Physical Activity Readiness Questionnaire and were excluded if they reported previous history of lower extremity injury or disorder. Approval for the use of Human Subjects was obtained from the Institutional Review Board prior to commencing the study. Informed consent forms were read and signed by each subject prior to data collection.

Warm-up prior to the test criteria via a sledge drop landing consisted of at least 3 minutes of low intensity work on a cycle ergometer. This was followed by static stretching including one exercise for each major muscle group with static stretches held from 12-15 seconds. Following the warm-up and stretching exercises, the subjects were allowed at least 5 minutes rest prior to beginning the test.

The subject was placed into the chair of the 30° inclined sledge (Figure1). Each subject was instructed to land on the platform from drop of a height of 70 cm (perpendicular to the sledge). Two practice jumps were performed to allow the subject to become accustomed to the test criteria. Following the two practice jumps, the subject was asked to perform the drop landings with four foot conditions with three landing locations; anterior, neutral, and posterior locations (Figure1). Foot conditions were barefoot, ski boots loose (none of hooks of the top two buckles of the ski boots were engaged), ski boots medium (one to two hooks were engaged), and ski boots tight (subjects were asked to engage as many hooks as possible). A total of twelve landings were performed by each subject: barefoot at anterior (BA), neutral (BN), and posterior (BP), ski boots loose at anterior (LA), neutral (LN), posterior (LP), the medium at anterior (MA), neutral (MN), posterior (MP), and tight at anterior (TA), neutral (TN), posterior (TP). The landings using ski boots were randomly ordered except the barefoot conditions were always performed first. A one minute rest interval was maintained between each landing.

A force platform (OR6-5-2000, AMTI, Watertown, MA, USA) was mounted on the 60° inclined landing plate (Figure 1). Ground Reaction Force data were collected at 2000 Hz to measure shock attenuation time, real time displayed and saved with the use of computer software (NetForce 2.0, AMTI, Watertown, MA, USA) for later analysis. The shock attenuation time was measured as time expressed in milliseconds between initial contact of the landing and the first minimal vertical reaction force following the peak vertical reaction force after initial contacts.

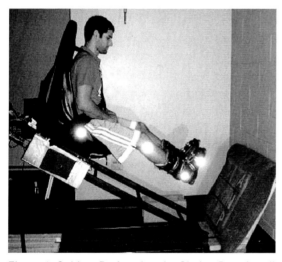

Figure 1: Subject Performing the Sledge Drop Landing.

Video of the landings was obtained at 60 Hz from the right side using 1 cm reflective markers placed on the seat near the greater trochanter, fibular head, lateral malleolus and the fifth metatarsal for the barefoot condition, the tip of the lowest buckle of the ski boot, near the fifth metatarsal for three ski boot conditions. Markers were digitized using automatic digitizing software Motus 8.5 (Vicon/Peak, Centennial, Co, USA) and minimal knee, and ankle joint angles were determined after data were smoothed using a fourth order Butterworth filter (Winter, 1990). Relative angles for knee and ankle were measured with a smaller joint angle value indicating more flexion of the knee and more dorsiflexion at the ankle.

Kinetic and kinematic data were matched using a synchronizing signal to determine the initial contacts. A cubic spline data interpolation program was written (using Matlab 6.5) to time normalize the data files.

Shock attenuation time and minimal knee and ankle joint angles were the dependent variables studied. A Two-Way 4 (type of boot condition) x 3 (foot landing condition) Repeated Measures Analysis of Variance (ANOVA) was used to analyze differences between means for the dependent variables. An alpha level of 0.05 was used to determine significance for all comparisons. In the case of significance, follow-up comparisons were made using the

Bonferroni adjustment. Statistical analysis was completed using SPSS version 16.0 (SPSS Inc., Chicago, IL).

RESULTS: As shown in Table 1, there were no differences (p>0.05) in the shock attenuation time. Table 2 shows that Min Knee angle at all landing locations, Anterior, Neutral and Posterior were different from each other (p < 0.05). As shown in Table 3, minimal (Min) ankle joint angle with ski boots loose condition at all three landing locations were significantly different (p < 0.05) from all other foot and other landing conditions.

Table 1 Shock attenuation time (millisecond), minimal (Min) joint angles (Mean ± SD) for the Knee, and Ankle (N=8).

	Barefoot	Loose	Medium	Tight
Anterior	559±216	519±220	550±186	517±208
Neutral	561±260	475±160	419±98	390±139
Posterior	476±123	393±34	402±37	372±55

Shock attenuation time was measured in

Table 2 Minimal (Min) Knee angle (degree) (Mean± SD) (N=8).

	Barefoot	Loose	Medium	Tight
Anterior [a]	104.8±19.93	94.3±16.09	95.7±15.08	92.5±15.92
Neutral [b]	110.8±24.18	108.3±10.19	111.6±15.66	117.8±17.05
Posterior [c]	127.4±14.96	127.1±18.38	135.5±9.01	136.2±10.55

[a] Indicates significant difference from Neutral and Posterior (p < 0.05).
[b] Indicates significant difference from Anterior and Posterior (p < 0.05).
[c] Indicates significant difference from Anterior and Neutral (p < 0.05).

Table 3 Minimal (Min) Ankle angle (degree) (Mean± SD) (N=8).

	Barefoot	Loose [a]	Medium	Tight
Anterior [b]	97.4±5.82	88.3±6.53	92.1±4.06	93.1±2.46
Neutral	87.3±6.44	81.4±3.21	87.6±3.93	89.3±3.44
Posterior	87.0±4.54	79.3±2.62	83.9±2.85	86.0±2.66

[a] Indicates significant difference from barefoot, medium, tight (p < 0.05).
[b] Indicates significant difference from Neutral and Posterior position (p < 0.05).

DISCUSSION: The first objective of this investigation was to study if the different ski boots' tightness can affect shock attenuation time in drop landings. Our results showed that there are no differences in the shock attenuation time between the different foot conditions with three landing locations. This indicates that ski boot tightness has no impact on the shock attenuation time during the landing phase in alpine skiing. This result could be explained by the lack of involvement or interactions between ski boots, bindings, and skis. Thus, this limitation of our study suggests further investigation of the effect of the ski boot tightness on the shock attenuation time should involve the ski bindings and skis.

Hame et al. (2002) and McConkey (1986) reported that hyperflexion of the knee with the additional tibial torque appear to be mechanisms of injury for the ACL in alpine skiers. The results of our study showed there are no differences in Min Knee angles between different landing locations. This indicates that ski boot tightness may not affect the range of motion of the knee during the landing in alpine skiing. However, Min Knee angle at the Anterior location differed from two other locations. It has been suggested that the ACL is at risk of rupture when the skiers fall backward with knees flexed (Gerritsen et al., 1996; McConkey, 1986; 1987). Since our results showed that subjects flexed their knees more at anterior location than at two other locations, this indicates that landings at anterior location may lead the skiers falling backward by increasing knee flexion.

Malliaras et al. (2006) found that patellar tendinopathy was associated with a reduced range of ankle dorsiflexion and restricted dorsiflexion may alter lower limb landing biomechanics. Our data suggested that engaging more hooks of the ski boot buckles restricts dorsiflexion of the ankles during the landing. Noé et al. (2007) found that wearing ski-boots induced changes in a skiers' postural strategy. Thus, tightening the ski boot buckle may influence balancing techniques during landings due to restriction of dorsiflexion at the ankles.

CONCLUSION: As suggested by previous researchers, backward falling may increase the chance of the ACL injuries. Our data indicates that landing at the Anterior location can increase knee flexion, which likely will result in backward falling. Although our results showed that different ski boot tightness does not affect the shock attenuation time, tightening ski boots can restrict dorsiflexion of the ankle during landing. Further research should involve ski bindings and skies to examine if the ski boot tightness with skis can affect shock attenuation time.

REFERENCES:
Gerritsen, K.G.M., Nachbauer, W., and Bogert, A.J. (1996). Computer simulation of landing movement in downhill skiing: Anterior cruciate ligament injuries. *Journal of Biomechanics, 29*, 845-854.

Geyer, M., and Wirth, C.J. (1991). A new mechanism of injury of the anterior cruciate ligament. *Unfallchirurg, 94*, 69-72. Cited in Hame et al. (2002).

Hame, S.L., Oakes, D.A., and Markolf, K.L. (2002). Injury to the anterior cruciate ligament during alpine skiing: A biomechanical analysis of tibial torque and knee flexion angle. *American Journal of Sports Medicine, 30*, 537-540.

Malliaras, P., Cook, J.L., and Kent, P. (2006). Reduced ankle dorsiflexion range may increase the risk of patellar tendon inury among volleyball players. *Journal of Science and Medicine in Sport, 9*, 304-309.

McConkey, J.P. (1986). Anterior cruciate ligament rupture in skiing: A new mechanism of injury. *American Journal of Sports Medicine, 14*,160-164.

McConkey, J.P. (1987). Mechanisms of knee ligament injuries in alpine skiing. *Canadian Journal of Sport Sciences, 12*, 163-169.

Noé, F., Amarantini, D., and Paillard, T. (2007). How experienced alpine-skiers cope with restrictions of ankle degree-of-freedom when wearing ski-boots in postural exercises. *Journal of Electromyography and Kinesiology, 19*, 341-346.

Pujol, N., and Blanchi, M.P.R. (2007). The incidence of anterior cruciate ligament injuries among competitive alpine skiers: A 25-year investigation. *American Journal of Sports Medicine, 35*, 1070-1074.

Winter, D.A. (1990). *Biomechanics and motor control of human movement* (2nd Ed). New York: Wiley Interscience.

Acknowledgement
Sponsored in part by a Northern Michigan University Faculty Research Grant and a Northern Michigan University College of Professional Studies Grant. The authors would like to thank Amanda Leonard for collecting the descriptive data of the subjects.

THE EFFECT OF WALKING SPEED AND SKILL LEVELS ON ELBOW FLEXION AND UPPER LIMB EMG SIGNALS IN NORDIC WALKING: A PILOT STUDY

Nicola Petrone[1], Manuel Orsetti[1] and Giuseppe Marcolin[2]

Department of Mechanical Engineering University of Padova, Italy [1]
Department of Anatomy and Physiology University of Padova, Italy [2]

Aims of the study were to evaluate the effect of the walking speed on the elbow's range of motion and the EMG activity levels on eight upper limb muscles when performing level Nordic Walking in outdoor sessions. The study involved both skilled Nordic Walking instructors and unskilled beginners to highlight the effect of a correct technical execution on the elbow's flexion angle and the EMG signals. All the subjects performed also level walking tests without poles at the same speeds of the NW tests: the EMG activation levels during walking were taken as control values of each subject to estimate the additional activation due to the poles.

KEY WORDS: Nordic Walking, EMG, elbow flexion-extension angle, walking speed, skill.

INTRODUCTION: Nordic Walking is becoming a popular physical activity due to its claimed advantages on the musculoskeletal and cardiovascular systems related to higher values of upper limb biomechanical involvement and full body energy consumption.

Despite its popularity, there are several publications focusing on lower limb biomechanics (Schwameder 1999, Willson 2001) or on the physiological effects on the full body (Schiffer 2006) but few works focused on the upper limb biomechanics and in particular on the EMG activation patterns of Nordic Walking (NW) compared to normal walking (W) in standard conditions.

In the present work, a protocol for the comparative evaluation of Expert and Beginner subjects walking with or without poles at different speeds was defined and applied to a small number of subjects in order to orient further wider works.

METHODS: Participants: Four subjects were involved in the study: two were National Instructors from the Italian Nordic Walking Association (ANWI) and two were students at the Exercise Science Faculty that had never performed NW and received a short verbal lesson about the NW technique before the tests. All subjects signed an informed consent before the tests.

Table 1 Data about the subjects involved in the study.

Subject	ID	Age	Height [cm]	Mass [kg]	Skilll	Pole length [mm]	Vmax [km/h]
E1	OM	28	170	77	Expert	1150	8,50
E2	ZL	42	180	68	Expert	1200	8,60
B3	GM	31	186	78	Beginner	1250	8,40
B4	ND	23	186	85	Beginner	1250	8,80

Data Collection: The research protocol was based on the possibility of recording outdoor the EMG signals on eight muscles together with the elbow flexion angle at controlled walking speeds. In order to account for the specificity of each subject's anthropometry, gender, skill and training state, the test speeds were referred not to absolute speed values but relative to the maximum sustainable walking speed selected by each subject in a pre-test with poles.

Data of each test trial were recorded by means of a 16 channels PocketEMG (BTS-Italy) portable PC placed on the back of the subject (145x95x20mm, mass 0.3kg). Eight channels were used to collect EMG data with bipolar surface electrodes placed on the muscle bellies at a distance between electrodes of 25mm. The reference electrode was applied on C7. The eight muscles involved in the measure were: triceps brachii caput longus (TBCL); deltoideus posterior (DP); latissimus dorsi (LD); pectoralis major sternal head (PECMSH); trapezius

transversalis (TRM), trapezio ascendens (TRS); obliquus externus abdominis (OEA); erector spinae (ERSL).
A Biometrics® goniometer was placed on the left elbow, and zeroed at fully extended arm: all data were synchronously recorded at 1Khz. A wrist Global Positioning System (GPS) Garmin Forerunner305® was used to measure the speed of each subject during each trial: to avoid any influence on the correct execution of walking, a second operator following closely the subject under test was reading the speed on the GPS and giving instructions for maintaining the walking speed at the preset value. Each trial was filmed with a commercial digital camera at 25Hz from the right side.
The tests were performed in summer on grass surface of a city park in Padova at sea level: a track of 60 m length was marked with cones and covered in the two directions for each speed: in a pre-test the maximum sustainable speed with poles v_{pmax} was selected by each subject. Then, for each trial, the track was walked in one way to familiarize with the speed: on the way back the recording was started. After the forward-backward trial with the poles, the subjects performed at the same speed a trial without poles, again recording the data on the way back. These trials were then performed at 80% of the maximum walking speed v_{pmax} and finally at 60% of the v_{pmax}. Eventually, still wearing the data acquisition system, the subjects performed their isometric Maximal Voluntary Contractions (MVCs) on restraint frames helped by operators.

Data Analysis: Analysis of trials focused only on the way back on the track and, in particular, on the last 30 m of the 60 m track length, to avoid speed settling of initial strides. The electrogoniometer signal at the elbow was used to define the gait cycle as the interval between two consecutive minimum values of the elbow flexion angles.
EMG raw signals of the eight muscles were first rectified, then integrated with a mobile window of 200ms, filtered with a 5Hz low pass Butterworth filter and finally normalized with respect to the Maximal Voluntary Contractions (MVCs) and to the cycle length. The final step after the time normalization was to average 10-15 strides performed in the last 30 m of the track to obtain the mean curves of the different analyzed values (Figure 1.a).
To evaluate the specific effect of using the NW poles, the difference between the peak EMG signal of trials performed with poles (NW) and the peak EMG signal of trials without poles (W) was calculated and used to estimate the increment in muscle activation due to the Nordic Walking with respect to normal Walking: this was named ΔEMG_{MUSCLE} and expressed as %MVC.

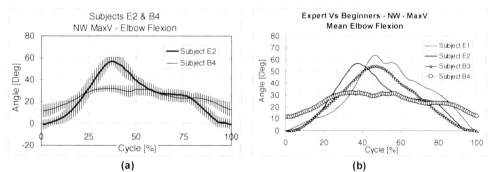

Figure 1: (a) Elbow flexion mean angles from 14 consecutive strides with SD band from E2 and B4. (b) Mean elbow flexion curves from the four subjects during Nordic Walking at maximum speed.

RESULTS: Mean curves of elbow flexion at maximum speed in Nordic Walking from the four subjects are presented in Figure 1.b: an initial evaluation of video and elbow flexion data performed in accordance with NW instructors induced the authors of the present pilot study to focus further comparisons on subjects E2 and B4. In fact, subject E2 was preferred to E1 for his smoother technique, whereas B4 was preferred to B3 due to the surprisingly positive response to verbal lesson of the latter. The elbow flexion mean curves from E2 showed similar peak values with increasing speed, whereas a significant decrease of peak flexion angle and flexion Range of Motion (R.O.M.) at increasing speed were found for subject B4 (Figure 2).

The normalized signals of EMG at the TBCL muscle from expert subject E2 when walking without poles (W) showed specific patterns with lower EMG peaks than walking with poles (NW) (Figure 3). This appeared consistently in the other muscles and justified the introduction of ΔEMG_{MUSCLE} in the data analysis. The walking speed effect on EMG signals was less evident in the expert subject E2 compared to beginner B4 for the TBCL muscle (Figure 4), as well as for other most activated muscles. Peak EMG values for all the measured muscles of two representative subjects E2 and B4 are reported in terms of percentage of MVC in Table 2 together with the corresponding ΔEMG_{MUSCLE} values expressing increments of NW relative to W.

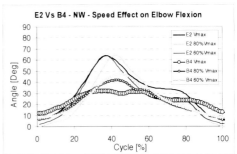

Figure 2. Comparison of the Elbow Flexion between Expert (E2) and Beginner (B4) at different Nordic Walking speeds.

DISCUSSION: The aim of the pilot study was to evaluate the effect of walking speed and skill levels on the elbow flexion and the EMG activations of eight upper limb muscles across compared W and NW paired trials. A major limitation of the work is the limited number of subjects and the variability of poling techniques during NW for beginners as shown in Figure 1.b for subject B3. On the contrary, small values of the intra-subject variability were represented by narrow SD bands for E2 and B4 (Fig. 1.a), encouraging future tests.

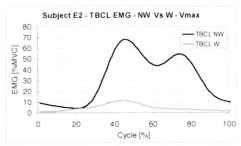

Figure 3: Comparison of the TBCL normalized EMG signal between Nordic Walking and Walking for expert subject E2 at maximum speed.

There was little effect of the walking speed on the elbow R.O.M. for the expert E2, whereas a speed increase reduced the elbow R.O.M. for beginner B4 (Figure 2). The opposite speed effect was evident in beginner B4 for the peakEMG signals that consistently increased with increasing speed: the expert E2 data (Table 2) showed lower peakEMG increments.

Most activated muscles in the expert E2 during NW were respectively LD, TBCL, ERSP, DP and OEA, as shown by peakEMG values expressed as percent of MVC in Table 2. On the other hand, the need of a paired differential analysis with control walking trials at each speed was confirmed by the evidence of non-zero EMG curves during walking (eg. Figure 3).

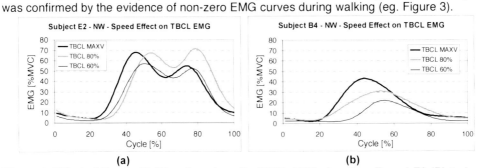

Figure 4: Effect of Nordic Walking Speed on the TBCL EMG signal on Expert E2 (Plot a) and Beginners B4 (Plot b).

Table 2. Table of results for comparison between Expert and Beginner analysed subjects.

Muscle	Values	Expert (E2)			Beginner (B4)		
		MAX V	80%	60%	MAX V	80%	60%
Elbow Angle	Max Flexion	+ 57°	+ 57°	+ 57°	+ 32°	+ 42°	+ 47°
	Min Flexion	- 01°	- 07°	- 05°	+ 12°	+ 05°	+ 07°
TBCL	NWpeakEMG [% of MVC]	68%	71%	57%	43%	31%	22%
	ΔEMG=(NW-W) [% of MVC]	+56%	+63%	+53%	+13%	+23%	+20%
DP	NWpeakEMG [% of MVC]	46%	37%	40%	76%	33%	20%
	ΔEMG=(NW-W) [% of MVC]	+33%	+25%	+30%	+40%	+18%	+12%
LD	NWpeakEMG [% of MVC]	90%	91%	80%	37%	22%	13%
	ΔEMG=(NW-W) [% of MVC]	+71%	+82%	+72%	+01%	+04%	- 10%
PECMSH	NWpeakEMG [% of MVC]	19%	20%	17%	26%	14%	06%
	ΔEMG=(NW-W) [% of MVC]	+13%	+15%	+14%	- 03%	+04%	+01%
TRM	NWpeakEMG [% of MVC]	14%	10%	08%	27%	13%	08%
	ΔEMG=(NW-W) [% of MVC]	+01%	- 01%	- 01%	- 04%	-06%	- 02%
TRS	NWpeakEMG [% of MVC]	07%	06%	06%	45%	40%	22%
	ΔEMG=(NW-W) [% of MVC]	+04%	+04%	+04%	- 10%	+16%	+05%
OEA	NWpeakEMG [% of MVC]	43%	36%	30%	15%	10%	04%
	ΔEMG=(NW-W) [% of MVC]	+06%	+05%	+04%	- 06%	+03%	00%
ERSL	NWpeakEMG [% of MVC]	51%	45%	38%	60%	51%	31%
	ΔEMG=(NW-W) [% of MVC]	+00%	+01%	+07%	-11%	+17%	- 02%

Furthermore, very low values of ΔEMG$_{MUSCLE}$ for muscles like ERSL and OEA suggested that NW has an incremental effect for well trained subjects only on specific muscles like LD, TBCL and DP. The beginner experienced much lower increments on the same muscles.

The proposed method seems appropriate for quantifying the effect of Skill and Speed on the elbow ROM and the EMG activity of selected upper limb muscles.

CONCLUSIONS: As supposed, the need for proper training of beginners in correct technical execution and the maximization of sport advantages was evident from these pilot tests. A skilled subject maintained its elbow R.O.M. at increasing walking speed and maximized the increment of EMG activation on specific muscles like LD, TBCL & DP. An unskilled subject reduced significantly its elbow R.O.M. at increasing walking speed and experienced much lower increments of EMG activation only on TBCL and DP muscles. Further tests may support statistically the results of the present work.

REFERENCES:

Schwameder H, Roithner R, Müller E, Niessen W, Raschner C, (2009), Knee joint forces during downhill walking with hiking poles, *Journal of Sports Science*, 17, 969-78.

Willson J, Torry MR, Decker MJ, Kernozek T, Steadman JR (2001) Effects of walking poles on lower extremity gait mechanics, *Medecine & Science in Sports & Exercise*, 33, 142-7.

Schiffer T, Knicker A, Hoffman U, Harwig B, Hollmann W, Strüder HK. (2006), Physiological responses to nordic walking, walking and jogging, *European Journal of Applied Physiology*, 98, 56-61.

ISBS 2009

Oral Session S16

Training Methods

GAIT DOES NOT RETURN TO NORMAL FOLLOWING TOTAL HIP ARTHROPLASTY: IMPLICATIONS FOR A RETURN TO ATHLETIC ACTIVITIES

Mario Lamontagne[1,2], Mélanie L. Beaulieu[1] and Paul E. Beaulé[3]

School of Human Kinetics, University of Ottawa, Ottawa, Canada[1]
Department of Mechanical Engineering, University of Ottawa, Ottawa, Canada[2]
Division of Orthopaedic Surgery, University of Ottawa, Ottawa, Canada[3]

The purpose of this study was to determine the effect of total hip arthroplasty (THA) on the biomechanics of the lower extremity during walking. Twenty THA patients and 20 healthy control participants performed several trials of level walking for which three-dimensional (3D) hip, knee and ankle angles, forces, moments and powers were recorded and calculated. Results revealed that the gait mechanics of THA patients do not return to normal following surgery, especially during the transition from double- to single-limb stance. These patients produced lower hip abduction moments that are perhaps a result of hip abductor weakness. Kinematic and kinetic adaptations at the distal joints were also found. Hip musculature deficiencies should be addressed in rehabilitation programs, especially if patients want to return to athletic activities.

KEY WORDS: total hip arthroplasty, hip abductors, gait, kinematics, kinetics.

INTRODUCTION: The characteristics of individuals undergoing total joint replacement have vastly changed over the past decades. Total hip arthroplasty (THA) patients are more physically active and live longer (Crowninshield et al., 2006), and thus, have higher expectations for the longevity and performance of their implant. Although pain relief continues to be one of the top-ranked expectations of THA candidates, the majority of these individuals also places great importance on their expectation for the surgery to improve their ability to exercise and play sports (Mancuso et al., 2003). The literature demonstrates that THA patients do indeed experience a remarkable relief in pain following surgery. They are not satisfied, however, with their ability to perform athletic activities and sports, such as swimming, tennis, golf and hiking, among others (Mancuso et al., 1997). Given that results from various studies show that one's gait does not return to normal following THA (Foucher et al., 2007; Vogt et al., 2004), this dissatisfaction may stem from a muscle weakness that is thought to be responsible for the abnormal gait patterns. Hence, muscle weakness may not only decrease THA patients' satisfaction of their athletic performance, but also increase their risk of injury during such activities and jeopardize the longevity of the implant. Although many researchers have performed gait analyses on THA patients post-operatively, most studies have ignored adjacent joints that are essential parts of the kinetic chain (e.g., knee and ankle joints) without performing a complete biomechanics analysis (Foucher et al., 2007; Vogt et al., 2004). And by completing such a thorough analysis, we might gain a better understanding of the cause(s) of these patients' deficiencies, which could be subsequently targeted during rehabilitation, and thus result in greater patient satisfaction during athletic activities. Consequently, the purpose of this study was to determine the effect of THA on mobility by comparing three-dimensional (3D) hip, knee and ankle joint angles, resultant inter-segmental joint forces and moments and powers during level walking of THA patients with those of healthy, matched control participants.

METHODS: Participants: Twenty THA patients (10 women, 10 men; age: 66.2 ± 6.7 yr; BMI: 27.2 ± 5 kg/m^2) and 20 healthy control participants, matched for gender, age and BMI (10 women, 10 men; age: 63.5 ± 4.4 yr; BMI: 24.9 ± 3.5 kg/m^2) were recruited on a voluntary basis. Exclusion criteria included bilateral hip replacement, hip replacement due to infection, fracture or failure of a previous prosthesis, concomitant surgical procedure during the surgery, as well as any past or present condition that could alter gait (e.g. stroke). Furthermore, all control participants had no current or history of serious lower limb injury or disease. All THA patients were operated by means of a lateral approach and were subsequently tested between 6 and 15 months postoperatively. Informed written consent, approved by the institutions' research ethics boards, was obtained from each participant.

Data Collection: A nine-camera digital optical motion capture system (Vicon MX, Oxford, UK) was used to capture, at 200 Hz, 45 retro-reflective markers (14 mm diameter) placed on various landmarks of the participants, according to a modified Helen Hayes marker set, while they executed the walking trials at a natural speed. Additionally, a force platform (AMTI, Model ORC-6-2000, Watertown, MA, USA) was used to record, at 1000 Hz, ground reaction forces during the stance phase of the gait cycle. Each participant performed six successful walking trials, three with their left foot and three with their right foot landing on the force platform. Walking trials during which the participant altered his/her gait to make contact with the force platform were discarded.

Data Analysis: The 3D marker trajectories were filtered using a Woltring filter (predicted Mean-Square Error value of 15 mm^2), whereas a low pass Butterworth filter (cut-off frequency of 6 Hz) was applied to the ground reaction forces. Following calculation of 3D hip, knee and ankle angles, the peak and range of these joint angles of the gait cycle were obtained. From the resultant inter-segmental joint forces and moments, as well as the joint powers, which were computed by means of the inverse dynamics approach, the peak joint kinetics during the stance phase of the cycle were extracted as variables of interest.

Statistics: Using SPSS statistical analysis software (SSPS for Windows, version 15.0, SPSS Inc., Chicago, USA), a series of one-way ANOVAs were executed to determine the presence of significant differences between the THA and control groups with regard to the 3D hip, knee and ankle kinematic and kinetic variables. Alpha levels of 0.0167 and 0.025 (corrected for multiple comparisons) were used to determine statistical significance of the kinematic and kinetic variables, respectively.

RESULTS: No significant differences between the THA and control groups were found with regard to age and BMI ($p > 0.05$). Results from one-way ANOVAs showed that a large portion of the statistically significant differences found between the experimental and control groups occurred at the time of transition from double- to single-limb stance (i.e., contralateral foot-off), as listed in Table 1. For this reason, as well as space limitation, only those variables will be presented and discussed, although several other significant differences were found between groups.

Table 1: Means (standard deviation) of the kinematic and kinetic variables found to be significantly different between THA patients and control participants during the transition from double- to single-limb stance of the gait cycle.

Type of variable	Joint	Variable	Value THA	Control	p-value
Angle (°)	Hip	Peak adduction	7.6 (2.5)	9.8 (2.2)	0.005
	Ankle	Peak external rotation	-10.5 (4.3)	-14.2 (3.9)	0.007
Joint Reaction Force (N/kg)	Hip	Peak anterior	-1.52 (0.81)	-2.31 (0.83)	0.004
	Knee	Peak proximal	9.27 (0.88)	10.04 (0.69)	0.004
	Ankle	Peak posterior	1.75 (0.43)	2.11 (0.32)	0.004
	Ankle	Peak proximal	9.64 (0.87)	10.37 (0.70)	0.006
Moment (Nm/kg)	Hip	Peak abduction	-0.76 (0.15)	-0.90 (0.11)	0.003
	Ankle	Peak internal rotation	0.09 (0.10)	0.03 (0.02)	0.010

DISCUSSION: It was found that the THA patients' gait mechanics during the task of walking on a level surface do not return to normal following surgery, especially during the transition from double- to single-limb stance. As most of the body weight shifted on the ipsilateral hip, the control participants produced a greater hip abduction moment of force while their hip was in a more adducted position in comparison with the THA patients (Figure 1).

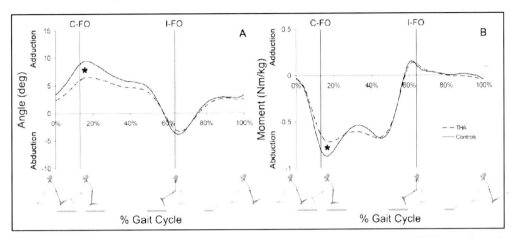

Figure 1: Average (and standard deviation represented by grey vertical lines) (A) hip angle and (B) hip moment of force in the frontal plane during level walking, time-normalized to the gait cycle. The asterisks (*) represent statistically significant differences between the THA and control groups.
C-FO = foot-off of contralateral limb; I-FO = foot-off of ipsilateral limb.

Similar findings have been previously reported in the literature (Foucher et al., 2007). It has been speculated that this altered gait pattern results from a weakness of the hip abductors. By adopting a mechanical strategy that places the hip in this less adducted position, THA patients require a smaller counteracting hip abduction moment of force to stabilize the pelvis in the frontal plane. Furthermore, several studies have demonstrated lower hip abductor strength in patients, both pre- and post-operatively, in comparison with control participants (Shih et al., 1994), although pre-surgery strength measurements may be questionable given that most THA candidates experience hip pain that can limit their ability to maximally contract their hip abductors. Several theories have been proposed as to the origin of this hip abductor weakness. Some researchers hypothesize that this weakness stems from disuse atrophy developed pre-operatively as a result of adopted gait patterns that limit contraction of the hip abductors in order to reduce hip loading, and thus reduce pain. Others believe that hip abductor weakness is an effect of surgery seeing that the lateral surgical approach to THA involves the detachment (and repair) of the anterior third of the gluteus medius. We suggest that the deficiency observed in the hip abductors of the THA patients is a result of both these occurrences. Given that Foucher and colleagues (2007) found that post-surgery hip abduction moments were not correlated with pre-surgery moments in THA patients performing the task of walking, pre-surgery disuse muscle atrophy cannot be entirely responsible for the lower hip abduction moments measured post-surgery in THA patients. Consequently, a portion of the group differences in hip abduction moments could be explained by the surgical approach utilized by the surgeons – the lateral approach, which disturbs the hip abductors. A comparison of the presented data with those from THA patients for which an anterior approach to surgery was used would potentially support our explanation regarding the origin of hip abductor weakness exhibited in THA patients. This is being currently addressed in an ongoing study. Nonetheless, a weakness of the musculature surrounding the hip joint may reduce the protection of the surface on which the implant is affixed, especially during athletic activities, and thus be detrimental to implant longevity. Given that athletic activities are generally more physically demanding than level walking, these gait deficiencies found to be present in THA patients may be amplified during sports, especially during those that require walking and/or running as the mean of locomotion It becomes imperative, therefore, for patients to follow a post-surgery rehabilitation program with a particular focus on muscle strengthening, especially of the hip abductors.

Furthermore, this strategy seemed to have coincided with an adaptation in transverse plane ankle mechanics. It seemed to demand a less externally rotated ankle – the result of an increase in ankle internal rotation moment produced by the THA patients. It is unclear, however, what consequences, if any, these greater ankle transverse plane moments may eventually have on the health of the musculoskeletal system. It was also found that the THA patients displayed lower hip, knee and ankle resultant inter-segmental joint forces than the control participants (Table 1). Hence, the former group exhibited a joint loading strategy that favoured unloading of the affected limb. Since these forces are highly dependent on those produced by the musculature surrounding the joints, THA patients displayed a reduction in not only hip, but also knee and ankle joint muscle activation, in comparison with their healthy, matched counterparts, as they transitioned from double- to single-limb stance of walking. This may be a result of the deficient frontal plane hip moments produced by the THA patients that propagated to the neighbouring joints. These differences in gait loading pattern at all three joints of the lower limb between the THA and control groups only reinforce a need for a rehabilitation program which addresses these muscle strength deficits. By neglecting such deficits, THA patients run the risk of prolonging their inferior ability to perform daily living and athletic activities, as well as the risk of jeopardizing the longevity of their implant.

CONCLUSION: Consequently, the results of the present study reveal that the gait mechanics of patients walking on a level surface do not return to normal following total hip arthroplasty. This was found to be particularly true as the THA patients transitioned from double- to single-limb stance. During this portion of the gait cycle, these patients produced lower hip abduction moments that are presumably consequential to hip abductor strength deficiencies or a preconditioning of the hip muscles. Moreover, gait mechanics at the distal joint were also affected. For these reasons, hip musculature deficiencies should be addressed in rehabilitation programs prior to and after THA, especially if patients want to return to athletic activities since these are generally more demanding than walking on a flat surface.

REFERENCES:
Crowninshield, R. D., Rosenberg, A. G., and Sporer, S. M. (2006). Changing demographics of patients with total joint replacement. *Clinical Orthopaedics and Related Research, 443*, 266-272.

Foucher, K. C., Hurwitz, D. E., and Wimmer, M. A. (2007). Preoperative gait adaptations persist one year after surgery in clinically well-functioning total hip replacement patients. *Journal of Biomechanics, 40*, 3432-3437.

Mancuso, C. A., Salvati, E. A., Johanson, N. A., Peterson, M. G., and Charlson, M. E. (1997). Patients' expectations and satisfaction with total hip arthroplasty. *Journal of Arthroplasty, 12*, 387-396.

Mancuso, C. A., Sculco, T. P., and Salvati, E. A. (2003). Patients with poor preoperative functional status have high expectations of total hip arthroplasty. *Journal of Arthroplasty, 18*, 872-878.

Shih, C. H., Du, Y. K., Lin, Y. H., and Wu, C. C. (1994). Muscular recovery around the hip joint after total hip arthroplasty. *Clinical Orthopaedics and Related Research* 115-120.

Vogt, L., Banzer, W., Pfeifer, K., and Galm, R. (2004). Muscle activation pattern of hip arthroplasty patients in walking. *Research in Sports Medicine, 12*, 191-199.

Acknowledgement
The authors would like to acknowledge the contribution of Drs. Kim, Feibel, and Giachino of The Ottawa Hospital, Dr Cranney of the OHRI, as well as Drs. Benoit and Li of the University of Ottawa. This research was supported, in part, by the Canadian Institutes of Health Research.

THE EFFECTS ON STRENGTH, POWER, AND GENU VALGUM FOLLOWING A FIVE WEEK TRAINING PROGRAM WITH WHOLE-BODY VIBRATION

Sarah Hilgers and Bryan Christensen

Department of Health, Nutrition, and Exercise Sciences
North Dakota State University, Fargo, ND, USA

KEYWORDS: whole-body vibration, resistance training, strength, power, genu valgum

INTRODUCTION: Human movements are controlled by the motor and sensory command centers of the body, and evidence indicates that voluntary activation of muscles is limited in force, power, and strength production (Ronnestad, 2004). The increase of motor unit activation through whole-body vibration (WBV) allows the muscles to contract and relax at a higher rate utilizing more muscle fibers to enhance athletic performance. The purpose is to examine the effects on strength, power, and genu valgum angle following a five week resistance training program either with or without whole-body vibration.

METHODS: There are two groups, resistance training (RT) and resistance training with WBV (RT-V). Both groups performed the same training program. Dumbbells were used for all exercises, except push-ups, with increasing weight by 2-5kg each week. The RT group trained on the gym floor, whereas the RT-V group performed all exercises on a vibration platform beginning at 30 hertz and increasing by five hertz every week. Pre- and post-test evaluation included maximum countermovement vertical jump (cm), standing long jump (cm), medicine ball toss (cm), 1RM (kg) of both chest press and leg press, and genu valgum measured at maximum genu valgum angle (degrees) at knee bend following a land and load maneuver.

RESULTS/DISCUSSION: The results of this study indicated greater improvement in both the countermovement vertical jump and leg press following the vibration training; however, the RT group displayed greater improvement in standing long jump, medicine ball toss, and chest press. One subject in the RT group and two subjects in the RT-V group exhibited genu valgum in one of their legs; all three subjects improved at the post-test.

CONCLUSION: Although this study showed improvement in only two areas following WBV, Bosco et al. (1999) found significant improvement of force-velocity, force-power, and vertical jump performance following ten, 60 second WBV treatments. Therefore, future research is needed to determine the appropriate frequency and training program for maximal performance.

REFERENCES:
Bosco, C., Colli, R., Introini, E., Cardinale, M., Tsarpela, O., Madella, et al. (1999). Adaptive responses of human skeletal muscle to vibration exposure. *Clinical Physiology, 19,* 183-187.

Ronnestad, B. R. (2004). Comparing the performance-enhancing effects of squats on a vibration platform with conventional squats in recreationally resistance-trained men. *Journal of Strength and Conditioning Research, 18,* 839-845.

Acknowledgments:
The authors would like to thank all of the individuals that agreed to participate in this study.

THE EFFECTS OF RECOVERY MODALITIES INCLUDING ICE BATH IMMERSION ON RECOVERY FROM EXERCISE AND SUBSEQUENT PERFORMANCE

Paula Fitzpatrick[1] and Giles Warrington[1]
Dublin City University, Dublin 9, Ireland[1]

The purpose of this study was to examine the effects of recovery modalities on recovery and subsequent performance. Ten trained male rugby union players were tested three times completing a different, randomly assigned recovery modality on each occasion. Each test began with a maximal aerobic endurance field test (20 metre shuttle test) followed by one of three recovery strategies – passive recovery, active recovery or ice bath immersion. Passive recovery involved lying on a recovery bed for 20 mins; active recovery entailed cycling for 20 mins at 50% heart rate reserve; and ice bath immersion required subjects to sit waist deep in an ice bath (5-8°C) for 3 x 30 s repetitions separated by one minute standing outside the bath. Following the 45 minute post-recovery strategy period, subjects completed 6 shuttles of a timed performance test (Illinois agility test). Plasma lactate concentrations and muscle soreness ratings were measured at various intervals throughout the testing. Analysis of the data revealed that active recovery resulted in significantly greater rates of lactate removal 5 mins into the recovery strategy when compared to passive recovery ($p = 0.01$). Muscle soreness was significantly lower for ice bath immersion than for active recovery immediately after the 20 minute recovery period ($p = 0.006$). No significant differences were observed for the subsequent performance test.

KEY WORDS: Recovery, ice bath immersion, subsequent performance.

INTRODUCTION: For decades various cryotherapies have been adopted for injury rehabilitation and more recently as a modality to accelerate the recovery process and enhance subsequent athletic performance. Many variations of cryotherapy are commonly used including cryo-chambers, ice pack therapy and ever more prevalently, ice baths, which are a focus of the present study. Due to the lack of definitive scientific evidence to support the efficacy of ice baths or any clear explanation of the possible mechanisms underlying their usage, there are currently no standardised procedures for their usage with a wide variety of recovery protocols being adopted. As a consequence, further research is required investigating ice bath immersion as a recovery strategy, which focuses on methodological issues such as specific temperature of the bath, repetitions, intervals between multiple repetitions, training effects associated with repeated use of ice bath immersion as a recovery strategy (Cochrane, 2004) and the effect of this recovery modality on subsequent performance. Recovery is an evolving area in sport. Recovery strategies are now not only being used extensively for treatment and rehabilitation from injury, but also to accelerate the recovery process from training and optimise subsequent athletic performance. The use of ice baths as a recovery strategy in particular has gained widespread popularity in recent times. Despite this there is a dearth of scientific evidence to support their usage. The aim of this study was to investigate the effects of three different recovery modalities (passive recovery, active recovery and ice bath immersion) both on recovery from exercise but also subsequent high intensity exercise performance.

METHODS: Data Collection: 10 healthy male subjects, aged between 18-30 years, volunteered to participate in the study. All experimental procedures were approved by the Dublin City University Research Ethics Committee. Subjects consisted of trained club rugby players of Junior 2 level or above and were recruited from Leinster rugby clubs. Subjects were restricted to backs position players to keep the groups as homogenous as possible. The descriptive and anthropometric data for the 10 subjects are presented in Table 1.

The study took place at the sports science facilities at Dublin City University (DCU). The subjects visited DCU on three separate occasions, with each visit separated by no less than a week. Each visit was divided into 2 stages separated by a 45 minute post-recovery period, during which nutrition, fluid intake and activity levels were monitored closely.

Table 1: Descriptive Anthropometric Data

Descriptive Anthropometric Data, n=10	(Mean(±SD))
Age (yrs)	21 (± 2.3)
Height (m)	1.76 (± 0.08)
Mass (kg)	73.3 (± 6.05)
Body mass index (kg.m^{-2})	23.5 (± 0.9)
Mean 20MST Score (mins elapsed)	12 (± 1.1)
Resting heart rate (bpm)	58 (± 13.7)
Target heart rate for active recovery (bpm)	130 (± 10.1)

The first part of the visit lasted approximately 35 mins and involved anthropometric measurements, a maximal endurance 20 metre shuttle run test (Leger and Lambert, 1982) and completion of one recovery strategy. Blood lactate measurements and muscle soreness ratings were recorded before testing began and at various intervals throughout each visit as outlined below:
- Sample A: Baseline resting measurement prior to 20MST
- Sample B: Immediately post-maximal test
- Sample C: 5 mins into the recovery strategy (equivalent to end of ice bath)
- Sample D: Immediately post recovery strategy (15 mins post ice bath)
- Sample E: Immediately pre-subsequent performance (after 45 minute period)
- Sample F: Immediately post-subsequent performance

Blood lactate measurements were obtained by drawing 1ml of blood from the earlobe onto a lactate strip and inserting into a blood lactate analyser (Lactate Pro™, Quesnel, Canada). Muscle soreness was rated using an 11 point scale (0-10) adapted from the differential descriptor scale (DDS) from Gracely et al. (1978).

Passive recovery involved the subjects lying supine on a recovery bed for a total of 20 mins with as little movement as possible. Active recovery consisted of the subjects cycling on a cycle ergometer (Monark™, UK) for 20 mins at a pace that corresponded to 50% of their heart rate reserve (HRR). The target heart rate (THR) was calculated using the Karvonen method involving the following equation: THR: ((HR $_{max}$ – HR $_{rest}$) x % intensity) + HR $_{rest}$. The ice bath recovery strategy was conducted by filling an ice bath (Symbol™, UK) with water to a level of 35cm and ice was used to keep the temperature constant in the range of 5-8°C using a 110mm immersion thermometer. The subject was then immersed in the bath to waist height with legs outstretched. Subjects completed three bouts of 30 s ice bath immersions with a 1 minute break in between each bout, in which the subject exited the bath. The subject was then instructed to remain seated for 15 mins after which time blood sample D and muscle soreness ratings were collected to coincide with the termination of the passive and active recovery strategies. Following each 20 minute recovery strategy period, subjects were given instructions regarding physical activity, hydration and nutrition for the 45 minute rest period before completing the subsequent performance test.

Following the 45 minute post-recovery period, subjects completed the second stage of the test which involved 6 shuttles, with 20 s recovery between each shuttle, of the Illinois agility test (Foran, 2001), which was used as the performance measure by which the recovery strategies were assessed. Muscle soreness ratings and blood samples E and F were taken immediately before and after the performance test respectively.

Data Analysis: SPSS for Windows v.15.0 (SPSS, USA) was used to conduct the statistical analysis. An analysis of variance (ANOVA) with repeated measures was carried out to determine any statistical significance between recovery groups at each of the six different time points. Post-hoc testing with Bonferroni adjustment was carried out along with tests of normality and Mauchly's test of sphericity. A probability of 0.05 was used to ascertain significance.

RESULTS: There were no significant differences in maximal endurance performance (20MST), baseline lactate levels and baseline muscle soreness ratings between subjects and within subjects across trials suggesting a large degree of homogeneity among the subjects. The active recovery modality exhibited significantly greater rates of lactate removal ($p = 0.01$) when compared to passive recovery at 5 mins into the recovery strategies (Figure 1).

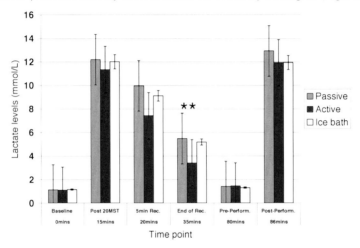

(** = $p<0.01$, * = $p<0.05$)
Figure1: Comparison of lactate levels for each recovery strategy.

The only other significant difference between trials was immediately following the 20 minute recovery period where ice bath immersion was associated with significantly lower muscle soreness ratings ($p = 0.006$) when compared to active recovery (Figure 2).

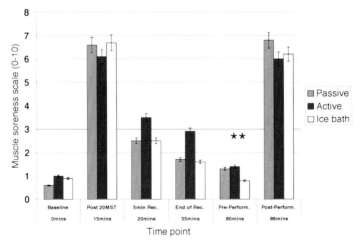

(** = $p<0.01$, * = $p<0.05$)
Figure 2: Comparison of muscle soreness ratings for each recovery strategy.

DISCUSSION: Ice bath immersion is one of the most popular recovery strategies used in team sports such as rugby. Despite this, research to support its effectiveness as a recovery modality is limited and contradictory (Barnett, 2006). In terms of the recovery indices measured, significant differences were observed for blood lactate clearance for active recovery compared to passive recovery ($p = 0.01$). However, no differences in subsequent performance were observed between the three recovery strategies investigated. This suggests that lactate removal rates may not be related to performance, in terms of muscle function in the current exercise trial. This finding is supported by Bond et al. (1991).

Upon completion of the 20 minute recovery strategy period muscle soreness ratings were found to be significantly lower for ice bath immersion when compared to active recovery ($p = 0.006$). The significantly greater muscle soreness associated with active recovery compared to ice bath immersion may be attributed to the intensity at which the active recovery strategy was implemented (50% heart rate reserve) compared to ice bath immersion which consisted of effectively 15 mins of passive rest following the 5 minute strategy. Another possible explanation relates to subjects' perception of the benefits of ice bath immersion as a recovery strategy. From the ice bath questionnaire completed prior to testing, 40% of the subject group envisaged a decrease in muscle soreness and an increase in performance with ice bath immersion. Alternatively, the reduced muscle soreness may be a delayed effect of the ice bath immersion recovery strategy.

The present study revealed that no significant differences were found in mean performance agility times, across trials for any of the three recovery strategies ($p = 1.000$).

The specific recovery protocols adopted in the study involved a relatively short post-exercise recovery period of 45 mins, in the anticipation that the effects of the recovery strategies would still be evident and thus, significant differences in performance may have been observed. However, as is shown in the results, lactate and muscle soreness levels had almost returned to baseline following the 45 minute post-recovery period and no significant differences were detected at this time point. Therefore, there are grounds for further research to incorporate even shorter post-exercise recovery periods. A period of 10-15 mins would typically replicate the half time interval in a team sport such as rugby union and as a result any findings may have practical connotations in this respect.

CONCLUSION: The findings of this study question the efficacy of ice bath immersion as an effective strategy to improve recovery and enhance performance. Furthermore ice baths appear to offer no additional advantage over more traditional recovery methods such as active recovery. To date much of the evidence to support the use of ice baths has been anecdotal with a distinct lack of scientific evidence. Therefore, despite their increasing popularity and use by many sports teams, none of the recovery strategies investigated in this study demonstrated any significant effect on performance.

REFERENCES:

Barnett, A. (2006). Using recovery modalities between training sessions in elite athletes, does it help? *Journal of Sports Medicine*. 36 (9): 781-796.

Cochrane, D.J. (2004). Alternating hot and cold water immersion for athlete recovery: a review. *Physical Therapy in Sport* (5): 26-32.

Foran, B (2001). High-Performance Sports Conditioning. *Human Kinetics, Champaign, IL:* 315.

Bond, V., Adams, R.G., Tearney, R.J., Gresham, K. and Ruff, W. (1991). Effects of active and passive recovery on lactate removal and subsequent isokinetic muscle function. *Journal of Sports Medicine and Physical Fitness* 31 (3): 357-361.

Gracely, R. H., McGrath, P. and Dubner, R. (1978). Ratio scales of sensory and affective verbal pain descriptors. *Pain*, 5: 5–18.

SPECIFIC TRAINING CAN IMPROVE SENSORIMOTOR CONTROL IN TYPE 2 DIABETIC PATIENTS

Thorwesten L[1], Eichler A[1], Sperlbaum C[1], Eils E[2], Rosenbaum D[2] and Völker K[1]

[1] Institute of Sports Medicine, University Hospital Münster, Germany
[2] Movement Analysis Lab, Orthopaedic Department, University Hospital Münster, Germany

Keywords: type 2 diabetic, polyneurpathy, proprioception

INTRODUCTION: Diabetes mellitus often is associated with proprioceptive and sensory deficits as a result of distal diabetic polyneuropathy (DPN). The aim of this prospective controlled longitudinal trial was to evaluate a specific sport intervention program regarding sensorimotor capabilities in type 2 diabetic patients compared to healthy controls. A higher incidence of fall-related injuries is given in the literature (Allet et al.2008; Allet et al 2009).

METHODS: 15 type 2 diabetic patients (7 female, 8 male; age 64 +/-10 years) whose disease is known since 9.5 years, and 14 healthy volunteers (7 woman, 7 men; age 56 +/-7 years) participated in this study. Using Semmes-Weinstein-Monofilaments (SMW, NorthCoast™) 5.07 log10(g) the diabetic patients were divided into diabetic PN as well as diabetic non PN group (non-DPN group n=9, DPN group n=6). 5 different measuring systems were used in test and retest: Sensibility threshold testing using SWM, postural balance testing with force plate (Kistler, type 9261A) with a sampling frequency of 40hz, active angle-reproduction test of the ankle using a custom made device with a Penny & Giles goniometer (ADU301), plantar foot pressure measuring (Emed ST, Novel) dynamic stability testing (Biodex-Stability-System, Biodex). A 90-minute special training program focusing on enhancing proprioceptive capabilities has to be completed once a week over an 8 month period. Statistics: Using SPSS v12.0, non parametric testing based on wilcoxon-test, mann-whitney-u-test, as well as kruskal-wallis-test were applied.

RESULTS: Comparing test-retest results of the sensibility threshold testing with SMW all plantar measuring points showed a slight increasing sensibility in the DNP group. Significant changes in threshold testing could be demonstrated in lateral forefoot and hallux area. Joint position sense measured with the angle reproduction test showed significant reduction of reproduction error in the diabetic group. For all other measuring methods no significant changes could be found.

DISCUSSION & CONCLUSION: A specific training can lead to positive changes in sensorimotor capabilities in type 2 diabetic patients. These results demonstrate the possibility of modulating proprioceptive deficits in diabetic patients during therapeutic intervention. Basic cause of this modulation could be found in an increasing cerebral processing using the given reafferent sensory input. This could result in a better management of ADL and could even be used as fall prevention because patients with diabetes are at higher risk of experiencing fall-related injuries when walking than healthy controls.

REFERENCES:
Allet, L.; et al. (2008): Gait characteristics of diabetic patients: a systematic review. Diabetes Metab Res Rev. Mar-Apr;24(3): 173-91
Allet, L.; et al (2009): Gait alterations of diabetic patients while walking on different surfaces. Gait Posture Apr; 29(3): 488-93

ISBS 2009

Scientific Sessions

Thursday AM

ISBS 2009

Oral Session S17

Variability

A COMPARISON OF VARIABILITY IN GROUND REACTION FORCE AND KNEE ANGLE PATTERNS BETWEEN MALE AND FEMALE ATHLETES

Sarah Breen[1] and Michael Hanlon[2]

Biomechanics Research Unit, University of Limerick, Limerick, Ireland[1]
Sport and Exercise Sciences Research Institute, University of Ulster, N Ireland[2]

The purpose of the present study was to compare the variability of movement and force production in males and females during a diagonal reaction task. Male (n=8) and female (n=8) subjects performed an unanticipated diagonal side cut task eight times with a 90s rest interval between trials. Variability of dominant limb knee angle and ground reaction forces were calculated for each subject over the eight trials. No significant differences were reported between genders for variability in any of the four parameters. This indicates that the variability of sagittal plane knee movement and ground reaction force patterns is not to be related to the increased incidence of anterior cruciate ligament injury in females.

KEY WORDS: Variability, Anterior Cruciate Ligament, Gender, Side cut

INTRODUCTION: Females are reported to be two to eight times more likely to sustain an anterior cruciate ligament (ACL) rupture than their male counterparts (Arendt and Dick, 1995). Despite the abundance of research in this area, there has been no definitive cause reported for the increased injury rate in females. Recent research (Pollard et al., 2005) has, however, reported that males demonstrate higher levels of movement variability in comparison to females, during a side cut task. This increased movement variability has been linked with a reduction in proneness to injuries such as ACL injury (Hamill et al., 1999; James et al., 2000; Mc Lean et al., 2004; Pollard et al., 2005). This variability-injury hypothesis states that increased variability may allow individuals to adapt to environmental perturbations experienced in dynamic sporting tasks, whereas individuals with reduced variability may have limited adaptation and increased incidence of injury (Hamill et al., 1999). Increased variability may provide a broader distribution of loads and allow a longer adaptation time, therefore reducing or slowing the ill-effects of repeated loading (James et al., 2000)

There is a need to further investigate this association between gender, variability and injury. The analysis of lower limb loading patterns would provide a new insight into the variability-injury hypothesis. This was undertaken in this investigation with the quantification of variability in ground reaction force (GRF) in the vertical (Fz), anterior-posterior (Fy) and medio-lateral planes (Fx) and variability in knee angle displacement patterns. It was hypothesised that females would demonstrate less variable force and movement patterns compared to males during an unanticipated side cut task.

METHODS: Subjects included eight male (age 21 ± 1 yrs; height 1.79 ± 0.06 m; mass 73.82 ± 13.47 kg) and eight female (age 20 ± 1 yrs; height 1.66 ± 0.1 m; mass 65.50 ± 6.94 kg) collegiate basketball and hockey players, with equal gender distribution between sports. Subjects were required to perform a diagonal side cut task to an unanticipated direction (left or right) at a 45° angle followed by running to a marker 2.5 m away as shown in Figure 1. Each subject began in an athletic ready position 0.4 m behind the force platform. The subject was then prompted by a green light to jump forward landing on both feet (dominant leg on the force platform) while following a directional cue (0.3 s after the green light) which indicated the direction of the cut. Practice trials were provided, to allow full familiarisation with the testing procedure. Variability was quantified for knee angle, Fx, Fy and Fz of the dominant leg for each subject.

Data Collection: A five-camera high-speed motion analysis system (Hawk; Motion Analysis Corp., Santa Rosa, CA) recording at 200Hz was synchronised with an AMTI single force platform system (1000 Hz) for kinematic and kinetic data collection, respectively. Three retro-reflective markers were secured on lateral malleolus, lateral femoral epicondoyle, and greater trochanter of the dominant leg for each subject for 3D kinematic data collection. Subjects performed 12 trials of the side cut task, eight of which involved a push off from the dominant leg and were utilised for analysis. A rest period of 90 s was provided between trials. Kinematic data was filtered using a Butterworth filter at 15 Hz (Winter, 1990). Vertical GRF (>10N) data were used to define the stance phase (Cowley et al., 2006). All kinematic and kinetic data were cropped accordingly and normalised to 1001 points. Kinetic data were normalised to body weight (expressed as percentage of body weight or %BW)

Data Analysis: Variability in all parameters was quantified for each subject by calculating standard deviation values around the mean ensemble curve (of eight trials) for each of the 1001 data points. The average of these 1001 standard deviation values provided a variability score for each parameter. Gender differences were assessed by parametric independent t-tests and non-parametric Mann-Whitney U test when data was not normally distributed. Statistical significance was set at alpha (α) < 0.05. Cohen's d values were reported as a measure of effect size, where 0.2, 0.5 and 0.8 represent a small moderate and large effect, respectively (Cohen, 1990).

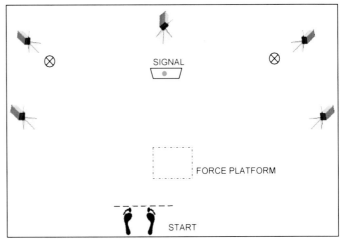

Figure 1 Experimental Set-Up

RESULTS: Variability scores for all parameters are presented in Table 1. No significant differences were found between males and females with regard to the variability of the knee angle or normalised GRF during the side cut.

Table 1 Average movement variability for each variable

Variable	Males	Females	Difference	Effect size	p-value
Knee Angle (°)	7.4° ±1.4	7.1° ± 1.4	0.3°	0.21	0.616
Fx (% BW)	7% ± 2	6% ± 2	1%	0.48	0.320
Fy (% BW)	7% ± 2	6% ± 1	1%	0.58	0.273
Fz (% BW)	18% ± 7	14% ± 5	3%	0.54	0.442

Figure 2 illustrates the pattern of variability for all 4 parameters. An increase in variability pattern for both genders can be seen in the GRF variability patterns (a, b, c) during the first ~15% of the stance phase. Post hoc analysis showed that the average and peak variability for Fx, Fy and Fz presented by the males during this initial phase was not significantly greater than the females, however larger effect sizes were found (Fx, Fy Fz effect sizes for average variability = 0.63, 0.73, and 0.83, while effect sizes for peak variability = 0.22, 0.53, and 0.75)

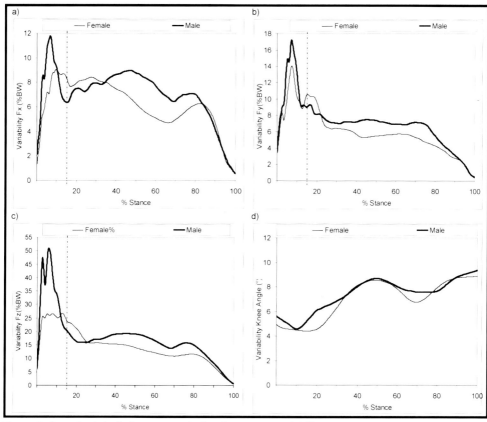

Figure 2: Standard deviation curves indicating mean variability within each gender for each parameter across the stance phase

DISCUSSION: The purpose of the present study was to compare variability for movement and force production in males and females during a diagonal reaction task. The variability in movement and force production patterns was not shown to differ significantly between males and females, and the hypothesis that females would be less variable was not supported.

The gender similarities do not support the proposed link between decreased variability in females and increased risk of ACL injury in accordance with the variability-injury hypothesis (Hamill et al., 1999; Pollard et al., 2005). The lack of differences in variability between genders also disagrees with the gender differences presented by Pollard et al. (2005) for intra-limb coupling variability. When this study is considered alongside the work of Pollard et al. (2005) it indicates similar variability in the resultant forces applied to the ground between genders, despite greater variability in the coordination of lower limbs in males. This implies that the higher movement coordination variability seen in males is somehow counterbalanced between limbs to result in more repeatable GRF profiles.

Although no statistically significant differences in variability were recorded, the effect sizes of differences did indicate moderate-large effects of gender. The effect sizes were particularly large for the first 15% of the stance phase. This is especially relevant in the case of ACL injury as the mechanism for an ACL injury is reported to occur just after foot contact during the first 20% of the stance phase (Boden, 2000). The higher levels of variability in male GRF patterns during the initial 15% of stance could depict a broader distribution of loads through the joints of the lower limb. Based on the large effect sizes, it is possible that future studies on a larger cohort of subjects would show this increased variability in males to be statistically significant. The reduction of inter-subject variability by controlling for exit velocity (running velocity when leaving the force plate) would assist in increasing the sensitivity of future research in this area. This future research is required to further investigate the role of movement variability in the gender imbalance of ACL injury risk.

CONCLUSION: No statistically significant gender differences were recorded in the variability of GRF or sagittal knee angle pattern data during an unanticipated side cut task. This indicates that despite previously reported increased variability in male lower limb coordination patterns, the resultant GRFs produced may not exhibit these same gender-related differences. Further research is necessary to investigate the relationship between variability in movement kinematics, coordination and GRFs and ACL injury risk with a view to the development of appropriate preventative training programmes.

REFERENCES:

Arendt, E. and Dick, R. (1995) Knee injury patterns among men and women in collegiate basketball and soccer NCAA data and review of literature. *American Journal of Sports Medicine,* **23**, 694-701.

Boden, B.P., Dean, G.S., Feagin Jr, J.A., Garrett Jr, W.E (2000) Mechanisms of anterior cruciate ligament injury *Orthopedics,* **23**, 573-578

Carlton, L.G. and Newell, K.M. (1993) Force variability and characteristics of force production in Newell, K.M. and Corcos, D.M. (eds.) *Variability and Motor Control,* Champaign: Human Kinetics.

Cohen (1990). *Statistical power analysis for the behavioural sciences.* (2nd edn.). Mahwah, NJ: Erlbaum.

Cowley, H.R., Ford, K.R., Myer, G.D., Kernozek, T.W. & Hewett, T.E. (2006) Differences in Neuromuscular Strategies Between Landing and Cutting Tasks in Female Basketball and Soccer Athletes. *Journal of Athletic Training.* **41**, 67-73.

De Vita, P., Skelly, W.A., (1992) Effect of landing stiffness on joint kinetics energetics in the lower extremity. *Medicine & Science in Sport and Exercise,* **24**, 108-115

Hamill, J. van Emmerick, E.A., Heiderscheit, B.C. and Li, L. (1999) A dynamical systems approach to lower extremity running injuries. *Clinical Biomechanics,* **14**, 297-308.

Hewett, T.E., Ford, K.R., Myer, G.D., Wanstrath, K. and Scheper, M. (2005) Gender Differences in Hip Adduction Motion and Torque during a Single-Leg Agility Maneuver. *Journal of Orthopedic Research,* **24**, 416-421

James, C.G., Dufek, J.S. and Bates, B.T. (2000) Effects of injury proneness and task difficulty on joint kinetic variability. *Medicine and Science in Sports and Exercise,* **32**, 1833-1844.

Landry, S.C., McKean, K.A., Hubley-Kozey, C.L., Stanish, W.D. and Deluzio, K.J. (2007) Neuromuscular and Lower Limb Biomechanical Differences Exist Between Male and Female Elite Adolescent Soccer Players During an Unanticipated Side-cut Maneuver. *The American Journal of Sports Medicine,* **35**, 1888-1899.

Mc Lean, S.G., Lipfert, S.W. and van Den Bogart, A.J. (2004) Effect of gender and defensive opponent on the biomechanics of sidestep cutting. *Medicine and Science in Sports and Exercise,* **36**, 1008-1016.

Pollard, C.D., Heiderscheit, B.C., van Emmerick, R.E.A. and Hamill, J. (2005) Gender Differences in Lower Extremity Coupling Variability during an unanticipated cutting maneuver. *Journal of Applied Biomechanics,* **21**, 143-152.

Acknowledgement

The authors would like to thank the following for their contribution to the data collection process: Brian Canty, Caroline Walsh, Derina Bourke, Eoin Kerin, Sinead Bruton, and Thomas Twomey. The Irish Research Council for Engineering and Science are also acknowledged for the funding provided which supported this work.

INTERSEGMENTAL COORDINATION DIFFERENCES BETWEEN BEGINNING PERFORMERS EXECUTING A BADMINTON SMASH FOR ACCURACY OR VELOCITY

H. Scott Strohmeyer, Cole Armstrong, Yuriy Litvinsky, Robert Nooney, Jaban Moore and Kevin Smith

University of Central Missouri, Warrensburg, Missouri USA

KEY WORDS: Intersegmental Coordination, Badminton Smash.

INTRODUCTION: Proper coordination is generally accepted as a very important process in skilful execution of many movement activities. It is also believed ballistic skills will exhibit a sequential intersegmental coordination pattern. Yet, what we know about how intersegmental coordination develops is relatively minimal. The purpose of this investigation was to examine a new skill (badminton smash) under two conditions (accuracy and velocity) to determine which condition would elicit the most theoretically correct intersegmental coordination pattern (Work in progress).

METHODS: Thirty-four individuals with no formal experience in badminton, were asked to execute the badminton smash. Teaching cues given, prior to participation in the study, were that the badminton smash should be executed: 1) with a high velocity; 2) a below horizontal trajectory angle; and 3) so it lands within the constraints of a legal court. Seventeen individuals were then tested executing the badminton smash with the net in place and the court boundaries marked (only those smashes meeting teaching cue #3 (above) were analyzed). The remaining subjects executed the badminton smash with no net or court boundaries. Three dimensional motion analysis using Vicon Motus™ was conducted for the best of three (highest velocity) smashes for each subject.

RESULTS: Initial data analyses point towards the subjects executing the badminton smash without a net are more sequential in their exhibition of intersegmental coordination patterns. Further, the timing of the sequences more closely matches the theoretical model developed by Morehouse and Cooper (1950) and further examined by Hudson, et al. (1991). Those individuals with the net in place are exhibiting a sequential pattern, however, the timing of segmental contributions is not as closely aligned to the theoretical model.

DISCUSSION: In 1991, Hudson et al. reported that interception skills add a level of complexity to ballistic tasks such that initial success in skill execution comes at the expense of proper mechanics, specifically intersegmental coordination. This statement does not take into account different ways of teaching movement skills. This investigation examines the effect of drill task on the intersegmental coordination of a skill that should be sequential in its exhibition. It will be important to determine which group appears to be "better" as future investigations will focus on which individuals will more quickly develop the movement patterns allowing them to be more proficient at ballistic skills.

CONCLUSION: Initial analysis has confirmed that the badminton smash exhibits a sequential intersegmental coordination pattern of movement. It is yet to be fully determined whether the accuracy group or the velocity group exhibit intersegmental coordination patterns that are more closely associated with the theoretically correct model.

REFERENCES:
Hudson, J., Bird, M., Strohmeyer, H., Horna, F., Hills, L., Rife, D., Clifton, R., & Walters, M. (1991). Coordination: Some questions and case studies. *9th International Symposium on Biomechanics in Sports*. Iowa State University, Ames, Iowa, USA, pp. 60-61.
Morehouse, L.E. & Cooper, J.M. (1950). *Kinesiology*. St. Louis: Mosby.

SKILL LEVEL AND PARTICIPATION OF UNIVERSITY STUDENTS IN RECREATIONAL SPORT

Philip Kearney[1,2], PJ Smyth[2] & Ross Anderson[1,2]

Biomechanics Research Unit, University of Limerick, Limerick, Ireland[1]
Department of Physical Education & Sports Sciences, University of Limerick, Limerick, Ireland[2]

This study investigated the relationship between throwing skill level and the engagement of college students in sports involving an element of throwing. There is a lack of knowledge about the throwing level of typical college students, and how this skill level influences students' participation in physical activity. 54 undergraduate students were qualitatively analysed performing the overarm throw and the volleyball serve, and completed questionnaires detailing their engagement in sports involving an element of throwing. Results indicated that college students are not proficient at throwing and that a higher throwing skill level is correlated with better serve form. Throwing skill level was not related to engagement in sports involving an element of throwing.

KEY WORDS: fundamental movement skill, overarm throw, physical activity

INTRODUCTION: An underlying concept in motor development is that the failure to develop proficient fundamental movement skills, such as throwing, will limit involvement in sports that utilise related skills, such as tennis (Gallahue & Ozmun, 2002). Proficiency is defined as the basic movement form that all learners can and should reach (Gallahue & Ozmun, 2002). A recent longitudinal study in Australia has reported that adolescent fitness levels, physical activity engagement and perceived competence may be positively influenced by the development of proficient object control skills (Barnett et al., 2008). Qualitative ratings of movement form at age 10 were significantly correlated to a questionnaire-based measure of physical activity taken 6 years later.

University education may have an important role to play in the development of an active population as it provides subsidised access to a wide range of sporting activities (Lunn, 2007). Although some research has investigated university students' physical activity (Staten et al., 2003; Suminski et al., 2002), the focus has been on factors such as the availability of facilities and physical activity promotion iniatives. No research was identified linking the effect of fundamental movement skill proficiency to sport participation in college students.

Limited research has looked at the fundamental movement skills of adults. In terms of throwing, Leme & Shambes (1978) qualitatively assessed 18 female adult students, selected on the basis of having a poor release velocity, and found all to have less than proficient throwing patterns. Rose & Heath (1990), utilising the same qualitative method as the present study, also found less than proficient throwing form typical of a sample of male and female college participants enrolled in tennis lessons. Without opportunity for practice and/or specialist instruction, it appears that low levels of throwing form will be maintained into adulthood (Haywood & Getchell, 2001). No quantitative studies were identified, however it is likely that as many children are not reaching fundamental movement skill proficiency, they will not be proficient as adults (Stodden et al., 2008).

Stodden et al. (2008) suggested that "If children cannot proficiently run, jump, catch, throw, etc., then they will have limited opportunities for engagement in physical activities later in their lives because they will not have the prerequisite skills to be active" (p. 291). The present study seeks to test this statement with regard to the throwing ability and engagement in sports involving throwing of a subset of Irish first year university students. Throwing was selected as it underlies a large number of sport skills (in volleyball, badminton, tennis, etc).

METHODS: Data Collection: 54 first year students volunteered to participate in the study (mean age 19.54 years). Ethical approval was obtained from the Physical Education & Sport Sciences department research ethics committee (PESSREC 61/07). Participants were free of injury assessed using standard pre-test questionnaire.

Procedures: Participants performed a self-directed general warm up. For the overarm throw and the volleyball serve similar procedures were observed. The task-specific protocol was explained. A pre-recorded demonstration of a proficient performer was observed. Five practice trials were performed followed by five recorded trials. A single camera (JVC Everio 50Hz) was placed perpindicular to the direction of the action on the throwing/striking side at a distance of 6m. Inter-trial interval was at least 30 seconds. Throws were towards an A3 sheet placed vertically on a wall 14m distant centred at a height of 1.80m. Participants were informed that accuracy was not being measured, and the goal of the task was to throw as fast as possible. A radar gun (Stalker™ ATS) was used to measure throwing speed. Before each throw participants were encouraged to throw "as hard as you can". Participants served towards a marked area ($3m^2$) in the centre-rear of the opposite court over a net of height 2.24m. Finally participants completed two questionnaires detailing both their present and past involvement in sports involving an element of throwing.

Table 1. Component actions for the volleyball overhead serve

Component	Level	Description
Foot action	1	Feet together or homolateral step taken
	2	Contralateral foot forward, no step taken
	3	Contralateral foot forward, step taken or pronounced weight shift
Trunk action	1	No rotation, flexion-extension only
	2	Blocked rotation, ball contact before front facing
	3	Blocked rotation, ball contact at or after front facing
Backswing action	1	Shoulder remains in a horizontally adducted position
	2	Shoulder horizontal abduction, shoulder abduction less than 90°
	3	Shoulder horizontal abduction, shoulder abduction greater than 90°
Forearm action	1	No forearm lag
	2	Forearm lag, maximum lag before front facing
	3	Delayed forearm lag, maximum lag after front facing
Contact point	1	Ball contacted behind shoulder or greater than one ball width ahead of shoulder
	2	Ball contacted above shoulder, elbow extension less than 150°
	3	Ball contacted above shoulder, elbow extension greater than 150°

Data Analysis: Video was analysed using the component approach (Roberton & Halverson, 1984). The lead author reduced the video data after establishing intra-individual objectivity following the procedure outlined by Roberton (1977). Four weeks after final data reduction 25 trials were randomly selected and categorised a second time. Intra-observer agreement ranged from 94-100%, in excess of the 80% agreement typically required (Roberton, 1977). For ease of comparison, participants were defined as Proficient (all components at proficient level), Moderate (all bar two components at proficient level) or Poor (more than two components less than proficient level) based on their summed component scores (O'Keeffe, 2001). The volleyball serve was analysed in a similar manner following a qualitative checklist (Table 1) devised following the main teaching points of the skill as identified by Viera & Ferguson (1996). The activity questionnaires asked the participants to detail which throwing-related sports they had engaged in (a) at any time previously, and (b) during the previous college semester. Participants who indicated participation were asked to (i) rate their perceived skill level on a Likert-type scale (1 = poor, 5 = excellent), and (ii) to identify how often they participated in the sport in question. Activity in throwing sport was classified as participating greater than or equal to four times per month in one or more throwing related activities. This figure was chosen as it represents the minimum frequency of training for throwing-related sports clubs within the university.

Statistical Analysis: Kendall's tau was used to investigate the relationship between throwing form and volleyball serve form as both data sets were ordinal in nature, and there were a moderate number of tied ranks. Fisher's exact test was used to examine the relationship between throwing skill level and engagement in sports involving an element of throwing. Independent t-tests examined the relationship between throwing speed and engagement in sports involving an element of throwing. Alpha was set to 0.05.

RESULTS: Of the 51 participants, only 3 (all male) demonstrated proficiency in the overarm throw (Figure 1). Kendall's tau indicated that individuals who scored higher on throwing form also scored higher on volleyball form, т (49) = 0.467, $p < .01$. 38% of respondents were classified as being active in throwing-related activities. 25.5% of participants had never engaged in throwing-related activity during the semester. There was no relationship between throwing form and being active in a throwing related sport, $p = 0.301$, Fisher's exact text. Active males (M=26.05, SE=1.29) threw faster than inactive males (M=23.88, SE=0.77), t(24)=-1.528, and active females (M=16.49, SE=1.03) threw faster than inactive females (M=14.84, SE=0.63), t(18)=-1.409, but these differences were not significant, $p>0.05$.

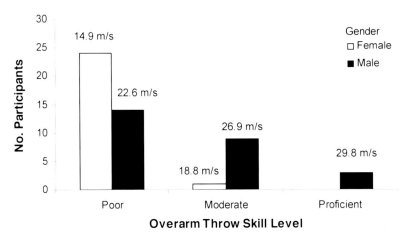

Figure 1. The number of participants attaining each skill level as defined by O'Keeffe (2001). The mean throwing velocities by skill level and by gender are listed over each subgroup.

DISCUSSION: Many authors (e.g., Gallahue & Ozmun, 2002; Haywood & Getchell, 2001) have argued that most children have the potential to be proficient in throwing by six to seven years of age. Only three participants in the current study demonstrated proficiency, as defined by Roberton & Halverson (1984). The mean throwing speeds obtained supported this finding: values were comparable to those previously reported for 13-year old children (Ehl et al., 2005) for males, and were even poorer for females.

Third level education is an important factor in the development of a sporting population, offering a wide range of subsidised opportunities to participate (Lunn, 2007). A less than proficient throwing pattern, as demonstrated by this population of college students, is suggested to present a barrier to their engagement in sports involving an element of throwing (Stodden et al., 2008). The present research did not support this suggestion. Participants with a higher throwing skill level were not more likely to participate in a sport involving an element of throwing than their less skilled peers. A number of factors may have contributed to this finding: (a) it is possible to participate recreationally in many sports that incorporate an element of throwing, such as badminton or volleyball, without utilising overarm strokes or without using a proficient action, (b) limitations in estimating physical activity from recall, and (c) participants demonstrating proficient throwing actions may not participate in throw-related sports due to a preference for alternative sports. General engagement in physical activity was not measured in this study. A further limitation was in the small number of proficient throwers located. The finding that participants who manifested higher form on the throw also tended to show higher form on the serve suggests that having a proficient fundamental movement skill may be an advantage when attempting to learn a novel sport skill, but that it's importance to continuing participation may be diluted with respect to other factors. Skill level is only one of a multitude of interacting factors affecting participation (Barnett et al., 2008). Future research must make an attempt to tease out the influence of these different factors.

CONCLUSION: The majority of a sample of young Irish adults had not reached proficiency in the overarm throw. Individuals who performed better on the overarm throw tended to also perform better on the volleyball serve. There was no relationship between throwing ability and engagement in throw-related sport. The role of fundamental movement skills in participation in sports involving an element of those fundamental skills may be better assessed through direct observation of performance rather than through questionnaire-based assessments of participation.

REFERENCES:

Barnett, L. M., Morgan, P. J., van Beurden, E., & Beard, J. R. (2008). Perceived sports competence mediates the relationship between childhood motor skill proficiency and adolescent physical activity and fitness: a longitudinal assessment. *International Journal of Behavioral Nutrition and Physical Activity, 5,* -.

Ehl, T., Roberton, M. A., & Langendorfer, S. J. (2005). Does the throwing "Gender gap" occur in Germany? *Research Quarterly for Exercise and Sport, 76*(4), 488-493.

Gallahue, D. L., & Ozmun, J. C. (2002). *Understanding Motor Development: Infants, Children, Adolescents, Adults* (5th ed.). Boston: McGraw-Hill.

Haywood, K., & Getchell, N. (2001). *Life Span Motor Development* (3rd ed.). Champaign, IL: Human Kinetics.

Leme, S. A., & Shambes, G. M. (1978). Immature Throwing Patterns in Normal Adult Women. *Journal of Human Movement Studies, 4,* 85-93.

Lunn, P. (2007). *Ballpark figures: Key Research for Irish Sports Policy.* Dublin: Irish Sports Council.

O'Keeffe, S. L. (2001). *The Relationship between Fundamental Motor skills and sport specific skills: Testing Gallahue's Theoretical Model of Motor Development.* University of Limerick, Limerick, Ireland.

Roberton, M. A. (1977). Stability of Stage Categorizations Across Trials: Implications for the 'Stage Theory' of Overarm Throw Development. *Journal of Human Movement Studies, 3,* 49-59.

Roberton, M. A., & Halverson, L. E. (1984). *Developing Children: Their Changing Movement.* Philadelphia: Lea & Febiger.

Rose, D. J., & Heath, E. M. (1990). The Contribution of a Fundamental Motor Skill to the Performance and Learning of a Complex Sport Skill. *Journal of Human Movement Studies, 19,* 75-84.

Staten, R. R., Miller, K. H., Noland, M. P., & Rayens, M. K. (2003). An application of an ecological model to understanding physical activity and dietary behaviors among college students. *Research Quarterly for Exercise and Sport, 74*(1), A19-A20.

Stodden, D. F., Goodway, J. D., Langendorfer, S., Roberton, M. A., Garcia, C., Garcia, L., et al. (2008). A Developmental Perspective on the Role of Motor Skill Competence in Physical Activity: An Emergent Relationship. *Quest, 60,* 290-306.

Suminski, R. R., Petosa, R., Utter, A. C., & Zhang, J. J. (2002). Physical activity among ethnically diverse college students. *Journal of American College Health, 51*(2), 75-80.

Viera, B. L., & Ferguson, B. J. (1996). *Volleyball: Steps to Success* (2nd ed.). Champaign, IL: Human Kinetics.

AUGMENTED KNOWLEDGE OF RESULTS FEEDBACK IN TENNIS SERVE TRAINING

Kieran Moran[1], Colm Murphy[1], Garry Cahill[2], Brendan Marshall[1]

[1] School of Health and Human Performance, Faculty of Science and Health, Dublin City University, Ireland; [2] Tennis Ireland, Dublin, Ireland

> The purpose of this study was to determine if augmented knowledge of results (KR) feedback, in the form of service speed, could enhance service speed after a six week training period. Twelve national standard junior tennis players completed six weeks of dedicated service training, three times per week; each training session consisted of 90 serves. Six players received augmented KR feedback and six did not. A Wilcoxon signed-rank test was used to compare pre- vs post- intervention results, and a Mann-Whitney test was used to examine if the differences between these results were larger in one of the groups ($\alpha = 0.05$). Results showed that both the augmented KR feedback group and the no feedback group significantly improved (0.8 ± 0.4 m.s^{-1} vs 0.2 ± 0.1 m.s^{-1}, respectively), but that the improvement was significantly larger for the augmented KR feedback group. These results suggest that it is beneficial to utilize augmented KR feedback, in the form of service speed, when undertaking a six weeks training program with elite tennis players.

KEY WORDS: augmented feedback, tennis, service speed

INTRODUCTION: The serve in tennis is commonly considered the most important stroke in the game because it is a high predictor of success (Knudson, 2006; Roetert and Groppel, 2001); its effectiveness is primarily dependent on ball speed (Elliot et al., 2003). It is therefore desirable to utilise learning strategies that help improve service technique. To this end knowledge of results [KR] (e.g. tennis serve speed) is an important type of feedback which athletes use, in combination with knowledge of performance [KP] feedback, to enhance technique (Adams, 1987; Salmoni et al., 1984). For some actions KR can be ascertained via internal sensors/systems (e.g. vision, hearing). However, a recent study (Moran et al., 2009) has shown that elite junior national tennis players can not differentiate the speed of consecutive serves, and in consequence it is not possible for them to use this information as an input in optimising technique. In light of this it could be recommended that augmented KR feedback from an external source (e.g. speed gun) may offer important additional information that may facilitate technique optimisation. External augmented KR feedback has been shown to have a positive effect on motor performance (Bilodeau et al., 1959; Kontinnen et al., 2004), however such studies have primarily involved a novel task. No studies could be found that examined the effect of providing feedback on service speed during training as a means of improving service speed. The present study aims to address this.

METHODS: Participants: Twelve junior national tennis players between the age of 13 and 18 (15.9 ± 1.7 years) volunteered to participate in this study. Players were free from any injury that would have prevented them from using maximum effort. All players were training between 20 and 26 hours as part of the Tennis Ireland National Squad programme. Approval by the Dublin City University Ethical Board was obtained in advance.

Participants were ranked by their senior coach in terms of his 'perception of how good their technique was'. Subsequently they were assigned to either the augmented KR feedback group (participants 1, 4, 5, 8, 9, 12) or the no augmented feedback group. A Mann-Whitney test indicated no significant difference between the augmented feedback group and the no feedback group for pre-intervention service speed (46.7 ± 4.7 m.s^{-1} vs. 46.5 ± 3.5 m.s^{-1}, respectively; $U = 17.0$, $p = 0.47$, effect size $r = 0.05$].

Data Collection: All sessions took place during participants' usual training times in the National Tennis Centre. Participants attended one pre-test session (to determine their baseline service speed), six weeks of training sessions (three times per week), and one post-intervention session. Each training session required the player to complete ninety serves; fifteen wide, fifteen body and fifteen T on both the Deuce and Advantage sides of the court. A 1.5 m x 1.5 m area was marked in the three areas of the court as a target. Serves were to be hit "as hard as possible with the aim of landing the serve in the service box". Feedback to the augmented feedback group was given immediately (< 1sec) after each serve via a large electronic display. Service speed was measured using a speed gun (StalkerPro, Stalker, USA) placed in line with the intended direction of serve (4m behind the baseline). Absolute errors for the speed gun are very small ($\pm$ 0.04 m.s^{-1}). It is acknowledged however that serves to either side of the sensor of the speed gun would contain inaccuracies; these were estimated to be up to 0.25%.

Initial baseline and post-intervention service speed were determined as the average of 15 serves to a 1.5 m x 1.5 m area of the T in the Deuce service box. A Wilcoxon signed-rank test was employed to examine pre- versus post- intervention results (α = 0.05). Where both groups exhibited a significant training effect a subsequent Mann-Whitney test was used to determine if there was a significant difference between the two groups in the magnitude of pre- to post- enhancement (α = 0.05).

Data Analysis: A Wilcoxen signed rank non-parametric test was employed (α = 0.05). The two variables used were (i) the number of correct differentiation (faster/slower) out of 10 and (ii) the number of correct differentiations expected due to chance (5).

RESULTS: For both groups there was a significant enhancement in service speed following 6 weeks of training [Z = -2.2, p = 0.03, effect size r = 0.64]. However, the augmented KR feedback group improved their service speed significantly more than the no augmented feedback group; 0.8 $\pm$ 0.4 m.s^{-1} vs. 0.2 $\pm$ 0.1 m.s^{-1}, respectively [U = 3.5, p = 0.01, effect size r = 0.68]. Average speeds for the augmented feedback and the no feedback groups pre-intervention were 46.7 $\pm$ 4.7 m.s^{-1} and 46.5 $\pm$ 3.5 m.s^{-1}, respectively; and post-intervention were 47.6 $\pm$ 4.7 m.s^{-1} and 46.7 $\pm$ 3.5 m.s^{-1}, respectively.

DISCUSSION: The present study showed that augmented KR feedback, in the form of service speed, resulted in a significantly greater improvement in service speed in national standard junior tennis players after a six week training period, in comparison to when no augmented KR feedback was provided. It appears that the augmented information of service speed is important because players at this level can not accurately determine the speed of their serve from trial to trial using intrinsic processes (e.g. vision, hearing) (Moran et al., 2009). As such they can not identify which of their attempts are able to produce the faster serves, which would allow them to possibly select those elements of the technique that optimise it. This supports the view that where there is no, or very limited, source of knowledge of results from intrinsic sources, performance will not improve without augmented feedback (Magill, 1998).

No previous studies appear to have investigated the effects of augmented KR feedback on improvement in tennis serve speed following a period of training; in fact no studies appear to have examined the effect of such feedback on any well learnt striking/throwing task in experienced athletes. The majority of previous studies on KR feedback have examined novel motor tasks or well learnt slow manual tasks. Bilodeau et al. (1959) demonstrated positive significant effects of receiving augmented KR feedback in trials of a linear positioning task. Kontinnen et al. (2004) also demonstrated the positive effects of KR feedback in psychomotor skill learning in precision shooting. However, in contrast to this, Buekers et al. (1992) showed no significant learning effect in a group receiving KR feedback for an anticipation timing task.

It has been postulated that augmented KR feedback may result in an enhancement in technique by: (i) facilitating the process by which elements of a technique from the present and previous attempts are selected for integration and utilisation (Magill, 1998), and/or (ii) motivating the athlete to maximise effort and for longer periods of time (Adams, 1987). No attempt was made in the present study to distinguish between them.

While the present study only examined the use of augmented KR feedback in tennis, it is worth noting that other sports have a similar situation where the path of the ball is impeded (either by a player or the environment) and therefore players can not judge how far the ball 'would have travelled'. Such information would likely provide sufficient information on ball speed because of the clear relationship between speed and distance (and trajectory). Sports facing similar conditions include table tennis, squash, penalty kicks in soccer.

CONCLUSION: Elite junior tennis players are able to improve their service speed using augmented KR feedback, in the form of speed of serve, more than players who receive no augmented feedback. The use of augmented KR feedback should be utilised, at least in short training cycles lasting six weeks.

REFERENCES:

Adams, J. A. (1987). Historical review and appraisal of research on the learning, retention and transfer of human motor skill. *Psychological Bulletin* 101, 41-47.

Buekers, M. J., Magill, R. A. & Hall, K. G. (1992). The effect of erroneous knowledge of results on skill acquisition when augmented information is redundant. *Quarterly Journal of Experimental Psychology.* 44A. 105-117.

Bilodeua, E. A., Bilodeau, I. M. & Schumsky, D. A. (1959). Some effects of introducing and withdrawing knowledge of results early and late in practice. *Journal of Experimental Psychology.* 58. 142-144.

Elliot, B., Fleisig, R. & Nicholls, R. E. (2003). Technique effects on upper limb loading in the tennis serve. *Journal of Sport Science and Medicine.* 6. 76-87. 27.

Knudson, D. (2006). *Biomechanical Principles of Tennis Technique.* Racquet Technical Publishing. 45

Magill, R. A. (1998). *Motor Learning, Concepts and Applications.* McGraw-Hill Companies. New York. 171-224.

Kontinnen, N., Mononen, K., Viitasalo, J. & Mets, T. (2004). The effects of augmented auditory feedback on psychomotor skill learning in precision shooting. *Journal of Sport and Exercise Psychology.* 26 (2). 306-316.

Moran, K., Murphy, C., Cahill, G., & Marshall, B. (2009) Can elite tennis players judge their service speed? Submitted to the proceedings of the *International Society of Biomechnics in Sports.* 17-21 August, Limerick.

Roetert, P. & Groppel, J. (2001). World Class Tennis Technique. Human Kinetics Publishers. Leeds. 207. 47.

Salmoni, A. W., Schmidt, R. A. & Walter, C. B. (1984). Knowledge of results and motor learning: A review and critical reappraisal. *Psychological Review.* 82. 225-260.

THE SPANNING SET AS A MEASURE OF MOVEMENT VARIABILITY

Michael Hanlon[1], Joan Condell[2] and Philip Kearney[3]

Sport and Exercise Sciences Research Institute, University of Ulster, N Ireland[1]
Computer Science Research Institute, University of Ulster, N Ireland[2]
Biomechanics Research Unit, University of Limerick, Limerick, Ireland[3]

The variability of an individual's movement pattern is an increasingly important focus of research in sport and exercise biomechanics. Inter-trial variability of a single variable is typically assessed using mean deviation or coefficient of variation, however, recent alternatives to these have been proposed such as the spanning set technique. This paper presents an investigation into the validity of the spanning set measure. Variability scores using the spanning set were compared against more traditional measures of variability (mean deviation, coefficient of variation and variance ratio). Results indicate that the spanning set is biased towards early-phase variability and may inaccurately describe the overall level of movement variability.

KEY WORDS: spanning set, mean deviation, variability, variance ratio.

INTRODUCTION: The interpretation of human movement variability has evolved from the historical standpoint, which viewed variability as noise or error in movement patterns, to the more recent viewpoint which considers the functional role of inter-trial variability and suggests both positive and negative relationships between variability and performance or health (Button et al. 2003; Crowther et al. 2008; James et al. 2000). While research does suggest direct links between performance and variability, the specific directional effects (beneficial or detrimental) of these links appear to be determined by complex interactions between skill type, performance variable and the level of performer (Button et al. 2003). Research which attempts to elucidate on these complex performance-variability interactions clearly requires valid and informative measures with which to quantify variability. The spanning set (SS) is one measure of inter-trial variability that has been recently recommended in the literature. Kurz et al. (2003) proposed the SS as an alternative and more sensitive measure than traditional variability measures such as mean deviation (MD) and coefficient of variation (CV). Subsequently, the SS technique has been used to quantify variability in running kinematics between different footwear types (Kurz & Stergiou, 2003) and in walking kinematics between control subjects and patients with peripheral arterial disease (Crowther et al. 2008). The mechanics of the SS approach for variability assessment are based on work by Lay (2000) and described in detail by Kurz and Stergiou (2004). In brief, the technique first involves fitting high-order polynomials to the standard deviation (SD) curves of a mean ensemble curve. The coefficients of each polynomial are then used to define the vectors of a spanning set between the two SD curves. The greater the difference between the two spanning set vectors (calculated as the root sum of squared differences between coefficient pairs) the greater the variability that is indicated in the mean ensemble curve.

Despite findings in favour of the SS technique (Kurz et al. 2003), no research work has strategically assessed its functionality in order to determine its validity as a measure of inter-trial variability. In reviewing the SS technique, there are indications that the mathematical procedures which underpin it are overly biased towards variability at the beginning of the movement cycle and far less sensitive to any variability occurring later in the movement cycle. Therefore, the purpose of this study was to assess the validity of the SS measure in a controlled manner using movement patterns with incidences of discrete phase-specific variability.

METHODS: Technique validation: Any measure of inter-trial variability should be equally sensitive to increases in variability at all phases of the movement cycle. Due to concerns about the phase-related sensitivity of the SS technique, it was assessed using four alternative phase-variability models, each involving variability at a different phase of the movement cycle (see Table 1). These variability models were applied to datasets from two diverse movement patterns: the sagittal plane knee angle in gait and the sagittal plane elbow angle during a basketball free throw. The mean movement patterns for each skill were based on actual data, with simulated variability added according to each variability model. These patterns were time-normalised to 101 data points (0-100% of cycle). Figure 1 illustrates the four phase variability models as applied to the knee angle during a complete gait cycle. The SS technique was conducted in accordance with the guidelines provided by Kurz and Stergiou (2004). Polynomial curve fitting was carried out using the least squares procedure in LabVIEW 8.2.

Table 1 Datasets used to assess phase-related sensitivity of SS technique.

Phase variability model	Knee angle (Gait)	Elbow Angle (Basketball)
1. Control	SD = 2^0 over complete cycle	SD = 5^0 over complete cycle
2. Variability Start	SD is 80% > than control during 0-30% phase of cycle	SD is 60% > than control during 0-30% phase of cycle
3. Variability Middle	SD is 80% > than control during 35-65% phase of cycle	SD is 60% > than control during 35-65% phase of cycle
4. Variability End	SD is 80% > than control during 70-100% phase of cycle	SD is 60% > than control during 70-100% phase of cycle

NOTE: The variability in the unaltered 70% of models 2-4 was equal to that of the control model.

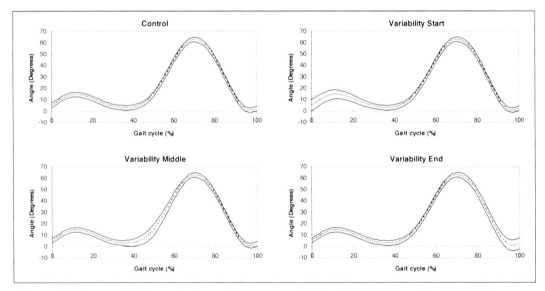

Figure 1: Phase-variability models used to assess validity of SS technique (applied to knee angle dataset)

Data Analysis: The SS scores were compared against three other recommended techniques for inter-trial variability: MD, variance ratio (VR) and CV (Hershler & Milner 1978; Kurz & Stergiou 2004).

RESULTS: Table 2 displays the variability scores (SS, MD, CV, and VR) for each dataset. As the relationship between variability models was identical for MD, CV and VR, only the MD results are plotted against the SS results in Figure 2. As the results trend was also identical between knee angle data and elbow angle data, only the knee angle results are illustrated in Figure 2. The addition of variability at the start of the movement resulted in an average increase in the SS score of 122% (knee +148%, elbow +96%), versus average decreases of 21% (knee -24%, elbow -18%) and 3% (knee -3%, elbow -2%) when variability was added at the middle and the end of the movement respectively. Conversely, the addition of variability resulted in changes in MD, CV and VR that were identical, regardless of the phase during which variability was added.

Table 2 Variability scores for SS and traditional variability measures in each dataset. Units for SS and MD are degrees, while CV is a % (VR is a unitless ratio).

	Knee angle (Gait)				Elbow angle (Basketball)			
	SS	MD	CV	VR	SS	MD	CV	VR
1. Control	4.00	2.00	9.27	0.010	10.00	5.00	6.14	0.051
2. Variability Start	9.91	2.48	11.48	0.017	19.55	5.89	7.23	0.074
3. Variability Middle	3.02	2.48	11.48	0.017	8.18	5.89	7.23	0.074
4. Variability End	3.89	2.48	11.48	0.017	9.76	5.89	7.23	0.074

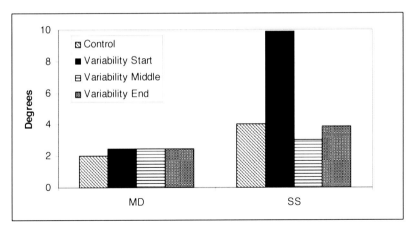

Figure 2: MD and SS scores for each variability model applied to the knee angle dataset.

DISCUSSION: This study assessed the phase-related sensitivity of the SS variability score in an effort to gauge its validity as a measure of inter-trial variability. It is evident from the results that the SS measure is heavily biased towards variability at the start of the movement (122% increase in scores) while being relatively insensitive to variability occurring at the middle and end of movement patterns (reductions of 21% and 3% respectively). This is in contrast to traditional measures of variability which show equal increases in their scores regardless of the phase during which variability is added. The fact that the SS is unequally weighted towards early-phase variability means that it is unsuitable for comparisons between subjects or between conditions (where variability could occur during any movement phase). Also, the erroneous finding of decreased variability by SS scores after mid-phase variability has been added to the movement is also a strong finding against the SS technique.

The combination of these results with a functional analysis of the SS technique indicates possible causes of the phase-related bias in variability scores. The SS score uses the calculated differences between coefficient pairs from polynomials mapping the upper and lower SD curves (Kurz and Stergiou 2004). As the largest coefficient value is typically the first coefficient, this has the greatest influence on the overall score. However, this first coefficient is also an indicator of the intercept value of the SD curve (i.e. the SD value at the start of the movement cycle), hence the bias towards early-phase variability.

These results prompt reinterpretation of the findings of previous studies using the SS technique. For example, the claim by Kurz et al. (2003) that the SS offered a more sensitive measure of movement variability between shod and barefoot running would appear to be solely the result of increased variability at the start of the movement cycle in barefoot versus shod conditions, rather than increased variability throughout the complete gait cycle. The suggested sensitivity of the SS technique is, therefore, an artefact of the calculation which only relates to early-phase variability. If the aim of a research study is to assess variability changes at specific phases of the movement cycle, then comparing the SD at these specific phases (rather than calculating the MD over the complete cycle) should allow the sensitivity of measurement required.

In considering alternatives to the spanning set, a strong note of caution should be issued in relation to the use of the CV. This quantity is seen as a useful way of normalising the SD so that variability can be compared between different conditions, individuals and variables. However, as pointed out by Mullineaux et al. (2001), the inclusion of the mean as the denominator can lead to imbalances between CV values and absolute SD values (e.g. when the mean is close to zero). This is further evidenced by the results of this study which show higher CV values for the knee angle data than the elbow angle data, despite the opposite trend being shown in MD values. The VR technique appears to offer a useful alternative to the CV value. This technique normalises the variation in curves to the average deviation from the overall single mean value, therefore accounting for the higher variability expected in more dynamic movements. This is supported by the findings in this study which show similar relationships between knee angle and elbow angle data for both the MD and VR techniques.

CONCLUSION: The present study shows that the SS method for assessing inter-trial movement variability is biased towards variability occurring at the start of a movement pattern and cannot be recommended as a valid measure. Researchers and practitioners seeking to understand the links between variability and performance should use the MD and VR measures instead of the spanning set, as these appear to offer greater accuracy in quantifying variability.

REFERENCES:

Button, C., Macleod, M., Sanders, R., & Coleman, S. (2003) Examining movement variability in the basketball free-throw action at different skill levels. *Research Quarterly for Exercise & Sport*, 74, 257-269.

Crowther, R.G., Spinks, W.L., Leicht, A.S., Quigley, F., & Golledge, J. (2008) Lower limb movement variability in patients with peripheral arterial disease. *Clinical Biomechanics*, 23, 1080-1085.

Hershler, C., & Milner, M. (1978) An optimality criterion for processing electromyographic (EMG) signals relating to human locomotion. *IEEE Transactions on Biomedical Engineering*, 25, 413-420

James, C.R., Dufek, J.S., & Bates, B.T. (2000) Effects of injury proneness and task difficulty on joint kinetic variability. *Medicine and Science in Sports and Exercise*, 32, 1833-1844.

Kurz, M.J., & Stergiou, N. (2003) The spanning set indicates that variability during the stance period of running is affected by footwear. *Gait and Posture*, 17, 132-135.

Kurz, M.J., & Stergiou, N. (2004) Mathematical measures of coordination and variability in gait patterns. In N. Stergiou (Ed.), *Innovative Analysis of Human Movement*. Champaign: Human Kinetics.

Kurz, M.J., Stergiou, N. & Blanke, D. (2003) Spanning set defines variability in locomotive patterns. *Medical & Biological Engineering & Computing*, 41, 211-214.

Lay, D.C. (2000) *Linear algebra and its applications* New York: Addison-Wesley.

Mullineaux, D.R., Bartlett, R.M., & Bennett, S. (2001) Research design and statistics in biomechanics and motor control. *Journal of Sport Sciences*, 19, 739-760.

ACUTE EFFECTS OF WHOLE-BODY VIBRATION ON KNEE JOINT DROP LANDING KINEMATICS AND DYNAMIC POSTURAL STABILITY

Eamonn Delahunt[1], Domenico Crognale[1], Brian Green[1], Chris Lonsdale[1], Denise McGrath[1]

School of Physiotherapy and Performance Science, University College Dublin, Dublin, Ireland[1]

KEY WORDS: squat, knee, landing, joint movement, balance.

INTRODUCTION: Whole-body vibration (WBV) is being increasingly utilized in addition to other training modalities in order to prevent and rehabilitate athletic injuries. Excessive knee joint movement has been reported to be a contributing factor to many traumatic and overuse knee joint injuries (Sigward et al., 2008). However the effects of WBV on sensorimotor function and consequent knee joint kinematics is unknown. Thus, the aim of the present study was to examine the effects of an acute WBV exposure on knee joint drop landing kinematics and dynamic postural stability in healthy participants. The null hypothesis was that acute WBV exposure would not influence lower limb drop landing kinematics or dynamic postural stability.

METHODS: 12 healthy male subjects with no history of previous knee joint injury volunteered to participate in this study. Each participant performed a series of 10 squats under three different counter-balanced conditions, with those conditions being: (1) over-ground, (2) 30 Hz vibration, (3) 50 Hz vibration. Immediately following each series of squats, each participant performed 3 single-leg drop landings from a height of 30 cm. Knee joint kinematics (peak valgus, peak flexion, and peak internal rotation) were assessed using a CODA 3D motion analysis system, while ground reaction forces were recorded using an AMTI force-plate, and subsequently used to calculate the dynamic postural stability index (DPSI).

RESULTS: Separate repeated-measures ANOVAs were conducted, using a Greenhouse-Geisser correction when violations of the sphericity assumption were observed. These analyses revealed no significant differences at $P < .05$ for peak knee valgus ($F_{1.07, 10.60} = .81$, $P = .69$, $\eta^2 = .08$, observed power = .07), peak knee flexion ($F_{2,20} = 2.44$, $P = .11$, $\eta^2 = .20$, observed power = .43), peak knee internal rotation ($F_{1.11, 11.12} = 1.12$, $P = .32$ $\eta^2 = .10$, observed power = .17) or the DPSI ($F_{1.30, 13.03} = 1.03$, $\eta^2 = .09$, observed power = .17).

DISCUSSION: In this study, acute vibration exposure showed no significant effect on knee joint drop landing kinematics or dynamic postural stability.

CONCLUSION: Further research with a larger sample is needed to determine whether vibration stimuli can enhance parameters of knee joint neuromuscular control in non-injured and injured subjects. Also the effect of a specific training programme conducted over a period of weeks requires consideration.

REFERENCES:
Sigward, S.M., Ota, S., and Powers, C.M. (2008). Predictors of frontal plane knee excursion during a drop landing in young female soccer players. *Journal of Orthopaedic and Sports Physical Therapy*, 38, 661-667.

ISBS 2009

Oral Session S18

Other Sports II

THE EFFECT OF MAXIMAL FATIGUE ON THE MECHANCIAL PROPERTIES OF SKELETAL MUSCLE IN STRENGTH & ENDURANCE ATHLETES.

Niamh Ni Cheilleachair and A.J. Harrison

Biomechanics Research Unit, University of Limerick, Limerick, Ireland.

The purpose of this study was to compare the effects of maximal fatigue on the mechanical performance of strength and endurance athletes. Ten strength trained athletes and nine endurance athletes performed a maximum fatigue protocol on a sledge and force plate apparatus followed by drop and rebound jumps at 15, 45, 120, 300 and 600 seconds post fatigue. Measurements of peak force, ground contact time, leg spring stiffness and height jumped were calculated prior to the fatigue protocol to establish baseline values, and also for each jump following the fatigue protocol. The fatigue protocol resulted in a significant reduction in peak force ($p<0.01$) and height jumped ($p<0.01$) in both groups while leg spring stiffness was also reduced in the strength athletes ($p<0.01$). In addition the endurance athletes indicated a potentiation effect with a significant increase in peak force ($p<0.01$) and leg spring stiffness ($p<0.05$) during the post fatigue jumps.

KEY WORDS: fatigue, potentiation, SSC, strength athletes, endurance athletes

INTRODUCTION: Fatigue is a complex, multidimensional phenomenon and is a reflection of the type of work that has been done. It has been described as "the transient decrease in performance capacity of muscles when they have been active for a certain time, usually evidenced by a failure to maintain or develop a certain force or power" (Asmussen, 1979). The nature and cause of fatigue depends on the type of exercise being performed. Differences in fatigue between strength and endurance athletes have been documented (Skurvydas et al, 2002; Edwards et al, 1977; Sahlin et al 1998) but are usually attributed to muscle metabolism. A number of studies, however, have reported that muscle fatigue is not always associated with metabolic changes (Edwards et al, 1977; Sahlin et al 1998). Decreases in muscle performance following exercise may be somewhat the result of the impaired utilization of muscle stiffness-mediated elastic energy (Avela & Komi, 1998). Komi et al (1986) reported similar findings when they concluded that repetitive impact loads may decrease the ability of the leg extensor muscles to maintain the necessary load and subsequently the muscle may lose its recoil ability. When active muscle is stretched, or when passively stretched muscle is suddenly activated, the muscle increases its tension and stores potential elastic energy in its series elastic component, which can then reappear during a subsequent shortening of the muscle. This phenomenon involving eccentric and concentric contractions is known as the stretch shortening cycle (SSC).

In coexistence with fatigue is the concept of potentiation. Post-activation potentiation (PAP) is an acute transient improvement in performance as a result of prior muscle activation. PAP initially exhibits a depression, most likely the result of acute fatigue, followed by a rapid rise in muscle function. Few studies have examined the effect of maximal SSC fatigue on the performance of subsequent SSC activities. Consequently the aim of this study was to investigate and compare the affect of maximum SSC fatigue on endurance and strength trained athletes and to also investigate the path of recovery with any subsequent PAP.

METHODS: Ten strength trained athletes (STA) (rugby players) and nine endurance trained athletes (ETA) (rowers) participated in this study. All 10 STA and 6 of the ETA were high level national athletes while 3 ETA were international athletes.

Table 1: Physical Characteristics of the Subjects.

	Age (years)	Height (cm)	Mass (kg)
ETA	25.9 ±6.3	187.7 ±9.1	82.2 ±11.4
STA	20.0 ±0.8	180.9 ±4.9	89.8 ±12.8

The nature of the study was explained to each participant and written informed consent was obtained. The protocol was approved by the University of Limerick research ethics committee. All participants attended the laboratory for one session to complete the testing. On arrival at the laboratory the drop jump (DJ) and rebound jump (RBJ) technique was explained and demonstrated to the participants and following a general warm up of jogging and stretching the participants performed practice jumps. The jumps were performed on a sledge and force plate apparatus with the sledge inclined at 30° and the AMTI OR5-6 force platform mounted at right angles to the sledge apparatus. For all jumps the participants were secured to the chair with a harness and waist belt. Instruction was given to the participants to keep both arms folded across the shoulders in order to minimise upper body movement during the jumps. The participants were also instructed to perform each jump maximally while attempting to minimise ground contact time (CT) and maximise jump height (JH). Prior to the fatigue workout sets of jumps comprised of one DJ followed immediately by a RBJ were performed to establish the participant's baseline values for each of the dependent variables: peak ground reaction force (pGRF), CT, JH and leg spring stiffness (k_{vert}). Testing commenced with four sets of DJ and RBJ with the participants being dropped from a predetermined height of 30 cm. Each set of jumps was immediately analysed using AMTI Bioanalysis software. Through use of the AMTI force plate instants of initial foot contact, take off and subsequent landing were obtained. From these ground reaction force traces, each of the dependent variables were calculated. Peak GRF was identified as the maximum force reading recorded from the ground reaction force traces from the force plate which was sampling at 1000Hz. CT was calculated as the difference between the time of initial foot contact and the time of take off. JH was calculated from the flight time (time difference between the take off and landing for jumps) and the use of the equation for linear motion $s = ut + 0.5at^2$. The calculation of k_{vert} involved the use of SVHS video recordings which were digitised using Peak Motus (Peak Performance Technologies, Colorado, USA) to identify the displacement from landing to full crouch. The pGRF was then divided by this displacement to calculate k_{vert} which is defined as the ratio of GRF to the displacement of a spring.

Following the analysis of each set of baseline jumps, the jump with the highest recorded JH was selected for further analysis. 90% of this maximum jump was calculated and this value was marked on sledge rails from a position where the participant was seated in the chair with the dominant leg fully extended. The fatigue workout then began with the participant being dropped from a height of 30cm for 1 DJ followed by repeated RBJ until the 90% mark was not reached on three consecutive jumps. 15, 45, 120, 300 and 600 seconds following the termination of the fatigue workout the participants were dropped from 30cm to perform one set of DJ and RBJ. From these recovery intervals each subject's minimum and maximum score for each dependent variable, irrespective of time, was identified. This allowed for identification of fatigue and any possible PAP without the interference of individual variation across recovery times.

Statistical Analysis: The software package SPSS (Version 16) was used to conduct all statistical analysis. A mixed effect split plot analysis of variance (SPANOVA) with repeated measures was used to evaluate differences between the average of the baseline scores and the minimum and maximum scores achieved during each recovery interval. The SPANOVA had 1 within-subjects factor namely Condition with 3 levels (baseline, minimum and maximum) and one between-subjects factor namely Group with 2 levels (ETA and STA)

RESULTS: Table 2 shows the mean ±SD of the number of jumps performed during the fatigue workout and the duration of the workout.

Table 2: Results of fatigue workout

	No of jumps	Duration of workout (s)
ETA	65.3 ±29.3	84.6 ±32.8
STA	55 ±24.9	69.3 ±21.2

The mean dependent variable scores for the baseline jumps were subtracted from the maximum and minimum scores for the jumps done after the fatigue workout and the results can be seen in figure 1. In this figure the baseline value is represented by the x-axis. The statistical results (GLM ANOVA) revealed a significant reduction in pGRF (p<0.01) and JH (p<0.01) in both groups of athletes following the fatigue workout and is illustrated by the "error" bars in the figure below. The STA also showed a significant reduction in k_{vert} (p<0.01), and while the ETA had a reduction of 6.38% in k_{vert} it was not a significant change. The difference between the baseline jumps and the maximum scores achieved during the 600 second recovery showed a significant increase in pGRF (p = 0.001), an increase of 8.44% and k_{vert} (p = 0.049), an increase of 23.75% for the ETA. The STA also showed increases in these dependent variables, a change in GRF of 7.34% and a change in k_{vert} of 11.62%, but neither was statistically significant.

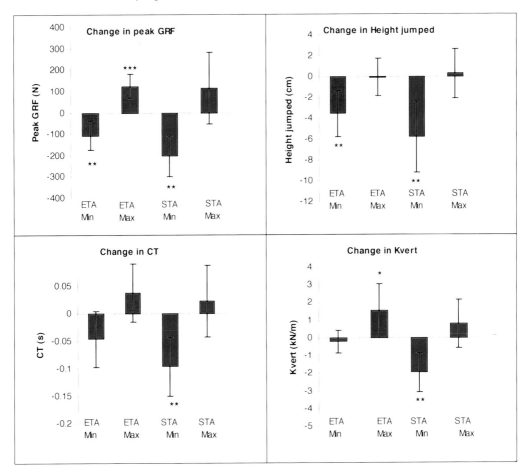

Figure 1: Mean ± 95% confidence interval pGRF, JH, CT and k_{vert} difference between the baseline jumps and the minimum and maximum values achieved following the fatigue workout. (***p<0.001; **p<0.01; *p<0.05))

DISCUSSION: The maximal fatigue workout resulted in significant reductions in pGRF and JH for both ETA and STA. STA also showed a significant reduction in k_{vert}. These findings are similar to studies which utilised submaximal SSC fatigue workouts (Avela & Komi,1998; Gollhofer et al. 1987 a,b; Nicol et al. 1991) and to a study which utilised maximal SSC fatigue workouts (Comyns, 2006). Comyns (2006) found a loss of efficacy in the SSC in a group of strength trained rugby players through a significant reduction in flight time and pGRF and an increase in ground contact time. Comyns (2006) also identified a reduction in k_{vert}, however it was not statistically significant.

Potentiation is known to co-exist with fatigue, whereby after an initial decline in performance following fatigue, there is a rapid rise in performance. While both ETA and STA in this study demonstrated an increase in performance following their initial decline, only the ETA demonstrated a significant improvement. The ETA showed a significant improvement in both pGRF (8.44% improvement) and k_{vert} (23.75% improvement). While ETA have been reported to withstand fatigue better and recovery quicker than strength athletes, (Hakkinen & Myllyla, 1990), there is a dearth of information surrounding the potential for ETA to benefit from the phenomenon of potentiation and several studies have suggested that ETA are indeed not capable of receiving potential benefits of potentiation. Comyns et al. (2005) reported a similar potentiation effect, to that found in this study, on the biomechanics of performance of a fast SSC exercise due to a prior contractile activity, except with strength trained athletes. The results of the present study, however, indicate that ETA have the ability to alter the biomechanics of rebound jumps and perform the jump with a stiffer leg action.

CONCLUSION: The results of the study indicate that ETA and STA fatigue significantly in pGRF generation and JH following a maximal fatigue workout. STA also show a significant reduction in k_{vert}, while ETA, despite showing a significant reduction in all other areas, appear to resist a significant reduction in k_{vert}. The results also indicate that during the recovery after a maximal fatigue workout, ETA have an ability to potentiate. This was evident from a significant increase in pGRF and k_{vert} above baseline values. These changes indicate that there is an enhancement in RBJ performance shortly after maximal fatigue as the RBJ is performed with a stiffer and more elastic leg spring action.

REFERENCES:
Asmussen. A. (179). Muscle Fatigue. *Medicine and Science in Sports.* **11(4),** 313-312
Avela, J., and Komi, P.V. (1998). Interaction between muscle stiffness and stretch-reflex sensitivity after long-term stretch-shortening cycle exercise. *Muscle & Nerve.* **21**, 1224-1227.
Comyns, T.M., Harrison A.J., & L.K. Hennessy (2006). The effect of a maximal stretch-shortening cycle fatigue workout on fast stretch-shortening cycle performance. *Proceedings of the XXIV International Symposium of Biomechanics*
Edwards, R., Hill, D., Jones, D. & Merton, P. (1977). Fatigue of long duration in human skeletal muscle after exercise. *J Physiol.* **272**, 769-778
Gollhofer, A., Komi P.V., Fujitsuka, N. and M. Miyashita. (1987a). Fatigue during stretch-shortening cycle exercises. II. Changes in neuromuscular activation patterns of human skeletal muscle. *International Journal of Sports Medicine.* **8** (Suppl.), 38-47.
Gollhofer, A., Komi P.V., Miyashita, M. and Aura, O. (1987b). Fatigue during stretch-shortening cycle exercises. I. Changes in mechanical performance of human skeletal muscle. *International Journal of Sports Medicine.* **8** (Suppl.), 71-78.
Hakkinen, K. & Myllyla, E. (1990). Acute effects of muscle fatigue and recovery on force production and relaxation in endurance, power and strength athletes. *J Sports Med Phys Fitn.* **30**, 5-12
Komi, P. (1986) Training of muscle strength and power: interaction of neuromotoric, hypertonic, and mechanical factors. *Inter J Sports Med.* **7**(suppl.) 10
Nicol, C. and Komi, P.V. (1991). Fatigue effects of marathon running on neuromuscular performance. I. Changes in muscle force and stiffness characteristics. *Scandinavian Journal of Medicine and Science in Sports.* **1**, 10-17.
Sahlin, K., Tonkonogi M. & Soderlund, K. (1998). Energy supply and muscle fatigue in humans. *Acta Physiol Scand.* **162**, 261-266
Skurvydas, A., Dudoniene, V., Kalvenas, A. & Zuoza, A. (2002). Skeletal muscle fatigue in long-distance runners, sprinters and untrained men after repeated drop jumps performed at maximal intensity. *Scand J Med Sci Sports.* **12**, 34-39

Acknowledgement
The researchers would like to thank IRCSET for providing funding for this research.

A STUDY OF SCULLING SWIMMING PROPULSIVE PHASES AND THEIR RELATIONSHIP WITH HIP VELOCITY

Raúl Arellano[1], Blanca de la Fuente[2] and Rocío Domínguez

Faculty of Physical Activity and Sports Sciences. University of Granada, Granada, Spain[1]
High Performance Centre, Sierra Nevada, Granada, Spain[2]

The purpose of this study was to identify the effect of sculling propulsive arm actions in displacement on the intra-cycle velocity of the hip. Four phases were defined (based on hand movements) prior to the development of the study: inward, pronation, outward and supination. A group of 9 international synchronized swimmers participated in the study. A displacement of 15 m was recorded using a velocimeter and underwater video cameras (bottom and sagittal views). Mean cycle velocity 0.548m/s, duration 0.828s, sculling frequency 1.220 sculling length 0,455 m and percentage of phase duration: inward (38.6%), pronation (10.3%), outward (33.5%) and supination (17.6%) were obtained. The mean body velocities were similar in the phases, while the durations where significantly different. The sculling propulsive action helps body displacement in the inward, outward and supination phases; while the pronation had a reduced contribution. Reversal stroke actions help to support the hand fixed it the water while the arm muscles are contracted helping the next propulsive phases (inward or outward) to move the hand an body forward.

KEY WORDS: unsteady propulsion, kinematics, velocimetry, vortex recapture

INTRODUCTION: Swimming stroke propulsion is composed of different hand and forearm propulsive actions that were classified as: outsweep, downsweep, insweep and upsweep (Costill, Maglischo, & Richardson, 1992). These propulsive actions can be studied from an steady or quasi-steady approach (Schleihauf, 1979) or considering unsteady propulsive mechanisms (Arellano, Terrés-Nicoli, & Redondo, 2006; Matsuuchi et al., 2009; Ungerechts, 2007). Sculling is a basic propulsive action applied in isolation in synchronized swimming or other aquatic sports, or that can be included in the pulling actions/trajectories of formal strokes. This simple propulsive action, at first sight, reveals an extraordinary complexity when it is carefully studied. Our previous 3D studies of supportive sculling while maintaining a vertical body position revealed a complex 3D path that is changed in length, inclination or peak velocities, under different vertical load conditions (Pochon & Arellano, 2007). Four phases can be defined in the sculling action when the unsteady propulsive mechanics that have been studied in similar biological propellers (Biewener, 2003): inward, pronation [stroke reversal 1], outward and supination [stroke reversal 2]. Supportive sculling (without body displacement) and propulsive sculling (where the body is displaced) were compared observing the pulling path and the wake produced in both conditions. Two pulling paths, "infinite" and "zigzag" shaped, were found (observing in this second case a continuous forward displacement of the hands). The wakes observed were clearly different with a vortex propulsive structure in the second case (Arellano & Pardillo, 2007). Hip velocimetry combined with sculling phase analysis seem a different procedure to analyse this propulsive action and it will be applied in our study. The aim of this study was to measure the effect of propulsive sculling action during horizontal body displacement, measured by the hip velocity and considering the sculling propulsive phases.

METHODS: Nine international synchronized swimmers participated as volunteers in the study. Their basic characteristics are described in Table 1. Each participant performed 3 trials executing propulsive sculling while keeping the body in a horizontal position. The trial with highest mean horizontal hip velocity was selected for the analysis. The swimmer was encouraged to perform as the highest velocity possible.

Table 1. Basic information about the participants

Variable	Mean	SD
Age (y)	20.9	1.4
Height (cm)	171.1	4.3
Mass (kg)	59.1	5.1

Data Collection: The hip velocity was obtained from a velocity–time data recorded using a position transducer, recording at 200 Hz. The apparatus consisted in a resistive sensor (which produced a resistance of 2.45N) with a coiled cable that was fastened to the swimmers' waists at the height of second and third lumbar vertebrae by means of a belt. An external box including a 12 bit analogue to digital converter plus and video processor to allow synchronization the recorded velocity with the video signal (50Hz) to combine both records and to save them in the computer through the USB 2.0 interface. Three views were combined in the video signals using a video mixer and a picture in picture video digital tool that is installed in the Altitude Training Centre of Sierra Nevada swimming pool (up 8 video signal can be combined in different frame positions). The bottom, sagital and lateral-posterior views were combined to observe in detail the initiation of each phase. At least three consecutive cycles were analyzed after the swimmer covered 8 m in the prescribed position.

Variables: The sculling actions were analyzing using the four phases durations (s), sculling frequency (Hz), body displacement by cycle (m), cycle mean velocity (m/s) and sculling phase mean velocity (m/s). To compare or produce coherent data analysis between participants, the durations were later normalized and expressed in percentages (0% was the initiation of the inward phase (I) and 100% was the end of the supination phase (IV).

Data analysis: Descriptive statistics were obtained mean and standard deviation plus ANOVA of repeated measures and post-hoc tests using standard statistical software.

RESULTS: The results are set out in Tables 2 and 3. This propulsive movement produced a limited body velocity; however, this velocity maintains a flat sinusoidal path throughout the cycle duration, with amplitude about ± 0.1 ms^{-1} related with the mean cycle velocity (0.548 ms^{-1}). The sculling frequency seems a little higher related to formal strokes, while the body length displacement is about one third or one quarter.

Table 2. Results (means and SD) obtained after records analysis per scull cycle.

Variables	Mean (SD) Total Cycle
Mean Velocity ms^{-1})	0.548 (0.097)
Duration (s)	0.828 (0.081)
Max. Velocity (ms^{-1})	0.639 (0.120)
Min. Velocity (ms^{-1})	0.475 (0.102)
Sculling Frequency (Hz)	1.220 (0.140)
Sculling Length (m)	0.455 (0.092)

Table 3. Results (means and SD) obtained after records analysis per phase of scull cycle

Variables	Phase I Inward	Phase II Pronation	Phase III Outward	Phase IV Supination
Velocity (m/s)	0.538 (0.106)	0.536 (0.100)	0.543 (0.090)	0.578 (0.106)
Duration (s)	0.322 (0.061)	0.086 (0.013)	0.278 (0.061)	0.143 (0.044)
Percentage (%)	38.58 (4.99)	10.35 (1.34)	33.47 (6.33)	17.6 (6.54)

The phase analysis enabled us to define specific durations related to the hand movements. The inward phase is about a 40% of the total duration, while the pronation is the shortest one. Outward is about a third of total cycle duration and supination is longer than pronation with higher variability. Significant differences have been found between all durantion's phases after apply a post-hoc test of ANOVA of repeated measures ($p<0.01$). However these differences have not been found in the mean velocities between phases. Peak velocities were more frequently obtained in the inward phase (5 cases of 9) and the supination phase (2 cases of 9), while the slowest velocities were more frequently produced at the inward phase (6 cases of 9) and while the swimmers performed the pronation (3 cases of 9).

DISCUSSION: Typical propulsive actions are described during the sculling phases in the inward phase. The path and hand position supported quasi-steady based propulsion with a possible lift and drag propulsive component involved. The velocities values obtained during stroke reversal phases (II & IV) were high enough to keep the body velocity similar or closer than typical propulsive phases (I & III). This stroke reversal is characterized by a complex wake interaction with propulsive hand that should produce enhanced lift [as is explained in several papers on animal biomechanics, see Biewener (2003). These effects added to the inertial influences that could be taken into all phases should explain all the phenomena. A clear forward displacement of the hand during the pulling trajectory, observed mainly through the bottom camera, in a zigzag shape, confirms the previous statements to explain the velocity of the body path.

CONCLUSION: The sculling propulsive action helps body displacement in the inward, outward and supination phases; while the pronation had a reduced contribution. Reversal stroke actions help to support the hand fixed it the water while the arm muscles are contracted helping the next propulsive phases (inward or outward) to move the hand an body forward. A wake analysis combined with the acceleration of the body will enable us to understand the sculling propulsive action in depth in the near future.

REFERENCES:
Arellano, R., & Pardillo, S. (2007). *Study of sculling actions during hovering and displacement, applying cinematic analysis, flow visualization and velocimetry (Abstract)* Paper presented at the European College of Sport Science Conference 2007, Jyvaskyla, Finland.
Arellano, R., Terrés-Nicoli, J. M., & Redondo, J. M. (2006). Fundamental hydrodynamics of swimming propulsion. *Portuguese Journal of Sport Science - Suppl. Biomechanics and Medicine in Swimming X, 6*(Supl.2), 15-20.
Biewener, A. A. (2003). *Animal locomotion* (1 ed. Vol. 1). Oxford: Oxford University Press.
Costill, D. L., Maglischo, E. W., & Richardson, A. B. (1992). *Swimming* (1 ed.). Oxford: Blackwell Scientific Publications Ltd.
Matsuuchi, K., Miwa, T., Nomura, T., Sakakibara, J., Shintani, H., & Ungerechts, B. E. (2009). Unsteady flow field around a human hand and propulsive force in swimming. *J Biomech, 42*(1), 42-47.
Pochon, A., & Arellano, R. (2007). Analysis of a 3d sculling path in a vertical body position under different load conditions. In J. A. S. M. Raúl Arellano Colomina, Fernando Navarro Valdivielso, Esther Morales Ortiz y Gracia López Contreras (Ed.), *SWIMMING SCIENCE I* (pp. 239-244). Granada: Universidad de Granada.
Schleihauf, R. E. (1979). A Hydrodynamical Analysis of Swimming Propulsion. In T. a. Bedingfield (Ed.), *SWIMMING III - Third Int.Symp.of Biomechanics in Swimming* (1 ed., pp. 70-109). Baltimore, Maryland (Estados Unidos): University Park Press.
Ungerechts, B. E. (2007). Unsteady mechanisms of swimming propulsion. In J. A. S. M. Raúl Arellano Colomina, Fernando Navarro Valdivielso, Esther Morales Ortiz y Gracia López Contreras (Ed.), *SWIMMING SCIENCE I* (pp. 31-40). Granada: Universidad de Granada.

Acknowledgement
The authors would like to thank all those participants' members of the Italian Synchronized Swimmers of the Italian National Team and particularly their coaches Laura de Renzis and Roberta Farinelli.

PROPOSAL OF AN ADDITIONAL-INERTIA TUNING METHOD TO VISCOUS LOAD FOR HIGH SPEED MOTION TRAINING

Kenji Shigetoshi[1], Sadao Kawamura[2] and Tadao Isaka[2]

Multimedia Center, Shiga University of Medical Science, Otsu, Shiga, Japan[1]
Department of Robotics, Ritsumeikan University, Kusatsu, Shiga, Japan[2]

This research proposes a new training system for high speed motions. In the proposed training system, a relatively small inertia is added to a viscous load in order to increase the load without decreasing the maximum speed. Moreover, in order to increase energy consumption of humans, we artificially make the additional inertia zero during motions. Consequently, the additional inertia is tuned to increase the load torque and the energy with keeping the maximum speed. The effectiveness of the proposed system is verified by some experimental results.

KEY WORDS: Muscle Strength Training, Mechanical Impedance, Variable Load, High Speed Training.

INTRODUCTION: Traditional muscle strength training devices are commonly classified into three types (inertia, viscosity, and elasticity). If inertia, viscosity or elasticity is individually used as a training load, there are basic problems. An Inertia load can give large load values to concentric muscles during accelerating, but during decelerating cannot. A viscous load can give load values in a full range of motions, but the load values become small at the beginning and the ending of motions. An elastic load cannot give large load values in the beginning of motions because this load increases in proportion to the position. Therefore each load has the merit and the demerit.

In this research, we consider training systems for high speed motions. If we base on training specificity, training systems should realize real high speed in order to improve athletic performance. It is reasonable to use viscous loads for high speed motion training because it is expected that viscous loads can easily guarantee safety in comparison with others. However, as mentioned above viscous loads can not make large load values in the beginning and ending of motions.

To solve this problem with viscous loads, this research proposes a new additional-inertia tuning method. In the proposed training system, a relatively small inertia is added to a viscous load in order to increase the load torque without decreasing the maximum speed. Moreover, in order to increase energy consumption of humans, we artificially make the additional inertia zero during motions. Consequently, the additional inertia is tuned to increase the load torque and the energy with keeping the maximum speed.

In recent studies, the effects of variable resistance training (VRT) were examined. However, the load combination patterns are limited and the load values can not be changed during motions because traditional inertia, viscosity, and elasticity devices were utilized (Wallace et al., 2006; Coker et al., 2006).

In order to exquisitely change inertia values, this research uses a variable mechanical impedance device, which was developed by the authors (Shigetoshi et al., 2008). By using this device, the additional inertia load can be suitably tuned in the beginning of motions and can be eliminated in the ending of motions. The effectiveness of the proposed system is verified by some experimental results for elbow flexion.

METHODS: Training System: This system consists of a direct drive motor with rotary encoder, a control computer and a lever arm with a handle grip. The initial angle of the lever arm sets at 0.785 rad from the downward vertical direction. For safety, a stopper is set in order to limit the lever arm rotation from 0 to 2.6 rad. To realize the variable mechanical impedance, this research applied a control method and mechanical impedance (inertia, viscosity, elasticity) could have variable values. More details are shown in (Shigetoshi et al., 2008).

Subject: One adult male (age, stature and body mass: 34 years, 1.78 m and 72.0 kg) participated in this study. Informed consent to participate in this study was obtained from this subject.

Experimental Conditions: We instructed the subject to generate the maximum muscle strength. Maximal voluntary elbow flexion from full extension to full flexion was measured. The upper arm was put on a stand. The centre of the elbow joint axis was aligned to the axis of the lever arm. Loading patterns we used are as follows.
1. Viscous Load: 0.5, 0.8, 1.2, 2.0, 3.0, 5.0 Nm/(rad/s) of viscous coefficients were used to obtain the standard data. In fact, it is impossible to perfectly eliminate the inertia and the elasticity because instability of a direct drive motor control occurs. Therefore, the inertia and the elastic coefficients was set 0.02 Nm/(rad/s^2) and 0.2 Nm/rad, respectively.
2. Additional Inertia: To increase the torque of the beginning of the motion and not to decrease the maximum speed, we investigated the suitable additional inertia value. As the result, we found that 0.2 Nm/(rad/s^2) was appropriate as the additional inertia.
3. Inertia Zeroing: The inertia coefficient 0.2 Nm/(rad/s^2) from 0 rad to $\pi/3$ rad of the joint angle was set. After $\pi/3$ rad, the inertia coefficient was set 0.02 Nm/(rad/s^2).

Data Analysis: Values are reported as mean ± SD. The experimental results for different loading were compared using One-way Factorial ANOVA. The level of significance was set $p \leq 0.05$.

RESULTS AND DISCUSSION:

Experimental Results of Additional Inertia Load: Figure 1 represents the experimental results of the viscosity case and the additional inertia case. The broken line shows the viscous load, and the solid line shows the additional inertia load in each graph. Figure 1 (A) shows the relationship between the angular velocity and the joint angle. It is observed that the maximum speed slightly downs in the case of the additional inertia load. On the other hand, the torque at the beginning of the motion significantly becomes large as seen in Figure 1 (B) and Figure 1 (C). Here, it should be noted that the negative torque is generated by inertia effect.

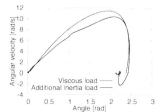 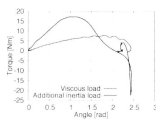 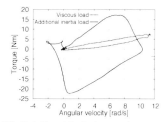

(A) Relationship between angular velocity and joint angle   (B) Relationship between torque and joint angle   (C) Relationship between torque and angular velocity

Figure 1: Viscous load vs. additional inertia load at low intensity load. Viscous load: inertia=0.02 Nm/(rad/s^2), viscosity=0.5 Nm/(rad/s), elasticity=0.2 Nm/rad. Additional inertia load: inertia=0.2 Nm/(rad/s^2), viscosity=0.5 Nm/(rad/s), elasticity=0.2 Nm/rad

Figure 2 represents the comparison of the viscous load with the additional inertia load for the maximum torque, the energy consumption and the maximum angular velocity. Figure 2 (A) shows the maximum torque on each load. From the figures, it is clear that the additional inertia load realizes much higher torque than the viscous load at 0.5 – 1.2 Nm/(rad/s) of viscous coefficient because the additional inertia becomes dominant at low viscous coefficients cases. Figure 2 (B) shows the energy consumption from the motion start to the end on each load. That was calculated by Eq. (1).

$$E = \int_0^T \tau_h(t)\dot{\theta}dt \qquad (1)$$

where T, $\tau_h(t)$, $\dot{\theta}$ denote the terminal time of motions, subject's exerted torque, angular velocity, respectively. From 0.5 to 3.0 Nm/(rad/s) of viscous coefficient, the energy consumption of the additional inertia load intends to be low as compared with the viscous load because negative torque is induced in the ending motion by the additional inertia. Figure 2 (C) shows the maximum angular velocity on each load. From 0.5 to 2.0 Nm/(rad/s) of viscous coefficient, the maximum angular velocity is significantly low as compared with the viscous load.

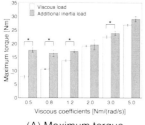
(A) Maximum torque
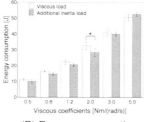
(B) Energy consumption
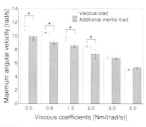
(C) Maximum angular velocity

Figure 2: Comparison between viscous load and additional inertia load for maximum torque, energy consumption and maximum angular velocity (* indicates significant difference at 0.05 level). Viscous load: inertia=0.02 Nm/(rad/s^2), viscosity=0.5-5.0 Nm/(rad/s), elasticity=0.2 Nm/rad. Additional inertia load: inertia=0.2 Nm/(rad/s^2), viscosity=0.5-5.0 Nm/(rad/s), elasticity=0.2 Nm/rad

Experimental Results of Inertia Zeroing: Figure 3 represents the characteristics of the inertia load. Figure 3 (A) shows a torque element induced by acceleration. Figure 3 (B) and Figure 3 (C) show angular velocity and power calculated by multiplying the torque and the angular velocity, respectively. This simply demonstrates the reason why the negative energy consumption is generated in deceleration phase.

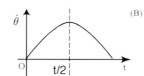
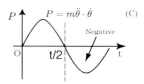

Figure 3: Characteristics of the inertia load

Figure 4 represents the experimental results of the viscosity case and the inertia zeroing case. The broken line shows the result of the viscous load and the solid line shows the result of the inertia zeroing load in each figure. Figure 4 (A) shows the relationship between angular velocity and joint angle. As seen in Figure 4 (A), the two peaks of speed are almost identical even though the phase of the peak is moved. It is understood that the angular velocity can increase after inertia zeroing. Figure 4 (B) and (C) demonstrate the effectiveness of the elimination of the negative torque.

Figure 5 represents the comparison of the viscous load with the inertia zeroing load for the maximum torque, the energy consumption and the maximum angular velocity. It is worth to compare Figure 2 with Figure 5. At first, the maximum torque can be effectively increased in both Figures 2 (A) and 5 (A). Next, the energy consumption can be significantly increased in low viscosity areas because of the elimination of the inertia as seen in Figures 2 (B) and 5 (B). Finally, the maximum angular velocity of the zeroing inertia (Figure 5 (C)) becomes closer to that of the viscosity load than that of the non-zeroing load (Figure 2 (C)).

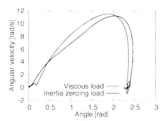 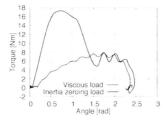 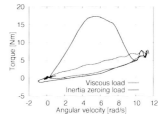

(A) Relationship between angular velocity and joint angle (B) Relationship between torque and joint angle (C) Relationship between torque and angular velocity

Figure 4: Viscous load vs. inertia zeroing load at low intensity load. Viscous load: inertia=0.02 Nm/(rad/s^2), viscosity=0.5 Nm/(rad/s), elasticity=0.2 Nm/rad. Inertia zeroing load: inertia=0.2 to 0.02 Nm/(rad/s^2), viscosity=0.5 Nm/(rad/s), elasticity=0.2 Nm/rad

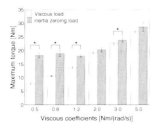 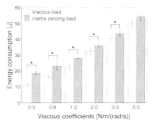 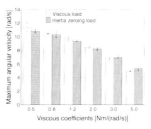

(A) Maximum torque (B) Energy consumption (C) Maximum angular velocity

Figure 5: Comparison between viscous load and inertia zeroing load for maximum torque, energy consumption and maximum angular velocity(* indicates significant difference at 0.05 level). Viscous load: inertia=0.02 Nm/(rad/s^2), viscosity=0.5-5.0 Nm/(rad/s), elasticity=0.2 Nm/rad. Inertia zeroing load: inertia=0.2 to 0.02 Nm/(rad/s^2), viscosity=0.5-5.0 Nm/(rad/s), elasticity=0.2 Nm/rad

CONCLUSION: In this paper, we proposed a new strength training system for high speed motions. This research investigated how inertia should be added to viscous load in order to increase torque and energy consumption and not to decrease maximum speed of motions. From the experimental results, simple additional inertia to viscous load can increase the maximum torque but the energy consumption and the maximum angular velocity decrease. Therefore, we introduced the inertia zeroing method. It was experimentally verified that the inertia zeroing method can increase the maximum torque and the energy consumption without decreasing the maximum angular velocity.

REFERENCES:

Coker, C. A., J. M. Berning & Briggs, D. L. (2006). A preliminary investigation of the biomechanical and perceptual influence of chain resistance on the performance of the snatch. *Journal of Strength and Conditioning Research*, 20(4), 887-891.

Shigetoshi, K., T. Isaka & S. Kawamura. (2008). A new evaluation system for dynamic muscular strength characteristics using isoviscous loading. *The Engineering of Sport 7*. M. Estivalet and P. Brisson. 2, 419-428.

Wallace, B. J., J. B. Winchester & McGuigan, M. R. (2006). Effects of elastic bands on force and power characteristics during the back squat exercise. *Journal of Strength and Conditioning Research*, 20(2), 268-272.

FACTORS IN UPPER EXTREMITY LOADING IN THE POWER DROP EXERCISE

Duane Knudson[1] and ChengTu Hsieh[2]
Texas State University, San Marcos, TX, USA[1]
California State University-Chico, Chico, CA, USA[2]

This case study examined the factors that were related to peak vertical force applied to a medicine ball in an upper body plyometric exercise. Sagittal plane video and force platform data were collected for two male athletes performing 30 power drop exercises with a 5 kg medicine ball. Force on the medicine ball, net joint torques, and several technique variables were analyzed with partial correlations. Drop height was related to the impulse of the exercise, but was not uniquely associated with higher peak forces measured by video or the force platform because of intercorrelations between joint torques. Peak forces on the medicine ball were 44 to 69% of the peak vertical ground reaction forces (600 Hz) and were not uniquely associated with drop height.

KEY WORDS: force, medicine ball, stretch-shorten cycle, plyometrics.

INTRODUCTION: Plyometrics are a common training strategy for improving performance in a variety of high-intensity and speed athletic events. Upper-body plyometrics (UBP) often use medicine balls, kettle bells, and instrumented Smith machines (Wilson, et al., 1993) to train stretch-shortening cycle (SSC) muscle actions. This ballistic throwing of a training mass results in greater muscle activation and larger percentages of the range of motion with positive acceleration, minimizing the negative acceleration phase in normal weight lifting (Newton et al., 1996, 1997). Forces applied to a medicine ball in UBP exercises are likely higher than the small weight of the resistance because of the large accelerations during the UBP movements (Knudson, 2001; Newton et al., 1996).

Previous research has attempted to estimate forces in UBP training for the power drop exercise by bouncing MB's off a force platform (Ebben et al., 1999). Knudson (2001) reported that this methodology was inaccurate because of the differences between a rigid platform and an exercising human. In fact, higher drop heights in the power drop did not always result in larger forces in some athletes (Knudson, 2001). Perceptual and technique variations may affect the duration and intensity of each power drop exercise as much as drop height (Knudson, 2001). Plyometric push-ups and power drop exercises have total contact times between 0.3 to 0.6 seconds (Jones et al., 1999; Knudson, 2001). Similar to lower extremity plyometrics (Bobbert et al., 1986), there is variation in how athletes perform UBP exercise that likely affects the training forces experienced in UBP (Knudson, 2001).

There is a need to understand the typical forces on the upper extremity in UBP exercises so that safe and effective training loads can be prescribed. The purpose of this case study was to examine the factors related to peak vertical force loading on the hands and peak joint torques in the power drop exercise for two experienced athletes. We also specifically examined how well the force platform method correlated with the more time intensive quantitative videography in documenting power drop loads. It was hypothesized that forces applied to the MB would vary according to interactions of exercise variables within each athlete.

METHODS: Two intercollegiate male athletes experienced in UBP exercise (94.6 and 88.4 kg) gave informed consent to participate in the study and attended a single testing session. Reflective markers were placed on the joint axes of the right arm. Following a warm-up the subjects performed 25 power drop UBP exercises with a 5 kg MB dropped from heights between 0.5 and 1.4 m. The heights were normally distributed and presented in a random order. There was approximately one minute of rest between each exercise. Subjects then performed 5 power drop exercises from the same height (0.8 m) to document the reliability of the dependent variables. The subjects were able to execute these exercises with the arms primarily in a sagittal plane.

Power drop exercises were performed in a supine position with flexed knees and hips on a small (100 by 33 cm) bench placed on top of a Kistler 9286 force platform. Vertical force data (600 Hz) was synchronized with kinematic data and analyzed with Kistler Bioware® software. To document the MB and upper extremity motion in the power drops, all trials were videotaped (60 Hz) in the sagittal plane. A two-dimensional rigid body model of the MB, hand, forearm, and upper arm of the left upper extremity was created. The center of the MB and the four markers were digitized from 10 fields before hand-ball contact to 10 fields after release using Vicon Motus® 9.2 software. All kinematic data were smoothed with a Butterworth digital filter using the automatic cut-off frequency selected by the system.

The loading variables examined were the vertical impulse (J) and peak vertical force measured by the force platform (PF_{FP}), and the peak net joint torques at the wrist (PT_W), elbow (PT_E), and shoulder (PT_S) from inverse dynamics with wrist extension, elbow and shoulder flexion defined as positive. Kinetics were calculated from the MB down the upper extremity. The peak vertical force on the MB (PF_V) was calculated from vertical acceleration measures of the MB and Newton's 2nd Law of motion, and was then compared to PF_{FP}. The technique variables examined in the study were the drop height, vertical hand velocity one field prior to contact, duration of contact with the MB, and the percentage of contact with negative vertical MB velocity. The association between the dependent variables and technique variables were analyzed within-subject with correlation and partial correlations with statistical significance accepted at the $p < 0.05$ level. Data reliability were documented with coefficients of variation of the five repeated trials with the same drop height. Descriptive data are reported as means (SD).

RESULTS AND DISCUSSION: The dependent variables showed good consistency, although peak joint torques showed more variability than forces or impulse. Mean coefficients of variation for the loading variables were 4% J, 5% PF_{FP}, 10% PF_V, 18 % PT_W, 26% PT_E, and 9% PT_S.

Partial correlation analysis showed the variables that were significantly and uniquely ($p < 0.05$) associated with drop height were specific to each athlete. Drop height also had significant partial correlations with J of the exercise ($r = 0.81$) and T_S ($r = 0.41$) for subject 1. Drop height was significantly and uniquely associated with J ($r = 0.80$) and percentage contact with negative vertical MB velocity ($r = -0.51$) for subject 2. The lack of an association between drop height and PF_{FP} was in agreement with several of the subjects studied by Knudson (2001).

The zero-order correlation ($r = 0.60$) between drop height and PF_V also disappeared in the partial correlation analysis for subject 1. This could be due to significant intercorrelations (|r| > 0.64) between PF_V and all peak joint torques. Significant partial correlations between joint torques in both subjects confirmed technique interactions between joints that likely confound associations between drop height and loading in power drop exercises. The peak torque at the elbow was significantly associated with the torque at the wrist ($r = 0.45$ and 0.77) and the shoulder ($r = 0.57$ and 0.37) for subjects 1 and 2, respectively. Both subjects also had significant partial correlations ($r = 0.50$ and 0.44) between vertical hand velocity prior to contact and the percentage of contact with negative vertical MB velocity.

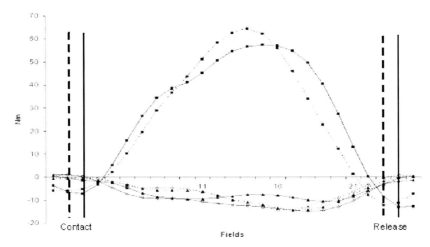

Figure 1. Typical patterns of net wrist ◊, elbow ▲, and shoulder ■ torques for both subjects (S1 – and S2 ---) in the power drop exercise. MB contact and release indicated by vertical lines.

Both subjects had similar patterns of net wrist and shoulder torques. There was, however, a different pattern of elbow torques between the subjects (Figure 1).It is possible the large variability of the elbow net torque is a result of the interaction of joint torques in this exercise for these subjects. These data support the hypothesis that loading in power drop exercise is not easily predicted by drop height because of complex interactions of technique variables. Descriptive data for the loading variables across all trials are reported in Table 1.

PF_{FP} was apparently correlated (r = 0.72 and 0.76) with PF_V for both subjects, but the partial correlations showed the unique associations were not statistically significant (r = 0.24 and 0.35, respectively). The interactions of technique variables and acceleration of upper extremity mass reduces the potential association between PF_{FP} and PF_V.

Table 1 Mean (SD) Loading Variables Over 30 Power Drop Exercises

	J	PF_{FP}	PF_V	PT_W	PT_E	PT_S
	Ns	N	N	Nm	Nm	Nm
Sub 1	85 (6)	548 (111)	250 (23)	-19 (3)	-15 (4)	70 (9)
Sub 2	84 (6)	494 (69)	235 (21)	-13 (4)	-11 (4)	66 (8)

* See method for abbreviations.

Knudson (2001) reported that estimating power drop exercise loads by dropping medicine balls on force platforms (Ebben et al., 1999) was inaccurate and unlikely to be useful in planning training. The results of the present study confirm and extend this observation. PF_{FP} over a 300 to 420 ms exercise would be different from the peak force of a MB bouncing off a rigid force platform in 40 to 50 ms. More importantly, PF_V were 44 to 69% of the PR_{FP} and were not significantly correlated for these two subjects. This overestimation using the force platform alone is likely related to not accounting for the acceleration of the mass of the arms and is consistent with research of this effect in bench press exercises (Rambaud et al., 2008).

Some of the limitations of this study were small sample size and the typical assumptions of 2D inverse dynamics. Despite these limitations, the subjects in this study confirmed the hypothesis of technique variations in power drop exercises that do not easily allow drop height or force platform measurements to predict athlete loading. This is consistent with research on indicating that desirable loading is difficult to establish because of interactions with machine restraints, duration, range of motion, and percentages of time that inertial resistances are accelerated and negatively accelerated (Frost et al., 2008a, 2008b; Newton et al., 1996).

CONCLUSION: Increasing MB drop height in power drop exercises for two experienced athletes was correlated with the vertical impulse of the exercise, but not peak vertical forces due to the interaction of technique factors acceleration of upper extremity mass. These data were consistent with previous research on ballistic exercises, and suggests that force platform data alone cannot be used to estimate loading in the power drop exercise.

REFERENCES:
Bobbert, M.F., Mackay, M., Schinkelshoek, D., Huijing, P.A., & van Ingen Schenau, G.V.. (1986). A biomechanical analysis of drop and countermovement jumps. *European Journal of Applied Physiology*, 54, 566-573.

Ebben, W.P., Blackard, D.O., & Jensen. R.L. (1999). Quantification of medicine ball vertical impact forces: estimating effective training loads. *Journal of Strength and Conditioning Research*, 13, 271-274.

Frost, D.M., Cronin, J.B., & Newton, R.U. (2008a). A comparison of the kinematics, kinetics, and muscle activity between pneumatic and free weight resistance. *European Journal of Applied Physiology*, 104, 937-956.

Frost, D.M., Cronin, J.B., & Newton, R.U. (2008b). Have we underestimated the kinematic and kinetic benefits of non-ballistic motion? *Sports Biomechanics*, 7, 372-385.

Jones, K, Hunter, G., Fleisig, G., Escamilla, R., & Lemak, L. (1999). The effects of compensatory acceleration on upper-body strength and power in collegiate football players. *Journal of Strength and Conditioning Research*, 13, 99-105.

Knudson, D. (2001). Accuracy of predicted peak forces during the power drop exercise. In J.R. Blackwell (Ed.) *Proceedings of Oral Sessions: XIX International Symposium on Biomechanics in Sports*, (pp. 135-138). San Francisco, CA: University of San Francisco.

Newton R.U., Kraemer, W.J., Hakkinen, K., Humphries, B.J., & Murphy, A.J. (1996). Kinematics, kinetics, and muscle activation during explosive upper body movements. *Journal of Applied Biomechanics*, 12, 31-43.

Newton R.U., Murphy, A.J., Humphries, B.J., Wilson, G.J., Kraemer, W.J., & Keijo, Hakkinen. (1997). Influence of load and stretch shortening cycle on the kinematics, kinetics and muscle activation that occurs during explosive upper-body movements. *European Journal of Applied Physiology*, 75, 333-342.

Rambaud, O., Rahmani, A., Boyen, B., & Bourdin, M. (2008). Importance of upper-limb inertia in calculating concentric bench press force. *Journal of Strength and Conditioning Research*, 22, 383-389.

Wilson G.J, Newton, R.U., Murphy, A.J., & Humphries, B.J. (1993). The optimal training load for the development of dynamic athletic performance. *Medicine and Science in Sports and Exercise*, 25, 1279-1286.

MECHANICAL EFFICIENCY IN BASEBALL PITCHING

Dave Fortenbaugh, Glenn Fleisig

American Sports Medicine Institute, Birmingham, AL, USA

Efficient pitching mechanics should maximize ball velocity while minimizing stress on the pitching arm. The purpose of this study was to quantify the relationship between ball velocity and upper extremity kinetics (UEKs) and define the kinematic patterns that achieve the most efficient pitching mechanics. Healthy collegiate and professional pitchers (n=147) threw maximal effort pitches from the wind-up. After determining the overall relationship between ball velocities and UEKs, two subgroups of pitchers were identified as efficient and inefficient. Efficient pitchers had significantly more ball velocity and similar or lower kinetic values. 10 of 23 kinematic variables were significantly different between the groups. It is recommended that coaches and researchers use the efficient group's mechanics as a point of reference when analyzing and teaching pitching biomechanics.

INTRODUCTION: Mechanical efficiency is the ratio of input energy to output energy. Loosely defined in terms of baseball pitching, the input energy can be seen as the mechanical stresses placed on the arm while the output energy is the ball velocity. The mechanical stresses (i.e. forces and torques) placed on the shoulder and elbow joints during baseball pitching routinely approach the limit that those structures can withstand with every pitch (Buchanan, Delp, & Solbeck, 1998; Morrey & An, 1983). To minimize arm strain and reduce the risk of injury, pitchers seek to maximize ball velocity through total body mechanical efficiency, bearing larger loads on the stronger leg and core muscles rather than the weaker arm muscles. Escamilla et. al. (2007) found a significant loss in ball velocity without a significant reduction in upper extremity kinetic (UEK) values as pitchers approached fatigue. This means that even though they are in a weakened state from being fatigued, pitchers continue to place extremely high loads on the shoulder and elbow joints. Previous research has confirmed that fatigue is a major risk factor for arm injuries that necessitate surgery (Olsen et. al., 2006).

At least six UEK parameters have been associated with ball velocity: shoulder anterior and proximal force, shoulder horizontal adduction and internal rotation torque, and elbow varus and flexion torque (Stodden et. al., 2005). Positive correlations can be seen in reported pitching biomechanics data between upper extremity kinetics (UEKs) and ball velocity (Fleisig et. al., 1999; Stodden et. al., 2005), though the exact nature of the relationship has yet to be determined. It is assumed that there are at least two main components that establish this relationship: one general and one specific. The general relationship stems from Newton's second law motion that states that force is directly proportional to acceleration given a constant mass. In the case of baseball pitching, this means that more force/torque must be applied to the body, and subsequently to the baseball, in order to accelerate the arm faster and release the ball with greater velocity. From this concept, one might expect to find a nearly perfect linear relationship between upper extremity kinetics and velocity. However, it is very unrealistic to assume that pitchers can apply force that perfectly. This leads to the second component of the relationship, which is what logically separates the efficient pitchers from the inefficient ones. Those who are able to best utilize the kinetic chain to maximize their ball velocity and simultaneously minimize UEKs are placing the least amount of stress on their arms for a given performance level. The first purpose of this retrospective study was to assess the linearity of the relationship between UEKs and ball velocity. It was hypothesized that strong linear correlations between them would be confirmed. The second purpose of the study was to develop a method for determining what constitutes an efficient pitcher and how the biomechanical profiles of mechanically efficient and inefficient pitchers can be differentiated.

METHODS:
Data were collected from healthy collegiate and professional baseball pitchers (n=147) tested at the American Sports Medicine Institute. The biomechanical testing procedures followed a previously reported protocol (Escamilla et. al., 1998). Each pitcher threw up to 10 fastballs with maximum effort at the regulation distance of 18.4m. All available trials were used for analysis, and each pitcher's data were derived from the average among trials. Players' height, mass, ball velocity, and biomechanical variables (23 kinematic, 6 kinetic, and 2 temporal) were measured. The first step in data analysis was to calculate Pearson r correlation values between ball velocity and each of the seven UEK parameters. Next, means and standard deviations (SDs) were calculated for ball velocity and the UEKs. For each variable of each pitcher, values of these parameters were classified as high (greater than one SD above the mean), average (within one SD of the mean), or low (more than one SD below the mean). Efficient pitchers were defined as those in the high velocity group with average or low UEKs and those in the average velocity group with low UEKs, while inefficient pitchers were those in the average velocity group with high UEKs and those in the low velocity group with average or high UEK values. Efficiency levels for each pitcher (efficient, normal, or inefficient) were initially determined for all six UEK values, but the focus was centered on four specific values (shoulder horizontal adduction torque and proximal force and elbow varus and flexion torque). This was done to isolate different phases of the delivery, reduce redundancy, and equally weight the effects of the risk of shoulder and elbow injuries. The pool of pitchers was ultimately reduced to two groups: those with three or four out of four efficient kinetics and no inefficient kinetics (n=16), and those with three or four out of four inefficient kinetics and no efficient kinetics (n=16). Independent t-tests were used to compare the two groups across all biomechanical and anthropometric variables, as well as ball velocity. To help protect against family-wise errors, for all tests, $\alpha=.01$.

RESULTS: The bivariate correlations between ball velocity and upper extremity kinetics for all pitchers with kinetic data (n=145) are shown in Table 1. All six kinetic values were significantly correlated with ball velocity ($p<.01$), with shoulder proximal force and elbow flexion torque showing correlation coefficients greater than or equal to 0.50.

Table 1. Coefficients of correlation between upper extremity kinetics and ball velocity

Upper extremity kinetic (UEK) parameter	Correlation coefficient
Shoulder Proximal Force (N)*	0.57
Elbow Flexion Torque (Nm)*	0.50
Shoulder Internal Rotation Torque (Nm)*	0.48
Elbow Varus Torque (Nm)*	0.48
Shoulder Horizontal Adduction Torque (Nm)*	0.35
Shoulder Anterior Force (N)*	0.22

* $p<.01$

The efficient pitchers had similar heights as the inefficient pitchers (192 cm to 189 cm), but the efficient pitchers had significantly less mass (88 kg to 96 kg) and threw with significantly greater ball velocity (39 m*s^{-1} to 35 m*s^{-1}). The comparisons of kinematic, temporal, and kinetic data between the efficient pitchers and inefficient pitchers are shown in Tables 2, 3, and 4 respectively. 10 of the 23 kinematic parameters were significantly different between groups, but neither of the temporal variables was significantly different. Efficient pitchers exhibited significantly greater maximum pelvis, upper trunk, and shoulder internal rotation velocities. They also had greater maximum shoulder external rotation, shoulder horizontal abduction at FC, and forward trunk tilt and elbow extension at BR. Finally, efficient pitchers maintained a shoulder abduction angle closer to 90 degrees at FC and BR and had less lateral trunk tilt at FC. Efficient pitchers had either significantly lower UEK values (shoulder anterior force and horizontal adduction torque) or statistically indifferent UEK values (shoulder proximal force and internal rotation torque, elbow varus and flexion torque) for all kinetic variables.

Table 2. Kinematic differences between efficient and inefficient pitchers.

Variable	Efficient	Inefficient
Stride length ratio (% ht)	83 ± 4	83 ± 3
Lead foot position (cm)	20 ± 10	19 ± 10
Lead foot angle (deg)	14 ± 8	12 ± 11
Lead knee flexion at FC (deg)	43 ± 8	42 ± 7
Pelvis rotation at FC (deg)	33 ± 11	32 ± 12
Pelvis to upper trunk separation at FC (deg)	53 ± 9	46 ± 9
Lateral trunk tilt at FC (deg)*	3 ± 5	9 ± 9
Shoulder abduction at FC (deg)	86 ± 11	96 ± 14
Shoulder horizontal abduction at FC (deg)*	30 ± 11	19 ± 12
Shoulder external rotation at FC (deg)	42 ± 25	53 ± 20
Elbow flexion at FC (deg)	90 ± 9	91 ± 17
Max pelvis rotation velocity (deg/s)*	621 ± 86	535 ± 68
Upper trunk rotation velocity (deg/s)*	1155 ± 71	1058 ± 72
Max shoulder external rotation (deg)*	189 ± 6	176 ± 9
Max shoulder horizontal adduction (deg)	12 ± 7	16 ± 8
Max elbow flexion (deg)	97 ± 10	96 ± 11
Max shoulder internal rotation velocity (deg/s)*	7773 ± 889	6732 ± 1225
Max elbow extension velocity (deg/s)	2346 ± 246	2202 ± 308
Lead knee flexion at BR (deg)	35 ± 14	37 ± 13
Forward trunk flexion at BR (deg)	38 ± 6	33 ± 7
Lateral trunk flexion at BR (deg)	24 ± 11	22 ± 12
Shoulder abduction at BR (deg)*	89 ± 7	99 ± 7
Elbow flexion at BR (deg)*	21 ± 4	27 ± 5

*$p < .01$

Table 3. Temporal differences between efficient and inefficient pitchers.

Variable	Efficient	Inefficient
Max pelvis rotation velocity (% pitch)	28 ± 15	22 ± 11
Max upper trunk rotation velocity (% pitch)	48 ± 8	46 ± 9

Table 4. Kinetic differences between efficient and inefficient pitchers.

Variable	Efficient	Inefficient
Shoulder anterior force (N)*	271 ± 47	334 ± 49
Shoulder proximal force (N)	1152 ± 200	1147 ± 189
Shoulder horizontal adduction torque (Nm)*	92 ± 17	109 ± 18

Shoulder internal rotation torque (Nm)	84 ± 19	96 ± 17
Elbow varus torque (Nm)	83 ± 179	93 ± 17
Elbow flexion torque (Nm)	47 ± 10	44 ± 13

*$p<.01$

DISCUSSION AND CONCLUSIONS: The purposes of this study were to define the relationship between ball velocity and upper extremity kinetic (UEK) values during baseball pitching and use these relationships to differentiate mechanically efficient pitchers from inefficient pitchers. The efficient pitching group displayed clear advantages by throwing significantly faster with similar or lower UEK values. They were also capable of accomplishing this despite having significantly less body mass. The efficient pitchers were able to generate greater pelvis, upper trunk, and shoulder internal rotation velocities. It is likely that this optimized use of the kinetic chain is what allowed them to achieve greater ball velocity. The increased upper trunk rotation probably accounted for some of the pitchers' ability to achieve greater maximum external rotation at the shoulder, another factor which has been linked to increased throwing velocity (Matsuo *et. al.*, 2001). They maintained more anatomical stability by keeping the shoulder closer to 90 degrees of abduction throughout the movement, flexed their trunk forward more at ball release to help drive the ball towards the plate, and had more elbow extension to maximize the "whip effect" of the arm in the throwing motion. Efficient pitchers likewise had greater horizontal abduction at foot contact to help activate the stretch reflex in the anterior shoulder and maintained a more neutral lateral trunk position to focus their body's energy toward the plate.

Baseball coaches, clinicians, and researchers are looking for the pitching mechanics that maximize performance while minimizing the risk of injury. Based on the results of this analysis, it is recommended that practitioners use the biomechanical profile of the efficient pitchers in this study as the referential group when assessing an individual's mechanics.

REFERENCES:
Buchanan TS, Delp SL & Solbeck JA (1998). Muscular resistance to varus and valgus loads at the elbow. *Journal of Biomechanical Engineering*, 120(5), 634-639.

Escamilla RF *et. al.* (1998). Kinematic comparisons of throwing different types of baseball pitches. *Journal of Applied Biomechanics*, 1, 213-228.

Escamilla RF *et. al.* (2007). Pitching biomechanics as a pitcher approaches muscular fatigue during a simulated game. *American Journal of Sports Medicine*, 35(1), 23-33.

Fleisig GS *et. al.* (1999). Kinematic and kinetic comparison of baseball pitching among various levels of development. *Journal of Biomechanics*, 32, 1371-1375.

Matsuo T *et. al.*, (2001). Comparison of kinematic and temporal parameters between different pitch velocity groups. *Journal of Applied Biomechanics*, 17, 1-13.

Morrey BF, & An, KN (1983). Articular and ligamentous contributions to the stability of the elbow joint. *American Journal of Sports Medicine*, 11(5), 315-319.

Olsen SJ *et. al.* (2006). Risk factors for shoulder and elbow injuries in adolescent baseball pitchers. *American Journal of Sports Medicine*, 34(6), 905-912.

Stodden DF *et. al.* (2005). Relationship of biomechanical factors to baseball pitching velocity: within pitcher v variation. *Journal of Applied Biomechanics*, 21, 44-56.

INTRA-LIMB JOINT COUPLING PATTERNS DURING THE USE OF THREE LOWER EXTREMITY EXERCISE MACHINES

Thomas J Cunningham, David R Mullineaux, Shaun K Stinton
University of Kentucky, Lexington, KY, USA

The purpose of this study was to preliminarily describe sagittal plane joint coupling patterns for a spectrum of common lower extremity exercises. Each participant performed 3, 10 second sessions on a stationary bicycle, elliptical and treadmill. Intra-limb coupling angles of the hip and knee for two recreational athletes were quantified using vector coding techniques on randomly selected cycles from each movement. Variability patterns within the same movements were repeatable within and between each participant while each movement's distinguishable variability pattern differed both spatially and temporally between pieces of exercise equipment. These findings suggest that each exercise machine studied is distinguishable characteristics in its variability pattern. Comparison of variability patterns might be a useful method in the design of functional training exercises to aid in optimally mimicking task kinematics.

KEY WORDS: biomechanics, dynamical systems, variability, vector coding, specificity.

INTRODUCTION: Running is a desirable skill utilized in sports, exercise and everyday life to maintain an active, healthy lifestyle. Unfortunately, running can be demanding on the body and has been shown to be attributed to many lower extremity injuries (Sutton, 1984) or be difficult to perform if an injury is present (Dauty, Potiron-Josse & Rochcongar 2003). Multitudes of training and rehabilitation protocols involve mimicking the running movement as closely as possible by performing lower extremity cyclic motions to train neuromuscular coordinative structures associated with this skill (Kilding, Scott & Mullineaux, 2007). A fundamental requirement to correctly simulate the running motion during a training protocol involves reproducing the lower extremity joint kinematics seen during running. Inability to adhere to this specificity principle can result in deficient motor learning patterns that can lead to possible future injuries or inefficient muscle recruitment patterns (Kilding et al., 2007).

Commonly used exercise devices that constrain the distal lower extremity to adhere to movements range from simple circular movements produced by a fixed length pedal crank in bicycles to a more sophisticated cyclical motion produced by a cam commonly referred to as an "elliptical" motion. Running is an open chain movement where the distal limb is not constrained, which introduces variances in movement that are beneficial for locomotion (Heiderscheit, 2000) but are difficult to reproduce for training. If the training standard is to mimic the lower extremity actions seen during running, it is necessary to establish kinematic patterns associated with currently used training interventions to gauge differences in seemingly similar motions. The ability to quantify differences in movements may help optimize development of functional training exercise movements or aid in the design of equipment to accurately facilitate movement patterns. It is therefore the aim of this study to compare the joint coupling patterns of both a simple cyclical movement (bicycle) and a supposedly more complex movement (elliptical) to the standard of running.

METHOD: Data Collection: Two recreational athlete, males participated in this study (1: age=28 yrs, height=192 cm, mass=94 kg, leg length=103 cm; 2: age=26 yrs, height=175 cm, mass=81 kg, leg length=79 cm). Individual retro-reflective markers were placed on the sacrum, bilateral ASIS, medial/lateral femoral epicondyles & malleoli with rigid clusters attached to the thigh and shank to describe rotations of the right hip and knee joint in the sagittal plane. Participants were asked to stand in an anatomically neutral stance while an anatomic static calibration was captured. Participants then performed 3, 10 second sessions of activity on a generic commercial treadmill (CS6.0; TRUE Fitness, St. Louis, MO, USA), elliptical machine (Prp350XL; Octane Fitness, Brooklyn Park, MN, USA) and stationary bicycle (Schwinn Evolution-SR; Nautilus, Vancouver, WA, USA). Speed was only controlled during the treadmill sessions which were performed at 3.8 m/s. Marker locations were captured using four digital motion cameras (Eagle-4; Motion Analysis Corp., Santa Rosa, CA) at a sampling frequency of 60 Hz with Cortex V1.0 software (Analysis Corp., Santa Rosa, CA, USA).

Data Analysis: Data were exported to Visual 3D (3.9, C-Motion, Germantown, MD, USA) for initial analysis. Data were filtered using a fourth-order Butterworth low-pass filter with a cutoff frequency of 6-Hz as determined by residual analysis (Winter, 1990). Joint angles were calculated for both the hip and knee for all trials with the hip being 0° and the knee 180° during standing posture with extension values being negative. Calculated parameters were further analyzed in Matlab R2007b (Mathworks, Natick, MA) where gait cycles were determined using thigh velocity. The start of each movement cycle for the elliptical and treadmill was defined as the maximum thigh position in the anterior direction relative to the body, while the start of the bike cycle was determined to begin at the thigh's most superior point. Data from each determined cycle were time normalized to 101 points representing 100% of the respective gait cycle. Three nonconsecutive trials for each condition were chosen randomly to represent each movement condition. Angle-angle diagrams were constructed to qualitatively describe the range and timing of joint movements. Vector coding techniques were then implemented in Excel (Microsoft Corp., Redmond, WA) to quantify intra-segmental coupling characteristics between all three movement conditions. (Heiderscheit, Hamill, & Emmerik, 2002)

RESULTS:

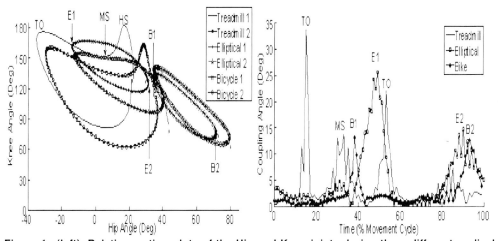

Figure 1a (left): Relative motion plots of the Hip and Knee joints during three different cyclical movements for two participants. Movement is counter-clockwise about each curve beginning at the asterisk (*). Heel-strike (HS), Mid-Stance (MS) & Toe-Off (TO) are visually distinguishable points during stance phase of the treadmill movement; E1 & E2 on the elliptical, B1 & B2 during the bike movement are abrupt changes in joint angle. Hip angle= 0° & knee angle =180° at standing posture.
Figure 1b (right): Flexion/Extension coupling angle variation plots between the Hip and Knee for the 1st participant obtained using vector coding. HS, MS, TO, E1, E2, B1 & B2 correspond to the same points labelled in Figure 1(left) and points of local maximums in variation.

After initial visual observation of constructed relative motion plots, it was determined by visual inspection that the within-participant variance between trials for each movement was considered negligible in the demonstration of the joint coupling patterns for each respective movement. Therefore, the ensemble average of the three trials chosen for each individual are considered to be adequate representations of the angular joint motions demonstrated during this study. These average relative motion plots are shown in Figure 1a. Movements are superimposed on the same figure to qualitatively compare inter-movement and intra-participant variation. The key gait determinants during stance phase (Heel-strike (HS), Mid-stance (MS) & Toe-off (TO)) are labelled on the treadmill plot for reference. Elliptical and bicycle movements were hypothesized to be more generic cyclical movements but did have key transition points in movement similar to that in running gait which are labelled; E1 & E2 for the elliptical and B1 & B2 for the bicycle.

A more quantitative tool for gauging differences in angular motions between movements at given points during each cycle can be seen in Figure 1b. Again, the same points are labelled in this figure as in Figure 1a for easy comparison. It should be noted that intra-movement variability was strikingly similar between participants (less than 3° SD & 1% SD) for all points in the gait cycle and therefore only one participant's coupling angle pattern is shown.

Treadmill running appeared to be the most complex movement with three distinct points highlighted in Figure 1b (HS, MS & TO). A fourth recognizable point associated during late swing phase is discernable at 84% of the gait cycle, however its magnitude in variability is comparatively less than expected (8°) given the large amount of joint movement associated at this point in the gait cycle and when compared to HS (36°), MS (15°) & TO (22°). Elliptical movement did experience similar magnitudes in variability as treadmill running most apparent at E1 (23°) which also was only separated temporally from TO by 5% of their movement cycles. Despite the similarity at these points, a large magnitude was seen at E2 (21°) but did not directly correspond temporally to a point on the treadmill gait cycle. The bicycle also had two discernable points (B1 & B2) which neither corresponded in magnitude nor timing to the treadmill. B2 (9°) did however seem to occur at similar time points as E2 but was somewhat less in magnitude. All other points not labelled were similar in magnitude (less than 5°) and did not seem to experience a noticeable amount of variation.

Timing was drastically different of labelled local variation peaks when compared to running. This could be a result of separating gait cycles using thigh velocity but most likely is evidence that the movements are substantially different. It appears as if E1 and E2 might be out of phase with Heel-Strike and Toe-Off by approximately 180°. However, the rate constant for the elliptical is substantially lower indicated by the width of the variation about E1 and E2. Magnitudes were comparable between these conditions which might correspond to the more similar ranges of motion the elliptical condition had with running than compared to the bicycle which had extremely lower magnitudes in variation.

DISCUSSION: Joint kinematic ranges and associated patterns observed in our study agreed with previous research that has established typical joint movement characteristics for the three presently studied movements (Horvais, Samozino, Textoris, Hautier, & Hintzy, 2008; Raasch & Zajac, 1999; Swanson & Caldwell, 2000). Key running gait determinants labelled and used for comparison were recognizable but associations to other movement's discernable points (E1, E2, B1 & B2) need to be compared with caution considering each movement's cycle was normalized using the same methods despite the movements clearly being different. Key points for both the elliptical and bicycle condition appear to have clear transition points but both the temporal and spatial characteristics of their variability would be subjective to change depending on the exact cam or pedal profile utilized by the piece of equipment. This is comparable to the differences observed in the relative motion plots between participants on the treadmill. Participant 1 had a substantially larger leg length than participant 2 (24 cm). This could infer that at the same speed participant 1 needed a relatively smaller joint angle range to maintain speed (Dillman, 1975). This is shown in the large shift of the angle-angle plot downward and to the right in the sagittal plane angle-angle state space indicating differences in leg length would inherently dictate a separate movement pattern for each individual. However; inter-participant variability was small and consistent in both magnitude and timing within each movement category. This indicates that despite the anthropometric heterogeneity of these participants, if replication of joint coupling patterns is the ultimate goal for intervention design, assumptions that the running pattern "standard" will not deviate to a large degree may be founded. This is consistent with previous literature that has shown repeatable variation patterns at specific gait points during stance phase (Heiderscheit, Hamill, & Emmerik, 2002). This observation also gives some evidence to support hypotheses that categorically different cyclical lower extremity movements each have their own distinguishable pattern with relatively small inherent variation between individuals. A future direction of research based on these preliminary observations might involve experiments which change movement parameters of exercise interventions such as cam radius or inclination angle which might temporally shift or alter the magnitude of variation in points similar to E1, E2, B1 and B2. Observing changes in variability by altering constraints to the kinematic environment suggests that subtly different categories of movement might appear similar but actually are measurably unique. Likewise, minimal inter-movement variation of variability patterns would indicate that ability to replicate a particular movement is measurable to an extent.

CONCLUSION: This preliminary study suggests that there are unique variability characteristics distinguishing each of these commonly used lower extremity exercise movements and that these patterns also allow for small variations between individuals. Measuring changes in both the magnitude and temporal phase of joint coupling variability patterns might be beneficial in the design and evaluation of optimal movement patterns for training. Reducing specific aspects of variability between a standard movement of running and functional training movements designed by coaches, trainers or equipment designers might provide a useful tool for specificity training.

REFERENCES:

Dauty, M., Potiron-Josse, M., & Rochcongar, P. (2003). Consequences and prediction of hamstring muscle injury with concentric and eccentric isokinetic parameters in elite soccer players. Translated. *Annales de Readaption et de Medecine Physique*, 46(9), 601-606.

Dillman, C. J. (1975). Kinematic analyses of running. *Exercise Sport Science Reviews*, 193-216.

Heiderscheit, B. C. (2000). Movement variability as a clinical measure for locomotion. *Journal of Applied Biomechanics*, 16, 419-427.

Heiderscheit, B. C., Hamill, J., & Emmerik, R. E. A. v. (2002). Variability of stride characteristics and joint coordination among individuals with unilateral patellofemoral pain. *Journal of Applied Biomechanics*, 18(2), 110-121.

Horvais, N., Samozino, P., Textoris, V., Hautier, C., & Hintzy, F. (2008). Biomechanical and physiological descriptions of the elliptical cycle locomotion. *Isokinetics and Exercise Science*, 16(1), 11-17.

Kilding, A. E., Scott, M. A., & Mullineaux, D. R. (2007). A kinematic comparison of deep water running and overground running in endurance runners. *Journal of Strength and Conditioning Research*, 21(2), 476-480.

Raasch, C. C., & Zajac, F. E. (1999). Locomotor strategy for pedaling: muscle groups and biomechanical functions. *Journal of Neurophysiology*, 82, 515-525.

Sutton, G. (1984). Hamstrung by hamstring strains: A review of the literature. *Journal of Orthopaedic and Sports Physical Therapy*, 5(4), 184-195.

Swanson, S. C., & Caldwell, G. E. (2000). An integrated biomechanical analysis of high speed incline and level treadmill running. *Medicine and Science in Sports and Exercise*, 32, 1146-1155.

Winter, D. A. (1990). *The Biomechanics and Motor Control of Human Movement (2nd ed.)*. New York: John Wiley & Sons.

ISBS 2009

Oral Session S19

Track & Field II

THE EFFECT OF EXHAUSTIVE RUNNING ON POSTURAL DYNAMICS

Mark Walsh[1], Adam Strang[2], Mathias Hieronymus[1], Josh Haworth[1]

Department of Kinesiology and Health, Miami University, Oxford, Ohio, USA[1]
Department of Psychology, Miami University, Oxford, Ohio, USA[2]

KEY WORDS: balance, fatigue, nonlinear dynamics.

INTRODUCTION: Recently, researchers have begun to use nonlinear measures such as the Lyapunovexponent (LyE) and Approximate Entropy (ApEn) to examine temporal structure in the continuous behavior of biological systems. When using these measures a higher score indicates lesser periodicity and greater chaotic behavior. A decrease in LyE and ApEn values have been shown in some cases to indicate pathological conditions. Studies of postural control have found that after a cerebral concussion, an athlete's center of pressure oscillations measured by ApEn are significantly decreased even up to 96 hours post-injury as compared with their preseason ApEn scores, even when the athlete appears steady (Cavanaugh, 2006). In regards to walking gate, local dynamic stability showed decreased variability when assessed amongst ACL reconstruction patients (Stergiou, 2004). The purpose of the current experiment is to use ApEn versus a set of traditional postural measure to evaluate a postural sway during upright stance prior to and following a bout of exhaustive running. In published balance studies only about 50% report significant improvements, possibly because traditional measures aren't capturing the improvements in postural control. Promising results from the above cited studies indicate that nonlinear measures may be measuring elements that the traditional measures don't detect.

METHODS: Following warm-up, participants (N=19) ran on a treadmill to exhaustion. Postural sway data was recorded via Center of Pressure (COP) obtained from a forceplate during upright stance for periods of 30s at six times during the experiment (baseline, post warm-up, post exhaustive running, and at 2, 5 and 10 minutes following the exhaustive run). From COP data researchers computed postural sway measures of COP path length, position variability, and ApEn for anterior-posterior and medial-lateral movement planes. It was predicted that measures of postural sway would exhibit a quadratic trend with measures deviating from baseline measures following bouts of exercise, returning to baseline levels with recovery. This would indicate that fatigued subjects would become more periodic and with recovery would return to a state similar to baseline values. To examine this hypothesis we performed a set quadratic contrast for each postural measure ($\alpha = 0.05$).

RESULTS: In both M-L and A-P planes measures of COP variability (p=0.025, p=0.022) and path length (p=0.048, p=0.002) displayed significant quadratic trends. Only ApEn values of COP data in M-L plane (p=0.002) exhibited the predicted trend. (A-P ApEn was p=0.078).

DISCUSSION: Our findings show that both sets of measures were successful detecting changes in postural control due to fatigue. These changes seem to imply that exhaustive running may compromise postural stability for a brief period of time.

CONCLUSION: We conclude that both traditional measures of postural sway and ApEn are effective tools for detecting changes in postural dynamics following exhaustive running. More research is needed in the area of nonlinear measurements and their application to analyzing human movement.

REFERENCES:
Cavanaugh, J.T., Guskiewicz, K.M., Giuliani, C.G., Marshall, S.M., Mercer, V.S., & Stergiou, N. (2006). Recovery of Postural Control After Cerebral Concussion: New Insights Using Approximate Entropy. *41*, 305-313.
Stergiou, N., Buzzi, U.H., Kurz, M.J., & Heidel, J (2004). *Innovative Analyses of Human Movement: Analytical Tools of Human Movement Research.* Champaign, IL:Human Kinetics.

GRAVITY'S ROLE IN ACCELERATED RUNNING - A COMPARISON OF AN EXPERIENCED POSE® AND HEEL-TOE RUNNER

Graham Fletcher[1], Marcus Dunn[2], Nicholas Romanov[3]

University of the Fraser Valley, 33844 King Rd, Abbotsford, BC, Canada[1]
Sheffield Hallam University, Sheffield, England[2]
Posetech, Miami, FL, USA[3]

The purpose of this study was to determine gravity's role in accelerated running using an experienced male Pose® and heel-toe runner as a comparison. A two-step accelerated run found that maximum horizontal acceleration of the centre of mass (COM) occurred before maximum horizontal ground reaction force (GRF). Maximum horizontal and angular acceleration of the arms and trunk occurred at or before maximum horizontal acceleration of the COM. At maximum horizontal GRF both participants' stance feet were vertically accelerated. It is suggested that acceleration of the COM occurs via a gravitational torque with GRF being the consequence of, not the cause of these movements. Therefore, practitioners might find this novel perspective helpful when applied to accelerated running.

KEYWORDS: Pose®, heel-toe, gravity, centre-of-mass, accelerated running

INTRODUCTION: A novel running technique, the Pose® method has made claims it is an effective way to run (Fletcher et al., 2008). The Pose® method of running teaches that movement occurs by changing from one support foot to the other while the centre of mass (COM) falls forward of the point of support (COP) via a gravitational torque, defined as $mg\ r\ \sin\theta$ (where m is mass, g is gravity, r is vector from COP to COM and θ is the angle between r and global vertical) (Romanov & Fletcher, 2007). This is achieved by pulling the support foot upwards from the ground toward the hip using the hamstring muscles as the body falls forward after mid-stance (Fletcher et al., 2008). The ipsilateral leg is not driven forwards during flight but allowed to fall to the ground under the COM to land in the next running Pose® (Romanov & Fletcher, 2007).

A recent critique (Brodie et al., 2008) asserted that during a complete running cycle, gravity does no net work and from mid-stance to terminal-stance actually retards the athlete in constant speed running. To date, research on Pose® running has focused on constant speed running, however accelerated running might provide a clearer explanation of gravity's role. Heel-toe (HT) runners encounter the same forces when running (ground reaction force (GRF), muscle force, gravity and muscle elasticity and air resistance) as Pose® runners. To accelerate, the runners' COM must experience a net external force, e.g. gravity and/or GRF. Therefore, the purpose of this study was to provide a comparison of accelerated running using an experienced HT and Pose® runner to further understand gravity's role in running.

METHODS: One male Pose® (age: 53 years, stature: 1.73 m, mass: 71.0 kg) and one male HT (age: 55 years, stature: 1.69 m, mass: 72.5 kg) runner participated in the current study. Both were experienced runners (>30 years) and considered to be exemplars of their respective techniques. Prior to participation, ethics approval for all procedures was obtained from Schriners Gait Laboratory, Vancouver and both participants provided written informed consent. Both participants used a two-step start for a fast acceleration run across a force platform measuring 0.40 × 0.50 m (AMTI, OR65), for 10 trials using a right foot contact. The force platform was integrated with an online, eight camera motion analysis system (Motion Analysis Corporation, Santa Rosa, CA), tracking the movement of 41 retro-reflective markers (NIH marker set). Three-dimensional kinematic and kinetic data were collected at 240 Hz and 1200 Hz respectively using EVaRT (5.0.4, Motion Analysis Corporation, CA, USA) and exported to C3D files for further analysis in Visual 3D (3.79, C-Motion, MD, USA). Kinematic and kinetic data were filtered using a second order, low-pass Butterworth bidirectional filter with cut-off frequencies of 10 and 50 Hz respectively and applied to a full-body kinetic model.

Calculated data, including full-body and segment COM position data, were exported to ASCII files for analysis in MATLAB (R2006b, The MathWorks, MA, USA). Linear and angular data were calculated relative to global and pelvis coordinate systems respectively, instantaneous velocity and acceleration were calculated as first and second time derivatives respectively.

RESULTS AND DISCUSSION: Running and in particular accelerated running is a component of many sports. Hence understanding whether the horizontal acceleration of a runner's COM occurs by pushing off of the ground by the foot or falling forwards via a gravitational torque while pulling the foot from the ground, is important for teaching running technique. Figure 1 shows that maximum horizontal acceleration of the runner's COM occurred before the maximum horizontal GRF in both runners. Usually, maximum horizontal GRF is associated with the foot pushing from the ground in order to increase the horizontal acceleration of the runner. Therefore, it was important to identify the runner's movements at maximum horizontal acceleration of the COM and at maximum horizontal GRF (Figure 1; Table 1).

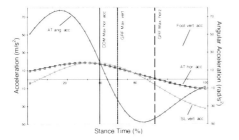 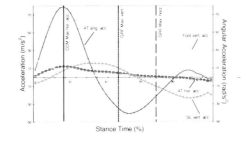

Figure 1: Pose® (left) and HT (right) runner's motion in stance (mean of ten trials)

At the instant of maximum horizontal acceleration of the COM (Figure 1: solid vertical line), the horizontal and angular acceleration of the HT runner's arms and trunk (AT) peak, whilst the angle of inclination of the COM (θ) is near vertical (Table 1). In contrast, the AT of the Pose® runner had just passed maximum horizontal acceleration and angular acceleration of the AT was approaching zero. At the same instant, the Pose® runners' stance foot had its lowest vertical acceleration, whereas vertical acceleration of the HT runners' stance foot was still decreasing. Also, the Pose® runners' swing leg (SL) achieved near maximum vertical acceleration whereas vertical acceleration of the HT runner's SL was still increasing.

Just before maximum horizontal GRF (Figure 1: dashed vertical line), the HT runners' support foot is accelerated superiorly. At this time, linear horizontal acceleration of the AT was approximately zero (velocity close to maximum). Minimum angular acceleration of the AT occurred at 55% of stance while vertical acceleration of the SL is zero at 60% of stance before the foot is then vertically accelerated. Minimum angular acceleration of the Pose® runner's AT occurred at 65% of stance while vertical acceleration of the SL passes through zero. At this time, the stance foot was vertically accelerated before maximum horizontal GRF (Figure 1: dashed vertical line) and horizontal acceleration of the AT passes through zero.

Therefore, at maximum horizontal acceleration of the COM, the support foot has minimal vertical acceleration and the AT has maximum horizontal acceleration in both runners. However, the angular velocity of the AT is near maximum in the Pose® runner but close to zero in the HT runner. At maximum horizontal GRF, the horizontal acceleration of the AT, horizontal acceleration of COM and vertical acceleration of the SL were close to zero in the Pose® runner as the support foot initiates its vertical acceleration. The HT runner is less coordinated at this time owing to negative vertical acceleration of the SL, but generally follows a similar movement pattern. At maximal horizontal acceleration of the COM, the HT runner experienced greater angular acceleration of AT by 67.9 rad/s^2 (13.1) owing to an increased forward lean of the upper body owing to θ being close to the vertical (Table 1).

Table 1 Pose® and HT movement variables at key instants during stance ($\bar{x} \pm s$)

Instant	Runner	GT (Nm)	AT hor. vel. (m/s)	SL ang. acc. (rad/s²)	COM hor. vel. (m/s)	θ (°)	Foot vert. vel. (m/s)
Initial contact	Pose®	-64.4 ± 16.0	3.74 ± 0.09	-84.6 ± 10.7	4.16 ± 0.09	-5.8 ± 1.5	-0.66 ± 0.30
	HT	-80.2 ± 22.0	2.93 ± 0.17	-64.5 ± 5.7	3.31 ± 0.13	-7.3 ± 2.0	-0.60 ± 0.11
Max COM hor. acc.	Pose®	122.9 ± 20.8	4.68 ± 0.21	0.83 ± 21.6	4.51 ± 0.16	11.2 ± 1.8	0.12 ± 0.05
	HT	-0.8 ± 21.3	3.19 ± 0.12	-48.0 ± 7.0	3.44 ± 0.14	-0.2 ± 2.0	-0.09 ± 0.07
Max vert. GRF	Pose®	165.8 ± 9.7	4.91 ± 0.09	36.6 ± 27.0	4.62 ± 0.11	14.9 ± 0.9	0.17 ± 0.01
	HT	131.0 ± 7.9	3.69 ± 0.13	8.1 ± 13.1	3.70 ± 0.12	12.0 ± 0.7	0.13 ± 0.01
Max hor. GRF	Pose®	268.6 ± 16.2	5.12 ± 0.13	82.9 ± 18.9	4.83 ± 0.15	23.0 ± 1.4	0.49 ± 0.10
	HT	240.5 ± 4.1	3.83 ± 0.09	65.5 ± 11.7	3.83 ± 0.11	21.0 ± 0.4	0.30 ± 0.05
Terminal stance	Pose®	391.1 ± 19.4	4.83 ± 0.13	99.5 ± 15.1	4.92 ± 0.13	31.0 ± 1.5	1.94 ± 0.12
	HT	380.3 ± 12.3	3.80 ± 0.11	80.9 ± 8.8	3.95 ± 0.12	30.4 ± 0.9	1.52 ± 0.10

GT = gravitational torque ($mg\,r\,\sin\theta$), AT = arms and trunk, SL = swing leg, θ = angle of vector COP-COM to vertical, hor. = horizontal, vert. = vertical, vel. = velocity, acc. = acceleration, ang. = angular.

However, at the same instant the Pose® runner experienced greater horizontal acceleration for the AT and the COM by 3.8 (1.8) and 1.8 (0.03) m/s respectively. It appears the Pose® runner was able to translate rotational movement into linear movement more successfully possibly owing to a greater gravitational torque because the COM was forward of the support foot. At maximum horizontal GRF the stance foot's vertical velocity was 61% greater in the Pose® runner highlighting a stable stance position which enables foot lift rather than SL drive. We briefly suggest several reasons for these similarities and differences between the two techniques. Recently, Chang et al. (2000) found that not only the vertical but also the horizontal GRF experienced by runners was affected by reductions in Earth's gravity. For example, when decreasing gravity by 75%, horizontal GRF impulse decreased by 53% whereas with a 30% increase in only the inertial force, there was only approximately a 9% increase in horizontal GRF impulse. They deduced from these data that differences in the horizontal impulses were due solely to gravity, but offered no explanation for the reduction in horizontal GRF. A vertical force, for example gravity, cannot affect a horizontal force. The horizontal force that resists the foot is friction (F) or horizontal GRF. The equations of motion (1-2) presented below, are for a runner with the rotational term reflecting motion of the COM about the support foot.

F (hor GRF) = $m\,dv_x / dt$ (1a)
N (vert. GRF) = $mg + m\,dv_y / dt$ (1b)
$I_{COM}\,d\omega / dt = N\,r\,\sin\theta - F r \cos\theta$ (2)

where m is mass of the runner, v_x and v_y are velocity components of the COM, r is the distance from the COM to the support foot's COP, I_{COM} is the moment of inertia about an axis through the COM, and angular velocity $\omega = d\theta / dt$

If F is less than the coefficient of friction (μ) and vertical GRF (N) then the runner does not slip and their body can then rotate about their support foot and the following equations are valid. Hence, $v_x = r\omega \cos\theta$ and $v_y = -r\omega \sin\theta$. Therefore,

$F = m\,r\,(\cos\theta\,d\omega / dt - \omega^2 \sin\theta)$ (3)
$N = mg - m\,r\,(\sin\theta\,d\omega / dt + \omega^2 \cos\theta)$ (4)

Equation 3 illustrates F is related to the angular acceleration of the runner about the support foot. Angular acceleration of the runner's COM is caused by a gravitational torque as the substitution of equations 3 and 4 into 2 yields,

$$\frac{d\omega}{dt} = \frac{d^2\theta}{dt^2} = \omega^2 \sin\theta \tag{5}$$

Equation 5 shows gravity's affect on GRF because equation 5 can also be derived by equating the gravitational torque $mg\ r\ \sin\theta$ about the support foot (Romanov & Fletcher, 2007). In support, Chang et al. (2000) found the resultant GRF vector at maximum horizontal GRF remained nearly constant between normal and 75% gravity. Hence, the changes in the magnitude of the vertical component of GRF were accompanied by proportional changes in the horizontal component of GRF to maintain the orientation of the resultant force vector. Therefore, gravity does affect F by virtue of the radius about the support foot and therefore its torque. The faster a runner rotates around their support foot, the greater the increase in F, owing to increased angular acceleration (equation 3) without the need for additional internal, muscle force to push-off against the ground. Horizontal acceleration of the COM occurs because of the rotation of the body via a gravitational torque ($v_x = r\ \omega\ \cos\theta$). Maximum F appears to be the optimal and stable time for the body to act as a support (see Zatsiorsky, 2002; angle of friction) to begin to pull the support foot from the ground as the body minimises angular and linear acceleration.

CONCLUSION: Findings indicate that both runners' bodies rotate about a near stationary support foot at maximum horizontal acceleration of the COM via a gravitational torque before the onset of maximum horizontal GRF. Accelerations of the AT and SL were zero close to maximum horizontal GRF for the Pose® runner except for the support foot, which was accelerated superiorly. The HT runner followed similar movement patterns to that of the Pose® runner but was less coordinated between the upper body and swing leg. Gravity completes no net work during stance in constant speed running, but achieves angular work via a gravitational torque accelerating the COM in both constant speed and accelerated running. The current study was limited with regard to sample size however, this research does enable practitioners to re-examine running technique from this novel perspective. Future research should consider these findings within larger groups.

REFERENCES:

Brodie, M., Walmsley, A. & Page, W. (2008). Comments on "Runners do not push off but fall forward via a gravitational torque". *Sports Biomechanics*, 7, 403-405.
Chang, Y., Wen, H., Huang, C., Hamerski, C.M., & Kram, R. (2000). The independent effects of gravity and inertia on running mechanics. The Journal of Experimental Biology, 203, 229–238.
Fletcher, G, Romanov, N.S. & Bartlett, R. M. (2008). Pose® method technique improves running performance without economy changes. *International Journal of Sports Science & Coaching*, 3, 365-380.
Fletcher, G. & Romanov, N.S. (2008). Comments on "Runners do not push off but fall forward via a gravitational torque". *Sports Biomechanics*, 7, 406 – 411.
Romanov, N. & Fletcher, G. (2007). Runners do not push off the ground but fall forwards via a gravitational torque. *Sports Biomechanics*, 6, 434-452.
Zatsiorsky, V. (2002). *Kinetics of human motion*. p. 77. Champaign: Human Kinetics.

Acknowledgment

The authors would like to thank Alec Black at Schriners Gait Laboratory, Vancouver.

ATHLETE-SPECIFIC ANALYSES OF LEG JOINT KINETICS DURING MAXIMUM VELOCITY SPRINT RUNNING

Ian N. Bezodis[1], Aki I.T. Salo[2] and David G. Kerwin[1]

[1]Cardiff School of Sport, University of Wales Institute, Cardiff, Cardiff, UK
[2]Sport and Exercise Science, University of Bath, Bath, UK

The effect of variations in joint kinetics on sprint performance in individual athletes is not yet known. To investigate biomechanical contributions to maximum velocity sprint running, data were collected from one elite male sprinter performing maximum effort 60 m sprints. High-speed video (200 Hz) and ground reaction force (1000 Hz) data were collected at the 45 m mark. Horizontal velocity and joint kinetics, via inverse dynamics, were calculated for two trials. The velocity of the step was closely linked to step length, knee angular velocity before touchdown, peak-to-peak centre of mass oscillation, hip extension moment during stance and ankle positive work before take-off. The study revealed the potential for athlete-specific, detailed biomechanical analysis and feedback to aid the technical work of athletes and their coaches across a range of sporting skills.

KEY WORDS: track and field athletics, intra-subject variation, inverse dynamics analysis

INTRODUCTION:

The study of joint kinetics can improve the understanding of the underlying causes of a movement (Winter, 2005). Biomechanical investigations of sprint running have studied the joint kinetics of the movement, but a comprehensive understanding of its causative mechanisms has not yet been achieved. To date, several studies have presented group-level analyses of the importance of joint kinetic factors to sprint performance (e.g. Mann, 1981; Johnson & Buckley, 2001; Belli et al., 2002). One possible approach to increasing understanding is to investigate the factors that relate to performance on a within-subject basis: examples from sprinting include Weyand et al. (2000) and Hunter et al. (2004), although these have only reported joint kinematics and ground reaction forces, and have not extended to joint kinetics. Recently, individual athlete results for joint kinetic variables have been presented (Bezodis et al., 2007; 2008), but a detailed analysis on an individual level in order to highlight the kinematic and kinetic variables that are most important to sprint performance has yet to be presented. Analysing and communicating these data sets in this manner would greatly help elite coaches with the development of specific individualised technical training programs that would allow athletes to focus on the development of targeted biomechanical variables with the explicit goal of improving sprint performance. The aim of this study, therefore, was to understand and summarise the individual biomechanical factors that contribute to changes in sprint performance between runs in an elite sprinter.

METHODS: Data collection: One elite male sprinter (height: 1.76 m; mass: 74.9 kg; 100 m PB: 9.98 s) with no recent injuries gave written informed consent to participate. Data were collected in the National Indoor Athletics Centre, Cardiff in late November. A force plate (9287BA, Kistler Instruments Ltd., Switzerland) operating at 1000 Hz was placed in a customised housing in the centre the track, and covered with a secured piece of the synthetic track surface. A high-speed camera (resolution 768 x 604 pixels; Redlake, MotionPro HS-1, USA) was placed perpendicular to the direction of the sprint, 25.0 m from the centre of the lane, with a 3.0 m field of view centred on the force plate. The high-speed camera was set up with a frame rate of 200 Hz, a shutter speed of 1/600 s, and was manually focussed. A 50 Hz digital video camera (DCR-TRV 900E, Sony, Japan) was located 3.5 m above the track surface, 6.3 m away from the centre of the running lane and 1.5 m before the centre of the force plate to give a field of view of 6.5 m in the direction of the running lane. The 50 Hz camera was set up with a shutter speed of 1/600 s, and was manually focused. Images of a 6-point sagittal plane calibration object were captured with each camera before the start of the running trials. A single synchronisation unit was used to link the cameras with the force plate. The area around the force plate was illuminated with 7600 W of floodlighting.

A customised starting check mark was located approximately 45 m before the force plate. This was used to aid the athlete in striking the force plate without the need to alter technique in the steps immediately preceding force plate contact (targeting). The athlete performed six 60 m sprints, consisting of a 30 m build up followed by a timed 'flying 30 m', within which the force plate was centred. A trial was deemed successful if the participant was able to strike the force plate at maximum velocity without noticeably or consciously altering his stride pattern. Two successful trials (labelled 1A and 1B) were achieved from the six runs.

Data Processing: Video data from the 50 Hz camera were imported into Target (Loughborough Innovations Limited, UK) and digitised using a 20-point model, comprising shoulder, elbow, wrist, fingertip, hip, knee, ankle, head of the second metatarsal and toe on each side of the body, and top of the head and base of the neck. Video data from the high-speed camera were imported into Peak Motus (v8.1.4.0, Peak Performance Technologies, Inc. USA), and digitised using a 5-point model, comprising head of the second metatarsal, and the ankle, knee, hip and shoulder joint centres on the side of the support (right) leg. All digitised coordinates were reconstructed using the 2D-DLT with lens correction (Walton, 1981). Trial 1A was digitised three times, on separate days, to examine the effect of digitising errors. Horizontal velocity, step length and step frequency of the step from the force plate in each trial were calculated using the information taken from the 50 Hz camera, as described by Bezodis et al. (2008). The step cycle was defined from the instant of touchdown on the force plate to the subsequent contact of the contra-lateral foot.

Vertical and horizontal ground reaction forces and coordinates of all digitised points from each camera for each successful trial were subjected to a residual analysis in order to determine optimum cut-off frequencies (Winter, 2005). Once filtered at the respective optimum cut-off frequencies, the ground reaction force data were matched to a video frame from the high-speed camera and were extracted at 200 Hz. However, the instant of touchdown was identified using the 1000 Hz force data. Body segment inertia parameters were taken from de Leva (1996) with the exception of the foot segment, for which data were taken from Winter (2005). The mass of a typical sprinting shoe (200 g) was added to the mass of the foot segment (Hunter et al., 2004). Joint moments, power and work were calculated by standard inverse dynamics equations, as presented by Winter (2005), then normalised (Hof, 1996). Three other well-trained sprinters from the same training group had also given written informed consent to take part in the same data collection and each achieved two successful trials. For interest, to facilitate the presentation of results to the athlete and coach all variables were calculated as a percentage of the group mean for each individual variable.

RESULTS: The athlete ran two steps at over 10 m/s, showing that this testing session was conducted at a true elite performance level. There was a difference in velocity between the two steps of 0.23 m/s. Results (in table 1) showed that the faster of the two trials yielded several variables that were markedly greater in magnitude than the slower of the trials. These included: step length, knee angular velocity before touchdown, peak-to-peak centre of mass oscillation, hip extension moment during stance and ankle positive work before take-off.

Pilot testing revealed that step variables could be measured to within the following RMS differences of a known criterion; 0.02 m/s for velocity, 0.01 m for step length and 0.01 Hz for step frequency. An error analysis of joint moments and work revealed intra-trial variability to be between three and twelve times greater than differences arising from repeat digitisations of a single trial. Figure 1 shows data as percentage of the group mean (the bars correspond to the respective variables from table 1).

Table 1 Selected kinematic and kinetic results for each trial.

Variable	Trial 1A	Trial 1B
Step velocity [m/s]	10.37	10.14
Step length [m]	2.25	2.17
Step frequency [Hz]	4.62	4.68
Knee angular velocity at TD [rad/s]	-4.80	-3.60
Ground contact time [s]	0.097	0.095
Maximum vertical force [BW]	4.41	4.39
Peak-to-peak CM oscillation [m]	0.057	0.047
Peak knee flexion moment after TD	-0.210	-0.212
Peak hip extension moment after TD	0.636	0.615
Peak hip extension moment in stance	0.299	0.276
Ankle positive work before take-off	0.064	0.040

Note: Joint kinetic variables have been normalised by body weight and height (Hof, 1996): TD = touchdown.

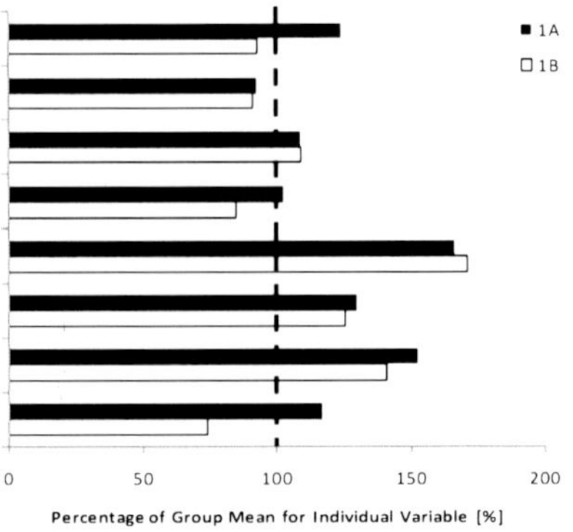

Figure 1 Selected results for each trial, presented as a percentage of training group mean.

DISCUSSION:

The running velocity achieved by the athlete in this study was greater than that of participants in previous studies of joint kinetics in maximum velocity sprinting (Mann, 1981; Belli et al., 2002). Step length values were lower and step frequency higher than previously reported findings from elite sprinters (Mann and Herman, 1985).

The aim of this study was to understand and summarise the individual biomechanical factors that contribute to changes in sprint performance between runs in an elite sprinter, and to present this information to the athlete and his coach. In order to represent complicated biomechanical variables in a manner that could be readily understood by the practitioners, specific variables other than step velocity, length and frequency were presented as a percentage of the mean value of that variable from the training group of four athletes. These relative values, along with an explanation of the meaningfulness of each specific variable facilitated understanding of complex technical data for the coach and his training group.

In trial 1A, the faster of the two measured, the knee flexion velocity at touchdown was over 20% above the group mean. This showed that the athlete was rapidly 'clawing' at the track and adopting a suitable position for the production of force during the stance phase (Mann and Herman, 1985). The combination of a below group average contact time and above group average maximum vertical force in the athlete studied displayed a method of generating impulses during stance that has been shown to be beneficial to sprint performance (Weyand et al., 2000). The vertical impulse during stance dictates the change in vertical velocity and therefore the peak-to-peak vertical oscillation of the centre of mass throughout the measured step. In trial 1B, the oscillation was 15% below the group mean, resulting in a reduced step length and therefore a reduced step velocity. The three joint moment values presented here ranged from 30-71% above group mean in both trials, giving a good reflection of the overall muscular strength that the athlete possessed. The athlete 'pulled' himself rapidly over the contact foot after touchdown, reducing the braking effect in early stance, by the large knee flexion and hip extension moments, which have been shown to be crucial to sprint performance (Mann and Sprague, 1980). The large hip extension moment later in the stance phase acted to drive the body over the contact foot and help to restore the forward momentum of the body. The magnitude of positive work performed at the ankle joint has previously been shown to be linked to sprint performance (Bezodis et al.,

2007), and in this athlete was over 40% of the group mean greater in trial 1A than 1B. This value gives a very good indication of the overall propulsion of the step, but occurs as a result of transfer of the propulsive forces from the larger proximal muscles by the bi-articular muscles of the leg.

CONCLUSION: This study has demonstrated athlete-specific analysis of joint kinetics during a sprint run that was previously suggested by Bezodis et al. (2007). The results could provide detailed biomechanical feedback to a coach and his athletes in a readily understandable manner that can facilitate the development of specific technical training programs designed to improve sprint performance. Whilst this study used two trials, a larger sample size for each athlete would improve the validity of the results, although this can be difficult to achieve in a practical setting. This athlete-specific analysis has the potential to improve the practitioner understanding of complex biomechanical concepts and could also be readily applied to a number of other sporting skills.

REFERENCES:

Belli, A., Kyröläinen, H. & Komi, P.V. (2002). Moment and Power of Lower Limb Joints in Running. *International Journal of Sports Medicine*, **23**, 136-141.

Bezodis, I.N., Salo, A.I.T. & Kerwin, D.G. (2007). Joint Kinetics in Maximum Velocity Sprint Running. In H.-J. Menzel and M.H. Chargas (Eds). *Proceedings of the XXVth Symposium of the International Society of Biomechanics in Sports*, pp 59-62. Ouro Preto, Brazil. ISBS.

Bezodis, I.N., Kerwin, D.G. & Salo, A.I.T. (2008). Lower-Limb Mechanics during the Support Phase of Maximum-Velocity Sprint Running. *Medicine and Science in Sports and Exercise*, 40, 707-715.

de Leva, P. (1996). Adjustments to Zatsiorsky-Seluyanov's segment inertia parameters. *Journal of Biomechanics*, **29** (9) 1223-1230.

Hof, A.L. (1996). Scaling gait data to body size. *Gait & Posture*, **4**: 222-223.

Hunter, J.P., Marshall, R.N. & McNair, P.J. (2004). Interaction of step length and step rate during sprint running. *Medicine and Science in Sports and Exercise*, **36** (2), 261-271.

Johnson, M.D. & Buckley, J.G. (2001). Muscle power patterns in the mid-acceleration phase of sprinting. *Journal of Sports Sciences*, **19**, 4, 263-272.

Mann, R.V. & Sprague, P. (1980). A kinetic analysis of the ground leg during sprint running. *Research Quarterly in Exercise and Sport*, **51**, 334-348.

Mann, R.V. (1981). A kinetic analysis of sprinting. *Medicine and Science in Sports and Exercise*, **13** (5), 325-328.

Mann, R.V. & Herman, J. (1985). Kinematic Analysis of Olympic Sprint Performance: Men's 200 Meters. *International Journal of Sport Biomechanics*, **1**, 151-162.

Walton, J. (1981). *Close-range cine-photogrammetry: A generalised technique for quantifying gross human motion*. Unpublished Doctoral Thesis. The Pennsylvannia State University.

Weyand, P.G., Sternlight, D.B., Bellizzi, M.J. & Wright, S. (2000). Faster top running speeds are achieved with greater ground forces not more rapid leg movements. *Journal of Applied Physiology*, **89** (5), 1991-1999.

Winter, D.A. (2005). *Biomechanics and Motor Control of Human Movement*. Hoboken: John Wiley and Sons, Inc.

Acknowledgements

The authors would like to thank UK Athletics Ltd. for financial support for this work, the coach and athletes for giving up their time to participate in the study, and Neil Bezodis, Tom Loney, Ollie Peacock and Paulien Roos for their assistance with data collection.

EFFECTS OF FATIGUE ON TECHNIQUE DURING 5 KM ROAD RUNNING

Brian Hanley and Laura Smith

Carnegie Research Institute, Leeds Metropolitan University, Leeds, UK

The purpose of this study was to investigate the effects of fatigue on kinematic parameters during a 5 km road race, and to establish how men and women fatigue differently. 17 highly competitive distance runners (9 male, 8 female) were videoed (50 Hz) as they completed the English National 5 km championships. Three-dimensional kinematic data were analysed using motion analysis software (SIMI, Munich). Data were recorded at 950 m, 2,400 m, and 3,850 m. Repeated measures ANOVA showed a significant decrease in speed ($p < 0.01$) which occurred due to both decreased step length and cadence ($p < 0.05$). Differences in speed, step length and contact time between men and women were found ($p < 0.05$). Athletes can reduce the risk of fatigue by using appropriate racing tactics.

KEY WORDS: fatigue, athletics, road running, gender differences.

INTRODUCTION: Great differences in running technique are evident between the earlier and latter stages of a race. Changes in technique are required by fatigued runners towards the end of a race to maintain speed and running economy (Anderson, 1996). This is often achieved through postural changes; however, such changes may be detrimental to actual running technique. Williams *et al.* (1991) suggested that with fatigue, runners should not markedly change their mechanics: runners that did exacerbated the effects of fatigue and subsequent technique deterioration. Research suggests step length decreases with fatigue due to irregularities of internal timing (Elliot and Ackland, 1981), changes in muscle firing patterns (Williams, 1990) and changes of the lower limb joint angles (Siler and Martin, 1991). Physiological factors may also contribute to the decrease in step length. Because of this, runners aim to attain an optimum technique that they can maintain in order to achieve race success. Elliot and Ackland (1981) proposed that identification and appropriate correction of deterioration in a runner's technique would be beneficial for enhancing performance. The aim of this study was to measure kinematic variables in 5 km distance running and investigate how they altered during the course of the run.

METHODS: Seventeen competitive distance runners (nine men, eight women) gave informed consent and the study was approved by the university's ethics committee. The participants' mean age was 26 years (± 4), stature 1.72 m (± 0.06) and mass 60 kg (± 7). The study was conducted at the National AAA 5K Road Championships in Horwich (GBR). All 17 participants completed the 5,000 m road race (three laps of 1,450 m, including an additional 650 m at the start). The race times ranged from sub-14 minutes to 17:30 (men sub-14 to 16:30 and women 15:45 to 17:30). Kinematic data were recorded at the same position on the course for the three laps of the race: 950 m (lap 1), 2,400 m (lap 2) and 3,850 m (lap 3). The race was filmed using two stationary DM-XL1 cameras (Canon, Tokyo), recording at 50 Hz. The reference volume was 5 m long, 2 m wide, and 2.16 m high; this ensured data collection of approximately three successive steps and allowed for 3D-DLT. Each camera was placed approximately 8 metres from the path of the runners. The video data were digitised using SIMI Motion (Munich) and filtered using a Butterworth 2nd order low-pass filter. Variables measured included joint angles, step length, cadence, and step time. In order to measure the effects of fatigue on the values obtained from the kinematic data and to compare the different effects between men and women, two factor mixed factorial repeated measures ANOVA was utilised, with post hoc pairwise tests using Bonferroni adjustments ($p < .05$).

RESULTS: The mean values presented in Table 1 show a decrease in speed over the three laps. Speed was defined as the average horizontal velocity during one complete stride cycle. The greatest decrease occurred between laps 1 and 2, from 20.40 km/hr to 19.38 km/hr: with a .04 km/hr decrease between laps 2 and 3. The decrease over the course of the race was significant ($p < 0.01$). Post hoc tests showed the values of lap 1 were significantly greater than both lap 2 and 3 ($p < 0.05$); values between lap 2 and 3 were not significantly different. For the men participants speed decreased constantly throughout the race, and this decrease was significant ($p < 0.05$). Post hoc tests established that the values at lap 1 were different to those at both laps 2 and 3. The women's mean speed decreased between lap 1 and 2, followed by a slight increase at lap 3. The values at lap 1 and 2, and lap 1 and 3 were different ($p < 0.01$). Men experienced a constant decrease with fatigue, whereas women decreased at a higher rate to begin with, followed by a slight increase. These differences between genders were significant ($p < 0.01$). Over the course of the race, men's speed decreased by 1.07 km/hr and women's by 1.05 km/hr, suggesting similar levels of fatigue.

Table 1: Mean (± SD) speed of the overall group, men and women participants.

	Overall (km/hr)	Men (km/hr)	Women (km/hr)
Lap 1	20.40 (± 1.25)	21.31 (± 0.96)	19.38 (± 0.55)
Lap 2	19.38 (± 1.57)	20.51 (± 1.29)	18.11 (± 0.53)
Lap 3	19.34 (± 1.48)	20.24 (± 1.49)	18.33 (± 0.52)
Mean (± SD)	19.71 (± 1.39)	20.81 (± 1.19)	18.63 (± 0.43)

Table 2 shows that the largest mean step length of all of the data was observed during lap 1 (1.78 m). Step length was defined as the distance the body travelled between a specific phase on one leg and the same phase on the other leg. The decrease in step length is consistent with the decrease in speed. Men's mean step length incurred the greatest decrease between lap 1 and 2; a slight decrease was exhibited from lap 2 to 3. This decrease was not significant. Women's mean step length decreased between laps 1 and 2, and then increased again slightly. Repeated measures ANOVA showed the differences were significant ($p < 0.01$). Post hoc tests showed lap 1's values were different from both lap 2 and 3's values. The men's mean step length was 0.17 m longer than the women participants'. Again, men and women athletes appeared to experience parameter changes differently over the course of the race. Men's step lengths decreased constantly, whereas an increase was evident for the women athletes following lap 2. The overall decrease in the men's length was 0.05 m, and the women's slightly greater at 0.07 m, suggesting women alter step length more as a result of fatigue. These differences were significant ($p < 0.05$).

Table 2: Mean (± SD) step length of the overall group, men and women.

	Overall (m)	Men (m)	Women (m)
Lap 1	1.78 (± 0.10)	1.84 (± 0.07)	1.71 (± 0.09)
Lap 2	1.72 (± 0.13)	1.80 (± 0.10)	1.62 (± 0.09)
Lap 3	1.72 (± 0.12)	1.79 (± 0.11)	1.64 (± 0.08)
Mean (± SD)	1.74 (± 0.10)	1.82 (± 0.09)	1.65 (± 0.10)

Mean cadence values for the entire sample followed a similar trend to step length and are shown in Table 3. Cadence was found by dividing speed by step length. The men's cadence decreased throughout the course of the race; no significant differences were found. The women's cadence decreased between lap 1 and lap 2 by 0.04 Hz, this was followed by a 0.01 Hz increase. Differences in cadence values were not statistically significant. The male participants exhibited a slightly larger mean cadence than the females. The men's cadences decreased constantly; women's cadences decreased from lap 1 to lap 2 but then increased in lap 3. However, no significant differences were found between genders. Cadence is determined by the completion time of each successive step. The mean step time for both men and women was 0.32 s. However, men had less contact time than women (0.17 and 0.19 s respectively) and more flight time (0.15 and 0.13 s respectively).

Table 3: Mean (± SD) cadence of the overall group, men and women.

	Overall (Hz)	Men (Hz)	Women (Hz)
Lap 1	3.18 (± 0.14)	3.21 (± 0.13)	3.15 (± 0.15)
Lap 2	3.14 (± 0.14)	3.16 (± 0.12)	3.11 (± 0.16)
Lap 3	3.13 (± 0.16)	3.14 (± 0.11)	3.12 (± 0.20)
Mean (± SD)	3.15 (± 0.14)	3.17 (± 0.12)	3.14 (± 0.17)

Leg joint angles are shown in Table 4. The hip was considered to be 0° at extension and therefore the angles at initial contact are a measure of hip flexion. The angle of the hip at toe-off (hyperextension) for all participants remained relatively constant at about 11°. Knee extension also remained constant during lap 1 and 2 followed by a 1° increase in lap 3. The ankle angle at toe-off remained relatively constant throughout the race. The angle of the hip at contact (hip flexion) was calculated as 28° during lap 1 and 2, and then decreased to 26°. No significant differences were found between any of the angular data between laps or between genders. Men and women appeared to alter their joint angles differently. The men's knee extension at toe-off decreased, whereas the women's angle increased. Males also exhibited greater ankle toe-off angles (117° as opposed to 112°). These differences however were not statistically significant.

Table 4 Lower limb joint angles for all participants (mean ± SD)

	Hip		Knee		Ankle	
	Initial contact (°)	Toe-off (°)	Initial contact (°)	Toe-off (°)	Initial contact (°)	Toe-off (°)
Lap 1	28 (± 5)	-11 (± 3)	151 (± 4)	154 (± 3)	103 (± 5)	115 (± 5)
Lap 2	28 (± 5)	-11 (± 3)	151 (± 5)	154 (± 3)	102 (± 4)	115 (± 6)
Lap 3	26 (± 4)	-12 (± 3)	152 (± 4)	155 (± 2)	103 (± 4)	115 (± 6)

The distance between the body's centre of mass and the foot's centre of mass at initial contact ('foot ahead') and toe-off ('foot behind') are shown in Table 5. The overall mean values for the distance of the foot behind remained constant at 0.41 m throughout the race. Following lap 2 the distance of the foot ahead decreased from 0.30 m to 0.29 m, however no significant differences were evident. The men participants' mean distance of the foot behind increased constantly throughout the race; an initial value of 0.40 m during lap 1, progressing to 0.41 m, and finally 0.42 m in lap 3. A similar increase was evident for the distance of the foot ahead in men: 0.28 m, 0.29 m and 0.30 m respectively. The women participants' mean distance of the foot behind the CM remained constant at 0.41 m throughout the course of the race. The distance ahead of the CM remained stable at 0.31 m up to lap 2, decreasing to 0.29 m during the final lap. The male and female participants exhibited similar mean foot ahead and foot behind lengths. No significant differences were found between genders.

Table 5 Position of the foot ahead and behind the body at contact and toe-off (mean ± SD)

	Men		Women	
	Foot ahead (m)	Foot behind (m)	Foot ahead (m)	Foot behind (m)
Lap 1	0.28 (± .03)	0.40 (± .04)	0.31 (± .03)	0.41 (± .02)
Lap 2	0.29 (± .03)	0.41 (± .04)	0.31 (± .04)	0.41 (± .03)
Lap 3	0.30 (± .03)	0.42 (± .04)	0.29 (± .04)	0.41 (± .02)

DISCUSSION: From the results it is evident that changes in technique become apparent as early as 2,400 m. Speed decreased for both men and women overall through the course of the race; men by 1.07 km/hr and women by 1.05 km/hr. However, whilst the men's speed decreased continually through the three laps, women's speed showed an increase from lap 2 to 3. This may have been due to the women slipstreaming behind the men during the final lap; it appeared from the video recordings that the women used the men as pace makers, possibly contributing to their increase in speed in the later stages of the race. Kyle (1979) estimated a 40% reduction in air resistance (consequently a 3% reduction in energy expenditure) when drafting 2 m behind another runner.

Results obtained in the present study support the notion that step length decreases partly due to changes in lower limb joint angles (Siler and Martin, 1991). Although step length decreased overall for men and women during the race (men by 0.05 m and women by 0.07 m), the changes were a reflection of the speed alterations; men decreased constantly whereas the women exhibited an increase following lap 2. Cadence decreased coinciding with increases in step and contact time, and changes in hip extension at toe-off.

It is possible that angles during the recovery leg's swing phase account for the decrease in step length. A decrease in knee flexion may occur due to muscular fatigue and an unconscious attempt to reduce stress on the hamstring muscles. As the hamstrings fatigue, this inhibits increased knee flexion, increasing the moment of inertia. The resultant increase causes less angular velocity, restricting the swing phase, and hence causing smaller steps.

Although not significant, men and women exhibited different joint angle adaptations with fatigue. The difference in knee angle at toe-off may reflect gender differences and running economy. These patterns are associated with better running economy for the respective gender. Changes in knee angle as fatigue ensued may have been an unconscious attempt to reduce energy expenditure.

Changes in landing distance affect acceleration and subsequently speed due to braking effects. The slight changes in landing technique, including the distance of the foot ahead of the CM and the ankle angles may reflect a compensatory strategy. No significant differences in the joint angles or the distances of the foot ahead and behind the CM were found. This suggests fatigue did not affect these variables.

CONCLUSION: The decrease in speed was a product of a decrease in step length and cadence. Step length decreased due to increased hip and knee angles at toe-off. Cadence however decreased due to increases in contact and step time, and hyperextension at toe-off. Consequently, fatigue had an effect on important gait parameters as early as 2,400 m. The women appeared to counter fatigue through technical strategies, such as slipstreaming to reduce air resistance and using other runners to pace themselves. Runners should consciously aim to run at a constant pace, or to have a negative time split (and within a group if possible), to delay the onset and effects of fatigue.

REFERENCES:

Anderson, T. (1996). Biomechanics and running economy. *Sports Medicine*, 22(2), 76-89

Elliot, B. & Ackland, T. (1981). Biomechanical effects of fatigue on 10,000 meter running technique. *Research Quarterly for Exercise and Sport*, 52, 160-166.

Kyle, C. R. (1979). Reduction of wind resistance and power output of racing cyclists and runners travelling in groups. *Ergonomics*, 22:387-397.

Siler, W. L. & Martin, P. E. (1991). Changes in running pattern during a treadmill run to volitional exhaustion: Fast versus slower runners. *International Journal of Sport Biomechanics*, 7, 12-28.

Williams, K. R., Snow, R., & Agruss, C. (1991). Changes in distance running kinematics with fatigue. *International Journal of Sport Biomechanics*, 7, 138-162.

Williams, K. R. (1990). Relationships between distance running biomechanics and running economy. In: Cavanagh, P. R. (Ed.), *Biomechanics of Distance Running*. Illinois: Human Kinetics, 271-306.

OPTIMIZATION OF DISCUS FLIGHT

Falk Hildebrand[1], Axel Schüler[1], Johannes Waldmann[2]
Institute of Applied Training Science, Leipzig, Germany[1]
Hochschule für Technik, Wirtschaft und Kultur (FH), Leipzig, Germany[2]

We use a 3-D model for men's and women's discus flight including initial discus flight path angle β_0, angle of attack α_0, pitch attitude δ_0 as well as release speed v_0 and initial spin rate p_0. We study in detail optimal release conditions depending on a constant wind velocity of v_0 =5 m/s blowing from different directions $\gamma=0°$, 10° up to 350°. Here $\gamma =0°$, 180°, 90°, and 270° correspond to tail wind, head wind, wind from the left, and wind from the right, respectively. The optimal wind for men is head wind from the right ($\gamma= 220°$). In this case optimal men's strategy at $v_0=25$ m/s is $\beta_0 = 33°$, $\alpha_0= 23°$, $\delta_0= 30°$ with a range r=74.80 m. Optimal wind for women is wind exactly from the right ($\gamma= 270°$). The optimal women's strategy at $v_0=24$ m/s is $\beta_0 = 41°$, $\alpha_0= 32°$, $\delta_0 = 30°$ with a range r=61.26 m. In all cases we assume an initial spin rate of $p_0=50$ rad/s. At the moment of release, the angle of attack α_0 of the discus symmetry plane should always be less than the flight path angle β_0. Also, we can show that a faster-spinning discus imparts greater gyroscopic stability and therefore achieves a better throw. We used evolutionary algorithms to perform the optimization.

KEY WORDS: Discus Flight, Release Conditions, Optimization, Evolutionary Algorithm.

INTRODUCTION: During the flight, drag and lift forces affect the discus. Due to lift, the discus often flies much further than the expected parabola distance. More than 10% of the throwing distance can be affected by aerodynamics, see Hildebrand (2001). Aerodynamic forces likewise generate a torque on the discus. Since the discus has an angular momentum at the moment of release, two classes of equations must be satisfied: First the equations of motion of the mass centre and secondly Euler´s gyroscopic equations. Recently, Hildebrand (2001), Hubbard (2002) and Hubbard and Cheng (2007) studied such a 3-D rotational model allowing the discus spin axis not to lie in a constant vertical flight plane.

The main purpose of this paper is to determine optimal release parameters depending on the wind direction. In all cases the modulus of wind velocity is assumed to be constant v_w = 5m/s.

Our objective is the following. The system of equations of motion were established and numerically solved by the first author's work, see Hildebrand (2001). The throwing distance (range) is then a function of the release parameters which are initial speed v_0, flight angle β_0, wind velocity v_w, spin rate p_0, angle of attack α_0 (in our notation this is the angle between the discus plane and the ground) and pitch attitude δ_0 (which is the angle between the thrower's arm and the ground). We used drag and lift coefficients from Tutevich (1976). The moments of inertia for men's and women's discuses were determined experimentally with a torsion pendulum, see Sommerfeld (1950). These values are listed below in Table 1. They are quite similar to those obtained by Hubbard and Cheng (2007). Varying the release conditions (α_0, β_0, δ_0) as well as the wind direction γ we are able to determine the maximum range. This provides useful orientations for coaches and athletes. In the last part we study dependence of the optimal range r of both the wind direction γ and the initial spin rate p_0. We consider nominal release speed v_0 =25 m/s.

Table 1 Parameters for men's and women's discuses

discus	mass [kg]	diameter [cm]	moments of inertia	
			I_a [kgm²]	I_t [kgm²]
men's	2	22.1	0.0147	0.0085
women's	1	18.2	0.0044	0.0022

Here I_a denotes the moment corresponding to the discus axis while I_t is the moment with respect to an axis in the discus plane.

METHODS: The discus is balanced on its rotational axis. Therefore we have gyroscopic equations of angular velocities about discus axes as follows. If M_1, M_2, M_3 are the moments of aerodynamic force corresponding to three orthogonal axes, then, see Hildebrand (2001):

$$I_t \cdot \frac{d\omega_1}{dt} - (I_a - I_t)\omega_2 \omega_3 = M_1, \quad I_t \cdot \frac{d\omega_2}{dt} - (I_a - I_t)\omega_3 \omega_1 = M_2, \quad I_a \cdot \frac{d\omega_3}{dt} = M_3.$$

The motion of the centre of mass is described by gravity, drag and lift forces as well as speed $\mathbf{v}$ and wind velocity $\mathbf{v}_w$. Let $\mathbf{w} = \mathbf{v}_w - \mathbf{v}$. The aerodynamic forces are determined by $\mathbf{F}_D = \frac{\rho}{2} A w^2 c_D \cdot \frac{\mathbf{w}}{w}$ and $\mathbf{F}_L = \frac{\rho}{2} A w^2 c_L \cdot \frac{\mathbf{w} \times (\mathbf{n} \times \mathbf{w})}{w^2}$. Here A denotes the area of the discus, ρ is the atmospheric density and w is the absolute value of $\mathbf{w}$. Note that drag and lift coefficients c_D and c_L depend on the angle between discus axis and direction of $\mathbf{w}$. These equations were numerically solved for a given set of nine initial parameters. These nine parameters form an individual in the evolutionary algorithm, see Weicker & Weicker (2003). At most four parameters are changed simultaneously in one evolution step. Usually wind direction and initial spin are fixed. A population is formed out of 50 individuals. We perform 20 mutations and 20 crosses in one evolution step and use a quadratic selection algorithm to choose the fittest 50 individuals for the next generation. Fitness function is the range. In general we let $β_0$ vary freely. It turns out that in all cases the optimal initial flight path angle $β_0$ is between 28° and 45°. We assume the following restrictions on $α_0$, $β_0$ and $δ_0$ to be satisfied, 0°≤ $β_0$ - $α_0$ ≤ 10° and 0°≤ $δ_0$ ≤ 30° which is justified by the abilities of any thrower. In all cases wind velocity is v_w = 5 m/s.

RESULTS: All tables are computed for right throwers. For left throwers take the values at γ'= 360°-γ. Figure 1 shows the optimal release parameters ($α_0$, $δ_0$, $β_0$) for a men's discus with initial spin rate p_0 = 50 rad/s and v_0= 25 m/s depending on wind direction. It turns out that optimal wind is head wind from the right (northeast); tail wind is worst. The maximal range in these cases differs by 10 m (r_1=74.60 m and r_2= 65.30 m). In all cases we obtain that the angle of flight direction (solid gray line) is about 10° bigger than the angle between discus and the ground (angle of attack, dashed line). For the sake of clarity, the pitch attitude $δ_0$ is drawn with a negative sign. It turns out that the range behaves quite sensitive to $δ_0$. In particular, changing $δ_0$ at γ=220° (northeast) from 0° to 30°, the range increases from 72.50 m to 74.60 m.

Figure 2 shows optimal release parameters ($α_0$, $δ_0$, $β_0$) for a women's discus with initial spin rate of p_0 = 50 rad/s and v_0= 25 m/s depending on wind direction. It turns out that optimal wind comes from the right (east, γ=270°); wind from the left is the worst. It is seen that the maximal range in these cases differs by 7 m (r_1=66.12 m and r_2= 58.82 m).

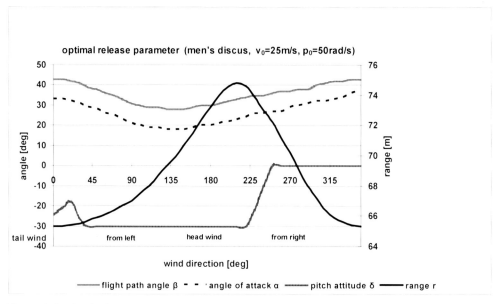

Figure 1: Optimal release parameter and maximal range depending on wind direction (men's discus)

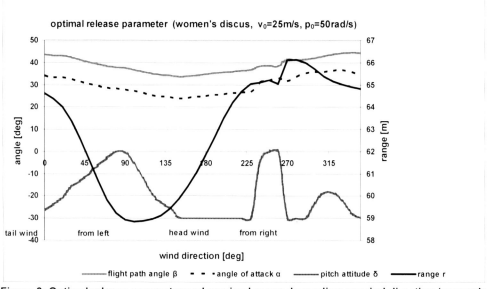

Figure 2: Optimal release parameter and maximal range depending on wind direction (women's discus)

In Figures 3 and 4 we show the optimal range for men's and women's discus depending on the initial spin rate and the wind direction. In general the best throw is obtained at the highest spin rate. The higher the spin the better is head wind. Head wind ($\gamma=180°$) is optimal for a men's discus with spin 90 rad/s, the corresponding range is r=77.36 m. Head wind from the right (Northeast, $\gamma=230°$) is optimal for women's discus with spin 90 rad/s, the corresponding range is r= 67.85 m. As the spin rate decreases, the optimal wind direction turns from north to east and the optimal range decreases, too. The worst wind for men is tail wind ($\gamma=0°$) while worst wind for women comes from the left ($\gamma=110°$).

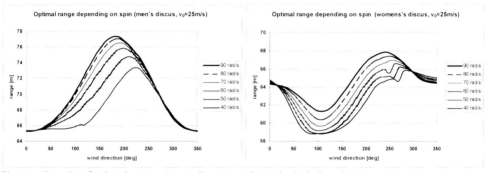

Figures 3 and 4: Optimal range depending on spin and wind direction

DISCUSSION: We have computed optimal release parameters (flight path angle, angle of attack, pitch attitude) for women's and men's discus depending on wind direction (v_w = 5 m/s) and initial spin rate. The release velocity was 25 m/s. Due to different mass and moments of inertia, we obtained different pictures for optimal range and optimal wind direction. However, the optimal flight path angles and angles of attack behave quite similar for both men and women. In contrast to the work of Hubbard & Cheng (2007) we discussed not only head and tail wind but all other wind directions, too. Similarly to Hubbard & Cheng (2007) our evolutionary algorithm shows that optimal release conditions must be wobble-free.

It was shown for a men's discus that optimal wind direction changes from head wind (γ=180°) to wind from northeast (γ=230°) when the initial spin rate decreases from 90 rad/s to 40 rad/s. In case of a women's discus the optimal wind direction changes from γ=240° (eastnortheast) to γ=290° (eastsoutheast) when the initial spin rate decreases from 90 rad/s to 40 rad/s.

CONCLUSION: The optimal release parameters from Figures 1 and 2 can immediately be applied by coaches and athletes. Knowing the wind direction the athletes can adjust flight path angle, angle of attack and pitch attitude. Other tables with different release velocity and different wind velocities had been computed. To get a more realistic view and to reduce the amount of computer time it is necessary to get smaller parameter spaces.

We used evolutionary algorithms to optimize release parameters. It turns out that this method is much more efficient up to factor ten compared with standard search algorithm.

REFERENCES:

Hildebrand, F. (2001). Modelling of discus flight. In: Blackwell, J. R. (Ed.), *Proceedings of Oral Sessions, XIX International Symposium on Biomechanics in Sports*, International Society of Biomechanics in Sports, San Francisco, 371-374.

Hubbard, M. (2002). Optimal discus release conditions including pitching-moment-induced roll. In: *Proceedings of World Congress of Biomechanics*, University of Calgary, Calgary.

Hubbard, M & Cheng, K. B. (2007). Optimal discus trajectories. *J. of Biomechanics* 40, 3650-3659.

Sommerfeld, A. (1950). Mechanics of Deformable Bodies. Lectures on Theoretical Physics, Vol. II. *Academic Press, Inc., New York, N. Y.*

Tutevich, V. N. (1976). Theorie der sportlichen Wuerfe. *Leistungssport* 7, 1-161.

Weicker, K. & Weicker, N. (2003), Basic principles for understanding evolutionary algorithms. *Fundam. Inform.* 55 (3-4), 387-403.

STABILISING THE HIP AND PELVIS DURING RUNNING: IS THERE AN EXPLOSIVE SOLUTION FOR UNINJURED ATHLETES?

Daniel L Meehan, Morgan D Williams, Cameron J Wilson and Elizabeth J Bradshaw

School of Exercise Science, ACU National, Melbourne, Australia

We compared the effectiveness of a conventional (slow-controlled) to a novel (explosive) eight week training program designed to improve lateral stability at the pelvis and hip during a running task. Parameters included: frontal-plane kinematics (500 Hz); electromyography recordings (1000 Hz) of gluteus medius (GM) and tensor fasciae latae (both sides); and oxygen kinetics. The groups were matched for hip and pelvis kinematics. After the training, reduction in peak angles at the hip and pelvis improved compared to baseline data regardless of group membership. Differences between groups were also found. Only the explosively trained group displayed changed GM onset times, where GM activation occurred earlier prior to ground contact when running. These differences in GM onsets support the notion of specific training adaptations, and the mechanism for hip and pelvis stability may not be the same for both groups. In addition, only the explosive group improved running performance (economy) further justifying this method of prescription.

KEY WORDS: Pelvic obliquity, Hip adduction, Running economy, Electromyography.

INTRODUCTION: Unstable frontal plane kinematics at the pelvis and hip during running have been linked with chronic/overuse injuries (e.g., Noehren, Davis & Hamill, 2007). The primary muscle used for lateral stability is the gluteus medius (GM). It prevents contra-lateral pelvic drop, and eccentrically controls hip adduction during the absorption phase. When instability is observed, the exercises prescribed to reduce joint amplitudes are based on increasing GM strength and control (e.g., Presswood, Cronin, Keogh & Whatman, 2008). Taken from a rehabilitation context, where patients are less mobile and experiencing pain, these exercises begin with non weight-bearing activities, and progress to "functional" weight-bearing activities. This conventional approach seems inappropriate for athletes. Generally, athletes who display instability at the hip and/or pelvis are pain free, moving freely and are only at risk of a chronic injury. Since, limited carryover above and below training velocity is expected from these slow-controlled exercises (Fleck & Kraemer, 1997), we consider this training less effective than more explosive exercises. We contend that explosive-strength exercises offer additional benefits to the athlete who is unstable but injury free. Evidence for this notion is currently not available, highlighting the purpose of this investigation.

METHODS: Participants: State Institute netballers (n = 8, age = 20.1 ± 2.1 years, stature = 178.2 ± 4.1 cm, mass = 69.1 ± 4.7 kg) all injury free took part in the study. Due to illness (unrelated to the study), one participant dropped out before the final trial. The testing protocol was repeated on three occasions over a period of 11 weeks. In week 1, baseline testing was performed, followed one week later by pre-intervention testing. These data were used to establish test-retest reliability. After an eight week training intervention, a final testing session was completed.

Data collection: At the start of each testing session anthropometric measures (Norton et al., 1996) were taken and used to estimate joint centres. Preparation of muscle sites and electromyography (EMG) electrode placement followed the Surface Electromyography for the Non-Invasive Assessment of Muscles initiative (Hermens, Freriks, Disselhorst-Klug & Rau, 1999). Bipolar Ag/AgCl surface electrodes (Bortec Bipole electrodes, Bortec Biomedical Ltd, Calgary, Canada) were placed on muscle sites, and the sacroiliac joint was used for the ground electrode. EMG data sampled at 1000 Hz was normalised against the participant's 100% maximum voluntary isometric contractions (MVIC) for the corresponding muscle. The

MVIC protocol is described elsewhere by Carcia and Martin (2007). EMG signals were processed and filtered using root mean square (RMS) smoothing (constant time windows = 100ms for MVIC and 25ms for running), a high pass filter with a 10 Hz cut-off and a low pass filter with 500 Hz cut-off. Retro-reflective ball markers (14mm diameter) identified the landmarks for the Plug-in-Gait lower limb marker set. Running kinematics was captured (500 Hz) using six cameras (model MX 13+) and Vicon MX motion analysis system (Vicon, Oxford, United Kingdom) with Vicon Nexus (version 1.3) software. All kinematic data were filtered using a Woltring filtering routine (Woltring, 1995). A metabolic cart (MOXUS Modular VO_2 system, MAX II Metabolic System, AEI Technologies, Pittsburgh, USA) was used to determine running economy (ml/kg/min). All running was performed on a high powered treadmill (H/P/Cosmos Sports and Medical GmbH, Pulsar 3p 4.0, Amsporplatz, Nussdorf-Traunstein, Germany) set to a 1% grade to reflect the energy cost of outdoor running. The running protocol included four minute stages at 8, 10, 12 and 14km/h. Data was collected during the final minute of running at the 10, 12 and 14km/h stages, including 3-D kinematics, EMG activity of the GM and tensor fasciae latae (TFL) muscles, and VO_2 kinetics for the calculation of running economy. Participants were matched based on the degree of frontal plane pelvic and hip instability demonstrated in the second testing session and assigned an eight-week training intervention designed to improve lateral hip/pelvic stability. Group 1 performed slow/controlled resistance exercises (Table 1). Group 2 performed explosive exercises (Table 2). Progressive overload was achieved for each exercise through either an increases in volume or intensity, or by increasing the complexity of the exercise performed. During the eight week training intervention period an average of 11 of the 16 planned training sessions were completed.

Table 1: Slow/controlled training intervention.

Exercise		Week 1	Week 2	Week 3	Week 4
1	Lying straight leg hip abduction	2 x 8	3 x 8	3 x 10	3 x 12
2	Split squats (static lunges)	2 x 8	3 x 8	3 x 10	3 x 12
3	Pelvic drops	2 x 8	3 x 8	3 x 10	3 x 12
		Week 5	Week 6	Week 7	Week 8
1	Reformer bed hip abduction (standing)	2 x 8	3 x 8	3 x 10	3 x 12
2	Single leg squats	2 x 8	3 x 8	3 x 10	3 x 12
3	Lateral lunges	2 x 8	3 x 8	3 x 10	3 x 12

Table 2: Explosive training intervention.

Exercise		Week 1	Week 2	Week 3	Week 4
1	Plyometric forward hops on gymnastics mat	2 x 6	2 x 6	3 x 6	4 x 6
2	Trampoline single leg hops (hands overhead*)	3 x 8	3 x 10	3 x 8	3 x 10
3	Split jerks (alternate forward leg)	4 x 4	4 x 4	4 x 4	4 x 4
		Week 5	Week 6	Week 7	Week 8
1	Plyometric lateral hops on floor (hip abduction)	2 x 6	2 x 6	3 x 6	4 x 6
2	Trampoline single leg hops (hands overhead*)	3 x 8	3 x 10	3 x 8	3 x 10
3	Split jerks (alternate forward leg)	4 x 4	4 x 4	4 x 4	4 x 4

* = overhead plate hold: 2.5kg in weeks 3-4; 5.0kg in weeks 5-6; 10.0kg in weeks 7-8

Data Analysis: Data analyses were performed using SPSS 15.0 for windows (Chicago, Illinois). Reliability using all participants' data (n=8) was assessed using dependent t-tests, typical error of measurement (TEM), coefficient of variation (CV%) and smallest worthwhile change (SWC). Non-parametric equivalents of *t*-tests were used to assess the effect of training and also group differences. A Wilcoxon signed-rank test, assessed the difference from pre- to post-intervention scores for the whole sample (n=7). A Mann-Whitney test, assessed group differences in baseline data and then group differences for changes in test variables from test 1 to test 2. Finally, corresponding effect size (r), median, minimum and maximum scores were also presented.

RESULTS: Reliability: At 12 and 14 km/h pace, test-retest reliability for all parameters was better than at 10 km/h. This was most evident for the EMG variables; TEM and CV scores were consistently higher when measured at 10 km/h (CV range 10.8 – 31.5%) compared to 12 km/h (5.4 – 23.8%) and 14 km/h (4.6 – 24.2%). Analysis was therefore performed on data collected at 12 and 14 km/h running velocities only. Despite showing CV scores > 5%, TEM for the EMG muscle onset time variables were judged as practically acceptable (SWC ranged between 1 to 4 ms). The TEM for the GM and TFL muscle onset variables, for example, ranged from just 5 to 8 ms.

Training effects: All kinematic data sets, except peak pelvic obliquity angle on the dominant side (P = 0.078, r = 0.50) decreased from pre- to post-intervention (P = 0.016-0.047, r = 0.54-0.63).

Group differences: Baseline scores did not differ between groups for all variables (P >0.050). In addition, no group differences for the change in scores in all kinematic variables were found. The explosive trained group did, however, display greater change in the onset of GM activation compared to the slow-controlled trained group at both 12 and 14 km/h, and on both sides. At 12 km/h, the change in dominant side GM muscle onset time was different (P=0.057, r=0.80), where the explosive trained group demonstrated a 22 (20 to 35) ms increase in muscle activation time prior to ground contact, compared to the slow-controlled group's median of 3 (1 to 6) ms. Similarly, non-dominant GM muscle activation time was different between groups (P=0.057, r=0.80). Again, a greater change in muscle onset time was found in the explosive trained group with a 16 (10 to 45) ms increase in activation time prior to ground contact, compared to the slow-controlled group's median of 0 (-1 to 2) ms, which resembled no change. Similar between group differences for GM onset times were found at 14 km/h (Figure 1). Running economy also improved from pre- to post- intervention in the explosive trained group, whilst a decrement was observed in two of the slow-controlled group (Figure 1). The changes in running economy were different between groups at both 12 km/h (P= 0.057, r = 0.80) and 14 km/h (P=0.057, r=0.80) velocities.

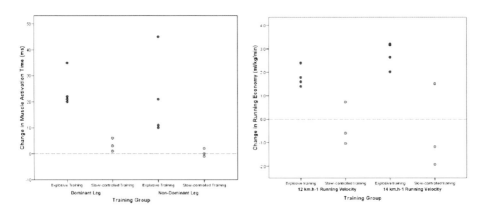

Figure 1: Group differences for changes in GM muscle onset at 14 km/h (left pane) and changes in running economy at 12 km/h and 14 km/h (right pane).

DISCUSSION: Prescribing exercises that encompassed 'specific' muscle contraction types, contraction velocities, ground contact times and loading to running improved hip and pelvis stability, but also running economy. Stability at the pelvis and hip were improved in both groups, however, the differences in GM onsets suggest that the mechanisms may differ and are specific to training. The change in GM pre-activity in explosive the trained group only was interpreted as an improvement in feed-forward neuromuscular control and possible muscle architectural changes in response to explosive training. By increasing the time GM is active before ground contact, joints become stiffer, taking advantage of the force-curve in anticipation of ground contact (Beard et al., 1993). Similar positive adaptations have been reported following plyometric training, but for landing tasks preceding a jump and not running (Myer, Ford, McLean & Hewett, 2006). Furthermore, all explosive group members increased economy, whilst, a decrement was observed in two out of three slow-controlled group members. Increased stiffness of the muscle-tendon system, and the resultant increase in elastic energy storage and return in the muscles of the trunk and legs may explain these improvements. Stiffer, less compliant muscles in the legs and lower trunk can enhance running economy via increased energy from elastic storage and return, which has no additional oxygen cost (Kyrolainen, Belli & Komi, 2001). Explosive-strength training increases the stiffness of the muscle-tendon system in this way in response to exposure of high eccentric loading and maximal concentric muscle contractions (Kyrolainen et al., 2001).

CONCLUSION: These findings support the notion that athletes, who are injury free, but at risk of chronic running related injuries, may benefit more from 'specific' explosive-strength exercises compared to the conventional (rehabilitation) approach. In particular, for coaches, where time available to prepare athletes is limited, eight weeks of explosive training was found to decrease hip and pelvis joint angles and also improve economy. The conventional approach of slow-controlled training only improved kinematics and not economy. Limited by sample size, the findings from this study warrant further investigation.

REFERENCES:

Beard, D. J., Kyberd, P. J., Fergusson, C. M., & Dodd, C. A. (1993). Proprioception after rupture of the anterior cruciate ligament. An objective indication of the need for surgery? *Journal of Bone and Joint Surgery, 75*, 311-315.

Carcia, C., & Martin, R. (2007). The influence of gender on gluteus medius activity during a drop jump. *Physical Therapy is Sport, 8*, 169-176.

Fleck, S. J., & Kraemer, W. J. (1997). *Designing resistance training programs*. Champaign: Human Kinetics Books.

Hermens, H. J., Freriks, B., Disselhorst-Klug, C., & Rau, G. (1999). The recommendations for sensors and sensor placement procedures for surface electromyography. In H. J. Hermens, B. Freriks, R. Merletti, G. Hagg, D. Stegeman, J. Blok & G. Rau (Eds.), *SENIAM 8: European recommendations for surface electromyography* (pp. 25-54). Enschede: Roessingh Research and Development.

Kyrolainen, H., Belli, A., & Komi, P. V. (2001). Biomechanical factors affecting running economy. *Medicine & Science in Sports & Exercise, 33*, 1330-1337.

Myer, G. D., Ford, K. R., McLean, S. G., & Hewett, T. E. (2006). The effects of plyometric versus dynamic stabilization and balance training on lower extremity biomechanics. *American Journal of Sports Medicine, 34*, 445-455.

Noehren, B., Davis, I., & Hamill, J. (2007). Prospective study of the biomechanical factors associated with iliotibial band syndrome. *Clinical Biomechanics, 22*, 951-956.

Norton, K., Whittingham, N., Carter, L., Kerr, D., Gore, C., & Marfell-Jones, M. (1996). Measurement techniques in Anthropometry. In K. Norton & T. Olds (Eds.), *Anthropometrica* (1st ed., pp. 25-76). Sydney: University of New South Wales Press.

Presswood, L., Cronin, J., Keogh, J., & Whatman, C. (2008). Gluteus medius: Applied anatomy, dysfunction, assessment, and progressive strengthening. *Strength and Conditioning Journal, 30*, 41-53.

Woltring, H. J. (1995). Smoothing and differentiation techniques applied to 3-D data. In P. Allard, I. Stokes & J. P. Blanchi (Eds.), *Three-dimensional analysis of human movement* (pp. 79-99). Champaign: Human Kinetics.

ISBS 2009

Scientific Sessions

Friday AM

ISBS 2009

Oral Session S20

Martial Arts & Dance

ANALYSIS OF PERFORMANCE OF THE KARATE PUNCH (GYAKU-ZUKI).

Edin K. Suwarganda[1], Ruhil A. Razali[1], Barry Wilson[1], Nick Flyger[1] and Arivalagan Ponniyah[2]

Centre for Biomechanics, National Sport Institute, Bukit Jalil, Malaysia[1]
National Coach, Karate Kumite, Bukit Jalil, Malaysia[2]

Variation in the movement sequence of the reverse punch (Gyaku-zuki) could affect kinematic variables such as punch time, distance and joint velocities. The reverse punches of nine elite Malaysian karate athletes were imaged in 3D at 150 Hz for two conditions (Jodan and Counter-chudan). Based on the linear resultant joint velocities of the shoulder and elbow two clusters are identified. One cluster is characterized by a more simultaneous movement sequence and the other by a more sequential movement sequence. The first cluster is mostly associated with female performances (87%) and the second cluster mostly with male performances (83%). It is found that the mostly male cluster achieved longer punch distance and higher peak linear resultant joint velocities for shoulder, elbow and wrist. Furthermore subgroups within the two clusters are identified and are associated with Jodan and Counter-chudan punches. The mostly female cluster achieves longer punch distance and higher peak linear resultant velocities in the subgroup associated with Jodan punches. However, the mostly male cluster achieves similar results for the subgroup associated with the Counter-chudan. Conclusion: the females tend to punch with a simultaneous sequence and men tend to punch with a sequential sequence with regard to the shoulder and elbow movements. Additionally women and men seem to have optimal performances in terms of punch distance and peak linear resultant joint velocities in different punching conditions.

KEY WORDS: reverse punch, elite karate athletes, cluster analysis, kinematic analysis.

INTRODUCTION: The Gyaku-zuki, also known as the reverse punch, is a technique commonly used in karate kumite, a form of competitive fighting (Emmermacher et al., 2005; Hofmann et al., 2008). The objective of a punch is to hit the opponent at a controlled distance in as little time possible (Emmermacher et al., 2005; Hofmann et al., 2008). Punching consists of the rapid execution of a sequence of body movements (Stull et al., 1988).

Only a few biomechanical studies have examined the technique of karate punching. Stull et al. (1988) investigated the relative timing sequence of peak joint velocities of the reverse punch by four different martial arts practitioners. They digitized multiple joint velocities (ankle, knee, hip, shoulder, elbow and wrist) from film (100 Hz) and found for example that the peak shoulder velocities of the shoulder occurred between 53% and 84% of total movement time. Variations in the sequence could affect kinematic variables such as total punch time, distance covered and joint velocities. Some researchers (Emmermacher et al., 2005; Hofmann et al., 2008 and Stull et al., 1988) measured one or more of these variables, but not in relation to any variation in the sequence of movements. Sforza et al. (2000) used 3 dimensional analyses of multiple joints to measure the repeatability (standard deviation) of karate punches (choku-tsuki and oi-tsuki). They found a larger standard deviation for horizontal motions than vertical motions.

The purpose of this research paper is to examine the reverse punch movements and variations in technique of elite Malaysian karate exponents. Two variants of the reverse punch or conditions were studied, the Jodan and Counter-chudan. Differentiating the variation in movement sequences was done by cluster analysis of the linear resultant joint velocities of the shoulder and elbow.

METHODS: The subjects were 8 full time Malaysian national karate athletes (4 males age 26.5±5.4 years; mass 65.0±11.5 kg and 4 females age 25.8±3.3 years; mass 55.7±7.3 kg). They had a minimum of 2 years experience in the national program. All subjects regularly practised the reverse punch, especially the Jodan variant where they punch the opponents head and the Counter-chudan variant where they punch an approaching opponent's trunk.

The punching movement was captured using 6 infra-red Motion Analysis Corporation Eagle cameras and a Kistler force plate. Force data were sampled at 1000 Hz and the camera frame rate was 150 Hz. Calibration and collection of data was done with the EVaRT 5.0 software application. The measurement error was less than 0.0005m.

Subjects completed a 10 minute warm up session before testing. Four reflective markers were placed on the hip (iliac crest), shoulder (acromion), elbow (lateral epicondyle) and wrist (ulnar styloid) of the preferred punching side. Subjects were instructed to position their back foot on the force plate when performing the punch. To simulate fighting a training partner acted as an opponent and stood at a desired distance set by the subject. Both the Jodan and Counter-chudan were executed 3 times each. There was a rest period of at least 15 seconds between each punch.

EVaRT was used to smooth the data with a butterworth filter at 12 Hz. After which the data was further analysed using Matlab 7.0 software. Computations of kinematic variables such as punch time, punch distance and linear resultant joint velocities were calculated. The athletes were allowed to move before executing the punch. Therefore the start of each punch is defined as the time of peak ground reaction force (GRF). The end of the punch is defined as the time of maximum elbow extension. Punch distance is defined as the horizontal position of the wrist at the end of the punch minus the horizontal position of the hip at the start of the punch.

One subject performed the Jodan only and 13 trials of the other subjects were excluded as incomplete trials leaving 39 trials for analysis. The shoulder and elbow linear resultant joint velocity data was normalized for time and passed into the statistical software R v2.6 (www.R-project.org). The data was clustered using the average linkage method within the hclust function from the R-Stat Cluster package. Groups were selected based on the Hubert-Γ score and p-values assigned to the clusters using a bootstrapping technique (Chow et al., 2008).

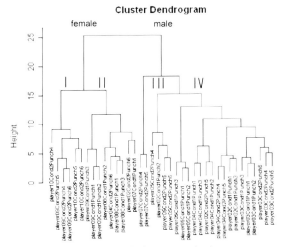

Figure 1: Dendrogram of cluster analysis.

RESULTS: Figure 1 shows that the cluster analysis identified two distinct groups (p=0.14). The first cluster included mostly trials from female subjects (87%) and the second cluster mostly trials from male subjects (83%). Table 1 contains the averaged peak values of kinematic variables of each cluster and the difference between the two clusters. It shows that cluster 2 has on average a longer punch distance and higher peak linear resultant velocities of shoulder, elbow and wrist. Examination of the ensemble average curves for the two clusters revealed that the mostly female group tended to show a simultaneous movement sequence and the mostly male group showed a sequential sequence with regard to movement of the shoulder and elbow. Examplar trials can be seen in Figure 2 and 3. The sequential sequence is characterized by the peak linear resultant elbow velocity occuring after the peak linear resultant shoulder velocity. The simultaneous sequence is characterized by the peak linear resultant elbow velocity occuring simultaneously with the peak linear resultant shoulder velocity.

Further examination identified two subgroups within each cluster (Figure 1). These four subgroups are mostly associated with the Jodan (subgroup II & IV) and Counter-chudan

Table 1 Average peak values of kinematic variables and the differences between cluster 1 & 2.

N=39	Cluster 1	Cluster 2	Difference
Punch time [s]	0.26 ± 0.04	0.27 ± 0.04	-0.01
Punch distance [m]	1.08 ± 0.16	1.29 ± 0.15	-0.21
Peak hip velocity [m/s]	2.42 ± 0.53	2.44 ± 0.57	-0.02
Peak shoulder velocity [m/s]	4.04 ± 0.58	4.61 ± 0.39	-0.57
Peak elbow velocity [m/s]	6.94 ± 1.10	7.36 ± 0.79	-0.42
Peak wrist velocity [m/s]	6.93 ± 1.12	7.65 ± 0.86	-0.72

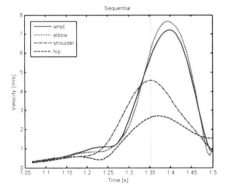
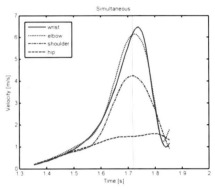

Figure 2 & 3: Exemplar linear resultant velocity curves of sequential (left) and simultaneous (right) movement sequences against time (not normalized).

Table 2 Average peak values of kinematic variables and the differences between the subgroups within cluster 1 & 2.

	Cluster 1			Cluster 2		
Subgroup (n)	Sub I (6)	Sub II (9)	Diff.	Sub III (6)	Sub IV (17)	Diff.
Punch time [s]	0.25 ± 0.04	0.27 ± 0.04	-0.02	0.28 ± 0.03	0.27 ± 0.05	0.01
Punch distance [m]	0.93 ± 0.08	1.19 ± 0.11	-0.26	1.38 ± 0.05	1.25 ± 0.16	0.13
Peak hip velocity [m/s]	2.02 ± 0.25	2.69 ± 0.51	-0.67	3.20 ± 0.33	2.18 ± 0.34	1.02
Peak shoulder vel. [m/s]	3.51 ± 0.36	4.40 ± 0.39	-0.89	5.14 ± 0.23	4.42 ± 0.22	0.72
Peak elbow vel. [m/s]	6.10 ± 0.78	7.49 ± 0.93	-1.39	8.11 ± 0.54	7.09 ± 0.70	1.02
Peak wrist vel. [m/s]	6.18 ± 1.04	7.43 ± 0.90	-1.25	8.52 ± 0.43	7.34 ± 0.76	1.18

(subgroup I & III). In subgroup II and IV respectively 78% and 65% are Jodan punches. In subgroup I and III respectively 100% and 83% are Counter-chudan punches. Table 2 contains the averaged peak values of kinematic variables of the subgroups and the difference between the subgroups within each cluster. It shows that subgroup II and subgroup III have a larger average punch distance and consistently higher averaged peak linear resultant joint velocities within the cluster.

DISCUSSION: The averaged peak linear resultant velocities are similar to those found in previous research by Emmermacher et al. (2005), Hofmann et al. (2008) and Stull et al. (1988). Summarizing the result, cluster 1 showed a more simultaneous movement between the shoulder and the elbow and cluster 2 showed a more sequential sequence. The kinematic results of cluster 2 does reveal larger values for the averaged peak linear resultant velocity of the hip, shoulder, elbow and wrist. Additionally, the average punch distance is larger in cluster 2. However this could be explained as an affect of gender differences. Therefore it is expected that the mostly male cluster have higher punch velocities and longer distance than the mostly female cluster. Logically, within the two clusters the ratio of Jodan and the Counter-chudan punches are alike. In the mostly female cluster 47% are Jodan punches and 53% Counter-chudan. For the mostly male cluster the percentages are 52% and 48% respectively. Interestingly the mostly female cluster achieve higher peak linear resultant joint velocity in the subgroup associated with Jodan whereas the mostly men cluster achieve higher peak linear resultant joint velocities in the subgroup that is associated with the Counter-chudan. However future research with more statistical power is needed to confirm these findings. Now only 8 subject participated and only 38 trials were included in the clustering analysis. Furthermore examining the effect of different punch conditions on kinematic variables could reveal important differences in techniques of male and female karate athletes.

CONCLUSION: The females in this study tend to show a more simultaneous movement sequence and the men show a more sequential sequence with regard to the shoulder and elbow movements. Additionaly their optimal performance is achieved in different punching conditions. The men punch optimal in terms of longer distance and peak linear joint velocities in the subgroup associated wth Counter-chudan punches whereas the women punch optimal in the subgroup associated with the Jodan punches.

REFERENCES:

Chow, J. Y., Davids, K., Button, C., & Rein, R. (2008). Dynamics of movement patterning in learning a discrete multiarticular action. *Motor Control*, 12, 219-240.
Emmermacher, P., Witte, K. & Hofmann, M. (2005). Acceleration course of fist push of Gyaku-zuki. In: Qing Wang (eds.), *Proceedings of XXIII International Symposium on Biomechanics in Sports Volume 2* (pp 884-887). Beijing: The People Sport Press.
Hofmann, M., Witte, K. & Emmermacher, P. (2008). Biomechanical analysis of fist punch Gyaku-zuki in karate. In: Kwon, Y.-H., Shim, J., Shim, J. K., & Shin, I.-S. (Eds.), *Scientific Proceedings of the XXVI International Conference on Biomechanics in Sports* (pp 576-579). Seoul, Korea: Seoul National University.
Sforza, C., Turci, M., Grassi, G., Fragnito, N., Pizzini, G. & Ferrario, F. (2000). The repeatability of choku-tsuki and oi-tsuki in traditional shotokan Karate: a morphological three-dimensional analysis. *Perceptual and Motor Skills*, 90 (Suppl. 3), 947-960.
Stull, R.A. & Barham, J.N. (1988). An analysis of movement patterns utilized by different karate styles in the karate reverse punch in front stance. In: Ellen Kreighbaum and Alex McNeil (eds.), *Biomechanics in Sports VI* (pp 233-243). Bozeman, Montana: Publishers.

ROUNDHOUSE KICK WITH AND WITHOUT IMPACT IN KARATEKA OF DIFFERENT TECHNICAL LEVEL

Valentina Camomilla[1], Paola Sbriccoli[1], Federico Quinzi[1], Elena Bergamini[1], Alberto Di Mario[2], Francesco Felici[1]

[1] Department of Human Movement and Sport Sciences, University of Rome "Foro Italico", Rome, Italy
[2] FIJLKAM – National Judo, Karate, Wrestling and Martial Arts Federation – Roma, Italy

The purpose of this study was to compare two different Karate roundhouse kicks performed by athletes of different technical level. The combination of high movement velocities and a high technical difficulty, qualify these actions as a good model to quantify the ability of a Karateka to execute complex movements. The first kick, directed to the face, entails a strong braking action to avoid the impact (NI), the other, directed to the chest, is concluded by an impact (IM). Technical aspects and the role of muscular co-activation as joint protector were investigated in six top level Karateka (KA) and six practicing karate amateurs (CO), by estimating joint kinematics and neuromuscular activity patterns. KA presented a faster execution for both tasks, prevalently due to a faster knee extension, supported by a low co-activation of the antagonist Biceps Femoris. This behaviour confirms that elite KA tend to lower the co-activation of antagonist muscles during fast movements, partially in contrast with the antagonists possible role in maintaining knee stability. The NI task, requiring higher technical competence and entailing a high target, is performed by KA athletes using a peculiar technique, based on a wide hip flexion-extension range, with a peak hip ab-adduction occurring earlier than in CO. A lower co-activation presented by CO during knee flexion is presumably due to their difficulty in mastering this complex kick.

KEYWORDS: Roundhouse kick, Karate, muscular co-activation

INTRODUCTION: Karate practice requires a fine control of movement associated to a great ability to perform the main technical actions as fast as possible (Zehr et al., 1997; Mori et al., 2002). While different studies dealt with front kick (Sørensen et al., 1996; Sforza et al.,2002), little attention has been devoted to roundhouse kicks (RK), despite their popularity and preferential use. Moreover, available doctoral theses and conference abstracts all focused on RK only in Taekwondo, whose practice does not include strong braking actions. Actually, these kicks seem to be a good model to quantify the ability of a karateka to perform complex actions since they combine high movement velocities with a high technical profile finalised to high precision. Thus, this work was designed to characterise top level karateka performance of two different roundhouse kicks: the first, directed to the face, entailing a strong braking action to avoid the impact (Mawashi Geri Jodan, No Impact - NI), the other, directed to the chest and concluded by an impact (Mawashi Geri Chudan, Impact - IM). Joint kinematics and neuromuscular activity pattern adopted by top level karateka and practicing karate amateurs were analysed to describe differences between the techniques. Muscular co-activation was analysed to investigate its role as joint protector in presence or absence of a braking action.

METHODS: Six top level Karateka (KA) (28±1yrs; 1.8±0.0m; 77±4kg) and six graduated students practicing karate as amateurs (Controls, CO) (25±1yrs; 1.78±0.03m; 73.8±4kg) were tested. After a visual trigger was delivered to the subjects, they performed with their preferred leg the two different roundhouse kicks, three times each: NI, directed to the face, and IM, directed to the chest and concluded by an impact on a punch bag.
Kinematic and electromyographic signals were obtained from the kicking leg. Surface electromyographic (sEMG) signals was recorded through a wi-fi transmission EMG amplifier (BTS Pocket EMG, Italy) from the Vastus Lateralis (VL), Rectus Femoris (RF), Biceps Femoris (BF), Gluteus Maximum (GM) and Gastrocnemious (GA) muscles. Magnetic field angular rate and gravity sensors (MTx, Xsens Motion Technologies, Enschede, The

Netherlands) were used to acquire pelvis, thigh, and shank orientation and angular velocity. The synchronization of the systems was provided by a switch connected to the visual trigger. Hip and knee angular kinematics were estimated first obtaining for each body segment, through a point-based anatomical calibration of the inertial sensors (Picerno et al. 2008), the time-invariant orientation of each anatomical frame with respect to the sensor technical frame and, then, combining this information with the time-variant orientation of the sensor technical frame with respect to global reference frame, measured by the inertial sensors. Joint angular kinematics was determined from the relative orientation of the proximal and distal anatomical frames using the Cardan convention (Grood and Suntay, 1983).

From the joint kinematics the following quantities were determined: the loading and kick phase, associated to knee flexion and knee extension, were identified in knee flexion-extension with an algorithm for peaks detection. Peak to peak knee flexion-extension (K_{fe}), hip flexion-extension, and ab-adduction (H_{fe}, H_{aa}) angles were determined within these two phases. The time interval from trigger to full knee extension (t_{task}) and the length of the knee extension phase (t_{KE}) were determined, as well as the time interval between the peak of hip ab-adduction and peak knee flexion-extension (t_{H-K}).

The raw EMG signal of each muscle was full-wave rectified and, for each subject and each muscle, normalized using the highest value within trials. Given the good signal to noise ratio of the EMG signal, a simple threshold-based method was used to detect the muscle activation onset (Staude et al., 2001). In each phase, a threshold was set equal to the average plus three times the standard deviation of the first 50 ms of the EMG signal; the onset was associated with the first sample of a 15ms window above threshold. The area underneath the EMG signal curve across the activation period was used to quantify the intensity of the contraction of a muscle from the onset, t_j, to the end of the above threshold period, t_f:

$$I_{muscle} = \sum \int_{t_j}^{t_f} EMG_{muscle}(t)dt \quad (1)$$

In order to quantify the activity of agonists and antagonists around the knee joint, a co-activation index (CI) was calculated as a percentage (0-100%) of the ratio between antagonist muscles and all muscle, with 100% indicating full coactivation (Kellis et al., 2003):

$$CI = \frac{I_{ant}}{I_{ago} + I_{ant}} \times 100\% \quad (2)$$

A CI index was calculated for knee extension, CI_E, and knee flexion, CI_F:

$$CI_E = \frac{I_{BF}}{I_{RF} + I_{VL} + I_{BF}} \times 100\% \qquad CI_F = \frac{I_{VL}}{I_{BF} + I_{VL}} \times 100\% \quad (3)$$

In CI_F the RF was excluded as antagonist due to its agonist role at the hip.

All data have been reported as Mean (Standard Errors, SE), in text, Table and Figure. Joint peak to peaks, K_{fe}, H_{fe}, H_{aa}, time intervals, t_{task}, t_{KE}, t_{H-K}, and co-activation indices, CI_E, CI_F were submitted to statistical analysis. Differences between groups of Karateka and Controls were tested through an unpaired T-Test. Differences between IM and NI kicks were investigated on the same variables using a paired T-Test. Statistical significance was set to p<0.05 for all tests performed. t_{H-K} was tested with a one-sample t-test, alpha = 0.05.

RESULTS: KA subjects presented shorter timings from the trigger to the full knee extension, t_{KE}, for both kicks, Table 1. This difference was more evident for the NI condition. When comparing kicks, no difference in time to task completion, t_{task}, was evidenced for KA subjects, while for CO subjects significantly higher times were necessary in the NI condition. For these subjects, the difference between IM and NI increased when focusing only on the length of the knee extension phase t_{KE}.

When observing joint kinematics, Table 1, no significant difference was evidenced for knee flexion extension range (KA: 138±9deg, CO: 118±8deg). Conversely, KA and CO subjects adopted different hip strategies. Hip flexion-extension range was wider in KA subjects, especially for NI kicks with respect to the IM condition (p = 0.0003). The hip ab-adduction range was not different between groups; however, it increased in both groups when changing from IM to NI condition (p<0.001). The hip ab-adduction peak happened consistently before maximal knee flexion for KA subjects (p < 0.04). CO subjects showed a consistent intra-subject behaviour, while presenting either an anticipation or a delay with respect to the peak knee flexion.

Table 1. Joint kinematics parameters and time intervals for both groups and kicks.

| | Joint kinematics peak to peak [deg] | | | | | | Time intervals [s] | | | | | |
| | K_{fe} | | H_{fe} | | H_{aa} | | t_{task} | | t_{KE} | | t_{H-K} | |
	IM	NI	IM	NI	IM	NI	IM	NI	IM	NI	IM	NI
KA	140 (9)	135 (10)	68 (12)	87 (18)	23 (3)	32 (5)	0.66 (0.03)	0.71 (0.06)	0.12 (0.02)	0.12 (0.02)	-0.04 (0.02)	0.00 (0.03)
CO	116 (8)	119 (8)	39 (3)	69 (10)	28 (4)	44 (6)	0.78 (0.02)	0.85 (0.04)	0.11 (0.00)	0.19 (0.02)	0.02 (0.02)	0.04 (0.02)

For KA subjects, a typical EMG activity pattern was observed, consisting of a biphasic and synergic activation of GM and VL muscles during both IM and NI kicks, the BF activation being shifted in time with respect to VL and GM muscles. A delayed activation of RF muscle was also observed. For CO subjects, no typical EMG activation pattern was observed.

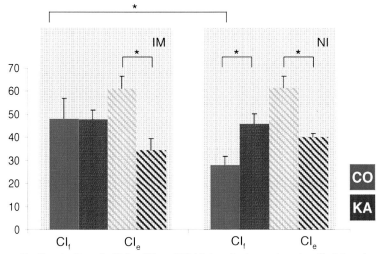

Figure 1: Co-activation indices in % for IM and NI kicks, during extension (full bars) and flexion (dashed bars) of CO (blue) and KA (magenta) athletes.

Co-activation strategies during flexion were different between groups only during NI kick, KA subjects presenting a positive 20% difference with respect to CO (p = 0.02). When comparing the kicks, no differences were evidenced for KA subjects, while a 17% decrease was evidenced in CO subjects when moving from IM to NI condition (p = 0.01). During extension, KA subjects had significantly lower CI_E in both conditions (p<0.004). No difference was evidenced between kicks.

DISCUSSION: Karateka presented an obvious faster execution for both tasks. Apart from presumably faster reaction times, not analysed in the present study, KA higher velocities with respect to CO subjects can prevalently be attributed to a faster knee extension. This velocity can be ascribed both to a higher extension activity of KA subjects or, as suggested by the CI_E behaviour, to a lower co-activation of the antagonists. This behaviour during knee extension confirms that elite KA tend to lower the co-activation of antagonist muscles during fast movements (Bazzucchi et al., 2008), partially in contrast with a possible role of antagonists in maintaining knee stability (Baratta et al, 1988). During knee flexion, conversely, CO subjects presented a lower co-activation with respect to KA when performing the NI kick. This result should be interpreted in the light of the following considerations. The IM task requires lower technical competence and entails a target positioned at a lower height. To accomplish the NI kick, a wider hip flexion-extension as well as ab-adduction range of movement is required with respect to the IM kick. KA athletes show a wider hip flexion-extension range and their peak hip ab-adduction occurs earlier than in controls, suggesting that KA use a different technique. Presumably, the lower proficiency of CO subjects in performing the NI technique leads to a lower hip angular velocity, accompanied a general lower co-activation strategy (Kellis and Katis, 2007).

CONCLUSION: This study provided insight in technical aspects typical of elite Karateka in the execution of Mawashi Geri Jodan and Mawashi Geri Chudan, different in target and as impact or no-impact tasks. Top elite athletes confirmed to perform fast ballistic movements with a co-activation strategy that favours the velocity to a little disadvantage in terms of safety of the knee joint. Further speculations will be formulated by analysing hip muscles co-activation strategy and observing the activity of each muscle involved.

REFERENCES:
Baratta, R., Solomonow, M., Zhou, B.H., Letson, D., Chuinard, R., D'Ambrosia, R. (1988). Muscular coactivation. The role of the antagonist musculature in maintaining knee stability. *Am J Sports Med.*, 16, 113-22.

Bazzucchi, I., Riccio, M.E., Felici, F. (2008). Tennis player show a lower coactivation of the elbow antagonist muscles during isokinetic exercises. *J Electromyogr Kinesiol.*, 18, 752-9.

Grood, E.S., Suntay, W.J. (1983). A joint coordinate system for the clinical description of three-dimensional motions: application to the knee. *J Biomech Eng*, 105, 136–44.

Kellis, E., Arabatzi, F., Papadopoulos, C. (2003). Muscle co-activation around the knee in drop jumping using the co-contraction index. *J Electromyogr Kinesiol*, 13, 229-38.

Kellis E., Katis A. (2007). Biomechanical characteristics and determinants of instep soccer kick. *J. Sports Science Med*, 6, 154-165.

Mori, S., Ohtani, Y. Imanaka, K.. (2002). Reaction times and anticipatory skills of karate athletes. *Hum. Mov. Sci.*, 21, 213-230.

Picerno, P., Cereatti, A., Cappozzo, A. (2008). Joint kinematics estimate using wearable inertial and magnetic sensing modules. *Gait Posture*, 28, 588–95.

Sforza, C., Turci, M., Grassi, G. P. Shirai, Y. F., Pizzini, G., Ferrario, V. F. (2002). Repeatability of mae-gaeri-keage in traditional karate: a three-dimensional analysis with black-belt karateka. *Percept. Mot. Skills*. 95, 433-444.

Sorensen, H., Zacho, M., Simonsen, E.B., Dyhre-Poulsen, P., Klausen, K. (1996). Dynamics of the martial arts front kick. *J Sports Sci.*, 14, 483-95.

Staude, G., Flachenecker, C., Daumer, M., Wolf, W. (2001). Onset Detection in Surface Electromyographic Signals: a Systematic Comparison of Methods. *EURASIP J Appl Signal Process.*, 2, 67–81.

Zehr, E. P., Sale, D. G., Dowling, J. J. (1997). Ballistic movement performance in karate athletes. *Med. Sci. Sports Exerc.*, 29, 1366-1373.

Acknowledgments
The experiments comply with the current Italian laws. Work partially founded by the Department of Human Movement and Sport Sciences, University of Rome "Foro Italico". A special thanks goes to all karateka for their active co-operation in the study.

A KINEMATIC ANALYSIS OF THE NAERYO-CHAGI TECHNIQUE IN TAEKWONDO

Michael Kloiber, Arnold Baca, Emanuel Preuschl, Brian Horsak

Department of Biomechanics, Kinesiology and Applied Computer Science, Centre of Sports Sciences and University Sports, University of Vienna, Vienna, Austria

The purpose of this study was to investigate the influence of selected kinematic parameters on the performance of the naeryo-chagi technique in taekwondo. Performance was quantified by the vertical velocity of the ankle at initial target contact (VIMP). METHOD: A sample of 19 competitive taekwondo athletes (17 males and 2 females) aged from 17 to 30 years (mean age = 19 ± 4), who were able to accomplish a correct naeryo-chagi technique, participated in this study. After warm up, participants were asked to perform several series of five naeryo-chagi kicks with their front leg at a kicking pad which was mounted on a frame at chin height. For data acquisition a motion tracking system comprising eight infrared cameras and a force plate were used. Only that series, which included the trial with the highest ankle velocity at initial target contact, was further processed. RESULTS: Significant differences between the best and worst performed kick of each athlete ($p = 0.025$) were found for the extension of the hip joint during the pull down phase (EHIP). No significant differences were found for the maximum ankle velocity during the strike out phase (AVSO; $p = 0.28$). Considering the best trials of each athlete only, Pearson correlation between EHIP and VIMP was significant ($r = 0.542$; $p = 0.017$), that between AVSO and VIMP was not ($r = 0.354$, $p = 0.137$). CONCLUSION: The magnitude of change of the hip flexion angle during the pull down movement seems to be an important factor for performing a kick featuring high velocity at initial target contact.

KEY WORDS: postural differences, motion analysis, kick-techniques, ankle velocity

INTRODUCTION:

Taekwondo is a combat sport, which originated out of Korea. It is especially well known for its fast, high and consecutive kicks. Since the year 2000 taekwondo is an official Olympic discipline. In competition, kicks and punches score points if the contact made to torso and/or head is sufficient to effect a displacement of body segments. A match can be won either by points or by knocking out the opponent. While punching on the head or kicking on the back head is not permitted, athletes can score by kicking torso, head or face or by punching torso. In taekwondo kicking techniques are most commonly used (Kazemi et al., 2006) and probably provide most efficiency.

The kicking technique naeryo-chagi is understood as a vertical taekwondo kick, which usually strikes the opponent's scull, face or clavicle. It can be described as follows: "The axe kick uses the rear of the heel to deliver a blow straight downwards. The kicking foot is swung up across the body until it is high in the air, then it is brought straight down onto the target." (Park, 1999). Aggeloussis et al. (2007) provide another description: "From the initial position, the knee is raised in an arc up and forward in front of the body, the leg is then extended and pulled down with the heel pointed downward. The arc can be performed in either a clockwise or counter-clockwise direction."

The first phase of the kick is initiated through a short preparation phase in which the athlete tries to move his centre of gravity forward for initiating the second phase. This phase is an upwards directed strike-out phase and ends with the top of stroke (reversal point of foot). During the downwards directed kicking phase the target is hit. The movement ends in the restoration of posture for either preparing a defence or attacking action.

For scoring in a competition the impact effects of kicks and punches are of high importance. Impact effect can physically be described by a linear momentum or via the kinetic energy. In both terms velocity takes a major role. It can be assumed that a higher maximum velocity at initial target contact gains more impact effect than punches or kicks with less velocity. Falco et al. (2009) investigated execution time and impact force in a roundhouse kick. Their findings approved significant differences in maximum impact force between expert and novice competitors. The purpose of this study was to analyse differences in selected kinematic variables with regard to the kick performance assessed by the ankle velocity at impact.

METHODS: Data Collection: A sample of 19 competitive taekwondo athletes (17 males and 2 females) aged from 17 to 30 years (mean age = 19 ± 4), who were able to accomplish a correct naeryo-chagi technique, participated in this study. After warm up, participants were asked to perform five naeryo-chagi kicks with their front leg onto a kicking pad which was mounted on a frame at chin height. A force plate was used to measure ground reaction forces. Athletes started from a normal standing position (supporting leg behind kicking leg). Each participant performed one to five series with five successive maximum-effort kicks onto the kicking pad. Athletes were allowed to choose the distance to the target. A motion tracking system (Vicon Motion Systems Limited, Oxford, UK) was used for analysis. The system consisted of eight infrared cameras (six cameras with a resolution of 1.3 Mega Pixels, and two cameras with a resolution of 4.0 Mega Pixels), an acquisition station system (Vicon MX Net) connected to a personal computer and 3D reconstruction software (Vicon Nexus 1.2 and Polygon 3.1). Data were collected with 250 Hz. After raw data acquisition, data were smoothed using a GCV-Auto Woltring filter routine, provided by Vicon Nexus 1.2 (see Woltring, 1986). The series which included the trial with the highest ankle velocity at initial target contact was selected for further processing.

Data Analysis: Movement was classified into three phases: start of the first phase was determined by loss of contact to the force plate by the kicking leg. The end was defined as the top of stroke (reversal point: ankle velocity in vertical direction = 0 m/s). The start of the second phase was defined by the reversal point and ended at initial target contact. Third phase started with initial target contact and ended with the first contact to ground of the kicking leg. Figure 1 illustrates the defined phases:

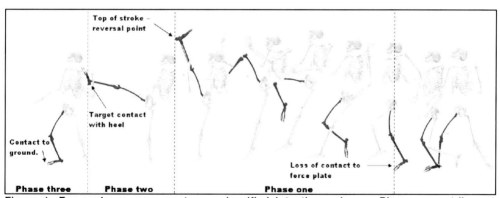

Figure 1: For analyses movement was classified into three phases. Phase one: strike out movement, phase two: target contact, phase three: restoration of posture.

Several kinematic parameters have been investigated on their influence on performance. Two of them will be discussed in more detail within this paper. Variable AVSO was defined as the maximum value of the vertical ankle velocity of the kicking leg during the phase one. The variable which represented EHIP was defined as the extension angle of the hip joint (for the kicking leg) during phase two. Values for both variables (ASVO and EHIP) were determined for the best and worst performed kick (best kick = maximum ankle velocity; worst kick = minimum ankle velocity at initial target contact (VIMP)) of each athlete within the series selected. Statistics were calculated using Student's t-test for dependent variables and the Pearson correlation coefficient considering $p < 0.05$ to be significant.

RESULTS: Figure 2 shows a typical curve progression of the ankle velocity and the hip angles of the kicking leg exemplarily plotted for one athlete.

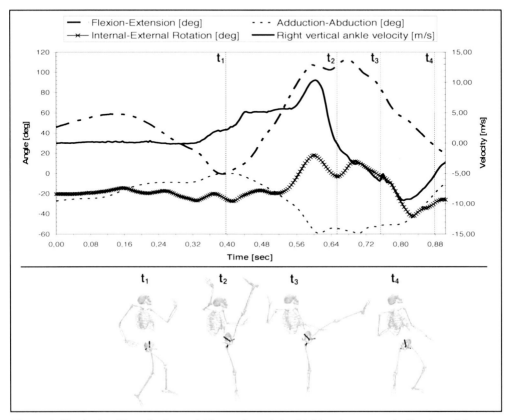

Figure 2: Vertical ankle velocity and hip angles of the kicking leg exemplarily plotted for one athlete: Flexion, adduction and internal rotation feature positive values; Extension, abduction and external rotation feature negative values; t_1: take-off from force plate; t_2: reversal point of ankle; t_3: initial target contact; t_4: contact to ground.

Table 1 shows descriptive statistics for variables AVSO, EHIP and VIMP:

Table 1 Descriptive statistics for variables AVSO, EHIP and VIMP.

BEST (n=19)				WORST (n=19)			
	AVSO [m/s]	EHIP [°]	VIMP [m/s]		AVSO [m/s]	EHIP [°]	VIMP [m/s]
Mean	10	13	8	Mean	10	9	7
± SD	1	12	1	± SD	1	13	1

Statistical analyses showed no significant differences between the best and worst performed kicks in their values for maximum ankle velocity during strike out (AVSO; $p = 0.28$). Significant differences ($p = 0.025$) were found for the extension of the hip joint (EHIP). A positive correlation between EHIP and VIMP could be found ($r = 0.542$; $p = 0.017$) whereas no significant correlation was found for AVSO and VIMP ($r = 0.354$; $p = 0.137$).

DISCUSSION: Two different variables were defined for statistical analyses: maximum ankle velocity during strike out (kicking leg) and range of hip extension before initial target contact. Although EHIP is not the only crucial factor for acceleration towards the target, it significantly ($r = 0.542$, $p = 0.017$) correlated with VIMP. Moreover, significant differences between the best and worst performed kicks could be found ($p=0.025$). An increase of acceleration distance seems to be a reliable predictor for an increase of kicking velocity. It should, however, be taken into account that longer acceleration paths go along with longer times to initial target contact thus raising the chance for the opponent to anticipate the movement.

Contrary to expectations no significant differences between the best and worst performed kicks could be found for AVSO ($p = 0.28$) and no significant correlation with VIMP could be shown ($r = 0.354$, $p = 0.137$). We would suggest that velocities of the kicking leg during the stretch phase above a certain individual threshold could have negative effects on the efficiency of the stretch-shortening cycle, thereby resulting in a reduced ankle velocity at target contact. Ishikawa et al. (2005) analysed the stretch-shortening cycle in short-contact drop jumps. They observed that the efficacy of elastic recoil decreased with increasing drop intensity. Their conclusions might also apply to the results of the present investigation. Less storage of elastic energy may occur at larger strain rates, the duration of the stretch-shortening cycle may not be enough for efficient elastic storage and recoil.

CONCLUSION: A significant difference between the best and worst kicks for the extension of the hip joint (EHIP) and a positive correlation for EHIP and the vertical velocity of the ankle at initial target contact (VIMP) was found. This indicates that the range of strike out movement is a crucial factor for performing the naeryo-chagi technique. On the other hand longer acceleration paths increase the risk of the opponent anticipating the attacker's actions. No significant difference and correlation were found for AVSO. It is possible that too high velocities during strike out effect negatively on athlete's performance as too low velocities do.

REFERENCES:

Aggeloussis, N., Gourgoulis, V., Sertsou, M., Giannakou, E. & Mavromatis, G. (2007). Repeatability of electromyographic waveforms during the Naeryo Chagi in taekwondo. *Journal of Sports Science and Medicine*, 6, 6-9.

Falco, C., Alvarez, O., Castillo, I., Estevan, I., Martos, J., Mugarro, F. & Iradi, A. (2009). Influence of the distance in a roundhouse kick's execution time and impact force in Taekwondo. *Journal of Biomechanics*, 42, 242-248.

Ishikawa, M., Niemelä, E. & Komi, P.V. (2005). Interaction between fascicle and tendinous tissues in short-contact stretch-shortening cycle exercise with varying eccentric intensities. *Journal of Applied Physiology*, 99, 217-223.

Kazemi, M., Waalen, J., Morgan, C. & White, A.R. (2006). A profile of Olympic Taekwondo competitors. *Journal of Sports Science and Medicine* CSSI, 5, 114-121.

Park, Y. (1999). *Taekwondo. The ultimate reference guide to the world's most popular martial art*. New York: Facts on file.

Woltring, H.J. (1986). A Fortran package for generalized, cross-validatory spline smoothing and differentiation. *Advances in Engineering software*, 8, 104-113.

A KINEMATIC ANALYSIS OF FINGER MOTION IN ARCHERY

Brian Horsak, Mario Heller, Arnold Baca

Department of Biomechanics, Kinesiology and Applied Computer Science, Centre of Sports Sciences and University Sports, University of Vienna, Vienna, Austria

This paper examines finger motion during the bow string release in archery. METHOD: Fifty-six shots from one athlete were captured with an infrared motion tracking system. Kinematics for index, third and ring fingers were calculated. Two different kinematic variables were defined, related to the proximal interphalangeal joint (PIP) of the third finger: maximum angular velocity (MAX) and minimum angular velocity (MIN). For statistical analysis shots were separated into two groups (very good shots: shots which hit the innermost score area and bad shots: score of 8 or less; shots which achieved a nine or a ten were excluded). A Mann-Whitney test was used. RESULTS: No significant differences were found in the variables MAX and MIN between very good and bad shots ($p > 0.05$). CONCLUSION: Findings in this study show that there are no significant differences in angular velocity (related to the PIP joint) between very good and bad shots, but that reproducibility of kinematic characteristics are possible crucial factors in archer's performance.

KEY WORDS: motion analysis, release, bow, finger joints

INTRODUCTION:

Coaches in archery often pay a lot of attention to the release phase of the shot. To avoid lateral deflection, the release phase of the fingers must be well balanced and highly reproducible (Ertan et al., 2003; Leroyer et al., 1993; Martin et al., 1990; Soylu et al., 2006). For investigation of finger movement several authors used surface EMG for analysing muscular activity, muscular coordination and different types of release strategies (Ertan et al., 2003; Martin et al., 1990 and Soylu et al., 2006). However, none of the investigations focused on the finger motion itself in its three-dimensional aspects. In the past few years motion analyses of finger and hand motion have gained large attention (e.g. Cerveri et al., 2007). Due to the relatively new opportunity of using motion capturing systems in hand motion analysis and due to the fact, that there has been hardly any three-dimensional kinematic research of finger motion in archery, this paper is going to demonstrate how an optoelectrical system can be used for three-dimensional analysis of the finger motion during the bow string release in Olympic recurve archery. The finger movement of a participant was analysed due to specific kinematic parameters. Statistical analyses were performed to investigate if there are any significant differences in selected kinematic values between "very good" and "bad" shots.

METHOD:

Data Collection: A motion tracking system (Vicon Motion Systems Limited, Oxford, UK) was used for analysis. The system consisted of eight infrared cameras (six cameras with a resolution of 1.3 Mega Pixels, and two cameras with a resolution of 4.0 Mega Pixels), an acquisition station system (Vicon MX Net) connected to a personal computer and 3D reconstruction software (Vicon Nexus 1.2 and BodyBuilder 3.6). Data were collected with 500 Hz. System accuracy was tested by tracking two markers mounted on a rigid object. Marker distance yielded differences in length of less than 0.2 mm. B-spline approximation was used for interpolation (2000 Hz) and differentiation. Kinematics in the finger joints were calculated using the VICON right hand model. This kinematic model calculated flexion/extension and abduction/adduction for all fingers (using the technique of Cheng & Pearcy, 1999) of the right hand and was programmed for BodyBuilder 3.6. A majority of competitive archers shoots using the three finger grip release. The finger release is defined as the point on which the bow string slips off from the finger tips. Due to this fact analysis in this paper was focussing on the three finger grip which included index, third and ring fingers. Semicircular markers with a diameter of 9mm were used and positioned on bony landmarks on the archer's right hand (Figure 1). The marker positions were based on the VICON right hand model: The investigated participant was a competitive archer from the Austrian B-National Team (age = 49 years, best FITA Indoor score = 1131 out of 1200 points) who took part in several national and international competitions (e.g. World Games 2007). Prior to participation, the subject read and signed a consent form, which was approved by Institutional Review Board at the Centre of Sport Science and University Sport of Vienna. The subject participated in three test sessions (one session included 30 shots, ten times three shots) in the Biomechanics Laboratory at the University of Vienna. Thirty shots at a distance of 18 meters (a FITA Indoor 40 cm vertical triple target face was used) were captured in one session. In sum, ninety shots were captured out of which, due to marker occlusion, fifty-six shots could be used for further analysis.

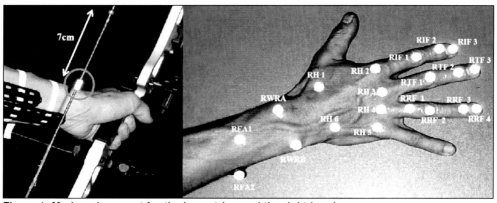

Figure 1: Marker placement for the bow string and the right hand.

Data Analysis: Analogue as performed in Keast & Elliot (1990) shots were corresponding to their achieved score separated into two groups: very good (score of x; x is the innermost score area of the target face) and bad (score of 8 or less) shots; shots which achieved a nine or a ten were excluded. Two different kinematic variables were defined (related to the PIP joint of the third finger): maximum (MAX) and minimum (MIN) angular velocity during the release of the bow-string (Figure 2). A Mann-Whitney test was used for statistical analysis.

For each shot, the ranges of motion of the metacarpophalangeal (MCP), proximal interphalangeal (PIP) and distal interphalangeal (DIP) joints during the shot itself were measured.

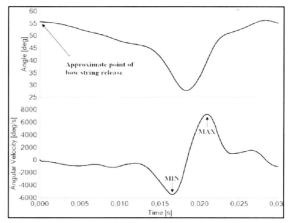

Figure 2: Kinematic variables: maximum (MAX) and minimum (MIN) angular velocity during the release of the bow-string (related to the PIP joint of the third finger).

The range of motion of each finger (FROM) was defined as the sum of ranges of motion of its three joints (JROM). The range of motion of each joint was then expressed as a percentage of the corresponding finger's range of motion (PCROM).

RESULTS: A Mann-Whitney test was used for analysis considering $p < 0.05$ to be significant. As values in Table 1 show, no significant differences were found.

Table 1 Descriptive statistics for very good and bad shots and results for statistic analysis (Mann-Whitney test for very good and bad shots).

Bad shots			Very good shots		
	MAX [°/s]	MIN [°/s]		MAX [°/s]	MIN [°/s]
Mean	687	-5123	Mean	6923	-4558
± SD	743	810	± SD	1063	443
Minimum	5675	-6394	Minimum	5495	-5193
Maximum	7895	-3943	Maximum	8392	-3656
	p-value for MAX	p-value for MIN			
Mann-Whitney t.	0.870	0.221			

Table 2 shows analysis for ranges of motion for all three fingers and appending joints. In sum the index finger achieved the highest value of FROM with 81 ± 5°. Most active joint was the DIP joint for the index finger (42 ± 5°), the PIP joint (32 ± 2°) for the third finger and the DIP joint (28 ± 4°) for the ring finger. Least movement was performed in the MCP joints (all fingers). The DIP joint showed a majority in PCROM for the index and ring finger (52 ± 6 and 57 ± 9 %). The PIP joint achieved a major PCROM for the third finger (55 ± 4 %).

Table 2 Results for range of motion analysis during the shot.

	FROM [°]		
Index finger	81 ± 5		
Third finger	58 ± 3		
Ring finger	48 ± 5		
	JROM [°]		
	MCP	PIP	DIP
Index finger	9 ± 28	30 ± 2	42 ± 5
Third finger	7 ± 2	32 ± 2	19 ± 2
Ring finger	6 ± 2	15 ± 2	28 ± 4
	PCROM [%]		
	MCP	PIP	DIP
Index finger	11 ± 2	37 ± 3	52 ± 6
Third finger	12 ± 3	55 ± 4	33 ± 4
Ring finger	13 ± 3	30 ± 5	57 ± 9

Note. FROM = range of motion of each finger; JROM = range of motion of each joint; PCROM = percent of the corresponding finger's range of motion.

DISCUSSION: No significant differences were found for the variables MAX and MIN (note that correlations between performance and variables, which were calculated in addition, showed no significant values, also). One possible reason might be numerical problems due to low accuracy of the second derivatives. Results for ranges of motion showed that especially the PIP and DIP joints are highly involved in the finger release. Least JROM was quantified for the MCP joints. Analysing the range of motion for each joint in respect to the range of motion for each finger (PCROM) the DIP joints showed highest values for the index and ring fingers. For the third finger the PIP joint showed the highest percentage of PCROM. Kinematic characteristics let presume that reproducibility is a possible crucial factor in archer's performance. Figure 3 and 4 show path-time diagrams of the MCP, PIP and DIP joints of the third finger. Three randomly selected shots out of the very good and bad group are exemplarily plotted. As it can be seen, the peak minimum values in both groups are almost similar for all three shots, but the characteristics of the graph before and after their minimum peak show hardly any differences for the "very good shots" and clearly more differences for the "bad shots". These findings probably support the assumption of Edelmann-Nusser (2005) that the motoric programme of arrow release in the manner of an open-loop movement is already initiated before clicker's fall.

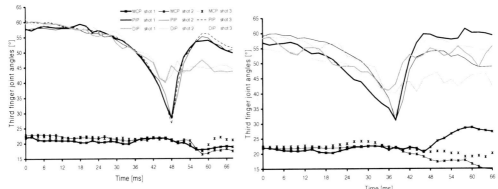

Figure 3: Time-path diagram of three randomly selected shots: "very good shots" (time-synchronized over minimum PIP peak).

Figure 4: Time-path diagram of three randomly selected shots: "bad shots" (time-synchronized over minimum PIP peak).

CONCLUSION: A measuring setup for three-dimensional motion analysis of hand motion in archery was developed and it was shown, that modern tracking systems can be used for research in finger motion analysis in archery. Time-path diagrams for kinematic data showed similar parameter values and graph-characteristics throughout the movement for "very good shots" and less similarity in graph-characteristics for "bad shots". These findings should be further investigated with a higher number of subjects.

REFERENCES:

Cerveri, P., De Momi, E., Lopomo, N., Baud-Bovy, G., Barros, R.M.L. & Ferrigno, G. (2007). Finger kinematic modeling and real-time hand motion estimation. *Annals of Biomedical Engineering*, 35, 1989-2002.

Cheng, P.L. & Pearcy, M. (1999). A three-dimensional definition for the flexion/extension and abduction/adduction angles. *Medical & Biological Engineering & Computing*, 37, 440-444.

Edelmann-Nusser, J. (2005). *Sport und Technik- Anwendungen moderner Technologien in der Sportwissenschaft*. Aachen: Shaker Verlag.

Ertan, H., Kentel, B., Tümer, T. & Korkusuz, F. (2003). Activation patterns in forearm muscles during archery shooting. *Human Movement Science*, 22, 37-45.

Keast, D. & Elliot, B. (1990). Fine body movements and the cardiac cycle in archery. *Journal of Sports Science*, 8, 203-213.

Leroyer, P., Van Hoecke, J. & Helal, JN. (1993). Biomechanical study of the final push-pull in archery. *Journal of Sports Sciences*, 11, 63-9.

Martin, P.E. & Heise, G.D. (1992). Archery bow grip force distribution: relationships with performance and fatigue. *International Journal of Sports Biomechanics*, 8, 305-319.

Martin, P.E., Siler, W.L. & Hoffman, D. (1990). Electromyographic analysis of bow string release in highly skilled archers. *Journal of Sports Sciences*, 8, 215-221.

Nishizono, A., Shibayama, H., Izuta, T. & Saito, K. (1987). Analysis of archery shooting techniques by means of EMG. In H. Ertan, Kentel, B., Tümer, T., & Korkusuz, F. (2005). Reliability and validity testing of an archery chronometer. *Journal of Sports Science and Medicine*, 4, 95-104.

Soylu, AR., Ertan, H. & Korkusuz, F. (2006). Archery performance level and repeatability of event-related EMG. *Human Movement Science*, 25, 767-74.

BALLET DANCER INJURIES DURING PERFORMANCE AND REHEARSAL ON VARIED DANCE SURFACES

Luke Hopper[1], Nick Allen[2], Matthew Wyon[3], Jacqueline Alderson[1], Bruce Elliott[1] and Tim Ackland[1]

School of Sport Science, Exercise and Health, University of Western Australia [1]
The Jerwood Centre, Birmingham, UK [2]
School of Sport, Performing Art and Leisure, University of Wolverhampton, UK[3]

Three dance surfaces regularly used by a professional touring ballet company (n=60) were quantified using standard sports surface testing apparatus. Surface sub-structure construction varied between surfaces and a range of surface force reduction values were reported. Injuries and associated variables occurring within the ballet company were recorded by the company medical staff. An injury was recorded if a dancer experienced an incident that restricted the dancer from performing all activities that were required of them for the period 24hrs after the incident. Injuries were delimited to those occurring in the lower limbs or trunk during reported non-lifting dance activity. Analysis of statistical significance was restricted due to a low injury data sample size. However certain trends in the injury data warrant future research. The surface with the highest variability in intra-surface force reduction was associated with the highest injury rates per week, lower limb injuries per week, mean days lost dancing per injury and likelihood of injury per performance day. Variability in intra-surface force reduction may have a stronger association with injury risk than mean surface force reduction magnitudes.

KEY WORDS: dance, injury, sports surface, landing

INTRODUCTION: The mechanical properties of floor surfaces used for athletic performance affect human landing mechanics (Ferris et al., 1999; Moritz et al., 2004) and it seems intuitive that certain floor surface types may have an associated injury risk. Yet epidemiological evidence investigating the relationship between surface properties and athletic injury is limited. Previous comparisons of floor surfaces and athletic injury have not quantified the mechanical properties of the investigated surfaces (Nigg and Segesser, 1988; Ekstrand et al., 2003; Nigg, 2003). Further, the relationship between sports surface and injury is difficult to establish due to confounding variables such as shoe/surface interactions and other environmental conditions (Shorten, 2007). Nonetheless, previous literature has identified that intra-surface variability may pose an injury risk to the athlete (Nigg, 2003) and human landing mechanics differ between known, unexpected or unknown surfaces (Moritz et al., 2004) but are highly adaptable to a known change in surface structure (Ferris et al., 1999).

Ballet dancers are highly trained athletes who demonstrate unique and highly developed motor functions (Imura et al., 2008). Dancers wear footwear during dance activity that provide little impact attenuation and perform in a well rehearsed and controlled environment. Therefore the injuries within a dance population may provide valuable insight into the potential injury risks associated with quantified dance surface mechanical properties.

The majority of dancer injuries are overuse in nature and occur in the lower limbs and trunk (Solomon et al., 1999; Nilsson et al., 2001). The interaction between a dancer and selected floor surfaces has been suggested to influence the incidence of lower limb and trunk injuries. Descriptors such as 'hard', 'stiff' or 'unsprung' have been used to describe floor properties that may present an injury risk to dancers (Bowling, 1989; Khan et al., 1995; Liederbach and Richardson, 2007). The actual mechanical properties of dance floors and how they relate to the incidence of lower limb and trunk injuries need to be quantified to assess any associated injury risks to the dancer. The aim of this study was to compare the lower limb and trunk injury rates within a professional ballet company to the mechanical properties of the surface on which the injuries occurred, as a means of identifying potential surface injury risk criteria.

METHODS: Data Collection: The mechanical properties of three stage surfaces used by a professional ballet company (n=60) were quantified using standard sports surface testing equipment, the advanced artificial athlete (Metaalmaatwork, NL). Advanced artificial athlete test protocols involve a drop mass consisting of a uni-axial accelerometer, a 20kg mass and a 2220N/mm spring with a circular (100mm diameter) impact surface, which is dropped from 55mm above the test surface. Residual analysis determined that the data be filtered at 92Hz with a low pass second order Butterworth filter. Acceleration data, captured at 9600Hz, were then converted to a force time series. The force reduction (FR) of each test point was calculated as the difference in the peak force recorded on the individual test points (F_t) to that on a rigid concrete surface (F_c) as a percentage of the peak force measured on a rigid concrete surface ($FR=((F_c-F_t)/F_c)100$). The test protocols followed were adopted as per the European standard; Surfaces for sports areas: Indoor surfaces for multi-sports use (BS EN 14808). Testing sessions were conducted in an ambient temperature of 18-21°C.

Test floors consisted of 20-30mm thick plywood boards covered by Harlequin Cascade Vinyl (Harlequin Pty Ltd., London, UK), a thin vinyl layer used to enhance the friction characteristics of the surface. Primary support structures for all surfaces were steel girders, which supported the surfaces above large storage areas. Wooden, metal and/or foam secondary support structures lay in between the plywood and the primary support structures. The orientation and spacing of support structures differed between surfaces. The primary support contact surface area for surfaces 1, 2 and 3 accounted for approximately 9.8%, 6.3% and 14.7% of the total surface area of each surface respectively. Test points were identified on each surface to ensure that FR values on and between floor supports were quantified. Support locations were only identifiable from below the stage and could not be identified by the performing dancers prior to landing.

FR data were compared to the company injury rates. Injury data were collated by the company medical staff over a two year period. An injury was recorded if a dancer experienced an incident that restricted the dancer from performing all activities that were required of them for the period 24hrs after the incident. Injuries were delimited to that occurring in the lower limb or trunk during non-lifting activities. All test venues were within one day's road travel from the home venue. When at a venue, all dance activity was conducted on the test surface.

Data Analysis: Descriptive statistics of the FR data were generated for each venue. Descriptive statistics of the injury frequencies were generated by grouping injuries into per week or per performance day categories. Total injuries per week were also separated into lower limb and trunk categories and the severity of the injuries were assessed by calculating the mean days missed dancing, as a result of injury. One week at a venue represented approximately 1620 dancer hours, based on 60 dancers, rehearsing or performing 6 days per week for 4.5 hours per day. Limited injury data sample sizes restricted analysis of statistical significance. FR values were compared with the European standard; Surfaces for sports areas: Indoor surfaces for multi-sports use (BS EN 14808).

RESULTS: Descriptive statistics of the FR values and injury data are presented in Tables 1 and 2 respectively. The lowest injury rates for four of the five injury variables occurred on surface 1, which was characterized by relatively high mean FR values across all support structures (45.90%) and moderate intra-surface variability (Std Dev 7.06; Range 21.93%). Surface 1 primary (46.45%) and on and between secondary support (45.46%) FR means differed by 0.99%. Injury rates on surface 2 were the second highest for all injury variables. Surface 2 was characterised by the lowest FR means and FR variability. A large difference in the mean FR values was recorded between surface 2 primary supports (12.88%) and on and between secondary supports (35.46%) but the primary support contact surface area was low relative to the total stage surface area (6.3%), which resulted in a low frequency of variable test points. Injury rates on surface 3 were the highest for four of the five injury variables. Surface 3 was characterised by the largest on and between secondary support FR variability and the largest distance between secondary supports. Surface 3 primary support mean FR (21.51%) was also markedly lower than the on and between secondary support mean FR (39.31%), but surface 3 was more variable than surface 2 due to a higher relative primary support contact surface area (14.7%). None of the surfaces complied with European standard; Surfaces for sports areas: Indoor surfaces for multi-sports use (BS EN 14808) (55-70% FR).

Table 1 Force reduction values for each surface
Primary Support Values (On supports only)

Surface	Test Points	Mean (%)	Standard Deviation	Range (%)	Maximum (%)	Minimum (%)	Distance Between Supports (m)
1	6	46.45	6.84	20.61	54.86	34.25	3.30
2	7	12.88	2.94	8.40	16.17	7.77	3.25
3	9	21.51	2.66	6.46	25.05	18.58	2.40

Secondary Support Values (On and between supports)

Surface	Test Points	Mean (%)	Standard Deviation	Range (%)	Maximum (%)	Minimum (%)	Distance Between Supports (m)
1	20	45.46	6.96	20.15	54.33	34.18	0.38
2	22	35.46	3.35	12.36	41.80	29.45	0.23
3	27	39.31	9.32	35.59	57.33	21.74	0.60

Table 2 Ballet company injury data on each surface

Surface	Weeks at Venue	Performance Days at Venue	Total Injuries per Week	Lower Limb Injuries per Week	Trunk Injuries per Week	Mean Days Lost Dancing per Injury	Performance Days Resulting in Injury
1	3	13	4.50	2.75	1.75	0.12	38%
2	4	20	5.67	4.00	1.67	0.35	50%
3	5	24	6.80	5.40	1.40	0.97	67%

DISCUSSION: This study assessed the lower limb and trunk injury rates within a professional touring ballet company during activity on three quantified performance venue floor surfaces. Assessed in multiple ways, injury rates were highest on the surface, which was characterised by (a) the largest FR variability on and between secondary supports; (b) the largest distance between secondary supports; (c) the primary support mean FR was markedly lower than the mean FR recorded on and between the secondary supports and (d) a high relative primary support contact surface area. As a result of these properties, surface 3 had a high frequency of variable FR values that were unidentifiable by the performing dancers prior to landing. Some support has therefore been provided for the findings of Nigg (2003) that surfaces with high intra-surface FR variability used for human physical activity may pose a greater injury risk to the individual than surfaces with low, uniform surface FR.

The adaptations of human landing mechanics to variations in surface FR are dependent on the individual's pre-landing perception of the impact surface and human landing mechanics change if surface FR is known, unknown or unexpected (Moritz et al., 2004, Ferris et al., 1999). Injury risks associated with these landing strategies are unknown, but the dependence of the neuromuscular system on post-impact sensory feedback for mechanical control on unknown surfaces may cause a delay in the necessary mechanical adaptations to the surface and pre-dispose the lower limb to injury.

The findings of this study are limited due to the low sample size. Nonetheless, the minimal footwear used by dancers and the controlled environment of professional dance delimits many of the confounding variables associated with previous sports surface epidemiological research. It should be noted that all surfaces did provide some FR and that dancer performance on a surface with a 0% FR value may be associated with an injury risk that has not been investigated. In response to these findings the ballet company now uses a portable performance floor on surfaces 1, 2 and 3 that complies with the European standards as quantified by the test protocols used in this study and future investigation of the injury rates on this surface will hopefully be conducted.

CONCLUSIONS: Ballet dancers demonstrate comparable skill and ability to that of elite athletes, but can be required to perform on surfaces that are sub-standard for elite performance. Variability in intra-surface FR, as identified in previous research, may present a greater risk of injury than low variability in intra-surface FR or low mean FR across the surface. Further investigation into the mechanics of human landing on surfaces of unknown FR is required to further investigate surface injury risk criteria.

REFERENCES:

Bowling, A. (1989). Injuries to dancers: prevalence, treatment and perception of causes. *British Medical Journal, 298,* 731-734.

Ekstrand, J., Timpka, T. and Hägglund, M. (2006). Risk of injury in elite football played on natural turf versus natural grass: a prospective two cohort study. *British Journal of Sports Medicine,* 40, 975-980.

Ferris, D.P., Liang, K. and Farley, C.T. (1999). Runners adjust leg stiffness for their first step on a new running surface. *Journal of Biomechanics,* 32, 787-794.

Imura, A., Iino, Y. and Kojima, T. (2008). Biomechanics of the continuity and speed change during one rotation of the Fouette turn. *Human Movement Science,* 27, 903-913.

Khan, K., Brown, J., Way, S., Vass, N., Crichton, K., Alexander, R., Baxter, A., Butler, M. and Wark, J. (1995). Overuse injuries in classical ballet. *Sports medicine, 19*(5), 341-357.

Liederbach, M. J., and Richardson, M. (2007). The importance of standardised injury reporting in dance. *Journal of Dance Medicine and Science, 11*(2), 45-48.

Moritz, C.T. and Farley, C.T. (2004). Passive dynamics change leg mechanics for an unexpected surface during human hopping. *Journal of Applied Physiology,* 97, 1313-1322.

Nigg, B.M. (2003). The stages of the Cirque du Soleil. In: B.M. Nigg, G.K. Cole & D.J. Stephanshyn (Eds.), *Sports Surfaces, biomechanics, injuries, performance, testing, installation.* (pp. 8-9). Calgary: Topline Printing.

Nigg, B.M., and Segesser, B. (1988). The influence of playing surfaces on the load on the locomotor system and on football and tennis injuries. *Sports Medicine,* 5, 375-385.

Nilsson, C., Leanderson, J., Wykman, A., and Strender, L. (2001). The injury panorama in a Swedish professional ballet company. *Knee Surgery, Sports Traumatology, Arthroscopy,* 9, 242-246.

Shorten, M. (2007). Sports surfaces and injury: The missing links. In: P. Fleming, C. Young, S. Dixon, I. James, M. Carré & C. Walker (Eds.), *Science, Technology and Research into Sports Surfaces.* Loughborough: Loughborough University.

Solomon, R., Solomon, J., Micheli, L.J. and McGray, E. (1999). The "cost" of injuries in a professional ballet company: A five-year study. *Medical Problems of Performing Artists, 14*(4), 164-169.

ISBS 2009

Oral Session S21

Jumping II

QUANTIFYING THE ONSET OF THE CONCENTRIC PHASE OF THE FORCE-TIME RECORD DURING JUMPING

R.L. Jensen[1], S.K. Leissring[1], L.R. Garceau[2], E.J. Petushek[2], and W.P. Ebben[2]
Dept. HPER, Northern Michigan University, Marquette, MI, USA[1]
Dept. Physical Therapy, Program in Exercise Science, Marquette University, Milwaukee, WI, USA[2]

Knowledge of the onset of concentric movement is needed to determine many variables including rate of force development. This study sought to identify the onset of the concentric phase of the force-time record of a variety of jumps. Twelve subjects performed a depth jump, single leg countermovement jump, and a counter movement jump. Vertical ground reaction force (GRF) obtained via force plate and video analysis of a marker placed on the hip, was used to estimate the time of onset of the concentric phase on the force time record. Repeated Measures ANOVA indicated differences in the onset of the concentric phase for the different jumps. Results can be used to assist in the calculation of outcome variables based on the determination of the onset of the concentric phase of the force-time record, when motion analysis is unavailable.

KEY WORDS: positive acceleration phase, eccentric, plyometric

INTRODUCTION: Force plates, and the resultant force-time records, have often been used in the analysis of jumping and plyometric exercises (Ramey, 1983), producing numerous potential outcome variables (Schieb, 1986). The calculation of some variables, such as time to maximum rate of force development (RFD), time to peak force, time to take off, RFD for the first 100ms, and RFD to peak requires identification of the point of onset of the concentric (also known as positive or upwards) phase of the jump (Ebben et al., 2007; Ebben et al., 2008; Harrison and Bourke, 2009; Wilson et al., 1995).
The onset of the concentric phase may be determined via the synchronization of motion analysis systems with a force plate. In the absence of motion analysis, the start of the concentric phase has been estimated as a point at where GRF readings were 10 N greater than the subjects static starting GRF, or the first point where the GRF of the force-time record exceeded body mass, after the eccentric phase in which GRFs fell below body mass (Ebben et al., 2007; Ebben et al., 2008; Harrison and Bourke, 2009). Additionally, the RFD has been determined from the point of the minimum GRF value attained at the end of the eccentric and presumably the beginning of the concentric phase (Wilson et al., 1995). Finally, a point near the peak of the GRF from the force-time record has been proposed to be the beginning of the concentric phase of countermovement jump (CMJ) (Bosco et al., 1982; Linthorne, 2001). Thus, a number of different methods of determining the onset of the concentric phase of the force-time record have been proposed or used, likely producing variability in the results. Therefore, the purpose of the current study was to estimate the onset of the concentric phase of the force-time record during a variety of jumping conditions.

METHODS: Twelve NCAA Division II athletes and recreationally active college students (seven female and five male; mean ± SD; age = 24.6±4.4 years, height = 175.0±7.9 cm; body mass = 69.7±10.8 kg, vertical jump = 45.2±7.9 cm) volunteered to serve as subjects for the study. Subjects completed a Physical Activity Readiness-Questionnaire and signed an informed consent form prior to participating in the study. Approval by the Institutional Review Board was obtained prior to commencing the study.
Warm-up prior to the study consisted of 3 minutes of low intensity work on a cycle ergometer. This was followed by dynamic stretching including one exercise for each major muscle group. Following the warm-up and stretching exercises, the subjects performed 5 repetitions of the CMJ followed by at least 5 minutes rest prior to beginning the test jumps. The test jumps included a depth jump (DJ) from a height equal to the subject's vertical jump, a single leg jump from the left leg (SLJ), and a counter movement jump with arm swing

(CMJ) (Potach, 2004). Three test jumps were performed for each condition and the mean of the three used as the criterion value. The order of each test jump was randomly assigned. For the DJ, subjects were instructed to drop directly down off the box and immediately perform a maximum vertical jump. For the SLJ and CMJ, subjects were instructed to jump for maximal height. A one minute rest interval was maintained between each trial. Subjects performed no strength or plyometric training in the 48 hours prior to the study.

The test jumps were performed on a force platform (OR6-5-2000, AMTI, Watertown, MA, USA). Ground reaction force data were collected at 1000 Hz, real time displayed and saved with the use of computer software (Net Force 2.0, AMTI, Watertown, MA, USA) for later analysis. Video of the exercises was obtained at 60 Hz from the saggital view using a 1 cm reflective marker placed on the greater trochanter. The marker was digitized using software (Motus 8.5, Vicon/Peak, Centennial, CO, USA) and data were smoothed using a fourth order Butterworth filter (Winter, 1990). The onset of the concentric phase was determined as the point where the marker moved vertically after the countermovement for the CMJ and SLJ or the landing of the DJ.

A signal was used to initialize kinetic data collection which also inserted as an audio tone in the video, in order to synchronize kinetic and kinematic data. Data were then combined into a single file and splined to create a file of equal length at 1000Hz.

The lowest point of vertical GRF during the countermovement for the CMJ and SLJ was determined from the force-time record. For the DJ the lowest point of vertical GRF was when contact was made with the force platform. In addition, the point where the vertical GRF equaled body mass following the countermovement was also determined. The time for each of these points on the force-time record was subtracted from the time determined from the onset of concentric movement as defined by upward movement of the marker on the greater trochanter. These times were used as the dependent variables for the study.

All statistical analyses of the data were carried out in SPSS © (Version 16.0). A repeated measures ANOVA was used to determine possible differences between trials. Bonferroni adjusted pairwise comparisons identified the specific differences between the test jumps. The criterion for significance was set at an alpha level of $p \leq 0.05$. Effect sizes were determined using partial eta squared (η_p^2).

RESULTS: Results of the ANOVAs are shown in Table 1. There was a significant main effect for the onset of the concentric phase as assessed by the point when vertical force equaled body mass ($p < 0.05$) with a medium effect size of $\eta_p^2 = 0.281$. However, post hoc Bonferroni comparisons found no significant differences between the different jumps ($p > 0.075$). For the onset of the concentric phase as assessed by the lowest point of the vertical force to the start of the vertical rise of the hip, there were significant main effects ($p < 0.05$) with a medium effect size of $\eta_p^2 = 0.391$. Post hoc analysis revealed the time of onset of the concentric phase for DJ was significantly less than the CMJ and SLJ ($p < 0.05$). The CMJ and SLJ did not differ ($p > 0.10$). Sample force-time records and approximate point of onset of the concentric phase for the DJ, SLJ, and CMJ are shown in Figures 1-3.

Table 1. Point of onset of the concentric phase of the force-time record, expressed in milliseconds after the point at which body mass is attained and after the eccentric phase of each jump. Data are expressed in ms. All data are expressed as mean ± SD for 12 subjects.

	Onset of concentric phase (ms after body mass achieved)	Onset of concentric phase (ms after the low point of vertical force)
DJ	248.4 ± 54.0	257.1 ± 55.1 [a]
SLJ	287.8 ± 52.8	445.6 ± 81.8
CMJ	306.3 ± 48.9	439.6 ± 64.4

[a] DJ was lower than CMJ and SLJ ($p < 0.05$)

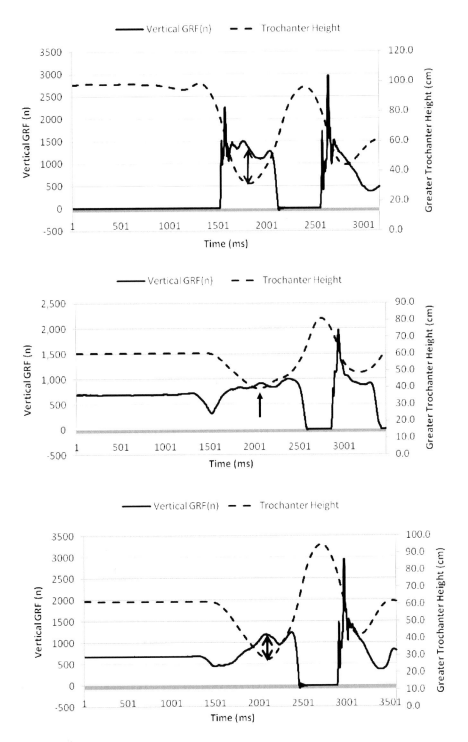

Figures 1-3. Figures 1-3 show the force-time record for the DJ, SLJ and CMJ with arrows identifying the approximate point of onset of the concentric phase.

DISCUSSION: This study demonstrates that the onset of the concentric phase of the force-time record occurs between 248.3 to 306.3 seconds after the force-time record reaches a value equal to body mass, subsequent to falling below body mass during the eccentric phase. Values were not statistically different between the three jumps, when assessed this way. Differences were found between the jumps when the onset of the concentric phase was assessed as the difference in time from the lowest point of the force-time record. Thus, the point of onset of the concentric phase may be dependent on the type of jump performed. The present study demonstrated that the onset of the concentric phase of the force-time record during jumping is different than the methods used in previous studies which attempted to approximate the start of the onset of the concentric phase from force-time records. These previous studies included estimations of this value as the point at where GRF readings were 10 N greater than the subjects static starting GRF, or the first point where the GRF of the force-time record exceeded body mass, after the eccentric phase in which GRFs fell below body mass (Ebben et al., 2007; Ebben et al., 2008; Harrison and Bourke, 2009; Wilson et al., 1995). Results of the present study are most similar to previous assertion that a point near the peak of the GRF from the force-time record was the beginning of the concentric phase of the CMJ (Bosco et al. 1982; Linthorne, 2001). Thus, the onset of the concentric phase may be estimated as the first GRF peak for the SLJ and the CMJ. However further research should be performed with additional subjects from a variety of heights for the DJ. Never-the-less, data from the present study can be used to approximate point of onset of the concentric phase of the force-time record, in order to calculate outcome variables that require the knowledge of this value in the absence of motion analysis.

CONCLUSION: Kinetic analysis provides useful information for the quantification of a variety of aspects of jumping. In the absence of motion analysis to determine the onset of the concentric phase, the results of this study provide an approximate measure of this value. Outcome variables that require knowledge of the onset of the concentric phase of the force-time record can be calculated based on this measure.

REFERENCES:

Bosco, C., Tihnayi, J., Komi, P.V., Fekete, G., and Apor, P. (1982) Store and recoil of elastic energy in slow and fast types of human skeletal muscle. *Acta Physiologica Scandinavica, 116*, 343-349.

Ebben, W.P., Flanagan, and Jensen. R.L. (2007) Gender similarities in rate of force development and time to takeoff during the countermovement jump. *J Exerc Physiol Online 10,* 10-18.

Ebben, W.P., Flanagan E.P., and Jensen R.L. (2008) Jaw clenching results in concurrent activation potentiation during the countermovement jump. *J Strength Cond Res 22,* 1850-1854.

Flanagan, E.P., Ebben, W.P., and Jensen, R.L. (2008) Reliability of the reactive strength index and time to stabilization during plyometric depth jumps. *J Strength Cond Res 22,* 1677-1682.

Harrison, A.J. and Bourke, G. (2009) The effect of resisted sprint training on speed and strength performance in male rugby players. *J Strength Cond Res 23,* 275-283.

Jensen, R.L. and Ebben, W.P (2007) Quantifying plyometric intensity via rate of force development, knee joint and ground reaction forces. *J Strength Cond Res 21,* 763-767.

Linthorne, N.P. (2001) Analysis of standing vertical jumps using a force platform. *Am. J. Phyysiol 69,* 1198-1204.

Ramey, M.R. (1983) The use of force plates for jumping research. In: *Proceedings of the Ist International Symposium of Biomechanics in Sports* 81-91.

Schieb, DA. (1986) The force plate in sports biomechanics research. In: *Proceedings of the IVth International Symposium of Biomechanics in Sports* 337-365.

Wilson, GJ, Lyttle, A.D., Ostrowski, K.J., and Murphy, AJ. (1995) Assessing dynamic performance: A comparison of rate of force development tests. *J Strength Cond Res 9,* 176-181.

Winter, D.A. (1990) *Biomechanics and motor control of human movement* (2nd Ed). New York: Wiley Interscience.

Acknowledgement: Funded by NMU Freshman Fellows and College of Professional Studies grants.

KNEE SEPARATION DISTANCE AND QUADRICEPS AND HAMSTRINGS STRENGTH DURING DROP VERTICAL JUMP LANDINGS

Orna Donoghue, Lorraine Steel, Rachel Collins, Harriet Young and Simon Coleman

Dept of Physical Education, Sport & Leisure Studies, Moray House School of Education, University of Edinburgh, Edinburgh, UK

KEY WORDS: ACL injury, functional strength ratio

INTRODUCTION: Non-contact anterior cruciate ligament (ACL) injury is common particularly in female athletes during jump landing tasks. Ligament dominance occurs when the muscles cannot control knee movement sufficiently thus increasing medial knee motion. Preferential use of the quadriceps during landing and greater strength compared to the hamstrings also increase the load placed on the ACL (Dugan, 2005). Noyes et al. (2005) measured knee separation distance during jump landings finding an increase after neuromuscular training. The aim of this study was to examine the relationship between knee separation distance during drop jump landing and hamstrings and quadriceps strength in female athletes.

METHODS: Ethical approval was obtained from the university ethics committee. Females were recruited from high risk i.e. volleyball, netball (n=18, mean ± SD: age 19.9 ±1.2 yrs, mass 69.3 ± 5.5 kg, height 1.72 ± 0.06 m) and low risk i.e. running, cycling (n=19, mean ± SD: age 21.1 ± 1.8 yrs, mass 63.8 ± 9.4 kg, height 1.67 ± 0.06 m) university clubs. Markers were placed on both greater trochanters, lateral femoral condyles and lateral malleoli. Participants performed two-footed drop vertical jumps from a height of 30 cm. A Canon digital camera recorded frontal plane marker coordinates at 50 Hz. Knee separation distance (absolute and normalised to hip separation distance) was calculated during landing using Ariel Performance Analysis System. Eccentric quadriceps (Qecc) and concentric hamstrings (Hcon) strength was measured at 60°/s and 180°/s using a Biodex isokinetic dynamometer; Hcon:Qecc ratios were subsequently obtained. A one-way independent t-test with $\alpha=0.05$ was carried out in SPSS v15. Pearson's correlations were calculated to examine relationships between knee separation distance and strength measures.

RESULTS: Concentric hamstrings strength in the dominant (p=0.001) and non-dominant (p=0.026) legs was significantly lower for participants in high risk sports. Both groups displayed similar knee separation distances. All correlations between knee separation distance and Qecc, Hcon or Hcon:Qecc ratio were low (r<0.4).

DISCUSSION: Female landing strategies in jumping based sports may result in preferential training of the quadriceps accounting for the current findings. However, as in Bennett et al. (2008), peak strength values were unable to predict dynamic loading during landing.

CONCLUSION: Hamstrings and quadriceps strength levels were not correlated with knee separation distances displayed during jump landings. Further analysis is required.

REFERENCES:
Bennett, D., Blackburn, J., Boling, M., McGrath, M., Walusz, H. and Padua D. (2008). The relationship between anterior tibial shear force during a jump landing task and quadriceps and hamstring strength. *Clinical Biomechanics*, 23, 1165-1171.
Dugan, S. (2005) Sports-related knee injuries in female athletes: what gives? *American Journal of Physical Medicine and Rehabilitation*, 84(2), 122-130.
Noyes, F.R., Barber-Westin, S.D., Fleckenstein, C., Walsh, C. and West, J. (2005). The drop-jump screening test difference in lower limb control by gender and effect of neuromuscular training in female athletes. *American Journal of Sports Medicine*, 33, 197-207.

LATERALITY IN VERTICAL JUMPS

Diana Lang, Gerda Strutzenberger, Anne Richter, Gunther Kurz and Hermann Schwameder

BioMotion Center, Department of Sport and Sport Science, Karlsruhe Institute of Technology (KIT), Germany

FoSS – Research Center for Physical Education and Sports of Children and Adolescents, Karlsruhe, Germany

KEY WORDS: laterality, vertical jump.

INTRODUCTION: Laterality is a widely investigated phenomenon in motor activities. Various studies deal with the functional dominance of one limb or one side of the body in sports (e.g. Fischer, 1988, Oberbeck, 1989). One common method for the identification of lateral differences in the lower limbs is the single-leg vertical jump (e.g. Stephens, 2005). In order to reduce the risk of injury and the coordinative demand Impellizzeri et al. (2007) and Newton et al. (2006) proposed a double-leg vertical jump force test for the assessment of bilateral strength asymmetry. The focus of these studies was set on the strength imbalance between the right and left leg using the maximum force as the relevant factor. Further parameters to describe lateral differences were neglected and still little is known about the coherence between laterality and jumping performance. Therefore, the aim of this study was to investigate laterality in established double-leg vertical jumps in performance diagnostics, such as the counter movement jump (CMJ), squat jump (SJ) and drop jump (DJ).

METHODS: 12 male and female athletic athletes (16.8 ± 2.9 yrs, 1.75 ± 0.08 m, 62.1 ± 6.6 kg) participated in this study. All subjects were experienced in vertical jumps. After a short warm-up consisting of jogging on a treadmill each subject performed six DJ from a height of 20 cm, six SJ and six CMJ with the arms akimbo. In all conditions the instruction was to jump as high as possible. Vertical ground reaction forces were measured with two, side-by-side, force plates (AMTI, 1000 Hz). Kinematic data were recorded using ten infrared-cameras for three-dimensional analysis (Vicon, 200 Hz). In addition to the singular dynamic and kinematic parameters, the process of the complete jump will be analysed separately for both legs. All results are described by a laterality index $LI = 100 \cdot (r - l)/(r + l)$.

RESULTS AND DISCUSSION: Regarding the vertical ground reaction forces, first results for the CMJ show lateral differences during the jump, e.g. in the maximum force (3.9 ± 2.1%), and in the previous stance phase. There is a small correlation between maximum force and jump height ($r = 0.54$). A higher minimum force of one leg leads to a higher explosive force and a higher momentum in the CMJ. Nearly all subjects show a small variability in their jumps with individual movement patterns. The kinematic data as well as the data of SJ and DJ is not yet available.

REFERENCES:

Fischer, K. (1988). *Rechts-Links-Probleme in Sport und Training*. Band 3: Reihe Motorik. Schorndorf: Hofmann.

Impellizzeri, F.M., Rampinini, E., Maffiuletti, N. & Marcora, A.M. (2007). A vertical jump force test for assessing bilateral strength asymmetry in athletes. *Medicine & Science in Sports & Exercise*, 39 (11), 2044-2050.

Newton, R.U., Gerber, A., Nimphius, S., Shim, J.K., Doan, B.K., Robertson, M., Pearson, D.R., Craig, B.W., Häkkinen, K. & Kraemer, W.J. (2006). Determination of functional strength imbalance of the lower extremities. *The Journal of Strength and Conditioning Research*, 20 (4), 971-977.

Oberbeck, H. (1989). *Seitigkeitsphänomene und Seitigkeitstypologie im Sport*. Band 68: Schriftenreihe des Bundesinstituts für Sportwissenschaft. Schorndorf: Hofmann.

Stephens, T.M., Lawson, B.R. & Reiser, R.F. (2005). Bilateral asymmetries in max effort single-leg vertical jumps. *Biomedical Sciences Instrumentation*, 41, 317-322.

MECHANICS OF AMPUTEE JUMPING – JOINT WORK

Marlene Schoeman, Ceri Diss and Siobhan Strike
School of Human and Life Sciences, Roehampton University, London, UK

The purpose of this study was determine if dynamic elastic response (DER) prostheses could absorb energy in the eccentric phase of a vertical jump performed by trans-tibial amputees phase and return this energy in the propulsive phase. Further, given the active nature of the ankle, the study aimed to determine the mechanisms required at the remaining joints to compensate for the pathological ankle. Six amputee (AMP) and 10 able-bodied participants (AB) performed maximal vertical jumps on two force plates which were synchronised with a 9-camera VICON infra red system. The amputees did not jump as high as the AB participants. Only minimal negative work was recorded at the prosthetic ankle in the eccentric phase which resulted in minimal positive work at the ankle in the concentric phase. The intact side produced greater work than the affected side in the concentric phase. The amputees generally adopted a hip strategy to generate positive work. The work recorded at the knee was reduced on the intact and affected side and indicates the prosthesis influences the movement on both sides. To enable amputees to participate in activities which require jumping, prostheses need to be developed and amputees need to be taught how to adjust their biomechanics to store and release energy in the prosthesis.

KEY WORDS: amputee, vertical, jump, joint work.

INTRODUCTION:

Vertical jumping is a skill required for many sports. It is frequently used to assess explosive strength and as a field test of performance capability. There is little research on amputee bilateral vertical jumping and on the compensations that result from amputation of the ankle.
dynamic elastic response (DER) prostheses have been developed in order to return energy to the system. A key concept of these prostheses is that elastic energy is stored when they compress under loading and this energy is returned later in the activity when the limb is unloaded. A key criteria for the effective use of this energy is that it is able to return the energy at the right time, frequency and location. This is difficult in a generic high activity prosthesis as the energy requirements vary depending on the activity being undertaken. Specific to the countermovement (CMJ), if the energy stored in the prosthesis in the eccentric phase can be returned in the concentric phase then the DER prostheses should be able to contribute to the total work required at the joints. However, if the energy is not stored, or is not returned effectively, then the remaining joints will have to compensate for the prosthesis. It is unclear if a DER prosthesis will be able to make a useful contribution and if the amputees will be able to compensate sufficiently at the other joints to be able to achieve a jump.
There is some debate about the relative contribution of the joints to the total positive work produced in the concentric phase leading up to flight. Hubley & Wells (1983) suggested that the knee was the main producer of work, followed by the hip and then the ankle. In contrast Fukashiro and Komi (1987) found that the hip was the most important contributor, followed by the knee and the ankle. Vanezis & Lees (2005) suggested that these discrepancies could be the result of the high variability in the data and identified two key strategies of jumping, with emphasis on the knee or the hip.
The primary aim of this study was to examine the work done in the prosthetic ankle in storage and return and to assess the effect of the passive foot in jumping. The secondary aim was to determine the work compensations at the other lower limb joints.

METHODS: Participants: Six unilateral transtibial amputees (5 males and 1 female) who were between 18 and 50 years, more than 12 months post-operative, with no secondary pathology and had an amputation of a traumatic nature were recruited. All the AMPs wore patellar tendon-bearing sockets with rigid pylons and their own prosthesis. Ten able bodied (AB) participants (9 male and 1 female) of the same age range with no pathology were used to facilitate the comparison of results. All participants (AMPs and AB) were recreationally active with similar proficiency in jumping and wore their own footwear (athletic trainers). All participants signed an informed consent form approved by the University and the National Health Services' Ethics Committee.

Data Collection: Data were collected in a single session. Following a 5 minute warm-up on a treadmill at a self-selected fast walking velocity, participants were given the opportunity to practise and familiarise themselves with the jumping criteria and laboratory conditions. Ten maximal bilateral countermovement jumps were performed with arms akimbo with 1 minute rest between each trial. The only instruction given was to jump as high as possible. The jumps were performed with each foot on a separate force plate. Trials were excluded if the participants used their arms or if they missed the force plates during landing. On average 13 trials were required to collect 10 successful trials. Data were collected using two Kistler (model 9581B, sampling at 1080Hz) force plates synchronized with a 9-camera Vicon (model 612, sampling at 120Hz) infra red system. Thirty four 25 mm diameter reflective markers were attached to specific anatomical landmarks according to Vicon's Plug-in-Gait full body gait model (Oxford Metrics). Measurements were taken for each individual according to the Vicon requirements for full body modelling of each dynamic CMJ trial.

Data Analysis: Jump kinetics and kinematics were calculated using Vicon Workstation software. Kinematic data were smoothed using a Woltering quintic spline (MSE = 15 mm) filter. The trial with the highest flight height was chosen for further analysis. Inverse dynamics using standard procedures were used to determine the net joint reaction components and the net joint moments at the ankle, knee and hip from the ground reaction force data associated with each foot. Joint power (the product of the net joint moment and joint angular velocity) and the work done in each phase (the time integral of the power production in the eccentric and concentric phases, as determined by the movement of the centre of mass (CoM) were calculated using standard procedures (de Koning and van IngenSchenau, 1994). All variables were normalised to body mass. To facilitate comparison with results in the literature, the magnitude of the results at each joint were summated to give an overall joint value. Flight height was defined by the CoM displacement (maximum height less height at take-off) as determined by the kinematic analysis. All other variables are presented for the AMPs as intact and affected limb separately. For the AB, the results are for the preferred and non-preferred jumping limb (and are presented under intact and affected for ease of analysis).

RESULTS:

Table 1 Flight height of the centre of mass and negative work done at the prosthetic ankle

Participant	FH (m)	Negative work at the prosthetic ankle ($W.kg^{-1}$)
AMP_1	0.24	-0.004
AMP_2	0.19	-0.002
AMP_3	0.17	-0.003
AMP_4	0.13	-0.005
AMP_5	0.10	-0.004
AMP_6	0.09	-0.005
$\bar{x}_{(AMP)}$ (± sd)	0.15±0.06	-0.004±0.001
$\bar{x}_{(AB)}$ (± sd)	0.31± 0.04	-0.1369±0.03

AMP flight height was lower compared to AB participants. There was little negative work done at the prosthetic ankle, indicating that little energy was absorbed in the eccentric phase (Table 1).

Positive work in the concentric phase at the intact ankle generally followed the trend of high to low, and was generally greater than for the AB participants. Very little work was done at the prosthetic ankle. The work done at both knees was similar and was lower than for the AB participants. There was no obvious trend at the intact or residual hip, however, the AMP participant with the highest jump achieved the greatest work at the hip and was symmetrical (Figure 1).

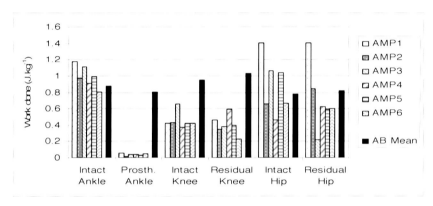

Figure 1: Positive work done at each joint in the concentric phase of jumping

The intact side dominated for overall positive work (Table 2). Every AMP produced more work at the intact compared with the prosthetic ankle (Table 2). When the intact and affected sides are added together to get an overall hip, knee and ankle work contribution, 5 out of 6 AMPs produced most work at the hips, followed by the ankles and then the knees. Amp 4 produced most at the hip, followed by the knee and then the ankle.

Table 2 Relative contribution of each joint to the total positive joint work in the concentric phase.

Participant	Intact contribution (%)				Prosthetic contribution (%)			
	Ankle	Knee	Hip	Total	Ankle	Knee	Hip	Total
AMP$_1$	24	9	28	61	1	9	28	39
AMP$_2$	30	13	20	63	1	11	26	37
AMP$_3$	32	19	31	82	1	11	6	18
AMP$_4$	30	12	15	58	1	20	21	42
AMP$_5$	29	12	30	71	1	12	17	29
AMP$_6$	29	15	24	68	2	8	22	32
$\bar{x}_{(AMP)}$ (± sd)	29± 3	13± 3	25± 6	67± 8	1± 0.4	12± 4	20± 8	33± 9
$\bar{x}_{(AB)}$ (± sd)	17± 3	18± 5	15± 6	50± 7	16± 3	20± 7	15± 7	50± 7

DISCUSSION: Vertical jumping is a fundamental skill common to numerous recreational activities and training strategies. It consist of clear phases, each with its own underlying performance criterion for successful execution. This makes it a valuable experimental model in assessing the cause and effect of human movement strategies for different population groups (Challis, 1998; Strike & Diss, 2005). As jumping is a multi-joint action that requires substantial muscular effort from the ankle, knee and hip joints, it was expected that biomechanical compensations would result from the amputation and that these would not be sufficient for amputees to jump effectively. The AMPs did not achieve flight heights equivalent to the AB participants who were matched for jumping experience. Only the AMP who jumped the highest reached a height which was similar to the lowest AB participant. It is clear that the prosthetic ankle did not sufficiently compensate for the intact ankle. Our first aim was to determine if the prosthesis was effective in performing negative work (energy absorption) in the eccentric phase and if this was returned effectively as positive work in the concentric phase. Although dynamic ankles are designed to store energy under loading and return this energy when the load is removed, the prosthetic ankle did not do this effectively in this movement. Only small amounts of negative work were recorded in the ankle in the eccentric phase. As a result there was little energy to be returned in the concentric phase, as indicated in the small positive work done at this joint. The dynamic nature of the prosthesis is not utilised effectively by AMPs in jumping. The second aim of the research was to determine how the other joints compensated for the ankle pathology. The other joints could not compensate effectively, with less work also recorded at the residual knee and hip compared to the AB participants. For the intact side the ankle and hip were the main joints at which compensations occurred, with the hip as the main source of work for all AMPs. When the intact and affected side work are added together, all AMPs except one produced most of the work at the hip, followed by the ankle and then the knee. Clearly the AMPs adopt the hip strategy as described by Venezis and Lees (2005). The lack of the biarticular gastrocnemius muscle clearly influences the knee mechanics on the affected side, but it also seems to influence them on the intact side. This is a result which requires further study.

CONCLUSION: In the absence of an intact ankle, AMPs did not reach the heights attained by experience matched AB participants. The work done by the prosthesis was not sufficient to replace the anatomical structure. There is a clear need for an accessible lower leg prosthesis for non-competitive amputees which will accommodate both everyday ambulation and more vigorous activities associated with a physically active lifestyle (Hafner et al, 2002). Continuous research into lower limb prostheses should aim to enhance amputee movement adapting the elastic energy storage and return properties of the prosthesis so that the magnitude, frequency and timing of the energy absorption and return is better suited to reduce the compensations required at the remaining joints and to enhance biomechanical performance.

REFERENCES:

Challis, J. H. (1998). An investigation of the influence of bi-lateral deficit on human jumping. *Human Movement Science, 17*, 307-325.
de Koning, J.J. and van IngenSchenau, G.J. (1994) On the estimation of mechanical power in endurance sports. *Sports Science Review, 3*, 34-54
Fukashiro, S. and Komi, P.V. (1987) Joint moment and mechanical flow of the lower limb during vertical jump. *International Journal of Sport Medicine, 8,* 15-21
Hafner, B. J., Sanders, J. E., Czerniecki, J., & Fergason, J. (2002). Energy storage and return prostheses: does patient perception correlate with biomechanical analysis? *Clinical Biomechanics, 17,* 325-344.
Hubley, C. L., & Wells, R. P. (1983). A work-energy approach to determine individual joint contributions to vertical jump performance. *European Journal of Applied Physiology, 50,* 247-254.
Strike, S., & Diss, C. (2005). The biomechanics of one-footed vertical jump performance in unilateral transtibial amputees. *Prosthetics and Orthotics International, 29,* 39-51.
Vanezis, A., & Lees, A. (2005). A biomechanical analysis of good and poor performers of the vertical jump. *Ergonomics, 48,* 1594 - 1603.

THE INFLUENCE OF SAND SURFACE ON THE KINEMATICS OF VOLLEYBALL SPIKE JUMPS

Dr. Markus Tilp, Institute of Sport Sciences, Karl-Franzens-University Graz
Dr. Herbert Wagner, Department of Sport Science and Kinesiology, University of Salzburg
Prof. Dr. Erich Müller, Department of Sport Science and Kinesiology, University of Salzburg

KEY WORDS: 3D Analysis, beach volleyball, volleyball

INTRODUCTION: Several types of sports, like soccer, European handball, and volleyball have expanded their classical fields of activity from indoor surfaces made of wood or synthetic and outdoor turf to sand surfaces. Previous studies reported differences in both body mechanics and energy demands during several types of movement on sand or similar compliant surfaces compared to hard surfaces. Therefore, we investigated the differences in kinematics during spike jumps performed on indoor or sand surface (Tilp et al., 2008).

METHODS: Eight male elite volleyball players performed spike jump movements on both surfaces. An eight camera motion capturing system (Vicon Peak, Oxford, UK, 250 Hz) was used to generate 3D kinematic data. Seven groups of variables representing the kinematics of the centre of mass, the countermovement, the approaching phase, and the angular amplitudes and maximal velocities of lower and upper limbs were tested for differences with Hotellings T_2^2.

RESULTS: Significant differences ($p<0.05$) were observed regarding the movement of centre of mass, the countermovement, the kinematics of the approaching phase, and the angular amplitudes of the lower limbs. However, no significant differences were found in the maximal angular velocities of the lower and upper limbs or in the amplitudes of the upper limb motion.

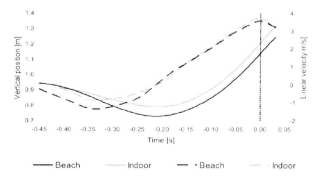

Figure 1 Position (solid) and Velocity (dashed lines) of centre of mass during spike jumps on indoor and beach surface

DISCUSSION: Results revealed significant technique adaptations associated with the changing of the movement surface. Movements on sand were significantly slower and led to lower jumping heights. The switch from eccentric to concentric work of the knee extensors was also significantly slower on sand. Athletes tried to compensate the disadvantageous properties of sand by altering approach techniques and amplifying lower joint amplitudes.

CONCLUSION: The altered kinematics of the movement implies altered demands on the neuromuscular system to reach high performance. This should be taken into account when planning fitness training.

REFERENCES:
Tilp, M., Wagner, H. & Müller, E. (2008) Differences in 3D kinematics between volleyball and beach volleyball spike movements, Sports Biomechanics, 7(3), 386-397.

ISBS 2009

Oral Session S22

Injury II

EFFICACY OF FUNCTIONAL KNEE BRACES IN SPORT: A REVIEW

Mario Lamontagne[1,2] and Nicholas Brisson[1]

School of Human Kinetics, University of Ottawa, Ottawa, Canada[1]
Department of Mechanical Engineering, University of Ottawa, Ottawa, Canada[2]

The purpose of this work was to review previously published studies on functional knee braces in an attempt to understand their functionality. A better comprehension of the effects of functional knee bracing could assist ACL-deficient athletes in deciding whether or not these devices could be beneficial in supporting their unstable knee joint during sports. Results showed that functional knee braces alter knee joint kinematics, kinetics and muscle activity. Their mechanism of action, however, might be attributed to the modification of proprioception and motor patterns of the lower limb rather than their mechanical stabilizing effect.

KEY WORDS: functional knee brace, anterior cruciate ligament (ACL) deficiency

INTRODUCTION: Knee injuries are caused either by direct contact or non contact movements and account for up to 50% of all sport injuries (de Loës, Dahlstedt & Thomée, 2000). The three main movements producing knee injury are planting and cutting, straight knee landings and rapid one-step stops (McNair, Marshall Matheson, 1990; Olsen, Myklebust, Engebretsen & Bahr, 2004). During these movements, the anterior cruciate ligament (ACL) tightens with extremes of extension and contraction, accompanied by internal or external rotation of the knee (Ebstrup & Bujsen-Moeller, 2000). Ligamentous tears to the knee, especially to the ACL, are the most debilitating of any sport injury and result in the longest time of absence from sport participation (Thacker et al., 2003). Functional knee braces (FKBs), devices designed to provide support for unstable knees, are frequently utilized to stabilize a ligament-injured knee during sports (Wirth & DeLee, 1990). They include medial and lateral vertical hinges, which may be uniaxial or polyaxial, and a mechanism to limit hyperextension (Martin & the Committee on Sports Medicine and Fitness, 2001). Although functional knee bracing has been extensively documented in literature, no clear consensus has been reached regarding their efficacy. The purpose of this work was to review previously published studies on FKBs in an attempt to assess their efficacy, thereby aiding ACL-deficient and ACL-reconstructed athletes in deciding whether or not a FKB could be beneficial in supporting their unstable knee joint during sports.

METHODS: A search of electronic databases (i.e., PubMed and Scopus from 1975 through March 2009) was performed using the keywords *knee brace, knee orthosis* and *anterior cruciate ligament (ACL)*. The search was then refined using the terms *functional knee brace* and *ACL deficiency*. Citations were further identified from the reference sections of the research papers retrieved. Studies were included in the review if they investigated the effects of FKBs on the knee joint. Abstracts, unpublished studies, papers not written in English and papers that addressed brace types other than FKBs, were excluded. In total, 96 articles were considered for the review. Retained studies consisted of biomechanical analyses on healthy knees, ACL-deficient knees and ACL-reconstructed knees, with and without FKB. In vitro studies on cadavers and knee joint models evaluated the effects of FKBs on the ACL. In these studies, the cadaveric knee or knee model was attached to a testing jig, which assessed torsional and antero-posterior loading via external force transducers and potentiometers (Wojtys, Loubert, Samson & Viviano, 1990; Liu, Lunsford, Gude & Vangsness, 1994). In vivo three-dimensional kinematic analyses of the knee joint were performed using intra-cortical pin implantation. Bone pins, affixed with target clusters, were inserted into the femur and tibia. Roentgen-stereophotogrammetric X-rays were taken with the implanted pins to record the position of the markers and to define the anatomical landmarks of the tibia and femur. Infrared cameras captured the markers about the knee, determining antero-posterior displacement of the tibia as well as tibial rotation of the knee joint (Ramsey et al., 2001). In addition, studies used ACL-implanted strain transducers to measure the displacement behavior of the ACL and calculate its strain response/elongation in vivo (Beynnon & Fleming, 1998; Cerulli, Benoit, Lamontagne, Caraffa & Liti, 2003). Other biomechanical studies examined knee joint kinematics and kinetics, with and without FKB, during quick stops, landings, lateral displacements and change of direction. Some studies have included in their methods surface and needle electromyography (EMG) to determine lower limb muscle activity with and without FKB (Wojtys, Kothari & Huston, 1996). Moreover, investigations conducted in a laboratory setting used dynamometers, such as isokinetic devices, to measure power, peak torques, peak forces and moments of force at the knee joint with and without FKB (Houston & Goemans, 1982). These studies, combined with our experience on biomechanical analyses of FKBs, were used to understand the functionality of FKBs.

RESULTS: Cadaveric studies have shown that FKBs can restrain anterior tibial displacement by 29-39% when the hamstring, quadriceps and gastrocnemius muscles are relaxed and by 70-85% when these muscles are contracted (Wojtys et al., 1990). Conversely, studies on mechanical knee surrogates have demonstrated that the aforementioned effects are present exclusively when the knee is subjected to static or low anterior shear forces; FKBs tend to fail in instances where high loads are present (France, Paulos, Jayaraman & Rosenberg, 1987). In vivo studies supported this finding by demonstrating that FKBs did not significantly reduce anterior tibial displacement during one legged jumps (Ramsey et al., 2001). Furthermore, braces have been shown to decrease peak load on the ACL in weightbearing and non-weightbearing knees (Beynnon et al., 1997) and to some extent control internal rotation of the knee under low loads. They cannot, however, control external rotation of the knee (Beynnon et al., 1992). Kinematic and EMG analyses showed that FKBs altered knee joint kinematics, force distribution characteristics as well as muscle activation patterns during the stance phase of running in normal and in ACL-deficient subjects (Knutzen, Bates, Schot & Hamill, 1987; Théoret & Lamontagne, 2006). In addition, EMG evaluations have found that brace application improved both quadriceps and hamstring muscle response times during anterior tibial translation stress testing and sports. On the contrary, both quadriceps and hamstring muscle reaction times were slowed by the knee braces (Wojtys et al., 1996; Németh, Lamontagne, Tho & Eriksson, 1997). Moreover, surrogate knee models were used to assess FKB composition and its effects on the transmission and absorption of low-level, repetitive impact forces at the knee. The results revealed that stiffer materials better resisted bending and therefore provided more protection to the knee joint given that the ligaments were required to provide a smaller proportion of the total resistance to the impact forces (Patterson & Eason, 1996). Others focused on distal migration of FKBs and concluded that misalignment of the brace does not modify lower limb mechanics during gait (Singer & Lamontagne, 2007). FKBs have been shown to produce several adverse effects on the lower limb. Markedly, EMG analyses demonstrated that knee bracing resulted in decreased hamstring and quadriceps activities during cutting maneuvers (Branch, Hunter & Donath, 1989). Isokinetic testing demonstrated that knee bracing decreased maximal torque output of the quadriceps by 12-30% and reduced maximal knee extension velocity by 20% (Houston et al., 1982). Furthermore, increased energy expenditure and premature muscle fatigue caused by regional muscle ischemia and lactic acid build-up during physical activity were concurrent with knee bracing (Styf, 1999). Similarly, clinical testing revealed that knee bracing reduced individuals' speed and agility (Marans, Jackson, Piccinin, Silver & Kennedy, 1991).

DISCUSSION: FKBs are commonly utilized to counteract the combining effects of antero-posterior displacement and internal rotation of the knee in ACL-deficient and ACL-reconstructed athletes. An ideal brace would synergize with the muscular and ligamentous knee stabilizers throughout the normal range of motion. Also, it would diminish negative effects of valgus, varus, rotational and translational forces exerted on the knee (Martin & the Committee on Sports Medicine and Fitness, 2001). Results revealed that functional knee braces do in fact modify knee joint kinetics and kinematics; however, their mechanisms of action might not necessarily be those initially speculated. Several authors have put forth the idea that perhaps FKBs do not stabilize the knee from a mechanical perspective and, as a result, research should focus on how FKBs alter proprioception and motor patterns of the lower limb. Modification of muscle activation, timing and coordination during physical activity could be beneficial in reducing strain on the ACL, reducing tibial rotation and limiting anterior tibial displacement (Németh et al., 1997; Ramsey, Wretenberg, Lamontagne & Németh, 2003).

CONCLUSION: This work reviewed previously published studies on FKBs in an attempt to assess their efficacy. FKBs seem to have some desirable effects on the knee joint. The use of a FKB can assist ACL-deficient or ACL-reconstructed athletes in continuing their sport practices, although performance may be compromised. These devices, however, should not replace a proper rehabilitation program or necessary surgery. Although numerous studies have investigated the effectiveness of FKBs, most of them focused on the mechanical stabilization effects of such devices. Future research is needed regarding the effects of FKBs on neuromuscular control as well as modification of locomotion patterns.

REFERENCES:

Beynnon, B.D. & Fleming, B.C. (1998). Anterior cruciate ligament strain in-vivo: A review of previous work. *Journal of Biomechanics*, 3, 519-525.

Beynnon, B.D., Johnson, R.J., Fleming B.C., Peura, G.D., Renstrom, P.A., Nichols, C.E., et al. (1997). The effect of functional knee bracing on the anterior cruciate ligament in the weightbearing and nonweightbearing knee. *American Journal of Sports Medicine*, 25, 353-359.

Beynnon, B.D., Pope, M.H., Wertheimer, C.M., Johnson, R.J., Fleming, B.C., Nichols, C.E. & Howe, J.G. (1992).The effect of functional knee-braces on strain on the anterior cruciate ligament in vivo. *Journal of Bone and Joint Surgery*, 74A, 1298-1312.

Branch, T.P., Hunter, R. & Donath, M. (1989). Dynamic EMG analysis of anterior cruciate deficient legs with and without bracing during cutting. *American Journal of Sports Medicine*, 17, 35-41.

Cerulli, G., Benoit, D.L., Lamontagne, M., Caraffa, A. & Liti, A. (2003). In vivo anterior cruciate ligament strain behaviour during a rapid deceleration movement: case report. *Knee Surgery, Sports Traumatology, Arthroscopy*, 11, 307-311.

de Loës, M., Dahlstedt, L.J. & Thomée, R. (2000). A 7-year study on risks and costs of knee injuries in male and female youth participants in 12 sports. *Scandinavian Journal of Medicine & Science in Sports*, 10, 90-97.

Ebstrup, J.F. & Bujsen-Moeller, F. (2000). Case report. Anterior crucial ligament injury in indoor ball games. *Scandinavian Journal of Medicine & Science in Sports*, 10, 114-116.

France, E.P., Paulos, L.E., Jayaraman, G. & Rosenberg, T.D. (1987). The biomechanics of lateral knee bracing. Part II – Impact response of the braced knee. *American Journal of Sports Medicine*, 15, 430-438.

Houston, M.E., & Goemans, P.H. (1982). Leg muscle performance of athletes with and without knee support braces. *Archives of Physical Medicine and Rehabilitation*, 63, 431-432.

Knutzen, K.M., Bates, B.T., Schot, P. & Hamill, J. (1987). A biomechanical analysis of two functional knee braces. *Medicine and Science in Sports and Exercise*, 19, 303-309.

Liu, S.H., Lunsford, T., Gude, S. & Vangsness, T. (1994). Comparison of Functional Knee Braces for Control of Anterior Tibial Displacement. *Clinical Orthopaedics and Related Research*, 303, 203-210.

Marans, H.J., Jackson, R.W., Piccinin, J., Silver, R.L. & Kennedy, D.K. (1991). Functional testing of braces for anterior cruciate ligament-deficient knees. *Canadian Journal of Surgery*, 34, 167-172.

Martin, T.J. & the Committee on Sports Medicine and Fitness. (2001). Technical Report: Knee Brace Use in the Young Athlete. *Pediatrics*, 108, 503-507.

McNair, P.J., Marshall, R.N. & Matheson, J.A. (1990). Important features associated with acute anterior cruciate ligament injury. *New Zealand Medicine Journal*, 103, 537-539.

Németh, G., Lamontagne, M., Tho, K.S. & Eriksson, E. (1997). Electromyographic Activity in Expert Downhill Skiers Using Functional Knee Braces After Anterior Cruciate Ligament Injuries. *American Journal of Sports Medicine*, 25, 635-641.

Olsen, O., Myklebust, G., Engebretsen, L. & Bahr, R. (2004). Injury Mechanisms for Anterior Cruciate Ligament Injuries in Team Handball: A Systemic Video Analysis. *American Journal of Sports Medicine*, 32, 1002-1012.

Patterson, P.E. & Eason, J. (1996). The Effects of Prophylactic Brace Construction Materials on the Reactive Responses of the MCL During Repetitive Impacts. *Journal of Athletic Training*, 31, 329-333.

Ramsey, D.K., Lamontagne, M., Wertenberg, P.F., Valentin, A., Engström, B. & Németh, G. (2001). Assessment of functional knee bracing: an in vivo three-dimensional kinematic analysis of the anterior cruciate deficient knee. *Clinical Biomechanics*, 16, 61-70.

Ramsey, D.K., Wretenberg, P.F., Lamontagne, M. & Németh, G. (2003) Electromyographic and biomechanic analysis of anterior cruciate ligament deficiency and functional knee bracing. *Clinical Biomechanics*, 18, 28-34.

Singer, J.C., Lamontagne, M. (2007). The effect of functional knee brace design and hinge misalignment on lower limb joint mechanics. *Clinical Biomechanics*, 23, 52-59.

Styf, Jorma. (1999). The Effects of Functional Knee Bracing on Muscle Function and Performance. *Sports Medicine*, 28, 77-81.

Thacker, S.B., Stroup, D.F., Branche, C.M., Gilchrist, J., Goodman, R.A. & Porter Kelling, E. (2003). Prevention of knee injuries in sports: A systemic review of the literature. *Journal of Sports Medicine and Physical Fitness*, 43, 165.

Théoret, D. & Lamontagne, M. (2006). Study on three-dimensional kinematics and electromyography of ACL deficient knee participants wearing a functional knee brace during running. *Knee Surgery, Sports Traumatology, Arthroscopy*, 14, 555-563.

Wirth, M.A. & DeLee J.C. (1990). The history and classification of knee braces. *Clinical Journal of Sports Medicine*, 9, 731-741.

Wojtys, E.M, Kothari, S.U. & Huston, L.J. (1996). Anterior Cruciate Ligament Functional Brace Use in Sports. *American Journal of Sports Medicine*, 24, 539-546.

Wojtys, E.M., Loubert, P.V., Samson, S.Y. & Viviano, D.M. (1990). Use of a knee-brace for control of tibial translation and rotation. A comparison, in cadavera, of available models. *Journal of Bone and Joint Surgery*, 72, 1323-1329.

ESTIMATION OF POTENTIAL RISK OF ACL RUPTURE IN FEMALE SOCCER PLAYERS AND EFFECTIVENESS OF A PREVENTION TRAINING PROGRAM

Thomas Jöllenbeck, Britta Grebe, Dorothee Neuhaus
Institute for Biomechanics, Klinik Lindenplatz, Bad Sassendorf, Germany

The purpose of this study was to make an estimation of the potential risk of ACL rupture in female soccer players by means of video based analysis and controlled by biomechanical testing as well as to relate to effects by carrying out an 8 week specific training program. Results show that video based screening seems to be suitable in order to achieve a good estimation of the risk for ACL injuries without large expenditure. The performed ACL prevention training program shows a slight reduction of the potential risk of ACL rupture.

KEY WORDS: ACL, prevention, soccer.

INTRODUCTION: Rupture of the anterior cruciate ligament (ACL) is a serious knee injury with a long phase of rehabilitation. Often there are fatal results for the athlete up to sports disability. Most frequent situations for ACL ruptures are landing from a jump, stopping and plant and cut manoeuvres. High risk sports are team handball, soccer and basketball (Petersen et al. 2005[1]). Approximately 70% of ACL injuries occur in non-contact situations. Studies have shown a considerably higher incidence of ACL rupture within female athletes. The rate is 2.4 to 9.5 times higher than in male athletes. Different possible factors of risk are discussed. As anatomical causes joint angle positions of lower extremity, laxity and condylar space are taken into consideration. Referring to hormonal factors female oestrogen, relaxin or menstruation are discussed. From biomechanical point of view muscle strengths, activation patterns, functional stability, propriozeption and neuromuscular control of the knee leading muscles are mentioned. Dynamic knee valgus is discussed as a major to the risk factor for ACL rupture. Additionally external conditions like outfit, shoes or playing ground are to be considered. Based on existing results programs were made to prevent ACL ruptures by means of explanation, propriozeption or jump exercises or consist of a combination of all three aspects. Sports specific programs for soccer and team handball were made out up to now (Petersen et al. 2005[2]).

The effectiveness of prevention programs is not yet regarded as secured. There still exists no screening test for the individual estimation of risk. Therefore it is the aim of the present study to make an estimation of the potential risk of ACL rupture in female soccer players by means of video analysis and based on biomechanical testing and additionally to relate the results to effects by carrying out of a specific training program.

METHODS: In the study participated 16 healthy female U17-soccer players (16.5 y, 63.0 kg, 170.2 cm) without injuries or former ACL ruptures. A pre-test and a post-test consisting of a jump test and a strength test were carried out at an interval of 8 weeks. Statistical analysis was performed for the data of 13 players; the data of 3 players had to be excluded from the study because of various injuries between pre- and post-test.

The jump test consisted of a series of drop, squat and counter movement jumps with 3 trials each. Drop jumps with a falling height of 39 cm were performed with regard to the dynamic knee valgus. Squat and counter movement jumps were performed with regard to the jump height. Two 3-dimensional force plates (Kistler) were used for recording of ground reaction forces and jump heights with a measuring frequency of 1000 Hz. Two video cameras (JVC, 50 Hz) in frontal and dorsal plane were used to determine dynamic knee valgus via knee angle and internal rotation in landing. Video and force plate data were synchronised and stored with the software Simi-Motion.

The strength test consisted of 3 exercises with 3 MVC's each of the abductors (ABD) as well as of the external rotators in a knee angle of 45 degrees (ARO45) and 90 degrees (ARO90). For this, a mobile force transducer (Biovision) was attached with adjustable straps between both thighs across the kneecap and fixed in a hip wide and parallel position. MVC's were registered on a PDA with a measuring frequency of 1000 Hz with the Software PLab. MVC of the abductor muscles was measured by abduction of a stretched against a standing leg in an upright standing position. MVC of the external rotators was measured sitting on a chair with fixed feet on the floor by abduction of the flexed legs. The height of the sitting position was adapted to the knee joint angles of 45 respectively 90 degrees.

Videos of the drop jumps were analysed for to estimate the potential risk for ACL injury with regard to the dynamic knee valgus between the first contact to ground and the deepest bending position. The qualitative risk factor (rf) was defined from a value of 0 (no risk, neither knee valgus nor inward knee movement) to a value of 4 (very high risk, knee valgus with knee contact, fig. 1). Additionally the horizontal distance of knees and feet at the moment of ground contact as well as of the deepest bending position was measured by 2D analysis. Between pre- and post-test the players performed the PEP (Prevent injury Enhance Performance) ACL prevention program (Mandelbaum et al. 2005, Grimm, K. & Kirkendall, D. 2007) 2 times a week instructed by a physiotherapist. PEP contains a series of exercises consisting of warm-up, stretch, strength, plyometrics, agilities and knee-bent with a duration of 15 minutes in all. The results were processed by the means of Simi-Motion, Simi-Onforce, PLab and Excel, SPSS was used for statistical analysis.

Figure 1 Example pictures for estimation of risk factor for ACL injury with regard to the dynamic knee valgus in deepest bending position, left: risk factor 0, right: risk factor 4, knee valgus with contact

RESULTS: Subjective video analysis in pre-test (pre) without knowledge of kinematic data shows a high potential risk for ACL injuries (pre rf: 3.0 ±1.0) respectively 10 from 13 players are rated with a risk-factor of 2.5 and higher (table 1). Risk is reduced significantly in the post-test (post rf: 2.5 ±1.2, p=.006), but 8 players remain in the high risk group.
Quantitative analysis shows a straight leg position at the moment of first ground contact and a significant horizontal approximation of the knees at the deepest bending position (pre 5.0 ±2.5cm, post 5.5 ±2.0 cm, p=.000). There is also a significant trend for an increasing knee distance at initial ground contact (pre 23.7 ±3.2cm, post 24.5 ±3.5 cm, p=.064). During landing there are no further significant changes, neither in horizontal distances nor in approximations of feet or knees between pre- and post-test. Subjective estimation of potential risk correlates significant with decreasing horizontal distance of the knees during landing at ground contact (pre r=-.74, p=.004, post r=-.67, p=.013), in the deepest point (pre r=-.95, p=.000, post r=-.89, p=.000) and concerning the approximation of knees (pre r=-.73, p=.005, post r= .74, p=.004). The vertical movement of the knee joint while landing is reduced significantly in the pre-test (pre 10.5 ±2.3cm, post 9.7 ±1.9cm, p=.047).
There is no significant change in the maximum jump heights in each kind of trial. The landing index (LI) of drop jump as the quotient of Jump maximum ground reaction force while landing and body weight is reduced significantly in the post-test (pre 4.4 ±0,5, post 3.8 ±0.6, p=.006). The maximum strengths of abductor muscles and external rotators at 45 degrees are significantly raised in the post-test between 7-13% (ABD: pre 217 ±36N, post 233 ±24N, p=.033, ARO45: pre 204 ±32N, post 231 ±42N, p=.011, external rotators at 90 degrees show a significant trend (ARO90: pre 222 ±29N, post 239 ±47N, p=.085).

Table 1 Main results

	Parameter	pre-test	post-test	t-test (p)
1	risk factor, subjective [0-no, 4-high risk]	3.0 ±1.0	2.5 ±1.2	.006
2	horizontal knee distance at initial ground contact [cm]	23.7 ±3.2	24.5±3.5	.064
3	horizontal knee distance at deepest bending position [cm]	18.6 ±4.4	19.0 ±4.3	.334
4	horizontal approximation of knee during landing [cm] * t-test between 2 and 3	-5.0 ±2.5 (p =.002)*	-5.5 ±2.0 (p =.000)*	
5	horizontal foot distance at initial ground contact [cm]	23.7 ±3.7	24.0 ±4.0	.375
6	vertical movement of knee during landing [cm]	10.5 ±2.3	9.7 ±1.9	.047
7	landing index (max ground reaction force / body weight)	4.4 ±0.5	3.8 ±0.6	.006
8	ABD - MVC abductor muscles [N]	217 ±36	233 ±24	.033
9	ARO45 - MVC external rotator muscles at 45° knee angle [N]	204 ±32	231 ±42	.011
10	ARO90 - MVC external rotator muscles at 90° knee angle [N]	222 ±29	239 ±47	.085

DISCUSSION: The results distinctly show a dynamic knee valgus while landing in drop jump discussed as a major risk factor for ACL rupture. There is a good agreement between video based estimation and biomechanical determined knee valgus motion. However, the video based screening appears to be influenced by two parameters. An increase of valgoid leg position at ground contact leads to a higher risk score as well as the approximation of knees up to the deepest bending position. Both parameters do not correlate with each other. However, they refer independent from each other to unfavourable motion patterns dependent on generated lever arms.
The ACL preventing training program (PEP) leads to a slight reduction of risk factor on principle, even if a high risk level remains. In addition to the subjective estimation, the landing index respectively the initial impact is also reduced significantly i.e. landing in post-test becomes softer than in pre-test. However, the vertical motion of the knees is also reduced ac-

cording to the training program. On the one hand, this may be interpreted as an optimization of the landing mechanism in connection with a softer landing. On the other hand, it is reported that a lesser bent knee while landing favours ACL injuries (Devita et al. 1992). The resulting effect remains uncertain. The raised strengths of abductors and external rotators may be speculated as a factor for stabilization and optimization of knee joint movement respectively for reduction of the risk situation. Finally it should still be mentioned, that team coaches report a clearly raised flexibility and agility of the players as a result of the PEP-program.

CONCLUSION: Video based screening seems to be suitable in order to achieve a good estimation of the potential risk for ACL injuries without large expenditure. But the results also show a remaining high potential risk in spite of an 8-week ACL preventive training program. Consequently, the effectiveness of the propagated prevention program needs some further investigation. From the biomechanical point of view the mechanisms promoting ACL rupture have to be cleared up. In particular the antagonistic coordination of quadriceps and hamstring muscles to stabilise the knee joint and to prevent an anterior glide of the tibia and condylar impingement favouring ACL injuries seems to be very important. The influences of muscular fatigue and the resulting changes in muscular coordination within long activity times to the incidence of ACL injuries are not yet investigated but may also be relevant. Further studies should deal with the activation patterns of thigh muscles in the recovered and fatigued state with female and male athletes with different expertise level and should improve knowledge for an optimization of training programs for neuromuscular control and activation to prevent respectively to reduce the potential for ACL rupture.

REFERENCES:

Devita P, Skelly WA: Effect of landing stiffness on joint kinetics and energetics in the lower extremity. Med Sci Sports Exerc 24 (1992) 108-115.

Mandelbaum BR, Silvers HJ, Watanabe DS. Effectiveness of a neuromuscluIar and proprioceptive training program in preventing anterior cruciate ligament injuries in female athletes: a 2 year follow up. Amer J Sports Med. 2005: 33(7) 1003

Petersen W., Rosenbaum, D., Raschke M.: Rupturen des vorderen Kreuzbandes bei weiblichen Athleten. Teil 1: Epidemiologie, Verletzungsmechanismen und Ursachen. Deutsche Zeitschrift für Sportmedizin, 56, 6, 2005, 150-156.

Petersen W., Zantop, T., Rosenbaum, D., Raschke M.: Rupturen des vorderen Kreuzbandes bei weiblichen Athleten. Teil 2: Präventionsstrategien und Präventionsprogramme. Deutsche Zeitschrift für Sportmedizin, 56, 6, 2005, 150-156.

Grimm, K., Kirkendall, D.: Gesundheit und Fitness für Fußballerinnen. FIFA (Fédération Internationale de Football Association) (ed.) rva, Altstätten, CH, 2007.

LEG PRESS EXERCISE IN PATELLOFEMORAL PAIN- A ONE-YEAR FOLLOW-UP STUDY

Chen-Yi Song[1] and Mei-Hwa Jan[1,2]

School and Graduate Institute of Physical Therapy, College of Medicine,
National Taiwan University, Taipei, Taiwan[1]
Physical Therapy Center, National Taiwan University Hospital, Taipei, Taiwan [2]

The purpose of this study was to investigate the short- and long-term effect of leg-press exercises in dealing with patellofemoral pain. Sixty subjects with patellofemoral pain participated. They were randomly assigned into leg-press exercise or control (no exercise) group. Training consisted of three weekly sessions for eight weeks. Measurements of pain (VAS), Lysholm scale score, morphology of vastus medialis obliquus (including cross-sectional area and volume by ultrasonography) were obtained before and after 8-wk treatment. Long-term follow-ups were carried out (on leg-press group only) at 6-month and 12-month later. Significant improvements in pain, functional score, and muscle hypertrophy were observed after leg-press intervention, but not in the control group. The good subjective and functional outcomes achieved immediately after exercise intervention were maintained at long-term follow-up. Since the short- and long-term prognoses of subjects who underwent leg-press exercise were relatively good, simple and convenient leg-press exercise was recommended in rehabilitation of patellofemoral pain.

KEY WORDS: knee, leg press, morphology.

INTRODUCTION: Patellofemoral pain is a common musculoskeletal problem of the knee that frequently affects both young and sporty populations. It was thought to occur with lateral malalignment of the patella where hypotrophy or atrophy of vastus medialis obliquus (VMO) muscle was one of possible cause that commonly seen in patients with patellofemoral pain. Since the VMO plays an important role in medial stabilization of patella, numerous rehabilitation protocols regarding quadriceps strengthening were described for dealing with this problem. Among them, the leg-press exercise was a common approach. However, the clinical evidence regarding the efficacy of this approach, especially on VMO morphology, was lacking. It is unclear if the improvement of subjective outcome, i.e. decrease of pain and increase of functional ability, is interrelated with VMO hypertrophy. The purpose of this study was to investigate the short-term (2-month) and long-term (6- and 12-month) effects of leg-press exercise. The morphology of VMO along with the pain and functional ability were measured.

METHODS: Data Collection: A total of 60 participants diagnosed with unilateral or bilateral patellofremoral pain were enrolled in this study. They were randomly allocated into leg-press exercise (LP) or control (no exercise) group. Three-weekly exercise interventions were carried out by one physical therapist. Four assessment sessions, at time of initial evaluation (pre-training), 2-month post-training, and 6-month, 12-month later, were performed by another physical therapist who was blinded to each patient's grouping. At long-term follow-up (6- and 12-month), only leg-press exercise group was evaluated.

Leg-press exercises training was performed unilaterally starting from 45° of knee flexion to full extension using leg-press machine (Enraf-Nonius B.V., Rotterdam, The Netherlands) since the patellofemoral joint stress was less in the functional range of knee motion. Prior to the beginning of exercise training, unilateral one-repetition-maximum (1RM) strength of the lower extremity was determined with repetition-to-fatigue testing. Patients were unilaterally trained at 60% of 1 RM for 5 sets of 10 repetitions. For the advancement of training resistance, the 1RM was re-measured every 2 weeks and the exercise intensity adjusted accordingly. A metronome was used to control the exercise pace at 2 second concentric and eccentric contractions from 45° of knee flexion to full extension. There was a 2 second break between each repetition and a 2 minute break between each set. Left and right limbs were alternatively trained between each exercise set. A hot pack was applied to quadriceps for 15 minutes before exercise. After exercise, participants were taught to stretch their quadriceps, hamstrings, iliotibial bands and calf muscle groups, and had a cold pack applied to their knee joints for 10 minutes. Self-stretches were maintained for 30 seconds and were repeated 3 times for each muscle group. Control group participants did not receive any exercise intervention, but were provided with health educational material regarding patellofemoral pain. During the intervention period, all participants were advised not to perform or receive any other exercise program or intervention. Neither tape nor brace was used. After that, 8-wk exercise program (the same with that of leg-press group) was then given to control group. The exercise intervention participants then received health education. In addition, simple home exercise programs (including general quadriceps strengthening exercise and lower-extremity stretching exercise) were taught, but they were not requested to keep up the program during follow-up period.

The outcome measures in this study included VAS pain assessment, and the worst pain experienced in the previous week was measured using the 100-mm VAS line. The functional ability was measured by Lysholm scale (0-100 point scale) where 100 point indicating maximal functions. Additionally, VMO morphology, including VMO cross-sectional area (CSA) on the patella-base level and VMO volume under the patella-base level, were assessed by ultrasonography (HDI 5000, Advanced Technology Laboratories, Bothell, WA) with a 5 to 12 MHz broadband linear-array transducer (38 mm). All ultrasonographic measurements were obtained while participants were lying on a bed, with both legs relaxed (feet were positioned in a frame to prevent leg rotation) and a thick padded towel placed underneath the knee to maintain resting position. The longitudinal length of the patella in mm was determined from the upper to the lower border with calipers. The VMO volume under the patella-base was approximated from a series of VMO CSAs using the trapezoidal rule (Lin et al., 2008). To obtain a valid calculation of VMO volume from the sonographic image, a custom-made holder was used to fix the probe (Lin et al., 2008). The holder was calibrated to quantify movement of the transducer by synchronizing with a scaled hub, which was turned in a full circle to mobilize the transducer by 1 mm from the proximal patellar-base toward the distal patellar-apex along a line perpendicular to the horizontal representing the upper border of the patella. The first VMO CSA was taken from the line passing through the patella-base level. Serial VMO CSAs were obtained every 2 mm, until the VMO image on the visual display faded (Lin et al., 2008).

Data Analysis: Data obtained from the most symptomatic knee were analyzed using SPSS 11.0 (SPSS, Inc., Chicago, IL). Data were subjected to an intention-to-treat analysis and included all drop-outs. The data of control group gathered at post-training evaluation was then used in long-term follow-ups for comparison with exercise group. Descriptive statistics (mean ± standard deviation, SD) were used to determine participant characteristics. Prior to statistical analysis, the Kolmogorov-Smirnov test was performed to assess the normality of continuous data. Normally distributed baseline demographic variables were compared independent t-test. Non-normally distributed variables were compared by Mann-Whitney test with an alpha 0.05. Gender and numbers of afflicted sides (bilateral vs. unilateral) were compared by Chi-square test with an alpha 0.05. For each outcome variable measured, a 4 (assessment time) × 2 (treatment groups) two-way mixed ANOVA was performed. When a significant two-way interaction was detected, post-hoc analysis was performed using Bonferroni adjustment.

RESULTS: The demographic data for both LP and control group participants was presented in Table 1. There were no significant between group differences, except symptom duration ($P= 0.025$), for the demographic variables.

Table 1 Demographic Data for Study Participants

	LP (n= 30)	Control (n= 30)	P value
Sex (Male : Female)	8 : 22	4 : 26	.197
Age (y/o)	40.2±9.9	43.9±9.8	.155
Height (cm)	161.3±8.4	159.7±5.2	.390
Weight (kg)	60.1±11.2	57.4±6.9	.259
BMI (kg/m^2)	23.0±3.0	22.5±2.1	.423
Involved side (Bilateral : Unilateral)	18 : 12	18 : 12	-
Duration of symptoms (month)	38.3±34.2	27.7±41.0	.025 *

During 8-wk intervention period, 8 participants dropped out of the study due to personal factors (not knee pain) or work commitment. Fifty-two participants completed the trial (27 in the LP exercise group and 25 in the control group). The follow-up rate was 0.90 in LP and 0.83 in control group at post-intervention evaluation. At 6- and 12-month, the follow-up rate was 0.83 for LP exercise group.

The main results of this study were summarized in Figure 1-4, with 4 assessment time in horizontal axis, where 0-month denoted pre-intervention, 2-month denote post-intervention, and 6-month and 12-month represent the follow-ups. There were no significant between-group differences in all outcome measures at baseline (pre-intervention). Significant decreases in pain, increases in functional score and VMO muscle hypertrophy were observed after LP intervention (all P< 0.05), but not in the control group. Only the good subjective and functional outcomes (VAS and Lysholm scale score) achieved immediately after exercise intervention were maintained at long-term follow-up (all P< 0.05 as compared to pre-intervention).

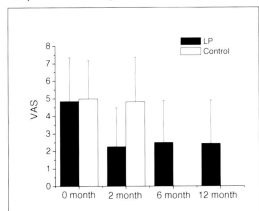

Figure 1: Comparison between Assessment Time Changes of VAS among Groups

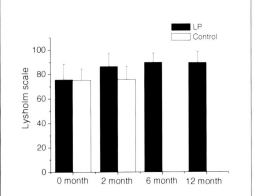

Figure 2: Comparison between Assessment Time Changes of Lysholm score among Groups

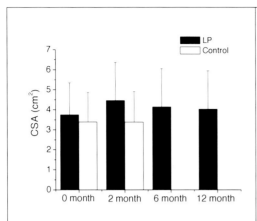

Figure 3: Comparison between Assessment Time Changes of VMO CSA among Groups

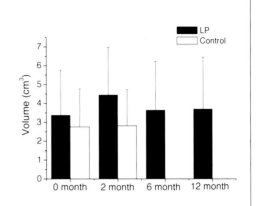

Figure 4: Comparison between Assessment Time Changes of VMO Volume among Groups

DISCUSSION: The results of present study showed decreased pain and increased function at both short-and long-term (1-yr, at least) follow-up which was consistent with previous study (Herrington, et al., 2007, Witvrouw, et al., 2000,2004). To our knowledge, we were the first to use ultrasonography to examine the therapeutic effect on VMO muscle morphology. The better result of VMO hypertrophy was found after 8-wk leg-press training, however, it was lost after 6 to 12-month follow-up time while patients remained less pain and better function compared to pre-training status. Restoration of quadriceps strength, flexibility and adjusted physical activity (life style) taught by health education may contribute to improvement in symptom and function. According to homeostasis theory proposed by Dye (Dye, 2005), patients could become pain-free once they function within the envelope of function. Further research regarding to what extent the VMO hypotrophy or atrophy may response for symptomatic patellofemoral pain will be of great interest.

CONCLUSION: The short- and long-term prognoses of subjects who underwent leg-press exercise were relatively good. Simple leg-press alone was recommended in clinical practice for dealing with patellofemoral pain.

REFERENCES:

Dye, S.F. (2005). The pathophysiology of patellofemoral pain. *Clin Orthop*, 436, 100-10.

Lin, Y.F., Lin, J.J., Cheng, C.K., Lin, D.H., Jan, M.H. (2008) Association between sonographic morphology of vastus medialis obliquus and patellar alignment in patients with patellofemoral pain syndrome. *J Orthop Sports Phys Ther*. 38:196-202.

Herrington, L., Al-Sherhi, A. (2007). A controlled trial of weight-bearing versus non-weight-bearing exercises for patellofemoral pain. *J Orthop Sports Phys Ther*, 37(4):155-60.

Witvrouw, E., Danneels, L., Tiggelen, D.V., Willems, T.M., Cambier, D. (2004). Open versus closed kinetic chain exercises in patellofemoral pain. A 5-year prospective randomized study. *Am J Sports Med*, 32(5):1122-30.

Witvrouw, E., Lysens, R., Bellemans, J., Peers, K., Vanderstraeten, G. (2000). Open versus closed kinetic chain exercises for patellofemoral pain. *Am J Sports Med*, 28(5):687-94.

Acknowledgement
We would like thank the support from Nation Science Council in Taiwan for this study.

ASSESSMENT OF THE EFFECT OF INJURY ON THE KINEMATIC DIFFERENCES IN THE SAGGITAL PLANE UPON LANDING

Amanda Clifford, Kieran O'Sullivan and Marie Tierney
Physiotherapy Dept., University of Limerick, Limerick, Ireland

KEY WORDS: Kinematics, Landing, Injury, Gaelic Football

INTRODUCTION: Landing on one leg is a common activity in Gaelic Football. Research has shown that the internal and external forces on the joints of the lower extremity can be modulated by changing kinematic patterns of lower limb function (Schmitz et al, 2007). Previous studies in other sports have suggested that uninjured limbs tend to land in more flexed positions (Ortiz et al, 2008). The purpose of this pilot study was to assess if injury had an effect on the kinematic pattern of a drop land in Gaelic footballers.

METHODS: Ethical approval was obtained for this study from the University of Limerick research ethics committee. This was a quantitative cross sectional study. 11 male college level Gaelic Footballers provided written informed consent and completed 5 single leg drop lands on each leg from a height of 0.6m while 3D kinematic data was simultaneously collected. The CODA motion analysis system was used to track 22 markers which were placed on specific anatomical landmarks, which allowed the measurement of lower-limb joint angle displacement during each drop land. An average of the five drop lands i.e. mean of the maximum joint angle achieved at the ankle, knee and hip during the drop were calculated and statistical analysis was performed on data using SPSS version 15.0. Limbs were classified as injured if they sustained an injury which prevented their participation in their sport for greater than 2 weeks. Injures were lower limb soft tissue injuries and did not include ruptures.

RESULTS AND DISCUSSION: The results of this study show that the uninjured ankle dorsiflexed more than the injured ankle (6.7°). This change is statistically significant (p=0.009). The other joints show that the injured limb flexed more, however these changes were not determined to be clinically or statistically significant (all p>0.05). This result differs from previous research in other sporting populations and may suggest a differing kinematic landing pattern among Gaelic Footballers. As the previous literature in this area is limited especially among Gaelic Footballers this study serves to add to the current research carried out on this population and this topic. The results of this preliminary study should be examined with caution considering its small sample size. However, it is hoped that this pilot will inform a future larger study in this area.

CONCLUSION: This study suggests that kinematics alterations may exist in Gaelic Footballers following injury. In addition, this study may indicate that Gaelic Footballers rehabilitation programmes may require modifying in order to optimise the function of the injured limb.

REFERENCES:

Negrete, R.J., Schick, E.A. and Cooper, J.P. (2007) Lower-Limb Dominance as a Possible Etiologic Factor in Noncontact Anterior Cruciate Ligament Tears, *Journal of Strength and Conditioning Research*, 21(1), 270-273

Ortiz, A., Olson, S., Libby, C.L., Trudelle-Jackson, E., Kwon, Y.H., Etnyre, B. and Bartlett, W. (2008) Landing Mechanisms Between Noninjured Women and Women with Anterior Cruciate Ligament Reconstruciton During 2 Jump Tasks *The American Journal of Sports Medicine*, 36(1): 149-157

Schmitz, R.J., Kulas, A.S., Perrin, D.H., Riemann, B.L., and Shultz, S.J. (2007) Sex differences in lower extremity biomechanics during single leg landings, *Clinical Biomechanics*, 22, 681-688

LOW BACK PAIN IN GOLF: DOES THE CRUNCH FACTOR CONTRIBUTE TO LOW BACK INJURIES IN GOLFERS?

Michael H. Cole[1] and Paul N. Grimshaw[2]

University of South Australia, Australia[1]
University of Adelaide, Australia[2]

KEY WORDS: golf injuries; lumbar lateral flexion; rotational velocity; trunk kinematics.

INTRODUCTION: Nearly 41% of low back injuries in golf occur around impact or during the early follow-through (McHardy et al., 2007). In view of these recent statistics, it is important to consider the significance of the crunch factor as a possible contributor to golf-related low back injuries. The crunch factor was described by Sugaya et al. (1997) as the instantaneous product of lateral trunk flexion (LFA) and axial trunk rotational velocity (ARV) and was based on the knowledge that both of these measures reach their peak close to impact. The authors reported that these factors would contribute to spinal degeneration and stated that the crunch factor could be useful to compare trunk mechanics in injured and healthy golfers. However, as only one earlier study (Lindsay & Horton, 2002) has examined the crunch factor in injured golfers, this work further considered the importance of this measure in low back pain golfers.

METHODS: Fifteen healthy golfers (NLBP) and twelve golfers with a mild or greater level of low back pain (LBP) were recruited. Each golfer performed 20 drives, whilst being filmed by three genlocked video cameras (50 Hz). Three-dimensional kinematics were derived for the best three swings using Peak Motus. The crunch factor was calculated as the instantaneous product of LFA and ARV, where LFA was the angle between the segments joining the mid-hip and mid-shoulder markers and the right and left hip markers minus ninety degrees and ARV was the first derivative of the hip to mid-trunk differential angle with respect to time. An ANCOVA controlling for age was used to assess for inter-group differences.

RESULTS: The crunch factor for both groups increased rapidly from the mid-point of the downswing through impact and into the follow-through, but the statistical results showed no significant difference between the groups with respect to the peak value. Similarly, peak lateral flexion and axial trunk rotational velocity did not differ between the golfers (Table 1).

Table 1: Peak crunch factor, lateral flexion and axial trunk rotational velocities.

	LBP		NLBP			
	Mean	SD	Mean	SD	p	Cohen's d
Peak Crunch (deg^2/s)	4879.7	2194.9	4920.2	2273.4	0.44	0.24
Peak Lateral Flexion (deg)	-19.1	5.6	-19.1	5.7	0.36	0.28
Peak Axial Trunk Rotational Velocity (deg/s)	-271.0	76.8	-260.4	50.3	0.36	0.33

DISCUSSION: This research showed no significant difference between the LBP and NLBP groups for peak LFA, ARV or the resulting crunch factor. These data were comparable to the peak crunch factors reported previously for six injured and uninjured golfers (Lindsay & Horton, 2002), but were greater than those presented for healthy golfers (Morgan et al.,1999). The non-significant findings together with small effect sizes suggest that the crunch factor is not a contributory factor in the development of low back pain in golfers.

REFERENCES:
Lindsay, D. M., & Horton, J. F. (2002). *Journal of Sports Sciences, 20*(8), 599-605.
McHardy, A., et al. (2007). *Journal of Chiropractic Medicine, 6*(1), 20-26.
Morgan, D. et al. (1999). *Science and Golf III*, pp.120-126. Champaign, IL: Human Kinetics.
Sugaya, H., et al. (1997). *22nd Annual Meeting of the AOSSM*, Sun Valley, ID.

ISBS 2009

Oral Session S23

Other Sports III

KINEMATIC ANALYSIS OF THE SLIDING STOP IN WESTERN RIDING AT THE MALLORCA WESTERN REINING TROPHY 2006

Maren Fröger and Christian Peham

Movement Science Group, Department for companion animals and horses, University of Veterinary Medicine, Vienna, Austria

The purpose of this study was to show the acceleration acting during the sliding stop. As data source we used a DVD of 10 finalists of the Mallorca Western Festivals 2006 (reining competition). These videos were analysed using the SIMI-Motion software. Additionally to the defined location on the horse and rider the reference points on the horse (saddle pad) and on the panel fence (advertising board) were digitised. With the help of the reference point the coordinates were determined and the acceleration was calculated. The maximum acceleration of the sliding stop in the running direction was mean=37.92 m/s² (SD=9.47). The vertical acceleration of the sliding stop at this time was mean=8.50 m/s² (SD=6.26). With an expected mass of horse and rider between 500 to 600 Kg, this acceleration will lead to a load between 11.6 KN and 37.6 KN. The conclusion is that the acting load during the sliding stop is comparable to load on the extremity during a gallop race. The question remains what are the effects of the sliding to the lower hind extremities of the horse and does it lead to injuries?

KEY WORDS: Sliding stop, western riding, acceleration, load, video analysis.

INTRODUCTION: Reining is a western riding competition for horses where the riders guide the horses through a precise pattern of circles, spins, and stops. All work is done at the lope and gallop; the fastest of the horse gaits. Sliding Stop: the horse goes from a gallop immediately to a complete halt, planting its hind feet in the footing and allowing its hind feet to slide several meters, while continuing to let its front feet "walk" forward. The back should be raised upward and hindquarters come well underneath (FEI, 2009). See figure 1.

Figure 1: Sliding stop during reining competition (Mallorca Western Reining Trophy 2006).

It is obvious, that the load during the sliding stop on the horse and on the rider is enormous. Especially to the horses' hind legs, therefore it is of importance to know the load to protect the hoofs with suitable horseshoes (skid boots, sliding plates). The aim of our study was to show the stress, expressed by acceleration (deceleration), acting in the centre of mass (Horse and rider), when performing a sliding stop.

METHODS:

Horses and Riders: 7 horses were Quarter horses and 3 Paint horses. The mean age of the horses was 6.8 years (SD 1.3). Riders and horses were top performers. Included were European champion, Futurity-champion, Derby-champion and two riders earned more than 1 Million US$ prize-money.

Data Collection: Videos of a DVD of the Mallorca Western Reining Trophy 2006 were the base material. Type of the camera is unknown to the authors. 10 finalists of the reining competition were analysed for this investigation. For obtaining a two-dimensional motion analysis, the recorded videos were imported into the motion analysis program SIMI-Motion. The next step was to calibrate each frame with the help of reference points. The calibration procedure was two folded. We used 2 reference points on the horse (saddle pad) to determine the effect of zooming (local coordinate system). The length of the saddle pad is known, 91 cm in all horses. Additionally 5 reference points on the panel fence were necessary to complete the calibration (Global Coordinate System). With help of the reference points the coordinates were calculated. The sampling rate of the videos was 50 frames per second.

The starting frame at digitising was the moment when the horse prepares for the sliding stop; i. e. the horse's hind quarters were under the horse's body in stretched position, ready to slide. The last frame was the moment when the horse stopped (Figure 2).

Figure 2 shows digitized points; the reference points and the estimated centre of mass

Data Processing: The data were smoothed using a moving average. Then the accelerations were calculated by differentiating the smoothed data twice. The error induced by the smoothing can be expressed through the goodness of fit (r^2). For all horses the goodness of fit was higher than 0.9 ($r^2 > 0.9$). The minimum acceleration was used to determine the forces (see figure 3). The Centre of mass was estimated under rider in the middle of the saddle pad (Buchner et al., 2000). The resulting forces have been obtained on the basis of mass of horses and riders between 500 kg and 700 kg.

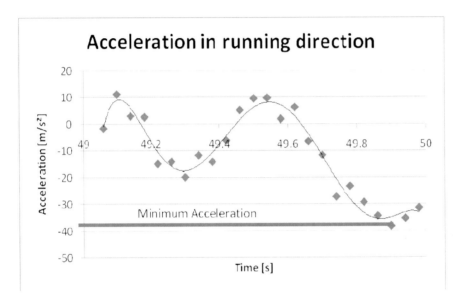

Figure 3 shows the horizontal acceleration of horse 1. The black line is the polynomial fit of the curve (6th order).

RESULTS: Table 1 shows the maximum acceleration of all horses (rider) in running and vertical direction. The main acceleration (deceleration) occurs in running direction. The values lay between -53.65 and -23.19 m/s².

Table 1 Maximum Acceleration of the Centre of Mass

Horse	Acceleration m.s^{-2}	
	Running Direction	Vertical
1	-38.15	3.65
2	-42.16	2.3
3	-35.36	8.95
4	-39	4.52
5	-38.1	12.01
6	-53.65	18.86
7	-33.73	8.94
8	-50.15	6.5
9	-25.69	1.01
10	-23.19	18.24
Mean	-37.92	8.50
SD	9.47	6.26

DISCUSSION: The usual load on the horses hoof is in the vertical direction. The sliding phase depends on the gait of the horse (walk, trot, gallop), but is commonly very short, compared to the sliding stop in Western Riding (Pardoe et al., 2001). It is obvious, that the acceleration in vertical direction is much smaller than in running direction. The mass of the analysed horses and rider were unknown to the authors. But if a mass of horse and rider between 500 and 700Kg is assumed, forces caused by the deceleration are in the range between 11.6 KN and 37.6 KN. This values are comparable to the load on the fore limbs when horse jump over a fence of one meter.

As it is shown in Figure 1, during the sliding stop the hind extremities are invisible. This limits this study, because the estimated stress to the hind limbs is inaccurate. Therefore question remains what are the effects of the sliding to the lower hind extremities of the horse and does it lead to injuries?

CONCLUSION: This study showed the enormous stress to rider and horse during the sliding stop; especially the hind limbs of the horse are used to decelerate. To avoid injuries for the horse a suitable training of horse and rider is essential; e. g. To start with lower speeds, use different grounds (lawn, sand) so that the horses will learn to adapt (Schöllhorn et al.,2009). Besides the training the appropriate equipment (Skid boots, sliding Plates) and the condition of the sand are crucial (Thomason and Paterson, 2008).

REFERENCES:

Fédération Equestere Internationale (FEI) (2009). *Rules for Reining Competitions*
http://www.fei.org/Disciplines/Reining/Pages/Rules.aspx
Pardoe CH, McGuigan MP, Rogers KM, Rowe LL, Wilson AM. (2001). *The effect of shoe material on the kinetics and kinematics of foot slip at impact on concrete.Equine Vet J Suppl. 33:70-73.*
Thomason JJ, Peterson ML (2008). *Biomechanical and mechanical investigations of the hoof-track interface in racing horses. Vet Clin North Am Equine Pract.;24(1):53-77.*
Schöllhorn WI, Mayer-Kress G, Newell KM, Michelbrink M. (2009). *Time scales of adaptive behavior and motor learning in the presence of stochastic perturbations. Hum Mov Sci. 2009 Jun;28(3):319-33.*
Buchner HH, Obermüller S, Scheidl M. (2000). *Body centre of mass movement in the sound horse. Vet J. Nov;160(3):225-34.*

TOWARDS UNDERSTANDING HUMAN BALANCE - ANALYZING STICK BALANCING

Manfred Vieten
University of Konstanz, Konstanz, Germany

In this study stick balancing serves as a role model for more demanding balancing tasks. The purpose is to detect the movement parameters important for stick balancing and their interrelations. Two tilt angles were defined and the relations with the stick coordinates and their derivations analyzed. The correlation between tilt angle and acceleration of the lower coordinate, the angular velocity and the same acceleration of the lower coordinate proved to be the most important relations. The relations were identified to serve as a guideline for establishing a computer simulation of stick balancing which is presented in a separate study. Four parameters were identified and the values determined for comparison with the results of a computer simulation.

Keywords: balancing, movement pattern, motion analysis, stability, correlations, Fourier analysis

INTRODUCTION: Balance is an ability to maintain the center of gravity of a body within the base of support with minimal postural sway (Shumway-Cook, Anson et al. 1988). It is an essential feature for achievement in most human movement tasks. The understanding of movement strategies and the mechanism behind human motion, if in sports or otherwise is of great importance. Good balance ensures top performance for example in gymnastics; it also avoids falling in human gait. Stick balancing represents a classical example for balancing. Cabrera and Milton, between 2002 and 2009, published a series of papers on different aspects of balance using this example. On-off intermittency and parametric noise was the topic of Cabrera and Milton (2002). Milton, Townsend et al. (2009) published a paper where the role of feedback strategies to maintain balance was the focus. Balance and human balance in specific is a topic with a multitude of aspects, and we are far from having a far reaching knowledge. Our long-term strategy of raising our understanding of balance is to develop a computer simulation and compare its results with measurements. Computer models depend on underlying rules that govern the simulation. Therefore, if the predictions and the measurements are in agreement, the underlying rules have a high probability to be valid. In a first attempt we seek some data from stick balancing and moreover its interrelations. To accelerate the developing process of the simulation we looked at data already available. The main task, depicted in this paper, was to analyse the data and identify the underlying relations between the stick's movement and the subjects' action/reaction. This subject-stick interplay mechanism is an important part in a computer simulation. The question was "What must happen for a subject to react?" We expected a deviation from the stick's upright position to trigger a reaction. Also, the stick's angular velocity might provoke the subjects' response. A subject's reaction must result in an accelerated movement of the finger respectively the lower end of the stick. We decided to look into the interplay of the parameters tilt angles (see Figure 1) and coordinate of the lower end of the stick, as well as into the combination of the first and the second derivatives of these parameters. Finally we identified four parameters, which can be used to compare measured data and simulation results.

METHODS: The experimental data were taken from a students' project. A stick (about 1 m of length) had to be balanced on the finger tip for one minute. Two active LED markers of a 3D Lukotronic digitizing system were glued onto the lower part of a stick at a distance of 0.328±0.001 m of each other. Nine sports students (7 males and 2 females) performed the task of balancing the stick for one minute. They repeated the task five times in sequence with short breaks of one to two minutes. The mass of the stick was not determined but this does not compromise the data, since within the equation of motion developed for the computer simulation (separate study) mass is not a parameter. We defined two tilt angles as in Figure

1 and equation (1.1). Tilt angles and derivatives of angles and coordinates were calculated from the raw data. Thereafter, each parameter was filtered using the low pass triple F filter (Vieten 2004). Before a residual analysis (Winter 2005) was performed to find the appropriate cut-off parameter. A correlation matrix was calculated containing the tilt angles, its derivatives angular velocity and angular acceleration and the stick's lower end coordinate

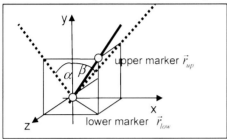

$$\alpha = \arctan\left(\frac{z_{up} - z_{low}}{y_{up} - y_{low}}\right) = \arctan\left(\frac{z}{y}\right)$$
$$\beta = \arctan\left(\frac{x_{up} - x_{low}}{y_{up} - y_{low}}\right) = \arctan\left(\frac{x}{y}\right)$$
(1.1)

Figure 1: Experimental setup and marker placement

as well as its velocity and acceleration. Cross-correlations were performed between the tilt angles and the stick's horizontal acceleration as well as between angular velocities and the stick's acceleration. We determined t_{shift}, the time lag for the highest correlation coefficient r to occur. For comparison between measured data and simulation results we chose the four parameters: 1. $|\overline{\beta}|$, the average of maximal $|\beta|$ at the vertical reversal points (all maxima with $|\beta| \geq 1°$ were included). 2. Correlation coefficient between α respectively β and its horizontal stick acceleration. 3. t_{shift}. 4. The frequency expectation as given in equation (1.2)

$$\langle v \rangle = \int_0^{v_s/2} v \cdot |F(v)| \cdot dv \bigg/ \int_0^{v_s/2} |F(v)| \cdot dv \qquad (1.2)$$

Here is v the frequency, v_s the sampling frequency, and $F(v)$ the Fourier transform of α respectively β.

RESULTS:

We calculated the cutoff frequency for the low pass F^3 filter to 10 Hz. All secondary parameters (angles and derivations) were calculated first, and then filtered. The average absolute value of the tilt angles at the reversal points averaged for all participants is 2.81° for

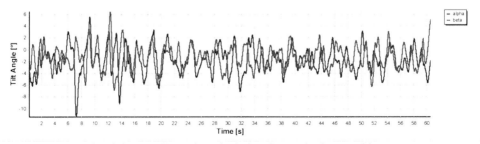

Figure 2: 60 seconds of the two tilt angles of a balancing task of one subject.

α with a standard deviation of ±1.81° and for β it is 2.38°±1.41°. A typical curve of α and β is depicted in **Error! Reference source not found.**. The correlation matrix between the tilt angles, lower marker coordinates and their derivatives for the subject CH4 are displayed in Table 1. The average

Table 1: Correlations between angles, coordinates and its derivatives (p<0.001) for the same subject and task as displayed in Figure 2. n.s. denote non significant results.

Subject CH4	alpha	d(alpha)/dt	d²(alpha)/dt²	beta	d(beta)/dt	d²(beta)/dt²
alpha	1.000	n.s.	-0.467	0.067	-0.048	n.s.
d(alpha)/dt	n.s.	1.000	n.s.	0.051	-0.031	n.s.
d²(alpha)/dt²	-0.467	n.s.	1.000	n.s.	-0.031	-0.244
beta	0.067	0.051	n.s.	1.000	n.s	-0.339
d(beta)/dt	-0.048	-0.031	-0.031	n.s.	1.000	n.s.
d²(beta)/dt²	n.s.	n.s.	-0.244	-0.339	n.s	1.000
x_low	0.050	-0.047	n.s.	-0.525	-0.038	0.090
d(x_low)/dt	0.119	n.s.	n.s.	n.s.	-0.869	n.s.
d²(x_low)/dt²	n.s.	n.s.	0.199	0.645	n.s	-0.747
y_low	-0.081	0.036	0.069	0.022	0.026	n.s.
d(y_low)/dt	-0.085	-0.394	0.197	-0.060	0.154	-0.064
d²(y_low)/dt²	0.125	-0.161	-0.253	-0.062	0.064	n.s.
z_low	-0.511	n.s.	0.099	n.s.	0.041	0.021
d(z_low)/dt	n.s.	-0.783	n.s.	-0.116	-0.029	n.s.
d²(z_low)/dt²	0.717	n.s.	-0.932	0.022	n.s	0.080

value of the correlation $\dot{\beta} \Leftrightarrow \ddot{x}_{low}$ for all subjects is 0.53±0.09. The cross-correlation between the angular velocity $\dot{\beta}$ of the tilt angle β and the respective acceleration of the lower coordinate $\ddot{z}_{low}$ are shown in Figure 3. The results for $\dot{\alpha}$ are skipped because they resemble

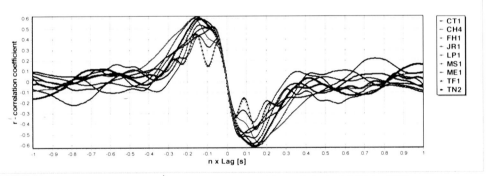

Figure 3: Cross-correlations $\dot{\beta} \Leftrightarrow \ddot{x}_{low}$ for the subjects labeled CT1 to TN2

very much to those received for $\dot{\beta}$. Cross-correlation analysis reveals a high correlation up to r = ± 0.6 for time shifted data in the range of ±0.1 to ±0.2 seconds. The mean shift time (n x Lag [s]) averaged over all subjects is 0.13±0.03 seconds. The expectation value (1.2) averaged over all the subjects is 7.82±3.32 Hz.

DISCUSSION:
Table 1 shows the correlations between tilt angles, lower marker coordinate and its derivatives for one subject. The actual results naturally vary between subjects but the magnitudes do not. A low but statistically significant correlation between $\ddot{\alpha}$ and $\ddot{\beta}$ indicates that the movement activation in x-direction has a moderate effect on the activation in z-direction. Also, it is a hallmark of the stick falling in an arbitrary direction and is not always in alignment with either x- or z-axis. The medium to high correlations between α and z as well as β and x and the relation between their first and second derivatives are just a reminder of their functional relationship and for this reason do not give much insight. A similar argument can be given for correlations between the tilt angles and the vertical component. However, it seems possible that subjects' different strategies to maintain the equilibrium are mildly reflected in the stick's change of the vertical coordinate. Fundamental are the strong correlat-

ions $\alpha \Box \ddot{z}$ and $\beta \Box \ddot{x}$ that are strongest for zero time shift. This is a direct indication of an immediate action once a deviation from the upright position occurs. Taking into account the human reaction time from receiving a stimulus to muscular enervation in the order of 0.1 to 0.2 seconds, anticipation must compensate for the reaction time. $\dot{\alpha} \Box \ddot{z}$ and $\dot{\beta} \Box \ddot{x}$ show significant correlation only after a time shift coinciding with the human reaction time.

CONCLUSION: The values of the maximal tilt angles show the magnitude of the deviations tolerable before the stick turns over. The most remarkable results however are that in order to control the stick, the angle as well as the angular velocity play substantial roles in a controlling strategy. Further more; we pinpointed the four parameters adequate for a comparison between measured data and a computer simulation as given in the method section. With these results we are well prepared for the development of a computer simulation of stick balancing. The outcome of such a simulation can be directly compared with the results of this paper. We did this analysis because we see stick balancing as a role model for more complicated tasks. Most balancing tasks important for humans such as standing and walking seem to be of different nature. In stick balancing an external object is balanced and the controlling force is applied from outside the balancing system. In most human balancing tasks however, actuation comes from within the balancing system, the human body. This does not mean we are talking about different classes of balancing tasks! We know from looking at the equations of motion that the force exchange between a balancing system and the outside is responsible for the main part of the controlling. Newton's third law teaches us that any force provokes a counter-force. As a consequence there is no principle difference in applying the force from outside or inside the balance system. Those cases having a much higher degree of freedom are different in the muscular interplay to provide the force exchange with the environment. But, the probability is high to find the same principle reaction type as described in this paper in other balancing systems as well.

REFERENCES:

Cabrera, J. L. and J. G. Milton (2002). "On-off intermittency in a human balancing task." Physical Review Letters 89(15).
Milton, J., J. L. Townsend, et al. (2009). "Balancing with positive feedback: the case for discontinuous control." Philosophical Transactions of the Royal Society a-Mathematical Physical and Engineering Sciences 367(1891): 1181-1193.
Shumway-Cook, A., D. Anson, et al. (1988). "Postural sway biofeedback: its effect on reestablishing stance stability in hemiplegic patients." Arch. Phys. Med. Rehabil. **69**: 395-400.
Vieten, M. (2006). Statfree: a software tool for biomechanics. In International Society of Biomechanics in Sports, Proceedings of XXIV International Symposium on Biomechanics in Sports 2006, Salzburg, Austria, University of Salzburg, 2006, p.589-593. Austria.
Vieten, M. M. (2004). Triple F (F³) Filtering of Kinemaitc Data. ISBS 2004, Ottawa, Canada, Faculty of Health Science University of Ottawa.
Winter, D. A. (2005). Biomechanics of human movement. New York, Wiley.

Acknowledgement
The raw data were taken during summer 2006 in a seminar on biomechanics. The author was lecturer for this class and supervised the student during the whole data acquisition. All numerical analyses within this study as well as the generation of the diagrams were done using the software StatFree (Vieten 2006). It is freely available on the internet at
http://www.uni-konstanz.de/FuF/SportWiss/vieten/Software/.

THE USE OF PASSIVE DRAG TO INTERPRET VARIATION IN ACTIVE DRAG MEASUREMENTS

Bruce R. Mason, Danielle Formosa and Vince Raleigh
Australian Institute of Sports Aquatics

This study investigated if a measure of mean passive drag could explain the huge differences in propulsive force required by different swimmers to swim at a similar high velocity. Nineteen elite male and female national freestyle swimmers were subjects. The subject's mean active and passive drag was measured at each swimmer's top swimming pace. Stepwise regression analysis was used in the analysis. Passive drag was accepted into the equation to calculate mean propulsive force, prior to velocity being rejected. The correlation coefficient for the relationship between mean propelling force and mean passive drag was 0.77. This was statistically significant at the $p<0.001$ level and explained 59% in the variance of mean propulsive force that swimmers need to produce to reach their top swimming pace. Clearly, a measure of the swimmer's passive drag does provide an explanation for the huge differences in propulsive force required by swimmers to be competitive in the free swim aspect of races.

KEY WORDS: Biomechanics, swimming, passive drag, active drag, propulsive forces

INTRODUCTION: The Australian Institute of Sport (A.I.S.) has developed protocols and equipment based on the velocity perturbation method developed in Russia by Kolmogorov et al. (2000) in which the active drag force that opposes the swimmer's motion may be computed at the swimmer's top swimming pace. In constant velocity unaided swimming, the propulsive force applied by the swimmer to the water is equivalent but opposite in direction to the active drag force applied by the water to the swimmer. Unlike the measurement of propulsive forces generated by athletes in many other sports to create and sustain movement and which can be measured directly, propulsive forces in swimming cannot be readily measured by a direct method. However, active drag measurement provides knowledge of the forces generated by swimmers to propel themselves through the water.

The A.I.S. method of assessing active drag as reported by Alcock et al. (2007) relies on towing the swimmer through water at a constant pace that is five percent greater than the top swimming velocity of the subject. The towing velocity is kept at a five percent higher constant velocity by a powerful dynamometer while the swimmer provides a maximum effort to swim with the tow. The dynamometer is mounted on a force platform that measures the force (total Y component of force from the force platform) applied to the swimmer by the dynamometer to maintain this higher velocity. Assuming equal power applied by the swimmer to the water for propulsion in both the unaided and the active drag towing trial, the propelling force at the swimmer's top swimming velocity may be computed using the towing force recorded from the force platform. While the propelling force so obtained is an oscillating parameter which is characteristic of the swimmer's stroke mechanics, it is more productive in this project to report the active drag or propelling force by a mean force over the whole stroke. The A.I.S. had been monitoring swimmers and obtaining the mean propulsive force generated by swimmers to attain their top swimming velocity. It became apparent that there were vast differences in the propelling force required by swimmers at similar swimming velocities. In an attempt to understand the reason for these differences, the A.I.S. entered a project to monitor both active and passive drag for swimmers at the swimmer's top swimming velocity, with the prospect to interpret such differences in mean propelling force from the mean passive drag values.

METHODS: Data Collection: Nineteen Australian swimmers (11 males and 8 females) were tested in the A.I.S. aquatics laboratory. The calibre of the swimmer was such that each had the ability to reach the final in at least one of the three freestyle events (50 m, 100 m and 200 m) at the Australian National Open Swimming Championships. All testing included only the Australian crawl swimming stroke. The testing included a number of tests sections.

The first part of the testing involved obtaining the top swim velocity of each swimmer. Here the subject was instructed to swim at their maximum velocity through a 10 m interval with a swim-in 10 m interval to build up to maximum pace, before reaching the start of the timing interval. The timing of each trial was performed utilising a video system (50 hertz) which included two cameras, with one focused on the start and the other at the end of the timing interval. Each camera's view included vision of an elapsed electronic timing clock, synchronised in each camera view and accurate to a hundredth of a second, to assess the time taken to swim the 10 m interval and hence enabled the calculation of the subject's mean swimming velocity. Three such trials were conducted on each swimmer and the trial with the quickest time was chosen to represent the swimmer's maximum pace.

The second part of the testing involved obtaining a measurement of passive drag for each subject at the swimmer's top swimming velocity. Prior to testing, the subject was instructed how to attain the streamline position, arms extended in front of the head (no breathing), that they were to adopt during the tow. The towing was performed with a device that attached to the second finger of the subject through a non stretch Kevlar cable linked to the dynamometer. The swimmer was thoroughly familiarised with the towing activity prior to any testing. Seven trials were conducted on each subject at a velocity equal to the swimmer's top swimming pace. The force required to tow each swimmer was sampled by computer over a 10 m interval. A trigger was used by the testers to indicate to the data capture computer and video timing system when the subject reached the start and also the end of the 10 m towing interval. Force data (total Y component from force platform) was collected by the computer on a 12 bit analogue to digital board at a sampling rate of 500 hertz. One second of kinetic data was collected prior to the starting signal and six seconds of data was captured after the signal. Both the start and end trigger signals were also collected by the computer as a separate channel of data, from the analogue to digital board, in order to enable the force data to be processed during only the 10 m timed interval. The mean value for the passive drag force over the 10 m interval was recorded for each trial. The two extreme mean trial scores were eliminated in the statistical analysis leaving five means for the calculation of the subject's mean passive drag force.

The third part of the testing involved obtaining a measurement of active drag for each swimmer at the subject's top swimming velocity. Each swimmer was familiarised with the active drag tow which was performed to obtain the swimmer's mean active drag at the subject's top swimming pace. The tow attachment to the swimmer was connecting to the belt worn around the waist and attached anterior to the body. Five trials were conducted with each swimmer at a tow velocity that was 5% faster than the subject's top swimming pace. The swimmer swam at top pace with the tow (no breathing). The first trigger for the collection of data occurred at the beginning of a stroke (right hand entry) and the kinetic data was captured for four complete strokes as denoted by a second trigger. The force data collected in these trials represented the additional force required to tow the swimmer at the subject's 5% higher velocity beyond that required at the swimmer's top pace. The active drag or propulsive force measurement for the swimmer's maximum speed was then able to be computed under the assumption that an equal effort was produced by the swimmer in both the free swim maximum effort unaided trials and in the active drag trials. The mean value for the active drag force over the four complete strokes was recorded for each of the five trials. The five mean trial scores were utilised to obtain the subject's mean propulsive force.

Data Analysis: The top swimming pace for each swimmer was obtained as the fastest of three free swim trials. The raw kinetic data in the seven passive tows was smoothed with a Butterworth digital low pass filter with a cut off frequency of 5 hertz. The mean passive drag force was computed over the 10 m interval for all seven trials and the two trials with extreme values were excluded from further analysis. The force obtained from the five active drag tows were smoothed as above and the active drag computed for each trial prior to obtaining a mean propulsive force value for each of the five trials. From the above five passive drag trials per subject, a single force value representing mean passive drag was obtained by averaging over the five trials. A single value representing mean propulsive force per subject was similarly obtained from the five active drag trials.

A stepwise regression SPSS package using propulsive force as the dependent variable and swim velocity and mean passive drag as the independent variables was carried out. A significance value of 0.05 for acceptance and 0.10 for rejection was chosen for the regression equation.

RESULTS:

Subject	Gender	Speed (m.s^{-1})	Mean Propulsive Force (N)	Mean Passive Drag Force (N)
1	M	1.90	161.7	77.2
2	M	1.92	226.4	81.9
3	M	1.92	151.0	85.3
4	M	1.85	235.7	71.3
5	M	1.91	256.5	88.3
6	M	1.89	302.2	84.5
7	M	1.92	325.3	84.6
8	M	2.02	237.2	100.0
9	M	1.98	204.9	81.6
10	M	1.83	184.3	64.8
11	M	1.89	286.4	73.7
12	F	1.72	163.7	49.9
13	F	1.76	127.4	52.5
14	F	1.71	77.5	47.5
15	F	1.74	164.6	51.0
16	F	1.69	171.3	43.9
17	F	1.61	95.3	37.9
18	F	1.64	89.3	40.7
19	F	1.64	100.1	38.4

The correlation coefficient for swim velocity with mean propulsive force was 0.738, for mean passive drag with mean propulsive force was 0.766 and correlation coefficient between the two independent variables (velocity with mean passive drag) was 0.977. The significance level for the both independent variables with the dependent variable was $p<0.001$.

In the regression equation mean passive drag was accepted into the equation first. Velocity was rejected from the equation. The reason for the rejection of velocity was probably because of the high correlation between mean passive drag and velocity. This indicated that there was relatively little further information that velocity could add to the equation.

DISCUSSION: It was revealed in A.I.S. active drag testing that male swimmers of approximately the same ability sometimes had vastly different mean propulsive force characteristics. In fact two male swimmers who were participating in the same testing session recorded identical top swimming velocities. These swimmers were tested in active drag analysis and the mean propulsive force over four whole strokes for one swimmer was found to be twice the value of the other swimmer's mean propulsive force. Such differences in effort for athletes that perform to the same level of performance are rarely seen in other sports unless there are huge size differences in the athletes. This then raised the question as to why one swimmer needed to exert twice the propelling force to swim at the same velocity as the other swimmer. Was it possible that a measure of some other parameter may be able to explain the difference? The obvious parameter to investigate was passive drag as it theoretically incorporated many of the anthropometric characteristics of the swimmer.

It is common knowledge that the force required to swim at a set velocity is a function of velocity squared. Therefore velocity itself must be highly related to the mean propulsive force produced by a swimmer. The researchers investigated the relationship between the dependent variable mean propulsive force and independent variables mean passive drag and swim velocity. Velocity was itself an additional independent variable that was known to relate to the mean propulsive force so that a comparison between the relationships of both independent variables and the dependent variable would be of interest. The regression analyses revealed that mean passive drag was slightly more highly correlated to mean propulsive force than was velocity. Mean passive drag was included in the regression equation to derive mean propulsive force but velocity was rejected because of the high correlation between velocity and mean passive drag. The project did however identify mean passive drag to be a slightly better predictor of mean propelling force than was swim velocity in a maximum swim effort. Mean passive drag was shown to account for close to 60% of the variance in the mean propulsive force generated by a swimmer at a maximum swim velocity.

CONCLUSION: This study identified that mean passive drag and mean velocity were both significantly related to mean active drag and hence mean propulsive force. Mean passive drag was in fact more highly related to mean propulsive force than was swim velocity. Therefore mean passive drag was identified to be a good indicator of why some swimmers need to generate much greater propulsive forces than do other swimmers that travel at relatively similar velocities. The differences in the mean propulsive force required by swimmers to travel at maximum velocity can to some extent be explained in the measurement of their mean passive drag at a similar velocity. Surely individuals that require considerably more force and power to swim at a set pace are at a greater disadvantage than those that require considerably less force and power. As passive drag reflects the level of propulsive force required for a swimmer to swim at top pace, then passive drag may be a good indicator of the future capabilities of a subject's swimming ability, as passive drag is to a limited extent dependent upon non adjustable anthropometric characteristics of the individual concerned.

REFERENCES:

Alcock.A., & Mason.B. (2007). Biomechanical Analysis of active drag in swimming, In Menzel.H & Chagas.M (Eds.), *Proceedings of the XXVth International Symposium of Biomechanics in Sports*, Brazil (pp212-215)

Kolmogorov,S., Lyapin,S., Rumyantseva,O., & Vilas-Boas,J. (2000). Technology for decreasing active drag at the maximal swimming velocity. In Sanders,R & Hong,Y. (Eds.), *Proceedings of XVIII International Symposium on Biomechanics in Sport Applied Program: Application of Biomechanical Studies in Swimming, Hong Kong* (pp 39-47).

SURFACE ELECTROMYOGRAPHY OF ABDOMINAL AND SPINAL MUSCLES IN ADULT HORSERIDERS DURING RISING TROT

Annette Pantall, Sharron Barton and Peter Collins
European School of Osteopathy, Maidstone, UK

The purpose of this study was to determine activation patterns of iliocostalis lumborum and rectus abdominis in horse riders during rising trot. Horse riders (n=10) of varying abilities (casual rider to Prix St George rider), aged from 20 – 57 were involved in the study. Electromyographic activity was recorded bilaterally from iliocostalis lumborum and rectus abdominis. In addition, movement at the lumbosacral joint was recorded using an electrogoniometer. The subjects were asked to walk and then perform rising trot on their horses. The electromyographic signals were full-wave rectifed and a moving filter was applied to the signal. Patterns of activation between sides and between muscles were determined. Coactivation of the right and left sides was present for all riders. The novice riders displayed coactivation of rectus abdominis and iliocostalis lumborum whereas the experienced riders had a phase shift between these two muscles. Rectus abdominis behaved as an agonist in the experienced rider, contracting as the rider made contact with the saddle on the outside diagonal. Controlled activation of the rectus abdominis is an important feature in the experienced rider. Training regimes for horse riders should incorporate specific exercising of the abdominal muscles.

KEY WORDS: horseback riding, electromyography, abdominal muscle, spine

INTRODUCTION: Horseback riding is a unique sporting activity in that it involves two living entities moving together in a synergistic fashion to produce a coordinated movement pattern. The movement of the horse is transmitted to the rider principally through forces acting on the saddle. In turn, the position of centre of mass of the rider's trunk relative to the horse will influence the gait of the horse. The position of the centre of mass of the rider's trunk is controlled mainly by contractions of the abdominal muscles and paravertebral muscles. Knowledge about the kinetics and kinematics of the rider is important for equestrian sports, in particular dressage. Dressage demands very precise sequencing of the horse's footfall, which implies equally even movement of the rider's trunk (Hodson et al., 1999). The type and frequency of movement of the rider will depend on the gait of the horse. A variety of gaits have been described, which include walk, tölt, pace, passage, trot, canter, rotary gallop and transverse gallop (Barrey, 1999). Trot is a two-time movement with the forelimb and contra-lateral hindlimb moving together. A full gait cycle during trot is composed of two stance phases and two swing phases. In rising trot, the rider will be seated during one of the stance phases and out of the saddle for the second stance phase. Rising trot requires the rider to control trunk movement accurately in order to minimize variability of the kinematics of the horse. Movement of the trunk is initiated primarily by contraction of the abdominal and paravertebral muscles. Surface electromyography (sEMG) provides a non-invasive method to analyze the activity of these muscles. However, few biomechanical studies have been undertaken on the horserider. The two main investigations on sEMG activity of riders' muscles have been performed by Terada (2000) and Terada et al. (2004). The former study reported a difference in muscle activation patterns between novice riders and advanced riders during walk, trot and canter. The advanced riders displayed cocontraction of rectus abdominis (RA) and the erector spinae and minimal involvement of adductor magnus. The study by Terada et al. (2004) investigated timings of muscle contraction of upper extremity muscles and RA in advanced riders during sitting trot. The purpose of this study was to investigate the contraction pattern of two trunk muscles to establish how the rider changes trunk position and whether the pattern differs between novice and advanced riders.

METHODS: Data Collection: Ten horse riders participated in the study. The age of the rider varied as did the height of the horse (Table 1). The riders were of mixed ability ranging from novice riders to experienced riders competing at international (Prix St. George) level. The sEMG of four trunk muscles was measured. The right and left RA and the right and left iliocostalis lumborum (IL) were selected for measuring. IL is the most lateral muscle of the erector spinae and therefore is a significant lateral flexor of the trunk. The Biometrics DataLOG (Biometrics Ltd, Gwent, UK) with sx232 bipolar electrodes was used to record the sEMG. The electrodes contained a high pass 3rd order filter, cut-off 25Hz, and a low pass 8th order filter, cut-off at 460 Hz. The DataLOG was attached to the rider's waist via a belt. The site of electrode placement was in the position recommended by the SENIAM report (Freriks & Hermens, 1999) for IL and by Cram and Kasman (1998) for RA. The skin area was cleaned with an alcohol wipe prior to attaching the electrodes. Sampling of sEMG was at 1000Hz in accordance with the Nyquist Theorem. Movement of the lumbar spine was measured in the sagittal and coronal planes using a twin axis goniometer (Biometrics Ltd, Gwent,UK). One end of the goniometer was fixed with surgical tape to the skin overlying the middle of the sacrum and the other end was attached to skin overlying the spinous process of the fifth lumbar vertebra. Data from the goniometer was collected at a sampling rate of 20Hz. All measurements took place in an indoor school at a livery stable in Kent. Riders were asked to initially sit still for 60 seconds then to perform a medium walk between 2 markers placed 18m apart and finally to do a working rising trot between these 2 markers. All riders rode in the anti-clockwise direction and sat down on the outside diagonal (right forelimb and left hindlimb in contact with the ground).

Table 1 Age, height, weight of rider, experience of riding and height of horse.

Subject	Age (years)	Height (m)	Weight (kg)	Number of years riding	Height of horse (m)
S1	42	1.65	76.4	15+	1.73
S2	47	1.68	73.2	15+	1.65
S3	23	1.68	58.5	15+	1.52
S4	28	1.79	63.6	10-15	1.68
S5	47	1.88	82.7	1-5	1.50
S6	36	1.57	57.0	15+	1.47
S7	20	1.60	58.2	10-15	1.52
S8	22	1.70	66.8	15+	1.50
S9	57	1.70	76.4	1-5	1.65
S10	43	1.62	54.1	1-5	1.57

Data Analysis: Analysis of the raw sEMG consisted first of full wave rectification of the signal and subsequent filtering of the data. Smoothing of the data was undertaken using the MATLAB (The Mathworks Inc., Natick, Mass.) forward / reverse *filtfilt* routine with a 100ms window to create a linear envelope. A 10-second mid-section of the filtered data was taken for each of the muscles during halt, walk and trot and the mean values calculated. Statistical tests were then applied to determine whether there was a change in mean value of the sEMG and lumbosacral angles from halt to walk, walk to trot and halt to trot. Fisher's F-test was applied to the data to determine whether there was equal variance and then the appropriate t-test was applied. The patterns of sEMG activity were observed to determine the presence of coactivation between RA and IL and the relationship of the sEMG to the lumbosacral angle in the sagittal plane.

RESULTS: Full sEMG and kinematic data were obtained for 7 out of the 10 subjects. No left RA data was available for subject S2, no right RA data for subjects S2 and S4 and no left IL was recorded from subject S1. There was increased activity of all muscles from halt to trot and increased activity of IL from walk to trot. The mean sagittal lumbosacral angle was greater from halt to trot but was less during trot than during walk. The lumbosacral angle showed a regular pattern of flexion and extension during riding trot for the 7 experienced riders with a frequency of 1.29 Hz to 1.88 Hz. A small amount of extension of the lumbosacral junction was observed as the rider's pelvis made contact with the saddle (Figure 1).

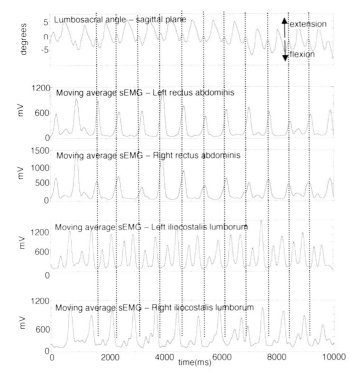

Figure 1: sEMG and L/S angles in rising trot for Subject S8. Dotted lines indicate times when contact is made with the saddle.

RA displayed peaks of activity at maximum lumbosacral flexion for 5 of the riders (S1, S3, S6, S8 and S10) and in one case maximum activity at extension (S4). In the remaining 4 cases, there was no clear correlation between lumbosacral angle and RA activity. There was a less obvious pattern for IL with 3 subjects (S8, S9 and S4) showing greatest activity at lumbosacral flexion and 2 subjects (S3 and S6) exhibiting peak activity at extension. All subjects displayed evidence of synergistic muscle activity between the left and right sides. Patterns of coactivation between RA and IL varied between riders. Four of the subjects (S2, S7, S9 and S10) displayed synergistic contraction of RA and IL whereas another group of 4 riders (S3, S4, S6 and S8) showed a phase shift in activity between RA and IL. Subjects S1 and S5 demonstrated irregular patterns of activity between RA and IL.

DISCUSSION: One would anticipate the frequency of the horserider's trunk movements to be similar to the frequency of the horse's footfall during rising trot. However, the frequency range of the lumbosacral angle in the sagittal plane was smaller than the frequency of working trot that has been calculated to be from 1.39 Hz to 1.56 Hz (Morales et al., 1998). This difference can be attributed to the varying abilities of the riders and sizes of the horses. Since all the riders were riding in the anti-clockwise direction, peak height from the saddle was reached during mid-stance of the inside diagonal. The lumbosacral angle will be at maximum extension during this phase of the horse's gait. During terminal stance of the inside diagonal, the rider descends relative to the saddle to sit during the outside diagonal. The lumbosacral angle is at maximum flexion during midstance of the outside diagonal. In the experienced riders, RA behaved as an agonist and was most active during lumbosacral flexion. This finding is consistent with that of Terada (2000) who reported that RA was maximal during mid-stance although this was measured during sitting trot.

The riders showed different patterns of activity for RA and IL. The phase-shifted pattern of activity of RA and IL activity was only evident in the more experienced riders whereas coactivation of muscles was observed in the less experienced riders (S9 and S10) and rider S2 who had undergone spinal surgery. An additional observation was that subjects S2 and S9 were two of the oldest riders. Coactivation of muscles has been associated with the ageing process (Enoka et al., 2003). Coactivation of muscles requires that the agonist must contract more strongly to overcome the force developed in the antagonist muscle. This pattern of muscle activity is therefore an energetically inefficient control system. The least experienced rider, subject S5, displayed a very varied pattern of sEMG activity and irregular changes in lumbosacral angle suggesting poor development of neuromuscular control.

CONCLUSION: Analysis of sEMG provided information regarding the ability of the horserider to adjust the position of their trunk and through this affect the movement of the horse. The more experienced rider used their RA as an agonist to the dominant movement of trunk flexion. This indicates more efficient control of muscle activation than in novice riders. Inexperienced riders coactivated their RA and IL, which utilizes a greater amount of energy. The study indicated that age may be an additional factor affecting the control strategy of the trunk muscles, with older riders less able to activate selectively their RA without IL contraction to produce trunk flexion. Further research is needed to investigate the relationship between the rider's control of muscle activity and how it is reflected in the horse's performance.

REFERENCES:

Barrey, E. (1999). Methods, Applications and Limitations of Gait Analysis in Horses. *The Veterinary Journal*, 157, 7-22.

Cram, J. & Kasman, G. (1998). *Introduction to surface electromyography*. Maryland: Aspen Publishers.

Enoka, R.M., Christou, E.A., Hunter, S.K., Kornatz, K.W., Semmler, J.G., Taylor, A.M. & Tracy, B.L. (2003). Mechanisms that contribute to differences in motor performance between young and old adults. *Journal of Electromyography and Kinesiology*, 13(1), 1-12.

Freriks, B. & Hermens, H.J. (1999). *European Recommendations for Surface Electromyography-SENIAM Project*, Enschede: Roessingh Research and Development.

Hodson, E.F., Clayton, H.M. & Lanovaz, J.L. (1999). Temporal analysis of walk movements in the Grand Prix dressage test at the 1996 Olympic Games. *Applied Animal Behaviour Science*, 62, 89-97.

Morales, J.L., Manchado, M., Cano, M.R., Miro, F. & Galisteo, A.M. (1998). Temporal and linear kinematics in elite and riding horses at the trot. *Journal of Equine Veterinary Science*, 18, 835-839.

Terada, K. (2000). Comparison of Head Movement and EMG Activity of Muscles between Advanced and Novice Horseback Riders at Different Gaits. *Journal of Equine Veterinary Science*, 11, 83-90.

Terada, K., Mullineaux, D.R., Lanovaz, J., Kato, K. & Clayton, J.M. (2004). Electromyographic analysis of the rider's muscles at trot. *Equine and Comparative Exercise Physiology*, I, 193-198.

ACCURACY OF A PORTABLE (PTZ DIGITAL) CAMERA SYSTEM DESIGNED FOR AQUATIC THREE-DIMENSIONAL ANALYSIS

Georgios Machtsiras and Ross H. Sanders

Centre for Aquatics Research and Education, PESLS, University of Edinburgh, Edinburgh, UK

KEY WORDS: swimming, biomechanics, 3D analysis.

INTRODUCTION: Three-dimensional (3D) motion analysis of aquatic activities such as swimming requires high accuracy throughout large volumes particularly when above and below water data are merged for full body analysis. The purpose of this study was to assess the accuracy of a recently developed portable camera system designed for 3D kinematic data collection of aquatic activities including swimming.

METHOD: Ten markers with known 3D coordinates in a 6.75 m^3 calibration frame were digitized from video clips recorded with 4 underwater cameras (Elmo PTC-450C, Elmo CO., Ltd, Nagoya, Japan) and the data input to a DLT programme. A different set of ten markers with known locations were digitized and their 3D coordinates were calculated using the DLT equations generated from the digitized data of the original set of markers. Concurrently the same procedure was conducted for data recorded by a second camera system for which high accuracy and reliability has been previously established (Psycharakis, Sanders and Mill, 2005).

RESULTS: The mean differences and the root mean square errors (RMS) in estimating the locations of the second set of markers relative to their known locations were quantified to assess reconstruction accuracy. Moreover, the accuracy of the new system was drawn based on the magnitude of the RMS errors with reference to the accuracy of the established system as shown in Table 1.

Table 1 Mean differences (mm) and RMS errors (mm)

Recording system	Mean differences (mm)			RMS errors (mm)		
	X	Y	Z	X	Y	Z
Assessed system	2.0	2.8	2.7	2.3	3.3	3.3
Established system	6.7	5.5	4.3	8.0	6.7	4.7

DISCUSSION: The results of this study showed that the mean differences were lower for the assessed system when compared to the established system. RMS errors represent 0.05% of the calibrated area for X axis, 0.2% for Y axis and 0.3% for Z axis. These results were considerably lower than the results of the second camera system used and other systems reported in the literature.

CONCLUSION: The camera system assessed in this study showed high accuracy when used for aquatic three-dimensional analysis. Considering the additional advantage of being portable, the tested camera system can be regarded a valuable research tool for swimming biomechanics.

REFERENCES:
Psycharakis S.G., Sanders R. & Mill F. (2005). A calibration frame for 3D swimming analysis. In: Wang Q (Ed.), *Proceedings of the XVII International Symposium on Biomechanics in Sports*. The China Institute of Sport Science, Beijing. Pp.: 901-904.

ISBS 2009

Poster Session PS 1

THE ROLE OF ANXIETY IN GOLF PUTTING PERFORMANCE

Ian Kenny[1], Áine MacNamara[2], Amir Shafat[2], Orla Dunphy[2], Sinead Murphy[2], Kenneth O'Connor[2], Tara Ryan[2] and Gerry Waldron[2]

[1]Biomechanics Research Unit, University of Limerick, [2]Department of Physical Education & Sport Sciences, University of Limerick

KEY WORDS: 3D, Anxiety, Accuracy

INTRODUCTION: Anxiety's influence on performance continues to be one of the main research interests for sport psychologists (Hanin, 2000). It is apparent, though, that there is a lack of empirical research characterising the multi-disciplinary effect of anxiety on sports performance. The current study aimed to ascertain biomechanical (accuracy, movement variability) and psychological (anxiety) markers to determine how anxiety affects golf putting.

METHODS: 22 healthy subjects (12 male, 10 female, 21.7±2.0 yrs, 175.3±8.1 cm, 76.4±10.0 kg, all data mean±SD) who had played golf recreationally previously but with no recorded handicap were recruited. Subjects performed thirty 3.05 m putts using their own putter under a control and anxiety condition. Anxiety was elevated using environmental cues (e.g. presence of spectators) and a competition scenario. Three-dimensional motion was tracked using a six camera Motion Analysis^{TM} system operating at 240 Hz. Final ball position from the hole was ascertained using overhead digital photogrammetry. Self reported anxiety was measured pre, during and post putting using standardised self-report anxiety questionnaires. The Competitive State Anxiety Inventory (CSAI) was used to measure state anxiety intensity and direction across three sub scales (i.e. cognitive anxiety, somatic anxiety and self-confidence). The shorter Mental Readiness Form (MRF) was used to obtain anxiety measures during performance.

RESULTS: Significant changes in self-reported anxiety were reported between the control and anxiety conditions. Student's t-test revealed that performance, as measured by distance from hole was not different in control (0.68±0.52 m) and anxiety conditions (0.56±0.33 m). Females, n=10, significantly worsened their performance under anxiety condition (1.02±0.45 m) compared to control (0.73±0.37 m). Movement analysis showed that swing tempo, represented as a ratio of backswing to downswing time increased significantly ($p<0.05$), from 0.57 to 0.65 from control to anxious conditions respectively. Total swing time increased by an average 0.16 s for anxious putts and left wrist angle was also more open at impact by 0.97 degrees.

DISCUSSION: Apparent reported anxiety did not affect overall putting performance. Results are supportive of suggestions in the literature that individuals may increase mental effort on the task to compensate for the negative influence of anxiety on performance (Wilson *et al.*, 2007). Changes in temporal aspects of the putt seemed to compensate for any movement variability which increased anxiety may have induced.

CONCLUSION: Results show anxiety did not cause significant change in putting performance overall.

REFERENCES:
Hanin, Y.L. (2000) Successful and poor performance and emotions, In: Emotions in sport. Ed: Hanin, Y. Champaign, IL: Human Kinetics. 157-188
Wilson M., Smith N.C., Holmes P.S. (2007) The role of effort in influencing the effect of anxiety on performance: testing the conflicting predictions of processing efficiency theory and the conscious processing hypothesis. *British Journal of Psychology*, vol.98 no.3, pp.411-428

FATIGABILITY OF TRUNK MUSCLES WHEN SIMULATING PUSHING MOVEMENT DURING TREADMILL WALKING

Yi-Ling Peng[1], Yang-Hua Lin[1] and Hen-Yu Lien[1]
Grad. Inst. of Rehabilitation Science, Chang Gung University, Taoyuan, Taiwan[1]

KEY WORDS: pushing, trunk muscles activity, muscle fatigue, median frequency

INTRODUCTION: Pushing is a common movement in moving objects, and it also related to about 9% to 20% low back injuries occurrence (Hoozemans et al., 1998). The purpose of the present study was to examine the effect of fatigue on trunk muscle activity during treadmill walking with and without a pushing movement.

METHODS: Twenty healthy young adults were asked to walk with and without a simulated pushing movement. Volunteers first performed walking with simulating pushing by their maximum pushing force. Stopping criterion was reaching subjective assessment of trunk muscle fatigue by Borg CR-10 Scale, and return to carry out walking without pushing movement one week later. The electromyography signal of erector spinea (ES), multifidus (MF), rectus abdominis (RA) and external oblique (EO) muscles at both sides were collected by surface electrodes. Median frequency (MDF) and root-mean-square (RMS) which provide information of muscle fatigability and activity were calculated in the initial, middle and final periods of both walking with and without pushing movements. Repeated measure ANOVA was used to compare the differences of MDF and RMS between the two conditions in three periods. Tukey HSD post hoc analysis was performed on all appropriate statistically significant main effects and interactions. Two-side significance was defined as $p<0.05$.

RESULTS: The mean time of walking with pushing was 19 min 32 sec ± 8 min 32 sec. MDF of ES, MF and RA decreased in both walking with and without pushing conditions, but only significant in without pushing condition (table 1). Normalized RMS of ES and, MF increased (range: 0.21-0.53) in walking with pushing condition, but decreased (range: 0.17-0.34) in walking without pushing condition. However, normalized RMS increased in both with and without pushing conditions.

Table 1 MDF change in with pushing and without pushing conditions during treadmill walking

MDF change (Hz)	Dominate/ Non-dominate	Erector spinae	Multifidus	Rectus abdominis	External oblique
With pushing	initial–middle	-3.5/ -2.1	-2.7/ -2.4	1.6/ -0.7	0.4/ 1.9
	initial–final	-7.5/ -4.0	-6.7/ -6.6	-1.8/ -0.1	1.8/ 3.8
Without pushing	initial–middle	-10.9*/-9.3*	-7.7*/ -15.3*	-11.7*/ -9.4	0.1/ 0.9
	initial–final	-7.6/ -7.9*	-11.1*/-13.59*	-8.9/ -2.5	8.9*/ 7.4*

* Significant difference of MDF change (Tukey HSD post hoc analysis)

DISCUSSION: MDF change had similar trends in both walking with and without pushing conditions. However, RMS change of back muscles increased in walking with pushing but decreased in walking without pushing. In these two different walking conditions, increased activity of back muscles may result in less fatigue in walking with pushing than without pushing.

CONCLUSION: More fatigability of back muscles was showed in walking without pushing than with pushing condition. Future research should investigate trunk muscle fatigability in subjects with LBP during pushing movement and treadmill walking compare to healthy subjects.

REFERENCE: Hoozemans, M. J & van der Beek, A. J. (1998). *Ergonomics, 41*, 757-81.
Acknowledgement : This work was supported by National Science Council (NSC96-2314-B-182-024)

A COMPARATIVE STUDY BETWEEN BLADES AND STUDS IN FOOTBALL BOOTS

J L Nutt, G P Arnold, S Nasir, W Wang, & R J Abboud

Institute of Motion Analysis & Research (IMAR), University of Dundee, TORT Centre, Ninewells Hospital & Medical School, Dundee DD1 9SY, Scotland

KEY WORDS: Football, Rugby, Injury, Blades, Studs, sEMG

INTRODUCTION: The incidence of non-contact injuries in football is high. A significant proportion of the blame is aimed at the footwear worn. Bladed design boots have attracted criticism, with high profile sports teams banning them amid fears of causing knee injury BBC, 2009). This study aimed to biomechanically compare a bladed boot design with a more conventional studded boot design to assess if either boot type produces a greater muscle response and suggest their potential for causing non-contact injury.

METHODS: 31 competitive football and rugby players were recruited to this study. Each participant was required to perform a standard running and sidestepping maneuver as commonly performed during match play. This was carried out along 18 meters of FIFA® approved synthetic turf. A TMSI Mobi sEMG device (TMS International, The Netherlands – www.tmsi.com) was used to measure peak muscle activity in four lower limb muscles closely related to the stability of the anterior cruciate ligament. A Vicon motion analysis system (Vicon UK, Oxford – www.vicon.com) was used to track the participants, and monitor stance phase (contact time) during the sidestep maneuver. Peak average sEMG values were recorded from all four muscles. To standardise the sEMG values recorded, percentage difference between the two boot types was calculated and used to test for any significant difference.

RESULTS: Data from 29 participants was suitable for analysis. Results showed that there is no statistically significant difference between peak average sEMG recordings when using bladed boots compared to studded boots. The four muscles: Rectus Femoris, Vastus Medialis/Lateralis and Semi Tendinosus, show very small percentage differences in peak average sEMG values between the two boot types.

DISCUSSION: The results show no difference in muscle activity when performing sidestepping maneuvers in bladed boots compared with studded boots. This suggests that the muscle forces acting across the joints of the lower limb are not affected by altering the shoe to ground interface. Specifically with relation to the muscles under investigation, there is no difference in the magnitude of muscle force acting across the knee joint.

CONCLUSION: Although there are many factors that influence the incidence of sporting injury. This study shows that with relation to muscle activity, there is no evidence to prove that bladed boots are more responsible for causing non contact lower limb injury than their studded counterparts. Neither boot could be concluded to be substantially more dangerous than the other and further in-depth investigation is required.

REFERENCE:
BBC (2008) BBC NEWS, BBC SPORT, Online UK Edition (2005). *Ferguson bladed boots ban*, 24 September 2005, Available from: http://news.bbc.co.uk/sport1/hi/football/teams/m/man_utd/4277722.stm [accessed 15.10.2008]

STEP HEIGHT EFFECTS ON LOWER LIMB BIOMECHANICS AND BODY CENTRE OF MASS MOTION DURING ELLIPTICAL EXERCISE

Yen-Pai Chen[1,2], Chu-Fen Chang[1], Hui-Lien Chien[1], Yi-Cheng Chen[2] and Tung-Wu Lu[1]

Institute of Biomedical Engineering, National Taiwan University, Taiwan[1]
Graduate Institute of Sport Equipment Technology, Taipei Physical Education College, Taiwan[2]

KEY WORDS: elliptical trainer, step heights, kinematics, kinetics, lower limbs

INTRODUCTION: Elliptical exercise (EE) has been shown to be beneficial for the development and maintenance of cardiorespiratory fitness. Despite these benefits, the feet are constrained by pedals to follow an elliptical trajectory, with the possibility of producing disadvantageous joint loads, body instability and potential musculoskeletal overuse injuries (Lu et al., 2007). Proper selection of step height during EE may help reduce these disadvantageous joint loads and instability. The purpose of the study was to study the effects of step height on the lower limb biomechanics and associated body center of mass (COM) motion during EE.

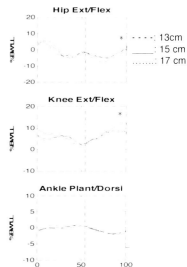

Figure1: Joint moments at three step heights. ($*$: $p<0.05$)

METHODS: Ten healthy male adults (23.7 ± 1.2 yr; 173.5 ± 3.9 cm; 72.2 ± 10.8 kg) performed EE at three step heights (13, 15 and 17 cm) with a pedal rate close to walking cadence and without workload while 3D kinematic data of the whole body and left pedal reaction forces (PRF) were measured using a motion analysis system and a six-component force transducer, respectively. The peak excursion of the body COM, PRF, lower limb joint angles and moments between step height conditions were analyzed using repeated measures analysis of variance with a significance level of 0.05.

RESULTS: With increasing step height, the subjects managed to keep the COM motion within a limited range ($p>0.05$), whereas the posterior component of PRF was significantly reduced in early stance ($p=0.003$). During stance phase, peak hip extension was decreased ($p=0.04$). During swing phase, peak hip and knee flexion angles were increased ($p=0.002$; $p=0.001$), while peak hip abduction angle and peak knee extensor moments were diminished ($p=0.02$; $p<0.001$). Peak hip extensor moments were increased and peak hip abductor moment decreased in late swing ($p<0.001$) (Fig. 1).

DISCUSSION & CONCLUSION: During EE, greater peak flexion at the hip and knee, and peak hip abduction during swing phase were used to compensate for the change of pedal trajectory and to keep the body stable when step height increased. With the more flexed posture in response to increased step height, the reduced posterior shear force shifted the line of action of the PRF more anterior to the hip joint center and less posterior to the knee joint center, leading to increased hip extensor and decreased knee extensor moments in late swing. The current results showed that proper selection of step heights during EE may help reduce harmful joint loadings, especially for the knee joints, and thus reduce potential risk of injuries.

REFERENCES:

Lu, T.-W., Chien, H.-L. and Chen, H.-L. (2007) Joint loadings in the lower extremities during elliptical exercise, *Medicine & Science in Sports & Exercise*. 39:1651-1658.

EFFECTS OF TAI-CHI CHUAN ON THE CONTROL OF BODY'S CENTRE OF MASS MOTION DRUING OBSTACLE-CROSSING IN THE ELDERLY

Tsung-Jung Ho[1], Sheng-Chang Chen[2], Chu-Fen Chang[2] and Tung-Wu Lu[2]

[1]College of Chinese Medicine, China Medical University, Taiwan
[2]Institute of Biomedical Engineering, National Taiwan University, Taiwan

KEY WORDS: gait analysis, inclination angle, centre of pressure

INTRODUCTION: Tripping over obstacles is a common problem among the elderly, which often leads to physical injuries. Previous studies have shown the positive effects of Tai Chi Chuan (TCC) exercises on muscle power, flexibility, endurance, dexterity, physical fitness, and balance (Wolf et al., 1996; Li et al., 2004). These effects are beneficial for the control of the body stability and prevention of falls. The aim of this study was to investigate the effects of Tai-Chi Chun and obstacle heights on the inclination angles of the center of mass (COM) relative to the center of foot pressure (COP) during obstacle-crossing.

METHODS: Ten TCC practitioners (age: 71±5.3 years, height: 163±6.7 cm, mass: 58.7±6.5 kg, TCC experience: 25±9.9 years) and fifteen healthy controls (age: 72±6 years, height: 160 ±5.7 cm, mass: 58±10.4 kg) crossed obstacles of heights of 10%, 20% and 30% of their leg lengths while the COM and COP position data were measured using a 3D motion analysis system and forceplates, respectively. COM-COP inclination angles at the instances when the swing toe was above the obstacle for both leading and trailing limb (T1 and T2) were extracted and statistically analyzed using a mixed analysis of variance with one between-subject factor (subject group) and one within-subject factor (obstacle height).

RESULTS and DISCUSSION: On the anterioposterior (A/P) COM-COP inclination angles, significant group effects were found at T1 ($p<0.001$) and T2 ($p<0.001$). Greater anterior COM-COP inclination angles in the control group indicated that they needed more effort to maintain A/P stability. Significant height effects were found at T1 ($p<0.001$) and T2 ($p<0.001$) and A/P COM-COP inclination angles were decreased with increasing obstacle height, except for the TCC group at T1. These results suggest that the non-TCC older people tended to keep their COM close to COP to increase the body's stability when the obstacle height increased. However, there were no effects on the mediolateral (M/L) COM-COP inclination angles at T1 and T2.

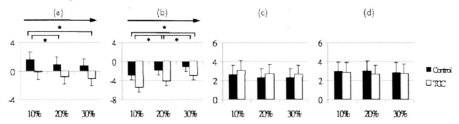

Figure 1 means and standard deviations of the A(+)/P(-) COM-COP inclination angles at T1 (a) and T2 (b), and these of the M(+)/L(-) angles at T1 (c) and T2 (d). (an arrow: a decreasing trend with increasing obstacle height; an asterisk: a significant difference between groups.)

CONCLUSION: The differences of the control strategies adopted during crossing between the TCC and control groups in the sagittal plane may be related to the strengthening effects of one leg standing commonly practiced in the TCC. The findings will be helpful for future studies on kinematics and control strategies adopted during obstacle-crossing.

REFERENCES:

Wolf et al. (1996).. *J Am Geriatr Soc.* 44, 489-97.

Li et al. (2004). *Med Sci Sports Exerc.* 36, 2046-52

KINEMATICS OF TACTICS IN THE MEN'S 1500 M FREESTYLE SWIMMING FINAL AT THE BEIJING 2008 OLYMPIC GAMES

Patrycja Lipinska and Włodzimierz S. Erdmann
Jedrzej Sniadecki Academy of Physical Education and Sport, Gdansk, Poland

The purpose of this study was to obtain a knowledge on tactics of long distance swimming at the highest professional level. Eight swimmers - men, finalists of 1500 m distance of freestyle of the Beijing 2008 Olympic Games were investigated. The distribution of velocity of swimming for the entire distance based on 50 m segments was analyzed. Partial, halves and tierces velocities, velocity indexes and linear regression equations were calculated. It was revealed that better swimmers (placed 1 – 4) had their distribution of swimming as ascending line and with very small difference between segmental velocities and that of the entire distance, while the rest of swimmers had descending velocity line and higher dispersion of partial values.

KEY WORDS: swimming, tactics, kinematics, men, Olympic Games, Beijing 2008.

INTRODUCTION: There are many research articles devoted to swimming. Within biomechanics there are a lot of different approaches to this sport discipline. The most often were investigated: technique of movement of free swimming, i.e. position of body parts, stroke length, rate, velocity, and indices, technique of start and turn, improving of propulsion, diminishing of resistance forces, structure of the entire distance divided into the start segment, free swimming, turn segments, the finish segment. The scientists who devoted their research work especially to biomechanics of swimming are, e.g.: Schleighauf (1979), Hollander et al. 1985, Sanders (2000). There is a lack of investigations on biomechanics of tactics of swimming especially within the last 10 years. The theory of effort says the lowest energy expenditure is obtained when the velocity of movement tends to be steady. The differences of velocity between segments of the distance and mean velocity of the entire distance should be minimized. Also the second part of the distance should be covered with higher velocity than the first one. This was already observed by scientists of the Centre of Locomotion Research in Gdańsk in such sports as: marathon running (Lipinska, 2006), swimming (Erdmann, 2008) and in other sports. The aim of this study was to obtain knowledge on tactics of long distance swimming at the highest professional (Olympic) level.

METHODS: Data Collection: Eight swimmers (men) were investigated. They participated in 1500 m of freestyle swimming final of the Beijing 2008 Olympic Games. Data on the entire time of swimming and also split times for every 50 m (30 segments) were obtained from the Official Website of the Beijing 2008 Olympic Games (results.beijing2008.cn).

Data Analysis: Data on velocities were calculated for the entire distance and for every 50 m segment. Data on velocities were calculated also for halves and tierces of the entire distance but here the first 50 m of the distance were not included since the first meters a swimmer covers in the air after release by jumping forward from the starting block and obtains much higher velocity than for the rest 50 m segments. The entire distance without the first 50 m was 1450 m and was named quasi-entire distance (QED). The difference between the entire distance (ED) and QED is just 3.3%.
In order to compare velocities of halves (H) and tierces (T) and to assess the tendency of velocity distribution through the entire distance indices were calculated. Their calculation was based on mean velocity of the QED. The quasi-half (QH) segment (the first one) had 700 m, and the half H (the second segment) had 750 m. The quasi-tierce (QT) segment (the first one) had 450 m, the rest T segments (second and third) had 500 m. Additionally linear equations of regression for each swimmer were defined.
Differences were calculated between velocities of segments and QED. Then squares of differences were calculated in order to obtain only positive values. Next, sums of differences' squares were calculated. They formed velocity differences' index (VDI) for

halves and tierces. The lower the VDI, the better. To compare steadiness of velocity of swimmers with different mean velocity, VDI was divided by QED. The results formed relative velocity differences' index (RVDI) for halves and tierces. To compare 50 m fragments analysis of variation was used.

RESULTS AND DISCUSSION: Looking at the velocities obtained by eight swimmers for the entire distance of 1500 m one can observe (see Figure 1) that the highest velocity was obtained for the first 50 m, then there were different tendencies of mid-distance swimming, and again higher velocity was obtained for the last 50 m segment.

Only the best four swimmers had ascending line of velocity, and the highest ascending line had the first swimmer at the finish. This is in concordance with the results obtained for the best sportspersons cited earlier. Next four swimmers had descending lines of velocity. They swam too fast at the beginning according to their possibilities. This is seen especially in a velocity line of swimmers no. 6 – 8.

Analysis of variation show substantial differences for velocities of 50 m segments, especially among one before last and last 50 m segment in comparison to the rest 50 m segments (exclusive of first 50 m).

Looking at the bars depicting mean velocities of QH and H segments (Figure 2A) one can see only the first four swimmers swam the second part of the distance faster than the first one. Looking at the bars depicting velocities of QT and T segments (Figure 2B) it is shown the winner and next three swimmers have a tendency of swimming with the middle segment the slowest but last segment is the fastest one. Again, one can see swimmers no. 6 – 8 swam the third part of the entire distance slower.

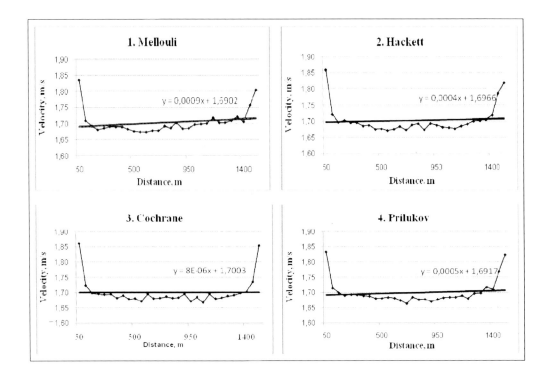

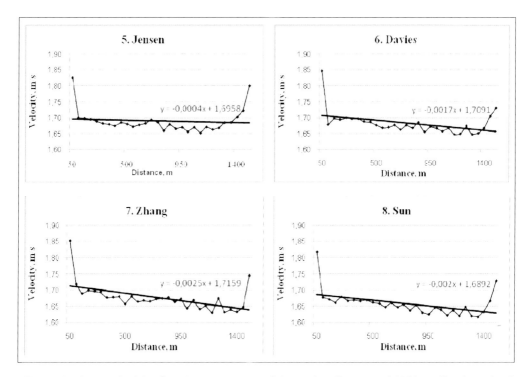

Figure 1: Mean velocities for 50 m segments of the entire distance of 1500 m. Numbers 1 – 8 are the consecutive places obtained in the competition

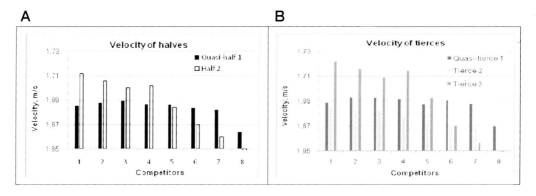

Figure 2: Mean velocities for halves (A) and tierces (B). The first half and the first tierce is without the first 50 m. Numbers 1 – 8 are consecutive places obtained in the competition

In Table 1 detailed results of mean velocities are presented. The data of indices can be assessed only with the knowledge of tendencies of velocities obtained. For example, swimmer no. 4 has indices for tierces similar to those of the winner, i. e. differences of segmental velocities according to mean velocity of the quasi-entire distance were small, and tendencies of velocity lines were similar.

Table 1 Result time (min:sec) for the entire distance and mean velocity (m/s) for the entire distance, halves and tierces and their indices: VDI - velocity differences' index; RVDI - relative velocity differences' index

No.	Name	Result time	Entire distance	Quasi-entire distance	Quasi-half 1	Half 2	Quasi-tierce 1	Tierce 2	Tierce 3
1	2	3	4	5	6	7	8	9	10
1	Mellouli Oussama	14:40.84	1.704	1.699	1.685	1.712	1.689	1.685	1.722
	VDI, RDVI				0.351	0.207	0.829	0.488	
2	Hackett Grant	14:41.53	1.703	1.697	1.688	1.706	1.693	1.682	1.716
	VDI, RDVI				0.167	0.098	0.612	0.361	
3	Cochrane Ryan	14:42.69	1.700	1.695	1.689	1.700	1.693	1.683	1.709
	VDI, RDVI				0.064	0.038	0.357	0.211	
4	Prilukov Yuriy	14:43.2	1.699	1.695	1.687	1.702	1.692	1.677	1.715
	VDI, RDVI				0.122	0.072	0.726	0.428	
5	Jensen Larsen	14:48.2	1.690	1.685	1.686	1.684	1.688	1.675	1.693
	VDI, RDVI				0.002	0.001	0.168	0.100	
6	Davies David	14:52.1	1.682	1.676	1.684	1.670	1.691	1.670	1.670
	VDI, RDVI				0.093	0.056	0.280	0.167	
7	Zhang Lin	14:55.2	1.677	1.670	1.682	1.660	1.688	1.669	1.657
	VDI, RDVI				0.241	0.144	0.510	0.305	
8	Sun Yang	15:05.1	1.658	1.653	1.664	1.642	1.670	1.646	1.643
	VDI, RDVI				0.247	0.149	0.411	0.249	

CONCLUSION: In order to approach sport performance with proper tactics sportspersons need to be taught how to achieve a good distribution of velocity of movement along the entire distance of swimming with general tendency to swim faster within the second half of distance in comparison with quasi-half 1 and swim faster in second and last tierce in comparison with quasi-tierce 1.

REFERENCES:

Erdmann, W. S. (2009) Kinematics of Tactics of Men's 1500 m Freestyle Swimming at 2008 U.S. Olympic Team Trials Finals. *Research Yearbook, Studies in Physical Education and Sport*, 14, 2, 92-98.

Hollander, A. P., Touissant, H. M., van Ingen Schenau, G J. (1985) Active drag and swimming performance. *New Zeal JSport Med*, 13, 110-113.

Lipińska, P. (2006) *Wielkosci kinematyczne i geometria trasy a taktyka biegu w maratonie [Kinematic quantities and geometry of the course and tactics of running in marathon]*. Doctoral dissertation, Gdansk: Sniadecki University School of Physical Education and Sport.

results.beijing2008.cn/WRM/ENG/INF/SW/C73A1/SWM015101.shtml#SWM015101 (2009.03), the Official Website of the Beijing 2008 Olympic Games.

Sanders, R. (2000) Beyond race analysis. In: R. Sanders, Y. Hong (Eds.), *Proceedings of XVIII International Symposium on Biomechanics in Sports. Applied Program: Application of Biomechanical Study in Swimming* (pp 3-11). Hong Kong: The Chinese University of Hong Kong.

Schleighauf, R. E. (1979) A hydrodynamic analysis of swimming propulsion. In: J. Terauds, E. Bedingfield (Eds.), *Swimming III* (pp 70-109). Baltimore: University Park Press.

EFFECT OF FATIGUE ON THE COORDINATION VARIABILITY IN ROWERS

Paul Talty & Dr. Ross Anderson

Biomechanics Research Unit, Department of Physical Education and Sport Sciences, University of Limerick, Limerick, Ireland

Keywords: rowing, continuous relative phase, dynamical systems theory

INTRODUCTION: According to the Dynamical Systems Theory (DST), movement variability is an essential feature of human motor behaviour. This theory of motor control has led to a paradigm shift in sport biomechanics research. Whereas previously, variability in sports biomechanics data was viewed as erroneous and in need of elimination, more recently the existence, amount and effect of variability on different aspects of sports biomechanics has been considered (Hamill et al., 1999). Numerous studies have also considered the effect of fatigue on coordination. Aune et al. (2008) found that fatigue induced a reduction in variability in segment positioning for highly skilled table-tennis players. Also a high level of performance was maintained by reducing racket velocity and thus altering the racket position at the point of racket-ball contact. In rowing, the athlete is seated throughout the event the back is continuously in a flexed position, this leads to high levels of lower back pain (LBP) among rowers (Perich et al., 2006). The current study will investigate coordination variability using an ergometer based rowing protocol designed to induce fatigue. The current research question is how does coordination variability react to fatigue and how does this relate to LBP.

METHODS: Three national level rowers (2 males, 1 female) involved in the winter training stage of their season participated in the study. The rowers were asked to row at a steady, challenging pace on a RowPerfect ergometer (CARE RowPerfect, Netherlands). Retro-reflective markers were placed bilaterally on the wrist, elbow, shoulder, knee, greater trochanter, mid-axilla line at the level of the iliac crest and used to identify the elbow and hip angles for each subject. An infra-red motion analysis system (Motion Analysis Corporation, Santa Rosa, CA) recorded the movements of each subject at 60Hz. Upon reaching a rating of 17 on the RPE scale the rower was instructed to stop (rowers 1, 2, 3 – 10, 10, 12 mins). All data (circa. 38000 frames per rower) were subsequently processed and analysed using customised software written on LabVIEW (National Instruments, Texas, USA); this software identified each stroke based on a kinematic event, normalised the data to 101 points, and calculated the continuous relative phase (CRP) relationship between the elbow and hip angle.

RESULTS: Initial CRP based results indicate that the variability of the rower's kinematics were relatively unaffected by the fatiguing protocol used here [rower 1 – mean baseline vs. mean fatigued – 3.35rads. vs 3.21rads.; rower 2 – mean baseline vs. mean fatigued – 3.4rads. vs. 3.53rads.; rower 3 – mean baseline vs. mean fatigued – 3.73rads. vs. 3.52rads.].

CONCLUSIONS: From the results it is clear that fatigue does not affect coordination variability amongst these rowers; this may be due to the ability of these rowers to perform skilfully under fatigued conditions during competition. The next stage of the research is to examine the link between these results and LBP.

REFERENCES
Hamill, J., et al. (1999) Clinical Biomechanics, 14, 297-308
Aune, T.K., et al. (2008) Perceptual and Motor Skills, 106, 371-386
Perich, D., et al. (2006) 'Low Back Pain in Adolescent Female Rowers and the Associated Factors', in Schwameder, H., eds., Proceedings of the XXIV Int. Symp. Biomech. Sports, Salzburg, Austria, 16th - 18th July, 2006

BIOMECHANICS AND POTENTIAL INJURY MECHANISMS OF WRESTLING

Tsong-Rong Jang[1], Sheng-Chang Chen[2], Chu-Fen Chang[2], Yang-Chieh Fu[2] and Tung-Wu Lu[2]

[1]Athletics Department and Graduate School, National Taiwan Sport University, Taiwan, R.O.C.
[2]Institute of Biomedical Engineering, National Taiwan University, Taiwan, R.O.C.

KEY WORDS: wrestling, injury mechanism, joint kinematics, centre of pressure

INTRODUCTION: Wrestling is one of the oldest and most popular competitive sports in the world. However, knowledge of the biomechanics of wrestling is not well established and the biomechanical risk factors of injuries remain unclear (Boden et al., 2002). The purpose of this study was to investigate the joint kinematics of the lower limbs and the center of pressure (COP) movements in Greco-Roman style (GR) and free style (FS) wrestlers during tackle defense.

METHODS: 18 male college wrestlers participated in this study: 10 majored in GR (height: 171.1±8.0 cm; weight: 73.9±11.5 kg) and 8 in FS (height: 169.0±5.2 cm; weight: 71.8±11.4 kg). The wrestlers received tackle attacks from the front (FD), left (LD) and right (RD) while their kinematic data measured by a 7-camera motion capture system (Vicon 512) at a sampling rate of 120Hz and ground reaction forces from two AMTI forceplates at a sampling rate of 1080Hz. Independent t-test was used for comparisons of the calculated variables between the GR and FS groups for each condition.

RESULTS and DISCUSSION: Compared to the GR group, the FS wrestlers tended to have greater anterior/posterior (A/P) excursions of the COP during FD and LD with greater knee flexion (Fig. 1 & Table 1). This flexed knee strategy may be related to the rule of the game and the training the FS wrestlers received. The FS group was also found to defend tackles with greater knee rotation. Increased joint angles in the transverse plane at the knee and ankle found in this study may subject the joints to a higher risk of ligament injuries commonly found in wrestlers, especially when resisting external force during practices or competitions.

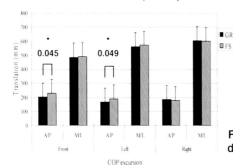

Table 1 Averaged peaks of knee and ankle joint angles

Degrees	FD		LD		RD	
	GR	FS	GR	FS	GR	FS
Knee flexion	70.40	94.53	61.49	88.31	78.62	88.09
Knee IR	0.63	11.33	3.70	2.68	8.61	4.05
Ankle IR	4.40	11.06	6.78	4.40	5.27	6.71

Fig. 1: A/P and medial/lateral (M/L) excursions of COP during FD, LD and RD in GR and FS group. ∗: $p<0.05$.

CONCLUSION: Differences between GR and FS wrestling were found in the resisting duration, joint angles and COP movements in the current study. The defending strategies adopted in the two styles of wrestling were quite different when facing a tackle attack. The current results support the high injury risk of the lower extremities among wrestlers. Strengthening of the muscles of the lower extremities is suggested for the improvement of the functional performance and prevention of lower extremities injuries.

REFERENCES:

Boden BP, Lin W, Young M, Mueller FO (2002) Catastrophic injuries in wrestlers. *American Journal of Sports Medicine.* 30, 791-795.

COMPARISON OF BALL-AND- RACKET IMPACT FORCE IN TWO-HANDED BACKHAND BETWEEN DIFFERENT DIRECTIONS OF STROKE

Chin-Fu Hsu[1], Kuo-Cheng Lo[2], Yi-Chien Peng[1], and Lin-Hwa Wang[1]

[1]Institute of Physical Education, Health & Leisure, National Cheng Kung University, Tainan, Taiwan
[2]Physical Education Office, Kun Shan University, Tainan, Taiwan

KEYWORDS: tennis, two-handed backhand, impact force

INTRODUCTION: Modern rackets have facilitated a modification in playing style from one of technique to one characterized by power (Miller, 2006). The aim of this report is compare the impact force and moment of upper extremity joint between the advanced and the intermediate group during two-handed stroke in across-court and down-the-line. Based on a few studies in backhand stroke, especially the two-handed backhand with different direction, this study is essential for understanding stroke characteristics.

METHODS: This study recruits six right-handed male tennis players. All subjects use the two-handed backhand stroke. This study adopted a 3-D motion analysis system including 8 cameras (500Hz) for recording subjects' upper extremities to trace their motion in cross-court and down-and-line, respectively. The inverse dynamics model was employed for the calculation of the forces of the upper segment joints.

RESULTS:

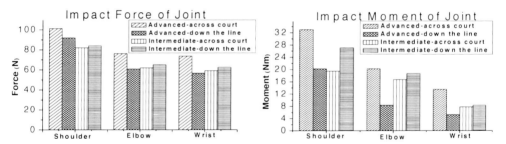

Figure 1: The joint impact force and moment between two directions of hit.

DISCUSSION: The result was found the force and moment of across-court was greater than down-the-line, it was possible that needed more trunk rotation (Reid & Elliott, 2002). In addition, the force and moment of intermediate with down-the-line were more than with across-court. It might apply more movement of elbow and wrist. Previous study found the lower level player had high risk of tennis elbow (Elliott, 2006), it was incidence with this study. One possible reason for the inconsistent result between advanced and intermediate groups may be the cause of the instable technique of intermediate group.

CONCLUSION: These findings confirm that the level of players and the different direction of stroke had difference of impact force and moment.

REFERENCES:
Elliott, B. (2006). Biomechanics and tennis. *Br J Sports Med, 40*(5), 392-396.
Miller, S. (2006). Modern tennis rackets, balls, and surfaces. *Br J Sports Med, 40*(5), 401-405.
Reid, M., & Elliott, B. (2002). The one- and two-handed backhands in tennis. *Sports Biomech, 1*(1), 47-68.

INFLUENCES OF THE MASS OF BOXING GLOVES ON THE IMPACT FORCE OF A REAR HAND STRAIGHT PUNCH

Genki Nakano, Yoichi Iino and Takeji Kojima

Department of Life Sciences, Graduate School of Arts and Sciences, the University of Tokyo, Tokyo, Japan

KEY WORDS: peak of the impact force, impulse, unfixed target, boxing glove

INTRODUCTION: The purpose of this paper was to investigate the influences of the mass of boxing gloves on the impact force of a rear hand straight punch thrown into an unfixed target whose mass (4.36kg) was similar to that of the typical human head.

METHODS: Nine male collegiate boxers whose varsity had been the title holder of Japan University Championship for the last two years threw rear hand straight punches into the unfixed target at full power wearing seven kinds of gloves (10, 12, 14 and 16oz gloves with no added weight and 10oz gloves with added lead weights of 2, 4 and 6oz: 10oz is official). The target was free to move straight horizontally without a noticeable frictional force. The orders of trials were randomized. Several punches were measured for each glove. A minimum of 1 minute rest was taken between two consecutive punches. Accelerations of the target were recorded using an accelerometer attached to the target at 10kHz, and then were low-pass filtered with a cutoff frequency of 500Hz. The peak of the impact force and its impulse were determined using the acceleration data and the mass value of the target: the acceleration became 0 at the end of the impact. The averaged data of the two trials with the largest impact forces for each glove for each boxer was used for later analysis. One-way repeated measures ANOVAs were used to investigate combined effects of the glove on the impact force using the gloves without added weights and to investigate effects of the mass of the glove on the force using the gloves with added weights. A post hoc test was performed using Holm's method if an ANOVA showed a significant effect. Comparisons between two same mass gloves (e.g. 12oz glove with no added weight and 10oz glove with 2oz lead weight) were also performed to test the effects of the cushioning on the impact force using Holm's method.

RESULTS: The duration of the impact force ranged from 18.5ms to 28.8ms. Table 1 shows the peak value of the impact force and its impulse for each glove. The effect of the gloves with no added weight on the peak impact force was significant ($p<0.01$) However, post hoc tests showed no significant differences between any pair of gloves. The effect of mass of the gloves on the impact force was not significant. The peak impact forces of the two same mass gloves were significantly different only between 16oz and 10+6oz ($p<0.05$).

Table1 The peaks and impulses of the impact forces.

		10oz	12oz	14oz	16oz	10+2oz	10+4oz	10+6oz
Peak of the impact force(N)	Ave.	2090	2140	1920	1900	2050	2100	2130
	SD	460	540	330	420	540	550	520
Impulse(N·s)	Ave.	18.6	18.7	19.0	19.1	21.6	21.9	21.8
	SD	2.1	1.8	1.2	1.9	4.2	1.3	1.6

DISCUSSION: The peak impact forces of the larger mass gloves with no added weight tended to be lower than the smaller ones. It is suggested that the peak impact forces of the gloves with no added weight depend on the cushioning rather than the mass of the gloves if the punching motions were the same irrespective of the masses of the gloves.

CONCLUSION: The difference in cushioning rather than mass of the glove would affect the magnitude of the peak impact force.

Acknowledgement: The authors thank Prof. Y. Yoshifuku who advised us about the experiment.

THE EFFECT OF VARYING CLUB HEAD MASS ON VELOCITY AND KINETIC ENERGY

Catherine B. Tucker, Ian C. Kenny, Derek J. Byrne and Ross Anderson

Biomechanics Research Unit, University of Limerick, Ireland

KEY WORDS: computer simulation, golf, swing velocity, timing.

INTRODUCTION: Typically, the standard club head mass of a driver is 0.2 kg approximately (White 2006). The golfing governing bodies do not stipulate driver club head mass. Theoretically, an increase in club head mass will lead to an increase in momentum transfer, but it is thought the increased mass leads to a reduction in club head velocity which is a more important determinant of how far the ball travels. The present study investigated the effect of increasing driver club head mass on club head velocity and total kinetic energy applied to the club at the grip by means of a computer simulation.

METHODS: Kinematic data for one subject (25 yrs, 91.3 kg, +2 handicap) was collected using a 5-camera set-up (Motion Analysis Corp). The subject performed eight shots with his own driver. A LifeMOD computer model was constructed with 42 degrees-of-freedom. Kinematic data collected using MAC was used to drive the model inverse dynamics and forward dynamics simulations. Validation for this model was carried out for club head velocity (r=0.999), kinematics (r=0.983) and kinetics (for more details see, Kenny et al., 2008).

RESULTS: Table 1 lists club head velocity and kinetic energy of the hand-club interface, at impact.

Table 1 Impact Velocity ms^{-1} and Total Kinetic Energy (J) Values at different club head masses

Mass (kg)	0.05	0.10	0.15	0.20	0.25	0.30	0.35	0.40	0.45	0.50	1.00
Velocity (m/s)	31.84	32.42	33.10	33.77	34.25	34.96	35.52	35.89	36.50	36.99	40.52
Kinetic Energy (J)	61.12	61.83	62.06	62.10	61.65	62.26	62.41	61.42	62.37	62.28	61.20

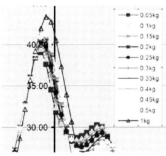

Figure 1: Club head velocities at impact – (y-axis indicates impact position)

DISCUSSION: The results show that for increased club head mass, there was an increase in club head velocity but no increase in kinetic energy applied by the hand on the club. Thus, these simulation results indicate that the increase in velocity was an equipment effect and not due to increased energy input from the subject.

CONCLUSION: This study identified that increasing club head mass can lead to increased club head velocity without increased energy input at the point where the hand grips the club. Further investigation is needed to ascertain if this holds true in "real" subjects.

REFERENCES:
White, R. (2006). On the efficiency of the golf swing. *American Journal of Physics*, 74, 1088-1094.
Kenny, I.C., McCloy, A.J., Wallace, E.S. and Otto, S.R. (2008). Segmental Sequencing of kinetic energy in a computer-simulated golf swing. *Sports Engineering*, 11, 37-45

Acknowledgement :The authors would like to acknowledge the Irish Research Council for Science Engineering and Technology for their support in this research.

RELIABILITY OF FORCES DURING VARIATIONS OF PLYOMETRIC EXERCISES

Sarah K. Leissring[1], William P. Ebben[2], Luke R. Garceau[2], Erich J. Petushek[2], Randall L. Jensen[1]

Dept. HPER, Northern Michigan University, Marquette, MI, USA[1]
Dept. Physical Therapy, Program in Exercise Science, Marquette University, Milwaukee, WI, USA[2]

Thirteen college students performed a drop jump from height equal to their peak vertical jump, single leg jumps from the left and right legs, and a counter movement jump. Vertical ground reaction force (GRF) obtained via an AMTI force plate and video analysis of markers placed on the hip, knee, lateral malleolus, and fifth metatarsal were used to estimate reaction forces on the knee joint. One-way Repeated Measures ANOVA indicated no differences for knee joint reaction forces relative to body weight or peak GRF for any of the jumps ($p > 0.05$). Average measures Intraclass Correlation Coefficients ranged from $r = 0.90$ to 0.97. Results indicate that peak GRF and knee joint reaction forces during the drop jump, counter movement jump, and single leg left and right leg jumps are reliable measures.

KEY WORDS: knee joint reaction forces, ground reaction forces, jumping

INTRODUCTION: Plyometric exercises have been demonstrated to improve athletic performance (Wilson et al., 1996) and enhance bone mass (Bauer et al., 2001). Ground reaction forces and knee joint reaction forces of plyometric exercises have been shown to vary depending on the type of jump performed or the height of a drop jump (Jensen & Ebben, 2002; 2007).

Evaluating the intensity of plyometric activity has been done using many variables (Bauer, et al., 2001; Ebben et al., 2008; Flanagan et al., 2008; Jensen & Ebben, 2002; 2007). However, the reliability of these variables while performing variations of plyometric exercises have not been extensively studied. Knowledge of a measure's reliability is important as it allows for comparisons across time as well as within various conditions. Without this knowledge it is difficult for researchers, practitioners, coaches, and/or athletes to know whether the measures of interest are changed via fatigue or training levels, or if they simply are not consistent. Therefore the purpose of the current study was to estimate the reliability of peak ground reaction forces and knee joint reaction forces while performing four variations of plyometric exercises.

METHODS: Thirteen active students (seven female and six male; mean ± SD; age = 24.3±4.3 years, height = 175.5±7.8 cm; body mass = 71.9±12.9 kg) volunteered to serve as subjects for the study. Subjects completed a Physical Activity Readiness-Questionnaire and signed an informed consent form prior to participating in the study. Approval for the use of human subjects was obtained from the institution prior to commencing the study. Subjects had performed no strength training in the 48 hours prior to data collection.

Warm-up prior to the plyometric exercises consisted of at least 3 minutes of low intensity work on a cycle ergometer. This was followed by dynamic stretching including one exercise for each major muscle group. Following the warm-up and stretching exercises, the subjects performed two trials of a maximal standing vertical jump (vertical jump = 46.2±8.4 cm) followed by at least 5 minutes rest prior to beginning the plyometric vertical jump tests. The order of plyometric exercises was randomly assigned and consisted of three trials each of drop jumps from a height equal to the subject's peak vertical jump (DJ), single leg jumps from the left and right legs (LLJ and RLJ respectively), and a counter movement jump (CMJ) with arm swing (Potach, 2004). For the drop jump subjects were instructed to drop directly down off the box and immediately perform a maximum vertical jump. For the other jumps they were asked to jump for maximal height. A one minute rest interval was maintained between each trial.

The plyometric exercises were performed by taking off from and landing on a force platform (OR6-5-2000, AMTI, Watertown, MA, USA). Ground Reaction Force (GRF) data were collected at 1000 Hz, real time displayed and saved with the use of computer software (Net Force 2.0, AMTI, Watertown, MA, USA) for later analysis. Peak GRF was the highest value attained during the movement and occurred during the landing.

Video of the exercises was obtained at 60 Hz from the sagittal view using 1 cm reflective markers placed on the greater trochanter, lateral knee joint line, lateral malleolus and the fifth metatarsal. Markers were digitized using automatic digitizing software (Motus 8.5 Peak Performance Technologies, Englewood, CO) and acceleration of the joint segment center of mass was determined after data were smoothed using a fourth order Butterworth filter (Winter, 1990).

To synchronize data a signal was used to initialize kinetic data collection which also inserted an audio tone in the video data. Data were then combined into a single file and splined to create a file of equal length at 1000Hz (see Figure 1). Knee joint reaction forces (KRF) were estimated according to Bauer *et al* (2001). Because GRF for the drop jump and counter movement jump would have been distributed across both feet (and therefore both knees) these GRF values were divided by two prior to calculation of KRF. Variables assessed were GRF, KRF/Body weight, and GRF/Body weight for the drop jump, counter movement jump, left leg jump, and right leg jump.

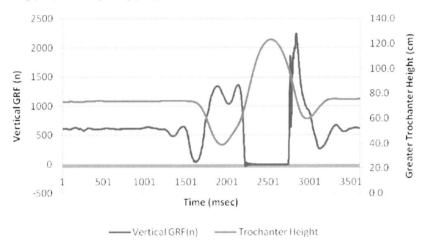

Figure 1. Illustration of vertical ground reaction forces and height of the greater trochanter marker relative to the force platform for the counter movement jump.

All statistical analyses of the data were carried out in SPSS © (Version 16.0). Trial-to-trial reliability analysis of recorded variables used both single (ICC_{single}) and average (ICC_{ave}) measures intra-class correlations. The ICC classifications of Fleiss (1986) (less than 0.4 was poor, between 0.4 and 0.75 was fair to good, and greater than 0.75 was excellent) were used to describe the range of ICC values A repeated measures ANOVA was used to determine possible differences between trials. The criterion for significance was set at an alpha level of $p < 0.05$

RESULTS: Table 1 displays the Mean ± SD of peak vertical ground reaction force, GRF/BW and knee joint reaction force relative to body weight variables measured during three trials of DJ, LLJ, RLJ, and CMJ. As shown there were no differences across the trials for any of the variables ($p > 0.05$).

The trial-to-trial reliability of all dependent variables measured during drop jump, left and right legged single leg jumps, and countermovement jump as depicted by the single (ICC_{single}) and average (ICC_{ave}) measures intra-class correlation coefficients is illustrated in Table 2. Values for ICC_{ave} ranged from r = 0.90 to 0.97, while ICC_{single} ranged from r = 0.76 to 0.91

Table 1. Peak ground reaction force (GRF), peak GRF relative to body weight (GRF/BW), and Knee joint reaction force relative to body weight (KRF/BW) (mean ± SD) across the three trials for the DJ, LLJ, RLJ, and CMJ (n=13).

	GRF (N)	GRF/BW (N·Kg^{-1})	KRF/BW (N·Kg^{-1})
DJ			
Trial 1	2710.7 ± 417.3	17.36 ± 2.70	4.67 ± 1.0
Trial 2	2734.2 ± 339.5	17.54 ± 2.30	4.81 ± 1.0
Trial 3	2650.3 ± 432.3	16.96 ± 2.70	4.93 ± 1.2
LLJ			
Trial 1	2134.0 ± 573.5	13.38 ± 2.20	5.86 ± 1.79
Trial 2	2240.2 ± 552.9	14.14 ± 2.41	5.70 ± 1.73
Trial 3	2256.4 ± 517.7	14.27 ± 2.26	5.70 ± 1.66
RLJ			
Trial 1	2097.3 ± 582.1	13.34 ± 3.35	7.30 ± 3.50
Trial 2	2255.8 ± 528.5	14.39 ± 3.09	7.21 ± 2.95
Trial 3	2139.0 ± 574.3	13.50 ± 2.69	6.90 ± 3.16
CMJ			
Trial 1	2500.9 ± 468.7	15.95 ± 2.09	3.52 ± 0.82
Trial 2	2502.8 ± 506.9	15.84 ± 1.95	3.64 ± 0.96
Trial 3	2432.3 ± 491.2	15.51 ± 2.45	3.57 ± 1.22

All trials within a specific jump were not different (p > 0.05)

Table 2. Intraclass Correlation Coefficient and 95% Confidence Interval (ICC:95% CI) for Peak ground reaction force (GRF), peak GRF relative to body weight (GRF/BW), and Knee joint reaction force relative to body weight (KRF/BW) across the three trials for the DJ, LLJ, RLJ, and CMJ (n=13).

	Average Measures (ICC:95% CI)				Single Measure (ICC:95% CI)			
	DJ	LLJ	RLJ	CMJ	DJ	LLJ	RLJ	CMJ
GRF	.94	.97	.97	.96	.84	.91	.91	.89
(N)	.85-.98	.92-.99	.92-.99	.91-.99	.66-.94	.80-.97	.78-.97	.76-.96
GRF/BW	.92	.91	.94	.91	.80	.78	.85	.77
(N·Kg^{-1})	.81-.97	.78-.97	.86-.98	.77-.97	.58-.93	.55-.92	.67-.95	.53-.91
KRF/BW	.90	.97	.95	.96	.76	.91	.87	.88
(N·Kg^{-1})	.76-.97	.92-.99	.88-.98	.89-.99	.52-.91	.78-.97	.71-.95	.73-.96

DISCUSSION: The major finding of the current study indicates that repeated measures of peak vertical GRF, peak GRF relative to body weight and KRF relative to body weight plyometric exercises can be reliably repeated on the same day for the exercises studied. The lack of difference across three trials and values for ICC_{ave} greater than r = 0.90 indicated a high level of reliability.

The findings for peak GRF are consistent with those of Stålbom et al. (2007) who found high reliability of forces in a single leg drop jump followed by a horizontal jump (ICC_{ave} r = 0.84). Though values for peak GRF (1880 ± 247 N) were less than those of the current study (2186 ± 554 N), subjects in the current study dropped from a mean height of 46 cm, while those of Stålbom and coworkers (2007) dropped from 20 cm. Similarly, Ford et al. (2007) found two legged drop jumps from the top of a 31 cm box were also highly reliable for kinematic variables, displaying values of ICC_{ave} r > 0.90. In addition, Flanagan and colleagues (2008) also found high reliability for the reactive strength index (height jumped/time spent on the ground between the landing and takeoff of a drop jump) when performing drop jumps (ICC_{ave} r >0.95).

CONCLUSION: The reliability of the current measures indicated that repeated measures of plyometric exercises can be reliably performed. Previous research has shown differences between the types of plyometric exercises (or jumps) for the studied variables (Jensen & Ebben 2002; 2007). Thus if researchers wish to compare plyometric exercises, the high reliability found in the current study would suggest that differences found between exercises would be due to the exercises themselves and not differences across the trials. As a result practitioners, coaches, and/or athletes may find these performance measures useful in assessing fatigue or training levels when performing drop, countermovement and single leg jumps.

REFERENCES

Bauer, J.J., Fuchs, R.K., Smith, G.A., and Snow, C.M. (2001) Quantifying force magnitude and loading rate from drop landings that induce osteogenesis. *Journal of Applied Biomechanics 17*, 142-152.

Ebben, W.P., Simenz, C., and Jensen, R.L. (2008) Evaluation of plyometric intensity using electromyography. *Journal Strength and Conditioning Research 22*, 861–868

Flanagan, E.P., Ebben, W.P., and Jensen, R.L. (2008) Reliability of the reactive strength index and time to stabilization during plyometric depth jumps. *Journal Strength and Conditioning Research 22*, 1677-1682.

Fleiss, J.L. (1986) *The Design and Analysis of Clinical Experiments*. New York, NY: Wiley.

Ford, K. R., Myer, G.D., and Hewett, T.E. (2007) Reliability of landing 3D motion analysis: Implications for longitudinal analyses. *Medicine and Science in Sports and Exercise 39*, 2021-2028.

Jensen, R.L. and Ebben, W.P. (2002) Effects of plyometric variations on jumping impulse. *Medicine and Science in Sports and Exercise 34*, S84.

Jensen, R.L. and Ebben, W.P (2007) Quantifying plyometric intensity via rate of force development, knee joint and ground reaction forces. *Journal Strength and Conditioning Research 21*, 763-767.

Potach D.H. (2004) Plyometric and Speed Training. In: Earle, R.W. and Baechle, T.R. (Eds) *NSCA's Essentials of Personal Training* Human Kinetics, Champaign, IL. 425-458.

Stålbom, M., Holm, D.J., Cronin, J.B., and Keogh, J.W.L. (2007) Reliability of kinematics and kinetics associated with horizontal single leg drop jump assessment. A brief report. *Journal of Sports Science and Medicine 6*, 261-264.

Wilson, GJ, Murphy, AJ, and Giorgi, A. (1996) Weight and plyometric training: effects on eccentric and concentric force production. *Canadian Journal of Applied Physiology 21*, 301–315.

Winter, D.A. (1990) *Biomechanics and motor control of human movement* (2nd Ed). New York: Wiley Interscience.

Acknowledgement: Funded by Northern Michigan University Freshman Fellows and College of Professional Studies grants.

AN INVESTIGATION INTO THE EFFECTS OF A SIMULATED EFFUSION IN HEALTHY SUBJECTS ON KNEE KINEMATICS AND LOWER LIMB MUSCLE ACTIVITY DURING A SINGLE LEG DROP LANDING

Garrett Coughlan[1], Rod Mc Loughlin[2], Ulrik McCarthy Persson[1], Brian Caulfield[1]

School of Physiotherapy and Performance Science, UCD, Dublin, Ireland [1]
O'Neill's Sports Injury Clinic, Sports Centre, UCD, Dublin, Ireland [2]

Arthrogenic muscle inhibition (AMI) is defined as an ongoing reflex inhibition of the musculature surrounding a joint following distension or damage to the structures of that joint [Hopkins and Ingersoll, 2000]. AMI following joint injury may affect movement and muscle recruitment which may impair rehabilitation and delay the return to activity. Knee angular displacement and velocity as well as lower limb EMG were measured in the period 250 milliseconds pre initial contact to 250 milliseconds post initial contact during a single leg drop jump in 8 healthy subjects before and after a simulated knee joint effusion of 60 millilitres. Repeated measures ANOVA and post hoc testing revealed no statistically significant differences in pre and post effusion in knee kinematic or lower limb EMG measures undertaken. A simulated knee effusion did not result in significant alterations to knee joint mechanics or lower limb muscle activation patterns during a single leg drop landing. The mechanism by which an effusion affects motor control during functional and dynamic weight bearing tasks warrants further investigation.

KEY WORDS: drop landing, knee, effusion, kinematics, EMG

INTRODUCTION: Arthrogenic muscle inhibition (AMI) is defined as an ongoing reflex inhibition of the musculature surrounding a joint following distension or damage to the structures of that joint *(Hopkins and Ingersoll. 2000)*. It is the natural response of the body to injury which may result in inadequate neuromuscular control in functional activities and a delay in the return to activity. Long term effects of inactivity following injury, potentially caused by AMI, can adversely affect muscles, bones, ligaments, and neural activity. Inhibition of the quadriceps is most likely the cause of strength loss, atrophy, and deficits in neuromuscular control after knee injury *(Hopkins et al. 2001)*. The presence of an effusion can lead to persistent quadriceps muscle weakness, resulting in knee instability as the capacity of the muscle group to respond to external loads generated by functional activity is compromised. Insufficient control and strength at the knee as a result of AMI can lead predispose an individual to reinjury as well as to the development of chronic degenerative conditions. An individuals post injury or pathological state cannot establish whether a change in movement or muscle recruitment patterns is a direct result of knee pathology or contributed to its etiology. The effusion model has the advantage of nullifying other factors of injury, pain and inflammation which are difficult to quantify. Recent investigations have assessed the effects of a simulated effusion on a range of measurements at the knee joint including muscle strength, postural control, proprioception and quadriceps H-reflex. However few studies have quantified the impact of this type of effusion on high speed dynamic weight bearing activity. Therefore the aim of this research was to quantify the effects of a simulated effusion on knee movement patterns and lower limb muscle activity in the period pre and post IC during a single leg drop landing in a sole testing session. An understanding of the influence that an effusion may have on this type of dynamic function could assist therapists in understanding the myriad of abnormalities associated with neuromuscular control due to injury and in the rehabilitation from lower limb injuries.

METHODS: Data Collection: Eight physically active subjects who gave informed consent participated in this study (24.6 years ± 4.3, 174.1 cm ± 0.1, 70.6 kg ± 12.5). Subjects stood on a 35 centimetre high platform adjacent to a rectangular force plate embedded in the laboratory floor with the test leg relaxed and non-weight bearing. The subject then used the contralateral leg to propel him/herself from the platform and stick the landing when impacting on the force plate. The motion analysis system (CODA, Charnwood Dynamics Ltd, Leicestershire, UK), force plate and EMG were manually triggered to simultaneously begin recording as the subject was given a verbal command to land on the platform. Surface EMG activity from the vastus medialis (VM), vastus lateralis (VL), biceps femoris (BF) and soleus (SOL) muscles were recorded on all subjects during the drop landing. Electrodes were placed on specific sites in accordance with the SENIAM research groups recommendations. Data was recorded on a Biopac MP100A (Biopac Systems Inc. Santa Barbara, CA, USA). In order to address potential measurement error in data recording, data was recorded in three measurement intervals throughout the testing session, twice prior to the effusion, Control 1 (C1) and Control 2 (C2), and once following the effusion, Post Effusion (PE). Following the control testing sessions, 2 ml of 2% Lidocaine was injected subcutaneously lateral to the knee joint line for anaesthetic purposes and 60 ml of saline solution (0.9% w/v Sodium Chloride Intravenous Infusion) was subsequently injected into the knee joint capsule. A ballotable patella test and an effusion wave test were performed to ensure that the effusion was within the knee joint.

Data Analysis: Sagittal and coronal knee angular displacement and velocity at initial contact (IC), and peak/trough values during the 250 ms period prior to and following IC were identified was identified for five drop landing trials for each subject. IC during the single leg drop jump was identified using the vertical component of the ground reaction force (GRF) using 15 Newtons (N) as a threshold for detection of impact. Forceplate data was collected at 200 Hz sampling rate. EMG signals were amplified (gain 300] and sampled at a rate of 1000 Hz. They were subsequently band pass filtered (Blackman 61 dB) at 20 Hz (low) and 500 Hz (high). The data was then full wave rectified and average over a 15 ms moving window. The selected time periods were the period 250 ms pre to cover the initial EMG pre activation prior to ground contact and 250 ms post ground contact as this corresponded with the initial weight acceptance phase. Statistical analysis was carried out using SPSS for Windows (Version 12.0.1; SPSS Inc, Chicago, IL, USA). We used a general linear model three factor repeated measures analysis of variance to analyse differences in kinematic/EMG variables at each of the test intervals. In each case the dependent variable was the kinematic/EMG variable in question and the independent variables were test interval (C1, C2 and PE). Post hoc paired t-tests were then carried out to test for differences in variables between individual pairs of test intervals (C1vC2, C2vPE, C1vPE). The alpha level was set at 0.05. Due to the potential for multiple comparison errors in the analysis, we used a Bonferroni adjustment to re-calculate the P value for the repeated measures and post hoc t-tests with adjusted critical P values of 0.001 and 0.002 respectively at a 95% confidence level for kinematics. Similarily, Bonferroni adjustments were also used for EMG calculations with adjusted critical P values of 0.002 and 0.003 at a 95% confidence level for the repeated measures and post hoc tests respectively.

RESULTS: Repeated measures ANOVA and post hoc testing revealed no statistically significant differences in pre and post effusion in knee kinematic (Fig 1) or lower limb EMG (Table 1) measures undertaken in 8 healthy subjects during a single leg drop landing task.

Table 1: EMG activity during the 250ms pre IC and 250ms post IC drop landing

Muscle	Variable	Measurement Interval			Repeated Measures ANOVA Level of Significance
		C1	C2	PE	
Vastus Medialis	250ms to IC	23.83 (6.93)	20.25 (4.60)	17.92 (6.57)	0.04
	IC to 250ms	60.24 (10.25)	58.40 (11.00)	49.31 (19.30)	0.09
Vastus Lateralis	250ms to IC	22.98 (4.43)	19.17 (5.28)	18.06 (6.65)	0.05
	IC to 250ms	60.91 (10.02)	59.37 (7.33)	49.95 (13.60)	0.15
Biceps Femoris	250ms to IC	27.20 (12.26)	26.24 (16.8)	19.13 (17.70)	0.07
	IC to 250ms	44.58 (16.21)	36.82 (12.96)	26.85 (10.43)	0.08
Soleus	250ms to IC	23.77 (7.04)	19.76 (5.79)	20.58 (8.15)	0.25
	IC to 250ms	40.96 (8.77)	39.23 (8.28)	38.40 (9.90)	0.63

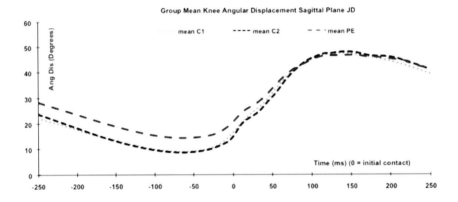

Fig 1: Knee angular displacement sagittal plane during the 250ms pre IC and 250ms post IC

DISCUSSION: We hypothesised that there would be a significant decrease in quadriceps muscle activity based on previous research (Torry et al. 2000 and 2005, Palmieri Smith et al. 2007). These muscles have a crucial role in stabilising the knee joint when landing from a jump as their eccentric contraction assists in dissipating large forces generated during this task. A decrease in their recruitment capability could reduce stability of the knee joint during a dynamic task which may predispose a patient to initial or repeated injury. The principal finding in this study was that a 60 ml simulated effusion of the knee joint does not result in alterations to knee kinematics or lower limb muscle activity in healthy subjects during single leg drop landing. The only other similar study in this area was conducted by Palmieri-Smith et al (2007) which used two levels of effusion (30 ml and 60 ml) and measured sagittal plane kinematics and lower limb EMG in the period 250 ms post IC. The results of that study demonstrated a decrease in knee flexion position at landing using an effusion of 60 ml, as well as a reduction in vastus medialis and vastus lateralis muscle activity between the high and low level effusions. The authors concluded that a reduction in muscle activity altered landing mechanics and therefore larger forces to be transferred through the knee as evidenced by an increased ground reaction force. The present study did not observe similar sagittal plane kinematics or EMG results. The lack of correlation between the findings of these two studies may be due to methodological differences, for example the methodology of the EMG analysis. EMG is normalised in studies to allow comparisons between conditions and/or subjects. The *Palmieri-Smith et al (2007)* study normalised their EMG by comparing their recording to that from a maximum voluntary isometric contraction [MVIC] and expressing the EMG as a percentage of maximal contraction. We choose to normalise our EMG data to mean or peak of an ensemble average as *Yang and Winter (1984)* have reported that normalising EMG with respect to MVIC is less reliable than normalising relative to a sub maximal contraction as well as been reported as have large inter subject variability. The disparity between the two studies may also have been due to the measurement protocol used in our investigation. Our data collection was conducted in a single testing session whereas the *Palmieri-Smith et al (2007)* study was conducted at four separate testing sessions on different days. In a reliability study, *Monaghan et al (2006)* observed limits of agreement of over eight degrees in sagittal plane gait variables during gait between testing sessions on different days. These limits may increase in a task such as drop landing as the patterns of movement are not as robust as that of gait. Coupled with this, in order to address the issue of test-retest variance that is inherent in kinematic studies we examined measures in three separate measurement intervals. We could have arrived at a different set of conclusions to those presented here had we employed a straightforward test-effusion-retest model and if we had not applied the Bonferroni adjusted p-values to our analysis to account for multiple comparisons. In particular a number of the EMG findings would have either bordered on or displayed statistical significance.

CONCLUSION: Despite these levels of effusion having an effect on patients in a clinical setting, they may not be sufficient to elicit major changes in movement patterns and muscle function during high velocity cyclical tasks in healthy subjects following a simulated effusion. Allied with this, the presence of a long term effusion with associated inflammation and pain in acute or chronically injured patients with adaptations to learned movement and muscle recruitment patterns may be responsible for the loss of proprioception and muscle inhibition as observed in clinical situations.

REFERENCES

Hopkins, J.T. & Ingersoll, C.D. (2000). Arthrogenic muscle inhibition: a limiting factor in joint rehabilitation. *Journal of Sport Rehabilitation.* 9, 135-59.
Hopkins, J.T., Ingersoll, C.D., Krause, A., Edwards, J.E. & Cordova, M.L. (2001) Effect of knee joint effusion on quadriceps and soleus motoneuron pool excitability. *Medicine in Science. Sports.and Exercise.* 33, 123-6.
Palmieri-Smith, R.M., Kreinbrink, J., Ashton-Miller, J.A. & Edward, M.W. (2007) Quadriceps inhibition induced by an experimental knee joint effusion affects knee joint mechanics during a single-legged droop landing. *American Journal of. Sports Medicine.*; 35, 1269-75.
Torry, M.R., Decker, M.J., Millett, P.J., Steadman, J.R. & Sterett, W.I. (2005). The effects of knee joint effusion on quadriceps electromyography during jogging. *Journal of. Sports Science and Medicine.* 4, 1-8.
Torry, M.R., Decker, M.J., Viola, R.M., O'Connor, D.D. & Steadman, J.R. (2000) Intra-articular knee effusion induces quadriceps avoidance gait patterns. *Clinical. Biomechanics.* 15, 147-59.
Yang J.F. & Winter D.A. (1984). Electromyographic amplitude normalisation methods: improving their sensitivity as a diagnostic tool for gait analysis. *Archives of Physical Medicine and Rehabilitation.* 65, 517-21.

Acknowledgement
Research funded by UCD Seed Funding Scholarships 2007

VARIABILITY OF STRIDE FREQUENCY AND PRONATION VELOCITY DURING A 16 DAY RELAY-RUN AROUND GERMANY – A CASE STUDY

Nina Gras, Thorsten Sterzing, Torsten Brauner, Doris Oriwol, Jens Heidenfelder, Thomas L. Milani

Department of Human Locomotion, Chemnitz University of Technology, Chemnitz, Germany

This case study analyzed stride frequency, represented by time of gait cycle (TGC), and maximum pronation velocity (MPV) for one subject running 350 km over 16 consecutive days. Data collection took place during a day-and-night team relay-race around Germany in 2008. TGC and MPV measurements were performed by a gyrometer incorporated in the subject's running shoe and recorded by a portable data logger. For data analysis TGC and MPV velocity were determined in 25 runs for altogether 112,532 steps of the right foot. Means and standard deviations of both parameters for complete runs and for all consecutive five-minute segments within each run were calculated. Results showed an increase of TGC and a decrease of MPV within single runs. Between run comparisons across all 25 runs showed no systematic change in TGC and MPV during the relay-race. Interestingly, during the unfamiliar night-runs TGC and MPV were increased compared to day-runs, potentially caused by altered biorhythm or limited vision at night.

KEY WORDS: running, pronation velocity, fatigue, mobile measurements

INTRODUCTION: Biomechanical research on distance running primarily focuses on injury prevention aspects, addressing those variables that characterize impact and rearfoot motion. Excessive pronation and pronation velocity are considered risk factors for achillodynia and knee injuries (Brüggemann et al., 2007; Grau & Horstmann, 2007). Thus, rearfoot stability characteristics are deemed an important aspect of injury prevention and rehabilitation.

Study designs addressing running style modification due to prolonged runs and fatigue include intervention studies with measurements right before and after a running intervention (Butler, Hamill & Davis, 2007) but also continuous measurement protocols that allow data collection at different times during the actual run (Brüggemann et al., 1995; Sterzing & Hennig, 1999; Derrick et al., 2002). The latter represent a more realistic running scenario. However, results of these studies are not uniform. Some studies report an increase in pronation and pronation velocity values (Derrick et al., 2002) while others do not observe this behaviour (Sterzing & Hennig, 1999; Butler, Hamill & Davis, 2007).

In addition to biomechanical variables, extreme mileages, and sudden increase of mileage are considered risk factors for running injuries. O'Toole (1992) reported that a relatively high increase in running mileage over a short period of time puts runners at risk for getting injured. Referring to these considerations, Heidenfelder et al. (2009) examined the running style of 11 runners one day before and one day after a 16-day team relay-race around Germany, which required each athlete to run approximately 300 to 350 kilometres. No systematic changes in any of the biomechanical impact or rearfoot motion variables were found for the group of runners or for individual athletes. It was concluded that the fundamental running style of runners is strongly determined and thus not subject to change fundamentally even due to extreme mileage interventions.

All of the referenced studies have in common that they may have induced artefacts not occurring during realistic outdoor running (laboratory, treadmill, instrumentation backpacks). Recently, there have been innovative efforts to implement mobile measuring devices to this research area. A miniature gyrometer attached to the shoe was shown to provide similar pronation velocity information as the established motion analysis and electrogoniometer approaches with a correlation coefficient for pronation velocity determination of 0.72 compared to a Vicon motion analysis system (Brauner, Sterzing & Milani, 2009a). These findings allow the study of pronation velocity in realistic running situations by an instrumented shoe not interfering at all with the runner. The measurement technique was shown to allow reliable recognition of gait cycle characteristics and determination of maximum pronation velocity within and across runners (Brauner et al., 2009b).

The purpose of this study was to examine time of gait cycle (TGC) and maximum pronation velocity (MPV) and corresponding variability in a one subject case study during a 16 day-and-night team relay-race around Germany in 2008 (www.lauf-kultour.de). Furthermore, as one half of the runs took place during day and the other half during night we looked at TGC and MPV values at full daylight compared to values at night during reduced visibility.

METHODS: One injury free male subject (27 yrs, 83 kg, 182 cm) participated in this study measured during a 16 day-and-night team relay-race. He was required to perform two runs per day (4am and 4pm) adding up to a total of 30 runs.

Running times were recorded and distances were determined using a high-sensitivity GPS receiver (*GARMIN, eTrex® H, Garmin International Inc., Olathe, KS, USA*). Mean duration of the runs was 59.65 ± 26.4 min with an average distance of 11.9 ± 4.3 km resulting in a mean running velocity of 3.4 ± 0.3 m s^{-1}.

During each of these runs the subject wore an instrumented running shoe (*Saucony Progrid Ride M 28025-3*) with an integrated gyrometer (*Murata ENC-03R, Murata Manufacturing Company, Ltd., Japan*) on his right foot. This instrumentation was developed for mobile pronation velocity measurements and its use showed high correlation to frontal plane kinematics measured with a Vicon 3D motion analysis system (Brauner et al., 2009b).In their laboratory study the authors were also able to demonstrate that the gyrometer could be used without any calibration process for pronation velocity measurements. Data was recorded with a data logger at 1000 Hz (*eLAS.net MultiLOG2, MSR-Electronics, Switzerland*). A digital high pass filter at 0.1 Hz (1st order Butterworth) eliminated the sensor inherent drift.

Figure 1: Subject/ measurement-setup: Instrumented shoe, integrated miniature gyrometer and data logger at a waist belt

Data of five runs of the first three days had to be eliminated due to a broken sensor, leaving the data of altogether 25 consecutive runs for gait cycle detection and MPV determination according to the previously developed algorithm (Brauner et al., 2009b). A total of 112,532 steps of the right foot (4,501 ± 1,465 per run) were identified, and each corresponding MPV was determined and used for further analysis. Furthermore, TGC was determined and used to interpret dependent stride frequency. Data analysis was divided in between-run and within-run analyses.

Within-runs analysis

For Within-runs analysis only the first 50 minutes of each of the 25 runs were analysed. These 50 minutes were divided into ten five-minute intervals. For each five-minute interval, mean and standard deviation of TGC and MPV were calculated. Furthermore, corresponding intervals were averaged over all 25 runs to investigate the effect of fatigue on TGC and MPV within runs (e.g. all 25 first five-minute intervals were averaged to represent the behaviour of the parameter in the first five minutes of a run, then all 25 second five-minute intervals and so on). Consequently, standard deviations were also calculated to represent variability for each calculated mean.

Between-runs analysis

For each of the 25 analysed runs means and standard deviations of TGC and MPV were calculated for all steps. This allowed analyzing TGC and MPV variability of single runs and comparing runs within the same day as well as between days.

For statistical analyses only descriptive methods were applied because this case study data shows a descriptive nature. The results can therefore not be used for generalization but they can provide a first impression into pronation behaviour of a single runner during prolonged field testing.

RESULTS AND DISCUSSION: Within-run analyses unveil a continuous increase of TGC and reduction of MPV over the time of each run (Figure 2). The reduction of MPV during the course of a run supports the findings of Sterzing & Hennig (1999) and Butler, Hamill & Davis (2007) but is in direct contrast to Derrick et al. (2002), who state increased pronation velocity during prolonged runs.

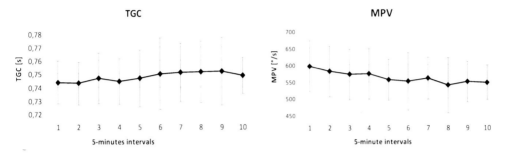

Figure 2: Within-run analysis: Averaged values over 25 runs for each five-minute interval

A fatigue effect over the complete period of the 25 runs was not observed, since TGC and MPV showed no systematic trend over all runs (Figure 3). Also, standard deviations of both variables were lower in single runs ($SD_{TGC}=0.046$, $SD_{MPV}=118.1$) than between runs ($SD_{TGC}=0.016$, $SD_{MPV}=58.9$), consolidating the observed effects. Astonishingly, variability of TGC was highly increased towards the end of the running event, as standard deviation of the last seven runs showed 2.9 times higher values compared to the first 18 runs.

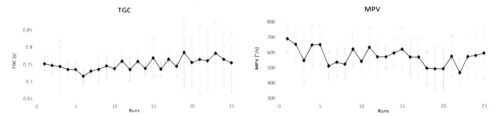

Figure 3: Between-run analysis: Mean values for each run

Interestingly, systematic differences occurred between day-runs and night-runs for both variables. Ten of eleven day-night comparisons (runs within 24 hours) unveiled higher MPV averages during night-runs in contrast to day-runs that took place within 24 hours (Figure 4).

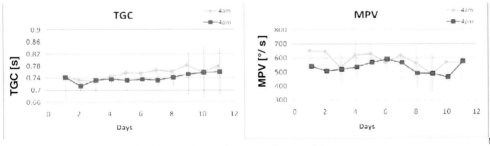

Figure 4: Day-night analysis: Mean values of runs at 4am and 4pm

subject reduced his stride frequency while increasing maximum pronation velocity during night runs. Night runs took place at unfamiliar biorhythm, limited vision and also during lower temperature compared to day runs, which are potential explanations for the observed effects.

CONCLUSION: Analysis of time of gait cycle and maximum pronation velocity allow deeper insight in athletes running performance characteristics. The methods used in our study allowed to observe these parameters during prolonged outdoor running. The presented decrease of maximum pronation velocity values within single runs for our subject contradicts the findings in the literature. However, this one subject observation needs to be interpreted with highest caution as this case study has no power to state general group effects of prolonged running. It rather marks the basis of a future set of studies aiming to solve the questions that arose during this study.

In future studies, with more subjects, the effects of prolonged running on time of gait cycle and maximum pronation velocity and their corresponding variability needs to be investigated. The proposed reasons (unfamiliar biorhythm, reduced vision, and lower temperature) for day and night run differences of the investigated variables need to be verified. This calls for systematically arranged follow-up studies approaching these issues.

The further enhancement of mobile measurement technology should provide better suited tools in order to deal with aspects aiming to link running characteristics to running injuries, a research area that is highly interesting but has not been able to come up with clear answers so far.

REFERENCES:

Brauner, T., Sterzing, T. & Milani, T.L. (2009a). Ankle frontal plane kinematics determined by goniometer, gyrometer and motion analysis system: A measurement device validation, *22. Congress International Society of Biomechanics*, Cape Town, South Africa.

Brauner, T., Sterzing, T., Oriwol, D. & Milani, T.L. (2009b). A single gyrometer inside an instrumented running shoe allows mobile determination of gait cycle and pronation velocity during outdoor running, *9. Footwear Biomechanics Symposium*, Stellenbosch, South Africa.

Brüggemann, G.P., Arndt, T., Kersting, U.G. & Knicker, A. J. (1995). Influence of fatigue on impact force and rearfoot motion during running, *15. Congress International Society of Biomechanics*, Jyväskylä, Finland, 132–133.

Brüggemann, G.P., Potthast, W., Lersch, C. & Segesser, B. (2007). Achilles tendon strain distribution is related to foot and shank kinematics and muscle forces, *Journal of Biomechanics*, 40(2), 139.

Butler R.J., Hamill, J. & Davis, I. (2007). Effect of footwear on high and low arched runners' mechanics during a prolonged run, *Gait & Posture*, 2, 219–225.

Derrick T.R., Derreu, D. & McLean, S.P. (2002). Impacts and kinematic adjustments during an exhaustive run, *Medicine and Science in Sports and Exercise*, 6, 1998–1002.

Grau, S. & Horstmann, T. (2007). Entwicklung eines Stabilitätslaufschuhs zur Prävention von Achillessehnenbeschwerden – Nike Air Cesium, *Sport-Orthopädie/Sport-Traumatologie – Sports Orthopedics/Traumatology*, 23(3), 179–184.

Heidenfelder, J., Brauner, T., Gras, N., Sterzing, T. & Milani, T.L. (2009). A 16-day running intervention did not influence biomechanical running variables, *22. Congress International Society of Biomechanics*, Cape Town, South Africa.

O'Toole, M.L. (1992). Prevention and treatment of injuries to runners, *Medicine and Science in Sports and Exercise*, 24(9) Suppl., 360-363.

Sterzing, T.F. & Hennig, E.M. (1999). Measurement of plantar pressures, rearfoot motion, and tibial shock during running 10 km on a 400 m track, In: E.M. Hennig and D.J. Stefanyshyn (Eds.) *Proceedings 4. Symposium on Footwear Biomechanics*, Canmore, Alberta, Canada, 88-89.

ACKNOWLEDGEMENT: We want to especially thank our subject Jörg Kersten for taking part in this research. This study was supported by Puma Inc., Germany.

EFFECT OF BILATERAL OR SINGLE LEG LANDING ON KNEE KINEMATICS

S. Abe, M.K.D. Lewis, K. Malliah, P.L. Malin, and R.L. Jensen

Department of Health Physical Education and Recreation, Northern Michigan University, Marquette, Michigan, USA

KEY WORDS: range of motion, knee angular acceleration, knee angles, gender differences

INTRODUCTION: Non-contact anterior cruciate ligament (ACL) injury has been shown to be more common in women than men, occurring at a rate of 6-8 times that of men (Hughes et al. (2008). Causes for this discrepancy are unknown, but differences in knee landing angle have been suggested as a possibility. Kinematic variables have been shown to be a major predictor in peak anterior tibial shear force (Schultz et al., 2009). Hughes et al. (2008) found that women display greater valgus angles and range of motion than men in a two legged landing. While Elvin et al. (2007) reported that knee contact angle was correlated to ground reaction forces and segment axial accelerations. Thus while it appears that knee kinematics are an important factor determining the stress on the knee in bilateral drop landings, in many cases, individuals cannot land on both legs and instead are forced to landing on a single leg. However, there is information lacking comparing knee joint angle and acceleration when landing on one versus two legs and these variables affect soft tissue forces on the knee. The purpose of this study was to compare knee joint range of motion and angular acceleration of the right and left leg during bilateral and single leg landings for males and females.

METHODS: Five male and six female recreationally active college students were recruited for this study. Subjects were excluded if they reported previous history of lower extremity injury or less than 60 minutes of physical activity per week. The study was approved by the Institutional Review Board and informed consent forms were signed by each subject.

Video of the landings was obtained at 60 Hz from six cameras using 1 cm reflective markers placed bilaterally on the iliac crest, greater trochanter, lateral knee joint line, lateral malleolus, second metatarsal and posterior heel. Markers were digitized using Motus 8.5 (Peak Performance Technologies, Englewood, CO) and knee joint angle and acceleration data were determined after smoothing with a fourth order Butterworth filter (Winter, 1990).

Subjects performed three drop landing trials each onto the right, left and both legs (bilateral) from a height that maintained their feet 33 cm above the landing surface. When instructed, they released their hands from a horizontal bar and dropped to the landing surface. Order of landing condition (right, left, and bilateral) was randomly assigned for each subject.

The dependent variables were knee angular acceleration, peak knee angle, and relative knee angle of the right and left legs. Means and standard deviations were calculated for the dependent variables. A 4×2 (legs × gender) mixed design ANOVA was used to evaluate the dependent variables, where the individual leg was the repeated measure.

RESULTS & DISCUSSION: Preliminary findings indicate that knee joint angles differ when landing with one leg compared to two legs, but angular acceleration of the knee does not differ across the landings. There was no difference between genders. Further study should focus on whether different knee angles alter forces on the knee during two and one leg landings and if these strategies can be learned.

REFERENCES:

Elvin, N.G., Elvin, A.A., Arnoczky, S.P., and Torry, M.R. (2007) The correlation of segment accelerations and impact forces with knee angle in jump landing. *J Appl Biomech 23:* 203-212.

Hughes, G., Watkins, J., Owen, N. (2008) Gender differences in lower limb frontal plane kinematics during landing. *Sports Biomech 7:* 333-341.

Schultz, S.J., Nguyen, A-D., Leonard, M.D., and Schmitz, R.J. (2009) Thigh strength and activation as predictors of knee biomechanics during a drop jump task. *Med Sci Sports Exerc 41:* 857-866.

LANDING STRATEGY MODULATION IN BACKWARD ROTATING PIKED AND TUCKED SOMERSAULT DISMOUNTS FROM BEAM

Marianne Gittoes[1], Gareth Irwin[1], David Mullineaux[2] and David Kerwin[1]

Cardiff School of Sport, University of Wales Institute, Cardiff, United Kingdom[1]
College of Education, University of Kentucky, USA[2]

The aim of this study was to develop understanding of the landing strategy modifications made when performing backward rotating piked (BP) and tucked (BT) dismounts from beam. Sagittal plane lower-body joint angular kinematic profiles were determined for four female gymnasts during the landing phase of BP and BT somersaulting dismounts. A common hip-biased landing strategy was employed by the four gymnasts in the dismounting skills. The more complex BP task was distinguished by the use of a more extended (3.7°) and flexed (5.0°) initial knee and hip joint configuration, respectively compared to the basic BT skill performed. Effective skill developments of backward rotating dismounts from beam may require modulation to the individual joint patterns defining the lower-body landing strategy.

KEY WORDS: strategy bias, lower-body control, kinematics, gymnastics

INTRODUCTION: The dismount is a critical element of a gymnastic routine. Complex dismounts evolve from gymnasts acquiring the more basic tucked version prior to the integration of more complex versions, which are distinguished by diverse spatial orientation objectives. In dismounts from the beam apparatus, an aerial phase comprising forward or backward rotating somersaults with or without twists about the longitudinal axis may be performed. Although, the backward tucked somersault may be considered a basic skill for competitive gymnasts to perform, the spatial and temporal constraints of the skill are complex (Davlin et al., 2001).

The entire dismount can be defined by separate aerial and landing phases, where the onset of each is established by the loss of contact with the apparatus and first contact with the landing surface, respectively (Gervais & Dunn, 2003). The mastering of a pre-programmed movement pattern in the landing phase of the 'simplistic' backward tucked (BT) dismount has been considered important in minimising point deductions and maximising safety in landing (Gervais & Dunn, 2003). Contacting the ground with an appropriate kinematic pattern i.e. body position and angular velocity, ensures a gymnast can achieve landing balance during dismounting (Sheets & Hubbard, 2007). While studies of simple drop landings have reported relatively invariant kinematic control strategies (McNitt-Gray, 1991), more complex gymnastic landings distinguished by diverse initial momentum conditions have been suggested to require a control strategy that uses a hierarchical relationship between more than one criteria (McNitt-Gray et al., 2001). The execution of a modified landing strategy capable of accommodating the requirements of more complex dismount skills from beam e.g. the backward piked (BP) version may subsequently be necessary to ensure skill-specific mastery and the achievement of a successful and 'safe' landing.

A mechanical understanding of control strategies used in basic and more complex skills has been advocated to be valuable in providing a mechanism for effective skill development (Irwin & Kerwin, 2007), enhancing performance, and minimising potential injury risk in landing. The aim of this study was to develop understanding of the kinematic strategy used in the landing phase of backward rotating dismounts from beam. The diverse spatial orientation objectives of the BP and BT skills were hypothesised to impose modulations in the lower-body joint kinematic strategies used in the landing phase.

METHODS: Four national level female gymnasts (mean ± SD height: 1.64 ±0.08 m, body mass: 59.0 ±6.9 kg) provided written informed consent to participate in the data collection, for which the protocol had been approved by the University's Research Ethics Committee. Each gymnast performed 10 successful backward single somersault dismounts from beam in a piked and tucked position (N = 80 trials). Successful performances were qualitatively judged by a national-level coach using the FIG Code of Points (2008).

Active markers were located on the lateral, right side of each gymnast at the metatarsalphalangeal (mtp) and on the ankle, knee, hip and shoulder joint centres. Co-aligned CODA CX1 motion analysis scanners (Charnwood Dynamics Ltd., Leicestershire, UK) were used to obtain marker locations (sample rate: 200 Hz; sample duration: 6 s) during each dismount routine. The three-dimensional coordinate data of each marker were reduced to two-dimensions (z-vertical and y-anterior-posterior) and low-passed filtered at 10 Hz.

The onset of the dismount phase was established as the instant at which the mtp z-displacement first exceeded the respective loaded displacement on the beam. The landing phase was defined as the duration between first ground contact, established by the descent of the vertical displacement of the mtp marker below the unloaded landing surface height and the instant at which the mtp joint maintained a stable, loaded position on the ground.

Ankle, knee and hip joint flexion-extension angular displacements and velocities were determined for each landing phase. Strategy bias, which was used by McNitt-Gray et al. (1994) to define drop landing strategies, was determined as the ratio of the knee-hip minimum in the joint angle profile of the landing phase. Skill- and phase-specific initial joint angle configurations, range of motion (ROM) and peak angular velocities, which were determined as the mean across all respective trials (BP: 40 trials; BT: 40 trials), were also used to describe the skill-specific landing strategy. Paired t-tests (15 tests) were performed to analyse between skill differences (n = 40 trials for each skill) in the phase duration, strategy bias, and each joints initial configuration, ROM and magnitude and time of peak angular velocity. The statistical significance level was set at $p<0.05$ for all analyses.

RESULTS: The BP landing phase duration (mean ±SD: 0.48 ±0.10 s) was significantly shorter than ($p<0.05$) the more basic BT skill (mean ±SD: 0.51 ±0.11 s). When expressed relative to the dismount duration, landing phase durations were similar between skills. Strategy bias measures consistently exceeding one indicated the use of a similar hip-biased strategy in the BP (mean ±SD: 1.56 ±0.26) and BT (mean ±SD: 1.57 ±0.35) landing phases.

As illustrated in Table 1, the strategies were differentiated ($p<0.05$) by the use of a 3.7° more extended and 5.0° more flexed initial knee and hip joint configuration, respectively in the BP compared to the BT skill. While similar knee and hip joint flexion ranges were employed, a significantly larger ($p<0.05$) ankle joint ROM and more rapid hip joint angular velocity were used across the landing phase of the BP compared to the BT skill. No significant difference between skills was found in the timing of the peak joint angular velocities.

Table 1 Mean [SD] landing phase kinematics for BP and BT somersault dismounts

	Initial Ankle θ (°)	Initial Knee θ (°)*	Initial Hip θ (°)*
BP	107.7 [9.3]	160.0 [4.9]	100.8 [7.6]
BT	106.4 [10.4]	156.3 [5.8]	105.8 [8.1]
	ROM Ankle θ (°)*	ROM Knee θ (°)	ROM Hip θ (°)
BP	34.1 [7.3]	60.2 [9.2]	43.2 [16.3]
BT	32.0 [7.3]	62.8 [9.8]	48.8 [4.6]
	Peak Ankle ω (rad.s^{-1})	Peak Knee ω (rad.s^{-1})	Peak Hip ω (rad.s^{-1})*
BP	-8.7 [2.1]	-11.5 [0.9]	-6.4 [1.1]
BT	-7.8 [2.3]	-11.5 [1.1]	-7.4 [2.3]
	Time of Peak Ankle ω (% phase)	Time of Peak Knee ω (% phase)	Time of Peak Hip ω (% phase)
BP	1.7 [2.4]	10.8 [2.6]	17.8 [3.9]
BT	2.2 [2.3]	11.1 [2.9]	16.9 [3.3]

*Significant difference between skills at p<0.05

The knee joint was associated with the largest joint angle changes across the landing phase profile compared to the ankle and hip joints. When expressed as a percentage in the BP joint angle range, the root mean squared difference (RMSD) between the landing phase joint angle-time profiles of the skills was 3.8 %, 8.8 %, and 6.4 % for the ankle, knee and hip joints (Figure 1a), respectively. As illustrated in Figure 1b, the knee joint contrastingly produced the most invariant joint angular velocity profile between skills (RMSD: 1.9 %), when compared to the ankle (RMSD: 5.4 %) and hip motions (RMSD: 3.7 %).

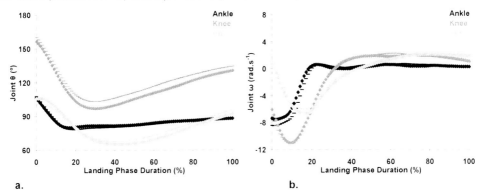

Figure 1: Group ensemble landing phase (a) joint angle [θ] and (b) angular velocity [ω] time profiles during the BP (thin) and BT (thick) dismount skills.

DISCUSSION: The aim of this investigation was to develop insight into the kinematic landing strategy modifications made when performing backward rotating piked and tucked dismounts from beam. Although the BP and BT skills are differentiated in complexity by diverse spatial orientation objectives in the aerial phase, a common hip-biased landing strategy was employed in the basic (BT) and more complex (BP) skill. The use of a hip-biased landing strategy has previously been suggested for gymnasts performing simplistic controlled drop landing task (McNitt-Gray, et al., 1994). The marginally shorter landing phase duration of the BP compared to the BT skill demonstrated that the more complex skill may have incurred slightly increased whole body momentum reduction demands on the gymnasts compared to the basic skill.

While a common global lower-body strategy was employed between skills, modulations in the local control strategy i.e. individual joint movement responses were made to accommodate the more challenging mechanical demands incurred with the more complex dismount skill. The BP landing strategy was characterised by individual joint movement patterns comprising a more extended knee at initial ground contact and simultaneously a more flexed hip than used in the BT skill. Discrepancies in the spatial orientation of the hip during the aerial phase of each skill may explain an earlier occurring, and more pronounced hip joint flexion in the BP compared to the BT landing phase, which was compensated for by a comparatively more extended knee joint at the onset of landing. The use of a relatively more rapid hip joint motion and greater ankle joint ROM during the subsequent ground contact phase of the BP skill may be a further compensatory response to the constrained contribution to whole body momentum reduction provided by the hip at the onset of landing. The more extended (3.7°) and flexed (5.0°) respective initial knee and hip joint configurations used in the BP compared to the BT skill suggested that lower-body and joint-specific strategy modulations may be achieved with increasing skill complexity. The diversity in the joint kinematic patterns used in the BP and BT skills investigated subsequently confirmed the study's hypothesis and supported previous suggestions that multi-joint control in landing tasks should consider local mechanical objectives (McNitt-Gray et al., 2001) and modified strategies may be required in more complex gymnastic dismounts (McNitt-Gray et al., 1994).

Between-skill comparisons in joint configuration profiles suggested that the increased dismount complexity was associated with greater modulations in the knee and hip joint configurations across the entire landing phase compared to the ankle joint. More successful landing performances in BT somersaults have previously been characterised by greater knee and hip joint excursions (Gervais & Dunn, 2003). The similarity in knee (BP: 60.2°; BT: 62.8°) and hip joint (BP: 43.2°; BT: 48.8°) ROM in the BP and BT landing strategies may however suggest that increasing dismount complexity does not compromise landing performance for the four gymnasts investigated. The association between the level of success of the landing performance and the associated joint kinematics was however not considered in this investigation. Future examination of successful and unsuccessful backward rotation dismounts using a larger sample size of gymnasts may therefore be warranted to gain further insight the landing performance constraints incurred by increasing skill complexity.

CONCLUSION: Whilst a common lower-body multi-joint control strategy was used in the skill development of backward rotating dismounts from beam, the landing objectives of a more complex dismount task was achieved through modulation of independent joint movement patterns. The development from a basic tucked to more complex piked dismount skill may be suggested to require a joint-specific modulation in the configuration at the onset of landing and across the landing phase profile.

REFERENCES:
Davlin, C.D., Sands, W.A. & Shultz, B.B. (2001). The role of vision in control of orientation in a back tuck somersault. *Motor Control*, 5, 337-346.
Federation International de Gymnastique. (2008). Code of Points, Artistic gymnastics for women. Switzerland: FIG.
Gervais, P. & Dunn, J. (2003). The double back somersault dismount from the parallel bars. *Sports Biomechanics*, 2 (1), 85-101.
Irwin, G., and Kerwin, D.G. (2007). Musculoskeletal work of high bar progressions. *Sports Biomechanics*. 6(3), 360-373.
McNitt-Gray, J.L. (1991). Kinematics and impulse characteristics of drop landings from three heights. *International Journal of Sport Biomechanics*, 7, 201-224.
McNitt-Gray, J.L., Hester, D.M.E., Mathiyakom, W. & Munkasy, B.A. (2001). Mechanical demand and multijoint control during landing depend on orientation of the body segments relative to the reaction force. *Journal of Biomechanics*, 34, 1471-1482.
McNitt-Gray, J.L., Yokoi, T. & Milward, C. (1994). Landing strategies used by gymnasts on different surfaces. *Journal of Applied Biomechanics*, 10, 237-252.
Sheets, A. & Hubbard, M. (2007). A dynamic approximation of balanced gymnastics landings. *Sports Engineering*, 10, 209-220.

THE EFFECT OF IMPACT CONDITION ON THE RELATIONSHIP BETWEEN LINEAR AND ANGULAR ACCELERATION

Philippe Rousseau, Evan Walsh, Scott Foreman, and T. Blaine Hoshizaki

Neurotrauma Impact Science Laboratory, University of Ottawa, Ottawa, Canada

KEY WORDS: Sports, concussion, biomechanics, linear acceleration, angular acceleration.

INTRODUCTION: Helmets are mandatory in many contact sports and are designed to prevent traumatic brain injuries. When assessing their performance, angular acceleration is not measured, as it is generally assumed to be highly correlated with linear acceleration (Pellman et al., 2003). Although being common, this assumption is not supported by strong data. The aim of this study was to establish the relationship between linear and angular acceleration.

METHODS: A 13.8 kg linear impactor was used to produce impacts at 5.5 m·s^{-1} to an instrumented Hybrid III headform. Five impact sites (Front, Front Boss, Side, Rear Boss, and Rear) were tested at four impact angles (Center of Gravity, Positive Azimuth, Negative Azimuth, and Positive Elevation).

RESULTS & DISCUSSION: Results show a moderate correlation between linear acceleration and angular acceleration ($R^2 = 0.401$). This is in part due to the front impacts through the centre of gravity, which clearly act as statistical outliers (Figure 1). Interestingly, this impact location is the most commonly used during helmet testing. This implies that adding the measurement of angular acceleration to current test protocols in order to predict mTBI is insufficient. New impact locations are necessary to characterize properly mTBI risk.

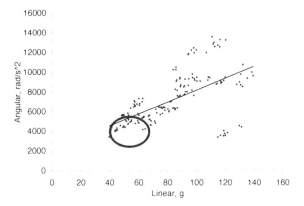

Figure 1: Relationship between linear acceleration and peak angular acceleration at 5.5 m/s. Front impacts through the centre of gravity are circled in blue.

CONCLUSION: The moderate correlation between linear and angular accelerations supports the inclusion of the latter in helmet performance tests. Doing so will allow for a better assessment of concussion protection in sport.

REFERENCES:

Pellman, E. J., Viano, D. C., Tucker, A. M., Casson, I. R., and Waeckerle, J. F. (2003). Concussion in professional football: reconstruction of game impacts and injuries. *Neurosurgery*, 53(4), 799-812.

Acknowledgement

The Authors would like to thank Xenith for supporting the laboratory.

JOINT KINEMATIC VARIABILITY IN THE AERIAL AND LANDING PHASES OF BACKWARD ROTATING DISMOUNTS FROM BEAM

Marianne Gittoes[1], Gareth Irwin[1], David Mullineaux[2] and David Kerwin[1]

Cardiff School of Sport, University of Wales Institute, Cardiff, United Kingdom[1]
College of Education, University of Kentucky, USA[2]

This study aimed to develop insight into the joint kinematic variability in backward rotating dismounts from beam. Two-dimensional lower-body coordinate data were obtained for ten backward piked (BP) and tucked (BT) dismounts performed by four gymnasts (N = 80 trials). The within-gymnast coefficient of variation (CV%) in the joint angle profiles was lower in the aerial-landing phase transition than the remaining dismount element. The CV% was consistently larger in the initial joint configurations of the BP aerial and landing phases than the more basic BT skill. Initial ankle and hip joint landing configurations produced the largest CV% difference between skills (ankle: 9.4 %, hip: 9.4 %). The development of complex dismounts from beam requires a pre-programmed control strategy allowing joint kinematic modulations at the onset of aerial and landing phases.

KEY WORDS: gymnastics, tucked, piked, coefficient of variation, skill development

INTRODUCTION: Dismounting is a crucial element in gymnastic routines and the mastering of fundamental dismounts has been considered beneficial in providing a foundation for the development of more complex skills (Takei et al., 1992). While the dismount element is commonly defined by an aerial and landing phase, diverse spatial orientation objectives in the aerial phase leads to a distinction in skill complexity. In dismounting from the beam apparatus, the backward tucked (BT) somersault can be considered a basic skill for competitive level gymnasts to master, while the development of a more complex dismount involves the acquisition of the back piked (BP) version. As dismount skill complexity increases, modified task objectives must be addressed by the gymnast's biological structures and varied movement patterns are potentially required to achieve effective skill development. Variability in human movement has traditionally been considered detrimental, and van Emmerik et al. (2005) highlighted that the majority of studies on motor learning have linked a decreased task performance variation with the learning process. In contrast, contemporary research employing a dynamical systems perspective has considered variability to have a functional role in locomotion and athletic tasks (Hamill et al., 1999; van Emmerik et al., 2005). Changes to the multi-joint movement strategy executed to achieve a desired performance outcome were recently considered necessary by Bradshaw et al. (2007) to accommodate the demands of a sprint-based athletic task. Skill development in triple-jump has similarly been associated with the presence of a more variable inter-limb coordination strategy (Wilson et al., 2008).

While insight into the movement variability in gymnastic skill development has been limited, Irwin & Kerwin (2007) recently advocated the importance of considering movement variability in understanding skill development due to discrepancies in joint kinematic variability in longswing progressions. Examination of the kinematic variability associated with dismounting has the potential to enhance insight into the control strategy modulation demanded for effective skill development in a commonly performed gymnastic task, and to further contribute to understanding of the role of movement variability in performance development. While between-gymnast comparisons have traditionally been employed to examine the characteristics of complex skills in gymnastics, Gervais & Dunn (2003) advocated the use of within-gymnast analyses for gaining insight into training and control strategies used in learning fundamental gymnastics dismounts. The aim of this study was to develop insight into the within-gymnast joint kinematic variability associated with the skill development of backward rotating dismounts from beam. Increased kinematic variability was hypothesised to be greater in the landing than aerial phase, and with the execution of a more complex skill.

METHODS: Four national level female gymnasts (mean ± SD height: 1.64 ±0.08 m, body mass: 59.0 ±6.9 kg) were recruited for the study and gave written informed consent. The experimental protocols were approved by the University's Research Ethics Committee. During the collection session, each gymnast performed ten successful backward rotating somersault dismounts from beam in piked and tucked positions (N = 20 trials for each gymnast). Successful performances were qualitatively judged by a national-level coach using the FIG Code of Points (2008). During the data collection session, active markers were located on the lateral, right side of each gymnast at the metatarsalphalangeal (mtp) and on the ankle, knee, hip and shoulder joint centres. Co-aligned CODA CX1 motion analysis scanners (Charnwood Dynamics Ltd., Leicestershire, UK) were used to obtain the active marker locations (sample rate: 200 Hz; sample duration: 6 s) during each dismount routine.

The three-dimensional marker coordinate data were subsequently reduced to two-dimensions (z-vertical and y-anterior-posterior) and low-passed filtered at 10 Hz. Separate aerial and landing phases were defined for each dismount routine. The aerial phase was defined as the duration between the instant at which the mtp z-displacement first exceeded the respective loaded displacement on the beam, and the time point prior to the instant at which the vertical displacement of the mtp marker descended below the unloaded landing surface height (ground contact). The landing phase was subsequently defined as the duration between first ground contact, and the instant at which the mtp joint maintained a stable, loaded position on the ground.

Sagittal plane ankle, knee and hip joint kinematic profiles were determined for the dismount duration using the filtered two-dimensional coordinate data and the phase-specific initial, range of motion (ROM), peak flexion and flexion angular velocity of each joint were identified. The within-gymnast coefficient of variation (CV%) in the lower-body kinematic measures were calculated as the percentage of the mean standard deviation across gymnasts (N = 4 gymnasts) relative to the group mean for the respective discrete kinematic measure. Joint angle CV% profiles were determined as the within-gymnast CV% at each time point in the respective aerial and landing phase. Paired t-tests (α level: 0.05) were conducted to examine between phase differences in the discrete measures CV% for the combined dismount routines.

RESULTS: Within-gymnast variability in the joint kinematic profiles (Figure 1) was typically lower at the completion of the aerial phase and the subsequent onset of landing compared to the remainder of the respective phases. Individual joint analyses demonstrated the hip joint angle CV% profile to be notably more prominent than the ankle and knee joint profiles in the latter stages of the landing phase.

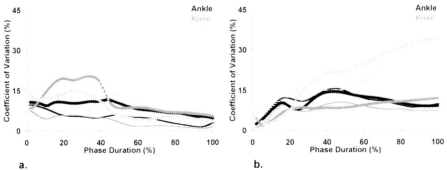

Figure 1: Within-gymnast CV% in joint angle profiles in the aerial (a) and landing (b) phase of BP (thin) and BT (thick) dismounts.

As illustrated in Table 1, the within-gymnast CV% was consistently larger for the initial joint configurations used in the aerial and landing phases of the BP dismount compared to the corresponding phases of the more basic BT skill. The ankle and hip joint configuration used at the onset of landing produced the largest CV% difference between skills (ankle: 9.4%, hip: 9.4%). In contrast, the BT skill was associated with a greater CV% in the knee and hip joint kinematic landing strategies than used in the more complex BP skill. Between skills CV% differences were greatest for the aerial phase hip joint ROM (22.9%) and the peak knee angular velocity (35.6%) of the landing phase.

Table 1 Within-gymnast CV% in lower-body kinematics of the aerial and landing phases of BP and BT dismount routines

	Aerial	Landing	Aerial	Landing	Aerial	Landing
	Initial Ankle θ (%)*		Initial Knee θ (%)*		Initial Hip θ (%)*	
BP	9.6	11.5	7.5	7.8	12.4	13.6
BT	2.4	2.1	1.6	2.1	3.6	4.2
	ROM Ankle θ (%)		ROM Knee θ (%)		ROM Hip θ (%)*	
BP	12.4	13.3	12.2	6.0	3.7	9.1
BT	11.9	10.5	9.9	13.8	26.6	24.9
	Maximum Flexion (%)*		Maximum Flexion (%)		Maximum Flexion (%)*	
BP	5.2	8.5	5.9	7.1	5.6	4.7
BT	4.8	4.4	6.6	10.8	11.5	15.7
	Peak Ankle ω (%)		Peak Knee ω (%)		Peak Hip ω (%)*	
BP	11.8	10.1	17.0	6.1	6.0	6.3
BT	9.8	9.0	4.9	42.6	12.2	22.6

*Significant difference between aerial and landing phase CV% at p<0.05 (N = 8 phases x 8 phases)

DISCUSSION: The within-gymnast kinematic variability associated with the execution of the aerial and landing phases of a basic and more complex backward rotating dismount from beam was examined. More invariant lower-body kinematic measures were typically associated with the aerial compared to the subsequent landing phase, which suggested that the gymnasts were more readily capable of replicating the movement patterns required to satisfy the task constraints of the flight routine. The dismount routines were commonly characterised by a relatively invariant joint kinematic profile at the end and onset of the aerial and landing phases respectively, when compared to the remaining dismount profile. The invariant lower-body kinematic profiles in the transfer from flight to ground contact suggested the need for a more constrained control strategy during the phase transition, which contradicted the dynamical systems perspective suggesting movement pattern variability has a functional role in allowing transitions between movement patterns (van Emmerik et al., 2005). The individual joint analyses conducted in this investigation may explain the discrepancy in findings from previous applications of the dynamical systems theory, which have frequently examined joint coupling control strategies. Future investigation of the coordination pattern variability in the joint strategy used may subsequently be suggested to extend insight into the phase-based demands of dismounting skills.

Increasing skill complexity was associated with greater within-gymnast modulations in the initial ankle, knee and hip joint angle configurations used in the BP and BT phase strategies, and was consistently associated with reduced variability in the landing phase knee and hip joint kinematics. In contrast, the landing phase variability in the discrete ankle joint measures was marginally larger in the BP skill compared to the more basic BT skill. The larger within-gymnast variability in the initial joint configurations used in the BP skill suggested that the execution of more complex dismount skills requires a control strategy that allows greater flexibility in the joint kinematic patterns used in initiating the aerial and landing phases. In contrast, the progression from a basic to more complex dismount skill may be considered to require a more constrained knee and hip joint control strategy during landing as evidenced by the more consistent BP knee and hip kinematic measures. The relatively invariant knee and hip joint and less consistent ankle joint kinematic measures associated with the complex BP landing strategy further supported previous suggestions (Irwin & Kerwin, 2005) that kinematic modifications in gymnastic skill development are specific to individual joints. Examination of the performance variability associated with increasing task complexity may be beneficial in furthering understanding of the control strategies used in successful dismount performances and may support the previously suggested (van Emmerik *et al.*, 2005) departure from traditional concepts that equate variability with inferior performance.

CONCLUSION: Backward rotating piked and tucked dismounts from beam are commonly defined by a flight-ground contact transition characterised by a constrained lower-body joint kinematic pattern. The development of more complex dismount routines from beam however, requires the use of a pre-programmed movement pattern that allows perturbations in the initial configurations of the lower-body joints at the onset of the aerial and landing phases. The effective development of more complex gymnastic dismounts is further characterised by independent modulations to individual lower-body joint patterns within the separate dismount phases.

REFERENCES
Bradshaw, E. J., Maulder, P.S. & Keogh, J.W.L. (2007). Biological movement variability during the sprint starts: Performance enhancement or hindrance? *Sports Biomechanics*, 6 (3), 246-260.
Federation International de Gymnastique. (2008). Code of Points, Artistic gymnastics for women. Switzerland: FIG.
Gervais, P. & Dunn, J. (2003). The double back somersault dismount from the parallel bars. *Sports Biomechanics*, 2 (1), 85-101.
Hamill, J., van Emmerik, R.E., Heiderscheit, B.C. & Li, L. (1999). A dynamical systems approach to lower extremity running injuries. *Clinical Biomechanics*, 14, 297-308.
Irwin, G. & Kerwin, D.G. (2005). Biomechanical similarities in the progressions for the longswing on high bar. *Sports Biomechanics*, 4 (2), 163-178.
Takei, Y. Nohara, H. & Kamimura, M. (1992). Techniques used by elite gymnasts in the 1992 Olympic compulsory dismount from the horizontal bar. *International Journal of Sport Biomechanics*, 8, 207-232.
van Emmerik, R.E.A., Hamill, W.J. & McDermott (2005). Variability and coordinative function in human gait. *Quest*, 57, 102-123.
Wilson, C. Simpson, S., van Emmerik, R.E.A. & Hamill, J. (2008). Coordination variability and skill development in expert triple jumpers. *Sports Biomechanics*, 7 (1), 2-9.

A COMPARISON OF RUNNING KINEMATICS BETWEEN TOP 6 AND HONG KONG ELITE TRIATHLETES IN 2008 ASIAN CHAMPIONSHIPS

Angus T. K. Lam, Danny P. K. Chu and P.M. Cheung
Hong Kong Sports Institute, Hong Kong SAR, China

KEY WORDS: Running, triathlon, kinematics.

INTRODUCTION: Triathlon is a multi-sport endurance event consists of swimming, cycling, and running in immediate succession between disciplines. A distinguishing feature of running in triathlon is the athlete may need to tolerate fatigue condition and lactate accumulation after exercising two disciplines for an extended period of time. Also, a sudden change of muscle group usage from cycling to running may cause a poor running form in the triathletes. Running is the last discipline in triathlon competition. According to the experience of the coaches, running performance had great effect on the overall result. In order to gather information on the performance of the athletes, on site data collection in high-level competition was conducted. The purpose of this study was to quantify the running kinematics of the triathletes and to determine the running kinematic difference between top 6 male triathletes in running event and HK elite triathletes in 2008 Asian Championships. This information could provide an updated reference on the performance of the Asian top triathletes. Based on the individual result, the strengths and weaknesses in each athlete were identified and specific training program could be provided for skill correction.

METHODS: Measurements were performed during the 2008 Asian Triathlon Championships at Guangzhou, China. Top 6 male triathletes in running event and 5 HK elite male triathletes were investigated at 0.9Km, 3.4Km, 5.9Km & 8.4Km of their 10Km race. The mean age of HK athletes was 26 year old. The average body height and weight were 175cm and 67Kg respectively. Their running motion in sagittal plane was videotaped at 50 Hz with a SONY DCR-TRV950E handycam. The camera was positioned perpendicular to the running plane of the triathletes. The video footages were analyzed by Dartfish TeamPro 4.5.2 software. A single running step was selected in each sampling point for analysis on the following parameters. Running speed (SP), step rate (SR), step length (SL), minimum left knee angle (KAng), left hip angle at touch down (TD) and left hip angle at take off (TO) were measured. TD was defined as an acute angle between heel-hip and the vertical. TO was an extension angle between trunk line and lower limb. SPSS 10.0 was used in statistical analysis. And Independent Sample t-test was employed to determine any significant difference of the data between two groups.

RESULTS and DISCUSSION: The SP of Top 6 and HK Elite were found to be 5.21 m/s and 4.63 m/s respectively. The SR of Top 6 and HK Elite were found to be 3.17 step/s and 2.99 step/s respectively. Moreover, the SL of Top 6 and HK Elite were found to be 1.65 m and 1.55 m respectively. Significant difference was found in SP ($p<0.01$), KAng ($p<0.05$), and TO ($p<0.01$). Although no significant difference was found in SR, the p-value 0.07 was closed to 0.05. The SR may be significant if the sample size could be increased. Significant difference was found in KAng and the mean values of the top 6 male triathletes and HK elite male triathletes were found to be 51° and 58° respectively. Decrease of KAng could reduce the moment of inertia of the leg with respect to the hip joint, thus reducing resistance to hip flexion. On the other hand, result showed that SL and TD were comparable between two groups.

CONCLUSION: This study provided information on the running characteristic of HK elite male triathletes. As SL of HK elite male athletes was comparable to top-level athletes, they should pay attention to their SR. Moreover, KAng and TO of HK elite male athletes should also be improved.

COMPARISON OF MALE AND FEMALE PEAK TORQUE USING A VARIABLE NUMBER OF REPETITIONS DURING A KNEE JOINT ISOKINETIC TEST

Barbara L. Warren, Sarah Moody and Deanna Malikie

University of Puget Sound, Tacoma, Washington, USA

The purpose of this study was to assess peak torque in males and females using a variable number of repetitions at five different isokinetic velocities. Sixty subjects, males and females, athletes and non athletes were tested on four separate occasions. Each testing session the subject executed a set number of repetitions, either 4, 6, 8, or 10, at velocities of 60, 120, 180, 240 and 300 $°\cdot s^{-1}$ with a 60 second rest period between each velocity set. The order of repetitions was randomly assigned. A 2 X 2 X 4 X 5 repeated measures ANOVA was used to analyze the data with $\alpha < .05$. The independent variables were gender, athlete, repetitions, and velocity with peak torque as the dependent variable. The following were found to be significant: interaction between gender and velocity; athlete and velocity; velocity; gender; athlete/non athlete; but no significant differences in number of repetitions were found. Therefore, peak torque was demonstrated equally regardless of the number of repetitions. However, by carefully reviewing the data for females it was evident that the women athletes did not reach peak torque at any of the five velocities when they executed four repetitions. In the female non athletes only at the higher velocities did they reach peak torque with four repetitions. It was concluded that females may need more repetitions to achieve peak torque than their male counterparts and that this should be taken into account when they are being tested for strength.

KEY WORDS: isokinetic, peak torque

INTRODUCTION: Considerable research has been conducted using isokinetic machines to assess strength, power and work of particular muscle groups. This has been accomplished using a variety of protocols including various rest periods (Longwell & Warren, 2008; Parcell, Sawyer, Tricoli, & Chinevere, 2002; Warren, 2007), velocities (Cotte & Ferret, 2003; Dauty, Dupre, Potiron-Josee, & Dubois, 2007; Ozcaldiran, 2008; Parcell et. al, 2002), and number of repetitions (Cotte & Ferret, 2003; Magalhaes, Oliveira, Ascensao & Soares, 2004; Pincivero & Campy, 2004). Rest periods between velocity sets have varied from 15 sec to 300 sec with some findings indicating that 60 sec is adequate (Longwell & Warren, 2008; Parcell et al., 2002). The number of velocities at which to test and the order varies from two to eight velocities with most tests being conducted with an ascending order although some have been randomly ordered (Arnold & Perrin, 1995; Dauty et al, 2007; Longwell & Warren, 2008; Magalhaes et al., 2004; Parcell et al., 2002). Many investigators who have assessed strength have used anywhere from three to five repetitions at more than one velocity, although others have used as many as 30 repetitions. Brown & Weir (2001) stated that no more than five repetitions are necessary when assessing strength. However, it has not been clear if these findings are similar in male and females subjects. The purpose of this study was to assess peak torque in males and females using a variable number of repetitions at five different isokinetic velocities. The desired outcome would be to recommend to future researchers a set number of repetitions to be used during isokinetic testing when assessing strength in both male and female populations.

METHODS: Data Collection: Thirty male and 30 female (15 athletes and 15 non athletes of each gender) apparently healthy, college aged students were recruited for this study. The athletes were those who were members of a university athletic team, while the non athletes were students who were not members of a university athletic team. Subjects were excluded if they had a previous knee injury. The study was approved by the University IRB and all subjects signed informed consent. The mean age, height, and weight of the males

respectively were 20.87 ± 1.25 yrs, 179.92 ± 49.7 cm, and 81.33 ± 14.04 kg., while the means for the females were 20 ± 1 years, 170.2 ± 5.58 cm, and 70.63 ± 12.90 kg.
A Cybex NORM isokinetic machine was used for all testing. For the present study gravity correction was integrated in all tests and the Cybex NORM was calibrated prior to collection of any data.
Subjects reported to the lab on five or six separate occasions, depending on their familiarity with the isokinetic machine. One or two were familiarization sessions and four were experimental testing sessions, all of which included a required five minute warm up on a bicycle ergometer at a self selected pace. The warmup on the isokinetic machine included four submaximal knee extensions at 60, 120, 180, 240 and 300 $°·s^{-1}$ with a 60 second rest period between each velocity. All isokinetic tests used a 90° range of motion.
During the familiarization sessions, subjects were fitted on the isokinetic system and settings were recorded to ensure the same positioning for all four experimental tests. After the warmup protocol, the subjects performed four maximal contractions at isokinetic velocities of 60, 180, and 300$°·s^{-1}$ with a 3-min rest between sets.
When experimental testing began, subjects were requested to abstain from maximal exercise bouts 24 hours prior to each session. During experimental testing at velocities of 60, 120, 180, 240 and 300 $°·s^{-1}$., subjects performed sets of maximal contractions of either four, six, eight or ten randomly assigned repetitions. Rest periods between velocity sets were standardized at 60 seconds Subjects were instructed to contract maximally during knee extension, while flexion velocity was set at 300 $°·s^{-1}$, which offered very little resistance.
Subjects were given both visual and verbal feedback during the tests. Each velocity tested was considered a set and the peak torque value for each velocity set was used for comparison.

Data Analysis: A 2 X 2 X 4 X5 repeated measures ANOVA was used to analyze the data with alpha < .05. The independent variables were gender, athlete, number of repetitions and velocity sets respectively, while the dependent variable was peak torque. For a two-tailed alpha=.05, power=.8, an effect size was calculated to be .7.

RESULTS: Analysis of the data revealed significant interactions between velocity and gender: (means and standard deviations are reported in ascending order of velocities) males 238.7±55.9, 189.7±46.2, 155.1±37.6, 131.8±30.9, 114.0±25.6 and females 151.2±37.8, 120.9±28.6, 98.7±23.1, 83.3±19.6, 71.0±16.5 (Figure 1); and between velocity and athlete: athletes 214.8±66.9, 168.7±55.1, 138.5±44.5, 118.1±36.9, 100.9±30.9, and non athletes 175.0±56.0, 142.0±44.1, 115.3±35.9, 97.0±30.57, 83.9±27.4 (Figure 2). As expected there was a significant difference in peak torque at different velocities (Figures 3-6), a significant difference in peak torque between genders, and a significant difference between whether one was an athlete or non athlete. However, there were no significant differences in peak torque at each velocity when using a different number of repetitions.

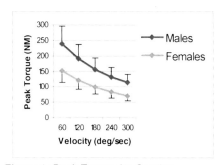

Figure 1: Peak Torque by Gender

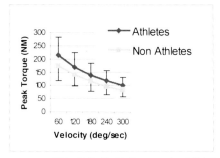

Figure 2: Peak Torque by Athlete

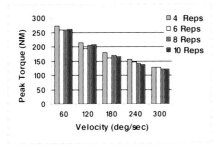

Figure 3: Male Athlete Peak Torque

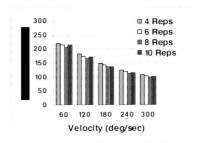

Figure 4: Male Non Athlete Peak Torque

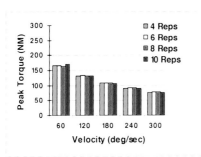

Figure 5: Female Athlete Peak Torque

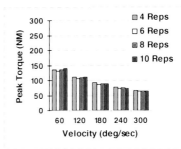

Figure 6: Female Non Athlete Peak Torque

DISCUSSION: The primary focus of this study was evaluating whether the number of repetitions executed during isokinetic testing would affect the amount of peak torque each subject could produce. Brown and Weir (2001) indicated that if strength was being measured there was no reason to use more than five repetitions. In fact, the majority of studies assessing strength as well as other variables using an isokinetic machine ask the subjects to execute between three and five repetitions (Aasa, Jaric, Barnekow-Gergkvist, Johansson, 2003; Arnold & Perrin, 1995; Cotte & Ferret, 2003; Longwell & Warren, 2008; Magalhaes et al., 2004; Ozcaldiran, 2008; Parcell et al., 2002; Warren, 2007). This study attempted to include a large number of subjects, athletes and non athletes as well as both genders. What was found about the male subjects is consistent with the findings of many of the studies cited above. Essentially, males can produce peak torque when executing four repetitions (Figures 3, 4). However, the female athletes' peak torque scores were not as high at any of the five velocities when using only four repetitions (Figure 5). They needed six or eight repetitions to achieve their peak torque. The female non athletes were unable to achieve peak torque with four repetitions at the lower velocities of 60 or 120 $°·s^{-1}$, while at the higher velocities they could produce peak torque with four repetitions (Figure 6). Although the peak torques between repetitions for women were not significant differences, it seems important to note they had difficulty producing peak torque with fewer repetitions. This observation would seem to be important for clinicians, trainers, and researchers who are assessing peak torque in the following situations: prior to a season, bilateral comparisons, evaluation of clients who are completing rehabilitation, or isokinetic testing for research purposes.

CONCLUSIONS: It was concluded that regardless of if one is an athlete; males seem able to produce peak torque with four repetitions, while females need at least six or eight repetitions to demonstrate peak torque. This is applicable for both clinicians and researchers who are using peak torque as an indicator of successful rehabilitation or performance, as women would not have achieved peak torque in most cases if only four repetitions are used for evaluation.

REFERENCES

Arnold, B. & Perrin, D. (1995). Effect of repeated isokinetic concentric and eccentric contractions on quadriceps femoris muscle fatigue. *Isokinetics and Exercise Science*, 5, 81-84.

Aasa, U., Jaric, S., Barnekow-Bergkvist, M., & Johansson, J. (2003). Muscle strength assessment from functional performance tests: Role of body size. *Journal of Strength and Conditioning Association*, 17, 664-670.

Brown, L. & Weir, J. (2001). ASEP procedures recommendation I: accurate assessment of muscular strength and power. *Journal of Exercise Physiology online*, 4 (3), 1-21.

Cotte, T., & Ferret, J. (2003). Comparative study of isokinetics dynamometers: CYBEX NORM vs. CON-TREX MJ. *Isokinetics and Exercise Science*, 11, 37-43.

Dauty, M., Dupre, M., Potiron-Josse, M., & Dubois, C. (2007). Identification of mechanical consequences of jumper's knee by isokinetic concentric torque measurement in elite basketball players. *Isokinetics and Exercise Science*, 15, 37-41.

Longwell, K., & Warren, B. (2008). The effects of rest interval on peak torque in females during isokinetic knee extension. *Medicine and Science in Sports and Exercise*, 40 (5) S1: S449.

Magalhaes, J., Oliveira, J., Ascensao, A., & Soares, J. (2004). Concentric quadriceps and hamstrings isokinetic strength in volleyball players. *Journal of Sports Medicine and Physical Fitness*, 44, 119-25.

Ozcaldiran, B. (2008). Knee flexibility and knee muscles isokinetic strength in swimmers and soccer players. *Isokinetics and Exercise Science*, 16, 55-59.

Parcell, A., Sawyer, R., Tricoli, V., & Chinevere, T. (2002). Minimum rest period for strength recovery during a common isokinetic testing protocol. *Medicine and Science in Sports and Exercise*, 34, 1018-1022.

Pincivero, D., & Campy, R. (2004). The effects of rest interval length and training on quadriceps femoris muscle. Part I: Knee extensor torque and muscle fatigue. *Journal of Sports Medicine and Physical Fitness*, 44, 111-8.

Warren, B. (2007). Minimum rest period for peak torque recovery during isokinetic testing. *Medicine and Science in Sports and Exercise*, 39 (5), pp. S302.

KINEMATIC ANALYSIS OF THE UPPER LIMB AT DIFFERENT IMPACT HEIGHTS IN BASEBALL BATTING

Takahito Tago[1], Michiyoshi Ae[2], Daisuke Tsuchioka[1], Nobuko Ishii[1], Tadashi Wada[3]

Tokushima Bunri University, Kagawa, Japan[1], University of Tsukuba, Ibaraki, Japan[2], Kokushikan University, Tokyo, Japan[3]

The purpose of this study was to investigate the change in the upper limb motion to three different hitting areas of the strike zone: high, middle, and low. Subjects were ten right-handed male skilled batters of a university baseball team. Data were collected using a three dimensional automatic motion analysis system (Vicon 612). The joint angles of the upper limbs were computed. Comparison of the hitting in the high area vs. low area revealed that to hit the ball in the low area the batter more extended his left elbow, and flexed more his both shoulders and horizontal adduction angle of the left shoulder was large at the phase of the Left upper arm parallel (LUP). At the impact phase he flexed his left elbow more, adduction angle of the left shoulder was small in the case of the high area than the case of the low area. The opposite tendency to the high area was observed in the case of the low area.

KEY WORDS: three dimensional motion analysis, angular kinematics, striking.

INTRODUCTION: Many investigations on baseball batting have analyzed the techniques by which a batter hits a ball in the middle in the hitting areas (McIntyre, 1982; Messier 1985). However, since the pitching course in actual game varies, the batter has to modify and change the batting swing so that he or she reacts to various courses. Little information of how a batter modifies the motion to various pitching courses have been reworded. Tago et al. (Tago, 2006) reported that in case of the high, middle hitting areas, the rotation of the shoulder at the impact phase were larger than the low hitting areas. The purpose of this study was to investigate the change in the upper limb motion to the different hitting areas.

METHODS: Subjects were ten right-handed batters of a university baseball team. Informed consent was collected after the explanation of the experiment procedure. Three different hitting areas were set in accordance to the rule of baseball. The batting tee commonly used during practice was used to modify hitting areas. The high areas for right-handed batters were defined as 1, 2, and 3 of Figure1, the middle areas as 4, 5 and 6 of Figure1, the low areas as 7, 8 and 9 of Figure1.The subjects were given the hitting areas in random order, and the position of non-stride leg was set as the same position at the beginning. The

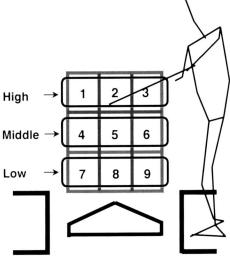

Fig.1 Hitting areas set in this study

coordinate axes were defined as follows: the Y axis was set as the direction to a pitcher, the X axis as the medio-lateral direction, and the Z axis as the perpendicular direction. Data were collected by using a three dimensional automatic motion analysis system (Vicon 612). Nine cameras operating at 250Hz were used to capture the players' motion. From several trials for each point, one trial of the fastest ball velocity and the best self-evaluation was chosen in each point and subject for analysis. For the analysis and description of data, the batting swing was divided by seven instants as follows : TBS···The phase at which the bat grip began to move toward a catcher (Start of take back). Toe-off···The phase at which the stride leg broke the contact with the ground. Knee-high···The phase at which the knee of the stride leg was in the highest position. Toe-on···The phase at which the tip of the foot of the stride leg contacted with the ground. SS···The phase at which the bat grip began to move toward a pitcher (Swing start). LUP···The phase at which the left upper arm of the batter was in parallel to the X-axis (L-upper arm parallel). IMP···The phase at which the bat contacted with the ball (Impact).
Angular kinematics computed were joint angles of the right and left elbows, and flexion-extension, adduction-abduction, horizontal adduction-abduction angle of the shoulders. Two-way ANOVA (three heights times three courses) was used to examine the difference in the angular kinematics of the phases mentioned above between hitting areas, setting significant level at 5%.

RESULTS AND DISCUSSION: Figures 2 and 3 show the average joint angles at seven phases during hitting the high and low hitting areas. Figure 2-1 shows the elbow joint angles and Figure 2-2 shows Flexion-Extension angles of the shoulder. In Figures, R indicates the right limb, L is the left limb, and one example is shown in the present study, and (1),(4),(7)

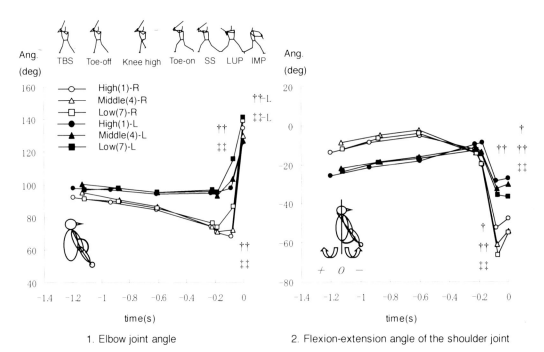

1. Elbow joint angle   2. Flexion-extension angle of the shoulder joint

Fig.2 Changes in the elbow and flexion-extension angle of the shoulder during batting in height hitting areas.
Significant difference (p<0.05) : High vs. Middle ; †, High vs. Low ; ††, Middle vs. Low ; ‡

indicates the hitting area (Refer to Figure.1). Significant differences are shown by a symbol (†, ††, ‡‡). And the definition of the each joint angle is shown in the picture in the graph. In Figure 2-1, only minor change in the left elbow angle was found before the phase of the left upper arm parallel(LUP), and then the elbow joint was abruptly extended toward the impact in both hitting of high and low hitting areas. The significant difference in the elbow was observed at the phase of the LUP and impact (IMP), i.e. the elbow joint angle of the low area was larger than that of the high area. In Figure 2-2, flexion angle of the right shoulder remained constant before the swing start (SS), after that the flexion angle of the right shoulder quickly increased toward the impact in high and low hitting areas. The significant difference in the right shoulder was observed at the phase of the LUP, i.e. the flexion angle of the right shoulder at the high area was smaller than that of the low area. Extension angle of the left shoulder gradually increased toward the swing start. After that the flexion angle of this joint suddenly increased toward the impact in both high and low areas. The significant difference was observed at the LUP and IMP, i.e. the flexion angle of the left shoulder at the low area was larger than that of the high area.

In Figure 3-1, abduction angle of the right shoulder was almost constant until the phase of the SS. After that the adduction angle of the right shoulder quickly increased toward the phase of the IMP. However, no significant difference was observed at the seven phases. Adduction angle of the left shoulder was almost constant until the phase of the SS. After that the abduction angle of the left shoulder quickly increased toward the phase of the IMP. The significant difference was observed at the IMP, i.e. the adduction angle of the left shoulder at

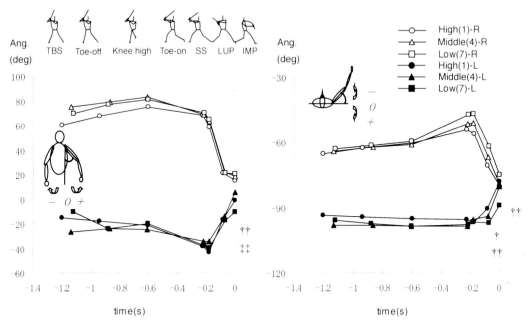

1. Adduction-abduction angle of the shoulder joint

2. Horizontal adduction-abduction angle of the shoulder joint

Fig.3 Changes in the adduction-abduction and horizontal adduction-abduction angle of the shoulder during batting in height hitting areas.
Significant difference (p<0.05) : High vs. Middle ; †, High vs. Low ; ††, Middle vs. Low ; ‡‡

the low area was larger than that of the high or middle areas.

In Figure 3-2, horizontal adduction angle of the right shoulder was almost constant until the instant of the Knee-high. After that the horizontal abduction angle of the right shoulder increased toward the phase of the SS. After that the horizontal adduction angle of the right shoulder quickly increased toward the phase of the IMP. However, no significant difference was observed at the seven phases. Horizontal adduction angle of the left shoulder was almost constant until the phase of the SS. After that the horizontal abduction angle of the left shoulder quickly increased toward the phase of the IMP. The significant difference was observed at the LUP and IMP, i.e. the horizontal adduction angle of the left shoulder at the high area was larger than that of the low area.

Comparing hitting the ball in the high area with the low area, we will be able to identify that hitting a low compared with a high ball was characterized by ; the batter more extended his left elbow, and flexed more his both shoulders and horizontal adduction angle of the left shoulder was large at the phase of the LUP. At the impact he flexed his left elbow more, adduction angle of the left shoulder was small in the case of the high area than the case of the low area. The opposite tendency to the high area was observed in the case of the low area. The significant differences in selected joint angles were observed after the commencement of the swing, which may imply that adjustments occur during the forwards swing period. It was suggested that the movement of a left arm from which a significant difference was seen in all the joint angles after the commencement of the swing be especially important.

CONCLUSION: Kinematic comparisons indicated that in hitting a ball at high and low areas, the batter adjusts the position of the bat by modifying shoulder and elbow angles, particularly at LUP and modifies the angles in the left upper limb just before impact.

REFERENCES:
McIntyre D.R., Pfautsch E.W. (1982). A kinematic analysis of the baseball batting swings involved in opposite-field and same-field hitting. *Res. Quart.*, 53,206−213
Messier S.P., Owen M.G. (1985). The Mechanics of Batting: Analysis of Ground Reaction   Forces and Selected Lower Extremity Kinematics. *Res. Quart.*, 56(2),138−143
Tago T., Ae M., Fujii N., Koike S., Takahashi K., and Kawamura T. (2006). Effects of the height of hitting point on joint angular kenematics in baseball batting. *Japanese journal of biomechanics in sport and exercise*, 10(1), 2-13

TREND ANALYSIS OF COMPLEX RELEASE AND RE-GRASP SKILLS ON THE HIGH BAR

Mark Samuels, Gareth Irwin, David Kerwin & Marianne Gittoes
Cardiff School of Sport, University of Wales Institute, Cardiff, Wales, UK

The purpose of this study was to examine the influence of the Code of Points on the frequency and type of Kovacs performed on the high bar by elite gymnasts. Video recordings of high bar performances were collected from the 2000 Sydney Olympic Games and 2006 Aarhus World Championships. Development of the Kovacs skill within the Code of Points was informed by a trend analysis of the type and frequency of release skills at the two international competitions. Kovacs variations contributed to 50% of the release skills at the Olympic Games and 40% post qualification round at the World Championships. The high frequency of the Kovacs was attributed to the increase in associated difficulty ranking, potential to increase difficulty through shape and twisting option and durability within the Code of Points. Biomechanical analysis is required for further understanding of the completion and development of the Kovacs.

KEY WORDS: Gymnastics, release and re-grasp skill, Code of Points.

INTRODUCTION: High bar performances consist of a unique movement, namely the release & re-grasp skill setting the high bar apart from the rest of the men's programme. Release skills emerged in the early 1950's, opening up a new avenue for skill development, resulting in an increase in innovative release skills at international competitions (Brüggemann et al., 1994). Release and re-grasp skills are mandatory in elite gymnastics high bar routines (Arampatzis & Brüggemann, 2001) and represent a direct means of accumulating high difficulty tariff. Gervais & Tally (1993) associated the performance of release skills in elite high bar routines with a potential to score highly and a small margin for error. Skill complexity is quantified by the FIG via the Code of Points (2008), using an ordinal system, where 'A' is the least difficult and 'F' is the most difficult. The Code of Points is subject to a review every four years, to encompass technological developments, coach and gymnast innovation and increasing levels of ability (Brüggemann et al., 1994). The ordinal scale utilised by the Code of Points has increased five times since 1950, to ensure skills are appropriately ranked and to remove redundant skills (Liang & Tian, 2003). The Kovacs is a highly ranked release and re-grasp skill consisting of a one and half backward somersaults, whilst travelling backwards over the high bar and re-grasping the bar. Six variation of the Kovacs release exist in the current Code of Points (2008) and it has emerged as a key requirement of elite high bar. The six variations of the Kovacs vary in difficulty rating, based on the shape and addition of twists in the flight phase (FIG, 2008). Previous biomechanical research within gymnastics has primarily focused on fundamental and commonly performed skills (Gervais & Dunn, 2003). Skill trends on high bar have moved toward the complex release skills in the last two decades (Brüggemann et al., 1994), as a result biomechanical research has focused analysis on the preparatory long swing and release parameters of these skills. Understanding the completion of these difficult skills is important, particularly in the learning phases, as it enables practitioners and coaches to educate their gymnasts safely and efficiently (Gervais & Tally, 1993). Biomechanical analysis of frequently performed skills with more than one variation can inform coaches regarding the development between variations (Holvoet et al., 2002; Irwin et al., 2007). Current skill trends in elite level high bar performances is therefore paramount to coaches and researchers, further developing the link between biomechanics and coaching practice. Therefore, the aim of this study was to investigate and quantify the trend in Kovacs release and re-grasp skills and to evaluate the development of the Kovacs within the Code of Points. The purpose of this study was to further the understanding of the use of the Kovacs and evaluate the influence the Code of Points since the introduction of the skill.

METHODS: Data collection: Elite male gymnasts from 12 and 43 nations performed high bar routines at the 2000 Sydney Olympic Games (OG) and the 2006 Aarhus World Championships (WC), respectively, on a standard 2.65m competition sprung high bar. Seventy-eight and two hundred and twenty six high bar performances were recorded across the qualification round, individual all-around, team and high bar finals from the OG and the WC, respectively. The camera was positioned approximately 35 m away from the high bar and 8 m above the high bar location at the OG. High bar performances from the WC were recorded using a digital camcorder located at floor level and positioned alongside the high bar at a 45° angle to the centre of performance.

Data Processing: High bar performances from the four competition rounds at the OG and the WC were reviewed via laptop (Toshiba Satellite PRO L100), to play back the performance data. Release and re-grasp skill trend analysis was carried out by an experienced gymnastics coach, to ensure successful and consistent skill recognition throughout high bar performances.

Data analysis: The analysis of the high bar performance data from the OG and WC occurred in two phases. Phase 1 of the analysis employed a hand notation system to analyse the trend in release skills. Analysis of the high bar performances began from when the gymnasts initially made contact with the bar, until the gymnast dismounted the high bar. Frequency and type of successful release skills performed at the international competitions was recorded. Release skills were divided into two separate groups based on the flight phases, namely travelling over the bar and non-travelling skills. Travelling over the bar skills were defined as, skills that released the bar and travelled over the bar and re-grasped on the opposing side. Non-travelling skills were defined as, skills that released the bar and re-grasped on the same side of the bar. Successful release skills were defined when the gymnast released and performed the required movements during the flight phase, re-grasped the bar and was able to carry on the routine uninhibited (Holvoet et al., 2002). Frequently performed release skills were identified in Phase 2 of the analysis, which used a review of the Code of Points (FIG, 2008). The review was conducted from the introduction of the frequently performed release skill into and up to the present Code of Points (FIG, 2008). Changes in difficult ranking and skill development were monitored over a 21 year time period (1985 – 2006).

RESULTS: Phase 1. The frequency and type of Kovacs at the OC and the WC are demonstrated in Figure 1. Travelling over the bar release skills represented 79% and 80% of the release skills performed at the OG and the WC, respectively. Four variations of the complex Kovacs release were performed at the OG and the contribution to the overall performance of release and re-grasp skills performed was vast, contributing to over 50% of the releases performed in each competition round. The tucked Kovacs was the most frequently performed Kovacs variation throughout the OG, contributing to 31% of the overall releases performed. The performance of the full twisting tucked Kovacs increased in the high bar final to 25%, equivalent to the tucked Kovacs, subsequently increasing the percentage performance of the Kovacs skills in the OG high bar final to 65%. The trend in performance frequency of the Kovacs continued at the WC post the qualification round. Post qualification, the four variations of the Kovacs contributed an average of 40% of the release skills performed in each round at the WC. The number of Kovacs variations increased with the performance of the full twisting straight Kovacs. The full twisting tucked Kovacs was performed in each competition round and was the most frequent release performed in the WC high bar final.

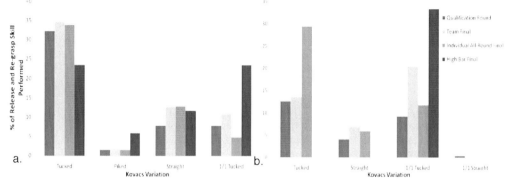

Figure 1: Type and frequency of release and re-grasp skill the Kovacs, performed during the qualifying round, team, individual all-around and high bar finals at the OG (a) and the WC (b)

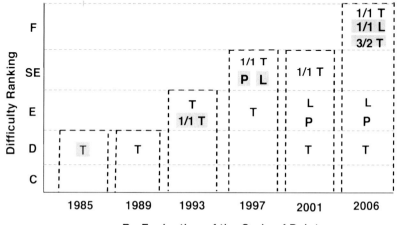

Figure 2: Evolution of the Kovacs within the Code of Points. Tuck (T), Pike (P), Straight (L), Full twisting (1/1) and one and a half twisting (3/2) Kovacs variations. Difficulty maximum of the respective Code of Points is marked by dashed line (FIG, 1985 – 2006). Highlighted Kovacs variation represents introduction to the Code of Points

Phase 2. The evolution of the Kovacs release skill on the high bar, in parallel with the Code of Points is illustrated in Figure 2. The Kovacs release skill has evolved from the basic tucked Kovacs introduced in to the Code in 1985 (FIG, 1985), with the introduction of the piked, straight, full twisting tucked and straight and the one and half twisting tucked Kovacs (FIG 1993; 1997; 2006). The new variations of the Kovacs entered the Code of Points in the top difficulty of the respective Code of Points. The tucked and full twisting adaptation of the Kovacs upon entry to the Code of Points remained in the top difficulty for three cycles (12 years) and to the current Code of Points, respectively. Post entry the piked and straight Kovacs were re-evaluated and placed in the E difficulty, simultaneously the tucked Kovacs was reduced to a D ranked skill. In the current Code of Point the full twisting straight and the one and half twisting tucked entered the top difficulty ranking.

DISCUSSION: Phase 1: The relatively high frequency of travelling over the bar releases compared to the non-travelling releases, could be due to the position of the gymnast at re-grasp, as a majority of non-travelling skills re-grasp below horizontal, therefore requiring the use of compensatory movements to continue the routine successfully (Cuk, 1995). The tucked Kovacs was the most frequently performed release and Kovacs variation at the OG and was the joint highest in the high bar final. The increase in full twisting tucked Kovacs during the OG high final could be associated to the high skill level of the gymnasts competing and therefore performing the most complex skills (Brüggemann et al., 1994). Kovacs releases made up 40% of the releases performed at the WC post qualification. Fluctuation in the frequency of releases in the qualification round could be associated with the ability level of the gymnasts, resulting in an increase in lesser ranked releases (Brüggemann et al., 1994). The full twisting tucked Kovacs was performed throughout and was the most frequent release performed in the WC high bar final. Kovacs variations increased at the WC with the development of the full twisting straight Kovacs, increasing the difficulty level of the Kovacs grouping. Increase in full twisting tucked and decrease in piked and straight Kovacs frequency, between the two international competitions could be related to alteration to the difficulty ranking of the skills within the Code of Point (2006).

Phase 2. The Kovacs has undergone expansion, in terms of skill development since introduction to the Code of Point in 1985. Alteration to shape and addition of twists has resulted in an attractive structured development pathway, consisting of six variations. All new variations of the Kovacs entered the Code of Points in the top difficulty ranking and demonstrated longevity, where other releases such as the Tkachev have decreased upon re-evaluation (FIG, 1997; 2001). Tucked and full twisting tucked Kovacs exhibited good stability within the Code of Points, maintaining top difficulty ranking for a number of years, which would prove beneficial to elite gymnasts, as career length increase. Historically, the Kovacs has offered a respectably high associated difficulty value for all variations, particularly in comparison to other release skills. This is evident in the current Code of Points, with all six variations, from the basic tucked to the full twisting straight, offering high difficulty ranking.

CONCLUSION: The findings of this study established the Kovacs as frequently performed and important release skill at the international level. The Kovacs and variations popularity stems from the development from the basic to the complex variation, associated high difficulty ranking of the releases and durability within the Code of Points. This study highlights the Kovacs release as a requirement of international male gymnasts and the need for biomechanical investigation, to further the understanding of the completion of the Kovacs variations and provide an insight into the development between variations of these skills. This research provides a platform for future trend analysis at major competition, in parallel with examination of the Code of Points to build on current findings and develop a greater understanding of the role of the Kovacs in elite gymnastics.

REFERENCES
Arampatzis, A. & Brüggemann, G.P. (2001). Mechanical energetic processes during the giant swing before Tkatchev Exercise. *Journal of Biomechanics*, **34**, 505-512.
Brüggemann, G. P., Cheetham, P., Alp, Y., & Arampatzis, D. (1994). Approach to a biomechanical profile of dismounts and release and regrasp skills of the high bar. *Journal of Applied Biomechanics*, **18**, 332-344.
Cuk, I. (1995). Kolman and pegan saltos on the high bar. In *ISBS 1995: Proceedings of XIII Symposium of Biomechanics in Sport* (edited by T. Bauer), pp. 118-122. Lakehead University: Thunder Bay, Ontario.
Federation Internationale de Gymnastique (1985, 1989, 1993, 1997, 2001, 2006, 2008). *Code of Points, artistic gymnastics for men*. Switzerland: FIG.
Gervais, P. & Dunn, J. (2003). The double back somersault dismount from parallel bars. *Sports Biomechanics*, **2**, 85 – 101.
Gervais, P. & Tally, F. (1993). The beat swing and mechanical descriptors of three horizontal bar release-regrasp skills. *Journal of Applied Biomechanics*, **9**, 66-83.
Holvoet, P., Lacouture, P. & Duboy, J. (2002) Practical use of airborne simulation in a release-regrasp skill on the high bar. *Journal of Applied Biomechanics*, **18**, 332-344.
Irwin, G., Kerwin, D.G. & Samuels, M. (2007) Biomechanics of the Longswing Preceding the Tkachev. In Proceedings of XXV International Symposium on Biomechanics in Sports (ed. H,-J. Menzel and M.H. Chagas) Ouro Petro, Brazil, 431-434.
Liang, C. & Tian, M. (2003). On Gymnastics Frontier Technical Creations, Unknown
Prassas, S., Kwon, Y.H. and Sands, W.A. (2006). Biomechanical research in artistic gymnastics: a review. *Sport Biomechanics*, **5**, 261-291.

EFFECT OF JUMP PERFORMANCE ON DIFFERENT COMPRESSION GARMENTS

S.Y. Liu 1, W.C. Chen2 and T.Y. Shiang2
Department of Physical Education, National Taiwan Normal University, Taipei, Taiwan1, Inst. of Exercise & Sport Science, National Taiwan Normal University, Taipei, Taiwan2

KEY WORDS: CMJ, muscle, RFD

INTRODUCTION: Recently, more people like to wear the compression garments in athletics and fitness activities. Previous studies showed that Lycra-type compression shorts could enhance athletics performance such as repetitive jump power (Kraemer et al, 1998), improve proprioception and increase endurance. However, there comes up a new type of compression garment with muscle support system composed of high elastic band bordered on garments. Therefore, the purpose of this study is to examine the jumping performance with varied kind compression garments.

METHODS: 8 women (height 163.1±5.9cm, age 22.75±2.49 years, body mass 55.76±5.55 kg) were recruited to randomly wear general sport shorts, compression garment (644290, Adidas) and one new type of compression garment (HXY-109,CW-X) with high elastic band along the quadriceps femoris muscle as executing countermovement jump (CMJ) with hands always on waist. Three parameters were analyzed including flight height of CMJ was measured the difference of the centers of mass (COM) from standing position to highest position in air by Vicon motion analysis system. The concentric impulse and rate of force development (RFD) were calculated from Kistler force plate. Impulse was the integral of the force from the lowest COM as squatting to takeoff, and the RFD referred to the speed at which force can be produced. One-way ANOVA was used to compare the difference of jumping performance with three different garments.

RESULTS: Jumping performance with CW-X is better than others shorts but without significance (table 1).

Table 1: The parameters of counter movement jumps wearing different compression garments.

	Height (mm)	Impulse (N*ms)	RFD (N/s)
general sport shorts	34.51 ± 3.87	119.36 ± 15.99	2158.0 ± 998.7
ADIDAS	34.65 ± 3.75	120.53 ± 16.51	2207.5 ± 753.5
CW-X	35.26 ± 4.05	121.69 ± 16.97	2593.4 ± 1293.5

DISCUSSION: The new type compression shorts with muscle system support based on anatomical research would reinforce the agonist strength by retaining material strain while subject doing stretching movement. However, the high elastic tension may increase the workload to squat among general women who don't have strong muscle strength to oppose that. Therefore, we attempt to recruit more female athletes to exam the subject effects as well.

CONCLUSION: This research is still undergoing, however, there has been a trend of better jumping performance with the compression garments with muscle system support so far.

REFERENCES
Brandon ,K.D., Young-Hoo K., Robert U.N., Jaekun S., Eva M..P., Ryan A.R., Lori R.B., Mike R. and William J.K. (2003) Journal of Sport Sciences ,21,601-610.
Kraemer, W.J., Bush, J., Newton, R.U., Duncan, N.D., Volek, J., Denegar, C.R., Canavan, P., Johnston, J., Putukian, M. and Sebastianelli, W. (1998) Sports Medicine, Training and Rehabilitation, 8, 163–184.

VARIABILITY IN COMPETITIVE PERFORMANCE OF ELITE TRACK CYCLISTS

Nick Flyger

National Sports Institute of Malaysia, National Sports Complex, Kuala Lumpur, Malaysia

This study calculated the individual variation in performance times for cyclists competing in international track cycling events as Typical Error and attempted to express that variation in terms of power. Performance times were collated from six international events during the 2005/06 UCI Track Season and log transformed. Typical Error was calculated via the back transformation of the RMSE from a two-way ANOVA excluding the interaction term. The average Typical Error over all events was 1.0% (0.8 – 1.3 95% CL). Theoretically when performance is expressed as average power, the variation is approximately 3%. Modelling of power output for typical male and female pursuit cyclists appears to confirm this relationship under typical race conditions. These results can be used to assess the suitability of a field-based aerodynamic test for measuring the smallest worthwhile performance enhancing change in drag, whether a cyclist has shown worthwhile improvements in power during a laboratory performance test or in performance time during a competitive season.

KEY WORDS: Cycling, Variability, Performance.

INTRODUCTION: Good coaches will monitor performance over time searching for trends due to training effects, the environment, injury etc..., which influence their athlete's performance. To judge whether a certain coaching or sport science intervention has been successful it is important to know how variable an athlete's performances are and the smallest worthwhile change (beneficial or harmful) in performance for that sport or event.

Variation in performance is calculated as the standard deviation of repeated performances by an individual athlete (within-subject standard deviation). Hopkins (1999) defines that individual variation of an athlete's competitive performance as the typical error in performance or simply the Typical Error. As sport scientists we are interested in Typical Error to quantify the smallest worthwhile change that is beneficial (or harmful) to the athlete. Hopkins (1999) has shown via statistical modelling that an improvement in performance equal to half an athletes Typical Error will result in around 10% more wins than usual.

Paton & Hopkins (2006) have previously studied variation in track cycling performance, in a broader study of cycling performance times. Using two years of the United States men's national Kilo (1km) time-trial series as their representation of track cycling they found that the Typical Error for time in the Kilo was 1.0% (0.8 – 1.4 95% CL). This suggests the smallest worthwhile change for a track athlete maybe as little as 0.5%. However their study was limited by the use of only one track event, which differs from other track events involving other distances, flying starts (200 m sprints), pacing strategies (pursuit) and teamwork (team pursuit). In addition the authors did not investigate the variation of female track cyclists nor cyclists of international quality.

This study aimed to quantify the Typical Error as a Coefficient of Variation (CV) in the performance times of internationally elite competitive track cyclists and expand on the findings of Paton & Hopkins (2006). The results will be related to power so coaches and sport scientists can determine the smallest worthwhile changes in power required to deem a training or sport science intervention as successful.

METHODS: Data Collection: Results from the four 2005/06 UCI World Cups, the 2006 World Championships and the 2006 Commonwealth Games were obtained from their respective websites (www.uci.ch and www.melbourne2006.com.au). Seven events were selected for analysis. They were: Men's and Women's Sprint qualification round (200m);

Women's 500m TT; Men's Kilo (1 km TT); Women's 3km Individual Pursuit; Men's 4km Individual Pursuit; and Men's 4km Team Pursuit. Cyclists who did not compete in three or more races were discarded from the analysis in order to keep the CV calculation exact as outlined in Hopkins et al. (2001).

Data Analysis: Once collated in MS Excel 2003 the times were log-transformed. This made it easier to express the resulting Typical Error as a percentage (CV) and accounts for skewness typically seen in timed athletic data. SPSS v13 was used to analyse the log times using a two-way ANOVA (Cyclist*Race). The ANOVA was not used to calculate significant differences. Rather the RMSE from the ANOVA table was back transformed to give a coefficient of variation of performance time as a percentage (Hopkins, 2000), that is, the average typical error for that event. The 95% confidence limit was calculated using the total degrees of freedom in the ANOVA and the Chi-squared distribution (Hopkins, 2000).

To express Typical Error as a percentage of Mean Power the relationship developed in Hopkins et al. (2001) was used. In that paper they derived that the typical error in mean power is equal to three times the typical error of cycling velocity. That derived relationship was validated by modeling power over a range of velocities for two 'virtual cyclists' riding at conditions typical to a dry warm indoor track at sea level. The model from Martin et al. (2006) was used to calculate power, where ρ is air density calculated from air temperature, humidity and pressure; C_{Drag} is the coefficient of aerodynamic drag for the cyclists; V is velocity; m is mass; g is gravity; C_{Roll} is the typical rolling resistance for a racing tubular tire on a wooden track

$$\text{Modeled Power (P)} = 0.5 \times \rho \times C_{Drag} \times V^3 + m \times g \times C_{Roll} \times V$$

The typical cyclists were modeled on data collected from two real male and female pursuit cyclists who have competed at UCI World Cups. Both cyclists had carried out an aerodynamic test to calculate C_{Drag} either in the field using SRMs and regression analysis or in a wind tunnel. The C_{Roll} value was an average value for a tubular tire chosen from the literature (Martin & Cobb, 2002). The velocities modeled represented the typical range of velocities by elite cyclists. For example the highest velocity used for the male and female cyclist would break their current respective Individual Pursuit world records by around 5 seconds. The lowest velocity would have been good enough to win gold in the respective Individual Pursuit at the 2007 South East Asian Games.

RESULTS: The cyclists used in this study all posted times typical of elite international cyclist. Figure 1 displays the CV (± 95% CL) for each event. The average Typical Error for all events over the 2005/06 UCI track season was 1.0%. The results of modeling percent change in power for an increase in cycling velocity suggest that a 1% increase in velocity is equal to 2.89 - 2.94% increase in mean power. There was however some evidence that this ratio increased as velocity increased through the typical race paces and when other factors of the model, such as air density and drag coefficients were modified. However the largest increase observed was about 0.03%.

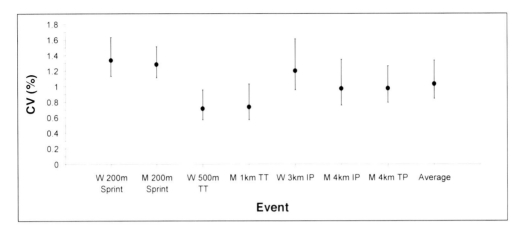

Figure 1: CV (±95% CL) for each event

DISCUSSION: The average Typical Error for all track events in this study (1%) is identical to the Typical Error for the Kilo race series calculated by Paton & Hopkins (2006). However, their Typical Error for the Mens Kilo is 1.4 (0.9 – 2.1 95% CL) times greater than the 0.7% (0.6 – 1.0 95%CL) observed for Mens Kilo in this study. Given that the cyclists in their study were nearly 5 seconds slower, on average, than those in the present study (64 vs. 69 seconds) the results may indicate that less talented cyclists are more variable in their performance. However it should be noted they used a mixed modeling procedure rather than the ANOVA approach I used. Further, the Typical Error measured in this study is a gross measure of within-cyclist variation over the entire season. Other sources of variation in the present study include: pacing strategies; changing fitness during the season; motivation for events like the World Champs; using different equipment during the season; altered riding position between events (changing aerodynamic drag). Future study is needed under controlled conditions to determine if less talented cyclists have more variation in their performance.

From the relationship derived by Hopkins (2001), a 1% change in performance time should equate to 3% change in mean power during a steady pace effort. This modeling approach assumes a constant power output, which is somewhat unrealistic and further validation is required using actual measured velocity profiles from real cyclists. While still anecdotal, by modeling a typical male and female pursuit cyclist over a range of race velocities it appears that the ratio of percentage change in power for a given percentage change in velocity (or time) is indeed close to 3 to 1.

The most important application of these findings is in assessing improvements in competitive performance and physiological tests or the success of changes made to the rider position through aerodynamic testing. When performance is measured in time we simply find the percentage difference between two most recent performances (or the most recent performance and the average performance). If that percentage difference is greater than half the Typical Error for that cyclist's event (from figure 1), then we assume there is some meaningful effect (beneficial or harmful) occurring (Hopkins, 1999). Where the performance is measured in average power we again calculate the percentage difference of power between two performances. However we assume there is some meaningful effect (beneficial or harmful) occurring when that percentage difference is greater than 1.5 times the Typical Error for that cyclist's event (from figure 1). Using 1.5 as a multiplier of Typical Error will convert the smallest worthwhile change from time to power.

In a physiological or aerodynamic test by knowing the smallest worthwhile change in power (half of the Typical Error) and the Typical Error of the test (also as a CV of power), we can judge whether the test is sensitive enough to detect the smallest worthwhile improvement in performance. From the average Typical Error of the track cyclists in this study if test error < 1.5% then it probably will detect the smallest worthwhile change. If test error ≈ 1.5% it may detect the smallest worthwhile change provided multiple trials are conducted. If test error > 1.5% then the test probably will not detect the smallest worthwhile change.

CONCLUSION:

- The average typical variation in the performance time of Elite Track Cyclists is 1.0% (0.8 – 1.3 95% CL).

- The typical variation in average power of Elite Track Cyclists during a race is around 3 times the typical variation of performance time or 3%.

- The smallest worthwhile enhancement in performance time is about 0.5% in the performance time of Elite Track Cyclists for various track events.

- The smallest worthwhile enhancement in average power of Elite Track Cyclists during a physiological or aerodynamic test (field or lab) is about 1.5%.

- A physiological or aerodynamic test (field or lab) should have a test-retest reliability CV < 1.5% to detect the smallest worthwhile change in power.

REFERENCES

Hopkins, W.G., Hawley J.A., and Burke, L.M. (1999). Design and analysis of research on sport performance enhancement. *Medicine & Science in Sports & Exercise.*, **31**, 472-485.
Hopkins, W.G. (2000). Measures of reliability in sports medicine and science. *Sports Medicine*, **30**, 1-15.
Hopkins W.G., Schabort E.J., and Hawley J.A. (2001). Reliability of power in physical performance tests. *Sports Medicine* **31**, 211-234
Martin, J., and Cobb, J. (2002). Bicycle Frame, Wheels & Tires. In *High Performance Cycling*. Ed. Jeukendrup, A. Chap 10, p 113 - 127
Martin, J., Gardner, A. S., Barras, M., and Martin, D. T. (2006). Modeling Sprint Cycling Using Field-Derived Parameters and Forward Integration. *Medicine & Science in Sports & Exercise.*, **38**, 592-597.
Paton, C., and Hopkins, W. (2006). Variation in performance of elite cyclists from race to race. *European Journal of Sport Science*, **6**, 25-31.

Acknowledgement
The author would like to thank Prof. Will Hopkins for advising on the appropriate statistical method and Dr Barry Wilson for commenting on earlier drafts.

DYNAMIC BALANCE IN ALPINE SKIERS

Tom Cresswell, Andrew Mitchell and Naomi Hewitt
Department of Sport Health & Exercise, University of Hertfordshire, England

KEY WORDS: proprioception, skiing, knee, injury.

INTRODUCTION: There are more than 200 million alpine skiers worldwide (Hunter, 1999) but currently ski equipment does not protect the knee as it does the rest of the lower leg and there has been a dramatic rise in knee ligament and meniscus injuries associated with alpine skiing in recent years (Pecina, 2002). Pujol *et al.* (2007) stated that 45-60% of knee injuries during alpine skiing involve the ACL. Balance is an important component of performance in skiing (Laskowski, 1999) and Natri *et al.* (1999) found that many skiers sustaining an injury believed they were only temporarily off balance and capable of regaining control. The three most common mechanisms of injury to the knee in skiing are linked to a loss of dynamic balance (Rossi *et al.*, 2003).

METHODS: Institutional ethical approval was granted and 9 males (age 22.4 ± 6.3 years; height 175.6 ± 8.7cm; mass 76.4 ± 7.8) and 13 females (age 20.3 ± 1.2; height 166.2 ± 7.2cm; mass 66.4 ± 8.6kg). Subjects wore ski boots on a newly developed ski specific balance apparatus with a 20° slope and strived to maintain the level position during three two-minute trials. Times spent in balance (0° +/- 5°) and in right (> 5°) and left deviation (< -5°) were recorded and analysed with a Biometrics Electrogoniometer and DataLog System.

RESULTS: Although this study is a work in progress and data collection is not complete, early signs suggest that the advanced group spent more time in balance than the beginners and more time was spent in left deviation than right by both the beginners and the advanced skiers.

DISCUSSION: At this time it appears that skill level demonstrates a performance effect suggesting that dynamic ski balance increases with skill level. If subsequent statistical analysis confirms that beginners demonstrate significantly worse dynamic balance than advanced skiers, this may put them at increased risk of skiing injury as observed by Keohle *et al.* (2002). Differences being observed in right and left deviation were not expected, and require further examination. Prehabilitation training has been shown to reduce the incidence of ski related ACL injury by up to 62% (Natri *et al.*, 1999) and Lephart *et al.* (1998) highlighted the importance of proprioceptive training in the rehabilitation from knee ligament injury. The apparatus developed in this study has potential uses in rehabiltiation from knee injury, proprioceptive prehabilitation, pre-season screening and talent identification. Further testing will be carried out to examine the contribution of lower limb muscles to dynamic ski balance using electromyography and to investigate the effect of a counterbalance measure as a rehabilitation tool.

REFERENCES:
Pujol N., Blanchi M. N. and Chambat P. (2007). The incidence of anterior cruciate ligament injuries among competitive alpine skiers. *The American journal of sports medicine;* **35(7):** 1070-1074
Koehle M. S., Lloyd-Smith R., and Taunton J. E. (2002), Alpine ski injuries and their prevention. *Sports Medicine;* **32:** 785-793
Laskowski, E. R., Newcomer-Aney, K., and Smith, J. (2000). Proprioception. *Physical Medicine and Rehabilitation Clinics of North America,* **11,** 323-340.
Lephart, S. M., Pincivero, D. M., and Rozzi, S. L. (1998). Proprioception of the ankle and knee. *Sports Medicine,* **25,** 149-155.
Natri, A., Beynnon, B. D., Ettlinger, C. F., Johnson, R. J. and Shealy, J. E. (1999), Alpine ski bindings and injuries. *Sports Medicine,* **28,** 35-48.
Pecina M. (2002), Injuries in downhill (alpine) skiing. *Croatian Medical Journal;* **43:** 257-260
Rossi M. J., Lubowitz J. H., and Guttmann D., (2003). The skier's knee. *The Journal of Arthroscopic and Related Surgery;* **19:** 75-84

COMPARISON OF TURN TECHNIQUES IN PERFORMING THE BASKET WITH HALF TURN TO HANDSTAND ON PARALLEL BARS

Tetsu Yamada[1], Daisuke Nishikawa[2], Yusuke Sato[2] and Maiko Sato[3]

Hyogo University of Health Sciences, Kobe, Japan[1], Nihon University, Tokyo, Japan[2], Japan Women's College of Physical Education, Tokyo, Japan[3]

KEY WORDS: gymnastics, parallel bars, basket, turn, Felge

INTRODUCTION: The basket with half turn to hand stand is performed in competition by male gymnasts on the parallel bars. The turn techniques in performing this maneuver are classified into two types. The early turn technique resembles a cast with half turn to support. The late turn technique looks like a basket to handstand on one rail followed by a half turn backward in a handstand. The former technique can improve the difficulty value by increasing the turn. The purpose of this study was to compare the different turn techniques in performing the basket with half turn to handstand on the parallel bars.

METHODS: Four senior male gymnasts (early turn: 2, late turn: 2) competing nationally were asked to perform the basket with half turn to handstand. The performances were videotaped using two digital video cameras (60 Hz) from a lateral view and a diagonal view of the front. Twenty-two body landmarks (right and left third MP, wrist, elbow, shoulder, toe, heel, ankle, knee and hip, and vertex, midpoint of tragions, suprasternale, lower end of thorax) were digitized. Three dimensional coordinates were synchronized and reconstructed using the method of Yeadon and King (1999). The coordinates were filtered with cut-off frequencies ranging from 3.4 to 5.4 Hz, and the center of mass and moment of inertia of each segment and of the whole body were estimated using the body segment inertia parameters of a Japanese athlete model (Ae, 1996).

RESULTS AND DISCUSSION: Figure 1 shows horizontal and vertical velocities of the center of gravity. Vertical lines in the graphs represent rotation angles of the line hand to center of gravity from the vertical position (handstand) and the instants of releasing and re-grasping the bar. Upward and downward peak velocities were similar in value. The late type of turn had a peak of backward velocity around 90 deg., but the early type of turn had a relatively constant backward velocity until 90 deg. Forward velocity of the early type of turn was greater than that of the late type of turn. This result shows that kinetic energy of the center of gravity, acting like a pendulum, was used more effectively in the early type of turn than in the late type.

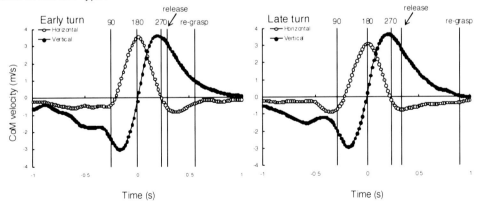

Figure 1 Horizontal and vertical velocities of the center of gravity.

THE EFFECT OF DIFFERENT EXTERNAL ELASTIC COMPRESSION ON MUSCLE STRENGTH, FATIGUE, EMG AND MMG ACTIVITY

Yu Liu, Wei-jie Fu, and Xiao-jie Xiong
School of Kinesiology, Shanghai University of Sport, Shanghai, China

The purpose of this study was to quantify the effects of three different compression conditions on (a) performance of muscle strength/power and fatigue in lower extremity, and (b) the responses of electromyography (EMG) and mechanomyography (MMG) of rectus femoris (RF) under repeated concentric muscle actions. All subjects (N=12) performed maximal voluntary contractions (MVC) and consecutive, maximal isokinetic knee extension movements at 60°/s & 300°/s velocities with three different compression conditions. The results indicated that local elastic compression of lower extremity, while not significant in improving isokinetic strength in short period, may have a positive effect on fatigue by helping maintain long-term force production through altering muscle activity in high-velocity of locomotion.

KEY WORDS: different compression, strength, fatigue, electromyography, mechanomyography.

INTRODUCTION: In high-intensive competitions or leisure sports, the use of compressive garments (e.g., tights, pants, and suits) has become increasingly widespread with the need to reduce muscle injury and maintain muscle function (Trennell et al., 2006; Wallace et al., 2006). Commercially available compression garments have been proposed to provide positive effects on athletes (Houghton et al., 2007). Mechanisms to explain the improved performance included changes in blood flow, improvement on muscle function and the damping of soft tissue vibrations (Herzog, 1993; Kraemer et al., 2001a; Coze and Nigg, 2008). Recently, there have been many publications demonstrating the benefits of compressive garments, however, there are still a quantity of studies from which the results do not support the positive effects claimed by some apparel manufacturers (Duffield and Portus, 2007; Sato et al., 2008). Moreover, far fewer rigorous scientific studies have been conducted to make a detailed investigation into the performance and muscle function under different compression conditions.

In addition, surface electromyography (sEMG) and mechanomyography (MMG) have been widely used to obtain biological information of muscle activities in sport science research field (Tarata, 2003). Specifically, the MMG amplitude is considered to reflect motor unit recruitment, whereas the EMG amplitude reflects both motor unit recruitment and firing rate (Beck et al., 2005; Orizio 1992). Thus, simultaneous measures of EMG and MMG provide additional insight with regard to the motor control strategies utilized by the electromechanical function of active skeletal muscle during a concentric fatiguing task.

Therefore the purpose of this study was to quantify the effects of three different compression conditions on (a) performance of muscle strength/power and fatigue in lower extremity, and (b) the responses of electromyography (EMG) and mechanomyography (MMG) of rectus femoris (RF) under repeated concentric muscle actions.

METHODS: Subjects: Twelve male students specialized in sport were recruited for this experiment (age: 21.2±1.4, mass: 67.1±6.4 kg, height: 1.78±0.05 m). All the participants were with 4-5 years of experience in track and field specializing in sprint or jump events and had no previous musculoskeletal injuries of the lower extremity half a year before this study.

Elastic Compression: Testing utilized adjustable compressive wrap which made of 43% polyamide, 42% cotton, 10% elastodiene, 5% elastane, and its tension was obtained by the force-length curve. The three conditions of testing were: medium loads, high loads and control condition (no compression). The covered area was from thigh to just above knee.

Isokinetic Strength Testing: Prior to the isokinetic testing, the strength of maximal isometric voluntary contraction (MVC) of the quadriceps was measured at 60° angles of knee flexion.

Isokinetic testing consisted of 1 set of 25 consecutive, maximal isokinetic knee extension movements at two selected velocities (60°/s and 300°/s) on a calibrated Contrex Isokinetic System (CMV AG Corp. Switzerland) for each condition.

Muscle Activity Procedures: The amplitude of EMG (Biovision, Wehrhaim, Germany) and MMG (a biaxial accelerometer, Biovision, Germany; bandwidth of DC–1000 Hz) of the rectus femoris (RF) under wearing different compression elastic textile were acquired with Dasylab 8.0 software both at a sampling frequency of 1000Hz. Furthermore, in order to observe isokinetic knee extension movements synchronously, an inclinometer was positioned with its values recorded along with EMG and MMG signals at the same time through a Biovision data acquisition system.

Signal Processing and Statistical Analyses: Date analysis for muscle activity was performed with custom programs written with Dasylab 8.0 software. The EMG and MMG amplitude (RMS) were calculated over the middle third of each repetition based on a total range of motion of 90° (approximately a 30°range of motion; 0.5s for 60°/s and 0.1s for 300°/s) (Ebersole et al., 2006). The EMG signals were bandpass filtered from 10 to 700 Hz, while the MMG signals were filtered with a pass band of 10–100 Hz prior to signal analysis.

The main variables discussed in this study for force production were peak torque, peak power and average power for the 1st five repetitions of knee extension, for fatigue performance were total work and decaying ratio of torque (k), and for muscle activity were EMG and MMG root mean square amplitude (rmsEMG/rmsMMG) of the rectus femoris for each of 1st five repetitions and total 25 repetitions. One-way ANOVAs were used for the analysis of different compression conditions using SPSS 13 (SPSS Inc., Chicago, IL). The significant level was set at $\alpha=0.05$.

RESULTS AND DISCUSSION: Torque and Power: The isokinetic results for force production (peak torque, peak power and average power) were shown in Table 1 (N=11). No significant differences were found among control, medium and high external elastic compression conditions at MVC and other two angular velocities.

Table 1: Force production of knee extensors at MVC and two angular velocities. Values are mean *(SD)*

Knee Extension	MVC			60°/s			300°/s		
	Control	Medium	High	Control	Medium	High	Control	Medium	High
Peak Torque(Nm)	223.0	212.8	213.6	175.7	179.1	174.4	140.0	140.6	130.7
	(36.7)	*(34.7)*	*(34.8)*	*(30.3)*	*(35.0)*	*(23.2)*	*(29.9)*	*(22.4)*	*(28.7)*
Peak Power(W)				182.7	186.7	181.6	616.0	614.4	590.5
				(31.6)	*(37.5)*	*(25.1)*	*(88.0)*	*(88.9)*	*(98.4)*
Ave. Power(W) for 1st five repetitions				107.4	110.2	103.7	233.9	238.3	231.1
				(14.3)	*(20.0)*	*(18.0)*	*(36.2)*	*(26.8)*	*(46.8)*

Fatigue Performance: No significant differences were found between elastic compression and control condition in total work and decaying ratio of torque k (Figure 1 and 2). However, it is worth noting that trends toward declined work and ascended k were observed in high loads compression at two angular velocities. That is to say, compared control condition with high loads, the total work of the latter one decreased approximately 4% both at 60°/s and 300°/s.

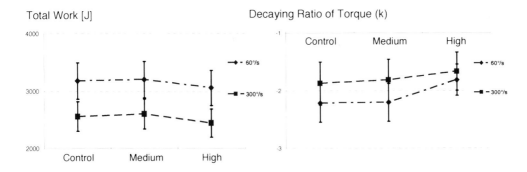

Figure 1 and 2: Influence of elastic compression conditions on total work and decay ratio of torque (k) at two angular velocities

EMG and MMG Amplitude: The amplitude (RMS) of EMG and MMG of RF under three different loads for MVC and the 1st five repetitions of two angular velocities was presented in Table 2 (N=12). No significant differences were found at MVC and 60°/s, but at 300°/s, the medium loads compression condition showed 21.5% and 37.9% increase in EMG and MMG ($p<0.05$) respectively when compared to the control condition.

Table 2: The RMS of EMG and MMG for the RF under three loads at MVC and 1st five repetitions

	MVC			60°/s			300°/s		
	Control	Medium	High	Control	Medium	High	Control	Medium	High
EMG	0.59±0.3	0.53±0.2	0.60±0.4	0.48±0.3	0.45±0.2	0.43±0.2	0.65±0.2	**0.79±0.3***	0.56±0.2
MMG	0.29±0.1	0.24±0.1	0.29±0.1	0.46±0.1	0.33±0.1	0.35±0.1	0.66±0.3	**0.91±0.3#**	0.79±0.4

* and # indicate a significant difference compared with control condition, $p<0.05$.

Results above indicated that if the compression and velocity of locomotion could not reach a certain range, the effect of external elastic compression for short-term force production may not be as distinct as we considered before. Meanwhile, compared with no or high compression in local area, medium loads might have a better ability of recruiting additional motor units, especially for fast twitch fibers (Beck et al., 2005), in helping improve short-term performance. This may to a certain extent provide indirect evidence of enhancing proprioception after using the compression (Kraemer et al., 1996).

For total 25 repetitions, there was an significant gradual decrease ($p<0.01$) in rmsEMG with its elastic compression enhanced (from zero to high loads) both at 60°/s and 300°/s, whereas no significant differences were found in rmsMMG (Figure 4 and 5).

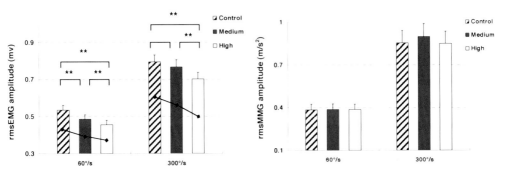

Figure 4 and 5: Influence of elastic compression conditions on the amplitude of EMG and MMG at two angular velocities for total 25 repetitions (**$p<0.01$).

Definitely, from EMG amplitude, elastic compression has an effect on muscle activity in isokinetic knee extension movements. These findings are similar to recent investigations reporting compression apparel decreased muscle pre-activation during running (Coza & Nigg, 2008). Considering with the results of declined work and ascended k showed above, it implies that these changes may have a positive effect on fatigue and performance. From rmsMMG, the effect in reducing muscle oscillation was insensitive to the changes in compression. This could be partly due to the pattern of muscle force production and the deficient range of motion which was not as extensive as running or jumping. However, this speculation requires further investigation.

CONCLUSION: The results of this study indicated that local elastic compression of lower extremity, while not significant in improving isokinetic strength in short period, may have a positive effect on fatigue performance by helping maintain long-term power output. Moreover, the amplitude (RMS) of EMG and MMG suggested that appropriate external elastic compression may be beneficial to recruiting additional motor units of rectus femoris in high-velocity of locomotion for short-term force production and have a positive effect on muscle activity in isokinetic knee extension movements for fatigue and performance. However, further work should focus on frequency-domain responses (e.g. mean power frequency) for more definite fatigue performance and optimal loads (press or compression) for different muscles to comprehend the benefits and mechanisms underlying the use of compression garments in athletes and healthy populations.

REFERENCES

Beck, T.W. et al. (2005). Mechanomyographic amplitude and frequency responses during dynamic muscle actions: a comprehensive review. *BioMedical Engineering OnLine*, 4, 1-27.

Coza, A., Nigg, B.M. (2008). *Compression apparel effects on soft tissue vibrations*, The 4th North American Congress on Biomechanics, University of Michigan, Ann Arbor, MI, USA.

Cramer, J.T. et al. (2002). Power output, mechanomyographic, and electromyographic responses to maximal, concentric, isokinetic muscle actions in men and women. *Journal of Strength and Conditioning Research*, 16, 399-408.

Duffield, R., Portus, M. (2007). Comparison of three types of full-body compression garments on throwing and repeat-sprint performance in cricket players. *British Journal of Sports Medicine*, 41, 409-414; discussion 414.

Ebersole, K.T. et al. (2000). The effects of leg angular velocity on mean power frequency and amplitude of the mechanomyographic signal. *Electromyography and Clinical Neurophysiology*, 40, 49-55.

Herzog, J.A. (1993). Deep vein thrombosis in the rehabilitation client: diagnostic tools, prevention, and treatment modalities. *Rehabilitation Nursing*, 18, 8-11.

Houghton, L.A. et al. (2009). Effects of wearing compression garments on thermoregulation during simulated team sport activity in temperate environmental conditions. *Journal of Science and Medicine in Sport*, 12, 303-309.

Kraemer, W.J. et al. (1996). Influence of compression garments on vertical jump performance in NCAA Division I volleyball players. *Journal of Strength and Conditioning Research*, 10, 180-183.

Orizio, C. (1992). Sound myogram and EMG cross-spectrum during exhausting isometric contractions in humans. *Journal of Electromyography and Kinesiology*, 2, 141-149.

Sato, K. et al. (2008). *The effect of compression tights and duration of testing on continuous jumping mechanical power*, ISBS Conference 2008, Seoul, Korea.

Tarata, M.T. (2003). Mechanomyography versus electromyography, in monitoring the muscular fatigue. *BioMedical Engineering OnLine*, 2, 3.

Trenell, M.I. et al. (2006). Compression garments and recovery from eccentric exercise. *Journal of Sports Science and Medicine*, 5, 106-114.

Wallace, L. et al. (2006). Compression garments: Do they influence athletic performance and recovery. *Sports Coach*, 28, 38-39.

COMPARISON OF SINGLE- AND MULTILAYER MATERIALS USED AS DAMPENING ELEMENTS IN KNEE-PROTECTORS

A.M. Kotschwar and C. Peham

Movement Science Group Vienna, University of Veterinary Medicine, Austria

The purpose of this study was to identify characteristics of protectors and materials used to assemble protectors, which can be used to create a ranking and proof that a protector has the effects wanted. Single layer neoprene of increasing material strength (n=7) was compared to prototype multilayer materials (n=18) and different commercially available knee protectors (n=18). The test object was attached to a realistic knee dummy, and a fall to the floor was recorded, both kinematically and kinetically. Maximum acceleration and pressure on a single sensor was calculated at the time of the impact, as well as the height of the first rebounce after impact. For single layer materials, results showed a linear correlation of material strength and all three measured parameters. While max. acceleration and pressure both decreased with growing material strength, bounce height increased. This behaviour cannot be observed in multilayer systems. For our test materials as well as fully assembled protectors, pressure values were almost identical, while bounce height varied in a wide range. Different protectors showed great difference in their effectiveness to reduce maximum acceleration.

KEY WORDS: knee, protection, falling, kinematic, pressure

INTRODUCTION: The frequency of falling in sport poses a high risk, for amateurs (Nugent 1974), as well as for professional athletes. Typical techniques e.g. in Volleyball or Handball make it very likely for the athletes to make ground contact with their knees Long recovery times due to joint injuries, and therefore financial loss are often a consequence. Especially knee injuries are known to possibly lead to permanent disability (Kujala et al. 1995) Protectors are sometimes, but not always used. Materials and configuration of these protectors are not standardized in any way, thereby possibly putting the health of the athlete at risk.

This study evaluates different materials and commercially available protectors concerning their dampening properties, to show up the importance of high quality products, not only in elite sports, but also in sport for the masses.

METHODS: Data Collection: A realistic dummy of the knee was molded by applying a polyurethane cast to a human knee in 90° flexed position. Then the cast was cut open to remove it, and reinforced with additional cast layers. The mass of the dummy was 2kg. This dummy was mounted onto a steel construct at the ankle area, allowing rotation around the horizontal axis. (See Figure 1) In the topmost position the height of the reference marker OL1 (boxed marker in Fig. 2) is 63.5cm.

Figure 1: Knee Dummy and Marker Position

8 reflecting markers were attached for the kinematic measurement, 4 at the upper leg part, and 4 at the lower leg part of the dummy. Measurement took place using a 5 camera system at 1000Hz (Motion Analysis). At the same time, data of a 4 by 4 cm pressure measurement sensor array (16 sensors, calibrated to a maximum of 200N/cm² each) (Pliance, Novel), positioned at the area of impact on the dummy, was collected. The two systems were set to be triggered simultaneously, to ensure synchronous data.

3 virtual markers were calculated, one at the assumed position of the kneecap, and one at the longitudinal axis of the lower leg and the upper leg. These were used for visualization purposes only (Segment axes in Fig.2).

The reference for all the following Data is the marker OL1 (see Figure 2,boxed) – due to its position it moves most and is least likely to be disturbed by noise. Rebound height was calculated in reference to the height of the resting position of the knee

Figure 2: Camera and Marker Setup

The kinematic data were used to calculate maximum acceleration at the moment of impact. Additionally, the height of rebounce was calculated..

From the kinetic data, maximum pressure onto a single sensor was derived.

Our samples are divided in 3 groups as shown in Table 1.

Table 1: Test Objects

Item	n
1. Single Layer Neoprene of 1.5,2,3,4,5,7 and 14 mm strength.	7
2. Prototype multilayer materials of varying composition and strength	18
3. Commercially available knee protectors	18

The samples were attached to the dummy, then the pressure sensors where set to zero, to remove bias from preload due to fixation. Each sample was measured 5 times in a row, without changing position.

Data Analysis: Kinematic and kinetic data were imported to Matlab 2008b (The Mathworks), smoothed (10thorder lowpass butterworth filter, 30Hz cutoff frequency), the time of impact calculated using self written code and the results saved to Excel 2007.

RESULTS: Single layer neoprene is the first group of tested materials.

At increasing material strength, maximum acceleration at the time of ground impact decreases from 102g(+/- sd 21.7) at 1.5mm, to 57.9g(+/- sd 2.3) at 14mm, as shown in Figure 3 (whiskers indicate standard deviation). The maximum pressure per sensor is also decreasing linearly from 76.25N/cm²(+/- sd 10.7) to 41.88N/cm²(+/- sd 1.72). One can easily see the high correlation between these two values. (Correlation of 87.5%)

At the same time, the height of rebounce after impact increases, from 71.4mm (+/- sd 4.7) to 172.9mm (+/- sd 3.6)

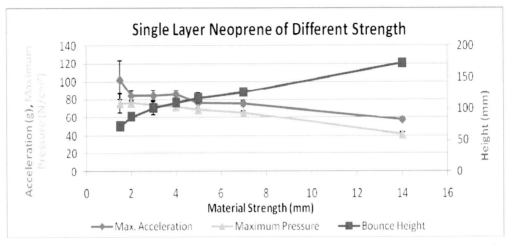

Figure 3: Max. Acceleration, Pressure and Bounce Height of single layer neoprene of increasing strength.

The characteristics of the tested multilayer materials are shown in Figure 4. Though max. acceleration differs greatly, max. pressure stays within a small value range. Bounce height doesn't show the inverse proportional behavior.

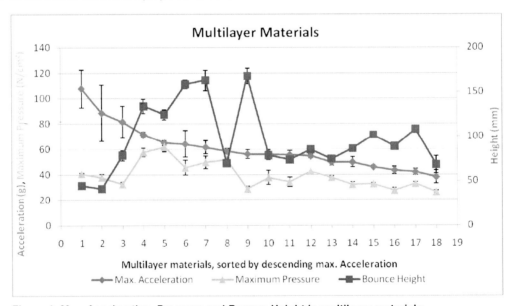

Figure 4: Max. Acceleration, Pressure and Bounce Height in multilayer materials

The same behavior can be seen in figure 5 for the tested knee protectors. Again, no differences in pressure can be seen, while bounce height doesn't behave proportional to the other parameters.

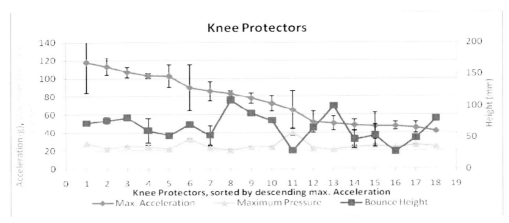

Figure 5: Max. Acceleration, Pressure and Bounce Height in knee protectors

DISCUSSION: We considered the use of a knee dummy essential for our investigation, since the conformation of the knee is quite complex. A simple flat surface could not give us an impression of what really happens at the knee. On the other hand, this uneven surface posed a big problem for our pressure measurement equipment. The Novel sensors are built to work best on flat surfaces, bending them while fixing to a surface leads to inaccurate measurements. Another effect is bridge building, where you get reasonable data on one sensor, but not on the next. Therefore only maximum pressure of a single sensor was evaluated, and no force calculations could be established. The investigation of the single layer neoprene shows the expected result of decreasing acceleration and pressure, and increasing bounce height when the material gets thicker (Figure 3). This proportional behavior between the measured parameters cannot be observed in multilayer materials (Figure 4) and protectors (Figure 5). Both groups show very consistent results in the maximum pressure. The less elastic layers of the materials make the contact area to the ground bigger and distribute the pressure evenly. Bounce height is a sign for the energy that is stored in the material and passed on to it right after the impact. In closed cell materials like neoprene, this effect is more pronounced than in open cell materials, where part of the compression is due to air pressed out of the pores. This can be a reason for the big variations in bounce height.

CONCLUSION: The materials used in building knee protectors highly influence the effect the protector has on maximum acceleration and pressure distribution at the moment of impact. Most protectors show a multilayer setup, and it is shown in our investigation, that the overall behavior of such a complex system is hard to predict.

Commercially available knee protectors show a big difference in effectiveness concerning the maximum acceleration that reaches the knee. Best types reduce it to one third of the least effective ones in our test (42.3g compared to 117.5g). It's therefore essential for the responsible athlete or trainer, to test and choose the right protector for the respective kind of sport.

REFERENCES

Kujala, U.M. et al., 1995. Acute injuries in soccer, ice hockey, volleyball, basketball, judo, and karate: analysis of national registry data. *BMJ (Clinical research ed.)*, 311(7018), 1465-8.
Nugent, R., 1974. Protective Equipment In Amateur Sport. *Canadian Family Physician*, 20(4), 73-76.

Acknowledgement
We would like to thank TSM-Bandagen, Germany, for supplying us with samples for our analysis, as well as for the financial support, that made this investigation possible.

LOWER LIMB JOINT STIFFNESS IN THE SPRINT START PUSH-OFF

Laura Charalambous[1], Gareth Irwin[1], Ian N. Bezodis[1], David G. Kerwin[1] and Robert Harle[2]

Cardiff School of Sport, University of Wales Institute Cardiff, Cardiff, UK[1]
Computer Laboratory, University of Cambridge, Cambridge, UK[2]

Previous studies have calculated joint stiffness (k_{JOINT}) during ballistic movements, but little has been reported on the sprint push-off. The aim of this study was to report lower limb k_{JOINT} during the first stance phase of the sprint start push-off. One sprinter performed 10 maximal sprint starts. An automatic motion analysis system (200 Hz) was synchronised with a force plate (1000 Hz) to collect kinematic profiles at the hip, knee and ankle and ground reaction forces for the first stance phase. Cluster markers defined the orientation of the segments of interest in true 3D, while reducing error from soft tissue artefacts. Kinematic and kinetic data were combined using inverse dynamics analysis to calculate joint moment (M). k_{JOINT} was calculated as change in M divided by change in joint angle ($k_{JOINT} = \Delta M / \Delta \theta$). The findings support the calculation of k_{JOINT} over separate phases of stance and not entire stance.

KEY WORDS: 3D inverse dynamics, sprint starts, joint kinetics

INTRODUCTION: In sprint running a powerful start is essential to reach a high level of performance (Mero, 1988). The importance of the stretch shortening cycle (SSC) has been recognised during the block (Mero et al., 1983) and second stance phases (Jacobs et al., 1996), where an increased utilisation of the SSC is associated with greater joint moment (M). Elastic structures contribute to contractions where contact time is 0.25 s and below (Komi, 1993). The first stance phase contact time is reportedly between 0.18 – 0.20 s (Coh et al., 2006), further indicating the SSC plays and important role in sprint push-off performance. The influence of musculoskeletal stiffness on the utilisation of SSC is well established (Latash & Zatsiorsky, 1993). Although the human body can regulate stiffness to meet performance requirements, many athletes do not fully utilise the SSC and modifying leg stiffness (k_{LEG}) by altering joint stiffness (k_{JOINT}) may be a method to achieve more efficient use of metabolic energy (Latash & Zatsiorsky, 1993). Essentially, k_{JOINT} represents lower limb stiffness at a joint level, whereas k_{LEG} considers the lower limb as a single 'spring'. Calculating k_{JOINT} as M change divided by joint angle change provides a measure of the angular torsion of this spring. Kuitunen et al. (2002) and Gunther & Blickhan (2002) found k_{LEG} during running to be mainly modulated by knee stiffness (k_{KNEE}). Conversely, Yoon et al. (2007) underlined the influence of ankle stiffness (k_{ANKLE}) on jumping performance and utilisation of the SSC at the ankle. Previous studies have agreed that high k_{ANKLE} results in lower contact times and thus enhances the mechanical efficiency of locomotion in sprinting (Kuitunen et al., 2002), hopping (Farley & Morgenroth, 1999) and jumping (Yoon et al., 2007). Stefanyshyn & Nigg (1998) analysed k_{ANKLE} during acceleration and Kuitunen et al. (2002) studied k_{ANKLE} and k_{KNEE} during maximum velocity sprint phases. However, no studies have assessed k_{JOINT} during the first stance phase of the sprint push-off. Furthermore, hip stiffness (k_{HIP}) has only been calculated during hopping tasks (Farley & Morgenroth, 1999). Stefanyshyn & Nigg (1998) showed that k_{ANKLE} increased with running speed whilst Kuitunen et al. (2002) found that it was k_{KNEE} that increased. These studies and Gunther & Blickhan (2002) agreed that k_{JOINT} changed with running speed and thus it has been speculated that increased k_{JOINT} results in greater running speed within an individual (Butler et al., 2003). The aim of this paper was to report lower limb k_{JOINT} during the first stance phase of the sprint push-off for repeated trials of a single athlete.

METHODS: Data Collection: An internationally competitive male sprint hurdler participated in the study (age 27 yrs, height 1.80 m, mass 74.4 kg, 110 m PB: 13.48 s). Four cx1 CODA scanners (Charnwood Dynamics Ltd, UK) were located around a force plate (Kistler Instruments 9287BA, Switzerland) for data collection. Kinematic data (200 Hz) and synchronised ground reaction force data (GRF, 1000 Hz) were captured during the first stance phase out of the blocks. 31 active markers were placed on the subject including three rigid clusters (anterior-lateral aspect of the thigh; lateral aspects of the shank and foot) on the first contact limb. A hip marker was located on the greater trochanter of the same limb. The 4-marker clusters defined the orientation of the segments in 3D, while reducing error from soft tissue artefact (Schache et al., 2008). Kinematic data collected during static trials, together with additional anatomical reference markers, were used to calibrate the athlete, before he completed 10 maximal sprint starts on a start signal. A successful trial was achieved when the athlete accelerated well beyond the measurement volume (>9 m), made first stance contact on the force plate and produced a start with no obvious deviation in technique.

Data Processing: Coordinate data were smoothed using a fourth-order Butterworth low-pass digital filter with a cut-off frequency of 8 Hz, determined by residual analysis of one trial (Winter, 2005). All outputs were normalised to 100% stance using multi-point cubic splines interpolation. The contact phase of the first stance was defined as the period where vertical GRF (F_z) exceeded 10 N. Angle (θ) and angular velocity (ω) of stance leg hip, knee and ankle during the contact phase were calculated from coordinate data. Net joint moments (M) at the hip, knee and ankle for the first stance phase were calculated from kinematic, GRF and anthropometric data using the inverse dynamics method presented by Winter (2005). Anthropometric data were taken from de Leva et al. (1996), with the exception of the foot segment (Winter, 2005). The mass of a typical sprinting shoe (0.2 kg) was added to the mass of the foot segment (Hunter et al., 2004). Mechanical power (P) at each joint was calculated as the product of M and ω. In accordance with Hof (1996), M and P were scaled to body weight and height and were thus dimensionless. k_{ANKLE}, k_{KNEE} and k_{HIP} were calculated from the gradient of the $M - \theta$ curves. k_{JOINT} for the eccentric and concentric P phases, defined from the P-time history, at the ankle and hip, and pre- and post-maximal extensor M at the knee were calculated separately at each joint.

RESULTS & DISCUSSION: Mean profiles across trials for k_{ANKLE}, k_{KNEE} and k_{HIP} during the entire stance and for eccentric and concentric phases are displayed in Figure 1. The graphs depict normalised M (Hof, 1996) plotted against respective joint angle. Previous studies reporting k_{JOINT} used non-normalised M, and so normalised and non-normalised k_{JOINT} values have been included in this study for comparison.

Latash & Zatsiorsky (1993) found that the efficiency of the SSC was improved when k_{JOINT} was greater. In this study k_{ANKLE} over the entire stance (6.84 ± 0.09 Nm/°) was lower than that of sprint acceleration (7.38 ± 1.08 Nm/°) found by Stefanyshyn & Nigg (1998) and maximum velocity sprinting (~8.5 Nm/°) analysed by Kuitunen et al. (2002). The differences between the first stance k_{ANKLE} reported in this study and the subsequent phases in the aforementioned studies may supplement previous reports that k_{ANKLE} increases during the acceleration phase of sprint starting (Stefanyshyn & Nigg, 1998). Further research is required involving k_{ANKLE} calculation in each phase of the sprint, within a subject group, before it can be concluded that k_{ANKLE} increases with the velocity of the task as shown in hopping tasks (Farley & Morgenroth, 1999). When separated into eccentric and concentric P phases the normalised k_{ANKLE} values were 0.019 and 0.006, respectively.

Compared to the pattern of k_{KNEE} reported by Kuitunen et al. (2002) a dissimilar pattern was found in this study. This is attributable to the different knee M pattern compared to the work of Kuitunen et al. (2002) and also the lack of knee flexion prior to extension exhibited by the athlete in this study. Consequently, k_{KNEE} over the entire first stance found in this study (0.016 Nm/°) was very different to the values reported by Kuitunen et al. (2002) across a range of running velocities (17 – 24 Nm/°).

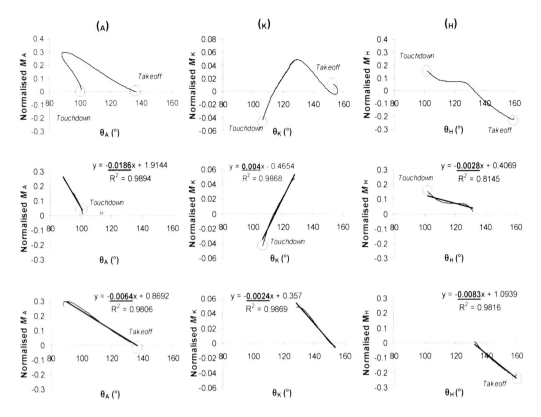

Figure 1: Time-normalised moment-angle plots of the ankle ($_A$), knee ($_K$) and hip ($_H$) for the first stance phase. The first graphs show the mean profile of k_{JOINT} over the entire stance. Stance is then separated into the first and second phases as highlighted in the method section. Linear regression is used to determine the gradient of each curve and thus k_{JOINT} (underlined value).

Further research is required to assess the measurement of k_{KNEE} over an entire stance. To gain a better understanding of k_{KNEE} and its relationship with performance it may be beneficial to separate the stance phase into individual power phases. This study separates k_{KNEE} into two phases. Firstly, the period from touchdown to peak knee extensor M (pre-max). Secondly, from peak knee extensor M to peak knee flexor M at around 85% of stance (post-max). When separated into these phases the normalised k_{KNEE} values were 0.004 and 0.002, respectively.

Although k_{HIP} (0.099) had not been previously reported in any study involving running movements, Farley & Morgenroth (1999) analysed single leg hopping and found changes in k_{HIP} to have little effect on overall k_{LEG} or performance. When stance is separated into phases of concentric and eccentric hip P, k_{HIP} is greater during the eccentric phase (0.008). The accuracy of subsequent studies may be enhanced by using these separate phases when investigating k_{HIP}. Further research is required to analyse whether k_{HIP} alters with velocity of the task and assess the effect of within subject k_{HIP} variation on sprint push-off performance.

CONCLUSION: This research was exploratory, based on the use of segmental clusters on repeated trials for a single elite subject and introduced measures of hip stiffness. A possible pattern between ankle stiffness and velocity of sprinting was identified, but further studies are needed to clarify this relationship. The research highlighted the importance of treating the 'push-off' as a separate phase. Furthermore, correlation coefficients close to ±1 for the eccentric and concentric phases indicated that sprint starting joint stiffness values should be analysed more like those for rebound jumping.

REFERENCES

Arampatzis, A., Bruggemann, G. & Metzler, V. (1999). The effect of speed on leg stiffness and joint kinetics in human running. *Journal of Biomechanics*, 32, 1349-1353.

Butler, R.J., Crowell, H.P. & Davis, I.M. (2003). Lower extremity stiffness: implications for performance and injury. *Clinical Biomechanics*, 18, 511-517.

Coh, M., Tomazin, K. & Stuhec, S. (2006). The biomechanical model of the sprint start and block acceleration. *Physical Education and Sport*, 4, 103-114.

de Leva, P. (1996). Adjustments to Zatsiorsky-Seluyanov's segment inertia parameters. *Journal of Biomechanics*, 29, 1223-1230.

Farley, C.T. & Morgenroth, D.C. (1999). Leg stiffness primarily depends on ankle stiffness during human hopping. *Journal of Biomechanics*, 32, 267-273.

Hunter, J.P., Marshall, R.N. & McNair, P.J. (2004). Segment-interaction analysis of the stance limb in sprint running. *Journal of Biomechanics*, 37, 1439-1446.

Hof, A.L. (1996). Scaling gait data to body size. *Gait Posture*, 4, 222-223.

Jacobs, R. &Ingen Schenau, G.J.van. (1992). Intermuscular coordination in a sprint push-off. *Journal of Biomechanics*, 25, 953-965

Komi, P.V. (1993). Stretch-shortening cycle. In P.V. Komi (Ed.), *Strength and Power in Sport* (pp. 169-179). Oxford: Blackwell Scientific Publications.

Gunther, M. & Blickhan, R. (2002). Joint stiffness of the ankle and the knee in running. *Journal of Biomechanics*, 35, 1459-1474.

Jacobs, R., Bobbert, M.F. & Ingen Schenau, G.J.van. (1996). Mechanical output from individual muscles during explosive leg extensions: the role of biarticular muscles. *Journal of Biomechanics*, 29, 513-523.

Kuitunen, S., Komi, P.V. & Kyrolainen, H. (2002). Knee and ankle joint stiffness in sprint running. *Medicine and Science in Sports and Exercise*, 24, 483-489.

Latash, M.L. & Zatsiorsky, V.M. (1993). Joint stiffness: myth or reality? *Human Movement Science*, 12, 653-692.

Mero, A. (1988). Force – time characteristics and running velocity of male sprinters during the acceleration phase of sprinting. Research Quarterly for Exercise and Sport, 59, 94 – 98.

Mero, A., Luhtanen, P., & Komi, P. V. (1983). A biomechanical study of the sprint start. *Scandinavian Journal of Sports Sciences*, 5, 20-28.

Schache, A.G., Baker, R. & Lamoreux, L.W. (2008). Influence of thigh cluster configuration on the estimation of hip axial rotation. *Gait and Posture*, 20, 60-69.

Stefanyshyn, D.J. & Nigg, B.M. (1998). Dynamic angular stiffness of the ankle joint during running and sprinting. *Journal of Applied Biomechanics*, 14, 292-299.

Winter, D.A. (2005). *Biomechanics & Motor Control of Human Movement*. New York: John Wiley & Sons.

Yoon, S., Tauchi, K. & Takamatsu, K. (2007). Effect of ankle joint stiffness during eccentric phase in rebound jumps on ankle torque at midpoint. *International Journal of Sports Medicine*, 28, 66-71.

A KINETIC COMPARISON OF RUNNING ON TREADMILL AND OVERGROUND SURFACES: AN ANALYSIS OF PLANTAR PRESSURE

Justin Wai-Yuk Lee, Mandy Man-Ling Chung, Kam-Ming Mok, Youlian Hong

Human Movement Laboratory, Department of Sports Science and Physical Education, Faculty of Education, The Chinese University of Hong Kong, Hong Kong SAR, China

The purpose of this study was to compare the plantar pressure in treadmill and overground running. It aimed to investigate whether treadmill is a suitable surface to carry out running shoe cushioning test. Fourteen male volunteers were recruited to run on four different running conditions i.e. treadmill, tartan, grass, and concrete with controlled running speed. A mobile plantar pressure measuring system was employed and peak pressure was measured. The results showed that the plantar pressure of treadmill running was different to that of overground running in total foot, medial midfoot, lateral midfoot and lesser toes.

KEY WORDS: overground, running, treadmill, plantar pressure, shoe testing

INTRODUCTION: In running shoe testing, treadmill is widely adopted by sports scientists. The speed and slope can be easily controlled and data from repeated running cycle could be obtained (Lavcanska, Taylor, & Schache, 2005). Several studies reported that there are differences in different biomechanical aspects, such as 3D kinematics, kinetics and electromyography (EMG) between treadmill running and overground running (Dixon, Collop, & Batt, 2000; Nigg et al., 1995). However, the results are often conflicting and inconclusive. Researchers carry out cushioning functional tests on treadmill and gather kinetics data without running on overground surfaces, of which the runners spend most of the training time on. The aim of this investigation was to find out the plantar pressure difference between different running surface with a controlled running speed and condition.

METHODS: Fourteen male recreational heel-toe runners (age: 22.8 ± 4.4 years; height, 169.2 ±4.78cm; weight, 62.7 ± 9.7kg) were requested to run on four different running surfaces i.e. treadmill, tartan, grass, and concrete with controlled running speed. Every subject wore a standard running shoe model (TN 600, ASICS, Japan) with size 41. A mobile plantar pressure measuring system (Novel GmbH, Munich, Germany) was employed. Kinetic parameter, peak pressure was measured. Six minutes of warm up and familiarization with treadmill was carried out by each subject. After warm up, subjects were instructed to run at a speed of 3.8m/s for 2 minutes on treadmill with the mobile measuring system. The data of the last minute were extracted for data analysis. In the overground running, standard tartan track, grass and concrete surfaces were chosen. Figure 1 illustrates the experimental setting of the run way. The speed was controlled any between 3.6 and 3.8 ms-1. An infra red timing system (Brower, US) was used to monitor the running speed of each trial, and each trial was regarded as finishing the 8m runway. Six trials of each overground surface ware taken. The testing sequence was randomized.

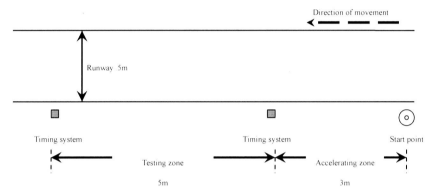

Figure 1: Experimental setup of overground running, i.e. tartan, grass and treadmill

All data were analyzed by Novel Pedar analyzing software (Germany). The only the dominant foot insole was used and divided into 9 recorded areas as shown in Figure 2. Using Novel Database-Pro software, peak pressure (PP) was extracted from each running step. A statistical tool SPSS 12 (SPSS, USA) was used. A repeated-measures analysis of variance (ANOVA) was performed. The assumption of sphericity was checked using Mauchly's test, and the LSD method was used to perform pairwise comparisons following a significant overall test result. The level of significance was set at an α level of 0.05 and data were presented as mean and standard deviation.

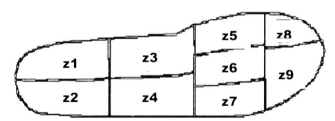

Figure 2: Zones of the plantar pressure surface, z1 (medial heel), z2 (lateral heel), z3 (medial midfoot), z4 (lateral midfoot), z5 (first metatarsal head), z6 (second metatarsal head), z7 (third, fourth and fifth metatarsal head), z8 (great toe), z9 (lesser toe).

RESULTS AND DISCUSSION: The aim of this study was to investigate the effect of different running surface on the plantar pressure in a controlled running speed. The results in table 1

showed that the plantar pressure of treadmill running was significantly lower than that of overground running in total foot, medial midfoot, lateral midfoot and lesser toes, $p<0.05$.

Table 1 - Mean and S.D. of peak pressure (PP) (kPa) in z1-z9 and total foot area. Only ANOVA tests with p value < 0.05 are shown in the table.

Zone	Treadmill	Tartan	Grass	Concrete	ANOVA F
Total foot	395.7 (86.3)	420.6 (92.3)	402.7 (75.9)	456.1(84.3)[a d]	3.5
z3	108.5 (21.4)	118.4 (36.8)	121.0 (31.2)[a]	124.5 (29.2)[b]	3.07
z4	137.3 (41.4)	148.0 (53.4)	142.1(50.2)	161.7(62.8)[a d]	3.76
z6	342.3(84.5)	338.2 (98.9)	329.6 (86.9)	379.0 (93.1)[d]	2.90
z7	245.2 (93.2)	261.2(83.3)	249.2 (72.1)	288.7 (78.4)[d]	3.01
z9	174.2(31.6)	198.6 (56.6)[a]	197.0 (52.3)[a]	219.0 (51.9)[b c]	7.54

a $p<0.05$ when compared with treadmill;   b $p<0.01$ when compared with treadmill;
c $p<0.05$ when compared with grass;   d $p<0.01$ when compared with grass.

In the total foot, PP concrete was found greater than PP treadmill. It's shown that a 15% greater plantar pressure was found when running on concrete compared with that of treadmill, $p<0.05$. With a more detailed analysis, in zone 3, 4 and 9, which represent the medial midfoot, lateral midfoot and lesser toes respectively, PP treadmill was found smaller than PP concrete. In the z3 lesser toes area, PP tartan, PP grass and PP concrete were significantly higher than PP treadmill, i.e. 14%, 13% and 26% higher respectively. The result was in consistent with the study from Baur et al. (2007). Fourteen runners ran on treadmill and a 400m track with a controlled speed. It's reported that PP overground of total foot area were significantly higher than that of PP treadmill. The difference was found mainly in forefoot area, in which PP overground is 25% higher than PP treadmill. They concluded that the muscular activity while running on the treadmill differs from that during overground running.

The plantar pressure difference in forefoot and midfoot areas might be caused by a different running mechanism between treadmill and overground running. According to Wank (1998), there was a lower electromyography (EMG) signal of vastus lateralis explained the less vertical displacement in treadmill running. And the higher EMG signal of biceps femoris in the take off phase might be caused by a greater forward lean of the trunk compared to overground running. It supported that running on treadmill might adopt a different running mechanism to that of overground running. More comprehensive analysis of kinematics and neuromuscular activity would provide further insight.

CONCLUSION: Based on the results of the study, we conclude that:
1) The total foot plantar pressure in treadmill running was found to be lower than that of

concrete running. After further investigation, the difference was mainly found in lateral and medial mid foot and lesser toes. In the forefoot lesser toes area, plantar pressure of treadmill running was significantly smaller than all the overground running i.e. tartan, grass and concrete.

2) To create a well-controlled testing environment, sports biomechanists carry out shoe cushioning test on treadmill instead of overground surfaces. Systematic errors may be introduced in the experiment design according to our finding. There might be a possibility that the absolute forefoot cushioning properties of the sports shoes are overestimated when the test is carried on treadmill. We suggest treadmill may not be a suitable running surface to carry those tests

REFERENCES

Baur, H., Hirschmuller, A., Muller, S., Gollhofer, A., & Mayer, F. (2007). Muscular activity in treadmill and overground running. *Isokinetics and Exercise Science*, 15, 165-171.

Dixon, S., Collop, A., & Batt, M. (2000). Surface effects on ground reaction forces and lower extremity kinematics in running. *Medicine & Science in Sports & Exercise*, 32(11), 1919-1926.

Lavcanska, V., Taylor, N. F., & Schache, A. G. (2005). Familiarization to treadmill running in young unimpaired adults. *Human Movement Science*, 24(4), 544-557.

Nigg, B., De Boer, R., & Fisher, V. (1995). A kinematic comparison of overground and treadmill running. *Medicine & Science in Sports & Exercise*, 27(1), 98-105.

Wank, V., Frick, U., & Schmidtbleicher, D. (1998). Kinematics and electromyography of lower limb muscles in overground and treadmill running. *Int J Sports Med*, 19(7), 455-461.

THE EFFECTS OF ADIDAS POWERWEB COMPRESSION SHORTS ON MUSCLE OSCILLATION AND DROP JUMP PERFORMANCE.

Russell Peters, Neal Smith, and Mike Lauder.
University of Chichester, Chichester, UK.

Adidas compression garment with PowerWeb technology was used in this study to explore the effects on athletic performance and influence on muscle oscillation during a drop jump task. Six male subjects performed 3 drop jumps under 2 conditions, bare leg (control condition) and PowerWeb compression shorts. Three dimensional kinematic data were collected using Vicon Motus software sampling at 500 Hz and force data with a Kistler force plate using Bioware software sampling at 500 Hz. Two-tailed paired sample t-test's were conducted to discover significant differences in muscle oscillation, maximum peak ground reaction force (GRF), peak vertical loading rate (PVLR) and jump height. Wearing PowerWeb compression shorts showed an improvement in jump height with an increase of 6.9 mm, although subjects experienced an extra 1 BW of force during landing.

KEY WORDS: Compression, drop-jump, muscle, oscillation, shorts.

INTRODUCTION: Compression technology is engineered to create a compression gradient within a body molded garment which involves wrapping muscles in tight-fitting fabric (Duffield and Portus, 2007). Recently this technique has been used to improve performance in sport and exercise and provide athletes with benefits such as increased comfort, improved performance and improved blood lactate removal (Kraemer et al., 1998; Doan et al., 2003). Research has suggested one of the benefits that a compression garment can have is focusing the direction of the muscle fibres by reducing muscle oscillation, which can hinder the alignment of muscle fibres when moving, reducing the function of the muscle (Kraemer et al., 1998). Therefore reducing the oscillation of a muscle could help improve technique and maximize the ability to recruit muscle fibres, which in turn can enhance performance (McComas, 1996).

Adidas PowerWeb compression garments aim to improve upon these suggested benefits by introducing powerbands at the rear of the garment. Thermoplastic Urethane (TPU) was selected for the bands as it was a material which would allow a spring like response but did not affect the comfort, or be detrimental to the performance of the garment. The theory is that as an athlete goes through a range of movement there is storage of energy in the elastic element (TPU), which is then delivered back to the athlete as they propel themselves along, as well as stabilising the muscles to reduce muscle oscillation to a greater degree than normal compression garments.

Studies to date have concentrated solely on the performance benefits, however, muscle mechanics suggests that changing the movement patterns of the muscles by wrapping them in tight fabric may have negative aspects. It is suggested that oscillation is a muscular mechanism to help dampen a force which has been applied to it (Herzog, 2000). Computer modelling has investigated this theory by comparing rigid segments model to segments with added wobbling mass. Both Gruber et al. (1998) and Pain and Challis (2006) have found that at the point of impact a wobbling mass independent from the skeleton plays an important role in dispersing energy with peak force significantly greater in the rigid model.

Therefore, the aim of this investigation was to explore the effects of PowerWeb compression shorts on athletic performance and influence on muscle oscillation using a drop jump task.

METHODS: Six males (age 28.67 ± 5.24 years; mass: 84.17 ± 6.88 Kg) volunteered for the study. All were free from injury and able to perform the task efficiently with no health problems. Informed consent was obtained and the subjects were free to withdraw from the study without prejudice at any time. The study had received University ethical clearance.

Subjects performed 3 drop jumps under 2 conditions, bare leg (control condition) and PowerWeb compression shorts. The compression shorts used were Adidas PowerWeb shorts, with the size of the shorts worn by the subjects decided by waist and inseam measurements as advised by Adidas. For the control condition ordinary gym shorts were used with the shorts of the limb being taped above the hip leaving the leg bare for analysis. Three retro reflective spherical markers (19 mm in diameter) were used with spotlights (Hedler). Figure 1 illustrates how the markers were attached to the dominant leg. Marker 2 was placed between the greater trochanter and lateral condyle of the knee. The girth was measured and then marker 1 and 3 were placed 12.5% to the posterior and anterior side of marker 2. This measurement system allowed there to be control over marker placement as anatomical landmarks were unable to be used. The measure of 12.5% also allowed the markers to be visible in both camera views during the performance.

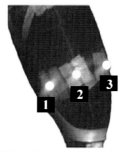

Figure 1: Marker system, 1) 12.5% to the posterior side of the mid-thigh marker, 2) mid-thigh and 3) 12.5% to anterior side of the mid-thigh marker.

All data acquisition was conducted on the same day with several familiarisation trials prior to performance. The drop jumps were performed by stepping off a box (60 cm) onto a 0.6 x 0.4 m piezoelectric force platform (Type 9851, Kistler, Alton, UK). The participants were instructed to keep their hands on their hips as a control construct. The subjects were asked to perform the jump naturally but also to be aware that the knee flexion should be approximately 90° and to keep trunk flexion to a minimum (Kollias et al., 2004) As the jumps were maximal each jump was separated by a 3 minute rest period.

The performed task was captured using 2 high speed (500 Hz) video cameras (TroubleShooter HR, Fastec Imaging, San Diego, US), genlocked, positioned 5 metres from the centre of the platform with an inter camera angle of approximately 110°. The cameras shutter speed was 1/5000 and each had a resolution of 640x480 pixels. The performance area was calibrated with a 17 point three dimensional calibration frame (Peak Performance Technologies, Englewood, USA).

Image digitisation and analysis were performed by Vicon Motus 9.2 software (Vicon, Los Angeles, California, USA). All trials were digitised at 500 Hz using a 3 point model. Acceleration data were chosen to analyse muscle oscillation as it provided greater sensitivity with regard to the movement velocity associated with the jump studied during this investigation. The velocity data were used to break the vertical jump into eccentric and concentric phases of the maximal jump. The present study was only interested in the concentric/propulsive phase of the jump as during this phase the muscles were contracting and it was at this point when a reduction in muscle oscillation could influence performance. This phase started when the velocity was at its lowest point, i.e. the bottom of the jump. Force data were collected with Bioware 3.21 software at a sampling rate of 500 Hz. This allowed the maximum peak ground reaction force (GRF), peak vertical loading rate (PVLR) and flight time to be calculated. Flight time was used to calculate the jump height. Once all data had been processed the 3 trials per condition for each subject were averaged. To establish whether there were any significant differences between the 2 conditions two-tailed paired sampled *t*-tests were used on the means with the statistical significance set at p≤ 0.05.

RESULTS AND DISCUSSION: Table 1 shows the results of the acceleration data. There were no significant differences in muscle oscillation accelerations between the conditions (mid thigh: $t_{(5)}$ = 1.220, p > 0.277; anterior: $t_{(5)}$ = 0.469, p > 0.659; and posterior: $t_{(5)}$ = .437, p > .680), which disagrees with the majority of the literature. However, although there was no significant difference at the mid-thigh marker there was a 7.9% decrease in muscle oscillation.

Table 1: Means (±SD) for peak accelerations and percentage difference.

Marker position	Bare Leg Condition	PowerWeb condition	% difference of PowerWeb condition compared to bare leg condition
Mid-thigh ($m.s^{-2}$)	85.8 ± 22.1	79.0 ± 22.6	↓7.9
Anterior ($m.s^{-2}$)	84.2 ± 24.7	82.7 ± 17.3	↓1.8
Posterior ($m.s^{-2}$)	86.6 ± 22.7	84.0 ± 18.4	↓3.0

The performance test of the garment with the PowerWeb condition produced on average a 0.69 cm higher jump compared to the bare leg condition (Table 2) ($t_{(5)}$ = -2.711, p < 0.042). The difference in height may appear minimal but to increase an athlete's performance by only a slight margin just by wearing a pair of shorts is noteworthy. Such findings are supported in the literature by Doan *et al.* (2003) who found that wearing compression shorts increased jump height by 2.4 cm. This does suggest that by wearing compression garments an athlete's performance maybe improved, however whether this due to muscle oscillation being reduced by allowing the muscle fibres to be optimally positioned or down to other factors such as the psychological influence wearing the shorts can have (Doan *et al.*, 2003) remains uncertain.

Table 2: Means (±SD) for Jump Height, Maximum Peak Ground Reaction Force (GRF) and Peak Vertical Loading Rate (PVLR). (Significant differences p<0.05 denoted by *).

Measure	Bare Leg Condition	PowerWeb condition
Jump Height (cm)	29.67 ± 4.95	*30.36 ± 5.08
Maximum Peak GRF (BW)	6.2 ± 2.2	7.2 ± 2.4
PVLR (BWs^{-1})	123.7 ± 80.8	128.2 ± 75.3

Maximum peak GRF and PVLR were derived from ground reaction force data. A suggested role of the PowerWeb shorts is to support and stabilise the muscle during movement and these force measures were used to indicate whether this caused any negative effect for the athlete. No significant differences were found between the 2 conditions (maximum peak GRF ($t_{(5)}$ = -0.578, p > 0.588) and PVLR ($t_{(5)}$ = -0.578, p > 0.588)), although were still of interest. The average maximum force for the bare leg condition was 6.2 BW compared to 7.2 BW in the PowerWeb condition. Also the PVLR shows the subjects on average experienced an extra 4.5 BW s^{-1} during the PowerWeb condition. If an athlete was wearing the garment during training and for competition this extra force, over time, may lead to injury for sports that experience heavy impacts (Candau et al., 1998). The results follow that of Pain and Challis (2006) who compared a simulated model with a wobbling mass compared to a rigid structure and found that when performing a drop jump the force at the ankle was 7.2 BW s^{-1} greater with the rigid model than that of the model with wobbling mass.

CONCLUSION: Adidas PowerWeb compression shorts provided a trend for increased jump height performance. Performance improvement could be down to the reduction of muscle oscillation although the present study found this not to be significant. However, further research of a larger population is required. This study has highlighted that there maybe a negative aspect to wearing the compression shorts as it has been suggested that the area covered by the garment may have its ability to disperse energy suppressed, which may be the reason for higher forces and loading rates being experienced while wearing the garment.

REFERENCES
Candau, R., Belli, A., Millet, G.Y., Georges, D., Barbier, B. and Rouillon, J.D. (1998). Energy cost and running mechanics during a treadmill run to voluntary exhaustion in humans. *European Journal of Applied Physiology and Occupational physiology*, **77**, 479-485.
Duffield, R. and Portus, M. (2007). Comparison of three types of full body compression garments on throwing and repeat-sprint performance in cricket players. *British Journal of Sports Medicine*, **41**, 409-414.
Doan, B., Kwon, Y., Newton, R., Shim, J., Popper, E., Rogers, R., Bolt, L., Robertson, M. and Kraemer, W. (2003). Evaluation of a lower-body compression garment. *Journal of Sport Sciences*, **21**, 601-610.
Gruber, K., Ruder, H., Denoth, J. and Schneider, K. (1998). A comparative study of impact dynamics: wobbling mass model versus rigid body model. *Journal of Biomechanics*, **31**, 439-444.
Herzog,W. (2000). Skeletal Muscle Mechanics: *From Mechanisms to Function*. Wiley. Chichester. UK.
Kollias, I., Panoutsakopoulos, V. and Papaiakovou, G. (2004). Comparing Jumping Ability Among Athletes of Various Sports: Vertical Drop Jumping From 60 Centimeters. *The Journal of Strength and Conditioning Research*, **18**, 546-550.
Kraemer, W.J., Bush, J.A., Bauer, J.A., Newton, R.U., Duncan, N.D., Volek, J.S., Denegar, C.R., Canavan, P., Johnston, J., Putukian, M. and Sebastianelli, W. (1998). Influence of a compression garment on repetitive power output production before and after different types of muscle fatigue. *Sports Medicine, Training and Rehabilitation*, **8**, 163-184.
McComas, A.J. (1996). Skeletal muscle. *Form and function*. Human Kinetics. Champaign. IL.
Pain, M.T.G. and Challis, J.H. (2006). The influence of soft tissue movement on ground reaction forces, joint torques and joint reaction forces in drop landing. *Journal of Biomechanics*, **39**, 119-124.

Acknowledgement
Thank you to Jason Lake and Hannah Gordon for their assistance during this study.

A KINEMATIC DESCRIPTION OF THE POST PUBESCENT WINDMILL SOFTBALL PITCHING MOTION

David W. Keeley, Gretchen D. Oliver, Priscilla Dwelly, and Hiedi J. Hoffman

Department of Health Sciences, Kinesiology, Recreation, and Dance
University of Arkansas, Fayetteville, USA

The purpose of this study was to describe the kinematics of the windmill softball pitch. Throughout the first three phases of the movement, both the pelvis and the trunk were rotated to a closed position while the throwing shoulder was flexed and externally rotated, and the throwing elbow was flexed. During the latter stages of the movement, the pelvis and torso opened up to face the plate, the throwing shoulder moved through an arc of hyperextension and was internally rotated while the throwing elbow extended. The kinematics identified may contribute to overuse injuries commonly reported by post pubescent softball pitchers. However, due to the limited data describing the windmill softball pitch, addition research is needed.

KEY WORDS: pitching, kinematics, post pubescent, softball, windmill.

INTRODUCTION: Similar to the baseball pitch, the windmill softball pitch has been labelled as a major contributing factor in upper extremity overuse injuries to pitchers (Hill et al., 2004; Maffet et al., 1997; Rojas et al., 2009; Werner et al., 2005; Werner et al., 2006). However, unlike with baseball pitching, there is limited research available that describes the windmill softball pitch in full. Since post pubescent softball pitchers often throw up to 2000 pitches in a weekend (Werner et al., 2006), it is important to understand their pitching mechanics from a biomechanical perspective. To date, the authors have identified only one study describing the kinematics of collegiate softball pitchers (Barrentine et al., 1986). Thus, the purpose of this study was to quantify the biomechanics specific to the pitching motion in post pubescent softball pitchers. In doing so, this study attempts to aid in the development of a fundamental basis for how post pubescent pitchers perform as well as how arm injuries may be sustained during performance.

METHODS: Data Collection: Four collegiate and three high school female post pubescent softball pitchers (age 17.7 ± 2.6; height 169cm ± 5.4; mass 69.1kg ± 5.4) participated in the current study. Data collection sessions were conducted at the University of Arkansas HPER building and testing protocols were approved by that institution's ethics board. Prior to testing each pitcher provided consent.

Kinematic data were collected using The Motion Monitor® system (Innovative Sports Training, Chicago IL) and calculated using the ISB recommendations of the international shoulder group (Wu et al., 2005). Prior to the conduction of test trials, the space in which the pitchers were to throw was calibrated using the following protocol. The origin of the world axes system was located on a wooden platform located 25.4 cm from the extended range transmitter used to generate the electromagnetic field. The orientation of the world axis system was similar to that described by Wu and Cavanaugh (1995) and was such that the world x-axis extended from the center of the pitching rubber toward the center of home plate, the world y-axis extended was orthogonal to the x-axis and extended vertically from the center of the pitching rubber. The world z-axis was orthogonal to both x and y, directed laterally to the right. To calibrate the space, a wooden stylus was attached to an electromagnetic sensor and placed at the world axes system origin, 15 cm from the origin along both the x and z axes, and at one random position above the origin per manufacturer recommendations. Following the establishment and calibration of the world axes, the root mean square error in calculating the three-dimensional location of markers within the calibrated space was determined to be less than 20 mm. In addition to kinematic data, force data were collected to identify when stride foot plant occured. To collect force data, a 40 x 60 cm Bertec force plate (Bertec Corp, Columbus, OH) was recessed into the platform at the location where stride foot plant was to occur.

Once set-up was complete and the system and space were calibrated, electromagnetic sensors were placed on the thorax, sacrum, distal throwing forearm, right and left mid-humerus, and right and left mid-shank of each subject and unlimited time was allotted for the participants to warm-up based on their normal routine. Following the warm-up, each participant threw fastball windmill style deliveries using an official softball (12 in. circumference, 0.17 kg) to a catcher behind the plate 12.2 m away. Both kinematic and force data were collected at a rate of 1000 Hz and were synchronized using Motion Monitor® (Innovative Sports Training, Chicago, IL). A total of five trials were recorded after they were deemed a successful strike and between trials, pitchers were allowed a 40-60 s rest period.

Data Analysis: After completion of the trials, positional kinematic data were filtered independently along the x, y, and z-axis using a Butterworth filtering techniques described by Werner et al. (2005) with a cut off frequency of 13 Hz. For analysis the movement was divided into the five phases described in Figure 1 and defined by Maffet et al. (1997). Although the softball pitch typically incorporates six phases, this study focused on all activity prior to ball release and at ball release. Throwing kinematics were calculated using the Internation Society of Biomechanics recommendations for reporting joint motion (Wu et al., 2005) and included forward and lateral flexion of the trunk, axial hip and trunk rotation, shoulder flexion, shoulder internal rotation, elbow flexion, and forearm pronation.

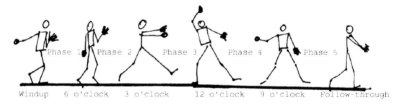

Figure 1: Windmill pitching phases.

RESULTS: During the pitch cycle, the hips close to a peak angle of -80° at 12 o'clock, before rotating open to an angle of -25° at ball release. Rotation of the upper torso follows a nearly identical pattern throughout the pitch cycle with the shoulder closing to an angle of -75° at 12 o'clock before rotating open to an angle of -17° at ball release. From 6 o'clock to 12 o'clock, the throwing shoulder is flexed to near 180° and externally rotated to -38°. From 12 o'clock through release, the throwing shoulder is moved through the near 180° arc of hyperextension back to an angle near 0°, as well as being internally rotated to an angle of -5°. In addition, the throwing elbow is initially hyper extended to an angle of -2° at 6 o'clock before being flexed to 26° at 12 o'clock. From here through release, the elbow was extended, reaching an angle of 4° at release. Also from 6 o'clock through 12 o'clock, the throwing forearm was pronated to 41° before being supinated back to an angle of 6° at release. The magnitudes for kinematic paramters at 6 o'clock, 12 o'clock, and release are presented in Table 1.

Table 1 Mean and standard deviation values for kinematic parameters at 6 o'clock, 12 oclock, and release

Parameter	6 o'clock	12 o'clock	release
hip rotation (°)	-12 ± 6	-80 ± 26	-25 ± 9
shoulder rotation (°)	-5 ± 7	-75 ± 22	-17 ± 11
shoulder flexion (°)	0 ± 2	164 ± 16	3 ± 4
shoulder internal rotation (°)	34 ± 16	-38 ± 15	-5 ± 8
elbow flexion (°)	-2 ± 4	26 ± 18	4 ± 6
forearm pronation (°)	18 ± 11	41 ± 29	6 ± 10

DISCUSSION: The sport of fast pitch softball is essentially a scaled down version of baseball which allows for comparisons across the two sports to be made. In terms of pitching, it has been the common perception that the windmill pitching motion in softball is a 'more natural movement' than the baseball pitching motion. This view has led to the opinion that the underhand motion is less stressful on the arm and may be the reason that softball pitching is less studied than baseball pitching. However, it has been found that the major kinematic and kinetic characteristics associated with pitching mechanics are similar across the two sports (Barrentine et al., 1998; Werner et al., 2006).

One common injury reported by post pubescent softball pitchers is anterior shoulder pain (Barrentine et al., 1998). The etiology of this pain may be related to both mechanical and musculoskeletal characteristics observed in female softball pitchers. First, during the initial phases (Phase 2 and 3) of the movement (i.e. 6 o'clock to 12 o'clock in softball) the throwing shoulder is both flexed and externally rotated. This high angle of shoulder flexion, coupled with the observed external shoulder rotation in softball pitchers may result in elevated anterior forces that may contribute to anterior/superior translation of the humeral head. This humeral translation has the potential to result in subacromial impingement injuries as well as posterior shoulder impingement injuries in post pubescent softball pitchers.

Another possible scenario for anterior shoulder problems in softball pitchers may result from a combination of both mechanical characteristic at the shoulder and elbow, and deficiencies in muscular strength in the bicep. Women typically exhibit less muscle mass and strength in the upper torso and arms when compared to their male counterparts (Miller, A.E. et al., 1993). As shown in Table 1, the elbow remains in a position near full extension throughout the windmill pitch. This, coupled with the large rotational arc of the arm throughout the movement may result in increased distraction forces at the shoulder (Barrentine et al., 1998). To resist this distraction the muscles of the rotator cuff, along with the biceps brachii fire to stabilize the head of the humerus against the glenoid fossa of the scapula (Glousman et al., 1988). As a result of this repeated increase in biceps activity, post pubescent softball pitchers may be at a greater risk of developing chronic tendonitis of the biceps and/or injury to the biceps labrum complex. It has also been reported that subacromial impingement may be a major factor in contributing to primary disease of the rotator cuff (Neer 1983). If the kinematics of softball pitching do contribute to subacromial impingement that resuls from incereased superior humeral translation, softball pitchers may experience repetitive microtrauma to the very muscles responsible for stablizing the humerus.

CONCLUSION: This study provides a kinematic description of the post pubescent windmill pitching motion. Although the data in the current study agree with previous reports (Barrentine, et al., 1998; Werner et al., 2006) in describing the actions occurring throughout the movement, they also identify some of the key factors that may be associated with the anterior shoulder pain commonly reported by post pubescent softball pitchers. In addition, this study, along with Barrentine et al., (1998) and Werner et al., (2006) shows that studies of the windmill softball pitch can be conducted in similar fashion to those for baseball pitching. Thus, because of the limited amount of literature currently available that describes the windmill softball pitch, further investigation into the etiology of injury, as well as the differences between genders and ages is needed.

REFERENCES:
Barrentine, S. Fleisig, G., Whiteside, J., Escamilla, R.F., & Andrews, J.R. (1998). Biomechanics of windmill softball pitching with implications about injury mechanisms at the shoulder and elbow. *Journal of Orthopedic and Sports Physical Therapy*, 28, 405-415.

Glousman, R., Jobe, F., Tibone, J., Moynes, D., Antonelli, D., & Perry, J., (1988). Dynamic electromyographic analysis of the throwing shoulder with glenohumeral instability. *Journal of Bone and Joint Surgery*, 70, 220-226.

Miller, A.E., MacDougall, J.D., Tarnopolsky, M.A., & Sale, D.G. (1993). Gender differences in strength and muscle fiber characteristics. *European Journal of Applied Physiology and Occupational Physiology*, 66, 254-262.

Neer CS: Impingement lesions. Clinical Orthopedics 173: 70–77, 1983

Rojas, I.L., Provencher, M.T., Bhatia, S., Foucher, K.C., Bach, B.R., Romeo, A.A., Wimmer, M.A., & Verma, N.N. (2009). Biceps activity during windmill softball pitching. *American Journal of Sports Medicine*, 37, (3), 558-565.

Werner, S.L., Guido, J.A., McNeice, R.P., Richardson, J.L., Delude, N.A. & Stewart, G.W. (2005) Biomechanics of youth windmill softball pitching. *American Journal of Sports Medicine*, 33 (4), 552-560.

Werner, S.L., Jones, D.G., Guido, J.A., & Brunet, M.E. (2006). Kinematics and kinetics of elite windmill softball pitching. *American Journal of Sports Medicine*, 34, (4), 597-603.

Wu, G., & Cavanagh, P.R. (1995). ISB recommendations for standardization in the reporting of kinematic data. *Journal of Biomechanics*, 28, 1257-1261.

Wu, G., van der Helm, F.C.T., Veeger, H.E.J., Makhsous, M., Van Roy, P., Anglin, C., et al. (2005). ISB recommendation on definitions of joint coordinate systems of various joints for the reporting of human joint motion – Part II: shoulder, elbow, wrist, and hand. *Journal of Biomechanics*, 38, 981–992.

Acknowledgement
The authors would like to thank Mr. Bob Carver for his support of the biomechanics and sports medicine programs at the University of Arkansas.

DEVELOPMENT OF A RECORDING SYSTEM TO EMPIRICALLY ANALYSE THE SHOOTING CHARACTERISTICS OF A CLAY PIGEON SHOOTER

Alan Swanton[1,2] and Ross Anderson[1]

Biomechanics Research Unit, University of Limerick, Limerick, Ireland[1]
Coaching Ireland, Limerick, Ireland[2]

KEY WORDS: real time acceleration measurement device, clay pigeon shooting, motion analysis.

INTRODUCTION: There is a dearth of published information with regard to the sport of clay pigeon shooting. The discipline involves the shooter firing a shotgun at a disc of clay, which is released within a known area but with an unknown trajectory. If the shooter hits the target, the clay breaks and the shooter receives instant feedback on the outcome of the shot. However, if the shooter misses the target the situation requires more analysis. The coach would hugely benefit from a method which gives empirical evidence outlining the kinematics of the shot. This project wil first attempt to identify the relavent characteristics i.e. timings within the shot, acceleration/movement of the gun. Once identified, the variability within these characteristics will be evaluated against the shot outcome results from a commercially available shooting simulator (Dryfire, Derby, UK).

METHODS: The preliminary stage of the project is to verify that the real time acceleration measurement device (RTAMD) correlates with acceleration data calculated from a motion analysis system (Motion Analysis Corporation, CA, USA). A reflective marker was placed directly on the RTAMD which was moved along the surface of a table.

RESULTS:

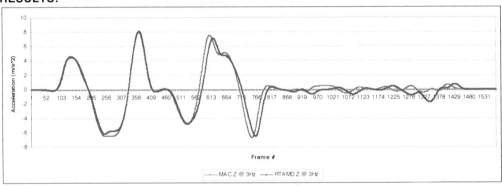

Figure 1 Comparison graph illustrating difference between acceleration (Z direction) calculated using a motion analysis system (MAC) and the RTAMD

Figure 1 illustrates the agreement between calculated acceleration in RTAMD and MAC. The root mean square error between the calculated accelerations is 0.565646 m/s/s. Preliminary results from a comparison between MAC and RTAMD are very promising.

CONCLUSION: This is a novel research project which attempts to develop a measurement scheme (coaching tool) for analysing the shooting characteristics of clay pigeon shooters, thus bridging the gap between lab based testing and in the field analysis.

Acknowledgement

I would like to thank Shane Corrigan (shooter) and Kevin Kilty, performance director within the Irish Clay Pigeon Shooting Association for their help through out this project.

COMPLEX TRAINING: AN EVALUATION OF POTENTIATION BETWEEN A 3RM BACK SQUAT AND A SQUAT JUMP

Darragh Graham and Andrew Harrison

Biomechanics Research Unit, University of Limerick, Limerick, Ireland

KEY WORDS: complex training, back squat, starting strength, rugby, sledge.

INTRODUCTION: Complex training (CT) is increasingly popular among strength and conditioning coaches. CT hypothesises that a near maximal muscle contraction will enhance the explosive capabilities of the muscle given the exercises are biomechanically similar (Docherty et al., 2004). Previous CT research has focused on intra-complex potentiation between near maximum exercises and a similar stretch-shortening cycle exercise. The effect of CT on starting strength (SS) has yet to be explored. The optimal rest interval between the loaded exercise and the explosive exercise is somewhat unclear. Comyns et al. (2006) investigated potentiation between a 5RM back squat (BS) and a counter movement jump. Results varied between individuals. The purpose of this study was to determine if a heavy loaded exercise (3RM BS) results in a performance increase on a SS exercise (squat jump).

METHODS: Twenty male rugby players, proficient with the technique of the BS and squat jump (SJ) participated in this study. All subjects were part of a professional (n=13) or a semi-professional rugby academy (n=7). Testing protocol consisted of a pre-test, 3RM back squat and a post-test. Pre and post tests consisted of 1 SJ every minute for 10 minutes. A pilot study confirmed 1 minute sufficient time for recovery between SJ's. The SJ starting position and 3RM BS depth was 90° flexion of the knee. The first post test SJ was performed one minute after the 3RM BS. All SJ's were performed on a sledge apparatus inclined at 30° as described by Harrison et al. (2004). Each SJ was recorded on an AMTI OR6-5 force platform mounted at right angles to the sledge apparatus sampling at 1000 Hz. For all SJ's subjects were guided into position through feedback from the experimenter, and a marker on the sledge. Once in the correct position subjects jumped approximately 2's thereafter. For each jump; height jumped, peak ground reaction force, rate of force development and SS were calculated.

RESULTS & DISCUSSION: Data from pre and post tests were compared. In figure 1, post-test mean max and min scores significantly changed from pre-test averages. Further analysis investigated individual potentiation and time to max potentiation indicated optimal rest intervals. Criterion for potentiation was post-test scores that surpassed the pre-test plus the typical error. Using this criterion, all subjects potentiated. Results from this study will aid coaches in designing weight training programmes that contain complex pairs.

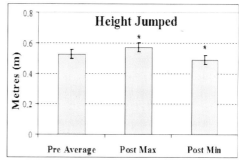

Figure 1: Group height jumped increases
(* = significant change from pre average, p<0.01)

REFERENCES:
Comyns, T.M., Harrison, A.J., Hennessy, L.K., and Jensen, R.L (2006). *The Optimal Complex Training Rest Interval for Athletes from Anaerobic Sports*. Journal of Strength and Conditioning Research. 20(3), 471–476.
Docherty, D., Robbins, D., Hodgson, M. (2004). *Complex Training Revisited: A Review of its Current Status as a Viable Training Approach*. National Strength and Conditioning Association. 26(6), 52-57.
Harrison, A.J., Keane, S.P., and Coglan, J. (2004). *Force-velocity relationship and stretch-shortening cycle function in sprint and endurance athletes*. Journal of Strength and Conditioning Research. 18(3), 473-479.

Acknowledgements: Joseph O'Halloran, Dr. Eamonn Flanagan and Dr. Tom Comyns

RELIABILITY OF DROP JUMP VARIATIONS IN PERFORMANCE DIAGNOSTICS

Gunther Kurz, Diana Lang, Anne Richter and Hermann Schwameder

BioMotion Center, Department of Sport and Sport Science, Karlsruhe Institute of Technology (KIT), Germany

KEY WORDS: drop jump, arm-swing, SSC

INTRODUCTION: In several sports athletes have to produce maximal power for accelerating the body and/or sport equipment (e.g. long jump, shot putt, basketball, etc.). The primary indicator used for estimating power output in drop jumps are the peak force rate (Fex) and the peak force (Fmax). These variables also characterise the reactive force ability of an athlete during a stretch-shortening-cycle (SSC). In order to use the accumulated energy produced in the eccentric phase during the concentric contraction, the contact please revise must not be longer than 200 ms (Komi 1984). Reactive force abilities are commonly tested in drop jumps (DJ). The accomplishment of DJ is variously described in the literature, so no standardized conditions are given (Baca 1999). One corresponding aspect is the arm position (free arm-swing vs. arms akimbo). Particularly, jumps with arm-swing require a higher grade of coordination (e.g. basketball) and are associated with an enhanced jump height. DJs with arms akimbo are often used in subjects with less sports experience or if the performance diagnostics primarily focuses on the power output in isolated leg extensions. Another aspect is the method of calculating the jump height from the measured force data. The calculation of the jump height from the momentum (based on the force-time-curve) might be inaccurate as the exact drop height is not known. Consequently, the jump height often is calculated from the flight time. This method, however, is also inaccurate if the landing position differs substantially from the take-off position. This problem might be solved performing a consecutive double DJ. Thus, the aim of the study is to investigate different DJ variations with respect to reliability.3

METHODS: 42 students experienced in DJ (Age=23.4 ± 2.6 yrs) participated in this study. All subjects were asked to perform three drop jumps in four different conditions: arm-swing - single contact (AS), arm-swing - double contact (AD), arm-akimbo - single contact (NAS), arm-akimbo – double contact (NAD). The measurements were repeated one week later for the same conditions. The subjects were asked to jump as high as possible. The ground reaction force was measured with an AMTI force plate (1000Hz). Flight time and ground reaction time were calculated from the recorded force-time curves. Data were analysed via Perason correlation.

RESULTS AND DISCUSSION: Test-retests correlated as follows: r_{AS}= 0.737; r_{AD}= 0.001; r_{NAS}= 0.727; r_{NAD}= 0.820. The correlation between single and double support conditions were: $r_{NAS-NAD}$= 0.836 and r_{AS-AD}= 0.852. Jump height with arm-swing was higher than jump height with arms akimbo (0.347 ± 0.067 m vs. 0.287 ± 0.053 m). This is in line with previous studies (Harman et al., 1990, Gerodimos et al., 2008). In conclusion, single contact jumps (AS, NAS) are sufficient for determine the reactive ability of an athlete. Only jump experienced athletes or specific movements in sports justify double contact jumps with arm-swing.

REFERENCES:

Baca, A. (1999). A comparison of methods for analyzing drop jump performance. *Medicine and Science in Sports and Exercise.* 31(3), 437-442

Gerodimos, V., Zafeiridis, A., Perkos, S., Dipla, K., Manou, V. & Kellis, S. (2008). The contribution of stretch-shortening cycle and arm-swing to vertical jumping performance in children, adolescents, and adult basketball players, Pediatric Exercise Science, 20, 379-389.

Harman, E.A., Rosenstein, M.T., Frykman, P.N. & Rosenstein, R.M. (1990). The effects of arms and countermovement on vertical jumping. *Medicine and Science in Sports and Exercise*, 22(6), 825-833.

Komi, P.V. (1984). Physiological and biomechanical correlates of muscle function: effects of muscle structure and stretch-shortening cycle on force and speed. In R.L. Terjung (ed.). Exercise and Sports Sciences Reviews. Vol 12, pp. 81-121. Collamore Press, Lexington, Mass.

TENNISSENSE: A MULTI-SENSORY APPROACH TO PERFORMANCE ANALYSIS IN TENNIS

Luke Conroy[1], Ciarán Ó Conaire[2], Shirley Coyle[2], Graham Healy[2], Philip Kelly[2], Damien Connaghan[2] Noel E. O'Connor[2], Alan F. Smeaton[2], Brian Caulfield[1], Paddy Nixon[1]

CLARITY Centre - University College Dublin[1] & Dublin City University[2], Ireland

KEY WORDS: Accelerometer, Tennis, Stroke, Coaching

INTRODUCTION: There is sufficient evidence in the current literature that the ability to accurately capture and model the accelerations, angular velocities and orientations involved in the tennis stroke could facilitate a major step forward in the application of biomechanics to tennis coaching (Tanabe & Ito, 2007; Gordon & Dapena, 2006). The TennisSense Project, run in collaboration with Tennis Ireland, aims to create the infrastructure required to digitally capture physical, tactical and physiological data from tennis players in order to assist in their coaching and improve performance. This study examined the potential for using Wireless Inertial Monitoring Units (WIMUs) to model the biomechanical aspects of the tennis stroke and for developing coaching tools that utilise this information.

METHODS: The TennisSense Technology Infrastructure comprises a UbiSense Spatial Localisation System and a Wireless Inertial Monitoring Unit (WIMU) worn on the forearm (racquet arm) of the player, that records acceleration, angular velocity and orientation in all three axes; using accelerometer, gyroscope and magnetometer sensors respectively. In addition, a total of nine cameras are positioned around the court, with pan, tilt and zoom capability. We recorded five separate training sessions of 20 minutes duration, using a single WIMU on the player's forearm, the UbiSense System and the full nine camera rig.

RESULTS & DISCUSSION: For this initial study, we looked solely at the accelerometer data from the WIMU that was synchronised with the camera data. The data was graphed using MATLAB® and compared to the video data. The most useful sequences of data were training drills; where the player attempts to *reproduce an identical shot each time*. It was clear from a manual inspection of these drill sequences; that significant similarities existed in the graphed accelerometer data between correctly executed shots of an *identical type*.

CONCLUSION: This examination of the results leads us to conclude that accelerometer data from 1 or more on-body sensors, could be readily used in a tool to automatically identify shot type, shot quality and subsequently the biomechanical determinants of correct/incorrect technique; based on a golden template of accelerometer data with appropriate matching criteria. The utility of such a tool is inferred from the literature examining the contributions of biomechanical factors to tennis stroke outcomes. Wrist flexion, radio-ulnar pronation and humeral internal rotation are identified as major contributors to speed of serve (Tanabe & Ito, 2007). Elbow extension and wrist flexion also appeared to be the major contributors to generation of racquet speed (Gordon & Dapena, 2006).

REFERENCES:
Tanabe & Ito (2007) A three-dimensional analysis of the contributions of upper limb joint movements to horizontal racket head velocity at ball impact during tennis serving *Sports Biomechanics* Sept 2007; 6(3): 418-433
Gordon & Dapena (2006) Contributions of joint rotations to racquet speed in the tennis serve *Journal of Sports Sciences* Jan 2006; 24(1): 31-49
Acknowledgement
This work is supported by Science Foundation Ireland under grant 07/CE/I1147.
This work is supported by the Tyndall National Institute under NAP Grant 209.

RELIABILITY OF TIME TO STABILIZATION IN SINGLE LEG STANDING

Mahendran Kaliyamoorthy and Randall L. Jensen
Department of Health, Physical Education and Recreation
Northern Michigan University, Marquette, Michigan, USA

The purpose of this study was to evaluate the reliability of time to stabilization in single leg standing. Time to Stabilization (TTS) is a measurement which can be used to analyze both static and dynamic stability of an individual. Twenty seven college students participated in this study. All the subjects were asked to perform a single leg standing task on the force platform placed on the floor. Five trials were performed by each subject to estimate the reliability of TTS. TTS was calculated as the time taken to reach body weight and stay within 5% of the body weight for one second. The results showed that the reliability was optimal with moderate correlation for the first two trials.

KEY WORDS: stabilization, posture, balance, dynamic stability.

INTRODUCTION: Time to Stabilization (TTS) is a measure recently proposed to evaluate both static and dynamic stability (Wikstrom et al, 2004). This measure incorporates both the sensory and mechanical systems to analyze one's balance. TTS as a measure of functional balance is used in various settings including the care of elderly (Ross et al, 2003) and in children (Cook et al, 2003). TTS is commonly used in measuring functional balance following ankle injury, or bracing of the ankle (Ross et al, 2005), in lower extremity muscle injury and fatigue (Wikstrom et al, 2004). Measures of TTS could also provide researchers with a mechanism to assess the importance of balance in an activity (Flanagan et al, 2008). TTS may also be a useful tool to evaluate one's balance and effect of balance on body posture. Activities like single leg balance are necessary in more demanding sports such as cross country skiing in which the skiers predominantly move and shift weight alternately on each leg during the gliding phase. Long duration activities may also cause fatigue which can alter the balance and affect performance. Hence we chose single leg standing as a task to assess TTS component using a force platform. The main purpose of this study was to estimate the reliability of TTS as a measure of dynamic stability in normal population. We hypothesized that there would be consistency in the time taken to reach a stable state during single leg standing for five trials.

METHODS: Twenty seven college students (Mean ± SD: age = 20.6 ± 3.0 years) were recruited for the study, 16 female and 11 male. Written informed consent was provided by all twenty seven participants for the study, which was approved by the University Institutional Review Board (# HS08-192).

Single leg stance was performed on a standardized force platform (FP) (OR6-5-2000, AMTI, Watertown, MA, USA) mounted on the ground. Ground Reaction Force (GRF) data were collected at 1000 Hz, real time displayed and saved with the use of computer software (NetForce 2.0, AMTI, Watertown, MA, USA) for later analysis.

The aim of the current study was to estimate TTS using a simple task of single leg standing on the FP, which involves stepping onto the FP and maintaining single leg balance for 30 seconds. Before the actual testing trials, participants performed several practice trials to practice the movement required for testing. Step Length was measured for all subjects before the test session. Subjects were asked to stand from one's step length distance away from the FP and asked to step on a line drawn in center of the FP (Jacobs et al, 2006). Subjects were instructed to step on the force platform and asked to achieve and maintain a stable state as soon as possible. Subjects were allowed to use arm movements to maintain their balance.

Time taken to attain stability was calculated from the acquired data (Flanagan et al, 2008) Time to stabilization was measured by the time taken for vertical ground reaction force to reach and remain within 5% of their own body weight for one second of duration. (Wikstrom et al, 2005)

All statistical analyses of the data were carried out in SPSS © (Version 16.0). Trial-to-trial reliability analysis of recorded variables used both single (ICC_{single}) and average (ICC_{ave}) measures intra-class correlations of absolute agreement. A one-way repeated measures ANOVA was used to determine possible differences between trials. Outliers were tested and subjects with data in excess of three standard deviations were eliminated from the analysis. The criterion for significance was set at an alpha level of $p \leq 0.05$.

Munro (2001) has suggested that Intraclass Correlation Coefficient (ICC) values be used to describe the degree of reliability with the following descriptors for ICC values: 0.00 to 0.25 = little, if any correlation; 0.26 to 0.49– low correlation; 0.50 to 0.69 – moderate correlation; 0.70 to 0.89 – high correlation and 0.90 to 1.00 – very high correlation. Thus high ICC were sought and deemed acceptable for comparison.

RESULTS: Data were obtained from all twenty seven subjects. Two subjects were removed due to data outliers and we performed One-way repeated measures ANOVA for twenty-five subjects. Thus to estimate the reliability of TTS, we performed a statistical analysis with twenty-five subjects. To estimate how many trials would be necessary to obtain acceptable reliability estimates of TTS we analyzed data using the first two trials, the first three trials, the first four trials and all five trials. The results of the data analysis are shown in the Tables 1 and 2. From the analysis no significant differences were found between the means of two, three, four or five trials with p>0.05

Table 1: Descriptive data for TTS trials: N=25

	Trial 1	Trial 2	Trial 3	Trial 4	Trial 5
Mean (s)	961.2	963.2	958.1	1000.5	914.4
Stand. Deviation (s)	311.7	353.1	263.3	267.5	254.8
Range (s)	1292	1545	1306	1006	1094

Table 2: Intraclass Correlation Coefficient and 95% Confidence Interval (ICC: 95% CI) for two, three, four and five trials of the TTS (n=25).

Number of Trials	ICC-single measures (95% CI)	ICC-Average measures (95% CI)
2	0.494 (0.131 – 0.740)	0.662 (0.232 – 0.851)
3	0.352 (0.105 – 0.603)	0.619 (0.260 – 0.820)
4	0.183 (0.007 – 0.420)	0.472 (0.029 – 0.744)
5	0.260 (0.097 – 0.478)	0.637 (0.350 – 0.821)

ICC- Intraclass correlation, CI-Confidence interval

ICC values for the various trials showed differences in repeatability of results for the TTS. $ICC_{average}$ measure for two (0.662), three (0.619) and five (0.637) trials had moderate correlations compared with four trials analysis (0.472) which was a low correlation. ICC_{single}

measures show all combinations of different trial analyses had low correlations (see Table 2). Since there was no significant difference among five trials we did not perform post hoc test for the data.

DISCUSSION: This study is the first that is known to the authors to estimate the reliability of TTS for single leg standing in an adult population. Results indicate that the reliability differed depending on the number of trials used to estimate reliability of TTS. From the data analysis no significant differences were found among the trials whether the comparisons consisted of two, three, four or five trials ($P > 0.05$).

Despite a lack of difference between the trials, the four trials analysis resulted in low $ICC_{average}$ (0.472) compared to two, three and five trial analyses. $ICC_{average}$ measures for two trials was $r = 0.662$, three trials $r = 0.619$ and five trials $r = 0.637$ indicating moderate correlation among these numbers of trials. ICC_{single} measure for two trials was $r = 0.494$, three trials $r = 0.352$, four trials $r = 0.183$ and five trials $r = 0.260$. Thus although average measures ICC is moderate for all but four trials, single measures ICC is low for all trials (Table 2).

The results showed that with two, three and five trials analysis the correlation stayed within moderate level of significance. Hence to estimate TTS for single leg standing it may be sufficient to perform just two trials instead up to five trials.

Subjects participated in a practice session to ensure they could reproduce the actual movement required for the study. Only a few subjects could not maintain their balance in the initial phase of stepping on the FP. Because all the subjects were healthy college students without any known balance problems, it is likely there was very little difference between the individuals and this may have decreased the variability throughout additional trials. The decreased variability would have allowed a greater crossover between subjects, thus lowering the ICC. Another possible reason for the difference among trials could have been due to attention demands during trials. Since all subjects performed several practice trials and testing, lack of attention in subsequent trials may be a factor causing difference among trials (Simoneau et al, 2006). Individual variation is also one of the factors to be considered in the low significance of reliability among five trials. Hence the most possible reason we could think about this variation could be attention demands of the subjects while performing the single leg standing on the FP.

CONCLUSION: In summary TTS displays moderate reliability based on the ICC and lack of difference across trials for 2, 3, or 5 trials. However, because two trials provide the same degree of reliability as performing more (up to five trials), we recommend the use of two trials when assessing standing single leg balance performance.

This may also be applicable to other tests used to measure TTS as well. Since single leg standing is one of the most common methods used to test the lower extremity balance, there needs to be consistency of results and reliability of single leg standing in assessing lower extremity balance. More studies are recommended to reproduce similar results and provide further information on the reliability of TTS in single leg standing.

REFERENCES:
Cook, A.-S, Hutchinson S, Kartin D, Price R and Wollacott M. (2003) Effects of balance training recovery in children's with cerebral palsy. *Dev Med & Child Neurol.* 45, 591-602
Flanagan E.P, Ebben W.P and Jensen R.L (2008). Reliability of the reactive strength index and time to stabilization during depth jumps. *J Strength Cond Res.* 22, 1677-1682
Jacobs N, Mokha M.B, Ludwig K and Krkeljas Z. (2006) Do ankle stabilizers influence dynamic stability in persons with functional ankle instability? In *Proceedings of the XXIV International Symposium of Biomechanics in Sports*, (H. Schwameder, G. Strutzenberger, V. Fastenbauer, S. Lindinger, E. Müller, editors) 383
Munro B.H. (2001) *Statistical Methods for Health Care Research*, fourth ed., Lippincott, Philadelphia.

Ross S.E and Guskiewicz K.M (2003). Time to stabilization: A method for analyzing dynamic postural stability. *Athl Ther Today.8*, 37-39

Ross S.E, Guskiewicz K.M and Yu B (2005). Single-leg jump landing stabilization times in subjects with functionally unstable ankle. *J Athletic Train.40*, 298-304

Simoneau M, Begin F and Teasdale N (2006). The effects of moderate fatigue on dynamic balance control and attention demands. *J Neuroeng and Rehabil.3*, 1-9

Wikstrom E.A, Powers M.E, Tillman M.D. (2004). Dynamic stabilization time after isokinetic and functional fatigue. *J Athletic Train.39*, 247-253

Wikstrom E.A, Tillman M.D and Borsa P.A (2005). Detection of dynamic stability deficits in subjects with functional ankle instability. *Med Sci Sports Exerc.37*, 169-175

EFFECT OF FATIGUE ON DYNAMIC BALANCE AFTER MAXIMUM INTENSITY CROSS-COUNTRY SKIING

Mahendran Kaliyamoorthy, Phillip B. Watts, Randall L. Jensen and James A. LaChapelle

Department of Health Physical Education and Recreation, Northern Michigan University, Marquette, Michigan, USA

KEY WORDS: XC Skiing, maximum exercise, balance, fatigue.

INTRODUCTION: Cross-country skiing stresses most of the joints, muscles and tendons in the body giving an overall workout. Skiing requires aerobic and anaerobic power, muscular strength and a variety of complex motor abilities including reaction time, agility, balance, coordination and attention demands (Emily & Arthur, 1989). Muscular fatigue is a key factor which can influence performance via impaired joint proprioception and postural control. Fatigue alters the force generation capacity of the muscle and ultimately leads to task failure (Mahyar et al, 2007). Injury risk increases as time duration of the skiing increases (Smith, Matheson & Meeuwisse, 1996). The maintenance of body posture and balance is an essential requirement for performance of daily tasks and sporting activities. Thus fatigue could affect a skier's performance through an effect on balance. Fast starts at the beginning of races and short intense efforts required for ascending hills could result in periods of fatigue that could affect balance and performance. Hence fatigue may either result in injury or affect the finish time of the skiers. The main purpose of this study was to evaluate how dynamic balance of the skiers can be influenced by fatigue states following maximum exercise

METHODS: Twelve experienced collegiate cross-country skiers participated in this study. All subjects were given oral and written instruction about the purpose of study and the methods used. An oversized treadmill was used for roller-skiing and a standardized force platform to measure dynamic balance. All subjects were asked to perform roller-skiing with a combination of V1 and/or V2 skating techniques on the oversized treadmill to the point of exhaustion to determine maximum oxygen uptake level. The dynamic balance was measured before and after the high intensity roller skiing test using a force platform (OR6-5-2000, AMTI, Watertown, MA, USA). All subjects were asked to step onto the force platform and maintain balance as soon as possible. The dynamic balance was measured using the time taken to reach stability i.e., from the point of the step onto the force platform until they achieved and maintained a steady state of vertical ground reaction force within 5 percent of their body weight. The data were recorded in AMTI NetForce 2.0 and were analyzed for the vertical force generation when the subjects stepped on to the force platform, using the body weight measured in Newton units. Trials post fatigue occurred within 60 s of fatiguing roller-ski exercise. Subjects performed three trials at 30 s intervals before and after roller-skiing.

RESULTS AND DISCUSSION: Data analysis revealed an interaction of trials and the fatigue state in balance measurement ($p<0.05$). Thus fatigue affected balance in skiers across the trials after high intensity exercise. Main effects differences across trials and fatigue state were not significant ($p = 0.635$ and $p = 0.059$ respectively). This may help to explain changes in performance with fatigue.

REFERENCES:

Emily MH and Arthur LD (1989). Characteristics of elite male and female ski racers. *Med Sci Sports Exerc 3*:153-158.

Mahyar S, Mojgan M, Ismaeil E and Amir MA (2007). Changes in postural stability with fatigue lower extremity frontal and sagittal plane movers. *J of Gait Posture 26*:214-215.

Smith M, Matheson GO and Meeuwisse WH (1996). Injuries in cross-country skiing: a critical appraisal of the literature. *Sports Med 21*:239-50

ISBS 2009

Poster Session PS2

IS PASSIVE PLANTAR FLEXION TORQUE DETERMINANT OF LOWER LIMB STIFFNESS?

Watanabe. K and Yanagiya. T

Laboratory for Biomechanics, Juntendo University, Chiba, Japan

KEY WORDS: passive plantar flexion torque, leg stiffness

INTRODUCTION: It has been supposed that leg stiffness affect repeated jump performance, such as hopping. Previous study indicated that ankle joint stiffness is important in leg stiffness (Arampatzis et al. 2001). Furthermore, it was demonstrated that Achilles tendon properties could contribute to the total mechanical work during hopping (Lichtwark and Wilson 2005). On the other hand, it is known that passive plantar flexion torque (PT) is determined by tendon properties (Kawakami et al. 2008). However, no study has been focused on the effect of PT on leg stiffness in hopping. The purpose of this study was to investigate the relationship between passive plantar flexion torque and leg stiffness.

METHODS: Ten female students participated as subjects and gave written informed consent prior to the study. They seated with the right knee extended and ankle positioned at every ten degrees from +30 degrees to -30 degrees (0 degrees; neutral position), and PT was measured simultaneously. Furthermore, they performed in-place hopping, matching metronome beats at 3.0 and 1.5Hz on a force platform. During hopping, subjects were filmed from the right side in sagittal plane with high-speed video camera at a sampling frequency of 300Hz. Leg stiffness was calculated as the change in ground reaction force divided by the vertical change in the center of mass during contact.

RESULTS: Table 1 represented physical characteristics, PT and leg stiffness. PT had no significant correlation with leg stiffness at 1.5Hz ($p > 0.05$), though significant correlation was found with leg stiffness at 3.0 Hz ($r=0.787$, $p<0.01$, Fig.2).

Table 1 physical characteristics, PT and leg stiffness

		mean ± SD
Hight (m)		1.63 ± 0.07
Weight (kg)		59.2 ± 9.0
Age (yrs.)		20.5 ± 0.8
Passive TQ (Nm)		7.1 ± 1.3
Leg Stiffness (kN/m)	1.5Hz	0.41 ± 0.12
	3.0Hz	0.84 ± 0.16

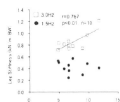

Fig.1 Relationship PT and leg stiffness

DISCUSSION: PT was significantly correlated with leg stiffness on faster counter movement. PT might have influenced the leg stiffness in hopping because it was related to the elasticity of Achilles tendon (Kawakami et al. 2008). If the contribution of the Achilles tendon increases along with the speed of the movement, this results suggest that PT may affect leg stiffness in hopping.

CONCLUSION: PT affected leg stiffness on fast repeated jump performance.

REFERENCES:

Kawakami Y, Kanehisa H, Fukunaga T. (2008). The relationship between passive ankle plantar flexion joint torque and gastrocnemius muscle and achilles tendon stiffness: implications for flexibility. *JOrthop Sports Phys Ther*.38(5), 269-76.

Arampatzis A, Schade F, Walsh M, Brüggemann GP. (2001) Influence of leg stiffness and its effect on myodynamic jumping performance *J Electromyogr Kinesiol*. 11(5):355-64
G. A. Lichtwark and A. M. Wilson. (2005). In vivomechanical properties of the human Achilles tendon during one-legged hopping. *J Exp Biol*. 208(24):4715-25

THE EFFECT OF FATIGUE ON REACTIVE STRENGTH IN ANTERIOR CRUCIATE LIGAMENT RECONSTRUCTED INDIVIDUALS

Eamonn Flanagan[1], Randall Jensen[2], Andrew Harrison[1] and Daniel Rickaby[2]

Biomechanics Research Unit, University of Limerick, Limerick, Ireland[1]
Department of Health, Physical Education, Recreation, Northern Michigan University, Marquette, MI, USA.[2]

This study examined the effect of fatigue on reactive strength in six subjects who had undergone surgical repair of the anterior cruciate ligament and returned to full sporting activity. Subjects' performance in rebound jumps on a force sledge apparatus was analysed before, during and after a maximal fatigue protocol. Flight time, contact time and reactive strength index were measured for each jump. No differences in reactive strength index were observed between legs in the non-fatigued or fatigued state. The data indicated that the subjects may recover reactive strength at a slower rate on the involved leg following maximal exercise. The data also indicates that reactive strength index is a highly suitable variable to examine during maximal SSC fatigue tasks.

KEY WORDS: Stretch shortening cycle, jumping, ground contact time.

INTRODUCTION: The impetus for this study was to address the gap in the literature on the status of reactive strength in athletes who have recovered from surgical repair of anterior cruciate ligament (ACL) rupture and subsequent rehabilitation. Injuries more often occur in the latter stages of sporting events when participants are in a fatigued state (Ostenberg & Roos, 2000; Zemper, 1989). At a basic level fatigue can be described as a loss of maximal force generating capacity, a loss of maximal power output (Vøllestad, 1997) or a failure to sustain further exercise at a required level (Strojnik & Komi, 1998). Fatigue can contribute to injury through a number of mechanisms. It is thought that fatigued muscles lose their strength and ability to act as a protective mechanism for anatomically weak joints. During vigorous activity, the musculature around a joint can fatigue at different rates due to different muscles containing varying proportions of fatigable and fatigue resistant muscle fibres. Since muscle fatigue leads to decreases in muscle force production (Vøllestad, 1997), force production around the joint can become unbalanced due to the relative fatigue state of each individual muscle. This can lead to abnormal or unnatural motions of the joints creating unbalanced and excessive stress distributions that contribute to injury (Kumar, 2001). Rozzi et al. (1999) observed that subjects exhibited deficiencies in proprioception and alterations in the muscular activity of the knee musculature in a fatigued state and concluded that fatigue predisposes athletes to an increased risk of knee ligament injury. Augustsson (2004) examined the effects of a pre-exhaustion exercise protocol on between-leg differences ACL reconstructed (ACL-R) subjects. In baseline testing, all subjects exhibited >90% leg symmetry compared with the fatigued condition in which two-thirds of patients showed abnormal leg symmetry (< 90%). Fatiguing exercise can increase the sensitivity and validity of assessment when examining the lower limb function of ACL-R subjects. Patients who are only examined in a non-fatigued state could be incorrectly cleared to participate in full activities despite retaining an increased injury risk in the fatigued state. The present study sought to examine the reactive strength status of rehabilitated ACL-R subjects who have returned to full activity in their chosen sports during a maximal stretch shortening cycle (SSC) fatigue protocol.

METHODS: Data Collection: ACL-R subjects, who had undergone post-surgery rehabilitation and returned to a level of physical activity comparable to their pre-injury status were recruited. Exclusion criteria included any episode of re-injury to the ACL following reconstructive surgery or any pathology or surgery in the hips, knees, ankles or feet of either leg within the last 6 months. Six adults participated in the study consisting of one male and five females. The group age was (mean ± S.D.) 25.5 ± 5 years; height 165.5 ± 8 cm and mass 65.3 ± 4 kg. Four of 6 reconstructions used a hamstring tendon autograft while 2 used a patellar tendon autograft. Mean time from surgery to participation in the study was 23 ± 12 months. Recruited subjects were from a variety of sports including soccer, martial arts, skiing and basketball. The University's research ethics board approved the study and subjects provided informed consent. Upon visiting the laboratory all subjects performed a standardised warm-up. Each subject's involved (INV) or uninvolved (UNINV) leg was selected at random to perform the testing protocol first. The testing protocol began with subjects performing one maximal set of rebound jumps (RBJ) in the force sledge apparatus. This set of jumps represent the pre-fatigue condition. In the RBJ protocol subjects were seated in the force sledge and winched to a height of 0.30cm from the force plate. The subjects were released and upon on landing performed four, single-legged, repeated maximal jumps. The 2nd, 3rd and 4th jumps in this set are considered RBJs and data to be analysed are derived from these jumps. This pre-fatigue data was analysed to find the average height jumped during each RBJ and represented the pre-fatigue condition. Ninety percent of this value was marked with a reflective marker on the sledge from a position where the subject was seated in the sledge chair with the leg fully extended at the hip and knee and with the ankle plantar flexed. Following this, subjects performed the fatigue protocol on the same leg. The fatigue protocol involved the subject being dropped from a height of 30cm and performing consecutive RBJs until they failed to reach the 90% level for three consecutive jumps. These three consecutive jumps below the 90% level represented the fatigued condition RBJs. An OMRON Opto-Switch (EE-SY410) was attached to the sledge chair and when it reached the reflective tape at the 90% level the light on the switch illuminated. It was clearly apparent whether or not the subject was reaching the 90% level on each jump. Komi (2000) has stated that these SSC fatigue protocols tax all the major elements of muscle function: metabolic, mechanical, and neural. These protocols cause disturbances in stretch reflex activation and consequently provide a strong basis for studying muscle function. Immediately following the fatigue protocol a further set of RBJs was performed. These jumps represented the post-fatigue condition. This set was performed an average of 50 seconds after the cessation of the fatigue protocol. Subjects were then given a 10 minute recovery period. Following this they performed the same series of protocols on the opposite leg. During all jumps ground reaction force was measured using an AMTI force plate sampling at 1000 Hz in the pre- and post-fatigue RBJs and at 100Hz during the fatigue protocol.

Data Analysis: Instants of take-off and landing were identified using the force traces for every jump performed. Flight time (FT) was calculated as the time between take-off and landing. Contact time (CT) was defined as the time between initial foot contact and take-off. RSI was calculated as the height jumped divided by CT, where, considering the 30° inclination of the force sledge apparatus, jump height was approximated as $(9.81 * FT^2)/16$.

Statistical Analyses: The three dependent variables analysed were FT, CT and RSI. Analyses focused on between-leg effects (INV and UNINV) and on within-legs between-condition effects (pre-fatigue, fatigued, post-fatigue). In the present study a small sample size was used (n = 6) and the pre- and post-fatigue data was not normally distributed so a GLM ANOVA was not deemed appropriate for use in the comparative analysis, between legs or between conditions. Means and standard deviations were reported and effect sizes were used to determine the magnitude of difference legs and between conditions. Effect sizes were calculated using η_p^2 and interpreted using the scale for effect size classification by Hopkins (2004). The behaviour of the lower limbs during the fatigue protocol was investigated using correlation analysis. The strength of the relationship between the fatigue process and the dependent variables was expressed using the variance explained statistic (r^2).

RESULTS: On both legs RSI diminished considerably from the pre-fatigue to the fatigued condition (see table 1). The reduction in RSI from the pre-fatigue to the fatigued condition was evident on both legs with very large effects sizes observed (η_p^2 = 0.881 and 0.907 for the INV and UNINV legs respectively). On both legs in the post-fatigue condition subjects'

RSI had returned toward levels comparable to that of the pre-fatigue condition but a deficit still remained on the INV leg with very large effect sizes observed between the pre- and post-fatigue conditions ($\eta_p^2 = 0.696$). No major differences in RSI were seen between legs across conditions. In the pre-fatigue and fatigued condition small effects sizes were in evidence between legs ($\eta_p^2 = 0.248$ and 0.063 for the pre-fatigue and fatigued conditions respectively). In the post-fatigue condition the subjects produced a moderately higher RSI on the involved leg than the uninvolved leg ($\eta_p^2 = 0.431$).

Table 1: RSI on the INV and UNINV leg, across conditions.

	Pre-Fatigue	Fatigued	Post-Fatigue
Involved Leg	0.506 (±0.20)	0.172 (±0.08)	0.459 (±0.17)
Uninvolved Leg	0.454 (±0.13)	0.189 (±0.10)	0.410 (±0.14)

On both legs FT reduced considerably from the pre-fatigue to the fatigued condition with large effect sizes observed ($\eta_p^2 = 0.997$ and 0.919 for the INV and UNINV legs respectively). On both legs CT increased with a large effect from the pre-fatigue to the fatigued condition ($\eta_p^2 = 0.918$ and 0.928 for the INV and UNINV legs respectively). Subjects' FT performance restored to pre-fatigue levels in the post-fatigue condition on both legs with small effect sizes observed between conditions ($\eta_p^2 = 0.241$ and 0.039 on INV and UNINV legs respectively). Moderate effect sizes in CT remained between the pre- and post- conditions with CT not fully returning to pre-fatigue levels ($\eta_p^2 = 0.561$ and 0.560 on the INV and UNINV legs respectively). A very strong negative correlation was observed between RSI and the fatigue protocol duration. This correlation was consistent between subjects and between legs. The average correlation coefficient on the involved leg was -0.942 (range: -0.893 to -0.975) and was -0.932 (range: -0.893 to -0.971) on the uninvolved leg (see figure 1). FT was strongly negatively correlated and CT was strongly positively correlated to the fatigue protocol duration between legs and between subjects.

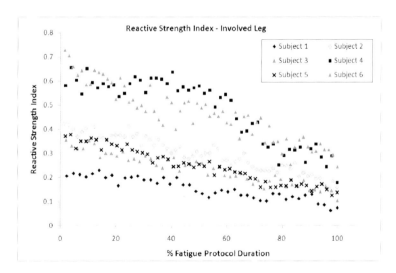

Figure 2: RSI during the fatigue protocol on the INV leg

DISCUSSION: Following reconstruction and rehabilitation no considerable differences in RSI were observed between the INV or UNINV legs in the pre-fatigue or the fatigued state. Subjects appeared to be able to restore reactive strength in ACL-R legs to levels comparable to that of their UNINV legs. Fatigue does not appear to degrade this capacity of the INV leg to a greater extent than it does on the UNINV limb. There was an indication of a deficiency on the INV leg in recovery following high intensity exercise as evidenced by a slower recovery of that leg to express RSI in the post-fatigue condition. This was evidenced by a large difference in RSI observed between the pre- and post-fatigue conditions on the involved leg ($\eta_p^2 = 0.696$).

To our knowledge, this is the first research study which has examined RSI throughout fatiguing exercise. An important finding of this study, in relation to the use of RSI as a marker of fatigue, was the very strong correlation between RSI and the fatigue protocol duration. The analysis of RSI throughout the fatigue protocol in this study reveals the importance of analyzing both jump height and ground contact time during fatiguing SSC exercise. A strong reduction in FT was observed in the fatigued condition on both legs. However, analysis of the mean FT and RSI data revealed that the reduction in RSI was twice as large as that of FT due to progressive elongation of CT in each jump. RSI is particularly important in the examination of fast SSC movements as the goal in such tasks is not just maximal jump height but also short ground contacts (Flanagan & Comyns, 2008). An individual may be able to attenuate declines in jump height but at the cost of elongating ground contact phases. In a sporting context, this would result in slower running velocities and reduced capacity to rapidly accelerate or change direction (Flanagan & Comyns, 2008). In the present study FT decreased, but CT increased in parallel. This had the net effect of drastically reducing subjects' RSI. Previous studies which have solely examined jump height during of after fatiguing SSC exercise may have underestimated the effect of fatigue on SSC function.

CONCLUSION: ACL-R subjects appear to be able to restore reactive strength in ACL-R legs to levels comparable to that of their UNINV legs and fatigue appears to degrade reactive strength function similarly on both legs. The data indicated that ACL-R subjects may have a possible deficiency on the INV leg in recovery following high intensity exercise as evident by a slower recovery capacity of that leg to express RSI 50 seconds after a maximally fatiguing exercise bout. RSI was very strongly negatively correlated with the duration of the fatigue protocol. The data presented outlines the suitability for using the RSI when monitoring fatigue during SSC exercise and its greater sensitivity during fatiguing exercise over FT or jump height alone.

REFERENCES:
Flanagan, E.P., Comyns, T.M. (2008) The use of ground contact times and the reactive strength index in optimizing training of the fast stretch shortening cycle. *Strength and Conditioning Journal*. 30: 32-38. 2008.
Hopkins, W.G. (2004) A new view of statistics. Online article: http://sportsci.org/resource/stats Date accessed: 7th February 2009.
Komi, P.V. (2000). Stretch-shortening cycle: a powerful model to study normal and fatigued muscle. *Journal of Biomechanics*, 33, 1197-1206.
Ostenberg, A. & Roos, H. (2000). Injury risk factors in female European football. A prospective study of 123 players during one season. *Scandinavian Journal of Medicine and Science in Sports*, 10, 279–285.
Strojnik, V. & Komi, P.V. (1998). Neuromuscular fatigue after maximal stretch-shortening cycle exercise. *Journal of Applied Physiology*, 84, 344-350.
Vollestad, N.K. (1997). Measurement of human muscle fatigue. *Journal of Neuroscience Methods*, 74, 219-227.
Zemper, E.D. (1989). Injury rates in a national sample of college football teams: a two-year prospective study. The Physician and Sportsmedicine. 17, 100-113.

Acknowledgement: The authors wish to thank the Irish Research Council for Science and Engineering Technology (IRCSET) for providing funding to support this research.

A WARM-UP INCLUDING A 5RM SQUAT PROTOCOL INCREASED BLOOD LACTATE, WITHOUT ALTERING THE SUBSEQUENT JUMP PERFORMANCE

Alexander J. Dinsdale, Athanassios Bissas and Sophie Reynolds
Carnegie Research Institute, Leeds Metropolitan University, Leeds, UK.

KEY WORDS: Potentiation, 5RM Back Squat, Acute Metabolic Changes, Vertical Jump.

INTRODUCTION: The execution of an acute resistance exercise has shown to enhance a subsequent explosive movement; this enhancement is termed Post Activation Potentiation (PAP). PAP exists in conjunction with both acute metabolic and neuromuscular fatigue mechanisms which still are not fully understood. This study aimed to identify the effect of a five repetition maximum (5RM) back squat on vertical jump performance and blood lactate.

METHODS: Six strength-trained male track and field athletes (21.50 ± 2.43 yrs, height 1.81 ± 0.08m, mass 83.05 ±10.29 kg and 1RM back squat strength 145.83 ± 9.17 kg) participated in this study. In brief, the study consisted of a single countermovement vertical jump (CMJ), a subsequent 5RM back squat and four successive CMJs, which were performed on a Kistler force plate (1000 Hz), at 1, 5, 9 and 13 minutes after completing the 5RM squat. Jump height and other mechanical variables (e.g. peak power) were calculated from the force-time curves. In addition, finger tip blood samples were collected in Lithium Heparin tubes and then analysed for lactate using the YSI 2300. The blood samples were collected at pre warm-up, post warm-up, after the pre squat CMJ and then at 15s, 3, 7, 11 and 15 minutes post squat.

RESULTS: A repeated measures ANOVA revealed significant increases in blood lactate in the post squat samples (15s – 11min) as compared with the pre squat samples ($p<0.05$), whereas the mechanical variables including the jump height did not change (graph 1).

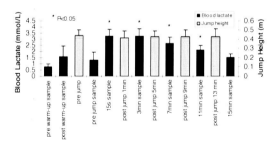

Figure 1: The effect of a 5RM back squat on both blood lactate and vertical jump height.

DISCUSSION: The 5RM squat induced acute metabolic changes, as this was observed in the elevated lactate levels, but it failed to affect positively the jump performance. It appears that the acute metabolic changes had no effect on the jump performance. The inappropriateness of the selected exercise to induce PAP through the necessary neuromuscular mechanisms and improve vertical jump performance has also been observed by Scott & Docherty (2004).

CONCLUSIONS: The integration of a 5RM back squat into a warm-up increased blood lactate but did not improve vertical jump performance in university level track and field athletes.

REFERENCES:
Scott, S. & Docherty, D. (2004) Acute effects of heavy preloading on vertical and horizontal jump performance. Journal of Strength & Conditioning Research. 18, 201-205.

Acknowledgment: We would like to thank the efforts of Laura Kirk and all the participants involved.

INFLUENCE OF ANKLE TAPING ON JUMP PERFORMANCE

Ian Kenny and Selva Prakash Jeyaram
Biomechanics Research Unit, University of Limerick

KEY WORDS: jumping, ankle taping, range of motion

INTRODUCTION: The present study aimed to investigate the influence of prophylactic ankle taping on jump performance, in the push off and the landing phase, for healthy subjects, for three types of jump. Ankle sprains represent from 38 to 50% of the total sport injuries (Jones et al., 2000). Functional taping and ankle braces are passive preventive measures frequently utilised in sports, however, studies on the influence of functional taping on sports tasks are scarce and most of them only analyse the passive ROM restriction (Hume and Gerrard, 1998).

METHODS: 12 healthy subjects (7 males, 5 female) were recruited. Following a subject-selected warm-up, subjects performed, taped and then un-taped, three each of countermovement jump (CMJ), 30 cm drop jump (DJ) and standing long jump (SLJ), using the arms, starting from/onto an AMTI force platform operating at 1000 Hz. A three-layer modified closed-basket inelastic taping technique was used on both ankles as shown in Figure one (Abian-Vicen et al., 2008). Ankle and knee active range of motion (ROM) was determined using an inclinometer before and after ankle taping. CMJ and DJ height was calculated via force platform flight time and SLJ distance was manually determined.

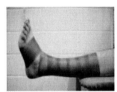

Figure1 Three-stage modified closed-basket ankle taping technique

RESULTS: Ankle taping produced no associated effect on ROM on knee extension or flexion, nor ankle dorsiflexion and eversion. Taping did significantly ($p<0.01$) reduce ankle plantar flexion (5.75°) and inversion (7.25°) ROM. While ankle taping brought about an average 16% and 6% increase in jump performance for DJ and SLJ respectively, results were not statistically significant.

DISCUSSION: The use of prophylactic ankle taping had no short-term influence on jump performance of healthy young subjects. Further analysis will examine variability of take-off and landing forces to ascertain whether any performance benefits found through the use of taping could be associated with increased landing forces thus injury potential.

CONCLUSION: Ankle taping has been found for dynamical exercise to not reduce performance and potentially may be of benefit to some individuals but dependent on the level of ROM.

REFERENCES:
Abián-Vicén, J. et al. (2008). Ankle taping does not impair performance in jump or balance tests. *Journal of Sports Science and Medicine*, 7, 350 – 356.

Hume, P.A. and Gerrard, D.F. (1998) Effectiveness of external ankle support. *Sports Medicine*, 25, 285-312.

Jones, D., Louw, Q. and Grimmer, K. (2000) Recreational and sporting injury to the adolescent knee and ankle: prevalence and causes. *Australian Journal of Physiotherapy*, 46, 179-188.

DOUBLE KNEE BEND IN THE POWER CLEAN

Laura-Anne M. Furlong[1,2], Gareth Irwin[1], Cassie Wilson[3], Huw Wiltshire[4] and David Kerwin[1]

[1]University of Wales Institute Cardiff, Cardiff, UK; [2]University of Limerick, Limerick, Ireland; [3]University of Bath, Bath, UK; [4]Welsh Rugby Union, Vale, UK

KEYWORDS: weight-training, stretch-shortening cycle, training specificity, sprinting

INTRODUCTION: The power clean is well established as the "gold standard" exercise for the development of lower extremity propulsive forces (Garhammer, 1982). The power clean has become a sprint specific strength and conditioning exercise, which is incorporated into periodised training programmes (Siff, 1992). Specifically the occurrence of a double knee bend (DKB) provides a mechanism to elicit a sprint specific stretch shortening cycle (SSC), maximising power output (Enoka, 1979). The aim of this exploratory study was to investigate whether the DKB occurred in power cleans as relative load increased.

METHODS: One elite male rugby player (age: 23 years; height: 1.72 m; mass: 85.5 kg), experienced in performance of the power clean, completed four lifts at each loading (60%, 70%, 80% and 90% of one repetition maximum lift (1RM)) in a randomized order over two days. Markers were placed on the 5th metatarsophalangeal joint, lateral malleolus, knee joint centre, greater trochanter and shoulder joint centre by the same researcher on both days. Kinematic data was recorded using four CODA CX1 scanners (Charnwood Dynamics, UK) sampling at 200 Hz. The DKB was defined as a local minimum in knee angle immediately prior to the catch phase of the power clean.

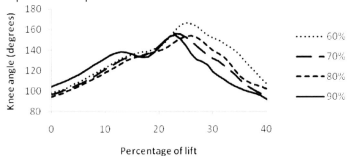

Figure 1. Knee angle during the first 40% of the power clean (A clear decrease in knee angle highlighting the occurrence of the DKB is shown during loading at 90% of 1RM)

RESULTS AND DISCUSSION: Figure 1 shows how the DKB does not occur until the bar is loaded at 90% of 1RM, as shown by a local minimum at 17% of the lift at 90% of 1RM. At low relative loads, bar velocity may have been high; decreased bar velocity with increased relative load increases time taken to complete the lift, allowing the DKB to occur.

CONCLUSION: The power clean appears to replicate the SSC seen in sprinting only when the bar is loaded at 90% of 1RM, but this is excessive for regular training. Further study with a larger sample of athletes would add power to this study and determine if this pattern is replicated.

REFERENCES:
Enoka, R.M. (1979). The pull in Olympic weightlifting. *Medicine Science and Sport*, 11, 131-137.
Garhammer, J. (1982). Energy flow during Olympic weightlifting. *Medicine and Science in Sport and Exercise*, 14(5), 353-360.
Siff, M. C. (1992). Biomechanical Foundations of Strength and Power training. In P.V.Komi (ed.), *Strength and Power in Sport* (pp. 103-139). Boston: Blackwell Scientific Publications.

STUDY ON DEVELOPMENTS OF BODY COMPOSITION AND PHYSICAL FITNESS FOR A YEAR IN JAPANESE ADOLESCENT TRACK AND FIELD ATHLETES

Aya Miyamoto[1], Joji Umezawa[1] and Toshio Yanagiya[2]

Graduate School of Health and Sports Science, Juntendo University, Chiba, JAPAN[1]

School of Health and Sports Science, Juntendo University, Chiba, JAPAN[2]

KEY WORDS: physical fitness development, adolescent athletes, longitudinal study .

INTRODUCTION: It is well known that physical fitness develops with the increased muscle thickness in adolescent boys especially from 16 to 18 years. Fukunaga et al. (1989) investigated the developments of body composition in Japanese boys and girls, and revealed that the muscle and fat cross sectional area increased with age from 16 to 18 years. Seefeldt et al. (1986) had reported that vertical jump height and maximal running velocity improved linearly with age from 5 to 18 years in boys who had no habitual physical training. However, no studies have investigated development of physical fitness of adolescent athletes. In addition, duration of measurement of longitudinal studies was a year basically. It is necessary to assess physical fitness several times within a year.The purpose of this study was to investigate developments of body composition and physical fitness in Japanese adolescent track and field athletes.

METHODS: Eleven boys (T1: 16.76 ± 0.69yrs., 170.5±4.9cm, 609.0±40.8N, T2: 17.19±0.68yrs., 170.7±4.9cm, 593.0±29.6N, T3: 17.49±0.65yrs., 170.7±5.0cm, 589.7±42.5N) participated in this study. Subjects belonged to track and field club of their high school. Body composition and physical fitness were determined 3 times a year; pre-season (T1), in-season (T2), and the off-season (T3). Measurements of body composition were height, body weight, the limb length, the limb girth, subcutaneous fat thickness(FT), and muscle thickness(MT). Anthropometry measurements, FT, and MT were the same methods described in Fukunaga et al. (1989). We applied test of physical fitness: sprint performance of 100m, jumping performance (counter movement jump, squat jump, drop jump), muscle power of trunk (medicine ball throw), single-joint isometric torque of knee and ankle. Dinn's multiple comparison procedure in corporating the Bonferroni correction was used to locate significant differences across season. Statistical significance was set at $p<0.05$.

RESULTS: No significant differences were found in height, body mass, the limb length, the limb girth except waist, subcutaneous fat thickness across seasons.On the other hands, significant difference was found in 100m time between T1 and T2.Moreover, there were significant differences in the torque of knee extension and knee flexion between T1 and T2.

DISCUSSION: Highest sprint performance was achieved in the in-season. Similar tendency was seen in knee extension and knee flexion torque. However, no significant difference was found in body composition for a year. These results indicated that improvement of sprint performance can be caused by the development of specific tension in lower limb muscles or improvement of sprint running skills.

CONCLUSION: Development of physical fitness was seen in the middle of track season in Japanese adolescent athletes.

REFERENCES:

Tetsuo FUKUNAGA et al. (1989). The Limb Composition in the Growing Generetion. Reprinted from *Journal of the Anthropological Society of Nippon.* 97 (1), 51-62

Seefeldt V (1986). Physical activity and wellbeing. Reston, American Allianace for Helth, Physical Education, Recreation and Dance.

A COMPARISON OF THE AEROBIC ENERGY DEMANDS OF TWO COMMERCIALLY AVAILABLE CYCLE ERGOMETERS IN TRAINED CYCLISTS

Gregory May & Dr. Giles Warrington

School of Health and Human Performance, Dublin City University, Dublin, Ireland & Clarity - Web Sensor Technologies, Dublin City University, Dublin, Ireland

The purpose of this study was to compare the energy demands of two cycling ergometers, (Velotron Dynafit Pro and Monark 834E) commonly used in the physiological monitoring of elite athletes. Eight trained male cyclists with a minimum 2 years training and racing experience participated in the study. Each subject completed an exercise trial involving a maximal incremental test. Testing was performed in a random order on either the Velotron or Monark cycle ergometer at the same time of day with no more than 14 days between each testing session. Subjects were requested to maintain their normal training and nutritional practices during the course of the study but to refrain from any intensive training 48 hours prior to each testing session. During the incremental testing significant differences for power output (PO), heart rate (HR), and oxygen uptake (VO_2) were found at both at fixed blood lactate (BL) reference points of; 2.5mmol l^{-1} (REF2.5mM) and at 4mmol l^{-1} (REF4mM). Overall the Velotron appeared to provide a more specific measure of cycling performance with significantly lower energy demands at fixed submaximal exercise intensities being observed as well as a significantly greater peak power output and time to exhaustion being attained, which may reflect the specific cycling position adopted. Further research is required to compare the findings of this study with actual cycling performance.

KEYWORDS: Ergometry, Cycling, Power, Physiology

INTRODUCTION: Ergometry is routinely used in a laboratory setting to evaluate physiological function and monitor changes in training status and performance of elite athletes from a range of sports. In road cycling, for laboratory testing to accurately replicate the physiological and biomechanical demands of the activity testing must be accurate, reliable and specific to the event. Traditionally physiological testing of cyclists is undertaken on a standard laboratory ergometer on which the athlete adopts a spurious non-specific riding position. In an attempt to address this issue a number of cycling ergometers have been developed which allow the accurate replication of the riding position that a cyclist adopts on their bicycle and hence measure performance predicting physiological variables with a much higher level of specificity. The purpose of this study therefore was to compare the aerobic energy demands of two commercially available cycle ergometers in trained cyclists.

METHODS: Eight trained male cyclists aged between 18 and 35 with at least 2 years racing and training experience were recruited to participate in this study. Subjects attended the Human Performance Laboratory located at the School of Health and Human Performance at DCU on two separate occasions. Each visit was separated by no more than 14 days with each test being performed at the same time of day. Prior to participating in each trial, the subjects were advised to maintain their normal training and nutritional practices, and to refrain from any strenuous activity in the preceding 48-hour period. Prior to undertaking any physiological tests, body mass and height were measured and recorded using a digital scales and stadiometer. Body Composition was measured and recorded using repeat skinfold measurements taken at 7 sites using a calibrated Harpenden skinfold calliper to estimate body composition. The Velotron Dynafit Pro (Racermate Inc., Seattle, U.S.A) was the sports specific cycling ergometer selected. It uses a magnetically braked flywheel, which is controlled from a PC. A preset protocol is loaded into the software and resistance is corrected automatically and independent of cadence. The ergometer arrives calibrated from the factory and no extra calibration is needed. It is possible to calibrate the Velotron via the AccuWattTM calibration checking procedure from the PC; this checks the original factory performance to verify that no change has occurred. The Monark 824 E Ergomedic (Monark Exercise AB, Vansbro, Sweden) generates a frictional force about the flywheel by tensioning a cord, which is attached to a hanging basket. This allows for weights to be added increasing the frictional force generated. Due to the nature of the resistive force the resistance is not independent of cadence. The flywheel is graduated to allow the usage of a photo-sensor to measure its rotational speed. This sensor connects to a PC via an ADC-11 data channel logger which, with software provided by Monark allows for calibration of the ergometer, and recording of data during testing. HR was measured continuously using a wireless Polar HR monitor (Polar Electro, Finland), Lactate measurement with a Lactate Pro analyser (Arkray, Japan); data were analysed using SPSS version 14. Significance was set at the p<0.05 level. Differences between means were analysed using a paired samples Student's t-test. Differences between ergometers were analysed using One Way ANOVA. Gas analysis was continuously measured using open circuit spirometry using an Innovision Innocor^{TM} (Innovision, Denmark) metabolic cart. The system was calibrated to manufacturer's specification using a 3L syringe and its own calibration procedure.

When each subject was using the Velotron, the ergometer was set up to replicate their preferred riding position by taking specific measurements from their own bicycle. The Velotron is adjustable along each of its major axes. Adjustments can be made to: saddle height; fore/aft position of saddle; vertical height of handle bars; horizontal position of handle bars; rotation of handle bars. Detailed measurements were taken of each subjects own bicycle so as to allow the cyclists own seating position to be replicated as closely as possible for each of the trials. Measurements were taken from: Centre of the stem top - Centre of the front hub; Centre of the bottom bracket – Saddle rails; Centre of handlebars – Centre of saddle rails; Horizontal measurement from centre of bottom bracket – Plum line at 90° dropped from centre of saddle rails. For the trial using the Monark bicycle, being a fixed frame, any adjustments were restricted to setting the saddle height to approximately that of the rider's own saddle. The participant's pedals were removed from their own bicycle and attached to the cranks of both ergometers. Warm-up and test were performed with the subject on the drops of the bars on the Velotron, and in any hand position on the Monark. During the maximal incremental test cyclists had to maintain a cadence of 80 - 100RPM, but not greater than 100RPM. The incremental test was set using 3-minute stages. HR, VO$_2$, and BL were taken at the end of every stage. The test started at 100W and increased by 50W for each stage until volitional exhaustion. The subject performed a 5-minute recovery at 100W with a self-selected cadence. Recovery BL was measured 1, 3 and 5 minutes post test.

RESULTS:

Table 1: Cardiorespiratoy Evaluation Analysis:

	Monark	SD +/-	Velotron	SD +/-
Power @ 2.5mmol	199.2	54.2	264.9**	33.1
Power @ 4mmol	253.3	33.1	302**	28.7
Heart Rate @ 2.5mmol	150	17.7	163**	14.8
Heart Rate @ 4mmol	167	10.6	173**	15.5
V02 @ 2.5mmol	34	6	37	5
V02 @ 4mmol	44	6	49**	5

= P<0.05, ** = P<0.01

Table 2: Maximal data:

	Monark	SD +/-	Velotron	SD +/-
VO2max (ml kg min^{-1})	60.0	6.4	58.7	11.8
Peak Power Output (W)	321.9	33.9	353.1*	31.2

* = P<0.05, ** = P<0.01

Significant differences were found for HR, BL and VO$_2$ during each stage of the incremental test between the two ergometers. Significant differences were found for PO and HR at both BL marker points; 2.5mmol l^{-1} (REF2.5MM) and 4mmol l^{-1} (REF4MM) respectively. However, corresponding significant changes for measured VO$_2$ occurred only at REF4MM and not at REF2.5MM.

DISCUSSION: When undertaking any laboratory based physiological and performance testing for any sport it is essential that the ergometry used, testing procedures and protocols adopted replicate the specific demands of the sport as closely as possible. In cycling the onset of recent technologies such as the development of sports specific ergometers including the Velotron Dynafit Pro and the SRM Cycling Ergometer, allow for a variety of positions to be adopted during testing, which therefore enable a cyclist to adopt their preferred training or racing position.

The age range of subjects in the present study was from 18 – 35 years and was representative of the typical age range of trained competitive cyclists. Results for height (1.75m +/- 0.04m) and body mass (68.9kg +/- 5kg) were very similar to those previously reported for professional road cyclists (Mujika & Padilla, 2001; Faria et al., 2005; Sallet, 2006). This indicates a possible natural propensity for a certain morphotype to become adept at road cycling (Jeunkendrup, 2002). Overall the sample group was in line with reported means, although significant differences exist between VO$_{2max}$ and PO$_{max}$ of the participants and professional riders; this is to be expected as the tested subjects were club level riders as opposed to full time training professional riders, and thus may not have as high a level of conditioning.

As the subjects were all trained road cyclists with a racing background all testing was performed with the cyclists using their own pedals and cycling shoes. Unlike traditional 'rat trap' pedals with straps and toe clips, a set of clipless pedals and shoes allow a cyclist to utilise a much more efficient stroke and increase the level of muscle recruitment during the stroke (De Koning & Van Soest, chapter 11; cited in Jeunkendrup, 2002). The cyclist is effectively able to scoop the pedal backwards while the foot is travelling in a positive vertical motion during the recovery section of the stroke and increase the time over which they are able to apply mechanical force (Burke, 2003). This allows a cyclist to pedal more effectively by eliminating dead spots at the top and bottom of the stroke, and cycle more economically due to greater power output to the same relative demand at sub-maximal intensities (Jeunkendrup, 2001). Testing in the race specific position was performed with the cyclist sitting on the drops of the road bars to further replicate the adopted race position on the Velotron. This position is adopted during racing for its aerodynamic properties which in comparison to the position adopted on the Monark would reduce typical drag area by ~1,000cm^2 and in real terms acts as to reduce the fatiguing effect of wind on a cyclist (Burke, 2003).

Analysis of the maximal data revealed no significant difference were found for VO$_{2max}$. In contrast significant differences were observed for PO$_{max}$ and time to exhaustion. Perhaps due to a training adaptation trained cyclists are able to produce a higher PO in a position they are used to. The same applies for assessing sub maximal power is scaled to another factor such as cadence or mass. This was shown to vary from ergometer to ergometer and implies that the position adopted on an ergometer will affect these performance determinants. With the length of the incremental test we also started to see the effects of muscle fatigue due to adopted body position during the Monark test. This may explain the increase in both BL and the VO$_2$ at REF2.5MM and REF4MM. As the body is unable to use the preferred and trained muscles from its normal cycling position, there may be a greater utilisation of type 2 muscle fibres in an attempt to compensate. A possible greater emphasis on the use fast glycolytic fibres might explain the elevated blood lactate levels at each exercise intensity on the Monark ergometer and significantly lower PO at REF2.5mM and REF4mM.

As cadence was maintained between 80 – 100RPM the effect of cadence on PO during the incremental test was not investigated. Previous studies have shown that increased pedal cadence during maximal trials can reduce the onset of fatigue and result in increase cycling economy (Lucia et al., 2004; Hagan RD, 1992). Furthermore, testing with drop bars on the Monark ergometer may have made more a specific position for the cyclist on the ergometer, due to changes in hip orientation and hence muscular recruitment, but as we were investigating the differences between the ergometers as per manufacturer's specifications it was not utilised. Further investigation into the effect of position on the cyclists ability to develop power at different parts of the cycle stroke could give a better insight into the effect of position on the different cycling ergometers.

CONCLUSION: Overall both ergometers may be valid for assessment of maximal capacity in cyclists. Both ergometers provided results with no significant differences in maximal physiological values or performance determinants during aerobic exercise. In contrast, sub maximal values were shown to be significantly different between the two ergometers indicating differences in the physiological demands which may in part be explained by the cycling position adopted and the ergometer itself. In conclusion, in order to apply any measures taken in a laboratory to road cycling situations it is essential to make sure all testing is performed in a similar position, repeated in the same manner, and as many factors expressed as either a sub maximal predictor, or as a predictor scaled to another factor. Hence investigation of sub maximal determinants of cycling performance may be as important as maximal predictors for cycling performance. Further research is required to compare the findings of this study with actual cycling performance.

REFERENCES:
Evangelisti M.I., Verde T.V, Andres F.F, Effects of Handlebar Position on Physiological Responses to Prolonged Cycling, *Journal of Strength and Conditioning Research*; 1995; 9(4): 243-246.

Burke E, High Tech Cycling (2nd edition), *Human Kinetics*, 2003

Faria EW, Parker DL, Faria IE, The Science of Cycling: Part 1, Journal of Sports Medicine 2005; 35 (4): 285-312

Faria EW, Parker DL, Faria IE, The Science of Cycling: Part 2, Journal of Sports Medicine 2005; 35 (4): 313-337

Hagan RD, Weis SE, Raven PB. Effect of Pedal Rate on Cardiovascular Responses During Continuous Exercise. *Medicine & Science in Sports & Exercise*. 1992; 24: 1088-95.

Jeukendrup A. & Martin J., Improving Cycling Performance, *Journal of Sports Medicine*, 2001; 31(7): 559-569.

Jeukendrup A, High-Performance Cycling, *Human kinetics*, 2002

Lucía A, Pardo J, Durántez A, Hoyos J, Chicharro JL, Physiological Differences Between Professional and Elite Road Cyclists, *International Journal of Sports Medicine*, 1998;19(5):342-8.

Lucia A, San Juan AF, Montilla M. In Professional Road Cyclists, Low Pedaling Cadences are Less Efficient. *Medicine & Science in Sports & Exercise*. 2004; 36: 1048-54.

Mujika I. & Padilla S., Physiological and Performance Characteristics of Male Professional Road Cyclists *Journal of Sports Medicine*, 2001; 31 (7): 479-487

Sallet P, Mathieu R, Fenech G, Baverel G, Physiological Differences of Elite and Professional Road Cyclists Related to Competition Level and Rider Specialization, *Journal of Sports Medicine and Physical Fitness*. 2006;46(3):361-5.

EFFECTS OF STEP LENGTH ON THE BIOMECHANICS OF LOWER LIMBS DURING ELLIPTICAL EXERCISE

Tung-Wu Lu[1], Chih-Hung Huang[2], Yen-Pai Chen[1], Chu-Fen Chang[1] and Hui-Lien Chien[1]

Institute of Biomedical Engineering, National Taiwan University, TAIWAN [1]
Human Computer Interaction Technology Center, ITRI South, TAIWAN [2]

KEY WORDS: elliptical trainer, stride lengths, kinematics, kinetics, lower limbs

INTRODUCTION: Elliptical exercise (EE) has been developed as a low-impact aerobic exercise modality with increased popularity in fitness training and clinical applications over the last decade. During EE, the feet are constrained by pedals to follow an elliptical trajectory, with the possibility of producing disadvantageous joint loads and potential musculoskeletal overuse injuries (Lu et al., 2007). Therefore, proper selection of step length during EE may be helpful for the reduction of these disadvantageous joint loads. The purpose of the study was to study the effects of three different step lengths on biomechanics of the lower limbs during EE.

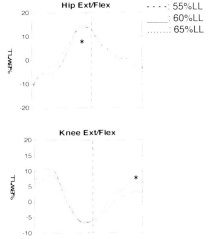

Fig1: Hip and knee joint moments at three step lengths. ($\ast$: $p<0.05$)

METHODS: Fifteen healthy male adults (23.7 ± 1.2 yr; 173.5 ± 3.9 cm; 72.2 ± 10.8 kg) performed EE on an elliptical trainer at three step lengths (55%, 60% and 65% leg length or LL), with a pedal rate close to their corresponding walking cadence and without workload. Three-dimensional kinematic data of the whole body and pedal reaction forces (PRF) were measured using a 7-camera motion capture system (Vicon 512) at a sampling rate of 120Hz and a six-component force transducer at a sampling rate of 1080Hz. Repeated measures analysis of variance was used to test the effects of step length on the peak excursion of the whole body center of mass (COM), PRF, joint angles and moments. The significance level was set at 0.05.

RESULTS and DISCUSSION: When the step length increased, significantly greater vertical motion of COM and increased vertical PRF were found during a period corresponding to the early stance of gait ($p<0.05$) while decreased ankle dorsiflexion and knee flexion were found at initial contact. Peak hip extensor moment at late stance and peak hip abductor moment at early stance increased as the step length increased, whereas the knee extensor moment at late swing decreased (Fig. 1). In the current results, subjects appeared to use the strategy of decreasing the ankle dorsiflexion and knee flexion to increase the length of lower limb for compensating for the changed pedal trajectory with increasing step length, and then altered joint kinematics further resulted in the observed greater vertical motion of COM. With greater vertical PRF, more flexed ankle and knee joints shifted the lines of action of the PRF to pass away from the hip joint center, leading to increased hip joint moments. Therefore, EE training should be used with smaller step length, preferably less then 55% LL, for patients with hip degeneration or pathologies.

CONCLUSION: This study showed that during EE increase of step length may increase harmful joint loadings, especially at the hip joint. The use of elliptical trainers for athletic and rehabilitative training should select appropriate pedal position and step length considering the user's joint function and muscle strength to avoid the potential risk of injury.

REFERENCES: Lu, T.-W., Chien, H.-L. and Chen, H.-L. (2007) Joint loadings in the lower extremities during elliptical exercise, *Medicine & Science in Sports & Exercise*. 39:1651-1658.

PREPARATORY LONGSWING TECHNIQUES FOR DISMOUNTS ON UNEVEN BARS

Dawn Tighe, Gareth Irwin, David G Kerwin and Marianne Gittoes
Cardiff School of Sport, University of Wales Institute, Cardiff, UK

The purpose of this study was to quantify the biomechanical differences between two methods of performing the preparatory longswing preceding the double layout dismount. Video images of 6 female Olympic level gymnasts performing the double layout dismount (3 = straddle preparatory longswing, 3 = dished preparatory longswing) were recorded using two synchronised 50 Hz digital cameras. 3D DLT reconstructed coordinates were combined with inertia values to define discrete release variables including vertical and horizontal velocity of the mass centre, release angle and angular momentum about the gymnast's mass centre. Joint angular kinematics at the hips and shoulders were contrasted with particular reference to the hip functional phase. Based on the reported release parameters the straddle longswing could be considered preferential.

KEY WORDS: angular momentum, functional phase, artistic gymnastics

INTRODUCTION: Successful performance of an uneven bars dismount is widely acknowledged to depend upon the preparatory longswing, and more specifically, actions at the hip and shoulder i.e. the functional phases (Irwin & Kerwin, 2007). The women's bar is less stiff than the men's high bar, and as a result, the hip and shoulder functional phases exhibited by female gymnasts may vary from those seen in men's preparatory longswings. The nature of the uneven bars means that, unlike their male counterparts, female gymnasts must swing past the low bar during descent of the longswing. The proximity of the low bar has previously been identified as acting to reduce the generation of the desired release parameters such as angular momentum during the preparatory backwards longswing (Arampatzis & Brüggemann, 1999). Consequently female gymnasts are reported to either straddle their legs, or increase the angle of hip flexion in order to pass the low bar whilst conserving the build up of required release parameters such as angular momentum (Hiley & Yeadon, 2005). There is a lack of research detailing the reasoning behind the choice of technique by elite female gymnasts, and the eventual biomechanical advantages of either technique prior to release. From a coaching perspective identification of the biomechanical advantages of either technique could lead to a better understanding of the technique to teach and consequently inform the coaching process through the effective selection of the preferred preparatory longswing for the successful execution of the double layout dismount (Irwin et al., 2005). The aim of this study was to identify the biomechanical characteristics of two longswing techniques favoured by elite female gymnasts as identified in a preliminary frequency analysis. The overall purpose is to identify if either technique provides an advantage for the gymnast in generating the release parameters required for successful completion of the double layout dismount.

METHODS: Subjects: Six female Olympic level gymnasts participated in the present study, three gymnasts performed the straddled preparatory longswing and three performed the dished preparatory longswing preceding the double layout dismount. Their mean ages, heights and masses were 18 ± 3 years, 1.50 ± 0.02 m and 40 ± 6 kg (straddled technique) and 17 ± 2 years, 1.47 ± 0.05 m and 36 ± 9 kg (dished technique).

Data Collection: Two digital video cameras (Sony Digital Handycam, DCR VX1000E) were positioned 30 m and 37 m from the horizontal bar and aligned so their optical axes intersected over the centre of the uneven bars. Images, recorded at 50 Hz, of a single vertical pole, with five spheres measuring 0.10 m in diameter, was positioned in six locations surrounding the apparatus at the Sydney 2000 Olympic Games. The resulting calibration volume was 3.2 m wide, 4.3 m long and 4.2 m high. Subsequently all the routines during the qualification rounds were recorded, from which six double layout dismounts were selected for analysis.

Data Processing: VICON PEAK MOTUS 9.0 (VICON PEAK, UK) motion analysis system was used throughout the digitising process. A 16 point model was used to represent the human performer during reconstruction, with the wrist, elbow, shoulder, hip, knee, ankle and toe on both sides of the body, the centre of the gymnast's head and the mid point between the gymnast's hands on the upper bar were digitised for each camera's view. Digitisation began ten images before the gymnast reached the handstand position at the start of the preparatory longswing and concluded ten images following the gymnast's touchdown on the landing mat.

Data Analysis: The digitised coordinates were synchronised following digitisation in accordance with the method outlined by Yeadon & King (1999). Reconstruction was conducted on the preparatory, flight and landing phases, a direct linear transformation (DLT) algorithm was used to acquire 3-D coordinates as described by Abdel-Aziz & Karara (1971). Matchcad^{14TM} (Adept Scientific, UK) was used to calculate discrete and continuous performance variables including angular momentum. Segmental inertia parameters for each gymnast were obtained using Yeadon's Inertia Model (1990) customised for each gymnast based on their reported mass and height values and scaled according to their limb dimensions determined from the reconstructed video data. Linear and quadratic functions were used to fit the centre of mass (CM) trajectory data from the instant immediately following bar release to the instant prior to touchdown. The resulting regression equations were differentiated and used to predict horizontal and vertical release velocity values (Vh and Vv). Circle angle (θ_{CM}) describing the angular position of the gymnast swinging about the bar was defined as the angle made by the line joining the mass centre of the gymnast to the neutral bar position with the vertical handstand position being 90°. Joint angles were represented by vectors joining virtual points, created by averaging pairs of digitised joint centre coordinates. The shoulder joint was defined by joining the mid-elbow, mid-shoulder and mid-hip, whilst the hip comprised vectors between mid-shoulder, mid-hip and mid-knee. The functional phases were determined using the methods reported by Irwin & Kerwin, (2005). Commencement of the functional phase was defined as the instant at which the joint angular velocity (ω) passed through 0 rad/s and concluded when the respective angular velocity returned to a negative value. Closing of the shoulder and opening of the hip joints were defined as positive angular velocities. The point of release was defined as the angular position at which the bar horizontal acceleration peaked on the upswing. In order for comparisons to be made between gymnasts of varying heights and masses, angular momentum values were normalised by dividing angular momentum by moment of inertia about the transverse axis through the CM in the anatomical position and 2π to convert to units of straight somersaults per second (SS/s). Normalised angular momentum (Ln) values were multiplied by flight time to give the equivalent number of straight somersaults in the subsequent flight phase (LnFT). This dimensionless value was thus a composite score based on CM velocity and normalised angular momentum at release. All values are reported as means (±sd) for gymnasts performing either the SLS and DLS technique.

RESULTS: The SLS technique showed the highest Ln values at release of 1.232 SS/s, which showed a percentage difference of 6% with regards to the DLS technique. The SLS technique resulted in Vv at release of 3.70 m/s and Vh at release of 1.59 m/s, which when compared to the DLS technique show a percentage increase of 7% and 23% respectively. The hip joint and circle angle ranges were greater (41% and 58%) in the DLS technique compared to the SLS technique. In contrast the shoulder joint and circle angle ranges were smaller for the DLS (39% and 22%) than the SLS technique (Table 2).

Table 1 Release parameters for the double layout dismount performed following a straddled (SLS n = 3) and dished (DLS n = 3) longswing

Release parameter	SLS	Standard Deviation	DLS	Standard Deviation
θ_{CM} (°)	333	7	345	7
Vv (m/s)	3.70	0.03	3.42	0.43
Vh (m/s)	1.59	0.66	1.23	0.43
Ln (SS/s)	1.232	0.080	1.162	0.082
ω_{CM} (rad/s)	7.10	0.18	7.90	0.20
LnFT (SS)	1.216	0.002	1.084	0.005

Table 2 Circle and joint angles at the start and end of the hip and shoulder (shd) functional phases for the SLS and DLS techniques.

Angle (°)	Phase	SLS	Standard Deviation	DLS	Standard Deviation
	Start	276	12	268	9
Circle (Hip)	End	325	5	345	7
	Change	49	9	77	8
	Start	276	12	268	9
Circle (Shd)	End	325	5	345	7
	Change	49	9	77	8
	Start	-23.7	9.6	-13.8	9.8
Hip	End	22.6	9.5	51.5	6.7
	Change	46.2	9.6	65.3	8.4
	Start	-5.2	2.9	-3.2	4.0
Shd	End	19.6	31.3	11.9	16.7
	Change	24.9	22.2	15.1	12.2

DISCUSSION: The present study aimed to identify mechanical differences between the straddle and dished preparatory longswings commonly performed by elite female gymnasts prior to the double layout dismount. The overall purpose being to quantify differences within the generation of advantageous release characteristics between the two longswings, subsequently facilitating effective skill development of the double layout dismount. Within a preparatory longswing accurate completion of the functional phases provides a means by which angular momentum can be maximised and provides the energy source required for successful execution of the longswing and succeeding dismount (Arampatzis & Brüggemann, 1999). The hip functional phase for the DLS is initiated earlier and occurs over an increased circle angle (58%) and hip angular range (41%) (Table 2) in comparison to the SLS, this could provide an explanation for why greater peak Ln values were seen for the DLS prior to release. This conclusion is supported by the findings of Hiley & Yeadon (2005), who reported increased hyper extension to flexion during this phase of the movement to result in increased angular momentum. Following the attainment of peak Ln values both techniques exhibit a decrease from peak values prior to release. The magnitude of this decrease is greater for the DLS resulting in Ln at release of 1.162 SS/s which is 6% less than the release Ln value for the SLS (Table 1). Sufficient levels of angular momentum must be generated during the preparatory longswing and maintained until the point of release in order that the gymnast possesses the capacity to complete the rotations in the dismount successfully. Thus the SLS could be considered advantageous due to increased levels of Ln prior to release. LnFT represents a weighting factor between the release parameters which govern success of the dismount. The SLS achieved a LnFT value 11% greater than that of the DLS indicating that the SLS may be preferential for the generation of the release parameters required for completion of the double layout dismount including Vv, Vh and Ln (Table 1). Brüggemann et al. (1994) identified the maintenance of sufficient height and rotation to be vital in the successful completion of a dismount. Height achieved is dependant upon Vv. Results showed that Vv was 7% greater for the SLS than the DLS, and as such indicates that the SLS could be considered preferential to dismount performance based upon potential to achieve a successful flight path following release (Table1). Brüggemann et al. (1994) also identified the need to create a flight path which allows the gymnast to travel safely away from the bar, this occurs through the achievement of adequate release Vh. The SLS produced 23% more Vh in comparison to the DLS, therefore the SLS could also be considered preferential in allowing the gymnast to travel safely away from the bar in flight (Table 1). Irwin et al. (2005) proposed that the biomechanical understanding of the movement pattern of a desired skill enhances the process of skill development. As such the present study identified biomechanical differences in the generation of desired release characteristics between the two favoured longswings, thus the findings of the present study have the potential to inform coaches about the biomechanics of the straddle and dished longswing techniques and thus allow for the development of the double layout dismounts on the uneven bars, with specific regard to the generation of Ln and Vh prior to release.

CONCLUSION: The straddle technique was identified as providing increased levels of Vv, Vh, Ln and LnFT. The straddle technique could therefore be considered preferential to the dished technique. The present findings have the potential to inform the coaching process and thus allow for the development of the double layout dismount. The angular kinematics reported in relation to the functional phases suggests that further kinetic analysis is required to provide an insight into the hip and shoulder moments. Further research should aim to incorporate other factors linked to the choice of longswing technique by gymnasts including coaching background and gymnast morphology. It would also be beneficial if future studies used a large sample size and performed repeated trials to account for inter and intra subject variability.

REFERENCES:
Abdel-Aziz, Y.I., and Karara, H.M. (1971). Direct linear transformation from the comparator coordinates into object space coordintes in close range photogrammetry. *In the proceedings of the Symposium on Close Range Photogrammetry.* Falls Church, VA: American Society of Photogrammetry.
Arampatzis, A. & Brüggemann, G-P. (1999). Mechanical energetic processes during the giant swing exercise before dismounts and fight elements on the high bar and the uneven parallel bars. *Journal of Biomechanics*, 32, 811-820.
Brüggemann, G.P., Cheetham, P.J., Alp, Y. & Arampatzis, D. (1994). Approach to a Biomechanical Profile of Dismounts and Release-Regrasp Skills of the High Bar. *Journal of Applied Biomechanics*,10, 291 – 312.
Hiley. M.J. & Yeadon, M.R. (2005). The Margin for Error When Releasing the Asymmetric Bars for Dismounts. *Journal of Applied Biomechanics*, 21, 223-235.
Irwin, G. & Kerwin, D. G. (2007). Musculoskeletal work of high bar progressions. Sports Biomechanics, **6**(3), 360-373.
Irwin, G. & Kerwin, D. G. (2005). Biomechanical similarities of progressions for the longswing on high bar. *Sports Biomechanics*, 4, 164–178.
Irwin, G., Hanton, S. & Kerwin, D. G. (2005). The conceptual process of skill progression development in artistic gymnastics. *Journal of Sports Sciences*, **23**(10), 1089-1099.
Yeadon, M.R. & King, M.A. (1999). A method for synchronising digitised video data. Journal of Biomechanics, 32 (9), 983-986.

PRACTICE AND TALENT EFFECTS IN SWING HIGH BAR INTER-JOINT COORDINATION OF NOVICE ADULTS

Albert Busquets[1], Michel Marina[1], Alfredo Irurtia[1]
and Rosa M. Angulo-Barroso[1,2]
INEFC, Universitat de Barcelona, Barcelona, Spain[1]
CHGD & School of Kinesiology, University of Michigan, Ann Arbor, MI (USA)[2]

This research describes changes in movement coordination after a two-month practice period of the swing on high bar in a novice cohort, which was divided by a-priory talent level into two groups: spontaneous-talented, ST, and non-spontaneous-talented, NST. Their performance was also compared with experienced gymnasts. Data were collected during pre- and post-practice sessions by two video cameras. Coordination between hip and shoulder joints was assessed. Results showed a similar practice effect in the swing enlargements in both novice groups. Interestingly, the ST group's inter-joint coordination variables on the downswing improved more than those of the NST group due to practice. Therefore, the two novice groups improved performance, but they showed diferent local coordination. Initial talent helped to improve both performance and coordination in the down-swing.

KEY WORDS: motor perceptual-learning, novel task, initial conditions, practice effect, gymnast.

INTRODUCTION: One of the main concerns a novice must resolve when facing a new motor task is to coordinate the participation of multiple joints and limbs (Bernstein, 1967; Hong & Newell, 2006). Coordination has been defined as the stable spatial-temporal relationship among movement system components (i.e. limb segments and joints) to achieve the task's goal (Irwin & Kerwin, 2007). Limb segments and joints are in-phase coordination when they move in synchrony with 0° phase lag between them, while the coordination is in anti-phase mode when segments or joints move in opposite directions with a phase lag of 180°. However, multiple intermediate out-of-phase coordination modes exist between in- and anti-phase coordinations. Moreover, movement performance and coordination could be impacted by an individuals' a-priori talent while learning a new task (Delignieres et al., 1998; Teulier & Delignieres, 2007). The aim of this research was to describe movement coordination changes after a two-month practice period of a novel task (longswing in high bar) in a novice cohort, which was divided in two groups considering the initial level of performance: spontaneous-talented versus non-spontaneous-talented. In addition, post-practice performance and coordination of the novices were compared to expert gymnasts. Given the task characteristics, we focused on the inter-joint coordination between the hip and shoulder in this study.

METHOD: Participants: Twenty-five novice participants who never performed a lonswing in high bar prior recruitment (fifteen males and ten females, age=20.2±2.2 years; height=1.70±0.1 m; body mass=66.5±11.7 kg) were divided by a-priory talent level (spontaneous-talented, ST, and non-spontaneous-talented, NST) using a k-means Cluster analysis on the basis of their best performed swing during the first trial in the first practice session. An additional expert group was made up of nine gymnasts from the national team (six males and three females, age=19.0±4.5 years; height=1.59±0.1 m; body mass=54.9 ±15.3 kg). All participants signed a consent form. The study was approved by the local ethics committee.

Data Collection: We defined three events independently for hip (H) and shoulder (S) angle joints (Fig. 1b): the smallest angle during downswing (P1H, P1S); the largest angle after P1 (P2H, P2S); and the smaller angle during upswing (P3H, P3S). Movement performance was defined as the total path of the swing (swing amplitude) measured from the maximum elevation on the downswing (Pi) and the maximum elevation on the upswing (Pf) (Fig. 1b).

Data were collected during pre- and post-practice sessions by two video cameras. Body position angle was defined as the angle formed by the line connecting the center of mass with the middle of the grasping hand and the vertical (z-axis) of the coordinate system (Arampatzis & Bruggemann, 1999) (Fig. 1a). Swing amplitude was normalized, as shown in Fig. 1a. In order to characterize the coordination between the hip and shoulder, we defined inter-joint phases (P1H-P1S, P2H-P2S, and P3H-P3S), and used continuous relative phase (CRP) to represent the phasing relationship between the actions of the two joints at every point. Each CRP was obtained by subtracting the phase angle of the distal joint (hip) from that of the proximal (shoulder), namely $\varphi_{shoulder-hip}$ (Clark & Phillips, 1993; Hamill et al., 1999). In turn, the phase angle (φ_i) was calculated from the normalized angular displacement (θ) and angular velocity (ω) using $\varphi=\tan^{-1}(\omega/\theta)$. In Figure 1c, the continuous relative phase graph of an expert gymnast was depicted for a longswing in high bar. Relative phase began around 0° indicating that the hip and shoulder moved in synchrony or did not move remaining the same angle values. This in-phase relationship became out-of-phase lead by hip flexion starting around -80%. Around 40% of the upswing the hip achieved the maximum extension (P3H) and the coordination changed faster to an out-of phase mode leaded by the shoulder's flexion with the same velocity as the hip initiated the extension slower. To quantify and statistically test differences between relative phase curves, the mean absolute relative phase (MARP) of each hip and shoulder phases were calculated (Stergiou et al., 2001).

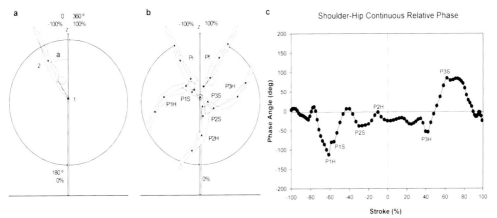

Figure 1: In (a), swing normalization and body position angle (a) defined by the z axis, middle grasping hand (1) and the body center of mass (2). In (b), initial position (Pi), final position (Pf) and swing events (P1, P2, and P3) from hip (H) and shoulder (S) movements. For simplicity, H and S events have been represented at the same instant of time for P1-P3. In (c), the continuous relative phase between the hip and shoulder joints during a longswing in an expert gymnast.

Data Analysis: To address the changes in performance and coordination between pre- and post-practice of the NST and ST groups, we used 2 (Group) X 2 (Time) mixed ANCOVAs with group as the between-participants factor, time (first and last trial) as the within- factor, and sex as the covariate. Planned comparisons between pre- and post-practice within each group were used. In addition, NST vs. ST group effects on the magnitude of change (post-minus pre-practice) were examined using students' t-tests for all variables. To address the secondary goal of this study, one-way ANCOVAs with Tukey multiple comparison post hocs were used for establishing differences between the NST, ST and Expert groups at the last trial. Sex (male and female) was included as the covariate in all statistical tests. Statistical significance was set at p<.05 level. P adjustments were conducted to control for multiple comparisons when appropriate.

RESULTS: We conducted 2x2 ANCOVAs for the swing amplitude, each inter-joint phase and MARP to investigate which parameters evolved differently between the groups. These analyses revealed that there were significant group ($F_{1,22}=8.44$, p=.008) and trial (time) ($F_{1,22}=11.51$, p=.003) main effects for the swing amplitude. In addition significant group by time interactions were found for P1H-P1S ($F_{1,22}=10.01$, p=.005) and P2H-P2S ($F_{1,22}=5.28$, p=.032). Indeed, simple main effects showed that the ST group improved P1H-P1S and P2H-P2S from pre- to post-practice (becoming closer to reference values, in this case values of the Expert group) while the NST group worsened despite practice (Table 1). When student t-tests compared the pre- and post-practice magnitude of change, significant differences existed between the NST and ST groups on P1H-P1S ($F_{1,22}=10.01$, p=.005) and P2H-P2S ($F_{1,22}=5.28$, p=.032). The sex covariate showed significant differences for the MARP P3H-Pf ($F_{1,22}=5.34$, p=.031) and MARP P3S-Pf ($F_{1,22}=8.29$, p=.009). Examining group differences at post-practice, the One-way ANCOVAs pointed out differences between expert and novice for the swing amplitude ($F_{2,30}=118.70$, p=.000) and the P2H-P2S inter-joint phase ($F_{2,30}=4.63$, p=.018). The post hoc test showed significant differences between both novice groups (ST and NST) and the experts for the swing amplitude, but there were not significant differences in paired group comparisons in P2H-P2S. The sex covariate was significant in One-Way ANCOVA time for P3H-P3S ($F_{2,30}=4.73$, p=.038).

Table 1 Participant characteristics and pre- and post-practice swing details,

	Pre		Post		
	NST (n = 15)	ST (n =10)	NST (n = 15)	ST (n = 10)	E (n = 9)
	Mean±sd	Mean±sd	Mean±sd	Mean±sd	Mean±sd
Swing Amplitude (%)	80.03±15.49	104.02±18.06	98.72±14.17	115.92±23.64	198.40±0.96
Inter-joint Phases [a]					
P1H-P1S	-4.37±12.12	-10.37±19.19	-15.80±18.61	-1.95±3.71	-6.45±20.15
P2H-P2S	-7.53±11.90	-7.24±15.91	-13.92±14.18	-2.44±10.05	1.09±12.77
P3H-P3S	-10.33±8.69	-10.33±17.80	-14.20±10.80	-13.14±12.67	-21.24±5.35
Swing MARP					
MARP P1H-P2H	42.41±25.06	37.28±18.47	47.84±22.96	49.76±16.35	41.86±23.85
MARP P2H-P3H	43.82±20.71	37.89±14.47	38.16±13.33	41.14±22.39	39.28±14.20
MARP P3H-Pf	42.04±23.16	45.98±17.14	49.63±20.30	49.92±22.75	46.84±25.41
MARP P1S-P2S	36.82±18.05	53.12±36.31	35.89±20.56	43.54±20.80	34.40±21.27
MARP P2S-P3S	39.16±22.93	43.14±21.96	42.20±17.66	33.94±12.26	37.58±14.69
MARP P3S-Pf	62.98±27.69	51.24±23.51	80.92±59.40	64.31±28.99	53.67±31.24

[a] All values in swing amplitude percentage

Overall, these results suggest novice participants improved their performance due to practice independent of group membership. However, novice group differences remained relevant across time. Group differences for the inter-joint phases mainly occurred in the downswing (-100-0%) while sex differences were revealed in the upswing (0-100%). When examining P1H-P1S and P2H-P2S, the ST group became closer to the experts by shortening these phases closer to an in-phase relationship between the hip and shoulder. In contrast, the NST group increased the phase durations closer to an out-of-phase coordination.

DISCUSSION: The main aim of this study was to describe coordination changes in the acquisition of a novel task (swing on high bar) due to practice and initial talent. It was hypothesized that within the novice cohort, participants with spontaneous talent would experience larger improvements than those less talented. However, our performance variable did not support our hypothesis while inter-joint phases variables yielded significant interaction effects. The P1H-P1S and P2H-P2S critically differentiated the two groups. The ST group reduced the time lag between the events of hip and shoulder during the downswing (-100-0%) becoming an in-phase coordination with closer values to experts. The NST group after practice worsened the two inter-joint phases with an out-of-phase coordination. These modifications could impact differently the goal achieved (swing amplitude) by the novice groups. Thereby, the ST group coupled the two movements in order to exploit the mechanical work of the gravity while the NST group separated the hip and shoulder events to increase the segmental velocities.

After the practice the novices achieved similar MARP values which did not differ from experts values. However, similar MARP values do not imply necessarily an improvement because different coordination modes could result in the same average. When we observed the continuous relative phase (CRP) graphics, the relationship between both joints the hip and the shoulder changed across 0° faster achieving larger values in novice groups, while the experts modified their coordination over a more expanded period of time.

CONCLUSION: In summary, our findings have shown that the novice participants improved the swing amplitude, but the coordination in the downswing was different for the ST and the NST groups. The ST group inter-joint phases during the down swing appeared more similar to the expert group. In contrast, the MARP values did not show differences between novices groups or expert group, perhaps due to the use of an average of different coordination modes. We suggest that the initial talent also helps to improve coordination in the downswing. Despite of talent level, upswing coordination will required more focused practice. Given the sex covariate was significant further exploration of these results are warranted.

REFERENCES:

Arampatzis, A., and Bruggemann, G.P. (1999). Mechanical energetic processes during the giant swing exercise before dismounts and flight elements on the high bar and the uneven parallel bars. *Journal of Biomechanics*, 32, 811-20.

Bernstein, N. (1967). *The coordination and regulation of movements.* New York: Pergamon Press.

Clark, J.E., and Phillips, S.J. (1993). A longitudinal study of intralimb coordination in the first year of independent walking: a dynamical systems analysis. *Child Development*, 64, 1143-57.

Delignieres, D., Nourrit, D., Sioud, R., Leroyer, P., Zattara, M., and Micaleff, J.P. (1998). Preferred coordination modes in the first steps of the learning of a complex gymnastics skill. *Human Movement Science*, 17, 221-241.

Hamill, J., van Emmerik, R.E., Heiderscheit, B.C., and Li, L. (1999). A dynamical systems approach to lower extremity running injuries. *Clinical Biomechanics*, 14, 297-308.

Hong, S.L., and Newell, K.M. (2006). Practice effects on local and global dynamics of the ski-simulator task. *Experimental Brain Research*, 169, 350-60.

Irwin, G., and Kerwin, D.G. (2007). Inter-segmental coordination in progressions for the longswing on high bar. *Sports Biomechanics*, 6, 131-44.

Stergiou, N., Jensen, J.L., Bates, B.T., Scholten, S.D., and Tzetzis, G. (2001). A dynamical systems investigation of lower extremity coordination during running over obstacles. *Clinical Biomechanics*, 16, 213-21.

Teulier, C., and Delignieres, D. (2007). The nature of the transition between novice and skilled coordination during learning to swing. *Human Movement Science*, 26, 376-92.

Acknowledgement
Support for this work was made possible by Secretaria General de l'Esport and the Departament d'Universitats, Recerca i Societat de la Informació (DURSI) of the Generalitat de Catalunya.

THE RELATIONSHIP BETWEEN GLUTEAL ACTIVITY AND PELVIC KINEMATICS DURING THE WINDMILL SOFTBALL PITCH

Hiedi J. Hoffman, Gretchen D. Oliver, David W. Keeley, Kasey B. Barber and Priscilla M. Dwelly

University of Arkansas, Fayetteville, AR. USA

KEY WORDS: electromyography, gluteus medius, gluteus maximus, kinematics

INTRODUCTION: Historically a number of pitch delivery methods have been used, with the windmill pitching motion becoming the most common and accepted pitch delivery method in recent years. Although millions of girls participate in softball, there are few studies investigating the mechanics of the windmill softball pitch. Thus, the purpose of this study is to quantify the activity of the gluteus maximus and gluteus medius on the dominant side; and to further determine if an increase in gluteal activity throughout each phase results in an increase of linear velocity of the pelvis at phase endpoint.

METHOD: Anthropometric data will be collected on softball pitchers' upper and lower extremity and torso specifically for this study: length, width, diameter, and circumference of the torso, dominant thigh, knee, shank, and ankle using a standard anthropometric kit (Rosscraft Innovator Inc). Electromyographic (EMG) data will be collected using a previously established protocol (Cram, & Kasman, 1998) with electrodes placed on both the gluteus maximus and gluteus medius. After placement of electrodes, manual muscle tests will be conducted to obtain maximum voluntary contractions so as to provide a baseline for each subject. Kinematic data will be collected using The Motion Monitor® system (Innovative Sports Training, Chicago, IL) and throwing kinematics will be calculated using the International Society of Biomechanics recommendations for reporting joint motion. The joint coordinate system will be defined by previously established protocol (Wu et al. 2005). Once the data are reduced, Pearson Product Moment Correlation Coefficients will be calculated to determine the strength of the relationship between gluteal activity and pelvic kinematics during the windmill softball pitching motion.

RESULTS: We will use the five phases of pitching based on Maffet et al. (1997) as illustrated in Figure 1.

Figure1. Phases and instances in time for the windmill softball pitch.

DISCUSSION/CONCLUSION: We anticipate that increases in dominant gluteus maximus and gluteus medius muscle activity during each phase will be positively correlated to the linear velocity of the pelvis at phase endpoint for each of the four instances in time (6 o'clock, 3 o'clock, 12 o'clock, and 9 o'clock).

REFERENCES:

Cram, J.R., & Kasman, G.S. *Introduction to surface electromyography.* Gaithersurg, Aspen Publishers, Inc, 1998.
Maffet, M.W., Jobe, F.W., Pink, M.M., Brault, H. & Mathiyakom, W. (1997). Shoulder muscle firing patterns during the windmill softball pitch. *American Journal of Sports Medicine*, 25, (3), 369-374.
Wu, G., van der Helm, F.C.T., Veeger, H.E.J., Makhsous, M., Van Roy, P., Anglin, C., et al. (2005). ISB recommendation on definitions of joint coordinate systems of various joints for the reporting of human joint motion – Part II: shoulder, elbow, wrist, and hand. *Journal of Biomechanics*, 38, 981–992.

HIP MOMENT PROFILES DURING CIRCLES IN SIDE SUPPORT AND IN CROSS SUPPORT ON THE POMMEL HORSE

Toshiyuki Fujihara and Pierre Gervais

Sports Biomechanics Laboratory, Faculty of Physical Education and Recreation, University of Alberta, Edmonton, AB, Canada

The purpose of this study was to analyze the hip moment profiles during Circles and to assess how gymnasts modulate their technique depending on different orientation on the pommel horse. Circles in side support and in cross support performed by six gymnasts were captured using Qualysis motion capture system. Hip joint moments were computed with the assumption that the total leg was a single rigid body. The results implied that the lateral-flexion movement at the hip joint was closely related to the important technique of both types of Circles. Cross-Circles were characterized by the greater flexor moment throughout a Circle and the smaller lateral flexor moment in the rear support phase.

KEY WORDS: Circle, Pommel horse, Hip joint moment

INTRODUCTION: Hip joint movements are one of the most critical factors of Circles, the most fundamental skill on the pommel horse in which a gymnast rotates his body horizontally alternating support from arm to arm (Fig.1). Although many textbooks describe that the whole body should be kept straight during Circles, a developing gymnast often bends his body at his hip joint to continue Circles, while keeping his knee joints straight. For this reason, the hip joint angle has been used as a discriminant variable that reflects a gymnast's level of expertise with Circles (e.g Baudry et al., 2006). Fujihara et al. (2009) asserted that lateral-flexion of the hip might be a technique unlike flexion considered a technical error. Mizushima (1998) discussed the hip rotation about the long axis, in relation to body twisting. As well as these kinematic studies, a kinetic study of the hip motions is necessary to further our analysis of Circle technique. Fujihara and Gervais (2008) observed that Circles in side support (Side-Circles, Figure1 top) and Circles in cross support (Cross-Circles, Figure1 bottom) showed different kinematic profiles due to the physical characteristics of the pommel horse. As an extension of this study, it was thought that conducting kinetic analysis using the same data set would provide a deeper insight of Circles technique. The purpose of the current study was, then, to analyze the hip moment profiles during Circles and to assess how gymnasts modulate their technique depending on the different orientation on the pommel horse.

Front support phase------Entry phase----------Rear support phase--------Exit phase---------- Front support phase

Figure 1: Circles in side support (top) and Circles in Cross support (bottom)

METHOD: Data Collection: Six gymnasts performed 3 sets of 10 Side-Circles and 3 sets of 10 Cross-Circles on a pommel horse. They were national or international level gymnasts. The first set of 10 Circles was randomly assigned as either Side-Circles or Cross-Circles, and in the subsequent sets, they performed two types of Circles alternatively. Our local ethics committee approved all experimental protocols. Prior to the experiment, each gymnast provided informed consent. Thirty seven retro-reflective markers were attached to the gymnast's body to estimate de Leva's (1996) suggested anatomical landmarks. Three-dimensional coordinates of these markers were acquired using 12 QUALISYS Proreflex motion tracking cameras operating at 120 Hz. Body mass and height were also measured.

Data Analysis: Three-dimensional coordinates data were smoothed using a fourth-order Butterworth digital filter at the optimal cut-off frequencies determined by automatic algorithm (3.67 Hz and 9.61 Hz) of Yokoi and McNitt-Gray (1990). The joint centres were estimated as the centres of two markers attached on the surface of each joint. To compute hip joint moments, all segments of the lower extremities—feet, shanks and thighs—were assumed to be a single rigid body. The moments of inertia of the total legs were computed using the parallel axis theorem based on the six segments. Note that under this assumption hip joint moments for each leg were not considered. Therefore, hip joint moment for adduction at each side of the hip, which was probably present to keep two legs together, was not taken into account. The hip joint moments were estimated by solving Euler's equations with two rigid bodies: the total leg and the lower trunk. Three local reference systems, two for the segments and one for the joint, were defined using vector cross products (Figure 2). In addition to the joint moment profiles, the joint power was computed as the product of the joint moment and the joint angular velocity, which was defined as the relative angular velocity of the distal segment with respect to the proximal segment. The joint moments and joint powers were normalized by each gymnast's body mass. For each set of 10 Circles, 7 Circles (3rd - 9th) were used so that the mean data for each variable were computed from the data of 21 Circles (7×3). The Wilcoxon matched-pairs test was used to compare Cross-Circles to Side-Circles. The effect sizes (ES) were computed as the differences divided by the standard deviations for Side-Circles.

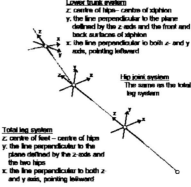

Figure 2: Definitions of the local reference systems

RESULTS: Figure 3 presents the angular velocities, joint moments, and joint powers at the hip joint during Side-Circles and Cross-Circles. For both types of Circles, the flexor moment was observed throughout a complete single Circle with mild peaks just before right-hand release, and during entry and exit phases. It should be reported, however, that the gymnast who seemed to perform the best quality Circles showed a small amount of extensor moment during the rear support phase for both Side- and Cross-Circles. In terms of flexion-extension movement, Cross-Circles was characterized by the greater flexor moments on average (0.43 vs 0.20 N·m/kg, $p = 0.028$, $ES = 2.11$), and the greater energy generation at the right-hand release (0.65 vs 0.10 W/kg, $p = 0.028$, $ES = 3.73$). In the lateral motions, the greater moments were observed for Side-Circles at the right hand contact (0.34 vs 0.10 N·m/kg, $p = 0.028$, $ES = 3.06$), and the left hand release (-0.64 vs -0.24 N·m/kg, $p = 0.028$, $ES = 2.20$). Both types of Circles showed the fast long-axial rotation of the total legs towards the right with respect to the lower trunk during the front support, but the magnitudes of the moments and powers were much smaller than those for flexion or lateral flexion.

DISCUSSION: On average, the hip flexor moment was observed throughout a Circle with several minor peaks. The patterns were not consistent among all gymnasts: each gymnast showed several peaks in different phases. Therefore, a more detailed analysis in relation to each individual technique is desirable to interpret the change in flexor moments. This is consistent with Fujihara and Fuchimoto (2006), who studied Side-Circles performed by 17 university

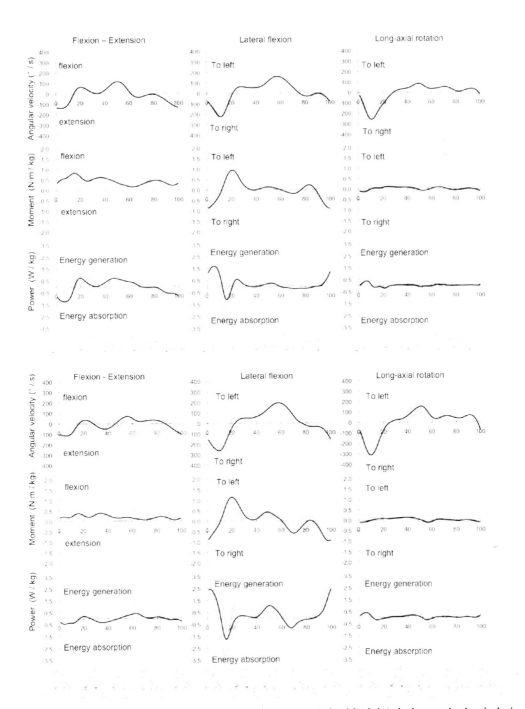

Figure 3: Angular velocities, moments, and powers at the hip joint during a single circle in cross and side support. The profiles are the averages (± one standard deviation) of 6 gymnasts (each performed 21 circles). The vertical lines and figures depict hand-contact and hand-release phases (left hand contact at 0 and 100%).

gymnasts, in the sense that no typical pattern of flexion-extension motion at the hip joint was found. As a general observation, however, the greater hip flexor moment was present in

Cross-Circles than in Side-Circles. Fujihara and Gervais (2008) reported the greater hip angle during the rear support for Cross-Circles and attributed it to the obstacle avoidance, namely, clearing the pommel horse. Although the hip flexor moment seemed to drop during the rear support, the greater moment in the preceding phases could explain the greater hip flexion in the rear support. On the other hand, a more consistent pattern was found in the lateral flexor moment among all gymnasts. Fujihara and Fuchimoto (2006) did report the general pattern of the lateral flexion motion. That is, gymnasts bended their body towards the non-supporting-hand side during the single hand support phases. Although it may not be as clear for Cross-Circles, the lateral flexor moment reached the major rightward peak at the left hand contact, the minor rightward peak at the right hand contact, the minor leftward peak at the left hand release, and the major leftward peak at the right hand release. Stated differently, the greater and smaller peaks occurred in the front support side and in the rear support side, respectively. Note that these peaks at the hand releases were eccentric work, and those at the hand contacts were concentric work. Fujihara et al. (2009) asserted that, during the double hand support phases, the upper body moved side-to-side to switch the supporting hands, and the body rotates mainly about its centre of mass. The lateral-side concentric work at the hand contacts seemed helpful to create the body rotation about the centre of mass during the side-to-side movement of the upper body. The greater side-to-side movement of the upper body for Side-Circles is perhaps one of the possible reasons for the greater magnitude seen in the lateral flexor moment.

The angular velocity of the long-axial rotation represents what is known as "twisting motion" of the hip. From the front support phase to the right hand release, a gymnast twists his hips rightward so that the front side of the legs faces up. Then, this twisted body turns in the opposite direction throughout a Circle. Both types of Circles showed this pattern, but the required moment for this motion was relatively small due to the shorter moment arm and smaller moment of inertia.

CONCLUSION: This study describes the hip moment profiles during two types of Circles. The results suggested that the lateral-flexion movement at the hip joint was an integral technique component of both types of Circles. Although more detailed analysis with individual technique is desirable, "piking" the body (flexion) appeared to be an individual adjustment to complete the skill (e.g. to keep dynamic balance or to avoid the collision with the pommel horse). It is recommended that coaches and gymnasts appreciate the importance of lateral hip motion for successful execution of Circles.

REFERENCES:

Baudry, L., Leroy, D., Thouvarecq, R. and Choller, D. (2006). Auditory concurrent feedback benefits on the circle performed in gymnastics. *Journal of Sports Sciences*, 24(2), 149-156.

de Leva, P. (1996). Adjustments to Zatsiorsky-Seluyanov's segment inertia parameters. *Journal of Biomechanics*, 29(9), 1223-1230

Fujihara, T. and Fuchimoto, T. (2006). Mechanical analysis in mechanism and technique of double leg circles on the pommel horse. *Japanese Journal of Biomechanics in Sports & Exercise*, 10(1), 27-41.

Fujihara, T., Fuchimoto, T. and Gervais, P. (2009). Biomechanical analysis of circles on pommel horse. *Sports Biomechanics*, 8(1), 22-38.

Fujihara, T. and Gervais, P. (2008). Kinematic comparison of circles in cross support and circles in side support. . In North American Congress of Biomechanics, August 5-9, 2008, hosted by the University of Michigan. Ann Arbor, Michigan. Abstract, American Society of Biomechanics and Canadian Society for Biomechanics.

Yokoi, T. and McNitt-Gray, J. L. (1990). A threshold to determine optimum cutoff frequency in automatic data smoothing using digital filter. Biomechanics: Proceedings for the 14th Annual Meeting, University of Miami, Miami, Florida, November 14-16, 1990 / American Society of Biomechanics. University of Miami, Miami, Florida, 209-210.

THE EFFECTS OF SPORTS TAPING ON IMPACT FORCES AND MECHANICAL BEHAVIORS OF SOFT TISSUE DURING DROP LANDING

Nyeon-Ju Kang, Woen-Sik Chae, Chang-Soo Yang[1] and Gye-San Lee[2]

Department of Physical Education, Kyungpook National University, Daegu, Korea
[1]Department of Martial Art, Incheon City College, Incheon, Korea
[2]Department of Physical Education, Kwandong University, Kangneung, Korea

KEY WORDS: impact force, mechanical behavior, drop landing, strain, stress

INTRODUCTION: Wakeling et al. (2002) proposed that resonant soft tissue vibrations caused by impact force may be potentially harmful. Bartold et al. (2009) reported that longitudinal strain of the plantar fascia was significantly reduced under the tape condition and it affected alleviating the symptoms of plantar fasciitis. Accordingly, if sports taping may develop the goal of reducing the strain and stress of muscle, it can be illustrated to help muscle tuning to minimize soft tissue vibration and prevent injury. Since the proven mechanical effects of sports taping not established, the purpose of this study was to determine how sports taping affects impact force and mechanical behaviors in the lower extremity.

METHODS: Twelve male university students (19.9±0.5 yrs, 175.2±5.5 cm, 646.6±40.9 N) were recruited. Kinematic data from a high speed camera (S.V.T., Motion Pro X3, 1000 frames/s) and GRF data from a force platform (AMTI OR6-5) were collected while subjects landed bilaterally from a drop height of 0.5 m with and without sports taping (Kinesio Tex, Japan) in random order. Twenty two reflective markers were attached to the right-hand side of the lower extremity to acquire strain and stress values calculated by using Marc 2005 (MSC software, USA). Strain and stress, peak VGRF and loading rate were determined for each trial. For each dependent variable a paired t-test was performed between the taping group (TG) and control group (CG) ($p<.05$).

RESULTS AND DISCUSSION: Strain is the amount by which a material deforms under stress. In the average and maximum values, longitudinal and principal strains in the thigh during the landing phase (LP) were reduced significantly in the TG. During the deceleration phase (DP) after impact, the maximum longitudinal strain in the TG was significantly decreased.

Table 1 Average and max strains for each phase

	TG (ave)	CG (ave)	p	TG (max)	CG (max)	p
LP longitudinal strain	1.036±0.021*	1.049±0.019*	.021	1.048±0.021*	1.063±0.018*	.010
LP principal strain	1.046±0.019*	1.058±0.016*	.030	1.061±0.019*	1.074±0.016*	.040
DP longitudinal strain	0.972±0.046	0.962±0.061	.424	1.046±0.022*	1.060±0.024*	.014

Lieber et al. (1993) showed that strain influences the amount of muscle damage. Since sports taping may not only increase the stability of lower limbs but also reduces excess stress and strain, it seems reasonable to suggest that sports taping is effective in preventing lower limb injuries. There were no significant differences in the shank during both phases. Since the strain was calculated by 2D analysis, it was too difficult to discreet real strain of the shank during the deceleration phase in particular. Peak VGRF and loading rate decreased in sports taping group but not significantly.

CONCLUSION: Wakeling et al. (2002) and Bartold et al. (2009) reported that resonant soft tissue vibrations caused greater movement of tissue compartment. The less strain values showed from sports taping may reduce the possibility of injury through minimizing movement of soft tissue at the landing event. To prove the effects of sports taping more accurately, future studies should examine 3D analysis of mechanical behaviors of the lower extremity during landing.

REFERENCES:
Bartold, S., Clarke, R., Franklyn-Miller, A., Falvey, E., Bryant, A., Briggs, C. and McCrory, R. (2009). The effect of taping on plantar fascia strain: A cadaveric study. *Journal of Science and Medicine in Sport*, 12(1), 74.
Lieber, R.L. and Friden, J. (1993). Muscle damage is not a function of muscle force but active muscle strain. *Journal of Applied physiology*, 74, 520–526.
Wakeling, J.M., Nigg, B.M. and Rozitis, A.I. (2002). Muscle activity damps the soft tissue resonance that occurs in response to pulsed and continuous vibrations. *Journal of Applied Physiology*, 93, 1093–1103.

FOOT FUNCTION IN SPRINTING: BAREFOOT AND SPRINT SPIKE CONDITIONS

Grace Smith[1,2] and Mark Lake[2]

Department of Sport and Exercise Sciences, University of Chester, UK[1]
RISES, Liverpool John Moores University, UK[2]

KEY WORDS: metatarsophalangeal joint, energy, sprinting, performance.

INTRODUCTION: The mechanical energy contribution of the metatarsophalangeal joint (MPJ) during sprinting has implications for improving performance. Mechanical properties of sprint spikes have been demonstrated to influence sprinting performance (Stefanyshyn and Fusco, 2004) but little work has examined foot function in relation to normal barefoot behaviour. This study investigated the effect of footwear on MPJ kinematics, kinetics and forefoot pressure distribution, comparing sprint spike conditions to barefoot sprinting.

METHOD: Trained sprinters performed maximal sprints on a 55 m indoor runway, contacting a force platform in the middle (Kistler, Switzerland, sampling at 1000 Hz). Kinematic data was also captured at 1000 Hz using 6 opto-electronic cameras (Qualisys Inc, Sweden). Plantar pressure distribution was also collected using an RSScan pressure mat. Four initial subjects ran barefoot and wearing their own sprint spikes. Kinematic and kinetic data was smoothed using a digital filter with a 100 Hz cut off frequency. The MPJ was modelled as having a single, oblique axis defined by markers on the 1st and 5th metatarsal heads.

PRELIMINARY RESULTS: Sprint spikes reduced the range of motion and energy lost at the MPJ (Table 1). Although energy was predominantly absorbed for both conditions, spikes tended to increase energy generation at takeoff but overall they reduced energy production during stance. Barefoot pressure results demonstrated that although lateral loading was evident at touchdown, overall loading was confined to the medial side of the foot during stance and progressed medially and distally for takeoff. In spikes, the loading transition was similar but loading was further concentrated on metatarsals 1, 2 and the hallux.

Table 1 Comparison of MPJ kinematics and kinetics in barefoot and sprint spikes (n=4)

	Sprint Spikes	Barefoot
Angular range of motion (°)	35.6 (± 3.8)	50.0 (± 4.6)
Energy absorbed during MPJ flexion (J)	-29.7 (± 5.7)	-38.6 (± 13.5)
Energy generated at touchdown (J)	5.1 (± 2.4)	9.4 (± 3.6)
Energy generated at takeoff (J)	0.6 (±0.5)	0.1 (±0.1)

DISCUSSION: Regardless of footwear, energy was mostly absorbed at the MPJ, agreeing with Stefanyshyn and Nigg (1997); however this energy loss appears to be reduced wearing sprint spikes. Loading occurred medially, concurrent with the notion that the MPJ axis is centred on metatarsals 1-3 in sprinting. The application of appropriate bending stiffness in relation to the MPJ axis, along with medio-lateral differences, dictated by pressure findings, could affect MPJ energetics and sprinting performance and warrants future investigation.

CONCLUSION: Preliminary findings suggest substantial changes in foot function and performance related parameters between barefoot and shod sprinting.

REFERENCES:

Stefanyshyn, D.J. and Fusco, C. (2004). Increased shoe bending stiffness increases sprint performance. *Sports Biomechanics*, 3, 55-66.

Stefanyshyn, D.J. and Nigg, B.M. (1997). Mechanical energy contribution of the metatarsophalangeal joint to running and sprinting. *Journal of Biomechanics*, 20, 1081-1085.

THE VELOCITY DEPENDENCE OF TECHNNIQUES COMMONLY LINKED WITH LOWER BACK INJURY IN CRICKET FAST BOWLING

Kane Middleton, Poonam Chauhan, Bruce Elliott and Jacqueline Alderson

The School of Sport Science, Exercise and Health
The University of Western Australia, Perth, Australia

The aim of this study was to examine the velocity dependence of shoulder alignment counter rotation, maximum hip-shoulder separation angle, maximum front knee flexion angle and maximum trunk lateral flexion. High-performance fast bowlers (n=17) were required to bowl multiple deliveries in a fast, normal and slower ball category. No statistical association was found between bowling velocity and maximum shoulder counter rotation or knee flexion. Significant associations were found between ball release velocity and trunk lateral flexion and maximum hip-shoulder separation angle. Significant differences were found between the bowling categories for separation angle and knee flexion. A regression analysis showed that trunk lateral flexion and separation angle only accounted for 11% of the ball velocity variance, for the normal delivery (31.3 ms^{-1}).

KEYWORDS: biomechanics, alignment, rotation, speed

INTRODUCTION: The fastest bowlers in the world attract large crowds due to the intensity they bring to the game. Mastery of the most appropriate technique, together with other aspects of bowling development are all essential if one is to reach full potential. Understanding what factors enhance bowling velocity without increasing the risk of injury will better equip and prepare bowlers to learn the art of fast bowling, and for coaches to detect and correct flaws in the bowling action. While the rewards of success are easily identified, the stresses and strains placed on the body during fast bowling are a major concern to the bowler. For an impact sport, this means large forces are transmitted through a variety of body tissues via the foot, ankle, knee, hip and various joints of the back. Often concurrently with these high loads, the trunk is flexing laterally and rotating in an endeavour to maximise the speed of the bowling-shoulder. A range of mechanical variables have been commonly linked with lower back injury and include, but are not delimited to: shoulder alignment counter rotation (CR), hip-shoulder alignment separation angle (SA), front knee flexion (KF) and trunk lateral flexion (TLF) (Foster et al., 1989; Burnett et al., 1995; Ranson et al., 2005). The purpose of the present study was to investigate the relationship between ball release velocity and the above mechanical variables.

METHOD: Seventeen male fast bowlers with a mean age 20.9 years (± 2.2), height 187.2 cm (± 6.3) and mass 81.4 kg (± 7.8), currently performing at first grade level or above within the Western Australian Cricket Association competition were recruited. Participants gave their written, informed consent prior to testing and procedures were approved by The University of Western Australia's Human Research Ethics Committee. Three-dimensional kinematics were captured using a 12 camera VICON MX 3-D motion analysis system (Oxford Metrics, Oxford, UK) operating at 250Hz. Static and dynamic calibration of the system was performed prior to the capture of dynamic trials. An abridged version of the UWA full body marker set described by Dempsey et al. (2007) consisting of 62, 12-millimetre retro-reflective markers, were affixed to the trunk, pelvis and the lower and upper limbs.

A cricket pitch was constructed with a rubber mat being laid down the length of the laboratory. Stumps were present at both ends and a bowling and popping crease were marked. Prior to the collection of the dynamic bowling trials, participants performed their personal warm-up. Participants bowled over the wicket and were instructed to aim to hit a target mark 0.3 m above and 0.3 m to the left of the top of the off stump, as viewed from the bowler's end. Each participant bowled five sets of six consecutive deliveries (an over), with a short break between each set in an attempt to replicate match conditions. Within each over, participants bowled four deliveries at their 'normal' match pace, while one delivery was bowled at 'very fast' pace, that is; the bowler's fastest possible delivery speed, and the other a 'slower' delivery. The delivery order was randomly allocated.

Vicon Workstation software (Oxford Metrics, Oxford, UK) was used to track, label and complete marker trajectories for each bowling trial. Data were filtered in Vicon Workstation using a Butterworth low pass filter at a cut-off frequency of 15 Hz. Filtered data were modelled using custom static and dynamic UWA models. The shoulder joint centre was calculated as the midpoint between a marker placed on the acromion process and markers on the anterior and posterior aspects of the shoulder overlying the glenohumeral joint with shoulder alignment defined as the line joining these points. The hip joint alignment was calculated as the line joining the left and right hip joint centres. SA was defined as the difference in the shoulder and hip alignments in the transverse plane. The knee joint centre was identified as the midpoint between the medial and lateral femoral condyles and the KF angle was calculated with 0° representing a straight limb. The maximum flexion angle between front foot impact (FFI) and ball release was recorded. TLF was calculated as the maximum relative angle between the vertical trunk axis and the horizontal pelvic axis in the frontal plane during the delivery action.

Means of the three 'very fast', three 'slower' and 12 'normal' deliveries from the 3rd, 4th and 5th overs of each bowler were used for analysis. Pearson's product moment correlations were calculated and used to identify associations between each dependent variable and ball velocity within each of the three velocity categories. A one-way ANOVA with repeated measures was used to determine whether there were differences between the bowling categories for each dependant variable. Where main effects were found, post-hoc tests with a Bonferroni correction were used to identify where these differences occurred. A stepwise linear regression was used for the pooled normal deliveries (n = 17 x 12 trials) to determine the influence of the independent variables, as predictor variables, on bowling velocity.

RESULTS: The mean value of the normal deliveries (n=17; 31.31 ms^{-1} ± 1.4) and the very fast deliveries (n=17; 32.46 ms^{-1} ± 1.26) were shown to be significantly different (t = 6.59, p < 0.001). No significant associations were found between ball release velocity and CR and KF in any of the ball velocity categories. There was a moderately significant positive relationship found between SA and ball release velocity in the slow bowling category (Table 1). Additionally, moderately significant negative relationships were found between TLF and ball release velocity in the normal and very fast bowling categories.

Table 1 Mean correlations (r) for ball velocity category and mechanical variables

Mechanical Variable	Bowlers (n)	Slower Delivery	Normal Delivery	Very Fast Delivery
CR (degrees)	10	0.162	-0.177	-0.114
Maximum TLF (degrees)	17	-0.400	-0.528*	-0.458*
Maximum SA (degrees)	17	0.453*	0.175	0.307
Maximum KF (degrees)	17	0.113	-0.164	-0.114

*Correlation is significant at the 0.05 level (1-tailed)

No significant differences were found for CR and TLF between the slow, normal and very fast bowling categories. For SA, significant differences were found between the slow and very fast bowling categories, as well as between the normal and very fast categories. Similarly for KF, significant differences were found between the slow and very fast bowling categories, as well as between the slow and normal bowling categories (Figure 1).

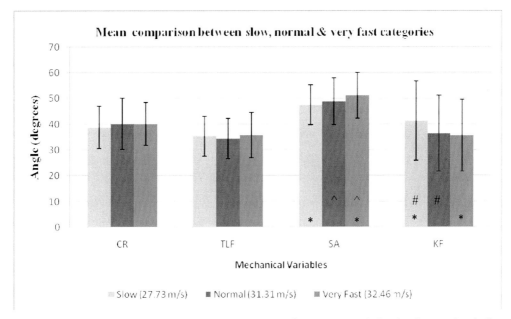

Figure 1 Mean comparison between slow, normal and fast category deliveries for mechanical variables. Matching symbols denote significant differences between pairs at the following confidence intervals: * $p \leq 0.001$, ^ $p = 0.031$, # $p = 0.01$

Only two variables remained in the stepwise linear regression as significant predictors of ball velocity. The results indicate that both TLF and SA together predict 11% of variance ($R^2=0.110$, SEE=1.623, $F(2, 197) = 12.114$, $p < .001$). That is, only 11% of the variance in ball velocity was explained by these two variables, for the normal delivery.

DISCUSSION: The results from this study, when combined with previous results (Portus et al., 2004; Salter et al., 2007), indicate that variations in bowling velocity are not significantly linked with an increase in shoulder alignment CR. Elliott et al. (2005) had previously shown that bowling over greater pitch lengths resulted in an increase in CR for young fast bowlers. Therefore there may be a trend for CR to be positively correlated with ball release velocity for young fast bowlers; however the results of this study suggest matured fast bowlers may produce extra ball velocity from other means when trying to bowl as fast as they can.

The significantly moderate negative correlation between ball release velocity and TLF in the normal and very fast categories seems to suggest bowlers who 'fell away' less were able to bowl at higher ball release velocities. This finding is in opposition to the results of Salter et al. (2007) who found a non-significant negative relationship between ball release velocity and ball release height ($r = -0.283$). The comparison between this study and that of Salter and colleagues is made with caution as ball release height is predominantly the summation of knee flexion, trunk flexion and upper arm abduction angle. Therefore arm abduction angle may have been a contributing factor to this negative relationship as it has been previously shown that lower arm abduction angles are significantly associated with ball release speed (Hanley et al., 2005).

The non-significant correlations between ball release velocity and SA in the normal and very fast categories suggest that there is no significant relationship between the two variables. However in the slower ball category, the positive correlation suggests that the greater velocity at which a bowler bowls a slower ball, the greater the SA. The significant difference in SA between the normal and very fast categories suggests that when bowlers attempt to volitionally bowl at a greater velocity, they are increasing the magnitude of a technique variable that has previously been related to lower back injury (Elliott, 2000).

As there was a significant difference between mean KF between the slow and normal bowling categories, bowlers may inadvertently increase this KF to slow their body's centre of mass when attempting to release the ball at a lower velocity. The non-significant difference between the normal and very fast bowling categories seems to suggest knee flexion may play more of a role in reducing peak impact forces over FFI (Foster et al., 1989) and may not influence bowling velocity. Given that only TLF and SA remained in the linear regression and only predicted 11% of variance, it may be reasonable to suggest that these two variables, along with CR and KF, are not useful when trying to predict ball release velocity. Therefore only one (SA) of the four mechanical variables that are commonly linked with lower back injury increase when bowlers attempt to volitionally bowl at higher velocities.

CONCLUSION: This study addressed the velocity dependence of techniques commonly linked with lower back injury in cricket fast bowling. For practical implications, only one (TLF) of the four variables, linked in the literature to lower back injuries, was shown to be significantly related to ball velocity for this sample of fast bowlers. As a negative correlation was found, decreasing TLF could potentially result in an increased ball release velocity as well as a decreased risk of sustaining a lower back injury. Of most interest, there was a significant increase in SA when bowlers attempted to volitionally increase their bowling velocity. Therefore bowlers who attempt to volitionally bowl at a greater velocity while increasing SA may be increasing their risk of sustaining a lower back injury, while not gaining any increase in ball release velocity from this variable. Additionally, bowlers should aim to restrict SA while bowling a slower ball, as this may be beneficial in reducing ball release speed as well as their risk of sustaining a lower back injury. A regression analysis showed that none of the four variables could strongly predict ball release speed and therefore cannot be confidently used in identifying potential fast-bowlers.

REFERENCES:

Burnett, A. F., Elliott, B. C. and Marshall, R. N. (1995). The effect of a 12-over spell on fast bowling technique in cricket. *Journal of Sports Sciences*, 13, 329-341.

Dempsey, A.R., Lloyd, D.G., Elliott, B.C., Steele, J.R., Munro, B.J. and Russo, K.A. (2007). The effect of technique change on knee loads during sidestep cutting. *Medicine & Science in Sports & Exercise*, 39, 1765-1773.

Elliott, B.C. (2000). Back injuries and the fast bowler in cricket. *Journal of Sports Sciences*, 18, 983-991.

Elliott, B., Plunkett, D. and Alderson, J. (2005). The effect of altered pitch length on performance and technique in junior fast bowlers. *Journal of Sports Sciences*, 23, 661-667.

Hanley, B., Lloyd, R. and Bissas, A. (2005). Annual Conference of the British Association of Sport and Exercise Sciences, *Journal of Sports Sciences*, 23, 93-223.

Foster, D., John, D., Elliott, B., Ackland, T. and Fitch, K. (1989). Back injuries to fast bowlers in cricket: A prospective study. *British Journal of Sports Medicine*, 23, 150-154.

Portus, M. R., Mason, B. R., Elliott, B. C., Pfitzner, M. C. and Done, R. P. (2004). Technique factors related to ball release speed and trunk injuries in high performance cricket fast bowlers. *Sports Biomechanics*, 3, 263-283.

Ranson, C., Kerslake, R., Burnett, A., Batt, M. and Abdi, S. (2005). Magnetic resonance imaging of the lumbar spine in asymptomatic professional fast bowlers in cricket. *Journal of bone and joint surgery*, 87, 1111-1116.

Salter, C. W., Sinclair, P. J. and Portus, M. R. (2007). The associations between fast bowling technique and ball release speed: A pilot study of the within-bowler and between-bowler approaches. *Journal of Sports Sciences*, 25, 1279-1285.

RELATIONSHIPS BETWEEN HIP AND SHOULDER ROTATION DURING BASEBALL PITCHING

David W. Keeley[1], Gretchen D. Oliver[1], Priscilla Dwelly[1], Hiedi J. Hoffman[1] & Christopher P. Dougherty[2]

University of Arkansas, Fayetteville, AR. USA[1]
Agility Center Orthopedics, Bentonville, AR. USA[2]

KEY WORDS: pitching, baseball, post pubescent

INTRODUCTION: Baseball pitching is one of the most dynamic movements in sports. Due to the violent nature of the pitching motion, baseball pitchers experience an extremely high rate of injury, particularly to the shoulder. One factor thought to be related to overuse injury at the shoulder is the high magnitude of shoulder external rotation. Because the pitching motion can be viewed as a kinematic chain in which the actions of individual segments have a direct impact on the actions of other segments, angles of internal/external rotation observed for the hip may be related to internal/external rotation observed at the shoulder. Thus, the purpose of the current project is to analyze internal/external rotation for both the shoulder and the hip in an attempt to identify a relationship between these two parameters. It is hypothesized that hip rotation will positively correlate with shoulder rotation.

METHOD: Kinematic data will be collected using The Motion Monitor® system (Innovative Sports Training, Chicago IL) and calculated using the ISB recommendations of the international shoulder group (Wu et al., 2005). The three-dimensional space at the pitching mound will be calibrated per manufacturer recommendations with the world axes system being oriented as described by Wu and Cavanaugh (1995). Once set-up and calibration is complete a set of 10 electromagnetic sensors will be placed at the following locations: the thorax (C7); the sacrum (L5/S1); the lateral epicondyle of the right and left humerus; the styloid process of the right and left ulna; the lateral epicondyle of the left and right femur; and the mid-point of the left and right shank. Following sensor placement, subjects will be allowed unlimited time for warm-up before throwing a series of maximal effort fastball pitches. Between trials, pitchers will be allowed a 40-60 s rest period. The internal/external rotation of both the right and left humerus and femur will be calculated throughout the pitching motion. Data will be reduced using established techniques. Data will be analyzed by calculating the Pearson product moment correlation coefficient for which hip rotation will be the independent variable and shoulder rotation will be the dependent variable.

DISCUSSION and CONCLUSION The identification of those factors associated with changes in the shoulder rotation of baseball pitchers is very important. As dynamic as the pitching motion is, the authors have identified no studies investigating the relationship between shoulder and hip rotation. If hip kinematics are related to shoulder kinematics, a better understanding of how the shoulder truly functions during baseball pitching can be developed through this study. In addition, this study has the potential to aid clinicians and sports medicine practitioners in understanding the underlying causes of shoulder injury in baseball pitchers by providing additional information about the relationship between the lower and upper extremities.

REFERENCES:

Wu, G. and Cavanagh, P.R. (1995). ISB recommendations for standardization in the reporting of kinematic data. *Journal of Biomechanics*, 28, 1257-1261.

Wu, G., van der Helm, F.C.T., Veeger, H.E.J., Makhsous, M., Van Roy, P., Anglin, C., et al. (2005). ISB recommendation on definitions of joint coordinate systems of various joints for the reporting of human joint motion – Part II: shoulder, elbow, wrist, and hand. *Journal of Biomechanics*, 38, 981–992.

ACUTE EFFECTS OF STATIC STRETCHING ON FORCE OUTPUT OF DORSI FLEXORS IN DANCE STUDENTS

Mayumi Kuno-Mizumura and Makiko Nagakura

Dept. of Performing Arts, Ochanomizu University, Tokyo, Japan

The purpose of this study was to investigate the effect of static stretching on maximal force production of the dorsi flexors in dance students who had been performing stretching exercise regularly. Eight female university students in the department of dance and dance education, and nine female university students without dance experience performed a maximal voluntary contraction of the dorsi flexors after six minutes of static stretching. After static stretching, maximal voluntary force of dorsi flexors decreased significantly in both groups, while the joint range of motion in plantar flexion was significantly higher in the dance students compared with the non-dancers. These results suggest that prolonged stretching exercise induces a similar deficit in force production after acute static stretching in both dancers and non-dancers.

KEY WORDS: dance, static stretching, force production.

INTRODUCTION: An acute bout of static muscle stretching can diminish force output or performance (Kokkoen et al., 1998, Avela et al., 1999) through two possible mechanisms: a change in active musculotendinous stiffness, and a depression of muscle activation. On the other hand, regular stretching training also inhibits the excitability of the stretch reflex induced by muscle stretching. Dance students perform stretching exercises regularly to improve their flexibility. The acute effects of static stretching on force output may therefore differ between dancers and non-dancers. We investigated the acute effect of static stretching on force output of dorsi flexors in dance students and students without dance experience. We hypothesized that acute static stretching would reduce force output to a smaller extent in dancers compared with non-dancers. The basis for this hypothesis on was that regular stretching induces adaptive responses in either musculotendinous architecture or neural pathways.

METHOD: Data Collection: Subjects were eight healthy female university students in the department of dance and dance education (mean age; 21.4 ±1.3 years, height; 1.58±0.06 m, body mass; 51.9 ±6.4 kg, %body fat; 21.4±1.3 %) and nine healthy university students without dance experience (mean age; 20.2 ±1.5 years, height; 1.59±0.05 m, body mass; 49.3 ±4.5 kg, %body fat; 21.4±2.6.%). Dance students had been taking classical ballet or modern dance classes more than two days per week. Their dance experience was 5.2±3.4 years. All subjects visited our laboratory on three separate days. On the first visit, subjects were asked to perform maximal dorsi flexion to be familiar with the testing protocol. Subjects sat on a test bench with hip joint at 90° flexion. The ankle joint was set at 0°, with the knee joint at full extension. The waist was secured by adjustable lap belts, and held in position. The foot was securely strapped to a footplate connected to the lever arm of the force platform (Kyowa dengyo, Japan). The force platform was attached to the sole of the subject's foot. The torque (Nm) during dorsi flexion with maximal effort was detected by the force platform. The range of motion in the ankle joint during plantar flexion and dorsi flexion were also measured at the first visit using a goniometer. A sit and reach test was also performed to evaluate each subject's overall flexibility. On the second or third visit, subjects performed maximal voluntary dorsi flexion before and after six minutes static stretching or six minutes rest with the same testing setup as the first visit. Subjects did not warm up before the six minutes of static stretching or at rest. During testing, room temperature was set at 25°C. Surface electromyogram (EMG) was recorded from the medial heads of gastrocnemius, soleus, and tibialis anterior muscles. Integrated EMG (iEMG) was calculated and normalized by the maximal voluntary contraction (%MVC). Electo-mechanical delay (EMD) was also calculated from the time series data of EMG and force output.

Data Analysis: Two-way ANOVA with repeated measures was used to analyse all variables. Post-hoc analysis of mean value was performed using Turkey's significant difference method. P<0.05 was accepted for statistical significance.

RESULTS: Range of motion (ROM) in plantar flexion and the results of sit-and-reach test were significantly higher in the dance students compared with the non-dancers, but ROM in dorsi flexion was similar between two groups.

After six minutes of static stretching, ROM of plantar flexion increased significantly in both groups (Figure 1).

Maximal voluntary force decreased significantly after six minutes stretching in both groups (Figure 2.), while there was no significant change in force output between before and after six minutes rest in the control trial. Integrated EMG also decreased after stretching. EMD did not change significantly before and after stretching, however EMD tended to increase more in the dance students before and after stretching compared with the non-dance students.

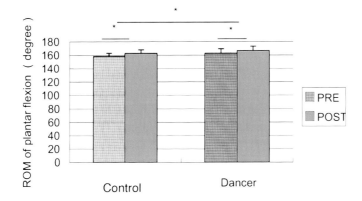

Figure 1 Range of motion in plantar flexion (dorsi flexors) for both groups before and after acute stretching

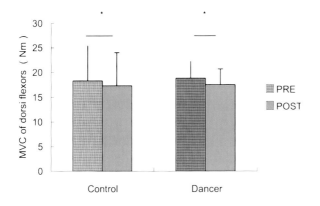

Figure 2 MVC force during dorsi flexion for both groups before and after acute stretching

DISCUSSION: The only adaptive response to regular stretching exercise in the dance students was greater flexibility in plantar flexion. The significant increase in ROM for plantar flexion after six minutes acute stretching in both groups indicates that muscle stretching increased joint mobility. In contrast to our hypothesis, maximal voluntary force and muscle activity in dorsi flexion decreased in both dance students and non-dance students. We propose that force deficit after acute stretching in this study would mainly due to changes in neural activity. In our previous studies, advanced dancers showed different responses in musculotendinous properties after stretching compared with the non-dancers. Future studies could assess the effects of force output individuals with great joint mobility, such as professional dancers.

CONCLUSION: Acute static stretching reduced force production and muscle activity during dorsi flexion in both dance students and non-dance students. Contrary to our hypothesis, the acute effect of passive muscle stretching was not dependent on joint ROM.

REFERENCES:
Kokkonen J., Nelson A.G. and Cornwell A. (1998). Acute muscle stretching inhibits maximal strength performance. *Research Quarterly in Exercise and Sport*, 69(4), 411-5.
Avela, J., Kyrolainen, H., & Komi, P. V. (1999) Altered reflex Sensitivity after repeated and prolonged passive muscle stretching. *Journal of Applied Physiology*, 86, 1283-1291.
Fowles, J. R., Sale, D. G. and MacDougall, J. D. (2000). Reduced strength after passive stretch of the human plantarflexors. *Journal of Applied Physiology*, 89, 1179-1188.

THE DIFFERENCE OF THE BALANCE ABILITIY BETWEEN THE FUNCTIONAL ANKLE INSTABILITY AND HEALTHY SUBJECTS

Yukio Urabe, Yuki Noda, Shinji Nomura, Takeshi Akimoto, Hiroe Shidahara and Yuki Yamanaka

Graduate School of Health Sciences, Hiroshima University, Japan

KEY WORDS: ankle instability, center of pressure, EMG activity, brace

INTRODUCTION: Ankle inversion sprains are one of the most common injuries occurring in sports activities. Repeated ankle sprains may lead to chronic ankle instability. In order to prevent the occurrence of ankle instability, it is necessary to understand the difference in the physiological characteristics of the subjects who have chronic ankle instability and those who do not. However, there is little known about the differences between the two. It has been reported that there are two types of ankle instability: mechanical ankle instability (MAI) and functional ankle instability (FAI) which is the disability to which patients refer when they say that their foot tends to "giving way". In this study, we have attempted to uncover the difference in the center of pressure (COP) and muscle activities during a single leg standing between FAI subjects and healthy subjects with or without an ankle brace.

METHODS: Nine FAI male subjects and 13 healthy male subjects participated in this study. FAI criteria were according to the Karlsson's Scoring scale of ankle instability. All FAI subjects had experienced an ankle inversion sprain and they felt that they had ankle instability. The COP was measured by a force plate UM-BAR (Unimec Co., Tokyo, Japan) and the total length of COP trajectory was calculated. In the EMG measurement, the use of a surface electrode system aided EMG measurements- Personal-EMG (Oisaka electric Co., Fukuyama, Japan). The EMG system was used to measure the Normalized integrated EMG (IEMG) of the Peroneus longus (PL), Gastrocnemius (GC), and Tibialis anterior (TA) muscles. Both FAI and healthy subjects wore a custom made ankle brace (Asics Co., Kobe, Japan). This brace has a single axis joint to allow the ankle to have free planter-dorsal flexion yet prevented ankle inversion movements. The COP and muscle activities while standing on a single leg were compared between the FAI subjects and healthy subjects with or without the ankle brace. We used the non-paired t test to quantitate the COP deviation between the FAI and the healthy. The paired t test was used between the COP with the brace and without the brace in the FAI and healthy subjects. The Statistical significance level was set as 0.05 in this study.

RESULTS: The length of COP trajectory was larger in the FAI as compared with the healthy subjects with open eyes ($p<0.05$). The subjects with the brace showed the smaller COP deviation than those without the brace in healthy subjects ($p<0.01$), while no significant difference in FAI subjects was observed. EMG activity showed no significant difference between the FAI and healthy subjects. However, EMG with the brace had larger tendency than without the brace both the FAI and the healthy.

DISCUSSIONS: There is little known about the physiological characteristic differences in the subjects who have chronic ankle instability. It was not influenced by treatment after the injury, age, gender, and type of sports. Only body weight was a risk factor that could lead to recurrent ankle sprains. This study demonstrated the differences in FAI and healthy subjects. The FAI showed large COP deviation.

CONCLUSIONS: We confirm the Karlsson's Scoring scale was a reasonable choice for categorizing FAI subjects. Recently, many athletes use the ankle brace to prevent reoccurring sprains. This study showed the ankle brace has the effect to maintain the stability of the ankle and that; repeated occurrences of ankle sprains might lead to chronic ankle instability. This study is presenting preliminary results, we will need to continue to measure more subjects and compare many varied physical conditions.

REFERENCE:
Karlsson, J.and Peterson, L. (1991). Evaluation of ankle joint function: the use of scoring scale. *The foot* 1(1):15-19

A BIOMECHANICAL COMPARISON OF THE LOWER EXTREMITY DURING FRONT AND BACK SQUATS IN HEALTHY TRAINED INDIVIDUALS

Erin C. Learoyd and Kathryn Ludwig

School of Human Performance and Leisure Sciences, Barry University, Miami Shores, Florida, USA

KEY WORDS: EMG, hip, knee, squat

INTRODUCTION: The two most common forms of the squat exercise are the back squat (BS) and the front squat (FS). While used interchangeably, there is little empirical evidence to inform the strength professional as to which variation may best benefit an athlete. Recently, Gullett et al. (2008) suggested that the lower compressive forces in the knee during a FS may make this variation the primary training choice. The purpose of this research is to compare the forces and moments at the hip, knee, and ankle during a BS and a FS in healthy trained participants.

METHODS: Currently, data have been collected from one individual, a healthy female who was trained in both the FS and BS variations. A 7-camera three-dimensional motion analysis system (Vicon, Centennial, CO) and AMTI force plate (Watertown, MA) were used to record data. Linear forces at the knee, moments at the hip, knee, and ankle, EMG of the rectus femoris (RF) and semitendinosis (ST) muscles, and kinematic data for the lower extremity were measured during a FS and BS movement at 65% of previously measured 1-RM loads.

RESULTS: Despite comparable relative loads, absolute loads differed between the FS and BS (52 kg and 55 kg, respectively). EMG measurements (%MVC) were higher in both RF and ST during the FS. Knee flexion and extension moments were similar between the squat conditions. This is contradictory to the results found by Gullett et al. (2008).

DISCUSSION: Further data should definitively support or contradict recent findings regarding FS and BS differences in the lower extremity. Linear and angular forces measured at the joints proximal and distal to the knee will help complete our understanding of lower-extremity biomechanics during these two squats types. The full results of all participants will be available at the August conference.

CONCLUSION: As technology advances, so does the ability of researchers to reexamine past practices for their safety, effectiveness, and efficiency. This enables the professional to make educated decisions based on empirical evidence regarding the nature of a prescribed training program. The back squat is a standard lower-body exercise and can be found in nearly any resistance training program. The front squat may elicit similar or higher muscle work output at a lower absolute load, resulting in comparable strength gains with lower joint stress. This would indicate that the front squat, while much less common, may in fact be safer and more efficient than the back squat, and if so, should be considered for use by strength and conditioning professionals and their peers.

REFERENCES:
Gullett, J.C., Tillman, M.D., Gutierrez, G.M. and Chow, J.W. (2008). A biomechanical comparison of back and front squats in healthy trained individuals. *Journal of Strength and Conditioning Research*, 23(1): 284-292.

QUALITY OF DINGHY HIKING: EFFECTS ON SPEED AND HEADING

R.N. Marshall

Eastern Institute of Technology, Taradale, Hawke's Bay, New Zealand

KEY WORDS: sailing, righting moment, technique,

INTRODUCTION: Dinghy hiking uses the sailor's bodyweight to produce a 'righting moment' which counteracts the moment of force caused by the wind on the sail. However, hiking is a physically demanding activity and maintaining quality hiking technique throughout a race is essential in competitive sailing. A land-based hiking simulator has been used to assess hiking technique, but no data were available to relate simulator results to on-water performance. This project evaluated the effect of decreased quality hiking positions on actual on-water dinghy speed and heading.

METHOD: GPS data on boat speed and heading as well as video of hiking positions were collected from two international-level dinghy sailors. Each sailed upwind in an International Laser, demonstrating three levels of righting moment while hiking. Data were collected for two trials of approximately sixty seconds for each hiking position for each sailor.

GPS data were averaged for each trial and across subjects. Single images from the video for each hiking position were analysed to calculate the sailor's centre of mass using a 4-segment model and righting moment about the fore/aft centreline of the boat. These data were averaged across subjects and trials for each hiking position. Average changes in righting moment were then plotted against average changes in speed and heading.

RESULTS: Changes in speed with hiking position can be seen in Figure 1. Concurrent with speed decreases were changes in heading. An average decreased righting moment of 119 Nm from 'Hard' to 'Rest' positions resulted in a speed decrease of 0.6 km/h and sailing 2° further off the wind. The 'Bunny' position resulted in an average 178 Nm righting moment decrease and a consequent speed drop of 1.7 km/h and sailing 6° off the wind.

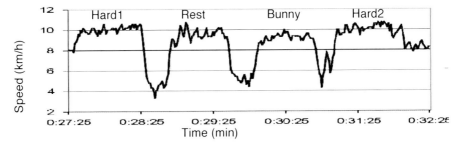

Figure 1 Speed vs hiking position

DISCUSSION: These preliminary data provide a guide to quantifying the effects of decreases in quality of hiking position which may be useful for both coaches and sailors. These data also permit calculation of the effects of speed and heading changes on the time it takes to sail a typical leg in a race. For example, sailing one half of a 3.3 km beat in the bunny position will take approximately 3 minutes and 20 seconds longer than sailing the entire leg in the fully extended ('Hard') position.

CONCLUSION: While coaches and sailors have known that poor hiking positions decrease boat speed and worsen sailing direction, this study provides data to quantify those effects in on-water sailing.

MECHANICAL COMPARISON BETWEEN ROUNDHOUSE KICK TO THE CHEST AND TO THE HEAD IN FUNCTION OF EXECUTION DISTANCE IN TAEKWONDO

Isaac Estevan[1], Coral Falco[1], Octavio Alvarez[2], Fernando Mugarra[3] and Antonio Iradi[4]

Faculty of Education and Sport Sciences, Catholic University of Valencia (Spain) [1], Cheste Sport Medicine Centre, Consell Valencia de l'Esport (Spain) [2], Faculty of Physics, University of Valencia (Spain) [3], Faculty of Medicine, University of Valencia (Spain) [4]

The purpose of this research was to examine and compare maximum impact force (MIF) and execution time (ET) in two different Taekwondo techniques, roundhouse kick to the chest (Bandal Chagui) and roundhouse kick to the head (Dolio Chagui) in terms of the execution distance and to analyse the relationship between maximum impact force and weight for two different groups and kicks. To measure the mechanical parameters, a model explained in Falco et al. (2009) was used. In this study, the 23 male taekwondo players participating were divided into two groups: medallists ($n = 12$) and non-medallists ($n = 11$). For the medallists' group no differences had been found in MIF or ET from either distance between roundhouse kick to the chest and to the head. However, significant differences were found in MIF in the non-medallists' group from all execution distances between roundhouse kick to the chest and to the head. For the non-medallists' group, weight significantly predicts MIF, but not in the medallists' group. In conclusion, medallist taekwondo players should perform roundhouse kick to the head instead to the chest, because it produces a better score in the same time.

KEY WORDS: Distance, impact force, execution time, kick, Taekwondo.

INTRODUCTION: Taekwondo is a combat sport with hits, in which leg techniques are the most commonly used in competition (Olivé, 2005). Among the leg techniques, there is roundhouse kick to the chest (Bandal Chagui), being the action with 27% of the total points are obtained in the combat and 10% of the KO, and there is also roundhouse kick to the head (Dolio Chagui) with which 11% of the points are obtained being the second technique to score KO (Olivé, 2005). One difference between Bandal and Dolio Chagui is the height of the target: to the chest and the face, respectively (O'Sullivan et al., 2008). In 2009, competition rules were changed: a kick to the head (i.e. Dolio Chagui) is worth three points while a kick to the chest is worth one point (i.e. Bandal Chagui) (WTF, 2009).
According to Kim et al. (2008), athletes choose techniques that score points, using less force and more efficiency. Roundhouse kicks are the actions which are easily adapted to execution distance (ED) (Kim et al., 2008). In addition to the ED (Hristovski et al., 2006), among the factors that condition the success of the kick, there are mechanical variables such as the maximum impact force (MIF) and the execution time (ET) that are relevant to score when kicking the opponent (Falco et al., 2009).
Among the studies analysing the differences in terms of the kind of roundhouse kick, Lee and Huang (2006) and Pedzich et al. (2006) point out that athletes carried out the more difficult kicks with a smaller MIF than the easier ones, influencing the kind of roundhouse kick in the MIF. This statement was confirmed by the research of O'Sullivan et al. (2008) who studied Bandal and Dolio Chagui and found higher impact force on the kicks whose objective was to hit to a smaller height ($p < .05$). On the other hand, Hong et al. (2000) found higher ET in the kicks directed at a greater height. In the study of Pedzich et al. (2006) the trajectory of the kick affecting the mechanical results of same was observed; in this line, Nien et al. (2006) found that in similar techniques there were differences in execution technique such as in the coordination of the lower limb segments what would be based on the principle of kinetic chain when each limb could be thought of as a chain consisting of rigid overlapping segments connected by a series of joints.

Estevan et al. (2009) and Falco et al. (2009) suggest that higher level athletes would base their execution on the principle of the kinetic chain, which supposes a better technical execution. Lee et al. (2005) point out that keeping the principle of the kinetic chain is essential to the effectiveness of the roundhouse kick. However, according to Pearson (1997) and Pedzich et al. (2006) athletes use their weight to increase MIF, these authors found a positive relationship between weight and MIF. In studies carried out using the same procedure as in the present one (three ED), Estevan et al. (2009) and Falco et al. (2009) found a positive correlation between weight and MIF in a lower level athletes' group, but not in a higher level athletes' group.

The overall purpose of this study was twofold: firstly, to examine the MIF and the ET in the Bandal Chagui and the Dolio Chagui from three EDs, in two different level groups. And secondly, to examine the relation between the MIF and the weight in terms of the kind of roundhouse kick and the athletes level, in three EDs.

METHOD: Participants: A convenience sample of 23 male taekwondo athletes gave informed consent to participate in the study; their average age was 26.78 years ($SD = 2.27$), weight of 73.80 kg ($SD = 12.60$) and height of 1.78 m ($SD = 0.07$). They were divided into two groups: the medallists' group ($n = 12$) and the non-medallists' group ($n = 11$) in terms of whether they could be medal winners in official national and international events or not. All the athletes had at least 4 years of competitive experience and trained for at least 3 hours per week. They accepted to participate anonymously and voluntarily in the research.

Procedure: In order to carry out the present study, a model explained in Falco et al. (2009), developing the same protocol (two trials for each of the three different EDs considering the length of the subjects' leg), was used to measure relevant parameters to kick performance and relating to the mechanical variables: MIF and ET.

RESULTS: The preliminary analysis (Kolmogorov – Smirnov) showed a normal distribution of all the considered variables. The Bonferrini correction was applied to reduce error accumulated in the 6 t tests carried out to assess potential differences in mechanical variables from three EDs between the Bandal and Dolio Chagui ($p < .01$), in the two different level groups. Statistical description (mean and standard deviation, minimum and maximum) are shown in Table 1.

For the medallists' group, no differences were found in the MIF and the ET between Bandal and Dolio Chagui. For the non-medallists' group, differences were found from ED_1 ($t = 3.00$; $p < .01$) and ED_2 ($t = 3.32$; $p < .01$), in the MIF of Bandal and Dolio Chagui, respectively.

In order to determine the relationship between the weight and the MIF, Pearson's correlation was calculated within both groups. For the medallists' group, there was no correlation in Bandal and Dolio Chagui. However, for the non-medallists' group, a significant positive correlation ($p < .05$) was found between weight and MIF from ED_3 ($r = .73$) in Bandal Chagui, and significant positive correlation was found from ED_1 ($r = .74$; $p < .01$), ED_2 ($r = .62$; $p < .05$) and ED_3 ($r = .64$; $p < .05$) in Dolio Chagui. So as to analyse the predictive power of weight on MIF, a regression analysis was made for the non-medallists' group, which showed that in ED_3 ($\beta = .73$) weight predicts 53.4% of the variation in MIF of Bandal Chagui ($p < .01$). Therefore, in ED_1 ($\beta = .74$; $p < .01$) weight predicts 54.7% of the variation in MIF of Dolio Chagui, in ED_2 ($\beta = .62$; $p < .05$) 38.9% of the variation in MIF of Dolio Chagui, and in ED_3 ($\beta = .64$; $p < .05$) 41.3% of the variation in MIF of Dolio Chagui.

Table 1. Comparative analysis between Bandal and Dolio Chagui in Maximum Impact Force and Execution Time in function of Execution Distance.

		Bandal chagui				Dolio Chagui				
	ED	M	SD	Min	Max	M	SD	Min	Max	t
		Medallist ($n = 12$)								
MIF	ED_1	1959.83	632.50	1258.50	3284.50	1934.13	573.82	1225.00	2660.50	.12
(N)	ED_2	2018.46	335.70	1491.50	2614.00	1889.00	401.23	1247.00	1405.00	.86
	ED_3	1852.29	381.09	1250.50	1431.50	1775.21	402.41	1278.00	2433.50	.49
ET	ED_1	.225	.038	.179	.312	.241	.030	.190	.285	-2.14
(s)	ED_2	.241	.020	.202	.267	.286	.100	.210	.580	-1.76
	ED_3	.329	.091	.236	.589	.313	.045	.227	.398	.63
		Non-Medallist ($n = 11$)								
MIF	ED_1	1910.82	677.18	664.00	2745.00	1457.10	537.71	665.00	2149.50	3.00*
(N)	ED_2	2040.68	307.45	1542.00	2436.50	1597.50	392.00	903.00	2390.50	3.32*
	ED_3	1871.64	370.61	1302.50	2391.00	1398.77	630.94	290.50	2580.00	2.74
ET	ED_1	.257	.095	.187	.472	.276	.015	.245	.290	-0.80
(s)	ED_2	.273	.043	.226	.378	.316	.072	.270	.520	-1.76
	ED_3	.344	.074	.218	.462	.370	.051	.292	.449	-1.18

Note. ED = Execution Distance (ED_1: short; ED_2: medium; ED_3: large); MIF = maximum impact force in Newtons (N); ET = execution time in seconds (s). * $p < .01$

DISCUSSION AND CONCLUSION: The main purpose of the present study was to examine whether the kind of roundhouse kick determines the impact force (MIF) and the execution time (ET). For the non-medallists' group, the influence of the kind of roundhouse kick on MIF on some the execution distances (ED) has been observed, that coincides with Lee and Huang (2006) and O'Sullivan et al. (2008) statements. Nevertheless, for the medallists' group, in any ED the kind of roundhouse kick does not condition the MIF. As for the ET, we observe that in any ED there are no differences between Bandal and Dolio Chagui for any group (medallists and non-medallists). In this way, for the medallists' group, the kind of roundhouse kick does not condition effectiveness of execution (same MIF and same ET), taking into consideration that Dolio Chagui is worth three points (two more than Bandal Chagui), we could guide the athletes to carry out more actions of Dolio Chagui in taekwondo combats to get more points.

Pearson (1997), Estevan et al. (2009) and Falco et al. (2009) analysed the relationship between the weight and the MIF in Taekwondo, they found a positive correlation between the weight and the impact force. In our research we have obtained results in the line of Estevan et al. (2009) and Falco et al. (2009) who found a relationship between the weight and the MIF, only for the non-medallists' group, that is, in the lower level athletes.

Estevan et al. (2009) point out that Dolio Chagui is a more complex kick than Bandal Chagui, which could explain for the non-medallists' group, the relationship between weight and MIF in Dolio Chagui in all EDs, instead Bandal Chagui that is only related in ED_3, explaining in Dolio Chagui higher percentage of impact force than in Bandal Chagui. On the other hand, the medallists' group would not use their weight to generate their MIF regardless of the roundhouse technique. Estevan et al. (2009) suggest that higher level athletes could achieve greater impact forces due to a better technical execution based on the principle of the kinetic chain.

MIF and ET in each ED was analysed in our study; to improve scientific bases of the efficiency of Taekwondo kicks future research should examine other mechanical parameters such as impact time or maximum impact force according to body mass, which may explain the differences depending on level. Furthermore, studies should analyse how reaction time influences each ED in mechanical parameters. Finally, it could be suggested that Dolio

Chagui is a more complex kick than Bandal Chagui, but for the medallists' group, the kick efficiency (with same force and same execution time) is not influenced by the kind of roundhouse kick. Therefore, coaches could guide the high level athletes to carry out more actions of Dolio Chagui in taekwondo combats to get a better score in the same time.

REFERENCES:

Estevan, I., Alvarez, O., Falco, C., Castillo, I., Mugarra, F. and Iradi, A. (2009). *Estimation of mechanical parameters influenced by execution distance in a roundhouse kick to the head in Taekwondo.* Manuscript submitted for publication.

Falco, C., Alvarez, O., Castillo, I., Estevan, I., Martos, J., Mugarra, F. and Iradi, A. (2009). Influence of the distance in a roundhouse kick's execution time and impact force in Taekwondo. *Journal of Biomechanics, 42* (3), 242-248.

Hristovski, R., Davids, K., Araújo, D. and Button, C. (2006). How boxers decide to punch a target: emergent behaviour in nonlinear dynamical movement systems. *Journal of Sports Science and Medicine, CSSI*, 60-73.

Hong, Y., Hing, K. L. and Luk, T. C. J. (2000). Biomechanical Analysis of Taekwondo Kicking Technique, Performance & Training Effects. *SDB Research Report, 2*, 1-29.

Kim, J. W., Yenuga, S. and Kwon, Y. H. (2008). The effect of target distance on trunk, pelvis, and kicking leg kinematics in Taekwondo round house kick. In Y. H. Kwon, J. Shim, J. K. Shim, & I. S. Shim (Eds.), *Proceedings of the 26th International symposium on biomechanics in sports* (p. 742). Seoul.

Lee, C. L., Chin, Y. F. and Liu, L. (2005). Comparing the difference between Front-Leg and Back-Leg Round-House Kicks attacking movement abilities in Taekwondo. In Q. Wang (Ed.), *Proceedings of the 23rd International symposium on biomechanics in sports* (pp. 877-880). Beijing.

Lee, C. L. and Huang, C. (2006). Biomechanical analysis of Back kicks attack movement in Taekwondo. In H. Schwameder, G. Strutzenberger, V. Fastenbauer, S. Lindinger, & E. Müller, (Eds.), *Proceedings of the 24th International symposium on biomechanics in sports* (pp. 1-4). Salzburg.

Nien, Y. H., Chang, J. S. andTang, W. T. (2006). The comparison of kinematics characteristics between single and successive Licking techniques for the taekwondo players with an olympic medal: a case study. *Journal of Biomechanics, 39*, S563.

Olivé, R. (2005). *Estudio de la cadera del practicante de taekwondo.* Unpublished doctoral dissertation. Universidad Autónoma de Barcelona, Barcelona.

O'Sullivan, D., Chung, C., Lee, K., Kim, E., Kang, S., Kim, T. and Shin, I. (2008). Measurement and comparison of Taekwondo and Yongmudo Turning Kick Impact Force for Two Target Heights. In Y. H. Kwon, J. Shim, J. K. Shim, & I. S. Shim (Eds.), *Proceedings of the 26th International symposium on biomechanics in sports* (pp. 525-528). Seoul.

Pearson, J. (1997). *Kinematics and kinetics of Taekwon-do turning kick.* Unpublished doctoral dissertation. University of Otago, Dunedin.

Pedzich, W., Mastalerz, A. and Urbanik, C. (2006). The comparison of the dynamics of selected leg strokes in taekwondo WTF. *Acta of Bioengineering and Biomechanics, 8* (1), 1-9.

World Taekwondo Federation. (2009). *Rules and Regulations.* Retrieved 3-14-2009 from http://wtforg.cafe24.com/wtf_eng/site/rules/file/Rules_and_Regulations_of_the_WTF_(as_of_Feb_3_2009).pdf.

Acknowledgement

This research has been partially supported by the Sports Service of the University of Valencia.

KINETIC AND KINEMATIC ANALYSIS OF THE DOMINANT AND NON-DOMINANT KICKING LEG IN THE TAEKWONDO ROUNDHOUSE KICK

Coral Falco[1], Octavio Alvarez[2], Isaac Estevan[1], Javier Molina-Garcia[1], Fernando Mugarra[3] and Antonio Iradi[4]

Faculty of Education and Sports Sciences, Catholic University of Valencia (Spain) [1], Cheste Sports Medicine Centre, Consell Valencia de l'Esport (Spain) [2], Faculty of Physics, University of Valencia (Spain) [3], Faculty of Medicine, University of Valencia (Spain) [4]

> The purpose of this study was to examine kinematic variables relevant to kick performance with the dominant and non-dominant leg, in a roundhouse kick measured from three execution distances. Forty-three taekwondo athletes that had competitive taekwondo experience participated in the study. A dependent t-test indicated that there were no differences between extremities at any distance ($p > .01$). Based on these results, competitive taekwondo players do not seem to reveal differences in limb kinematics. The results also showed the influence of the distance from which the dominant leg explains a larger percentage of variance in reaction time (24%), execution time (20%) and total response time (60%) of the non-dominant leg from a short distance, whereas regarding impact force (22%), this higher percentage is explained from a long distance.
>
> **KEY WORDS:** combat sport, impact force, reaction time, total response time, symmetry.

INTRODUCTION: In Taekwondo, like other sports, competitive athlete training starts with analysing the activity. In a kicking sport where techniques with the preferred and non-preferred leg alternate, the roundhouse kick with the back leg is widely used. According to Tang et al. (2007) symmetric skills are important to be among the elite in taekwondo. In general, athletes have preference for one particular foot to kick the target during training (Tang et al., 2007), and to date, it is unknown if this preference makes this sport as asymmetric. Pedzich et al. (2006) found statistically significantly higher stroke force values on the right side when the right limb was dominant in a yop and dwit chagi, respectively. Furthermore, the impulse stroke of the side kick (yop chagi) was found to be slightly higher (9%) for the left limb than for the right one. On the contrary, during dwit-chagi the impulse stroke was higher (9%) for the right limb than for the left one. In the same line, Peng (2006) found that in a roundhouse kick, the dominant leg was faster than the non-dominant one ($p < .05$). In contrast, Tang et al. (2007) found no significant differences between the preferred and non-preferred leg in movement time ($p > .05$) in this same type of kicking.

To date, there is still a lack of research on symmetries in taekwondo. With these investigations in mind, the purpose of this study was to examine if the dominant leg (D) produced better results than the non-dominant one (ND) in maximum impact force (MIF), reaction time (RT), execution time (ET) and total response time (TT) in a taekwondo roundhouse kick measured from 3 different execution distances (ED_1, ED_2, ED_3).

METHODS: In order to carry out the present study, a model based on the work of Falco et al. (2009), using a mannequin, contact platform, force platform, and adding a LED (see figure 1) to measure the reaction time and developing the same protocol; that is two trials for each of the three different distances considering the subjects' leg length (ED_2 or medium distance), 1/3 above (ED_3 or long distance) and 1/3 below (ED_1 or short distance), was used to measure parameters relevant to kick performance and related to mechanical variables: maximum impact force (MIF), reaction time (RT is initiated by switching on the LED on the chest of the mannequin until the athlete raises the kicking foot from the contact platform), execution time (ET starts when the athlete raises the foot of the kicking leg from the contact platform and stops when the athlete's foot impacts on the force platform reaching the maximum impact force) and total response time (TT is defined as reaction time plus execution time).

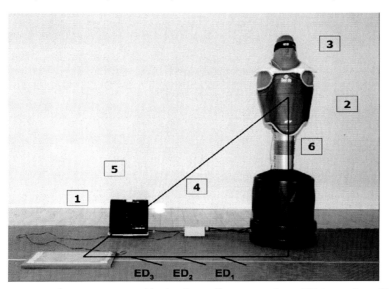

Figure 1. System set up with the three distances: short (ED_1), medium (ED_2) and long (ED_3). 1: contact platform, 2: force platform, 3: signal light, 4: microcontroller, 5: Pc, 6: mannequin

A sample of 43 taekwondo players (31 men and 12 women) aged from 13 to 38 years ($M = 24.49$; $SD = 5.94$), weighing from 46 to 101 kg ($M = 70.91$; $SD = 12.91$) and with a height ranging from 1.53 to 1.93 m ($M = 1.73$; $SD = 0.15$), kicking from distances between 0.69 (ED_1) and 1.37 (ED_3) metres ($M = 1.03$; $SD = 0.07$) (ED_2) were selected to participate in the study. All of them had been doing taekwondo for at least 4 years and had competitive taekwondo experience (including athletes that had won a medal in a national or international championship). Informed consent was given.

2-tailed, paired t tests were performed to assess the variation of the kinematical measures (MIF, RT, ET & TT) over execution distances (ED_1, ED_2, ED_3) through the dominant (D) and non-dominant (ND) leg. The Bonferroni correction was applied to reduce the accumulated error in the 12 t tests performed to assess potential differences in mechanical variables from three EDs between the dominant and non-dominant leg ($p < .01$). A Pearson product moment correlation was also performed to evaluate the relationship among paired variables between dominant and non-dominant leg. Moreover, a regression analysis was done to analyse the predictive power of the dominant leg on the non-dominant leg in the variables under study (MIF, RT, ET & TT) at each distance (ED_1, ED_2, ED_3). The criterion level set as $p < .01$ was considered significant for all analyses.

RESULTS: The preliminary analysis (Kolmogorov –Smirnov) showed a normal distribution in all the considered variables. Descriptive statistics (mean and standard deviation) and t coefficients are presented in Table 1. The results indicated that there were no differences between the dominant and non-dominant leg from any distance ($p > .01$).

Table 1. Descriptive statistics and *t* coefficients for dominant and non-dominant legs (N = 43)

		Dominant		Non-Dominant		
		M	SD	M	SD	t
MIF (N)	ED₁	1317.64	727.91	1024.11	519.73	2.52
	ED₂	1032.62	630.34	966.40	473.72	.65
	ED₃	960.71	515.01	866.97	494.74	1.18
RT (s)	ED₁	.50	.09	.50	.07	-.54
	ED₂	.52	.11	.53	.09	-.67
	ED₃	.61	.11	.62	.13	-.31
ET(s)	ED₁	.27	.07	.27	.07	-.33
	ED₂	.32	.10	.32	.11	-.23
	ED₃	.36	.10	.38	.13	-.72
TT (s)	ED₁	.76	.08	.78	.09	-1.68
	ED₂	.83	.12	.85	.11	-1.14
	ED₃	.96	.10	1.00	14	-1.73

Note. ED= Distance (1= short; 2= medium; 3= long); MIF = maximum impact force; RT = reaction time; ET = execution time; TT = total response time *$p < .01$

Significant positive correlations ($p < .01$) between the dominant and non-dominant leg were found in MIF from the ED_1 ($r = .34$), ED_2 ($r = .31$) and ED_3 ($r = .47$), in RT from the ED_1 ($r = .49$), in ET from ED_1 ($r = .45$) and ED_2 ($r = .31$) and in TT from the ED_1 ($r = .78$), from the ED_2 ($r = .36$) and from the ED_3 ($r = .40$). Also, the regression analysis results to predict the power of the dominant leg over the other leg into variables under study are showed in Table 2.

Table 2. Regression Analysis for all variables in the dominant over non-dominant leg (N = 43)

Variable	B	SE B	β	t	R²
IF$_D$ED$_1$- IF$_{ND}$ED$_1$	.24	.11	.34	2.25	.12
IF$_D$ED$_2$- IF$_{ND}$ED$_2$	.23	.11	.30	2.05	.09
IF$_D$ED$_3$- IF$_{ND}$ED$_3$	.45	.13	.47	3.42	.22
RT$_D$ED$_1$- RT$_{ND}$ED$_1$	.35	.10	.49	3.48	.24
ET$_D$ED$_1$- ET$_{ND}$ED$_1$	.42	.13	.45	3.11	.20
ET$_D$ED$_2$- ET$_{ND}$ED$_2$	.33	.16	.31	2.05	.09
TT$_D$ED$_1$- TT$_{ND}$ED$_1$	.81	.11	.78	7.56	.60
TT$_D$ED$_2$- TT$_{ND}$ED$_2$	.13	.32	.13	2.41	.13
TT$_D$ED$_3$- TT$_{ND}$ED$_3$	.57	.21	.40	2.73	.15

Note. ED= Distance (1= short; 2= medium; 3= long); D= dominant leg; ND = non-dominant leg; MIF = maximum impact force (N); RT = reaction time (s); ET = execution time (s); TT = total response time (s) *$p < .05$

DISCUSSION: The purpose of this study was to examine the mechanical variables maximum impact force, reaction time, execution time and total response time between the dominant and non-dominant leg, in a sample of 43 taekwondo athletes. In line with previous studies (Tang et al., 2007), results revealed that taekwondo athletes achieved a symmetrical sport style while performing a roundhouse kick. In contrast with other studies (Pedzich et al., 2006; Peng, 2006), the absence of differences exhibited in our study might be due to the technique under study if we take into account that yop and dwit chagi are considered more difficult to perform (Pedzich et al., 2006) while the roundhouse kick is the most useful technique. Perhaps as athletes´ experience and practice increase, the differences between limbs decrease. Thus, differences founded by Peng (2006) could have been due to the fact that the taekwondo athletes who performed roundhouse kicks were teenagers. As pointed out Tang et al. (2007) symmetrical kicking skills are important at the elite level of taekwondo.

In our study, in line with Falco et al. (2009), regression analysis also showed the influence of the distance in a roundhouse kick. This influence is witness to the major variance explained from the short distance from the dominant on non-dominant leg in reaction time, execution time and total response time, while in impact force this variance is better explained from the long one. Finally, a limitation of the study is the potency of this. In this way, future studies should examine the inter-relations between kinematical variables within different distances and into different level and age groups and be performed with a greater number of participants.

CONCLUSION: This study could suggest the absence of asymmetry during a roundhouse kick in a sample of taekwondo competitive athletes. Based on these results, we suggest that taekwondo players did not seem to show mechanical differences between the dominant and non-dominant leg. Perhaps symmetries in taekwondo kicks are important in order to reach excellence in this sport, but more studies are needed to confirm this comparing the variables by level and gender as well as using other techniques within the same groups. In addition, the results showed the influence of the distance to determine/predict impact force, reaction time, execution time and total response time. The measurements from the non-dominant leg can be better explained by the short distance except for impact force, which is better explained in the long one.

REFERENCES:

Falco, C., Alvarez, O., Castillo, I., Estevan, I., Martos, J., Mugarra, F. and Iradi, A. (2009). Influence of the distance in a roundhouse kick's execution time and impact force in Taekwondo. *Journal of Biomechanics, 42* (3), 242-248.

Pedzich, W., Mastalerz, A. And Urbanik, C. (2006). The comparison of the dynamics of selected leg strokes in taekwondo WTF. *Acta of Bioengineering and Biomechanics, 8* (1), 1-9.

Peng, C.T. (2006). The difference of strength and the speed, balance between the dominant and non-dominant leg during the roundhouse kick of tae kwon do athletes. Unpublished doctoral dissertation. National College of Physical Education. Taiwan.

Tang, W. T., Chang, J. S. and Nien, Y. H. (2007). The kinematics characteristics of preferred and non-preferred roundhouse kick in elite taekwondo athletes. *Journal of Biomechanics, 40* (S2), 780.

Acknowledgement

This research has been partially supported by the Sports Service of the University of Valencia.

RECONSTRUCTION ACCURACY FOR VISUAL CALIBRATION METHOD

Marc Elipot[1,2], Nicolas Houel[2], Philippe Hellard[2], Gilles Dietrich[1]

1- ECI – LAMA, Université Paris Descartes, Paris, France
2- Département Recherche Fédération Française de Natation, Paris, France

KEY WORDS: camera calibration algorithm, reconstruction error, motion analysis

INTRODUCTION: Motion analysis is a common technique in biomechanics and sport studies. Since Abdel-Aziz and Karara (1971), the direct linear transformation (DLT) is the most widely used method to analyse human movements. More recently, Drenk et al. (1999) proposed a modified DLT method, called DLT double-plane method (DLT DP), which involves two parallel control planes (rather than a whole 3D structure). With these two calibration methods, a set of coefficients is calculated. This set summarise indirectly the internal and external parameters of the camera. Kwon et al. (2002) and Elipot et al. (2008) have respectively shown that, in aerial and underwater conditions, the DLT DP method can reduce the reconstruction error.

Bouguet (1999) presented an alternative method to calibrate camera. In this method, called visual calibration (VC), camera internal and external parameters are directly calculated. Bases on the pinhole camera model, the visual calibration aim to solve the following equation:

$$\begin{bmatrix} u \\ v \\ 1 \end{bmatrix} \sim \begin{bmatrix} f_u & \alpha & U_0 \\ 0 & f_v & V_0 \\ 0 & 0 & 1 \end{bmatrix} \cdot R \cdot \begin{bmatrix} X \\ Y \\ Z \\ 1 \end{bmatrix} + T \qquad (1)$$

where

$\sim$: up to a non zero scale factor

$\begin{bmatrix} u \\ v \\ 1 \end{bmatrix}$: the image plane coordinate vector (pixel) of a point P , $\begin{bmatrix} X \\ Y \\ Z \\ 1 \end{bmatrix}$: the space coordinate vector of the same point P

R : A 3×3 rotation matrix ; T : A 3×1 translation matrix

$\begin{bmatrix} f_u \\ f_v \end{bmatrix}$: the focal length in pixel ; $\begin{bmatrix} U_0 \\ V_0 \end{bmatrix}$: the principal point coordinates ; α : The skew coefficient defining the angle between the pixel axes x and y

Nevertheless, reconstruction accuracy of the visual calibration has never been identified. The aim of this study is to identify the reconstruction accuracy of the visual calibration and to compare this calibration method to DLT and DLT DP for the reconstruction of points placed inside (without extrapolation) or outside (with extrapolation) of the calibrated space.

METHODS: Two mini-DV cameras, with a sampling frequency of 25 Hz, have been used in this study and placed as shown in the figure 1.

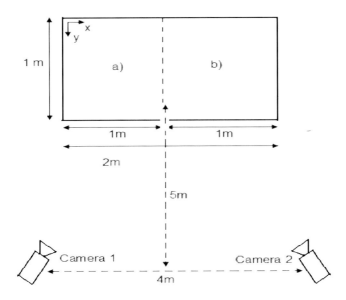

a) first set of calibration points
b) second set of calibration points
c) a + b : full set of calibration points

Figure 1: Experimental set up

The calibration procedure has been realised for the DLT and DLT DP methods using a calibration frame of $2 \times 1 \times 1$ m with 69 calibration points. The calibration procedure for the VC has been realized in two steps: 1- the internal parameters of the cameras (f_u, f_v, u_0, v_0 and $α$) have been inferred from images of the planar calibration rig (a checker board pattern). 2- the external parameters of the cameras (R and T) have been inferred from an image of the calibration frame.

Different configurations of calibration points have been used (depending on the algorithm used). These configurations are summarised in the table 1.

Space reconstruction accuracy has been calculated from equations 2 and 3 given by Kwon and Casebolt (2006):

$$\varepsilon = \sqrt{(Xk - Xr)^2 + (Yk - Yr)^2 + (Zk - Zr)^2} \quad (2)$$

$$\varepsilon_{RMS} = \sqrt{\frac{1}{n}\sum \varepsilon^2} \quad (3)$$

where (X_k, Y_k, Z_k) are the known object-space coordinates, and (X_r, Y_r, Z_r) are the reconstructed object-space coordinates, ε is the reconstruction error for a given control point, and ε_{RMS} is the overall reconstruction error. As shown in the table 1, for the condition without extrapolation (i.e. 0% extrapolation), reconstructed points are inside the calibrated space. For the condition with extrapolation, reconstruction accuracy is checked with points placed 25, 50, 75, 100 cm (on the x axis) outside of the calibrated space (i.e. from 25% extrapolation to 100% extrapolation).

Table 1: Configuration of the calibration points used

	Algorithm	Configuration of calibration points
Without extrapolated points	DLT	full set of points (c.)
	DLT DP	set c. split in two planes (front and back plane)
	VC	set c.
	VC DP	set c. split in two planes (front and back plane)
With extrapolated points	DLT	points in a. are used to calibrate points in b. (outside of the calibrated space) are used to check the reconstruction accuracy
	DLT DP	points in a. are used to calibrate (split in two planes) points in b. (outside of the calibrated space) are used to check the reconstruction accuracy
	VC	points in a. are used to calibrate points in b. (outside of the calibrated space) are used to check the reconstruction accuracy
	VC DP	points in a. are used to calibrate (split in two planes) points in b. (outside of the calibrated space) are used to check the reconstruction accuracy

RESULTS: All the results are summarised in the table 2.

Table 2: RMS and maximum reconstruction errors for the four algorithms

Algorithm	Extrapolation	RMS Error (mm)	Max Error (mm)	Max-to-RMS-ratio (%)
DLT standard	*0%*	*9*	*26*	*290.2*
	25%	9.8	12.5	128
	50%	14.1	20.5	145.3
	75%	18	20.6	114
	100%	19.9	28.8	144.5
DLT double-plane	*0%*	*5*	*10.9*	*214.6*
	25%	12.5	17.7	141.7
	50%	12.3	18	146.6
	75%	14.8	17	114
	100%	14.3	18	144.5
VC standard	*0%*	*9.7*	*18.9*	*193.8*
	25%	9.2	11.4	123.6
	50%	10.9	13.9	128.2
	75%	12.1	15.2	125.5
	100%	16	21.9	136.6
VC double-plane	*0%*	*6*	*10.7*	*179.3*
	25%	5.6	7.0	126
	50%	8.4	11.4	135.6
	75%	11.1	15.5	139.9
	100%	19.2	19.2	134.6

For the condition without extrapolation: The results show that, for the two algorithms and for the condition without extrapolation, the double plane method (DLT DP and VC DP) score smaller calibration errors for the RMS values and for the maximum errors values. Results also show that, without extrapolation, RMS errors for the DLT DP and DLT are respectively slightly smaller than the RMS errors for the VC DP and the VC. Nevertheless, maximum error observed for the VC and the VC DP are respectively smaller than those observed for the DLT and DLT DP. The max-to-RMS ratios are also smaller for the VC and VC DP.

For the condition with extrapolation: The results of the conditions with extrapolation show that, for every level of extrapolation, except for 100%, the RMS and maximum values are smaller for the VC DP than for the VC. Also, the RMS and maximum values are respectively smaller for the VC DP and the VC than the values observed for the DLT and the DLT DP.

DISCUSSION: The present data (Table 2) supports the conclusion that double plane methods provide more accurate reconstruction of points within the calibration space (0% extrapolation). The VC DP and the DLT DP allow an improvement of the reconstruction accuracy of respectively 41% and 43%. These results agree with those of Kwon et al. (2002) and Elipot et al. (2008).

Reconstruction errors increase for the DLT and DLT DP method if the reconstructed points are placed outside of the calibrated space. More especially, the DLT DP appears to be much more inaccurate (191% more inaccurate) when the extrapolation is fixed at 75%. These results agree with those of Kwon et al. (2002) and Kwon and Lindley (2000).

The VC DP may have slightly improved RMS reconstruction accuracy compared to the other methods, but it also create a more homogeneous space calibration. VC DP also improves reconstruction accuracy for points placed outside of the calibrated space. So the VC method is particularly convenient to study large movements or when it is impractical to build a calibration frame large enough to prevent any extrapolation.

REFERENCES:

Abdel-Aziz, Y.I., and Karara, H.M. (1971). Direct linear transformation from comparator coordinates into object space coordinates in close range photogrammetry. In ASP Symposium on Close Range Photogrammetry (pp. 1-18). Falls Church: American Society of Photogrammetry.

Bouguet, J.Y. (1999). Visual methods for three-dimensional modeling. Ph. D Thesis. Pasadena, California Institute of Technology.

Drenk, V., Hildebrand, F., Kindler, M., and Kliche, D. (1999). A 3D video technique for analysis of swimming in a flume. In R.H. Sanders and B.J. Gibson (Eds.), Proceedings of the XVII International Symposium on Biomechanics in Sports (pp. 361-364). Perth: Edith-Cowan University.

Elipot, M., Houel, N., Hellard, P., and Dietrich, G. (2008). Comparaison de méthodes de calibration de camera pour l'analyse du mouvement en conditions sous-marines. In M. Sidney, F. Potdevin, & P. Pelayo (Eds.), 4èmes journées spécialisées de natation (pp. 127-128). Lille : Université de Lille.

Kwon, Y.H., Ables, A., and Pope, P.G. (2002). Examination of different double-plane camera calibration strategies for underwater motion analysis. In K.E. Gianikellis (Ed.), Proceedings of the XXth International symposium on biomechanics in sports (pp. 329-332). Caceres: Universidad de Extramadura.

Kwon, Y.H., and Casebolt, J.B. (2006). Effects of light refraction on the accuracy of camera calibration and reconstruction in underwater motion analysis. *Sports biomechanics, 5*, 95-120.

Kwon, Y.H., and Lindley, S.L. (2000). Applicability of four localized-calibration methods in underwater motion analysis. In R. Sanders and Y. Hong (Eds.), Proceedings of the XVIII international symposium on biomechanics in sports. Applied program: Application of biomechanical study in swimming (pp. 48-55). Hong Kong: The Chinese University of Hong Kong.

Acknowledgements:
The authors would like thank Caroline Moreau and Lou Counil for their contribution during the data acquisition process.

MOTOR UNIT FIRINGS DURING VOLUNTARY ISOMETRIC RAMP AND BALLISTIC CONTRACTIONS IN HUMAN VASTUS MEDIALIS MUSCLE

Shinji Mizumura[1], Yoshihisa Masakado[2], Katsuhiko Maezawa[3], Kelly F McGrath[1]

School of Arts and Letters, Meiji University, Tokyo, Japan[1], School of Medicine, Tokai University, Kanagawa, Japan[2] and School of Medicine, Juntendo University, Tokyo, Japan[3]

Intra-muscular electromyographic (EMG) signals in vastus medialis muscle were decomposed into their constituent motor unit action potential trains using a specially designed quadrifilar wire electrode during voluntary isometric ramp and ballistic contractions. Five male adults participated in our experiments as subjects and performed ramp trapezoidal, ramp triangular, and ballistic contractions. By using a newly developed wire electrode, intra-muscular EMG signals were successfully decomposed into the individual motor unit action potential trains. The firing behaviors analyzed by the decomposition technique were consistent with previous studies on small muscles. This new quadrifilar wire electrode is potentially a useful tool for detecting intra-muscular electromyographic signals in large limb muscles such as the vastus medialis.

KEY WORDS: Motor unit, recruitment threshold, firing frequency

INTRODUCTION: A technique for analyzing motor unit (MU) action potentials called "Precision Decomposition" was previously developed by De Luca and his colleagues (De Luca 1993; De Luca & Adam 1999; LeFever & De Luca 1982; LeFever et al. 1982). This technique allows identification of action potentials from individual motor unit firings during isometric contractions by using a quadrifilar needle electrode. By using this technique, many important findings have been reported on motor unit firing behavior in humans. Recently we developed a new electrode using fine wire instead of a needle. The purpose of the present study was to demonstrate our new electrode design records reliable and stable intra-muscular EMG signals from a large proximal limb muscle such as the vastus medialis, allowing these signals to be decomposed into individual motor unit action potential trains during maximal isometric ramp and ballistic contractions.

METHODS: Subjects and apparatus: Five healthy male adults (mean $\pm$ SD; age: 30.4 $\pm$ 6.3 years; height: 174.4 $\pm$ 4.4 cm; weight: 66.4 $\pm$ 5.4 kg) participated in the present experiments as subjects. Informed consent was obtained from all subjects before their participation. The subjects sat on an experimental chair with their right knee flexed at 90 degrees. Isometric knee extension force was recorded using a force transducer (LSM-100KBSA67; Kyowa, Tokyo, Japan) positioned at the distal end of the right leg (Fig. 1).

Tasks: The subjects were asked to produce isometric knee extension force in three types of force exertion: ramp trapezoidal, ramp triangular and ballistic exertions. The ramp trapezoidal exertion task consisted of increasing, maintaining, and decreasing force phases and was used to test whether the new wire electrode could detect reliable signals during a slow and low force exertion task. Subjects were required to increase force gradually up to 25% MVC in 2.5 seconds (10% MVC /s), then maintain that force level for 5 seconds, and relax gradually in 2.5 seconds (-10% MVC/s). In the ramp triangular exertion task, to test whether the electrode could detect reliable signals during high force exertion, subjects were required to increase force up to 100% MVC in 5 seconds (20% MVC/s), and then relax as quickly as possible. The purpose of the ballistic exertion task was to assess the electrode in a rapid force exertion task, during which subjects produced a brief force, as fast as possible, up to the target force set at approximately 50% MVC.

Recording of intra-muscular EMG signals: Intra-muscular EMG signals were obtained from the right vastus medialis muscle. The original precision decomposition technique used a quadrifilar needle electrode (De Luca 1993) to obtain the intra-muscular EMG signal. This electrode has four detection surfaces, each 50 μm in diameter, 200 μm apart on the corners of the port of a square configuration, located on the side of a needle. In the present experiment, intra-muscular EMG signals were obtained using a specially designed quadrifilar 'wire' electrode whose detection surfaces were the same as the original quadrifilar needle (Unique Medical, Tokyo, Japan, Fig. 2). The EMG signals were differentially amplified and digitized at a sampling rate of 51.2 kHz together with the knee extension force signal sampled at 500 Hz (Counterpoint; Dantec, Skovlunde, Denmark).

Analysis of motor unit firings: A precision decomposition technique (Counter point, Dantec, Denmark) was used to analyze intra-muscular EMG signals in vastus medialis muscle. The intra-muscular EMG signals were band-pass filtered from 1 to 10 kHz, and then decomposed into individual motor unit action potential trains by decomposition algorithms: template matching, template updating, superposition resolution and firing statistics. Mean firing frequency was calculated every 400 ms time bin in the ramp trapezoidal and ramp triangular exertion tasks. In the ballistic exertion task, the frequency was calculated as a reciprocal of instantaneous inter-firing intervals.

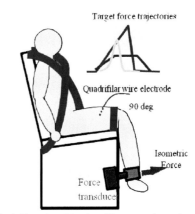

Fig. 1. Experimental setup of the present experiments

RESULTS: A total of 13 motor units (5 in the trapezoidal, 4 in the triangular and 4 in the ballistic tasks) from 5 subjects were successfully decomposed. Figure 3 shows the raw intra-muscular EMG signal (A), bar plot of the decomposed MU firings (B), mean firing frequency (C) and inter-firing interval time (D) for one trial in the ramp trapezoidal exertion task. In this case, three motor units were identified over the entire contraction. Their recruitment threshold forces were 7.5, 16.5 and 26.0% of MVC (Fig. 3B). Figure 3C shows the time courses of the mean firing frequencies of each motor unit. The mean firing frequency of the earlier recruited motor unit was higher than that of the later recruited MU. Figure 3D shows the time interval between continuous spikes. The motor unit with a higher recruitment threshold force showed a larger variation of the inter-firing interval.

Figure 4 shows motor unit firing behaviors in the ramp triangular task. In these trials, relatively high threshold motor units were identified (53.0, 57.0% MVC). The relationship between the recruitment

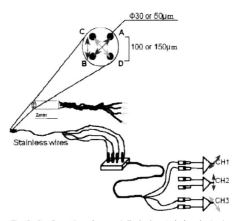

Fig. 2. Configuration of a specially designated wire electrode used in the present experiment

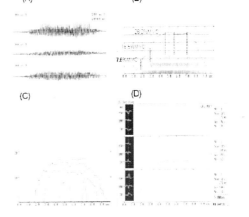

Fig. 3. Raw intra-muscular EMGs (A), bar plots (B), mean firing frequency (C) and inter firing interval(D) of a sample subject.

threshold force and the variation of the inter-firing interval in this task was the same as in the trapezoidal exertion task; the higher the recruitment threshold, the larger the variation of the inter-firing interval.

Figure 5 shows bar plots of the motor unit action potential trains in one trial of the ballistic exertion task. In this trial, the peak force value was 46% of MVC and four motor units were identified. All the motor units identified were recruited at approximately 0% MVC and Table 1 shows their mean firing frequencies. A remarkably high mean firing frequency was observed in motor unit 1. The mean firing frequencies of earlier recruited motor units were higher than those of later recruited MUs (Table 1).

DISCUSSION: The main finding of the present investigation was that by using a quadrifilar wire electrode specially designed by the authors, intra-muscular EMG signals of a large proximal limb muscle, such as the vastus medialis, were successfully decomposed into their individual motor unit action potential trains, when subjects performed large and rapid knee extension force production. Although the number of motor units that were analyze in the present study were too small to effectively describe the features of the motor unit firing behavior due to technical difficulties, our results still provided useful information on motor unit firing behavior of large muscles exerting high forces.

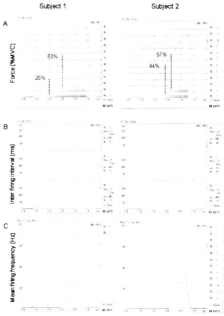

Fig. 4. Bar plots of inter firing interval(A), inter firing interval times(B), and mean firing frequencies (C) in the triangle task.

Recruitment threshold of motor unit: Recruitment threshold of the motor unit has been defined as a force value at which that motor unit begins to fire during voluntary isometric ramp contraction. It is well known that in human finger muscles, such as first dorsal interosseus muscle, which require a fine control of force magnitude, most motor units fire at less than 50% of maximal voluntary contraction (Basmajian and De Luca 1985, Masakado et al. 1995). On the other hand, in the more proximal limb muscles, such as the deltoid and tibialis anterior muscles, motor units fire at approximately 70% of MVC (Basmajian and De Luca 1985; Erim et al. 1996). In the present experiment, intra-muscular EMG signals were analyzed from the vastus medialis, a large proximal limb muscle. The highest recruitment threshold of the triangular task was 57.0% of MVC (MU2 of Subject 2 in Fig. 4). This suggests that the motor units with high recruitment threshold could also exist in the vastus medialis muscle.

Firing frequency of motor unit: The earlier recruited motor units always fired at greater frequencies compared to the later recruited

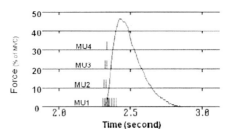

Fig. 5. Sample bar plots of motor unit action potential trains in the ballistic task.

Table 1. Mean firing frequencies of motor units identified for one trial identified in ballistic task.

MUs	Number of firings	Mean firing frequency (Hz)
MU4	1	—
MU3	3	42.6
MU2	2	52.6
MU1	10	96.8

motor units during the voluntary isometric contractions. Erim et al. (1996) called this property "*Onion Skin Phenomenon*". In the present experiment, the onion skin was clearly shown during the ramp trapezoidal (Fig. 3C), triangular (Fig. 4C) and ballistic tasks (Table 1). Therefore, our results are consistent with the observations from previous studies. These results suggest that the control of firing frequency of motor units in the vastus medialis muscle is similar to other limb muscles. To the knowledge of the authors, this reported recording of MU firing frequency during ballistic contraction of a large limb muscle is the first in the motor unit research field. A firm conclusion cannot be made since only one trial was successfully decomposed for the ballistic exertion task, but the mean firing frequency of all three MUs successfully decomposed were greater than 40Hz, and these values were much higher than those (20-30 Hz) for the ramp exertion tasks. Whether these low-threshold high-frequency MUs are the same MUs as those firing in the ramp exertion, or different from those recruited in the ramp exertion is a further question to be answered.

Quadrifilar wire electrode: This type of electrode affects the results of motor unit firing behaviors. In our preliminary experiments, we used a conventional quadrifilar needle electrode (De Luca 1993) to record intra-muscular EMG signals in the vastus medialis muscle. However, a reliable signal could not be obtained when using the needle electrode during isometric ramp and ballistic contractions as described above. In the present experiments, we used a specially designed wire electrode (Fig. 2). As a result we could obtain reliable signals. This suggests that the thick tip of the quadrifilar wire electrode plays a role like an anchor of a ship, so that the detection surface moves with the surrounding muscle fibers throughout the ramp and ballistic contractions.

CONCLUSION: In the present study a quadrifilar wire electrode was used to record reliable and stable intra-muscular EMG signals from the vastus medialis muscle, a large proximal limb muscle, during maximal isometric ramp and ballistic contractions. The signals were then decomposed into individual motor unit action potential trains. The quadrifilar wire electrode developed is potentially a valuable tool for recording intra-muscular EMG signals, especially in large limb muscles. Continued improvement in the rate of electrode sampling and decomposing of motor unit firing will serve to enhance our understanding of human muscle force control mechanisms.

REFERENCES:
Basmajian, J.V. and De Luca, C.J. (1985). *Muscles alive*. William & Wilkins, Baltimore.
De Luca, C.J. (1993). Precision decomposition of EMG signals. *Methods in clinical Neurophysiology*, 4, 1- 28.
De Luca, C.J., LeFever, R.S., McCue, M.P., and Xenakis, A.P. (1982). Behavior of human motor units in different muscles during linearly varying contractions. *J Physiology*, 329: 129- 142.
Erim, Z., De Luca, C.J. Mineo, K. Aoki, T. (1996) Rank-ordered regulation of motor units. *Muscle & Nerve*, 19: 563- 573.
LeFever, R. S. and De Luca C. J. (1982). A procedure for decomposing the myoelectric signal into its constituent action potentials Part I: Technique, Theory and Implementation. *IEEE Trans Biomed Eng*, Vol. BME-29, No. 3, 149-157.
LeFever, R. S., Xenakis, a. P. and De Luca C. J. (1982). A procedure for decomposing the myoelectric signal into its constituent action potentials Part II: Execution and Test for Accuracy. *IEEE Trans Biomed Eng*, Vol. BME-29, No. 3, 175-188.
Masakado, Y., Akaboshi, K., Nagata, M., Kimura, A. and Chino, N. (1995) Motor unit firing behavior in slow and fast contractions of the first dorsal interosseous muscle of healthy men. Electroencephalogra & Clin Neurophysiology/Electromyogra & Motor control, 97 (6), 290-295.

Acknowledgement
This study was funded by a Grant-in-Aid for Scientific Research.

EXAMINATION OF GLUTEAL MUSCLE FIRING and KINETICS OF THE LOWER EXTREMITY DURING THE WINDMILL SOFTBALL PITCH

Gretchen D. Oliver, Priscilla Dwelly, David Keeley, and Hiedi Hoffman
University of Arkansas, Fayetteville, Arkansas, USA

The purpose of this study was to examine the muscle activation and kinetic data of the lower extremity during the windmill softball pitch. Limited research has documented the lower extremity in the windmill softball pitch. Seven female post-pubescent softball players volunteered for the study. Pitchers were analyzed with surface electromyography and motion analysis software. There was no relationship observed between gluteal activation and ground reaction forces at SFP and maximum force. Only one investigation prior to this study reported ground reaction forces of windmill softball pitchers, however they did not examine the EMG activity.

KEY WORDS: ground reaction forces, sEMG, pitching mechanics

INTRODUCTION: In 2008, the Amateur Softball Association reported that 2.5 million fast-pitch players registered and of those, 1.3 million were reported as female post pubescent between the ages of 12 and 18 (Rojas et al., 2009). Regardless of the sport of softball being increasingly more popular, there has been limited research in the area of biomechanics and injury implications. The sport of baseball has been researched much more extensively. When comparing the two sports, the main difference is that softball is a scaled down version of baseball. Example, the softball pitcher throws from level ground 12.19 m from the batter and displays more of a leap in the delivery, whereas a baseball pitcher throws from an elevated mound 18.44 m from the batter and produces more of a stepping action for the delivery.

Previously, shoulder distraction forces have been examined for both baseball and softball pitchers. Distraction forces of 70% to 90% body weight at the shoulder have been reported in windmill softball pitchers (Barrentine et al., 1998; Werner et al. 2006). These distraction forces are comparable to those of baseball pitchers. It is the repeated application of these distraction forces that eventually implicate overuse injuries similar to those experienced in baseball pitching (Feltner, & Dapena, 1986; Fleisig, Andrews, Dillman, 1995; & Werner et al. 1993). To the researchers' knowledge only one study has been presented discussing the implication of lower extremity biomechanics during the windmill softball pitch.

Recently, stride foot ground reaction forces have been observed in windmill softball pitchers, where the findings were different from those of reported in baseball (Werner et al., 2005). It is believed that softball pitchers exhibit greater ground reaction forces because they are throwing from flat ground and exhibiting an abrupt deceleration stop upon ball release (Werner et al.,2005), whereas baseball pitchers pitch from an elevated mound. The lower extremity musculature must balance the core and upper extremity musculature in attempt to maintain the fluid motion of the pitch. The gluteal muscles are primary stabilizers of the core and therefore when the gluteals are not able to provide sufficient stability and energy transference the athlete is predisposed to injury. It was our purpose to examine muscle activation and kinetic data of the lower extremity during the windmill softball pitch. We anticipated that as the pitcher completes the delivery then ground reaction forces would increase, and gluteal muscle activation would be activated to control the lower extremity throughout the deceleration phase.

METHODS: Data Collection: Four collegiate and three high school female post pubescent softball pitchers (age 17.7 y $\pm$ 2.6; height 169 cm $\pm$ 5.4; mass 69.1kg $\pm$ 5.4) consented to participate. The study was granted Institutional Review Board approval. Surface electromyography (sEMG) data were collected on the muscle bellies of the dominant leg (non-stride leg) gluteus maximus and medius muscles using Myopac Jr 10 channel amplifier (RUN Technologies Scientific Systems, Laguna Hills, CA).

To determine maximum voluntary isometric contraction (MVIC) for the gluteal muscle group, manual muscle tests were performed three times for five seconds. The MVICs determined for both the gluteus maximus and medius were based on the work of Kendall et al (1993). The first and last second of each MVIC trial were removed from the data in attempt to obtain steady state results for each of the muscle groups. The manual muscle testing provided a base line reading for which all EMG data were based.

After unlimited time was allotted for the participants to warm-up based on their normal routine, each participant threw fastball windmill style deliveries using an official softball (30.48 cm, circumference, 170.1 g) to a catcher behind the plate 12.2 m away. The pitching mound was positioned so that the participants stride foot would land on top of the 40 X 60 cm Bertec force plate (Bertec Corp, Columbus, Ohio) which was anchored into the floor. A total of five trials were recorded after they were deemed a successful strike. Both sEMG and force plate data were collected at a rate of 1000 Hz. Force plate, kinematic, and sEMG data were synchronized using Motion Monitor® (Innovative Sports Training, Chicago, IL).

Data Analysis: After completion of the trials, positional kinematic data were filtered independently along the x, y, and z-axis using a 2nd order Butterworth filter (10 Hz) (Werner et al., 2005). The sEMG signals were preamplified (x 1200) near the electrodes and were band pass filtered between 10 and 500 Hz and sampled at a rate of 1000 Hz (Rojas et al., 2009). Surface EMG enveloped data were assessed through mean maximum sEMG reference values that were calculated for each muscle from stride foot plant (SFP) to maximum force production. Stride foot plant to maximum force production encompassed final aspect of phase 4 and phase 5 of the five phases of the windmill pitch as defined according to Maffet et al. (1997) illustrated in Figure 1. Stride foot plant phase was determined by the first recording of force on the force plate through the maximum force delivered on the force plate. According to Werner et al. (2005) maximum ground reaction force occurs just prior to ball release, thus after maximum force is achieved, the ball is released and the pitcher begins their follow-through phase. Five trials of sEMG data for each subject were analyzed to determine average peak amplitudes for all muscles during the SFP.

Figure 1. Windmill pitching phases.

RESULTS: *Surface EMG:* The gluteal muscles activation during SFP and maximum force or ball release are presented in Figure 2.

Kinetics: Breaking/propulsive, medial/lateral, and vertical ground reaction forces for the stride foot plant phase are presented in Figure 3. The force component Fx is in the direction of the pitch and can be interpreted as anterior/posterior force in reference to the body. The force Fy is the vertical component reflecting body weight support. Landing vertical forces peaked at ball release. The force Fz is the medial/lateral with lateral (positive) directed towards 3rd base for the right handed pitcher.

Pearson correlation results indicated weak and non-significant correlation between gluteus maximus activity and Fx ($r=-0.38$, $p=0.18$, $r^2=0.14$), Fy ($r=0.22$, $p=0.46$, $r^2=0.05$), and Fz ($r=0.08$, $p=0.76$, $r^2=0.01$). There were also weak and non-significant relationships between gluteus medius activity and Fx ($r=-0.001$, $p=0.1$, $r^2<0.001$), Fy ($r=-0.12$, $p=0.72$, $r^2=0.01$), and Fz ($r=0.27$, $p=0.36$, $r^2=0.07$).

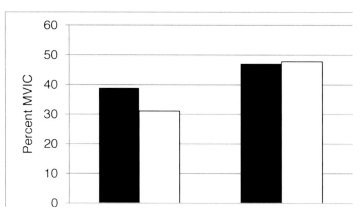

Figure 2: Gluteal activation during stride foot plant (SFP) and max ground reaction force (Max)

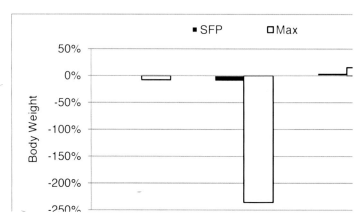

Figure 3. Representative (Fx) anterior/posterior forces, (Fy) vertical ground reaction forces, and (Fz) medial/lateral. Legend: SFP, stride foot plant; Max; maxium force.

DISCUSSION: The musculature of the lower extremity is important based on the theory of sequentiality (Putnam, 1993). According to Putnam (1993), proximal segments of the legs and trunk work sequentially in effort to accelerate the shoulder for optimal production in the upper extremity. During Phase 4, as the pitcher was attempting to abruptly decelerate for ball delivery on the stride leg, the dominant gluteus medius was attempting to hold the hip upright, while the pitcher was balanced on the stride leg. It is important that throughout the delivery that the gluteus medius is active to support the hip. If the gluteus medius is not active then the hip with drop predisposing the individual for muscular compensations and ultimately resulting in an overuse injury due to the breakdown in the kinetic chain (Kibler et al., 2006). Therefore, as more weight was applied to the stride leg, the gluteus medius increased activity. Kinetic data revealed that Fx was positive at SFP indicating the stride leg was attempting to maintain balance in order generate force in the direction of the ball. After SFP, the stride foot applied a breaking/posting force (Fy) in attempt to transfer energy to the upper extremity for ball release. This deceleration force was evident in a vertical ground reaction force of 235.96% body weight.

It has been shown that landing forces at the time of ball release correlated with wrist velocity during baseball pitching (MacWilliams et al., 1996). Thus, in throwing where forces are generated from the lower extremity to the upper extremity, leg drive is an important aspect. In addition, Werner (2005) found that peak deceleration forces of the stride were different from that of baseball pitching. In the current study the magnitude of peak vertical force was 235.96% of body weight.

CONCLUSION: We were able to identify muscle activation and kinetic data of the lower extremity during the windmill softball pitch in post-pubescent females. Our findings agreed with our postulation of the pitcher completing the pitch delivery with increased ground reaction forces as were their increases in the activation of the gluteal muscle group. Further investigations need to examine forces and torques of both the stride leg and non-stride leg during the pitching motion. Joint loads at the shoulder are known for producing overuse injuries for pitchers. However, more studies need to focus on the joint loads of the lower extremity in order to address the overuse injury potentials and their relationships to upper extremity injuries, especially concerning gluteal activation and core control.

REFERENCES:
Barrentine, S. Fleisig, G., Whiteside, J., Escamilla, R.F., and Andrews, J.R. (1998). Biomechanics of windmill softball pitching with implications about injury mechanisms at the shoulder and elbow. *Journal of Orthopedic and Sports Physical Therapy*, 28, 405-415.

Fletner, M.E. and Dapena, J.J. (1986). Dynamics of the shoulder and elbow joint of the throwing arm during the baseball pitch. *International Journal of Sport Biomechanics*, 2, 235-259.

Fleisig, G.S., Andrews, J.R., and Dillman, C.J. (1995). Kinetics of baseball pitching with implications about injury mechanisms. *American Journal of Sports Medicine*, 23, 233-239.

Hill, J.L., Humphries, B., Weidner, T. and Newton, R.U. (2004). Female collegiate windmill pitchers: influences to injury incidence. *Journal of Strength and Conditioning Research*, 18, (3), 426-431.

Kendall, F.P., McCreary, E.K., Provance, P.G., Rodgers, M.M., and Romani, W.A. *Muscles: Testing and Function.* Fourth edition. Baltimore, Williams & Wilkins, 1993.

Kibbler, W.B., Press, J., and Sciascia, A. (2006). The role of core stability in athletic function. *Sports Medicine*, 36(3), 189-198.

MacWilliams, B.A., Choi, T.,Perezous, M.K., Chao, E.Y.S., and McFarland, E.G. (1998). *American Journal of Sports Medicine*, 26, 66-71.

Maffet, M.W., Jobe, F.W., Pink, M.M., Brault, H. and Mathiyakom, W. (1997). Shoulder muscle firing patterns during the windmill softball pitch. *American Journal of Sports Medicine*, 25, (3), 369-374.

Putnam C.A. (1993). Sequential motions of body segments in striking and throwing skills: description and explanations. *Journal of Biomechanics*, 26, 125-135.

Rojas, I.L., Provencher, M.T., Bhatia, S., Foucher, K.C., Bach, B.R., Romeo, A.A., Wimmer, M.A., and Verma, N.N. (2009). Biceps activity during windmill softball pitching. *American Journal of Sports Medicine*, 37, (3), 558-565.

Werner, S.L., Fleisig, G.S., Dillman, C.J., and Andrews, J.R. (1993). Biomechanics of the elbow during baseball pitching. *Journal of Orthopaedic and Sports Physical Therapy*, 17, 274-278.

Werner, S.L., Guido, J.A., McNeice, R.P., Richardson, J.L., Delude, N.A. and Stewart, G.W. (2005) Biomechanics of youth windmill softball pitching. *American Journal of Sports Medicine*, 33 (4), 552-560.

Werner, S.L., Jones, D.G., Guido, J.A., and Brunet, M.E. (2006). Kinematics and kinetics of elite windmill softball pitching. *American Journal of Sports Medicine*, 34, (4), 597-603.

RESULTS OF BIOMECHANICAL STUDIES OF TWO WAYS TO EXECUTE THE GOLF SWING

Ferdinand Tusker[1] and Florian Kreuzpointner[2]

Institute for Biomechanics in Sport, Technische Universtät München, Munich, Germany[1, 2]

The purpose of this study was to identify the biomechanical distinctions of two different ways to learn the Golf Swing. We compared the classical method with a new method we call "carving-golf". The study compared the dynamic analyses of 12 carving-golf swings and 32 classic swings. The horizontal and vertical forces of the persons standing on a Kistler platform were analysed. We found significant differences in impact and time parameters. While "carving-golf" teaches not to start with a countermovement the evaluation of the horizontal forces showed that a countermovement took place. The mean value of this impact was only 25% of the mean value the classic swing shows. There was no significant change in the forward impact but the velocity the whole system generated towards the target course was diminished by 33%.

KEY WORDS: carving golf; forces; learning; motor skills.

INTRODUCTION: The Golf Swing is a complex movement. There are rotation and translation and we have to handle a tool that should be moved precisely to a specific point. This movement requires coordinated muscle activation. The more complex it is the more difficult it is to learn (McHardy and Pollard 2005). We also know, that movements with less complexity are more easy to remember if they get close to the function of everyday movements (Dijkstra, K.; MacMahon, C.; Misirlisoy, M, 2008).

To make the movement of the swing easier carving golf forced only to balance the weight to the front foot, not to do any active rotation and not to do a translation in the way of a countermovement. The registered ground reaction forces of the golfers who dominate this movement have been compared with ground reaction forces of golfers who play the classic technique.

METHOD: Data Collection: The ground reaction forces of 42 Golf Players were measured by Kistler force platform (type 9287A and Type 9286B). Analogue Data was collected with 1000 Hz by a 12bit AD-converter and a range of ±5V. The average handicap of 32 players playing the classic style was 10 and 12 players were beginners who learned to play "carving golf".

After a warm up each person has to do 5 swings with their feet on the force platform. The golf ball was placed for the personal needs on an artificial turf mat. The ball was beaten with iron 7. The turf mat was also placed on a force platform, so that we controlled a fast change in the force when the ball was hit.
We analyzed the horizontal ground reaction forces of the last five seconds before impact.

Data Analysis: Figure 1 shows the horizontal force of a classic swing for the line direction of the ball. It also gives us the information how the collected data was calculated. It is:

- t_1 – time when movement starts;
- t_2 – time when horizontal force turned algebraic sign first time;
- t_3 – time when horizontal impulse comes to zero;
- t_4 – time when horizontal force turned algebraic sign second time;
- t_5 – moment of impact to the golf ball.

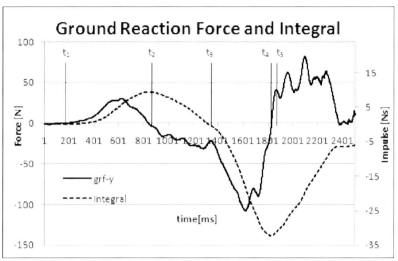

Figure 1: Horizontal ground reaction force in line direction of the golf ball and its integral with the recorded points of time

These time markers represent the following periods:

- $t_{minus} = t_2 - t_1$; the time period the force vector shows in countermovement direction;
- $t_{amort} = t_3 - t_2$, the time period the maximum velocity in countermovement direction comes to zero;
- $t_{produce} = t_4 - t_3$ as the time period the maximum impulse of the whole system in the line direction of the golf ball is generated;
- $t_{lack} = t_5 - t_4$ as the time period from maximum impulse of the whole system and the impact to the golf ball.

Additional we calculated:

- p_1: the maximum impulse of the whole system in countermovement direction;
- p_2: the maximum impulse of the whole system in line direction of the golf ball;
- v_1: the velocity of the centre of gravity in countermovement direction;
- v_2: the velocity of the centre of gravity in line direction of the ball.

To calculate the velocities we had to register the mass of our test persons.

The computed data were statistically tested by t-test for independent samples.

RESULTS: Table 1 shows the mean values of the parameters and significant differences. The duration when force in countermovement direction is produced is significant lower while playing carving golf. Classic golf players produce 3.8 times the impulse of carving golfers in countermovement direction and even more velocity in the same direction. The time to stop this movement shows no significant difference. The period classic golfers produce impulse in the line of the direction of the golf ball is significant shorter than carving golfers needed. Nevertheless is the impulse in the line of the direction of the golf ball for the classic players more than 25% higher than for the carving golfers and even more is the velocity. The mean value of the time period between the maximum impulse and the moment of impact is significant shorter for the carving golfers.

Table 1 Mean values of the parameters

Variable	Type	N	Mean	SD	Significance
$t_{minus}[s]$	Classic	32	.593	.154	$p<0.05$
	Carving	12	.441	.199	
$p_1[kg*m/s]$	Classic	32	11.09	4.60	$p<0.05$
	Carving	12	2.92	1.44	
$t_{amort}[s]$	Classic	32	.338	.106	$p>0.05$
	Carving	12	.308	.163	
$t_{produce}[s]$	Classic	32	.53	.144	$p<0.05$
	Carving	12	.657	.237	
$p_2[kg*m/s]$	Classic	32	-34.66	9.02	$p<0.05$
	Carving	12	-25.46	3.49	
$t_{lack}[s]$	Classic	32	.0469	.0571	$p<0.05$
	Carving	12	.00892	.0181	
$v_1[m/s]$	Classic	32	.142	.0608	$p<0.05$
	Carving	12	.0344	.0154	
$v_2[m/s]$	Classic	32	-.437	.105	$p<0.05$
	Carving	12	-.294	.0384	

DISCUSSION: Figure 2 shows the force time curve that was expected if you teach a player to do the first move in line direction of the golf ball. Under the condition that there was no move when we start to record the force values we realize that there is no countermovement. Our results depending on a larger number of players show such a movement with about 25% of the impulse classic golf players have. The perception of the carving golfers and the instruction to the golf players is different. Guadagnoli M et al. (2002) prove the necessity of video or verbal instruction to support the teaching of the golf swing. The analysis of ground reaction forces are additional information on the causes of a movement

To produce impulse in the line direction of the ball the classic group needed less time and induce a higher impulse. That is only possible with higher forces. The carving golf method was invented to avoid these higher forces. Higher forces are often the reason for a lack of accuracy. There is a lack of research literature regarding this coherence for golf game. Articles are focussing on putting when dealing with accuracy or they look for higher forces to motivate longer distance for the shot (Hume et al. (2005); Fletcher and Hartwell (2004); Doan et al. (2006)).

Assuming that the best performance for the length of the shot is given when the highest impulse is produced, it looks like that the carving golfers have advantages over the classic golf players. The delay between maximum velocity of the whole system and the impact to the golf ball is significantly shorter regarding the carving golfer opposite the classic golfer. However, we did not measure the speed of the club. Maybe there is a necessity for a larger delay between maximum horizontal impulse and impact to the golf ball to generate a higher performance.

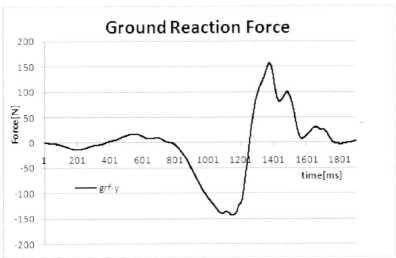

Figure 2: The expected horizontal ground reaction force in line direction of the golf ball when doing a carving golf movement.

CONCLUSION: This study identified mechanical parameters evoked by 2 different possibilities to realize a golf shot. It shows that even if a special movement is forced the player starts with a countermovement. This countermovement is not realized by the players.

Using the techniques of force platforms it is possible to distinguish between different kinds of movements while the golf swing. It provides with fundamental basics for moving techniques in golf. It also can assist to evaluate the quality of a golf swing movement.
Carving golf is based on a young training method to teach the golf swing. Controlling the procedure of learning by using biomechanical measurements can help us to improve our knowledge about the complexity of learning and playing golf.

REFERENCES:
Dijkstra K., MacMahon C., and Misirlisoy M.: The effects of golf expertise and presentation modality on memory for golf and everyday items. *Acta Psychol (Amst)*. 2008 Jun; 128 (2): 298-303, Epub 2008 Apr 25.
Doan BK, Newton RU, Kwon YH, and Kraemer WJ.: Effects of physical conditioning on intercollegiate golfer performance. *Journal of Strength and Conditioning Research*. 2006 Feb; 20(1):62-72.
Fletcher IM, and Hartwell M.: Effect of an 8-week combined weights and plyometrics training program on golf. *Journal of Strength and Conditioning Research*. 2004 Feb;18(1):59-62.
Guadagnoli M., Holcomb W., and Davis M.: The efficacy of video feedback for learning the golf swing. *Journal of Sports Science*. 2002 Aug; 20(8): 615-22.
Hume P.A., Keogh J., and Reid D.: The role of biomechanics in maximising distance and accuracy of golf shots. *Sports Medicine*. 2005 35(5): 429-49.
McHardy A., and Pollard H.: Muscle activity during the golf swing. *British Journal of Sports Medicine*. 1005 Nov; 39(11): 799-804; discussion 799-804.

DETERMINATION OF ARM AND LEG CONTRIBUTION TO PROPULSION AND PERCENTAGE OF COORDINATION IN BUTTERFLY SWIMMING

Morteza Shahbazi-Moghadam

School of Physics, University of Tehran & Centre for Aquatic Research and Education, the University of Edinburgh Scotland, UK

The Indirect Measurement of Active Drag (IMAD) was used to study the contribution of the legs and arms to propulsion in butterfly swimming. Contrary to MAD (Measuring of Active Drag) system, the IMAD can be used for all strokes and therefore enabled us to study the butterfly swim to estimate not only the percentage of leg and arm contribution to propulsion but also the percentage of swimmers' arms and legs co- ordinations. The method revealed that the best coordination was 78.% and that the contribution of arms and legs in propulsive force were 92% and 66% and in velocities were 98% and 88% respectively, showing that the swimmers received arm contribution better than leg contribution in propelling and velocity.

Keywords: legs and arms contribution, percentage of coordination, butterfly swim

INTRODUCTION: Few researchers dedicated research on determination of arm and leg contribution to propulsion and percentage of coordination in butterfly swimming. It is well known that the butterfly is the fastest style regulated by FINA. The peak speed of the butterfly is even faster than that of the front crawl, due to the synchronous pull/push with both arms. Yet since speed drops significantly during the recovery phase, it is overall slightly slower than the front crawl. Butterfly swimmers have a top speed of 2.18 m/s (4.87 mph), slightly under freestyle at 2.35 m/s (5.25 mph), over backstroke at 2.04 m/s (4.57 mph), and well over breaststroke at 1.84 m/s (4.11 mph). In butterfly swimming hands play the main role in propulsion however, it is unclear how many percentage legs may cause an increase in swimming speed. Shahbazi, (2007 and 2008) and Shahbazi et al., 2006 studies, by using the indirect measurement of active drag (IMAD), reported well these percentages in front and back crawl and breaststroke swims. Butterfly is a difficult stroke to swim as it needs both stamina and style.

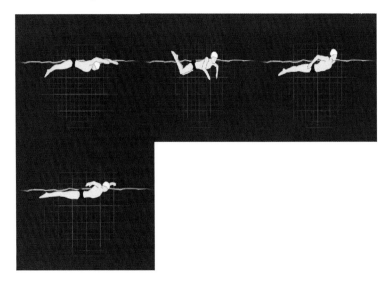

Figure 1. The four main phases in butterfly swim; start position, pre-thrust, thrust and flying phases in which the maximum drag is produced in third phase.

The butterfly stroke has three major parts, the pull, the push, and the recovery. These can also be further subdivided. From the initial position, the arm movement starts very similarly to the breast stroke. The pull movement follows a semicircle with the elbow higher than the hand and the hand pointing towards the center of the body and downward. The push pushes the palm backward through the water underneath the body at the beginning and at the side of the body at the end of the push. The swimmer only pushes the arms 1/3 of the way to the hips, making it easier to enter into the recovery, making the recovery shorter and making the breathing window shorter. The recovery swings the arms sideways across the water surface to the front, with the elbows straight. The arms should be swung forward from the end of the underwater movement, the extension of the triceps in combination with the butterfly kick will allow the arm to be brought forward relaxed yet quickly. In contrast to the front crawl recovery, this arm recovery is a ballistic shot. The leg movement is similar to the leg movement in the front crawl, except the legs are synchronized with each other, and it uses a wholly different set of muscles. The shoulders are brought above the surface by a strong up and medium down kick, and back below the surface by a strong down and medium up kick. A smooth undulation fuses the motion together. The feet are pressed together to avoid loss of water-pressure. The feet are naturally pointing downwards, giving downwards thrust, moving up the feet and pressing down the head. The aim of present study was to determine the contribution of arms and legs and also the percentage of coordination in butterfly swim.

METHODS: Seven male swimmers at national level (aged 18 ± 1 yr; weight 66.68 ± 10.89 kg; height 175.59 ± 14.35 cm) volunteered for this study. The mean best time for the subjects in the 100-m butterfly stroke, short course, was 62.5 ± 2.45 sec. The subjects swam butterfly under three conditions: (a) arms only with no bounding in legs, (b) legs only, and (c) full stroke. At a constant speed and using the arms only, the mean propelling force equals total drag at any given speed. In IMAD method (Shahbazi and Sanders 2002, 2004; Shahbazi et al., 2006; Shagbazi, 2007 and 2008), there is no special system but a tape-meter, a start-stop watch and appropriate formulae extracted from theoretical mathematical modeling.

The swimmers were requested to start swimming a 10 m long distance from still position by whistling as fast as they could and then at the end of the 10 m distance, again by whistling, they ceased swimming but gliding as far as possible. The time of 10m swim and the glided distance were used in the formulae (Shahbazi and Sanders, 2002, 2004) in order to estimate the propulsive force resulted from arms only, legs only, and the full stroke. In each step, swimmers swam three times with enough time of rest in between. The mean propulsive force is given as:

$$F_P = (C_1 V_L + C_2 VL^2) \qquad (1)$$

V_L is the maximum velocity that the swimmer can reach in 10 m swim; C_1 and C_2 are the hydrodynamic coefficients to be obtained by:

$$C_1 = 2MV/(X+Vt) \qquad (2)$$

X is the glided distance, V is the average velocity in 10 m swim, and

$$C_2 = X/M \qquad (3)$$

The maximum velocity (limit velocity) can be obtained by:

$$V_L = 0.5\{C_1/C_2 + [(C_1/C_2)^2 + (4MV/C_2 t)]^{1/2}\} \qquad (4)$$

RESULTS AND DISCUSSION: By measuring time of 10m swim with a precision of 10^{-2} s. and the glided distance with a precision of 10^{-2} m and using above formulae, the individual values for maximum swimming speed, hydrodynamic coefficients, drag force, and the relation between these variables for all subjects were obtained. In the second, third, and forth columns of the Table 1 the full stroke, arm only (with no leg support), and leg only forces, applied by subjects are presented. In column 5 of the Table 1 the sum of the arm and leg only forces is presented as theoretical force. In fact we considered as if these two forces were applied in the same direction (direction of velocity). In column 6 the difference between theoretical and real forces are presented. In column 7 of Table 1 the percentage of force which has not been used for increasing the swimmer velocity is presented. From these data the percentage of the arms and legs coordination can easily be achieved and is presented in column 8.

In columns 2, 3, and 4 of Table 2 the mean velocities of full stroke, arms and legs only are presented. In columns 5 and 6 the percentage of arms and legs are presented using their velocities and in column7 and 8 the percentage of arms and legs contributions are presented by using IMAD method. As is indicated in Table 1, IMAD method is capable of yielding the arms and legs forces separately, therefore the percentage of the contribution of arms and legs are calculated. Our results suggest that the whole leg force does not aid propulsion directly and therefore it follows from the present results that partly; an amount of ΔF (in Table 1) is used in stabilizing the trunk in the full stroke. Subject No.3 (75.2 kg) had the highest coordination (78.3%) and stabilizing in full stroke swim. Subject No.1, (85 kg), although 10kg heavier, showed significant coordination (73.8%). On the other hand, with less full stroke force he had significant mean velocity.

Table1. Mean $\pm$ SD of full, arm, and leg forces and the percentage of coordination

Subjects	Full-Stroke F_F (N)	Arms only F_A (N)	Legs only F_L (N)	Theoretical (F_A+F_L) (N)	Difference ΔF (N)	Loss $\Delta F/(F_A+F_L)$	Coordination %
1	72.1 $\pm$ 1.65	66.54 $\pm$ 3.85	31.09 $\pm$ 2.53	97.64 $\pm$ 2.85	25.54	26.2%	73.8%
2	72.32 $\pm$ 1.57	65.82 $\pm$ 3.47	43.79 $\pm$ 1.84	113.6 $\pm$ 3.92	41.30	36.4%	63.6%
3	68.86 $\pm$ 3.92	56.23 $\pm$ 2.57	30.33 $\pm$ 2.64	75.06 $\pm$ 3.33	16.28	21.7%	78.3%
4	55.77 $\pm$ 3.6	51.09 $\pm$ 1.25	26.92 $\pm$ 1.45	78.05 $\pm$ 3.95	22.26	28.6%	71.5%
5	44.76 $\pm$ 1.48	41.1 $\pm$ 0.92	21.05 $\pm$ 2.07	62.34 $\pm$ 2.33	17.54	28.2%	71.9%
6	49.08 $\pm$ 2.66	48.09 $\pm$ 1.59	36.46 $\pm$ 3.52	83.56 $\pm$ 2.96	34.46	41.2%	58.8%
7	41.92 $\pm$ 2.24	37.54 $\pm$ 1.12	24.52 $\pm$ 1.42	62.04 $\pm$ 1.95	20.09	32.4%	67.6%

Table 2. Mean $\pm$ SD of full, arm, and leg only velocities and their % of contributions

Subjects	Full-Stroke V_F (m/s)	Arms only V_A (m/s)	Legs only V_L (m/s)	V_A/V_F %	V_L/V_F %	F_A/F_F %	F_L/F_F %
1	1.41 $\pm$ 0.03	1.35 $\pm$ 0.02	0.95 $\pm$ 0.04	95.8%	67.4%	92.3%	43.1%
2	1.42 $\pm$ 0.05	1.4 $\pm$ 0.04	1.14 $\pm$ 0.03	95.6%	78.6%	91%	66.1%
3	1.44 $\pm$ 0.07	1.32 $\pm$ 0.03	1.23 $\pm$ 0.05	84.3%	76.9%	69.2%	58.5%
4	1.35 $\pm$ 0.04	1.29 $\pm$ 0.02	1.21 $\pm$ 0.03	95.6%	70.4%	91.6%	48.3%
5	1.33 $\pm$ 0.10	1.26 $\pm$ 0.03	1.15 $\pm$ 0.06	94.7%	70.0%	91.6%	47.6%
6	1.49 $\pm$ 0.06	1.46 $\pm$ 0.03	1.33 $\pm$ 0.12	98%	88.0%	96%	74.3%
7	1.39 $\pm$ 0.05	1.32 $\pm$ 0.02	1.25 $\pm$ 0.04	95%	78.4%	89.5%	58.4%

Our results showed that in butterfly swimming the arm forces were significantly higher than leg forces whereas the arms only velocities were predominant in butterfly swimming. Figure 2 shows that there are high correlations between arm forces and swimmer mass (82%) and full stroke (93.8%), while there were no significant correlations with leg forces.

Unfortunately, our subjects were not butterfly swimmers but still the results are satisfactorily acceptable. The method is reliable and simple to use, therefore other researchers can use this method for all other strokes and get fantastic results.

CONCLUSION: The IMAD method has been used to determine the contributions of arms and legs in propulsion and swimmers' velocity. The study revealed that there were significant correlations between swimmers' mass and arm forces, while it was not the case for leg forces. This meant that the swimmers' kicking was mostly used for body stabilizing and swimmers' mass was not much correlated with legs. Arms forces were significantly related with full stroke force. The IMAD reliably and easily revealed the swimmers parameters which could not be achieved with MAD.

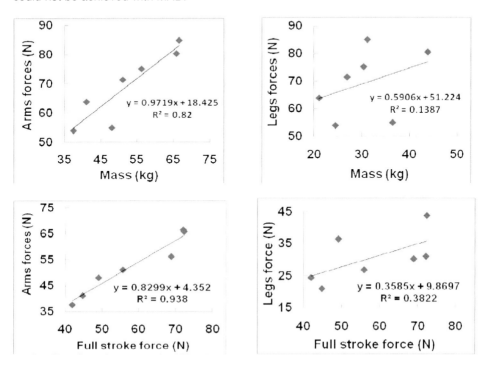

Figure 2. There are high correlations between arm forces and swimmer mass (82%) and full stroke (93.8%), while there were no significant correlations with legs forces.

REFERENCES:
Shahbazi, M.M. (2008). Determination of Arms and Legs Contribution to Propulsive and Percentage of Coordination in Breaststroke Swimming. Proceedings; XXVI International Symposium on Biomechanics in Sports. Seoul, Korea, pp. 228-231.
Shahbazi, M.M. (2007). Determination of Arms and Legs Contribution to Propulsive and Percentage of Coordination in Backstroke Swimming. Proceedings; XXV International Symposium on Biomechanics in Sports. Menzel, H.J. and Chagas, M. H. (eds.). Ouro Preto, Brazil. Pp208-211.
Shahbazi, M.M., Ravassi, A.A., Taghavi, H. (2006). Use of a New Indirect Method in Determining the Contribution of Legs and Hands to Propulsion in Front Crawl. Proceedings; XXIV International Symposium on Biomechanics in Sports. Schwameder, H., Struzenberger, G., Fastenbauer, S., Lindinger, S., Muller, E. (eds). P. 71-74. University of Salzburg- Austria.
Shahbazi, M.M., and Sanders, R.H. (2004). A Biomechanichal Approach to Drag Force and Hydrodynamic Coefficient Assessments. Proceedings; XXIV International Symposium on Biomechanics in Sports. Lamotagne, M. Gordon, G., and Sveistrup, H. (eds). Faculty of Health Sciences, University of Ottawa, Canada. P. 225-228.
Shahbazi, M.M., and Sanders, R.H. (2002). Kinematical Approaches to Hydrodynamic Force Assessments. Pakistan Journal of Applied Sciences. 2(9) 895-902.

THE DETERMINATION OF DRAG IN THE GLIDING PHASE IN SWIMMING

Daniel A. Marinho[1,2], Filipe Carvalho[2,3], Tiago M. Barbosa[2,4], Victor M. Reis[2,3] Francisco B. Alves[5], Abel I. Rouboa[6], António J. Silva[2,3]

Department of Sport Sciences, University of Beira Interior, Covilhã, Portugal[1]
Research Centre in Sports, Health and Human Development, Portugal[2]
Department of Sports, Health and Exercise, University of Trás-os-Montes and Alto Douro, Vila Real, Portugal[3]
Department of Sports Sciences, Polytechnic Institute of Bragança, Portugal[4]
Faculty of Human Kinetics, Technical University of Lisbon, Lisbon, Portugal[5]
Department of Engineering, University of Trás-os-Montes and Alto Douro, Vila Real, Portugal[6]

KEY WORDS: CFD, 2-D models, Hydrodynamics.

INTRODUCTION: The hydrodynamic drag forces produced by the swimmer during the sub aquatic gliding have been analyzed appealing to experimental investigation methods (e.g., Lyttle et al., 2000). However, the obtained results varied, which can translate some of the main inherent difficulties involved in the experimental studies. Thus, through application of a numerical method of Computational Fluid Dynamics (CFD), we intended to study the hydrodynamic drag forces, created during the displacement of the swimmer in different gliding positions, attempting to address some practical concerns to swimmers and coaches.

METHODS: Two-dimensional models of the human body in steady flow conditions have been studied. For the achievement of this study, two-dimensional virtual volumes, had been created in CAD software: i) a ventral position with the arms extended at the front of the body; ii) a ventral position with the arms placed alongside the trunk; iii) a lateral position with the arms extended at the front of the body; iv) a dorsal position with the arms extended at the front of the body. The drag coefficients (Cd) and the drag forces (Fd) had been calculated between speeds of 1.6 m/s and 2 m/s (with 0.05 m/s increments) in a two-dimensional Fluent® steady flow analysis.

RESULTS/DISCUSSION: For all the studied positions the Cd values decreased with speed whereas the Fd values increased with the flow speed, as expected (e.g., Bixler et al., 2007). The positions with the arms extended at the front presented hydrodynamic drag values lower than the position with the arms aside the trunk. The lateral position was the one in which the drag was lower. The ventral and the dorsal positions presented similar values.

CONCLUSION: The positions with the arms extended at the front seem to allow a higher reduction of the negative hydrodynamic effects of the human body morphology: a body with various pressure points due to the large changes in its shape. Thus, this position (perhaps performed in a lateral position) must be the one adopted after starts and turns. For instance, considering the breaststroke turn, the first gliding, performed with the arms at the front, must be emphasized in relation to the second gliding, performed with the arms along the trunk. Nevertheless, this concern demands further research using three-dimensional CFD models.

REFERENCES:
Bixler, B., Pease, D. and Fairhurst, F. (2007). The accuracy of computational fluid dynamics analysis of the passive drag of a male swimmer. *Sports Biomechanics*, 6, 81-98.
Lyttle, A., Blanksby, B., Elliot, B. and Lloyd, D. (2000). Net forces during tethered simulation of underwater streamlined gliding and kicking technique of the freestyle turn. *Journal of Sports Sciences*, 18, 801-807.

Acknowledgement
This work was supported by the Portuguese Government by Grants of the Science and Technology Foundation (SFRH/BD/25241/2005; POCI/DES/58872/2004).

ASYMMETRY IN FRONTCRAWL SWIMMING WITH AND WITHOUT HAND PADDLES

Mike Lauder and Rebecca Newell
University of Chichester, Chichester, West Sussex, UK.

The aim of this study was to determine whether asymmetry exists in underwater front crawl stroking patterns with and without hand paddles. Six senior national level male swimmers performed trials at 100m race pace, with and without large (480cm^2) hand paddles. Underwater motions for both right and left arms were filmed from the front and sides using three gen-locked video cameras and the video recordings were digitised at 50Hz to give three-dimensional coordinate data for a three segment model of the arm. The use of hand paddles significantly altered key temporal and kinematic features of the front crawl arm stroke for the left and right sides of the body. Specifically, the paddles increased time to complete the upsweep phase of the stroke on both sides. The paddles significantly reduced backward hand displacement on the left and right sides and altered the depth and lateral displacement of the stroke on the right side. Depth of stroke and elbow angle was also different without paddles on the right side and indicated asymmetry in technique, perhaps related to preferred breathing side.

KEY WORDS: Asymmetry, swimming, hand paddles, kinematics

INTRODUCTION: Specificity should be a key consideration in any strength or power development program (Payton and Lauder, 1995). Hand paddle swimming is seen by many as a specific form of strength training and enhancement of technique. The movement and speed specificity of hand paddles training has received limited attention in the literature. General conclusions indicate changes in these variables generated by paddles may compromise the training of specific muscles used in free swimming (Monteil and Rouard, 1992, 1994; Payton and Lauder, 1995). Regarding technique, the few studies that have been published are unanimous in their findings that paddles increase the stroke rate time from between 10% to 22% (Stoner and Luedtke, 1979; Monteil and Rouard, 1994; Payton and Lauder, 1995), thus, questioning the assumption that hand paddles are a specific form of training to enhance technique in front crawl swimming.

Previous studies in swimming have only evaluated one half of the body with presumed symmetry of the opposing side. However a symmetrical stroking pattern is assumed in elite swimmers and to date, no study has considered the effects of a possible imbalance between right and left arm pulling patterns. In other cyclic sports research has shown that imbalances can lead to overuse injuries (E.g. Kayaking: Lovell and Lauder, 2001). Therefore in light of the present review, the aims of this study were to determine whether asymmetry exists in underwater front crawl stroking patterns and to determine whether the use of hand paddles play a negative or positive role in the variation of stroke parameters with respect to training specificity.

METHODS: Participants: Participants for this study were six highly trained male competitive swimmers (mean age 21.5 ± 2.3 years). Best 100m front crawl performances ranged from 53.1s to 56.0s (mean time 54.9s ± 0.93s). Each subject was swimming approximately 5000 m a week with the hand paddles used in this study at the time of filming. All subjects' preferred breathing side was to the right.

Filming Procedure: After an appropriate warm-up and habituation to the experimental conditions, each subject swam front crawl trials through a performance volume that had been previously calibrated with a reference frame measuring 0.85m x 1.5m x 0.75m. The frame contained 52 control points of known location and distribution throughout the volume. Three cameras were used during filming. The first was positioned front-on to the swimming direction the second and third cameras were positioned to capture two separate halves of the calibration frame from the side view. The front-on camera filmed from an underwater

housing. Cameras 2 and 3 filmed from behind underwater windows. All cameras were aligned with the optical axis of the camera perpendicular to the glass/water interface.
Each subject swam a series of 25yard (22.9m) trials, with and without paddles, at the race pace. The hand paddles used were large WIN© paddles, which had a plan area of 480cm². The order of trials was randomised and subjects were given adequate time between trials to minimise the effects of fatigue.
Four trials per subject were selected for analysis; right hand with (RP) and without (RNP) paddles; left hand with (LP) and without paddles (LNP). Trials were selected on the criteria that the arm pull was executed naturally within the calibrated volume, no breath was taken whilst swimming through the volume and each swimmer achieved 95 ± 3% of their personal best time for the 100m front crawl sprint.
A three-segment model of the right and left arms: hand (including paddle), forearm, upper arm were defined by four body landmarks on each arm: the gleno-humeral joint centre, elbow joint, wrist joint and distal point of the third phalange. The estimated locations of these landmarks were manually digitised, at a sampling frequency of 50Hz, using a Panasonic VCR (NV- F75 HQ) editing system and an Archimedes 440 Computer equipped with Arvis Digitising card and KINE software (Bartlett and Bowen, 1993). Prior to filming, the skin overlaying these landmarks had been marked with black pen to help estimate their location.
Image coordinates were transformed to 3-D object space coordinates using a Direct Linear Transformation Algorithm correcting for linear lens distortion. These were then smoothed and differentiated using cross-validated quintic splines.

Definition of variables
The complete underwater motion of the arm, from hand entry to hand exit, was subdivided into four phases (Figure 1)
1&2 Downsweep phase from hand entry (A) to the most lateral position of the hand (B)
3 Insweep phase from (B) to the most lateral position of the hand (C)
4 Upsweep phase from (C) to hand exit (D)

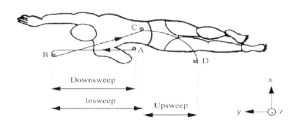

Figure 1. Swimming phases (Payton and Lauder, 1995)

The following variables were used to describe the kinematics of the underwater pull (Payton and Lauder 1995): Pull Width - Medial displacement (x-axis) of the hand during the Insweep phase; Pull Length - Backward (y-axis) displacement of the hand from its most forward position to its most backward position relative to the water; Pull Depth - Vertical displacement (z-axis) of the hand from entry to its' deepest point; Lateral Displacement - Entry (x-axis) to the widest point laterally; Elbow Angle - Maximum flexion during the Insweep Phase; Hand Velocity - Maximum velocity taken at the most distal point (third phalange) during Phase 3 and Phase 4

Statistical Analysis
Due to the small sample size, all variables were assumed to be normally distributed. To reasonably determine any significant differences in technique between the right and left

underwater hand pulling patterns (arm condition) and differences in technique between free and hand paddle swimming (paddle condition), Paired Samples T-Tests were employed with a two-tailed level of significance set at 0.05, as the validity in running more complex parametric tests in relation to the power of those tests for the sample size would be questionable.

RESULTS AND DISCUSSION: Significantly longer pull times were shown for the phase 4 with and without paddles irrespective of sides and a significantly longer total pull time was shown without paddles for the right side (Figure 2). The findings of longer pull times in phases 4 (upsweep) are in line with previous research (Payton and Lauder, 1995), however it would appear that without paddles the swimmers pull for a longer time on the right side.

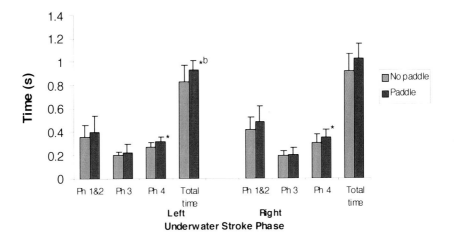

Figure 2. Durations of underwater phases for left and right sides, with and without paddles (mean ± SD). Significant at $P=0.05$ between: *with paddles and without paddles; [a]left and right with paddles; [b]left and right without paddles.

Right side differences were also shown for pull displacements (Figure 3). Between with and without paddle conditions, significantly shorter stroke lengths were shown on both the left and right sides. This finding is again in line with previous research where reduced stroke length has been shown to be as much as 28% when wearing paddles (Monteil and Rouard, 1992 &1994; Payton and Lauder, 1995). With paddles, the right side showed a deeper pull depth but less lateral displacement. Previous research has indicated that the use of paddles indicates a drag dominated approach to propulsion and it would appear that the kinematics for the right side support this. Without paddles the pull depth was deeper on the right side compared to the left. These findings may indicate that the right side is the stronger side and that the stroke kinematics reflects this strength difference.

Similarly, the results showed significantly slower pull velocities with paddles in phase 3 with paddles for both sides (Left NP: 2.53 ± 0.45 m.s^{-1}; Left P: 1.56 ± 1.05 m.s^{-1}; Right NP: 3.12 ± 1.06 m.s^{-1}; Right P: 1.46 ± 1.09 m.s^{-1}). No differences were shown for the velocity data when comparing left and right sides with paddles and left and right sides without paddles.

Elbow angle results showed only a significantly greater flexion of the elbow with paddles on the right side compared to the left (Right NP: $112.1 \pm 6.7°$; Right P: $105.7 \pm 10.5°$; Left NP $110.9 \pm 7.6°$; Left P: $111.4 \pm 7.9°$). This may indicate a greater degree of trunk motion or 'body roll' when wearing paddles as this finding does not support the earlier findings of a deeper pull on the right side when wearing paddles. This may also be related to the swimmers preferred breathing side as this was the right side for all the swimmers tested.

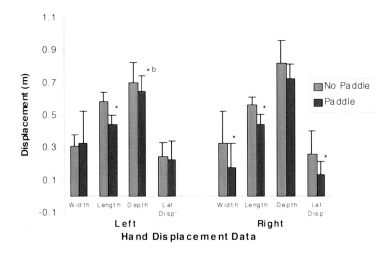

Figure 3. Hand displacement data for left and right sides, with and without hand paddles. Significant at $P=0.05$ between: *with paddles and without paddles; [a]left and right with paddles; [b]left and right without paddles.

CONCLUSION: The use of hand paddles significantly altered key temporal and kinematic features of the front crawl arm stroke for the left and right sides of the body. Specifically, the paddles increased time to complete the upsweep phase of the stroke on both sides. The paddles significantly reduced backward hand displacement on the left and right sides and altered the depth and lateral displacement of the stroke on the right side. Depth of stroke and elbow angle was also different without paddles on the right side and indicated asymmetry in technique, perhaps related to preferred breathing side.

REFERENCES:

Bartlett, R.M. and Bowen, T. (1993) *KINE system*. Manchester Metropolitan University: Alsager.

Lovell, G. and Lauder, M. (2001) Bilateral strength comparisons among injured and non-injured competitive flatwater kayakers. *Journal of Sport Rehabilitation*. 10(1), pp. 3-10.

Monteil, K.M. and Rouard, A.H. (1992) Influence of the size of the paddles in front crawl stroke. In *Biomechanics and Medicine in Swimming: Swimming Science VI* (eds. D. MacLaren, T. Reilly and A. Lees), E and FN Spon: London. pp. 99-104.

Monteil, K.M. and Rouard, A.H. (1994) Free swimming versus paddles swimming in front crawl. *Journal of Human Movement Studies*, 27: 89-99.

Payton, C.J. and Lauder, M.A. (1995) The influence of Hand paddles on the kinematics of front crawl swimming. *Journal of Human Movement Studies*, 28, 175-192.

Stoner, L.J. and Luedtke, D.L. (1979) Variations in the front crawl and back crawl arm strokes of varsity swimmers using hand paddles. In *Proceedings of the Third International Symposium of Biomechanics in Swimming. Swimming III.* (eds. J. Terauds and E.W. Bedingfield). University Park Press: Baltimore. 281-288.

DIFFERENCES IN SEGMENTAL MOMENTUM TRANSFERS BETWEEN TWO STROKE POSTURES FOR TENNIS TWO-HANDED BACKHAND STROKE

Lin-Hwa Wang[1,2], Kuo-Cheng Lo[3], Yung-Chun Hsieh[4], and Fong-Chin Su[1]

Institute of Biomedical Engineering[1], Institute of Physical Education, Health & Leisure Studies[2], National Cheng Kung University, Tainan, Taiwan
Physical Education Office, Kun Shan University, Tainan, Taiwan[3]
Institute of Physical Education, University of Tainan, Tainan, Taiwan[4]

KEYWORDS: biomechanics, athletes, racket

INTRODUCTION: Tennis stroke force depends on momentum transfer from racket to ball during ball-racket impact. Previous researchers study backhand stroke mechanics, focusing on comparison of one-handed and two-handed backhand stroke biomechanics (Reid & Elliott, 2002). This study investigated linear (LM) and angular momentum (AM) transfer from the trunk and upper extremities to the racket in open (OS) and square stances (SS) for different skill levels of players in the two-handed backhand stroke.

METHODS: 6 advanced (AG) and 6 intermediate (IG) players were recruited in this study. 21 retro-reflective markers were placed on each subject's upper extremities, trunk and racket and a 3-D 8-camera motion analysis system was adopted for recording their movements at sampling rate 500Hz at open and square stances, respectively. LM is the product of segment mass and velocity at the gravitational segment centre of mass position. AM is defined as the product of the principal moment of inertia and angular velocity in the segment coordinate system. A two-way ANOVA with repeated measures with a significance level of 0.05 was used.

RESULTS & DISCUSSION:

Table 1. Significant differences of the LM and AM between skill groups and stroke stances

Linear momentum (kg-m/s)			Angular momentum (kg-m²/s)		
Direction	Result	p	Direction	Result	p
Trunk back/fore	SS(6.26±0.79) > OS(0.93±0.16) IG(17.58±4.37) > AG(6.26±0.79)	.007 .009	Trunk L/R bending	IG(0.11±0.15) > AG(0.05±0.14)	.020
Trunk right/left	IG(7.78±2.32) > AG(0.17±0.30)	.004	Shoulder IR/ER	OS(0.07±0.01) > SS(0.05±0.01)	.047
Trunk up/down	IG(3.52±0.77) > AG(1.72±0.79)	.000	Wrist P/S	OS(0.14±0.04) > SS(0.10±0.04) AG(0.14±0.04) > IG(0.06±0.02)	.043 .043
Upper arm back/fore	SS(1.16±0.19) > OS(0.59±0.24) IG(2.27±0.66) > AG(1.16±0.19)	.018 .002			
up/down	IG(0.52±0.27) > AG(0.18±0.16)	.043			

The trunk produced larger backward, leftward and upward linear momentum in the IG group than in the AG group (Table 1). These LM components didn't help the stroke and might increase body instability and waste energy during the stroke for the IG. This study also found significant shoulder external rotational AM in the acceleration phase, significantly larger in OS than in SS. In term of ground reaction force transition, the shoulder joint plays a pivotal role. Enhancing the strength of shoulder rotator cuff muscles contributes to efficient racket momentum generation, particularly in OS.

CONCLUSION: The AG reduces trunk LM to keep stable and applies trunk and linkage segment rotation to generate backhand stroke power. The AG also has a quick backswing for increasing acceleration and maintains longer in the follow through phase for shock energy absorption. The SS has better LM transfers than OS. However, the OS generates larger shoulder rotational AM.

REFERENCES:
Reid, M. & Elliott, B. (2002). The one- and two-handed backhands in tennis. *Sports Biomechanics*, 1(1), 47-68.

Acknowledgement – Partially supported by National Science Council grant 89-2413-H-006-004, Taiwan

BIOMECHANICAL ANALYSIS OF ROUNDED OUTSOLE DESIGN SHOE DURING WALKING

Seungbum Park[1], Sangkyoon Park[1], Jungho Lee[1],
Kyungdeuk Lee[1], Daewoong Kim[1], Kookeun Seo[2] and Haksoo Shin[3]

Footwear Industrial Promotion Center, Busan, S. Korea[1]
Pusan National Univ., Busan, S. Korea[2], Daegu Univ., Daegu, S. Korea[3]

KEY WORDS: rounded outsole, barefoot, walking, knee angle, knee moment, immediate effect

INTRODUCTION: Studies have investigated biomechanical benefits of functional walking shoes such as unstable (Nigg, et al., 2006) or rounded outsole design shoes. However, it is still unclear how they influence lower extremity biomechanics during movement. The purpose of this study was to determine biomechanical differences and its mechanism of newly designed power walking shoes with lifted toe design compared to control shoe and bare foot conditions for middle-age females.

METHODS: Ten healthy females (43.6±2.99 years, 55.3±5.23 kg, 158.7±4.64 cm, shoe size: 235 mm) participated in this study. Kinematic and kinetic data were collected for the right leg of each subject during walking with 3 different shoe conditions (Figure 1). Subjects walked on a 10 m walkway at self-selected speed monitored by a motion capture system (Motion Analysis Corp., USA) and a force platform (AMTI, USA). The kinematic and kinetic data were filtered using a fourth order low-pass Butterworth filter with a cut-off frequency of 6 Hz and 100 Hz, respectively. Knee moment was calculated using a standard inverse dynamics approach while knee angle was calculated using a segment coordinate system (SCS).

Figure 1: Tested shoe model (A: Control, B: Power walking and C: Bare foot)

RESULTS: There were statistical differences in knee moment and ROM between type B shoe and type C (P<0.05). Subjects tended to have higher knee moment (abduction) when wearing type B shoes during walking compared to type A and type C shoes.

Table 1 Knee joint angle (unit: deg)

	Frontal	Sagittal	Transverse
A Type	9.05±3.30	63.20±3.56	16.64±5.40
B Type	9.41±3.98	**70.03±4.18**	19.46±5.66
C Type	9.10±3.74	**60.68±3.45**	18.80±6.64

Table 2 Knee joint moment (unit: Nm/kg*m)

	Frontal	Sagittal	Transverse
A Type	0.302±0.095	0.577±0.250	0.019±0.012
B Type	**0.305±0.076**	0.576±0.233	**0.018±0.014**
C Type	**0.261±0.078**	0.555±0.213	**0.023±0.017**

Bolded number inidates significant differences at α=0.05

DISCUSSION & CONCLUSION: As range of knee movement and knee moments increased with Type B shoes it may have a positive training effect as an exercise tool during walking. Long-term effect on biomechanical factors is unclear; kinetic and kinematic and EMG will be needed to be examined. This research concludes that the shoe B type has an immediate effect on biomechanical variables during walking.

REFERENCES: Nigg B. M., Hintzen S. and Ferber, R. (2006). Effect of an unstable shoe construction on lower extremity gait. *Clinical Biomechanics*, 21(1), 82-88.

PREPARATORY LONGSWINGS PRECEDING TKACHEVS ON UNEVEN BARS

Michelle Manning, Gareth Irwin, David Kerwin and Marianne Gittoes
Cardiff School of Sport, University of Wales Institute, Cardiff, Wales, UK

KEY WORDS: gymnastics, functional phases, angular momentum.

INTRODUCTION: The preparatory longswing on uneven bars is fundamental to the development of more complex skills in women's gymnastics. The preceding longswing governs the release parameters that in turn determine the success of the straddle Tkachev (Arampatzis & Brüggemann, 2001). Of the many longswing variations currently in use in women's gymnastics, this study aims to investigate differences in the biomechanics of three distinctive preparatory longswings used in performing the straddle Tkachev. The long term purpose is to increase understanding of these skills and potentially improve the effectiveness of coaching.

METHODS: Twin video image data from the 2000 Olympic Games were reconstructed using 3D DLT techniques. Pike (n=3), straddle (n=3) and arch (n=2) preceding longswings (LS) were analysed. The start and end of the hip and shoulder functional phases (FP), defined according to Irwin and Kerwin (2005), and their respective joint angle changes were calculated. Release angle (Θcm) and vertical and horizontal centre of mass (CM) release velocities (Vv and Vh), together with normalised angular momentum in straight somersaults per second (SS/s) of CM about the bar (Ln_b) and about gymnast CM (Ln_c) were calculated.

RESULTS & DISCUSSION: In all preparatory LS techniques, the FP at the hips precedes the shoulders. Between the three techniques the start of the FP occurred earliest for the arch, then straddle and finally pike. Ln_b was lowest in the pike technique resulting in a reduction in Vh and Vv. Ln_c for the pike LS was smallest with the corresponding values for the arch and straddle being approximately two and three times larger (Table 1).

Table 1 Mean [±SD] release parameters for the pike, straddle and arch longswing preceding the straddle Tkachev on uneven bars

	Θcm (°)	Vv (m/s)	Vh (m/s)	Ln_b (SS/s)	Ln_c (SS/s)
PIKE	54 [6]	1.39 [0.57]	-1.91 [0.30]	1.885 [0.079]	-0.098 [0.075]
STRADDLE	75 [10]	1.43 [0.35]	-2.00 [0.30]	2.162 [0.065]	-0.302 [0.147]
ARCH	74 [11]	1.62 [0.59]	-2.25 [0.49]	2.297 [0.122]	-0.213 [0.070]

The straddle and arch techniques each resulted in greater angular momentum about either the CM or the bar respectively. Each therefore appeared to benefit from the preceding longswing. Both angular momenta were lowest for the pike technique indicating that the reduced radius of rotation and delayed FP may limit the subsequent generation of angular momentum.

CONCLUSION: This study aimed to investigate how the biomechanics of the female longswing preceding the straddle Tkachev changes as a function of technique. From a coaching perspective, as the pike LS results in the lowest velocity and angular momentum at release it may be the least effective technique to adopt. Current work is focussed on examining a wider range of preceding longswing techniques across different competitions.

REFERENCES:

Arampatzis, A. and Brüggemann, G.P. (2001). Mechanical energetic processes during the giant swing before the Tkatchev exercise. *Journal of Biomechanics*, 34, 505-512.

Irwin, G. and Kerwin, D.G. (2005). Biomechanical similarities of progressions for the longswing on high bar. *Sports Biomechanics*, 4, 163 -178.

RELIABILITY OF INVERSE DYNAMICS OF THE WHOLE BODY IN THE TENNIS FOREHAND

Yoichi Iino and Takeji Kojima

Graduate School of Arts and Sciences, University of Tokyo, Tokyo, Japan

KEY WORDS: kinetics, inertial parameter, joint moment.

INTRODUCTION: Reliability of joint moment calculation using inverse dynamics is critical for evaluation of joint function and has been investigated for locomotion and lifting tasks, but not for sport movements in which the trunk can not be assumed to be rigid. The tennis forehand was studied in this paper because many biomechanical studies on the movement have been performed (Elliott et al., 1989) and the trunk twists substantially in the forehand. The purpose of this study was to investigate the reliability of the inverse dynamic analysis of the whole body in a tennis forehand using different segment inertial parameter (SIP) sets.

METHODS: Six high speed video cameras and two force plates were used to determine the moment acting on the pelvis during closed stance tennis forehands performed by six male tennis players. The difference between the pelvic moments determined by the top-down and bottom-up approaches was determined to evaluate the reliability (Plamondon et al, 1996). The effects of the different SIP sets (Zatsiorsky et al., 1983) on the RMS differences in the pelvic moments between the two approaches were tested using one-way repeated measures ANOVA ($p<0.05$). MC was adjusted by Dumas et al. (2007) and ZA was by de Leva (1996).

RESULTS: The RMS differences in the lateral flexion moment for AE and ZA were significantly smaller than that for MC (Table 1). The RMS difference in the axial rotation moment for AE was significantly smaller than the differences for MC and ZA.

Table 1 RMS differences in the pelvic moment components between the top-down and bottom-up approaches using the different SIP sets (Nm)

	AE		MC		ZA	
	Mean	s	Mean	s	Mean	s
Lateral flexion	20.7	4.9	46.9	7.2	23.8	7.8
Extension/Flexion	29.1	10.1	43.3	12.8	26.4	6.5
Axial rotation	16.6	3.2	23.7	2.5	20.2	3.9

DISCUSSION: The RMS differences were not negligible for all components of the moments determined by all SIP sets while overall AE provided better results than MC and ZA.

CONCLUSION: The reliability of the pelvic moment in a tennis forehand determined in this study was not so high for any SIP set used.

REFERENCES:

de Leva P. (1996) Adjustments to Zatsiorsky-Seluyanov's segment inertia parameters. *Journal of Biomechanics*, 29, 1223-1230.
Dumas R., Cheze L., and Verriest J.P. (2007) Adjustments to McConville et al. and Young et al. body segment inertial parameters. *Journal of Biomechanics*, 40, 543-553.
Elliott B., Marsh T., and Overheu P. (1989) A biomechanical comparison of the multisegment and single unit topspin forehand drives in tennis. *International Journal of Sport Biomechanics*, 5, 350-364.
Plamondon A., Gagnon M., and Desjardins P. (1996) Validation of two 3-D segment models to calculate the net reaction forces and moments at the L(5)/S(1) joint in lifting. *Clinical Biomechanics*, 11, 101-110.
Zatsiorsky, M. and Seluyanov, V.N. (1983) The mass and inertia characteristics of the main segments of the human body. *Biomechanics VIIIB*, Human Kinetics, Champaign, Illinois, 1152–1159.

Acknowledgement

This work was supported by Grant-in-Aid for Young Scientists (B), 19700497 from the Ministry of Education, Science, Sports, and Culture and Technology of Japan.

THE INFLUENCE OF EXTRA LOAD ON TIME AND FORCE STRUCTURE OF VERTICAL JUMP

Frantisek Vaverka[1], Zlatava Jakubsova[2] and Daniel Jandacka[3]

Faculty of Physical Culture, Palacky University, Olomouc, Czech Republic[1]
VSB-Technical University of Ostrava, Ostrava, Czech Republic[2]
Pedagogical Faculty, University of Ostrava, Ostrava, Czech Republic[3]

KEY WORDS: extra load, force-time structure, vertical jump, ground reaction force.

INTRODUCTION: Extra load is very often used in the training for development of explosive strength. Some research works are focused on a problem of take-off activity in relationship to extra load (Nelson & Martin, 1985; Bosco et al., 1984 and others). The main aim of this study was to find how the extra load influences time and force curve of the vertical jump.

METHODS: The analysis of the ground reaction force during vertical jump (Kistler force plate 9281CA, BioWare v3.2.6) using procedure described in Vaverka (2000) provided us with times of preparation, braking and accelerating phases and the total time of take-off duration, average force in braking and accelerating phases, the peak force in accelerating phase and braking and accelerating force impulses. The height of jump was computed from accelerating force impulse and body and external load weight. Two groups of university students (men, n = 18; women, n = 18) were asked to execute counter-movement vertical jump with arms excluded in four variants: without load and with extra load equal to 10%, 20%, and 30% of their body weight. Resulting 17 variables were statistically analyzed (ANOVA, Statistica v8).

RESULTS: We found the same tendency in results for both men and women. Statistically significant difference was found only in one variable – the height of jump, which decreased with increasing size of extra load (men: 38.3 – 34.4 – 30.9 – 27.6 cm, women: 29.7 – 26.0 – 23.3 – 20.2 cm). We also found slight and statistically non-significant increase in the total duration of take-off (men, 0.98–1.05 s, women 0.78-0.81 s) and in accelerating force impulse (men, 212–233 Ns, women 140–150 Ns).

DISCUSSION: Only small and statistically not significant differences in analyzed time and force variables of the Fz(t) were found. Increasing load (human body and external load) in combination with almost identical accelerating force impulses produced in different variants of jumps resulted in significantly decreasing height of jump.

CONCLUSION: Increased load resulted in small and statistically non-significant changes in measured time and force variables, except for the height of jump. We found only small extension in the total time of the take-off caused by longer time of braking and accelerating phase, and in the magnitude of accelerating force impulse. The research proved that increasing external load up to 30% of body weight does not significantly change the force-time structure of the vertical jump.

REFERENCES:

Bosco, C. et al. (1984). The influence of extra load on the mechanical behavior of skeletal muscle. *European Journal of Applied Physiology*, 53, 149-154.

Nelson, R.C. and Martin, P.E. (1985). Effects of gender and load on vertical jump performance. In D.A. Winter et al. (Eds.), *Biomechanics IX-B* (pp. 429-433). Champaign, IL: Human Kinetics Publishers.

Vaverka, F. (2000). Vertical jump – a suitable model for problems in biomechanics and motorics. In F. Vaverka and M. Janura (Eds.), *Proceedings of the conference BIOMECHANICS OF MAN 2000* (pp. 213-216). Olomouc: Univerzita Palackého.

Acknowledgement
Supported by the grant of the Grant Agency of the Czech Republic (GACR) No. 406/08/0572.

ESTIMATION OF HORSE LEG MUSCLES FORCE DURING JUMPING

Morteza Shahbazi-Moghaddam and Narges Khosravi

School of Physics, University of Tehran and Biomechanics Group, School of Physical Education of the University of Edinburgh Scotland, UK

The purpose of the present study was to estimate horses' leg muscle forces in jump height during jumping a spread fence of different heights. A digital camcorder was used (25 Hz) along with Ulead Studio program in order to obtain time, muscle lengths at rest, compression, extension, jump distance, and various angles in horse's legs data. The total jump distance and time of flight for each horse were measured with a precision of 10^{-2} m and 10^{-2} s respectively. Biomechanical formulae have been established in order to evaluate the muscles stiffness coefficients. Three groups of leg muscles; serratus ventralis, biceps brachii, and radial carpal extensor were considered in this study and their forces were successfully estimated.

Key words: horse, serratus ventralis, biceps brachii, radial carpal extensor, force, jumping

INTRODUCTION: So far few studies have been evaluating horses kinematics and kinetics; Shahbazi and Khosravi (2007, 2008), the CG kinematics of horses jumping over relatively small fences (=1m high) reported by Powers and Harrison (1999, 2000), and jumping over a water jump (=4.5m wide) reported by Clayton et al., (1996). An early study (Clayton and Barlow, 1989) examined the effect of fence dimensions on the limb placement of jumping horses, but no analysis was conducted on the CG kinematics and estimation of muscles stiffness. Shahbazi (2004) and Shahbazi and Erfani (2005) modeled human legs and reported reasonable stiffness coefficient values and thereafter estimated legs muscles forces in sprinters. The take-off kinematics of jumping horses in puissance competition was investigated and reported by Powers (2005). The body position and kinematics of a horses' centre of gravity at take-off are important factors determining the jump outcome. Unlike human athletes, horses are unable to significantly alter their body positions during jumping and therefore need to raise their CGs substantially, in order to clear the fence. The take-off is crucial to the jump outcome. Jumping requires the horse to raise its centre of mass high enough for all of its body parts to successfully clear the height and width of a fence. The jump should be viewed as an increased part of the suspension phase or an elevated canter stride as it occurs between the stance phase of the fore and hind limbs, for this reason jumping is mostly performed whilst cantering. So far there was no biomechanical investigation of horses' muscles kinetics in jumping and no leg muscles forces estimates were reported. The main aim of this study was to model and estimate horse's leg muscles forces in the linear CG kinematics and kinetics of take-off through establishing biomechanical relationships. These muscles were chosen because they were touchable to put landmarks on with visible variation in length to measure for our mathematical model.

METHODS: Video recordings (25 Hz) were obtained of eight top horses jumping over a fence of 1.0 m height. A single Sony camera was set up at 10m from the fence and perpendicular to it. The field of view measured about 6m wide and encompassed one full approach stride and the take-off phase. Video recordings were then transferred into ULead Studio program in order to measure the different angles of horse's leg's joints (angles between fibula-femur and pelvis) in two distinct positions and time sequences. The muscles lengths were then measured in all three phases; at rest, compression, and extension with Matlab programs. The riders average mass was (54±1.5 kg) and the horses average mass was 540±43 kg.

Formulae:
Shahbazi and Khosravi (2008) established the necessary formulae for muscles stiffness coefficients as followings:

$k_3 = -M [v^2 A (\Delta l_2)^2 + 2A\eta g - \lambda_1 (a_x + a_y)] / (\lambda_1 C + \lambda_3 A)$
(1)
$k_1 = M [v^2 C (\Delta l_2)^2 + 2C\eta g - \lambda_3 (a_x + a_y)] / (\lambda_1 C + \lambda_3 A)$
(2)
$k_2 = [k_1 \Delta l_1 \cos\theta_1 + k_3 \Delta l_3 \cos\theta_3] / (\Delta l_2 \cos\theta_2)$
(3)

In which λ_1, η, λ_3 and A and C are defined as follows:

$(\Delta l'_1)^2 (\Delta l_2)^2 - (\Delta l_1)^2 (\Delta l'_2)^2 \equiv \lambda_1$, $(\Delta l'_3)^2 (\Delta l_2)^2 - (\Delta l_3)^2 (\Delta l'_2)^2 \equiv \lambda_3$,   and $h'_{cm} (\Delta l_2)^2 - h_{cm} (\Delta l'_2)^2 \equiv \eta$
(4)

$\Delta l'_1 \sin(\theta'_1 + \theta'_2) + \Delta l_1 \sin(\theta_1 + \theta_2) \equiv A$ and $\Delta l'_3 \sin(\theta'_3 + \theta'_2) + \Delta l_3 \sin(\theta_3 + \theta_2) \equiv C$
(5)

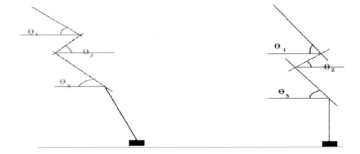

Figure1. One of the horses jumping; muscles groups at contraction (right) and in extension (left). The sticks showing the different angles in two phases.

Δl_1, Δl_2, Δl_3 are the serratus ventralis, biceps brachii, and radial carpal extensors muscle groups variations at compression phase respectively and $\Delta l'_1$, $\Delta l'_2$, and $\Delta l'_3$ are the variations in extension phase. θ_1, θ_2, θ_3, and θ'_1, θ'_2, and θ'_3 (Fig.1) are the angles which these muscle groups make with each other in the two phases, Shahbazi and Khosravi (2008).

RESULTS AND DISCUSSION: Eight horses jumped the fence at 1 m height, three times each. The procedures of finding the different parameters; $\Delta l_1, \Delta l_2, \Delta l_3, \Delta l'_1, \Delta l'_2, \Delta l'_3$, and $\theta_1, \theta_2, \theta_3, \theta'_1, \theta'_2, \theta'_3$ were the same as were recited in two previous papers, Shahbazi and Khosravi (2007, 2008). In order to get precise muscles lengths ($\Delta l_1, \Delta l_2, \Delta l_3$, and $\Delta l'_1, \Delta l'_2, \Delta l'_3$) in two phases, Matlab programming was successfully used.

Table 1. Values of different angles and variations of muscles lengths groups in two phases

Parameters	1	2	3	4	5	6	7	8
θ_1 Deg.	52	49	37	44.5	51	43.5	45.5	47.5
θ_2 Deg.	55.5	50	57	55	39	39.5	39	35
θ_3 Deg.	54	47	37	45.5	51	43	48.5	49.5
θ'_1 Deg.	30	35	38	23	35	27	33.5	30.5
θ'_2 Deg.	69	74	66	72.5	60	62.5	59	57
θ'_3 Deg.	42	47	46	43.5	52	44.5	46.5	46.5
Δl_1 cm	0.0175	0.065	0.07	0.04	0.05	0.03	0.0545	0.0525
Δl_2 cm	0.0375	0.055	0.11	0.015	0.07	0.0225	0.0335	0.0325
Δl_3 cm	0.0175	0.06	0.02	0.04	0.03	0.0175	0.028	0.025
$\Delta l'_1$ cm	0.0575	0.04	0.06	0.01	0.03	0.04	0.044	0.045
$\Delta l'_2$ cm	0.0625	0.09	0.14	0.08	0.09	0.05	0.090	0.095
$\Delta l'_3$ cm	0.025	0.01	0.04	0.02	0.03	0.08	0.033	0.035

Table 2. Values of muscles stiffness coefficients and corresponding forces

Muscles		1	2	3	4	5	6	7	8
serratus ventralis	k1	7.32E+06	3.69E+07	3.02E+06	6.27E+07	1.63E+07	5.91E+07	5.08E+05	5.28E+05
biceps brachii	k2	6.54E+06	1.55E+06	7.05E+04	1.51E+05	1.12E+04	6.78E+05	5.14E+04	5.44E+04
radial carpal	k3	6.67E+06	3.95E+07	1.14E+07	6.23E+07	2.76E+07	1.01E+08	1.28E+06	1.58E+06
serratus ventralis	F1	1.28E+05	2.40E+06	2.12E+05	2.31E+06	2.51E+06	1.77E+06	2.37E+04	2.77E+04
biceps brachii	F2	2.80E+03	8.51E+04	7.75E+05	2.78E+03	2.26E+03	1.53E+04	1.47E+03	1.77E+03
radial carpal	F3	1.17E+05	2.37E+06	2.29E+05	2.14E+06	2.49E+06	1.77E+06	3.65E+04	3.95E+04

In Table 1, values of different angles along with variations of muscle lengths groups in two phases for eight horses are given. In Table 2, muscles stiffness coefficients and corresponding forces are given. As can be seen, the values for the same angle vary remarkably from one horse to another, this means that the values depended remarkably on the rider technique and skill for jumping. The serratus ventralis muscle group force varied between; 2.4×10^4 to 2.5×10^6 N. and for biceps brachii muscle group force varied between; 2×10^3 to 7×10^5 N. and finally for radial carpal extensors muscle group varied between; 3.6×10^4 to 2.5×10^6 N.

CONCLUSION: The mechanical modeling suggested by Shahbazi and Khosravi (2008) has been used to evaluate the force of horse's leg muscle groups at take-off in jumping over a fence of 1m height. Mathematical formulae have been used to calculate these muscle forces. By using this method, the kinetic variables at take-off were found and riders, trainers and the owners of horses would be able to recognize their horse's muscle force and strength and can take necessary actions to improve their horse in jumping. The model needs filming and then measuring the time of flight of CM, the different leg joint angles in two positions; and the muscles measurements in different positions. With the help of Matlab programming and finally the use of the established formulae one can achieve horse leg muscles group forces.

REFERENCES:
Clayton, H.M., and Barlow, D.A. (1989). The effect of fence height and width on the limb placements of showjumping horses. Journal of Equine Veterinary Science, **9**, 179-185.
Clayton, H.M., Colborne, G.R., Lanovaz, J. & Burns, T.E. (1996). Linear kinematics of water jumping in Olympic show jumpers. Pferdeheilkunde, **12**, 657-660.
Powers, P.N.R., and Harrison, A.J. (2000). A study on the technique used by untrained horses during loose jumping. Journal of Equine Veterinary Science, **12**, 845-850.
Powers, P.N.R., Harrison, A.J., and Storey, N.B. (1999). Kinematic analysis of take-off parameters during loose jumping in young untrained horses. In: R.H. Sanders and B.J. Gibson (Eds). Proceedings of the XVII International Symposium on Biomechanics in Sports 101-104. Perth, Western Australia: School of Biomedical and Sports Science.
Shahbazi, M.M., and Brougeni, N. (1998). A new biomechanical aspect for assessing mechanical parameters in long jump. Proceedings of the XVI International Symposium on Biomechanics in Sports 324-327, Riehle, H., and M. Vieten (Eds), Konstanz University, Konstanz-Germany.
Shahbazi, M.M. (2004). A Mechanical Model for Leg Stiffness and Average and Maximum Force Estimations in Sprinters.
Proceedings of XXII International Symposium on Biomechanics in Sports. Pp. 370-374.
Shahbazi, M.M., Erfani, M. (2005). A Compter Simulation for Stiffness in Maximal Sprinting. Proceedings of XXIII International Symposium on Biomechanics in Sports. Pp.607-228.
Shahbazi, M.M., and Khosravi, N. (2007). A New Biomechanical Method for Investigating Horses Jumping Kinematics.
Proceedings of XXV International Symposium on Biomechanics in Sports. Pp.579-583.
Shahbazi, M.M., and Khosravi, N. (2008). A Mechanical Model to Estimate Legs Muscles Stiffness Coefficients in Horse During Jumping.
Proceedings of XXVI International Symposium on Biomechanics in Sports. Pp.624-627.

ACKNOWLEDGMENT:
The authors are thankful to the Iranian Horse Riding Federation for providing their best horses and riders to pursue the present study. The first author is also thankful to Sinclair Graham for helping with manuscript.

THE ANALYSIS OF PEDALING FORCE AND LOWER EXTREMITY EMG USING DIFFERENT PEDALING RATES AND LOADS

Cheng-Shuan Chang and Tzyy-Yuang Shiang

National Taiwan Normal University, Taipei, Taiwan

KEY WORDS: pedaling force, pedaling rate, load

INTRODUCTION: In cycling, the pedaling rate and load will affect the rider's performance and the enjoyment of riding. Previous studies usually analyzed the lower extremity EMG with different pedaling posture and pedaling rate, but mostly by professional cyclists (Neptune & Hull, 1999). However, most riders are recreational riders, therefore the results of previous studies are not suitable for the untrained persons, there were no results of lower extremity EMG and pedaling force in the study of pedaling rate and load. The purpose of this study was to analyze the effect of the different pedaling rates and loads on pedaling force and lower extremity EMG.

METHODS: Twelve untrained and healthy subjects (24.5±0.9 years; 174.2±4.3 cm; 73±8.3 kg) were recruited. All subjects completed the maximum exercise test on Lode ergometer by Vmax 29 in first stage, and setting up 3 exercise intensities (50%; 65%; 80% VO_2 max) for each subjects. In stage 2, all subjects completed 9 exercise tests [3 intensity/3 velocity (60; 75; 90 rpm)]. Pedaling force (left pedal) and lower extremity EMG (VM; RF; VL; BF; TA; GM; GL) were captured via 1000 Hz sampling rate using Vicon motion analysis system. A two-way ANOVA within repeated measures was used to perform statistical analysis.

RESULTS: The following figures show results of the IEMG and impulse of pedaling force.

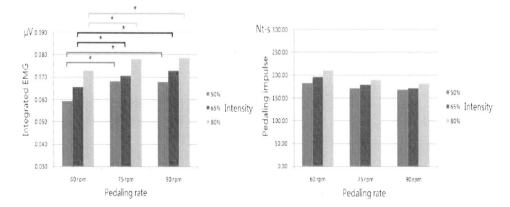

Fig 1: Lower extremity IEMG   Fig 2: Impluse of pedaling force

DISCUSSION: The IEMG in 60 rpm were significantly lower than 75 rpm and 90 rpm in three exercise intensities (Fig 1), and had no significant difference between 75 rpm and 90 rpm. EMG showed that the optimum cadence of untrained person was 60 rpm, which is much slower than 90 rpm by profession cyclist indicated in previous studies (Neptune & Hull, 1999).

CONCLUSION: The optimum cadence is 60 rpm on untrained person by analyzing IEMG. Optimum cadence for untrained person shouldn't be analyzed only by EMG, but also by adding more factors, such as pedaling force.

REFERENCES:

Neptune, R. R., and Hull, M. L. (1999). A theoretical analysis of preferred pedaling rate selection in endurance cycling. Journal of Biomechanics, 32(4), 409-415.

ISBS 2009

Poster Session PS3

ON-BOARD AND PRE-FLIGHT MECHANICAL MODEL OF YURCHENKO ONE TWIST ON VAULT: IMPLICATIONS FOR PERFORMANCE

Gabriella Penitente[1], Franco Merni[2] and Silvia Fantozzi[2]

Division of Sport and Exercise Science, University of Abertay Dundee, Scotland [1]
Faculty of Exercise and Sports Science, University of Bologna, Italy[2]

The aim of this study was to point out the biomechanical explanation of the judges' detection of scores relative to the on-board and pre-flight phases of the Yurchenko vault with one twist on (table). In an attempt to identify the weakness of technique and then to diagnose the likely causes of a poor performance, an extensive analysis was undertaken using a deterministic model. The 4 female gymnasts performing YU vault one twist on during the 2006 Italian Championship for Clubs were filmed by three cameras operating at 100Hz. Spearman's correlation coefficient was used to establish the strength of the relationship between the mechanical variables of the model and the judges' detection of points. Significant correlations indicated that the loss of credit depended mostly on angular variables. Firstly, low angular velocity of the center of mass (CM) at the impact of the board, then the small angular displacement of CM and high shoulder angular velocity produced on board and finally, a smaller hip extension and a larger shoulders extension at the take off from the board. In addition, other vertical variables determined a worst result: the lack of height of the CM at takeoff from the board, the decreased displacement of the CM on the board and the loss of the vertical velocity on the board.

KEY WORDS: Yurchenko vault, mechanical model, gymnastics

INTRODUCTION: In the Yurchenko (YU) family vaults – from round off entry – the on-board and pre-flight are both two critical phases because of the transitional action in which the gymnast's orientation change from forward to backward. The YU one twist on (table) is the most advanced evolution in this family of vaults. It is characterized by a full twist in the pre-flight that can negatively affect the linear velocity and angular momentum during the successive phases of the handspring on the table and the post-flight. Because of this, and the high risk of scores deductions, only few skilled gymnasts perform YU one twist on vaults in competition. In fact, the scoring system introduced in the 2005 provides that during the vault tournament gymnasts are awarded scores by two panels of judges. The first one is the difficulty score (D-score) represented by the specific value assigned to every vault in the Code of Points. The D-score is the same for every gymnast performing a similar vault. The second panel is the execution score (E-score) that evaluates the quality of the performance and this is the most important one. The base E-score is 10 and judges deduct points for errors in form, technique, execution and landing. Most of the studies conducted so far on YU (Nelson, Gross & Street, 1985; Elliott & Mitchell, 1991; Known et al, 1990; Koh, Jenning & Elliott 2003; Koh et al 2003) provided a biomechanical profile of this kind of vault performed on the old horse. The results carried out from several studies on vault (Koh, Jennings and Elliott, 2003; Know, Fortney & Shin, 1990; Penitente et al 2007) revealed that faults occurred in the latter phase of the skill are caused by the poor performance of the earlier phases. In particular, the landing is influenced by the post-flight. This in turn, is governed by the pre-flight and the on-board phases. The aim of the present study is to determine the weakness of performance and then to diagnose the likely causes of poor execution of the on board and pre-flight phases of the YU vault one twist on. The analysis was undertaken using a deterministic model arranged on the handspring vault model, developed by Takei (1998) and based on the method of Hay and Reid (1988).

METHODS: Data Collection: The 4 Yurchenkos performed by elite gymnasts during a women team competition in the 2006 Italian Championship for Clubs, were recorded for 3D motion analysis. Three high-speed synchronized cameras (BASLER 610, 3CCD, 1Mpixel) were used, each filming at a nominal rate of 100Hz, with the angle between their optical axes set at approximately 120°. The x-axis of the 3D reference system was directed along the runway, the z-axis was orthogonal to the floor, and the y axis oriented orthogonal to the x-z plane. Each trial included the following phases: snap-down phase of the round-off, on-board and pre-flight. The SIMI Motion System software was used to digitize approximately 50 frames of the movies from the impact on the board (BIMP) to the take-off from the table. The gymnasts' body was characterized by a 14-segment model identified by nineteen body points (head centre, tip of the nose, neck (7th C), shoulders, elbows, wrists, hands (3rd finger base), hips, knees, ankles, feet tip). The data were filtered with a low-pass second-order filter with a cut-off of 6Hz. The location of the center of mass (CM) was computed using the anthropometric parameters of Dempster (1955).

Data Analysis: Based on Takei's model (1998) and Hay and Reid's method (1988), a six-level deterministic model was developed. This included the linear and angular variables crucial for the on-board and pre-flight phases of the vault (Figure 1). The model depicted was used to analyze the correlations (Spearman's rank correlation coefficient) between the selected mechanical variables and the deduction of E-scores assigned by judges. In order to analyze the good form and technique in the early phases of the movement the deduction assigned for errors in landing have not been considered. Under the supervision of a certified international judge, after an accurate review of the movies, the landing point deductions have been quantified and subtract from the E-score awarded. The partial E-score deductions obtained has been used for the correlation analysis.

RESULTS: Figure 1 depicts the correlations of pre-flight factors with the partial E-score points deducted by judges. The thickness of the lines linking variables indicates the magnitude of the relationship. The significant correlations are indicated by bold boxes.

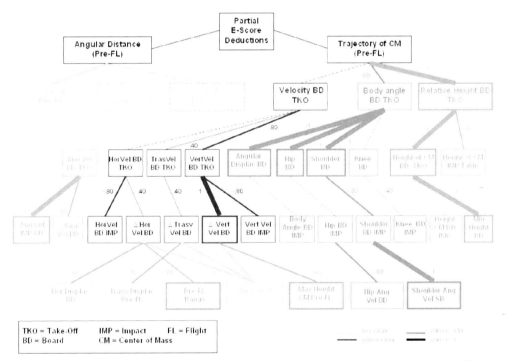

Figure 1 – Deterministic model of on board and pre-flight phases of the Yurchenco one twist on vault

In order to detect the influence of the mechanical quality relative to the early phases of the skill on the overall performance of the YU one twist on vault two variables were identified in the second level of the model; the angular distance and the trajectory of the CM during the pre-flight. Among the variables determinant for the angular displacement of the CM only the angular velocity of the CM at the impact of the board, located at the 5th level of the model, was significantly correlated with the partial deductions assigned by the judges (rho=-1). This negative correlation meant that the decrease of the angular velocity at the impact with the board increase the judges' penalty. Regarding the variables that determine the trajectory of the pre-flight, the score is negatively correlated between the vertical displacement achieved by the CM at the instant of take off from the board (rho=-1) (3rd and 4th level) and during the board phase (rho=-1) (5th level). This meant that these two variables were equally important because they accounted for similar variances with the judges' deductions. The less relative height of the CM at the take off from the board and the decreased displacement of the CM during the board phase increased the penalties. The reduction of the vertical velocity on the board was positively correlated with the score deduction, so the loss of velocity in this phase determined a worst score. The significant correlations found with the segmental angular variables in the 3rd level of the model revealed that a reduced hip angle (rho=-1) and a larger shoulder angle (rho=1) at the take off from the board determined a lower score. In addition, the angular velocity of the shoulder on the board was positively correlated with the score deductions (rho=1). The correlation of the angular displacement of the body on the board with the score deductions in the 3rd level, was significant (rho=-1). This meant that increased angular displacement on the board decreases the penalty assigned by judges.

DISCUSSION: The results from this study represented an analysis of the kinematic faults which occurred in the earlier phase of the YU one twist on vault. The deterministic model developed shows how a relatively large number of linear and angular variables during the on board and pre-flight phases can explain the penalty assigned by judges in competition. For the vault analyzed in the present study, the mechanical objectives of the on board phase is to generate the optimal pattern between lift, travel and biaxial rotation (somersault and twisting) in the pre-flight phase. The model displays the mutually interdependence between these three requirements and overviews the effect of the linear variables upon the angular ones and vice versa. The characteristics of the take-off from the board appeared to be very similar to those of the full twist backward somersault on the floor (George, 1980) in which gymnasts used an on ground twist technique to initiate the rotation about the longitudinal axis. The model showed that a too small blocking angle of the body, associated with a poor extension of the hip and an excessive extension of the shoulder at take off, were causes of deduction of points. Accumulation of these angular factors leads the gymnasts to over-initiate the twist on the board. This is caused by twisting the upper body about the lower body too early on the board creating an extended arched body shape at the take off. The high magnitude of the correlations of most of the variables in the last level of the model suggests that the gymnasts should spend more time on the board. This enables to travel through a larger blocking angle, to initiate the twist as later as possible and to increase the impulse. Developing a large angular displacement on the board would also achieve a higher quantity of lateral axis rotation and increase the quality of both the somersaulting and the travelling action. It was possible to detect that, for the YU one full twist on vault a too great angular velocity of the shoulder in the somersault direction affected the performance negatively, contrary to the traditional strategy used to improve the backward somersault rotation on board where a rapid swing of the upper body was desired to enhance the angular momentum (Koh et al 2003 and 2004). This is because, in the YU one full twist the arms are more involved in the generation of the longitudinal rotation. The somersault's momentum has to be generated developing a technically perfect blocking angle of the body instead of freely throwing the arms backwards. In comparison with other studies conducted on the YU vaults (Takei, 1998) that identified the maintenance of the horizontal velocity and the gain of the vertical velocity of the take off board as a crucial performance factor, the present study showed some dissimilarity. The horizontal component appeared to have just a marginal rule to better the score; in fact no correlations between the horizontal variables and the penalties were significant. On the other hands, the vertical component, both in terms of displacement and velocity of the CM, appeared to be an important factor for a successful performance of the YU one twist on vault.

The results show that the gymnasts can improve their score by, minimizing the descendent motion on the board and enhancing the vertical velocity during the on board phase, in order to achieve the desired lift of the CM during the pre-flight.

CONCLUSION: In conclusion, the deterministic model of the YU one full twist on vault drawn in the present study provides considerable evidence that the pre-flight full twisted phase makes this YU vault very hard to perfect score during competition. The bi-axial rotation in the early phase of the vault leads gymnasts to make kinematics faults that affected the general performance causing points deduction during later stages of the vault. The loss of credit resulted, is associated with an erroneous body shape and a low development of the vertical motion during the on board phase. These disadvantageous kinematic faults affected first, the optimal achievement of the pre-flight multi-tasks (lift, travel and rotate) thus, the successive phases and the general performance of the vault. In addition, this study has highlighted the importance of the different rule of the arms action in the generation of the angular momentum on the board between this kind of YU vault and the traditional YU vault style previously analyzed.

REFERENCES:
Nelson R.C., Gross T.S., and Street G.M. (1985) Vaults performed by female Olympic gymnasts: a biomechanics profile. *International Journal of Sport Biomechanics*, Vol. 1, pp 111-121.
Know Y., Fortney V.L., Shin I. (1990) 3-D Analysis of Yurchenko vaults performed by female gymnasts during the 1988 Seoul Olympic Games. *International journal of sport biomechanics*, Vol. 6, pp 157-176.
Koh M., Jenning L., Elliott B., and Lloyd D. (2003) A predicted optimal performance of the Yurchenko layout vault in women's artistic gymnastics *Journal of applied biomechanics,* Vol. 19, pp 187-204.
Koh M., Jennings L., and Elliott B. (2004) Role of joint torques generated in an optimised Yurchenko layout vault. *Sport Biomechanics*. Vol. 2(2), pp 177-190.
Elliott B., and Mitchell J.. (1991) A biomechanical comparison of the Yurchenko vault and two associated teaching skills. *International journal of sport biomechanics*, Vol. 7, pp 91-107.
Penitente G., Merni F. Fantozzi S., and Perretta N. (2007) Kinematics of the springboard phase in yurchenko-style vaults. 25th International Symposium on Biomechanics in Sport, pp 36-39.
Takei Y. (1998).Three-dimensional analysis od handspring with full turn vault: deterministic model, coaches' belief, and judges' scores. *Journal of applied biomechanics*, Vol 14, pp 190-210
Hay J.G., and Reid J.G. (1988) Anatomy, mechanics and human motion. Englewood Cliff,NJ: Prentice Hall.
Dempster W.T. (1955) *Space requirements of the seated operator.* (WADC Technical report 55-159). Wright-Patterson Air Force Base, OH: Wright Air Development Center.
George G.S. (1980) Biomechanics of women's gymnastics. Ed1. Englewood Cliff, Prentice-Hall, pp 53-58.

GROUND REACTION FORCE OF BASEBALL FLAT GROUND PITCHING

Chun-Lung Lin and Chen-fu Huang

Department of Physical Education, National Taiwan Normal University, Taipei, Taiwan

The purpose of this study was to describe the characteristics of the ground reaction force (GRF) of baseball flat ground pitching, and compares the characters with previous research which pitched on pitching mound. Fourteen division I college pitchers participated in this study. A VICON Motion capture system (10 cameras) and two force platforms were used to collect 3-D kinematic data (500Hz) and GRF data (1000Hz). Three successful trials for each subject were analyzed. The result shows the pivot foot anterior/posterior (AP) propulsive force was larger on flat ground, and the leading foot AP force was larger on pitching mound. The other two components GRF were similar in these two ground situations. The three components of GRF had low correlation with ball velocities. Comparing the peak GRF in three components between pitcher with fast and slow ball velocity groups, the fast velocity group produced a larger leading AP braking force. The leading foot AP braking force may be an important variable for identify the fast and slow pitching ball velocities.

KEY WORDS: pitching, baseball, biomechanics

INTRODUCTION: Pitcher plays an important role in a baseball game. The game result was often affected by pitcher performance, so the pitching training is very important. Pitching on flat ground is a common practice for training. This method is thought to be a training method with less pressure. But we don't really know why it is less pressure. MacWilliams et al. (1998) measure GRF from 6 collegiate and 1 high school pitcher. They found that when pitcher throw on the mound, ball velocity have high correlation with three components of peak ground reaction forces (GRF). (pivot foot AP forces $r^2=.82$, ML forces $r^2=.74$, vertical $r^2=0.76$. leading foot AP forces $r^2=0.86$, ML force $r^2=0.70$, vertical $r^2=0.88$). However, there was lack of research focus on flat ground pitching. Gottschall et al. (2004) in their research in running found when running from flat to downhill, the AP braking force increased, the AP propulsive force decreased. According to these two researches, we made two hypotheses. First, pitching on flat ground, the three components of GRF will have high correction with ball velocity. Second, when pitching on flat ground, the pivot foot AP propulsive force would be larger, but the leading foot AP braking force would be smaller than pitching on the mound.

METHODS: 14 healthy collegiate (12 right hand, 2 left hand) who plays in Chinese Taipei university baseball division I volunteered to participate in present research (age 19±1.1yr; height 172.8±6.6cm; body mass 74.2±8.0kg). All subjects were informed of the experimental procedures and gave their consent before participating. This study was approved by the local medical ethics committee. A VICON Motion capture system (Vicon Peak, Lake Forest, CA) with ten digital cameras (MX13) and two force platforms (Kistler model 9281, 9287) were used to collect 3-D kinematics data (500Hz) and force data (1000Hz). Thirty eight markers (8 mm in radius) were attached to the subject according VICON Plug-in model. Four markers were attached on the ball to compute ball velocity (resultant velocity). The anatomical nature position data were collected in the first trial. Subjects threw to a target (40*60cm,80cm high from ground.) attached to the safety net which is 3 m in front of the force platforms. The succeed trial was defined as the ball hit the target. Ten successful trials were collected. Kinematic data were computed by Visual3d. The GRF data were transformed into the frequency domain by Fourier transform and the first 32 harmonics coefficients were used for analysis (MacWilliams et al., 1998). The mean result of three trials with fastest ball velocity of each subject was used to compute correlations and t test. The correlations between GRF and ball velocities and t test were computed with SPSS 11.

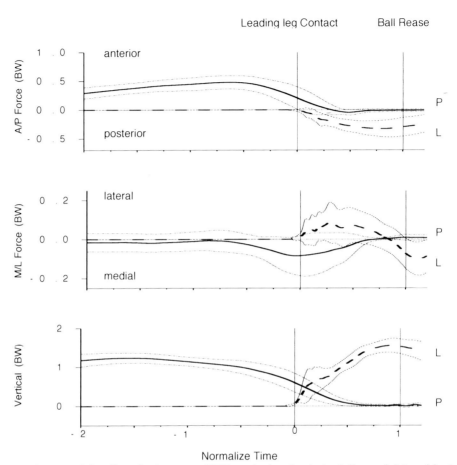

Figure 1. Pivot and leading foot mean GRF and standard deviation of 14 subjects (pivot foot ___ , leading foot _ _). Every subject's data were computed for mean of three trails. Force data are normalized with body weight.

Table 1. Comparison of GRF on fast and slow ball velocity groups

	Fast Speed (BW) (n=6)		Low Speed (BW) (n=6)		t	p
	Mean	Std	Mean	Std		
Pivot Foot						
A/P Force	0.49	0.12	0.55	0.10	-1.01	0.34
M/L Force	-0.19	0.04	-0.16	0.04	-1.59	0.14
Vertical	1.28	0.12	1.33	0.09	-0.91	0.38
Leading Foot						
A/P Force	-0.27	0.11	-0.43	0.15	2.26	0.05*
M/L Force	0.22	0.06	0.14	0.03	2.92	0.02*
Vertical	1.60	0.16	1.60	0.15	0.05	0.96

*$p<.05$

RESULTS: The mean ball velocity was 31.1±3.1m/s. The mean peak wrist velocity before ball release was 17.9±1.9m/s, maximum is 21.7 m/s, and minimum is 19.2m /s.

The pitcher produced pivot foot AP propulsive force when the leading leg started to stride, and AP propulsive force reached a peak 0.5BW before leading leg contact the ground. After contact, the pivot foot AP force rapidly decreased. The leading foot AP force appeared and increased after foot contact and reached a peak 0.3 BW just before ball release.

In medial/lateral (ML) direction, right side is lateral side for right-handed pitchers. For left-handed pitchers, it is opposite. The ML forces of the two feet were all small. Pivot foot medial force reached a peak 0.1BW just about leading leg contact. After the leading leg contact, the leading foot produced a lateral force, then switch to medial force after weight transfer and reached a peak 0.1BW after ball release.

The pivot vertical force maintained about 1.2BW, and decreased just before leading leg contact. Then the weight started to transform to the leading foot. The vertical force of leading foot appeared after it contact to the ground and increased to a peak 1.6BW just before ball release.

In this research, the coefficient of correlation between GRF and ball velocity were very weak. (pivot foot AP forces r^2=.14, ML forces r^2=.00, vertical r^2=0.20. leading foot AP forces r^2=0.15, ML force r^2=0.10, vertical r^2=0.00). For making the relationship between GRF and ball speed more clear, we took 2 subject of mid ball velocity out and compared the GRF peaks between the 6 pitchers with faster ball velocity and 6 with slower ball velocity. Only the leading foot AP and ML forces had significant difference. The result shows the pitcher with faster ball velocity produced more AP braking force.

DISCUSSION: Compare this result with previous research (MacWilliams et al., 1998), The wrist velocity was similar. The MP and vertical forces have similar value. But the AP forces were different. The pivot foot AP force in this study was larger but the leading AP force was smaller than the previous research. When compare with level running, Gottschall et al. (2004) found the result AP braking peak force was greater for downhill running, but AP propulsive peak force was greater for flat. The same phenomenon appeared on baseball pitching. Pivot foot produce more AP propulsive force when pitching on flat ground and leading foot produced more AP braking force on the mound.

MacWilliams et al. (1998) found that pitchers should train to develop powerful pivot leg drives as a normal part of the throwing motion, but they should not attempt to overpush to gain extra velocity. In their study pitching on the mound need more AP braking force than propulsive force. If pitchers overpush, they will not have enough AP braking force to maintain balance. So control propulsive force is an important factor. But on flat ground, the parallel braking force is smaller than propulsive force. No matter how hard the pitcher pushed off, he can produce enough force to braking. So it is easy to pitch on the flat ground and the pitcher can focus on other parts of the mechanic. The weak correlation between GRF and ball velocity might come from subjects' variation. The large different of AP forces show when faster ball was pitched greater braking force produced. The different correlation in this study and previous might reflect that pitching on the mound and flat ground is using different way to transferred the

Table 2 GRF data of previous study and present study. (The mean and standard of previous study are measure from paper's figure.)

	MacWilliams et al (BW)	present study (BW)
A/P force		
Pivot foot	0.35±0.07	0.48±0.10
Leading foot	0.72±0.08	0.33±0.14
M/L force		
Pivot foot	0.10±0.02	0.85±0.10
Leading foot	0.10±0.02	0.86±0.10
Pivot Force		
Pivot foot	1.00±0.03	1.2±0.14
Leading foot	1.50±0.05	1.6±0.21

energy to trunk and upper limbs. required further research to clarify the mechanic sequences of how energy produced and transferred.

CONCLUSION: when pitching on flat ground, the pivot foot AP propulsive force is larger, but the leading foot AP braking force is smaller. That makes the pitcher easy to balance. This study finds low correlation between GRF and ball velocity, and the leading foot produces greater braking force when faster ball was pitched. According this, pitching on flat ground can develop upper limb mechanic with less chance of injure. It is a better practice for rehabilitated pitcher for develop upper limb mechanic. When move from ground to the mound, they should pay attention to control their pivot foot force to avoid overpush. If pitcher loses balance in pitching process, the force produced from lower limb might hurt their arm or trunk again.

REFERENCES:
MacWilliams, B. A., Choi, T., Perezous, K. M., Chao, S. E., and MacFarland, G. E. (1998). Characteristic ground-reaction force in baseball Pitching. *American Journal of sports medicine*, 26, 66-71.
Gottschall, J. S., and Kram, R. (2004). Ground reaction forces during downhill and uphill running. *Journal of Biomechanics*, 38, 445-452.

NON-LINEAR CAMERA CALIBRATION FOR 3D RECONSTRUCTION USING STRAIGHT LINE PLANE OBJECT

Amanda P. Silvatti[1] Marcel M. Rossi[1] Fábio A. S. Dias[2] Neucimar J. Leite[2] and Ricardo M. L. Barros[1].

Faculty of Physical Education[1], Institute of Computing[2], University of Campinas, Campinas, Brazil.

KEY WORDS: non-linear camera calibration, 3D analysis.

INTRODUCTION: One critical aspect to 3D reconstruction of human motion using videogrammetry is related to the need for an accurate calibration of large volumes. Most of calibration methods used in biomechanics requires the construction, transportation and measurement of rigid structures and this is more difficult when larger volumes are involved. Recently, alternative approaches have been proposed to overcome this critical aspect (Cerveri et al., 1998; Zhang, 2000). This work presents preliminary results of the proposition and evaluation of a non-linear camera calibration method for 3D reconstruction using a plane object containing straight lines.

METHODS: The non-linear calibration method proposed in Zhang (2000) and implemented as a camera calibration toolbox for Matlab was adapted for video analysis and tested in this paper (named Chess Method). The method uses just a few points (12) distributed over a small volume (2.7x0.9x1.0 m^3) to obtain extrinsic parameters and a model plane contains a pattern of 5x8 squares (chess board) that defines straight orthogonal lines (100 mmx100 mm with 54 corners). The chess board was moved through all the acquisition volume to calculate the intrinsic calibration parameters and distortion parameters. We used four Basler cameras (60 Hz) with wide angle lens (4 mm) covering an acquisition volume of approximately 5x2x2 m^3. The chess board was tracked automatically in 700 frames. We analyzed the accuracy of the method using a rigid bar test. The accuracy was defined by the mean absolute errors of the curves of the distances between two markers (expected value=285.4 mm) obtained in function of time and we also calculated the mean error and the standard deviation of the curves of the distances. The results were compared to the literature.

RESULTS: Table 1 show the variables values: mean error, standard deviation (SD), accuracy found in the literature (Chiari et al., 2005) and using the Chess Method.

Table 1 Variables values compared with found in literature (mm). NA: data not available.

	Elite Plus	Vicon 370	Peak 5	Kinemetrix 3D	Ariel Apas	Chess Method
Mean error	NA	2.3	5.3	3.0	NA	1.1
SD	0.3	1.2	4.2	3.8	5.4	3.7
Accuracy	0.5	2.3	5.3	3.3	11.6	3.0

DISCUSSION: According to Chiari et al. (2005), accuracy in commercial system ranged from 0.5 mm to 11.6 mm. Our preliminary results showed compatible values. However, in our work the rigid bar was moved in a larger volume.

CONCLUSION: The preliminary results of the chess method revealed to be an applicable alternative with good accuracy for non-linear camera calibration and three-dimensional reconstruction using straight line plane object in larger volumes.

REFERENCES:

Zhang, Z. (2000) A flexible new technique for camera calibration, *IEEE Transactions on Pattern Analysis and Machine Intelligence*, 22, 11, 1330-1334.
Cerveri, P., Borghese, N.A., and Pedotti, A. (1998). Complete calibration of a stereo photogrammetric system through control points of unknown coordinates. *Journal of Biomechanics*, 31, 10, 935-940.
Chiari et al., (2005) Human movement analysis using stereophotogrammetry: Part 2: Instrumental errors. *Gait & Posture*, 21, 2, 197-211.
Research supported by FAPESP (00/01293-1, 06/02403-1), CNPq (451878/2005-1, 309245/2006-0, 473729/2008-3) and PRODOC-CAPES (0131/05-9).

THE CONTRIBUTION OF LOWER TORSO, UPPER TORSO AND UPPER LIMBS SEGMENTAL MOTION TO HAMMER HEAD VELOCITY DURING ACCELERATION PHASE

Hiroaki FUJII[1], Keigo OHYAMA BYUN[2], Mitsugi OGATA[2], Norihisa FUJII[2]

[1]Doctoral Program of Health and Sports Sciences University of Tsukuba and
[2]Institute of Health and Sport Sciences, University of Tsukuba, Japan

INTRODUCTION: In hammer throw event, the distance of the hammer throw is mainly determined by hammer head resultant velocity at release. During turn phase, hammer head resultant velocity is increased gradually with four turns and is increased during double support phase. The purpose of this study was to investigate the contribution of the motions in the lower torso, upper torso, upper limbs and hammer segment to the hammer head tangential velocity during double support phase.

METHODS: Six male hammer throwers with records in official competitions of 65.9~77.6m participated. Hammer throw with four turn technique was recorded by two high speed cameras (250Hz). The hammer head and 23 points of the body were digitized and coordinate data were three-dimensionally reconstructed using a DLT method. The coordinate data were smoothed with a Butterworth digital filter at optimum cut-off frequencies (6.0~8.25Hz) determined by a residual error method. Hammer head velocities were resolved into several components, namely the velocity caused by the segmental rotation at lower torso, upper torso, upper limbs and hammer head segment. The velocity of each component were calculated as the vector cross-products between the respective relative segmental angular velocity vector and the respective relative displacement vector from each joint to center of hammer head, by means of a mathematical model suggested by Springings et al (1994).

RESULTS AND DISCUSSION: Figure1 shows the averaged tangential velocities obtained from each motion component on all subjects. In each double support phase, the tangential velocity obtained from lower torso rotation (LTV), upper torso rotation (UTV), upper limbs rotation (ULV) contributed to hammer head tangential velocity mainly. Furthermore, at the beginning of the double support phase, UTV was increased. Although ULV reached the peak in middle of double support phase, Finally, LTV reached the primary component of hammer head tangential velocity during last half of double support phase. Because ULV was considered as a passive motion caused by the upper torso motion, these results suggested that lower torso rotation and upper torso rotation play KEY roles to accelerate hammer head. It was inferred from these results that technique of lower limbs motion that caused torso rotation and torsion were the critical factor of successful acceleration of hammer head.

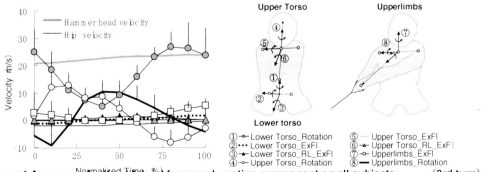

Figure1 Average velocity obtained from each motion component on all subjects (3rd turn).

REFERENCE: Springings et al. (1994): A three-dimensional kinematic method for determining the effectiveness of arm segment rotations in producing racket-head speed. J .Biomechanics. 41 :23-37

ASYMMETRIC LOADING DURING THE HANG POWER CLEAN - THE EFFECT THAT SIDE DOMINANCE HAS ON BARBELL POWER SYMMETRY

Jason Lake, Mike Lauder and Neal Smith
University of Chichester, Chichester, UK.

The vertical ground reaction force (GRF) of both feet and bar end kinematics were recorded using force platforms and high speed video simultaneously during hang power clean (HPC) performance with typical training intensities, in order to determine whether perceived handedness and ground kinetic asymmetry influenced bar end kinematics. There were significant differences between the GRF when side dominance was determined from GRF asymmetries ($p \leq 0.05$), but not when determined by perceived handedness. Similarly, there were significant differences between bar end power outputs when they were determined according to bar end asymmetries but not when determined by perceived handedness or GRF asymmetry. These results suggest mechanisms other than ground kinetic asymmetries influence bar end power output symmetry.

KEY WORDS: side dominance, force, resistance exercise, weightlifting.

INTRODUCTION: Measures of resistance exercise power output (RPO) are routinely used to monitor the affects of resistance exercise training programs (Hori et al., 2007). Recent research findings have shown that during controlled bilateral resistance exercise healthy individuals tend to favour a dominant side that may not necessarily correspond with the side they perceive to be dominant by as much as 10% (Flanagan & Salem, 2007; Newton et al., 2006). However, little is known about how ground kinetic asymmetries affect barbell kinematics. This could have important implications to strength and conditioning practitioners because measures of RPO tend to be based on the movement kinematics of one end of the barbell. Lake et al. (2008) recently reported that barbell kinematic symmetry was not influenced by ground kinetic asymmetries during dynamic lower-body resistance exercise. However, the effect that ground kinetic asymmetries have on barbell kinematics during ballistic lower-body resistance exercise has not been studied. This could have important implications to the strength and conditioning practitioner because the most common method of estimating RPO relies on the position-time data of one barbell end. With the above in mind this study set out to investigate the effect that side dominance had on the symmetry of hang power clean bar power output.

METHODS: Participants: Following University of Chichester ethical approval and a thorough explanation of the experimental aims and procedures, nine healthy males with a minimum of one year's hang power clean (HPC) experience volunteered to participate in this investigation. Their mean (±SD) physical characteristics were age: 28.8 (±8.5) years, mass: 84.1 (±18.9) Kg, height: 1.79 (±0.04) m, HPC 1RM: 70.8 (±18) Kg.

Test Procedures: Participant maximum HPC strength (1 RM) was established during a visit to the laboratory that occurred at least 48 hours but no more than one week prior to the asymmetry testing session and which followed a procedure that was similar to that outlined and used by (Kawamori et al., 2005).
During asymmetry testing each participant performed two single lifts with 80% of their 1RM (repetition maximum), with a minimum of one minute and a maximum of three minutes recovery between each lift (Reiser et al., 1996). Of the two lifts, the lift with the greatest bar end power output was selected for later analysis (Kawamori et al., 2005). Participants were instructed to lower the bar under control to above the knee and perform the positive phase as explosively as possible (Kawamori et al., 2005). The 80% 1RM load was selected to represent typical training loads (Reiser et al., 1996).

Measurements: The vertical GRF of HPC performance was recorded from both feet individually by two 0.4 by 0.6m Kistler 9851 force platforms (Alton, UK) at a sampling frequency of 500 Hz. The analog GRF signals were amplified by two type 9865E 8-channel charge amplifiers before they were digitally converted.

Three cameras (Basler A602fc-2, Germany) were positioned on rigid tripods approximately 5m from the centre of the area of interest around the right hand side of the participant, with an inter-camera angle of ~120 degrees recorded HPC performance at 100 Hz. The HPC GRF and movement footage data collection was synchronised using an MX Ultranet control unit (Peak Performance Technologies Inc., Englewood, Colorado).

All successful trials were digitised at 100 Hz using Peak Motus 9.2 software. The digitisation of reflective markers on the bar ends enabled the calculation of three dimensional spatial coordinates using the direct linear transformation procedure. Raw co-ordinate data were smoothed using a low pass filter with a cut off frequency of 6 Hz. The positive phase of HPC performance was determined from the velocity-time curve, and bar end velocity and acceleration used to calculate bar end power output in accordance to the methods outlined by Hori et al. (2007). The positive phase barbell power and GRF data were then averaged for further analysis. This is an approach that was recently used by Flanagan & Salem (2007), who suggested that peak performance data may not accurately represent the behaviour of parameters of interest over a selected period of time. Side dominance was reported according to perceived handedness (LRSD or left right side dominance) (Flanagan & Salem, 2007; Newton et al., 2006), and left and right foot positive phase GRF dominance (GRFSD or ground reaction force side dominance) (Flanagan & Salem, 2007; Newton et al., 2006). The bar end power outputs and GRF were then grouped according to their dominant (D) and non-dominant (ND) sides for analysis.

Statistical analysis: The different dominant (D) and non-dominant (ND) side GRF and bar end power outputs were compared using paired t tests. All statistical calculations were performed using SPSS version 16.0 for Windows (SPSS, Inc., Chicago, IL) and an alpha value of $p \leq 0.05$ was used to determine statistical significance.

RESULTS AND DISCUSSION: The mean (±SD) and confidence intervals (CI) for D and ND side positive phase average GRF and bar end power outputs are presented in Table 1.

Table 1 Mean (±SD; CI) mean HPC positive phase GRF and bar end power output

	Mean GRF (N)				Mean Bar Power (W)			
	LRSD		GRFSD		LRSD		GRFSD	
	D	ND	D	ND	D	ND	D	ND
Mean	668.0	648.1	691.0	625.1	673.5	667.6	672.2	668.9
SD	110.5	145.4	128.4	121.0	152.1	151.1	156.4	146.7
CI	2.3	3.0	2.7	2.5	3.2	3.2	3.3	3.1

*Note: LRSD: left right side dominance; GRFSD: GRF side dominance; D: dominant side; ND: non-dominant side.

There was a non-significant difference of -5.2% ($t(8)= 0.588$, $p=0.573$) between the mean left and right side GRF. This can be seen graphically in Figure 1 and was in agreement with the results reported by Newton et al. (2006) for movement asymmetries during back squat performance. The high standard deviation (see Table 1) may offer explanation for the lack of significant difference. Further analysis highlighted that 6 of the 9 participants did not display side dominance that was consistent with their perceived handedness, with individual participant differences ranging from -49.2% to 11.7%. This finding was in good agreement with the results recently published by Flanagan & Salem (2007).

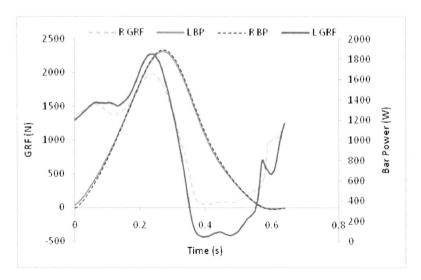

Figure 1: Typical GRF and bar end power output curve showing good movement symmetry.

A significant difference of 10.4% (t(8)=2.587, p=0.032) was established between the mean dominant and non-dominant side GRF. This is graphically presented in Figure 2. The mean difference was slightly greater than that recently reported for movement asymmetries during back squat performance (Flanagan & Salem, 2007; Newton et al., 2006). However, this may have been a consequence of the explosive nature of the HPC. It is worth noting that all 9 participants reported being right side dominant but that 6 of them demonstrated left mean GRF side dominance. This finding warrants further research as it suggests that there may be mechanisms other than ground kinetic side dominance that may underpin movement asymmetry.

There was a minimal and non-significant (0.8%; t(8)=0.834, p=0.428) mean difference between the mean left and right side bar end power outputs. There were no differences (t(8)=0.445, p=0.668) between the GRFSD mean dominant and non-dominant side bar end power outputs.

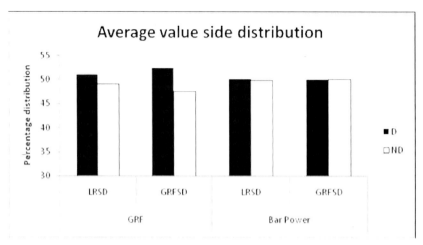

Figure 2: Mean LR and GRF side dominance D and ND GRF and bar end power distributions.

Further analysis of the bar end power output data showed that the side differences ranged from -5.9% to 6.1% for the left and right side bar end power output differences and -6.5% to 5.5% for the GRF dominant and non-dominant side bar end power output differences, indicating variability that was not a consequence of ground kinetic asymmetries. When side dominance was determined by bar end power output, that is the highest power output of the left and right bar end, power outputs became the dominant and the other side non-dominant, a mean difference of 2.8% was found, which was significant (t(8)=4.383, p=0.002). This finding supported the contention that mechanisms other than ground kinetics or perceived handedness underpin movement symmetry. This finding is a unique aspect of this study as these are the first asymmetry data to be determined by both ground kinetics and bar end kinematics.

CONCLUSION: It is apparent that any difference between the mean left and right side and GRF side dominant and non-dominant GRF does not necessarily influence the kinematics of the barbell. This is in good agreement with the findings recently presented by Lake *et al.* (2008) for asymmetric back squat performance but is the first time that ground kinetic and bar kinematic symmetry data from HPC performance have been presented. The results of this study support the integrity of methodologies that rely on the position-time data of one barbell end to estimate RPO. However, they also add to the growing body of research evidence that shows that healthy individuals tend to favour a dominant side may not necessarily correspond with the side they perceive to be dominant during controlled bilateral resistance exercise.

REFERENCES:
Flanagan, S., and Salem, G (2007). Bilateral differences in the net joint torques during the squat exercise. *Journal of Strength and Conditioning Research*, 21, 1220-26.

Garhammer, J. (1993). A review of power output studies of Olympic and Powerlifting: Methodology, performance prediction, and evaluation tests. *Journal of strength and Conditioning Research*, 7, 76-89.

Hori, N., Newton, R.U., Andrews, W.A., Kawamori, N., McGuigan, M.R. and Nosaka, K (2007). Comparison of four different methods to measure power output during the hang power clean and weighted jump squat. *Journal of Strength and Conditioning Research*, 21, 314-20.

Kawamori, N., Crum, A.J., Blumert, P.A., Kulik, J.R., Childers, J.T. and Wood, J.A (2005). Influence of different relative intensities on power output during the hang power clean: Identification of the optimal load. *Journal of Strength and Conditioning Research*, 19, 698-708.

Lake, J., Lauder, M. and Smith, N (2008). Squat asymmetry: A case study. *Journal of Sports Sciences*, 26(S1), 86.

Newton, R.U., Gerber, A., Nimphius, S., Shim, J.K., Doan, B.K. and Roberston, M (2006). Determination of functional strength imbalance of lower extremities. *Journal of Strength and Conditioning Research*, 20, 971-77.

Reiser, R.F., Smith, S.L. and Rattan, R (1996). Science and technology to enhance weightlifting performance: The Olympic program. *Strength and Conditioning*, 18, 43-51.

Acknowledgement
The authors would like to thank Russ Peters of the University of Chichester for his technical support during this project.

FOOT-TO-BALL INTERACTION IN PREFERRED AND NON-PREFERRED LEG AUSTRALIAN RULES KICKING

Jason Smith, Kevin Ball and Clare MacMahon
School of Sport and Exercise Science, Victoria University, Melbourne, Australia

Kicking is an integral skill within Australian Rules Football (ARF) and the ability to kick with either foot is essential at the elite level. A principal technical factor in the kicking skill is the nature of impact between the kicking foot and ball (Ball, 2008a). This study compared characteristics of foot-to-ball interaction between preferred and non-preferred kicking legs in Australian Rules football (ARF). Eighteen elite ARF players performed a maximum distance kick on their preferred and non-preferred legs. From high-speed video (6000Hz), two-dimensional digitised data of seven points (five on the kick leg, two on the ball) were used to quantify parameters near and during impact. The preferred foot produced significantly larger foot speed, ball speed, work done on the ball, ball displacement while in contact with the foot and change in shank angle during the ball contact phase.

KEYWORDS: Football, Kicking, Ball impact

INTRODUCTION: Australian Rules football (ARF) is one of the most popular sports in Australia. As the only form of disposal allowed for goal-scoring in ARF is kicking, the ability to kick accurately and effectively is integral to ARF players. In addition, as approximately 20% of kicks in Australian Football League (AFL) games are performed with the non-preferred foot, and this percentage can be as high as 45% for some individuals (Ball, 2003), the ability to kick effectively with either leg is an essential attribute for a player. This ability provides a player with greater disposal options and makes it more difficult for the opposition to defend.

The nature of foot-ball impact is a technical factor that is important to the kicking skill (Tsaousidis and Zatsiorsky, 1996; Dorge et al., 2002; Nunome et al., 2006a; Ball, 2008a). For ARF kicking, Ball (2008a) found that foot speed before impact, ball speed after impact, time in contact between foot and ball and ball displacement during contact were important parameters during foot-ball impact. From the results, Ball (2008a) argued that work was done on the ball during foot-ball impact which is important for the nature of foot-ball impact as momentum equations would be inappropriate for describing the foot-ball impact. Tsaousidis and Zatsiorsky (1996) found similar results for soccer kicking.

In spite of the importance of preferred and non-preferred leg kicking in ARF, there have been few studies examining these kicks and no studies focussing on the important aspect of nature of impact. The aim of this study was to compare ball to foot characteristics for preferred and non-preferred foot kicking in ARF.

METHODS: Eighteen Australian Football League (AFL) players (Age: 22.8 ± 4.2 years) kicked a Sherrin Australian Rules football (used in AFL competition, pressure range of 67-75 psi) with the preferred and non-preferred leg. Reflective markers were placed on the kicking leg prior to kicking. Players performed their preferred run-up approach and kicked for maximum distance through marker cones placed 40 metres from a line indicated by two markers. All players kicked until a 'good' kick (defined as a kick that went between the cones and was considered by the player and kicking coach to be a good maximal kick) was executed. All kicks were videoed at 6000 Hz using a Photron Fastcam APX-RS high speed camera (Photron Ltd, San Diego) placed perpendicular to the line of the kick with one good kick for each participant and each leg stored to disk.

Data Analysis: From the video for each kick, ten frames immediately prior to the initial contact between foot and ball (initial impact) and ten frames immediately after the end of contact between foot and ball (release) were identified for analysis. For each frame, seven points (head of fibula, lateral malleolus, heel of boot, head of the 5th metatarsal, toe of boot, top point of ball, bottom point of ball) were digitised using Silicon Coach Analysis tools

(Silicon Coach Ltd, NZ). The digitised XY coordinates were transferred to Microsoft Excel to enable the calculation and analysis of seven parameters identified as important during football impact (Table 1).

Table 1. Definition and calculation of measured parameters.

Parameter	Definition
Foot speed (ms^{-1})	Foot speed was defined as the average speed of the centre of the foot prior to initial foot-ball impact. X and Y coordinates of four points on the kick leg (ankle, heel, head of 5th metatarsal, toe of boot) were averaged to approximate the centre of the foot. Foot speed was calculated in the X and Y-directions between each frame, then averaged across all the digitised frames prior to initial foot-ball impact. The resultant foot speed was then calculated using quadrature summation.
Ball speed (ms^{-1})	Ball speed was defined as the average speed of the centre of the ball across all ten digitised frames after release. X and Y coordinates of two points (bottom of ball, top of ball) used to approximate the centre of the ball. Ball speed was calculated in the X and Y-directions between each frame, then averaged across all the digitised frames after foot-ball release. The resultant ball speed was then calculated using quadrature summation.
Ball:foot speed ratio	Ball:foot speed ratio was defined as the average ball speed at release divided by average foot speed at initial impact.
Time in contact (ms^{-1})	Period of contact between foot and ball from initial impact to release. Time in contact was calculated using the timer function supplied by Silicon Coach. The timer function was used to determine the number of frames in which the foot was in contact with the ball. Once the number of frames was determined, this was then divided by the frame rate (6000 Hz) to give time in contact between foot and ball.
Ball displacement (m)	Ball displacement defined as the change in displacement between the centre of the ball at initial impact and the centre of ball at release. The X and Y co-ordinates of the two points on the ball (top of ball, bottom of ball) were averaged to determine the position of the centre of the ball. Ball displacement was then calculated by subtracting the coordinates of the centre of the ball at impact from the coordinates of the centre of the ball at release.
Change in shank angle (°)	Difference in shank angle (angle between the horizontal axis and line between the head of fibula and ankle of the kick leg) between initial impact and release. The horizontal angle function supplied by Silicon Coach was used to find the shank angle by digitising the head of the fibula and ankle. Shank angle at impact was then subtracted from the shank angle at release to determine change in shank angle.
Work done on the ball (J)	Calculated using the formula, Work = mass x acceleration x displacement. Approximated using the mass of the ball (450g), ball acceleration during foot-ball impact (calculated from change in ball speed during foot-ball impact divided by time in contact) and ball displacement. Change in ball speed (used in ball acceleration) was defined as the difference between average ball speed before impact and average ball speed after release.

Statistical Analysis: Paired t-tests were conducted for each kicking parameter. Statistical significance was set at $p < 0.007$ after Bonferroni adjustment. Effect sizes (large: $d > 0.8$, medium: $d > 0.5$, small: $d > 0.2$) as defined by Cohen (1988) were also conducted.

RESULTS: Table 2 reports the mean and standard deviation values for the measured parameters as well as the results of statistical tests comparing preferred and non-preferred leg kicks.

Table 2. Mean and standard deviation of measured parameters for preferred and non-preferred kicking legs.

		Preferred Leg	Non-preferred Leg	t-test (p-value)	Effect size (d)
Foot speed (ms^{-1})	Mean	26.5	22.6	<0.001*	1.77
	SD	2.5	1.7		Large
Ball speed (ms^{-1})	Mean	32.6	27.0	<0.001*	1.32
	SD	4.4	3.8		Large
Ball:foot speed ratio	Mean	1.23	1.20	0.055	0.25
	SD	0.11	0.13		Medium
Time in contact (ms^{-1})	Mean	11.53	12.05	0.01	0.37
	SD	1.25	1.48		Medium
Ball displacement (m)	Mean	0.22	0.19	<0.001*	1.50
	SD	0.02	0.02		Large
Change in shank angle (°)	Mean	13	12	<0.001*	1.38
	SD	1	1		Large
Work done on the ball (J)	Mean	225.0	156.2	<0.001*	1.53
	SD	45.0	42.3		Large

* Significant difference ($p<0.007$) after Bonferroni adjustment.

Preferred and non-preferred legs differed significantly for five of the seven parameters analysed. In all cases, preferred leg kicks produced the greater values with a large effect size. No significant difference existed between kicking legs for ball:foot speed ratio and time in contact although for both, a small effect existed.

DISCUSSION: Foot speed recorded for the preferred leg in this study (26.5 ms^{-1}) was similar to values reported for elite ARF players (26.4 ms^{-1}, Ball, 2008b). No foot speed data exists for ARF kicking on the non-preferred leg, though the values in this study (22.6 ms^{-1}) were comparable to those found for soccer kicking (20.6 ms^{-1}, Nunome et al., 2006b). Ball speeds for preferred (32.6 ms^{-1}) and non-preferred (27.0 ms^{-1}) legs were similar to those reported by Nunome et al. (2006b) for soccer players (preferred: 32.1 ms^{-1}; non-preferred: 27.1 ms^{-1}). The values for ball:foot speed ratio, both preferred (1.23) and non-preferred leg (1.20), recorded in this study were within the range of values found for soccer kicking (Preferred leg: 1.06, Asami and Nolte, 1983; 1.35, Nunome et al., 2006b; Non-preferred leg: 1.19, Dorge et al., 2002). The mean time in contact for the preferred leg (11.53 ms^{-1}) was longer than the values reported by Ball (2008a) for ARF kicks (30 m kick: 9.8 ms^{-1}; 50 m kick: 10 ms^{-1}). However, the value in this study lay between values reported for soccer kicking (9.1 ms, Nunome et al., 2006a: 16 ms, Tsaousidis and Zatsiorsky, 1996). Preferred leg ball displacement (0.22 m) lay within the range of values reported by Ball (2008b) for elite ARF players (0.19-0.24 m).

Differences existed between preferred and non-preferred legs at foot-ball impact. Values for preferred leg kicking were significantly greater for foot speed, ball speed, ball displacement, change in shank angle and work done on the ball. No difference was found for ball:foot speed ratio and time in contact between preferred and non-preferred leg kicking.

Ball and foot speeds were greater for the preferred leg compared to the non-preferred leg, while no difference was found for ball:foot speed ratio. This may indicate that foot speed is the influential factor for ball speed. Strong correlations between foot and ball speed (Preferred leg, $r = 0.79$, $p < 0.001$; non-preferred leg, $r = 0.64$, $p = 0.005$) support foot speed being the influencing parameter for ball speed. However, the testing in this study limited kicks to 'good' kicks and the nature of impact between foot and ball might be more influential for 'mis-kicks' where the ball is not struck with an optimal orientation and/or position on the foot. This is an important useful future direction.

Change in shank angle was greater for the preferred leg, indicating that the shank moved through a greater range for the preferred leg compared to the non-preferred leg. More work was done on the ball for the preferred leg kick. In addition, a large significant effect existed for ball displacement, with the preferred leg moving the ball further than the non-preferred leg. However, no significant difference existed between kicking legs for time in contact. This indicates that greater power existed for the preferred leg, as may be expected.

Work was done to the ball during contact with the foot for both the preferred and non-preferred legs. Players produced significantly higher values for work done to the ball when kicking with the preferred leg compared due to both a greater average force being applied to the ball and greater displacement over which the force was applied. While the values for ball:foot speed ratio in this study may suggest that foot speed is the main influencing factor for ball speed, work done on the ball may also influence ball speed due to force being applied to the ball during foot-ball impact. Ball (2008a) suggested that muscular force generated at the hip and/or knee can be applied to the ball during foot-ball contact due to large values for change in shank angle, work done on the ball and ball displacement during foot-ball impact. The values in this study suggest a similar finding. As the values for work done on the ball, change of shank angle and ball displacement were all greater when kicking with the preferred leg, this indicates that the preferred leg was able to apply greater muscular force during foot-ball impact.

CONCLUSION: Differences exist in the nature of ball-foot impact between preferred and non-preferred foot kicks in ARF. Preferred foot kicks produced greater foot and ball speed, ball displacement, change in shank angle and work done to the ball. The greater amount of work being done to the ball during preferred leg kicking was due to both a greater average force being applied to the ball and greater displacement over which force was applied. Further research should focus on examining the source of work done to the ball. Suggestions that muscular force is responsible for the work done on the ball (Tsaousidis and Zatsiorsky, 1996; Ball, 2008a) are still to be determined, yet may provide important scientific information for coaching and strength and conditioning techniques of ARF players.

REFERENCES:
Asami, T., and Nolte, V. (1983). Analysis of powerful ball kicking. In: H. Matsui, and K. Kobayashi (Eds.).*Biomechanics VIII-B* (pp. 695-700). Champaign: Human Kinetics.
Ball, K. (2003). Profile of kicking in AFL football. Technical report for Fremantle football club. Perth.
Ball, K. (2008a). Foot interaction during kicking in Australian Rules Football. In: T. Reilly, F. Korkusuz (Eds.), *Science and football VI* (pp. 36-40). London: Routledge.
Ball, K. (2008b). Biomechanical considerations of distance kicking in Australian Rules football. *Sports Biomechanics*, **7**, 10-23.
Cohen, J. (1988). *Statistical Power Analysis for the Behavioural Sciences* (2nd Ed.). Hillsdale: Lawrence Erlbaum Associates.
Dorge, H., Bull Andersen, T., Sorensen, H., and Simonsen, E. (2002). Biomechanical differences in soccer kicking with the preferred and the non-preferred leg. *Journal of Sports Sciences*, **20**, 293-299.
Nunome, H., Ikegami, Y., Kozakai, R., Apriantono, T., and Sano, S. (2006a). Segmental dynamics of soccer instep kicking with the preferred and non-preferred leg. *Journal of Sports Sciences*, **24**, 529-541.
Nunome, H., Lake, M., Georgakis, A., and Stergioulas, L. (2006b). Impact phase kinematics of instep kicking in soccer. *Journal of Sports Sciences*, **24**, 11-22.
Tsaousidis, N., and Zatsiorsky, V. (1996). Two types of ball-effector interaction and their relative contribution to soccer kicking. *Human Movement Science*, **15**, 861-876.

THE STUDY AND APPLICATION OF THE INERTIA GAIT

A. S. Abdalla Wassf Isaac and A. Kamal

Department of Physics and Mathematics Engineering
Faculty of Engineering, Port Said, Egypt

KEY WORDS: The Inertia Gait, Measurement Apparatus, Study and Application.

INTRODUCTION: The aim is to develop a system of measurement suitable for both experimental (during sports) and clinical use which could be operated simply and with minimal disturbance of gait in situations outside the biomechanics laboratory. The use of accelerometers for the measurement of human gait parameters has not been widely accepted in biomechanics. However, several bioengineers and biomechanics in sports are interested in this field. Such readily obtainable information provides an index by which patient progress along the continuum of functional impairment may be monitored. The measurement of acceleration with subsequent integration can estimate velocity and position with reduced noise as Ladin et al.,1989).

METHODS: The gait analysis system consists of inertia accelerometer, electro-goniometer, amplifier-power supply interface, computer system and walkway, the laboratory walkway used for gait analysis. The subject wore the measurement shoe and attached goniometers. The initial output of each accelerometer was calibrated. In this subjects, there were 51 men (average age 29 year, mean weight 64 kg) and 42 women (average age 25.5 year, mean weight 54 kg). The five patients were male (average age 33.5 year, mean weight 65 kg).

RESULTS: The peak accelerations in each direction are presented in terms of the acceleration unit g=9.81m/s and the timing of these peak accelerations are presented as a percentage of the gait cycle, we can show it in table 1. The mean cycle time 1.39 s for normal subjects and 1.87 s for the patients. Compared to normal subjects, the five patients with incomplete paraplegia showed that the peak acceleration in the fore-aft direction was lowered ($\prec 0.5 g$).

Table 1 Peak acceleration and corresponding time for one location

	Peak acceleration (g)	Timing (o/o cycle)
Fore-aft direction	-2.0900 ± 0.580	68.79 ± 2.90
	$+1.8900 \pm 0.580$	99.00 ± 1.39
Up-down direction	$+1.0020 \pm 0.290$	68.50 ± 4.09
	-2.0010 ± 0.890	97.80 ± 2.25
Right-left direction	$+0.7700 \pm 0.210$	64.70 ± 4.89
	-1.0003 ± 0.303	94.40 ± 1.69

DISCUSSION: The results show that the acceleration of the lower leg during walking on level ground in 93 normal subjects varied with different phase, different direction and each measurement point. In the fore-aft direction, a negative peak acceleration appeared at toe-off and acceleration phase (65-70 %). In the vertical direction a positive peak acceleration appeared at early swing phase and a negative peak at the end of swing phase, at same time, there were two negative peaks and two positive peaks, for three different positions.

CONCLUSION: In this article, although this work does not attempt to analyse fully the parameters of gait. The inertial gait analysis apparatus provided an objective measure of acceleration of the lower leg and it is hoped to use the technique more extensively in clinical work and developed the method for use in the study of the sports movement.

REFERENCES: Ladin, Z., Flowers, W. C. and Messner, W., (1989). A quantitative comparison of a position measurement system and accelerometry. J. Biomechanics, vol. 22:295-308.

LOWER LIMB BIOMECHANICAL ADAPTATIONS TO TOTAL HIP ARTHROPLASTY EXIST DURING SITTING AND STANDING TASKS

Mélanie L. Beaulieu[1], Mario Lamontagne[1,2], Daniel Varin[1] and Paul E. Beaulé[3]

School of Human Kinetics, University of Ottawa, Ottawa, Canada[1]
Department of Mechanical Engineering, University of Ottawa, Ottawa, Canada[2]
Division of Orthopaedic Surgery, University of Ottawa, Ottawa, Canada[3]

The purpose of this study was to determine the effect of total hip arthroplasty (THA) on lower limb mechanics during the tasks of sit-to-stand and stand-to-sit. Twenty THA patients and 20 control participants performed three trials of sit-to-stand and stand-to-sit. Three-dimensional (3D) hip, knee and ankle angles were calculated. Forces, moments and powers were obtained with an inverse dynamics approach. THA patients exhibited lower joint forces and moments, as well as lower hip flexion and higher abduction angles, near seat-on and seat-off. These results indicate that THA patients were able to adopt a strategy that allowed them to reduce loading at the operated lower limb joints. Although such a strategy may be desirable given that higher loads can increase friction and accelerate wear of the prosthesis, reduced loading may be an indication of inadequate muscle strength that needs to be addressed.

KEY WORDS: total hip arthroplasty, sit-to-stand, stand-to-sit, kinematics, kinetics.

INTRODUCTION: Total hip arthroplasty (THA) is known to be a successful joint replacement procedure given that most patients experience significant pain alleviation, as well as an improvement in their capability to perform essential daily activities (Mancuso et al., 1997). In spite of these positive outcomes, the literature reveals that numerous biomechanical adaptations persist post-operatively during the execution of a number of daily activities (Foucher et al., 2008; Talis et al., 2008). For example, Talis et al. (2008) found that THA patients, when asked to stand up from a chair, displayed asymmetrical lower limb loading patterns that favoured the non-operated limb (i.e., unloading operated limb) – a strategy that may eventually elicit a joint disorder in the non-operated lower limb. However, kinematic and kinetic data of the lower limb joints were not obtained, thus preventing the researchers from elucidating the source of this asymmetry. To our knowledge, no other published studies have investigated the effect of THA on sit-to-stand biomechanics, despite the fact that it is a demanding, yet essential task that one must perform daily to maintain his/her independency. Furthermore, the sit-to-stand, as well as the stand-to-sit, tasks are also important elements of certain athletic activities, such as sitting in and standing out of a golf cart, canoe or kayak in order to enjoy a round of golf or paddling sports. In fact, THA patients have expressed dissatisfaction with their ability to perform athletic activities and sports (Mancuso et al., 1997). As proposed by Talis et al. (2008), their inability to perform standing and sitting manoeuvres may stem from prolonged muscle weakness, movement patterns adopted pre-operatively or adaptive behaviour as a result of surgery. Interestingly, it was found that sit-to-stand biomechanics of healthy adults were most affected by a reduction in hip muscle strength in comparison with strength of the muscles acting on the knee joint (Gross et al., 1998).

As for the biomechanics of the stand-to-sit task, a paucity of research has been published, even with regard to healthy individuals (Dubost et al., 2005). Consequently, a thorough biomechanical analysis of THA patients performing the sit-to-stand and stand-to-sit tasks is needed to elucidate the causes of these patients' deficiencies and dissatisfactions. Moreover, the results of this study could potentially be utilized to modify existing rehabilitation programs accordingly. The purpose of the present study was to determine the effect of THA on lower limb mechanics during the tasks of sit-to-stand and stand-to-sit by comparing three-dimensional (3D) hip, knee and ankle joint angles, joint reactions forces, moments and powers of THA patients with those of healthy, matched control participants.

METHODS: Participants: To achieve the purpose of this study, a total of 40 participants were recruited – 20 patients having undergone THA by means of a lateral surgical approach in the past 6-15 months (10 women, 10 men; age: 66.2 ± 6.7 yr; BMI: 27.2 ± 5 kg/m^2) and 20 healthy control participants, matched for gender, age and BMI (10 women, 10 men; age: 63.5 ± 4.4 yr; BMI: 24.9 ± 3.5 kg/m^2). THA patients were excluded if they had undergone hip replacement surgery for the contralateral hip joint, hip replacement due to an infection, a fracture or a failure of a previous prosthesis or hip replacement during which a concomitant surgical procedure was performed. Potential participants were also excluded if they suffered from any former or current condition that could alter their gait (e.g., stroke) or serious lower limb injury or disease (with the exception of the hip implant for the experimental group). Prior to participating in the study, which was approved by the institutions' research ethics boards, all participants signed an informed written consent.

Data Collection: Three-dimensional kinematics of the sit-to-stand and stand-to-sit tasks were collected at 200 Hz by means of a nine-camera digital optical motion capture system (Vicon MX, Oxford, UK), as well as 45 retro-reflective markers (14 mm diameter) placed on various landmarks of the participants according to a modified Helen Hayes marker set. Furthermore, 3D ground reaction forces were recorded at 1000 Hz with two force platforms (AMTI, Model ORC-6-2000, Watertown, MA, USA) positioned side-by-side and 10 cm in front of a height-adjustable bench. The height of the bench was adjusted to correspond to the height of the participant's tibial plateau. The participants performed three trials of each task – sit-to-stand and stand-to-sit – without the assistance of their arms and with each foot positioned on a force platform approximately shoulder-width apart, facing anteriorly.

Data Analysis: To remove noise from the data, a Woltring filter (predicted Mean-Square Error value of 15 mm^2) and a low pass Butterworth filter were applied to the 3D marker trajectories and ground reaction forces, respectively. The peak and range of the joint angles during the entire task, as well as the peak joint kinetics after seat-off for the sit-to-stand trials and before seat-on for the stand-to-sit trials were extracted from the calculated 3D hip, knee and ankle angles, joint reaction forces, moments and powers. These data were obtained from the operated limb in the THA group and from the dominance-matched limb in the control group.

Statistics: A series of one-way ANOVAs were performed, by means of SPSS statistical analysis software (SSPS for Windows, version 15.0, SPSS Inc., Chicago, USA), to determine the presence of significant differences between the THA and control groups with regard to all dependant variables. Alpha levels of 0.0167 and 0.025 (corrected for multiple comparisons) were used to determine statistical significance of the kinematic and kinetic variables, respectively.

RESULTS: Results from statistical analyses indicate that the THA and control groups were indeed matched for age (p=0.142) and BMI (p=0.092). It was also found that the majority of the statistically significant differences between the groups occurred near the time of seat-off and seat-on for the sit-to-stand (Table 1) and stand-to-sit (Table 2) tasks, respectively. For this reason, in addition to the constraint of limited space, only those variables will be presented and discussed.

DISCUSSION: Results clearly indicate that the mechanics of the sit-to-stand and stand-to-sit tasks have not returned to normal after total hip arthroplasty. Interestingly, most of the variables that significantly differed between groups for the sit-to-stand task were also found to differ for the stand-to-sit task. These abnormalities were most noticeable at seat-off and seat-on. With regard to kinematics, THA patients placed their operated hip in a more abducted position

Table 1. Means (standard deviation) of the significantly different kinematic and kinetic variables between THA patients (operated leg) and control participants (matched leg) near seat-off (SOF) of the sit-to-stand task.

Type of variable	Joint	Variable	Value THA	Value Control	p-value
Angle (°)	Hip	Peak flexion	81.6 (9.2)	90.3 (5.3)	0.001
	Hip	Peak abduction	-10.3 (6.8)	-4.7 (5.8)	0.008
	Knee	Flexion near SOF	73.9 (9.9)	83.6 (8.2)	0.002
	Ankle	External rotation near SOF	-16.8 (6.6)	-22.2 (6.3)	0.012
Joint Reaction Force (N/kg)	Hip	Peak anterior	-2.48 (0.49)	-3.33 (0.50)	0.000
	Knee	Peak proximal	4.20 (0.51)	4.98 (0.52)	0.000
	Ankle	Peak proximal	4.86 (0.52)	5.69 (0.49)	0.000
Moment (Nm/kg)	Hip	Peak extension	-0.48 (0.18)	-0.65 (0.19)	0.005
	Sum*	Peak extension	-0.96 (0.25)	-1.20 (0.18)	0.001

*This variable is defined as the sum of the sagittal plane hip, knee and ankle moments.

Table 2. Means (standard deviation) of the significantly different kinematic and kinetic variables between THA patients (operated leg) and control participants (matched leg) near seat-on (SON) of the stand-to-sit task.

Type of variable	Joint	Variable	Value THA	Value Control	p-value
Angle (°)	Hip	Peak flexion	81.5 (8.1)	91.5 (5.9)	0.000
	Hip	Abduction near SON	-9.1 (6.4)	-2.7 (7.3)	0.005
	Ankle	External rotation near SON	-16.2 (6.2)	-21.5 (6.6)	0.012
Joint Reaction Force (N/kg)	Hip	Peak anterior	-2.18 (0.72)	-3.09 (0.38)	0.000
	Knee	Peak proximal	3.56 (0.78)	4.33 (0.41)	0.000
	Ankle	Peak proximal	4.51 (0.83)	5.39 (0.43)	0.000
Moment (Nm/kg)	Hip	Peak extension	-0.47 (0.25)	-0.69 (0.16)	0.002
	Sum*	Peak extension	-0.80 (0.30)	-1.04 (0.15)	0.002

*This variable is defined as the sum of the sagittal plane hip, knee and ankle moments.

than the control participants. This phenomenon was accompanied by a more externally rotated ankle joint complex – most likely a result of this increased hip abduction. Such kinematics resulted in reduced joint reaction forces at the hip, knee and ankle in comparison with the control participants. Unfortunately, to our knowledge, no other study has analyzed the three-dimensional kinematics and kinetics of either sit-to-stand or stand-to-sit tasks in hip replacement patients, hence preventing us from comparing these results. THA patients also performed these tasks with lower moments of force about the hip joint. Similar results have been reported with total knee arthroplasty patients (Mizner & Snyder-Mackler, 2005), where low quadriceps strength and activity level were related to low hip and knee moments of force in the sagittal plane.

Our results revealed that the THA patients of the present study adopted a movement pattern that reduced forces and moments at all lower limb articulations. This could result from either the surgery, a pre-operative adaptation to alleviate pain or a strategy to prevent dislocation, as sit-to-stand is a task prone to posterior dislocation (Nadzadi et al., 2003). This strategy may gradually lead to muscle atrophy and consequently, a reduction in muscle strength. In such a case, patients would not be able to produce adequate muscle torque after hip replacement, regardless of the absence of pain. Muscle activity during sit-to-stand and stand-to-sit should be investigated in relation to kinematics and kinetics in future studies.

Some authors have expressed concerns about this unloading pattern of the operated hip joint (Talis et al., 2008). They found limb loading to be asymmetrical and thus assumed overloading of the healthy hip. Hence, they cautioned that such a loading pattern could cause premature wear of the non-operated hip. However, although not within the scope of, and therefore not reported in, the present paper, preliminary observations indicated that kinematic and kinetic data of the non-operated leg were similar to those of the control group. This suggests that THA patients adopted a strategy that allowed them to reduce the load on their operated lower limb without overcompensating with the non-operated side. Such a strategy could be desirable in the long term as higher loading could increase friction and accelerate wear of the prosthesis (Williams et al., 2006). On the other hand, this strategy may be an indication of, or may lead to, muscle weakness given that joint reaction forces and moments are highly dependent on those produced by the musculature surrounding the joint. Consequently, regaining muscle strength and being able to adequately load the new joint remains important in order to participate in recreational and athletic activities. Augmenting the strength of the muscles surrounding the new hip joint will ensure adequate muscle synergies, increase prosthesis stability and enable THA patients to better perform daily functional tasks and engage in their favourite recreational activities without limitations.

CONCLUSION: Results obtained from this study indicate that the THA patients displayed sit-to-stand and stand-to-sit kinematics and kinetics that differed from those of the healthy, control participants. These patients used a strategy that allowed them to reduce loading on the prosthesis without compromising the non-operated hip. Although such a strategy may be desirable given that higher loads can increase friction and thus decrease prosthesis longevity, reduced loading may be an indication of inadequate muscle strength. Such muscle weakness needs to be addressed to restore proper muscle synergies and thus, protect the surface on which the implant is affixed. Further investigations are needed to evaluate lower limb joint mechanics of THA patients performing more difficult tasks such as sit-to-stand from a lower position and/or with a lateral lower body displacement in order to better mimic certain tasks found in recreational activities (e.g., getting out of a kayak).

REFERENCES:
Dubost, V., Beauchet, O., Manckoundia, P., Herrmann, F., and Mourey, F. (2005). Decreased trunk angular displacement during sitting down: an early feature of aging. *Physical Therapy*, 85, 404-412.
Foucher, K. C., Hurwitz, D. E., and Wimmer, M. A. (2008). Do gait adaptations during stair climbing result in changes in implant forces in subjects with total hip replacements compared to normal subjects? *Clinical Biomechanics*, 23, 754-761.
Gross, M. M., Stevenson, P. J., Charette, S. L., Pyka, G., and Marcus, R. (1998). Effect of muscle strength and movement speed on the biomechanics of rising from a chair in healthy elderly and young women. *Gait and Posture*, 8, 175-185.
Mancuso, C. A., Salvati, E. A., Johanson, N. A., Peterson, M. G., and Charlson, M. E. (1997). Patients' expectations and satisfaction with total hip arthroplasty. *Journal of Arthroplasty*, 12, 387-396.
Mizner, R. L., and Snyder-Mackler, L. (2005). Altered loading during walking and sit-to-stand is affected by quadriceps weakness after total knee arthroplasty. *Journal of Orthopaedic Research*, 23, 1083-1090.
Nadzadi, M. E., Pedersen, D. R., Yack, H. J., Callaghan, J. J., and Brown, T. D. (2003). Kinematics, kinetics, and finite element analysis of commonplace manoeuvres at risk for total hip dislocation. *Journal of Biomechanics*, 36, 577-591.
Talis, V. L., Grishin, A. A., Solopova, I. A., Oskanyan, T. L., Belenky, V. E., and Ivanenko, Y. P. (2008). Asymmetric leg loading during sit-to-stand, walking and quiet standing in patients after unilateral total hip replacement surgery. *Clinical Biomechanics*, 23, 424-433.
Williams, S., Jalali-Vahid, D., Brockett, C., Jin, Z., Stone, M. H., Ingham, E., et al. (2006). Effect of swing phase load on metal-on-metal hip lubrication, friction and wear. *Journal of Biomechanics*, 39, 2274-2281.

Acknowledgement

The authors would like to acknowledge the contribution of Drs. Kim, Feibel, and Giachino of The Ottawa Hospital, Dr Cranney of the OHRI, as well as Drs. Benoit and Li of the University of Ottawa. This research was supported, in part, by the Canadian Institutes of Health Research.

FRONT AND PIPE SPIKES IN FEMALE ELITE VOLLEYBALL PLAYERS: IMPLICATIONS FOR THE IMPROVEMENT OF PIPE SPIKE TECHNIQUES

[1]Masanao MASUMURA, [1]Walter Q MARQUEZ, and [2]Michiyoshi AE

[1]Doctoral Program of Health and Sport Sciences, [2]Institute of Health and Sport Sciences, University of Tsukuba, Japan

KEY WORDS: volleyball, pipe spike, female, trunk twist, 3D motion analysis.

INTRODUCTION: The pipe spike was recently incorporated in offensive combinations increasing the complexity of volleyball tactics and making games more sophisticated. Such combinations are now seen in female games as a strategy of feinting to overcome the block more effectively. Although some investigations analysed front spike techniques, no study has been done on the pipe spike motion for female elite players in official games. Therefore, the purpose of this study was to compare the kinematics of front and pipe spikes of female elite volleyball players, and obtain insights into the techniques for pipe spike motion.

METHODS: The front and pipe spikes of four elite players (181.6±7.1 m, 69±3.9 kg) who frequently used front and pipe spikes in games were recorded by two high-speed VTR cameras (250 Hz, 1/2000 s) during six official games of the Volleyball World Grand Prix 2008, in Japan. The ball and 25 points on the body were digitized and three dimensional coordinate data were reconstructed using a DLT method. The coordinate data were smoothed with a Butterworth digital filter at optimum cut-off frequencies (5~12Hz) based on a residual method. The angles and angular velocities of the right shoulder were the primary variables analysed. The velocity of the ball and right hand, relative velocity between segment endpoints, and kinematics of the trunk and whole body centre of gravity (CG) were also computed.

RESULTS: The horizontal velocity of CG at the toe-off tended to be larger in the pipe than in the front spike, while the ball velocity and jump height were likely to be larger in the front than in the pipe spike. Figure 1 shows forward and lateral lean, and twist angles of PI and MK's front and pipe spikes, as typical examples. PI largely twisted the trunk in both types of spike, while MK used small trunk twist. MK's ball velocity of the pipe spike was much smaller (65.5 ms^{-1}) than that of the front spike (79.8 ms^{-1}).

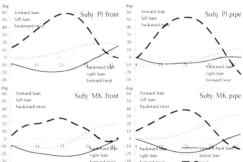

Figure1 Changes in the trunk angles during the airborne phase

DISCUSSION: Masumura et al. (2008) reported the horizontal velocity of CG at the toe-off of pipe spike was larger than that of front spike in all elite male players studied. In the present study, a similar trend was observed in the female players. This may have occurred due to jumping forward to cover a large distance and hit the ball. Furthermore, PI relied on a large trunk twist to apply force to the ball in both spikes, while MK relied more on forward and lateral leans. Based on these results, as it was already seen in males (Masumura et al. 2008), the trunk twist in the airborne phase seems to contribute to the ball velocity also in female players.

CONCLUSION: It was presumed that a fast forward trunk twist is more effective than trunk leaning motion to produce a powerful pipe spike.

REFERENCES:

Masumura, M., Marquez, W. Q., and Ae, M. (2008). A biomechanical analysis of pipe spike motion for elite male volleyball players in official games. *Proceedings of XXVI ISBS Congress*, 723-726.

SECURITY OF RUNNING OF COMPETETIVE COURSE IN ALPINE SKIING ACCORDING TO ITS GEOMETRY OF SETTING

Piotr Aschenbrenner, Wlodzimierz S. Erdmann

Sniadecki University of Physical Education and Sport, Gdansk, Poland, EU

In alpine ski disciplines there are plenty of incidents. Sometimes up to 40 % of competitors do not finish the run. The reasons are wrong preparation of skiers and wrong tactics of running. But there are also some errors in setting of gates of a course. The aim of this paper is to present some examples of those errors. One of them is a long intergate distances with small angles of deviation where skier achieves high velocity and just after that a big angle of deviation of a course with small intergate distance. The other is positioning of one gate above a downcast and the other just below a downcast. In Kvitfjel 2007 up to 10 skiers ran off the track in such a configuration.

KEY WORDS: alpine skiing, security, incidents, course, geometry

INTRODUCTION: Sport result in alpine skiing depends on fitness preparation, technique, equipment, environment conditions. At the highest level (Olympic Games, World Championships, World Cup) one can say they should not be a source of incidents (accidents, hitting a pole, missing a gate). Nevertheless there are plenty of incidents during running an alpine course. In 2006 Federation Internationale de Ski (FIS) established Injury Surveillance System in order to collect data on all incidents and to analyze a source of them.

Several authors (e.g. Blitzer et al., 1984; Ekeland et al., 1985, 1996, 1997; Ellman et al., 1989; Shealy & Ettinger, 1996, Erdmann & Giovanis, 1998, Bere et al., 2008) described classification of accidents and orthopedic treatment. Unfortunately there are small amount of papers dealing with reason and with prophylaxis against accidents. Bergstrøm et al. (2001) investigated a security problem during Junior Alpine Ski FIS Championships, Bavallen, Norway, 1995. He stated that in 998 runs (for male and female) in all alpine disciplines there were 4 accidents that ended with contusions. The most sever accident happened during the training downhill run. After that accident the position of one of the turning poles was changed. During Championships only 57 % of all runs were accomplished up to the finish. As much as 141 competitors did not finish their runs because of missing a gate or of a fall. There was no correlation between the number of accidents and a level of skiers. The most contusions happened during downhill.

Erdmann & Giovanis (1997) and Erdmann & Aschenbrenner (2002) presenting some connections between geometry of the alpine course and velocity of running revealed that more incidents happened while there was specific configuration of gates. Especially existence of a long intergate distances where skier achieved high velocity and just after that a big angle of deviation of a course with small next intergate distance was a source of the incidents.

The aim of the research was assessment of a risk of an incident of running a course in four alpine ski disciplines at the highest world level in the light of geometry of setting a course.

METHODS: Material: Competitions of four alpine ski disciplines of FIS World Cup – downhill (DH), super giant (SG), giant slalom (GS), and slalom (SL) were investigated. Altogether 12 competitions (each discipline was investigated three times) were under consideration. Within half of them (for slalom and giant slalom) two runs (two legs) were investigated. Research work was done since December 2006 until March 2007 at the following places: Val Gardena / Gröden and La Villa (Alta Badia) in Italy, Hinterstöder (Austria), Garmisch-Partenkirchen (Germany), Kranjska Gora (Slovenia), Lillehammer-Kvitfjel (Norway). The detailed data of the number of participating skiers and their incidents are presented in a Table.

Table 1. Alpine ski disciplines investigated, number of skiers appeared at the start, and number of incidents; DNF – do not finished, DSQ - disqualified

Discipline	Downhill			Super Giant		
Place of competition	Val Gardena	Garmish-Parten-kirchen	Kvitfjel	Val Gardena	Hinter-stöder	Kvitfjel
At the start	57	55	62	61	63	66
DNF	4	6	3	7	9	10
DSQ	0	0	0	2	0	0
DNF+DSQ %	7.0	10.9	4.8	14.8	14.3	15.2

Discipline	Giant Slalom			Slalom		
Place of competition	Alta Badia	Hinter-stöder	Kranjska Gora	Alta Badia	Garmish-Parten-kirchen	Kranjska Gora
At the start	100	100	93	112	104	105
DNF	8	20	22	28	16	27
DSQ	1	0	0	3	0	1
DNF+DSQ %	9	20	23.7	27.7	15.4	26.7

Geometry of setting of alpine courses, i.e. positioning of turning poles of the gates was obtained through Differential Global Positioning System (DGPS). Testing the accuracy of DGPS straight line was recorded during ascending while riding a lift. During reconstruction a ride again a straight line was achieved. The runs shown at the big screen at the finish area were recorded with video camera. Times achieved by skiers between gates were obtained. Having displacement and time intergate velocity was calculated. The distribution of velocity is important factor in security of running but this will be described in separate paper.

RESULTS AND DISCUSSION: There were big differences between geometry of courses of particular disciplines, e.g. length of a downhill was about 3.5 km (Figure 1 and 2), while SG was about 3/5, GS 1/2, and SL 1/5 of the length of DH. Vertical drop of DH from Figure 2 was about 840 m, and vertical drops of other disciplines were about 3/5 for SG, 1/2 for GS and 1/4 for SL. There were about between 50 and 60 gates situated at the courses. Some of them were positioned in a wrong manner. This plus wrong distribution of an effort during a run which was already revealed in other papers (Aschenbrenner, 2002; Giovanis, 1998) was a reason of incidents.

Figure 3 presents examples of fragments of geometry of courses where some of the incidents happened. As it was known from our previous papers one of the worst gate configurations was that there were long distances between gates with small angles of deviation and then small distance with big angle (Figure 3 A). Another example was positioning of a gate before a downcast and then positioning the next just below the downcast. When a setter puts poles for establishing a gate he holds his body erected. In this situation he can see a gate below a downcast. While a skier is running he/she is in a squat position so his/her eyes have problem to see what is below the downcast. In particular example spotted in Kvitfjel (Figure 3 B) as much as 10 skiers ran off the course at the same gate situated below a downcast.

CONCLUSION: In order to have alpine ski runs done in more security way it is postulated that the organizers of a competition will provide geometry of a course with gates. This information would be available for all participants of a competition, including setter and a referee for checking the correctness of a setting. Up to now competitors and coaches have one hour to check the course personally but without running. It looks like it is not enough since there are so many incidents.

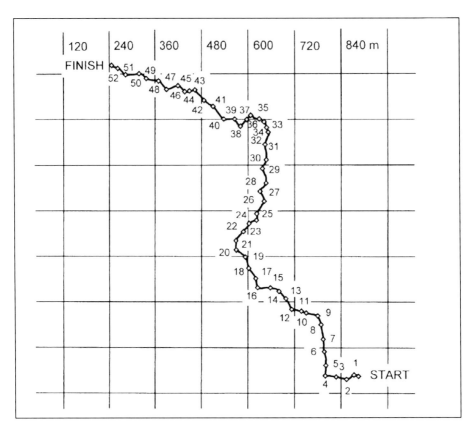

Figure 1. An example of a view perpendicular to the slope of setting of direction poles in giant slalom (Alta Badia, Italy).

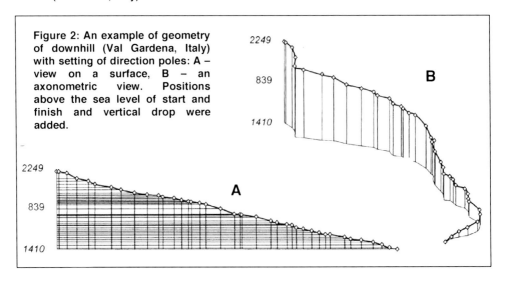

Figure 2: An example of geometry of downhill (Val Gardena, Italy) with setting of direction poles: A – view on a surface, B – an axonometric view. Positions above the sea level of start and finish and vertical drop were added.

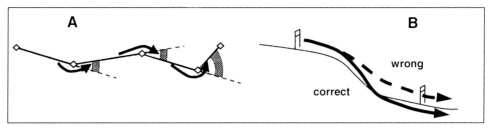

Figure 3: Examples of difficult fragments of an alpine course being a source of incidents: A – few long intergate distances with small angles of deviation and then short distance with big angle of deviation, B – one gate before a downcast, the next one just after it is a source of misguiding.

References:
Aschenbrenner, P. (2002) *Kinematics of competitors' running in the course of alpine ski giant slalom and its geometrical parameters* (in Polish). Doctoral dissertation, Gdansk: Sniadecki University School of Physical Education and Sport.

Bere, T.T., Florenes, T.W., Nordsletten, L., and Bahr, R. (2008) Injuries among world cup alpine skiers, *Book of Abstracts, 2nd World Congress on Sports Injury Prevention*, 26–28 June, 2008, Tromsø, Norway.

Bergstrøm, K.A., Bergstrøm, A., and Ekeland, A. (2001) Organisation of safety measures in an Alpine World Junior Championship, *British Journal of Sports Medicine*, 2001; 35:321–324

Blitzer, C.M, Johnson, R.J, Ettlinger, C.F, et al. (1984) Downhill skiing injuries in children. *Am J Sports Med*,12:142–7.

Ekeland, A., and Holm, A. (1985) Injury and race completion rates in alpine competitions. In R.J. Johnson, J.R. Mote (Eds.) *Skiing trauma and safety: 5th International Symposium*, ASTM STP 860, Philadelphia: American Society for Testing and Materials, p. 293–301.

Ekeland, A., Dimmen, S., Lystad, H. *et al.* (1996) Completion rate and injuries in alpine races during the 1994 Olympic Winter Games. *Scand J Med Sci Sports,* 6:287–90.

Ekeland, A., Nordsletten, L., Lystad, H., *et al.* (1997) Previous skiing injuries in alpine Olympic racers. In R.J. Johnson, J.R. Mote J.R., A. Ekeland (Eds.) *Skiing trauma and safety: 11th International Symposium*. Philadelphia: American Society for Testing and Materials, p. 17–13.

Ellman, B.R., Holmes III, E.M., Jordan, J., *et al.* (1989) Cruciate ligament injuries in female alpine ski racers. In R.J. Johnson, J.R. Mote J.R., M.H. Binet (Eds.) *Skiing trauma and safety: 7th International Symposium,* ASTM STP 1022. Philadelphia: American Society for Testing and Materials, p. 105–11.

Erdmann, W.S., and Giovanis, V. (1997) Investigations on kinematics of giant slalom's tactics in alpine skiing. In M. Miyashita, T. Fukunaga (Eds.) *Proceedings, XVI Congress of the International Society of Biomechanics*, University of Tokyo, Tokyo, Japan, p. 79.

Erdmann, W.S., and Giovanis, V. (1998) Incidents in alpine skiing giant slalom. In H.J. Riehle, M.M. Vieten (Eds.) *Proceedings, XVI Symposium of the International Symposium on Biomechanics in Sports*, University of Konstanz, Konstanz, Germany, p. 311-312.

Erdmann, W. S., and Aschenbrenner, P. (2002) *Geometry and running of the alpine ski FIS World Cup giant slalom. Part three – velocity as a function of geometry.* In K.E. Gianikelis (Ed.) *Scientific Proceedings, XX Symposium of the International Symposium on Biomechanics in Sports*. University of Extremadura, Caceres, Spain, p. 101

Giovanis, V. (1998) Kinematics of slalom tracks' running in alpine skiing and problem of traumatology (in Polish), Doctoral Dissertation, Academy of Physical Education, Cracow 1998.

Shealy, J.E., and Ettlinger, C.F. (1996) Gender-related injury patterns. In J.R. Mote, R.J. Johnson, W. Hauser, *et al.* (Eds.) *Skiing trauma and safety: tenth volume*, ASTM STP 1266. Philadelphia: American Society for Testing and Materials: 45– 57.

LOWER LIMB LANDING BIOMECHANICS ON NATURAL AND FOOTBALL TURF

Philippa Jones[1]; David Kerwin[1]; Gareth Irwin[1]; Len Nokes[2] and Rajiv Kaila[3]

Cardiff School of Sport, University of Wales Institute, Cardiff, United Kingdom[1]
School of Engineering, Cardiff University, United Kingdom[2]
The Royal National Orthopaedic Hospital, Stanmore, London, United Kingdom[3]

The aim of the study was to investigate variation in lower limb kinematics during jump landings on natural (NT) and artificial Football Turf (FT). One footballer performed 30 single leg jump landings, following a ball heading movement on NT and FT and immediately continued into a two-step forward run. Landing limb kinematics were recorded (200Hz) using CODA™ and cluster markers. There were similar knee and ankle touchdown kinematics and differing joint angle profiles throughout. FT landings showed greater knee flexion, adduction and internal rotation and reduced ankle eversion. During early impact, the ankle showed a tendency for greater plantar-flexion and inversion using FT compared to NT. These observations highlight a potential for altered lower limb kinematics on NT and FT which may be exaggerated during more demanding tasks and warrant further investigation.

KEY WORDS: artificial turf, injury, landings.

INTRODUCTION: Football Turf (FT), as described by the Fédération Internationale de Football Association (FIFA), is a third generation artificial surface developed specifically for football performance. Alterations to the Laws of the Game in 2004 acknowledged FT as an official surface for competitive football. The use of FT surfaces in professional competition continues to rise, but its suitability remains in question by the football community, based largely on the limitations of previous generations of artificial turf (Baker, 1990). Research in running has highlighted the potential for surfaces to impact on movement technique, altering performance, joint loading and subsequently injury risk (Dixon et al., 2000). Epidemiological investigations have highlighted a high incidence of lower limb injuries in professional football, often linked to landings (Murphy et al., 2003) and so differences in lower limb kinematics during landings as a function of playing surface may influence injury potential. This preliminary investigation aimed to identify whether lower limb kinematics differed when performing game specific jump landings on natural turf (NT) and FT.

METHODS: Data Collection: A female footballer (age 24 yrs, height 1.64 m and mass 56.8 kg) provided informed consent to participate in the study, with all procedures approved by the University's Ethics Committee. The participant was deemed experienced on natural turf having played competitive senior football for 10 years whilst inexperienced on FT having never played a competitive match on this novel surface. The An automated motion analysis system CODA, (Charnwood Dynamics Ltd, Leicestershire, UK) was used to collect the trajectories of 24 active LED markers at 200 Hz for 5 seconds per trial. Cluster marker sets were utilised with additional markers placed on anatomical landmarks of the lower limb for 10 static trials (2-6 s duration). The anatomical reference markers were removed for the movement trials. Ground reaction force data were collected for the landing leg using a force plate (Kistler 9287BA, Switzerland) sampling at 1000 Hz. Turf samples, housed in purpose built metal trays (900 mm x 600 mm x 50 mm) were mounted on the force plate and changed between trials. The participant performed 15 landings on each surface in a randomized order, wearing her own football boots (Predator Pulse II FG, Adidas). Trials were separated with a 5 minute interval to reduce the effects of fatigue on performance. Each trial comprised a single step approach into a jump to head a suspended size 5 official football followed by a single leg landing on turf and a two-step forward run (Figure 1).

  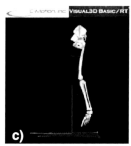

Figure 1: a) Participant performing experimental protocol and b) cluster set used to enable c) skeletal representation of lower limb during movement using Visual 3D software

Data Processing: The kinematic data were processed using motion analysis software (Visual 3D, C-motion Inc., USA) which enabled three dimensional kinematics and skeletal illustrations for each movement trial (Figure 1). The local coordinates for the thigh, shank and foot segments of the landing leg were defined from the static trial data and assigned to the 30 movement trials. Each trial was normalised to 100% from the instant when vertical ground reaction force (Fz) exceeded 10 N and the instant when Fz dropped below 10 N. The raw data were filtered at a cut off frequency of 19 Hz, determined by a Residual Analysis and the rotations about the three axes (flexion-extension, adduction-abduction and internal-external rotation) of the hip and knee joint and the two axes (plantar-dorsiflexion and inversion-eversion) of the ankle joint were exported from Visual 3D.

Data Analysis: Mean (±SD) curves were calculated for knee flexion/extension, adduction/abduction and internal/external rotation under NT and FT conditions, with percentage root mean squared differences (%RMSD) calculated for each joint angle profile. Variability was quantified by calculating %RMSD between the standard deviation profiles under each condition. Mean (±SD) joint angles were calculated for the instant of touchdown under each condition (Table 1).

RESULTS: The two greatest overall percentage difference (%RMSD) between the average NT and FT angle profiles for the hip, knee and ankle were found in ankle inversion-eversion and knee adduction-abduction (Table 1). The greatest variability between conditions was found in knee flexion-extension (Table 1). Percentage difference found between movement variability at the hip, knee and ankle were greater than the differences found between the mean angle profiles (Table 1). Similar joint angles were demonstrated under NT and FT at touchdown (Table 1).

Table 1 Overall percentage difference between the mean and the standard deviation angle profiles between NT and FT conditions

	Movement	Mean profile RMSD (%)	SD profile RMSD (%)	Mean (±SD) Angle at Touchdown NT (°)	Mean (±SD) Angle at Touchdown (FT)
Hip	Flexion-extension	4	45	12 (5)	10 (3)
	Adduction-abduction	18	35	9 (3)	9 (4)
	Internal-external rotation	5	49	-10 (4)	-10 (5)
Knee	Flexion-extension	9	82	-14 (5)	-13 (4)
	Adduction-abduction	22	40	-6 (3)	-5 (2)
	Internal-external rotation	13	34	15 (5)	15 (4)
Ankle	Plantar-dorsiflexion	5	9	-20 (10)	-26 (9)
	Inversion-eversion	25	33	0 (7)	-2 (4)

The angle profiles for the hip, knee and ankle when landing on NT and FT are presented in Figure 2. At touchdown under both surface conditions the hip was flexed, abducted and internally rotated, the knee was flexed, adducted and externally rotated and the ankle was plantar-flexed (Figure 2). A tendency towards greater knee flexion, adduction and internal rotation was observed under NT compared to FT (Figure 2). A tendency was also noted for additional plantar-flexion and inversion at the ankle joint during early contact and less ankle eversion throughout the landing under NT compared to FT.

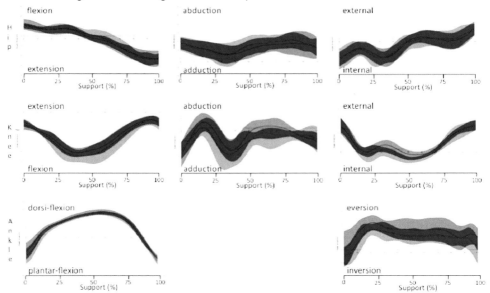

Figure 2: Mean (n=15) ±SD joint angle profiles for hip, knee and ankle under NT (black) and FT (red). Blue areas indicate overlapping standard deviation whilst the black indicates variability in NT alone and red indicates variability in FT alone

DISCUSSION: This preliminary investigation aimed to identify whether lower limb kinematics differed when performing game specific jump landings on NT and FT. Similar hip, knee and ankle angles at the instant of touchdown suggest similar preparatory landing strategies and force generation at touchdown under both surface conditions (Decker et al., 2003). However, differences in knee and ankle joint kinematics highlighted potential alterations in landing strategy when performing on NT and FT which warrants further investigation.

Knee Landing Strategy: The tendency towards additional knee flexion under FT compared to NT (Figure 2) suggests an increased ability to dissipate force (Decker et al., 2003), however, no difference was found in peak Fz between conditions. Previous authors (McNitt-Gray et al., 1993) noted greater knee flexion during landings as a function of additional surface stiffness, suggesting differing surface characteristics between NT and FT which may require further investigation. Landings under FT produced a tendency towards greater adduction and internal rotation of the knee compared with NT (Figure 2). Additional lateral movements of the knee may have induced similar lateral movements at the ankle (Andrews et al., 1996) providing rationale for the apparent tendency towards reduced ankle eversion under FT (Figure 2).

Ankle Landing Strategy: The participant demonstrated a tendency towards greater ankle plantar-flexion and inversion during early contact under FT compared to NT (Figure 2), with both movements believed to be key ankle injury mechanisms when landing (Renstrom and Konradsen, 1997). Landing in an extended ankle position is thought to improve force dissipation (Self and Paine, 2001) and increase lateral ankle ligament exposure (Caulfield and Garrett, 2004). Whilst this tendency appears minimal, a potential exaggeration may be evident during more demanding tasks. This concept may be particularly relevant to the suitability of FT due to the increased ankle injury incidence noted when performing on FT in comparison to NT (Ekstrand et al., 2006).

Knee Movement Variability: Substantial differences were reported in movement variability (MV) under NT and FT (Table 1), with landings on FT producing greater MV at the knee joint in comparison to NT. Optimal MV is believed to stem from experience (Wilson et al., 2008), which was believed to occur under NT conditions, reducing repetitive loading and increasing adaptability to perturbations (Hamill et al., 1999). In contrast, the greater MV produced at the knee under FT was deemed to be outside the functional limits of variability and coincided with the participant's lack of experience on this surface. The participant may have been attempting to acquire the appropriate characteristics of the landing movement, resulting in greater MV under FT (Wilson et al., 2008).

CONCLUSION: The present study has investigated the kinematics of the lower limb during football specific jump landings. Whilst being a preliminary investigation, tendencies towards greater knee flexion and internal rotation combined with additional lateral movements at the knee and ankle may highlight important differences in landing technique when performing on a novel FT surface. Future studies need to investigate how performers adapt to performing on FT incorporating greater sample sizes, inferential statistics and inter-subject comparisons across a range of turf types and sizes.

REFERENCES

Andrews, M., Hewett, T. and Andriacchi, T. P. (1996). Lower limb alignment and foot angle are related to stance phase knee adduction in normal subjects: A critical analysis of the reliability of gait analysis. Journal of Orthopaedic Research 14, 289-295.

Baker, S. W. (1990). Performance standards for professional soccer on artificial turf surfaces. *Journal of Sports Turf Research Institute*, 66, 83-92.

Caulfield, B. and Garrett, M. (2004). Changes in ground reaction force during jump landing in subjects with functional instability of the ankle joint. Clinical Biomechanics, 19, 617-621.

Decker, M. J., Torry, M. R., Wyland, D. J., Sterett, W. I. and Steadman, J. R. (2003). Gender differences in lower extremity kinematics, kinetics and energy absorption during landing. *Clinical Biomechanics*, 18, 662-669.

Dixon, S. J., Collop, A. C. and Batt, M. E. (2000). Surface effects on ground reaction forces and lower extremity kinematics in running. *Medicine & Science in Sports & Exercise*, 32, 1919-1926.

Ekstrand, J., Timpka, T. and Hagglund, M. (2006). Risk injury in elite football played on artificial turf versus natural grass: a prospective two-cohort study. *British Journal of Sports Medicine*, 40, 975-980.

Hamill, J. R. E., van Emmerik, B. C., Heiderscheit, B. C. and Li, L. (1999). A dynamical systems approach to lower extremity running injuries. *Clinical Biomechanics*, 14, 297-308.

McNitt-Gray, J. L., Yokoi, T. and Millward, C. (1993). Landing strategy adjustments made by female gymnasts in response to drop height and mat composition. Journal of Applied Biomechanics, 9, 173-190.

Murphy, D. F., Connolly, D. A. and Beynnon, B. D. (2003). Risk factors for lower extremity injury: a review of the literature. *British Journal of Sports Medicine*, 37, 13-29.

Renstrom, P. A. and Konradsen, L. (1997). Ankle ligament injuries. British Journal of Sports Medicine, 31, 11-20.

Self, B. and Paine, D. (2001). Ankle biomechanics during four landing techniques. Medicine & Science in Sports & Exercise, 33, 1338-1344.

Wilson, C., Simpson, S. E., van Emmerik, R. E. A. and Hamill, J. (2008). Coordination variability and skill development in expert triple jumpers. Sports Biomechanics, 7, 2-9.

Acknowledgement
This work was funded by UWIC Bursary Awards for Research Students and University of Wales Bursary.

FALL AND INJURY INCIDENCE RATES OF JOCKEYS WHILE RACING IN FRANCE, GREAT BRITAIN AND IRELAND

Manuel A. Forero Rueda[1] Walter L. Halley[2] & Michael D. Gilchrist[1]

School of Electrical, Electronic & Mechanical Engineering, University College Dublin, Belfield, Dublin 4, Ireland[1]
The Turf Club, The Curragh, Co. Kildare, Ireland[2]

KEY WORDS: Jockey; Horse racing; Equestrian; Head injury; Concussion.

INTRODUCTION: The objective of this study is to provide quantitative details of the frequency and severity of injuries, especially head injuries, sustained by jockeys while racing. There is a lack of worldwide equestrian injury data and a lack of uniformity in the data that is available.

METHODS: All available epidemiological injury data were collected and provided to the authors for the racing seasons of 1999-2006 by the senior Medical Officers of both the Irish Turf Club and France Galop. Corresponding data have also been collated and published previously by the medical personnel of the British Horseracing Authority (Balendra et al. 2007, McCrory et al. 2006). Incidence rates and proportions were compared between racing types and racing jurisdictions using Incidence Rate Ratios, Poisson Regression and two-proportion t-tests.

RESULTS: Of all jockeys, it is amateur jump racing jockeys that fall most frequently in Ireland, France and Britain. Jump racing also has the highest rates of injury/ride amongst both amateur and professional jockeys. Flat racing, however, has the highest rates of injuries/fall (34-44%). While there is a paucity of worldwide equestrian injury data and a lack of uniformity in the data that are available, it would appear that 15% of all injuries in both jump and flat racing populations of amateur and professional jockeys are concussive and more than half of these involve loss of consciousness (LOC).

DISCUSSION: Differences in fall rates between amateur jump jockeys from different countries could be due to different racing environments, such as the presence of jumps, field conditions and racing speed. Why there are relatively high incidences of concussion in equestrian racing is not clearly known. The prevalence of mildly concussive injuries to professional jockeys and the high chances of LOC in the event of a concussion suggest that the performance of current helmet designs (BSI 1997) would merit future investigation.

CONCLUSION: This study will help to motivate future efforts into designs for personal protective equipment including helmets, reconstructions of documented falls involving injury, and identifying fundamental differences between horse racing types and jurisdictions.

REFERENCES:

Balendra G, Turner M, McCrory P, Halley W. (2007). Injuries in amateur horse racing (point to point racing) in Great Britain and Ireland during 1993-2006. *British Journal of Sports Medicine; 41*, 162-6.

McCrory P, Turner M, LeMasson B, Bodere C, Allemandou A. (2006). An analysis of injuries resulting from professional horse racing in France during 1991-2001: A comparison with injuries resulting from professional horse racing in Great Britain during 1992-2001. *British Journal of Sports Medicine; 40*, 614-8.

BSI. EN 1384:1997 (1997). Specification for helmets for equestrian activities. *London, British Standards Distribution.*

Acknowledgement

We would like to thank the Turf Club for financing this project. Dr Benoit LeMasson of France Galop kindly provided the authors with French jockey injury data.

ACUTE EFFECTS OF WHOLE-BODY VIBRATION ON ELASTIC CHARGE TIME IN TRAINED MALE ATHLETES

Laura-Anne M. Furlong and Andrew J. Harrison

Biomechanics Research Unit, University of Limerick, Limerick, Ireland

Whole-body vibration (WBV) has been shown to increase jump height, power and strength but the mechanisms behind these changes are not fully understood. The aim of this study was to investigate the influence of WBV on elastic charge time, a surrogate measure of tendon and aponeurosis stiffness. 7 trained males were exposed to 10 vibrations at 30 Hz ± 4 mm with 60 seconds rest between each exposure. Pre and post-tests were conducted immediately, 5, 10, 15, 20, 30 and 40 minutes following vibration exposure. A significant increase in elastic charge times of both vibrated (p=0.004) and control (p=0.024) limbs suggest whole-body vibration decreases tendon and aponeurosis stiffness, possibly due to a warm-up effect of the lower limbs. Further study of the muscle stiffness response to vibration will improve understanding of the mechanisms behind performance improvements following vibration exposure.

KEYWORDS: electromechanical delay, tendon, aponeurosis, performance enhancement

INTRODUCTION: Whole-body vibration (WBV) is a novel training modality involving vibration of the body on a surface vibrating at a frequency between 20-50 Hz. It has gained popularity in recent years, due to reported acute and long-term improvements in performance. WBV has been shown to improve knee extensor strength and torque (Delecluse et al., 2003; Jacobs and Burns, 2009), jump height (Bosco et al., 1999; Cormie et al., 2006), acceleration, step length, step rate (Paradisis and Zacharogiannis, 2007) and power output during the back squat (Rhea and Kenn, 2009). The mechanisms explaining improvements following vibration exposure are not yet fully understood. The main neuromuscular mechanism stimulated is the tonic vibration reflex, which is a tonic contraction of the muscles as a result of sensitivity of the primary muscle spindle endings to stretch (Griffin, 1990). It is logical to assume however, that WBV also influences other musculoskeletal structures, such as tendon and aponeurosis. Improvements following WBV are particularly evident in outcome variables involving a stretch-shortening cycle (SSC), which has been shown to be related to muscle and tendon stiffness prior to the concentric phase of movement (Anderson, 1996). To date, the influence of WBV on these variables is unknown. Winter and Brookes (1990) proposed the stiffness of the musculotendinous structures can be estimated by determining electromechanical delay (EMD) in a simple heel raise movement. EMD is further subdivided into force development time (FDT), which represents the delay between muscle activation and force registration and elastic charge time (ECT) which describes the delay between the first registration of force and movement of the heel. ECT has therefore been proposed as a surrogate measure of tendon and aponeurosis stiffness (Winter and Brookes, 1991). There is a need to establish whether WBV produces acute changes in ECT of the triceps surae as it may explain more about the mechanism by which WBV affects performance in SSC activities. The purpose of this study was to improve understanding of the structures influenced by WBV. The aim of this study was to determine the acute effects of WBV on ECT in trained male athletes, as a possible mechanism for enhanced performance seen following vibration exposure.

METHODS: Data collection: Following university ethics committee approval, 7 trained males (mean ± SD: age: 22.1 ± 1.4 years; height: 1.84 ± 0.06 m; mass: 87.0 ± 7.5 kg) volunteered as subjects for this study. All competed at an elite/sub-elite level in their sport (sprinting=1, rugby=1, boxing=1, Gaelic football=2, hurling=2) and were injury free for at least 6 weeks prior to testing. Subjects completed a Physical Activity Readiness Questionnaire and signed an informed consent form prior to participation. None had previously used WBV as a regular training modality. ECT was determined using an adapted electromechanical delay technique (Winter and Brookes, 1991). Subjects sat on a plastic chair with the knee flexed at a 90° angle. The ball of the foot rested on a force plate (AMTI Technologies Inc., USA) sampling at 1000 Hz and the heel rested on a footswitch. The fixed position of the footswitch ensured the subject consistently placed their foot on the same place on the forceplate. A 3 cm x 6 cm piece of Perspex was taped to the heel to eliminate error from soft tissue movement over calcaneus. Following shaving and preparation of the skin with alcohol wipes (PDI Alcohol Prep Pads, Professional Disposables Inc., USA) two electrodes (Kendall Meditrace 100, Kendall Medical Supplies, USA) were placed on distal soleus, 2 cm apart at a point ⅔ of the distance between the lateral femoral condyle and lateral malleolus. EMG was recorded at 1000 Hz using a Powerlab system (ADI Instruments Powerlab 4/25T). After a '3, 2, 1' countdown, subjects were instructed to raise the heel as fast as possible.

A contra-limb control leg design was used, with the preferred kicking leg being exposed to vibration. Control and vibrated leg performed 5 heel lifts each immediately prior to vibration exposure. Subjects were exposed to 10 vibrations at 30 Hz ± 4 mm with 60 seconds rest between each exposure using a Vibrogym platform (Vibrogym Inc., Haarlem, The Netherlands) located beside the force plate. The knee of the vibrated leg was flexed to 110° and the knee of the control leg was flexed so it did not touch the vibration plate. Markings on the vibration plate ensured consistent foot placement. All vibrations were completed barefoot to remove confounding from footwear or socks. A post-test was conducted immediately after vibration exposure. Post-tests were also completed 5, 10, 15, 20, 30 and 40 minutes post-exposure and involved the vibrated and control leg performing 5 heel lifts each.

The instant of foot plantar flexion force was detected from force platform records and heel movement was detected by the footswitch. ECT was defined as the time interval between the registration of force on the force platform and initial movement of the heel and was determined by visual inspection of graphs in Microsoft Excel 2007 (Microsoft Inc., USA).

Data analysis: All statistical analysis was carried out in SPSS (Version 17, SPSS Inc., USA). A Shapiro-Wilk's test was used to check normality of data sets. ANOVA with repeated measures was used to determine if significant differences existed between pre-vibration ECT and maximum and minimum ECT post-vibration. Alpha was set at $p < 0.05$.

RESULTS: All data satisfied the Shapiro-Wilk test for normality ($p > 0.05$) with the exception of the control limb measures at 40 minutes ($p = 0.025$). Figure 1 presents the results of the effects of WBV on ECT times in trained males in both the vibrated and control leg.

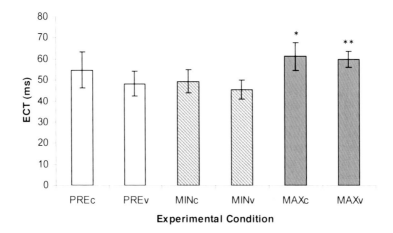

Figure 1: ECT in trained males in both the control and vibrated limb. ** indicates a significant difference compared with pre-test scores ($p < 0.01$); * indicates a significant difference compared with pre-test scores ($p < 0.05$)

DISCUSSION: The results of this experiment showed that following WBV, ECT increased significantly compared with pre-test ECT of both control (p=0.024) and vibrated (p=0.004) limbs. Increased ECT is thought to be indicative of decreased tendon and aponeurosis stiffness following vibration compared to ECT measures prior to vibration exposure, although it should be pointed out that ECT does not provide a direct measure of tendon stiffness and this limitation should be recognised. This provides a further explanation for the mechanisms behind performance improvements observed following WBV. Cochrane et al. (2008) reported an increase in muscle temperature and power output following WBV, suggesting a possible warm-up effect on the musculotendinous structures of the lower limbs which subsequently enhances performance. The warming effect of the test protocol was not measured in this study, but during vibration the subjects involved commented on feelings of increased warmth in the vibrated limb. Sweating was also observed, which suggests the protocol used here may have induced a warm-up effect on the vibrated limb. Due to the systemic nature of the body, this may also explain why a similar decrease in stiffness was seen in the control limb. The violation of the assumption of normality in data related to the control limb 40 minutes following vibration was considered trivial due to the robust nature of the statistical test used, and was not considered to influence overall study results.

CONCLUSIONS: The results of this study suggest WBV decreases tendon and aponeurosis stiffness in trained male athletes, possibly due to a warm-up effect. This may explain, in part, improvements in performance following WBV. Direct measurement of the tendon stiffness response and study of the muscle stiffness response to WBV will further enhance understanding of the mechanisms explaining performance improvements seen following WBV.

REFERENCES:
Anderson, T. (1996). Biomechanics and running economy. *Sports Medicine*, 22, 76-89.
Bosco, C., Colli, R., Introini, E., Cardinale, M., Tsarpela, O., Madella, A., Tihanyi, J. and Viru, A. (1999). Adaptive responses of human skeletal muscle to vibration exposure. *Clinical Physiology*, 19, 183-187.
Cochrane, D.J., Stannard, S.R., Sargeant, A.J. and Rittweger, J. (2008). The rate of muscle temperature increase during acute whole-body vibration exercise. *European Journal of Applied Physiology*, 103, 441-448.
Cormie, P., Deane, R.S., Triplett, N.T. and McBride, J.M. (2006). Acute effects of whole-body vibration on muscle activity, strength and power. *Journal of Strength and Conditioning Research*, 20(2), 257-261.
Delecluse, C., Roelants, M. and Verschueren, S. (2003). Strength increase after whole-body vibration compared with resistance training. *Journal of Strength and Conditioning Research*, 35(6), 1033-1041.
Griffin, M.J. (1990). *Handbook of Human Vibration*. London, UK: Academic Press.
Jacobs, P.L. and Burns, P. (2009). Acute enhancement of lower-extremity dynamic strength and flexibility with whole-body vibration. *Journal of Strength and Conditioning Research*. 23(1), 51-57.
Paradisis, G. and Zacharogiannis, E. (2007). Effects of whole-body vibration training on sprint running kinematics and explosive strength performance. *Journal of Sports Science and Medicine*, 6, 44-49.
Rhea, M.R. and Kenn, J.G. (2009). The effect of acute applications of whole-body vibration on the iTonic platform on subsequent lower-body power output during the back squat. *Journal of Strength and Conditioning Research*, 23(1), 58-61.
Winter, E.M, and Brookes, F.B.C. (1990) Electromechanical response times and muscle elasticity. *Journal of Physiology (London)*, 429, 106.
Winter, E.M., and Brookes, F.B.C, (1991) Electromechanical response times and muscle elasticity in men and women. *European Journal of Applied Physiology*. 63, 124-128.

AN INVESTIGATION INTO THE IMPACT FORCE EXPERIENCED BY DIFFERENT TYPES OF FOOTBALLS

Yo Chen[1], Jia-Hao Chang[1], Tai-Yen Hsu[2]

[1]Department of Physical Education, National Taiwan Normal University, Taipei, Taiwan
[2]Department of Physical Education, National Taichung University, Taichung, Taiwan

KEY WORDS: soccer, ball velocity, impact.

INTRODUCTION: The impact force of kicking varies with different materials and the estimated force in powerful (maximal velocity) instep kick was 1100N (Tsaousidis and Zatsiorsky, 1996). The force may cause injuries and stress accumulated on the foot especially in novices, due to unfamiliar skill. Rubber is durable, cheap but stiffer; TPU (Thermoplastic Polyurethane) and PU (Polyurethane) material has higher elasticity and impact absorbability. The purpose of this study was to identify the impact force, max velocity and travelling distance with different material footballs.

METHODS: Four footballs (inflated pressure 0.6 bar), molten FVA 5000 (PU), Adidas Final 8 (PU), Adidas Sportivo (TPU), and Spalding 61-731 (Rubber), were used. Impact force was measured by 2.5 m ball drop test on KISTLER 9187 force plane (90*60 cm^2, 1000 Hz, Unit: kg), each drop test was proceeded 3 trials. Two elite football players (subject A: 175 cm height and 80 kg weight; subject B: 172 cm height and 72 kg weight) without lower extremity injuries were recruited. A radar gun (sampling at 300 Hz) was used to measure peak ball velocity with powerful instep kick, and a measuring-wheel were used to measure ball travelling distance with powerful goal kick. For the reliability and repeatability each kicking task was repeated 3 times in the windless environment, and the kicking data required to exceed 90 $km.h^{-1}$ were identified.

RESULTS: The impact force, velocity and travelling distance of the footballs are shown in Table 1. Adidas Final 8 is a FIFA approved ball for UEFA Champions cup, and it has lower impact force and higher velocity and travelling distance in this experiment. In the drop test, ball and force plane contact time was 7-10 $m.s^{-1}$, and the impact force greater than 1000 N was maintained 2 $m.s^{-1}$.

Table 1. The impact force, velocity and travelling distance

	Molten FVA5000	Adidas Final 8	Adidas Sportivo	Spalding 61-731
Impact force (N)	1068	1058	1077	1145
Velocity (km/h)	94(5.0)	100(3.2)	98(4.2)	98(4.6)
Traveling distance (m)	49(2.9)	50(2.3)	48(2.8)	47(3.2)

DISCUSSION & CONCLUSION: The result 94 to 100 $km.h^{-1}$ ball speed is close to related research of elite player (Nunome et al.,2006). Adidas Final 8 had lower impact force and higher ball velocity and travelling distance in the test, and Sportivo had similar data with Final 8. The higher impact force was presented at Spalding rubber ball, TPU and PU material had lower impact force. Rubber balls are durable, but it is not appropriate in powerful kick or heading task for novices. Non-FIFA approved Adidas TPU ball also reach the standard level, it is suit to practice and competition.

REFERENCES:

Nunome H., Lake M., Georgakis A. and Stergioulas L. (2006). Impact phase kinematics of instep kicking in soccer. *Journal of Sports Sciences*, 24,11-22.

Tsaousidis N. and Zatsiorsky V. (1996). Two types of ball-effector interaction and their relative contribution to soccer kicking. *Human Movement Science*, 15, 861-876.

THE SYMMETRY IN GAIT KINEMATICS OF ADOLESCENTS' CASE STUDIES

Matilde Espinosa-Sánchez

Unidad de Investigación en Cómputo Aplicado,
Universidad Nacional Autónoma de México.
Zona Cultural; Ciudad Universitaria; México, D. F. 04510

A gait cycle of 20 adolescents has been analyzed, representing each of them an age-gender group within a period of five years. The basic step values, the flexion-extension angle values between trunk and thighs (hips), thighs and calves (knees), calves and feet (ankles), and the angle between the horizontal and thighs were calculated. The symmetric movements of the lower limbs of each individual can be observed by means of angle-angle diagrams and the statistical results indicate the significant differences between right and left joints. After the analysis it is possible to infer that the symmetry in walking is developed during the latest years of the adolescence, but there are significant differences between right and left angular displacements.

KEY WORDS: gait symmetry, adolescents, angular displacement

INTRODUCTION: During the childhood period the abilities of the movement depend on the development and the maturation of muscles, bones and nerves. The physical activities of the developing child influence the size, the form and the alignment of the muscle skeletal system (Skinner, 1994). Much of the human variability appears to be related to the answers to the stressors, and there are diverse answers: cultural, growth and of development. Also there are variations because the bilateral symmetry in the structure of the human body. The variation in the left and right side is random and also independent. There exists variability in the time required for the maturation of the individual's biological systems; the individual grows while she or he matures (Malina, 1991). The fast growth during adolescence is a phenomenon in all the individuals, and varies in intensity and duration from a child to another one (Tanner, 1966). The commonly accepted duration of adolescence is almost five years (Comas, 1976; Stang, 2008).

The case study, this is a single subject study, is a necessary and sufficient method for certain important tasks of investigation (Flyvbjerg, 2006). This means investigation of single case study or multiple case studies, that can include quantitative evidences are trustworthy sources of multiple evidences and are beneficiaries of previous theoretical development (Yin, 2002). It is not desirable to summarize and to generalize case studies but, it must be emphasized that the method contributes to the knowledge development (Flyvbjerg, 2006).

A comfortable speed means that the rate of movement is free, that each individual walked at the speed that he or she wants (Kerrigan et al., 1998). It has been studied that free walking (comfortable speed) was far less variable than forced walking varied with speed. It has been reported that when walking at freely chosen step rate, there is an invariant relationship between step length and step rate regardless of walking speed. Many observations have confirmed that the rhythmic stage of human walking is very consistent and is directly related to the optimal efficiency for an individual (Sutherland et al., 1994).

When walking and/or running the knee flexion-extension is related to the thigh rotation (forward and backward) of the same lower member, by means of angle-angle diagrams (Cavanagh and Grieve, 1973M; Enoka et al., 1982). This graphical representation is sensitive to details of movement and useful to show the range of movement magnitude (ROM), the differences between the right step and the left step, and therefore, these graphs indicate the magnitude of the individuals' gait asymmetry.

The calculations of the three-dimensional kinematics, by means of photogrametric and videogrametric procedures, need at least two simultaneous recordings in video from different views. Additionally, the DLT procedures could be used for the individuals' movement 3D reconstruction in a computer (Woltring, 1980; Shapiro, 1978; Espinosa, 1999). Commonly the human body representation with 14 linear segments is the Chandler inertial model (Chandler et al., 1975). The movement sequence analysis is for a gait cycle considered of the double support and the single support or oscillation phases (Vaughan et al., 1992).

There is the necessity to know how is developed the kinematics of walking while the adolescence period passes. The objective of this research has been to find in the analyzed sample, the asymmetries and differences in the angular displacements of the gait. This work can be useful in the considerations that the physical educator, the sport trainer and/or the physician, could do with respect to their criteria of education and diagnosis.

METHODS: Participants. This study is a cross-sectional one carried out in Mexico City with a population belonging to the middle socioeconomic layer. The sample of analyzed individuals (Table 1) is constituted by 20 case studies representing each of them an age-gender group within a period of five years that is the accepted duration of adolescence. **Data collection.** Three camcorders (Panasonic AG-EZ30 DV) were installed in a tripod separated to each other by an angle of 120 degrees, each with its focus line aiming towards the same target at a distance of 10 m that is the center of the path by which the adolescents walked. The individuals were asked to walk barefoot at a comfortable speed through the path of about 10 m long and 1.5 m wide, the more centered gait cycle was chosen. The source data were the coordinates of 19 anatomical points that were located and extracted from the video images. No markers were used to locate the anatomical points. The 3D data was approximated to 120 calculated frames per second by cubic beta-spline interpolation. **Data analysis.** The 3D angular displacements and the basic values of walking of the 20 individuals were calculated by means of computer programs developed for this purpose. The differences of the relation of the knee flexion-extension and the thigh rotation were observed in angle-angle diagrams. The Student's t-test was used to compare the performance between right and left lower limbs.

Table 1. Age and gender groups and individual cases.

FEMALE	GROUP	10f	105f	11f	115f	12f	125f	13f	135f	14f	145f
AGE	[years]	10.23	10.59	11.05	11.68	12.11	12.58	12.95	13.73	13.85	14.52
MALE	GROUP	12m	125m	13m	135m	14m	145m	15m	155m	16m	165m
AGE	[years]	11.97	12.66	13	13.73	14.09	14.34	15.06	15.62	15.83	16.72

RESULTS AND DISCUSSION: The basic step values: duration, length, rate and velocity, were calculated and are shown in Table 2. In average the male individuals use a more time to execute a gait cycle, they have a longer step length, and their step rate and velocity are slower. The differences between the female and male individuals and between the right and the left side are a few hundredths for most of the cases.

The angle-angle diagrams indicate how much symmetric are the knee-thigh relations, right side and left side. In Figure 1 the differences can be identified visually, the figure shows that the most marked asymmetries are those of the youngest individuals: female 10f, 105f, 11f and 115f, and male 12m, 125m, 135m and 145m. Although, making an only visual consideration it is possible to say that the knee flexion-extension and the thigh rotation ranks of movement are analogous in both genders.

The Student's t-test ($p<0.05$) was applied to analyze the differences in the angular displacement between right joints and left joints, and both thighs inclination of each one of the 20 individuals. There are significant differences at a 0.05 level of significance if the t* score of the statistic student lies outside the range -1.96 to 1.96. The Table 3 shows for the female individuals: the 100% of hips, the 70% of knees and of ankles, and the 20% of thighs have differences; for the male individuals: the 100% of hips, the 60% of knees, the 90% of ankles, and the 20% of thighs have differences.

Table 2. Right and left step duration, length, rate and velocity for both female and male individuals.

	ID	10f	105f	11f	115f	12f	125f	13f	135f	14f	145f
Right step	t [s]	0.53	0.49	0.42	0.4	0.48	0.41	0.48	0.44	0.55	0.46
	length [m]	0.69	0.64	0.82	0.67	0.87	0.72	0.62	0.7	0.62	0.66
	rate [steps/s]	1.89	2.04	2.38	2.5	2.08	2.44	2.08	2.27	1.82	2.17
	v [m/s]	1.3	1.31	1.95	1.68	1.81	1.76	1.29	1.59	1.13	1.43
Left step	t [s]	0.52	0.47	0.41	0.38	0.46	0.4	0.45	0.42	0.52	0.49
	length [m]	0.62	0.62	0.8	0.68	0.75	0.65	0.65	0.72	0.6	0.66
	rate [steps/s]	1.92	2.13	2.44	2.63	2.17	2.5	2.22	2.38	1.92	2.04
	v [m/s]	1.19	1.32	1.95	1.79	1.63	1.63	1.44	1.71	1.15	1.35
	ID	12m	125m	13m	135m	14m	145m	15m	155m	16m	165m
Right step	t [s]	0.49	0.55	0.6	0.44	0.43	0.59	0.5	0.48	0.5	0.4
	length [m]	0.7	0.65	0.76	0.7	0.8	0.65	0.7	0.74	0.7	0.79
	rate [steps/s]	2.04	1.82	1.67	2.27	2.33	1.69	2	2.08	2	2.5
	v [m/s]	1.43	1.18	1.27	1.59	1.86	1.1	1.4	1.54	1.4	1.98
Left step	t [s]	0.47	0.59	0.49	0.44	0.41	0.62	0.44	0.47	0.51	0.4
	length [m]	0.66	0.67	0.68	0.68	0.84	0.6	0.62	0.74	0.61	0.7
	rate [steps/s]	2.13	1.69	2.04	2.27	2.44	1.61	2.27	2.13	1.96	2.5
	v [m/s]	1.4	1.14	1.39	1.55	2.05	0.97	1.41	1.57	1.2	1.75

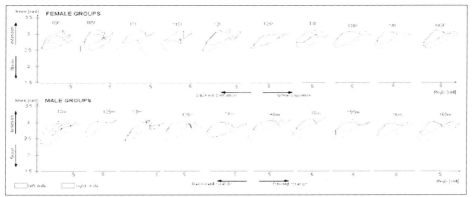

Figure 1. Angle – angle diagrams. The relation between the knee flexion - extension and the forwards and backwards thigh rotation.

Table 3. The Student's t-test statistic comparing right and left joints

Joints	10f	105f	11f	115f	12f	125f	13f	135f	14f	145f
HIPS	22.1*	-11.6*	-20.8*	-14.2*	-6.6*	-5.2*	-20.2*	-12.6*	-7.9*	-8.0*
KNEES	-7.9*	-6.4*	-21.7*	-12.6*	-3.8*	2.3*	-4.4*	-0.1*	-1.9*	-1.9*
ANKLES	-1.4*	3.7*	11.7*	-0.6*	0.8*	19.2*	8.7*	13.2*	3.9*	3.0*
THIGHS	1.1*	1.1*	3.6*	3.8*	1.8*	-0.6*	-0.4*	-0.7*	0.8*	0.4*
Joints	12m	125m	13m	135m	14m	145m	15m	155m	16m	165m
HIPS	-8.4*	-16.0*	2.1*	-13.6*	-4.3*	-16.7*	-14.1*	-8.4*	-11.5*	-11.2*
KNEES	-7.1*	-4.7*	-1.9*	-4.1*	-0.4*	-4.3*	-4.4*	1.1*	-1.9*	-3.7*
ANKLES	-3.9*	9.0*	0.5*	14.8*	12.1*	8.3*	10.5*	7.3*	9.5*	7.2*
THIGHS	3.1*	-1.7*	0.8*	-2.1*	-1.4*	0.3*	-0.1*	-0.9*	0.3*	0.3*

*$p<0.05$

CONCLUSION: The results of this study indicate that during the adolescence period, there are slightly differences in the basic step values between genders. The ranges of movement in the knee flexion-extension and thigh rotation relation, inferred that the symmetry in walking is developed during the latest years of the adolescence. Comparing the joints of the lower limbs, there are about 70% of significant differences between the right angular displacements and the left angular displacements.

It will be necessary to analyze a significant group of individuals for each age and gender group in order to know if the angular kinematic parameters provide the sufficient information to establish behavior patterns.

REFERENCES
Cavanagh,P.R. and Grieve,D.W. (1973). The graphical display of angular movement of the body. *Br. J. Sports Med*, 7, 121-136.
Chandler, R.F., Clauser, C.E., McConville, J.T., Raynolds, H.M., Young, J.W. (1975). *Investigation of inertial properties of the human body*. Final report: AMRL-TR-74-137. Aerospace Medical Research Laboratory. Wright-Patterson Air Force Base, Ohio. 162 p.
Comas, J. (1976). *Manual de antropología física*. Instituto de Investigaciones Antropológicas, Publishers. Universidad Nacional Autónoma de México. 652 p.
Enoka, R.M., Miller, D.I., Burgess, E.M. (1982). Below-Knee AmputeeRunning Gait. *American Journal of Physical Medicine*, 61, 70.
Espinosa-Sánchez, M. (1999). A3D coordinate approach to walking analysis. *Proceedings of the XVII International Symposium on Biomechanics in Sports*, Edith Cowan University, School of Biomedical and Sports Science Publisher, Australia (pp. 155-158).
Flyvbjerg, B. (2006), Five Misunderstandings About Case Study Research. *Qualitative Inquiry*, 12 (2), 219-245. .
Kerrigan, D.C., Todd, M.K., Della Croce, U., Lipsitz, L.A., Collins, J.J. (1998). Biomechanical gait alternatives independent of Speedy in the healthy elderly: evidence for specific limiting impairments. *Archives of Physical Medicine and Rehabilitation* 79(3), 317-322.
Malina, R.M., Bouchard, C. (1991). *Growth, Maturation and Physical Activity*. Human Kinetics.
Shapiro, R. (1978). Direct linear transformation method for three-dimensional cinematography. *Research Quarterly*, 49, 197-205.
Skinner, S. (1994). Chapter 6: Development in gait. In *Human walking*. Edited by Rose J, Gamble JG. Second Edition. Baltimore: Williams & Willins. p. 123-38.
Stang, J. and Story, M, (Eds) (2008). Guidelines for Adolescent Nutrition Services
http://www.epi.umn.edu/let/pubs/adol_book.shtm, Chapter 1 *Adolescent Growth and Development* Regents of the University of Minnesota, mar/2009
Sutherland, D.H., Kaufman, K.R., Moitoza, J.R. (1994). Kinematics of Normal Human Walking. *In Human Walking*. Edited by Rose J, Gamble JG. Second Edition. Baltimore: Williams and Wilkins.
Tanner, J.M. (1966). *Educación y desarrollo físico*. Siglo XXI Editores, S.A. México.
Vaughan, C.L., Davis, B.L., O´Connor, J.C. (1992). *Dynamics of human gait*. Human Kinetics.
Woltring, H.J. (1980). Planar control in multi-camera calibration for 3-D gait studies. *Journal of Biomechanics*, 13, 39-48.
Yin, R. (2002). *Case Study Research. Design and Methods*. Third Edition. Applied social research method series Volume 5. Sage Publications. California.

Acknowledgement
The author would like to thank my former tutor Johanna Faulhaber for her confidence in my effort.

ANKLE JOINT LOADING DURING THE DELIVERY STRIDE IN CRICKET MEDIUM- FAST BOWLING

Kathleen Shorter, Neal Smith, and Mike Lauder
University of Chichester, Chichester, UK

To date, biomechanical research investigating the aetiology of cricket injuries has studied the kinetics and kinematics of associated movements in isolation. The aim of this study was to apply inverse dynamics to investigate ankle joint forces during the delivery stride using four Basler 200 Hz cameras synchronised to two Kistler 9581B force plates with Peak Motus 9.2. Although peak ankle joint moment in the sagittal plane was greater for the front foot (mean: 3.21 ± 1.71 Nm·Kg^{-1}) in relation to the back foot (mean: 1.70 ± 0.87 Nm·Kg^{-1}); average rate of joint loading was 246% greater in the frontal plane for the back foot (mean: 1.11 ± 0.82 Nm·Kg^{-1}·s^{-1}) compared to the front foot (mean: 0.45 ± 0.20 Nm·Kg^{-1}·s^{-1}). Findings would suggest that whilst the front foot is prone to acute injuries, the back foot may be more susceptible to overuse injuries such as lateral ankle instability.

KEY WORDS: kinematics, force, joint moment, injury

INTRODUCTION: Cricket injury epidemiology studies have established 11% of injuries afflicting fast bowlers involve the foot and ankle with no distinction made between the front and back foot (Orchard et al., 2002). Smith (1999) reported a high incidence of posterior talar impingement afflicting elite South African bowlers, which was associated with rapid force application and ankle plantar flexion during front foot impact. Hurrion et al. (2000) established that impact forces during the delivery stride are in the region of 2.37 times body weight (BW) vertically and 0.94 BW horizontally for the back foot and 5.7 BW vertically and 3.5 BW horizontally during front foot impact, which may be modified by footwear (Shorter et al., 2008). Whilst it is assumed that the front foot would be at greater risk of injury due to larger impact forces, contrasting foot movement during the delivery stride as depicted in Figure 1 would suggest that the nature and susceptibility of injuries may vary between the back and front foot.

Back Foot

Front Foot

Figure 1: Pictorial description of back and front foot contact from heel strike to toe-off during the delivery stride

Cricket studies investigating the aetiology of injuries have reported the kinematics and kinetics of the delivery stride in isolation. Whilst informative, the multi-factorial nature of injuries supports the need to investigate the causation of injuries in a similar manner. The susceptibility of a joint to injury is dependent on both the forces acting upon it, combined with the position of the joint. The aim of this study was to apply inverse dynamics as described by

Winter et al. (1990), to quantify the ankle joint forces experienced during the delivery stride to provide greater insight into the pathomechanics of associated ankle injuries.

METHODS: After gaining University ethical approval eight medium-fast amateur bowlers of mixed ability and technique (age: 22.38 ± 2.50 years, height: 1.80 ± 0.06 m, mass: 83.03 ± 16.10 Kg) were recruited and provided informed consent. All bowlers wore standardised footwear with which they had become habituated.
Testing was conducted indoors with adequate movement space to allow for a standard run-up. Smartspeed light gates (Fusion Sport, Queensland, Australia) were set two metres apart, two metres from the popping crease to monitor approach velocity prior to the delivery stride. Foot impact during the delivery stride was captured using two Kistler 9581B force plates (600 Hz) synchronised with four 200 Hz Basler digital cameras using Peak Motus 9.2 software. Lower limb kinematics during the delivery stride were reconstructed using the CAST technique (Cappozzo et al, 1995). For each limb eight retroflective markers: femoral epicondyle, tibial cluster comprised of three markers, lateral malleolus, posterior calcaneous, 2nd metatarsal head and 5th metatarsal head, were used to define the shank and foot segments with a marker subset used whilst bowling.
Prior to data collection, participants underwent a self-selected warm-up. Participants were required to bowl 10 successful balls for the foot under investigation. A successful ball was classified as one when full foot interaction with the force plate occurred whilst also complying with the laws of cricket as determined by the researcher.
All trials were digitised using Peak Motus 9.2 software, with kinematic data filtered using quintic spine processing. Kinematic and kinetic data were extrapolated to 200 Hz and exported into Visual3D for analysis. Foot contact for each foot was defined from the first instant vertical force exceeded 20 N, with toe off subsequently defined by the moment vertical force equated zero. Ground reaction force (GRF) data were normalised to participant body weight and analysed to determine peak force in both vertical and horizontal planes. Ankle joint kinetics of the back and front foot during the delivery stride were defined by peak and average joint moments in each of the cardinal planes and normalised to body weight. To further describe the forces acting on the ankle joint throughout foot contact, the average rate of joint loading in each plane was calculated by dividing the average rectified joint moment by the duration of foot contact.

RESULTS AND DISCUSSION: Similar to the findings of Hurrion et al. (2000), ground kinetics between the front and back foot during the delivery stride differed. Figure 2 depicts a representative combined back and front foot GRF trace during the delivery stride. Peak GRFs for the front foot (3.02 ± 0.87 BW vertically, 1.52 ± 0.48 BW horizontally) were greater in magnitude compared to the back foot (2.16 ± 1.43 BW vertically, 0.50 ± 1.12 BW horizontally). Disparity in peak force generation may partly be attributed to prolonged front foot contact (0.50 ± 0.18 s), which was 1.67 times longer in duration compared to the back foot (0.30 ± 0.05 s). Duration of foot contact during the delivery stride between feet has not been previously reported, however the role of the front foot to provide a stable base of support during ball release may necessitate prolonged foot contact.
Ankle joint moments for both the back and front foot during the delivery stride were found to closely reflect the typical movement pattern associated with pace bowling (refer to Figure 1). Figure 3 depicts a representative combined back and front foot moment-time trace where differences in joint kinetics were evident reflecting the different functional requirements of each foot.

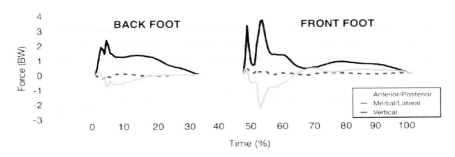

Figure 2: Representative combined back and front foot GRF trace during the delivery stride

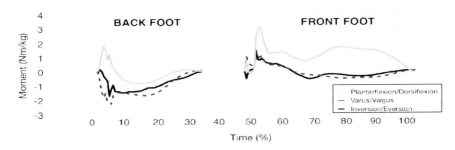

Figure 3: Representative combined back and front foot ankle joint moment trace during the delivery stride

Back foot contact was typified by a lateral forefoot strike corresponding with peak ankle joint moments. The peak eversion moment (mean: 0.96 ± 0.74 Nm·Kg^{-1}) closely coincided with the peak plantarflexion moment (mean: 1.70 ± 0.87 Nm·Kg^{-1}), in order to stabilise the foot through resisting ankle inversion. As foot contact progressed, sagittal plane torque decreased into dorsiflexion when the centre of pressure shifted towards the heel. Throughout foot contact a valgus moment (mean peak: 1.68 ± 1.11 Nm·Kg^{-1}) was evident due to the perpendicular position of the foot in relation to the body, combined with the upper body pivoting around the lower limb in preparation for ball release. As toe off commenced, sagittal plane torque began to increase but remained minimal as the bowler relies on momentum gained during the run-up to terminate foot contact rather than actively pushing off.

Ankle joint moments during front foot contact highlight the largely planar motion of the foot, with the peak moment in the sagittal plane (mean: 3.21 ± 1.71 Nm·Kg^{-1}) far greater in magnitude compared to those generated in both the frontal (mean: 2.37 ± 1.63 Nm·Kg^{-1}) and transverse (mean: 0.98 ± 0.36 Nm·Kg^{-1}) planes. Similar to the back foot, peak joint moments were found to closely coincide with heel contact. The main ankle moment experienced by the front foot during the delivery stride occurred in the sagittal plane, reflective of the functional demands placed upon it. Following heel strike, the sagittal plane moment decreased during foot flat but then rapidly ascended into plantarflexion as the bowler actively pushed off in order to increase the height of ball release. After ball release, ankle joint moments stabilised, reflective of the need for the front foot to provide a base of support during the remainder of the bowling action.

Disparity in ankle joint moments between the back and front foot were established (Table 1). Whilst peak ankle joint moments followed the trend of reported ground kinetics, with the front foot exhibiting greater magnitudes in comparison to the back foot, this trend did not pertain to both average joint moments and average rate of joint loading. Peak moments such as the large peak plantarflexion moment observed with the front foot may be associated with acute injury formation such as that reported by Smith (1999). Whilst reporting of peak values aids in the understanding of acute injuries, both average moments and average rate of joint loading provide greater insight into the overall strain placed on the ankle joint throughout foot contact. Average rate of joint loading in both the sagittal and transverse planes were similar between the front and back foot, however, loading in the frontal plane for the back foot (mean: 1.11 ± 0.82 $Nm \cdot Kg^{-1} \cdot s^{-1}$) was 246% greater compared to the front foot (mean: 0.45 ± 0.20 $Nm \cdot Kg^{-1} \cdot s^{-1}$). High frontal plane loading when combined with loading in the other cardinal planes places strain on the supportive structures of the ankle as they must stabilise the ankle from multidirectional forces. Findings from this study would suggest that whilst the front foot is prone to acute injuries, the back foot may be more susceptible to overuse injuries such as lateral ankle instability.

Table 1: Ankle joint kinetics (mean ± SD) during the delivery stride

Joint Kinetics	Back Foot	Front Foot
Peak joint moment ($Nm \cdot Kg^{-1}$)		
Plantarflexion/Dorsiflexion	1.70 ± 0.87	3.21 ± 1.71
Varus/Valgus	1.68 ± 1.11	2.37 ± 1.63
Inversion/Eversion	0.96 ± 0.74	0.98 ± 0.36
Average joint moment ($Nm \cdot Kg^{-1}$)		
Plantarflexion/Dorsiflexion	0.43 ± 0.29	1.10 ± 0.47
Varus/Valgus	0.55 ± 0.34	0.79 ± 0.54
Inversion/Eversion	0.32 ± 0.22	0.21 ± 0.08
Rate of joint loading ($Nm \cdot Kg^{-1} \cdot s^{-1}$)		
Plantarflexion/Dorsiflexion	1.64 ± 1.40	2.28 ± 1.05
Varus/Valgus	2.04 ± 1.50	1.72 ± 1.29
Inversion/Eversion	1.11 ± 0.82	0.45 ± 0.20

CONCLUSION: The aim of this study was to quantify ankle joint moments experienced by the front and back foot during the delivery stride to provide greater insight into the aetiology of associated ankle injuries. High peak sagittal moments indicate that the front foot is more prone to acute injuries, whilst the back foot may be more susceptible to overuse injuries due to the multidirectional forces placed upon it throughout foot contact. Findings from this study suggest that rather than utilising kinematic and kinetic data in isolation to investigate pathomechanics of injury, methodology such as inverse dynamics can be successfully applied to provide a greater holistic understanding.

REFERENCES:

Cappozzo, A., Catani, F., Croce, U. D., & Leardini, A. (1995). Position and orientation in space of bones during movement: Anatomical frame definition and determination. *Clinical Biomechanics*, 10, 171-178.

Hurrion, P., Dyson, R., Hale, T. (2000). Simultaneous measurement of back and front foot ground reaction forces during the same delivery stride of the fast-medium bowler. *Journal of Sports Sciences*, 18, 993-997.

Orchard, J., James, T., Alcott, E., Carter, S., & Farhart, P. (2002). Injuries in Australian cricket at first class level 1995/1996 to 2000/2001. *British Journal of Sports Medicine*, 36, 270-5

Shorter, K., Smith, N., Dyson, R. (2008). Kinetic and kinematic changes associated with altering heel height during cricket medium-fast cricket bowling. Proceedings of International Society for Biomechanics in Sports. p. 279.

Smith, C. (1999). Ankle injuries in fast bowlers: posterior talar impingement syndrome: the South African experience. *5th IOC World Congress on Sport Sciences*, p. 245. Canberra, ACT: Sports Medicine Australia.

Winter, D. (1990). *Biomechanics and motor control of human movement*. John Wiley and Sons, New York, NY.

EMG ANALYSIS OF THE LOWER EXTREMITY BETWEEN VARYING STANCE SQUAT WIDTHS IN BASEBALL CATCHER THROWING

Yi-Chien Peng[1], Kuo-Cheng Lo[2], Hwai-Ting Lin[3], and Lin-Hwa Wang[1]

Institute of Physical Education, Health & Leisure, National Cheng Kung University, Tainan, TAIWAN[1]
Physical Education Office, Kun Shan University, Tainan, TAIWAN[2]
Faculty of Sports Medicine, Kaohsiung Medical University, Kaohsiung, TAIWAN[3]

KEY WORDS: baseball athlete, squat position, electromyographic analysis

INTRODUCTION: In baseball, a catcher has to squat for long periods of time and complete ball throwing using a squat stance. Individual difference is often observed in each catcher's squat posture, and stance squat width, which affects the catcher's performances and the muscle activity of lower limbs, has seldom been studied. The purpose of this study will be to investigate muscle activity in lower extremity when throwing is performed with varying stance squat widths.

METHODS: Six collegiate male catchers who had catching experience for more than ten years participated in this study. Surface EMG system (1000 Hz) and an eight-camera Eagle System (500 Hz) were used. The EMG system was also used synchronously to record the muscle activity of 8 muscles (bilateral tibialis anterior (TA), bilateral gastronemius medialis (GM), bilateral rectus femoris (RF) and bilateral biceps femoris (BF).

RESULTS:

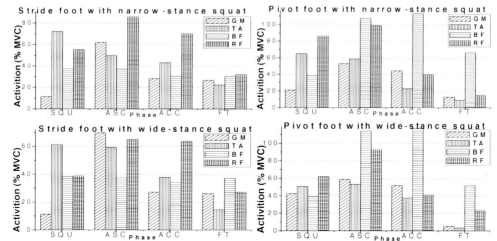

Figure 1: The muscle activity of stride and pivot foot between varying stance squat widths. (SQU:Squat, ASC:Ascending, ACC:Acceleration, FT:Follow-throught)

DISCUSSION: The results of wide-stance squat shown in figure 1 indicated that the wide-stance squat probably reduced the activity of RF in squat phase. Muscle activity of RF and BF in ascending and acceleration phases of throwing were different between stride and pivot foot.

CONCLUSION: The data provided in the article indicate that the using of varying stance squat widths has had a significantly effect on the activity of RF.

REFERENCES:
Escamilla, R. F., Fleisig, G. S., Lowry, T. M., Barrentine, S. W., & Andrews, J. R. (2001). A three-dimensional biomechanical analysis of the squat during varying stance widths. *Med Sci Sports Exerc, 33*(6), 984-998.

RATE FORCE DEVELOPMENT DURING BENCH PRESS IS ONLY RELATED TO THROWING VELOCITY WHEN USING LIGHT LOADS

Daniel A. Marinho[1,2], Ricardo Ferraz[1,2], Roland Tillaar[2], Aldo M. Costa[1,2], Victor M. Reis[2,3], António J. Silva[2,3], Juan J. González-Badillo[4], Mário C. Marques[1,2]

Sport Sciences Department, University of Beira Interior, Covilhã, Portugal[1]
Research Centre in Sports, Health and Human Development, CIDESD, Portugal[2]
Sport Department, University of Trás-os-Montes and Alto Douro, Vila Real, Portugal[3]
Sport and Informatics Depart., University Pablo de Olavide, Seville, Spain[4]

KEY WORDS: team handball, training, performance.

INTRODUCTION: The bench press is a widely used movement to develop strength and power in the upper-body in team handball players. Although bench press has been extensively used, data about kinematics factors in light free weights is limited (Marques & González-Badillo, 2006). Few studies have examined the relationship between ball throwing performance in elite team handball players with power or rate of force development, and bar velocity during muscle contractions of the upper-extremity in concentric only bench press exercise. The aim of this study was to examine the relationship between ball throwing velocity during a 3-step running throw and strength parameters in each force-time curve against three different light free weights.

METHODS: 13 semi professional team handball players (age 27.7 ± 3.5 yr, mass 83.1 ± 11.3 kg, height 1.82 ± 0.09 m; mean ± SD) were measured during a concentric only bench press test with 25, 35, and 45kg (40%, 55%, and 70% of the group mean 1RM). A linear transducer (Isocontrol, JLML, Madrid, Spain) attached to the barbell was used to sample displacement-time during the attempts. The ball throwing velocity was determined using a Doppler radar gun (Sports Radar 3300, Sports Electronics Inc.). These tests were performed during the in-season.

RESULTS: The results of this study clearly indicated significant associations between ball velocity and time at maximum rate of force development (r= 0.66; p≤0.05) and rate of force development at peak force (r= 0.56; p≤0.05) only with 25kg load.

DISCUSSION: This research indicated that ball velocity was moderately associated with maximum rate of force development when light loads were used. Indeed, the analysis of the rate force development during these strength tasks was demonstrated, as suggested by others (e.g. Wilson et al., 1995). Nevertheless, few data has been focus on this subject in specific throwing performance.

CONCLUSION: A training regimen designed to improve ball-throwing velocity in elite male team handball players should emphasize bench press movement using light loads.These findings should be interpreted with caution since correlations provide only associations and do not represent causation, so additional research is required to elucidate if improvements in upper body strength as a result of resistance and/or plyometric training will indeed improve maximal throwing velocity in elite team handball athletes.

REFERENCES:

Marques, M.C. & Gonzalez-Badillo JJ. (2006). In-season resistance training and detraining in professional team handball players. Journal of Strength and Conditioning Research, 20, 563-571.

Wilson, G.J., Lyttle, A.D., Ostrowski, K.J. & Murphy, A.J. (1995). Assessing dynamic performances: A comparison of rate force development tests. Journal of Strength and Conditioning Research, 9(3), 176-181.

WHAT ARE VALUES OF SHOELACES IN RUNNING?

Jing Xian Li[1], Lin Wang[2], Youlian Hong[2]
[1]School of Human Kinetics, University of Ottawa; [2]Department of Sports Science and Physical Education, The Chinese University of Hong Kong[2]

KEY WORDS: planter pressure distribution, rearfoot motion, perceived comfort

INTRODUCTION: Shoelaces are widely thought to make running shoes more comfortable and well-fitted. Studies on the effects of lacing patterns of running shoes on plantar pressure distribution, shock attenuation, and rearfoot motion found that shoe lacing patterns had a remarkable influence on foot-shoe coupling in running (Hagen & Henning, 2008). Their results showed that tighter shoelaces or lacing these closer to the ankle joint resulted in better use of running shoe features. Recently, a type of no-lace running shoes, in which an elastic material is used to replace the traditional lace structure, has been designed by a shoe company. However, there is no available information on the effects of shoes with or without lace structures on foot biomechanics during running. As such, the purpose of this study was to examine if there were differences between running shoes with and without the lace system on perceived comfort, plantar pressure distribution, and rearfoot motion control in running.

METHODS: 15 male experienced runners of 21.44±1.97 years old (60.70±4.76kg body weight, 1.69±0.03m body height) participated in the study. Two pairs of running shoes with the same design, one with the lace structure and another without it, were provided to each participant. For the comfort evaluation, questionnaires containing nine questions and featuring the visual analogue scale were completed by the participants after running 450m on a track at their preferred running speed. The rearfoot movement was filmed using a video camera (9800, JVC Inc. Japan) positioned posteriorly to the participant, with a sampling frequency of 200Hz when the participants ran on the treadmill at 3.8m/s. The APAS system was used in analyzing the video images and calculating the maximum rearfoot pronation angle. Plantar pressure signals during treadmill running were recorded and analyzed by an in-shoe force sensor system (Novel Pedar System, Germany). Paired t-tests were employed to determine if there were any differences in the measurements between the shoes with and without laces. Statistical significance was set at $p<0.05$.

RESULTS: A comparison of the measurements between two shoe conditions revealed that in the condition of the shoes without laces, perceived comfort ratings in shoe length, width, heel cup fitting, and forefoot cushioning were significantly lower. The peak plantar pressures on the lateral side of the midfoot and the 3rd to 5th metatarsal heads were higher, while the contact areas of the 3rd to 5th toes to the insole were significantly larger (204.83±46.58 Kpa vs 223.52±53.38Kpa, p=0.005). Maximum rearfoot pronation angle during running wearing shoes without laces was significantly larger than the conditions in which shoes with laces were wore (13.47±2.68° vs 14.15±3.40°, p=0.013).

DISCUSSION: The study demonstrated that shoelaces make shoes more comfortable and well-fitted during running. The evidences of higher pressure distribution on the lateral side of the midfoot and the 3rd to 5th metatarsal heads in the shoes without laces coincided with the results of comfortable evaluation. Larger maximal pronation in the shoes without laces suggests less function in terms of motion control.

CONCLUSION: A shoelace structure must be a necessary feature of running shoes toward achieving improved comfort and motion control during running.

REFERENCES
Hagen, M. & Henning, E.M. (2008). The influence of different shoe lacing conditions on plantar pressure distribution, shock attenuation and rearfoot motion in running. *Clin Biomech.* 23. 673 - 674.

CHANGES IN SCAPULAR MOTION DURING A SIMULATED BASEBALL GAME

Daisaku HIRAYAMA, Norihisa FUJII, Michiyoshi AE, Sekiya KOIKE

Institute of Health and Sport Sciences, University of Tsukuba, Ibaraki, Japan

KEY WORDS: baseball, pitching, scapula (shoulder girdle).

INTRODUCTION: The importance of role of the scapula in throwing motion is reported in many clinical situations. Therefore, stability of the scapula in relationship to the entire moving arm is key point in the throwing motion. However, the scapular motion is often ignored in conventional biomechanical research. The purpose of this study was to investigate the changes in scapular motion during a simulated baseball game.

METHODS: One male college baseball pitchers threw 15 pitches in an inning for 9 innings (135 pitches) in an indoor pitcher's mound. Rest time between innings was 6 minutes. Three-dimensional positions of 47 reflective markers attached to subject were tracked by an optical motion capture system (Vicon Motion System 612, Oxford Metrics) with eight cameras (250Hz). For each subject 75 fastball pitches (1st, 3rd, 5th, 7th, and 9th innings) were chosen for analysis. Kinematic parameters were analyzed by simple linear regression analysis ($p<0.05$). The scapula (shoulder girdle) coordinate system was defined according to the clavicle coordinate system and the thorax coordinate system. The scapular motion was scapular abduction/adduction (forward/backward) and scapular elevation/depression (upward/downward) (Figure 1).

RESULTS & DISCUSSION: The scapular adduction (backward direction) angle at stride foot contact was decreased with increasing the number of pitches. The scapular abduction-adduction range of motion during stride foot contact to ball release decreased with increasing the number of pitches (Figure 2). Decrease in the scapular range of motion is considered as decrease in the scapular adduction angle at stride foot contact.

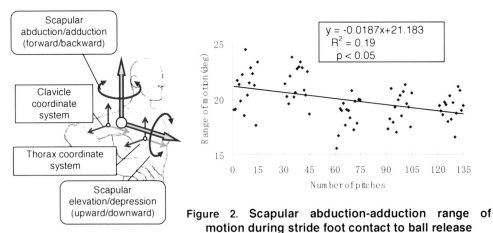

Figure 1. Scapular motion

Figure 2. Scapular abduction-adduction range of motion during stride foot contact to ball release

CONCLUSION: This study shows that increasing the number of pitches decreases the scapular range of motion.

REFERENCES:

Kibler, W.B. (1998). The role of the scapula in athletic shoulder function. *The American Journal of Sports Medicine*, 26(2), 325-337.

THE MEASUREMENT OF KINETIC VARIABLES IN RACE WALKING

Brian Hanley, Andi Drake and Athanassios Bissas
Carnegie Research Institute, Leeds Metropolitan University, Leeds, UK

The purpose of this study was to measure kinetic variables during race walking. Forty national and international race walkers walked either 5 km or 10 km at a pace equivalent to 105% of their season's best time. Junior athletes walked 5 km, while senior athletes (mostly 20 km walkers) walked 10 km. Kinetic data were collected using a Gaitway treadmill (1000 Hz). Data were collected at the 2.5 km point. Men had longer step lengths than women and walked faster as a result. There was little difference in cadence. Average flight times for each group of athlete were approximately 0.04 s. Senior athletes showed more 'typical' race walking vertical force patterns than the juniors; this may be linked to quantity of training experience and gait efficiency. Athletes are advised to develop muscular strength endurance to cope with loading rates upon initial contact.

KEY WORDS: race walking, force, athletics, treadmill.

INTRODUCTION: Previous research in race walking has predominantly focussed on kinematic data without reference to ground reaction forces (GRF). Payne (1978) measured the GRF for one international race walker and found that mediolateral forces were much greater than in normal walking with only small increases in vertical and anterioposterior forces. Fenton (1984) conducted studies on seven athletes of varying ability. Lower vertical impact peaks occurred in elite athletes compared to less skilled walkers. A second, weight-bearing peak then occurred, and a much smaller decrease in force to the point of midstance compared to normal walking. The final propulsive peak was smaller in race walking compared to normal walking as a proportion of the first peak (in race walking the push-off peak was approximately 0.5 BW below the weight-bearing peak). A limitation of measuring kinetic data by means of in-dwelling force plates is the difficulty with which multiple meaningful trials are possible. The accuracy of data collection can be affected by athletes targeting the force plates rather than walking naturally and ensuring that athletes walk at an appropriate, realistic pace. To combat this, Ávila et al. (1998) used a treadmill with in-built force plates (Gaitway, Traunstein) to compare normal walking GRFs with those in race walking; the four female athletes tended to display less variability in vertical forces when race walking. Neumann, Krug & Gohlitz (2008) used a similar treadmill to analyse four junior female race walkers at multiple points during a 6 and a 10 km walk, and found that the coefficient of variation for temporal variables tended to increase with distance walked. These studies focussed particularly on variability and not the importance of the kinetic parameters to race walking per se. Kinetic data collection has thus been negligible resulting in a lack of understanding of the mechanisms behind important kinematic parameters, particularly in women and junior athletes. The purpose of this study was to measure and provide a description of kinetic variables in male and female junior and senior race walkers.

METHODS: Data collection: Forty national and international race walkers gave informed consent and the study was approved by the university's ethics committee. The group of forty athletes comprised fifteen senior men (both 20 km and 50 km walkers), seven senior women, ten junior men, and eight junior women. The senior men had a mean age of 31 yrs (± 11), stature 1.81 m (± .08), and mass 72 kg (± 7). The senior women had a mean age of 26 yrs (± 7), stature 1.67 m (± .03), and mass 56 kg (± 8). The ten junior men had a mean age of 17 yrs (± 1), stature 1.80 m (± .06), and mass 65 kg (± 11) and the eight junior women had a mean age of 16 yrs (± 1), stature 1.67 m (± .04), and mass 56 kg (± 4). All participants were free from injury. Each senior athlete walked for 10 km on a treadmill (Gaitway, Traunstein) at a pace that resulted in a walking time equivalent to 105% (± 2) of their season's best time. Junior athletes who normally raced over 10 km walked on the treadmill for 5 km. Each athlete walked at a constant pace for the duration of the test. Kinetic data were recorded using the Gaitway treadmill, which has two in-dwelling force plates (Kistler, Winterthur). The sampling rate was 1000 Hz. Data were collected for thirty seconds at 2.5 km. As well as kinetic data, the associated software (Gaitway, Traunstein) gave values for step length, cadence, and temporal data.

Data analysis: Because of low sample sizes for each specific age and gender group, Pearson's product moment correlation coefficient was used to find associations within all male walkers ($N = 25$) and all female walkers ($N = 15$). Comparisons between groups (using one-way ANOVA) have not been undertaken due to the small sample sizes per group.

RESULTS: Table 1 shows the mean values of speed, step length and cadence for each group of athletes. The overall range of values for step length was from 0.96 m to 1.22 m for men, and from 0.93 m to 1.09 m for women. With regard to cadence, the overall range was from 2.80 to 3.37 Hz for men, and from 3.00 to 3.26 Hz for women. The mean step length value is also shown expressed as a percentage of the athletes' statures. It can be seen that in general, athletes had step lengths measuring approximately 60% of their standing heights. Walking speed is the product of step frequency (cadence) and step length. In both male and female groups, speed was positively correlated with step length when expressed in both absolute (m) and relative terms (%) ($p < .01$). Men's standing height was positively correlated with step length ($p < .01$) and negatively with cadence ($p < .01$). No significant correlations with stature were found in the women athletes. Speed was correlated with cadence in the men's group ($p < .05$) but not the women's ($p = .10$).

Table 1 Speed, step length, and cadence data for each group (mean ± SD)

	Speed (km/hr)	Step length (m)	Step length (%)	Cadence (Hz)
Senior men	12.74 (± 0.64)	1.13 (± 0.08)	62.2 (± 3.4)	3.13 (± 0.15)
Senior women	11.59 (± 0.58)	1.01 (± 0.05)	60.8 (± 2.9)	3.17 (± 0.07)
Junior men	11.99 (± 0.85)	1.09 (± 0.06)	60.6 (± 2.5)	3.05 (± 0.18)
Junior women	11.08 (± 0.56)	1.00 (± 0.05)	59.9 (± 3.1)	3.08 (± 0.05)

Cadence is determined by the time taken to complete each successive step; the shorter the step time, the higher the cadence. In turn, step time can be broken down into two components: contact time and flight time. The values for each of these variables, as well as the percentage of step time spent in contact, are shown in Table 2. Similarly to cadence in Table 1, there are only small differences between groups for step time values. Although strictly speaking there should be no flight time in race walking because of IAAF rule 230.1, most groups in this study had flight times of approximately 0.04 s, with junior women having flight times of 0.05 s. Flight time and contact time were negatively correlated with each other ($p < .01$). Within the male athletes, both step time and contact time were negatively correlated with speed ($p < .05$ and $p < .01$ respectively) and flight time was positively correlated with speed ($p < .05$). In the women walkers, only contact time showed a (negative) correlation with speed ($p < .05$). Flight time was positively correlated with step length in both male and female groups ($p < .05$). The negative correlation found between men's heights and cadence was more related to their contact times ($p = .06$) rather than to their flight times ($p = .87$).

Table 2 Step time, contact time, and flight time data for each group (mean ± SD)

	Step time (s)	Contact time (s)	Flight time (s)	Contact time (%)
Senior men	0.32 (± 0.02)	0.28 (± 0.02)	0.04 (± 0.02)	87.9 (± 4.7)
Senior women	0.32 (± 0.01)	0.28 (± 0.02)	0.04 (± 0.01)	87.1 (± 4.3)
Junior men	0.33 (± 0.02)	0.29 (± 0.03)	0.04 (± 0.01)	87.9 (± 4.8)
Junior women	0.33 (± 0.01)	0.27 (± 0.01)	0.05 (± 0.01)	84.3 (± 3.3)

Table 3 shows the forces at the impact, first (loading), midsupport and active (push-off) peaks. Impact peak was defined as the highest recorded force during the first 70 ms of contact with the treadmill; midsupport force was defined as the minimum force value recorded between the first and second peak forces. Only vertical ground reaction forces are displayed as it is not possible to record shear forces with the treadmill. All forces are shown as normalised data. The only correlations found between these data and speed was a positive correlation with midsupport force in men ($p < .05$) and with first peak force in women ($p < .01$). Furthermore, step length was also correlated with first peak force in women ($p < .05$) and cadence with second peak force in women ($p < .05$). Contact time was negatively correlated with both first peak force and midsupport force in both groups of athletes ($p < .05$ for men and $p < .01$ for women). In addition, flight time was positively correlated with first peak, midsupport, and second peak forces in both groups ($p < .01$ for men and $p < .05$ for women).

Table 3 Force data for each group (mean ± SD)

	Impact peak (BW)	First peak force (BW)	Midsupport force (BW)	Second peak force (BW)
Senior men	1.48 (± 0.21)	1.77 (± 0.13)	1.30 (± 0.28)	1.57 (± 0.11)
Senior women	1.56 (± 0.21)	1.80 (± 0.19)	1.41 (± 0.32)	1.55 (± 0.10)
Junior men	1.40 (± 0.18)	1.79 (± 0.10)	1.36 (± 0.28)	1.61 (± 0.09)
Junior women	1.56 (± 0.19)	1.73 (± 0.11)	1.34 (± 0.20)	1.68 (± 0.05)

Weight acceptance is the slope of the force curve during the loading phase, taken from the point of 10% of the impact peak force to the point of 90%; while the push-off rate is the slope of the force curve during unloading, taken from the point of 90% of push-off peak force to the point of 10%. Impulse was recorded in the vertical direction, while base of support is the average distance between one foot's mediolateral centre of pressure and the next opposing foot's mediolateral centre of pressure for each foot strike. Table 4 shows the results for these variables. In women, impulse was negatively correlated with weight acceptance rate ($p < .01$) and positively with push-off rate ($p < .05$). In men, there was a positive correlation between base of support and weight acceptance rate ($p < .05$) and a negative one between base of support and midsupport force ($p < .05$). Impulse was negatively correlated with speed in the men's group ($p < .05$).

Table 4 Loading rates, impulse, and base of support for each group (mean ± SD)

	Wt. Acceptance (BW/s)	Push-off rate (BW/s)	Impulse (BW.s)	Base of support (mm)
Senior men	29.4 (± 5.5)	16.1 (± 3.9)	0.32 (± 0.02)	37 (± 23)
Senior women	32.0 (± 5.9)	15.1 (± 3.2)	0.31 (± 0.01)	44 (± 18)
Junior men	30.8 (± 4.8)	13.3 (± 2.8)	0.33 (± 0.02)	30 (± 11)
Junior women	30.8 (± 4.5)	18.9 (± 3.6)	0.32 (± 0.01)	28 (± 13)

DISCUSSION: Both male groups were faster than the female groups and this was predominantly due to their longer step lengths. Very little difference was found between men and women for cadence values, although it could be assumed that faster walkers will have a balanced combination of both long step lengths and high walking cadences. One important factor for athletes and coaches to consider is the length of the athletes' stride in relation to their overall standing height. Consideration of this will allow for better comparison with other walkers.

Both groups of senior athletes had slightly shorter step times compared to the junior athletes, and the junior women had slightly longer flight times at 0.05 s. It is unlikely that this duration of flight (or the 0.04 s duration for other groups) would be visible to the naked eye when judging. Nonetheless, such flight times should be minimised as much as possible in order to adhere to the rules of the event and reduce the risk of disqualification.

The impact peaks for both senior and junior men were lower compared to both groups of women. This may suggest that the male athletes were more skilled than the females, as suggested by Fenton (1984). The men had an overall greater mean age and this may reflect more training experience. Midsupport forces in normal walking tend to drop below 1 BW; a range between 1.30 and 1.46 BW was found here. Furthermore, all groups had lower propulsive peak values than loading peaks, although the largest decrease was 0.25 BW. This value is less than that found by Fenton (1984) who found a 0.5 BW decrease, but the overall vertical force pattern suggested by him was mostly replicated here. The junior women showed the smallest decrease between these two values, and this may be a result of their relative inexperience in the event. Both groups of senior athletes showed relatively typical race walking GRF patterns.

All four groups of athletes had similar weight acceptance rates (approximately 30 BW/s); these values show the large loading rates experienced by the race walkers upon initial contact and the need for preventive exercises to prevent injury, particularly in the lower leg. Push-off rate was highest in the junior women although it must be remembered that this was a vertical force and therefore not necessarily conducive to forward propulsion.

A limitation of using the Gaitway treadmill is its inability to record shear forces; thus it is not possible to measure important variables which may be important (e.g. propulsive anterio-posterior forces). Although most athletes were experienced at walking on treadmills, it is an artificial setting in comparison with track or road walking and therefore it is not certain that the athletes adopted their normal walkign technique. Further studies investigating the validity of using treadmills to analyse race walking are warranted.

CONCLUSION: The pattern of vertical GRF in race walkers differs from normal walking in an attempt to maintain high speeds and prevent lifting of the centre of mass. Typical race walking patterns are not shown in less-experienced junior athletes. Coaches and athletes are advised to develop muscular endurance and work towards developing efficient race walking gait in order to develop efficiency of movement and reduce the risk of disqualification due to lifting.

REFERENCES:

Avila, A., Klavdianos, A., Manfio, E., Viollaz, F., Nasser, J., Fonseca, J., & Andrade, M. (1998). Racewalking and normal walking analysis. In *Proceedings of the XVI International Symposium on Biomechanics in Sports* (edited by Vieten, M.) Konstanz: University of Konstanz.

Fenton, R. M. (1984). Race walking ground reaction forces. In J. Terauds, *et al.*, (Eds.), Research Center for Sports, Del Mar, California, 61-70.

Neumann, H. F., Krug, J., & Gohlitz, D. (2008). Influence of fatigue on race walking stability. In *Proceedings of the XVII International Symposium in Sports* (edited by Kwon, Y.-H., Shim, J., Shim, J. K., and Shin, I.-S.), Seoul: Seoul National University, 428-431.

Payne, A. H. (1978). A comparison of ground reaction forces in race walking with those in normal walking and running. In *Biomechanics*, Vol. VI-A (edited by Asmussen, E. and Jorgensen, K.), 293-302. Baltimore: University Park Press.

IN-SHOE PLANTAR PRESSURE MEASUREMENTS DURING GOLF SWING PERFORMANCE WEARING METAL AND ALTERNATIVE SPIKED GOLF SHOES

Paul Worsfold*, Neal Smith and Rosemary Dyson.

*University of Chester, Chester, Cheshire, UK
University of Chichester, Chichester, West Sussex. UK.

In-shoe plantar pressures were assessed when two different golf shoes were worn. One shoe design incorporated 7 metal spikes and the other 7 alternative spikes. Eighteen male golfers (mean handicap ±SD: 12.4 ± 7.8) played 5 shots with a 7 iron while wearing each shoe on a natural grass surface. Pressures in 9 standardised foot regions were analysed to yield maximal, average and ball impact values. The greatest pressures were at the lateral regions of the front-foot from the point of ball impact when wearing the metal spike shoes. Significantly greater maximal pressures occurred when wearing the metal spike shoe (front foot: lateral 114 kPa; back foot: lateral heel 40 kPa, medial central 63 kPa; p<0.05). In the alternative spike shoe, significantly greater maximal (76 kPa) and average pressures (41 kPa) occurred on the back foot at the central toe region.

KEY WORDS: Footwear, golf, pressure, shoe, spike.

INTRODUCTION: Previous ground reaction force studies (Worsfold *et al.*, 2006, 2007) have identified the forces, torques and interactions between the golf shoe sole interface and a natural turf ground surface. To further assess the functional properties of golf shoe interfaces, an understanding of the shoe-foot interaction is required. McPoil & Cornwall, (1995) stated that in-shoe pressure measurements can provide a greater understanding of the effects of specialised footwear, as well as assist in their modification to maximise their benefit to the user. Wallace *et al.*, (1994) reported foot pressures during the golf swing of six right-handed male golfers wearing two types of shoe, one with metal spikes on the outer sole, and another with a rubber moulded sole. The golfers played shots with their own driver off a grass-covered tee-box outdoors while wearing shoes containing eight piezo-electric film transducers sampled at 400 Hz. The limited understanding of the relationship between within shoe plantar pressures and golf shoe sole traction designs does not allow prediction of the effects of modified golf shoe design features such as alternative spikes. Alternative spikes have been introduced in response to concerns about the damage to golf courses by metal spike shoes (Hammond & Baker, 2002). This research analysed, during the golf swing, in-shoe pressures to identify any plantar pressure differences associated with the wearing of a golf shoe fitted with metal spikes compared to a golf shoe fitted with alternative spikes.

METHODS: Eighteen right-handed male (mean ±SD age: 29 ± 2.2 years; mass: 81.4 ± 2.8 kg; height: 179.9 ± 1.9 cm) golfers volunteered for the study. The golfers' handicaps ranged between 24 and for the better golfers 0 (mean 12.4 ± 7.8). Two types of golf shoes with identical uppers but different sole interface designs were tested. One shoe incorporated alternative spikes (Figure 1A), and one metal spikes (Figure 1B), with seven spikes in total on both shoes in similar locations on the shoe sole

Figure 1: Golf shoe alternative spike (A) and metal spike (B) fitted to improve traction

Plantar pressure insoles (Footscan RSscan International, Belgium) were fitted inside both left and right shoes. The shoe insole size was selected to fit comfortably within each individual's shoe. Subjects were given time (walking and playing shots) to become accustomed to wearing the shoe and insole combination. Five shots were played on a natural turf surface with the subject's own 7-iron while wearing each shoe type. The order of shoe testing was randomized for each participant. A Footscan data logger was attached to the back of the golfer's waist to prevent any movement restrictions. During each golf swing, data from the insoles within the front foot (closest to the direction of the shot), and the back foot were recorded by the data logger for 8 seconds with sampling at 500Hz. Foot movements and ball impact were captured using a 200 Hz High Speed Peak Systems Camera (Peak Performance Technologies inc. Englewood, Colorado USA).

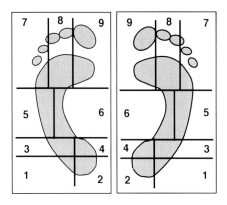

Analysis involved the setting of nine plantar foot regions using the dynamic regional analysis feature, which adapted for individual foot size and form by screening the foot from different directions. The Footscan software (version 2.33) placed a mask on top of the footprint which divided it into nine regions, which were proportional for foot length and width (Figure 2). Maximum pressure, average pressures during the whole swing process and pressures at ball impact were analysed for the nine regions of both the left and right feet. Mean and standard deviation values were calculated for all data. One way analysis of variance with repeated measures and a Tukey HSD post hoc test were applied.

Figure 2: The left (front) and the right (back) foot dynamic regional analysis identifying the nine foot regions (R1-R9)

RESULTS:

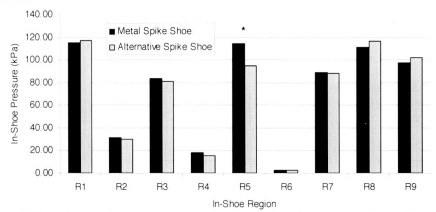

Figure 3: Front-foot maximum pressures during the whole swing process when using a 7-iron.
Note: * Denotes significant difference between shoes p<0.05.

In-shoe pressure analysis identified that wearing the metal spike shoe was associated with significantly greater ($p<0.05$) maximal plantar pressure at R5 (mid-lateral) of the front-foot and at ball impact (Figures 3 & 4). At the back-foot the metal spike shoe was associated with the production of significantly greater ($p<0.05$) maximal and average in-shoe pressures at the mid-medial arch position R6 (Figure 5). Significantly greater ($p<0.05$) metal spike maximal pressures were also identified at R1 at the outer heel of the back-foot. The alternative spike shoe was associated with significantly greater ($p<0.05$) maximal and average pressures at R8 (mid toe region) of the back-foot (Figure 6). Average pressures at the front foot were similar for both shoes.

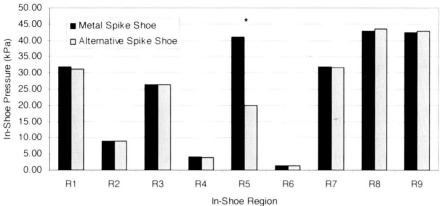

Figure 4: Front foot maximal pressures at time of ball impact when using 7-iron
Note: * Denotes significant difference between shoes p<0.05.

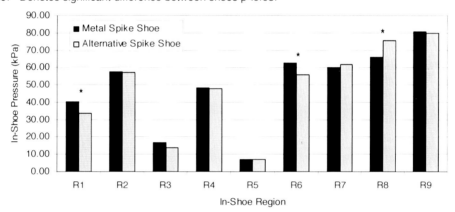

Figure 5: Back foot maximum pressure during the whole swing process when using 7-iron.
Note: * Denotes significant difference between shoes p<0.05.

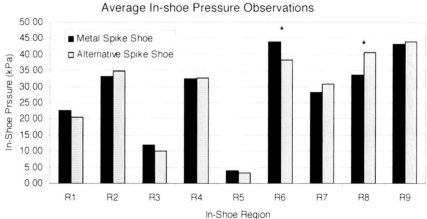

Figure 6: Back foot average pressures during the whole swing process when using 7-iron
Note: * Denotes significant difference between shoes p<0.05.

DISCUSSION: Pressures of up to 120 kPa were recorded inside the front foot shoe, while maximal pressures at the back foot peaked at approximately 80 kPa. These findings are in general accord with previous research (Worsfold *et al.*, 2007) which identified greater maximal vertical ground reaction force with the 7-iron at he front foot (1.1BW) than the back foot (0.82 BW), and the within shoe pressure study of Wallace et al. (1994). Significant differences in the pressures recorded under both the front and back feet within the shoe linked to shoe design outer sole features. At the front foot shoe the greatest pressures occurred at the lateral heel, lateral side of the foot and at the 1st, 2nd and 3rd metatarsals. The finding of the greatest maximal pressures at the 1st metatarsal in the metal spike shoe was in accord with earlier research reported by Wallace et al. (1994). The maximal pressure on the lateral side of the front foot was greater by 18% in the metal spike shoe, and at ball impact the pressure on the lateral side of the front foot was twice as great in the metal spike shoe compared to the alternative spike shoe (Figure 4). Such large significant differences identified during this experiment, when good front foot shoe traction was important to allow the club and body to decelerate around the front leg at impact and in the follow-through, lend support to the concept (e.g. Frederick, 1986) of human perception influencing physical muscular performance and foot action.

The alternative spike shoe was associated, notably, with the production of significantly greater maximal pressure over the back foot 2nd and 3rd metatarsal (R8, Figure 5), which was supported by average pressure analysis (Figure 6). This region was important during weight transfer at impact and in the follow-through when the back foot heel raised and stance stability was dependent on good forefoot traction.

CONCLUSION: Shoe design features on the outer sole of the golf shoe influence within shoe plantar pressures, and thus must be perceived by the player who adapts muscular control of the body and lower extremity during the golf swing. Such findings lend support to the concept (e.g. Frederick, 1986) of human perception influencing physical muscular performance and foot action. The importance of ensuring players familiarise with sporting tasks while wearing appropriate performance footwear was supported.

REFERENCES:
Hammond, L.K.F. & Baker, S.W. (2002). The effect of metal and alternative golf spikes on visual turf quality and playing characteristics of Fustuca Rubra/Agrostis Capillaries/Poa Annua putting surfaces. *Journal of Turfgrass and Sports Surface Science,* 76, 46-58.

Frederick, E.C. (1986). Kinematically mediated effect of sport shoe design: a review. *Journal of Sports Sciences,* 4, 169-184.

McPoil, T., Cornwall, M., & Yamada, W., (1995). A comparison of two in-shoe plantar pressure measurement systems. *Lower Extremity,* 95-103.

Wallace, E., Grimshaw, P., & Ashford, R. (1994). Discrete pressure profiles of the feet and weight transfer patterns during the golf swing. In *Science and Golf II. Proceedings of the World Scientific Congress of Golf.* (Ed. Cochran, A.J. and Farrally, M.R.). E and FN Spon. p.27-32.

Worsfold, P.R., Smith. N.A. & Dyson, R. (2006). Force generation at the golf shoe sole to grass turf interface occurring with alternative spike designs. In *Proceedings of XXIV International Symposium on Biomechanics in Sports* (Ed: Schwameder H, Strutzenberger G, Fastenbauer V, Lindinger S, Muller, E.). Austria: University of Salzburg p.502 – 505.

Worsfold, P., Smith. N.A. & Dyson, R.J. (2007). A comparison of golf shoe designs highlights greater ground reaction forces with shorter irons. *Journal of Sports Science and Medicine.* 6, 484-489

Acknowledgement
Footwear for this research was provided by Adidas and financial support from University of Chester.

STATIC AND DYNAMIC ANALYSIS OF THE FOOT IN SOCCER PLAYERS SUSTAINING PROXIMAL 5TH METATARSAL STRESS FRACTURE

Moshe Ayalon [1], Iftach Hetsroni [2], David Ben-Sira [1]

Biomechanics Laboratory, Zinmann College of Sports & Science, Wingate Institute, Israel [1]
Orthopaedic Department, Meir Medical Center, Kfar Saba, Israel [2]

KEY WORDS: Stress fracture, 5th metatarsal, Plantar pressure.

INTRODUCTION: Stress fracture of the proximal 5th metatarsal (MT) is a well recognized entity among athletes. Identification of specific risk factors for this injury may play an important role injury prevention. Lateral oveloading in rigid cavus foot have been suggested as contributing factors (Williams, 2001). The purpose of our study was to characterize static variables of foot structure and dynamic variables of foot function in soccer players which sustained 5th MT stress fracture.

METHODS: 10 injured soccer players who regained full professional activity following a unilateral proximal 5th MT stress fracture, and 10 control uninjured soccer players participated in this study. Static variables of arch height and ankle flexibility were measured. In addition, plantar pressure was assesed during barefoot walking on an EMED platform. Plantar pressure variables were analayzed for four stance cycles. Because some of the variables did not fulfill the assumption of normality, a non-parametric approach was applied using the Wilcoxon signed ranks test for bilateral comparisons and the Mann-Whitney test for between samples comparisons.

RESULTS: Static foot parameters were not different in the injured foot either from the sound foot of the injured group or from the control group. Peak pressure (PP) under 5th MT was significantly lower in the injured limb compared to the sound limb in the injured group. PP was significantly lower in the injured limb compared with the control group in both 4th MT and 5th MT and higher compared to the control group in the 1st MT . PP of the sound limb in the injured group was significantly lower compared to the control group in both 3rd and and 4th MTs. Normalized pressure time integral was significantly lower in the 5th MT and higher in the 1st MT for the injured limb compared with the control group. As for the sound limb, this variable was significantly lower in the 4th MT compared with the control group.

DISCUSSION: In contrast to the lack of association between static and passive indices and 5th MT stress fracture, the dynamic evaluation revealed several general trends, which differentiate injured and uninjured feet. Contrary to the expectations, in both feet, injured subjects demonstrated a relative mean lateral forefoot unloading and relative mean medial forefoot increased loading compared with the control group. Moreover, peak pressure in the 5th MT of the injured limb was noticeably reduced compared with the sound limb as well as the control group. The findings can be interpreted either as representing an inherent loading characteristic, which may function as a causative risk-related factor, or as an adaptive foot function consequent to the injury. Stress reduction over a long period of time may result in a relatively weakened infrastructure (Wolff's law), and increased susceptibility to a stress injury.

CONCLUSION: Athletes who sustained proximal 5th MT stress fractures demonstrate a unique loading pattern of the forefoot. We recommend that future studies focus on understanding the dynamic function of the foot and on isolation of dynamic risk factors and disregard static measurements.

REFERENCES:
Williams, DS 3rd, McClay, IS. & Hamill J. (2001). Arch structure and injury patterns in runners. *Clinical Biomech* ,16, 341-347.

BIOMECHANICAL CHARACTERISTICS OF GRINDING IN AMERICA'S CUP SAILING

Simon Pearson[1], Patria Hume[1], John Cronin[1,2], David Slyfield[3]

Institute of Sport and Recreation Research New Zealand, AUT University[1]
School of Exercise, Biomedical and Health Sciences, Edith Cowan University[2]
Emirates Team New Zealand[3]

KEY WORDS: kinematics, torque-angle analysis, electromyography (EMG).

INTRODUCTION: Understanding the biomechanics of a sporting movement and what aspects of the movement technique most influence performance can facilitate more specific training, both in terms of strength and conditioning and technical improvements. This study was undertaken to describe the kinetic, kinematic, and muscular activation characteristics of the grinding movement in America's Cup sailing, a high intensity constrained cyclic movement (similar to bicycling) performed with the upper limbs.

METHODS: Ten male America's Cup sailors (33.6±5.7 years, 97.9±13.4 kg, 186.6±7.4 cm) who performed grinding regularly as part of their on-board role participated in this study.
Each sailor performed eight maximal grinding performance tests, of eight-second duration, on a custom-built grinding simulator (Dynapack, New Zealand). Performance tests were conducted under four conditions: forward and backward grinding at both moderate and heavy loads, with two tests completed for each condition (eight tests in total). In addition to power output from the grinding simulator, which was used as a performance measure, the following biomechanical data was also collected:

- Full body (nine segment) 2D sagittal plane kinematics, digitised using APAS (Ariel Dynamics, USA).
- Torque-angle analysis from the grinding handles using an adapted SRM (Schoberer Rad Messtechnik) Powermeter system with torque analysis module (SRM, Germany).
- EMG (Bortec AMT-8 system; Bortec, Canada) on seven upper body sites.

RESULTS: Analysis of all data has yet to be completed; however, findings from preliminary results are as follows:

- Peak torque application typically occurs through 60-200° for forward grinding and 300-40° for backward grinding (0° = crank handle vertically upwards).

- Shoulder angles (relative to the trunk, 0° = upper arm adjacent) at peak torque were similar for forward and backward grinding, at 83.5±10.3° and 83.0±7.1° respectively.

- Variation in torque application throughout the grinding cycle was negatively associated with forward grinding performance (r = -0.60; lower 90% CL = -0.88, upper 90% CL = -0.02) but positively associated with backward grinding performance (r = 0.48; CL = -0.15, 0.83).

DISCUSSION: Although results are not yet complete and currently inconclusive, preliminary results do indicate areas in which adaptations to aid performance enhancement may be implemented. One potential area is that the stimulus during strength training could be altered to more specifically target the key joint angles relating to torque production in grinding, in particular for backward grinding where it appears that the majority of work is performed through specific sectors of the cycle. In contrast, the negative association between variability of torque application and forward grinding performance could mean that technical adaptations to maintain consistent handle force through the cycle may be beneficial.

ANALYSIS OF GROUND REACTION FORCE DURING FASTBALL AND CHANGE-UP SOFTBALL PITCHES

Jia-Hao Chang[1] and Wei-Ming Tseng[2]

[1]Department of Physical Education, [2]Institute of Exercise and Sports Science, National Taiwan Normal University, Taipei, Taiwan

KEY WORDS: kinetics, push-off, landing.

INTRODUCTION: The lower extremity provides stability and balance when we exercise. The roles what lower extremity plays should be clarified during pitching. Therefore, the purpose of this study was to observe the differences of ground reaction force for fastball and change-up windmill softball pitching.

METHODS: Ten collegiate female windmill pitchers (age: 19.5±2.0 years, height: 167.5±4.8 cm, and weight: 64.3±8.6 kg) participated in the study. Force platforms (9281, Kistler) sampling at 1000 Hz were used to collect the push-off and landing forces in the laboratory. Three pressure sensors placed on throwing finger, tiptoe, and heel of each subject were used to define the pitching cycle. The force data were normalized with subject's body weight. Time was normalized with pitch duration. Paired t-test was used to assess the differences of peak force and $p < .05$ was considered as significance.

RESULTS & DISCUSSION: The ground reaction forces of push-off and landing during fast ball and change-up pitching were similar (Figure 1 and Figure 2), and no differences of peak force were found in all directions ($p>.05$). Peak forces of push-off and landing in vertical direction resulted from this study were similar to baseball pitching in the literatures.

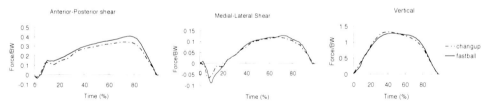

Figure 1. Push-off force based on the mean values of the ten subjects. Force data are normalized with body weight.

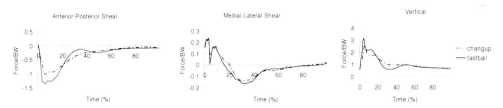

Figure 2. Landing force based on the mean values of the ten subjects. Force data are normalized with body weight.

CONCLUSION: The similar ground reaction forces in vertical, anterior-posterior, and medial-lateral directions were found during fast ball and change-up pitching, which suggests that the lower extremity provides the supports for stability and balance in pitching.

REFERENCES:

MacWilliams, B.A., Choi, T., Perezous, K.M., Chao, S.E., & MacFarland, G.E. (1998). Characteristic ground-reaction force in baseball pitching. The American journal of sports medicine, 26, 66-71.

THE INFLUENCE OF EXPERIENCE ON FUNCTIONAL PHASE KINEMATICS OF THE LONGSWING

Genevieve Williams, Gareth Irwin and David Kerwin
Cardiff School of Sport, University of Wales Institute, Cardiff, UK

KEY WORDS: gymnastics, high bar, joint kinematics

INTRODUCTION: The biomechanics of successful longswings are well understood, however, the influence of experience on execution is not well defined. This study aims to explore functional phase (FP) kinematics during repeated longswings performed by an experienced (E), inexperienced (I) and novice (N) participant.

METHODS: Three participants performed five sets of five longswings on a high bar. Data were collected using an automated motion analysis system (CODA CX-1), sampling at 200Hz. Circle angle (θ_C) was defined by the mass centre to bar vector with respect to the horizontal. Kinematics of FP's, defined by maximum shoulder flexion to extension (θ_{CS}) and hip extension to flexion (θ_{CH}), were analysed during swing three and four in each set.

RESULTS AND DISCUSSION: Variability of the start (1) and end (2) of the functional phases (θ_{CS12} and θ_{CH12}) were larger for participant I however, the range of the FP remained similar to that of participant N. These results suggest that the timing of FP initiation may well be important in the development of this particular skill (Figure 1).
The joint angle changes within the respective functional phases were smallest for E, and largest for N, where θ_{CS12} and θ_{CH12} occurred earlier. It is likely that these findings reflect the mechanical result of an early onset FP with increased shoulder and hip angles, indicating that the action and timing of the functional phases is critical in facilitating the mechanical demands of successful longswing performance (Irwin & Kerwin, 2005).

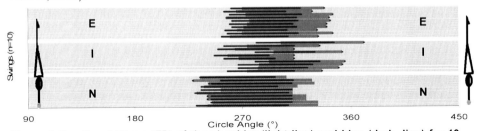

Figure 1 Functional Phase (FP) of the shoulder (light line) and hips (dark line) for 10 swings of experienced (E), inexperienced (I) and novice (N) participants

Table 1 Mean [±sd] shoulder (θ_S) and hip (θ_H) angles, and changes in corresponding joint angles ($\Delta\theta$), at the start (1) and end (2) of the functional phases for experienced (E), inexperienced (I) and novice (N) participants

	θ_{S1}(°)	θ_{S2}(°)	$\Delta\theta_{S12}$(°)	θ_{H1}(°)	θ_{H2}(°)	$\Delta\theta_{H12}$(°)
E	6 [2]	-27 [2]	33 [3]	17 [4]	-23 [2]	49 [4]
I	3 [2]	-33 [4]	37 [4]	22 [3]	-57 [4]	79 [6]
N	17 [3]	-32 [7]	49 [8]	33 [5]	-67 [4]	100 [7]

CONCLUSION: These results show that kinematic analysis throughout the development of the longswing can be a valuable tool in identifying key variables required for skill progression. Current work is examining more participants of varying experience.

REFERENCES: Irwin, G. and Kerwin, D. G. (2005). Biomechanical similarities of progressions for the longswing on high bar. *Sports Biomechanics*, 4, 164-178.

CONTRIBUTION OF THE SUPPORT LEG TO THE VELOCITY CHANGE IN THE CENTRE OF GRAVITY DURING CUTTING MOTION

Yuta Suzuki[1], Michiyoshi Ae[2] and Yasushi Enomoto[3]

Doctoral Program in Physical Education, Health and Sport Sciences, University of Tsukuba, Japan[1]
Institute of Health and Sport Sciences, University of Tsukuba, Japan[2]
Faculty of Education, Kyoto University of Education, Japan[3]

The purposes of this study were to investigate how the support leg contributes to the velocity change in the whole body center of gravity (CG) during cutting motion. Twenty male university students ran 30 m zigzag runs with five cuts of 90 degrees incorporating side step (SS) and cross step (CS) technique. The third cut of each trial was videotaped with two digital video cameras operating at 60 fps for a three-dimensional analysis of cutting motion. The thigh in the first half and the foot in the second half of support phase contributed to increase the velocity of the whole body centre of gravity. While the foot greatly contributed to the medial acceleration of the whole body center of gravity in cross step, the shank was contributor around the mid-support in SS. These results suggest that anterior lean of the thigh and ankle plantarflexion played an important role in cutting motion. In CS, the foot contributed to change the direction of the CG velocity with leaning the shank and directing the toe inward in the support phase. In SS, turning the foot to outward in the support phase may help increase the medial relative accelerative force of the shank with the ankle plantarflexion and knee extension.

KEY WORDS: motion analysis, side step cutting, cross step cutting, relative accelerative force

INTRODUCTION: In many ball games, players frequently encounter to change the direction of running as well as running straight. Andrews et al. (1977) classified techniques for changing running direction into 1) the side step cutting technique, in which the direction is changed by planting one foot opposite to the new direction, and 2) the cross step cutting technique, in which after planting one foot, the other leg crosses in front of the body to accelerate in the new direction. In the cutting motion, the player's center of gravity (CG) undergoes a horizontal deceleration and acceleration in the tangential direction and a normal acceleration to change the running direction. During the support phase in the cutting motion, the support leg plays an important role to generate force to change the running direction. Although studies on the cutting motion have investigated the CG velocity change, relationships of sprint running ability and plyometric ability to the cutting motion, there is little information of the techniques of the cutting. The purposes of this study were to investigate how the support leg contributes to the velocity change in the whole body CG during cutting motion, and to get insights into the techniques of cutting motion.

METHODS: Data Collection and Processing: Twenty male university players of soccer, basketball, rugby, and handball (age 19.8±1.3 yrs, height 1.73±0.04 m, body mass 66.0±5.4 kg) participated in this study. They performed 30 m zigzag runs with five sequential cuts of 90 degrees by the side step (SS) or the cross step (CS) technique. The third cutting motion of five sequential cuts was videotaped with two digital video cameras operating at 60 fps for the three-dimensional analysis. Twenty-five body landmarks were digitized and their three-dimensional coordinate data were reconstructed by a DLT method. The coordinate data were smoothed by a Butterworth digital filter with cut-off frequencies of 2.4 to 8.4Hz decided by a residual analysis. The CG coordinate data estimated after the Japanese athletes body segment parameters (Ae, 1996) were differentiated to obtain the velocity and acceleration of the CG.
Data Analysis: The relative momenta generated by the foot, shank, and thigh of the support leg were calculated using equations after Ae et al. (1985), assuming that the velocity of CG

for a proximal segment was a summation of velocities of CG and the joints relative to more distal segments and joints.

$$RM_{th} = (m_{ar} + m_t + m_{fl})V_{h/k} + m_{th}V_{th/k} \quad (1)$$

$$RM_{sh} = (m_{ar} + m_t + m_{fl} + m_{th})V_{k/a} + m_{sh}V_{sh/a} \quad (2)$$

$$RM_f = (m_{ar} + m_t + m_{fl} + m_{th} + m_{sh})V_a + m_fV_f \quad (3)$$

where RM = momentum generated, mi = segment mass, Vi/j = relative velocity of segment i to joint j, ar = arm, t = trunk, th = thigh, sh = shank, f = foot, fl = free leg, h = hip, k = knee, a = ankle. The relative accelerative forces (RAFs) were calculated as a derivative of the momenta generated. The RAFs were transformed to the CG coordinate system, i.e. the first (antero-posterior) axis was directed to the horizontal velocity of the whole body CG, the second was the vertical, and the third (medio-lateral) axis was defined as a cross product of the first and second axes. The antero-posterior lean angle of the leg segment was defined as an angle between the leg segment and the vertical axis on a plane composed by the antero-posterior and vertical axes, and the medio-lateral lean angle of the leg segment was defined as an angle between the leg segment and the vertical axis on a plane composed by the medio-lateral and vertical axes. The horizontal angle of the leg segment was defined as an angle between the leg segment and the antero-posterior axis on the horizontal plane. The RAFs were normalized by time of the support phase and the body mass of the subjects. The support phase was divided into the first half from the foot strike to the instant of minimum horizontal velocity of the whole body (mid-support), and the second half from the mid-support to the toe off.

RESULTS: Figure 1 presents the antero-posterior RAFs of the support leg segments and the acceleration of the whole body CG in the support phase for SS (the left figure) and CS (the right figure). In SS and CS, the acceleration of the whole body CG was negative (posterior) in the first half and positive (anterior) in the second half. The foot generated the posterior RAF at the foot strike until about 30% of the support phase. While the shank generated the posterior RAF followed by the foot, the thigh generated the anterior RAF in the first half of the support phase. In the second half, the foot generated the anterior RAF, but the thigh generated the posterior RAF.

Figure 2 presents the medio-lateral RAFs of the support leg segments and the acceleration of the whole body CG in the support phase for SS and CS. In SS and CS, the medio-lateral acceleration of the whole body CG was positive (medial). In CS, the foot generated large medial RAF through the support phase, and the RAF of the thigh was small. Although the foot generated the medial RAF in the support phase in SS, the shank generated remarkably larger medial RAF around the mid-support, compared to that of CS.

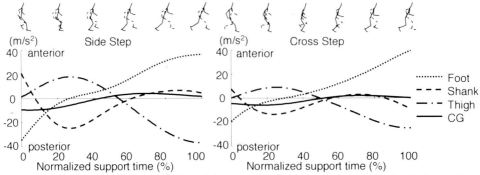

Figure 1 Averaged anterior-posterior relative accelerative force generated by the foot, shank, and thigh and the acceleration of the CG.

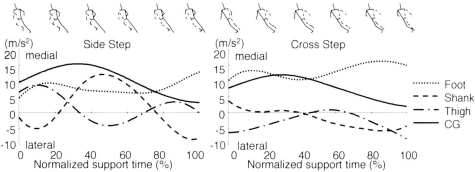

Figure 2 Averaged medial-lateral relative accelerative force generated by the foot, shank, and thigh and the acceleration of the CG.

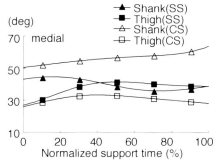

Figure 3 The medial lean angle of the shank and thigh.

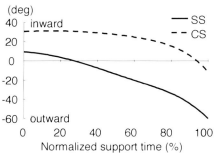

Figure 4 The horizontal angle of the foot of the support leg.

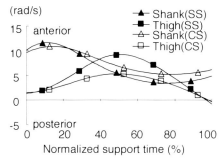

Figure 5 The shank and thigh angular velocities of the support leg.

Figure 3 shows the medio-lateral lean angle of the shank and thigh. In SS and CS, the shank and thigh leaned medially through the support phase. While the lean angle of the shank increasing in CS, it in SS decreased to 70% of the support time, and slightly increased. In SS and CS, there was a positive correlation between maximal medial lean angle of the shank and the medial velocity change by the foot (r=0.682, p<0.01; r=0.650, p<0.01). In SS and CS, the ankle was dorsiflexed in the first half, and plantarflexed in the second half of the support phase. There was a significant positive correlation between the acceleration of whole body CG and the extension angle of the ankle joint in the second half (r=0.525, p<0.05; r=0.754, p<0.01).

Figure 4 shows the horizontal angle of the foot. The angle zero means the support foot directed to the CG velocity vector on the horizontal plane. Although the foot in CS directed to inward except just before the toe off, it was outwarded after 20% of the support phase in SS. There were a significant negative correlation between the horizontal angle of the foot at the foot strike and the medio-lateral velocity change of the whole body CG in CS (r=-0.448, p<0.05), and a positive correlation between the horizontal angle the foot at the toe off and the medio-lateral velocity change of the whole body CG in SS (r=0.646, p<0.01).

Figure 5 shows the antero-posterior lean angular velocities of the shank and thigh. In SS and CS, the shank reached maximum lean angular velocity immediately after the foot strike, and slowed down until the toe off. The thigh angular velocity reached the peak around the mid-support. In SS and CS, there were a negative correlation between the deceleration of the whole body CG and the anterior lean angular velocity of the thigh at the foot strike (r=-0.478, p<0.05; r=-0.700, p<0.01), and a positive correlation between the acceleration of whole body CG in the second half and the maximal forward lean angular velocity of the thigh (r=0.636, p<0.01; r=0.664, p<0.01).

DISCUSSION: In CS, the thigh contributed to accelerate the whole body CG in the first half, the foot indicated a great contribution in the second half of the support phase. The anterior lean angular velocity of the thigh and the ankle's plantarflexion in the support phase seems to be related to the change in the magnitude of the CG velocity. These results indicate that leaning the thigh increases the anterior RAF in the first half, and that the ankle plantarflexion helps increase the anterior RAF in the second half. The findings on the segmental contributions in the straight running (Ae et al., 1985), that an accelerative force was generated by the shank, followed by the thigh, and then the foot, implied that a technique for the velocity change of the whole body CG in cutting motion was similar to the case of straight running. In CS, the foot greatly contributed to the medial acceleration of the whole body CG in the support phase. The horizontal angle of the foot was inward to the CG velocity vector at the foot strike until just before the toe off in CS. When the toe directed inward, the medial RAF of the foot increased by using the ankle plantarflexion moment. Since the shank medial lean angle affected the medial RAF of the foot, leaning the shank and directing the toe inward makes the plantarflexion of the ankle be effective to increase in the medial RAF of the foot.

As in CS, the thigh in the first and the foot in the second half contributed to the acceleration of the whole body CG, implying that the anterior lean of the thigh and the plantarflexion of the ankle played an important role to increase anterior RAF in SS. In contrast to CS, the shank contributed to the medial acceleration of whole body CG around the mid-support in SS. After 20% of the support phase, the horizontal angle of the foot was medial in SS. The foot horizontal angle affected the medial velocity change of the whole body CG. When the toe was directed outward in the support phase, the medial RAF of the shank increased by using the ankle plantarflexion and knee extension moments in SS. Sigward and Powers (2006) indicated that the great knee valgus moment during SS may increase the risk of the knee injury. Therefore, directing the toe to the outward may be a good technique not only to increase the medial RAF of the shank but also to prevent the knee injury in SS.

CONCLUSIONS: The thigh in the first half and the foot in the second half of support phase contributed to increase the velocity of the whole body CG. The thigh anterior lean and ankle plantarflexion played an important role in cutting motion. In CS, the foot contributed to change the direction of the CG velocity with leaning the shank and directing the toe inward in the support phase. In SS, turning the foot to outward in the support phase may help increase the medial RAF of the shank with the ankle plantarflexion and knee extension.

REFERENCES:
Ae M.(1996) Body segment inertia parameters for Japanese children and athletes. *Japanese Journal of Sports Sciences, 15, 155-162.*
Ae M., Miyashita K., Shibukawa K., Yokoi T., and Hashihara Y.(1985) Body segment contributions during the support phase while running at different velocities. *BIOMECHANICS IX-B, pp 343-349.*
Andrews J.R., McLeod W.D., Ward T. and Howard K.(1977) The cutting mechanism. *The American Journal of Sports Medicine, 5(3), 111-121.*
Sigward S.M. and Powers C.M. (2006) The influence of gender on knee kinematics, kinetics and muscle activation patterns during side-step cutting. *Clinical Biomechanics, 21(1).*

TIBIAL ROTATIONS DURING STEP UP EXERCISE DO NOT CHANGE KNEE EXTENSOR ACTIVITY IN THE PATELLOFEMORAL PAIN SYNDROME

Juliana Moreno Carmona[1], Cristina Maria Nunes Cabral[2], Amélia Pasqual Marques[1]

[1] University of Sao Paulo, Sao Paulo SP, Brazil
[2] University of the City of Sao Paulo, Sao Paulo SP, Brazil

The purpose of this study was to compare the electromyographic activity of vastus medialis (VM) and vastus lateralis (VL) muscles during step up exercises at different tibial rotations, between individuals with and without patellofemoral pain syndrome. The electromyographic activity of the VM and VL muscles during step up exercises was evaluated in three tibial rotations: without rotation, medial and lateral rotation. Thirteen patients with patellofemoral pain syndrome and 16 control subjects performed this exercise three times. The results showed a significant lower electromyographic activity of VM muscle in both groups, compared to the electromyographic activity of VL muscle ($p<0.05$). Moreover, a significant difference in electromyographic activity between muscles for all rotations and groups ($p<0.001$) was also found. The results of the present study suggest that tibial rotations performed during step up exercise did not change the electromyographic activity of VM and VL muscles.

KEY WORDS: patellofemoral pain syndrome, EMG activity, step up exercise.

INTRODUCTION: Patellofemoral pain syndrome (PFPS) is one of the most common joint complaints in the orthopaedic practice (Powers, 1998; Davlin et al., 1999) and is frequently diagnosed in athletes (Miller et al., 1997; Tyler et al., 2006). It represents 25% to 40% of all knee problems found in sports medicine centers (Fagan et al., 2008) and affects 25% of total population (Fagan et al., 2008; Harrington et al., 2005). PFPS patients report symptoms of anterior or retropatellar pain, aggravated by functional activities as ascending and descending stairs. Considering all the etiologic factors involved, tibial rotation seems to be an important one.

In one of the first studies about the relationship between tibial rotation and quadriceps femoris muscle activation, Wheatley and Jahnke (1951) determined that lateral tibial rotation increases the vastus medialis (VM) function. On the other hand, more recent studies found that medial tibial rotation increases the activity of vastus medialis oblique muscle (VMO) and that tibial or femural rotation could contribute to decreased contact area of patellofemoral joint (PFJ). Salsich et al. (2007) concluded that the PFJ contact area can be increased by the regulators of tibial and femoral rotation, decreasing pain in PFPS patients. Miller et al. (1997) reported that VMO/vastus lateralis (VL) ratio during step up/down stairs and wall slide exercises with lateral tibial rotation was minor than in other rotations, which means that the VMO activity is bigger in this specific situation.

Thus the purpose of this study was to compare the electromyographic (EMG) activity of VM and VL muscles during step up exercises at three tibial rotations – without rotation, medial rotation and lateral rotation, between individuals with and without PFPS.

METHODS: Subjects: The population studied consisted of 13 patients with PFPS diagnosis (PFPS group – PG), 5 male and 9 female, and 16 individuals without any musculoskeletal injury in lower limbs (control group – CG), 7 male and 9 female.

The institutional review board (CAPPesq, the ethics committee for analysis of research projects of the Clinical Hospital of the University of Sao Paulo Medicine School) approved the study.

Patients were included if aged between 18 and 35, and showed a positive patellar compression test (Magee, 2002). Other inclusion criteria included symptoms of PFPS for at least six months, anterior or retropatellar pain during or after at least two of the following activities: prolonged sitting, stepping up or down stairs, squatting, kneeling, running and jumping, and an insidious onset of symptoms unrelated to a traumatic incident (Cowan et al., 2002). Participants were excluded if they showed signs or symptoms of any other knee pathology or injury (Thomeé, 1997).

Procedures: Before the evaluation, the subjects signed an informed consent form. The EMG activity of VM and VL muscles, of the injured leg in PG and the right leg in CG, was detected using an 8-channel EMG equipment with an analogical-digital converter (EMG System do Brasil) and resolution of 12 bits, interfaced with a computer and data collection software – AqDados 5.0 (Lynx Electronics Technologies), with an acquisition frequency of 1000 Hz per channel and band pass of 20-500 Hz; active differential surface electrodes (EMG System do Brasil); and self-adhesive electrodes (Meditrace). An electrogoniometer (EMG System do Brasil) was also used.

The electrodes were placed on the muscle according to SENIAM recommendations (Hermens et al., 2006). Self-adhesive electrodes were positioned on the muscle belly, 2 cm apart, and fixed with adhesive tape. The EMG activity of the VM and VL muscles was measured during three repetitions of step up a stair with 19.5 cm-height in three tibial rotations – without rotation (WR), with 20° of medial rotation (MR) and 20° of lateral rotation (LR), controlled by positioning the foot through a piece of wood with the determined angle, in a self-selected speed. The movement amplitude during step up was controlled by an electrogoniometer, which permitted to identify the exact EMG amplitude during the exercise. An isometric knee extension contraction with the knee flexed at 90 degrees and anterior resistance applied at the distal tibia, with the subject in a sitting position, was also collected during four seconds and the volunteers were asked to extend their knee as much as possible.

EMG data analysis: The EMG signal collected during isometric contraction was processed by means of the software Origin 6.0 routines and transformed into root mean square (RMS) values as follows: the wave acquired was rectified and filtered through a 5 Hz band pass filter to obtain a linear envelope. After a visual inspection of the envelope, the 1-second period showing least variation and maximum EMG activity was selected and the RMS value of the rectified signal calculated with a band pass filter of 20 to 500 Hz in the selected period. The mean of the three attempts was analyzed. The EMG signal collected during step up exercise was then normalized by the RMS values obtained during isometric contraction.

Statistical analysis: The significance level was set at $p<0.05$. All the variables were analyzed as to normality using the Kolmogorov-Smirnov test and variance homogeneity by Levene test. Demographic variables were compared using T test for independent samples (weight, height, body mass index) and Mann Whitney test (age). EMG data presented normal distribution. In this way, three ANOVA two-way were performed, the first compared the effect of tibial rotations in the activity of VM muscles between groups, the second compared the effect of tibial rotations in the activity of VL muscles between groups and the last compared the muscles within different rotations. Tukey was the post hoc test used.

RESULTS: Comparing demographic variables between groups, a statistically significant difference in age ($p=0.0178$) was observed, while other data did not differ (age: GC- 20 (20-24.5), PG- 28 (22-30) years; weight: GC- 65±7.5, PG- 66±10.5 kg; height: GC- 1.7±0.1, PG– 1.7±0.1 m; body mass index: GC- 21.7±3, PG- 22.6±0.4 kg/m^2).

Results obtained during the evaluation showed significant difference in EMG activity between muscles for all rotations and groups ($p<0.001$). Table 1 presents the means (standard deviation) of normalized RMS values by the isometric contraction of VM and VL muscles during step up exercise in all three rotations – WR, MR and LR, which were obtained in both groups. Interactions were not observed in the two-way ANOVA, considering groups, muscles and rotations ($p>0.05$).

Table 1: Means (standard deviation) of normalized RMS values collected, in %.

Rotation	Muscles	PFPS Group (n=13)	Control Group (n=16)
Without rotation	VM	144.7 (160.4)[1]	132.1 (47.2)[4]
	VL	232.8 (221.8)[1]	190.1 (87.6)[4]
Medial rotation	VM	146.7 (159.4)[2]	134.6 (60.5)[5]
	VL	233.2 (222.7)[2]	173.8 (79.5)[5]
Lateral rotation	VM	143.8 (158.9)[3]	109.3 (138.3)[6]
	VL	233.3 (223.1)[3]	183.3 (91.3)[6]

[1] Significant difference between muscles (ANOVA p<0.001; Tukey p<0.001)
[2] Significant difference between muscles (ANOVA p<0.001; Tukey p<0.001)
[3] Significant difference between muscles (ANOVA p<0.001; Tukey p<0.001)
[4] Significant difference between muscles (ANOVA p<0.001; Tukey p<0.0068)
[5] Significant difference between muscles (ANOVA p<0.001; Tukey p<0.0477)
[6] Significant difference between muscles (ANOVA p<0.001; Tukey p<0.0305)

DISCUSSION: The results showed no significant difference in EMG activity of VM and VL muscles in the step up exercise with different tibial rotations between groups. The mean RMS values for VM and VL muscles in CG and PG were similar, despite previous results found in the literature (Wheatley and Jahnke, 1951; Miller et al., 1997).

One of the reasons for this difference can be the location of electrodes. In the present study, the SENIAM protocol was used, so the whole VM muscle had its EMG activity measured, not only its oblique fibers. These differences in the location of electrodes may lead to the difference of EMG data observed between this study and others found in the literature. On the other hand, Miller at al. (1997) reported EMG activity of VMO muscle, which was the dynamic stabilizer of the patella, and found fewer EMG activity of this muscle. Besides that, the authors analyzed the VMO/VL ratio, while the present study compared the normalized RMS values of VM and VL muscles separately.

Another methodological difference is that EMG data was acquired during a self-selected speed activity because our goal was to reproduce a natural movement that would not occur if the speed was controlled. This methodology was used in previous studies (Sacco et al., 2006).

Surprisingly, EMG activity of VL is bigger than VM muscle in both groups and rotations. Our supposition is that the stair height could provoke this result, because it is well known that if flexion angle of hip and/or knee increases, quadriceps femoris muscle activation is changed. When flexion angle of knee is low, the EMG activity of quadriceps femoris is low too (Cabral, 2001; Sheehy et al., 1998). But the finding about a greater activity of VL muscle in the CG needs more investigation.

Another possible reason for the differences found in this study according to the literature, is that our PG subjects perform regular sports activities, which could give more functional structure to muscles and change the personal EMG activity standard.

CONCLUSION: The results of the present study suggest that tibial rotations performed during step up exercise did not change the EMG activity of VM and VL muscles. Further studies are necessary to clarify the knee extensor function role in PFPS patients, since this study found that EMG activity of VL muscles is bigger than of VM, even in healthy subjects, suggesting that the knee extensor EMG activity may not be sufficient to predict the presence of PFPS.

REFERENCES:

Cabral CMN. (2001). Efeito dos exercícios em cadeia cinética fechada realizados no step na atividade elétrica dos componentes medial e lateral do músculo quadríceps femoral (dissertação de mestrado). São Paulo.

Cowan SM, Hodges PW, Bennell KL, Crossley KM. (2002). Altered vastii recruitment when people with patellofemoral pain syndrome complete a postural task. *Archives of Physical Medicine and Rehabilitation*, 83: 989-95.

Davlin CD, Holcomb WR, Guadagnoli MA. (1999). The effect of hip position and electromyographic biofeedback training on the vastus medialis oblique : vastus lateralis ratio. *Journal of Athletic Training.* 34 (4): 342-49.

Fagan V, Delahunt E. (2008). Patellofemoral Pain Syndrome: a review on the associated neuromuscular deficits and current treatment options. *British Journal of Sports Medicin.* Jul; 14.

Harrington L, Malloy S, Richards J. (2005). The effect of patella taping on vastus medialis oblique and vastus lateralis EMG activity and knee kinematic variables during stair descent. *Journal of Electromyography Kinesiology.* Dec; 15 (6): 604-7.

Hermens HJ, Freriks B. SENIAM. Enschede: Surface ElectroMyoGraphy for the Non-Invasive Assessment of Muscles. *http://www.seniam.org*, disponível em três de dezembro de 2006.

Magee DJ (2002). *Avaliação musculoesquelética.* São Paulo: Manole.

Miller JP, Sedory D, Croce RV. (1997). Leg rotation and vastus medialis oblique/vastus lateralis electromyogram activity ratio during closed chain kinetic exercises prescribed for patellofemoral pain. *Journal of Athletic Training.* Jul; 32 (3): 216-220.

Powers CM. (1998). Rehabilitation of patellofemoral joint disorders: a critical review. *Journal of Orthopaedic and Sports Physical Therapy.* Nov; 28 (5): 345-54.

Sacco ICN, Konno GK, Rojas GB, Arnone AC, Pássaro AC, Marques AP, Cabral CMN. (2006). Functional and EMG responses to a physical treatment in patellofemoral syndrome patients. *Journal of Electromyography and Kinesiology.*16:167-74.

Salsich GB, Perman WH. (2007). Patellofemoral joint contact area is influenced by tibiofemoral rotation alignment in individuals who have patellofemoral pain. *Journal of Orthopaedic and Sports Physical Therapy.* Sep;37(9):521-8.

Seehy P, Burdett RG, Irrgang JJ, VanSwearingen J. (1998). An electromyographic study of vastus medialis oblique and vastus lateralis activity while ascending and descending steps. *Journal of Orthopaedic and Sports Physical Therapy.* Jun;27(6):423-9.

Thomeé R. (1997). A comprehensive treatment approach for patellofemoral pain syndrome in young women. *Physical Therapy*, 77(12): 1690-703.

Tyler TF, Nicholas SJ, Mullaney MJ, McHugh MP. (2006). The role of hip muscle function in the treatment of patellofemoral pain syndrome. *American Journal of Sports Medicine.* 34 (4): 630-36.

Wheatley and Jahnke. Electromyographic study of the superficial thigh and hip muscles in normal individuals. (1951). *Archives of Physical Medicine and Rehabilitation.* Aug;32(8):508-15.

Acknowledgment: I would like to thank the Conselho Nacional de Desenvolvimento Científico e Tecnológico (CNPq) for the financial support given to this study and all my laboratory group, especially Cabral CMN and Marques AP for all time spent before, during and after the execution of this project.

BASIC PRINCIPLE OF RIDING ON A SNAKEBOARD

Alexander S. Kuleshov[1]

Department of Mechanics and Mathematics, Moscow State University, Moscow, Russia[1]

This paper describes the mathematical model of a derivative of a skateboard known as the snakeboard. Equations of motion of the model are derived and their analytical and numerical investigations are fulfilled assuming harmonic excitation for the angles of rotation by feet and a torso of the rider. The possibility of the forward motion for the snakeboard is analyzed.

KEY WORDS: snakeboard, dynamics, analysis of motion.

INTRODUCTION: The Snakeboard (Figure 1) is one of the modifications of a well-known skateboard. It allows the rider to propel himself forward without having to kick off the ground. The motion of the snakeboard becomes possible due to specific features of its construction and due to the special coordinated motions of the rider's feet and body. The first snakeboard appeared in 1989 and from this moment many fans among the amateurs of extreme sports it has found. Soon after the invention of a snakeboard the first attempts to describe the basic principles of motion have been made. The basic mathematical model for the snakeboard has been proposed by Lewis et al. (1994). In this paper we give the rigorous mathematical proof of the conditions for the forward motion of a snakeboard.

Figure 1: The Snakeboard. (Reproduced from Smith et al., (1991)).

The snakeboard consists of two wheel-based platforms upon which the rider is to place each of his feet. These platforms are connected by a rigid crossbar with hinges at each platform to allow rotation about the vertical axis. To propel the snakeboard the rider first turns both of his feet in. By moving his torso through an angle, the snakeboard moves through an arc defined by the wheel angles. The rider then turns both feet so that they point out and moves his torso in the opposite direction. By continuing this process the snakeboard may be propelled in the forward direction without the rider having to touch the ground.

METHODS: MATHEMATICAL MODEL AND EQUATIONS OF MOTION: The mathematical model of the snakeboard considered in this paper is represented in Figure 2. We assume that the snakeboard moves on the xy plane and let Oxy be the fixed coordinate system with origin at any point of this plane. Let x and y be the coordinates of the system centre of mass (point G) and θ is the angle between the central line of the snakeboard and the Ox-axis. In the basic model treated by Lewis et al. (1994) the platforms could rotate through the same angle in opposite directions with respect to a central line of the snakeboard (in other words, for the model described by Lewis et al. (1994) we have $\varphi_f = -\varphi_b = \varphi$, see Fig. 2). We suppose that platforms can rotate independently and their positions are defined by two independent variables φ_f and φ_b. The motion of the rider is modelled by a rotor. Its angle of rotation with respect to the crossbar is denoted by δ.

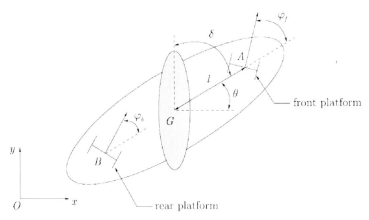

Figure 2: The Mathematical Model of the snakeboard. (Image modified from Lewis et al. (1994)).

Further we describe the motion of platforms using new variables ψ_1 and ψ_2 connected with variables φ_f and φ_b by relations:

$$\psi_1 = \varphi_f - \varphi_b, \qquad \psi_2 = \varphi_f + \varphi_b.$$

If $\varphi_f \neq \varphi_b \neq 0$ then there is a point on the line passing through a crossbar which has a zero lateral velocity and hence only the velocity along the crossbar. We denote this velocity by V. Control of the Snakeboard is realized by rotations of the platforms through φ_f and φ_b and by rotation of the rotor through δ. Therefore we assume that the variables δ, ψ_1 and ψ_2 are known functions of time t i.e.

$$\delta = \delta(t), \quad \psi_1 = \psi_1(t), \quad \psi_2 = \psi_2(t).$$

These variables are the controlled variables in this problem.
Parameters for the problem are:
m_b : the mass of the crossbar;
m_r : the mass of the rotor;
m_p : the mass of every platform (we assume the platforms are identical);
$m = m_b + m_r + 2m_p$: the total mass of the system;
J_b : the moment of inertia of the crossbar;
J_r : the moment of inertia of the rotor;
J_p : the moment of inertia of every platform;
ℓ : the length from the board's center of mass to the location of the wheels;

$$k^2 = \frac{J_b + J_r + 2J_p + 2m\ell^2}{m\ell^2}, \quad d_1 = \frac{J_r}{m\ell}, \quad d_2 = \frac{J_p}{m\ell}.$$

Equations of motion of the considered model of a snakeboard have the form (Kuleshov, 2007):

$$\dot{x} = V\cos\theta - \frac{V\sin\psi_2 \sin\theta}{\cos\psi_1 + \cos\psi_2}, \quad \dot{y} = V\sin\theta + \frac{V\sin\psi_2 \cos\theta}{\cos\psi_1 + \cos\psi_2}, \quad \dot{\theta} = \frac{V\sin\psi_1}{(\cos\psi_1 + \cos\psi_2)\ell}, \quad (1)$$

$$P_1(t)\dot{V} + P_2(t)V = Q(t), \quad (2)$$

$$P_1(t) = 1 + \frac{k^2 \sin^2\psi_1}{(\cos\psi_1 + \cos\psi_2)^2} + \frac{\sin^2\psi_2}{(\cos\psi_1 + \cos\psi_2)^2}, \quad Q(t) = -\frac{(d_1\ddot{\delta} + d_2\ddot{\psi}_2)\sin\psi_1}{\cos\psi_1 + \cos\psi_2},$$

$$P_2(t) = \frac{(\dot{\psi}_2 \sin\psi_2 \cos\psi_2 + k^2\dot{\psi}_1 \sin\psi_1 \cos\psi_1)}{(\cos\psi_1 + \cos\psi_2)^2} + \frac{(\dot{\psi}_1 \sin\psi_1 + \dot{\psi}_2 \sin\psi_2)(k^2 \sin^2\psi_1 + \sin^2\psi_2)}{(\cos\psi_1 + \cos\psi_2)^3}.$$

Equation (2) determines the dependence of the velocity V on the controlled variables $\delta(t)$, $\psi_1(t)$ and $\psi_2(t)$. Suppose that $V(0) = V_0 = 0$. Then the solution of equation (2) can be written as follows:

$$V(t) = -\frac{(\cos\psi_1(t) + \cos\psi_2(t))}{\sqrt{k^2 \sin^2\psi_1(t) + (\cos\psi_1(t) + \cos\psi_2(t))^2 + \sin^2\psi_2(t)}} \times \int_0^t \frac{\sin\psi_1(\tau)(d_1\ddot{\delta}(\tau) + d_2\ddot{\psi}_2(\tau))d\tau}{\sqrt{k^2 \sin^2\psi_1(\tau) + (\cos\psi_1(\tau) + \cos\psi_2(\tau))^2 + \sin^2\psi_2(\tau)}}. \quad (3)$$

Having the expression for $V(t)$ we obtain from the third equation of the system (1)

$$\theta(t) = \theta(0) + \int_0^t \frac{V(\tau)}{\ell} \frac{\sin\psi_1(\tau)}{(\cos\psi_1(\tau) + \cos\psi_2(\tau))} d\tau. \quad (4)$$

Using this formula we can obtain from the first two equations of the system (1):

$$x(t) = x(0) + \int_0^t V(\tau)\left(\cos\theta(\tau) - \frac{\sin\psi_2(\tau)\sin\theta(\tau)}{\cos\psi_1(\tau) + \cos\psi_2(\tau)}\right)d\tau,$$

$$y(t) = y(0) + \int_0^t V(\tau)\left(\sin\theta(\tau) + \frac{\sin\psi_2(\tau)\cos\theta(\tau)}{\cos\psi_1(\tau) + \cos\psi_2(\tau)}\right)d\tau \quad (5)$$

Thus, the problem of a Snakeboard dynamics at arbitrary controlled variables $\delta(t)$, $\psi_1(t)$ and $\psi_2(t)$ is completely solved in terms of integrals (3)-(5). However the calculation of these integrals for given controlled variables and the analysis of the exact solution is a rather complicated problem. Below we assume the harmonic excitation for the controlled variables.

RESULTS: Observations of actual snakeboard riders suggest that sinusoidal inputs provide a good starting point for our investigations:

$$\delta = a_0 \sin(\omega_0 t), \quad \psi_1 = a_1\varepsilon \sin(\omega_1 t), \quad \psi_2 = a_2\varepsilon \sin(\omega_2 t).$$

Here ε is a parameter. We assume that ε is sufficiently small such that for the angles ψ_1 and ψ_2 the following approximate formulae

$$\sin\psi_1 \approx \psi_1, \quad \sin\psi_2 \approx \psi_2, \quad \cos\psi_1 \approx 1 - \frac{\psi_1^2}{2}, \quad \cos\psi_1 \approx 1 - \frac{\psi_1^2}{2}$$

are valid. In other words, we will neglect the terms of order higher than the second on the parameter ε. This assumption is completely justified by the snakeboard construction.

The snakeboard is assumed to have its initial condition at the origin in the space state, i.e.

$$x(0) = 0, \quad y(0) = 0, \quad \theta(0) = 0.$$

Taking into account all these assumptions we have the following simplified formula for the velocity $V(t)$:

$$V(t) \approx \frac{d_1 \varepsilon a_0 a_1 \omega_0^2}{2}\int_0^t \sin(\omega_0\tau)\sin(\omega_1\tau)d\tau + \frac{d_2 \varepsilon^2 a_1 a_2 \omega_2^2}{2}\int_0^t \sin(\omega_1\tau)\sin(\omega_2\tau)d\tau.$$

This integral will be a periodic function except the case

$$\omega_0 = \omega_1 = \omega_2 = \omega.$$

In this case the velocity $V(t)$ is a linear function of time:

$$V(t) = \frac{(d_1 a_0 + d_2 a_2\varepsilon) a_1 \omega\varepsilon}{4}(\omega t - \sin(\omega t)\cos(\omega t)).$$

Thus we can formulate the main principle of a Snakeboard dynamics: to propel the snakeboard forward the rider should rotates his torso and his feet so that the frequencies of rotation by a torso and feet would be equal.
From the third of equations (1) an approximate formula for $\theta(t)$ has the form:

$$\theta(t) = \frac{d_1 a_0 a_1^2 \varepsilon^2}{8\ell}\left(\sin(\omega t) - \omega t \cos(\omega t) - \frac{\sin^3(\omega t)}{3}\right).$$

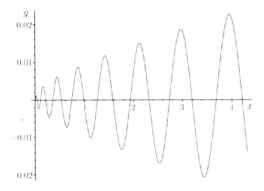

Figure 3: Trajectory of the snakeboard's center of mass.

Obviously, the function $\theta(t)$ as well as $V(t)$ is a linearly growing function of time. Therefore in a time interval of order $1/\varepsilon$ the angle θ remains proportional to ε. Thus in this time interval we can consider θ as a small angle. For small values of θ we have for $x(t)$ and $y(t)$:

$$x(t) = \frac{(d_1 a_0 + d_2 a_2 \varepsilon) a_1 \varepsilon}{8}\left(\omega^2 t^2 - \sin^2(\omega t)\right),$$

$$y(t) = \frac{d_1 a_0 a_1 a_2 \varepsilon^2}{4}\left(\sin(\omega t) - \omega t \cos(\omega t) - \frac{\sin^3(\omega t)}{3}\right).$$

Fig. 3 shows the trajectory of the snakeboard's center of mass in the considered case.

CONCLUSION: In this paper we found the condition for the forward motion of a snakeboard. We have proved that the rider propels the snakeboard forward using the 1:1 resonance between two frequencies: the frequency of rotation by a torso and the frequency of rotation by feet. These two frequencies should be equal.
Other snakeboard gaits have been investigated numerically in (Lewis et al., 1994) and analytically in (Kuleshov, 2007). It can be proved that every snakeboard gait can be achieved by a corresponding choice of resonant condition between the frequencies of rotation by a torso and feet. These results will be helpful for all people who makes the first steps in a snakeboard riding.

REFERENCES:
Kuleshov, A.S. (2007). Further Development of the Mathematical Model of a Snakeboard. *Regular and Chaotic Dynamics*, 12 (3), 321-334.
Lewis, A.D., Ostrowski, J.P., Murray, R.M. & Burdick, J.W. (1994). Nonholonomic Mechanics and Locomotion: the Snakeboard Example. *Proceedings of the IEEE ICRA*, 2391-2400.
Smith, E.O.M., Fisher J., King S. (1991). Skateboard. United States Patent USD338253.
http://www.freepatentsonline.com/D338253.pdf

LUMBAR SPINE IN SENIOR AND ELITE LEVEL ROWERS – A COMPARISON WITH THE LOW BACK PAIN POPULATION

Caroline MacManus[1] and Kieran O'Sullivan[1]

Department of Physiotherapy, University of Limerick, Limerick[1]

KEY WORDS: rowing, lumbar spine.

INTRODUCTION: Low back pain (LBP) is very prevalent in the sport of rowing. Elite rowers miss on average 24 days from training in rowing related injuries per year (Bernstein et al 2002). A wireless posture monitor (Sels Instruments, Belgium) has recently been developed, which can measure lumbar spine posture in real time. The monitor has good face validity and laboratory bench testing has established the accuracy of the monitor. The aims of this investigation were to; (1) determine the reliability of this novel method of monitoring lumbar spine posture during rowing, and (2) determine if there are differences in lumbar spine posture between rowers with/without LBP during an incremental "step-test". These original pilot studies will inform further development of the monitor prior to final validation studies, and use of the monitor in larger studies.

METHODS: Ethical approval was attained. The monitor is based on a strain gauge where increasing elongation of the monitor's strain gauge occurs with spinal flexion. The posture monitor was placed on the subjects pelvis (2nd sacral spinous process) and lumbar spine (3rd lumbar spinous process) determined by palpation. For the reliability study (n=7), subjects (6 males, 1 female, mean ± SD, age 29.4 ± 7.6 years, height 1.84 ± 0.57 m, weight 85.9 ± 9.8 kg) performed stage 1 of the "step-test" on 3 occasions. To compare LBP and non-LBP groups, 11 male subjects were recruited; LBP=6; controls =5; (mean ± SD, age 29.2 ± 5.6 years, height 1.88 ± 0.67 m, weight 86.3 ± 8.4 kgs). LBP subjects had self-reported "non-specific" LBP. They performed an incremental "step-test" which consisted of six 4-minute stages on a Concept 2 Model D ergometer.

RESULTS: The posture monitor demonstrated high reliability (all ICC values > 0.84, with low mean differences of <4% ROM). Comparing the LBP and non-LBP groups, a one-way ANOVA (with repeated measures for the step-test stages) revealed there was no significant interaction effect (p=0.541). On average across the two groups (LBP, non-LBP) there was no significant main effect for time (p=0.257) or group (p=0.620).

DISCUSSION: The posture monitor demonstrates high reliability. No significant differences between groups were noted within the "step-test". There was a non-significant trend for the LBP group to exhibit more extension than non-LBP group.

CONCLUSION: The study established that the posture monitor demonstrates high reliability and could be a useful tool in the field. The validity of the monitor as a measure of lumbar posture and movement compared to video fluoroscopy and another laboratory-based motion analysis system is currently being planned. When these studies are complete, larger studies using this device will be performed. This will include using more homogenous subgroups of LBP subjects, in addition to prolonged rowing.

REFERENCES:

Bernstein, I. A., Webber, O. & Woledge, R. (2002) An ergonomic comparison of rowing machine designs: possible implications for safety. *British Journal of Sports Medicine,* 36, 108-112.

Acknowledgement
The authors would like to thank all rowers who participated in the study.

EFFECT OF A SPECIFIC STRENGTH TRAINING ON THE DEPTH SQUAT WITH DIFFERENT LOAD: A CASE STUDY.

Ciacci S., Pecoraioli F.

Faculty of Exercise and Sport Sciences, University of Bologna, Italy
Dept. of Histology, Embryology and Applied Biology, University of Bologna, Italy

The purpose of this study is to verify if a specific strength training program can reduce the transition time from descent to ascent phase in squat exercise. The kinematic and temporal data of 3 depth squats with different load (35%, 40%, 45% of 1RM) during 2 test sessions, before and after a specific strength training program, have been collected.
The Track and Field 400 m Italian Champion have been analysed during the depth squat through an optoelectronic stereophotogrammetric system for tridimensional motion analysis. The collected data were: knee and hip angles, time of eccentric and concentric phase (descent and ascent times), the transition time (from descent to ascent phase).
The results show that a specific strength training can develop the speed of movement and, particularly, decrease the transition phase time at heavy load. This method can be useful to search an optimal depth squat load and to develop the Stretch Shortening Cycle (SSC) speed.

KEY WORDS: strength training, transition phase, 3d kinematic analysis, SSC time, sprint ability.

INTRODUCTION: The strength training is a fundamental component for the optimal performance in several sport (Rahmani 2001, Gollhofer 2003, Wisloff 2004, Harris 2008), especially in the sprint (Sleivert 2003, Harris 2008, Holm 2008). The squat is the most investigated strength exercise and the most used movement in strength training for the leg. The researchers studied this exercise in order both to explain the sport performance (Wisloff 2004, Harris 2008) and to prevent injuries (Escamilla 2001).
The squat parameters that the researchers studied, were: kinematics data (knee and hip angles and their angular velocity), temporal values (time of SSC), kinetics (joint moments) and EMG data (Rahmani 2001, Manabe 2007, Harris 2008).
The special strength training can improve the load used during a squat exercise or its stretch-shortening cycle speed (Rahmani 2001, Sleivert 2003, Manabe 2007, Harris 2008). The SSC speed, especially the short transition phase from eccentric to concentric phase, are fundamental to improve the powerful push off (Gollhofer 2003) and the sprinting performance (Sleivert 2003). Previous studies demonstrated that a powerful push off is possible only if the time of the SSC and especially of the transition phase are short (Manabe 2006, Gollhofer 2003). The purpose of this study is to verify if a specific strength training can improve the speed of depth squat exercise and, particularly, its transition time.

METHODS: The Track and Field 400 m National Italian Champion (183 cm high, 71 kg weight, personal best on 400 m, 46.30 s) have been analysed through an optoelectronic stereophotogrammetric system for tridimensional motion analysis. (VICON 460, Oxford Metrics, UK). The data were collected using six infrared 1.3 megapixel cameras, with a 100 Hz sampling frequency. A modified Helen Hayes marker set was used with the addition of a marker on the head of the fifth metatarsal and one marker on the trunk.
The subject performed 2 depth squat at 3 different loads (30%, 35%, 40% 1RM) all at maximum speed, before and after a special strength training program (4 training week). This program included 3 workouts for each week, except for the 4th (only 2 workouts). During each workout the athlete performed 4 set of 3 movements of depth squat with loads included between the 30% and 60% of 1RM, combined with 4 set of countermovement or drop jump.
During the test and the training, to standardize the exercise the barbell's movement have been checked through 2 marks on the bar supports.

Data collected were: angles and angular flexion/extension velocities of knee and hip joint and the SSC time (descent/eccentric, ascent/concentric an transition phase time).
The descending phase was defined as the period of movement from maximum bar height before the descent to the lowest point of the squat; the ascending phase was defined as the period from the lowest point of the squat to maximum bar height after ascent. The transition phase was defined as the period of the movement from the last millimetre of knee and hip markers in vertical and horizontal axis during the descent phase to the first millimetre of the same markers during the ascending phase, according to the system accuracy. During the transition phase the knee angle supposedly stays the same (Bosco et al. 1981).

RESULTS: In table 1 are reported the time of the different phases (Eccentric=ECC, Transition =T.P. and Concentric phase=CONC) at each load and the Range of Motions (ROM) of the knee joint during both test session (Pre-test and Post-test, before and after the training, respectively). The results show the speed development of the movement during the Post-test (Table 2). Particularly, the speed development is notable at 87 and 97 Kg of load, while at 77 kg the data appear very similar.

Table 1:

KNEE	Pre-test					Post-test				
30% 1RM	ECC. 1/100s	T.P. 1/100s	CONC. 1/100s	TOT 1/100s	ROM Deg.	ECC. 1/100s	T.P. 1/100s	CONC. 1/100s	TOT 1/100s	ROM Deg.
Average	59	38	62	121	102	59	33	61	120	98
SD	4	3	1	4	4	3	8	3	3	5
	Pre-test					Post-test				
35% 1RM	ECC. 1/100s	T.P. 1/100s	CONC. 1/100s	TOT 1/100s	ROM Deg.	ECC. 1/100s	T.P. 1/100s	CONC. 1/100s	TOT 1/100s	ROM Deg.
Average	63	48	72	135	101	55	40	67	122	100
SD	3	3	2	2	1	1	3	1	1	3
	Pre-test					Post-test				
40% 1RM	ECC. 1/100s	T.P. 1/100s	CONC. 1/100s	TOT 1/100s	ROM Deg.	ECC. 1/100s	T.P. 1/100s	CONC. 1/100s	TOT 1/100s	ROM Deg.
Average	66	61	78	144	98	62	51	66	128	95
SD	6	1	5	2	1	4	4	1	5	4

In table 2 are reported the differences between the data before and after the training. The grey colour underline the developments obtained in the post-test session.

Table 2:

KNEE JOINT DIFFERENCES					
% 1RM	ECC. 1/100s	T.P. 1/100s	CONC. 1/100s	TOT 1/100s	ROM Deg.
30%	0	-6	-1	-1	-4
35%	-8	-8	-5	-13	-1
40%	-4	-10	-12	-16	-3

The ROM of the knee results are very similar in both session tests and are comparable with the same data of others studies (Domire Z.J. 2007), confirming that the developments have been obtained in the speed of movement. A similar study (Manabe 2007) show the knee and hip joint ROM shorter than the same data of this research and consequently also the time of movement are shorter.
The same values of hip joint are collected in Table 3. Also about this joint, the greater developments of the speed have been notable in the post-test session. Table 4 shows the better performance obtained from the athlete during the post-test session, especially in the heaviest load squat. It's interesting to notice that at the lightest load (30% of 1RM), the subject at post-test make the movement more quickly than pre-test, even if the ROM is greater.

Table 3

HIP	Pre-test					Post-test				
30%1RM	ECC. 1/100s	T.P. 1/100s	CONC. 1/100s	TOT 1/100s	ROM Deg.	ECC. 1/100s	T.P. 1/100s	CONC. 1/100s	TOT 1/100s	ROM Deg.
Average	54	53	73	127	66	56	45	68	124	75
SD	4	2	9	10	3	4	2	12	11	5
	Pre-test					Post-test				
35%1RM	ECC. 1/100s	T.P. 1/100s	CONC. 1/100s	TOT 1/100s	ROM Deg.	ECC. 1/100s	T.P. 1/100s	CONC. 1/100s	TOT 1/100s	ROM Deg.
Average	59	66	75	134	68	52	52	71	123	63
SD	4	3	3	2	3	0	2	1	1	13
	Pre-test					Post-test				
40%1RM	ECC. 1/100s	T.P. 1/100s	CONC. 1/100s	TOT 1/100s	ROM Deg.	ECC. 1/100s	T.P. 1/100s	CONC. 1/100s	TOT 1/100s	ROM Deg.
Average	60	77	83	143	63	59	57	69	128	66
SD	4	4	2	3	1	3	4	3	3	3

Table 4

	HIP JOINT DIFFERENCES				
% 1RM	ECC. 1/100s	T.P. 1/100s	CONC. 1/100s	TOT 1/100s	ROM Deg.
30%	2	-8	-5	-3	8
35%	-7	-14	-4	-11	-5
40%	-1	-20	-14	-15	4

Also the angular velocities of the Flexion and Extension Peak at pre-test (FP1 and EP1 respectively) and the same data of post-test (FP2 and EP2 respectively) confirm the best performances obtained during the post-test (Table 5 and 6).

Table 5

°/sec	KNEE				HIP			
30%1RM	FP1	EP1	FP2	EP2	FP1	EP1	FP2	EP2
Average	321	324	355	376	294	210	332	248
SD ±	59,8	19,5	87,3	5,9	26,3	12,1	32,5	11,1
35%1RM	FP1	EP1	FP2	EP2	FP1	EP1	FP2	EP2
Average	395	334	418	360	321	205	348	222
SD ±	116,9	23	21,7	12,4	32,9	14,7	25,6	9,96
40%1RM	FP1	EP1	FP2	EP2	FP1	EP1	FP2	EP2
Average	357	319	358	345	285	212	312	216
SD ±	86,8	15,4	120,9	12,4	25,4	25,8	44,1	5,8

Table 6

°/sec	Knee Differences		Hip Differences	
% 1RM	FP	EP	FP	EP
30%	-33,8	-51,7	-38,1	-38,1
35%	-23,3	-25,3	-27,5	-16,4
40%	-1,1	-25,6	-26,8	-4,0

DISCUSSION: The results show that the kinematic and temporal data of depth squat obtained in the post-test (after the specific strength training program) are better than the same collected in the pre-test session. The time of eccentric and concentric phase is shorter at each loads of post-test. Particularly is shorter the transition phase time; the development of this time was the most important aim of this study.

All the enhancements are more noticeable at the heavier loads. In fact the improvement at 30% of 1RM seems to be not very important; this can to mean that the load is too easy to influence positively the strength training and consequently the results of the test before and after the training are similar. Especially about the knee joint time the data show not important differences (table 2).

Whereas at 30% of 1RM, it is interesting to notice the improvement of the flexion and extension angular velocity of the knee and hip joint.

Instead at the 35% and 40% of 1RM loads the improvement are considerable about all parameters analysed. Especially the transition and concentric phase time are shorter in post-test than the same data in pre-test session.

Also the flexion and extension angular velocities of knee and hip joint are faster. These data underline how it is very important in the strength training the choice of the right load to improve a specific performance parameter.

CONCLUSION: The results show that the specific strength training can be useful to improve the speed of depth squat exercise. Especially the aim of the specific training used in this research was to improve the velocity of the transition phase from descent/eccentric to ascent/concentric phase. In fact the others studies assert that if the transition phase is slower or delayed, no enhancement of the concentric phase occurs (Gollhofer 2003) with a consequent loss of power. The results of this study showed that this specific strength program resulted helpful to develop the transition phase of the squat in the observed athlete.

However in this study only 1 subject was analysed, therefore the conclusions are limited. To generalize the results of this research, it is necessary to analyze a sample including more athletes. But after this research seems to be possible to confirm the others studies of the literature that assert that the shorter SSC and the correct load in the squat exercise are decisive elements to improvement the special strength and the sprint performance (Rahmani 2001, Sleivert 2004, Harris 2008).

REFERENCES:

Bosco C., Komi, P.V., Ito, A. (1981) Prestretch potentiation of human skeletal muscle during ballistic movement Acta Physiol Scand, 111, 135-140.

Domire, J.Z., Challis, J.H. (2007) The influence of squat depth on maximal vertical jump performance. *Journal of Sport Sciences*, 25(2), 193-200.

Escamilla, R.F. (2001) Knee biomechanics of the dynamic squat exercise. *Med. Sci. Sports Exerc*, 33,1,127–141.

Gollhofer, A., Bruhn, S., (2003). The biomechanics of jumping. In: Gollhofer, A., Bruhn, S. (Eds.), *Handbook of Sports Medicine and Science: Volleyball. Blackwell Science Pub Publisher*, 3, 18-28.

Harris, N.K., Cronin, J.B., Hopkins, W.G., Hansen, K.T. (2008) Squat jump training at maximal power loads vs. Heavy loads: effect on sprint ability. *J Strength Cond Res* 22(6), 1742-1749.

Holm, D.J., Stalbom, M., Keogh, J.W.L., Cronin, J. (2008) Relationship between the kinetics and kinematics of a unilateral horizontal drop jump to sprint performance. *J Strength Cond Res* 22(5), 1589-1596.

Manabe, Y., Shimada, K., Ogata, M. (2007) Effect of slow movement and stretch-shortening cycle on lower extremity muscle activity and joint moments during squat. *J Sports med Phys Fitness*, 47,1-12.

Rahmani, A. Viale, F., Daleau, G., Lacour, J.R. (2001) Force/velocity and power/velocity relationships in squat exercise. *Eur J Appl Physiol*, 84, 227-232

Sleivert, G., Taingahue, M. (2004) The relationship between maximal jump-squat power and sprint acceleration in athletes. *Eur J Appl Physiol*, 91, 46-52.

Wisloff, U., Castagna, C., Helgerud, J., Jones, R., Hoff, J. (2004) Strong correlation of maximal squat strength with sprint performance and vertical jump height in elite soccer players. *Br J Sports Med*; 38:285–288.

EVALUATION OF SADDLE HEIGHT IN ELITE CYCLISTS

Marlene Mauch[1] and Andreas Goesele[1]
crossklinik, Swiss Olympic Medical Center, Basel, Switzerland[1]

KEY WORDS: cycling, seating position, anthropometrics.

INTRODUCTION: Proper bike fit is essential to prevent injuries and improve performance. Especially recreational cyclists, often short on experience, rely on professional adjustment for seating position. Common bike fitting methods use formulas based on rider anthropometrics, e.g. the LeMond method (pubic symphysis height [PSH] x 0.883) (LeMond, 1988) or the Hamley method (PSH x 1.09) (Hamley et al., 1967). Pruitt (2006) recommends the correct saddle height [SH] within a knee angle [KA] of 25-35°. The purpose of this study is to verify these methods in elite men and women cyclists for today's application.

METHODS: Anthropometrics and SH of 31 (8 women, 23 men; 27.2 ±4.1 years) elite cyclists were measured using a water-level ruler and measuring tape. Differences between nominal (calculated SH based on LeMond and Hamley method) and actual (measured) SH were analysed. Additionally, 25 of these cyclists were recorded using a video camera (25Hz) (Sony®) and KA were analysed using Templo® (Contemplas Inc., Germany). Independent t-test was used to determine the differences between men and women.

RESULTS: PSH averages 84.4 ±4.3 cm, KA 41.8° ±4.4° (Table 1). There were significant differences between women and men for both the LeMond ($p<0.01$) and Hamley methods ($p<0.05$): on average women sat 1.1 cm lower and men 0.4 and 0.1 cm higher than the calculated value recommended (Table 2).

Table 1 Knee angle (°)

Sex	Min	Max	Mean	SD
Women (n=7)	38.7	45.7	41.9	2.56
Men (n=18)	32.2	51.9	41.8	5.03

Table 2 Nominal/actual value comparison (cm)

Sex	LeMond-Method Difference		Hamley-Method Difference	
	Mean	SD	Mean	SD
Women (n=8)	-1.12	± 1.55	-1.12	± 1.56
Men (n=23)	0.43	± 1.18	0.10	± 1.25

DISCUSSION: The results indicate that elite cyclists, especially women, sit lower than recommended in current literature (Pruitt, 2006). This lower position could enable them to generate more power or achieve a better aerodynamic position. Sex-related differences could relate to different soft tissue characteristics or preferences in seating position.

CONCLUSION: In recreational cyclists, preventing injuries and overuse syndromes should have a high priority to allow for enjoyable and long-lasting activity. A low SH is often accompanied by injures (Pruitt, 2006). Further (prospective) studies should focus on the relationship of biomechanical factors and aerodynamics as well as the incidence of injuries to develop and validate a method in bike fitting applicable for both professionals and amateurs.

REFERENCES:
Hamley, E. & Thomas, V. (1967). Physiological and postural factors in the calibration of a bicycle ergometer. *Journal of Physiology*, 191, 55-57.
Pruitt, A. & Matheny, F. (2006). *Andy Pruitt's medical guide for cyclists*. Boulder, Colorado: VeloPress.
Le Mond, G. & Gordis, K. (1987). *Greg LeMond's complete book of bicycling*. New York: Perigee Books.

Acknowledgement
This study was partly supported by Cervélo Inc., Toronto, Canada.

LOWER LIMB JOINT KINETICS IN THE SPRINT START PUSH-OFF

Laura Charalambous[1], Ian Bezodis[1], Gareth Irwin[1], David Kerwin[1] and Robert Harle[2]

Cardiff School of Sport, University of Wales Institute Cardiff, Cardiff, UK[1]
Computer Laboratory, University of Cambridge, Cambridge, UK[2]

Previous studies have analysed lower limb joint kinetics during sprint performance, but not addressed the earliest contact out of the blocks. The aim of this study was to report lower limb joint moments and powers during the first stance phase of the sprint push-off. One competitive male sprinter performed 10 maximal sprint starts. An automatic motion analysis system (CODA, 200 Hz) with synchronised force plate data (1000 Hz) were used to collect kinematic profiles at the hip, knee and ankle and ground reaction forces for the first stance phase. Cluster markers defined the orientation of the lower limb segments in 3D. Knee and hip kinetics differed to the later phases of sprint, whereas similarities were found at the ankle. This study highlights the need for the push-off phase to be considered separately from both research and practical perspectives.

KEY WORDS: 3D inverse dynamics, joint moment, joint power

INTRODUCTION: Sprinting success relies on performance of a fast start followed by achievement and maintenance of the highest possible running velocity. Consequently, sprint performance has been separated into several distinct phases (Delecluse et al., 1995). A powerful start is essential to reach a high level of performance (Mero, 1988). Research has analysed joint kinetics during the block (Mero et al., 2006), second stance (Jacobs & van Ingen Schenau, 1992; Jacobs et al., 1996), acceleration (Johnson & Buckley, 2001), and maximal velocity phases (Bezodis et al., 2008; Kuitunen et al., 2002). This study concentrates on the first stance out of the blocks and extends the kinematic analysis reported by Coh et al., 2006. The aim of this study was to improve the understanding of lower limb joint kinetics during the first stance phase following the sprint start.

METHODS: Data Collection: An internationally competitive male sprint hurdler participated in the study (age 27 yrs, height 1.80 m, mass 74.4 kg, 110 m PB 13.48 s). Four cx1 CODA scanners (Charnwood Dynamics Ltd, UK) were located around a force plate (Kistler Instruments 9287BA, Switzerland) for data collection. Kinematic data (200 Hz) and synchronised ground reaction force data (GRF, 1000 Hz) were captured during the first stance phase out of the blocks. 31 active markers were placed on the subject including three rigid clusters (anterior-lateral aspect of the thigh; lateral aspects of the shank and foot) on the first contact limb. A hip marker was located on the greater trochanter of the same limb. The 4-marker clusters defined the orientation of the segments in 3D, while reducing error from soft tissue artefact (Schache et al., 2008). Kinematic data collected during static trials, together with additional anatomical reference markers, were used to calibrate the athlete, before he completed 10 maximal sprint starts on a start signal. A successful trial was achieved when the athlete accelerated well beyond the measurement volume (>9 m), made first stance contact on the force plate and produced a start with no obvious deviation in technique.

Data Processing: All data were processed in Visual 3D™. Coordinate data were smoothed using a fourth-order Butterworth low-pass digital filter with a cut-off frequency of 8 Hz, determined by residual analysis (Winter, 2005). All outputs were normalised to 100% stance phase (i.e. when F_z >10N). GRF in the vertical (F_z) and horizontal (F_y) directions were normalised to body weight units (BW). Angles (θ) and angular velocities (ω), and joint moments (M) at the ankle (M_A), knee (M_K) and hip (M_H) for the first stance phase were calculated from kinematic data, GRF and anthropometric data using standard inverse. Anthropometric data were taken from de Leva et al. (1996), with the exception of the foot segment where the value of Winter (2005) was used. The mass of a typical sprinting shoe

(0.2 kg) was added to the mass of the foot segment (Hunter et al., 2004). Mechanical power (P) at the hip (P_H), knee (P_K) and ankle (P_A) were calculated as the respective products of M and ω. In accordance with the recommendations of Hof (1996), M and P were scaled to body weight and height. Due to the critical motions of sprint performance, analysis was focussed on lower limb flexion and extension, with the latter being positive for ω and M. Mean values (±sd) for the 10 trials have been reported for hip, knee and ankle during the first stance out of the blocks.

RESULTS & DISCUSSION: To facilitate comparison, all values from previous research have been scaled according to Hof (1996). The entirely plantar-flexor M_A (Figure 1) was similar in pattern and magnitude (0.30 ± 0.02) to previously reported data obtained from later in sprint runs; Stefanyshyn & Nigg (1998) and Bezodis et al. (2008) found peak normalised values of 0.21 – 0.24 and 0.25, respectively. However, the peak M_A is somewhat higher than during the second contact phase peak of approximately 0.17 (Jacobs & van Ingen Schenau, 1992). The current values are also similar to those reported by Farley and Morgenroth (1999) during maximum height hopping (0.33). This element, in addition to a similar movement pattern at the ankle, may support the use of ballistic hopping in sprint push-off training. According to the electromyographic (EMG) analysis of Jacobs and van Ingen Schenau (1992), the increase in M_A during the first half of stance resulted from increased triceps surae and decreased tibialis anterior force. The subsequent decrease in M_A was explained by a decreased moment arm and increased contraction velocities of the plantar-flexor muscles.

The eccentric action (peak P_A = -0.23 ± 0.05) of the plantar-flexors until around 50% of stance, followed by concentric action indicates that the ankle absorbed energy during the first half of stance and generated energy during the second half. The eccentric phase is necessary to control the collapse of the lower limb and prevent the limb moving across the ground too quickly. The concentric phase has been shown to result from transported P from proximal muscles and P liberated by the plantar-flexors (Jacobs et al., 1996). The elastic energy stored and released by the plantar-flexors contributes to the high peak in concentric P_A (0.88 ± 0.06). This may support the importance of P generation and the SSC of the plantar-flexors to the sprint push-off (Cavagna et al., 1968). Although revealing a similar eccentric-concentric pattern, the magnitude of P_A differed from those reported during the later sprint phases. There was a considerably lower proportion of eccentric contraction and the magnitude of the eccentric phase was relatively small in this study. For example, Bezodis et al. (2008) found eccentric and concentric peaks of -1.26 and 1.01, respectively. The differences underline the importance in considering the push-off and maximum velocity phases separately and may also confirm that the transfer of energy mechanism described by Jacobs et al. (1996) produces the relatively high peak concentric P_A during the first stance phase.

The main P_K phase was concentric (power generating) extensor action, which began around 15% and continued until 80% of stance. The increasing net extensor M_K (peak = 0.05 ± 0.01) may be due to increasing activity of the extensors or decreasing activity of the flexors. The initial flexor M_K (-0.04 ± 0.01) after touchdown has been shown to be caused by the contraction of the hamstrings (Jacobs & van Ingen Schenau, 1992). Despite the torque caused by this contraction, (Figure 1), the knee joint still extends thus creating eccentric knee flexion (peak = -0.10 ± 0.03).

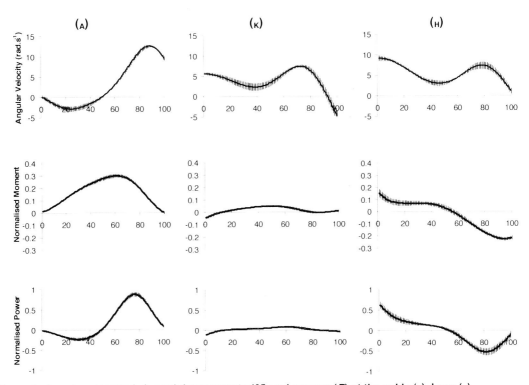

Figure 1: Angular velocity (ω), net joint moments (M) and powers (P) at the ankle ($_A$), knee ($_K$) and hip ($_H$) for the first stance phase. Values are the mean for all trials and expressed as a percentage of stance. Vertical bars indicate ±1sd.

A co-contraction of knee flexors and extensors prevented an early increase in ω (ω decreased from touchdown to 40% of stance) and thus prevented the knee from fully extending too early in stance, as observed previously by Jacobs et al. (1996). An early leg extension would result in an increase in vertical velocity of the CM that would contradict the aim of the task, to increase horizontal velocity of the CM. This increase and subsequent decrease in extensor M results in a corresponding increase and decrease in ω at the knee. Knee kinetics were similar to those reported for the second stance phase (Jacobs & van Ingen Schenau, 1992) but differ from the irregular and undulating patterns reported during later phases of sprinting (Bezodis et al., 2008; Johnson & Buckley, 2001). These observations further support the need for separate analysis of the push-off phase.

Due to constant extension of the hip, the extensor M_H during the first half of stance (peak = 0.15 ± 0.03) results in a P generating concentric phase (peak = 0.62 ± 0.09) and the flexor M_H (peak = -0.23 ± 0.01) in the second half of stance results in a P absorbing eccentric phase (peak = -0.53 ± 0.09). M_H and P_H were similar in both pattern and magnitude between the present study and second stance phase findings (Jacobs & van Ingen Schenau, 1992; Jacobs et al., 1996) but differ compared to hip kinetics found in the later phases (Bezodis et al., 2008; Johnson & Buckley, 2001). During the acceleration phase Johnson & Buckley (2001) recognised the horizontal GRF acted in the posterior direction and thus a hip flexion M was required to prevent premature extension of the hip. Thus, since the hip continued to extend despite this M, an eccentric P was produced during the acceleration phase. In this study, due to F_y acting in the anterior direction for the majority of the sprint push-off, there was no flexor M_H early in stance. In the present study M_H and P_H are similar both in pattern and magnitude to those reported by Jacobs & van Ingen Schenau (1992) and may indicate the mechanisms explaining the hip movement during the second stance also control the first.

Although these similarities could be a characteristic of the push-off phase, they may be an attribute of the subject used in this study. Further analysis of other athletes is required to validate these observations.

CONCLUSION: Joint kinetics for the first stance out of the blocks in the sprint start were found to be similar to those previously reported for the second stance. Differences, particularly in knee and hip kinetics, were evident between the push-off and data reported for the later stages of sprint performance. This underlines the importance of considering the push-off as an individual phase of sprint performance, from both research and practical perspectives. The similarity between ankle kinetics in hopping and the sprint push-off supports the use of ballistic hopping techniques in sprint training. Cluster markers facilitate greater scope than the location of joint centre method used by previous sprint studies and will enhance future research.

REFERENCES:
Arampatzis, A., Bruggemann, G. & Metzler, V. (1999). The effect of speed on leg stiffness and joint kinetics in human running. *Journal of Biomechanics*, 32, 1349-1353.

Bezodis, I.N., Kerwin, D.G. & Salo, A.T. (2008), Lower-limb mechanics during the support phase of maximum-velocity sprint running. *Medicine and Science in Sports and Exercise*, 40, 707-715.

Cavagna, G.A., Dusman, B. & Margaria, R. (1968). Positive work done by a previously stretched muscle. Journal of Applied Physiology, 24, 21-34.

Coh, M., Tomazin, K. & Stuhec, S. (2006). The biomechanical model of the sprint start and block acceleration. *Physical Education and Sport*, 4, 103-114.

Delecluse, C.H., van Coppenolle, H., Willems, E., Diels, R., Goris, M., van Leemputte, M. & Vuylsteke, M. (1995). Analysis of 100 meter sprint performance as a multidimensional skill. *Journal of Human Movement Studies*, 28, 87-101.

de Leva, P. (1996). Adjustments to Zatsiorsky-Seluyanov's segment inertia parameters. *Journal of Biomechanics*, 29, 1223-1230.

Farley, C.T. & Morgenroth, D.C. (1999). Leg stiffness primarily depends on ankle stiffness during human hopping. *Journal of Biomechanics*, 32, 267-273.

Hof, A.L. (1996). Scaling gait data to body size. *Gait Posture*, 4, 222-223.

Jacobs, R., Bobbert, M.F. & Ingen Schenau, G.J.van. (1996). Mechanical output from individual muscles during explosive leg extensions: the role of biarticular muscles. *Journal of Biomechanics*, 29, 513-523.

Jacobs, R. & Ingen Schenau, G.J.van. (1992). Intermuscular coordination in a sprint push-off. *Journal of Biomechanics*, 25, 953-965

Johnson, M.D. & Buckley, J.G. (2001). Muscle power patterns in the midacceleration phase of sprinting. *Journal of Sports Science*, 19, 263–72.

Kuitunen, S., Komi, P.V. & Kyrolainen, H. (2002). Knee and ankle joint stiffness in sprint running. *Medicine and Science in Sports and Exercise*, 24, 483-489.

Mero, A. (1988). Force – time characteristics and running velocity of male sprinters during the acceleration phase of sprinting. Research Quarterly for Exercise and Sport, 59, 94 – 98.

Mero, A., Kuitunen, S., Harland, M., Kyrolainen, H. & Komi, P.V. (2006) Effects of muscle – tendon length on joint moment and power during sprint starts. *Journal of Sports Science*, 24, 165-173.

Stefanyshyn, D.J. & Nigg, B.M. (1998), Dynamic angular stiffness of the ankle joint during running and sprinting. *Journal of Applied Biomechanics*, 14, 292-299.

Winter, D.A. (2005), *Biomechanics and Motor Control of Human Movement*. New York: John Wiley and Sons, Inc.

TEMPORAL CHARACTERISTICS OF THOMAS FLAIRS ON THE POMMEL AND FLOOR

S. Prassas[1], G. Ariel[2] & E. Tsarouchas[3]
[1]California State University East Bay, Hayward, CA USA; [2]Coto Research Center, Coto De Caza CA USA; [3]Olympic Athletic Sports Institute, Athens, Greece

INTRODUCTION: To perform successfully on any apparatus, gymnasts must execute skills with creativity and virtuosity. Whereas creativity is demonstrated by introducing new skills, combining existing ones, or adapting skills to different apparatuses, virtuosity is expressed by executing skills with exceptional technique (Prassas et al. 2006). The Thomas Flairs (Fig. 1), originally introduced and performed on the pommel horse, have been adapted on other apparatuses including the floor. Understanding the timing of the different phases of the skill and what effect the different physical characteristics of the two apparatuses may impose on that timing, would be valuable to coaches and gymnasts seeking to improve performance, judges evaluating gymnastic routines, and scientists studying motor skills.

Figure 1. Thomas Flairs on Floor (top) and Pommel (bottom) (Prassas et al., 2006)

METHODS: Thomas Flairs performed by two skilled gymnasts were videotaped utilizing two 60 Hz cameras. Viewed from above, both gymnasts performed the skill rotating clockwise. In each apparatus, Flairs were analyzed utilizing the Ariel Performance Analysis System (APAS). Temporal and kinematic data for one full circle beginning with and ending at the right hand contact with the floor or pommel were examined and compared. The feet, hips, shoulders, and elbow joints, and both hands were digitized. Position data were smoothed by digital filtering at 7 Hz.

RESULTS AND DISCUSSION: Temporal results are presented in Table 1. The results suggest that Tomas Flairs in both the pommel horse and floor are executed with similar "tempo". Specifically, gymnasts spent more time in front support than in rear and—for gymnasts rotating clockwise when seen from above—less time in right support than in left. It remains to be seen if these trends can be generalized.

Table 1. Temporal Characteristics of Thomas Flairs on Floor and Pommels

Variable	Floor		Pommels	
	Subj. 1	Subj. 2	Subj. 1	Subj. 2
Total Time (sec)	1.13	1.08	1.1	1.13
Time in Front Support (%)	17.6	13.8	16.7	17.6
Time in Rear Support (%)	14.7	10.8	13.6	11.8
Time in Right Support (%)	32.4	33.8	31.8	32.4
Time in Left Support (%)	35.3	41.5	37.9	38.2

REFERENCES:
Prassas, S., Ariel, G., Ostarello, J. & Tsarouchas, E. (2006). Thomas Flaires on the pommel and floor: a case study. *Proceedings of XXIV International Symposium on Biomechanics in Sports*, Volume 1 (p 262), Department of Sport Science and Kinesiology, University of Salzburg, Austria.

ISBS 2009

Poster Session PS4

REAL WORLD HEAD IMPACT DATA MEASUREMENTS ON JOCKEYS

Manuel A. Forero Rueda and Michael D. Gilchrist

School of Electrical, Electronic & Mechanical Engineering, University College Dublin, Belfield, Dublin 4, Ireland

KEY WORDS: Instrumentation; Head injury; Equestrian; Horse racing.

INTRODUCTION: A novel instrumentation system previously implemented on American football helmets (Greenwald et al. 2008) has been adapted and validated against a Hybrid III head and neck for use in an equestrian environment. This has been used to determine the forces applied to jockeys' heads during fall impacts in competition racing.

METHODS: Twelve professional Irish National Hunt jockeys participated in data collection of real life head impacts. Instrumented helmets were fitted to the jockeys to gather head impact data in case of a fall during a race. These helmets were fitted with sets of six linear accelerometers that, when appropriately aligned and used in conjunction with postprocessing techniques, provided the linear and angular acceleration values for the jockey's head defined along cartesian axes, the origin of which is at the centre of gravity of the jockey's head.

RESULTS: Fall/ride frequency was of 5%, ride average for the participating jockeys was of 300 rides/year. Four confirmed head impact events during racing with video data have been collected to date from different jockeys. The values for maximum linear and angular acceleration are shown in Table 1. None of these impact events led to head injury.

Table 1 Maximum linear and angular acceleration results for each fall

Fall Case	Linear Acceleration (g)	Angular Acceleration (rad/s^2)
1	31.3	1799.5
2	71.4	4403.0
3	47.7	3122.1
4	98.5	2428.3

DISCUSSION: Linear acceleration values are just below the accepted injury thresholds derived from the WSTC, where injury is expected above 100g for moderate duration impacts; therefore they were not expected to be related to injurious cases. From previous instrumented helmet data collected from American Football players (Greenwald et al. 2008), the data collected to date has a very low probability of concussion (<0.3).

CONCLUSION: This study is the first time that head impact linear and angular accelerations have ever been directly measured in equestrian activities in a real racing environment. The data can now be used to reconstruct head impacts through the use of video analysis and computational techniques. A larger data set will allow us to correlate linear and angular accelerations to head injury in jockeys. More actual fall measurements should be done in a more controlled environment to improve the frequency of positive results, and further technical improvements on the measurement system should be implemented to simplify the measurement procedure to optimize data collection in a horse riding environment.

REFERENCES:
Greenwald R, Gwin J, Chu J, Crisco J (2008). Head Impact Severity Measures for Evaluating Mild Traumatic Brain Injury Risk Exposure. *Neurosurgery*, 62(4), 789-98.

Acknowledgement
We would like to thank The Turf Club for financing this project.

RELATIONSHIP BETWEEN REDUCED OF MEDIAL LONGITUDINAL ARCH HEIGHT AND KNEE VALGUS

Takeshi Akimoto, Yukio Urabe, Yuki Yamanaka, Natsumi Kamiya and Shinji Nomura

Graduate School of Health Sciences, Hiroshima University, Hiroshima, Japan

KEY WORDS: knee valgus, foot arch, jump-landing.

INTRODUCTION: Knee valgus during jump landing is considered to be one of the situations that may cause injury to the anterior cruciate ligament (ACL). Decreased muscle torque during hip abduction is also reported to be a causative factor for knee valgus, as is pronation of the ankle joint (Joseph M et al. 2008). Further, knee valgus may occur when a reduction in the height of the medial longitudinal arch causes ankle pronation, leading to tibial inclination angle to the inside. In this study, motion analysis was performed to investigate the relationship between a reduction in the height of the medial longitudinal arch and knee valgus.

METHODS: This study involved 15 healthy subjects who were not experiencing any pain in the lower limbs at the time of measurement. Prior to the experiment, the distance from the navicular tuberosity to the floor was measured in seated position and in weightbearing position to determine reduction of the medial longitudinal arch height. The leg-heel angle was also measured. The subjects were asked to perform single leg jump landing as a trial task. Twelve markers were attached to each subject. Four high speed video cameras (FKN-HC200C. 4Assist, Inc., Japan, 200 Hz) were used for recording the trial task. The video sequences were digitized. A direct linear transformation method was performed to establish the three-dimensional coordinates of the femur and tibia during landing by a 3D motion analyzer (Frame-DIAS II, DKH Co., Japan). The angles of flexion, knee valgus, and tibial inclination angle to the inside were calculated using the joint coordinate system proposed (Grood ES et al. 1983). When knee flexion angle was 45° during the single-leg jump-landing task, the correlation between the reduction in the medial longitudinal arch height and (1) the knee valgus angle, (2) the tibial inclination angle to the inside, and (3) the leg-heel angle was examined using Pearson's correlation analysis. The statistical significance level was set at less than 5%.

RESULTS: A significant relationship was identified between a reduction in the medial longitudinal arch height and the knee valgus angle ($r=0.54$, $p<0.05$). A significant relationship was also recognized between the tibial inclination angle to the inside and the knee valgus angle ($r=0.59$, $p<0.05$). No significant correlation was observed between the leg-heel angle and the knee valgus angle.

DISCUSSION & CONCLUSION: Knee valgus angle is considered to be influenced not only the knee joints but also hip and ankle joints. The results of this study revealed that a reduction in the height of the medial longitudinal arch might trigger tibial inclination to the inside and increased the knee valgus angle. Thus we consider it necessary to assess foot function in order to prevent ACL injury. Interestingly, no relationship was identified between the leg-heel angle (the pronation angle of the calcaneus) and the knee valgus angle; the importance of this finding is made clear in our future studies. Our results indicated that a reduction in the height of the medial longitudinal arch increases the knee valgus angle consequently increasing the risk of ACL injury.

REFERENCES:
Joseph M et al. (2008). Knee valgus during drop jumps in National Collegiate Athletic Association Division I female athletes: the effect of a medial post. *Am J Sports Med*, 36(2), 285-289
Grood ES et al. (1983). A joint coordinate system for the clinical description of three dimensional motions: application to the knee. *J Biomech Eng*, 105(1), 136-144

A COMPARISON OF YOUTH PITCHING KINEMATICS ACROSS PREPUBESCENT AND PUBESCENT PITCHERS

David W. Keeley, Gretchen D. Oliver, Priscilla Dwelly, and Hiedi J. Hoffman
University of Arkansas, Fayetteville, AR. USA

To reduce injuries in youth baseball pitchers coaches teach proper mechanics at a young age. Unfortunately, the mechanics taught to beginning pitchers are based on data from adolescent pitchers and may result in techniques that could injure younger pitchers. Thus, the purpose of this study was to identify differences between the pitching mechanics of prepubescent and pubescent baseball pitchers. Of the 20 parameters analyzed in the study, 7 were observed to be different between the two groups. The findings of this study indicate that the mechanics currently being taught to youth pitchers may not be appropriate for all ages and that furthur study is needed to help identify what mechanics are correct for all ages of pitchers. The data produced in this study may help clinicians appreciate the mechanical differences between pitchers of various ages and better understand the etiology of pitching injuries as they relate to age.

KEY WORDS: pitching, kinematics, pubescent, prepubescent.

INTRODUCTION: Youth baseball pitchers suffer injury at extremely high rates, with most of the injuries thought to be the result of overuse (Sabick et al., 2004). The underlying cause of these injuries is often attributed to the use of improper mechanics during pitching which result in repeated, unnecessary stresses being placed on the throwing arm. Because of this it is currently thought that the best practice for reducing these stresses is to teach proper pitching mechanics at an early age (Fleisig et al., 1999). Although skill development for young pitchers may be most critical at the beginning of their career, the mechanics currently being taught to these pitchers are often based on the results of biomechanical analyses of the pubescent pitching motion. Previously, biomechanical studies incorporating subjects over the age of ten (Fleisig et al., 1999; Sabick et al., 2004; Nissen et al., 2007) did not take into account the age related differences between pubescent and prepubescent pitchers. Thus, the purpose of this study was to quantify pitching kinematics in both prepubescent and pubescent pitchers in an attempt to identify differences between the various age groups. It was hypothesized that prepubescent pitchers would exhibit mechanics that were significantly different than those observed in pubescent pitchers.

METHODS: Data Collection: Eighteen right hand-dominant baseball pitchers assigned to two separate groups participated in the study (9 prepubescent and 9 pubescent). Data collection sessions were conducted in the Human Motion Research Lab, Texas A&M University-Commerce. Testing protocols were approved by that institution's ethics board, and prior to testing each pitcher and their parent(s)/guardian(s) all provided consent.

Prior to testing, a 3.38 m^3 calibration cube with 18 calibration points was suspended above the pitching mound using techniques described by Escamilla et al. (1998) and the three-dimensional space was calibrated. Following calibration the root mean square error in calculating the three-dimensional location of markers within the calibrated space was determined to be less than 10 mm. Reflective markers were attached bilaterally to each subject on the greater trochanter of the hips and the lateral-superior tip of the acromions, the medial and lateral epicondyles of the throwing elbow, and the radial and ulnar styloid process of the throwing wrist. Each subject then performed their own specified warm-up routine before throwing three maximal effort fastballs for strikes toward a catcher located 13.4 m from the pitching mound. To be considered a successful trial, a pitch was required to pass through a strike zone ribbon suspended 0.4 m above home plate and encompassing an area of 0.2 m^2. In addition, the velocity of all successful trials was required to be within 3 mph of the velocity of the fastest strike thrown for each subject. Each pitch was digitally recorded by three synchronized high speed digital video cameras (Basler Vision Technologies, Germany) recording at 120 Hz and that were arranged to capture the movement from the dominant side. Between trials each subject was allowed a 40-60 s rest period.

Data Analysis: Reflective marker locations were digitized in each frame from maximum knee lift through maximum internal rotation. Following digitization, the three-dimensional location of each marker was calculated using Direct Linear Transformation (Abdel-Aziz & Karara, 1971), and then filtered independently in the X, Y, and Z axis using a 2nd order Butterworth filter set at a cut-off frequency of 13 Hz (Sabick et al., 2004). Previously established techniques were then used to calculate the torso and throwing arm kinematics defined in Figure 1 (Dillman et al., 1993; Fleisig et al., 1999). For each subject, mean and standard deviation was calculated for: 1) shoulder abduction, 2) shoulder horizontal abduction, 3) shoulder internal rotation, anc 4) elbow flexion. In addition to these parameters, the rate of axial torso rotation was calculated as the cross-product of the upper torso vector and its first derivative (Feltner and Dapena, 1989).

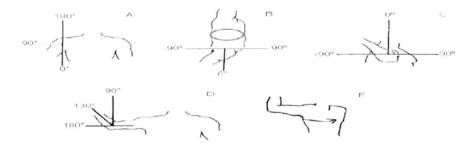

Figure 1: Definition of joint angles: A) shoulder abduction, B) shoulder horizontal abduction, C) shoulder external rotation, D) elbow flexion, and E) axial torso rotation. Adapted from Flesig et al., 1996.

Statistics: For each subject, mean and standard deviation were calculated for each kinematic parameter. Prior to testing for mean differences the nature of the distribution was analyzed, and after the data were deemed to be normally distributed paired independent sample *t*-test were used to compare mean values between the prepubescent and pubescent groups at the following intervals: 1) stride foot contact; 2) maximum shoulder external rotation; 3) ball release; and 4) maximum shoulder internal rotation. For each of the analyses, age was the independent variable and the kinematic parameter being analyzed was the dependent variable. Because the data were analyzed at four independent intervals, the level of significance for kinematic data was adjusted and set at alpha = 0.01. In addition, to discuss any differences identified between the groups in terms of standard deviation units, the effect size (d) was calculated for all parameters at all intervals.

RESULTS: The results of kinematic analyses are shown in Table 1 and of the 20 position parameters analyzed, 7 were different between groups. Previous reports indicate that pitching mechanics do not vary greatly with age and support the notion that proper pitching mechanics can be taught at an early age (Fleisig et al., 1999). The results of these studies must be interpreted with caution as they do not incorporate the youngest of pitchers. This study, by including younger subjects, identified several mechanical

Table 1 Kinematic differences between prepubescent and pubescent pitchers

	Prepubescent (n=9)	Pubescent (n=9)	Sig.	d
Stride foot contact				
Horizontal adduction (°)	22 ± 7	17 ± 9	*	0.08
Elbow extension (°)	101 ± 41	90 ± 43	*	0.19
Maximum shoulder external rotation				
Torso rotational velocity (°/s)	701 ± 440	1120 ± 414	*	1.60
Ball release				
Torso rotational velocity (°/s)	450 ± 312	772 ± 464	*	1.64
Maximum shoulder internal rotation				
Torso rotational velocity (°/s)	285 ± 272	676 ± 379	*	2.33
Horizontal adduction (°)	7 ± 9	34 ± 18	*	0.59
Elbow extension (°)	169 ± 8	148 ± 12	*	0.47

Note: * significant difference between groups ($p < 0.01$).

differences between prepubescent and pubescent baseball pitchers. Thus the experimental hypothesis of the study was retained. Of the 20 kinematic variables analyzed at various instances throughout the pitching motion, 7 differed significantly between groups. Although differences in the magnitude of various parameters were identified, the movement patterns observed were generally similar between the groups. Thus, to better compare the differences observed between the two independent groups, the effect size (d) was caluculated so that those differences might be discussed in terms of standard deviation units.

The results of the current study support the findings of previous studies (Aguinaldo et al., 2007) that indicate young pitchers have difficulty in controlling the rate of axial torso rotation thoughout the pitch cycle. However, it is often thought that more skilled pitchers are able to better control their torso rotation throughout the pitching cycle. Our results contradict this as we show that pubescent pitchers increase the velocity of their torso rotation early in the pitch cycle ("opening up early") with their rate of axial torso rotation consistenlty being 1.5 to 2 standard deviation units higher than the prepubescent group. Opening up early in the pitch cycle often means that the torso is rotating prior to proper positioning of the scapula and humerus, which could ultimately result in excessive horizontal abduction, or hyperangulation. It has also been speculated that slight changes in timing could result in reduced output and potentially harmful joint loads at the throwing shoulder (Fleisig et al., 1996). It may ultimately be the increased rate of torso rotation observed in pubescent pitchers that contributes to the high rate of shoulder soft tissue injuries in this younger group.

Unlike torso rotation, the magnitude of the observed differences between groups for the other paramters were quite small in terms of the effect size. Of interest however was that 3 of the 7 differences between prepubescent and pubescent pitchers were observed during the deceleration phase of the motion. The deceleration phase acts primarily as a mechanism for injury prevention by slowing the tremendous velocities generated during arm acceleration. Thus, it is important to understand how the differences observed in deceleration kinematics relate to number of injuries often experienced by pubescent pitchers. Pubescent pitchers often experience inflammation of the rotator cuff at a higher rate than prepubescent pitchers. During deceleration, the rotator cuff works to control internal rotation of the humerus as well as horizontal adduction of the arm across the torso (Fleisig et al., 1996). The large range of horizontal adduction observed in pubescent pitchers may result in increased activity of the rotator cuff musculature, ultimately placing pubescent pitchers at an increased risk for rotator cuff injury. This problem may compound as a youth pitchers become less capable of controlling internal rotation through activation of the rotator cuff. If damage to the rotator cuff results in a decreased ability to control internal rotation during deceleration, pubescent pitchers may be forced to compensate by increasing the already large range of horizontal adduction of the arm across the torso in order to decelerate the throwing arm.

CONCLUSION: Although previous reports have indicated that pitching mechanics do not vary greatly with age and have supported the notion that proper pitching mechanics can be taught at an early age, the results of the current study indicate that pitching mechanics may vary based on age. By including younger subjects, this study identified several mechanical differences between prepubescent and pubescent baseball pitchers, with the most dramatic differences being observed for axial torso rotation. The findings of this study indicate that the mechanics currently being taught to youth pitchers may not be appropriate for all ages and that furthur study is needed to help identify what mechanics are correct for all ages of pitchers. The data produced in this study may help clinicians appreciate the mechanical differences between pitchers of various ages and better understand the etiology of pitching injuries as they relate to age. For instance, one finding of the current study which may relate to overuse tendonopathy at the subscapularis in prepubescent pitchers was a brief period of shoulder abduction prior to maximum external rotation. As the arm is both abducted and externally rotated during late cocking, anterior shoulder stresses also increase. This increase may typically be reduced through an increase in the activity of the subscapularis. (Glousman et al., 1988). If a lack of muscular strength in prepubescent pitchers renders them unable to compensate for these increased stresses, a scenario becomes possible where the subscapularis is required to repeatedly work beyond its capacity, resulting in overuse tendonopathy and/or a loss of muscular integrity.

REFERENCES:

Abdel-Aziz, Y.I., & Karara, H.M. (1971). Direct linear transformation from computer coordinates into object coordinates in close-range photogrammetry. In *Proceedings ASPUI symposium on close-range photogrammetry* (pp.1-19). Falls Church, VA: American Society of Photogrammetry.

Aguinaldo, A.L., Buttermore, J., & Chambers, H. (2007). Effects of upper trunk rotation on shoulder joint torque among baseball pitchers of various levels. *Journal of Applied Biomechanics*, 23, 42-51.

Dillman, C.J., Fleisig, G.S., and Andrews, J.R., (1993). Biomechanics of pitching with emphasis upon shoulder kinematics. *Journal of Orthopaedic and Sports Physical Therapy*, 18, 402-408.

Escamilla, R.F., Fleisig, G.S., Barrentine, S.W., Zheng, N., & Andrews, J.R. (1998). Kinematic comparison of throwing different types of baseball pitches. *Journal of Applied Biomechanics*, 14(1), 1-23.

Feltner M.E., & Dapena J. (1986). Dynamics of the shoulder and elbow joints of the throwing arm during a baseball pitch. *International Journal of Sport Biomechanics*, 2, 235-59.

Fleisig, G.S., Barrentine, S.W., Escamilla, R.F., & Andrews, J.R. (1996). Biomechanics of overhand throwing with implications for injuries. *Sports Medicine*, 21(6), 421-437.

Fleisig, G.S., Barrentine, S.W., Zheng, N., Escamilla, R.F., & Andrews, J.R. (1999). Kinematic and kinetic comparison of baseball pitching among various levels of development. *Journal of Biomechanics*, 32, 1371-1375.

Glousman, R., Jobe, F., Tibone, J., Moynes, D., Antonelli, D., & Perry, J., (1988). Dynamic electromyographic analysis of the throwing shoulder with glenohumeral instability. Journal of *Bone and Joint Surgery, 70, 220-226.*

Nissen, C.W., Westwell, M., Ounpuu, S., Patel, M., Tate, J.P., Pierz, K., et al. (2007). Adolescent baseball pitching technique: A detailed three-dimensional biomechanical analysis. *Medicine and Science in Sports and Exercise*, 39, 1347-1357.

Sabick, M.B., Torry, M.R., Lawton, R.L., & Hawkins, R.J., (2004). Valgus torque in youth baseball pitchers: A biomechanical study. *Journal of Shoulder and Elbow Surgery*, 13, 349-355.

Acknowledgement

The authors would like to thank Dr. Jason Wicke, Director of the Human Motional Analysis Lab, Texas A&M University – Commerce.

PREVENTION & REHABILITATION OF DIVERS' CERVICAL VERTEBRAE INJURIES FROM THE PERSPECTIVE OF BIOMECHANICAL BALANCE

Renbo Qiao

Department of Physical Education and Sports, Jiangsu University, Zhenjiang, China

In this paper, the theory of biomechanical balance of the cervical vertebrae (CV) from medical studies was used to help explain the causes of CV injuries suffered by divers. By re-analysing the data of CV injuries of 312 divers, we concluded that CV injuries may be caused by damage of the biomechanical balance of the CV. Based on this theory, combined with some medical knowledge - suitable exercises and appropriate therapeutic intervention for coaches, therapists and trainers were suggested to treat and prevent CV injuries in sporting events.

KEY WORDS: biomechanical balance, cervical vertebra injuries, prevention, rehabilitation.

INTRODUCTION: Cervical vertebrae (CV) injuries are one of the most common injuries suffered by divers. Most research is done on the clinical investigation and prevention of it (He, 2005, Zhou, 2005), but seldom is conducted on the relationship between the biomechanical balance of the CV and its related injuries. Biomechanics of cv balance state that the CV are not in abnormal strain under physiological loads, and they have no excessive abnormal activities of functional spinal units. Medical research (Jiang et al. 2002, Ran, 2003) has shown that CV diseases were closely related to the biomechanical imbalance of the CV - not only from its etiology and pathology, but also from its treatment and prevention. Accordingly, some medical studies focus on treating CV diseases by restoring equilibrium to the biomechanical balance of the CV. It is both logical and intuitive to assume that this theory and its methods can be applied to CV injures. This paper studies the mechanism, prevention and rehabilitation of divers' CV injuries from the angle of biomechanical balance of the CV.

METHOD: Paying extra attention to the anatomical characteristics of the CV and the biomechanical balance principle from medical studies, we reanalysed data collected from this field in order to find out the causes of divers' CV injuries and their categories. Clinical checkups were conducted on a sample of 312 Chinese divers, which comprised of 165 male and 147 female divers whose ages are in the range 9 - 24 with 16 being the average. It found out the rate of cervical vertebra injuries and the main causes of them. At the same time, an image check was conducted on 173 randomly chosen divers. Through X-ray analyses, the conditions of cervical vertebrae and the categories of these injuries were clear. It also contained an analysis of the physical characteristics of diving - which may damage the biomechanical balance of the CV. Lastly, it outlines measures and methods that can be taken to prevent CV injuries in the perspective of biomechanical balance and by the analysis of principles of sports injury.

RESULTS: The conditions of biomechanical balance of CV: Ordinary people's CV balance and stability are maintained by two parts. First is endogenous balance: which includes centrum, accessories, intervertebral disk and connected ligament which forms static balance. Second is exogenous balance: this mainly concerns regulation and the control of muscles which form dynamic balance as the original source of energy. If any organ contained among these two systems does not work well with others, the normal biomechanical balance of the CV will be lost. As Jiang et al (2002) proposed - that at static balance in the CV, the inter-vertebral disk is the key part in the bearing system. The ligament helps maintain spinal stability, especially for the upper part of the CV (Weiss, 2002). From the neck muscle mechanical analysis (Zhang, 1986), it finds out C_{4-5} muscles are weak and have the worst stability in neck curved peak. The CV are often in dynamic balance during movement and its power-line movement will generate torque. So the divers need to keep their muscles moving in order to counteract this torque and maintain a balance. Enhancing the moving strength of the divers' neck muscles is very important for the dynamic balance of the CV.

The results of the checkups of the divers show that 95 out of 321 divers presented a history of neck injuries (30.45%), of which 42 divers' injuries are caused by a single reason, and the remaining 53 are caused by a variety of reasons, such as unskilled technique at the commencement of training, false entry work, wrong rip entry, incorrect angle entry and insufficient strength in cervix when entering water. The 173 X-rays show that only 85 cases are completely normal, 43 cases are of exercise-induced adaptive changes, and 45 cases are of CV injuries. The categories of the above injuries are listed as atlantoaxial joint dislocation, CV degenerative, anterior dislocation of atlas, congenital fusion of CV and cervical fatigued fractures (see table I, II).

Table I Causes of Cervical Vertebra Injuries

Causes of CV Injuries	N (Total = 95)	%
Unskilled technique at the commencement of training	87	91.6%
Incorrect rip entry	84	88.4%
Incorrect angle of entery	61	64.2%
Lack of strength in neck	65	68.4%

Table II Types of Cervical Vertebra Injuries

Type of CV Injuries	N (Total = 45)	%
Atlantoaxial joint dislocation,	20	44.4%
Cervical degenerative changes	15	33.3%
Anterior dislocation of atlas	5	11.1%
Congenital vertebral fusion	4	8.9%
Cervical fatigued fractures	1	2.2%

DISCUSSION AND RECOMMENDATIONS: The survey finds that the rate of divers' CV injuries is as high as 35.45%. These 173 X-ray samples show that 45 divers have CV injuries in minor or medium degrees and a minority have severe injuries. These injuries are mainly caused by the water pressure formed during the entry process. The categories of these injuries are listed as atlantoaxial joint dislocation, cervical vertebra degenerative, anterior dislocation of atlas. The X-rays show that atlantoaxial joint dislocation and anterior dislocation of the atlas all belong to abnormal atlantoaxial dislocation, indicating the laxity of the atlantoaxial ligament and joint and the over-range of motion as well as the destruction of endogenous balance of cervical vertebrae. The reasons are related to the previous ones, particularly in the entering process. When divers perform head flexion and extension, it is easy for them to have joint instability and thus form atlantoaxial joint dislocation. In the survey, 15 cases are of cervical vertebra degenerative, in which there are varying degrees of vertebral body hyperosteogeny. The intervertebral space of C_{4-5}, C_{5-6} is narrow and shows typical degenerative changes of cervical spondylosis. Because of the worst stability of cervical exogenous C_{4-5} and the trauma or chronic soft tissue injuries, plus muscle cramps caused by imbalance, the CV injuries inevitably occur in the central point of C_{4-5}.

As we know, the head and upper limbs make contact the water firstly during diving. Even if the entry into the water is performed correctly, the largest enduring force is 16 to 25 times that of the body weight (Stevenson, 1985). According to the '*moment balance principle*', if the angle into the water is not vertical and thus forms an acute angle between the CV curve and the horizontal axis i.e. the water, the pressure placed on the cervix will be greatly increased. When the atlas makes contact with the water, the resulting bend or extended state will add strain to atlanto-axial ligament, and cause avulsion or possibly a fracture. When the external force makes the CV exceed its natural range, it will shift one side of the small joints too much and the other undergoes a similar shift in the opposite direction. Thus, it results in unilateral joint dislocation. The above cases are all caused by the instant imbalance between the muscle strength and the external force, so that the power balance of the CV is perturbed, resulting in acute traumatic CV injury. This catalyst accelerates the detrimental effect on intervertebrae discs, small joints, vertebral body and other parts of the body. Day by day, this impact will damage the endogenous balance of CV factors, and cause chronic damage to CV. Divers should rotate their necks in order to complete the aerial somersault and twist. This kind of rotating movement will easily cause injuries to the ligaments, muscles and joints of the neck.

Maintaining the biomechanical balance of CV includes two aspects. First, improve self-ability in maintaining CV balance (including muscle strength and joint-ligament flexibility). Second, reduce the outside force to the CV during diving (including the angle of entry, the technique of rip entry, movements of the neck and head during the diving process). So the specific preventative measures for CV injuries are summarised as follows:

- Those who have congenital CV malformations are likely to suffer from CV injuries due to their abnormal anatomical structure and defects in their endogenous stability. These individuals are not suitable candidates for diving exercises.

- Further study should be undertaken regarding the techniques of rip entry and eradicate the incorrect aspects of their water entry.

- Strengthen the neck ligaments and exercise muscle groups in order to overcome the pressure caused by the water to the neck and head. Two methods are listed below which are suggested to help with this.

Method one: Undertake static training for neck muscles. Assume the prone position, then extend your chest and remain balanced for one minute. Then turn your head slowly in a controlled manner from right to left with six repetitions for each set. In this type of exercise, sports joint capsule, ligaments and muscles around the joints are continuously stretched so as to improve muscle coordination.

Method two: Assume the prone position, then place your hands close to your thighs which are in a lateral position, with feet closed together and your head and legs raised as high as possible. Do this for 5 to 10 seconds each time, and repeat 5 to 8 times daily. This exercises the neck muscles and ligaments, which helps improve the stability of the CV, while also helps strengthen the neck and back muscles through weight-lifting exercises at the same time.

- Focus on the preventive cervical traction and relaxation and then undergo an acu-point massage after training. e.g. use two thumbs or a middle finger to press and smear Fengchi, Jianjing, Chien Wai Shu, Waikuan, Meeting Valley and the tender point. Through this massage, blood circulation is improved and the local flow of lymph is promoted so as to add nutrition to body tissues. It should be obvious that when these tissues become soft and flexible, their functions are improved.

- Perform regular inspection of the CV. Once CV injuries happen during competitive training, medical workers can use traction, physical therapy and other methods to improve the microcirculation status around the divers' neck.

Undertaking these methods will strengthen and promote the divers' nutrient metabolism and muscle, so as to help restore (and improve) the dynamic and static balance in the CV.

CONCLUSION:

Understanding the biomechanical balance conditions of the CV and the causes of this kind of imbalance will be helpful in the design and implementation of correct training methods that will help prevent CV injuries. Following the advice that this paper has mentioned, implementing the preventive methods and the performing the exercises which improve neck strength in daily training will be easier than recovery training and it's a good way to prevent injuries and save valuable training time. As the adage says - "Prevention is better than cure!"

REFERENCES:

Jiang, H., Miao, Z.Y. & Wang, Y. J. (2002). Cervical power balance and the prevention and treatment of cervical spondylosis. *Journal of Traditional Chinese Medical Bone Setting*, 12(3), 49-50.

Weiss, J.A. & Cardiner, J.C. (2002). Bonifasi-listaic. Ligament material behaviour is nonlinear viscoelasic and rate independent under shear loading. J Biomech, 35, 943-950.

Zhou, H., Shi, H., Li, F., Xu, Z., & Ren, Y. (2006). Clinical investigation and prevention of cervical vertebra injuries among 312 divers. *Chinese Journal of Sports Medicine,* 11(6), 714-716.

Zhang Changjiang. (1986) Cervical Disease Prevention and Treatment of Chinese Medicine. China's Ancient Books Publishing House. 29-182

DETERMINANT OF LEG SPRING STIFFNESS DURING MAXIMAL HOPPING

Hiroaki Hobara[1], Kouki Gomi[2], Kazuyuki Kanosue[3]

**National Rehabilitation Center for Persons with Disabilities, Saitama, Japan[1]
Faculty of Sport Sciences, Waseda University, Saitama, Japan[2]**

KEY WORDS: Spring-mass model, stiffness regulation, hopping.

INTRODUCTION: Understanding stiffness of the lower extremities during human movement may provide important information for developing more effective training methods during sports activities. It has been reported that leg stiffness (K_{leg}) during submaximal hopping depends primarily on ankle stiffness (Farley & Morgenroth, 1999), but the way stiffness is regulated in maximal hopping is unknown. The aim of the present study was to investigate a major determinant of the leg stiffness during maximal hopping.

METHOD: Ten well-trained male athletes performed two-legged hopping in place with a maximal effort. Hopping was repeated 15 times and five consecutive hops from the sixth to the tenth were used in the analysis. Based on the spring-mass model, we determined the K_{leg} (the ratio of peak vertical ground reaction force to peak leg compression at midstance). Similarly, based on the torsional spring model, hip, knee and ankle stiffness (K_{hip}, K_{knee}, and K_{ankle}, respectively) was calculated. Multiple regression analyses were performed using K_{leg} as a dependent variable and K_{hip}, K_{knee}, and K_{ankle} as an independent variable. Within the multiple regression analysis, a standardized partial regression coefficient (β) was used to determine the relative importance of joint stiffness to K_{leg}.

RESULTS: Average values of hopping frequency, contact time and flight time were 1.272 ± 0.185 Hz, 0.185 ± 0.020 s, and 0.497 ± 0.056 s, respectively. Table 1 shows that the multiple regression model accounted for 84% of the variance of K_{leg} (Adjusted R^2 = 0.836, $p<0.05$). Further, the model revealed that only K_{knee} was significantly correlated with K_{leg}, and that the β coefficient of K_{knee} (0.644, $p<0.05$) was higher than K_{ankle} (0.371) and K_{hip} (-0.055).

Table 1. β coefficient of K_{ankle}, K_{knee} and K_{hip}.

K_{ankle}	0.371
K_{knee}	0.644*
K_{hip}	-0.055
Adjusted R^2	0.836*

DISCUSSION & CONCLUSION: Our results contrast with the previous study which stated that leg stiffness during submaximal hopping depends primarily on ankle stiffness (Farley and Morgenroth, 1999). On the other hand, some studies showed that for running velocities ranging from 2.5 to 9.73 m/s the leg spring stiffness is influenced by changes in knee joint stiffness (Arampatzis et al., 1999; Kuitunen et al., 2002). Thus, the present study indicates that knee stiffness is a major determinant of leg stiffness during not only running and sprinting, but also maximal hopping.

REFERENCES:

Arampatzis, A., Bruggemann, G.P. & Metzler, V. (1999). The effect of speed on leg stiffness and joint kinetics in human running. *Journal of Biomechanics* 32, 1349-1353.

Farley, C.T. & Morgenroth, D.C. (1999). Leg stiffness primarily depends on ankle stiffness during human hopping. *Journal of Biomechanics*, 32, 267-273.

Kuitunen, S., Komi, P.V. & Kyröläinen, H. (2002). Knee and ankle joint stiffness in sprint running. *Medicine and Science in Sport and Exercise* 34, 166-173.

SURFACE MARKERS VERSUS CLUSTERS FOR DETERMINING LOWER LIMB JOINT KINEMATICS IN SPRINT RUNNING

Timothy Exell, David Kerwin, Gareth Irwin & Marianne Gittoes
Cardiff School of Sport, University of Wales Institute, Cardiff, United Kingdom

The purpose of this study was to compare lower limb joint angle time histories using surface markers and segmental clusters. An athlete completed three single leg standing trials whilst moving the joints of the free leg from maximum flexion to maximum extension followed by seven maximal sprint runs. Trials were tracked by a three-dimensional CODA system. For standing trials, mean timing differences were greatest in maximum extension at the ankle and hip (0.01 s). Angle differences ranged from 2° (knee flexion) to 11° (ankle extension). Timing differences in sprinting were greatest in extension (hip 0.03 s) with joint angle differences in maximum flexion and extension 7 & 9° (ankle), 3 & 6° (knee) and 23 & 4° (hip) respectively. When comparing results from surface markers and clusters, a good level of agreement was found in the continuous knee flexion-extension profile, and the discrete timings for all joints.

KEY WORDS: flexion-extension, range of motion, precision.

INTRODUCTION: Calculation of key kinematic data is an important part of most biomechanical analyses of sprint running. Kinematic data offer a quantitative description of athletes' movements and are a required input for the calculation of many kinetic variables, for example, when using an inverse dynamic approach. It is important that kinematic values are calculated accurately and that they truly represent the movement being analysed (Challis & Kerwin, 1996). Joint angles at specific event times (i.e. maximum flexion/extension, touchdown) are commonly reported values associated with successful sprint performance (Mann & Herman, 1985).
Many previous studies have analysed sprinting using digitised markers on joint centres (Mann & Herman, 1985; Bezodis et al., 2008). Gait research studies often utilise three-dimensional (3D) kinematics and kinetics using segmental clusters (Cappozzo & Cappello, 1997; McClay & Manal, 1999), giving more information on the orientation of segments. This allows internal segmental rotations to be calculated. Whilst the calculation of 3D kinematic data is beneficial, there are inevitably situations, such as in competition, when collection of the required data is not possible and kinematic data calculated from joint centre locations is the norm. To enable direct comparisons between results calculated from clusters and surface markers, differences between the two methods should be quantified.
The aim of this study was to compare surface anatomical markers located on joint centres with segmental clusters for the calculation of temporal and angular characteristics of lower limb joints. Findings will be of use when comparing results calculated using more contemporary cluster methods with those collected using traditional or field based methods.

METHOD: Data Collection: Data collection took place on the 110 m straight of the National Indoor Athletics Centre, Cardiff. One male athlete participated in the study (age 24.1 years, body mass 70.5 kg, height 1.84 m). A CODA Motion Analysis System (Charnwood Dynamics, UK) was set up operating four scanners (CX1) at a sampling rate of 200 Hz (capture time: 8 s). Scanners were positioned at a height of 1.3 m above the track surface, in pairs 6 m from the centre of the lane and at a separation of 5 m along the lane. This gave a bilateral field of view of 10 m, between 30 and 40 m from the start line. The system was aligned according to manufacturer's guidelines. Active CODA markers (21) were attached to the athlete (Figure 1) as surface markers located on joint centres or as part of four-marker clusters (Charnwood Dynamics, UK). Prior to testing, four additional markers were attached to the athlete whilst a static calibration trial was captured, allowing orientations of clusters to be determined relative to joint centres. The foot cluster and individual markers were attached using double-sided adhesive tape, reinforced with PVC tape. Shank and thigh clusters were attached by means of silicon friction pads and Coban Self-Adherent Wrap (3M, USA).

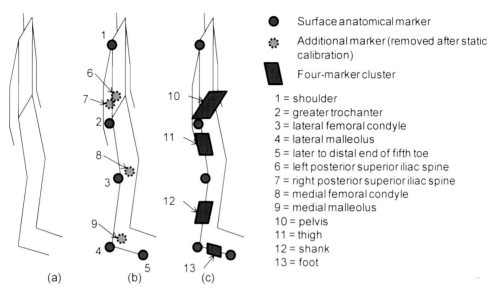

Figure 1: Stick figure representation of athlete (a) showing locations of surface anatomical markers (b) and four-marker clusters (c)

Data collection occurred in two phases; standing trials where the athlete stood on his left leg and moved the joints of his right leg through a full range of motion in the sagittal plane. Ten cycles of maximum flexion to extension were collected for each joint. In the second phase data were collected during seven maximal sprint runs. The athlete accelerated maximally from the start line through the data collection volume to a finish line 10 m beyond.

Data Analysis: Coordinate data were filtered via a low-pass Butterworth digital filter with a 17 Hz cut off frequency determined from Challis' (1999) autocorrelation method. Joint angles were calculated in two ways: for surface markers, three-point vector angles were calculated; for clusters, angles between adjacent segments' Cartesian locations were calculated. During sprint trials, right foot touchdown was calculated using the vertical acceleration method of Bezodis et al. (2007). Root mean squared differences (RMSD) between angles and event times calculated from clusters and surface markers were calculated for maximum flexion and extension. For sprint trials angular data were compared across one complete stride (right foot touchdown to subsequent right foot touchdown) and at instants of touchdown.

RESULTS: Mean RMSDs between joint angles calculated using the two methods are presented in Table 1. Discrepancies in event times were consistently low throughout all trials. A range of values were seen for mean differences between maximum flexion and extension angles, the lowest being for maximum knee flexion (<2°) and the largest being maximum ankle extension (~11°). During sprint trials, the least difference was at the knee with a RMSD of ~4° across all strides. The difference in hip flexion angle was surprisingly large (23°), however this magnitude did not occur throughout the range, shown by the difference in extension angle (~4°). Figure 2a shows time profiles for one standing trial. These show similarity of knee results, consistent differences at the ankle and inconsistency between methods for the hip. Figure 2b illustrates mean results for one stride from all sprint trials. Results are similar to standing trials, with consistency between angle differences for the ankle and knee, but not for the hip. Ankle differences were similar throughout the stride (~10°); increasing at approximately 70% of the stride cycle (~15°). Differences at the knee were small throughout the stride, with the largest differences occurring around maximum flexion. There was less consistency in differences at the hip joint with the largest occurring between 70 and 80% of the stride (>30°).

Table 1 Group mean [± SD] RMSD between lower-limb joint kinematic measures obtained using cluster and surface marker protocols during standing and sprint trials

Standing Trials	Flexion		Extension		RMSD	
	Time (s)	Angle (°)	Time (s)	Angle (°)	Angle (°)	
Ankle	0.01 [0.01]	9.5 [0.4]	0.01 [0.02]	10.7 [1.4]	10.2	
Knee	0.00 [0.00]	1.5 [0.4]	0.00 [0.00]	6.2 [0.6]	3.9	
Hip	0.01 [0.01]	6.0 [2.6]	0.01 [0.01]	8.2 [8.5]	12.0	
Sprint Trials	Flexion		Extension		RMSD	Mean TD
	Time (s)	Angle (°)	Time (s)	Angle (°)	Angle (°)	Angle (°)
Ankle	0.00 [0.00]	7.1 [1.4]	0.02 [0.02]	9.1 [1.4]	9.8	8.2 [0.9]
Knee	0.00 [0.01]	2.5 [2.2]	0.00 [0.00]	5.7 [0.7]	3.8	4.7 [0.9]
Hip	0.02 [0.01]	23.0 [7.4]	0.03 [0.02]	4.2 [3.9]	12.9	22.7 [20.6]

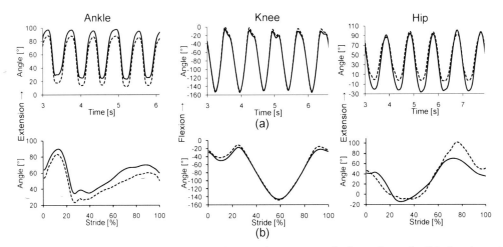

Figure 2: Joint profiles calculated using surface anatomical markers (solid lines) and segmental clusters (dashed lines) for (a) sample standing trial and (b) mean sprint trials (n=7)

DISCUSSION: This study's aim was to compare surface markers with segmental clusters for the calculation of temporal and angular characteristics relating to lower limb joint angles, as these are key variables that have been related to successful performance (Mann & Herman, 1985). Results showed that event times calculated using both methods are similar. Event timings at the knee were least different (0.00 s); ankle and hip values were higher at 0.01 and 0.02 s respectively. These were within precision levels possible for comparable data obtained from 50 Hz video (0.02 s). For standing trials, knee angle differences (~4°) were smaller than for the other joints. Ankle angles calculated from surface markers were consistently larger (~10°) than those from clusters, however there were smaller differences in ranges of movement between maximum flexion and extension (~2°). The largest differences were at the hip (RMSD = 12°), however, unlike ankle results, the magnitude and polarity of the differences varied throughout trials. Results for sprint trials followed a similar trend to standing trials with the knee showing the least differences in event times (0.00 s) and angles (~4°). Ankle differences were consistently ~10° throughout the stride. Hip results were least comparable with differences ranging from ~4° at maximum extension to ~23° for maximum flexion and touchdown angle (Figure 2b). One source of inconsistency in hip angle differences could be angle definition when using clusters and surface markers. Cluster hip angles were calculated relative to the pelvis, discounting upper body movement. However, for surface markers, the shoulder marker is typically included, therefore, upper body movements affect hip angle. The consistent difference at the ankle may be due to cluster angle definition from the static calibration trial. This introduces a systematic offset, relating to the athlete's stance in the static calibration, which may be correctable using offset normalisation (Mullineaux, 2004).

Care should be taken when comparing ankle angles calculated with the two methods; however, the range of movement and temporal characteristics were similar for both methods indicating that direct comparisons are possible for those variables. Knee angles were similar for both methods of calculation. Discrepancies were evident between hip angles, which may be due to shoulder motion affecting hip angle when calculated from surface markers.

The study has highlighted key performance related variables that can be compared across methods, allowing comparison between data collected using newer methods with previous data and data from competition. Further work could investigate effects of upper body movement on hip angle and whether this could be isolated when comparing results calculated using surface marker methods, as hip angle was highlighted by Mann and Herman (1985) as one of the most consistent success factors. Future work will benefit from a larger sample size and the comparison of cluster data with video based surface marker data.

CONCLUSION: Initial findings, based on single subject trials, indicate that surface markers and segmental clusters can be used to detect similar key event times of maximum flexion and extension of the ankle, knee and hip angles during maximal sprint running. When comparing kinematic measures derived from surface markers and clusters, a good level of agreement was found in the continuous knee flexion-extension profile, and discrete timings for all joints. The level of agreement in the hip joint angle range was lower than achieved for the ankle and knee.

REFERENCES:

Bezodis, I. N., Thomson, A., Gittoes, M. J. R. & Kerwin, D. G. (2007). Identification of instants of touchdown and take-off in sprint running using an automated motion analysis system. In H. J. Menzel & M. H. Chargas (Eds). *Proceedings of the XXVth Symposium of the International Society of Biomechanics in Sports* (pp 501-4). Ouro Preto, Brazil.

Bezodis, I. N., Kerwin, D. G. & Salo, A. I. T. (2008). Lower limb mechanics during the support phase of maximum-velocity sprint running. *Medicine and Science in Sports and Exercise.* 40, 707-15.

Cappozzo, A. & Cappello, A. (1997). Surface-marker cluster design criteria for 3-D bone movement reconstruction. *Transactions on Biomedical Engineering.* 44, 1165-74.

Challis, J. H. (1999). A procedure for the automatic determination of filter cutoff frequency for the processing of biomechanical data. *Journal of Applied Biomechanics.* 15, 303-17.

Challis, J. H. & Kerwin, D. G. (1996). Quantification of the uncertainties in resultant joint moments computed in a dynamic activity. *Journal of Sports Sciences.* 14, 219-31.

Mann, R. V. & Herman, J. (1985). Kinematic analysis of Olympic sprint performance: Men's 200 meters. *International Journal of Sport Biomechanics.* 1, 151-62.

McClay, I. & Manal, K. (1999). Three-dimensional kinetic analysis of running: significance of secondary planes of motion. *Medicine and Science in Sports and Exercise.* 31, 1629-37.

Mullineaux, D. R. (2007). Sample size and variability effects on statistical power. In C. J. Payton and R. M. Bartlett (Eds) *Biomechanical Analysis of Movement in Sport and Exercise: Biomechanics Testing Guidelines* (pp162). London: Routledge.

Acknowledgement
This work was funded by EPSRC grant number EP/D076943.

MEASUREMENT OF THIGH MUSCLE SIZE USING TAPE OR ULTRASOUND IS A POOR INDICATOR OF THIGH MUSCLE STRENGTH

Kieran O'Sullivan[1], David Sainsbury[1] and Richard O'Connor[1]

[1]Physiotherapy Department, University of Limerick, Ireland

KEY WORDS: ultrasound, isokinetics, muscle, size, strength, tape measurement

INTRODUCTION: Although commonly used, the validity of measures of thigh circumference and ultrasound measurements of muscle thickness as indicators of muscle strength is unclear (Maylia et al 1999). This study aimed to determine the relationship of these two simple measures of muscle size to the concentric isokinetic quadriceps and hamstrings strength of a group of Irish Gaelic footballers.

METHOD: Thigh circumference was measured using a tape measure. Linear ultrasound measurements of quadriceps and hamstring muscle thickness were also obtained. A pilot study on 15 subjects was performed in advance to determine the reliability of these measures. Twenty five senior and intermediate male gaelic footballers were recruited (mean age 20.74 years, mean BMI 24.78 kg/m^2) Concentric muscle strength was determined using the Biodex system 3 isokinetic dynamometer. Three commonly used speeds were used; 60°/sec, 180°/sec, and 300°/sec. Reliability was assessed using intra-class correlation coefficients and Bland & Altman methods. Muscle size measurements were correlated with torque values using Pearsons correlation. The alpha level was set at $p < 0.05$.

RESULTS: Both tape and ultrasound demonstrated moderate to excellent reliability at measuring quadriceps and hamstrings muscle size (ICC's 0.69-0.99). Hamstrings were significantly stronger on the dominant limb at 60°/sec ($p=0.046$) and 180°/sec ($p=0.005$), but not at 300°/sec ($p=0.092$). There was no significant difference in quadriceps strength at any speed (all $p>0.05$). Dominant limbs were significantly larger using the tape measure (10cm level: $p=0.005$, mid-thigh level: $p=0.003$). The dominant hamstrings ($p<0.001$), but not the dominant quadriceps ($p=0.399$), were significantly larger on ultrasound. There were statistically significant correlations between muscle strength and muscle size measurements, especially for the tape measurements. However, despite the fact that the dominant limb muscles were both stronger and larger, the strength of these correlations was only weak to moderate ($r = 0.176-0.526$).

DISCUSSION: Although reliable, gross anthropometric measures of thigh muscle thickness using tape and ultrasound correlate poorly with isokinetic muscle strength. This may be related to the strong influence of neural factors on muscle strength (Deschenes et al 2002).

CONCLUSION: Simple measures of muscle size are poor indicators of muscle strength.

REFERENCES:

Maylia, E, Fairclough, J, Nokes, L & Jones, M. (1999) Can Muscle Power Be Estimated From Thigh Bulk Measurements? A Preliminary Study. *Journal of Sports Rehabilitation* 8, 50-59.
Deschenes, MR, Giles, JA, McCoy, RW, Volek, JS, Gomez, AL & Kraemer, WJ (2002). Neural factors account for strength decrements observed after short-term muscle unloading. *American Journal Of Physiology. Regulatory, Integrative And Comparative Physiology* 282, R578-83.

PEAK VELOCITY OF NORDIC SKI DOUBLE POLE TECHNIQUE: STAND-UP VS. ADAPTIVE SIT-SKIING.

Jodi L. Tervo and Randall L. Jensen
Dept. HPER, Northern Michigan University, Marquette, Michigan, U.S.A.

KEY WORDS: cross country skiing, disabled sports, biomechanics.

INTRODUCTION: One of the event styles in cross-country skiing is the classic technique in which the skis move in groomed tracks. Double poling is a technique used under the classic skiing style, and is defined as when the upper body provides most of the propulsion via bilateral pole pushes. Double poling during classic cross-country skiing has become more popular in the past twenty years. It has also been shown to have strong correlations with increased race speed (Smith, Fewster, & Braudt, 1996).

Maximal velocity considers an overall velocity of the movements, but does not specify at which point during the poling phase that peak velocity occurs. By breaking a movement down into its components one may be able to critique technique more specifically. This study examined the point at which peak linear velocity occurred during the double poling cycle time in Nordic stand-up and sit-down skiing.

METHODS: Four female and two male collegiate athletes participated in stand-up and sit-skiing, and one experienced male sit-skier participated as a subject to reference for the analysis portion of this study. All subjects arrived at an outdoor cross-country ski venue with skis, boots, and poles. Ethical approval (#HS09-242) was received prior to conduction of the study, and signed an informed consent and PAR-Q questionnaire were completed prior to participating. A Canon Digital Video Camcorder NTSC Optura 20 (Canon Inc., Japan) was placed perpendicular to the ski line approximately 4.5 meters away. A 1/1000 shutter speed was used, with a 60 Hz camera. Each subject had markers placed on their ski binding to allow digitization of movement. They were asked to mimic a race start using the double pole technique. The video captured a recording of at least one complete cycle of double poling. One cycle is defined from pole plant to pole plant. The Peak Motus System version 8.5 (Vicon Motion Systems Inc., Centennial, CO) was used to analyze the data. One poling cycle lasted approximately one second. Using MATLAB each subject's trial was standardized to one second. This allowed the researchers to calculate the percentage of poling cycle (%PC) for comparisons.

RESULTS & DISCUSSION: Paired T-Test indicated that the point at which peak velocity occurs in Nordic sit-skiing (mean ± SD: 0.350 ± 0.066 %PC) and stand-up skiing is (0.231 ± 0.031 %PC) ($p = 0.017$). A One-sample T-Test comparing the experienced male sit-skier (0.350%PC) to the subjects found no difference ($p = 0.995$). Results of the study indicate that there is a difference in the point at which peak velocity occurs between Nordic sit-skiing and stand-up skiing. Furthermore, it may be possible to use experienced stand-up skiers as subjects in sit-skiing research. A larger subject base in the future may allow more extensive conclusions to be made.

REFERENCES:

Smith, G. A., Fewster, J. B., & Braudt, S. M. (1996). Double poling kinematics and performance in cross-country skiing. *Journal of Applied Biomechanics, 12*, 88 – 103.

HIGH- AND LOW-ARCHED ATHLETES EXHIBIT SIMILAR STIFFNESS VALUES WITHIN THE LOWER EXTREMITY

Douglas Powell[1] and Carrie Albright[2]

Biomechanics Laboratory, University of Texas of the Permian Basin, Odessa, TX, USA[1]
Department of Kinesiology, Towson University, Towson, MD, USA[2]

KEY WORDS: Arch, Kinetics, Landing

INTRODUCTION: Abnormal foot function has been associated with an increased propensity of injury (Kaufman et al., 1999). Both high- (HA) and low-arched (LA) athletes suffer a greater incidence of injury (Kaufman et al., 1999). Previous research has shown that HA compared to LA runners exhibit greater lower extremity stiffness and greater stiffness within the lower extremity joints (Williams et al., 2004). The purpose of the current study was to examine lower extremity stiffness as well as hip, knee and ankle joint stiffness in HA and LA athletes. It was hypothesized that the HA athletes would have greater stiffness values than the LA athletes within the lower extremity.

METHODS: Ten HA (arch index > 0.356) and 10 LA (arch index < 0.290) female athletes participated in the current study. Each subject performed five trials of landing from a 30cm box while three-dimensional kinematics (240Hz, Vicon Motion Systems Ltd., Oxford, UK) and ground reaction forces (1200, OR-6, AMTI, Watertown, MA) were collected simultaneously. Lower extremity stiffness was calculated as the peak vertical ground reaction force divided by the change in vertical height of the pelvis. Lower extremity joint stiffness was calculated as the joint moment divided by the joint angle. A 2x3 analysis of variance was used to determine significant differences between the HA and LA groups.

RESULTS & DISCUSSION: The HA and LA athletes exhibited similar lower extremity stiffness values ($p=0.061$). Additionally, no differences in hip ($p=0.912$), knee ($p=0.869$) or ankle stiffness ($p=0.198$) were observed between the HA and LA athletes. These data show that the HA and LA athletes exhibited similar stiffness values suggesting that unique injury patterns during dynamic tasks are not caused by differences in lower extremity stiffness.

CONCLUSIONS: Lower extremity injury patterns are different between HA and LA athletes, however these differences are not due to differences in lower extremity stiffness. Further research should examine differences in lower extremity joint moments between HA and LA athletes which may provide greater insight into these different injury patterns.

REFERENCES:
Kaufman, K.R., et al (1999). The effect of foot structure and range of motion on musculoskeletal overuse injuries. *Am J Sports Med*, 27(5), 585-93.

Williams, D.S., et al (2004). High-arched runners exhibit increased leg stiffness compared to low-arched runners. *Gait Posture*, 19(3), 263-9.

ARCHERY TRAINING IMPROVE POSTURAL CONTROL IN YOUNG CHILDREN

Alex, J. Y. Lee [1], Yu-Chi Chiu [1], Ying-Fang Liu [2], Wei-Hsiu Lin [3]

Department of P.E., National HsinChu University of Education, Taiwan [1]
Division of P.E., HsinSheng College of Medical Care and Management, Taiwan [2]
Department of P.E., National ChiaYi University, Taiwan [3]

KEY WORDS: exercise training, balance control, center of pressure

INTRODUCTION: Exercise training might be beneficial for postural control (PC) during exercise and daily activities. The purpose of the study was to examine the effects of an eight-week archery training program on PC in young children during the standard, aiming, and archery shooting standing posture.

METHOD: Thirty-two young children were recruited and grouped as experimental group (E, 16 children) or control group (C, 16 children). The E group underwent archery training for forty minutes a time, three times a week, for eight weeks, and the C group did not received any regular physical training. PC was evaluated by the measurement of COP displacement on a force platform as the COP radius, speed and sway area during each testing position (Lee & Lin, 2008). The training program was planned by expert archery coaches and based on the training of basic archery skill and coordination. Data were analyzed by the mixed design, two-way ANCOVA to determine if any difference in PC between groups before and after training during the standard posture. Repeated measures, dependent t test was also used to determine if any difference in PC before and after training during the aiming and shooting posture in experimental group. The significant level for all statistical tests was set at $p < .05$.

RESULTS: The archery performance was increased from 5.6 points before training to 8.2 points after training per shooting. The COP radius and sway area were significantly decreased after archery training in E group during the standard (0.41 ± 0.04 vs. 0.31 ± 0.03 mm/s, $t = 3.26$; 1.59 ± 0.22 vs. 1.03 ± 0.19 mm^2, $t = 2.30$) and shooting posture (1.86 ± 0.16 vs. 1.12 ± 0.08 mm/s, $t = 4.41$; 22.26 ± 3.56 vs. 7.19 ± 0.82 mm^2, $t = 4.27$).

DISCUSSION: During the archery training, subjects were asked to concentrate and focus on the target, therefore, have to control the balance and the movement during shooting practice. It might be the reason for these improvements on PC in COP radius (24-40 %) and sway area (35-68 %) after regular archery training.

CONCLUSION: This study demonstrated that regular eight-week archery training not only improved the performance of archery shooting, but also improved the PC during the standard, aiming, and shooting posture. The results of this study could provide a practical training regimen for archery coaching and understand the possible relationships between the archery performance and PC.

REFERENCES:
Lee AJY. and Lin W.H. (2008). Twelve-week biomechanical ankle platform system training on postural stability and ankle proprioception in subjects with unilateral functional ankle instability. *Clinical Biomechanics*, 23, 8, 1065-1072.

Acknowledgement
This research was supported by grants from National Science Committee (NSC 96-2413-H-134-008 & 97-2410-H-134-022) and National HsinChu University of Education, TAIWAN.

KNEE AND ANKLE JOINT KINEMATICS IN KENDO MOVEMENT AND THE REPEATABILITY

Sentaro Koshida, Tadamitsu Matsuda, and Kyohei Kawada
Ryotokuji University, Department of Health Science, Urayasu, Japan

KEY WORDS: martial arts, Achilles tendon injury, motion analysis

INTRODUCTION: Kendo, a Japanese martial art of sword fighting, comprises strike and thrust motion against a specific target of a part of an opponent's body. Among kendo athletes, chronic Achilles tendon (AT) injuries are frequently observed only on the left side probably because a main power source lies in the left side of the lower body for the repetitive forward-back bounding steps in the strike motion. It has been reported that the AT injury may be linked to abnormal joint biomechanics among runners. In order to make better clinical decision on the AT injury treatment in kendo athletes, therefore, we should focus on joint biomechanics of the left side of the lower extremity in kendo movement. The purpose of the study was to demonstrate the joint kinematics of knee and ankle in the kendo movement and quantify the repeatability.

METHOD: Ten male collegiate kendo athletes (mean($\pm$SD) age 20.5$\pm$1.1 years; height 172.8$\pm$4.5 cm; weight 76.1$\pm$9.7 kg; the kendo experience 12.1$\pm$2.4 years) attended the study and five of the participants were also tested a different day approximately at intervals of one week. After the informed consents were given, we instructed the participants to execute three sets of the kendo strike motion with a single forward step toward the target object. We obtained the joint kinematics data of the left ankle flexion-extension, rearfoot inversion-eversion, knee flexion-extension during the single support phase in the motion by an eight-camera Mac3D motion analysis system (Motion analysis corp., Santa Rosa, CA, USA). The coefficient of multiple correlation (CMC) was used to assess the repeatability of joint angle waveforms within a test day and between test days. Statistical significance was set as $P \leq 0.05$ in this study.

RESULTS AND DISCUSSION: Figure1 illustrated the representative data of the joint angle curve during the single stance phase. The CMC within a test day was from 0.986 to 0.999 in ankle flexion-extension angle, from 0.890 to 0.998 in rearfoot inversion-eversion angle, and from 0.930 to 0.999 in knee flexion-extension angle in the participants. The CMC between the test days were from 0.966 to 0.992 in ankle flexion-extension angle, from 0.710 to 0.996 in rearfoot inversion-eversion angle, and from 0.716 to 0.993 in knee flexion-extension angle. The current result suggests that it may be reasonable to make a clinical decision by using the result of single kendo motion analysis for experienced kendo athletes.

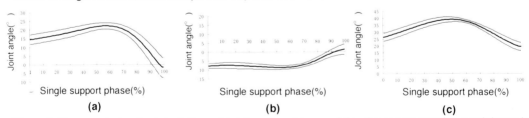

Figure1. Mean and standard deviation of joint motion (degrees) for the representative participant in ankle flexion-extension, (b) reafoot inversion-eversion, and (c) knee flexion-extension during the single stance phase of left leg in the kendo movement

CONCLUSION: The knee and ankle joint kinematics in the kendo motion may be quite repeatable both within a test day and between test days.

EFFECTS OF FATIGUE ON THE GROUND REACTION FORCES AND LEG KINEMATICS IN ALL-OUT 600 METERS RUNNING

Hirosuke Kadono[1] Michiyoshi Ae[2] Yuta Suzuki[1] and Kazuhito Shibayama[1]

Doctoral Program in Physical Education Health and Sports Sciences, University of Tsukuba, Tsukuba, Japan[1]
Institute of Health and Sports Science, University of Tsukuba, Tsukuba, Japan[2]

The purpose of this study was to investigate effects of fatigue on the ground reaction forces and leg kinematics during all-out 600m running, which was performed by eight male middle-distance runners. Their running motion was videotaped (300Hz) and the ground reaction forces were measured (500Hz) at the 150m and 550m marks of the 600m running. From the 150m to 550m mark, running speed significantly decreased ($p<0.001$) while the 2nd half of the support time significantly increased ($p<0.01$). During the 2nd half of the support phase, the horizontal impulse ($p<0.05$) and the average force ($p<0.001$) of the ground reaction force significantly decreased. Furthermore the average angular velocity of the support shank significantly correlated with the horizontal average force ($r=-0.811$, $p<0.001$), the ratio between the vertical and horizontal average force ($r=0.803$, $p<0.001$). Therefore it is likely to be one of the important techniques to maintain large forward lean and angular velocity of the support shank during the support phase in the final stage of the middle-distance running races.

KEY WORDS: support time, horizontal force, shank

INTRODUCTION: In the 400m sprint and 800m middle-distance running, which the running times are lesser than two minutes, the peak running velocity is achieved in the initial stage of the race, and the running velocity progressively decreases toward the end of the race (Gajer et al., 2007; Abbiss & Laursen, 2008). To achieve a high performance in these events, runners have to reach a large running velocity as quickly as possible and to maintain it as long as possible There are some studies on effects of fatigue on kinematics and kinetics for the sprint running (Chapman, 1982; Sprague & Mann, 1983; Nummela et al., 1996). But there is less information of the changes in kinematics and kinetics, especially the ground reaction forces (GRF) due to fatigue during the middle-distance running. In official 800m races, the final 200m is likely to be spent to decide a winner of the race with different running kinematics of stride frequency in defiance of fatigue witch accumulates in the preceding 600m. Therefore, we assumed that the 600m running with a full effort would be suitable to asses the effects of fatigue on the GRF during a large portion of the 800m race without unnecessary suffering runners. The purpose of this study was to investigate effects of fatigue on GRF and leg kinematics during an all-out 600m running, which was simulated the middle-distance race.

METHODS: Data Collection: Eight male middle-distance runners (height 1.76 ± 0.06m, body mass 64.3 ± 5.5 kg and 800m personal best record 1 min 49 s 77 ± 1 s 49) participated in this study. Subjects were asked to perform an all-out 600m running with a positive pacing strategy (Abbiss & Laursen, 2008) that the running speed at the initial stage of the race was larger. A pace maker's bicycling ahead of the subject was provided to keep a previously determined velocity. The subjects were videotaped (60 Hz) to determine the average running speeds at every 50 m intervals. The subject's running motion was videotaped (300 Hz) over one full running cycle and GRF were measured (500 Hz) at the marks of 150m and 550m of the 600m running.

Data Analysis: Twenty-three body landmarks were digitized at 150 Hz and reconstructed in real coordinate data. The real coordinates were smoothed by a Butterworth digital filter at cut off frequencies ranging from 6.0 to 7.5 Hz, which were decided by a residual method. The angle and angular velocity of the support leg segments, foot, shank and thigh were calculated from the smoothed coordinate's data. The running motion was divided into support and non support phases. The support phase was defined as the phase from the foot contact to the toe-off, and the non support phase was at the toe-off to next the foot contact. The support phase was further divided into 1st and 2nd half based on the instant of zero crossing of the anterior-posterior GRF. The impulses of the 1st and 2nd halves of the support phase were calculated by integration of GRF, the average forces were calculated by dividing the impulses by the support times, and a ratio of average vertical force to horizontal force calculated by dividing the average vertical force by the average horizontal one. The paired t-test was used to assess the significant differences between variables for the 150m and 550m mark. Pearson's correlation coefficient was calculated to examine the relationships between variables. The level of significance was set at $p<0.05$.

RESULTS: The average time of 600m running was 1 min 21 s 13 $\pm$ 1 s 61. Figure 1 shows that the averaged and individual patterns of the running speed change. The running speed increased after the start, rose to the peak at 50-100m interval, and gradually decreased toward the finish. Table 1 shows the running velocity, stride length and step time at the 150m and 550m marks. Running velocity at the 150m mark was significantly higher than the 550m mark ($p<0.001$). Stride length ($p<0.001$), support distance ($p<0.05$) and non support distance ($p<0.01$) at the 150m mark were significantly larger than the 550m mark. Step time, 1st half and 2nd half of support time at the 150m mark was significantly shorter than the 550m mark ($p<0.01$). Figure 2 shows the impulses and average forces of GRF. The vertical impulse in the 2nd half of the support phase at the 550m mark was significantly larger than the 150m mark ($p<0.05$), but the average force of the 2nd half of the support phase at the 150m mark was significantly larger than the 550m mark ($p<0.05$). In the horizontal component, the average force of the 1st half of the support phase at the 150m mark was significantly larger than the 550m mark. In the 2nd half of the support phase, the impulse and the average force in the 150m mark were significantly larger than the 550m mark. Figure 3 shows changes in the thigh and shank angles of the support leg, which were normalized by the support time and averaged. The shank angle was significantly smaller in 150m than 550m marks from

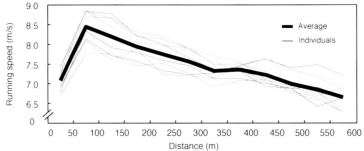

Figure 1 The averaged and individual patterns of running speed in the all-out 600m running.

Table 1 The running speed, stride length and stride frequency at the 150m and 550m mark of all-out 600m running.

			150m		550m		Difference
Running velocity		(m/s)	8.22	(0.43)	6.77	(0.33)	p<0.001
Stride length		(m)	2.17	(0.08)	1.97	(0.09)	p<0.001
Support distance		(m)	1.01	(0.06)	0.96	(0.04)	p<0.05
Non support distance		(m)	1.16	(0.08)	1.01	(0.09)	p<0.01
Step time		(s)	0.265	(0.014)	0.292	(0.017)	p<0.01
Support time	1st half	(s)	0.062	(0.003)	0.071	(0.006)	p<0.01
	2nd half	(s)	0.064	(0.006)	0.074	(0.005)	p<0.01
Non support time		(s)	0.139	(0.011)	0.147	(0.012)	ns

Figures in parentheses are standard deviations.

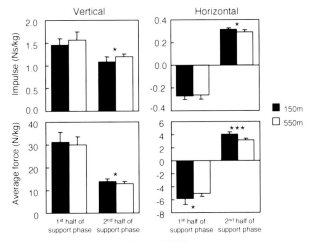

Figure 2 The Impulse and average force of GRF.
*, ** and *** represents a significant difference between 150m and 550m mark, p<0.05, p<0.01 and p<0.001.

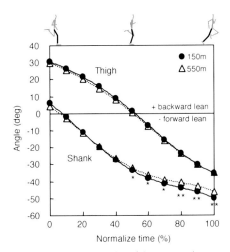

Figure 3 The thigh and shank angles of the support leg.
* and ** represent a significant difference between 150m and 550m mark, p<0.05 and p<0.01.

50% to 100% normalized time. During the 2nd half of the support phase, there were significantly relationships between the average angular velocity of the support shank and the horizontal average force (F_h, Figure 4, r=-0.811, p<0.001), and the ratio of the vertical average force (F_v) to the horizontal one (F_h), (F_v/F_h, Figure 5, r=0.803, p<0.001).

DISCUSSION: The 2nd half support time was significantly increased (p<0.01) while the running speed decreased (p<0.001) from the 150m to 550m mark. This change was similar to the results of the 400m sprint (Chapman, 1982; Sprague & Mann, 1983; Nummela et al., 1996). The average forces at the 550m mark were smaller than the 150m mark (Figure 2). This indicated that the subjects were unable to exert the large force during the support phase of the final stage of the race. Although the horizontal impulse and average force during the 1st half of the support phase decreased from the 150m to 550m mark, those also decreased during the 2nd half of the support phase. This indicates that the decrease in the running speed in the final

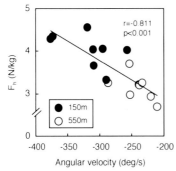
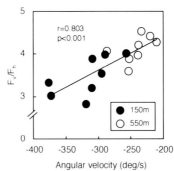

Figure 4 Relationships between the average angular velocity of the support shank and the horizontal average force (F_h) during the 2nd half of the support phase.

Figure 5 Relationships between the average angular velocity of the support shank and the ratio between the vertical and horizontal average force (F_v/F_h) during the 2nd half of support phase.

stage was caused by the decreased acceleration force rather than the increased deceleration force of the GRF and implies that the motion in the 2nd half of the support phase should be investigated to see effects of fatigue. Kadono et al. (2008) indicated that in the positive 800m races, the average shank angular velocity of the support leg during the 2nd half of the support phase decreased with the decrease in running speed. Accordingly, in the present study, during the 2nd half of the support phase, the shank at the 150m mark was leaned more forward than the 550m mark (Figure 3). From these and the correlation results (Figure 4, 5), it may be thought that the subjects who rotated the shank forward in a lower angular velocity could not direct the GRF more horizontally. Therefore, maintaining large forward lean and higher angular velocity of the support shank in the 2nd half of the support phase is likely to be an important technique during the final stage of the 800m middle-distance running races.

CONCLUSION: It was concluded that the decrease in the running speed in the 800m race was caused by the decreased acceleration component rather than the increased braking component, and that the forward lean of the support shank in the final stage of the 800m race become smaller and slower than the initial stage of the race. Therefore it is likely to be one of the important techniques to maintain large forward lean and fast angular velocity of the support shank during the support phase in the final stage of the 800m middle-distance running.

REFERENCES:

Abbiss, C.R. & Laursen, P.B. (2008) Describing and understanding pacing strategies during athletic competition. *Sports Medicine*, 38 (3), 239-252.

Chapman, A.E. (1982) Hierarchy of changes induced by fatigue in sprinting. *Canadian Journal of Applied Sport Sciences*, 7 (2), 116-122.

Gajer, B., Hanon, C. & Mathieu, C.T. (2007) Velocity and stride parameters in the 400 metres. *New Studies in Athletics*, 22 (3), 39-46.

Kadono, H., Ae, M., & Enomoto, Y. (2008) Effects of pacing strategies on the running motion of male 800 meter runners. *Proceedings of the XXVI International Conference on Biomechanics in Sports* (pp 677-680). Seoul, Korea. ISBS.

Nummela, A., Gundersen, J.S. & Rusko, H. (1996) Effects of fatigue on stride characteristics during a short-term maximal run. *Journal of Applied Biomechanics*, 12, 151-160.

Sprague, P. & Mann, R.V. (1983) The effects on muscular fatigue on the kinetic of sprint running. *Research Quarterly for Exercise and Sport*, 54 (1), 60-66.

IMPROVED MUSCLE ACTIVATION IN PERFORMING A BODY WEIGHT LUNGE COMPARED TO THE TRADITIONAL BACK SQUAT

Priscilla Dwelly[1], Gretchen Oliver[1], Heather Adams-Blair[2], David Keeley[1], Hiedi Hoffman[1]

University of Arkansas, Fayetteville, AR, USA[1]
Eastern Kentucky University, Richmond, KY, USA[2]

The purpose of this study was to examine the muscle activity during the body weight lunge and back squat at three depths, 90°, 60°, and 30°. Eight female collegiate athletes volunteered for the study. Each participant performed 5 squats at each depth and the body weight lunge. Surface EMG data and video were recorded to analyze the motion. The lunge activated most of the lower muscles greater than the squat. Sport performance is often initiated from a ready stance, similar to that of the squat however, upon movement the athlete typically steps into a lunged position. Therefore, with the similarities in sport motion to that of the lunge, and equal if not greater activation of the lower extremity, the lunge is an ideal activity to train athletes.

KEY WORDS: sEMG, quadriceps, hamstring.

INTRODUCTION: The back squat is a common exercise for performance strength training and rehabilitation with focus on quadriceps and hamstring activity. Contrary to common clinical practice, variation in stance width during the squat does not affect the isolation of quadriceps musculature (McCaw & Melrose, 1999; Signorile et al, 1995). With the increased focus on the hamstring musculature for injury prevention, especially the anterior cruciate ligament, exercises such as the lunge are more commonplace. Thus, the purpose of this study was to examine the gluteal, quadriceps and hamstring muscle activation at three different squat depths and during the body weight lunge.

METHOD: Eight healthy, female intercollegiate athletes (mean age 20.8 + 3.9 y; mean height, 177.8 + 10.9 cm; mean mass, 67.3 + 9.9 kg) consented to participate in the study. The study was approved by the Institutional Review Board. Prior to testing, adhesive 3M Red-Dot bipolar surface electrodes were placed over the muscle bellies on the subject's dominant side according to method of Basmajian and Deluca (1985) with an interelectrode distance of 25 mm (Hintermeister et al.1998). The muscles targeted were the following: rectus femoris, vastus lateralis, vastus medalis obliques, medial hamstring (semimembrinosus and semitendinosus), biceps femoris, gluteus medius and maximus.
Manual muscle tests were performed through maximum isometric voluntary contractions (MVIC) based on the work of Kendall et al. (1993). Three manual muscle tests were performed for a total of five seconds for each muscle group. The first and last second of each MVIC trails were removed from the data in attempt to obtain steady state results for each of the muscle groups. Each subject then performed several warm-up squats to assure proper technique and proper depth prior to each trail recording. Each subject performed five weighted squats of 70% of their body weight to 90°, 60°, and 30° of knee flexion respectively. Each subject was allotted three minutes of rest between the different depths. During the trials subjects were instructed on proper posture through verbal cues. After the squats were completed the subject also performed five body weight lunges with body weight only. In addition to EMG data, video data were also collected from a 90° lateral view to assure appropriate technique as well as to event mark trials. All trials were event marked for concentric and eccentric phases.
A Myopac Jr 10 channel amplifier (RUN Technologies Scientific Systems, Laguna Hills, CA) transmitted the all EMG raw data at 60 Hz via a fiber optic cable to the receiver unit. The EMG unit has a common mode rejection ratio of 90 dB. The gain for the surface electrodes was set at 2000. EMG data were recorded, stored, and analyzed with the analog data acquisition package of Peak Motus Software (version 9.0; Peak Performance, Englewood,

CO). EMG enveloped data were assessed. Mean maximum EMG reference values were calculated for each muscle within the phase. Five trials of EMG data for each subject were analyzed to determine average peak amplitudes for all muscles during each concentric and eccentric phase of the exercise.

Data Analysis: Data from each muscle were normalized as a percent of the contribution of electrical muscle activity of the MVIC. The Levene's test was performed to determine homogeneity of the variables, all variables except rectus femoris violated the Levene's statistic. Therefore, nonparametric Kruskal-Wallis tests were used to observe if the exercises (lunge and the squat at 90°, 60°, and 30°) had a main effect on the muscles involved (rectus femoris, vastus lateralis, vastus medialis oblique, semitendinosus, biceps femoris, gluteus maximus, and gluteus medius). Where exercise affected the muscle activity, Mann-Whitney tests identified specific differences between each exercise and each muscle. The level of significance was set at p<0.05 and all tests were performed using SPSS 15.0 (Chicago, IL).

RESULTS: Table 1 presents the summed ranks for rectus femoris, vastus lateralis, vastus medialis oblique, semitendinosus, biceps femoris, gluteus maximus, and gluteus medius for the 90°, 60°, 30° squats, and the lunge. Figure 1 illustrates the rank differences between the three different squat depths and the lunge. Kruskal-Wallis tests revealed a significant main affect for exercise type on the rectus femoris (χ^2=15.706, p=0.001); vastus medialis oblique (χ^2=8.767, p=0.03); vastus lateralis (χ^2=15.169, p=0.002); semitendinosus (χ^2=8.775, p=0.03); biceps femoris (χ^2=14.258, p=0.003); and gluteus medius (χ^2=10.387, p=0.02). Exercise type did not have a main affect on gluteus maximus muscle activity (p=0.05).

Specifically as illustrated in Figure 1 post-hoc Mann Whitney tests revealed that the rectus femoris had higher activation during the 90° compared to the 60° (U=6.5, p=0.007) or 30° squat (U=2.0, p=0.002) and during the lunge (U=6.0, p=0.006) compared to the 30° squat. The vastus medialis obliques had higher activation during the 90° squat compared to the 30° squat (U=11.5, p=0.03) and during the lunge compared to the 30° squat (U=7.0, p=0.009). The vastus lateralis had higher activation during the 90° squat compared to the 60° squat (U=9.5, p=0.02), and the 30° squat (U=5.5, p=0.005) and during the lunge compared to the 60° squat (U=7.5, p=0.01), and compared to the 30° squat (U=4.0, p=0.003). The semitendinosus/semimembrinosus had higher activation during the 90° squat compared to the 30° squat (U=4.5, p=0.004) and during the lunge compared to the 30° squat (U=11.0, p=0.03). The biceps femoris had higher activation during the 90° squat compared to the 30° squat (U=8.5, p=0.01) and during the lunge compared to the 90° squat (U=9.5, p=0.02), 60° squat (U=11.5, p=0.03), and compared to the 30° squat (U=2.0, p=0.002). The gluteus medius had higher activation during the lunge compared to the 90° squat (U=6.0, p=0.006).

Table 1 Summed ranks for each muscle based on exercise type.

	RF	VMO	VL	SEMI	BF	MAX	MED
90° Squat	24.88	18.88	22.31	20.56	17.56	14.44	16.19
60° Squat	12.63	16.50	11.75	16.5	14.56	14.31	16.38
30° Squat	8.06	8.69	8.56	8.5	8.25	12.88	9.19
Lunge	20.44	21.94	23.38	20.44	25.63	24.38	24.25
Chi-Square	15.706	8.767	15.169	8.775	14.258	7.791	10.387
Significance	0.001	0.033	0.002	0.032	0.003	0.051	0.016

Legend: Rectus femoris (RF), vastus medialis oblique (VMO), vastus lateralis (VL), semitendinosus and semimembrinosus (SEMI), biceps femoris (BF), gluteus maximus (MAX), and gluteus medius (MED).

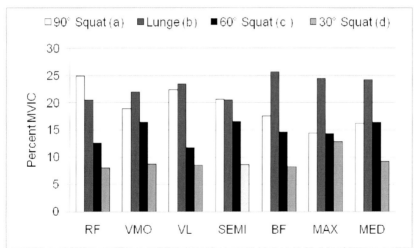

Figure 1. A comparison of the summed ranks of the percent muscle activation for the 90°, 60°, 30° squats and the lunge.
Legend: Rectus femoris (RF), vastus medialis oblique (VMO), vastus lateralis (VL), semitendinosus and semimembrinosus (SEMI), biceps femoris (BF), gluteus maximus (MAX), and gluteus medius (MED).

DISCUSSION: The results demonstrated that there was no greater muscle activation when performing any of the squat depths to that of the body weight lunge. It was revealed that the body weight lunge did indeed produce more activation in the majority of all muscles analyzed when compared to the three squat depths. Although two muscles, the rectus femoris and semitendinosus/semimembrinosus, displayed greater activation during the 90° squat when compared to the lunge, however this difference was not significant.

Thus, in attempt to enhance athletic performance, we want to focus on training athletes functionally for their sport not necessarily functional for the weight room. Functionality comes into question when one assesses how the athlete resumes their athletic position. If an athlete is in competition and is placed in a position where the knees are flexed to 90° there is a much greater chance that the athlete is either going to fall to the ground or walk out of the position. Majority of athletes move from a base position, which mimics the squat however upon moving the athlete naturally has to step forward, backward, or side to side. This step is crucial to transfer energy from potential to kinetic, therefore with the increased activation of the lower extremity muscles and sports function clinicians should train athletes from a lunge position.

CONCLUSION: By training the athletes in the body weight lunge, they can obtain the same results of that of squat to 90° training. The lunge allows the athlete to be in a more sport functional position. From the basic lunge position of the knees flexed to 90° we can begin to train explosive recovery moves, which would transfer over to competition. Ideally, athletes should be training their kinetic chain fluidly and dynamically; the more dynamic the activity the more fluid the athlete's movement and posture will be in competition. As a coach, personal trainer or athletic trainer, we should begin to worry when movement gets ridged because of the susceptibility to injury.

As an athletic trainer conditioning for rehabilitation or a coach training for performance, the data reveals that athletes do not need any type of equipment to train optimally and functionally. Using the athlete's own body weight and proper lunging instruction, they can maintain optimal activation of the primary muscles involved. Therefore, for those teams who do not have the weight room facility, the athletic trainer or coach can instruct the athletes on strength training and rehabilitation exercises on the actual playing surface, whether it be field or court.

REFERENCES:

Basmajian, J.V., & DELUCA, C.J. (1985). Apparatus, detection, and recording techniques. In: Butler JP, editor. Muscle alive, their functions revealed by electromyography. Baltimore: Williams & Wilkins, 19-64.

Hintermeister, R.A., Lange, G.A., Schultheis, J.M., Bey, M.J., & Hawkings, R.J.(1998) Electromyographic activity and applied load during shoulder rehabilitation exercises using elastic resistance. *American Journal of Sports Medicine*, 26, 210-220.

Kendall, F.P., McCreary, E.K., Provance, P.G., Rodgers, M.M.,& Romani, W.A.(1993) Muscles: Testing and Function. Fourth edition. Baltimore, Williams & Wilkins.

McCaw, S.T. & Melrose, D.R. (1999) Stance width and bar load effects on leg muscle activity during the parallel squat. *Medicine Science in Sports Exercise*. 31 (3), 428-436.

Signorile, J.F., Kwaitiwski, K., Caruso, J.F., & Robertson, B. (1995) Effect of foot position on the electromyographical activity of the superficial quadriceps muscles during the parallel squat and knee extension. *Journal of Strength and Conditioning Research*, 9 (3), 182-187.

THE EFFECT OF STRENGTH TRAINING ON THE KINEMATICS OF THE GOLF SWING

Amy C. Scarfe[1], Francois-Xavier Li[1] and Matthew W. Bridge[2]

School of Sport and Exercise Sciences[1],
School of Education[2], University of Birmingham, Birmingham, UK

KEY WORDS: Golf, Strength training, kinematics.

INTRODUCTION: Despite the wide acceptance of resistance training in other sports, use of such programs in golf is a relatively new concept; consequently there is a lack of research addressing the effects of these programs on golfing performance. The few studies addressing this issue have used golf-specific programs (Lephart *et al*, 2007) these can require specialist equipment and can be more difficult to follow. Outcome measures in these studies have shown that strength training improves performance measures i.e. distance the ball is carried yet neglected details of the coordination of the golfer themselves (Fletcher and Hartwell, 2004).

The aim of this study was to identify how a simple combined general resistance and plyometric program effects golf swing kinematics.

METHOD: Twelve category 1 golfers (mean handicap 3.51 ± 2.32) were assigned to two groups, Exercise and Control. Pre and post intervention kinematic data were collected, via a Vicon motion analysis system (250 Hz), of 10 swings with a 6Iron and Driver.

During the 5 week training period the exercise group carried out a strength and conditioning program based on that used by Fletcher and Hartwell (2004), 3 times a week, while the Control group did not perform any resistance training.

RESULTS: Strength training data showed improvements of between 25% and 809%, indicating that the intervention was successful and participants in the exercise group significantly improved their strength. There were significant increases for both clubs in exercise group wrist peak extension angular velocity (4% Driver; 16% 6Iron), maximum X-Factor (10% Driver; 8% 6Iron) and maximum X-factor velocity (20% Driver; 5% 6Iron) but no changes were observed for the control group. Exercise group peak driver club head velocity was also found to increase by 9% this increase was not evident in the 6Iron.

DISCUSSION: As in previous studies (Fletcher and Hartwell, 2004), the data indicated that a general strength training program can increase the club head velocity. Additionally this study also highlighted the improvements in key kinematic variables that are associated with increased club head velocity. The improvement in performance is even more remarkable considering the high standard of the participants and that the strength training program was less golf-specific than previous studies.

CONCLUSION: It is evident that after completing a 5-week strength training program a number of kinematic variables of the golf swing of skilled players can be improved. Exploration of the effects in less skilled golfers and different types of programs (golf-specific or non-specific) warrants further investigation.

REFERENCES:

Lephart, S.M., Smoliga, J.M., Myers, J.B., Sell, T.C. and Tsai, Y.S. (2007). An Eight-Week Golf-Specific Exercise Program Improves Physical Characteristics, Swing Mechanics, and Golf Performance in Recreational Golfers. *Journal of Strength and Conditioning Research*, 21, 860-869.

Fletcher, I. and Hartwell, M. (2004). Effects of an 8-Week Combined Weights and Plyometrics Training Program on Golf Drive Performance. *Journal of Strength and Conditioning Research*, 18, 59-62.

TRUNK KINEMATICS DURING THE TEE-SHOT OF MALE AND FEMALE GOLFERS

Michael H. Cole[1] and Paul N. Grimshaw[2]
University of South Australia, Australia[1]
University of Adelaide, Australia[2]

KEY WORDS: Golf; Shoulder Movement; Pelvic Movement; Hip to Shoulder Differential.

INTRODUCTION: While females comprise 20% of the golfing population in some Western countries (e.g. Australian Bureau of Statistics, 2007), previous research has typically assessed populations that are exclusively comprised of male golfers (e.g. Cheetham et al., 2008). However, the overall prevalence of golf-related injuries is reported to be similar for males and females (McHardy et al., 2006) and thus, it is of interest to assess whether the kinematics of the female golf swing are similar to those demonstrated by male players. This is important, as this knowledge will ensure that any changes that are made by coaches to improve performance and/or reduce the risk of injury in these golfers are appropriate.

METHOD: Hip, shoulder and trunk kinematics of a group of male (n = 5) and female (n = 5) golfers were compared during the performance of the tee-shot. Whilst performing 20 tee-shots, participants were filmed by three genlocked video cameras (50 Hz) and 3D kinematics were derived using Peak Motus 2000. The 3D kinematics were used to calculate; 1) *Hip angle:* the angle formed between the inter-trochanter vector and a theoretical line parallel to the y-axis between the tee and the target (Transverse plane); 2) *Shoulder angle:* the angle formed between the inter-acromion vector and the same theoretical line used to calculate hip angle (Transverse plane); 3) *X-Factor:* the differential angle between the inter-trochanter and the inter-acromion vectors (Transverse plane). To examine for statistically significant differences between the groups, the Mann-Whitney *U* test was used.

RESULTS: Female golfers showed an increased hip rotation range of motion (ROM) during the backswing (BS) and increased hip and shoulder rotation ROM during the downswing (DS). Given the concomitant differences in hip and shoulder rotation, it is not surprising that the groups did not differ with respect to X-Factor values throughout the swing (Table 1).

Table 1: Hip and shoulder ROM during the BS and DS and the peak X-Factor. * $p<0.05$; ᵛ $p<0.1$

		Male (n = 5)		Female (n = 5)	
		Mean	SD	Mean	SD
Hip Angle ROM (deg)	BS	37.8*	10.2	49.9*	4.4
	DS	62.4*	10.2	82.3*	11.3
Shoulder Angle ROM (deg)	BS	97.7ᵛ	6.1	109.7ᵛ	11.4
	DS	97.6*	8.0	124.5*	22.3
Peak X-Factor (deg)		56.6	9.1	55.9	12.6

DISCUSSION: The findings of this research are important, as coaches have often believed the female golf swing to be a slower version of the male golf swing. This research demonstrates that the golf swings of male and female golfers differ significantly with respect to the patterns of hip and shoulder rotation, which may suggest that female golfers could benefit from different coaching strategies to aid improved performance and reduce injury risk.

REFERENCES:

Australian Bureau of Statistics (2007). *Participation in sport & physical recreation* (4177.0). Canberra.
Cheetham, P. (2008). *Science and Golf V,* pp. 30-36. Mesa, AZ: Energy in Motion.
McHardy, A., Pollard, H., & Luo, K. (2006). *Sports Medicine, 36*(2), 171-187.

ANALYSIS OF BILATERAL ASYMMETRIES BY FLIGHT TIME OF ONE LEG COUNTERMOVEMENT JUMP

Hans-Joachim Menzel, Silvia Ribeiro Araújo, Mauro Heleno Chagas
Federal University of Minas Gerais, Belo Horizonte, Brazil

KEY WORDS: lower limbs' asymmetries, vertical jump, flight time.

INTRODUCTION: Lateral differences concerning the magnitude of strength are frequently found in the lower limbs (LL) and may be related to preference (dominant and non-dominant) and skill performance. For jump tests the lateral differences greater than 15% are considered clinically significant (Noyes et al., 1991). Since Countermovement Jumps (CMJ) on a force platform are an adequate method for the identification of lateral asymmetries (Menzel et al., 2006), the objective was to verify if lateral asymmetries of CMJ performance can also be identified by the flight time of single leg CMJ on a contact mat.

METHODS: 29 male physical education students without history of knee injury (mean age 20,56 ± 2,25 years) performed three maximal CMJ on a double force platform where the force-time histories of the vertical ground reaction force were measured separately for each leg. Three one leg CMJ with each leg were also performed on a contact mat where the jump height was determined by flight time. During the CMJ the hands remained fixed on the hips. The jumps were randomly ordered and a rest interval of 2 min was maintained between the jumps. The highest jump was used for further analysis. In order to verify the correspondence of lateral differences identified by the vertical impulse determined by two-leg CMJ on a force platform and one leg CMJ on a contact mat, the contingency coefficient was calculated.

RESULTS: The numbers of identified bilateral asymmetries by the different methods are shown in table 1. A significant contingency coefficient (0,617; p = 0.01) was found, which indicates a rather good concordance of diagnostic information between the methods.

Table 1 Crosstabulation of prevalent laterality determined by CMJ on force platform and on contact mat.

Contact mat (Flight time)	Force platform (Impulse)	
	Symmetry	Asymmetry
Symmetry	23	4
Asymmetry	0	2

DISCUSSION: Using a double force platform seems to be the more sensitive method for the identification of bilateral differences, since this method identified 6 cases with relevant differences, whereas one leg CMJ on a contact mat only identified 2 cases. Considering the double force platform a valid criterion, the contact mat method did not identify false asymmetries.

CONCLUSION: Though one leg CMJs on a contact mat, which is a less expensive and simpler method than the use of a force platform, do not identify false asymmetries, this method seems to be only a restricted alternative for the identification of bilateral asymmetries of the lower limbs since asymmetries are only partially identified.

REFERENCES:
Menzel, H.J.; Chagas, M.H. & Cruz, G.H. (2006). Identification of bilateral asymmetries in lower limbs of soccer players by vertical jumps on a double force platform. Salzburg, Austria: *XXIV ISBS*, 727-730.
Noyes, F.R., Barber, S.D. & Mangine, R.E. (1991). Abnormal lower limb symmetry determined by function hop tests after anterior cruciate ligament rupture. *American Journal of Sports Medicine*, 19, 513-518.

ACUTE STATIC STRETCH EFFECTS ON MULTIPLE BOUTS OF VERTICAL JUMP

Chelsea C. Walter & Michael Bird

Truman State University, Health & Exercise Sciences, Kirksville, Missouri, USA

The purpose of this study was to evaluate the acute effects of static stretching on vertical jumping and the ground reaction force kinetics when stretching was implemented between multiple performance bouts. Fifty-two young adults were randomly assigned to a control or stretch condition, each group performing four sets of three jumps. After the initial jump series, subjects in the stretch condition performed a set of four unilateral lower extremity stretches, holding each stretch for 30 s on each leg. Vertical jump height and time had a significant interaction effect ($p<0.05$), the stretch group had a significant decrease in jump height only after the first jump. Vertical jump kinetics was not significantly affected from one series to another or between the two conditions. Stretching between exercise bouts had a small, significant effect on the product of the jump, but no effect on the ground reaction force kinetics. No cumulative effects were found across jumps after subsequent stretching.

KEYWORDS: jump height, ground reaction force, rate of force development, impulse force

INTRODUCTION: Deciphering the best way to incorporate a stretch protocol with respect to performance has become increasingly important for athletes. Stretching has long been part of exercise routines, often performed before the exercise bout. Many coaches, athletes, and researchers have advocated the use of stretching for potential benefits in performance and decreased injury risk. While stretching has been hypothetically thought to enhance performance, there is little scientific evidence to back this claim. Moreover, recently researchers have found pre-performance stretching to have a detrimental influence on movement outcomes in various measures including power production (Manoel et al., 2008; Samuel et al., 2008), vertical jump performance (Knudson et al., 2001; Unick et al., 2005), force production (Brandenburg, 2006; Knudson & Noffal, 2004), rate of force development, and ground reaction forces (Young & Elliot, 2001). As stated by Nelson et al. (2008), "this paradox between accepted dogma and current research raises the dilemma over prioritizing performance and safety for athletes" (p. 338).

Some researchers have found small, yet significant, negative effects of pre-performance stretching on movement (Bradley et al., 2007; Brandenburg, 2006; Knudson & Noffal, 2005; Young & Elliot, 2001; Nelson et al., 2005). Others have found pre-performance stretching to have no impact on performance (Knudson et al., 2001; Manoel et al., 2008; Samuel et al., 2008; Unick et al., 2005). The dilemma of pre-performance stretching will continue to be relevant as researchers continue to strive toward best preparing the athlete for success. For example, little research has been conducted to assess the effect of stretch when it is executed within a series of performance bouts (Brandenburg, 2006; Samuel et al., 2008; Unick et al., 2005).

The timing of stretching with respect to performance could play an important role in competitions where athletes perform multiple bouts with time lapse between events such as sprinting, field events in track, and Olympic lifting. Knowledge of how stretching may impact performance, and whether the effects are cumulative may be valuable to both coaches and athletes as they plan the time between competition bouts. The purpose of this study was to evaluate the acute effects of static stretching on vertical jumping and the ground reaction force kinetics when stretching was implemented between multiple performance bouts.

METHODS: Data Collection: Fifty-two recreational and varsity athletes from the university population provided informed consent to participate in this study. Subjects consisted of 19 males and 33 females with ages ranging from 19-23 years. The height and mass of the subjects were 1.71±0.09 m and 74.6±16.0 kg, respectively. Subjects were randomly assigned to either a stretch (n=26) or no-stretch (control, n=26) condition and placed in testing groups of five to six individuals. All participants followed a standardized warm-up protocol consisting of sub-maximal cycling for five minutes, using Monark cycle ergometers at 120 W. This was immediately followed by four series of three vertical jumps with a six minute time lapse between each series. Jumps were performed with both hands on hips to isolate the lower extremities and to eliminate potential confounding arm coordination; subjects were instructed to jump as high as possible for all jumps. A Kistler force plate (model 9286AA) was used to measure the ground reaction force data (200 Hz sampling rate).

After the initial jump series, the stretch condition subjects completed four static stretches that targeted the gluteal, hamstring, quadriceps, and calf muscle groups. The same protocol was used after subsequent jump series (based on Knudson et al., 2001; Samuel et al., 2008; and Unick et al., 2005) and the no stretch condition subjects were allowed to rest. Each stretch was performed once unilaterally and held for a timed thirty seconds, as described in the National Strength and Conditioning Association guidelines. Adequate stretch was defined as committing to full range of motion at each desired joint until a slight discomfort, but not pain, was achieved. Stretching techniques were demonstrated to the subjects prior to data collection to ensure subject understanding of proper technique, and also monitored during experimental stretching routine. Each unilateral stretch was completed on both sides of the body.

Stretch protocol

Unilateral Gluteus Stretch: Subjects sat with knees flexed and their feet flat against the floor. After crossing one leg over the thigh of the other leg, they grasped the back of that same thigh with both hands. Subjects pulled their legs towards their torso to stretch.

Unilateral seated hamstring stretch: Subjects sat with an anterior tilt of the pelvis. The leg being stretched remained outstretched while the uninvolved leg was flexed in a figure-four position. Subjects then were instructed to lean forward, flexing at the hip, and reach with their hands towards their toes.

Unilateral standing quadriceps stretch: Subjects stood on one leg with a posterior pelvic tilt and one hand against a wall for balance. Subjects grasped non-weight bearing foot, bringing the knee into flexion as far as possible while keeping the knee perpendicular to the floor.

Unilateral standing calf stretch: Subjects stood with both hands placed against a wall in front of them (arms outstretched, elbows straight). While keeping left knee slightly bent, the toes of right foot slightly turned inward, subjects were to move right foot back one or two feet and place right heel and foot flat on the floor.

Data Analysis:

Variables and Statistics: Vertical jump height was calculated from the time off the force plate. Peak vertical force was the maximum vertical value prior to take off. Rate of force development was the difference between the first peak vertical force and body weight divided by the time between those two events. Total impulse force was calculated as the total area under the force curve during the eccentric and concentric phases. Each kinetic variable was normalized to body weight. Based on intraclass correlations of jump heights within each series ranging between 0.97 and 0.99, the first jump of each series was used for analysis. All variables were statistically evaluated using a 2X4 (group by jump trial) repeated measures ANOVA. Post-hoc tests for significant main effects were completed with a Bonferroni correction. The alpha level for significance was 0.05.

RESULTS: Mauchly's test of sphericity was significant ($p<0.05$), so Greenhouse-Geisser adjustments were used. There was no significant relationship between gender and time or three-way interaction (trials x gender x group), allowing for consequent analyses to collapsed data across gender. Vertical jump height had a significant main effect for trials ($p < 0.05$) and a significant interaction effect ($p < 0.05$) for trials and group (see Figure 1). The first jump was significantly greater than the second and third jumps, but not greater than the fourth jump. Peak vertical force had a main effect across jump trials, but no interaction effect between trial and group. Neither rate of force development nor impulse force produced any significant results.

Table 2. Kinetic variable results (Mean ±SD) are all relative to body weight. No significant differences or interactions were found. NS and S indicate no stretch and stretch conditions.

Group	Variable					
	Peak Vertical Force (BW)		Rate of Force Dev. (BW/s)		Total Impulse Force (BW•s)	
	NS	S	NS	S	NS	S
Jump 1	2.40 ±0.34	2.37 ±0.29	6.10 ±3.56	5.34 ±2.94	0.28 ±0.07	0.35 ±0.10
Jump 2	2.29 ±0.27	2.37 ±0.36	5.74 ±2.70	5.75 ±3.10	0.29 ±0.08	0.35 ±0.10
Jump 3	2.23 ±0.21	2.32 ±0.34	5.39 ±2.11	5.28 ±2.61	0.29 ±0.08	0.35 ±0.10
Jump 4	2.24 ±0.22	2.29 ±0.34	5.81 ±2.62	5.50 ±2.87	0.29 ±0.08	0.35 ±0.09

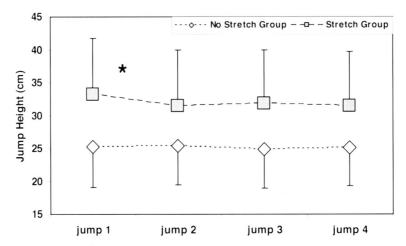

Figure 1. Jump height with SD (cm) across all four trials for each group. The * indicates the significant interaction effect and is evident as the stretch group values fall between jumps one and two. There was also a main effect between jumps as the total mean for the first jump was significantly greater than jumps two and three.

DISCUSSION: The purpose of this study was to determine how acute static stretching would affect vertical jump performance and ground reaction force kinetics when a stretch protocol was implemented between multiple jump bouts. The stretch group jump height significantly decreased ($p<0.05$) between the first and second jump, while the control group did not change. This small, 5.5% decrease was consistent with previous literature, where decreases in performance after stretching ranged from 3% to 24% (Bradley et al., 2007; Brandenburg, 2006; Rubini et al., 2007; Knudson & Noffal, 2005; Young & Elliot, 2001; Nelson et al., 2005). As the vertical jump is often used to reflect somatic power, these results differed from Manoel et al. (2008), who found no muscle power decrement after stretching. The significant interaction between jump and group (see Figure 1) had an effect size of 0.063, which reflects a very small effect. Indeed, the significant impact of stretch between the first and second jumps was not experienced by every subject, as 19 (73%) decreased in jump height, 4 (15%) changed by less than 2%, and 3 (12%) increased in jump height. In the control group, 14 (54%) decreased in jump height, 3 (12%) had little change, and 9 (35%) increased between the first and second jumps. Subsequent jump heights were not different and no cumulative effect of stretch was evident, this differed from Knudson & Noffal (2005), who found somewhat cumulative affects of stretch on strength.

No significant differences were found in the ground reaction force kinetics of the jumps. Normalized values for peak vertical force, rate of force development, and total impulse force were not significantly different across jump bouts or groups. Other studies have found strength or force differences after stretch (Brandenburg, 2008; Nelson et al., 2005; Knudson & Noffal, 2005). However, other studies have not (Young & Elliott, 2001; Samuel et al., 2008). Interestingly, Knudson et al. (2001) found no kinematic differences as a result of stretch. Within the limitations of this heterogeneous subject population and use of primarily ground reaction force data, the impact of pre-performance stretching remains tenuous and elusive, clearly the overall influence on the kinetics and kinematics requires more investigation.

CONCLUSION: Previous recommendations to remove stretch from pre-exercise routines (Brandenburg, 2008) may be considered for between performance stretching in multiple bout events. If stretching is desired, it may be best to complete it immediately after the performance to minimize any potential impacts on performance, especially if the time between bouts is more than a few minutes (Bradley et al., 2007). Stretching as part of the post-performance routine may offer the best benefits for flexibility and performance improvement for some athletes. No cumulative effect of stretch was found for vertical jump heights or ground reaction force kinetics.

REFERENCES:

Bradley, P., Olsen, P., & Portas, M. (2007). The effect of static, ballistic, and proprioceptive neuromuscular facilitation stretching on vertical jump performance. *Journal of Strength and Conditioning Research, 26*, 223-226.

Brandenburg, J. (2008). Duration of stretch does not influence the degree of force loss following static stretching. *Journal of Sports Medicine and Physical Fitness, 46*, 526-534.

Knudson, D., Bennett, K., Corn, R., Leick, D., & Smith, C. (2001). Acute effects of stretching are not evident in the kinematics of the vertical jump. *Journal of Stretch and Conditioning Research, 15*, 98-101.

Knudson, D., & Noffal, G. (2005). Time course of stretch-induced isometric strength deficits. *European Journal of Applied Physiology, 94*, 348-351.

Manoel, M., Harris-Love, M., Danoff, J., & Miller, T. (2008). Acute effects of static, dynamic, and proprioceptive neuromuscular facilitation stretching on muscle power in women. *Journal of Strength and Conditioning Research, 22*, 1528-1534.

Nelson, A., Kokkonen, J., & Arnall, D. (2005). Acute muscle stretching inhibits muscle strength endurance performance. *Journal of Strength and Conditioning Research, 19*, 338-343.

Rubini, E., Costa, A., & Gomes, P. (2007). The effects of stretching on strength performance. *Journal of Sports Medicine, 37*, 213-224.

Samuel, M., Holcomb, W., Guadagnoli, M., Rubley, M., & Wallmann, H. (2008). Acute effects of static and ballistic stretching on measures of strength and power. *Journal of Strength and Conditioning Research, 22*, 1422-1428.

Unick, J., Kieffer, H., Cheesman, W., & Feeney, A. (2005). The acute effects of static and ballistic stretching on vertical jump performance in trained women. *Journal of Strength and Conditioning Research, 19*, 206-212.

Young, W., & Elliott, S. (2001). Acute effects of static stretching, proprioceptive neuromuscular facilitation stretching and maximum voluntary contractions on explosive force production and jumping performance. *Research Quarterly for Exercise and Sport, 72*, 273-279.

Acknowledgements

The authors would like to recognize the significant efforts of Rebecca Pollock, Anna Mattlage, Ally Eberle, and Benjamin Riebl toward the data collection and processing of this study.

DYNAMIC STABILIZATION IN COLLEGIATE FEMALE VOLLEYBALL PLAYERS: EFFECTS OF LEG DOMINANCE AND OFF-SEASON

Kimitake Sato[1], Gary D. Heise[1], and Kathy Liu[2]

[1]University of Northern Colorado, Greeley, CO. USA
[2]University of Delaware, Newark, DE. USA

KEY WORDS: time to stabilization, hopping task, volleyball players.

INTRODUCTION: Adequate dynamic stabilization can be a key factor in preventing non-contact lower extremity injuries, especially in sports which require agile movements, such as volleyball. Individuals with functional ankle instability (FAI) took longer to stabilize in static and dynamic tasks when examining anterior/posterior (AP) and medial/lateral (ML) responses (Ross & Guskiewicz, 2004). In an effort to detect FAI in athletes, differences in time to stabilization (TTS) between post- and pre-seasons and between dominant and non-dominant legs should be identified. The purpose of the study was to identify those differences (post- vs. pre-season; leg dominance) across various hopping directions. It was hypothesized that the pre-season test and dominant leg conditions exhibit greater stability (i.e., shorter TTS) in all hopping directions.

METHODS: After a sufficient warm-up, collegiate female volleyball players ($N=8$) hopped onto a force plate (AMTI, Watertown, MA, USA), landing one-legged, from four different directions. The force data were collected for 10 sec, at 200 Hz. TTS was calculated in accordance with the procedures of Colby et al. (1999). Two (AP & ML) 2x2x4 mixed-design, repeated measures ANOVA examined the effects of season, leg dominance, and four different hopping directions (medial, lateral, 50% and 100% of leg-length forward hops) on TTS ($\alpha=.05$).

RESULTS: No main effects were found for season and leg dominance, but there was a main effect for hopping direction in both AP and ML forces (AP: ($F(3,21)=274.99$, $p<.001$), ML: ($F(3,21)=122.79$, $p<.001$)). No interaction effects were identified in Season*Leg Dominance, Leg Dominance*Hopping direction, Season*Hopping tasks, and Season*Leg Dominance*Hopping direction.

DISCUSSION & CONCLUSION: The results did not support the hypotheses that differences would be identified for all effects of season, leg dominance, and hopping direction. Between post-season and pre-season, the participants did not receive a specific treatment to improve the stability of their lower extremity, which may explain the lack of season main effect. During the season, two lower extremity injuries were reported (ACL tear & ankle sprain) among those tested. The hopping direction main effect indicates that TTS varies depending on the task; this is consistent with results from Wilkstrom et al. (2008). Building on this preliminary work, future studies should consider larger samples, different athletic populations, and specific treatment protocols to determine influences on dynamic stability.

REFERENCES:
Colby, S. M., Hintermeister, R. A., Torry, M. R., & Steadman, J. R. (1999). Lower limb stability with ACL impairment. *Journal of Orthopedic Sports Physical Therapy, 29*, 444 – 451.
Ross, S. E., & Guskiewicz, K. M. (2004). Examination of static and dynamic postural stability in individuals with functionally stable and unstable ankles. *Clinical Journal of Sport Medicine, 14*(6), 332 – 338.
Wikstrom, E. A., Tillman, M. D., Schenker, S. M., & Borsa, P. A. (2008). Jump-landing direction influences dynamic postural stability scores. *Journal of Science and Medicine in Sport, 11*, 106 – 111.

KINEMATIC ANALYSIS IN TEAM-HANDBALL JUMP THROW

Herbert Wagner[1,2], Michael Buchecker[1,2], Erich Müller[1,2]

Department of Sport Science and Kinesiology, University of Salzburg, Austria[1]

CD-Laboratory "Biomechanics in Skiing", University of Salzburg, Austria[2]

KEY WORDS: team-handball, jump throw, 3D-analysis

INTRODUCTION: The purposes of our study were to determine the proximal-to-distal sequence of the linear joint and angular velocities and to measure the influence of maximal angular velocities and performance level to ball release speed of the jump throw, which is the most applied throwing technique in team-handball (Wagner et al., 2008).

METHODS: 3-D kinematic data were analyzed via the Vicon MX 13 (8 cameras, 250 fps) from 26 male team-handball players of different performance levels (body height: 181.2 ± 7.6cm; body weight: 76.9 ± 11.3 kg; age: 21.2 ± 5.0 years; training experience: 6.8 ± 5.2 years). The performance level of the participants based on their experience and performance in competition and was rated from one (elite) to level six (novice) (level 1 [n=5], 2 [n=4], 3 [n=3], 4 [n=4], 5 [n=4], 6 [n=6]). The participants were instructed to throw the ball onto a target at 8m distance, and to hit a square of 0.5×0.5m at about eye level (1.75m), with maximum ball release speed and throwing precision. To calculate joint center positions, linear joint velocities, joint angular velocities and ball release speed, we used a three-dimensional model (Plug-In Gait Model, Vicon Peak, Oxford, UK). The measurement accuracy of our model is described in Tilp et al. (2008). For statistical analysis a one-way ANOVA were used to calculate the differences in the proximal-to-distal sequence and Pearson Product Moment correlation coefficient tests (two tailed tests) were calculated to determine association between ball release speed and maximal angular velocity, angular velocity at ball release and performance level.

RESULTS: Two different proximal-to-distal sequencings of the joints were found, separated into a classical and team-handball specific technique. We found correlation between ball release speed (20.0 ± 2.8m/s) and maximal trunk forward tilt angular velocity (392 ± 128°/s, $r=0.62$, $P<0.01$), the trunk forward tilt angular velocity (188 ± 109°/s, $r=0.53$, $P<0.01$), the trunk side tilt angular velocity (115 ± 139°/s, $r=-0.52$, $P<0.01$), the shoulder internal rotation angular velocity at ball release (3919 ± 1258°/s, $r=0.53$, $P<0.01$), the maximal elbow extension angular velocity (1537 ± 283°/s, $r=0.51$, $P<0.01$) and performance level (3.6 ± 1.9, $r=-0.80$, $P<0.001$).

DISCUSSION: Results of our study suggest that experienced team-handball players execute throws with better efficiency and movement coordination, trunk positioning, and arm movement velocity compared to less experienced players. Most team-handball players use both classical (complete proximal-to-distal sequence) and team-handball specific technique (incomplete proximal-to-distal sequence) depending on the game situation. The incomplete proximal-to-distal sequence described in our study were also found in team-handball standing throw by Fradet et al. (2004) and baseball throw by Hong et al. (2001).

CONCLUSION: We conclude that team-handball players need optimal movement coordination and the ability to produce greater force to perform jump throws with high ball release speed. The results of this study and those of recent studies in team-handball (Gorostiaga et al., 2005) suggest that specific strength and coordination training may increase ball release speed in the team-handball jump throw; however, additional training studies in team-handball players are warranted.

REFERENCES:
Paper truncated by editorial board; for a full list of references please contact the author.

INFLUENCE OF BODY WEIGHT ON JOINT LOADING IN STAIR CLIMBING

Gerda Strutzenberger, Anne Richter, Diana Lang and Hermann Schwameder

BioMotion Center, Department of Sport and Sport Science, Karlsruhe Institute of Technology (KIT), Germany

FoSS – Research Center for Physical Education and Sports of Children and Adolescents, Karlsruhe, Germany

KEY WORDS: joint loading, stair climbing, obesity

INTRODUCTION: Exercise is an essential treatment in childhood obesity. Due to the low impact on joint loading exercise recommendations are aerobic exercise such as swimming, cycling and walking (Hassink et al, 2008). Little is known though about the effect of adiposity on the function of the locomotor system (Wearing et al., 2006). Only limited research has been done on obese gait in children (Nantel et al. 2006) and even less is known about other weight bearing tasks such as climbing stairs. Therefore, the aim of this study was to examine the influences of obesity on the load pattern of the lower extremity joints of obese children while ascending and descending stairs.

METHODS: 17 normal weight children (10.4 ± 1.3 yrs, 143 ± 9 cm, 36.7 ± 7.5 kg) and 18 obese children (10.5 ± 1.5 yrs, 148 ± 10 cm, 56.6 ± 8.39 kg) participated in this study. A staircase with 6 steps (17cm x 28 cm per step) was built. Two force plates (AMTI, 1000 Hz) were embedded in the 3rd and 4th step. The kinematic data was collected using 10 infrared cameras (Vicon, 200 Hz). The children performed 3 valid trials walking up- and downstairs with a given speed of 110 steps/min. Dynamic data was normalized to body weight and time-normalized to stance phase. Inverse dynamics were calculated and mean peak values of ankle, knee and hip joint moments were identified. Independent t-tests were used to check for differences between the two groups.

RESULTS: The analysis of this study is still in progress. First results of 9 subjects (5 normal weight, 4 obese) can be reported (Table 1). Due to the low number of subjects no statistical analysis was performed. The transverse plane shows slightly higher peak moments in all joints. Additionally changes of the load pattern in the hip and knee while descending appeared in that plane.

Table 1 Mean peak moments of the hip, knee and ankle in sagittal and transverse plane.

	Hip M_{flex}	Hip M_{add}	Knee M_{flex}	Knee M_{varus}	Ankle $M_{dorsalext}$	Ankle M_{pron}
upstairs: max obese (Nm/BW)	0.81±0.24	0.61±0.15	0.85±0.24	0.50±0.15	1.61±0.32	-0.29±0.12
upstairs: max normal weight (Nm/BW)	0.93±0.21	0.49±0.04	0.97±0.21	0.42±0.04	1.38±0.15	-0.21±0.08
downstairs: max obese (Nm/BW)	0.32±0.24	1.01±0.20	0.96±0.22	0.69±0.18	1.40±0.10	-0.26±0.08
downstairs: max normal weight (Nm/BW)	0.60±0.39	0.98±0.12	0.93±0.30	0.61±0.10	1.67±0.26	-0.21±0.07

DISCUSSION: The differences of joint loading parameters between the two groups are small, but should not be neglected considering the higher body weight of the obese group. Therefore, weight bearing tasks challenge the obese musculoskeletal system, and could overload it when done too excessively. Exercise and sport performed by obese children should hence focus on training in load reduced conditions.

REFERENCES:
Hassink, S.G. et al. (2008). Exercise and the obese child. *Progress in Pediatric Cardiology*, 25, 153-157
Nantel, J.,et al. (2006). Locomotor Strategies in Obese and Non-obese Children. *Obesity*, 14, 1789-1794.
Wearing, S.C., et al. (2006). The biomechanics of restricted movement in adult obesity. *Obesity Reviews*, 7, 13-24.

ANALYSIS OF HUMAN MOTION WITH METHODS FROM MACHINE LEARNING

W. Seiberl[1], M. Karg[2], K. Kühnlenz[2], M. Buss[2] and A. Schwirtz[1]

Department of Biomechanics in Sports[1], Institute of Automatic Control Engineering[2], Technische Universität München, Germany

KEY WORDS: Motion analysis, biomechanics, machine learning.

INTRODUCTION: Usually, predefined kinematic parameters are investigated in biomechanical studies of human motion. In recent years, techniques of machine learning have been added to this field of research (Chau, 2001). In this study different dimension reduction methods like Principal Component Analysis (PCA) and Fourier Transformation (FT) are investigated as an alternative to common biomechanical approaches in motion analysis.

METHODS: Human gait in different variations of physical (full-body exhaustion) or psychological states (emotional states: happy, sad, angry, neutral) was tracked using a 6-camera VICON-system (240Hz). We chose a feature selection method which does not require information about an individual's normal gait. Instead of investigating a set of predefined features, we extracted structural and dynamical cues by PCA and FT. The procedure is leant on eigenpostures (EP) and eigenwalkers as proposed by Troje (2002). In addition nonlinear extensions, like Kernel LDA and Kernel PCA, are studied.

RESULTS: Results in recognition and classification of exhaustion (69%) and emotion (58-83% for individuals, Ø 65%) in human gait could be achieved significantly above chance. Extra success comparing to a random predictor is 36% and 52%, respectively. Applying the procedure to a statistically preselected pool of kinematic parameters leads to emotion recognition between 70-100% for individual subjects.

DISCUSSION: Applying barely feature extraction without involving expert knowledge leads to recognition rates clearly above chance level for emotional states or exhaustion in human walking. Hence, the explored algorithms give in means information which is at least 19% above guessing. Additionally, recognition rates can be improved if expert knowledge is integrated, e.g. by a previous statistical analysis which determines significant joint angles. This allows reducing the number of trained parameters in the machine learning algorithms and improves classification. Central task in future work is to enhance differentiation between hardly distinguishable motions by methods from machine learning which provide estimated probabilities for different motions involving also unknown causalities of unexpected parameters.

CONCLUSION: Techniques of machine learning showed abilities for recognition and classification of human gait just above chance. Results can be improved if expert knowledge is integrated. In a next step force-time curves of ski-jumping, measured with force plates applied to the jump-of platform (Oberstdorf, Germany), will be implemented in the explored algorithms to see whether this can enhance knowledge of technique and classify different types of performances.

REFERENCES:

Chau, Tom (2001). A review of analytical techniques for gait data. Part 1: fuzzy, statistical and fractal methods. *Gait & Posture, 13*(1), 49-66.

Troje, N. F. (2002). Decomposing biological motion: a framework for analysis and synthesis of human gait patterns. *Journal of Vision, 2*(5), 371-387.

Acknowledgement

This work is supported within the DFG excellence initiative research cluster "cognition for technical systems - CoTeSys", see also www.cotesys.org

POSTURAL CONTROL IN ELITE ARCHERS DURING SHOOTING

Wei-Hsiu Lin [1], Guo-Tang Huang [2], Ping-Kun Chiu [3], Alex, J. Y. Lee [2]

Department of P.E., National ChiaYi University, Taiwan [1]
Department of P.E., National HsinChu University of Education, Taiwan [2]
Department of Sports Training Science, National Taiwan Sport University [3]

KEY WORDS: center of pressure, static balance, dynamic balance.

INTRODUCTION: Archery is described as a static sport requiring fine movement control and proper endurance strength of the upper body (Soylu, Ertan, & Korkusuz, 2006). To investigate the differences of postural control (PC) between elite and general collegiate archers during static and shooting conditions.

METHODS: Nineteen archers were recruited as elite archers (EA, nine archers, FITA scores: 1210.1 ± 19.1 points, age: 20.2 ± 1.6 years, height: 170.0 ± 6.1 cm, mass = 81.2 ± 25.6 kg) and general archers (GA, ten archers, FITA scores: 1122.5 ± 47.3 points, age: 20.1 ± 1.0 years, height: 173.3 ± 9.6 cm, mass: 70.1 ± 14.9 kg) according to the scores of single round, International Archery Federation (FITA). PC was measured with a portable three-axis force plate which sample rate was set at 100 Hz as mean radius, velocity, and the sway area of the center of foot pressure (COP) during different testing conditions (Lee & Lin, 2008). The static testing posture included single/double limb(s) standing with open/closed eyes. The shooting testing posture included two times 6 arrows archery shooting for 50 meters. An independent t-test analysis was used to examine the differences between two groups in each PC parameters during the static and the shooting condition. The statistic significance was set at $p < .05$.

RESULTS: No significant differences were found in static testing condition between groups. However, EA showed significant smaller COP sway velocity and area in shooting condition than GA (3.53 ± 0.75 vs. 4.42 ± 0.93 mm/s, $t = -2.31$; 2.98 ± 1.43 vs. 6.50 ± 4.16 mm^2, $t = -2.41$). Furthermore, the COP sway area was also significant smaller during the better, high-scores shooting arrows than poor, lower shooting scores arrows (4.49 ± 3.70 vs. 5.30 ± 4.29 mm/s, $t = -2.48$).

DISCUSSION: Significant better PC during archery shooting in EA implied that physical training enhances bipedal equilibrium control and reduces body oscillations during static and dynamic equilibrium tests. In addition, elite gymnasts highlight good stability when completing a unipedal dynamic equilibrium task due to their physical training which develops equilibrium specifically.

CONCLUSION: Elite collegiate archers showed better PC during shooting conditions and high-scores shooting arrows than GA which demonstrated the close relationship between high archery performance and good postural control ability.

REFERENCES:

Lee AJY. and Lin W.H. (2008). Twelve-week biomechanical ankle platform system training on postural stability and ankle proprioception in subjects with unilateral functional ankle instability. *Clinical Biomechanics*, 23, 8, 1065-1072.
Soylu AR, Ertan H, and Korkusuz F. (2006). Archery performance level and repeatability of event-related EMG. *Human Movement Science*, 25(6), 767-74.

Acknowledgement

This research was supported by grants from National Science Committee (NSC 96-2413-H-134-008 & 97-2410-H-134-022) and National HsinChu University of Education, TAIWAN.

EVALUATION OF MECHANICAL POWER OUTPUT MEASUREMENT IN A BENCH PRESS EXERCISE UNDER VARIABLE LOAD

Daniel Jandačka[1] and František Vaverka[2]

Human motion diagnostic center, University of Ostrava, Czech Republic[1]
Department of Biomechanics and Engineering Cybernetics, Faculty of Physical Culture, Palacky University, Olomouc, Czech Republic[2]

KEY WORDS: bench press, muscle power output, validity.

INTRODUCTION: The study is aimed to investigate evaluation of mechanical power output measurement during bench press exercise by methods which are used at training practice. As the criterion of power output measurement evaluation we selected a method which estimates the output by means of empirical 3D mechanical model (work in progress).

METHODS: This study was performed on ten untrained middle-aged men. The power output was measured at following loads: 18.0; 26.5; 39.2 and 47.7 kg. The indirect power output method (indirect – used in training practice) employs indirect power output measurement which was performed on FitroDyne Premium (FDP, Slovakia) equipment. The output (**P**) is calculated from changes of vertical speed vector (**v**) of a dumbbell at time (**t**) and weight of a dumbbell (**m**) in the following way:

$$P = m\left(\frac{dv}{dt} - g\right)v(t)$$

where (**g**) is a vector of gravitational acceleration. At combined method of mechanical power output measurement we needed to obtain a dumbbell position in space and time, so we used a kinematic analysis of movement, system Qualisys (Sweden). For direct power measurement a dynamometer AMTI was used. 3D mechanical model will be created on the basis of a dynamic analysis of bench press exercise in Visual 3D software.

RESULTS: An average mechanic power output measured by the indirect method was significantly lower at loads 18.0 and 26.5 kg, than the power output measured by the combined method (size effect = 1.41 and 0.59). With higher loads the output under-estimation of the indirect method subsides.

DISCUSSION: Our findings are in compliance with a research performed by Hori et al. (2007), who show that the indirect method under-estimates measured power output in comparison with a combined method - which was by Cormie et al. (2007) established as a criterion. The limitation to load movement speed, without reference to movement of body elements, can be a source of significant inaccuracies in mechanical power output measurement.

CONCLUSION: Methods used in a training practice for estimation of mechanical muscle power output, which are based on indirect power measurement, under-estimates mechanical muscle power output in a bench press exercise. This under-estimation can influence the load optimization at power training. Future research will be necessary to create 3D mechanical models which will help to estimate more accurately the applied power and speed of gravity centre movement of body elements system and the load applied during individual exercises.

REFERENCES:
Hori, N., Newton, U. R., Andrews, A. W., Kawamori, N., McGuigan, R. M., & Nosaka, K. (2007). Comparison of four different methods to measure power output during the hang power clean and the weighted jump squat. *Journal of Strength and Conditioning Research*, 21, 314-320.
Cormie, P., McBride, M. J., & McCaulley, O. G. (2007). Validation of power measurement techniques in dynamics lower body resistance exercises. *Journal of Applied Biomechanics*, 23, 103-118.

Acknowledgement
The authors would like to thank Professor Joseph Hamill. This research was supported by the Grant Agency of the Czech Republic (No. 406/08/0572).

RESPONSE TIME AND JAB FORCE PUNCH OF THAI FEMALE AMATEUR BOXERS: A PRELIMINARY STUDY

Rat Tongaim, Weerawat Limroongreungrat, Sirirat Hirunrat, Duangjun Phantayuth & Sumethee Thanangkul[1]

College of Sports Science and Technology & Biomedical and Instrumentation for Research and Development Center[1], Salaya, Mahidol University, Thailand

KEY WORDS: response time, jab force, boxing

INTRODUCTION: The jab an important punch in amateur boxing used to interrupt an opponent's rhythm and to score points. To jab successfully, a boxer must respond quickly and hit a target with sufficiently high force. Luangtrakul et al. (2002) investigated response time alone during jab training but not force. Moreover, female boxers have not been studied. Understanding response time and jab force can be used to train boxers. The purpose of this study was to examine the effects of target choice response times and forces of the jab punch of female Thai national amateur boxers.

METHODS: Eight female Thai amateur boxers volunteered in the study. Participants jabbed as quickly as they could at a target with their lead hand when the light came on. Six different target positions were mounted on the same plane which consisted of the head, the chest, the right and left shoulders and lower abdomens. The orders of the targets were randomly assigned. Each target was connected to a uni-axial load cell to measure force and response time. Three trials were run and averaged. A repeated measure ANOVA and post hoc test were performed ($p < .05$).

RESULTS: The response time and force of jab for 6 targets are shown in Fig. 1 and 2.

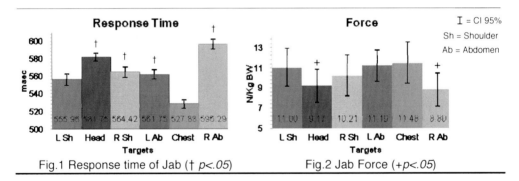

Fig.1 Response time of Jab († $p<.05$)   Fig.2 Jab Force (+$p<.05$)

DISCUSSION: The response time and jab forces were fastest and highest at the chest. This may be because the jab distance to the chest is the shortest and is directly aligned with the direction of jab. Moreover, the force to this target is almost perpendicular to the plane of load cell.

CONCLUSION: The study showed that the fastest response time and the highest force of Thai female boxers occurred with the body target. This may suggest that coaches focus on the targets that have poor response times and low force when training boxers.

REFERENCES:
Luangtrakul, Keawsri & Saksitwiwattana. (2002). The creation of a prototype for measuring and training response time. Sports authority of Thailand.

A PLANE-BASED CALIBRATION PROCEDURE FOR THE 3D ANALYSIS OF VIDEO RECORDINGS IN DISCUS THROWING DURING COMPETITION

Volker Drenk

Institute for Applied Training Science, Leipzig, Germany

Standard procedures for the calibration for 3D video measurements in sports biomechanics are the recording of a spatial calibration frame or the wand calibration. Both methods require access to the competition site, which often is not allowed in international championships. Therefore, alternative calibration routines are needed that utilize the geometric conditions of the competition site. For the calibration of 3D video recordings in discus throwing a new method is introduced. It is based solely on given coordinates of spatial control points in the background and the known interior camera parameters. After introducing the solution steps, the method is validated by comparing it to established standard methods using model data. The applicability of this method is demonstrated by analysing discus competition recordings.

KEY WORDS: videogrammetry, calibration, 3D, discus throwing.

INTRODUCTION:

The atmosphere of international championships often motivates athletes to achieve their top-performances. Obviously, detailed knowledge of these performances is of particular interest. A common method to gather such information is the analysis of video recordings. A typical problem for the calibration of the cameras results from the regulations for international championships: it is usually not allowed to enter the competition area. Therefore, the most common routines for camera calibration based on spatial control points or sets of well distributed homologous points (Direct Linear Transformation (DLT), wand calibration) cannot be applied. Alternative methods utilize geometric constraints of the sports field. Ariel et al. (1997) previously used such geometric information as well as anatomical landmarks of the athletes for calibration. For example, the geometry of the discus throwing facilities (figure 1) can provide markers for the spatial position in a two-dimensional plane whereas the vertical component can be generated by the height of the athletes. Based on these spatial control points the DLT method can be applied. It has been shown by Drenk & Hildebrand (2002) that the reconstruction of 3D-data can also be solved by multiple

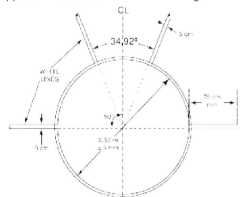

Figure 1: Layout of discus throw circle

views of the same plane from different panning angles.
Numerous studies of alternative calibration methods have been published in the field of computer vision. Here the reconstruction of the orientation parameters of the camera from planar geometric structures is a well studied problem. A solution presented by Zhang (1999) is based on the recording of a well known planar pattern from different perspectives. Sturm & Maybank (1999) allow an arbitrary number of views and calibration planes and the integration of existing interior parameters. Our approach aims at a geometric interpretation of orientation parameters as in Gurdjos et al. (2002). They calculate the centre circle from the homography between the object and image plane and show the connection to the Poncelet

theorem. Their method assumes the interior parameters to be unknown, so that several views are required. Moreover, in their cost function the distances of the projection centres to the centre planes defined by them have to be minimized. Therefore the application of the PTZ (Panning, Tilting, Zooming)-algorithm (Drenk 1994) for iterative solution finding is a new feature.

METHODS:

Figure 2 and 3 present the geometric relations and the recording setup for validation. Figure 3 shows the orientation of the calibration cube from the view of the camera. The model scene was constructed like this: The object plane is at z=0. The optical axis is tilted by 10° relating to the negative y-axis, its projection centre C is (6.08, 8.10, 1.48). The rolling angle is 0°. The image area in pixels is (768, 576), aspect ratio=1. The principal point is the image centre. The calibration frame is a cube with dimensions 1.50 m, translated from the origin by (5.5, 3,0).

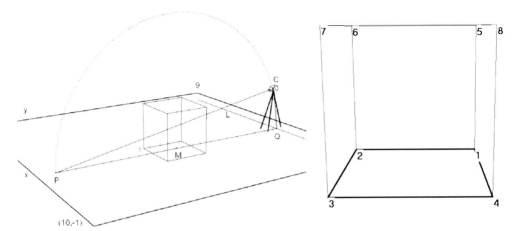

Figure 2: View of the model scene   **Figure 3: Cube seen from the camera**

It is assumed that there is a camera image of a planar scene with four or more control points. Aspect ratio and position of the principal point of the camera must also be available. Normally, these parameters can be determined easily in advance. In addition, the camera configuration is assumed to be stable and well-known. The solution consists of the following steps:

1. Determination of the 2D DLT parameters a1-a8 by evaluating the control points in the object plane z=0, i.e. in the bottom plane of the cube. The following holds:

$$x' = \frac{a_1 x + a_2 y + a_3}{a_7 x + a_8 y + 1} \qquad y' = \frac{a_4 x + a_5 y + a_6}{a_7 x + a_8 y + 1} \qquad (1)$$

2. Calculation of the following parameters in the object plane:
   - The intersection line L between the plane parallel to the image plane through the projection centre and the object plane. It is determined by the line equation in the denominator of (1).
   - The point P in the object plane that is assigned to the principal point is calculated by the evaluation of (1) for the principal point.
   - Perpendicular point Q from P to L

3. The half circle (Thales circle) over $\overline{PQ}$ in the plane perpendicular to the object plane is the geometric locus of the projection centre. This follows from the optical axis being perpendicular to the image plane.

4. Modelling of a camera with these specifications:

- Principal point and aspect ratio as given
- Angle between axes of image coordinate system 90°
- The optical axis is line CP

The remaining parameters are noncritical and can be determined heuristically.

5. The projection centre passes now through a half circle. For each local point the PTZ algorithm can be applied. This algorithm was expanded by a rotation about the optical axis (rolling). The PTZ algorithm answers the following question: how the camera has to be rotated in the centre of rotation C and how the focal length has to be adapted to locate the projection of the control points in the image plane as close as possible to the measured position (cost function: sum of squared Euclidean distances)?
6. The search area will be reduced to a plausible section of the circle arc. The step width is chosen in a way that the projection centre can be calculated at designated accuracy (1 cm). The position of the projection centre with minimum distances is required, the camera calibration results from the output parameters of the applied PTZ procedure.

The Thales circle has been used earlier by Drenk & Hildebrand (2002) to calculate the projection centre. Due to the wide-ranging area of control points an alternative method has been used to determine the projection centre on the circle arc. The locus of the panning axis was reconstructed from multiple tripod based views with different panning angles and constant interior orientation. Additionally, the centre of the circle in the object plane covered by the intersection lines L was calculated. Then, the projection centre is located with good approximation on the panning axis.

RESULTS AND DISCUSSION:

The evaluation of steps 1-6 of our model scene resulted in the following:

Table 1 Validation results

	DLT	Plane-based calibration	Model
projection centre x y z	6.09 8.06 1.47	6.08 8.07 1.48	6.08 8.10 1.48
control point	image coordinates of control points		
	complete	lower plane	
1	494 398	494 398	
2	211 398	211 398	
3	143 509	143 509	
4	537 509	537 509	
5	499 110		
6	202 110		
7	126 108		
8	549 108		

Considering that the measurements were made without sub-pixel accuracy, the result is very satisfying. Geometric information about the orientation of lines at the discus throw circle can be derived from the competition regulations (figure 1). By simple intersection relations a sufficient number of control points can be generated. The discus throw circle is embedded in a quadratic platform of 4 m length. Instead of the reconstructable points of the throwing circle the vertices of this platform were used, because they cover a larger area of the object plane (figure 4). Therefore a higher accuracy for the calculation of the 2D DLT parameters can be expected. Video recordings were done by DV cameras with clearly defined principal point (360, 288) and aspect ratio (0.9375). The search area was delimited for the z-interval (0.5 m, 2m).Using the plane-based calibration first the projection centres (left: camera: (2,85.,-4.55,1.07), right camera: (9.74,1.96,1.15) and then the complete calibrations were calculated. The resulting positions of the projection centres are very plausible.

Figure 4: Screenshot of the implemented measurement program

The implemented measurement program allows to measure simultaneously in both views by displaying epipolar lines. If the epipolar line from the second view is collinear with the object point from the first view, the correct calibration of both cameras is confirmed. This was valid for the whole evaluated sequence. In figure 4 right the top of the left foot was selected. The analysis of this competition succeeded.

CONCLUSION:

Spatial calibration frames are part of standard equipment of biomechanic measurement teams. Their application guarantees a high accuracy of the measurements based on video recordings. It has been shown in this paper that 3D-analyses are also possible provided these calibration frames cannot be used. Despite reduced information on planar control points the orientation of the cameras can be calculated accurately. The spectrum of established calibration routines for sports (biomechanic) measurements is completed reasonably by the method presented in this paper. It has been applied additionally in canoeing and swimming. Further application areas are ball sports (e.g. volleyball, tennis), where the playground marks provide accurate planar information.

REFERENCES:

Ariel, G. B., Finch, A., & Penny, M. A. (1997). Biomechanical analysis of discus throwing at the 1996 Atlanta Olympic Games. *Proceedings of the XV International Symposium on Biomechanics in Sports*. Texas Women's University, Denton, Texas, June 20-28, 1997, 365-371.

Drenk,V. (1994). Photogrammetric evaluation procedures for pannable and tiltable cameras of variable focal length. *Proceedings of the XII International Symposium on Biomechanics in Sports*. Budapest-Siofok, July 2-6,1994, 27-30.

Drenk, V. & Hildebrand, F. (2002). Plane-based camera calibration for 3D-videogrammetry for canoeing and rowing. *Proceedings of the XX International Symposium on Biomechanics in Sports*, 349. Caceres: Universidad de Extremadura.

Gurdjos, Pierre, Crouzil, A. & Payrissat, R. (2002). Another way of looking at plane-based calibration: The Centre Circle constraint. Heyden et al. (Eds.)*: ECCV 2002, LNCS 2353*, pp. 252–266. Springer-Verlag Berlin / Heidelberg.

Sturm, P. & Maybank,S. (1999). On plane based camera calibration: A general algorithm, singularities, applications. *Proceedings CVPR*, 432-437.

Zhengyou Zhang (1999). A flexible new technique for camera calibration. *Technical Report MSR-TR98 -71, Microsoft Research*, 1999. Updated version of March 25, 1999.

UPPER AND LOWER EXTREMITY MUSCLE FIRING PATTERNS DURING THE WINDMILL SOFTBALL PITCH

Gretchen D. Oliver, Priscilla Dwelly, David Keeley, and Hiedi Hoffman
University of Arkansas, Fayetteville, AR, USA

The purpose of this study was to describe the activity of both the upper and lower extremity muscles during the windmill softball pitch. Seven female post-pubescent softball pitchers volunteered for the study. Pitchers were analyzed with surface electromyography, and motion analysis software. The muscle firing patterns were described during five phases of the windmill softball pitch.

KEY WORDS: sEMG, female, mechanics of pitching

INTRODUCTION: With millions of girls participating in high school and collegiate softball, there is limited research available. The baseball pitch has been investigated intensively, and continues to be investigated. Barrentine et al. (1998) observed similar torques in softball as that of baseball, and therefore with the risk of injuries in windmill softball pitching becoming as paramount as those in baseball, the mechanics of the motion of the pitch are imperative to understand (Maffet et al., 1997; Werner et al., 2005).

However, it is also known that the body is a kinetic link model, which describes the body as interdependent segments, thus contribution of the entire body during sport activities is essential (McMullen & Uhl, 2000). The proximal segments of the legs and trunk work sequentially in effort to accelerate the shoulder for optimal force production in upper extremity activities (Putnam, 1993). Furthermore, the large muscles of the hips and trunk help position the thoracic spine to accommodate appropriate motions of the scapula which allow for functional shoulder motion. Adequate firing of the gluteal muscle group is vital in proximal to distal sequencing in ballistic/dynamic movements such as the windmill softball pitch.

Previously the research has focused solely on the upper extremity muscle function with the windmill softball pitch. Maffet et al. (1997) examined the activation patterns of eight muscles of the upper extremity and Rojas et al. (2009) examined the biceps during the five phases of the windmill pitch. However, there is no study to date that examines both the upper and lower extremity muscle-firing patterns throughout the phases of the windmill softball pitch. Therefore, the purpose of this study was to examine and describe the muscle firing patterns of three upper extremity muscles (biceps, triceps, and rhomboids [scapular stabilizers]) and two lower extremity muscles (gluteus maximus and medius) during the five phases of the windmill softball pitch.

METHODS: Data Collection: Four collegiate and two high school female post pubescent softball pitchers (age 17.7 y ± 2.6; height 169 cm ± 5.4; mass 69.1 kg ± 5.4) consented to participate. Participants were recruited from the local high school and University. The study was granted Institutional Review Board approval. None of the participants had any previous or current musculoskeletal injury. Surface electromyography (sEMG) data were collected on three muscles of the throwing (dominant) arm, and stride leg (non-dominant leg) as per previously described protocols of Maffet et al. (1997) and Rojas et al. (2009). Surface EMG electrodes were placed on the muscle bellies of the biceps, triceps, rhomboids (scapular stabilizers), gluteus maximus and gluteus medius using Myopac Jr 10 channel amplifier (RUN Technologies Scientific Systems, Laguna Hills, CA).

To assure proper electrode placement, a certified athletic trainer (P.D.) performed manual muscle tests through maximum isometric voluntary contractions (MVIC) based on the work of Kendall et al. (1993). In addition, all sEMG data were performed by the certified athletic trainer (P.D.). Manual muscle tests were performed on each muscle three times for five seconds. The first and last second of each MVIC trails were removed from the data in attempt to obtain steady state results for each of the muscle groups. The manual muscle testing provided a base line reading for which all EMG data were based.

In addition to sEMG data, kinematic data were collected simultaneously in attempt to identify the different phases of the pitch. Kinematic data were collected using The Motion Monitor® system (Innovative Sports Training, Chicago IL) and throwing kinematics were calculated using the International Society of Biomechanics recommendations for reporting joint motion (Wu et al., 2005). Electromagnetic sensors were placed on the thorax, sacrum, dominant distal-forearm, right and left distal-humerus, and right and left mid-shank. Both sEMG and force plate data were collected at a rate of 1000Hz. Force plate, kinematic, and sEMG data were synchronized using The Motion Monitor®.

After unlimited time was allotted for the participants to warm-up based on their normal routine, each participant threw fastball windmill style deliveries using an official softball (30.48 cm. circumference, 170.1 g.) to a catcher behind the plate 12.2 m away. Unlimited warm-up was allowed to account for individual differences in throwing preparation that would be similar to the participant throwing in a game situation. Five trials were recorded after they were deemed a successful strike.

Data Analysis: After completion of the trials, positional kinematic data were filtered independently along the x, y, and z-axis using a 2nd order Butterworth filter (10 Hz) (Werner et al. 2005). The sEMG signals were preamplified (x 1200) near the electrodes and were band pass filtered between 10 and 500 Hz and sampled at a rate of 1000 Hz (Rojas et al., 2009). Surface EMG enveloped data were assessed through mean maximum sEMG reference values that were calculated for each muscle during each of the five phases of the pitch. The five phases were defined according to Maffet et al. (1997) and are illistrated in Figure 1. In the softball pitch there are typically six phases, however, this study focused on all activity prior to ball release and at ball release, excluding the follow-through phase. Five trials of sEMG data for each participant were analyzed to determine average peak amplitudes for all muscles during the first five phases of the pitch. Phase 1 was described as the windup or from the initial movement to the 6 o'clock position. Phase 2 was from the 6 o'clock position to the 3 o'clock position. Phase 3 was from 3 o'clock to 12 o'clock. Phase 4 was from 12 o'clock to 9 o'clock and Phase 5 was from 9 o'clock to ball release.

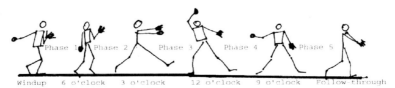

Figure 1: Windmill pitching phases.

RESULTS: The gluteus medius muscle had the greatest activity throughout the entire pitch. The gluteus maximus muscle and then the rhomboids (scapular stabilizers) followed with their activity. Results are graphically summarized in Figure 2.

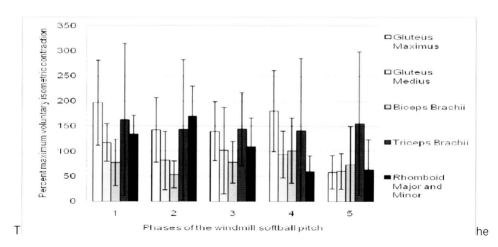

Figure 2: Mean and Standard Deviations (error bars) of Muscle Activation During the Windmill Softball Pitch

DISCUSSION: The wind-up or Phase 1 displayed greater muscle activity in the gluteus medius and maximus than the upper extremity muscles. And based on the weight shift during this phase, activation of the gluteals is required. Phase 2, where the arm was elevated to 90^0 and take off occurred for the stride leg, the gluteus medius acted to stabilize and generate torque of the pelvis; the rhomboids had increased their firing during this phase in attempt to stabilize the scapula throughout arm elevation in the scapular plane. Previous investigations have noted that prior to fatigue overhead throwers have increased upward rotation of the scapula compared to non-overhead throwers, indicating altered movement of the scapula (Myers et al, 2005). However in a separate investigation, after pitching in a regular collegiate event/fatigue, the scapula exhibited decreased upward rotation and external rotation (Birkelo et al, 2003); after a swimming event, investigators noted similar findings in altered scapular motion (Scibek and Borsa, 2003). An unstable scapula or inefficient movement of the scapula during such a dynamic movement would predispose the glenohumeral joint to migrate superiorly, which is associated with impingement syndrome (Deutsch et al, 1996).

During Phase 3 the activity of the gluteus medius increased, and where the humerus was not only being elevated but also externally rotated the triceps brachii activity remained consistent. Phase 4 displayed a continuation of the triceps brachii activity, as well as decreased activation of the scapular stabilizers.

Contrary to baseball mechanics the biceps brachii is most active during the acceleration phase during the windmill softball pitch compared to the deceleration phase (Fleisig et al, 1999). In Phase 4 as the pitcher was attempting to "post" for ball delivery on the stride leg, the dominant gluteus medius must hold the dominant hip upright, while the pitcher is balanced on the stride leg. During Phase 5 the triceps brachii experienced high activation while the core musculature of the gluteus maximus and medius decreased in activation. Throughout Phases 1-3 the rhomboids stayed consistent to stabilize the scapulae, as the arm was dropping below 90^0 of elevation and the humerus was internally rotating the rhomboids decreased in activity. The triceps brachii had the most variability during the windmill softball pitch, this may suggest different abilities to control the acceleration and deceleration phases between the pitchers involved. Future investigations with a larger sample size may look into a differences and relationships between experience level and muscular activation throughout the phases.

It is known that softball is the same game as baseball, but on a smaller field. It has been found that the upper extremity distraction forces during pitching are very similar between the two sports (Barrentine et al., 1998). However, the major apparent difference is the pitching surface from which the pitchers throw. In baseball, the pitchers throw from a mound that allows gravity to assist with the movement, while in softball pitchers throw from a level surface without the assistance of gravity. The windmill softball pitcher has to 'post' during Phase 4 of the pitching cycle and throughout ball release. The posting activity is not assisted by the force of gravity or 'falling' from a pitching mound; however the softball pitcher leaps forward to gain momentum. The requirement of balance is displayed in the evidence of gluteal activation during the last phases of the pitching cycle, where the dominant gluteus medius is highly active.

CONCLUSION: We were able to identify muscle activation for the upper and lower extremity during the windmill softball pitch in post-pubescent females. It should be noted that our sample size was small, however the protocol performed has been previously validated (Maffet et al., 1997; Rojas et al., 2009) and the certified athletic trainer was sufficiently trained in sEMG data collection. Further investigations need to not only address a different population group, such as pre-pubescent or professional, but also examine the activation of the scapular stabilizers. As this is the only investigation of our knowledge looking at the rhomboids throughout the windmill softball pitch, we are not able to generalize on the functionality of the rhomboids throughout the windmill pitch. In addition further investigations are needed on the lower extremity. An investigation of both dominant and non-dominant lower extremity and core musculature would provide insight to the dynamic balance required to perform a windmill softball pitch.

REFERENCES:

Barrentine, S. Fleisig, G., Whiteside, J., Escamilla, R.F., & Andrews, J.R. (1998). Biomechanics of windmill softball pitching with implications about injury mechanisms at the shoulder and elbow. *Journal of Orthopedic and Sports Physical Therapy*, 28, 405-415.

Birkelo, J.R., Padua, D.A., Guskiewicz, K.M., & Karas, S.G. (2003). Prolonged overhead throwing alters scapular kinematics and scapular muscle strength. *Journal of Athletic Training*, 38: S10-S11.

Deutsch, A., Altchek, D.W., Schwartz, E., Otis, J.C., & Warren, R.F. (1996). Radiologic measurement of superior displacement of the humeral head in the impingement syndrome. *Journal of Shoulder Elbow Surgery*, 5 (3) 186-193.

Fleisig, G.S., Barrentine, S.W., Zheng, N., Escamilla, R.F., & Andrews, J.R. (1999). Kinematic and kinetic comparison of baseball pitching among various levels of development. *Journal of Biomechanics*, 32, 1371-1375.

Kendall, F.P., McCreary, E.K., Provance, P.G., Rodgers, M.M., & Romani, W.A. *Muscles: Testing and Function*. Fourth edition. Baltimore, Williams & Wilkins, 1993.

Maffet, M.W., Jobe, F.W., Pink, M.M., Brault, H. & Mathiyakom, W. (1997). Shoulder muscle firing patterns during the windmill softball pitch. *American Journal of Sports Medicine*, 25, (3), 369-374.

McMullen, J., & Uhl, T.L. (2000). A kinetic chain approach for shoulder rehabilitation. *Journal of Athletic Training*, 35 (3), 329-337.

Meyers, J.B., Laudner, K.G., Pasquale, M.R., Bradley, J.P., & Lephart, S.M. (2005). Scapular position and orientation in throwing athletes. *American Journal of Sports Medicine*, 33 (2), 263-271.

Putnam C.A. (1993). Sequential motions of body segments in striking and throwing skills: description and explanations. *Journal of Biomechanics*, 26, 125-135.

Rojas, I.L., Provencher, M.T., Bhatia, S., Foucher, K.C., Bach, B.R., Romeo, A.A., Wimmer, M.A., & Verma, N.N. (2009). Biceps activity during windmill softball pitching. *American Journal of Sports Medicine*, 37, (3), 558-565.

Scibek, J.S., Borsa, P.A. (2003). Swimming practice significantly reduces scapular upward rotation. *Journal of Athletic Training*, 28: S11.

Werner, S.L., Guido, J.A., McNeice, R.P., Richardson, J.L., Delude, N.A. & Stewart, G.W. (2005) Biomechanics of youth windmill softball pitching. *American Journal of Sports Medicine*, 33 (4), 552-560.

Wu, G., van der Helm, F.C.T., Veeger, H.E.J., Makhsous, M., Van Roy, P., Anglin, C., et al. (2005). ISB recommendation on definitions of joint coordinate systems of various joints for the reporting of human joint motion – Part II: shoulder, elbow, wrist, and hand. *Journal of Biomechanics*, 38, 981–992.

A CASE STUDY OF THE EFFECTS OF INSTRUCTION USING MOBILE PHONE'S ANIMATION FEEDBACK ON THROWING KINEMATICS

Kengo Sasaki, Souishi Shimizu, Ami Ushizu, Kenji Kawabata, Takahiko Sato,
*Kazuhiro Matsui, *Yu Nakashima and *Hiroh Yamamoto

Biomechanics Lab., Graduate School of Ed., Kanazawa Univ., Kanazawa, Japan
* Biomechanics Lab., Fac. of Ed., Kanazawa Univ., Kanazawa, Japan

KEY WORDS: Animation Feedback, Throwing events, Mobile phone, Motion Analysis

INTRODUCTION: During recent years, the evolution of the mobile phone is highly active. This evolution transcends the framework of telephone. Japanese mobile phones have evolved as unique and Japan is called the Galapagos Islands of mobile phone. These mobile phones have included high quality camera and higher quality digital cameras. In conjunction with these facts, the mobile phone is life multi-tool that has positive possibilities for instruction of sports. The aim of this study was to obtain the data on the effects of instruction using the mobile phone's animation feedback on throwing kinematics during shot put and discus throw events.

METHODS:
Subjects: The two male and one female throwers of Track and Field team in the Kanazawa University were used as subjects in this study. The all subjects were right handed and signed an informed consent. Subject 1 is a female middle grade Discus thrower, Subject 2 is a male beginner Shot-put and Discus thrower and Subject 3 is a male middle grade Shot putter.
Feedback: The feedback test was designed by reference to previous study (James et.al., 2005). The feedback test consists of two sessions. The 1st session, the Base-line test (BT), Feedback and Performance test (PT) of Shot-put and Discus throw were conducted on the same day. The 2nd session, the Retention test (RT) of both events were conducted after 7 days from 1st session. Each test was consisted of 5 trials. The trial intervals were established two minutes. In Shot-Put trial, all subjects used the glide technique. The Base-line test was conducted without any technical instruction for subjects. After 5 BT Shot-put trials, subjects were given 20 minutes of rest. During rest, all subjects were given and watch the expert model animation and self 5 BT trials animation that were recorded by mobile phone data. After complete feedback, the 5 performance test trials were conducted. After complete Shot-put 1st session, the Discus throw 1st session (5 BT trials, Feedback and 5 PT trials) was carried out in a similar manner of Shot- put in same day. Additionally, the 2nd sessions (both Shot-put and Discus throw 5 RT trials) were conducted at 7 days later (Figure1).

As the expert model, both Shot-put and Discus throw animations of an abroad male player who has grate recode was able to download to mobile phones via internet was used in this study. For conform to both shot put and discus throw expert model animations, the BT and PT animations were recorded from Y axis using mobile phone's Cameras.

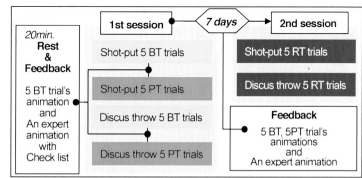

Figure1. Feedback test design

Additionally, the check list was used to provide feedback. The check list include some instruction as follows: (1) increase Trunk Tilt angle at Rear foot touch down, (2) decrease Hip-Shoulder Separation angle at Front Foot Take Off- Rear Foot Touch Down, (3) decrease Hip-Shoulder Separation angle at Front Foot Take Off in discus and (4) decrease Knee-Flexion angle and Hip-shoulder separation angle at Rear foot touch down, (5) decrease trunk-tilt angle and Knee-Flexion angle at Front foot touch down-Release in shot-put throw.

Data collection: The three VHS video cameras (DCR-TRV50: Sony) were used to record the throwing motion for analysis at a rate of 30 Hz (Figure2). And the three mobile phones (W31S, W42S: Sony, W51T: Toshiba / au by KDDI) were used to record the subject's throwing motion of BT and PT for feedback at a rate of 15 Hz. To investigate the relationship between quality of feedback and parameter displacement, the amount and type of feedback in one day was observed in one week from 1st session to Retention Test. The successfully 3 trials were extracted from every 5 trials of each test, the 27 Shot-put trials, and 27 Discus throw trials were used to digitize. The critical instants of Rear (right) Foot Take Off /RFTO, Front (left) Foot Take Off /FFTO: discus only, Rear Foot Touch Down /RFTD, Front Foot Touch Down /FFTD, and Release /Re of the Discus throw and Shot -put were identified from each video camera for every trial to calculated critical parameters (Young and Li, 2005, Leigh and YU, 2007).

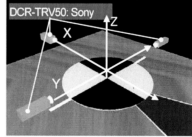

Figure 2.The Location and Axis

The three parameters (Figure 3) were calculated and investigate it's displacement through 3 tests (BT, PT and RT) in this study. The recorded animations by three VHS video cameras were into a personal computer (Versa VY12: NEC). And the 3-D motion analysis system (Frame DIAS Ver.3 for Windows: DKH) was used to calculate parameters. Ten (right and left shoulder, hip, knee and foot) body landmark were manually digitized in each instant.

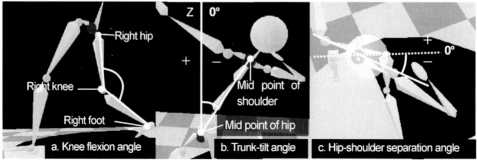

Figure 3.Paramaters

RESULTS: The average amounts of feedback in a day were shown table1. The subject 1 and subject 3 were watched more expert animation than self animation. In addition, the Subject 3 was less watched self animations (BT and RT) than other subjects. On the other hand, the subject2 was no difference between animation types.

Table 1. Amount of Feedback/ a day

Type of animation	Shot-put			Discus throw		
	Subj. 1	Sbuj. 2	Subj. 3	Subj. 1	Subj. 2	Subj.3
Before Feedback (BT)	5.0	2.0	0.4	4.7	2.0	0.4
After Feedback (PT)	3.0	2.3	1.0	3.7	2.6	0.4
Expert	7.7	3.4	4.9	15.7	3.4	1.1

The kinematics parameters and distances on Discus throw were shown Figure 4. On RT, the distance of subject 2 was increased concurrently with angular displacement of HSS from RFTD to Release was larger than other tests. The suject1 and subject 3 were not increase distance. On RT of subject 1, the angular displacement of HSS from RFTD to Release was larger than BT and PT. At the same time, angular displacement of TT on RT from FFTD to Release was smaller than other tests. Additionally, on RT of subject 3, the angular displacement of TT from RFTD to Release was larger than other tests. Concurrently with angular displacement of HSS on RT was small than other tests.

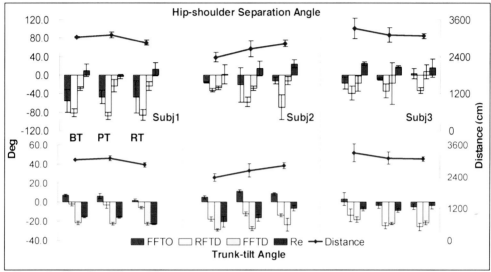

Figure 4. Results of Discus Throw

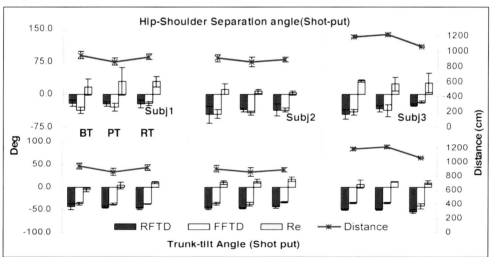

Figure 5. Results of Shot-Put

The distances and kinematics parameters of Shot-put were shown Figure 5. The distance and angular displacement of HSS from RFTD to Release on RT of subject3 (middle grade shot put player) were smallest in three tests. The distances on RT of subject1 and subject2 were lowest each other test, and distances of RT were higher than performance test. In all subjects, the angular displacement of TT on RT from FFTD to Release was larger than other tests.

The knee flexion angle at any instants of all subjects has negligible displacement during test sessions in each event.

DISCUSSION: In both Discus throw and Shot-put, the increased distance of subjects on RT may be due to improved angular kinetics by increased angular displacement of TT and HSS. The cases of without increase distance, angular displacement indicated increase. Despite these case, distance on RT was not increased may be cause of other angular displacements that are different from increased were became smaller than BT or RT. In Discus throw, The HSS at RFTD-FFTD and TT at release were determined parameters that indicated inverse correlation with distance (Leigh and YU, 2007). For this reason, the changes of angular displacement on RT of this study were appearing to improvement of parameter including without increase distance.

In the Shot-put, the increased angular displacement of TT from RFTD to Release on RT was appearing to cause of increasing distance. However, in previous study of Shot-put kinematics (Young and Li, 2005) was not indicate correlated distance with Trunk-Tilt angle. For this reason, the increasing TT appeared to not directly but indirectly effect of increasing distance such as improve angular kinetics on Shot-put in this study. Additionally, decreased subject 3's distances on RT may be due to the angular displacement of HSS from RFTD to Release on RT was smallest in three tests. At the same time, the subject 3 had a fewer self animation (BT and PT animations) feedback. This study was not intended to obtain statistic analysis. However the amounts of feedback may be correlated with parameter displacement. The feedback test of previous study (James et.al., 2005) found that the use of self or combination (self and expert animation augmented) videotape feedback was most useful for increasing kinematics and reducing kinetics during landing. For that founding, a few amount of self animation feedback in subject 3 was not enough to affect improve motion kinematics similar Discus throw. This means that the most cause of decreased subject3's distance on RT was appear to be another factor different from feedback effect.

In this study, the distance and parameters in some of subjects were indicated bit improve trend on RT. This result appears to suggest that animation augmented feedback is effective as a training tool. Additionally, the mobile phones were used to video feedback in this study. That appears to suggest the mobile phone such as life multi tool has positive possibility to use for instruction of sports. In addition, all subjects were inexperienced jump-landing test in previous study. And jump-landing test was more simply motion than throwing event's motion (James et.al., 2005). For these factors different from this study, the feedback effect may be difference by difference level of athletes and complexity of motion.

CONCLUSION: The feedback using mobile phone's animation for throwing event may allow kinematics to improve but we can not determine clearly the effect of feedback. In future studies, more subjects are needed. Investigating the difference of feedback effect from skill level of athletes and complexity of motion are needed. Multiple analyses of kinetics and kinematics were needed.

REFERENCES:
James A. Oñate, Kevin M. Guskiewicz, Stephen W. Marshall, Carol Giuliani, Bing Yu and William E. Garrett (2005).Instruction of Jump-Landing Technique Using Videotape Feedback, *Am J Sports* Med ,33, 831-842
Leigh, Steve and YU, Bing (2007).The associations of selected technical parameters with discus throwing performance: A cross-sectional study. *Sports Biomechanics*, 6, 269-284
Leigh, Steve, Gross, Michael T., Li, Li and Yu, Bing (2008).The relationship between discus throwing performance and combinations of selected technical parameters. *Sports Biomechanics*, 7, 173-193
Mont Hubbard, Neville J. de Mestre, John Scott (2001). Dependence of release variables in the shot put. *Journal of Biomechanics*, 34, 449–456
Young, Michael and Li, Li (2005).Athletics', Determination of Critical Parameters among Elite Female Shot Putters. *Sports Biomechanics*, 4, 131-148
Yu, B., Broker, J., and Silvester, J. L. (2002). A kinetic analysis of discus throwing techniques. *Sports Biomechanics*, 1-46

AN ANALYSIS OF THE IMPACT FORCES OF DIFFERENT MODES OF EXERCISE AS A CAUSAL FACTOR TO THE LOW BONE MINERAL DENSITY IN JOCKEYS

Sarah Jane Cullen[1], Giles Warrington[1], Eimear Dolan[1] and Kieran Moran[1]
Dublin City University, Dublin, Ireland[1]

The purpose of this study was to investigate the forces placed on the lower limbs of jockeys during riding and to determine whether these were comparable to the impact forces associated with traditional weight bearing activities such as walking and running. Evaluation of these forces will allow isolation of the key causes of the previously reported high incidence of low bone mineral density (BMD) associated with this population and indicate as to whether a lack of weight bearing exercise is a causative factor in this phenomenon. Eight apprentice jockeys completed 6 different activities including walking, running and riding (walk, trot, canter, gallop), where accelerometry data was collected to determine the amount of impact loading applied to the lower limbs. The impact accelerations of the lower limbs in horse riding were significantly lower than those seen in running ($p<0.05$). An individual walking appears to have no significant lower limb acceleration difference compared to trotting on a horse ($p<0.05$). However lower limb accelerations during walking are significantly higher to walking on a horse, and lower to cantering and galloping. The relatively non-weight bearing nature of the different riding trials compared to running suggests that jockeys may not receive adequate loading required to gain a sufficient osteogenic effect in order to optimise and maintain adequate BMD levels. Further research is required to validate the finding that, the lack of sufficient loading is a potential contributory factor to the low BMD observed in this population.

KEY WORDS: jockey, bone, musculoskeletal loading

INTRODUCTION: Osteoporosis, which is characterised by low bone mass and deterioration of bone tissue, is normally associated with an aging population and is most prevalent in post-menopausal (Reginster et al., 2006). More recently, it has been reported that professional male jockeys have abnormally low levels of mineralised bone tissue in a given area (BMD) compared to age and size matched subjects (Warrington et al., 2006, 2009). This high incidence of a low BMD observed in jockeys is of particular concern, in the context of the high incidence of falls and fractures recorded in horse racing (McCrory et al., 2006). To date, the primary causative factors which explain this pattern of low BMD in jockeys has not been determined. It has been suggested that chronic energy deficit due to restricted dietary intake, in order to make the strict weight requirements necessary to compete, may affect normal bone formation in jockeys (Leydon et al., 2002). Lack of appropriate weight bearing exercise is another modifiable risk factor which affects the formation of bones (Welton et al., 1994). The loading nature of horse racing has not yet been established, although it has been suggested that horse riding may not provide a sufficient osteogenic stimulus (Alfredson et al., 1998). The purpose of this study was to evaluate the forces placed on the lower limbs of the horse racing jockey during riding and compare this with data from traditional weight bearing activities.

METHODS:

Data Collection: Eight apprentice jockeys participated in this study (age 17±1 yrs; height 1.66 ±0.11 m; body mass 57.04 ±6.9 kg). Each subject was instrumented with 2 lightweight tri axial wireless accelerometers (Crossbow, type CXL100HF3, sensitivity 10 mVG, and range ±100 g), attached to the lateral side of the tibial tuberosity and the base of the lumbar vertebrae (L4). These chosen positions provided a bony prominence for good contact to be achieved, with minimal interference from the riding boot, saddle, horse or any clothing. Accelerometers were aligned along the longitudinal axis of both the tibia and spine whilst subjects were standing. Measurements from the base of the foot to the sensor were recorded for each subject; ensuring correct replacement of the sensor. Subjects completed

6 different trials, in a random order, involving walking, running and riding (walk, trot, canter and gallop). The walking and running trials were completed inside the gymnasium with the cones set up as in the diagram on the wooden floors (Figure 1). Subjects wore their own training shoes. Each subject was instructed to move at a constant pace, aiming to pass a cone by the time a whistle signalled. For walking, the whistle signalled every 6secs ensuring an average walking speed of 6 km/h (1.67 m/s); for running every 3 s for an average running speed of 12 km/h (3.33 m/s).

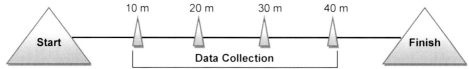

Figure 1: Apparatus used for the walking and running trial

The riding trials were completed in the indoor training arena with the test set up as in the diagram (Figure 2). Subjects wore their own riding boots and two different horses of the same training level were used. Moving clockwise around the arena, each subject was instructed to complete the 4 different gaits of locomotion on the horse (walk, trot, canter, and gallop), adjusting to the selected gait on arrival to the black cone.

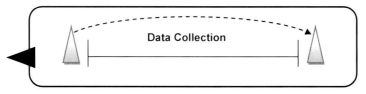

Figure 2: Apparatus used for the riding trial

Data Analysis: For this study, attention was focused on the vertical axis (Y), the direction of the shock waves transmitted vertically up through the bones. Five complete strikes were analysed for all conditions and all subjects. As a result, an average value was gathered for the accelerations applied to the tibia and the corresponding lower back value which was then used for statistical analysis (SPSS 16.0 for windows). Descriptive statistics were found for each dependent variable for each task; statistics included means and standard deviations. One way repeated measures ANOVA was performed on each of the group means to establish if a significant difference exists ($p \leq 0.05$) in the impact loading placed on the lower limbs of the body among the different tasks. A pair wise comparison identified the location of the significant differences ($p \leq 0.05$).

RESULTS: Acceleration magnitudes for both the lower leg and lower back were dependent on the mode of activity undertaken (Table 1).

Table 1: Impact Accelerations during the Different Activities

Activity	Lower Leg Acceleration (g)	Lower Back Acceleration (g)
Walk	2.25 ± 0.82^{e}	1.04 ± 0.14^{f}
Run	13.18 ± 6.99^{b}	3.54 ± 1.04^{c}
Walk – Horse Ride	$1.21 \pm 0.52^{a\,e}$	$0.68 \pm 0.33^{b\,f}$
Trot – Horse Ride	2.94 ± 1.12^{e}	0.75 ± 0.42^{f}
Canter – Horse Ride	$2.96 \pm 0.55^{a\,e}$	$1.64 \pm 0.58^{a\,e}$
Gallop – Horse Ride	$3.64 \pm 0.82^{b\,e}$	$1.79 \pm 0.69^{a\,d}$

Data presented as mean ± SD, $^{a}p \leq 0.05$; different to walking, $^{b}p \leq 0.01$; different to walking, $^{c}p \leq 0.001$; different to walking, $^{d}p \leq 0.05$; different to running, $^{e}p \leq 0.01$; different to running, $^{f}p \leq 0.001$; different to running.

The effect of the individual activities on lower leg and back impact accelerations are shown (Figures 3 and 4).

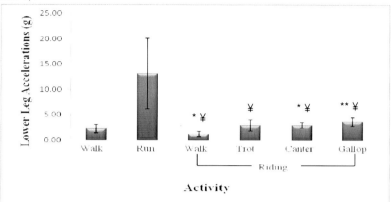

Figure 3: Activity dependent accelerations in the lower leg. The significant difference of walking and running to each of the riding activities is shown (*p≤0.05 for walking; **p≤0.01 for walking; ¥p≤0.01 for running).

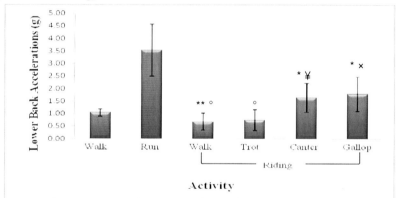

Figure 4: Activity dependent accelerations in the lower back. The significant difference of walking and running to each of the riding activities is shown (*p≤0.05 for walking; **p≤0.01 for walking; ˣp≤0.05 for running; ¥p≤0.01 for running; °p≤0.001 for running).

DISCUSSION: Based on a thorough analysis of current literature, no other studies were found that investigated the lower limb accelerations in jockeys during riding; therefore no comparative data is currently available. Depending on the horse riding gait chosen, noticeable differences in lower limb accelerations were observed. The lower limb accelerations in this study were significantly lower to all other accelerations while the subject was walking on the horse. Trotting resulted in impact accelerations quite similar to those for an individual walking on the ground. The riding gait that resulted in the greatest impact accelerations of the lower leg and lower back was galloping, however, a horse would not be able to remain galloping for a long period of time, so for the majority of training, the gait of cantering would be used. Although these riding accelerations during cantering were still significantly lower than those produced when the individual was running, observation of the acceleration patterns suggest that the impacts in riding were produced at a much greater frequency than those of the individual walking or running. To optimise bone health, it has previously been suggested the exercise includes mechanical loading strains of a high rate and magnitude, distributed in an unusual manner (Bailey et al., 2008). In one study of cyclists it was reported that the non-weight bearing nature of the activity coupled with the reasonably fixed body position adopted provokes a repetitive muscular strain pattern of moderately low magnitude and regular or even distribution (Nichols et al., 2003). In the current study, although a greater frequency in impact accelerations was observed in cantering compared to walking and running, the cycling theory may also be applied to the sport of horse racing such that the jockey does not 'sit' on the horse during cantering and galloping, but rather grips the horse with knees, ankles, and thighs. The relatively low impact nature of the sport of horse racing in addition to the even distribution may mean that jockeys do not receive sufficient weight bearing activity needed to maintain and increase BMD. Alfredson et al. (1998) reported that no significant difference in BMD existed between a group of female horse riders and the non actives at any site measured, suggesting the impact forces during riding are not sufficient to create osteogenic strains on the skeleton. Furthermore, conflicting evidence regarding the extra exercise jockeys partake in leaves doubts as to whether sufficient impact loading is attained elsewhere (Labadarios et al., 1993; Leydon et al., 2002).

CONCLUSION: This study aimed to evaluate the impact forces placed on the lower limbs of the horse racing jockey during riding and compare this with data from traditional weight bearing activities, including walking and running. Results indicated that the chosen gait of the jockey determined the impact acceleration. Lower limb accelerations during walking on legs are similar to trotting on a horse, greater than walking on a horse, however significantly lower to cantering and galloping. Running resulted in much greater lower limb accelerations compared to all other activities. The relatively non-weight bearing nature of the sport of horse riding compared to running suggests that jockeys may not receive sufficient loading required to optimize and maintain adequate BMD levels. It appears that in addition to nutritional factors, the lack of sufficient loading in horse riding is a causal factor to the low BMD observed in jockeys. Further research is required to validate the findings of this study.

REFERENCES:

Alfredson, H., Hedberg, G., Bergstrom, E., Nordstrom, P. & Lorentzon, R. (1998). High Thigh Muscle Strength but Not Bone Mass in Young Horseback-Riding Females. *Calcified Tissue International.* 62, 497-501.
Bailey, C.A. & Brooke-Wavell, K. (2008). Exercise for optimising peak bone mass in women. *Proceedings of the Nutrition Society.* 67, 9-18.
Labadarios, D. (1993). Jockeys and their Weight Practices in South Africa. *South African Medical Research Council.*
Leydon, M.A. & Wall, C. (2002). New Zealand jockeys' dietary habits and their potential impact on health. *International Journal of Sport Nutrition and Exercise Metabolism.* 12, 220-237.
Nichols, J.F., Palmer, J.E. & Levy, S.S. (2003). Low bone mineral density in highly trained male master cyclists. *Osteoporosis International.* 14, 644-649.
McCrory, P., Turner, M., LeMasson, B., Bodere, C. & Allemandou, A. (2006). An analysis of injuries resulting from professional horse racing in France during 1991–2001: a comparison with injuries resulting from professional horse racing in Great Britain during 1992–2001. *British Journal of Sports Medicine.* 40, 614-618.
Reginster J.Y. & Burlet N. (2006). Osteoporosis: A Still Increasing Prevalence. *Bone.* 38: S4-S9.
Warrington, G.D., MacManus, C., Griffin, M.G., McGoldrick, P.A. & Lyons, D. (2006) Bone mineral density and body composition characteristics of top level jockeys. *Medicine and Science in Sports and Exercise.* 38 (5), 246.
Warrington, G., McGoldrick, A., Dolan, E., McEvoy, J., MacManus, C., Griffin, M. & Lyons, D., (2009) Chronic weight control impacts on physiological function and bone health in elite jockeys. *Journal of Sports Sciences* (In Press).
Welten, D.C., Kemper, H.C., Post, G.B., Van Mechelen, W., Twisk, J., Lips, P. & Teule, G.J. (1994). Weight-bearing activity during youth is a more important factor for peak bone mass than calcium intake. *Journal of Bone and Mineral Research.* 9 (7), 1089-1096.

DAILY KNEE JOINT LAXITY IN FEMALES ACROSS A MENSTRUAL CYCLE AND MALES ACROSS A CALENDAR MONTH

Daniel Medrano Jr[1], MS; Darla R. Smith[1], PhD; Dayanand Kiran[2], PT; Mary E. Carlson[2], PT, PhD

Kinesiology, University of Texas at El Paso, El Paso, TX, USA[1]
Physical Therapy, University of Texas at El Paso, El Paso, TX, USA[2]

Increased athletic participation of females has resulted in a high occurrence of anterior cruciate ligament (ACL) injuries. Excessive knee joint laxity during hormonal peaks of endogenous sex hormones during the follicular, ovulatory, and luteal phases of the menstrual cycle has been associated with ACL injury risk. The purpose of this study was to determine the effect of gender and menstrual phase on knee joint laxity over the full course of a normal menstrual cycle in females and across a calendar month in males. A repeated measures ANOVA revealed no interaction effect between gender and phase, no main effect for gender, but a statistically significant main effect for phase. Since male participants demonstrated a similar inclination between phases, the significance of this trend should be interpreted as a possible random occurrence.

KEY WORDS: anterior cruciate ligament, knee laxity, menstrual cycle.

INTRODUCTION:

National Collegiate Athletic Association (NCAA) Injury Surveillance System data taken from 15 male and female collegiate sports across 1988-2004 have revealed significant annual increases in ACL injuries over time. These results are representative of a 1.3% average annual rate increase, p=0.02 in ACL occurrence (Hootman, Dick, & Agel, 2007). Rates of ACL injuries in female athletes were found to be 3 and 4 times higher than male athletes participating in soccer and basketball, respectively (Mihata, Beutler, & Boden, 2006). Conservative costs for surgical repair and rehabilitation have been estimated at $17,000 to $25,000 per injury (Hewett, Myer, & Ford, 2006). These expenses do not include costs or considerations of possible long-term degeneration of the knee joint or development of degenerative arthritis (Lovering & Romani, 2005).

In vitro findings of estrogen and progesterone receptor sites on the male and female ACL, have a growing number of researchers investigating the effects hormonal fluctuations associated with the menstrual cycle on knee joint laxity. A number of researchers have hypothesized that females would exhibit changes in knee laxity during hormonal peaks. Since, the ACL primarily serves to resist anterior tibial translation; researchers have used anterior knee joint laxity as a measure of ACL integrity (Pollard, Braun, & Hamill, 2006). The purpose of the study was to determine the effect of gender and menstrual phase on knee joint laxity over the full course of a normal menstrual cycle in females and across a calendar month in males.

METHODS:

Ten apparently healthy, normally menstruating females (mean age [y] 21.00 ± 1.56, mean height [cm] 160.89 ± 5.89, mean weight [kg] 56.79 ± 4.71) and twelve males (mean age [y] 21.83 ± 2.33, mean height [cm] 175.99 ± 9.46, mean weight [kg] 76.4 ± 15.23) volunteered to participate. Selection was based on the following inclusion criteria: 1) 18 - 28 years of age, 2) normal menstrual cycle, 3) no history of pregnancy, 4) no use of oral contraceptives for six months prior to testing, and 5) no history of knee ligament injury treated by surgery. Male participants were subject to inclusion criteria 1 and 5 listed above. Women who reported more than 3 days variation in the lengths of the 3 menstrual cycles prior to testing, missed menstrual cycles, and/or amenorrhea were considered to have irregular menstrual cycles and were not allowed to participate. All female participants included in the study were not on birth control or hormone replacement therapy and were not under the influence of any prescribed medication at the time of testing. Physical activity for all participants was restricted to 10 hrs of recreational and/or leisure activity (walking, bicycling, swimming, sporting games, etc.) per week. Participants were instructed to refrain from all physical activity 3 hours prior to testing. Knee joint assessments for participants were performed at the same time each day for the entire course of each individual testing period. All data were collected by the principal investigator.

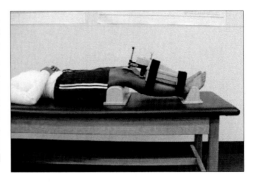

Figure 1: KT-1000 Knee Ligament Arthrometer

Knee Joint Laxity: Daily anterior knee joint laxity testing was performed on both the right and left knee using the KT-1000 Knee Ligament Arthrometer (MEDmetric Corporation, San Diego, CA USA). The participant was placed in a supine position with legs resting on a thigh support and feet positioned in a foot support to ensure bilateral symmetry of the participant's limbs during testing. Knees were bent at thirty degrees to assure uniform right/left and test-retest angles of flexion. The KT-1000 arthrometer was placed onto the anterior aspect of the tibia of the test limb. The apparatus was secured at the level of the gastrocnemius and the lower portion of the extremity above the medial and lateral malleolus by Velcro® straps as illustrated in Figure 1. A passive drawer test was conducted at a displacement load of 133N (30lb). This load was chosen because it is a commonly reported measure that allows for general comparisons across previous research (Shultz, Kirk, Johnson, Sander, & Perrin, 2004). The displacement load was applied through a force handle located 10 cm distal to the joint line of the knee. The mean of three trials was calculated and recorded as the participant's overall anterior knee laxity measure. If consecutive measurements differed by more than 0.5 mm, testing was halted and redone. Right and left leg measurements of anterior tibial translation were collapsed together and averaged into a single daily value for each participant. Intra-tester reliability was determined using an intraclass correlation coefficient of laxity measured from 12 participants over two consecutive days of testing. Results revealed a high degree of reproducibility between day 1 (5.30±1.86mm) and day 2 (5.18±1.81mm); ICC=0.96, $F(1,11)=0.36$, $p=0.56$.

Menstrual Cycle: Menstrual cycle length can vary considerably from one female to the next with an average length generally around 28 days (Wojtys, Huston, Boynton, Spindler, & Lindenfeld, 2002). ACL laxity in female participants was normalized to 28 days using a proportional scaling method. For example, if the participant's menstrual cycle was found to be 30 days long, each day would be multiplied by the fraction 30/28 (Belanger, Moore, Crisco III, Fadale, Hulstyn, & Ehrlich, 2004). An average of like days was then calculated to reduce the cycle into 28 days. Measurements of the female participants were started on the first day of menses (self-report) and continued until the onset of the subsequent menstrual cycle. Male participants were tested for 28 consecutive days.

Data Analysis: All data were analyzed using Minitab® Statistical Software Package. The independent variables for the study were gender (2 levels – male/female) and menstrual cycle phase (3 levels – follicular/ovulatory/luteal). The follicular phase represented days 1 - 9, the ovulatory phase represented days 10 - 14, and the luteal represented days 15 to end of phase. The dependant variable was anterior knee joint laxity. A 2 x 3 (repeated measures) ANOVA with Tukey's post hoc was used to analyze the data for significant differences. Alpha was set at 0.05 level of significance.

RESULTS:

Anterior knee joint laxity data across the follicular, ovulatory, luteal phases of the menstrual cycle for female participants and across 28 consecutive days in male participants are presented in Figure 2. The interaction effect between gender and phase was not statistically significant, F(2,40)=0.44, p=0.65. The main effect for gender, F(1,20)=0.88, p=0.36, also did not reach statistical significance. However, there was a statistically significant main effect for phase, F(2,40)=6.53, p<0.01. Post-hoc comparisons indicated that the mean laxity during the follicular phase (M=5.78mm) was significantly different from the ovulatory phase (M=6.11mm); (d=0.21) and the luteal phase (M=6.06mm); (d=0.17).

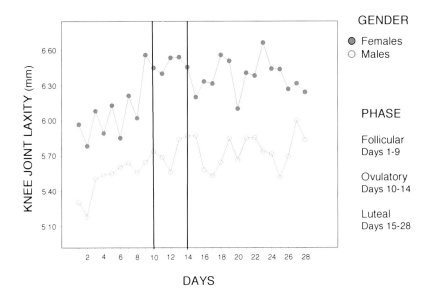

Figure 2: Mean knee joint laxity in males and females across a 28 day cycle

DISCUSSION:

Observed changes in knee joint laxity from the follicular to the ovulatory phase in females is suggestive of possible hormonal influences on ACL tissue and resulting laxity measurements. Findings partially supported results from Park, Stefanyshyn, Hart, Loitz-Ramage, and Ronsky, (2007) who found that maximum manual knee joint laxity values in females were significantly higher in the ovulatory and luteal phase when compared to the follicular phase. Results conflicted with Beynnon, Berstein, Belisle, Brattbakk, Devanny, Risinger, and Durant (2005) who found that females had significantly greater knee laxity across all data collection periods. These results should be placed contextually with the data collected for the male participants. Since male participants demonstrated a similar inclination between phases, the significance of this trend should be interpreted lightly or as a random occurrence. Due to the absence of a gender difference in knee joint laxity, it may be plausible that sex hormone concentrations have little to no effect on laxity. The lack of significance in the interaction for gender by phase further strengthens this assertion. It is also possible that the inter-subject variability was the reason for the lack of a significant gender difference. For example, within the follicular phase, the range of values for men was 3.11mm – 8.43mm, while the range for women was 4.51mm – 7.71mm.

Differences in previous research findings have been attributed to possible miscalculations of hormonal peaks and limited testing of anterior knee joint laxity during dates that may not have coincided with exact hormonal milieus during the menstrual cycle (Pollard, Braun, & Hamill, 2006; Shultz, Sander, Kirk, & Perrin, 2005). A number of studies have also used scaling methods to normalize menstrual cycle length to 28 days to contrast knee joint laxity across the follicular, ovulatory, and luteal phase. This may have additional implications on research findings since the length of regular menstrual cycles varies considerably.

CONCLUSION:

Although the results found within this study do not explain the effect sex hormones have on ACL tissue, the current research provides a clear continuous knee joint laxity profile for both females and males not found in previous literature. These findings may be of relevance when considering the number of data collections necessary to accurately report events that affect knee joint laxity in females and males. The erratic nature of documented joint laxity profiles warrants additional research that includes monitoring of daily circulating hormone concentrations and knee joint laxity over consecutive months. This may identify individual patterns of hormonal influence or render data that are observably random. While the exact mechanism for ACL injury cannot be determined conclusively from this study, structural, neuromuscular, and biomechanical factors may have a greater influence on a female athlete's susceptibility to ACL injuries than joint laxity.

REFERENCES:

Belanger, M.J., Moore, D.C., Crisco III, J.J., Fadale, P.D., Hulstyn, M.J., & Ehrlich, M.G. (2004). Knee laxity does not vary with the menstrual cycle, before or after exercise. *The American Journal of Sports Medicine, 32*(5), 1150-1157.

Beynnon, B.D., Bernstein, I.M., Belisle, A., Brattbakk, B., Devanny, P., Risinger, R., & Durant, D. (2005). The effect of estradiol and progesterone on knee and ankle joint laxity. *The American Journal of Sports Medicine, 33*(9), 1298-1304.

Hewett, T.E., Myer, G.D., & Ford, K.R. (2006). Anterior cruciate ligament injuries in female athletes: Part 1, mechanisms and risk factors. *The American Journal of Sports Medicine,* 34(2), 299-311.

Hootman, J.M., Dick, R., & Agel, J. (2007). Epidemiology of collegiate injuries for 15 sports: Summary and recommendations for injury prevention initiatives. *Journal of Athletic Training,* 42(2), 311-319.

Lovering, R.M. & Romani, W.A. (2005). Effect of testosterone on the female anterior cruciate ligament. *The American Journal of Physiology – Regulatory, Integrative, and Comparative Physiology,* 289, 15-22.

Mihata, L.C.S., Beutler, A.I., & Boden, B.P. (2006). Comparing the incidence of anterior cruciate ligament injury in collegiate lacrosse, soccer, and basketball players: Implications for anterior cruciate ligament mechanism and prevention. *The American Journal of Sports Medicine,* 34(6), 899-904.

Park, S.K., Stefanyshyn, D.J., Hart, D.A., Loitz-Ramage, B., & Ronsky, J.R. (2007). Influence of hormones on knee joint laxity and joint mechanics in healthy females. *Journal of Biomechanics, 40,* S2.

Pollard, C.D., Braun, B., & Hamill, J. (2006). Influence of gender, estrogen, and exercise on anterior knee laxity. *Clinical Biomechanics,* 21(10), 1060-1066.

Shultz, S.J., Kirk, S.E., Johnson, M.L., Sander, T.C., & Perrin, D.H. (2004). Relationship between sex hormones and anterior knee laxity across the menstrual cycle. *Medicine and Science in Sports and Exercise,* 36(7), 1165-1174.

Shultz, S.J., Sander, T.C., Kirk, S.E., & Perrin, D.H. (2005). Sex differences in knee joint laxity change across the female menstrual cycle. *The Journal of Sports Medicine and Physical Fitness,* 45(4), 594-603.

Wojtys, E.M., Huston, L.J., Boynton, M.D., Spindler, K.P., & Lindenfeld, T.N. (2002). The effect of the menstrual cycle on anterior cruciate ligament injuries in women as determined by hormone levels. *The American Journal of Sports Medicine,* 30(2), 182-188.

Acknowledgement

Advice on statistical analysis was provided by Dr. Julia Bader at the Statistical Consulting Laboratory at the University of Texas at El Paso, which is supported by a grant (5 G12RR008124) from the Research Centers in Minority Institutions, National Center for Research Resources, NIH.

DEVELOPMENT OF A CRITERION METHOD TO DETERMINE PEAK MECHANICAL POWER OUTPUT IN A COUNTERMOVEMENT JUMP

N. Owen[1], J. Watkins[1], L. Kilduff[1], D. Cunningham[1], M. Bennett[2] & H. Bevan[1]

Sport and Exercise Science Research Centre, Swansea University, UK[1]
Welsh Rugby Union, Cardiff, UK[2]

KEY WORDS: power, force platform, countermovement jump

INTRODUCTION: The ability of players to repeatedly generate high levels of muscular power is a key determinant for success in many sports. Variations of the countermovement vertical jump (CMJ) have long been used as a means of measuring lower body power (LBP) (Fox and Mathews 1972). The criterion method of measuring of LBP is based on performance in a CMJ off a force platform (FP) (Hatze, 1999). Instantaneous power is determined from the product of the vertical ground reaction force (VGRF) and the velocity of the whole body centre of gravity, velocity being derived by the integration of the resultant VGRF. However, there seems to be no published standard protocol for the criterion method. The purpose of this study was to establish a standard protocol for the criterion method.

METHODS: The variables necessary to define a reliable CMJ method were:

1. Vertical force range and resolution, determined per individual corner force transducer.
2. Force sampling frequency and resultant force integration frequency.
3. Method of integration of the resultant force.
4. Determination of body weight (BW).
5. Determination of the initiation of the CMJ.

Fifteen professional male rugby players (mass = 102.5 ± 13.3 kg) performed a maximal CMJ off a FP. The five variables were then optimised, for each CMJ, to maximise the reliability and validity of the measure of peak mechanical power.

RESULTS AND DISCUSSION: The results of the investigation are summarised in table 1.

Table 1 Criterion method specification for the measurement of instantaneous power in a CMJ using a force platform

Variable	Criterion method specification
1	Six times BW (total vertical force range) at 16 bit A/D resolution.
2	1000 Hz.
3	Simpson's rule or the trapezoidal rule.
4	Mean VGRF for I second of quiet standing immediately prior to jump signal.
5	10 ms before the instant that BW ± 5 SD is exceeded after the jump signal.

Uncertainties (95% CI) for variables 2, 3, 4 & 5 were ±0.5%, ±0.4%, ±0.1% & ±0.5% respectively. Peak power output was most sensitive to variables 4 and 5.

CONCLUSION: This study has established a reliable standard protocol for the criterion method of measuring peak power in a CMJ using a FP. As all other estimates and less reliable methods of determining peak power in a CMJ rely on the FP method for calibration, it is recommended that this protocol be used for all future criterion measures using a FP.

REFERENCES:

Fox, E.L. & Mathews, D.K. (1972). Interval training: conditioning for sport and general fitness. Philadelphia: W.B.Saunders, pp.257-258.

Hatze, H. (1998). Validity and reliability of methods for testing vertical jump performance. Journal of Applied Biomechanics, 14, 127-140.

ACUTE EFFECTS OF HOPPING WITH WEIGHTED VEST ON VERTICAL STIFFNESS

Orna Donoghue and Lawson Steele

Dept of Physical Education, Sport & Leisure Studies, Moray House School of Education, University of Edinburgh, Edinburgh, UK

KEY WORDS: stretch-shortening cycle, spring mass model

INTRODUCTION: Stiffness is defined as the resistance of a body to deformation (Brughelli & Cronin, 2008). It influences how the body interacts with the ground in terms of mechanics and joint kinematics (Farley & Morgenroth, 1999). Optimal stiffness is important in sprinting and jumping as it relates to efficient use of the stretch-shortening cycle. Weighted vests are used to overload the muscles in warm-up and training and have been found to enhance subsequent jumping performance (Faigenbaum et al., 2006). The purpose of this study was to examine the acute effects of wearing a weighted vest on vertical stiffness during hopping in place using a simple spring mass model.

METHODS: Ethical approval was obtained from the university ethics committee. Thirty active males were randomised into experimental (age: 21 ±1.4 years; mass: 85.6 ±10.1 kg; height: 1.83 ±7.3 m) and control groups (age: 21 ±1.1 years; mass: 85.5 ±9.5 kg; height: 1.85 ±7.6 m). Markers were placed on the sacrum, anterior superior iliac crests, greater trochanters, femoral condyles, tibial tubercles, lateral malleoli and 5th metatarsals. Participants performed 3 trials of double leg hopping on a Kistler force plate operating at 500 Hz. Each trial lasted 10 s with 4 min recovery between trials. Participants hopped for maximum height in time to the beat of a metronome at 2 Hz. The experimental group wore a vest weighted with 10% body weight during the second trial. Three-dimensional kinematics were obtained simultaneously using 6 Qualisys cameras, operating at 200 Hz. Vertical stiffness (k_{vert}) was calculated by dividing peak vertical ground reaction force (GRF) by vertical displacement of the sacrum marker during ground contact. GRFs and k_{vert} values were normalised to body weight for all participants. A repeated measures two-way ANOVA with 1 between-subjects factor (group) and 1 within-subjects factors (trial with 3 levels) was carried out using SPSS v.15.

RESULTS: Statistical analysis showed no significant group, trial or group x trial interaction effects for absolute or normalised k_{vert} or GRF ($p>0.05$).

DISCUSSION: Existing research has found that k_{vert} varies with surface and task demands. This suggests that there may be acute changes when wearing a weighted vest but this was not supported by the results. Brughelli & Cronin (2008) recommended that future research examine training practices that may affect stiffness and subsequent running performance.

CONCLUSION: The results indicated that wearing a weighted vest had no acute effects on vertical stiffness or ground reaction force. Future analysis will examine how wearing this device affected joint kinematics during this task.

REFERENCES:

Brughelli, M. & Cronin J. (2008). Influence of running velocity on vertical, leg and joint stiffness. *Sports Medicine*, 38, 647-657.
Faigenbaum, A.D., McFarland, J.E., Schwerdtman, J.A., Ratamess, N.A., Kang, J. and Hoffman, J.R. (2006). Dynamic Warm-Up Protocols, With and Without a Weighted Vest, and Fitness Performance in High School Female Athletes. *Journal of Athletic Training*, 41, 357-363.
Farley, C.T. & Morgenroth, D.C. (1999). Leg stiffness primarily depends on ankle stiffness during human hopping. *Journal of Biomechanics*, 32, 267-273.

GENDER DIFFERENCES IN INSTRUMENTED TREKKING POLE USE DURING DOWNHILL WALKING

Julianne Abendroth[1] Greg Dixon[1], and Michael Bohne[2]
Willamette University, Salem, Oregon USA[1]
Utah Valley University, Orem, Utah USA[2]

This study examined gender differences when hiking downhill with and without trekking poles. Fourteen men and thirteen women were recruited who had hiking and poling experience. Integrated and peak GRF and braking forces (BF), integrated EMG, and trekking pole forces were collected and analyzed. A MANOVA using mean gain scores examined statistical significance (p=.05). Moderate correlations were noted for pole forces and the dependent variables, but no statistical significance was found for the mean gain scores between gender. Trends were noted for peak Fz and BF between gender, with men demonstrating a greater reduction in forces. Men on average also generated greater pole loads, even when normalized for body mass. Four distinct patterns of pole use effectiveness were observed posthoc, but crossed gender lines. Overall, pole loading may be a contributing mechanism to a reduction in forces and muscle activity for men more so than women, but high subject variability limits the strength of this conclusion.

KEY WORDS: gait, hiking, forces, electromyography

INTRODUCTION: The benefits of hiking have been well established, but can be reduced by the incidence of injury or pain (Heggie & Heggie, 2004). Walking downhill causes changes in gait that increase the forces and moments acting on the body, including increased muscle activity of the lower extremity (Kuster, Sakurai & Wood, 1995) which increase the risk for overuse injury (Knapik, Harman & Reynolds, 1996). Trekking poles are thought to counteract the potentially harmful effects by reducing forces on the lower extremity. The effects of trekking pole use have been well established including significant decreases in foot plant forces (GRF, BF) and muscle activity of the lower extremity (Bohne & Abendroth, 2007; Willson, Torry, Decker, Kernozek & Steadman, 2000); however, the mechanism behind these effects is not well understood. Theoretically, with pole use, a portion of the force acting on the body is transferred to the upper extremity, thus decreasing the forces acting on the lower extremity (pole loading). No study has determined whether the effects seen with hiking pole use are the result of pole loading or some other mechanism that occurs during pole use. Research trends also indicate that men may be using poles more effectively than women when walking downhill (Abendroth, Benson & Bohne, 2003). Therefore, the purpose of the current study was to compare poling and gait forces between men and women during downhill walking. Specific goals were to first examine the relationship between pole load and un-weighting. That is, an increase in pole load was expected to correlate with a decrease in foot plant kinetics (GRF, braking forces (BF)) and muscle activity. The second goal was then to examine gender differences in the relative decrease in foot plant and pole forces, and muscle activity between the poling conditions.

METHODS: Twenty-seven healthy volunteers with previous hiking and pole use experience were recruited, and all signed informed consents (14 men, 13 women: age 39 ±12; 43 ±13 and mass 82.2 ±6.5; 61.2 ±7.1 kg, respectively). Approval for the study was obtained from the University IRB. Participants were instructed to wear their preferred hiking shoes, and then assigned to walk in a predetermined, counterbalanced order which included a no pole (NP) and a pole (P) condition. Participants walked at a self-selected pace, although walking speed was held constant between conditions using a Brower timing system. Participants completed 10 successful trials (complete force plate contact during "natural stride") in each condition.

A wooden ramp with ascending and descending 20 degree slopes was used to simulate a hiking experience. A Bertec force plate (1000Hz) mounted flush in the down slope portion of the ramp was used to collect ground reaction and braking forces (GRF/BF). Bertec instrumented Leki trekking poles (1000 hz) were used to measure pole load. Surface electromyographic (EMG) data of the anterior tibialis (TA), gastrocnemius (GA), vastus lateralis (VL), and biceps femoris (BiF) muscles of the participant's dominant leg were measured (Bortec, Inc.). A foot switch in the participant's shoe measured foot stride time. Force plate and EMG data were collected using DATAPAC 2K2 software. Pole forces were collected via Bertec Acquire software. All data were collected simultaneously over 3 second intervals during the downhill portion of the walk. Data collected during the stance phase (when the participant's foot was in contact with the force plate), were used to perform analyses between pole use and gender.

GRF were converted to vertical GRF (VGRF) since the force plate was mounted at a 20 degree angle. Peak and integrated VGRF and BF were averaged over the ten trials for each condition. EMG data were rectified and integrated through a low-pass Butterworth filter at 10 Hz (IEMG). The individual muscles were averaged over the ten trials for each condition, and then summed for overall muscle use. The relative changes in conditions were calculated using mean gain scores (no poles - poles) for all variables. Positive mean gain scores indicated a greater force/activity was generated in the no-pole condition, while a negative mean gain score indicated a greater force/activity was generated for the pole condition. Pole loads were filtered at 500 Hz, and synchronized with the stance phase of the force data. The loads generated for both poles during this stance phase were integrated, averaged, and summed together.

Statistical analyses were performed using SPSS. Correlations between pole loading and the mean gain scores were made using Pearsons r. Positive correlations indicated that with more pole loading, a greater difference was also noted in the change in forces between the no pole and pole conditions (mean gain score). Comparisons between men and women for the mean gain scores between NP/P conditions were performed using a MANOVA. Statistical significance was set at an alpha = .05 for comparisons. Statistical power was calculated to be 80%, based on the selected sample size and an effect size of 0.7(CHECK THIS).

RESULTS: Small to moderate correlations were noted for pole forces to the mean gain scores (no pole – pole) for integrated VGRFs and peak VGRFs for both men and women, as well as for integrated BF and peakBF (see Table 1). However, no statistical significance was noted. Men demonstrated relatively stronger correlations for peak forces, while women demonstrated relatively stronger correlations for integrated forces. Correlations for IEMG were small, although stronger for the women.

Table 1. Correlations of pole forces to mean gain scores (no pole – pole) for integrated vertical ground reaction and braking forces (IVGRF & IBF), peak GRF and BF, and muscle activity.

	IVGRF	peakVGRF	IBF	peakBF	Summed IEMG
men	0.37	0.41	0.20	0.35	0.18
	(p = .09)	(p = .07)	(p = .24)	(p= .11)	(p=.27)
women	0.41	0.31	0.43	0.28	0.39
	(p = .08)	(p = .15)	(p = .07)	(p= .18)	(p=.09)

Mean gain scores for pole use between gender are shown in Figure 1. Women, on average, demonstrated more or similar forces and muscle activity when using poles in comparison to no poles, except for peakFz. Men demonstrated decreased peak forces in both VGRF and BF, but the integrated forces were greater during the poling condition. Less muscle activity was observed, on average, for men when using poles.

The MANOVA revealed no statistically significant differences for any of the force measurements or summed muscle activity, between men and women (Wilkes Lambda F = 7.7, df = 7, p = .37). Pole forces, even when normalized for body mass, was greater for men than women, although also not statistically significant, as part of the MANOVA. Stride time was the timing of a complete stride onto or off of the force plate.

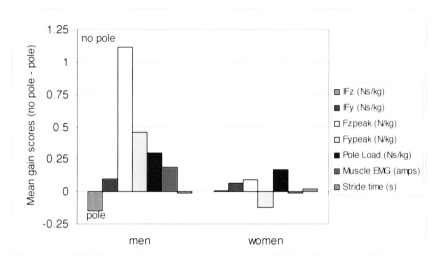

Figure 1. Mean gain scores for no pole vs. pole condition. Positive scores indicate greater force/activity without poles, while negative scores show greater force/activity while using poles.

Effect sizes (ES) were examined between gender for practical significance. Integrated VGRF and BF effect sizes were small (.31 and .15, respectively). Peak forces demonstrated greater ES (.65 and .72, for VGRF and BF, respectively). These differences included an average reduction of peak VGRF of 1.1 N/kg for men, and only 0.09 N/kg for women. For a typical 70 kg person, these peak forces would represent a 77 N reduction for a man, and a 6 N reduction for a similar sized woman. The braking forces were reduced for men by an average of 0.46 N/kg while women increased their braking forces with poles, by 0.12 N/kg. Likewise men reduced muscle activity by 18.7amps, while women increased muscle activity by 1.39 amps, when using poles. A moderate ES was noted at .42.

High variability within gender was noted. The data was reexamined posthoc, for effective pole use, which was defined as a reduction in IVGRF and/ or muscle activity. Four distinct patterns were noted. Eight participants (4 men, 4 women) demonstrated reductions in both forces and muscle activity, which would be the most effective use of poles. The average pole load for this group was greatest. A second group reduced forces overall, but not muscle activity (6 women, 1 man), while a third group reduced muscle activity but not forces, overall (2 women, 6 men). Both of these groups demonstrated similar pole forces. The final group (3 women, 3 men) demonstrated no reductions with pole use, and the pole load was small as well.

DISCUSSION: The primary purpose of this study was to examine the relationship between trekking pole load and the mean differences in foot plant forces and muscle activity for both men and women. Due to the moderate correlations between pole load and the relative decreases in foot plants forces and muscle activity observed, it is surmised that pole loading is partially responsible for the kinetic effects observed during trekking pole use, although the relationships are similar between men and women.

Pole use did not reproduce the expected results of reducing forces and muscle activity, on average, for men or women. While peak forces did decrease on average, with pole use, integrated forces for men increased, perhaps due to a technique difference, since the stride time was not different between conditions. A large variability within gender may have served to confound the results. From a practical significance, some gender differences do seem to exist. Men load the poles greater than women, even when normalized. When grouped by effective pole use, this distinction held true, with the men in the most effective group of lesser forces and lesser muscle activity demonstrating greater pole loading on average, than the women in the same group. The second pattern of pole use, with lesser forces but greater muscle activity, was noted in more women. This may support the idea that women are not using to poles predominately to unweight, but rather may be using a combination of pole loading with more muscle activity to produce the unweighting effect. This effect may highlighted by the increase in the peak braking force for women when using poles, where the poles are not being used to slow their forward momentum. More men were in the third group of lesser muscle activity, but greater forces with pole use. These participants may be gaining some effect of pole use in terms of lessening muscle activity may lessen fatigue, even though the unweighting effect is missing. The final group, equal numbers of men and women, appear to gain no benefit from pole use. The four distinct pole use techniques appeared to be independent of any particular demographic that could have been predicted or controlled including age, gender, and hiking/pole use experience.

CONCLUSION: Gender alone does not appear to a predominant indicator in effective pole use. While some trends in gender differences were noted with pole use while hiking downhill, no definitive conclusions were drawn. When grouped by like-results, the most effective group who were able to reduce forces and muscle activity with pole use, also produced the greatest pole load, whereas the least effective group (greater forces and more muscle activity, with pole use) also produced practically no pole load, indicating that pole load may in fact be a contributing mechanism behind the effects of trekking pole use. However, as pole load was not strongly correlated to relative decreases in foot plant forces and muscle activity across the entire subject population, it cannot be concluded that pole loading is the only or primary mechanism behind the established effects of trekking pole use.

REFERENCES:

Abendroth-Smith, J., Benson, A., & Bohne, M. (2003) Kinetic patterns of seasoned downhill hikers using none, one, or two hiking poles. *Med Sci Sport Exer, 35(5), s98.*

Bohne, M., & Abendroth-Smith, J. (2007). Effects of hiking downhill using trekking poles while carrying external loads. *Med Sci Sport Exer, 39(1), 1-7.*

Heggie, T.W., & Heggie, T.M. (2004). Viewing Lava Safely: An Epidemiology of Hiker Injury and Illness in Hawaii Volcanoes National Park. *Wilderness and Env Med,* 15, 77-81.

Knapik, J. Harman, E., & Reynolds, K. (1996). Load carriage using packs: a review of physiological, biomechanical and medical aspects. *Applied Ergonomics,* 27(3), 207-216.

Kuster, M., Sakurai, S., & Wood, G.A. (1995). Kinematic and kinetic comparison of downhill and level walking. *Clin Biomechanics,* 10(2), 79-84.

Willson, J., Torry, M.R., Decker, M.J., Kernozek, T., & Steadman, J.R. (2000). Effects of walking poles on lower extremity gait mechanics. *Med Sci Sport Exer,* 33(1), 142-147.

Acknowledgement
The authors would like to thank Bertec Corp. for their efforts in designing and building the instrumented hiking poles.

DURABILITY OF RUNNING SHOES WITH EVA AND PU MIDSOLE

Youlian Hong[1], Yau Choi Ngai[1], Lin Wang[1], Jing Xian Li[2]
Department of Sports Science and Physical Education, The Chinese University of Hong Kong, Hong Kong, China[1]
School of Human Kinetics, University of Ottawa, Ottawa, Canada[2]

KEY WORDS: durability, cushioning, midsole material, shoe testing.

INTRODUCTION: Running shoes may play an important role in preventing injuries by absorbing external shock due to ground impact (Cook et al., 1990; Verdejo and Mills, 2004). Shoe age maybe an important factor in running injuries. One prospective study showed that running injury was associated with shoe age (Taunton et al., 2003). In recent years, different types of foam materials have been developed for running shoe midsoles. Two common types of foam materials, Ethylene Vinyl Acetate (EVA) and Polyurethane (PU), are now widely used in running shoe midsoles. The purpose of the present study was to examine the durability of running shoes with common types of EVA and PU midsole materials.

METHODS: Three types of running shoes, with different midsole materials EVA, PU1 (newly developed material - density and hardness close to EVA), and PU2 (current material with higher density and hardness), were worn by human subjects and shoe cushioning characteristics (peak force and energy return in the impact test) at the heel were measured using a commercial impact tester every 50 km running distance. The mechanical impact test maked equal comparisons across all shoes and every 50 km running distance.

RESULTS: Change of cushioning characteristics was as below:
Peak force: EVA and PU1 shoes had a lower peak force than PU2 shoes at all running distances. The changes of the peak force at 500 km with reference to 0 km were EVA +4.8%, PU1 -2.6%, and PU2: -5.0%.
Energy return: EVA shoes had higher energy return than PU1 and PU2 shoes at all running distances. The changes of energy return at 500 km with reference to 0 km were EVA -0.5%, PU1 +5.5%, and PU2 -1.1%.

DISCUSSION: The benefit of this study is that it provided true information about the durability of current EVA, PU1, and PU2 materials used in conventional running shoes under normal use. The change of cushioning characteristics was smaller when compared with other previous studies that shoes were tested by machine simulated or human subjects. The EVA and PU midsoles didn't deteriorate so much as reported by other previous studies. It maybe due to the improvement in midsole materials and manufacturing processes in recent years provides better cushioning and durable EVA and PU midsoles for running shoes.

CONCLUSION: As the running distance increased, the cushioning characteristics of midsole materials changed continuously. EVA, PU1, and PU2 showed different patterns of positive or negative changes. The change of peak force at the 500 km running distance was only between -5% and +5%. These findings provided useful information to runners about the durability of conventional running shoes with an EVA and PU midsole.

REFERENCES:
Cook, S.D., Brinker, M.R, Poche, M. (1990). Running shoes. Their relationship to running injuries. *Sports Med.*, 10(1):1-8.
Taunton JE, Ryan MB, Clement DB, McKenzie DC, Lloyd-Smith DR, Zumbo BD (2003). A prospective study of running injuries: the Vancouver Sun Run "In Training" clinics. *British J Sports Med* 37:239-44.
Verdejo, R., Mills, N.J. (2004). Heel-shoe interactions and the durability of EVA foam running-shoe midsoles. *J. of Biomechanics* 37; 1379-1386.

Acknowledgement: Not applicable

ISBS 2009

Poster Session PS 5

Effect of Kinetic mechanisms of lower limbs on torso motion in baseball batting for different ball speeds

Tokio Takagi[1] Norihisa Fujii[2] Sekiya Koike[2] Michiyoshi Ae[2]

[1]Doctoral Program in Physical Education, Health and Sport Sciences,
[2]Institute of Health and Sport Sciences, University of Tsukuba, Japan

KEY WORDS: kinetics, baseball batting, different speed ball, lower limbs

INTRODUCTION: It is known that baseball batters adjust movements of the torso, and lower and upper limbs to hit the ball when ball course and speed vary. However, few studies investigated the effect of kinetic mechanisms of lower limbs on torso motion for different ball speeds. Therefore, the purpose of this study was to investigate effects of kinetic mechanisms of lower limbs on torso motion in baseball batting for different ball speeds.

METHODS: Twenty nine university baseball players hit baseballs thrown by a machine at SLOW (80-85km/h), MEDIUM (100-105km/h), and FAST (125-130km/h) speeds. Three dimensional kinematic data were collected using Vicon 612 system (250Hz), and ground reaction forces were collected with two force platforms (500Hz). Segment torque powers (STP) and joint torque powers (JTP), and joint force powers (JFP) were calculated for the lower body.

RESULTS and DISCUSSION: In our previous study (Takagi et al, 2008), batters reduced the displacement of the center of gravity (CG) towards the machine and rotated the upper torso earlier in FAST than in SLOW. Figure 1 shows hip JFP of the pivot leg in the lower torso averaged for every ball speed conditions. Hip JFP of the pivot leg was slightly less in FAST than in SLOW from CGmin to SR, mainly because of the less velocity of the hip joint. The other hip JFP and STPs of both legs were almost zero. Therefore, it was suggested that less hip JFP of the pivot leg was one of the reason of less displacement of the torso in FAST than in SLOW. Figure 2 shows hip JTPs of the both legs averaged for every ball speed conditions. Batters generated the hip JTPs earlier in FAST than in SLOW just after SR, and these powers were transferred to the upper torso. Therefore, it was concluded that the powers generated earlier in the hip joints helped the batters to rotate the upper torso earlier in FAST.

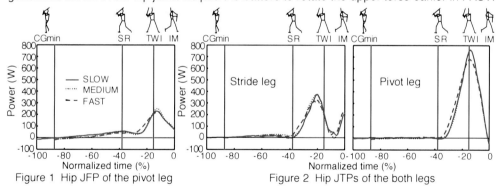

Figure 1 Hip JFP of the pivot leg   Figure 2 Hip JTPs of the both legs

CGmin: Minimum CG displacement against machine, **SR**: Start of rotation of lower torso, **TWI**: Maximum torso twist, **IM**: Impact. The averaged time from CGmin to IMPACT was normalized for every ball speed conditions (-100% : CGmin of SLOW). Vertical lines indicate CGmin of FAST and MEDIUM, SR of every ball speed conditions, TWI of every ball speed conditions in order.

REFERENCES:

Takagi, T., Fujii, N., Koike, S., Ae, M. (2008). The kinematic characteristics of swing motion to the different speed balls in baseball batting. *Journal of the Society of Biomechanisms (in Japanese)*, 32(3), 158-166.

MODERATION OF LOWER LIMB MUSCULAR ACTIVITY DURING JUMP LANDING BY THE APPLICATION OF ANKLE TAPING

Ashleigh Dodd, Rosemary Dyson and Russell Peters
University of Chichester, Chichester, UK

Ground reaction forces were measured using a Kistler force platform when 8 female subjects, wearing their own shoes, landed as they received a netball chest pass. Simultaneously electromyography recordings were made of three ankle muscles. The peak impact (total 2.3 BW; vertical 2.2 BW) and landing (total 1.8 BW; vertical 1.8 BW) forces were similar whether the ankle was taped or not ($p>0.05$). Statistical analysis of the electromyography data indicated that, for the subject group as a whole, the peak muscular activity of the gastrocnemius medius ($p=0.031$), tibialis anterior ($p=0.018$) and peroneus longus ($p=0.019$) muscles was significantly reduced during landing when the ankle was taped using zinc oxide tape.

KEY WORDS: ankle, electromyography, force, netball, taping.

INTRODUCTION: Ankle taping is commonly used within sport to provide ankle support and to prevent injury (MacDonald, 1994). The aim of this study was to determine the effect of ankle taping upon the muscle activity of three ankle muscles during jump landing from a netball chest pass.

METHODS: Eight female netball players from the 1st or 2nd University teams volunteered for the study and wore their own netball shoes. All participants had been free of injury for 12 weeks, had no current musculoskeletal lower limb problems, and provided written informed consent. The University gave ethics approval. Participants were required to catch a netball chest pass from a feeder by forward jumping and landing on the dominant leg on a 9851B Kistler force platform which was sampled at 500 Hz for 4 seconds using Provec 5.0 software. The jump distance was 1.25 times the participants' leg length (Hopper et al., 1999). Three jumps were performed with landing on the dominant limb without any taping and then another three jumps after ankle support had been applied using Transpore zinc oxide tape. The applied taping was a combination of a Gibney basketweave and figure of 8 continuous heel locks as described by Rarick et al. (1962) consisting of 2 anchor strips above the achilles, and 2 at the tarsal arch of the foot. Three vertical, and 3 horizontal horsehoe-shaped strips were used as stirrups. The vertical stirrups were pulled from the medial to the lateral side of the foot, slightly everting the foot (Davies,1977) which was found to aid with prevention from the inversion sprain mechanism. Figure-of-eight heel locks were wrapped under the foot, behind the heel, and the lower leg, which were repeated once more, and adding 2 figure of 6 tapes (Hopper et al.,1999). Prior to testing in each condition the participants completed a 3 minute warm up on a Monark cycle ergometer. Subjects were allowed up to 3 practice jumps before recording began. Peak total impact and landing forces and peak vertical impact and landing forces were normalised for each subject's body weight (BW). Taped ankle range of motion was measured after application and after the completion of testing (Physio Med International standard goniometer). Electromyography (EMG) activity of the dominant leg was recorded from the ankle plantar flexor gastrocnemius medius, dorsiflexor and invertor tibialis anterior, evertor and assisting plantar flexor peroneus longus using a patella reference electrode. Medicotest blue sensor electrodes were applied following skin cleaning and mild abrasion using disposable products. An MIE radiotelemetry system was used to record EMG data at 500 Hz for 4 seconds with Myodat 3.0 software and later to perform linear envelope analysis. Results from the two most representative jumps for each participant in each condition were averaged. Statistical analysis was performed using paired two-tailed t tests.

RESULTS: Upon landing the ground total force trace usually displayed an initial greater impact peak with a following second peak as the foot came into full contact with the floor as shown in Figure 1.

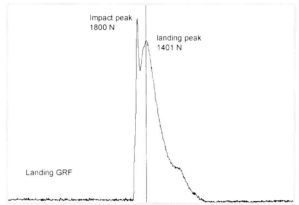

Figure 1: A representative example of total ground reaction force (reflecting the vertical, sagital and mediolateral planes) after receiving a netball pass when ankle taping had been applied.

Impact total forces were higher than landing total forces for both the control condition and when ankle taping was applied as shown in Figure 2. When compared to the control condition there were similar impact and landing total forces recorded with ankle taping (Figure 2). These peak total forces ($\sqrt{(Fx^2 + Fy^2 + Fz^2)}$) represent the integrated response of the foot in the horizontal plane (sagittal and mediolateral) and were adopted as the foot was not always in line with the force platform horizontal axes on landing, and consequently separate measures would be misleading. When taped, the ankle range of movement was similar before and after jump testing (p=0.655). The peak vertical forces at impact were similar in the control (2.25 ± 0.31 BW) and taped condition (2.21 ± 0.16 BW; p>0.05). Similar peak vertical landing forces were recorded with and without taping (control 1.84 ± 0.09 BW; taped 1.78 ± 0.07 BW). Mean peak total forces were slightly greater than the vertical forces reflecting the contribution of the integrated horizontal ground reaction force components

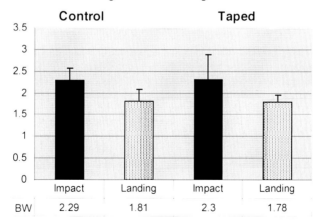

Figure 2: Total forces, reported for jump landing (±SD) after catching a netball chest pass when the ankle was not taped (control) and when taping was applied (n =8).

Figure 3 shows a recruitment sequence of the three muscles during the take off and landing. At take off, the gastrocnemius and peroneus muscles simultaneously reached peak activity as they acted to plantarflex the ankle. After air time, the gastrocnemius acted to plantarflex the ankle for the toe contact landing and tibialis anterior to gradually dorsiflex the ankle. The peroneus activity increased to aid in counteracting the secondary inversion action of the tibialis anterior by eversion.

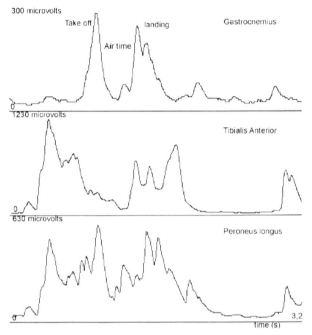

Figure 3: Linear enveloped EMG activity of ankle muscles during take off and landing to receive a netball chest pass.

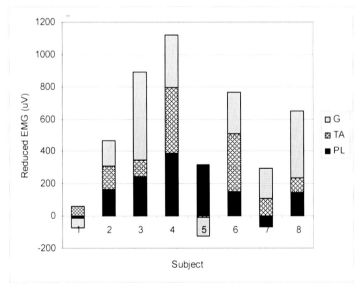

Figure 4: Reduction in muscular activity recorded from the gastrocnemius medius (G), tibialis anterior (TA) and peroneus longus (PL) muscles when the ankle was taped before taking and landing from a netball pass, relative to the control ankle condition without any tape application.

For the whole group, the muscle activity of the gastrocnemius medius (p=0.031), tibialis anterior (p=0.018) and peroneus longus (p=0.019) muscles was significantly reduced when landing from receiving the netball pass when ankle taping had been applied. Figure 4 summarizes the reduction in muscular activity, determined by subtraction of the peak values with taping from those without taping (Mitchell et al., 2008). The reduction in peroneus longus peak activity when the ankle was taped in slight eversion was evident. Peroneus longus has an important counteracting eversion protection action in the avoidance of ankle inversion injuries. However within subject reductions in activity of the ankle dorsiflexor and invertor tibialis anterior is also evident together with reduced gastrocnemius activity.

DISCUSSION: In review, the role of ankle taping is not just of avoiding injury by increased mechanical stability, but also of improving proprioception in foot orientation and athletic confidence (Robbins & Waked, 1998). Most ankle sprains are inversion injuries, and in netball most injuries occur when landing from a jump (Hopper et al., 1999). Within netball, taping has relevance as each episode of play lasts 15 minutes, less than the 20 minutes research has indicated that taping efficacy is maintained (Robbins & Waked, 1998) albeit more than the 10-15 mins reported during research review (Kadakia & Haddad, 2003). Hopper et al. (1999) reported similar EMG peak activity when tape was applied in contrast to this research. Maximal voluntary contraction normalisation has not been used in this research area due to the dynamic nature of the jumping and landing task, and the predictable effect of weakening the taping applied. Robbins & Waked (1998) considered that reduced EMG peak activity induced by taping might increase the injury risk. Study of induced EMG activity with graded ankle range of movement restriction could be worth investigation to identify the contributions of the slight eversion and restriction. Total forces (reflecting the forces in the frontal, saggital and vertical planes of the foot), and vertical forces of peak impact and landing were similar whether the ankle was taped or not. This is in accord with general reports in the reviews of Robbins & Waked (1998) and Bot et al. (2003), and for netball by Hopper et al. (1999). In this research peak vertical forces were less than the 3.3 BW reported by Hopper et al. (1999). Larger studies would clarify if this is due to dvances in sport shoe design over 20 years.

CONCLUSION: Maximal vertical and total ground forces were 2.3 BW when landing from a netball chest pass, and were similar if ankle taping was applied. However, the maximal EMG activity of the evertor peroneus longus, and also the gastrocnemius medius and tibialis anterior muscles was significantly reduced when the ankle was taped in slight eversion with zinc oxide tape.

REFERENCES:

Bot, S.M., Verhagen, E.A.L.M., & Van Mechelen, W. (2003). The effect of ankle bracing and taping on functional performance: A review of the literature. *International SportMed Journal*, 4, (5). http://www.esportmed.com

Davies, G. J (1977). The ankle wrap: Variation from the traditional. *Athletic Train, Journal of National Athletic Trainers' Association*, 12, 194-197.

Hopper, D.M., McNair, P.M. & Elliott, B.C. (1999) Landing in netball:effects of taping and bracing the ankle. *British Journal of Sports Medicine*, 33, 409-413.

Kadakia, A.R. & Haddad, S.L. (2003). The role of bracing and taping in the secondary prevention of ankle sprains in athletes. *International SportMed Journal*, 4, (5).http://www.esportmed.com

MacDonald, R (1994). *Taping techniques*. Butterworth Heineman. England:London p 27-29.

Mitchell, A., Dyson, R., Hale, T. & Abraham, C. (2008). Biomechanics of ankle instability part 1: Reaction to simulated ankle sprain. *Medicine and Science in Sport and Exercise*, 40, 1515-1521.

Rarick, G.L, Bigley, G., Karst, R., & Malina, R.M (1962). The measureable support of the ankle joint by conventional methods of taping. *Journal of bone joint surgery*, 44, 1183-1190.

Robbins, S. & Waked, E. (1998). Factors associated with ankle injuries: Preventative measures. *Sports Medicine*, 25, 63-72.

KINEMATIC ANALYSIS OF THE TRADITIONAL BACK SQUAT AND SMITH MACHINE SQUAT EXERCISES

Anthony Gutierrez[1] and Rafael Bahamonde[2]

Department of Kinesiology, University of Wisconsin, Madison, USA[1]
Biomechanics Laboratory, Indiana University, Indianapolis, USA[2]

The purpose of this study was to compare the kinematics of the traditional back squat (TBS) and the Smith Machine Squat (SMS). The squat exercise is a common exercise in strength and conditioning programs as well as in rehabilitation settings. Eight experienced college age weight lifters performed both TBS and SMS. Three dimensional video analyses were used to analyze the motions. Lower extremity joints and trunk angular motions were computed and compared using Paired T-test. The TBS generated larger ROM than the SMS in all the joints measured. Due to the linear restriction of the bar motion along the vertical axis in the SMS the subjects positioned their feet forward to enable bar lowering. This feet placement positioned the bar farther away from the knee at the instance of maximal knee flexion.

KEY WORDS: Back Squat, Smith Machine, Kinematics.

INTRODUCTION:

The traditional back squat (TBS) exercise is primarily associated with strength training in athletes at all levels of competition. In addition, due to its classification of a closed kinetic chain exercise (Escamilla, Fleisig et al. 1998), it is often used in rehabilitative settings. The Smith machine (SM) is a type of equipment used for squats and other exercises that is commonly available in health clubs and recreation centers. It consists of a barbell that is fixed within rails, so that it can only move vertically, although new variations allow some forward and backward movement (See Figure 1). Because of its fixed motion it decreases the need for balancing the bar and weight plates which increase the safety of the exercise. In the fitness industry the SMS has been suggested as an alternative exercise to TBS (Glenn, 2009) but it has also been labelled as one of the least effective exercise to perform (Russi, 2008). Although several studies have analyzed the biomechanics of the (TBS) (Escamilla, Fleisig et al., 1998; Escamilla 2001; Escamilla, Fleisig et al., 2001; Zink, Perry et al., 2006; Flanagan and Salem 2007; Robertson, Wilson et al., 2008; Sahli, Rebai et al., 2008; Gullett, Tillman et al., 2009) only a handfull of studies have looked at the mechanics of the SM squat (SMS) (Escamilla, Fleisig et al., 1998; Abelbeck 2002; Jacobson 2003; Cotterman, Darby et al., 2005). Abelbeck (2002), developed a mechanical saggital plane model of the SMS that allowed for variations in foot placement. This model showed a decrease in knee moment and an increase in hip moment as the model's feet were positioned anterior to the body. In contrast, in a study by Jacobson (2003) the SMS generated greater anterior-posterior and compressive knee forces than the TBS. More recently significant differences were found between the 1RM using the SMS and TBS (Cotterman, Darby et al. 2005). Subjects were able to generate greater 1RM using the SMS. Therefore the aim of his study was to compare the kinematics of the TBS and the SMS. It was hypothesized that the fixed vertical motion of the bar during the SMS leads to kinematic changes at the joints and bar motions which may also affect the joints kinetics.

Figure 1

METHODS:

Data Collection: Eight, experienced, college age lifters (3 male and 5 female) with a mean body height (♂= 1.74 ± 0.16 m ♀= 1.64 ± 0.10 m) and a mean body mass (♂= 86.5 ± 11.8 Kg ♀= 64.4 ± 14.1 Kg) were recruited for the study. The TBS used an Olympic weight bar (201.6 N) while the SMS bar weighted (89.6 N). To reduce any variations in technique and to normalize the load, the subject's lifted only 50% of their body weight and were instructed on how to perform the squats according to guidelines of the National Strength and Conditioning Association (NCSA) (Robertson, Wilson et al., 2008). Each subject warmed up 10 min prior to exercise and performed three repetitions of each exercise. StreamPix Video Capture Software (Norpix, Inc., Quebec, Canada) and two Basler A602 cameras (50Hz) were used to record the motions. Video analysis was done using KWON-3D Motion Analysis Software (Visol, Korea).

Data Analysis: The squat motion was divided into two phases, the descending and ascending phases. The end of the descending phase and the start of the ascending phase were determined from the time of maximum knee flexion (MKF). Because of differences in trial length and for comparison purposes the combined times of the ascending and descending phase was normalized to 100%. Joints centers, 12 body landmarks and the weight plates were marked, digitized, and used to define the rigid segments representing the bar, trunk, thighs, shanks and feet. Three dimensional coordinate data were smoothed using a 4th degree Butterworth-low pass filter with a smoothing factor of 6 Hz. Due to the symmetric nature of the movements only the left side of the subjects was used for comparisons. Linear and angular kinematics were computed using Kwon 3D. The joint motions computed were: trunk flexion (relative to the vertical axis), hip flexion/extension, knee flexion and extension, and ankle dorsiflexion-plantar flexion.

Paired t-tests (Sigma Stat, Systat Software, Inc.) were performed on the following parameters: ROM of the hip, knee, ankle and trunk angles, angular position at MKF, and horizontal position of the bar relative to the knee at MKF. Effects size was calculated according to Cohen (1977).

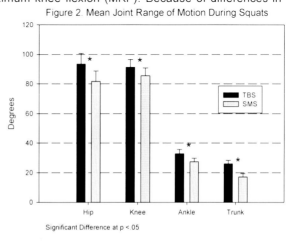

Figure 2. Mean Joint Range of Motion During Squats

Significant Difference at p < .05

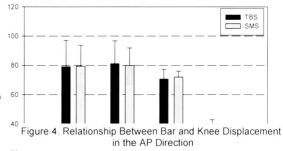

Figure 3. Joint Angles at Maximum Knee Flexion

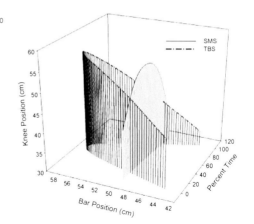

Figure 4. Relationship Between Bar and Knee Displacement in the AP Direction

RESULTS:

Mean MKF occurred at about the same normalized time for both type of squats techniques (53% and 51% for TBS and SMS respectively). Figure 2 shows the mean ROM values for hip and knee flexion/extension, ankle dorsiflexion and plantar flexion and trunk flexion. All subjects exhibited larger ROM in all the joints during the TBS. Trunk and hip flexion ROM values were considerably greater in the TBS than in the SMS. Although there were no significant differences between joint angles at MKF for the two types of squat techniques (See Figure 3), there was greater trunk flexion (ES = 1.49) during the TBS at MKF and less ankle dorsiflexion (ES = .54) during the SMS. Figure 4 shows the relationship between the bar and knee positions with respect to time during the squat motions. Because of the linear motion of the SMS the bar is restricted in the AP (anterior-posterior) direction (Figure 4 - Gray plane), whereas in the TBS the bar and the knee move forward during the descent phase. Figure 5 shows the horizontal distance of the bar relative to position of the knee. Although non-significant (ES =.50), the bar was positioned further back from the knee during the SMS at the instance of MKF. In the TBS the bar moved closer over the knee joint center.

DISCUSSION:

The SMS is a common weight training exercise because it is safer to perform, as it removes the necessity of balance and it can be done without spotters. It has been the assumption that the SMS has similar joint motions than the TBS, although some studies have shown that the SMS generates greater 1RM and different knee joint moments and forces than the TBS. There has not been a kinematic comparison of the two types of squats and there is some controversy regarding the knee kinetics. One study suggests that the SMS enhances the 1RM by reducing the knee torques due the position of the feet in front of the bar (Cotterman, Darby et al., 2005) which reduces the stress at the knee. This theory was supported by the biomechanical model presented by Abelbeck (2002) in which the knee moments decreased and the hip moments increased as the feet were positioned anterior to the hip joint (from the standing position), but contradicts the research by Jacobson (2003) in which the SMS generated greater compressive and AP forces. Our subjects had greater ROM at the joints when using the TBS. The joints ROM during the TBS were similar to those reported in previous studies (Escamilla, Fleisig et al., 1998; Robertson, Wilson et al., 2008). At the instance of MKF there was great deal of variability in the knee and hip joint angles, it was clear that there was less trunk flexion and less ankle dorsiflexion. To achieve the desired ROM in the SMS the subjects had to position the feet farther forward which increased the horizontal distance from the bar to the knee. In the TBS the bar moved forward as the trunk flexed forward to position the load closer to the knee joint. These kinematic changes can lead to more stress to knee joint during the SMS (Abelbeck, 2002; Jacobson, 2003) but further research is needed to confirm this hypothesis.

Figure 5. Horizontal Distance Between Bar and Knee Position at MKF

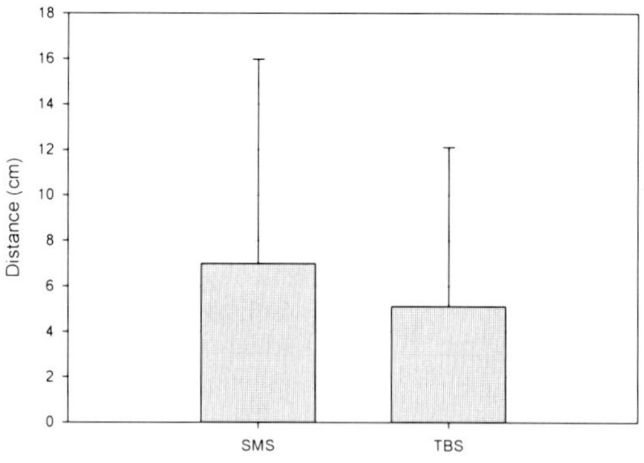

CONCLUSION:

The motions of the SMS are slightly different that the motions of the TBS. The restricted linear motion of the bar affects the ROM of the joints and alters the foot placement in relation to the bar. These changes may affect the kinetics of the lower extremity joints (Abelbeck, 2002; Jacobson, 2003). Knowledge of this information could be beneficial in testing, training, and rehabilitation programs and could have implications for prescribing exercise modalities. Caution should be used to assume that the SMS produces similar motions to the TBS.

REFERENCES:

Abelbeck, K. G. (2002). Biomechanical model and evaluation of a linear motion squat type exercise. *Journal of Strength Conditioning Research,* 16(4), 516-24.

Cohen, J. (1977). *Statistical Power Analysis for Behavioral Science.* New York, Academic Press.

Cotterman, M. L., Darby, L. A. et al. (2005). Comparison of muscle force production using the Smith machine and free weights for bench press and squat exercises. *Journal of Strength Conditioning Research,* 19(1), 169-76.

Escamilla, R. F. (2001). Knee biomechanics of the dynamic squat exercise. *Medicine & Science in Sports Exercise,* 33(1), 127-41.

Escamilla, R. F., Fleisig, G. S. et al. (2001). A three-dimensional biomechanical analysis of the squat during varying stance widths. *Medicine & Science in Sports Exercise,* 33(6), 984-98.

Escamilla, R. F., Fleisig, G. S. et al. (1998). Biomechanics of the knee during closed kinetic chain and open kinetic chain exercises. *Medicine & Science in Sports Exercise,* 30(4), 556-69.

Flanagan, S. P. and Salem G. J. (2007). Bilateral differences in the net joint torques during the squat exercise. *Journal of Strength Conditioning Research,* 21(4), 1220-6.

Glenn, L. (2009) Smith Machine Squat. WWW.MuscleMagfitness.com. Retrieved June 5, 2009, from http://www.musclemagfitness.com/bodybuilding/exercises/smith-machine-squat.html

Gullett, J. C., Tillman, M. D. et al. (2009). A biomechanical comparison of back and front squats in healthy trained individuals. *Journal of Strength Conditioning Research,* 23(1), 284-92.

Jacobson, T. R. (2003). Biomechanical Analysis of Barbell and Smith Machine Squats. Exercise and Sport Sciences. Gainsville, University of Florida. Masters.

Robertson, D. G., Wilson, J. M. et al. (2008). Lower extremity muscle functions during full squats. *Journal of Applied Biomechanics,* 24(4), 333-9.

Russi, B. (2008) 9 least effective exercises. WWW.WEBMD.com, Retrieved June 5, 2009, from http://www.webmd.com/fitness-exercise/features/9-least-effective-exercises

Sahli, S., Rebai, H. et al. (2008). Tibiofemoral joint kinetics during squatting with increasing external load. *Journal of Sport Rehabilitation,* 17(3), 300-15.

Zink, A. J., Perry, A. C. et al. (2006). Peak power, ground reaction forces, and velocity during the squat exercise performed at different loads. *Journal of Strength Conditioning Research,* 20(3), 658-64.

Acknowledgment

The authors would like to thank Dr. Alan Mikesky for his input in the paper and the Diversity Scholars Research Program at Indiana University Purdue University Indianapolis for its funding.

DO PEOPLE WITH UNILATERAL CAM FAI FAVOUR THEIR SYMPTOMATIC LEG DURING MAXIMAL DEPTH SQUATS?

Matthew J Kennedy[1], Mario Lamontagne[1,2], Paul E Beaulé[3]

School of Human Kinetics, University of Ottawa, ON, Canada[1]
Department of Mechanical Engineering, University of Ottawa, ON, Canada[2]
Division of Orthopaedic Surgery, University of Ottawa, ON, Canada[3]

Cam Femoroacetabular Impingement (FAI) is caused by an abnormally convex femoral head-neck junction and can damage the peripheral acetabulum in activities requiring a large hip range of motion (ROM). This study analyzed the three-dimensional (3D) ground reaction forces (GRF) and moments (GRM) and the resultant GRF of the symptomatic and asymptomatic legs in participants with unilateral cam FAI during a maximal depth squat. Seventeen participants with unilateral cam FAI performed 5 maximal depth squats with each leg on a separate forceplate. No significant differences were found between the two legs. These results indicate that participants with cam FAI do not favour their affected leg during maximal depth squats.

KEY WORDS: Femoroacetabular impingement, Biomechanics, Squat, Kinetics

INTRODUCTION: FAI is a cause of hip pain in young active adults and is believed to cause osteoarthritis (OA) of the hip (Ganz et al., 2003; Wisniewski and Grogg, 2006). Up to 24% of highly athletic males may have cam FAI (Murray and Duncan, 1971) and elite athletes with FAI report decreased sport performance (Philippon et al., 2007). Cam FAI is caused by an abnormal bony ridge on the femoral head which is driven into the peripheral acetabulum at the limits of hip ROM (Ito et al., 2001). This repetitive contact particularly during rapid athletic movements can cause the articular cartilage to be sheared off the acetabular rim and can result in labral tears (Beck et al., 2005). The high prevalence of FAI in athletes and the serious damage it can cause make it imperative for us to increase our understanding of this deleterious condition.

Cam FAI causes hip pain in movements which require large hip flexion such as deep sitting, and athletic activities (Wisniewski and Grogg, 2006; Laude et al., 2007), and reduces hip function. Both peak hip ROM and maximal squat depth are reduced in participants with cam FAI as compared to healthy controls (Kennedy et al., 2009 (In Press); Lamontagne et al., 2009). Furthermore, Philippon et al. (2007) found that people with unilateral FAI have reduced passive hip ROM in the symptomatic leg compared to the contralateral leg. Maximal depth squat is a demanding but controlled movement requiring near maximal hip flexion angles in participants with FAI (Flanagan et al., 2003; Kennedy, 2009 (In Press)), making it a good movement to isolate biomechanical differences between symptomatic and asymptomatic legs in unilateral FAI. The purpose of this study was to determine whether or not participants diagnosed with unilateral cam FAI favour their symptomatic leg during strenuous closed chain activities such as maximal depth squats. We postulated that participants with unilateral cam FAI would have a larger resultant GRF in the asymptomatic leg compared to the symptomatic leg.

Figure 1. Maximal depth squat resultant ground reaction force between the asymptomatic and symptomatic legs of participants with unilateral cam FAI. A very similar peak resultant GRF was generated by the symptomatic and asymptomatic legs (p = 0.786)

METHODS: Data Collection: Seventeen participants (7 females; 10 males) diagnosed with unilateral cam FAI participated in this study. All participants were fit and otherwise healthy, with an average age of 35.5 (± 10.6) years, and average BMI of 23 (± 2.3) kg/m^2. Participants were all diagnosed by the same clinician (PB) having a positive impingement test, visible cam morphology and an α angle indicative of FAI as determined by radiographs. Participants were excluded if OA was visible on radiographs. All participants signed an informed consent approved by the Ottawa Hospital and the University of Ottawa Ethics Boards.

After a preliminary stretch and warm-up, participants performed five maximal depth squats with one foot on each of two forceplates (AMTI OR-6, Watertown, MA, USA). Participants stood with their feet shoulder-width apart, their arms extended anteriorly and maintained heel contact throughout the squat. Maximal squat depth was controlled by an adjustable bench set to 1/3 tibial height, and positioned behind the participant. Participants were instructed to squat as low as possible and to ascend back to standing while maintaining control. Maximal depth was attained by touching their buttocks to the bench.

Data Analysis: 3D GRF, GRM and the resultant GRF generated by the asymptomatic and symptomatic legs of each participant were measured for all five trials. These variables were then averaged across trails and ensemble averaged across participants and compared between the two legs using one-way between-group ANOVAs (α = 0.05) for each dependent variable. A one-way between-group ANOVA (α = 0.05) was also run comparing the resultant force between the dominant and non-dominant legs to ensure that leg dominance was not a confounding variable.

RESULTS: There were no significant differences found between the 3D GRFs, GRMs, or in the resultant GRF (Figure 1) between the asymptomatic and symptomatic legs.

The peak 3D GRFs and resultant GRF were very similar between the two legs, with p values ranging from 0.869 to 0.765 for the 3D GRFs, and 0.786 for the resultant GRF. There were also very little differences between the symptomatic and asymptomatic legs for the peak GRMs, with p values ranging from 0.995 to 0.389.

DISCUSSION: There have been no previous kinetic studies comparing symptomatic and asymptomatic legs in participants with unilateral FAI. However, as mentioned previously, people with unilateral FAI have reduced hip ROM in their symptomatic hip compared to their asymptomatic hip (Philippon et al., 2007). Furthermore, people with FAI experience pain in their symptomatic hip(s) during deep sitting (Laude et al., 2007) – which is very similar to deep squats. Both of these findings indicate that people diagnosed with unilateral FAI have limitations in their symptomatic leg compared to their asymptomatic leg.

According to a study conducted in our lab using a similar population to this study, people with unilateral cam FAI cannot squat as low as matched controls. Only 33% of FAI participants could attain the lowest squat depth compared with 91% of the controls (Lamontagne et al., 2009). So few FAI participants being able to squat to the lowest depth indicates that squatting is a demanding activity for this population. Since participants with unilateral FAI have limited passive hip ROM in their symptomatic leg compared to their asymptomatic leg, and experience pain in deep sitting it seems logical that they would in turn favour this leg during strenuous activities requiring large hip mobility such as maximal depth squats. This however was not the case.

Surprisingly, in contradiction to our hypothesis unilateral FAI participants had no differences between the resultant GRF of their symptomatic and asymptomatic legs during maximal depth squats ($p = 0.786$). Furthermore there were no differences in any of the 3D GRFs or GRMs between the two legs. This indicates that this population does not favour their symptomatic leg during maximal depth squats.

One might assume that leg dominance could contribute to the kinetic symmetry between the symptomatic and asymptomatic legs. However Hesse et al. (1996) reported that although asymmetric weight distribution during sit-to-stand is common, it is not related to leg dominance. Furthermore, there was no significant difference in the resultant GRF between dominant and non-dominant legs ($p = 0.286$).

Based on the results from Lamontagne et al. (2009) which used a similar study population, the FAI group squatted using two different strategies, one with hip adduction and the other with hip abduction. This resulted in large kinematic variability in the frontal plane, which would likely be accompanied by large kinetic variability. Squat strategy was not restricted to ensure a natural movement representing participant's usual lower-limb kinetics. Although this may have masked differences in the individual GRFs and GRMs between the symptomatic and asymptomatic legs it would not affect the resultant GRF - our primary measure - since only its magnitude was compared between the two legs without considering the angle of application.

No differences were found in GRFs and GRMs during maximal depth squats indicating that participants with unilateral cam FAI generate the same net force in both legs during strenuous closed chain activities. Whether or not this uniform net GRF is generated using the same muscle contributions from each leg cannot be determined with certainty from this data, limiting its clinical relevance. However, the fact that the resultant GRF and all of the 3D GRF and GRM components were very similar between the two legs suggests similar joint kinetics. Although these are interesting novel results, it should be noted that these findings do not necessarily transfer to the rapid open chain and ballistic activities common in most sports.

CONCLUSION: Since FAI is a serious medical condition which primarily affects active adults and athletes, increasing our understanding of it is very important for sport scientists. Previous research has shown that FAI limits lower limb functionality and sports performance. Furthermore, in unilateral cases it has been shown that the symptomatic hip has a lower passive ROM compared to the asymptomatic hip. Although it was postulated that people with unilateral cam FAI would favour their symptomatic leg during maximal depth squats, this hypothesis was disproven. Unilateral cam FAI participants had very similar GRFs and GRMs between the symptomatic hip and the asymptomatic hip during maximal depth squats. This indicates that people with this condition do not favour their symptomatic hip during strenuous closed chain activities.

REFERENCES:

Beck, M., Kalhor, M., Leunig, M., & Ganz, R. (2005). Hip morphology influences the pattern of damage to the acetabular cartilage: femoroacetabular impingement as a cause of early osteoarthritis of the hip. *The Journal of Bone and Joint Surgery, 87*, 1012-1018.

Flanagan, S., Salem, G. J., Wang, M. Y., Sanker, S. E., & Greendale, G. A. (2003). Squatting exercises in older adults: kinematic and kinetic comparisons. *Medicine and Science in Sports and Exercise, 35*, 635-643.

Ganz, R., Parvizi, J., Beck, M., Leunig, M., Notzli, H., & Siebenrock, K. A. (2003). Femoroacetabular impingement: a cause for osteoarthritis of the hip. *Clinical Orthopaedics and Related Research*, 112-120.

Hesse, S., Schauer, M., Jahnke, M. (1996). Standing up in healthy subjects: symmetry of weight distribution and lateral displacement of the centre of mass as related to limb dominance. *Gait & Posture, 4*, 287-292.

Ito, K., Minka, M. A., 2nd, Leunig, M., Werlen, S., & Ganz, R. (2001). Femoroacetabular impingement and the cam-effect. A MRI-based quantitative anatomical study of the femoral head-neck offset. *The Journal of Bone and Joint Surgery, 83*, 171-176.

Kennedy, M., Lamontagne, M, Beaule, PE. (2009 (In Press)). The Effect of Cam Femoroacetabular Impingement on Hip Maximal Dynamic Range of Motion. *Journal of Orthopedics*.

Lamontagne, M., Kennedy, M. J., & Beaule, P. E. (2009). The effect of cam FAI on hip and pelvic motion during maximum squat. *Clinical Orthopaedics and Related Research, 467*, 645-650.

Laude, F., Boyer, T., & Nogier, A. (2007). Anterior femoroacetabular impingement. *Joint Bone Spine, 74*, 127-132.

Murray, R. O., & Duncan, C. (1971). Athletic activity in adolescence as an etiological factor in degenerative hip disease. *The Journal of Bone and Joint Surgery, 53*, 406-419.

Philippon, M., Schenker, M., Briggs, K., & Kuppersmith, D. (2007). Femoroacetabular impingement in 45 professional athletes: associated pathologies and return to sport following arthroscopic decompression. *Knee Surgery, Sports Traumatology, Arthroscopy, 15*, 908-914.

Philippon, M. J., Maxwell, R. B., Johnston, T. L., Schenker, M., & Briggs, K. K. (2007). Clinical presentation of femoroacetabular impingement. *Knee Surgery, Sports Traumatology, Arthroscopy, 15*, 1041-1047.

Wisniewski, S. J., & Grogg, B. (2006). Femoroacetabular impingement: an overlooked cause of hip pain. *American journal of physical medicine & rehabilitation / Association of Academic Physiatrists, 85*, 546-549.

EFFECTS OF SHORT-TERM SLED TOWING AND UNLOADED SPRINT TRAINING ON LEG POWER AND STIFFNESS

Pedro E. Alcaraz[1], José L.L. Elvira[2], José M. Palao[3], and Vicente Ávila[3]

Department of Sport Sciences. Alfonso X el Sabio University, Madrid (Spain)[1]
Physical Education and Sport Area. UMH, Elche, Alicante (Spain)[2]
Department of Physical Activity and Sport Sciences. UCAM, Murcia (Spain)[3]

KEY WORDS: athletics, strength, specificity, resisted sprint training, stiffness.

INTRODUCTION: The relationship between stiffness and athletic performance is of great interest to the sport and research communities. Unfortunately, there are no longitudinal studies that have investigated the effects of strength or power training on mechanical stiffness in sprinters. The aim of the study was to examine the effect of resisted and unloaded sprint training programs on sprint time, leg power, and stiffness.

METHODS: Eight female and 14 male athletes (100 m PB: 10.5-11.5 s for men, 12.0-13.0 s for women; > 8 training years) participated. After 3 weeks of standardized training, they were randomly assigned to a resisted (RG) or control (CG) group. They performed the same sprint-specific training program for 4 weeks, which included 2 days of strength training and 2 days of sprints per week: 30 m flying and 50 m maximum intensity. The difference in the RG was the use of sled towing (~8% body mass) for the 30 m fly sprints. The pre and post-tests were: 1) 50 m sprints, where interval times were measured with photocells: 0-15, 15-30, and 30-50 m. A 2D photogrammetric analysis of the run at 40 m from the start was done. Leg and vertical stiffness (k) were estimated based on the modelling of the force-time curve from the kinematic variables (Morin et al., 2005); 2) Squat jump (90°) (SJ); and 3) countermovement jump (CMJ) on a force plate. A two-way factorial ANOVA with one between-subjects factor (training group) and one within-subjects factor with repeated measures (times in pre and post test) were used to determine the interaction group x time ($p \leq 0.05$).

RESULTS: Power, sprint time, leg/vertical k of each training group are shown in Table 1.

Table 1. Results of the resisted training and control groups in the pre and post test.

		$SJ_{pow/bw}$ (W·kgf^{-1})	$CMJ_{pow/bw}$ (W·kgf^{-1})	T_{0-15} (s)	T_{15-30} (s)	T_{30-50} (s)	k_{leg} (kN·m^{-1})	$k_{vertical}$ (kN·m^{-1})
Resisted	Pre	55.0 ± 7.6	57.3 ± 8.0	2.32 ± 0.17	1.78 ± 0.15	2.28±0.22	13.4±3.1	156.9±37.9
	Post	55.0 ± 7.8	57.6 ± 8.1	2.24 ± 0.11	1.74 ± 0.14*	2.25±0.19	13.2±3.2	155.8±38.2
Control	Pre	54.5 ± 3.4	52.3 ± 6.3	2.36 ± 0.16	1.70 ± 0.10	2.20±0.14	10.7±2.8	125.6±31.6
	Post	51.8 ± 7.7	54.0 ± 7.3	2.26 ± 0.09	1.68 ± 0.09	2.16±0.11*	11.4±3.0	136.9±33.5

Pow/bw- power relative to body weight; T- sprint time in the three phases of the run; k_{leg} and $K_{vertical}$- leg and vertical stiffness while running. *Significantly different in the main effect Time (p<0.05). No significant differences were found in the interaction group x time.

DISCUSSION: The traditional sprint training improved performance in the maximum velocity phase. The resisted training improved performance in the transition phase. No interaction between the group factor and improvements was found, so we cannot determine that the effect was caused by the training. SJ/CMJ power and k were not increased with any of the methods used. The reasons for these results could be the years of training experience, the short duration of the treatment, and the low volume and/or inadequate load of sled training.

CONCLUSION: The effect of the proposed short-term resisted training demonstrated no difference with the unloaded sprint training in the measured variables. The appropriateness and the way of training with sled towing for trained sprinters should be studied more in depth.

REFERENCES:
Morin, J. B., Dalleau, G., Kyrolainen, H., Jeannin, T., and Belli, A. (2005). A simple method for measuring stiffness during running. J. Appl. Biomech., 21, 167-180.

EFFECTS OF UPHILL RUNNING ON SPRINTING TECHNIQUE IN FOOTBALL PLAYERS

José L.L. Elvira[1], Vicente Ávila[2], José M. Palao[2] and Pedro E. Alcaraz[3]

Physical Education and Sport Area. UMH, Elche, Alicante (Spain)[1]
Department of Physical Activity and Sport Sciences. UCAM, Murcia (Spain)[2]
Department of Sport Sciences. Alfonso X el Sabio University, Madrid (Spain)[3]

KEY WORDS: uphill, strength, specificity, resisted sprint training.

INTRODUCTION: The success in many actions in team sports is determined by the player's ability to develop high speed and acceleration. There are many resisted methods for training the strength within the specific running technique, each one with different application according to the characteristics of the overload. Uphill sprinting is one of these methods. A criticism related to the use of resisted methods is that the athletes may use a modified running technique and so subsequently could alter their movement pattern if repeated in time (e.g. Alcaraz et al. (2008) showed that some methods modify the body lean). The purpose of this work was to clarify the effects of uphill sprinting on variables related to the running performance and technique in football players.

METHODS: Sixteen male football players of a team competing in the 3rd Spanish Football League participated in the study. They performed 30 m horizontal and uphill (4% slope) sprint running, over artificial grass with their daily training shoes. Interval times were measured with photocells (0-10, 10-20, and 20-30 m). A 2D photogrammetric study was conducted in each condition at the 20-30 m interval. Two trials were recorded and the one with the best time was analysed. A T-test for related samples was applied ($p \leq 0.05$) to compare both running situations.

RESULTS: The measured variables for both conditions are shown in Table 1.

Table 1. Results of the performance and sprinting technique variables.

	S_{0-10} (m/s)	S_{10-20} (m/s)	S_{20-30} (m/s)	Stride Length (m)	Stride Rate (Hz)	Body Lean (°)	Recovering Thigh (°)
Horizontal	5.45 ± 0.20	7.75 ± 0.30	8.32 ± 0.39	1.85 ± 0.11	1.14 ± 0.09	21 ± 4	117 ± 4
	*	*	*	*			*
Uphill	5.19 ± 0.20	7.29 ± 0.25	7.53 ± 0.37	1.81 ± 0.13	1.12 ± 0.08	20 ± 4	120 ± 4

S_{0-10}- mean speed in the 0 to 10 m; S_{10-20}- mean speed in the 10 to 20 m; S_{20-30}- mean speed in the 20 to 30 m; Body Lean and Recovering Thigh- recorded at the take-off instant.
*statistically significant differences ($p \leq 0.05$) between horizontal and uphill running.

DISCUSSION: Our results support that uphill running reduces the sprint time and stride length while keeps constant the stride frequency as does the sled towing (16% body mass) (Alcaraz et al., 2008). Nevertheless, an interesting effect of the uphill running is that it keeps constant the body lean angle, but increases the elevation of the thigh when recovering the leg, which is assumed to be beneficial for increasing the stride length.

CONCLUSION: Uphill running appears to be an interesting resisted method to overload the athlete to train for increasing stride length while introducing minimal changes in the body lean.

REFERENCES:

Alcaraz, P. E., Palao, J. M., Elvira, J. L. L., and Linthorne, N. P. (2008). Effects of three types of resisted sprint training devices on the kinematics of sprinting at maximum velocity. J. Strength Cond. Res., 22, 890-897.

The Development of Low Cost Sensor technology to provide Augmented Feedback for On-Water Rowing

DJ Collins[1&2], Ross Anderson[1], Derek O'Keeffe[2]

1 Biomechanics Research Unit, PESS Department, University of Limerick
2 Bioelectronics Research Laboratory, ECE Department, University of Limerick

KEY WORDS: rowing, wireless, feedback, sensor development

INTRODUCTION Investigations to identify ways to improve stroke technique of rowers are on going (Anderson et al 2005). However, the majority of these studies are laboratory based and thus neglect the effects of the oar and water have on the rowing stroke. Traditionally on-water testing has tended to concentrate on the forces produced by the rower at the oar handle, foot stretcher and on boat by the rower but not the actual rowing technique (Hill, 2001). The purpose of this study was to design and develop a system that would provide quantitative feedback of the rowers' technique on the water. This was accomplished through the integration of low cost electronic sensors, specific software interface and use of wireless technology (Anderson & Collins 2004).

METHODS: A full-scale model of a rowing station of a coxless four sweep rowing boat was constructed. This model allowed for the testing and validation of the sensors using a 3D motion analysis system (MotionAnalysis, USA). All sensors were to be fixed to the model as not to hinder the rowers' performance. A rotary potentiometer, fixed to the swivel of the gunwale, was chosen to measure the oar angle. From this position the oar's angular velocity and acceleration is calculated through differentiation. A string based potentiometer was used to measure the blade angle (Celesco, USA). In order to measure the blade angle, the potentiometer was fixed to the oar collar and to the oar. To measure the seat speed and position a second string potentiometer (Celesco, USA) was fixed at the end of the slide rails and attached to the back of the seat. A laptop with a DAQ card (DAQCard-6024E - NI Corporation, USA) acquired the data from the sensors. The information was then transmitted across a wireless network. The received data was then processed using NI LabView 8.2 software (NI Corporation, Texas, USA) to calculate and display the acquired data. The interface allows the coach to select what aspect of the stroke they wish to view in real time. The system was tested in full in a rowing tank at the University of Limerick Boat House before being tested out on the water.

RESULTS: Initial laboratory based results have proved successful (Linear R^2 values of >0.990 when individual sensor data are compared to the Motion Analysis system data) and testing is currently at the rowing tank stage.

CONCLUSIONS: The study will conclude having developed a complete wireless augmented feedback system for on-water rowing.

REFERENCES

Anderson, R. Harrison, A. Lyons, G.M. (2005) Accelerometry-based feedback – can it improve movement consistency in rowing. Journal of Sports Biomechanics 4(2):179-195

Anderson, R. and Collins, D.J. (2004) Are wireless technologies the future for augmented feedback? In: ISEA Conference Proceedings Vol 2, 561-567

Hill, H. (2001) Dynamics of coordination within elite rowing crews: evidence from force pattern analysis. Journal of Sports Sciences 20: 101-117

QUANTITATIVE ANALYSIS ON THE MUSCULAR ACTIVITY OF LOWER EXTREMITY DURING WATER WALKING

Koichi KANEDA[1], Daisuke SATO[2], Hitoshi WAKABAYASHI[3], Yuji OHGI[1] and Takeo NOMURA[4]

Graduate School of Media and Governance, Keio University, Fujisawa, JAPAN[1]
Niigata University of Health and Welfare, Niigata, JAPAN[2]
Kyushu University, Fukuoka, JAPAN[3]
NPO Tsukuba Aqua Life Research Institute, Tsukuba, JAPAN[4]

This study compared water walking (WW) with land walking (LW) in order to evaluate the muscular activities of the lower extremity. Nine young healthy subjects performed WW at voluntary slow, normal and fast speeds for 8 seconds with two repetitions. On the LW condition, subjects performed two trials at normal pace. Surface electromyography electrodes were placed on the tibialis anterior (TA), medial gastrocnemius (GAS), rectus femoris (RF) and biceps femoris (BF). As for WW, each muscular activity patterns at different speeds had moderate or high correlation with LW in cross correlation function (r = 0.53-0.90). The mean value of the muscular activity of GAS at slow speed condition during WW were lower than that of LW. At the fast speed condition, TA, RF and BF activities in WW were higher than that of LW. It was considered that WW was able to simulate LW at any levels of speeds and stimulate thigh muscles and TA sufficiently even in slow speed WW.

KEY WORDS: water, walking, electromyography, similarity.

INTRODUCTION:

Water exercise has become more popular for both the fitness enhancement and the rehabilitation training. In water, buoyancy acts on the body to reduce the gravitational stress at the joints, while water viscosity requires the subject to exert greater force than when moving on land (Miyoshi et al., 2005). Though many studies about muscular activity of water walking (WW) have been conducted and often compared with land walking (LW), almost all the studies were mainly focused on the difference between LW and WW (Kato et al., 2002; Barela et al., 2006). From the viewpoint of rehabilitation training, the similarity of LW and WW should be important. It might be able to suggest the utility of WW as a walking practice. The purpose of this study was to analyze quantitatively muscular activities of the lower extremity during WW compared with LW.

METHODS:

Subject: Nine healthy young males participated in this experiment. Their mean age, height, weight and %fat were 24.9 ± 2.2 yr, 172.0 ± 3.8 cm, 69.3 ± 3.7 kg, and 19.4 ± 4.1%, respectively. Ethics Committee of the Institute of Health and Sports Sciences in the University of Tsukuba approved this study.

Experimental design: Subjects performed LW at normal (comfortable) speed. On the other hand, for the WW, they walked voluntary slow, normal and fast speed for 8 seconds with two repetitions. The water temperature was kept 27 ± 2°C. Depth of the pool was 1.1 m which corresponds to approximately the level of the navel.

Electromyogram recordings: The left lower extremity muscular activity of the tibialis anterior (TA), medial gastrocnemius (GAS), rectus femoris (RF) and long head of the biceps femoris (BF) were measured using a surface electromyography (EMG) electrodes (Mini Ag/AgCl Skin Electrode, NT-511G; Nihon Kohden Corp., Japan) at 1.5 cm intervals of two electrodes. A reference electrode was placed on the clavicle. The skin cuticle was removed carefully using a blood lancet (Blood Lancet; Asahi Polyslider Co. Ltd., Japan) and cleaned with alcohol wipes so that the inter-electrode impedance was less than 20 kΩ. The electrodes were covered with transparent film and putty for waterproofing. Subjects were instructed to perform isometric maximal voluntary contraction (MVC) for 5 seconds for each muscle on land in advance. The raw EMG signals were collected by using the eight-channel multi-telemeter system (WEB-5500 multi-telemeter system; Nihon Kohden Corp., Japan) at a time constant of 0.03 seconds and with 2 kHz sampling rate.

Data analysis: A digital video camera which was synchronized with the EMG was placed on the left side of the subject; it allowed coverage of one cycle from the latter part of the trials at 30 Hz frame rate. Ball-marker with 3cm diameter was attached to the top of the head, inorder to analyze the walking speed. One single cycle EMG data which was defined by the interval between two sequential heel contacts were recorded. The raw EMG data were filtered using 4th-order band-pass filters with cut-off frequencies of 500 Hz and 10 Hz, and subsequently full-wave rectified and low-pass filtered with moving average at 5 Hz to obtain the linear envelope. The muscular activity level was presented at percentages of peak 1-second values of the maximum voluntary contraction (%MVC) for each muscle.

Statistics: Every single cycle duration was normalized into 0% to 100% with a step of 1%, and its data was presented as mean ± standard deviation (SD). A cross correlation function was adapted to compare the muscular activity patterns during one cycle. An one-way repeated measures analysis of variance (ANOVA) with the Dunnet post hoc test was used to compare the mean values of the data. All comparison was conducted between the LW and each speed of the WW. Statistical significance was inferred at $p < 0.05$.

RESULTS:

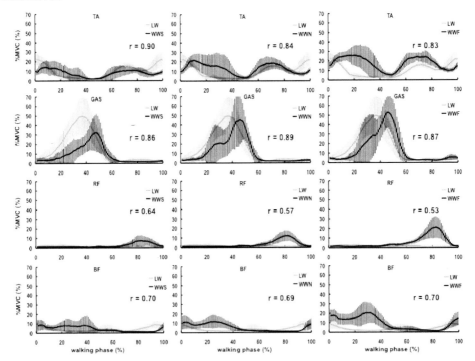

Figure 1. Each muscular activity pattern during one single cycle on LW and WW trials.
Values are mean ± SD.
r: cross correlation coefficient values.
LW: land walking, WWS: water walking at slow speed, WWN: water walking at normal speed, WWF: water walking at fast speed. TA: the tibialis anterior, GAS: medial gastrocnemius, RF: rectus femoris, BF: long head of the biceps femoris.

Muscular activity pattern: Figure 1 shows each muscular activity pattern during one single cycle for LW and WW. The cross correlation coefficients between LW and each speed of the WW were moderate in the RF and BF (r = 0.53 - 0.70), and were high in the TA and GAS (r = 0.83 - 0.90). All coefficient values were significant (p < 0.05).

Mean values of muscular activity: Table 1 represents the mean values of the %MVC during one single cycle for each trial. Significant lower muscular activities of the GAS were found in slow and normal speed WW than that of LW (p < 0.05). The muscular activities of the TA in the normal and fast speed WW were significantly higher than that of the LW (p < 0.05). The muscular activities of the RF were also significantly higher in the normal and fast speed WW than the LW (p < 0.05). The BF activity in the fast speed WW was significantly higher than that of the LW (p < 0.05).

Walking speed: The walking speeds were 1.26 ± 0.14 m/s at the LW, 0.44 ± 0.08 m/s at the slow speed WW, 0.57 ± 0.07 m/s at the normal speed of the WW, 0.72 ± 0.06 m/s at the fast speed WW. All speeds of the WW were significantly slower than that of the LW (p < 0.05).

Table 1. The mean values of the muscular activity during one cycle.

	LW	WWS	WWN	WWF
TA (%)	9.9 ± 2.1	8.1 ± 3.5	12.7 ± 5.0 *	17.5 ± 3.8 *
GAS (%)	15.6 ± 5.4	8.6 ± 3.9 *	11.7 ± 5.1 *	13.8 ± 5.6
RF (%)	1.8 ± 0.8	2.4 ± 1.2	3.5 ± 1.4 *	5.5 ± 2.2 *
BF (%)	4.0 ± 2.2	4.5 ± 2.8	5.5 ± 2.3	9.3 ± 4.2 *

Mean ± SD.
* ; significant difference between LW and each WW at p < 0.05.
LW: land walking, WWS: water walking at slow speed, WWN: water walking at normal speed, WWF: water walking at fast speed. TA: the tibialis anterior, GAS: medial gastrocnemius, RF: rectus femoris, BF: long head of the biceps femoris.

DISCUSSION:

Muscular activity pattern: The muscular activity patterns of the LW and each speed of the WW in the present study were very similar to the previous study which investigated the activitys of TA, GAS, BF, vastus lateralis, erector spine and rectus abdominis during normal speed of WW (Barela et al., 2006). The peak activities of the GAS in WW were relatively 10% later than that of in LW. On the other hand, the TA was activated mainly during the swing phase, and BF was activated during the support phase. Statistical analysis revealed that the significantly moderate to high cross correlation coefficients were obtained in comparison between the LW and the each speed of the WW. The moderate to high cross correlation coefficient indicate that the muscle activity pattern of the WW were similar to that of the LW even in the slow speed WW and fast speed WW. Especially, the correlation coefficients of the RF and BF were moderate, whereas those of the TA and GAS were high at the every speed on the WW. The reason of that in the RF was due to the higher muscular activity in the WW than the LW during the latter phase of one cycle to overcome water resistance for forward swing of lower extremity (Kato et al., 2002). As for the BF, the higher muscular activity in the WW than the LW during the former part of one cycle, which was due to the generation of high propulsive force to overcome water resistance (Barela et al., 2006), would affect to the moderate cross correlation values. In addition, the authors found that the higher muscle activity of the TA in WW than that of on LW. It was considered that subjects should to walk with stable posture during WW (Kato et al., 2002).

Mean values of the muscular activity: The mean values of the muscular activity of the TA and RF were significantly higher in the normal and fast speed WW than that of the LW, and in the BF, the muscular activity was significantly higher in the fast speed WW than LW. The reason of that was to overcome water resistance during forward swing and to generate high propulsive force during stance phase (Kato et al., 2002; Barela et al., 2006) with stable posture (Kato et al., 2002). The mean values of the muscular activity of the GAS were significantly lower in the slow and normal speed WW than the LW. Kato et al (2004) found that the GAS activity during WW was higher than LW in the same walking speed. However, the maximal walking speed of the present study during WW was slower than that of during LW. It was considered that the extremely slow speed in the WW was affected to the GAS activity for generating propulsive force by ankle plantar flexion (Miyoshi et al., 2006). However, there was no significant difference between the slow speed WW and LW among the TA, RF and BF, and between the fast speed WW and LW in the GAS. It was considered that the thigh muscles and TA was able to stimulate sufficiently even in slow speed WW, however GAS could stimulate insufficiently except at the fast speed WW.

Availability of the water walking: Water exercise has become widely used for rehabilitation training for people with lower extremity arthritis and frail elderly people, because of the effect of the buoyancy (Suomi et al., 2000; Sato et al., 2007). From the viewpoint of the rehabilitation training, it must be important to clarify both the similarity and the difference between WW and LW. On one hand, many previous studies have focused on the difference between LW and WW, our present study focuses on the similarity of the muscular activity. As a result, the WW was able to simulate the LW highly regardless of the speed of the WW, and stimulate the thigh muscles and TA sufficiently even in the slow speed WW. It was suggested that the slow speed WW also effective as a walking practice on muscle training especially in thigh muscles and TA with less risk of injury and increased overall safety (Devereux et al., 2005).

CONCLUSION:

This study investigated muscular activity during LW and WW, and compared with the LW and the WW for these patterns and mean values of the muscular activity. As a result, the walking patterns of the WW, even in the slow and fast speed conditions, were highly simulated to that of the LW. Additionally, the slow speed WW could stimulate thigh muscle and TA sufficiently, because there was no difference in the mean values of the muscular activities. Therefore, the WW has a possibility to adopt as a walking practice on muscle training for those who aims to re-acquire walking ability in daily life, such as, frail elderly receiving nursing care.

REFERENCES:

Barela, A. Stolf, S. & Duarte, M. (2006). Biomechanical characteristics of adults walking in shallow water and on land. *Journal of Electromyography and Kinesiology.* 16(3), 250-259.

Devereux, K. Robertson, D. & Brila, K. (2005). Effects of a water-based program on women 65 years and over: a randomized controlled trial. *Australian Journal of Physiotherapy.* 51(2), 102–108.

Kato, T. Sugagima, Y. Koeda, M. Fukuzawa, S. & Kitagawa, K. (2002). Electromyogram activity of leg muscles during different types underwater walking. *Advances in Exercise and Sports Physiology.* 8(2), 39-44.

Miyoshi, T. Shirota,T. Yamamoto, S. Nakazawa, K. & Akai, M. (2005). Functional roles of lower-limb joint moments while walking in water. *Clinical Biomechanics*, 20, 194-201.

Miyoshi, T. Nakazawa, K. Tanizaki, M. Sato, T. & Akai, M. (2006). Altered activation pattern in synergistic ankle plantarflexor muscles in a reduced-gravity environment. *Gait and Posture.* 24(1), 94-99.

Sato, D. Kaneda, K. Wakabayashi, H. & Nomura, T. (2007). The water exercise improves health-related quality of life of frail elderly people at day service facility. *Quality of Life Research.* 16, 1577-1585.

Suomi, R. Koceja, D.M. (2000). Postural sway characteristics in women with lower extremity arthritis before and after an aquatic exercise intervention. *Archives of Physical Medicine and Rehabilitation.* 81(6), 780-785.

STUD LENGTH AND STUD GEOMETRY OF SOCCER BOOTS INFLUENCE RUNNING PERFORMANCE ON THIRD GENERATION ARTIFICIAL TURF

Clemens Müller, Thorsten Sterzing, Thomas L. Milani

Department of Human Locomotion, Chemnitz University of Technology, Chemnitz, Germany

The purpose of this study was to evaluate the influence of different stud lengths and stud geometries of soccer boots on soccer specific running performance. The study involved performance testing by running through two functional traction courses and corresponding subjective testing. Variables of this study were objectively measured running times and perception ratings of running performance. 15 experienced soccer players participated in the study. Players run slower when performing with shorter studs ($p<0.01$). Here, measured running times were reflected by players' perception ($p<0.01$). For stud geometry, bladed studs were more supportive with regard to objectively measured running performance compared to elliptic studs ($p<0.05$). In contrast, no differences were found with regard to players' perception ($p=0.19$). In conclusion, longer stud length provides better traction resulting in better performance. Bladed studs provide a functional traction advantage compared to elliptic studs with respect to running performance.

KEY WORDS: soccer, stud configuration, traction, running time, perception.

INTRODUCTION:

In soccer the player-surface interaction is a central performance aspect, which is influenced by the sole configuration of the boot and the surface. Dependent on external factors like surface conditions suited soccer shoes may vary in stud length. Longer studs are suited for wet and deep surfaces, shorter studs for dry and hard surfaces. These variations are important when playing on natural grass. Additionally, different stud positioning, stud geometries and also different numbers of studs are available for soccer shoes. All these factors may affect playing (running) performance and also injury prevention. Thus, the sole configuration of a soccer boot plays an important role with regard to the functionality of traction properties. Sole configurations that provide unsuitable low traction properties increase the risk of slipping. Sole configurations may also provide unsuitable high traction, thus increasing injury potential. In several studies it was shown that different sole/traction configurations effect running speed (Krahenbuhl 1974, Müller et al. 2008, Sterzing et al. 2009). Krahenbuhl (1974) showed a performance benefit for soft ground design compared to multicleated design and tennis shoes on natural grass. Müller et al. (2008) found out, that plane and not so aggressive stud configurations enable players to run faster on artificial turf. In a series of eight studies Sterzing et al. (2009) summarized performance differences for various sole configurations on different surface conditions.

Since 2004 high quality types of third generation artificial turf have been approved for official game play by the FIFA (FIFA 2007). However, the influence of stud lengths and stud geometries with regard to performance benefit on artificial turf is not well understood and thus need to be investigated.

Therefore, the objective of this study was to examine the effect of different stud lengths and different stud geometries on traction properties during running and to evaluate corresponding perception of players. Furthermore, players were asked to assess the traction suitability of the different shoe conditions on the given surface.

METHODS:

15 experienced soccer players (age: 22.8 ± 1.5 years; height: 175.9 ± 4.2 cm; weight: 70.8 ± 3.4 kg) participated in both studies. The artificial turf used in both studies was *LigaTurf 240 22/4 RPU brown* (Polytan, Burgheim/Germany), which is considered to be state of the art 2-Star artificial soccer turf.

Figure 1: Shoe conditions (a) **Figure 2: Shoe conditions (b)**

(a) Stud length
The basic shoe model was the *Nike Mercurial Vapor II*. Three different stud lengths were investigated: 100% stud length (NM 100), which marked the regular stud length, 50% stud length (NM 50) with studs shortened to half of the regular length and 0% stud length (NM 0) with studs completely removed (no studs). Reduction of stud length was performed by an orthopaedic shoe technician who abraded the studs (Figure 1).

(b) Stud geometry
The basic shoe model was the *Nike Tiempo Premier*. One version had elliptic studs and the other one had bladed studs (Figure 2). A specification of the stud geometry is displayed in table 1.

Table 1: Stud specification of the shoe conditions

	NM 100		NM 50		NM 0		Elliptic		Bladed	
	rearfoot	forefoot	rearfoot	forefoot	rearfoot	forefoot	rearfoot	forefoot	rearfoot	forefoot
Numbers of studs	4	8	4	8	0	0	4	8	4	9
$\sum$ Medial edged contact surface of the studs [cm^2]	8.2	8.4	4.1	4.2	0.0	0.0	3.6	4.6	4.9	4.8
$\sum$ Lateral edged contact surface of the studs [cm^2]	8.2	9.8	4.1	4.9	0.0	0.0	3.6	6.2	4.9	6.4

The shoe upper was equal for shoe conditions of both studies. Both studies incorporated a performance testing and a subjective testing protocol.

1. Running Performance Testing:
For running performance testing the subjects ran through two different soccer-specific functional traction courses, slalom and acceleration, as fast as possible (Figure 3). The slalom course had a total length of 26 m containing 11 cutting and 12 acceleration movements. Subjects had to go through the course three times in each shoe condition. Shoe order was randomized and shoes were changed after each single run. A rest of two minutes was mandatory after each run in order to prevent the subjects from getting fatigued. The acceleration course had a total length of 6 m requiring the subjects to perform maximum acceleration. The testing procedure was the same as for the slalom course. Running times were measured by double light barriers (TAG Heuer, Marin-Epagnier/Switzerland). After performing all the required runs of each running course subjects were asked to give a speed ranking of the investigated shoe condition for the respective course (fastest performance to slowest performance).

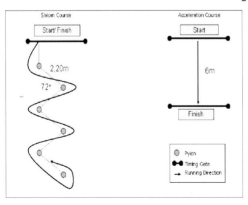

Figure 3: Functional traction courses

2. Subjective Testing of Traction Suitability:
For subjective testing of traction suitability the subjects performed soccer specific movements with the different shoe conditions, e.g. maximum straight accelerations, complete stops, cutting movements to the left and right, and turning movements. The subjects then rated their perceived traction suitability of each soccer shoe model on a nine-point perception scale. The traction suitability testing was categorized in intensity of traction (1-high to 5-neutral to 9-low) and liking of traction (1-good to 5-neutral to 9-bad) (NSRL 2003).
Means and standard deviations for all running time variables were calculated. These were analyzed using a one-way repeated measure ANOVA for comparing shoe-surface characteristics. Post-hoc analyses were applied when appropriate according to Fisher's LSD. The level of significance was set to $p<0.05$ and $p<0.01$. For subjective variables mean ranks (Friedman-test) and medians were calculated.

RESULTS AND DISCUSSION:
(a) Stud length
In both courses subjects run slowest ($p<0.01$) when wearing the NM0 condition (Figure 4). In addition, NM0 was clearly perceived to exhibit the slowest running times in the slalom and in the acceleration task. In the NM100 subjects ran faster ($p<0.01$) which was also clearly perceived by the subjects ($p<0.01$).

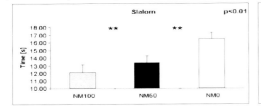 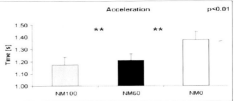

Figure 4: Slalom and acceleration running times - stud length (** $p<0.01$)

The NM100 was rated to be the best suited traction design with regard to traction intensity and traction liking ($p<0.01$) whereas the NM0 was rated to be the least suited traction design among the three shoes in this study.

Table 2: Speed ranking (1-fastest to 3-slowest), Traction rating (1-high/good to 9-low/bad)

		NM100	NM50	NM0
Speed ranking slalom	[mean rank]	1.07	1.93	3.00
Speed ranking acceleration	[mean rank]	1.13	1.87	3.00
Traction intensity	[median]	1	5	9
Traction liking	[median]	2	5	9

(b) *Stud geometry*
With the bladed design the subjects ran faster in the slalom course ($p<0.05$) compared to the elliptic design. In the speed ranking the subjects perceived statistically no differences between the two types of shoes. In acceleration task no statistical differences were found between the two types of shoes with regard to running time and speed ranking.

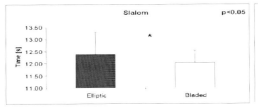 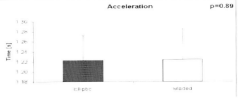

Figure 5: Slalom and acceleration running times - stud geometry (* $p<0.05$)

The subjects rated the bladed design to be the better suited traction design with regard to traction intensity and traction liking (p<0.01).

Table 3: Speed ranking (1-fastest to 2-slowest), Traction rating (1-high/good to 9-low/bad)

		Elliptic	Bladed
Speed ranking slalom	[mean rank]	1.69	1.31
Speed ranking acceleration	[mean rank]	1.70	1.30
Traction intensity	[median]	5	2
Traction liking	[median]	5	3

Lowering the regular stud length on artificial turf influenced subjects' running performance negatively. This is most likely due to the increased risk of slipping during dynamic acceleration and cutting movements. Therefore, it is assumed that subjects performing with NM0 showed more cautious movement behaviour. This movement adaptation strategy resulted in weaker performance (longer running times). The bladed design had a positive effect on running performance compared to the elliptic design. This might be due to the bigger contact surface (Table 1) of the bladed studs allowing more dynamic sideward propulsion. Sterzing & Hennig (2005) also demonstrated a traction benefit for bladed studs compared to elliptic studs. With regard to traction perception, the objectively measured running times are reflected by subjects perception.

CONCLUSION:
Performance and perception testing showed differences between the investigated shoe designs. Stud length affects traction properties of soccer shoes with regard to running performance. This shows that lowering the original stud length while maintaining the regular number of studs decreases performance. In further studies stud lengths between 50% and 100% should be investigated or the length of studs may be even increased in order to quantify the range of stud length that offers best traction. Especially the latter approach needs to have extraordinary caution as injury potential with increase of stud length will increase too. Stud geometry also affects running performance. In the slalom course the bladed stud design shows better performance which is reflected by players' perception. Thus, the geometry of studs is an important factor to develop high level traction conditions. Further enhancement of lateral and medial contact surfaces of bladed studs may result in better performance.

REFERENCES:
FIFA (2007). FIFA – laws of the game 2007/2008. www.fifa.com, (29.11.2007).

Krahenbuhl, G.S. (1974). Speed of movement with varying footwear conditions on synthetic turf and natural grass. *Research Quarterly* 45 (1).

Müller, C., Sterzing, T. & Milani, T.L. (2008). Comprehensive evaluation of player-surface interaction on artificial soccer turf during cutting movements. *26. Symposium International Society of Biomechanics in Sports*, Seoul, Korea.

Nike Sport Research Lab (2003). Product testing and sensory evaluation. Sport Research Review 2.

Sterzing, T., Müller, C., Hennig, E.M. & Milani, T.L. (2009). Actual and perceived running performance in soccer shoes - a series of eight studies. *Footwear Science* (accepted, in press).

Sterzing, T. & Hennig, E. (2005). Stability in soccer shoes: The relationship between perception of stability and biomechanical parameters, *Science and Football V*, T. Reilly, J. Cabri, D. Araujo, London/New York, Routledge Taylor and Francis Group.

EFFECTS OF CONCENTRIC VERSUS ECCENTRIC TRAINING ON MUSCLE STRENGTH AND NEUROMUSCULAR ACTIVATION

Chun-Han Tseng[1], Yu-Ru Kuo[1], Chi-Huang Huang[1] and Heng-Ju Lee[2]

Graduate Institute of Athletic Training & Health Science, National Taiwan Sport University, Taoyuan, Taiwan[1]

Department of Physical Education, National Taiwan Normal University, Taipei, Taiwan[2]

KEY WORDS: isokinetic muscle contraction, electromyography, strength training

INTRODUCTION: Eccentric contraction (EC) involves fewer motor units but produces more tension than concentric contraction (CC) (Kay, 2000). Both EC and CC training can stimulate strength gain (Miller, 2006). However, it is not clear whether one method is more effective than the other and the effect of each training on motor units recruited after training. The purpose of this study was to compare the effects of EC and CC isokinetic training exercises on quadriceps muscle strength and neuromuscular activations.

METHODS: Sixteen healthy subjects (8 males/8 females) were randomly divided into two groups: EC training group (ECTG, mean age 23.3 ± 2.9 yrs) and CC training group (CCTG, mean age 22.8 ± 3.0 yrs). Isokinetic muscle strength training was performed on right knee at 120°/s, 3 sets of 10 reps at 80% maximal efforts, 3 days a week for 6 weeks. Surface EMG electrodes were placed on rectus femoris of each subject. Isokinetic dynamometer (Biodex systems 3) and surface EMG system (MP150) were used to collect data simultaneously at 1000 Hz. Before and after 6 weeks training period, maximum voluntary isometric knee extension torque was obtained at 60° of knee flexion, and maximum isokinetic CC/EC torque was obtained at 120°/s. Knee extension torque and EMG signals were collected total range of motion from 10°-100° of knee flexion (0° = full extension). Data were analyzed from 30°-70° range during maximal isokinetic contraction. EMG signals were digitally filtered (bandwidth 10–450 Hz) and full wave rectified. Torque data were normalized by body weight and EMG data were normalized by EMG signals collected during MVC. Two way ANOVA was used to compare torques and EMG data between groups, and between pre and post-test. (α=.05).

RESULTS: Significant differences were found for both contraction types after CCTG and ECTG except EC after CTG (Table 1). Normalized EMG data demonstrated no significant results through different training modes and contraction types.

Table 1 : Normalized Torque (Nm/kg) of CCTG and ECTG at Pre/Post Test for CC and EC Test Modes; △%: represents reduction rate between pre and post

Group	CC			EC		
	pre	post	△%	pre	post	△%
CCTG	1.32 ± 0.39	1.77 ± 0.43*	44%	2.37 ± 0.37	2.63 ± 0.53	11%
ECTG	1.30 ± 0.33	1.70 ± 0.28*	36%	2.03 ± 0.53	2.76 ± 0.35*	42%

DISCUSSION: The difference between CCTG and ECTG only occurred during EC. For CCTG, it only demonstrated greater torque production when contracted at the same training type. The reason for no significant difference in EMG data might be that both groups adopted the same amount of motor units recruitment, and therefore the efficiency of producing force was better after training (Seger & Thorstensson, 2005).

CONCLUSION: The level of motor units recruited was not increased after 6 weeks isokinetic training. Muscle strength increased significantly after both training modes, but concentric muscle strength was only enhanced in CCTG.

REFERENCES:
Kay, D., St Clair Gibson, A., et al. (2000). Different neuromuscular recruitment patterns during eccentric, concentric and isometric contractions. *Journal of Electromyography and Kinesiology, 10*, 425-31.
Miller, L.E., Pierson, L.M., et al. (2006). Knee extensor and flexor torque development with concentric and eccentric isokinetic training. *Research Quarterly for Exercise & Sport, 77*, 58-63.
Seger, J.Y., & Thorstensson, A. (2005). Effects of eccentric versus concentric training on thigh muscle strength and EMG. *International Journal of Sports Medicine, 26*, 45-52.

ASSISTED AND RESISTED SPRINT TRAINING MAY REDUCE ACTIVE DRAG IN SWIMMERS IN AN AEROBIC TRAINING PHASE

Per-Ludvik Kjendlie[1] og Tommy Pedersen[1]

[1]Norwegian School of Sport Sciences, Department of Physical Performance, Oslo, Norway

KEY WORDS: Swimming, biomechanics, active drag, assisted sprint, resisted sprint

INTRODUCTION: Active drag is one of the most important performance determining factors of swimming. Often, swimmers and coaches use assisted and resisted sprint training to enhance performance and one hypothesis is that sprinting with supramaximal velocity may, over time, help the swimmer to learn drag reducing mechanisms. The aim of this study was to investigate whether assisted and resisted sprint swimming may influence active drag.

METHODS: A randomised, matched-control intervention design was used, including 9 swimmers in the assisted – resisted sprint (SP) and 9 swimmers in the control group (CON). The subjects were national level swimmers of both genders, aged 18.0 ±1.5 years. The perturbation method of Kolmogorov & Duplisheva (1992) was used to measure the active drag. Sprint swimming velocity was measured using a swimming speedometer attached to the swimmers hip, measuring instantaneous velocity, one trial swimming freely and one trial towing the perturbation buoy. The velocity difference between the two conditions and the drag characteristics of the perturbation buoy was used to calculate active drag. The subjects were trained regularly for 8 weeks in an aerobic phase of the season. SP group did additional sprinting 3 times a week using surgical tubing to resist and assist (achieving supramaximal velocities) their swimming. CON trained identically, and sprinted without assistance or resistance from the surgical tubing.

RESULTS: A mean difference in relative change between pre and post testing active drag coefficient were 6.9% (±23.7%) for the SP and -12.8% (±32.1%) for CON. However, there was no statistical difference between the two groups mean pre-post difference (p=0.20), probably due to large variations within each group, and a relatively small sample.

CONCLUSION: Although not statistically different, the numerical trend of our data may indicate that using additional assisted and resisted sprint training may reduce the active drag of swimmers in an 8 week long aerobic training phase.

REFERENCES:
S. V. Kolmogorov and O. A. Duplisheva. Active drag, useful mechanical power output and hydrodynamic force coefficient in different swimming strokes at maximal velocity. J.Biomech. 25 (3):311-318, 1992.

DIFFERENCES BETWEEN CONCENTRIC AND ECCENTRIC CONTRACTION INDUCED MUSCLE FATIGUE

Chun-Han Tseng[1], Yu-Ru Kuo[1], Chi-Huang Huang[1] and Heng-Ju Lee[2]

Graduate Institute of Athletic Training & Health Science, National Taiwan Sport University, Taoyuan, Taiwan[1]

Department of Physical Education, National Taiwan Normal University, Taipei, Taiwan[2]

KEY WORDS: MPF, eccentric contraction, EMG.

INTRODUCTION: Studies of neuromuscular activation often evaluated through isometric contractions. However, this type of contraction may not truly represent muscle actions during activities. EMG analysis is not only used to determine motor unit activations, but also used to determine muscle conduction velocity by transforming signals into frequency spectrum. Studies have shown that fatigue mucles produced a relativly slower conduction velocity measured by mean power frequency (MPF). Therefore, the purpose of this study was to compare the effects of muscle fatigue generated by two different types of contraction. We hypothesized that muscle fatigue generated by concentric contractions (CC) would cause gretaer muscle contraction frequency reduction than eccentric contractions (EC).

METHODS: Seventeen healthy male subjects (aged 18-31 y/o) were recruited for this study. Each subject required to attend two test sessions (CC/EC) that were separated by one week. During each test section, subjects were seated and right forearms were fixed onto an isokinetic dynamometer (BIODEX Medicine System). Each test section included two measurements (pre and post fatigue test) and one fatigue protocol (CC or EC). For each test, measurements were taken before (pre) and immediately after (post) the fatigue protocol. At the pre-test, isometric MVC torque of elbow flexor and biceps EMG signals at 90° of elbow flexion were be collected simotaneously. During the fatigue protocol, subjects were asked to execute elbow flexors concentricaly or eccentricaly throught 30-120° of elbow flexion at angular velocity of 45°/s. The definition of muscle fatigue was joint torque reduced to 50% of MVC torque. All data were synchonized and collected by BIOPAC MP150 system. EMG signals were digitally filtered (bandwidth 10–450 Hz) and transferred to power spectrum to calculate mean power frequency (MPF). Paired t-test was used for statistical analysis.

RESULTS: Elbow flexion torque of EC had a siginificant greater reduction than CC. However, CC had a significant greater reduction of bicpes brachii MPF than EC(Table 1).

Table 1 Joint torque (Nm) and MPF (Hz) at pre and post test of CC and EC protocol. △: represents difference between pre and post test

	CC			EC		
	pre	post	△Pre-post	pre	post	△Pre-post
Torque	55.4±10.1	39.8 ± 8.9	15.6±2.9	53.7 ±10.0	33.8 ± 8.5	19.9 ± 5.7*
MPF	111.9±11.4	87.6 ± 9.1	24.3±8.9*	112.2 ±12.9	100.7 ± 11.6	11.4 ±13.6

DISCUSSION: EC fatigue protocol could cause more torque reduction than CC, which would possibly explain that fatigue muscle strained more frequently during eccentric contraction. The greater reduction of MPF caused by CC fatigue protocol could possibly result from reduction of neuromuscular conduction velocity (Kay, D., et al., 2000).

CONCLUSION: Muscle fatigue generated by CC or EC demonstrated two different effects on torque production and muscle contraction frequency measured by MPF. Both torque reduction and slower MPF were often used to be the indicator of muscle fatigue. However, data from the current study revealed that different types of muscle contractions could cause different results of muscle fatigue indicators.

REFERENCES:
Kay, D., St Clair Gibson, A., Mitchell, M. J., Lambert, M. I., & Noakes, T. D. (2000). Different neuromuscular recruitment patterns during eccentric, concentric and isometric contractions. *J Electromyogr Kinesiol*, 10(6), 425-431.

AN ORIGINAL INVERSE KINEMATICS ALGORITHM FOR KAYAKING

Vincent Fohanno[1], Mickaël Begon[2], Floren Colloud[1] and Patrick Lacouture[1]

Laboratoire de Mécanique des Solides, Université de Poitiers, Poitiers, France[1]
Département de Kinésiologie, Université de Montréal, Montréal, Canada[2]

KEY WORDS: Inverse Kinematics, Global Optimization, Kayak.

INTRODUCTION: For some sports, e.g. kayaking and rowing, ecological conditions represent a challenge for collecting three-dimensional kinematics. The lower-limbs are partially hidden by the boat and motion analysis systems used in laboratory are not suitable for outdoor and on-water measurements. An inverse kinematics (IK) approach has been proposed where a few tasks would be measured by inertial sensors or inclinometers (Begon & Sardain, 2007). However, due to the lack of actual kinematics, the authors could not assess its reliability. The purpose of this study is to assess the accuracy of this IK algorithm by comparison with a standard algorithm of global optimization.

METHODS: A kayaker equipped with 88 markers performed six trials of paddling on an ergometer. The kinematics were collected with a 10-cameras motion analysis system at 300 Hz (T40, Vicon, Oxford, UK). A 17-segment 42-degree of freedom (*dof*) chain model was defined and personalized for the kayaker. The IK algorithm was composed of two parts to determine the joint kinematics based on both upper-limb with paddle and lower-limb closed loop constraints. The lower-limbs had as much *dof* as tasks (n=7). The tasks were the ankles positions and the pelvis rotation in the horizontal plane. Using an iterative procedure, the inversion of the Jacobian matrix (**J**) gave the configuration which satisfied the task equation. For the upper-limbs, the control space was larger than the task space (22 *dof* versus 7 tasks). The tasks were the rotation of the scapular girdle in the horizontal plane, the paddle position and orientation. The algorithm involved a damped pseudo-inverse of **J** and an optimization term to keep the joint angles as far as possible of the joint limits by projecting its gradient into the null space of **J**. The Root mean square differences (*RMSd*) for each joint angle time history between the IK algorithm and a standard algorithm of global optimization were calculated and expressed in degrees.

RESULTS and DISCUSSION: The average *RMSd* were 6°, 10° and 8° for the trunk, upper-limb and lower-limb *dof* respectively. These results correspond to errors reported in locomotion (Reinschmidt et al, 1997). In paddling, kayakers fully extend elbows and knees. These configurations close to singularities were handled by a pseudo-inverse of **J** and the optimization term. Due to the large control space, a lexicographic method should improve the kinematic agreement (Marler, 2005). The next step will be to collect the tasks using inertial and position wireless sensors for a population of elite kayakers.

CONCLUSION: This IK method can be applied to many sports in order to estimate the athlete's kinematics using a few tasks measured by wireless sensors.

REFERENCES:

Begon, M. & Sardain, P. (2007). Simulation of paddling an kayak ergometer. *Mathematical modelling in sport*, IMA2007 – Manchester 24-26 June 2007.

Reinschmidt, C., et al (1997). Effect on skin movement on the analysis of skeletal knee joint motion during running. *Journal of Biomechanics*, 30, 729-32.

Marler, R.T. (2005). A study of multi-objective optimization methods for engineering applications. *Thesis*, University of Iowa, Iowa city, IA.

Acknowledgement:
The financial support of Région Poitou-Charentes and EU (CPER 2007-2013) is gratefully acknowledged.

ESTIMATION OF KNEE EXTENSION MOMENT CONSIDERING VELOCITY EFFECT AND MUSCLE ACTIVATION USING TENDON SLACK LENGTH OPTIMIZATION

Hyun Woo Uhm[1], Woo Eun Lee[1] and Yoonsu Nam[3]

The Department of Mechatronics Engineering, Graduate School, Kangwon National University, Chun-Cheon, Republic of Korea[1]
School of Mechanical and Mechatronics Engineering, College of Engineering, Kangwon National University, Chun-Cheon, Republic of Korea[3]

This study presents a model to estimate knee extension moment considering muscle velocity effect and muscle activation. The muscle tendon force is very sensitive to the tendon slack length. To predict tendon slack length, exact muscle parameters of a human are needed. But it is difficult to measure all of the muscle parameters from human body. So we propose the algorithm which finds the tendon slack length of quadriceps for more accurate estimation of knee extension moment. Finally considering muscle velocity effect and muscle activation, knee extension moment is estimated. Algorithm was embodied by MATLAB optimization toolbox. And it is evaluated by using an experimental data.

KEY WORDS: tendon slack length, optimization, EMG, knee joint moment.

INTRODUCTION: This research is the basic research of developing gait orthosis for rehabilitation of a handicapped person. Orthosis should grasp the intention of human movement. By analyzing this intention, proper force will be estimated and delivered to human. For this process, accurate muscle skeleton model parameters are needed to estimate proper force, i.e. joint moment. Also muscle-tendon length which is changed by joint movement and moment arm should be searched. But those parameters can be hardly found for a specific person. S. L. Delp referred reported data until 1990, and defined muscle-tendon parameters of 43 muscles related to movement from young cadavers. Using these data he developed SIMM(Software for Interactive Musculoskeletal Modeling) which is muscle-skeleton movement analysis program. L. L. Menegaldo using this program, he proposed a function expressed by muscle length and variation of joint moment arm with the angle of joint.

But these parameters can not be a method of prediction for joint moment because every human has different values of muscle parameters. When we change tendon slack length, l_t^s and maximum isometric muscle force, F_0^m which are the most sensitive parameters related to muscle tendon force(F^t), it is expected that more accurate analysis will be executed.
In this paper, to solve this problem we will discuss about the method to decide tendon slack length and maximum isometric muscle force which is the most important element, using specified person's knee joint isometric extension moment in MVC(Maximum Voluntary Contraction) condition. And finally using optimized tendon slack length, optimized scale factor for maximum isometric muscle force, velocity effect and muscle activation, isokinetic extension moment can be estimated.

METHOD: Data Collection: Using a dynamometer, CON-TREX MJ system, isometric knee extension moment and isokinetic concentric knee moment were measured during the experiment. Isometric extension moment was collected in MVC (Maximum Voluntary Contration) condition. The process of isometric moment experiment was that a subject put forth his strength 2 times to extensional direction for 5 seconds of each one. Isometric knee extension moment was measured at $10°$, $30°$, $50°$, $60°$, $70°$, $80°$, $90°$ and $100°$.

The isokinetic moment was collected at 30, 60 and 180 deg/s. Surface type EMG sensors are on the thigh to RF(Rectus Femoris), VM(Vastus Medialis) and VL(Vastus Lateralis) during the whole session. So muscle activations, a(t) can be calculated from processing those EMG signals. The subject sitting on the dynamometer at 85° hip angle had executed experiment only changing knee joint angle. And each maximum moment of each joint angle was selected for estimating knee extension moment.

Data Analysis: The Hill type muscle model was used for estimating joint moment.

$$F_t = \{f_{act}(a(t), \tilde{l}_m) f(v) + f_{psv}(\tilde{l}_m)\} F_0^m \cos\phi \quad (1)$$

Where a(t) is muscle activation, f_{act} is active muscle force term, f_{psv} is passive muscle force term, $f(v)$ is muscle force related to muscle contraction velocity, F_0^m is maximum isometric muscle force, $\tilde{l}_m$ is normalized muscle length and ϕ is pennation angle.

Equation (1) can be derived into equation (2) considering isometric contraction condition which is that muscle activation is 1 in MVC (Maximum Voluntary Contraction) condition and muscle velocity is 1 due to no movement during the isometric contraction.

$$\overline{F^t} = \{f_{act}(\tilde{l}_m)|_{a(t)=1} + f_{psv}(\tilde{l}_m)\} \cos\phi \quad (2)$$

$$\overline{F^t} = \begin{cases} 1480.3\varepsilon^2 & 0 < \varepsilon < 0.0127 \\ 37.5\varepsilon - 0.2375 & \varepsilon \geq 0.0127 \end{cases} \quad (3)$$

$$\varepsilon = \frac{l^t - l_s^t}{l_s^t} = \frac{l^{mt}(\theta) - l_o^m \tilde{l}_m \cos\phi - l_s^t}{l_s^t} \quad (4)$$

Where l^t is tendon length
l_s^t is tendon slack length
l^{mt} is muscle tendon length
l_o^m is optimal muscle length

For the reliable estimated knee joint moment, model parameter values for each muscle are needed. Also Menegaldo's formula have to be applied to variation of muscle length and moment arm. But changes of muscle-tendon length due to muscle contraction or relaxation or each muscle character parameters are not measurable. However if the structure of muscle-skeleton is similar to each other, appropriate size scaling can help using data from S. L. Delp or L. L. Menegaldo. The optimization algorithm was built for finding tendon slack length and scale factor k which is minimizing the differences of calculated knee extension moment from experimental data by MATLAB optimization toolbox. The algorithm is designed on the assumption that muscle force characteristics of each individual can be expressed by controlling values of tendon slack length(l_s^t) of each muscle and maximum isometric force(F_0^m). To minimize the number of optimization variables, just one scale factor (k) is used for RF, VM and VL instead of optimizing F_0^m of every muscle. Delp's data of tendon slack length and k=1 were used for initial values of optimization. Each optimized tendon slack length and k from the optimization algorithm in figure 1 are listed in table 1.

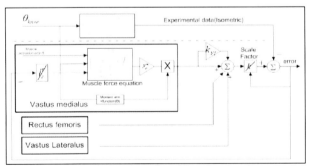

Figure 1: An optimization algorithm for tendon slack length and moment scale factor.

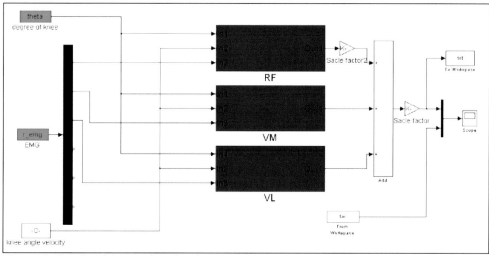

Figure 2: An algorithm estimating knee extension moment

RESULTS: After getting each optimized tendon slack length of RF, VM and VL and optimized scale factor k, to estimate isokinetic extension moment, input parameters of l'_s, k, a(t), knee angle, muscle velocity and muscle parameters were allocated in the algorithm in figure 2.

The result of estimated knee isokinetic extension moment after performing the algorithm in figure 2 is showed in figure 3

Although the RF, VM, VL and VI(Vastus Intermedius) is the most influential muscle to generating knee extension moment, only RF, VM and VL were used due to not attaching an EMG sensor on VI. Assuming that PCSA (Physiological Cross Sectional Area) ratio of each muscle related knee joint extension will be invariable for each individual, a scale factor for VI considering PCAS ratio is multiplied to RF because of RF having similar patterns of the EMG signal to VI.

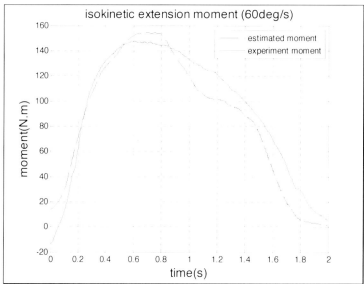

Figure 3: Estimated knee isokinetic extension moment in 60deg/s

Table 1 optimized tendon slack length and scale factor

	A: Initial values (Delp)	B: optimized values	(A-B)/A(%)
l'_{sRF}	0.346	0.3239	6.387
l'_{sVM}	0.126	0.1299	-3.095
l'_{sVL}	0.157	0.1490	5.096
k	1	1.4361	-43.61

In table 1, l'_s and k which were calculated after the optimization process by using MATLAB optimization toolbox in figure 1 are indicated. Searching range for tendon slack length is ±15 and ±50 for scale factor. The initial tendon slack length values of each muscle were refered to the paper of S. L. Delp.

DISCUSSION:

To estimate joint moment, precise musculotendon parameters are needed. However it is pretty difficult to know accurate musculotendon parameters of specified person. Optimization algorithm for estimating joint moment was designed with finding values of tendon slack length and scale factor k. Change of searching range of tendon slack length and k makes considerable differences between experimental moment and estimated moment. The best matching factors are selected for the optimization.

CONCLUSION: Tendon slack length is the most sensitive and important parameter to determine muscle tendon force. So adjusting tendon slack length of the muscles and scale factor k, expected knee extension moment is approximately matched with experimental data. Muscle activation derived from processing EMG signals and muscle velocity also used for algorithm of estimating knee extension moment.

This study can help analyzing the intention of movement by using EMG signals and kinematic information to help handicapped person with gait orthosis.

REFERENCES:

Fleischer, C. & Hommel, G. (2006). Torque Control of an exoskeletal knee with EMG signals, *Proceedings of the Joint Conf. on Robotics: ISR 2006 and Robotik 2006.*
Delp, S. L. (1990). Surgery simulation: A computer graphics system to analyze and design musculoskeletal reconstructions of the lower limb, *Ph.D. dissertation, Stanford University.*
Menegaldo, L. L., Fleury, A. de T. & Weber, H. I. (2004). Moment arms and musculotendon lengths estimation for a three-dimensional lower limb model. *Journal of Biomechanics*, Vol.37, pp.1447-1453.
Buchanan, T. S., Llyod, D. G., Manal, K., & Besier, T. F. (2004). Neuromusculoskeletal modeling: estimation of muscle forces and joint moments and movements from measurements of neural command. *Journal of applied biomechanics*, Vol. 20, pp. 367-395.
Anderson, D. E., Madigan, M. L. & Nussbaum, M. A. (2007). Maximum voluntary joint torque as a function of joint angle and angular velocity: model development and application to the lower limb. *Journal of Biomechanics.*
Yoonsu Nam & Woo Eun Lee. (2008). Developing a model to estimate knee joint moment in MVC condition, *Journal of biomedical engineering research.* Vel. 29 no.3, pp.222~230

Acknowledgement
This work was supported by Korean Science & Engineering foundation (R01-2008-000-20375-0).

EVALUATION OF HOCKEY HELMET PERFORMANCE BY FINITE ELEMENT MODELING

Andrew Post[1], Blaine Hoshizaki[1] and Michael Gilchrist[2]

[1]Neurotrauma Impact Science Laboratory, University of Ottawa, Ottawa, Canada
[2]University College Dublin, Dublin, Ireland

KEY WORDS: Finite element analysis, Modeling, Hockey injury, Concussion

INTRODUCTION: Since the advent of helmet use in ice hockey the incidence of traumatic brain injury (TBI) has decreased, however the prevalence of mild traumatic brain injury (mTBI) has not (Wennberg and Tator, 2003). Recently finite element modeling (FEM) has been used in an attempt to identify mTBI thresholds from an impact using shear stress strain (SSS) and other parameters to aid in reducing these injuries (Zhang et al., 2004). The following study employs the University College Dublin Brain Trauma Model (UCDBTM) to evaluate the ability of vinyl nitrile (VN) and expanded polypropolene (EPP) hockey helmets to reduce the risk of brain injury.

METHODS: A helmeted and unhelmeted hybrid III headform/neck, instrumented with a 3-2-2-2 accelerometer array, was impacted three times at 5.3 ± 0.05 m/s (Padgaonkar et al., 1975). The location of impact was front boss, with a negative azimuth, the weight of the impactor was 16.9kg. The results, in x, y and z components of linear and angular acceleration were inputted into the UCDBTM.

RESULTS AND DISCUSSION: The results indicated that while the general location of peak SSS did not change, its magnitude differed between the three conditions. The no helmet condition had a peak of 0.17 ± 0.001 mm/mm, the EPP helmet peak was 0.17 ± 0.002 mm/mm, and the VN helmet peak was 0.13 ± 0.005 mm/mm (Figure 1).

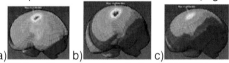

a)   b)   c)

Figure 1: Impact to hybrid III headform for, a) no helmet, b) EPP helmet, and c) VN helmet.

While the SSS area was reduced between the no helmet and helmet condition, there was a reduction in peak SSS for the VN helmet where the EPP and no helmet remained the same. It should be noted that the peak linear and angular acceleration inputs into the model did not differ significantly, however, the slope of the curve to peak was steeper for the EPP helmet over the VN helmet. This would support the research that SSS is rate dependent (Zhang et al., 2004).

CONCLUSION: The UCDBTM was successful at showing how the use of a helmet can decrease the chance of brain strains. The results also suggest that there is a difference in how VN and EPP function in relation to actual brain injury, separate from peak acceleration values.

REFERENCES:
Wennberg, R.A. and Tator, C.H. (2003). National Hockey League Reported Concussions, 1986-87 to 2001-2002. *Canadian Journal of Neurological Science,* 30 (3), p206-209.

Zhang, L., Yang, K.H., & King, A.I. (2004). A Proposed Injury Threshold for Mild Traumatic Brain Injury. *Journal of Biomechanical Engineering,* 126, 226-236.

Padgaonkar, A.J., Kreiger, K.W., and King, A.I. (1975). Measurements of Angular Accelerations of a Rigid Body using Linear Accelerometers. *Journal of Applied Mechanics,* 42, p 552-556.

A KINEMATIC COMPARISON OF RUNNING ON TREADMILL AND OVERGROUND SURFACES

Kam-Ming MOK, Justin Wai-Yuk LEE, Mandy Man-Ling CHUNG, Youlian HONG

Human Movement Laboratory,
Department of Sports Science and Physical Education,
Faculty of Education,
The Chinese University of Hong Kong, Hong Kong SAR, China

Treadmills are often used for research in sport running shoes and physical training. The purpose of this study was to compare the kinematics in treadmill and overground running, and to investigate if the shoe testing on treadmill can reflect the performance of the running shoes on overground surfaces. Thirteen male subjects were recruited to run on treadmill, tartan, grass and concrete surfaces. Effective vertical stiffness, temporal and kinematic parameters were measured. The results showed that the running patterns within overground surfaces were not significantly different, while the significant differences were found between treadmill and overground running.

KEY WORDS: treadmill, shoe testing, overground, kinematics

INTRODUCTION: Treadmills are often used for research in sport running shoes and physical training. Human performing similar running pattern on treadmill compared to overground is necessary. Otherwise, the change of the running pattern may affect the research findings and training effects. Therefore, the increase in the usage of treadmill on scientific investigations and physical trainings raises the discussion on the difference in running pattern between treadmill and overground running. Previous studies (Riley et al., 2008; Schache et al., 2001; Novacheck, 1998; Wank et al., 1998; Nigg et al., 1995) reported the body kinematics on treadmill running and overground running. Those studies showed similar results on hip and knee kinematics, however, contradictory results on truck lean angle. Although the change of truck lean angle was preliminarily explained by the external drag force from the belt on treadmill (Wank et al., 1998), the mechanism is still unclear. The ankle kinematics is important information on sport running shoe testing. Previous studies reported ankle kinematics on sagittal plane; however, the rearfoot motion is not well-investigated. On the other hand, previous studies only included hard surfaces such as tartan and concrete as overground. The difference between treadmill running and soft overground running, such as grass surface, is not well known. The aim of the present study is to compare the kinematics in treadmill and overground running. Overground surfaces are tested, including tartan, grass and concrete, to investigate if the shoe testing on treadmill can reflect the performance of the running shoes on overground surfaces. This study provides information to judge the validity on human running simulation on treadmill in all aspects.

METHODS: Thirteen male subjects aged 22.4 ± 3.9 years (body height, 1.70 ± 0.06 m; body mass, 63.6 ± 9.2kg) were recruited from The Chinese University of Hong Kong. All subjects were heel-toe runners. Each subject wore a standard running shoe model (TN600, ASICS, Japan). Six minutes warm up and familiarization session (12 km/hr) on treadmill was provided to subjects before testing. They were tested on a treadmill (6300HR, SportsArt, US), tartan, grass and concrete. Tartan, grass and concrete were overground surfaces in different surface stiffness. Figure 1 shows the experimental set up on field. For the sagittal kinematic analysis, markers were attached to acromion process, greater trochanter, lateral fermoral condyle, lateral malleolus, heel and first metatarsal head. For rearfoot, the Achilles tendon and centre of calf were marked. Two posterior markers on shoe formed a vertical line in unloading condition. The running speed on treadmill was 3.8m/s. The acceptable running speed on overground surfaces was 3.6~4.0 m/s for each trial. An infra red timing system (Brower, US) was used to monitor the running speed on overground surfaces of each trial. In the testing, the average running speed of subjects on overground running was 3.85 m/s. Two high-speed cameras (DVL9600, Panasonic, Japan) with 50 Hz captured the motion from sagittal and rear view. Regarding each subject, five running trials from each of four different surfaces were processed. Temporal and kinematic parameters were calculated and outputted by motion analysis system (APAS, Ariel Dynamics, US). The kinematic data were smoothes using a fourth-order low-pass Butterworth filter with cut-off frequency of 8 Hz. The events of initial foot contact and toe off were defined from the trajectory of heel marker and first metatarsal head marker respectively, with the aid of visual identification on video recordings. The temporal parameter includes stride time, stance time. For sagittal view, the kinematic parameters included the range, the maximum value in one stride, the minimum value in one stride, the value at initial foot contact (IFC) and the value at toe off (TO) of trunk angle, hip angle, knee angle and ankle angle. For rear view, touch down angle, total range of motion, relative maximum pronation and peak pronation velocity were measured. Effective vertical stiffness was calculated from subject body mass and stance time by the method suggested by Cavagna (1988).The parameters were tested with a one-way repeated measure ANOVA (P<0.05), post-hoc test Tukey. The data were analyzed by SPSS statistical software (SPSS 15.0, SPSS Inc., US).

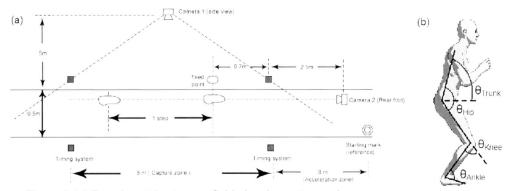

Figure 1: (a) Experimental set up on field, drawing not to scale
(b) Definition of kinematic parameters

RESULTS: Statistical analysis showed that most of the kinematic parameters in sagittal plane were found to be significantly different (P<0.05) between treadmill running and overground running. No significant difference was found between tartan, grass and concrete surface. Statistical analysis showed that there was no significant difference on the rearfoot motion parameters. Table 1 shows the temporal and kinematic parameters with significant difference.

DISCUSSION: Significant differences were found on the stride length and temporal parameters between treadmill and overground running. Stride length, stride time and stance time were smaller in treadmill running when compared to overground running. The results were consistent to previous studies (Schache et al., 2001; Wank et al., 1998) although our study was limited by 50Hz video motion analysis. In treadmill running, the backward moving belt provided an external backward drag force on the foot. Then, that external backward drag force assisted the runner to complete the stance phase. As a results, the stance time decreases and the stride time was shortened.

Table 1 Mean and S.D. of temporal and kinematic parameters (stance phase in % of stride; duration in seconds; angle in degrees; length in metres; vertical stiffness in kN/m; N=13)

Parameters	Treadmill	Tartan	Grass	Concrete
Stride Length	2.38(0.47)	2.74(0.16)[a]	2.76(0.15)[a]	2.78(0.11)[b]
Stride Time	0.67(0.027)	0.71(0.025)[b]	0.70(0.026)[b]	0.71(0.030)[b]
Stance Time	0.196(0.142)	0.225(0.020)[b]	0.221(0.021)[b]	0.228(0.167)[b]
Stance Phase	29.55(2.73)	31.72(3.26)[a]	31.99(2.64)[a]	32.63(2.74)[b]
θ_{Trunk} Range	10.9(3.50)	19.3(5.55)[b]	18.8(5.33)[b]	20.1(5.00)[b]
θ_{Trunk} Max	88.8(4.02)	84.9(4.95)[b]	85.4(5.08)[b]	86.6(4.84)[b]
θ_{Trunk} Min	77.9(3.18)	65.5(7.11)[b]	66.5(7.59)[b]	66.4(5.67)[b]
θ_{Trunk} IFC	87.3(3.66)	74.3(7.70)[b]	75.0(8.15)[b]	74.4(6.41)[b]
θ_{Trunk} TO	78.8(3.23)	66.3(6.83)[b]	67.4(7.30)[b]	67.2(5.29)[b]
θ_{Hip} Range	48.9(4.88)	56.4(5.49)[b]	58.1(6.52)[b]	55.2(6.44)[b]
θ_{Hip} Max	185.9(5.16)	181.5(6.53)[b]	183.0(6.81)[a]	180.8(6.27)[b]
θ_{Hip} Min	137.0(5.67)	125.0(8.23)[b]	124.8(8.68)[b]	125.5(5.58)[b]
θ_{Hip} IFC	145.7(5.07)	128.4(9.39)[b]	128.7(9.72)[b]	128.2(7.16)[b]
θ_{Knee} Range	80.5(9.81)	89.2(11.9)[a]	89.9(9.89)[b]	88.0(10.6)[a]
θ_{Knee} Max	106.2(8.50)	113.2(10.2)[a]	113.2(9.58)[a]	109.4(9.04)
θ_{Knee} IFC	28.9(3.81)	41.0(7.38)[b]	41.3(7.57)[b]	38.6(7.03)[b]
θ_{Knee} TO	41.7(5.55)	28.1(5.38)[b]	27.3(4.37)[b]	27.0(5.89)[b]
θ_{Ankle} Range	42.1(7.50)	51.7(9.06)[b]	51.9(11.6)[b]	48.4(7.12)[b]
θ_{Ankle} TO	100.9(4.32)	113.2(10.6)[b]	111.6(7.56)[b]	112.6(7.84)[b]
Effective Vertical Stiffness	65.18(12.30)	50.49(11.16)[b]	50.45(10.37)[b]	47.15(8.05)[b]

[a] $P<0.05$ when compared with Treadmill
[b] $P<0.01$ when compared with Treadmill

The ankle angle at toe off was found to be less in treadmill running when compared to overground running. It can be explained that the external backward drag force from the treadmill belt assisted the foot to complete the take off motion with less plantarflexion. This change in running pattern leads to different plantar pressure distribution. A study on the difference in plantar pressure distribution between treadmill and overground running is needed.

Significance differences were found on the parameters of trunk angle between treadmill running and overground running. Less forward lean of trunk was found on treadmill running. In treadmill running, center of gravity of the body (CG) is not required to be moved forward. Novacheck (1998) suggested that greater forward trunk lean would move the CG forward of the support foot in stance. A greater horizontal GRF could be exerted against the running surface (Novacheck, 1998). However, the CG of runner is not required to move forward in treadmill running, and thus less horizontal GRF is required. Therefore, trunk was less forward leaned in treadmill running compared to overground running. It implied that runners who frequently trained on treadmill may have their running pattern changed to less truck forward lean because they used to not moving the CG forwards.

Stiffness during running has been related to the risk for bony injuries such as knee osteoarthritis and stress fractures. (Butler et al., 2003; Granata et al., 2002, Grimston et al., 1991).Increased stiffness is typically associated with reduced lower extremity excursions and increased peak force (Butler et al., 2003). The effective vertical stiffness values on overground running in this study were similar to those reported by other study (Arampatzis et al., 1999). Effective vertical stiffness in treadmill running was found to be significantly higher than that in overground running. It appears that there is a potential injury risk in treadmill running and should be studied in the future.

CONCLUSION: The running pattern changed significantly between treadmill running and overground running. It appears that treadmill running is not able to simulate overground running. Further investigation on trunk motion and plantar pressure in treadmill and overground running is needed. Moreover, the effective vertical stiffness was found to be higher in treadmill running. Research on the impact of the higher effective vertical stiffness in treadmill running is valuable.

REFERENCES:

Arampatzis, A., Brüggemann, G.P. & Metzler, V. (1999). The effect of speed on leg stiffness and joint kinetics in human running. *Journal of Biomechanics*, 32, 1349-1353.

Butler, J.R., Crowell III, H.P. & Davis, I.M. (2003). Lower extremity stiffness: implications for performance and injury. *Clinical Biomechanics*, 18, 511-517.

Cavagna, G.A., Franzetti, P., Heglund, N.C. & Willems, P. (1988). The determinants of the step frequency in running, totting and hopping in man and other vertebrates. *Journal of Physiology*, 399, 81-92.

Granata, K.P., Padua, D.A., & Wilson, S.E. (2002). Gender differences in active musculoskeletal stiffness. Part II. Quantification of leg stiffness during functional hopping tasks. *Journal of Electromyography and Kinesiology*, 12, 127-135.

Grimston, S.K., Ensberg, J.R., Kloiber, R. & Hanley, D.A. (1991). Bone mass, external loads, and stress fractures in female runners. *Journal of Applied Biomechanics*, 7, 293-302.

Nigg, B.M., De Boer, R.W. & Fisher, V. (1995). A kinematic comparison of overground and treadmill running. *Medicine and Science in Sports and Exercise*, 27, 98-105.

Novacheck, T.F. (1998). The biomechanics of running. *Gait Posture*, 7, 77-95.

Pinnington, H.C. & Llord, D.G. (2005). Kinematic and electromyography analysis of submaximal differences running on a firm surface compared with soft, dry sand. *European Journal of Applied Physiology*, 94, 242-253.

Riley, P.O., Dicharry, J., Franz, J., Croce, U.D., Wilder, R.P. & Kerrigan D.C. (2008). A kinematics and kinetic comparison of overground and treadmill running. *Medicine & Science in Sports & Exercise*, 40, 1093-1100.

Schache, A.G., Blanch, P.D., Rath, D.A., Wrigley, T.V., Starr, R. & Bennell, K.L. (2001). A comparison of overground and treadmill running for measuring the three-dimensional kinematics of the lumbo-pelvic-hip complex. *Clinical Biomechanics*, 16, 667-680.

Wank, V., Frick, U. & Schmidtbleicher, D. (1998). Kinematics and Electromyography of Lower Limb Muscles in Overground and Treadmill Running. *International Journal of Sports Medicine*, 19, 455-461.

Acknowledgements
We wish to express gratitude to Mr. Philip Ngai, Mr. Lin Wang, Miss. Sally Tsang and Mr. Ricky Sung for data collection and technical supports.

EFFECTS OF STATIC STRETCHING, PNF STRETCHING, AND DYNAMIC WARM-UP ON MAXIMUM POWER OUTPUT AND FATIGUE

Joshua Aman[1] and Bryan Christensen[2]

[1]University of Minnesota, Minneapolis, MN, U.S.A
[2]North Dakota State University, Fargo, ND, U.S.A

The purpose of this study was to determine the effects of static stretching, PNF stretching, and dynamic warm-up on maximum power output and fatigue. Ten participants were recruited to perform a vertical jump test at 3 minutes and 20 minutes post-treatment for all treatments until voluntary fatigue. Participants performed a standard protocol including one of the stretching/warm-up treatments followed by two repeated, counter-movement, vertical jump tests. Results of the study showed no statistically significant differences in maximum power output although the dynamic warm-up group resulted in a 10% and 9% higher average output compared to the control group. Results also showed no statistically significant differences in percent decline in power output as well as time to voluntary fatigue, although there was up to a 6 s difference between treatments and the control group. Although this study concluded with no statistical significance, an argument could be made for applicable significance.

KEY WORDS: PNF, static stretching, dynamic warm-up, power output, fatigue.

INTRODUCTION: Pre-physical activity stretching and/or warm-ups are used in almost every activity or sporting event. Many recent studies have claimed that stretches such as static and proprioceptive neuromuscular facilitation (PNF) stretching are hindering, rather than aiding, to maximum power output (Behm et al., 2006; Cramer et al., 2005, Marek et al., 2005). Dynamic warm-up does not appear to increase range of motion (ROM) as much as a stretching exercises, however, beneficial effects of performing an active warm-up prior to supramaximal exertion have been previously shown (Bergh & Ekblom, 1979; O'Brien et al.,1997). Two primary hypotheses have been developed to explain the stretching-induced strength deficit: (1) mechanical factors, such as alterations in the viscoelastic properties of the muscle that may affect the length/tension relationship, and (2) neural factors, such as decreased motor unit activation, firing frequency, and/or altered reflex sensitivity (Cramer et al., 2005; Marek et al., 2005). Since the tendon has viscoelastic properties, it is able to change in length without immediately 'springing' back to its original length. The compliance of the tendon has been viewed as both positive and negative in that a more compliant tendon may aid in reducing injuries, where as a less compliant tendon may aid in increase performance (Åstrand et al., 2003). To maximize performance as well as ensure maximal possible safety of those involved, stretching procedures should be studied to identify the appropriate techniques to prepare properly an individual for physical exertion. Therefore, the purpose of this study was to determine the effects of static stretching, PNF stretching, and dynamic warm-up on maximum power output and fatigue. The following research hypotheses were tested: (1) there will be a difference in power output following the static stretching, PNF stretching, and dynamic warm-up treatments when compared to the no treatment condition, (2) there will be a difference in fatigue following the static stretching, PNF stretching, and dynamic warm-up treatments when compared to the no treatment condition, (3) there will be a difference in power output and fatigue when comparing values of the three-minute post-treatment tests to the 20-minute post-treatment tests.

METHODS: Ten participants, five males and five females, with average age of 19.8 yrs (±1.8), were recruited to participate in this study. Participants had an average mass of 75.6 kg (±15.5 kg). Each participant completed a control trial (no stretching) as well as each treatment in random order, each on a separate day. Treatments consisted of a pre-testing protocol utilizing static stretching, PNF stretching, dynamic warm-up, or no treatment (control). The order of each session was as follows: (1) Perform a warm-up on a stationary bicycle for three minutes instructed as a "slow, easy peddle," (2) perform one of the treatment protocols or no treatment (control), followed by a three-minute rest, (3) perform a repeated, counter-movement, vertical jump test until voluntary fatigue, (4) sit quietly until 20 minutes post-treatment, and (5) repeat vertical jump test. Duration of the treatment protocol lasted approximately three minutes. During the control trial, participants sat quietly in place of the stretching or dynamic warm-up exercise.

Protocols: The static and PNF stretching protocols being used have been established as an appropriate amount of stretch time to increase ROM (Bandy et al., 1997; Shrier & Gossal, 2000). Both the static and PNF stretching routines were designed to stretch the knee flexors, hip extensors, knee extensors, hip flexors, and plantar flexors. During the static and PNF stretching protocols, each muscle group was stretched with the participant in either the supine or prone position. For the static stretch protocol, each stretch was performed twice and held for 30 s with a 10 s rest between stretches.

In the PNF stretching protocol, a contract-relax method was used. Participants were stretched by the researcher for 15 s, followed by a concentric contraction of the stretched muscle against the resistance of the researcher for 6 s, followed by a second passive stretch of the participant by the researcher.

The dynamic warm-up protocol used was established by the literature to effectively increase muscle temperature (Bergh & Ekblom, 1979; Sargeant, 1987; Stewart et al., 2003). Drills performed were in the following order: high knee drill (slow jog), gluteus kicks (slow jog), stationary body squats (x10), high knee lunge (walking), carioka (slow jog), forward and lateral leg swings (x10 each).Each warm-up drill was performed twice for approximately 10 s, with a 10 s rest between sets.

Data Analysis: Maximum power output and time to voluntary fatigue were recorded during the vertical jump performance using a customized portable force plate (AMTI Inc., Boston, MA) amplified at 200 Hz. Percent decline in power output was also used to measure fatigue and was calculated by determining their change in power output, from their max power output to minimum power output. All values recorded for both the three-minute and 20-minute post-treatment performance tests were compared between each treatment using a repeated mesures ANOVA test followed by a Tukey post hoc test. SPSS Version 14.0 statistical software was used to analyze data. An alpha level of ≤0.05 was accepted as significant.

RESULTS: The following figures (1-3) show means and standard deviation bars of all four treatments as well as trial 1 versus trial 2 for maximum power output (Fig. 1), percent decline of maximum power output (Fig. 2), and time to voluntary fatigue (Fig. 3). There were no statistical differences between any of the treatments in trial 1 nor in trial 2 for maximum power output, percent decline of power output, and time to voluntary fatigue.

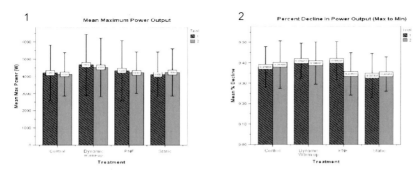

Fig. 1: Average maximum power output for all four treatments in trial 1 and trial 2.
Fig. 2: Average percent decline of power output for all four treatments in trial 1 and trial 2.

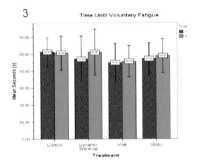

Fig. 3: Average time to voluntary fatigue for all four treatments in trial 1 and trial 2.

Data analysis of the results also showed no statistically significant differences between trial 1 and trial 2 for each treatment. Therefore, all three hypotheses were rejected.

DISCUSSION: Results of this study indicate that different stretching and warm-up techniques had no statistically significant influence on power output nor fatigue, three or 20 minutes post-treatment. This study did not concur with many previous studies that reported statistically significant changes due to stretching and warm-up protocols. Some results, however, were similar to previous studies in a practical sense. Bergh and Ekbolm (1979) showed a correlation between an increase in muscle temperature and an increase in vertical jump and peak power output. In this study, the dynamic warm-up treatment, which has been shown to increase muscle temperature, had an average max power output of 4672 W three minutes post-treatment. After 20-minutes post-treatment, which would most likely result in a muscle temperature cool-down, max power output dropped to 4520.9 W. The control group had an average maximum power of 4201 W (10% lower than the dynamic warm-up condition) in trial 1 and an average maximum power output of 4113 W (9% lower than the dynamic warm-up condition) in trial 2. According to the equation of Sayers et al. (1999) for estimating peak power, an increase of 470 W of power output, similar to the difference between the dynamic warm-up and control group in trial 1, would increase a 90 kg person's vertical jump from 76 cm to approximately 84 cm, a dramatic increase in performance for an athlete.

In this study, all treatments in trial 1 produced quicker times to fatigue when compared to the control group by at least 4 s. Specifically, the PNF treatment was 6.38 s shorter in duration when compared to the control group in trial 1. In trial 2, the PNF group was 4.96 s quicker to fatigue than the control group. Although not statistically significant in this study, in any given anaerobic physical activity, 4-6 s is a long period of time. Therefore, this data set may indicate PNF stretching has a detrimental effect on performance and this effect may last up to 20 minutes post-stretch. Just as well, the static stretching treatment resulted in a time to fatigue of 4.06 s shorter than the control group in trial 1. This may indicate that static stretching may also have a detrimental effect on performance immediately following stretching. Also noteworthy was the difference from trial 1 to trial 2 in time to fatigue for the dynamic warm-up treatment. The quicker time to fatigue in trial 1 (by 4 s) may indicate that a higher muscle temperature, and ultimatley a dynamic warm-up, may be related to a quicker time to fatigue.

The PNF condition showed the highest percent decline of power output in trial 1 at 41%. Interestingly, the PNF treatment also showed the largest change from trial 1 to trial 2, resulting in a 34.5% decline in trial 2. Percent decline in the control group remained relatively unchanged. This may indicate that PNF stretching had an effect on performance at three minutes post-stretch. The static treatment showed the lowest percent decline in power output in both trials (~34% for trial 1 and trial 2). Overall, although not statistically significant, an argument could be made that all treatments had an influence on fatigue, particularly in trial 1, and that the PNF stretching treatment may have had a larger impact on fatigue than static stretching or dynamic warm-up.

CONCLUSION: Results of this study may have applicable relevance to practitioners including athletes and coaches. The data set, although not statistically significant, indicates that performing a dynamic warm-up prior to an activity may allow for a greater maximum power output performance during that activity compared to performing a stretching routine or no routine at all prior to an activity. However, the data set also indicates a dynamic warm-up, PNF, and static stretching may all lead to a quicker time to fatigue when compared to the control group. The dynamic warm-up treatment also showed the greatest difference in time to fatigue when comparing trial 1 to trial 2. This may indicate that allowing a rest period (20 minutes or greater) would certainly be beneficial for time to fatigue, but it may have a detrimental effect on power output. Based on trends of this study data, recommendations could be made for performing a dynamic warm-up prior to any activity reliant on maximum power output, and stretching routines be reserved for post-activity. Future research may benefit from having a set amount of time each individual will test for and then measuring percent decline of power output that occurred in that set amount of time. To receive a more accurate measure of stretching, future research may need to measure angles of joints when stretching participants so as to limit some of the variability in relying on the participants feeling of "slight discomfort." Increasing the number of participants would also give rise to data with more statistical power.

REFERENCES:

Åstrand, P. O., Rohahl, K., Dahl, H. A., & Strømme, S. B. (2003). Fatigue. In M. S. Bahrke, M. Schrag, K. Bernard, D. Campbell, J. L. Davies, J. Anderson, & K. Bojda (Eds.), *Textbook of work physiology: Physiological bases of exercise* (4th ed., pp. 454-477). New York: McGraw-Hill.

Bandy, W. D., Irion, J. M., & Briggler, M. (1997). The effect of time and frequency of static stretching on flexibility of the hamstring muscles. *Physical Therapy, 77*(10), 1090-1096.

Behm, D. G., Bradbury, E. E., Haynes, A. T., Hodder, J. N., Leonard, A. M., & Paddock, N. R. (2006). Flexibility is not related to stretch-induced deficits in force or power. *Journal of Sports Science and Medicine, 5*, 33-42.

Bergh, U., & Ekblom, B. (1979). Influence of muscle temperature on maximal muscle strength and power output in human muscles. *Acta Physiologica Scandinavica, 107*, 33-37.

Cramer, J. T., Housh, T. J., Johnson, G. O. Ebersole, K. T., Perry, S. R., & Bull, A. J. (2000). Mechanomyographic and elctromyographic responses of the superficial muscles of the quadriceps femoris during maximal, concentric isokinetic muscle action. *Isokinetic Exercise Science, 8*, 1826-1831.

Marek, S. M., Cramer, J. T., Fincher, A. L., Massey, L. L., Dangelmaier, S. M., Purkayastha, S., et al. (2005). Acute effects of static and proprioceptive neuromuscular facilitation stretching on muscle strength and power output. *Journal of Athletic Training, 40*, 94-103.

O'Brien, B., Payne, W., Gastin, P., & Burge, C. (1997). A comparison of active and passive warm-ups on energy system contribution and performance in moderate heat. *Australian Journal of Science and Medicine in Sport, 29*, 106-109.

Sayers, S. P., Harackiewicz, D. V., Harman, E. A., Frykman, P. N., & Rosenstein, M. T. (1999). Cross-validation of three jump power eqations. *Medicine & Science in Sports & Exercise, 31*, 572-577.

Sargeant, A. J. (1987). Effect of muscle temperature on leg extension force and short-term power output in humans. *European Journal of Applied Physiology, 56*, 693-698.

Shrier, I., & Gossal, K. (2000). Myths and truths of stretching: Individualized recommendations for healthy muscle. *The Physician and Sportsmedicine, 28*(8), 57-63.

Stewart, D., Macaluso, A., & De Vito, G. (2003). The effect of an active warm-up on surface EMG and muscle performance in healthy humans. *European Journal of Applied Physiology, 89*, 509-513.

Acknowledgements
The authors would like to thank Athletic Republic for supplying the equipment for testing purposes as well as those athletes who volunteered for testing.

KINEMATICS ANALYSIS OF TWO STYLES OF BOW FOR MARTIAL ARTISTS AND AVARAGE STUDENTS

Kenji Kawabata, Yusuke Miyazawa, Soichi Shimizu, Takahiko Sato, Yoshinori Takeuchi[1], Ryoichi Nishitani[1], Naotoshi Minamitani[2] and Hiroh Yamamoto[1]

Biomechanics Lab., Graduate School of Ed., Kanazawa Univ., Kanazawa, Japan
Biomechanics Lab., Fac. of Ed., Kanazawa Univ., Kanazawa, Japan[1]
Center of Development for Education, Hokuriku Univ., Kanazawa, Japan[2]

> The aim of this study was to conduct operating analysis of two styles of bow (RITSUREI and ZAREI) for martial artists and average students, and to compare their kinematics characteristic of. Subjects of this study were healthy men (n=33) who were martial artist group (MA, n = 18) and average student group (AS, n = 15). Subjects conducted RITSUREI and ZAREI three times each. Results of paired t-test, on angular degree parameter of RITSUREI, MA made smaller angular degree of the upper limbs and angular degree of the neck than that of AS. regarding time parameter components, the total time of ZAREI, from beginning of it to stopping, from stopping of it to end, and from beginning of RITSUREI to stopping were significantly shorter than that of AS. MA made several characteristic shown in text. BUDO bow is regarded as practical bow even in social rule when bow teaching is conducted

KEY WORDS: RITSUREI, ZAREI, kinematics analysis

INTRODUCTION: Bow has been done frequently not only in daily life but also in all sports scene. In Japan, athletes tend to bow at the beginning and ending of all sports scene because it is quite natural to bow at both beginning and ending in BUDO (martial arts) rule. Because they have an martial art consciousness "start with bow and finish up with bow" since old days in Japan, and so Japanese people have more occasions than foreign people to perform bow. Recently, the way the bow is done is taught to social freshman, is learning social etiquette, and many children to let them know social moral.

However, how to bow is different from person like how to walk and throw. It seems that sports experience which each person has is making this kind of difference. Especially, martial artists regard bowing as important thing, so it is considered that their bow is different from that of other sports players. The aim of this study was to conduct operating analysis of RITSUREI (standing bow) and ZAREI (sitting bow) of martial artists and average students, and to compare their kinematics characteristic.

METHODS: Subjects of this study were healthy men (n=33) who consisted martial artist group (MA, n=18) and average student group (AS, n=15). MA consisted of JUDO expert, KENDO expert and AIKIDO expert. Informed consent was obtained from subjects. Subjects were attached 10 markers on top of head, ear, Seventh Cervical spine(C7), shoulder, elbow, Anterior Superior Iliac Spain(ASIS), center of Posterior Superior Iliac Spain (PSIS), great trochanter, knee and lateral malleolus to operating analysis on sagittal plan(Figure 1). Markers were attached on only the right side of the body except for top of head and C7. Subjects conducted six trials that consisted of three times each RITUREI and ZAREI after rehearsal. (Figure 2, 3). This experiment was conducted using two-dimensional analysis. Front-back direction was regarded as the X-axis, and vertical direction relative to the ground was regarded as Y-axis.

On completion of the experiment, it was expected that bow angular degree and how long bow is done would differ between subjects. Therefore, in this study, object of bow was regarded as person who was respected to unify experiment condition. Factors that were composed of total bow time of each subject, time from beginning to stopping, stopping time, time from stopping to end, angular degree of the upper limbs, angular degree of the neck, and how far great trochanter moved when standing. Comparisons were carried out by unpaired t-test with Welch corrections. Post hoc coefficient correlation was calculated by Pearson product-moment correlation coefficient. Significant level was set at $p<0.05$.

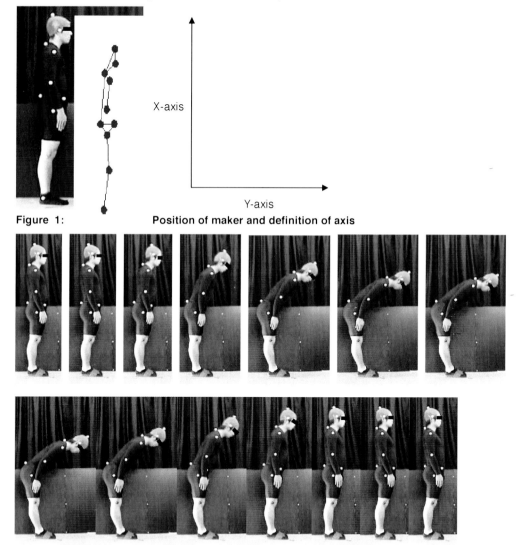

Figure 1: Position of maker and definition of axis

Figure 2: Posture change with sagittal plane of right in RITSUREI

Figure 3: Posture change with sagittal plane of right in ZAREI

RESULTS: Regarding angular degree parameter of RITSUREI, MA made smaller angular degree of the upper limbs and angular degree of the neck than that of AS. In the angular parameter of ZAREI, there was no significant difference (Table1). In time parameter, components of total time of ZAREI, from beginning of it to stopping, from stopping of it to end, and from beginning of RITSUREI to stopping were significantly shorter than that of AS (Table2). In all components, average moving backward distance (±) at standing:12.9 ±3.9cm by AS, 12.1 ±3.3cm by MA. There was no significant difference between AS and MA in this parameter.

Table 1: Average amount and standard deviation of angular degree parameter in AS and MA

	AS(n=18) mean	S.D.	MA(n=15) mean	S.D.
ZAREI				
Maximum angular degree of upper limb(deg)	84.4	6.6	83.8	8.7
Maximum angular degree of neck(deg)	35.0	10.2	30.0	15.3
RITSUREI				
Maximum angular degree of upper limb(deg)	60.0	11.1	45.3 *	12.6
Maximum angular degree of neck(deg)	21.6	11.7	12.9 *	11.3

Table 2: Average amount and standard deviation of time parameter in AS and MA

	AS (n=18)		MA (n=15)	
	mean	S.D.	mean	S.D.
ZAREI				
Total (s)	4.89	0.86	3.96*	0.49
Biginning – Stopping (s)	1.80	0.44	1.28*	0.14
Stopping (s)	1.16	0.46	0.96	0.28
Stopping – End (s)	1.95	0.32	1.72*	0.20
RITSUREI				
Total (s)	3.55	0.82	3.04	0.66
Biginning – Stopping (s)	1.21	0.41	0.84*	0.34
Stopping (s)	0.95	0.35	0.93	0.22
Stopping – End (s)	1.38	0.32	1.27	0.29

DISCUSSION AND CONCLUSIONS: Regarding angular degree parameter, it is important to keep flexion from 15 degree to 30 degree, to keep direction of eyes constant, not to flex only head, and not to flex too much (edo et al., 1985. son, 1987). In this study, average maximum angular degree of loin conducted by MA was bigger than that of one shown in text. However, this typical characteristic of bow in BUDO made their flexion lower than that of NS. It is said that NS tended to make deeper-flexion when they bow (Morishita et al., 1985). This is one kind of reason that there was significant difference.

In the time parameter, how long bow was conducted by MA was obviously shorter than that of NS. This is because MA provided for bow sooner compare for NS. In business rule, it is one way to provide for bow sooner. This is because this way is easier to show one's feeling of appreciation, respect and deference. That is reason why BUDO bow is regarded as practical bow even in social rule when bow teaching is conducted.

In displacement parameter, it is regarded as better to keep our mode of behavior natural in BUDO rule. This is reason why it is thought that bow conducted by MA tended to make low angular degree. However, this hypothesis was not established because there was no significant difference in two groups. From medical perspective, BUDO bow is less stressful for loins than that of conducted by NS, and it is expected that that kind of bow is suitable for person who have prior history of back pain.

REFERENCES:
T. Edo, K. Tsumura, K. Sizawa, H. Yano and K. Watanabe. Actual fight of KENDO. 1985
H. Morishita, N. Iwashita. Motion, business-bow, compliments. Science of physical education. 1985; 35(11): 823-826
N. Minamitani, Yoichi K. A study of Terminological History on" SEIZA". Journal of Hokuriku University. 2006; 29: 97-114

KINEMATICAL ANALYSIS OF 110M HURDLES
- FOCUSING ON THE STEP LENGTH

Kazuhito SHIBAYAMA[1], Norihisa FUJII[2] and Michiyoshi AE[2]

1) Doctoral Program in Physical Education, Health and Sport Sciences, University of Tsukuba, Japan
2) Institute of Health and Sport Sciences, University of Tsukuba, Japan

KEY WORDS: 110m hurdles, running velocity, step length

INTRODUCTION: The adaptations to each of the four steps are required for good performance in 110m hurdles (McDonald, 2002). However, the motions related with both running velocity and step length are not studied yet. Thus the purpose of this study was to investigate kinematic characteristics of 110m hurdlers with reference to step length.

METHODS: Twenty nine male hurdlers (Height: 1.84 ± 0.05m, Mass: 74.6 ± 6.9kg, Time in analyzed race: 13.77 ± 0.45 s) participated in this study. The motions from the 6th to 7th hurdles (1cycle) were videotaped during world and Japanese top class official competitions. The kinematic parameters primarily analyzed were: 1) running velocity (RV), which was defined as the average of the horizontal CG velocities during 1cycle; 2) vertical velocity of the CG at the takeoff of the 2nd step; 3) step length, which was defined as the distance between the toe at the 2nd and 3rd steps; 4) step frequency, which was defined as the inverse of the duration of the 2nd step; 5) thigh angle; 6) landing distance (LD), which was defined as the distance from the hurdle to the toe at the touchdown; 7) support time, which was defined as the duration from the foot contact to the toe off; and 8) air time, which was defined as the duration from the toe off to the foot contact. A multiple regression analysis was conducted with RV and LD as independent variables with P at .05.

RESULTS: Table 1 shows the result of multiple regression analysis with kinematical parameters at the 2nd step as dependent variables. Step frequency positively correlated with RV. Support time, air time and vertical velocity of the CG at the takeoff of the 2nd step were negatively correlated with RV. Figure 1 shows the thigh angle of trail-leg during 1 cycle, which was normalized at 100%, with significant differences of RV and LD. In the 2nd step (2on-3on, 6-27%), the thigh angle was negatively correlated with RV (5-10%), and positively correlated with RV (16-22%). There was no significant difference with LD.

Table 1 The Effect of running velocity and landing distance.

Dependent variables (mean + SD)	Independent variables (β)		R^2
	RV (m/s)	LD (m)	
Step length (2.00 + 0.08 m)	-.026	-.279	.074
Step frequency (4.30 + 0.23 s⁻¹)	.784***	.137	.562***
Support time (0.13 + 0.01 s)	-.699***	-.066	.462***
Air time (0.11 + 0.01 s)	-.493*	-.029	.235*
Vertical velocity of CG at takeoff (0.25 + 0.16 m/s)	-.581**	-.158	.301**

*: $p<.05$, **: $p<.01$, ***: $p<.001$

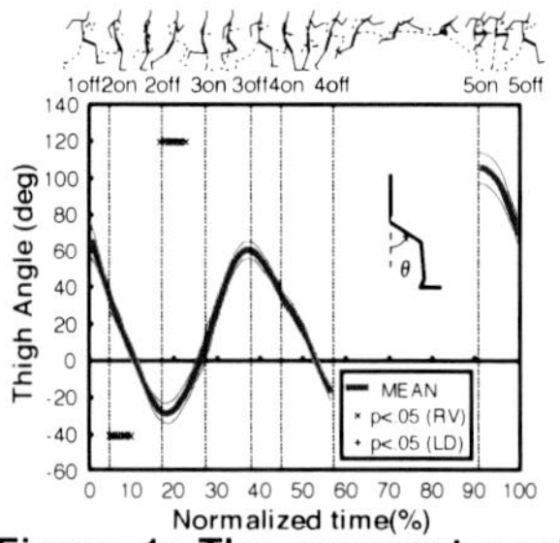

Figure 1: The segment angle of thigh of trail-leg during 1

DISCUSSION: It is thought that faster hurdlers shortened support time of the 2nd step by a toe-contact nearer the CG at touchdown. The small range of motion of the thigh during the 2nd step was effective for reducing vertical velocity of CG at takeoff and shortened air time. Faster hurdlers achieved shortened support and air times and higher step frequency of the 2nd step.

CONCLUSION: The results of this study revealed that the motions to achieve higher step frequency during the 2nd step were effective for good performance of 110m hurdles.

REFERENCES: McDonald, C. (2002). Hurdling is not sprinting. *Track Coach 161*, 5137-5143.

LANDING DEVELOPMENT: A FIRST LOOK AT YOUNG CHILDREN

Pamela J. Russell[1], Jean Eckrich[2] and Madison Hawkins[2]
Bridgewater State College, Bridgewater, MA, USA[1]
Colby-Sawyer College, New London, NH, USA[2]

The purpose of this study was to examine sagittal and frontal views of children (n=14) aged 4-9 landing from a maximal effort vertical jump to begin a description of landing development. Video records (collected at 30 frames/sec) of the jump and landing were viewed frame by frame with Windows Movie Maker and analyzed with a simple scoring system validated to detect improper movements during landing. Findings indicated that this stop-landing task challenged balance as most landings included a step, straddled foot position, and a wide stance. Mechanisms for force absorption (knee and hip flexion) tended to occur more often in landings of older children, but incidences of knee valgus also increased with age. Further investigation may establish developmental expectations for landing and help coaches and physical educators correct potentially harmful patterns as children age and pursue more competitive sport.

KEY WORDS: landing evaluation, motor development, pedagogy, youth sport

INTRODUCTION: Anterior cruciate ligament (ACL) injury risk increases when landings include little knee flexion, knee valgus (i.e., outward tending of the tibia distal to the knee), and too much tibial rotation (e.g., Hewett et al., 2006). Physical training programs can improve upon these potentially harmful patterns. In a large (n=1435) randomized controlled trial an alternate warm-up routine (Prevent Injury and Enhance Performance (PEP)) appeared to decrease ACL injury risk for college-level women soccer players by improving neuromuscular control (Gilchrist et al., 2008). However, a critical aspect of injury prevention may be instruction in proper movement techniques and correction of poor movements, for a less select group such as recreational athletes. DiStefano et al., (2009) advocates this approach. In her recent work, intervention programs targeted to correct strength and flexibility imbalances and postural malalignment when landing appeared successful for 10-17 year-old recreational youth soccer players (n=173). ACL injury risk increases about age 11-12 in soccer, so landing technique instruction prior to this age may benefit more sport participants.

In formal physical education (PE) classes, most school-aged children receive instruction that promotes competence with many movements, such as jumping and force absorption during landings. Teachers use multiple tools, including established developmental progressions that delineate quality movements (e.g., Wickstrom, 1977), to provide appropriate movement feedback. However, there is no developmental progression for landings. The physical educator does not know what lower extremity positions to expect at each age, making developmentally appropriate landing instruction impossible. The intent of this study was to begin a description of landing skill development in children (ages 4 to 9) though examination of knee and foot positions and movements when landing from a maximal effort vertical jump (MVJ). Educators and youth sport coaches may use knowledge of expected landing qualities at varied developmental stages to positively influence landing skills in young participants. Perhaps facilitating neuromuscular habits associated with decreased injury risk will better prepare young sport participants for landings later challenged by their own physical growth.

METHODS: Participants: Parents associated with a small New England college were informed of the study to gather a volunteer sample of 14 healthy children without lower extremity musculoskeletal injuries or disabling conditions. Three similar-aged subgroups were created (i.e., 4-5 yrs (n=6), 6-7 yrs (n=5), 7-8 yrs (n=3)) to permit developmental observations. College approved consent/assent forms were signed by a parent and the child(ren) prior to data collection. For each child, a parent indicated activity level and youth sport participation.

Data Collection: Data were collected in two sessions in a gymnasium. Prior to subject arrival two digital video cameras (Panasonic PV-DV400; Panasonic PV-GS55) were leveled and positioned (1.13 m high; 2.31 m from target: 0.94 m high; 1.53 m from target) to view the frontal and sagittal planes, respectively, of a MVJ and stop-landing task. Participants wore shorts, a short-sleeved shirt, and sneakers for data collection. Each child completed 5 min of lower extremity static stretching and jogging or ball kicking to warm-up. A research assistant determined each child's height, weight, standing reach, and MVJ height estimate. To estimate MVJ height, participants jumped as high as possible, once, to touch the wall marked with height increments. Next, each participant moved to the filming area and practised jumping to a target (a balloon (0.64 m circumference) hung from a yardstick positioned parallel to the floor and clamped to an adjustable Vertec® (Sports Imports, Inc. Columbus, OH)). Participants started from a line 0.39 m behind the target and completed 3-5 practise trials. Verbal instructions indicated for them to jump as high as possible, touch the balloon with both hands, and land balanced facing the camera. Starting with the estimated MVJ height, the target height was adjusted during practise to challenge each child, but still allow him/her to touch the balloon with both hands. Target height was recorded as MVJ height. After the practise, five MVJs and landings were filmed from the frontal and sagittal views.

Data Reduction: Digital video recordings were transferred to Windows Movie Maker (v. 5.1, Microsoft Corporation, 2007) for single frame viewing at 30 frames/sec. Developmental characteristics (e.g., Gallahue & Ozmun, 2006) of sagittal view MVJ trials were examined for maturity to ascertain MVJ task difficulty and help explain potential landing differences, as jumping maturity influences landing ability (Ayalon et al., 1987). Frontal and sagittal view landing characteristics were described using portions of the Lower Extremity Scoring System (LESS) (Padua et al., 2004), a clinical movement analysis tool assessed for validity and reliability to detect improper movement patterns during jump-landing tasks (e.g., Distefano et al., 2009; Padua et al., 2007). Tables 3 and 4 list the landing qualities observed; timing of foot contact and subsequent balance, stance width, toe position, foot placement, and knee and hip positions and motions at initial contact (IC) and maximum knee flexion (MxKFx). Two researchers reviewed these qualities in multiple trials to ensure evaluation criteria consistency. One researcher analyzed all of the data. Trials were excluded if the subject did not touch the target. Analyses used a binary technique (evidence for or against the presence of each jumping and landing characteristic was simply tallied). Number of occurrences was expressed as a percent of the number of jumps and landings. Simple descriptive analysis techniques were purposely selected to mimic tools that coaches or physical educators might use in their professional settings.

RESULTS AND DISCUSSION: With age participants' height, weight, jump height, and seasons of youth sport participation increased (Table 1). Parent survey responses showed that 78.6% of the children were moderately active after school and over the weekends (i.e., spent *more free time* doing activities that *kept the heart rate elevated*, {e.g., tag, basketball, soccer, scooters, in-line skates, free-play, biking, organized youth sport} than time in activities that did not elevate the heart rate much {e.g., video games, television, board games, reading, arts and crafts}).

Sagittal jumping observations showed that all subjects consistently used forceful extension at the hips, knees, and ankles from a 2-foot take-off coupled with upward head tilt focusing the eyes on the target. These qualities reflect mature jumping (Gallahue & Ozmun, 2006). In addition, the oldest participants' jumping patterns indicated 60°-90° of knee flexion in preparation (93% of jumps), no exaggerated trunk lean, no leg tucking during flight, and simultaneous coordination of the arm swing (33% of jumps) (Table 2). Other preparatory and jumping motions varied with age but the tendency, not surprisingly, was for jumping ability to mature with age in this sample (Table 2).

Table 1: Subject Characteristics (Means and Standard Deviations)

Age Groups	Age (mos)	Height (m)	Weight (kg)	Jump Height (m)	PE Classes (#/week)[1]	Youth Sport Seasons[2]	Activity Instruction Classes[3]
4-5 yrs (n = 6)	65.0 (5.48)	1.1 (.04)	19.6 (.76)	0.21 (.03)	0.3 (.52)	0.5 (.84)	1.8 (2.14)
6-7 yrs (n = 5)	85.8 (8.96)	1.2 (.06)	21.9 (1.07)	0.21 (.05)	1.4 (.89)	2.0 (1.87)	2.2 (1.48)
8-9 yrs (n = 3)	107.7 (4.16)	1.3 (.03)	23.3 (.47)	0.25 (.04)	1.0 (0.0)	6.7 (4.93)	1.0 (1.41)
Total (n = 14)	81.6 (18.16)	1.2 (0.10)	21.2 (1.72)	0.22 (.04)	0.9 (.77)	2.4 (3.32)	1.9 (1.73)

[1] Structured physical education classes in a private or public school setting. [2] Number of seasons [3] Number of class units.

Table 2: Sagittal View Differences in Jumping Characteristics Across Age Groups

Jumping Characteristic	Mature Level	All Ages (%) (69 landings)	4-5 yrs (%) (29 landings)	6-7 yrs (%) (25 landings)	8-9 yrs (%) (15 landings)
Knees > 90°	Low	43	52	60	0
Knees 60°-90°	High	57	48	40	93
Too much trunk lean	Low	12	14	16	0
Arms aid unequally	Low	88	100	92	60
Arms coordinated	High	10	0	8	33
Leg tuck	Low	9	21	0	0

Landing observations from the sagittal plane indicated difficulty with the task (Table 3). Only 6% of all landings included a simultaneous touch of both feet and then a stop. The oldest age group accomplished this task most often. In 80% of all landings, simultaneous touch of both feet was followed by a step before stopping. All age groups stepped back more often than stepping forward. Other stepping patterns and directions (e.g., 2 steps forward) were evident but not tallied. These findings suggest that simultaneous control of balance and force absorption during landing was difficult. Balance was maintained as none of the participants fell. Landings consistently (99%) showed toe to heel contact, with foot placement mostly straddled (left (L) to right (R) or R to L) as opposed to side-by-side. Asymmetric foot placement may promote unequal force distribution between R and L legs. Less than 50% of all landings showed at least 30° of knee flexion (FX) at IC, with little variation across age groups. However, with age, a greater percentage of landings showed at least 45° of knee FX at MxKFx. Few landings (33%) showed hip FX at IC, but most landings (61%) showed increased hip FX from the time of IC to MxKFx. More knee and hip flexion allows more time to absorb landing forces. This positive landing trait may be more prevalent as children age.

Table 3: Sagittal View Landing Characteristics Across Age Groups

Landing Characteristics	All Ages (%) (69 landings)	4-5 yrs (%) (29 landings)	6-7 yrs (%) (25 landings)	8-9 yrs (%) (15 landings)
Simultaneous: 2-ft touch – stop	6	3	0	20
Simultaneous: 2 ft touch - 1 step[a,b]	80	86	92	60
[a]Step forward	17	17	28	7
[b]Step back	54	55	64	53
Toe-to-heel contact	99	100	96	100
Straddled (L-R)/(R-L)	46/29	38/44	56/24	47/7
IC: Knee 30° FX	43	41	48	40
MxKFx: Knee 45° FX	72	59	76	93
IC: Hip some FX	33	38	32	27
MxKFx: Hip > FX	61	45	76	67

Frontal plane observations showed that in each age group foot placement was at or greater than shoulder width (SW) in the majority of landings (Table 4). Children may use this placement coupled with a straddled foot position to control side-to-side and front to back balance. In only 30% of landings were one or both toes turned in or out at IC. This could be a positive trait as toeing in/out rotates the tibia, stressing the knee joint. In each age group, toeing in/out most often affected just one foot, promoting landing asymmetry. Only 35% of landings demonstrated knee valgus (VAL) at IC, but the oldest age group had the most instances of knee VAL at IC (73%). By the time of MxKFx, 39% of landings showed knee VAL for one leg, but with increasing age instances of knee VAL increased (28% to 40% to 60%). This finding is troubling as knee valgus may increase ACL injury risk in adolescence.

Table 4: Frontal View Landing Characteristics Across Age Groups

Landing Characteristics	All Ages (%) (69 landings)	4-5 yrs (%) (29 landings)	6-7 yrs (%) (25 landings)	8-9 yrs (%) (15 landings)
Stance: $\geq$ SW	68	79	52	73
Stance: <SW	32	21	48	27
IC: Toe out/in (1 foot)	28/29	38/24	24/24	13/47
IC: Toes out/in (2 feet)	9/14	3/21	20/8	0/13
IC: Knee VAL (1 or 2)	35	24	24	73
MxKFx: Knee VAL(1)	39	28	40	60
MxKFx: Knee VAL(2)	26	31	16	33

CONCLUSION: Sagittal and frontal view examination of children aged 4-9 jumping then landing indicated that the MVJ task presented similar levels of difficulty across age groups, despite a tendency for more mature jumping patterns with age. Landings did not seem to improve with age. The stop-landing task challenged balance as most landings included a step, straddled foot position, and wide stance. Knee and hip flexion mechanisms for force absorption tended to occur more often in landings of older children, but incidences of knee valgus also tended to increase with age. Findings are preliminary, yet highlight the need for further investigation to establish developmental expectations for landing that might allow potentially harmful mechanics to be corrected as children age. Future findings could benefit physical educators and youth sport coaches and positively influence landing skill development in young children.

REFERENCES:

Ayalon, A., Ben-Sira, D., & Leibermann, D. Characteristics of landing from different heights and their relationships to various fitness and motor tests. *Physical Fitness and the Ages of Man: Proceeding of the Symposium of the International Council for Physical Fitness Research*, 1987; Jerusalem.

DiStefano, LJ, Padua, DA, DiStefano, MJ, & Marshall, SW. (2009). *Influence of Age, Sex, Technique, and Exercise Program on Movement Patterns After an Anterior Cruciate Ligament Injury Prevention Program in Youth Soccer Players*, American Journal of Sports Medicine, 37: 495-505.

Gallahue, DL & Ozmun, JD. (2006). *Understanding motor development: infants, children, adolescents, adults*. Boston: McGraw Hill.

Gilchrist, J, Mandelbaum, BR, Melancon, H, et al. (2008). A randomized controlled trial to prevent noncontact anterior cruciate ligament injury in female collegiate soccer players. . *American Journal of Sports Medicine*, 36, 1476-1483.

Hewett, TE, Myer, GD, & Ford, KR. (2006). Anterior cruciate ligament injuries in female athletes: part 1, mechanisms and risk factors. *American Journal of Sports Medicine*, 34, 299-311.

Padua DA, Marshall SW, Beutler AI, Boling MC, Thigpen CA. (2007). *Differences in jump-landing technique between ACL injured and non-injured individuals: a prospective cohort study*. Journal of Athletic Training, 42, S85.

Padua DA, Marshall SW, Onate JA, et al. (2004) Reliability and validity of the Landing Error Scoring System: implications on ACL injury risk assessment. *Journal of Athletic Training*, 39, S110.

Wickstrom, RL. (1983). *Fundamental movement patterns*. Philadelphia: Lea & Febiger.

A COMPARISON OF THE PERFECT PUSH-UP™ TO TRADITIONAL PUSH-UP

Michael Bohne, Jason Slack, Tim Claybaugh, and Jeff Cowley
Utah Valley University, Orem, Utah, USA

KEY WORDS: push-ups, electromyography, biomechanics.

INTRODUCTION: The push-up has traditionally been used to help improve arm and shoulder girdle strength and endurance (ASGSE) (Baumgartner, Oh, Chung, and Hales, 2002). The Perfect Push-Up™ was designed to help improve the work-out that could be obtained while performing push-ups. While the claims from the manufacturer seem appealing there is little to no research to support their claims. The purpose of the current study is to examine the muscle activity and arm kinematics while using the Perfect Push-Up™ compared to traditional push-ups, thus addressing the claims of an improved work-out.

METHODS: Thirty-four healthy male volunteers (age 19-30) performed 10 push-ups in both, in a counter balanced order, the traditional method and using the Perfect Push-Ups™. Approval was obtained from University IRB. Surface electromyography (SEMG) was recorded for 7 muscles, 20 seconds per condition, using a Delsys Bagnoli 8-channel system (Boston, MA, USA). Those muscles include: Pectoralis Major (PM), Anterior Deltoid (AD), Rectus Abdominus (RA), Biceps Brachii (BB), Triceps Brachii (TB), Latisimus Dorsi (LD) and External Oblique (EO). Additionally, participants were videotaped at 60 Hz to examine the arm kinematics. Participants were given basic instructions on how to use the Perfect Push-Up™, additionally they were instructed to be consistent during both push-up conditions. SEMG data were analyzed using average rectified EMG amplitude (RMS). Kinematic data were analyzed using MaxTraq motion analysis software (Ann Arbor, MI, USA). Correlational data for elbow range of motion were collected between conditions to ensure the same technique was employed between conditions. Statistical data were analyzed using a paired t-test (alpha=0.05) for each muscle and arm kinematics by SPSS 16.0 (Chicago, IL, USA)

RESULTS: There was a significant increase in the activity for five of the seven muscles while using the Perfect Push-Up™. A significant increase was observed for PM ($p<0.01$), AD ($p<0.05$), BB ($p<0.01$), TB ($p<0.05$) and EO ($p<0.05$). Additionally, there is evidence of a significant increase in the elbow range of motion while using the Perfect Push-Up™ ($p<0.01$). Also, there was strong correlation in the elbow range of motion ($r=0.871$).

DISCUSSION: Examining the elbow range of motion during the push-up shows evidence that the participants were consistent in how they performed the push-up. An increase in the muscle activity should lead to greater gains in ASGSE. The increase in the ROM provides the possible explanation for these results. However, participants travel deeper than some previous research suggest is safe (Cooper Institute for Aerobics Research, 1992), which could lead to add to the increase in muscle activity.

CONCLUSION: Increases in muscle activity and elbow range of motion were observed while using the Perfect Push-Up™ possibly confirming manufacturer's claims. However further research is required to pinpoint the exact cause of these increases beyond range of motion.

REFERENCES:
Baumgartner, T.A., Oh, S., Chung, H. and Hales, D. (2002) Objectivity, Reliability, and Validity for a Revised Push-Up Test Protocol. *Measurement in Physical Education and Exercise Science*, 6(4) 225-242.

Cooper Institute for Aerobics Research (1992) *The Prudential Fitnessgram: Test administration manual*. Dallas, TX.

EFFECT OF RELAY CHANGEOVER POSITION ON SKATING SPEED FOR ELITE SHORT TRACK SPEED SKATERS

Conor Osborough and Sarah Henderson

School of Science and Technology, Nottingham Trent University, Nottingham, UK

KEY WORDS: winter sports, speed skating, performance.

INTRODUCTION: Relay changeovers in short track speed skating take place when the incoming skater pushes the outgoing skater at the start of the straight. Time can be gained or lost during these changeovers, depending on how effectively they are executed. The position on the track where initial contact between skaters is made is thought to be a critical factor in an effective relay changeover (Riewald et al., 1997). The aim of this study was to determine how the position of the relay changeover on the track would affect skating speed (SS).

METHODS: Eight elite male short track speed skaters (21.8 ± 2.4 yrs), who were members of the British Short Track Speed Skating Team, consented to participate in this study. Participants completed six relay trials during two separate testing sessions, under simulated race conditions. For each relay, participants were instructed to initiate their changeover in one of three positional zones: Early; Middle; Late (i.e. a distance of 3.5 m, 4.0 m or 4.5 m from the start of the straight respectively). On the second date, the trials were counterbalanced. Trials were recorded, at 50 Hz, from the stands using three digital video camcorders. Forty-nine relay changeovers were selected for analysis. SS was determined from the positional change of each skater's hip joint centre, which was digitised from the video-footage over the length of the straight. For each positional condition, the SS of each changeover was plotted and the equation of the trendline found. Using this, the SS of the second and sixteenth changeover was determined. Changes in mean SS of the changeover straight were analysed for the three positional conditions (1 x 3 ANOVA). The level of statistical significance was set at $p < 0.05$.

RESULTS: There was a change in skating speed between the second and the sixteenth changeover, for each of the three positional conditions (Early: 11.01 $m·s^{-1}$ vs. 11.20 $m·s^{-1}$; Middle: 11.18 $m·s^{-1}$ vs. 11.14 $m·s^{-1}$; Late: 11.21 $m·s^{-1}$ vs. 10.86 $m·s^{-1}$). Mean changeover straight SS (11.23 ± 0.67$m·s^{-1}$) using the middle condition was non-significantly higher, when compared to the other two conditions (Early: 11.06 ± 0.25 $m·s^{-1}$; Late: 11.09 ± 0.30 $m·s^{-1}$).

DISCUSSION: Although there were non significant findings, there was some evidence to suggest that initiating the relay changeover in the middle zone produced the fastest SS during the changeover straight, when compared to the Early or Late zones. Furthermore, SS appeared to be most consistent between the start and end of the relay when the changeovers were initiated in the middle zone.

CONCLUSION: The results from this study suggest that small performance gains (in the region of 1% to 1.5%) might be found if short track relay changeovers are consistently initiated between 3.5 m and 4.0 m from the start of the straight. Such improvement gains are of value in an event where a team's success is usually decided by fractions of a second.

REFERENCES:
Riewald, S., Broker, J., Smith, S. & Otter, J. (1997). Energetics and timing of relay exchanges in short track speed skating. *Medicine and Science in Sports and Exercise, 29*(5), 45.

Acknowledgement
The authors would like to thank the skaters for their participation and British Speed Skating for their support in this project.

MATURATION EFFECTS ON LOWER EXTREMITY KINEMATICS IN A DROP VERTICAL JUMP

ChangSoo Yang[1], InSik Shin[2], Gye-San Lee[3], and BeeOh Lim[2]

Department of Martial Art, Incheon City College, Incheon, Korea[1]
Sports Science Institute, Seoul National University, Seoul, Korea[2]
Department of Physical Education, Kwandong University, Kangneung, Korea[3]

KEY WORDS: maturation, lower extremity, kinematics.

INTRODUCTION: As children increase in biologic age, body height and weight increase, and subsequent maturation of the nervous, endocrine, muscular, and cardiovascular systems leads to alterations in neuromuscular performance (Naughton et al., 2000). It is important to understand the effects of growth and development on sports performance and sports injuries. The purpose of this study were to investigate maturation effects on lower extremity kinematics in a drop vertical jump.

METHOD: Three-dimensional video graphic data were collected for 11 prepubescent females (7~10 age), 11 pubescent females (13~16 age), and post pubescent females (21~24 age) performing a drop vertical jump. A total of 21 reflective markers were then placed in pre-assigned positions. The subject was instructed to drop off the box, leave both feet at the same time, land, and then immediately perform a maximum vertical jump. Three trials were performed from a wooden box of each subject's relative knee lateral epicondyle height. In this controlled laboratory study, a one-way ANOVA experimental design was used for the statistical analysis ($p<.05$). Post-hoc tests with Tukey's correction were used.

RESULTS: Pubescent and post pubescent females demonstrated significantly decreased knee flexion angles (21.3 ± 6.7, 22.9 ± 5.6, 28.4 ± 7.1, respectively) ($p<.05$) and hip flexion angles (27.5 ± 8.4, 28.7 ± 7.7, 41.3 ± 9.5, respectively) ($p<.05$) at initial ground contact during the landing of a drop vertical jump task than prepubescent females.

DISCUSSION: The results of this study supported previous study that Hass et al. (2003) showed a significant decrease in knee flexion angles during the landings of post pubescent subjects in stride jump landing task in comparison to prepubescent subjects. Landing with small knee flexion angles in a drop vertical jump task may increase the load on the ACL. Previous study on knee anatomy and biomechanics have shown that the anterior shear force applied on the tibia by the quadriceps muscles increases as the knee flexion angle decreases because the patellar tendon-tibial shaft angle increases as the knee flexion angle decreases (Buff et al., 1988). The results of this study provide significant information for research on the prevention of non-contact anterior cruciate ligament injuries.

CONCLUSION: Pubescent and post pubescent females have decreased knee and hip flexion angles at initial ground contact and decreased knee and hip flexion motions during the landing of a drop vertical jump task than prepubescent females. These age differences in knee and hip flexion motion patterns occur after the onset of puberty.

REFERENCES:

Buff et al., (1988). Experimental determination of forces transmitted through the patello-femoral joint. J Biomech. 21, 17-23.

Hass, et al., (2003). Lower extremity biomechanics differ in prepubescent and postpubescent female athletes during stride jump landings. *J Appl Biomech.* 19, 139-152.

Naughton, et al. (2000). Physiological issues surrounding the performance of adolescent athletes. *Sports Med.* 30, 309-325.

THE EFFECTS OF AGING ON THE HIP AND SPINAL MOTIONS IN THE GOLF SWING

Clive Lathey, Siobhan Strike and Raymond Lee
School of Human and Life Sciences, Roehampton University, London, UK

The purpose of this study was to determine the effect of aging on the kinematics of the lumbar spine and hips in the golf swing. Aging was shown to alter the joint coupling but not the magnitude of joint loading.

KEY WORDS: spine and hip motion, spine and hip coupling, golf.

INTRODUCTION: There are two primary motions in the golf swing, body rotation and arm movement. Body rotation occurs predominantly in the spinal and hip joints. As an individual ages, biological changes gradually affect the mechanics of the human body (Adams, 2006). The resulting decrease in muscle strength, endurance and a reduction in flexibility may compromise spinal functions. Therefore a natural consequence of the aging process is a decrease in golf swing performance. The spine-hip coupling may compensate for the reduced spinal flexibility. Therefore the aim of this research is to determine how the kinematics of the spine and hips and their coupling change with increased age.

METHODS: Data Collection: Data was collected on healthy golfers (handicap 10-20) in two age groups: young (age=18-30) and older (age=46+). Participants must not have had back pain for 6 months prior to data collection and no previous surgery. All participants signed an informed consent form approved by the University Ethics Committee. Following a warm-up, four electromagnetic motion sensors (Fastrak, Polhemus Ltd) were placed on T12, L5/S1, and the lateral sides of the right and left thigh. Participants completed 5 maximal shots with their own 7 iron, and 3D motion data were collected with respect to the previously recorded neutral positions. The best shot, defined by distance and accuracy, was chosen for analysis.

Data Analysis: To assess the effect of aging on the range of motion, ANOVA was conducted between the 2 groups for the 3-D motions at the top of the backswing. Cross-correlation of the movement-time curves was conducted to study spine-hip coupling.

RESULTS: It was shown that the back swing was mainly accomplished by forward flexion, side bending and axial rotation of the spine, (43.7 ±5.7°, 32.5 ±5.0° and 42.6 ±2.8° respectively) and flexion and axial rotation of the hips (35.0 ±11.2° and 41.1 ±11.2°). There was minimal abduction or adduction of the hips. There was strong coupling between the movements of the spine and hip. Cross correlation analysis showed that the coefficient was significantly reduced in the older age group (mean r=0.81 for the young group and r=0.64 for the older group). However, aging did not appear to affect the range of motions.

DISCUSSION and CONCLUSION: The results indicate that aging affects the kinematics of the golf swing. It does not appear to alter the magnitude of joint motions, but the coordination between different joints. This may affect golf performance. Coaches should be aware of the change in coordination in aged golfers as this cannot be corrected by altering the magnitude of motion of the spine and hips.

It is concluded that aging affects coordination of motions of spine and hips during a golf swing, and this needs to be taken into account in designing training programme.

REFERENCES:
Adams, M, Bogduk, N, Burton, K & Nolan, P (2006) *The biomechanics of back pain* (2nd ed). Churchill Livingstone

DEVELOPMENT AND VALIDATION OF A SYSTEM FOR POLING FORCE MEASUREMENT IN CROSS-COUNTRY SKIING AND NORDIC WALKING

Bortolan Lorenzo, Pellegrini Barbara and Federico Schena

Research Centre for Bioengineering and Motor Sciences, Rovereto, Italy

The purpose of this study was to describe and validate a force transducer system specifically designed to measure the force exerted through the poles in cross-country skiing and Nordic walking. It is constituted by a custom built load cell and by a mounting system that allow to minimise cross talk effects. The system is applicable to standard carbon racing shafts to ensure the standard stiffness of the pole. The reliability of the system has been tested performing different static and dynamic tests. The comparison with the reference load cell has shown a good measurement linearity in the range of typical values for poling propulsion and a sensitivity only to the force axially applied to the shaft. The test performed on a 2D platform and with a motion capture system for the measurement of pole inclination, demonstrated the possibility to obtain a reliable measure of the vertical, longitudinal and lateral components of the force exerted by the subject. The accuracy, the portability of the system and their applicability to different shafts allow evaluation of poling action in both laboratory and field conditions, providing important information in cross-country skiing and Nordic walking biomechanical research.

KEY WORDS: Nordic walking, cross-country skiing, force, poles, validity.

INTRODUCTION: The increasing number of studies published in recent years highlights the growing interest in understanding the contribution of upper limbs in different forms of locomotion such as cross-country skiing and Nordic walking. Early studies investigated the forces applied on the poles during cross country skiing (Pierce, 1987). In the last few years, the interest in Nordic walking as health-enhancing physical activity has also increased the interest in the analysis of the propulsive role of the upper limbs in non purely sporting contexts The quantification of the forces applied through the poles can be obtained using traditional and special force platforms, or force transducers mounted on the poles, (Komi, 1987, Holmberg et al., 2005). A limitation of force platform is that the measurements are obtained on just on one or few consecutive poling cycles. Furthermore, this measure can be collected during a limited range of conditions (Komi, 1987) To overcome these shortcomings, the use of force transducers mounted below the poles handgrip has been developed and proposed (Holmberg et al., 2005). This method is more advantageous in terms of portability and increases the range of conditions that can be investigated. Recently, commercial handgrips for both cross-country skiing and Nordic walking poles have been modified by replacing the old wrist strap with a wrapper cuff. This new solution reduces the hand control on the grip and moves away the force point application from the axis of the pole. To minimize the cross talk due to bending moment a new system has been developed. The aims of this study were to describe the system and to examine its accuracy and reliability in acquiring the force exerted through the poles during skiing and Nordic walking.

METHODS: Force measurement system: A mono axial custom made load cell (Deltatech, Italy) was mounted between the upper end of the shaft and the handgrip. The transducer was calibrated using a specific calibration apparatus with 6 standard weights (5-30 kg). The hand-grip has been modified by inserting an aluminium pipe that permits a free longitudinal movement of the handle respect to the shaft (Figure 1). In this way only the axial force is transmitted by the load cell from the handgrip to the section of the pole, while the transverse forces are unburden by the coaxial aluminium pipe. The movement of aluminium pipe and hand grip is limited by a rubber band that maintains the cell preloaded. The load cell is retained in the correct position by a light pipe cemented with the cell and that contains the amplifier. The load cell and amplifier power dissipation is about 200 mW . To allow the individual selection of the poles length and to ensure the standard stiffness of the pole, the traditional carbon racing shafts (CT1; Swix Sport, Norway; Diamond 10 Max; One Way,

Finland) with different lengths (115 cm-170 cm, 2.5 cm steps) were used. The weight of the load cell is 30 g and the entire measurement system adds 50 g to the original pole mass. A further increment in the pole mass is due to the special carbide tip (49 g) that is used when the tests are performed in the lab to allow a suitable friction with the rubber surface of the treadmill. The position of the centre of mass does not change after applying the force measurement system and the special tip (64.5 vs. 64.7% of pole length), while rises until the 70.8% of pole length with standard tip.

Experimental testing and data analysis: Different statical and dynamical tests were performed in order to assess the reliability of the force measurement system. All signals were acquired by means of a data acquisition board (NI DAQ-PAD-6016, 16 bit; National Instruments, USA), while the inclination and the motion of the pole in the different setups were collected at 200 Hz by means of an optoelectronic motion capture system (6 cameras MCU240, ProReflex; Qualisys, Sweden). Five reflective markers were axially adhered along the pole at a distance of 20 cm between each other. Data collection was triggered by a digital signal in order to ensure the synchronization between force and kinematic signals.

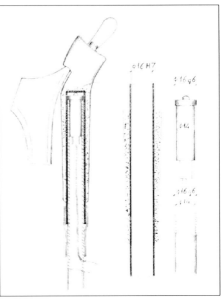

Figure 1: Drawing of the load cell and its position under the handgrip

1. A static test was conducted to analyse the measures obtained when a non-axial force is applied. The test was performed suspending a mass of 4 kg to the strap and inclining the pole at different tilt angles. The measured values were compared with the theoretical values calculated as the product of the weight force by the sine of the tilt angle.
2. A validation of the measurement system linearity was performed by axially pushing the instrumented pole against a reference load cell (546QD; DSEurope, Italy) simulating 15 poling cycles. Both signals were sampled at 200 Hz.
3. Five poling imitation at every five different pole inclinations were performed on a two axial force platform (P114-BIAX-S-A/5000N; Deltatech, Italy) maintaining the pole on the XZ plane. The pole force has been factorized in longitudinal (X) and vertical (Z) force as product of pole force by the cosine and the sine respectively of the angle between the line through the 1st and the 5th marker and horizontal line. The calculated longitudinal and vertical forces have been then compared with the respective force components acquired by the force platform.
4. The accuracy of the device in dynamic conditions was examined measuring the resonant frequency of the dynamometric system. The load cell mounted on the specific calibration apparatus and the whole system were impacted with a hard plastic hammer on the upper side of the hand grip. Force signal was acquired at 10 kHz.
5. To test the response of the pole during force action, a double-poling exercise was performed on a treadmill using roller skis. The flexion of the pole was determined by measuring the camber of the shaft under loading condition. An athlete skied at 13 km·h^{-1} at 3° of slope. A 30 seconds recording period was acquired at 200 Hz by the kinematic system. The camber was evaluated analysing the maximal displacement between the line through the 1st and the 5th marker and 3rd degrees polynomial curve that fits all pole markers.

RESULTS:

1. The difference between the measured and the reference values is reported as absolute and percentage difference respect to the full scale at different pole inclinations.

Table 1: Results of static measurement at different pole inclination

Inclination [°]	78.1	63.1	53.7	49.2	36.3
Measured [N]	38.18 ±0.17	35.25 ±0.15	31.26 ±0.14	28.74 ±0.12	23.98 ±0.10
Theoretical [N]	38.40	34.99	31.62	29.70	23.23
Abs error [N]	-0.22	0.25	-0.37	-0.96	0.75
Error/FSD [%]	-0.04	0.05	-0.07	-0.19	0.15

2. Coefficient for the linear regression between reference load cell value and pole force measurement is 0.999 ($p<0.001$) while the constant coefficient of the equation for the linear regression is -4.025 (Figure 2).

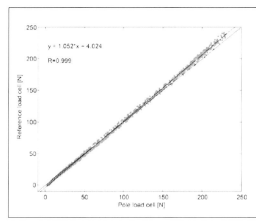
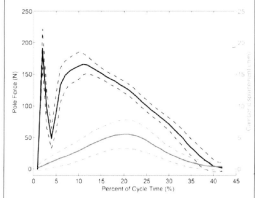

Figure 2: Regression between pole load cell and reference load cell.

Figure 3: Typical curve for poling force for double poling (black line) and measure of the camber of the shaft (grey line).

3. The difference between the force measured by the pole force transducer and force platform was calculated as percentage with respect to full scale for both vertical (Z axis) and longitudinal (X axis) components. A relation between the magnitude of the error and the absolute values was found for the vertical component. Specifically, the lowest the pole angle the highest the difference between the two measures.

Table 2: Results of poling imitation test at different pole inclination

Inclination [°]	88.2 ±0.65	64 ±0.71	47.5 ±1.66	41.5 ±0.78	34.9 ±0.90
Error/FSD Z [%]	1.01 ±0.75	1.13 ±1.19	1.23 ±1.44	1.83 ±2.35	3.03 ±2.74
Error/FSD X [%]	-0.03 ±0.46	-0.45 ±0.62	-0.74 ±1.00	-0.31 ±1.77	0.61 ±2.55

4. No resonant frequency has found for the load cell transducer while a 39.6Hz resonant frequency was found for the complete system (measurement equipment and pole).

5. A small but significant flexion of the pole during the poling action was found (Figure 3). The maximal camber is 5.5 ±2.2 mm occurring at about 20% of the poling cycle time.

DISCUSSION: The results of this study support the validity of a new force transducer system for measuring the force exerted through the poles. The insertion of the force measurement system to the traditional racing pole increased the original weight but did not change its centre of mass. This is particularly appreciated by the elite athletes that referred their technical action was not modified by the modified poles. The static test showed that the transducer inserted on the pole is sensitive only to the force axially applied to the shaft, indicating that the measurement is minimally corrupted by cross-talk effect. This result was confirmed in the dynamical situation. Indeed, the test performed on the 2D platform showed that almost all the force applied along the pole was accounted by the longitudinal and vertical force components. This also demonstrated that with a system that can measure the inclination of the pole, it is possible to obtain a reliable measure of the vertical, longitudinal and lateral components, and not only the total component of the force exerted by the subject. This decomposition is particularly important because it allows to identify the components that effectively contribute to the forward propulsion. The comparison with the reference load cell has shown a good measurement linearity in the range of typical values for poling propulsion (Holmberg et al., 2005). However, the offset resulting from the regression equation suggested that the assembling of the transducer and the handgrip using the rubber caused a preload of the load cell. To take in account this preload that could be different for each system, we suggest to perform a calibration before every test. No resonance frequency is associated to the transducer while an oscillation has been seen for the complete system probably due to the shaft vibration. However the results of double poling test on the treadmill showed no effects of vibration during typical poling force exertion. A slight flexion during double poling actions of the shaft was detected during the test performed in specific condition. Further studies are needed to examine if camber and vibration are related to pole length, pole stiffness and/or ground hardness. The low energy consumption and the range of the output signal of this load cell are well suited for a portable pole measurement system since it is only necessary to add small batteries and a portable datalogger.

CONCLUSION: This study demonstrated the accuracy and reliability of this new force transducer system specifically designed to measure poling force. The main advantage of this system is to minimize the cross talk effect due to the bending moment allowing the measure of the force exerted through the pole. Furthermore, combining kinematic data, it allows to determine longitudinal force to better understand the contribution of the poling action in cross-country skiing and Nordic walking biomechanical research. The portability of the system and their simple applicability to the different type of shafts allow to evaluate poling contribution in both controlled (laboratory) and ecological (field) conditions.

REFERENCES:
Holmberg,H.C., Lindinger,S., Stoggl,T., Eitzlmair,E., and Muller,E. (2005). Biomechanical analysis of double poling in elite cross-country skiers. *Med.Sci.Sports Exerc.*, 37, 807-818.

Komi,P.V. (1987). Force measurement during cross-country skiing. *International Journal of Sport Biomechanics*, 3, 370-381.

Pierce,J.C., Pope,M.H., Renstrom,P., Johnson,R.J., Dufek,J., and Dillman,C. (1987). Force Measurement in Cross-Country Skiing. *International Journal of Sport Biomechanics*, 3, 382-391.

EFFECTS OF DIFFERENT JUMP-LANDING DIRECTIONS ON LOWER EXTREMITY MUSCLE ACTIVATIONS

Yu-Ming Li and Heng-Ju Lee

Department of Physical Education, National Taiwan Normal University, Taipei, Taiwan

KEY WORDS : jump-landing, single-leg landing, muscle activations

INTRODUCTION: When performing a jump-landing task in sports, we may jump and land at different directions instead of one direction. During landing period, single-leg landing is generally considered at a higher risk of injury than double-leg landing. However, there were few studies on discussing about musculoskeletal responses when landed at different directions. Therefore, the purpose of this study was to analyze the effects of different jump-landing directions on lower extremity muscle activations when performing single-leg landing.

METHODS: Six healthy subjects [3 males/3 females (24 ±3.3 years old; 63.5 ±12 kg; 168 ±7.3 cm)] were recruited for this study. All subjects required to jump with double-leg at three different directions (forward, diagonal and lateral) and land with single-leg (dominant leg only). During jump-landing task, subjects were asked to touch an object which was placed at 50% of their maximum jumping height with both hands. When landed with single-leg, subjects were asked to maintain their balance for 3 seconds. If the trial which subject was unable to perform requirements above, it would be considered as 'failed trial'. EMG data were collected by BIOPAC MP150 system with surface electrodes. Target muscles for EMG measurements were vastus lateralis (VL), biceps femoris (BF), tibialis anterior (TA) and medial portion of gastrocnemius (MG) on the dominant leg. All EMG data were normalized by signals collected during MVC. One way ANOVA within repeated measures were used to compare muscle activations at different jump-landing directions.

RESULTS: There were no significant differences of muscle activation among three directions. However, forward jump-landing protocol had higher lower extremity muscle activations than diagonal and lateral directions during initial landing (0-1s)(Figure 1). During stability period (1-3s), all lower extremity muscle activations dropped significantly than initial landing(Figure 2).

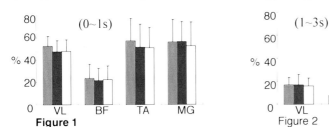

Figure 1 Figure 2

DISCUSSION: Muscles are often considered as dynamic stabilizers of a joint. During landing, greater muscle activations could enhance the function of dynamic stabilizer and lead to better joint stability. Compared to forward jump-landing, lateral and diagonal protocol demonstrated less muscle activations. Less muscle activation might cause less dynamic joint stability support and increase risk of injuries. Therefore, higher muscle activations could reinforce joint stability to prevent injuries.

CONCLUSION: In the current study, diagonal and lateral jump-landing protocol excited less lower extremity muscle activations which could lead to joint instability and result in injuries.

REFERENCES:
Wikstrom, E. A., Tillman, M. D., Schenker, S. M., & Borsa, P. A. (2008). Jump-landing direction influences dynamic postural stability scores. *J Sci Med Sport, 11*(2), 106-111.
Olsen, O. E., Myklebust, G., Engebretsen, L., & Bahr, R. (2004). Injury mechanisms for anterior cruciate ligament injuries in team handball: a systematic video analysis. *Am J Sports Med, 32*(4), 1002-1012.

BIOMECHANICAL ANALYSIS OF STANDING LONG JUMP: A 3D STUDY

Chen-fu Huang, Ray-hsien Tang
National Taiwan Normal University, Taipei, Taiwan

KEY WORDS : kinematics, torque, loading jump

INTRODUCTION: Many studies had investigated 2D standing long jump, and some indicated that when jump with optimal load, the jumping performance would be improved. But till now, no study reported the 3D kinetic and kinematic data in standing long jump. The purpose of the study was to compare the kinetic and kinematic difference between restricted arm jump, normal jump, and loaded standing long jump.

METHODS: Six male junior high school students (age: 15.3±0.4 years, height: 177.2±6.11 cm, weight: 66.4 ± 5.87 kg) participated the study. All subjects did 10-mimute warm-up and performed restricted arm jump (RAJ), normal jump (NJ), and load 3.9kg standing long jump (LJ). Ten Vicon Mx13 cameras (200 Hz) were synchronized with a Kistler force platform (1000Hz) to collect the 3D biomechanical data. Sixty- nine reflective markers were placed on subjects' body and 15-segment model were used to calculate the variables. The raw data were smoothed by second order Butterworth low-pass filter with 6 Hz cut off frequency, and kinetic and kinematic variables were calculated using Visual 3D software.

RESULTS AND DISCUSSION: The LJ increased the jump distances (2.47m > 2.31m > 1.94m), the result is the similar to previous studies (Minetti & Ardigo,2002). The horizontal CG takeoff velocity was enhanced (3.57m/s > 3.49m/s > 3.09m/s), too. The vertical CG takeoff velocity was enhanced in RAJ (1.93m/s > 1.71m/s > 1.56m/s), but jump distance was decreased in RAJ. The peak GRFx was decreased in RAJ (646.6N < 796N), the result is the similar to previous studies (Ashby & Heegaard, 2002). The LJ had less hip extension (20.55 Nm < 27.00 Nm) and external rotation torque (10.24 Nm < 16.33 Nm), and less knee flexion torque (-19.50 Nm < -25.34 Nm).

Table1 : Kinematic variables and peak GRF of jumping performances

	RAJ	NJ	LJ
Distance (m) *	1.936±0.12	2.309±0.15	2.466±0.18
H Velocity (m/s) *	3.085±0.14	3.492±0.30	3.569±0.33
V Velocity (m/s) *	1.929±0.17	1.707±0.22	1.557±1.11
Peak GRFx *	646.6±102.1	796.0±143.9	716.2±63.5
Peak Hip Tx (Nm) *	21.63±2.70	27.00±3.59	20.55±4.62
Peak Hip Tz (Nm) *	13.95±3.86	16.33±3.58	10.24±3.38
Peak Knee Tx (Nm) *	-18.33±3.84	-25.34±4.49	-19.50±4.85

*$p<.05$ X direction: flexion (-) and extension (+); y direction: adduction (-) and abduction (+); z direction: internal rotation (-) and external rotation (+).

CONCLUSION: The LJ increased the jump distances, but RAJ decreased. The horizontal CG takeoff velocity was enhanced in LJ. The vertical CG takeoff velocity was enhanced in RAJ.

REFERENCES
Ashby, B. M. & Heegaard, J. H. (2002). Role of arm motion in the standing long jump. *Journal of Biomechanics, 35,1631*-1637.
Minetti, A. E.; Ardigo, L. P. (2002). Halteres used in ancient Olympic long jump. *Nature, November 2002; 420(14)*: 141-142.

WRIST POSITION AFFECTS HAND-GRIP STRENGTH IN TENNIS PLAYERS

Davide Susta and David O' Connell
School of Health and Human Performance, Dublin City University, Dublin, Ireland

In tennis the wrist is required to be in different degrees of orientation at ball impact depending on the stroke and type of shot being hit. To date, little is known about the interplay between wrist position and grip strength, despite the fact that hitting the ball and firmly holding the racket when the wrist is flexed has been suggested as factor predisposing tennis players to lateral epicondylitis. The aim of this study was to investigate the effect of different wrist positions on isometric grip strength at self-selected grip size. Thirty-seven tennis players performed three isometric contractions at each of the following wrist positions: neutral, extension, flexion, ulnar deviation and radial deviation. Maximal isometric grip force was measured at each wrist position with the use of a hand-held grip dynamometer and then the highest value at each position selected for analysis. Our results are as follows: at neutral the force exerted was 80.2 ± 22.07 (mean ± sd) kg, at wrist extension 56.99 ± 18.40 kg, at wrist flexion 33.96 ± 9.47 kg, at radial deviation 56.26 ± 19.39 kg and at ulnar deviation 56.64 ± 17.60 kg. Our findings show that, compare to the position defined as neutral, the maximum isometric force exerted by the fingers' flexor muscles is significantly affected (lowered) by wrist position ($p<0.001$).

KEY WORDS: Tennis, grip strength, wrist position

INTRODUCTION: Tennis strokes are performed by holding the racket while the wrist is in different degrees of orientation and depending on the stroke and type of shot being hit players have to manage grip forces and racket control at ball impact. While there are only six main strokes in the game of tennis (serve, forehand/backhand groundstroke, forehand/ backhand volley and smash) there are multiple variations in technique within these strokes. Technique, as well as wrist position, will vary depending on the individual playing style, the type of ball received and the type of shot a player wishes to produce. The two major factors affecting wrist position on ball strike are 1) impact point and 2) grip technique chosen by a player.
As pointed out by Rettig (2002) at ball contact on the serve the wrist is in extension and ulnar deviation, while during the forehand groundstroke and volley the wrist is usually extended and in ulnar deviation on impact with more advanced players showing a greater degree of extension compared to beginner players. In addition, players using a 'western grip' technique have a greater degree of ulnar deviation on contact with the ball compared to those using an 'eastern grip'. On the backhand groundstroke and volley the wrist of experienced players is in extension and 2-5° radial deviation at ball impact compared to novice players who tend to contact the ball in wrist flexion, a faulty and potentially harmful backhand groundstroke technique (Blackwell 1994).
Grip forces exerted by fingers flexor muscles are isometric and accordingly to the force-length relationship a tennis player should hit the ball when the forearm muscles are at a quasi-optimal length in order to reduce the relative strength used, increase control and minimise muscles and soft tissues overload. Surprisingly, the amount of information available from the scientific literature is limited and little is known about the relation between wrist position and grip strength in tennis players. Therefore, aim of the present study was to measure maximum isometric grip strength at five different wrist positions: neutral, extension, flexion, ulnar deviation and radial deviation.

METHODS: Thirty-seven healthy tennis players (male, age range 16 to 52 years) with no current or recent (< 6 months) hand, wrist or elbow injury participated in the study. Subjects tested were tennis players of various playing levels ranging between international (10), advanced (13) intermediate (12) and beginner (2) standards. The instrumentation used to measure isometric grip force was a hydraulic hand-held dynamometer with a dual scale readout displaying grip forces in kg and lb. The dynamometer we used has only discrete adjustable sizes so, as additional inclusion criterium participants self-selected grip size had to be 4½ inches. Subjects were tested in five different wrist positions which included neutral, maximal extension, maximal flexion, maximal ulnar and maximal radial deviation. For each wrist position subjects were instructed to move their wrist into the "end active range of motion" position, while holding the hand-dynamometer. They were then required to produce three two second maximal isometric grip efforts. Before each trial, subjects were asked to further move their wrist, while keeping the end range of motion position, to check the position was maximal and the trial recorded only when subjects were unable to further move the joint. We then selected the highest value out of three trials.

Data Analysis: A one-way repeated-measures ANOVA and a Bonferroni post-hoc test were performed to compare neutral versus other wrist positions grip strengths. An alpha level of 0.05 was used to denote statistical significance.

RESULTS: At neutral, fingers flexor muscles isometric grip force was 80.2 ± 22.07 kg (mean $\pm$ sd). At wrist extension hand grip force was 56.99 ± 18.40 kg, at wrist flexion 33.96 ± 9.47 kg, at radial deviation 56.26 ± 19.39 kg and at ulnar deviation 56.64 ± 17.60 kg.
The results show that wrist position affects the isometric force exerted ($F = 80.12$, $p<0.001$)
Post-hoc analyses revealed that the force the fingers' flexor muscles are able to exert isometrically, i.e. holding a hand grip of the same size of self-selected tennis hand grip, is lower at extension ($p<0.001$), flexion ($p<0.001$), ulnar deviation ($p<0.001$) and radial deviation ($p<0.001$) compared to the force exerted at neutral position. Wrist flexion was the position at which the highest force exerted was maximally lowered.

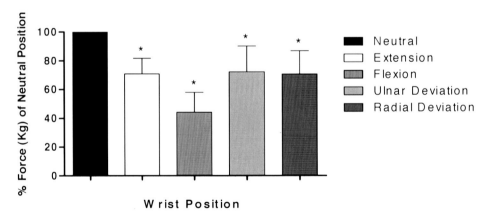

Figure 1: Grip strength at different wrist positions expressed as percentage of the maximum force produced with the wrist joint in neutral position.

DISCUSSION: Overuse of the extensor carpi radialis brevis (ECRB) muscle, one of the muscle determining wrist joint extension, can result in lateral epicondylitis by repetitive overloading microtraumas (Kraushaar, 1999). In a tennis player lateral epicondylitis is believed to result from the force of repetitive impacts between the racket and the ball, combined with a lack of wrist stability during the backhand stroke (Hatch, 2006). In the present study grip strength of male tennis players of various standards was significantly lower at all the wrist positions compared to neutral. The main possible explanation of these findings is that, taking for example wrist flexion, the position at which the grip strength showed the lowest values, the fingers flexor muscles are at a length which is not optimal to develop the highest force. Additionally, the wrist and fingers extensor muscles could play a role in limiting the force exerted by fingers flexors due to their length: when the wrist joint is in maximal flexion (and fingers flexed) wrist and fingers extensor muscles are almost at their physiologically maximal length and the fingers' flexors have to overcome this passive resistance to develop force. It's is likely that such a reduced strength could result in poor racket control and grip stability, which could be key factors in predisposing novice tennis players to repeated overloads whilst, for example, performing one-hand backhand strokes. Therefore, it is reasonable to suggest that at ball impact, being 'extreme' wrist positions associated with higher vulnerability, wrist position should be close to neutral, the position allowing to produce the highest hand grip force, in order to minimise soft tissues overloading. For these reasons, the results of the present study may have implications for a better understanding of the mechanism of wrist and forearm soft tissues injuries in tennis. Our findings adds further weight to previous studies which have proposed faulty backhand technique, i.e. with the wrist in flexion at ball impact, as a possible mechanism determining a sudden ECRB elongation, and provides a useful information to plan a training programme aimed at improving strength (and racket control) when the player has to hold the racket with the wrist in flexion, extension, radial deviation and ulnar deviation.

CONCLUSION: Our results showed that wrist position significantly affects the hand grip force that tennis players are able to produce and this reduced strength might play an important role in predisposing tennis players to repeated overloads, by decreasing racket stability and control. The information provided by our study is especially useful to novice tennis players or to those with poor backhand technique, i.e. prone to hit the ball when the wrist is in a position which is not neutral. Our findings also suggest that a specific training aimed at increasing forearm muscles strength at different wrist positions could be helpful and should included in tennis conditioning programmes to reduce the risk of soft tissues overloading.
Further research should investigate the effects of a different grip size on the hand grip strength exerted to hold and control a tennis racket.

REFERENCES:

Blackwell J, Cole K. Wrist kinematics differ in expert and novice tennis players performing the backhand stroke: Implications for tennis elbow. (1994) Journal of Biomechanics; 27:509–16

Kraushaar BS, Nirschl RP. Tendinosis of the elbow (tennis elbow): clinical features and findings of histological, immunohistochemical, and electron microscopy studies (1999). Joiurnal of Bone Joint Surgery America; 81:259-278

Hatch, G, Marilyn M. Pink, Karen J. Mohr, Paul M. Sethi and Frank W. Jobe. The Effect of Tennis Racket Grip Size on Forearm Muscle Firing Patterns. (2006) American Journal of Sports Medecine: 34; 1977-1989

Nirschl RP, Ashman ES. Elbow tendinopathy: tennis elbow. (2003) Clinical Sports Medecine;22: 813-836.

Rettig A.C. (2002). Hand and wrist injuries in tennis. In: Renström P.A.F.H (Editor), Handbook of sports medicine and science: Tennis, IOC, Blackwell Publishing, 223-232.

BIOMECHANICAL ANALYSIS OF THE HANDBALL IN AUSTRALIAN FOOTBALL

Lucy Parrington, Kevin Ball, Clare MacMahon and Simon Taylor
School of Sport and Exercise Science, Victoria University, Melbourne, Australia

The handball pass in Australian Football has become increasingly important in recent years. However, important technical elements of handballing have not been identified in the scientific literature. The purposes of this study were to provide a descriptive analysis of the handball through the evaluation of a player considered to have good technique, to compare handballs for maximal speed and accuracy, and to compare preferred and non-preferred hands. Three-dimensional data were collected from one elite level Australian Football player using Optotrak Certus. The player performed three handballs for maximal speed and three handballs for accuracy with both the preferred and non-preferred hand. Linear hand speed, linear shoulder speed, shoulder angular velocity and elbow angular velocity were larger in the maximal speed condition. Differences in the development of hand speed were found for preferred and non-preferred hands.

KEY WORDS: handball, Australian football, three-dimensional analysis.

INTRODUCTION: Within Australian football (AF), the two methods of passing the ball are kicking and handballing. Handballing permits quick movement to assist in catching the opposition off guard (McLeod and Jaques, 2006) and is becoming an increasingly important skill in AF. The handball involves gripping the ovoid shaped ball (similar to American football/ rugby ball) with a stationary platform hand; punching the ball from the platform hand using the area between the thumb and forefinger of the punching hand; connecting with the ball near the back point where the laces meet; stepping toward the target to gain power and distance; and to follow-through with the punching hand in a motion upwards and towards the target (McLeod and Jaques, 2006). Figure 1 displays a pictorial sequence of the handball.

Figure 1: Handball sequence

Game trends over the past ten years indicate increasing use and importance of the handball in the game of AF. Since 1998, the total number of handballs per Australian Football League season has increased by 45%, from 224 to 325 per game. Additionally, the ratio of kicks to handballs has changed, with handballs contributing 44% of disposals (i.e. kick or handball) in 2008 compared to only 35% in 1999 (Champion data, 2008).

Handballs are performed for both accuracy and speed of pass, and with both hands. Game-based handball analysis (Parrington et al., 2008) demonstrated both left and right hands were used for handballing (41% and 59% for the left and right hands respectively). Further, handballs were produced over short (less than 2 m) and long distances (greater than 10 m); and were performed under different time constraints with some passes predominantly requiring accuracy only while others additionally required greater speed of ball flight to reach the intended target before an opponent could intercept the pass. As such, the ability to dispose the ball effectively with either the left or right hand is an important skill (B. Gotch, AFL Development Coach, personal communication). Additionally, there are increased options when the ball can be passed over both short and long distances.

In spite of its increased importance, there have been no scientific studies performed on handballing with biomechanical research largely focused on kicking (e.g. accuracy kicking, Dichiera et al., 2006; distance kicking, Ball, 2008). This study aimed to evaluate handball technique, and identify technical differences that exist between distance and accuracy passes and between preferred and non-preferred hands.

METHODS: Data Collection: One male elite AF player (19 years, 192cm, 80kg) considered by the coaches at his club to have good handball technique participated in this study. Handpasses were performed using a Sherrin football (used in competition) for maximal speed/ distance and accuracy. Six trials were captured for maximal speed (three on each hand) and six trials were captured for accuracy (three on each hand). For maximal trials, the participant was instructed to perform a handball with the aim of passing the ball to another player 10 m away as fast as possible by propelling the ball with maximal speed. For accuracy trials, the participant was instructed to attempt to hit a 0.5 m x 0.3 m rectangle (6.5 m from the test area with the bottom edge positioned at a height of 1.5 m).

Prior to testing, rigid body markers composed of clusters of light emitting diodes [Optotrak Certus (Northern Digital Inc., Ontario, Canada)] were placed on the upper extremity (upper arm, forearm and hand), lower extremity (shank, thigh and foot) and torso (upper trunk and pelvis). Elastic neoprene wrap assisted the attachment of rigid bodies to each segment and sports strapping tape was used in addition to tightly secure each placement.

Joint centres were digitised at the shoulder, elbow, wrist, metacarpophalangeal joints two and five, hip, knee, ankle, and metatarsophalangeal joints two and five. For each trial, a three tower Optotrak motion analysis system, sampling at 100Hz captured 3D coordinates of these markers during handball execution from the initial forward step until after the end of follow through. The player landed his front foot on an AMTI OR6-5-1 force plate (Advanced Mechanical Technologies, Inc, Massachusetts, USA). Two dimensional video was used to determine the time between front foot heel strike and ball contact. One researcher recorded target accuracy, with a score of zero allocated to a direct hit, and scores of one, two and three as the ball moved in sections 0.3 m, 0.6 m and 0.9 m away from the central square.

Data Analysis: Three-dimensional trajectory values were reconstructed using Visual3D (C-Motion, Inc., Maryland, USA). Raw data were interpolated and smoothed using a lowpass Butterworth filter (10Hz cutoff frequency), then were transferred to Microsoft Excel. Data were normalised from the onset of downswing (defined as the first downward motion of the hand after backswing) to ball contact (100%). Mean, standard deviation and effect sizes (Cohen's *d*) were calculated to provide additional information. However, effect size values should be viewed with caution due to the small number of trials analysed.

RESULTS: Parameter definitions and mean values are presented in Table 1. Differences existed between the maximal pass (MP) condition and the accuracy pass (AP) condition. All linear and angular velocities, step length and 3D shoulder angle were larger for the MP condition, while larger values were recorded for the AP for hand height and trunk angle. Differences also existed between the preferred hand (P) trials and non-preferred hand (NP) trials. Step length, 3D shoulder angle, linear velocities, trunk angle and forearm angular velocity were all larger for the P, while 3D elbow angle, shoulder angular velocity and upper arm angular velocity were larger for the NP. Greater shoulder range of motion (ROM) was matched with smaller elbow ROM for the NP trials. Accuracy was better in the P than NP.

DISCUSSION: Speed of the distal segment at ball contact is a major determinant of ball speed in kicking (Ball, 2008), required for maximal based passes. Hand speed was thus used as an indicator of performance, along with accuracy scores in this study. Based on the magnitude of the values, shoulder angular velocity appears to be a major contributor to hand speed. Elbow angular velocity contributes comparatively less, and although values were small (between 0.5 and 1.2m/s) shoulder speed appears to play a role.

Table 1 Kinematic Variables (Parameters refer to punching hand at ball contact unless stated).

Parameter	Definition	Condition	Preferred		Non-preferred	
			mean	SD	mean	SD
Step length (m)	Distance between the left and right foot. Measured from the fifth metatarsal head.	Speed	1.07	0.04	0.96	0.02
		Accuracy	0.94	-	0.89	0.05
Hand height (m)	The vertical distance from the hand to the ground	Speed	0.77	0.02	0.76	0.01
		Accuracy	0.82	0.01	0.83	0.02
Hand speed (m/s)	Speed of the hand segment	Speed	10.4	0.2	10.1	0.3
		Accuracy	9.8	0.6	9.4	0.4
Shoulder speed (m/s)	Linear speed of the shoulder	Speed	1.2	0.1	0.7	0.1
		Accuracy	1.0	0.01	0.5	0.2
3D Shoulder angle (°)	Angle measured between the trunk and upper arm segment	Speed	36	1	31	5
		Accuracy	32	2	30	3
3D Elbow angle (°)	Angle between the upper arm and the forearm segment	Speed	96	1	99	3
		Accuracy	95	0.3	100	6
Trunk angle (°)	Angle of the trunk in the 2D sagittal plane	Speed	55	2	48	4
		Accuracy	57	1	53	5
Shoulder angular velocity (°/s)	Angular velocity of the shoulder	Speed	268	71	565	103
		Accuracy	230	-	503	123
Elbow angular velocity (°/s)	Angular velocity of the elbow	Speed	-39	4	-35	13
		Accuracy	-32	-	-33	3
Shoulder ROM (°)	Difference between minimum and maximum shoulder angle measured between the onset of down swing and ball contact	Speed	31	12	37	9
		Accuracy	20	-	42	18
Elbow ROM (°)	Difference between minimum and maximum elbow angle measured between the onset of down swing and ball contact	Speed	12	-	5	-
		Accuracy	17	-	2	-
Accuracy	Lower values represent hits closer to the target	Accuracy	2		4	

*Some standard deviations are not reported. Due to marker occlusion, values could not be calculated for some parameters for all trials (i.e. N < 3).

Greater hand speed was found in the MP condition (P, d=1.2, NP, d=1.8). Similarly, the MP condition revealed greater values for shoulder speed (P, d=3.4, NP, d=1.6), angular shoulder velocity (P, d=0.8, NP, d=0.5) and elbow angular velocity (P, d=0.7, NP, d=0.2) than the AP condition. The data presents similar achievement of hand speed across hands. However, there was a difference in the contribution of the shoulder speed and shoulder angular velocity in the development of this speed. Shoulder speed values for P (1.2m/s and 1.0m/s) were approximately double that found for NP (0.7m/s and 0.5m/s respectively). In comparison, shoulder angular velocity in the P (268°/s and 230°/s) was just under half of the NP value (565°/s and 503°/s respectively). This demonstrates a similar outcome of hand speeds were achieved through different movements for the P and NP for this player.

The larger step length for the MP (NP, d=1.7) supported the coaching literature that states stepping toward the target is important in order to gain power and distance (McLeod and Jaques, 2006). For this player, a larger step length was evident in the MP condition for both P and NP. The increase in step length may assist in developing speed of the body towards the target, demonstrated by larger shoulder speed for the MP task. In turn the greater shoulder speed might contribute to the larger hand speed also displayed in the MP condition. Competency to perform handballs with either hand is important in elite competition. In comparing performance, accuracy was greater for the P, and while hand speed between hands was quite similar, it was greater in the P (MP, d=1.1, AP, d=0.7). As discussed above, differing contributions of shoulder speed and shoulder angular velocity provide similar performance of hand speed. Although hand speed performance was similar across hands, results suggest a less developed motor skill for the NP. Greater shoulder speed (MP, d=7.4,

AP, d=3.8) and elbow angular velocity (for MP; d=0.5), and lower shoulder angular velocity (MP, d=3.4) in the P trials suggests the development of hand speed occurs through the contribution of a number of parameters in the preferred arm. In comparison, it appears that the development of hand speed in the NP occurs mainly through greater shoulder angular velocity. The higher contribution to the movement from a number of parameters in the P trials indicates a releasing of the degrees of freedom in the preferred arm (Vereijken et al., 1992). In comparison, the arm of the NP locks the elbow (only 2 and 5 degrees of movement) during down swing, prior to ball contact. Moreover, the less skilled arm appears to compensate for a lack of range of the elbow joint, by increasing ROM at the shoulder. Releasing degrees of freedom has been shown to occur in the later stages of learning (Vereijken et al., 1992). Greater accuracy in the P trials also indicates a more refined skill in the preferred arm.

Differences between MP and AP relate to the player developing greater hand speed at ball contact in the MP condition. Larger values were found for shoulder speed, angular shoulder velocity and elbow angular velocity. These differences may be attributed to the attempt to swing faster in the MP to attain maximum speed/ distance; and/or a slower controlled swing in the AP in order to establish control and stabilise the hand at ball contact for accuracy. Nonetheless, similarities in the pattern of values between the MP and AP indicate a similar, but scaled movement pattern for this player.

The main limitations of this study relate to the positioning of the motion capture towers and the target in the AP condition. First, positioning of the cameras made tracking of the non-punching arm difficult. Modifications to tower locations may allow greater capture of both arms during data collection. Second, the positioning of the body in relation to the target is needed to gain an understanding of segment orientations. Future directions for study could involve looking at segment orientations in reference to the target, and parameters in the non-punching arm and the lower extremities. Further study with more participants would be beneficial to establish whether any values reported show significant differences or trends.

CONCLUSION: This study explored the kinematics of handballing in Australian football through the comparison of maximal and accuracy passing, and preferred and non-preferred hands. All linear speed and angular velocity values were greater for the MP. Different contributions of shoulder speed and shoulder angular velocity were used to develop similar hand speed values across hands, while accuracy was greater for the P. Findings require further investigation with more participants to establish any significant differences or trends.

REFERENCES:

Ball, K. (2008). Biomechanical considerations of distance kicking in Australian Rules football. *Sports Biomechanics*, 7(1), 10-23.

Champion Data. (2008). Statistical analysis of 2008 AFL game data. Technical Report for AFL and AFL clubs. Melbourne: Champion Data.

Dawson, B., Hopkinson, R., Appleby, B., Stewart, G., and Roberts, C. (2004). Player movement patterns and game activities in the Australian Football League. *Journal of Science and Medicine in Sport*, 7(3), 278-291.

Dichiera, A., Webster, K., Kuilboer, L., Morris, M., Bach, T., and Feller, J. (2006). Kinematic patterns associated with accuracy of the drop punt kick in Australian Football. *Journal of Science and Medicine in Sport*, 9, 292-298.

McLeod, A., and Jaques, T. (2006). *Australian football: steps to success* (2nd ed.). Champaign, IL: Human Kinetics.

Parrington, L., MacMahon, C., and Ball, K. (2008). Game-based handball analysis. Technical Report for the Western Bulldogs Football Club.

Vereijken, B., Whiting, H., Newell, K., and van Emmerik, R. (1992). Free(z)ing degrees of freedom in skill acquisition. *Journal of Motor Behavior*, 24(1), 133-142.

Acknowledgement
This study was made possible by the support of the Western Bulldogs Football Club and its players.

EFFECTS OF INDEPENDENT CRANK ARMS AND SLOPE ON PEDALING MECHANICS

Saori Hanaki-Martin, David R. Mullinaeux and Stacy M. Underwood

Department of Kinesiology and Health Promotion, University of Kentucky Lexington, Kentucky, USA

The aim of this study was to identify the effects of independent crank arms and slope on pedaling kinetics during an anaerobic maximal-effort cycling bout. After undergoing 6 weeks of training with independent crank arms, each of 6 male cyclists completed four 30 s Wingate tests under different cycling conditions of: fixed crank arms on level surface; fixed crank arms on a slope; independent crank arms on level, and; independent crank arms on a slope. Two-dimensional pedal forces recorded using instrumented pedals were used to derive pedaling effectiveness, work distribution and power output. The effects of the crank arms and the slope were minimal, but highly effective and consistent pedaling force (90% effectiveness, 70% work and effective force of 155±6 N) was observed between 45-135° of the crank cycle in all experimental conditions.

KEYWORDS: cycling, efficiency, exercise, power, work

INTRODUCTION:

Road cycling has become more competitive as it has gained popularity. Specifically, ability to perform well on incline is thought to be associated with success in competition. More innovative tools have been adapted by cyclists to become successful. Independent crank arms (IND) are one of these tools that are intended to improve cycling technique. These crank arms move independently, so the rider is prohibited to rely on one leg's action to move the other. The manufacturers claim that the benefit of IND would result in promoting more active recruitment of muscles that are not typically used with the conventional crank arms (FIX) during the upstroke (PowerCranks Science, 2006). Empirical studies have provided inconclusive findings, where one study reported improved gross efficiency after a 6-week training period with IND (Luttrell & Potteiger, 2003) others reported that their effects on physiological functions and power output were minimal (Lucia et al., 2004 & Santalla et al., 2002). A comparison between two independent groups that underwent a short 5-week training period with the IND versus FIX reported no difference in power output, but a modified work distribution pattern was observed in the IND group (Bohm et al., 2008). Cyclists anecdotally report changed cycling techniques while cycling with the IND (Luttrell & Potteiger, 2003) while effects of such crank arms had not been investigated. Examining pedaling patterns while using the IND may clarify the uncertainty in their training effects reported previously. Additionally, it is beneficial to investigate pedaling techniques on incline, as hills in competitive cycling are unavoidable. In obtaining better understanding of pedaling kinetics, examining work distribution as well as the conventional index of pedaling efficiency has been suggested as it describes contributions of different pedaling phases to revolution (Bohm et al., 2008). Therefore, the purpose of the present study was, after 6 weeks of familiarization training, to investigate the effects of the IND and an incline on the pedaling power, effective force and indices of pedaling effectiveness and work distribution.

METHODS: Six male cyclists (24.7 ±3.8 years, 1.80 ±0.04 m, 73.1 ±4.5 kg) who were members of a local cycling team volunteered in the study. In accordance with the study protocol, all subjects were 18-30 years of age, and had been regularly cycling and were free of injury or illness at the time of study. For 6 weeks, each subject trained 3 times a week with a road bike equipped with a pair of IND (PowerCranks, Walnut Creek, CA, USA) that was mounted on a fluid-resisted cycle trainer (Minoura, Gifu, Japan). The subjects were instructed to gradually increase the amount of time spent riding with the IND setting during the training. Each training session lasted 60 minutes. During the 7th week, all subjects performed 4 separate Wingate tests that were 24 hours apart. The 4 testing conditions were:

1) FIX-level (FL); 2) FIX-slope (FS); 3) IND-level (IL), and; 4) IND-slope (IS). The order of tests was randomly assigned.

The load for the Wingate test was set using a front to rear cog ratio of the subject's body mass (BM kg) of BM*1.00 to 14 (actual rear cog size) for the level (0% incline), and BM*1.12 to 14 for the slope (17.6% incline). The ratio for the 0% grade was determined from pilot testing, and the ratio for the 17.6% grade was derived mathematically as the increased workload required for the increased grade equates to the product of the cyclist's body weight and tangent of the grade (Mognoni & di Prampero, 2003). The gear ratio was controlled using the virtual gear function of a stationary bike with an electro-magnetic brake (Velotron Elite, RacerMate, Inc., Seattle, WA, USA). Retro-reflective markers were placed on the pedals, crank arms, crank axis, and the cyclist's foot to track the motion of the pedal and the crank arms. After a 10-minute warm up, the subject increased the pedaling cadence to their maximum and the test was started. During the 30-second testing, positions of the crank arms and pedals were recorded using 12 high-speed cameras and Cortex software v1.0 (Motion Analysis Corp., Santa Rosa, CA, USA) at 200 Hz. A pair of custom instrumented pedals (Newmiller et al., 1988) mounted on the stationary bike were used to record the vertical and antero-posterior pedal forces at 1000 Hz. The motion and the force data were synchronized and collected simultaneously.

In the present study, only the right pedal data were included. The dependent variables derived from the collected data were: 1) the single leg power (SLP); 2) effective force (F_{Eff}) that is defined as the pedal force perpendicular to the crank; 3) the index of pedal effectiveness (IE, the ratio of the useful force to cause crank torque to the total force applied to the pedal (Coyle et al., 1991)) for the complete pedal cycle (IE_{360}) and for 4 sectors of the crank cycle (IE_{down}=45-135°; IE_{back}=136-225°; IE_{up}=226-315°; IE_{fore}=316-45°), and; 4) percent work for the 4 sectors (%W_{down}; %W_{back}; %W_{up}; %W_{fore}) for each revolution (Bohm et al., 2008). All variables were averaged over all pedal cycles performed during the 30 second trial. After the normality distribution of the variables was tested, the effects of the crank type (FIX/IND) and slope (level/slope) on each variable were examined by repeated measures ANOVA (for normally distributed data) or Friedman test (for non-normally distributed data) at α = 0.05. The means, 95% confidence intervals, and the effect sizes of the variables were determined with Bonferroni adjustment when appropriate (SPSS, v.17, SPSS Inc., Chicago, IL, USA).

RESULTS AND DISCUSSION: Both IND and slope were associated with lesser numbers of revolutions during the 30-second Wingate test [FIX v. IND: 40.2 - 42.2 v. 37.4 - 41.2, F =

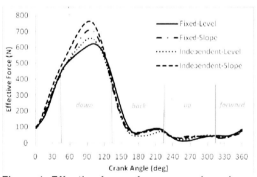

Figure 1. Effective forces for one crank cycle of a single subject during maximal effort for 4 cycling conditions

Table 1. Single leg power and effective force means (SD) during a 30-s maximal effort for 4 cycling conditions

	Single Leg Power [W]	Effective Force [N]
FL	363 (27)	215 (16)
FS	364 (20)	231 (23)
IL	336 (38)	207 (16)
IS	345 (36)	224 (17)

FL: fixed-level; FS: fixed-slope; IL independent-level; IS: independent-slope

8.00, p = 0.037, η_p^2 = 0.62; level v. slope: 39.9 – 42.8 v. 37.4 – 41.2, F = 49.71, p < 0.01, η_p^2 = 0.91]. The decreased cadence on incline has been previously reported and the reduction appears to contribute to an increase in crank torque (Caldwell et al., 1998). The greater F_{Eff}

in slope conditions in the present study [$\chi^2(3) = 11.0$, $p = 0.01$] (Table 1) might indicate differences in crank torque as F_{Eff} is directly related to crank torque. Although the source of the difference was not analyzed, a typical F_{Eff} curves from one of the subjects (Figure 1) showed greater force magnitudes between about 70-110° for slope conditions, an implication of a raise in crank torque associated with incline. However, the average SLP did not differ across 4 experimental conditions [$\chi^2(3) = 6.20$, $p = 0.10$] (Table 1). These results suggest that the cyclist maintains power output by increasing F_{Eff} to compensate the lower pedaling cadence while cycling on an incline.

IE_{360} did not differ across conditions [crank: $F = 0.89$, $p = 0.39$, $\eta_p^2 = 0.15$; slope: $F = 0.69$, $p = 0.47$, $\eta_p^2 = 0.11$] that indicated no change in overall pedaling effectiveness while using IND on the level and incline. However, when the IE was determined separately for difference sectors of the pedal cycle, it appeared that the cyclists used different techniques to accommodate different riding conditions (Figure 2). IE_{back} was influenced by the slope indicating that the crank torque was generated more effectively between 135° and 225° on incline [level v. incline: 21.0% - 55.1% v. 32.1% - 64.8%, $F = 63.44$, $p < 0.01$, $\eta_p^2 = 0.93$]. This might be related to the position of the cranks relative to the gravitational force. Due to the offset of the angle by the incline, the gravity acted differently on the limb, pedal, and crank. Brown et al. (1996) reported that how gravity acts on the body alone affected muscle activation pattern. Therefore, it is possible that the gravitational force might have influenced muscle recruitment in this and other sectors. The line of action for the gravitational force relative to the bike could explain the difference observed in the forward sector IE across the conditions [$\chi^2(3) = 11.4$, $p < 0.01$]. IE_{up} varied greatly between subjects. With FIX, 2 of the subjects exhibited negative IE_{up} values. Though insignificant, the mean IE_{up} increased with use of IND and made between-subject variability smaller [FIX: 45.4±19.8% v. IND: 49.3±9.8%]. Unlike IEs in aforementioned sectors, IE_{down} was consistently high (average 89% - 91%) with a low deviation across all conditions. IE_{down} has been shown to be consistent at a wide range of power outputs (Bohm et al., 2008). In the present study, it was shown that a 10° shift in crank angle (i.e. slope) did not affect it. This consistency of IE in down phase might be because the cyclists had already learned to optimize the down phase from their cycling experience. It might also imply that the muscles function more favorably and are minimally affected by the gravity within this particular range of crank cycle. The minimally affected total and sectional IEs associated with IND might suggest that the IND do not affect the pedaling effectiveness. However, the incline appeared to improve the pedaling effectiveness during the backward section regardless of the crank arm type.

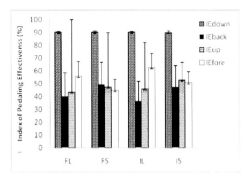
Figure 2. Index of pedaling effectiveness (mean±SD) over 30 s of maximal effort for 4 sectors for different cycling conditions
(FL: fixed-level; FS: fixed-slope; IL independent-level; IS: independent-slope)

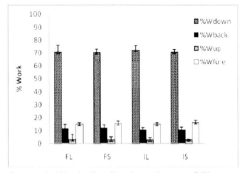
Figure 3. Work distributions (mean±SD) over 30 s of maximal effort for 4 sectors for different cycling conditions
(FL: fixed-level; FS: fixed-slope; IL independent-level; IS: independent-slope)

The work distribution in the downward phase was also consistently high and was not affected by the crank or slope conditions [$\chi^2(3) = 1.40$, $p = 0.71$] (Figure 3). The $\%W_{up}$ was not affected by the experimental conditions. Additionally, the %W_{up} was positive in all conditions. A previous study investigated pedaling work distributions observed negative work during the upward sector (Bohm et al., 2008).That observation might be as a result of different testing protocol. Since this study involved a short, all-out trial without cadence restriction, the subjects were able to actively pull up their feet quickly during the up-stroke throughout the trial. The only difference observed in the work distribution was the %W_{back}. IND had reduced the amount of work contributed to this phase [FIX v. IND: 9.2% - 14.2% v. 8.6% - 12.4%, F = 17.67, p = 0.008, η_p^2 = 0.78]. This could be because the leg has no need to drive the contra-lateral leg to clear the top dead center (i.e. 0°).

CONCLUSION: Work distribution and IE changes resulted from independent crank arms and slope during a short anaerobic cycling bout were limited to modified effectiveness in the back and forward sectors. There was no change in overall pedaling effectiveness. The power output was maintained primarily by sustaining the effectiveness during the downward sector of the pedaling cycle. Both the independent crank arms and the slope were associated with lesser number of pedal revolutions that appeared to be related to an improvement in effective force production. These minimal changes are based on a short cycling session, and the effects of independent crank may differ in a longer cycling session. Investigations including more subjects, longer training, and different cycling ability may yield different results. To obtain the comprehensive understanding of the effects of the independent crank arms, other measurements, such as joint moments and electromyography should also be considered in future investigations.

REFERENCES:
Bohm, H., Siebert S., & Walsh M. (2008). Effects of short-term training using SmartCranks on cycle work distribution and power output during cycling. *European Journal of Applied Physiology*, 103, 225-232.
Brown, D. A., Kautz, S. A., & Dairaghi, C. A., (1996). Muscle activity patterns altered during pedaling at different body orientations. *Journal of Biomechanics*, 29, 1349-1356.
Caldwell, G.E., McCole, S.D., Hagberg, J. M. & Li, L. (1998). Pedal and crank kinetics in uphill cycling. *Journal of Applied Biomechanics*, 14, 245-259.
Coyle, E.F., Feltner, M. E., Kautz, S. A., Hamilton, M. T., Montain, S. J., Baylor, A. M., Abraham, L. D., & Petrek, G. W. (1991). Physiological and biomechanical factors associated with elite endurance cycling performance. *Medicine and Science in Sports and Exercise*, 23, 93-107.
Lucia, A., Balmer, J., Davison, R.C.R., Perez, M., Santalla, A., & Smith, P.M. (2004). Effects of the rotor pedalling system on the performance of trained cyclists during incremental and constant-load cycle ergometer tests. *International Journal of Sports Medicine*, 25, 479-485.
Luttrell, M.D., & Potteiger, J.A. (2003). Effects of short-term training using powercranks on cardiovascular fitness and cycling efficiency. *Journal of Strengthening and Conditioning Research*, 17, 785-791.
Mognoni, P., & di Prampero P.E. (2003). Gear, inertial work and road slopes as determinants of biomechanics in cycling. *European Journal of Applied Physiology*, 90, 372-376.
Newmiller, J., Hull, M.L., & Zajac, F.E. (1988). A mechanically decoupled 2 force component bicycle pedal dynamometer. *Journal of Biomechanics*, 21, 375-386.
PowerCranks, (2006). Science: Some science behind PowerCranks and cycling improvements. Retreived April 10, 2009, from http://www.powercranks.com/sports/Studies/cycling%20science.htm
Santalla, A., Manzano, J.M., Perez, M., & Lucia, A. (2002). A new pedaling design: the Rotor-effects on cycling performance. *Medicine and Science in Sports and Exercise*, 34, 1854-1858.

Acknowledgement
The authors would like to thank PowerCranks Inc. for providing the independent crank arms used in this study.

THE MOTION ANALYSIS OF WALK ON RACE WALKING PLAYERS

Yoshinori Takeuchi, Yoshinari Oka, Tatsuya Nishimura*, Kenji Kawabata*, Kengo Sasaki*, Ami Ushizu*, Yu Nakashima and Hiroh Yamamoto

Biomechanics Lab., Fac. of Ed., Kanazawa Univ., Kanazawa, Japan
*Biomechanics Lab., Graduate School of Ed., Kanazawa Univ., Kanazawa, Japan

The purpose of this study was to determine what characteristic race walking players have in their daily walking motion. The healthy four female race waking players and healthy four female university students were determined as subjects in this study. The 13m walking road was set. Each group conducted three Normal Walking trials (NW) and three Brisk Walking trials (BW). After that, the Race Walking group conducted three Race Walking trials (RW). The 3D motion analysis system (Frame Dias②ver.3.0 for windows) was used to calculate parameters. As to NW knee segment parameters, the maximum knee extension angle, range of motion and average knee angles of race walking group were significantly different from that of control group ($p<0.05$). The most important discovery of this study was that race walking player's daily walking motion was influenced by their race walking characteristic.

Key Words: Motion Analysis, Race Walking

INTRODUCTION: The walking motion of race walking in track and field is greatly-different from that of daily walking. So, we think that race walking players maintain their race walking characteristics in daily walking motion. However, there are no reports that investigate biomechanics difference of normal walking motion between race walking players and general people. The purpose of this study was to determine what characteristic race walking players have in their daily walking motion.

METHODS: Subjects: The subject of this study were healthy four female race waking players (Race Walking group : W) and healthy four female university student (Control group : C). Informed consent was obtained from all subjects.
Design: The 13m walking load was set. The Race Walking and Control group conducted three Normal Walking trials (NW) and three Brisk Walking trials (BW). After that, the Race Walking group was conducted three Race Walking trials (RW).
Data Collection: The definition of calculated these kinematics parameter were shown Figure 1-3. Observed from above, as to the θ3, left ASIS was regarded as plus when it goes forward. Observed from forward, as to the θ4, when horizontal ASIS was horizontal against ground, it was regarded as standard. In addition, condition that it turns down for left side was defined as plus. The definition of kinematics parameter was shown Table 1. Reflection markers were attached on C7, right and left shoulder, ASIS, knee, foot, heel and toe (M5) for calculation of parameter. The 3D motion analysis system (Frame Dias②ver.3.0 for windows) was used to calculate parameters. The synchronized four video cameras were used to record motion.

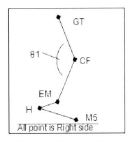
Figure 1: Knee angle

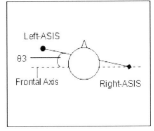
Figure 2: Pelvis revolution

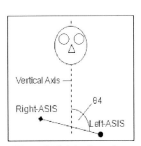
Figure 3: Pelvis tilt angle

Table 1: Define Kinematics parameters

Parameters	Define
Maximum Knee Extension Angle	Maximum θ_1
Maximum Knee Flexion Angle	Minimum θ_1
Range of Knee Motion	Maximum θ_1 - Minimum θ_1
Average Knee Angler	Average θ_1 in 1 cycle
Range of pelvis revolution	|Maximum θ_3| |Minimum θ_3|
Maximum pelvis left tilt angle	Maximum θ_4
Maximum pelvis right tilt angle	Minimum θ_4
Range of pelvis tilt angle	Maximum θ_4 - Mimimum θ_4
Average pelvis tilt angle	Average θ_4 in 1 cycle

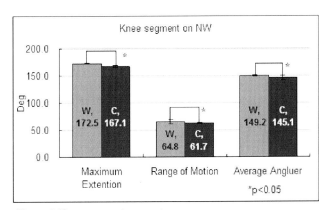
Figure 4: Knee segment parameters

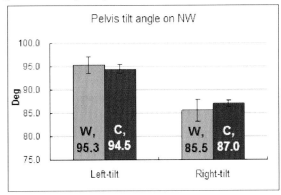
Figure 5: Pelvis parameters

Data Analysis: Paired t-test was used to determine differences between conditions. Significant level was set at 5%.

RESULTS: The results of comparing Race walking and Control group on Normal Walk were shown Figure4-6.
As to knee segment parameters (Fig.4), the maximum knee extension angle, range of motion and average knee angles were significantly different between race walking and control group (p<0.05). And the race walking group was wider than control group.

As to pelvis, left and right tilt angles were not significantly different between race walking and control group (Fig. 5). The pelvis tilt range of angles was significant different between race walking group and control group (Fig. 6). And the race walking group was wider than control group.

DISCUSSION: The subjects of race walking group who were trained as specialist of race walking tends to extend overly in 1 cycle walking in normal walking compared to control group. And, race walking group tended to tilt left and right pelvis deeper than that of control group, when race walking group walk BW. The race walking players who are subject of this study walked with their knee and pelvis mainly used at normal walking that is never fast. It is speculated that their walking way was reflected by their walking motion. The W subjects who were trained as race walking specialist, in both NW and BW compared to C, tended to extend knee widely to make their motional range wide. In addition, race walking group tilted pelvis wider than that of control group from when race walking group walk by 1.4(m/s). And, they convoluted their pelvis more actively when they walk by 2.0(m/s).

CONCLUSION: In this study, female university race waking players were mainly determined, but we never focused on the number of subjects. At further study, the larger numbers of subjects need to be determined including high school student and civilians. Sexual distinction needs to be considered and data of this study need to be compared to that of male race walking players. Finally, in further study, side motion like inversion and extroversion needs to be considered because, in this study, only longitudinal motion was focused.

REFERENCES:

Arne Nagel, Frauke Fernholz, Carolin Kibele, Dieter Rosenbaum. Long distance running increases plantar pressures beneath the metatarsal heads.　Gait&Posture,　2008 ; 27 : 152-155
Gabor Barton, Paulo Lisboa, Adrian lees, Steve Attfield. Gait quality assessment using self-organising artificial neural networks.　Gait&Posture,　2008 ; 25 : 374-379
Matthew K. Seeley, Brian R. Umberger, Robert Shanpiro. A test of functional asymmetry hypothesis in walking. Gait&Posture, 2008 ; 28 : 24-28
Murray T Vanderpool, Steven H Collins, Arthus D Kuo. Ankle Fixation need not increase the energetic cost of human walking. Gait&Posture, 2008 ; 28 : 427-433
5. Sadeghi H, Allard P, Prince F, Laabelle H. Symmetry and limb Dominance in able-bodied gait. Gait&Posture,　2000 ; 12 : 34-45
Sasaki K, Neptune R. Differences in muscle function during walking and running at the same speed. Journal of Biomechanics,　2006 ; 39 : 2005-2013
Wendy J, Terese L. Differences in normal and perturbed walking kinematics between male and female athletes. Clinical Biomechanics, 2004 ; 19 : 465-472

PRELIMINARY EVALUATION OF A PRECISE STARTING SENSOR FOR SHORT DISTANCE ATHLETIC SPORTS BELOW 400 m

Jeong-Tae Lee[1], Han-Wook Song[1*] and Cheong-Hwan Oh[2], Chan-Ho Park[2], Eun- Hye Huh[2], Jin Lee[2]

[1]Div. of Physical Metrology, KRISS, Daejeon, South Korea
[2]Chungnam National University, Daejeon, South Korea

The importance of starting a race in short distance athletic sports below 400 m was rarely considered to the extent that it should be. The main research theme in this field has mainly been the relationship between the starting signal and the response speed of leg muscles. The records in short distance athletic sports have been improved through training athletes to increase their response speed. However, the improvements in records have also been due to the starting time speed; thus, there is another way to improve times, that is, through the starting speed. The starting speed related to the kicking force against the starting blocks at the start of a race. The objectives of this research were to present a method for analyzing forces acting upon a starting block at the start of a race and to optimize the starting conditions for each athlete. To achieve these objectives, a starting block with Wheatstone bridge type strain gauges which could measure, in normal and horizontal directions, the repulsive forces acting on the starting blocks at a starting point in real-time, was developed. The use of this block was expected to correct the posture of each athlete and record the sports dynamics data for each athlete.

KEY WORDS: starting block, strain gauge, athletic.

INTRODUCTION: The importance of starting in short distance athletic sports below 400 m has begun to be recognized for a long time. Currently, starting blocks are used as tools to make starting easier in every race of short distance athletic sports and, in effect, to improve records. The starting block helps increase the kicking force of athletes' feet. It is possible to control the lengths and angles of starting blocks according to individual athletes to improve their starting speed. Due to the structure of starting blocks, before track athletes start a race, no forces are applied to the footboard. When the athletes start, they push the footboard with their feet. Kicking forces to the footboard are distributed vertically to the longitudinal direction of the footboard. Also, forces are exerted on connection parts that support and fasten the footboard. Forces that are exerted on the connection parts are distributed horizontally to the surface, (Aron et al, 2003; Bret et al 2003). This study presents a method to analyze forces that affect starting blocks at the start of a race and which will, ultimately, help each athlete to determine his or her optimal starting conditions. In short distance track races, the start is the most important point of the race. For this study, existing starting blocks were used. These starting blocks help athletes reach their maximum speed by using the greatest amount of power and physical balance along with the smallest amount of energy in the shortest period of time. Repulsive force sensors were developed and attached to the starting blocks according to their sizes. These sensors will correct the postures of short distance track athletes at the starting line and measure the repulsive forces of each athlete, thus contributing to record improvements. The purpose of this study is to provide equipment that can measure the exact starting time of an athlete by installing a sensor to the parts of starting blocks where forces are applied and where they change.

METHODS: As shown in Figure 1, starting blocks are composed of the footboard that an athlete's feet meet and fixed ends that fix the footboard to the ground by the anchor. The vertical direction force (F_t) to the footboard occurs orthogonally to the plane of the footboard. This vertical direction force (F_t) is measured by a normal sensor. The connection part receives the horizontal direction force (F_x) from the kicking force to the footboard. The horizontal direction force (F_x) is measured by a horizontal sensor.

By measuring this vertical direction force (F_t) and horizontal direction force (F_x), the gravitational direction force (F_z) that allows an athlete to stand can be calculated. The angle (θ) that is formed by vertical direction force (F_t) and horizontal direction force (F_x) is the same as the one formed by the orthogonal direction (θ) of the plane of the footboard, the ground of the support, and the horizontal direction. Therefore, according to Pythagoras's theorem, gravitational direction force (F_z) is calculated as Formula (1), shown below.

$$F_t^2 = F_x^2 + F_z^2 \qquad (1)$$

The footboards consist of Board One, where the right foot is placed, and Board Two, where the left foot is placed. The components of the footboards are a connection part in the bottom, control equipment that adjusts the angle between the footboard and the ground, another connection part, and a normal sensor installed in the footboard. There are holes in footboards so the angle can be adjusted by inserting pins in the holes and grooves.

The binding equipment is installed in the upper part of the connection part, the salient part, and in the bottom of the connection part. The binding equipment consists of a thread, and a bolt connects the connection part with the binding equipment.

The normal sensor and the horizontal sensor consist of strain gauges. Eight strain gauges are needed for the normal sensor. The eight strain gauges consist of Compression Strain Gauge One, Two, Three, and Four, which are all located on the top, and Tension Strain Gauge One, Two, Three, and Four, which are located on the bottom. The compression strain gauge is located on the top corner, and the tension strain gauge is located at the bottom corner (Figure 2).

There are four strain gauges of the horizontal sensor in the connection part. The four strain gauges consist of the first tension strain gauge, positioned in a part of the connection part; the first compression strain gauge; the second tension strain gauge, positioned in the other part of the connection part; and the second compression strain gauge. The tension strain gauge is installed on the part that receives tension through forces that are applied horizontally to the connection part, and a compression strain gauge is installed to the part that is compressed by forces that are applied horizontally.

Figure 1: Starting block with sensors   Figure 2: Normal sensor

Each strain gauge is connected as a part of a Wheatstone bridge circuit. In the Wheatstone bridge circuit of the normal sensor, the first compression strain gauge, the second compression strain gauge, the third compression strain gauge, and the fourth compression strain gauge, along with the first tension strain gauge, the second tension strain gauge, the third tension strain gauge, and the fourth tension strain gauge on the other hand are connected serially. Specifically, the first compression strain gauge and the second compression strain gauge are connected to the third and fourth compression strain gauges face to face. Meanwhile, the first and the second tension strain gauges are connected with the third and fourth strain gauges face to face.

In the Wheatstone bridge circuit of the horizontal sensor attached to the connection part, the first and the second compression strain gauges are connected face to face and the first and the second tension strain gauges are connected face to face as well.

The signal process equipment is composed of an amplifier connected to a galvanometer, a display part that is connected to the amplifier with connection equipment that shows output voltage, and analysis equipment that detects the starting time by analyzing display parts.

RESULTS AND DISCUSSION: A galvanometer measures the output voltage from the normal sensor attached to the footboard, and the output stage of the Wheatstone bridge circuit connected to the connection part. At the start, when forces are applied, the output voltage changes, and the amplifier amplifies this output voltage. The amplifier is connected to the display part with the connection equipment. Horizontal to the connection part, and a compression strain gauge is installed on the part that is compressed by forces that are applied horizontally (Figure 3).

As shown in the analysis equipment, the time when forces are applied is found and starting time is established. The display part shows output voltage and thus, when an athlete uses force to start a race, output voltage drastically changes. By measuring the change in time, the exact starting time of the athlete can be determined.

Figure 3: Horizontal sensor   Figure 4: Calibration procedures

The starting block described above is anchored. As shown in figure 4, the produced sensor is anchored by a deadweight force standard machine and weight. All equipment is compressed and fixed using "the Standard Calibration Procedure of Electric Force Measuring Devices" and the Relative Expanded Uncertainty of the standard was within 0.005 % (a confidence level of approximately 95 %, k=2) [4]. The relative expanded uncertainty (REU) of each piece of equipment was calculated as shown in tables 1 and 2, which REU was defined as $W_p = k \cdot w_c (\%)$, k=2 (95%) [5]. By using this equipment, any data that applies to the athlete may be acquired.

Table 1 Data of the relative expanded uncertainty (left foot)

(a)F_n Compression			(b)F_x Compression		
Real weight (N)	Relative expanded Uncertainty(REU) (W_i, %)	Uncertainty into force (Column 1 * column 2) (N)	Real weight (N)	Relative expanded Uncertainty(REU) (W_i, %)	Uncertainty into force (Column 1 * column 2) (N)
0	0	0	0	0	0
300	2.267	6.67	50	3.456	1.69
500	1.870	9.17	100	3.031	2.97
1000	1.205	11.81	150	2.189	3.22
1200	1.126	13.24	200	2.066	4.05
1500	1.060	15.59	250	1.526	3.74
2000	0.705	13.81	300	1.017	2.99
2500	0.318	7.78	350	0.688	2.36
3000	0.104	3.05	400	0.472	1.85

Table 2 Data of the relative expanded uncertainty (right foot)

(a) F_n Compression			(b) F_x Compression		
Real weight (N)	Relative expanded Uncertainty(REU) (W_i, %)	Uncertainty into force (Column 1 * column 2) (N)	Real weight (N)	Relative expanded Uncertainty(REU) (W_i, %)	Uncertainty into force (Column 1 * column 2) (N)
0	0	0	0	0	0
300	1.930	5.68	50	3.920	1.92
500	1.618	7.93	100	3.405	3.34
1000	1.586	15.55	150	2.440	3.59
1200	1.183	13.92	200	1.972	3.86
1500	0.800	11.76	250	1.695	4.15
2000	0.436	8.55	300	1.014	2.98
2500	0.238	5.82	350	0.538	1.84
3000	0.073	2.13	400	0.148	0.59

CONCLUSION:

Through this study, a sensor system that uses Wheatstone bridge type strain gauges was created to measure repulsive forces in vertical and horizontal directions that are applied to starting blocks at the start of short distance athletic races. The equipment produced shows a maximum of 4% repeatability and uncertainty in each of the operating forces. It is expected that the actual results of athletes will be acquired using the proposed equipment and that these results will help to determine the optimal stance of each athlete.

REFERENCES:

Aron, J. M, Robert, G. L., & Aaron, J. C.(2003) Kinematic Determinants of Early Acceleration on Field Sport Athletes, *J. Sports Sci. Med.*, 2(3), 144-150.

Bret, C, Rahmani, A., Dufour, A. B., Messonnier, L. & Lacour, J. R.(2002) Leg Strength and Stiffness as Ability Factors in 100m Sprint Running, *J. Sports Med. Phys. Fit.*, 42, 274-281.

GUM (Guide to the Expression of Uncertainty in Measurement), ISO, 1993

Harrison, A. J., Keane, S. P., & Coglan, J.(2004) Force-Velocity and Stretch-Shortening Cycle Function in Sprint and Endurance Athletes, *J. Str. Condres*, 18(3), pp. 473-479.

KRISS, Standard Calibration Procedure of Electric Force Measuring Devices (c-07-1-0040-2002), KRISS, 2002

ACKNOWLEDGEMENT

This study was made possible through support from the Korea Research Institute of Standards and Science (KRISS) for the article "The Establishment of Measurement Standards for the Human Body Medical Industry" and support from the Small and Mid-Size Business Innovation Development Project for "The Development of a Starting Block Precision Sensor for the Improvement of Performance in Short Distance Athletic Races.".

A PROPOSED METHODOLOGY FOR TESTING ICE HOCKEY HELMETS

T. Blaine Hoshizaki, Evan Stuart Walsh, Philippe Rousseau, and Scott Foreman
Neurotrauma Impact Science Laboratory, University of Ottawa, Ottawa, Canada

KEY WORDS: Ice hockey, mild traumatic brain injury, angular acceleration.

INTRODUCTION: Ice hockey helmets were initially created to reduce the incidence of major traumatic brain injuries and have proven successful for this task. Unfortunately, helmets have not been as efficient in reducing mild traumatic brain injuries (mTBI) (Flick, Lyman & Marx, 2005). Standards regulating ice hockey helmets only measure linear acceleration, at a single inbound velocity, which may not be adequate to assess their ability to mitigate mTBI (King et al., 2003). The purpose of this study was to propose a new methodology incorporating both angular and linear accelerations, and various velocities, to quantify helmet performance.

METHODS: A 13.8 kg linear impactor was used to produce impacts at 5.5, 7.5, and 9.5 m·s^{-1} to an instrumented Hybrid III headform. The eight impact conditions tested were identified in a previous study as having a high-risk response to injury criteria.

RESULTS AND DISCUSSION: The results indicated that the tested ice hockey helmet was effective in keeping peak linear accelerations below a 50% risk of mTBI (Zhang, Yang & King, 2004), yet this was not the case for peak angular acceleration. This is of concern due to the link established in the literature between angular acceleration and mTBI (King et al., 2003).

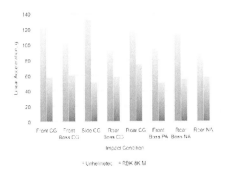 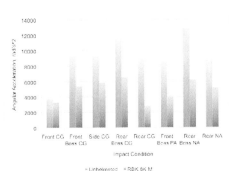

Figure 1: A comparison of peak linear (left) and angular (right) accelerations between a bare and a helmeted Hybrid III headform at 5.5 m/s.

CONCLUSION: A new methodology was devised to assess the ability of ice hockey helmets to manage linear and angular accelerations. The test was performed on a certified ice hockey helmet, which showed a poor ability at reducing angular acceleration.

REFERENCES:
Flik, K., Lyman, S., & Marx, R.G. (2005). American collegiate men's ice hockey: an analysis of injuries. *The American Journal of Sports Medicine*, 33(2), 183-187.

King, A. I., Yang, K. H., Zhang, L., & Hardy, W. G. (2003). Is head injury caused by linear or angular acceleration? *Paper presented at the 2003 IRCOBI Conference*, Lisbon, Portugal.

Zhang, L., Yang, K. H., & King, A. I (2004). A proposed injury threshold for mild traumatic brain injury. *Journal of Biomechanical Engineering*, 126(2), 226-236.

Acknowledgement: The authors would like to thank Xenith for supporting the Neurotrauma Impact Science Laboratory.

EFFECTS OF BALANCING HAMSTRING AND QUADRICEPS MUSCLE TORQUE ON RUNNING TECHNIQUE.

Doug Rosemond[1], Peter Blanch[2], Ross Smith[3] and Tudor Bidder[4]

Biomechanics & Performance Analysis[1], Physical Therapies[2], Strength & Conditioning[4] and Track & Field[4] Dept's., Australian Institute of Sport

KEY WORDS: hamstring, quadriceps, running, technique.

INTRODUCTION: It has been suggested that balancing the isokinetic strength of quadriceps (Q) and hamstring (H) muscles can reduce hamstring injuries during running (Croisier et al 2008). The efficacy of this type of intervention has been previously explored. To further the knowledge of the H: Q relationship we have examined the intervention's affect on running technique as presented here.

METHODS: An elite male 400m runner (age: 20 yrs, mass: 78.1 kg) performed a maximal isokinetic muscle test (Kin Com Cahttanga, TN) to determine peak concentric and eccentric hamstring and quadriceps torques ($H_{ecc}, H_{conc}, Q_{ecc}, Q_{conc}$) at a specific range of angular velocities (Croisier et al 2008). The athlete also ran at 9.6m/s for a full kinematic and kinetic analysis of overground running technique (8 force platforms, 22 cameras and 16 EMG channels). The tests were repeated after 10 weeks of periodised eccentric strength training.

RESULTS: The second isokinetic testing results showed a reduction of bilateral asymmetries to below 15%. H_{ecc}/Q_{conc} ratios increased for left leg (0.51 to 0.72) and right leg (0.54 to 0.70). H_{ecc}/Q_{conc} improvements are a result of increases in H_{ecc} peak torques of 19% (right leg) and 40% (left leg) at 30°/s. Significant changes in running technique were noted (see Table 1).

Table 1 Running technique variables pre and post intervention focused on correcting hamstring and quadriceps imbalances. Trials /session = 6. Steps/trial = 4.

Run technique variable	Pre Test (mean)	SD	Post Test (mean)	SD	T-Test (P value)	Effect
Step length (m)	2.20	0.04	2.25	0.04	0.02	1.21
Step frequency (steps/s)	4.4	0.1	4.0	0.7	0.09	0.80
Horizontal foot speed prior to contact (m/s)	2.3	0.2	1.8	0.2	0.00	2.93
Net vertical impulse (N.s)	90.45	4.33	96.21	2.99	0.02	1.55

DISCUSSION: The increased effectiveness of running technique was associated with foot activity prior to and during initial foot contact where hamstring EMG activity was at its highest. Changes to running technique are likely to be the result of factors related to improved hamstring strength. As post intervention values for H_{ecc}/Q_{conc} ratios were below the critical value (0.87) (Croisier et al 2008), further improvements in running technique variables could be expected with future strength gains. The athlete has since incurred a hamstring strain providing an opportunity to investigate the outcomes of his recovery in relation to running technique and leg strength.

CONCLUSION: Improving hamstring function to reduce the likelihood of hamstring injury has been shown, in this case, to improve effectiveness of foot contact during running, however the level of improvement was not sufficient to prevent a hamstring injury.

REFERENCES:

Croisier, J. L., Ganteaume, S., Binet, J., Genty, M., & Ferret, J. M. (2008). Strength imbalances and prevention of hamstring injury in professional soccer players: a prospective study. *The American Journal Of Sports Medicine, 36*(8), 1469-1475.

COMPARISON OF VERTICAL FORCES BETWEEN A PRESSURE MEASUREMENT SYSTEM AND A FORCE PLATE

Nicholas Brisson and Marshall Kendall
School of Human Kinetics, University of Ottawa, Ottawa, Canada

KEY WORDS: vertical force, gait, pressure mapping, force plate.

INTRODUCTION: Force plates and pressure measurement systems are commonly used for clinical and experimental applications. The MatScan (Tekscan Inc.), a lightweight, portable and cost-efficient pressure mapping system, can serve as an alternative to a force plate for measuring vertical forces. There has been much debate, however, as to the accuracy and precision of pressure mapping systems when compared to force plates, which have been considered the gold standard in force measurement (Hsiao, Guan & Weatherly, 2002). Although much research has been conducted using different force and pressure measurement systems, there is virtually no literature that discusses the comparisons of measured forces between these types of systems. The goal of this study was to compare the vertical forces measured by the MatScan to those measured by an AMTI force plate.

METHODS: Three participants were recruited for this study. Each participant executed a total of 10 walking trials at a natural, self-selected pace (5 trials with left foot and 5 trials with right foot landing on the MatScan/force plate). The AMTI force plate (OR6-6-2000) was zeroed with the MatScan fixated on top of it. The MatScan was calibrated to subjects' weight (standing two feet on the mat). For each trial, the MatScan and force plate recorded data simultaneously. All trials were averaged and normalized to 100% of the stance phase.

RESULTS: Mean vertical forces (N) measured by the AMTI force plate and the MatScan during the stance phase for subject 1 (left foot on the MatScan/force plate) are shown in Figure 1. The absolute mean force difference (%) between the gait patterns obtained from the MatScan and the force plate was 12.7%.

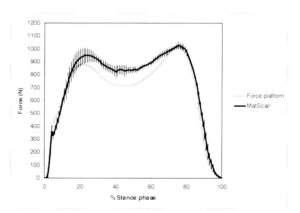

Figure 1 Mean vertical forces (N) measured by the AMTI force plate and the MatScan during the stance phase for subject 1, walking at a natural speed with left foot on the MatScan/force plate.

DISCUSSION & CONCLUSION: The MatScan significantly overestimated vertical forces during natural walking when compared to the AMTI force plate. Hence, the MatScan and the AMTI force plate do not produce similar force values in dynamic conditions. The difference between the forces measured by the two systems may be due to the calibration procedure utilized. Furthermore, the MatScan's encasing may have absorbed some of the force, thereby reducing the amount of force applied to the force plate. More research should be conducted on the MatScan's ability to accurately and precisely measure vertical forces.

REFERENCES:

Hsiao, H., Guan, J. & Weatherly, M. (2002). Accuracy and precision of two in-show pressure measurement systems, *Ergonomics*, 45, 537-555.

THE DETERMINATION OF NOVEL IMPACT CONDITIONS FOR THE ASSESMENT OF LINEAR AND ANGULAR HEADFORM ACCELERATIONS

Evan Stuart Walsh, Philippe Rousseau, Scott Foreman, and T. Blaine Hoshizaki
Neurotrauma Impact Science Laboratory, University of Ottawa, Ottawa, Canada

KEYWORDS: Neurotrauma, impact, injury, angular acceleration, sports helmets

INTRODUCTION: Sports helmets, albeit very effective at preventing traumatic brain injury, have not mitigated the risk of mild traumatic brain injury in sport (Flik, Lyman, & Marx, 2005). Current protocols utilized in sports helmet testing incorporate only impact vectors through the center of mass, eliciting primarily linear accelerations. Angular acceleration has been suggested to be a better predictor of diffuse head injury than linear acceleration (Holbourn, 1943); therefore, the objective of this study was to develop a protocol capable of producing and measuring both forms of acceleration for future implementation into sports helmet standards.

METHODS: A 13.8 kg linear impactor was used to produce impacts at 5.5 m/s to an instrumented Hybrid III headform. This velocity was selected to match existing protocols (NOCSAE). Five impact sites (Front, Front Boss, Side, Rear Boss, and Rear) were tested at four impact angles (Center of Gravity, Positive Azimuth, Negative Azimuth, and Positive Elevation). Resultant linear and angular headform accelerations were analyzed with respect to published injury thresholds (Zhang, Yang, & King, 2004).

RESULTS AND DISCUSSION:

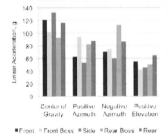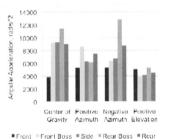

Figure 1: Peak linear and angular accelerations for multiple impact conditions at 5.5 m/s.

Three off-center impact conditions (Front Boss Positive Azimuth, Rear Boss Negative Azimuth, and Rear Negative Azimuth), along with the five center of gravity conditions, were determined to be above the published 80% injury risk prediction thresholds. A protocol using the above eight impact conditions could ameliorate mild traumatic brain injury prediction.

CONCLUSION: A new impact protocol employing both linear and angular accelerations for the prediction of mild traumatic brain injury was created. This protocol can be used towards the development of safer head protection for sport.

REFERENCES:

Flik, K., Lyman, S., & Marx, R.G. (2005). American collegiate men's ice hockey: an analysis of injuries. *The American Journal of Sports Medicine*, 33(2), 183-187.

Holbourn, A.H.S. (1943). Mechanics of Head Injury. *The Lancet*, ii, 438-441.

Zhang, L., Yang, K. H., & King, A. I (2004). A proposed injury threshold for mild traumatic brain injury. *Journal of Biomechanical Engineering*, 126(2), 226-236.

Acknowledgement
The authors would like to thank Xenith for supporting the Neurotrauma Impact Science Laboratory.

A CROSS-SECTIONAL STUDY OF GENDER DIFFERENCES IN PULLING STRENGTH OF TOW FOR JAPANESE ELEMENTARY SCHOOL CHILDREN

Takahiko Sato, Saki Sodeyama*, Ryuji Nagahama*, Soichi Shimizu, Kengo Sasaki, Ryoichi Nishitani*, Cao Yulin* and Hiroh Yamamoto*

Biomechanics Lab., Graduate School of Ed., Kanazawa Univ., Kanazawa, Japan
*Biomechanics Lab., Fac. Of Educ., Kanazawa Univ., Kanazawa, Japan

The aim of this study was to obtain the data of gender differences of pulling strength during experimentally executed TOW for Japanese elementary school children. In mean back strength, gender difference was small from 1st grade to 4th grade, but on 5th and 6th grade, gender difference became large. In mean pulling strength, gender difference was large in 5th and 6th grade. But no tendency was found from 1st grade to 4th grade. In male children, sum of pulling strength increases substantially when the grade changes from 4th to 5th. But pulling strength tended to grow constantly. On the other hand, in female children, sum of pulling strength increases substantially when the grade changes from 2nd to 3rd. And from 4th to 6th, sum of back strength and rope tension were very close to each other. Results suggested that though male children get grow for muscles, female children get motor function more than male children.

KEY WORDS: gender difference, tug of war, elementary school

INTRODUCTION: Pulling at a rope has been performed ritually all over the world. Today, it is enjoyed as a sport called tug-of-war; TOW. The appeal of TOW is to cooperate with any other teammates to aim for win. Although TOW is commonly sport for us, 'pulling for maximal effort and step backward', 'hold for maximal effort against opponents pull' and 'step forward unable to hold' in TOW are complicated movements and not usual.
TOW is constructed for three types of phase; Drop phase, Hold phase, and Drive phase (Masahiro et al, 2005). Drop phase is from start to run out the slack. Hold phase is hold against opponent or step forward unable to hold. Drive phase is step backward. From previous studies the agonists in Hold phase are backside muscles, and in Drive phase are abdominal muscle and backside muscles (Shigeki et al, 1988).
In TOW, power and endurance of muscles is very important. Before 10 years old no gender difference in development of muscles is found out, but after 10 years old males start to grow more than females. Furthermore, in 10 years old, the muscle development rate normalized by peak is 35%. On the other hand, developments of motor function is dependent on motor learning; growing muscles and neural system, usually experience. Starting to learn walking forward is 3 years old, running forward is 7 years old, walking and running backward is 11 years old (Saki et al, 2005). That is to say elementary school children have been in developmental stage.
In most previous studies, athletes, adults, and students at collage had performed and had been measured. On the other hand, there are few studies for elementary school children. So, the purpose of this study was to obtain the data of gender differences of pulling strength during experimentally executed TOW for Japanese elementary school children.

METHODS:

Subjects of this study were healthy elementary school children. 8 male and 8 female children in each grade participated (male: 8×6=48, female: 8×6=48). Table1 shows the physical characteristics of subjects. In a preliminary session, subjects were measured for back strength and pulling strength. Both parameters were measured by load cell (TCLP-2000KA, Tokyo Sokki Kenkyujo, Co., LTD Japan). The subjects performed 1

Table 1: Pysical characteristic

grade	gender	height(cm)	weight(N)	age(years)
1	male	122.5±4.4	233.4 ± 33.3	7.2±0.3
	female	119.0±2.6	217.7 ± 26.5	6.7±0.4
2	male	127.1±2.0	257.9 ± 18.6	8.0±0.3
	female	123.1±5.8	245.2 ± 45.1	7.7±0.1
3	male	130.4±6.5	271.7 ± 42.2	9.0±0.3
	female	126.3±3.8	241.3 ± 23.5	8.6±0.3
4	male	135.0±4.8	306.0 ± 46.1	9.9±0.2
	female	138.3±7.9	340.3 ± 104.0	9.6±0.3
5	male	140.3±3.8	348.1 ± 66.7	11.1±0.3
	female	143.1±6.7	324.6 ± 62.8	10.9±0.3
6	male	154.8±3.1	472.7 ± 108.9	11.5±0.6
	female	145.4±7.5	355.0 ± 37.3	11.6±0.4

trial for each parameter. The 1 trial was 5 seconds and designed for Figure1 and Figure2. Those parameters were defined as average in 5 seconds. And as test session, the 8 students in each grade were split into 2 groups (group A: n=4, group B: N=4) and performed mini game (4:4) to be measured the rope tension. The sum of back strengths in each group were equalized as possible based on preliminary session. The mini game was set for 30 seconds and performed 3~5 games for each pairing (Figure3). This difference was caused by the fatigue of subjects. The rope tension was defined as average in all trials.

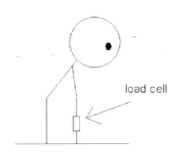

Figure1: Back strength trial

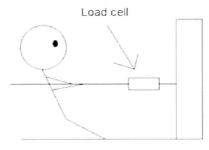

Figure2: Pulling strength trial

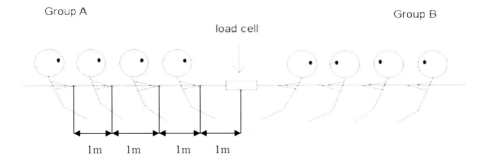

Figure3: Mini game

RESULTS:

Table 2 shows the measurements in preliminary session. In mean back strength, gender difference was only small from 1st grade to 4th grade, but on 5th and 6th grade, gender difference became large. In mean pulling strength, gender difference was large on 5th and 6th grade. But no tendency was found from 1st grade to 4th grade.

Table2: Mean back strength and pulling strength (±SD)

grade	gender	Back strength (N)	Pulling strength (N)
1	male	119.6 ± 47.0	176.4 ± 41.2
	female	101.9 ± 20.6	131.3 ± 21.6
2	male	186.2 ± 53.9	193.1 ± 30.4
	female	180.3 ± 59.8	149.9 ± 31.4
3	male	214.6 ± 76.4	183.3 ± 43.1
	female	260.7 ± 43.1	203.8 ± 12.7
4	male	255.8 ± 26.5	184.2 ± 36.3
	female	234.2 ± 72.5	192.1 ± 49.0
5	male	345.9 ± 87.2	254.8 ± 58.8
	female	248.9 ± 30.4	209.7 ± 34.3
6	male	390.0 ± 63.7	296.9 ± 43.1
	female	293.0 ± 66.6	229.3 ± 33.3

The developments of rope tension and sum of pulling strength were shown for Figure 4 and Figure 5. Sum of pulling strength in those figures was defined as intermediate value between group A's and group B's. In male children, sum of pulling strength got grows largely when the grade changes from 4th to 5th. But pulling strength was tended to grow constantly. On the other hand, in female children, sum of pulling strength got increases substantially when the grade changes from 2nd to 3rd. And from 4th to 6th, sum of back strength and rope tension were very close to each other.

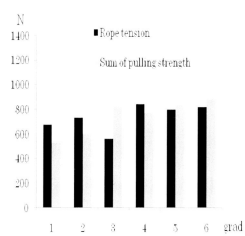

Figure 4: The developments of rope tension and sum of pulling strength in male children

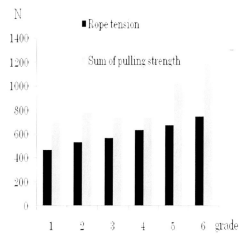

Figure 5: The developments of rope tension and sum of pulling strength in female children

The difference between rope tension and sum of pulling strength in male children was larger than female's. Because of this tendency, it remains possible that female children take advantage of muscle power to perform any complicated movements more than males. Thus, it is suggested that though male children develop for muscles, female children develop motor function more than male children.

Through this study, the data of pulling strength during experimentally executed TOW for Japanese elementary school children has been obtained and gender differences in tendency of developments of muscle and motor function have been shown.

In further investigation, it needs to increase the number of subjects and find out the statistical significance. On the other hand, it is very important to address not only a cross-sectional study like this but a longitudinal study to certify the muscles and motor function developing.

REFERENCES:

Masahiro Nakagawa et al. (2005). CHARACTERISTIC OF PULLING MOVEMENT FOR JAPANEASE ELITE TUG OF WAR ATHLETES. XXIII International Symposium on Biomechanics in Sports, Beijing.2005 Proceedings, pp475-478

Shigeki Kawahara, Hiroh Yamamoto (1988). The biomechanical analysis on Tug of war, Seoul Olympic Scientific Congress, pp356-357

Saki Sodeyama et al. (2005). THE CARACTERISTICS OF BACKWARD WALKING AND BACKWARD RUNNING IN PRIMARY SCHOOL CHILDREN. XXIII International Symposium on Biomechanics in Sports, Beijing.2005 Proceedings: pp539-542

POSTURAL EFFECTS ON COMPARTMENTAL VOLUME CHANGES OF BREATHING BY OPTOELECTRONIC PLETHYSMOGRAPHY IN HEALTHY SUBJECTS

Rong-Jiuan Liing[1], Kwa-Hwa Lin[1], Tung-Wu Lu[2], Sheng-Chang Chen[2]

[1]Graduate Institute of Physical Therapy, National Taiwan University, Taiwan
[2]Institute of Biomedical Engineering, National Taiwan University, Taiwan

KEY WORDS: Postural effect, compartmental volume changes, optoelectronic plethysmography

INTRODUCTION: Breathing pattern was an important factor to affect the performance of sports for athletes. Optoelectronic plethysmography (OEP) was a new method to evaluate breathing pattern by measuring compartmental volume (upper thorax (UT), lower thorax (LT), and abdomen (AB)) freely without limitation. Previous study already investigated the swimmers had better breathing pattern measured by OEP (Karine et al., 2008) in sitting posture. Swimming, such as backstroke, is perfromed in supine posture, but previous study did not consider the postural effect on breathing pattern. This study explored the compartmental volume changes of healthy subjects in different postures.

METHOD: 7 healthy male subjects were recruited in this study (age: 41.4±12.9 yrs, height: 170.7±4.3 cm, body weight: 77.3±12.2 kg). To compare the volume changes in supine and sitting, subjects performed breathing in sitting posture with a back support. Passive markers were applied in front and lateral side of chest wall as figure 1. For posterior part of chest wall, the model used the reference plane of the back of chair and bed. 45 markers on the chest wall identified three compartments during breathing from functional residual capacity to maximal inspiration in 2 postures. A 2-way repeated ANOVA was performed. ($\alpha=0.05$).

RESULTS: The results indicated that there was interaction between compartmental volume

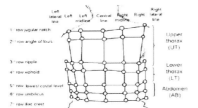 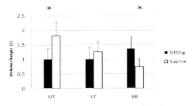

Figure1: Geometric diagram of chest wall with 3-compartment model.

Figure2: The compartmental volume changes between 2 postures. (*: $p<0.05$)

changes and postures ($p<0.001$). The post hoc comparison showed significant differences in UT and AB between different postures ($p = 0.015$ and 0.001 perspectively) in figure 2.

DISCUSSION: The results show that posture can affect the breathing pattern to increase the volume of UT and decrease the AB volume. Gravity would help the diaphragm descend in sitting. For supine posture, this position did not benefit the diaphragm movement. The main volume change was contributed by upper thorax (Estenne et al., 1985). The further work will recruit backstroke swimmers to discuss the postural effect on breathing pattern. This study provided a feasible method to assess the breathing pattern in sitting and supine postures and useful information about breathing patterns of healthy male subjects in different postures.

CONCLUSION: The breathing pattern was influenced by posture in healthy subjects, especially in upper thorax and abdomen.

REFERENCES:
Karine et al.,(2008). Coordination between ribs motion and thoracoabdominal volumes in swimmers during respiratory maneuvers. *Journal of Sports Science and Medicine*, 7, 195-200.

Estenne et al.,(1985). Rib cage and diaphragm compliance in humans: effects of age and posture. *Journal of Applied physiology*, 59, 1842-48.

AN INVESTIGATION OF THE ACTIVATION OF THE SUBDIVISIONS OF GLUTEUS MEDIUS DURING ISOMETRIC HIP CONTRACTIONS

Catriona O' Dwyer[1], David Sainsbury[1] and Kieran O'Sullivan[1]
Department of Physiotherapy, University of Limerick, Limerick, Ireland [1]

KEY WORDS: gluteus medius, functional subdivisions, electromyography

INTRODUCTION: Gluteus medius is involved in movement and stability of the hip and gluteus medius dysfunction is commonly implicated in many lower limb pathologies (Fredericson et al 2000). It is proposed that functional subdivisions exist within the gluteus medius muscle (Conneely and O'Sullivan 2008). There is however a lack of empirical evidence examining the role of the subdivisions of gluteus medius. This study compared the muscle activation of these subdivisions (anterior, middle and posterior) during isometric contractions of hip abduction, internal and external rotation in normal subjects.

METHODS: A single-test design was undertaken. Standardised electrode locations were determined in advance. Three surface electromyography (sEMG) electrodes were placed on each subject (n=15) to record muscle activity of each muscle subdivision. Subjects performed three maximal voluntary isometric contractions for hip abduction, internal and external rotation on the Biodex System 3 Isokinetic Dynamometer with simultaneous recording of sEMG activity of the subdivisions of gluteus medius (Motion Lab System multi-channel EMG system). The average root mean square of the sEMG was calculated. Data was analysed using a one-way ANOVA for muscle segment, with repeated measures on isometric contraction direction (post-hoc Bonferroni).

RESULTS: There was a significant main effect for subdivision ($p<0.001$) and isometric contraction direction ($p<0.001$), and a significant interaction between subdivision and direction ($p<0.005$). The anterior gluteus medius had a significantly greater activation across all three isometric contractions than the middle and posterior gluteus medius (both $p<0.001$). Furthermore, there was a significantly greater activation of all three functional subdivisions during abduction and internal rotation when compared to external rotation (both $p<0.001$).

DISCUSSION: The results suggest that gluteus medius is more active during abduction and internal rotation than in external rotation, in line with previous research (Earl 2004). The results support the hypothesis that muscle activation is not homogenous throughout the entire gluteus medius muscle and that functional subdivisions exist within the muscle.

CONCLUSION: Functional subdivisions exist within the gluteus medius muscle. Muscle activation is effected by both subdivision and isometric contraction direction with the greatest activation found in the anterior gluteus medius subdivision, and during abduction and internal rotation. Future studies should examine the role of the functional subdivisions in subjects with lower limb pathologies.

REFERENCES:

Conneely, M. and O'Sullivan, K. (2008). Gluteus maximus and gluteus medius in pelvic and hip stability: isolation or synergistic activation? *Physiotherapy Ireland*, 29 (1), 6-10.

Earl, J. (2004). Gluteus Medius Activity during three Variations of Isometric Single-Leg Stance. *Journal of Sports Rehabilitation*, 13, 1-11.

Fredericson, M., Cookingham, C., Chaudhari, A., Dowdell, B., Oestreicher, N. and Sahrmann, S. (2000). Hip Abductor Weakness in Distance Runners with Iliotibial Band Syndrome. *Clinical Journal of Sport Medicine*, 10 (3), 169-175.

Acknowledgement

I would like to acknowledge the contribution of Sharon Smith to this research project.

APPLICABILITY OF OPERATIONS RESEARCH AND ARTIFICIAL INTELLIGENCE APPROACHES TO NON- CONTACT ANTERIOR CRUCIATE LIGAMENT INJURY STUDIES

Nicholas Ali[1], Gordon Robertson[1], Gholamreza Rouhi[2]

School of Human Kinetics, University of Ottawa, Ottawa, Canada[1]
Department of Mechanical Engineering, University of Ottawa, Ottawa, Canada[2]

Numerous problems in biomechanics can be tackled using optimization methods. The primary objective of this paper is to explore the applicability and illustrate the importance of two major optimization approaches, namely, operation research (OR) and artificial intelligence (AI) to studies in non-contact anterior cruciate ligament (ACL) injury biomechanics. This paper focuses on the applicability of the two approaches to bring attention to the enabling capabilities that can be offered to address challenges faced in non-contact ACL injury studies. The differences and similarities, as well as, advantages and disadvantages of these two approaches are discussed. Some of the key techniques covered in the two different approaches are highlighted. As well, the area in which there is a common ground for both approaches is outlined. It was determined for a small search space and highly tailored problems, classical exhaustive OR methods usually suffice; however, for large search spaces an AI technique must be employed. Thus, an AI technique is better suited to tackle the challenges faced in non-contact ACL injury studies especially given the multifaceted nature of such problems.

KEY WORDS: ACL injury, optimisation, operations research, artificial intelligence

INTRODUCTION: Operations research (OR) is concerned with the quantitative specifications of problems and the use of mathematical techniques to solve these problems. Artificial intelligence (AI) is concerned with making computers capable of emulating intelligent behaviour (Holsapple, Jacob et al. 1994). Since the ability to solve problems is one of the hallmarks of human intelligence, a major AI focus is on computer-based techniques for solving problems (Carter and Price 2001). Historically, OR and AI research communities worked in relative isolation from one another. This separation is difficult to understand, since both disciplines share many similarities deeply concerned with questions of human problem solving and decision-making. In addition, both approaches are highly computer dependent, share a common conceptual framework, and require methods that effectively cope with uncertainty and imprecision (Kanal, Lemmer et al. 1987; Pearl 1988). Despite these similarities both approaches are highly distinct from each other. For instance, AI has a strong foundation in logic with methods that emphasizes automatic theorem proving, while OR puts emphasis on the mathematics of optimization (Holsapple, Jacob et al. 1994). In addition, OR has been found useful for decision-making, while AI pursuit deals in automatic decision-making. As well, OR focuses on the management of procedural knowledge, while AI tends to focus on the management of reasoning knowledge. There are few studies on the applications of AI and OR techniques to improve our understanding of non-contact injury to the ACL. To study ACL mechanics classical optimization techniques such as the Newton-Raphson search method or mathematical programming cannot be employed because of the absence of a polynomial type function. More importantly, these techniques cannot handle many design parameters in a large domain. Monte Carlo and mathematical programming method are the dominant OR techniques used in the literature to investigate non-contact ACL injury (Blankevoort and Huiskes 1996; McLean, Su et al. 2004). These two techniques are used primarily in applications to evaluate the probability of random outcomes of human movement (McLean, Su et al. 2004) and to solve a series of mathematical equations (Abdel-Rahman and Hefzy 1993; Blankevoort and Huiskes 1996) respectively. Monte Carlo simulation is an attractive tool since it allows researchers to study and predict risk of sustaining an injury before injury occurs. Non-contact ACL injuries typically occur when several extreme conditions or risk factors happen at the same time. Monte Carlo simulation can estimate the probability when multiple extreme conditions are satisfied, which then allows one to predict the risk of injury. Nonetheless, Monte-Carlo methods are algorithms that randomly generate and retain the best solutions before going to the next search iteration. Rather than presenting an historical review of what has been done, this paper takes a more explicit approach, discussing the work in three fields: 1) overview of AI and OR approaches; 2) present challenges in non-contact ACL injury studies; and 3) application of AI and OR to enable us to address some of these challenges encountered in non-contact ACL injury studies.

METHODS: This article reviewed the relevant literature on non-contact ACL injury mechanisms, OR and AI optimization techniques in the PubMed electronic database using MEDLINE (1966 to 2007) and Applied and Complementary Medicine Database (AMED) on Ovid (1985 to Sept. 2007). Unpublished data and abstracts were excluded. Our search was supplemented by reviewing the bibliographies of retrieved articles as well as hand searching scholarly journals.

RESULTS: Overview of Optimization: Any problem in which design parameters have to be determined, (assuming such parameters exist), given certain constraints, can be treated as an optimization problem (Carter and Price 2001). Typically optimization methods seek to improve rather than create new ideas. However, the ability of optimization methods to mold new ideas should not be ruled out, since optimized solutions usually lead to new designs that are in some cases unique.

Operations Research (OR) Perspective: From an OR perspective there are specific problems and explicit procedures for solving such problems. It is the wording or problem statement that helps us to decide which OR method to use. Hence, a given procedure requires a certain type of problem statement as a precondition for execution. This has somewhat made OR procedures unattractive for solving complex problems. Mathematical programming appears to be the cornerstone of OR techniques and is arguably the most widely used and visible extension of the field. To gain a more thorough understanding of OR, available OR techniques, as well as, details on coding ones own technique, please consult (Winston 1994).

An Artificial Intelligent (AI) Perspective: For many practical problems encountered, the only way to be sure of finding an optimal solution is to search completely through the whole set of possible solutions. The time required to carry out such an exhaustive search is, although finite, far greater than most institutions can afford. The pretext then is to find shortcuts that will allow one to organize the search process so that it is no longer a complete search over all possible solutions, but rather it becomes an affordable search that is likely to find optimal or near optimal solutions. These methods are called artificial intelligent, heuristic or soft computing techniques. To gain a more thorough understanding of AI, available AI techniques, as well as, details on coding ones own technique, please consult (Pham and Karaboga 2000).

Present Challenges in Understanding non-contact ACL Injury: There has been much interest in quantifying the ACL loading *in vivo* during activity (Beynnon, Fleming et al. 1995). This is motivated by the high incidence of ACL injury, lack of understanding of ligament mechanics, and frequent requirements for surgical treatment. Despite the advances in research to understand knee mechanics, little is known about the mechanism of ACL injuries. There is also no clear consensus that implicated factors for non-contact ACL injury are indeed risk factors.

Presently, *in vitro* investigations using cadavers dominate in the arena of ACL injury mechanics. Here the ACL is initially studied when intact, after the ACL is sectioned, and finally after the ACL is reconstructed. *In vitro* studies have provided tremendous knowledge on the function of the ACL, but they do not appreciably help us increase our insight and understanding on the mechanism of ACL injury.

The vast majority of studies do not address the effects of hip and ankle kinematics on ACL loads. Torso, hip, and foot motions were suggested to modulate 3 of the 4 quadriceps muscles, the gastrocnemius, and the hamstrings via their length tension relationship, which has been shown to affect risk of ACL injury (Withrow, Huston et al. 2006). In addition, the vast majority of the studies in the literature do not account for the effects of whole body movement on ACL loads, despite its distinct effect (Peña, Calvo et al. 2007).

Other overarching challenges in non-contact ACL injury studies include the difficulty entailed when comparing results from one study to another due to the tremendous heterogeneity between studies. As well, there is large variability in biological tissues material properties reported in the literature. In addition, extracting soft tissue three dimensional (3D) geometries remains a challenge, especially for those that are intra-articular like the ACL.

AI and OR Applied to Non-contact ACL Injury Studies: Numerous real-world problems entail the fusion of many disciplines. This is due to the recognition that the design and development of complex systems can no longer be done in subsections with each discipline isolated from each other. The needed for a more comprehensive strategy is imperative given the increase level of complexity within disciplines, and the need to extract the advantages of a synergistic design process (Hajela 1999). Multidisciplinary design optimization (MDO) has recently emerged as a field of research and practice that brings together many previously disjointed disciplines and tools of engineering. Typically MDO involves many design variables, many constraints, and analysis from various contributing disciplines that are not independent.

Many studies utilizing optimization in biomechanics have been concentrated in the orthopedics arena (Yoon and Mansour 1982). AI techniques have received tremendous attention over the years compared to OR techniques primarily because they do not require gradient information, they are more robust in handling both continuous and discrete design variables, and above all, they share an enhanced ability to locate the global optimal solution.

Clinical studies, interviews with athletes, and video analyses are non-contact ACL injury study approaches that provide mostly qualitative data, and as such, they are not well suited for obtaining a clear and comprehensive understanding ACL injury mechanism. On the other hand, computational modeling, musculoskeletal modeling, and experimentation are quantitative ACL injury study approaches that all have their own advantages and disadvantages. The authors' view is that the seamless integration of all three quantitative methods, i.e. computational modeling; musculoskeletal modeling; and experimentation, is a more robust study approach to better predict non-contact ACL injuries. In this approach the three qualitative approaches, i.e. clinical studies; interviews with athletes; and video analyses, can be used to aid validation. More importantly, an AI technique is employed to combine all six study approaches, as well as, to facilitate search and optimization. This type of study approach process is absent from the literature. The author's believe this method can enable us to simultaneously determine knee kinematics and kinetics, muscle activation and loading patterns, joint compressive stresses, and ground reaction forces (GRF) at which the ACL is at high risk of injury. To elucidate, one possible approach is a 3D musculoskeletal-driven finite element model of the lower extremity that is verified and validated with experiment, clinical studies, interviews with athletes, and video analysis studies. In this approach to address the numerous variables and constraints involved, the immense variability in material properties, uncertainty and unknowns such as inciting position of ACL injury, a genetic algorithm can be utilized. The genetic algorithm also functions to tie all disciplines together thereby facilitating an automatic or semi-automatic transfer of data.

Key Findings: It was determined that application of AI and OR approaches to research centered on non-contact ACL mechanics is primarily limited to OR techniques. A possible study approach that can enable researchers to overcome some of the challenges faced in non-contact ACL mechanics studies was presented. It was emphasized that despite scarcity of AI application to ACL injury mechanics studies, AI is very much well suited for addressing many of the limitations of existing ACL mechanics study approaches. This can provide opportunities to researchers in this field to take benefit of powerful AI techniques.

CONCLUSIONS: A brief overview of optimization was presented. The two main areas of optimization namely operations research and artificial intelligence were discussed in the context of its application to study non-contact ACL injury. Some of the key gaps in knowledge in the literature pertaining to non-contact ACL injury studies were also presented. Finally, the capabilities of AI and OR techniques were covered, and recommendations put forth on how existing non-contact ACL injury study approaches can be united with an AI technique to reinforce our understanding of ACL mechanics. Even though OR techniques applied to study biomechanics problems is prevalent they lack the robustness required to investigate complex problems such as non-contact ACL injury. So the nature and complexity abound non-contact ACL injury requires an AI approach.

REFERENCES:
Abdel-Rahman, E. and M. S. Hefzy (1993). A Two-Dimensional Dynamic Anatomical Model of the Human Knee Joint. Journal of Biomechanical Engineering **115**(4A): 357-365.
Beynnon, B. D., B. C. Fleming, et al. (1995). Anterior Cruciate Ligament Strain Behavior during Rehabilitation Exercises In Vivo. The American Journal of Sports Medicine **23**(1): 24-34.
Blankevoort, L. and R. Huiskes (1996). Validation of a three-dimensional model of the knee." Journal of Biomechanics **29**(7): 955-961.
Carter, M. W. and C. C. Price (2001). Operations Research: A Practical Introduction, CRC Press.
Hajela, P. (1999). Non-gradient methods in multidisciplinary design optimization-status and potential. Journal of Aircraft **36**(1): 255-265.
Holsapple, C. W., V. S. Jacob, et al. (1994). Operations Research and Artificial Intelligence, Intellect Books.
Kanal, L. N., J. F. Lemmer, et al. (1987). Uncertainty in artificial intelligence. IEEE Transactions on Systems, Man and Cybernetics **17**(2): 332-332.
McLean, S. G., A. Su, et al. (2004). Development and Validation of a 3-D Model to Predict Knee Joint Loading During Dynamic Movement. Journal of Biomechanical Engineering **125**: 864-874.
Pearl, J. (1988). Probabilistic reasoning in intelligent systems: networks of plausible inference, Morgan Kaufmann.
Peña, E., B. Calvo, et al. (2007). Computational Modeling of Diarthrodial Joints. Physiological, Pathological and Pos-Surgery Simulations. Arch Comput Methods Eng **14**: 47-91.
Pham, D. T. and D. Karaboga (2000). Intelligent optimisation techniques. Secaucus, Springer New York.
Winston, W. L. (1994). Operations research: applications and algorithms. Boston, MA, Duxbury Press.
Withrow, T. J., L. J. Huston, et al. (2006). The Relationship between Quadriceps Muscle Force, Knee Flexion, and Anterior Cruciate Ligament Strain in an In Vitro Simulated Jump Landing. American Journal of Sports Medicine **34**(2): 269-274.
Yoon, Y. S. and J. M. Mansour (1982). The passive elastic moment at the hip. Journal of Biomechanics **15**(12): 905-10.

ISBS 2009

Applied Sessions

ISBS 2009

Applied Session AS1

Coaching Biomechanics

QUALITATIVE BIOMECHANICS FOR COACHING

Duane Knudson[1], Jacque Alderson[2], Rafael Bahamonde[3], Michael Bird[4]

Texas State University, USA[1]
University of Western Australia, Perth, Australia[2]
Indiana University-Purdue University at Indianapolis, USA[3]
Truman State University, USA[4]

Session Information:

Coaches must apply principles of biomechanics in their qualitative judgments of the technique used by athletes. These judgments can have a major influence on performance and injury risk. This session will focus on the most effective use of qualitative biomechanical analyses and video replay software. Several scholars who have experience teaching qualitative biomechanical analysis to future coaches will present, followed by a question and answer session.

Schedule of Presentations:

Dr. Knudson will introduce the session and provide a brief overview of qualitative biomechanical analysis. 11:00 – 11:15

Dr. Alderson will present sport injury models as they apply to assessment, intervention and rehabilitation of common injuries in cricket, tennis and running. Relevant qualitative and quantitative 2D features of *SiliconCoach* that can be utilised by a coach to potentially reduce injury incidence will be presented. 11:15 – 11:45

Dr. Bahamonde will present how qualitative analysis can be used to teach biomechanics concepts to physical education and coaching students. Movement examples from tennis, soccer and track field and meaningful features of *Hu-m-an* software will be illustrated. *Hu-m-an* is unique in that it was developed with a specific teaching-learning focus. 11:45 – 12:15

Dr. Bird will present biomechanical core concepts as a "common language" to evaluate and improve all human movements. The core concepts are visually observable, but meaningful features of *Dartfish* will be illustrated to enhance what is seen by both the coach and the mover. Movement examples from golf, resistance training, basketball, and other sports will be presented. 12:15 – 12:45

Discussion: 12:45 – 13:00

This will provide an opportunity for delegates to ask specific questions relating to any of the presenters.

ISBS 2009

Applied Session AS2

Rowing Biomechanics

Rowing Applied Session @ ISBS 2009

Aim
The aim of the Rowing Applied Session is to see how an understanding of rowing biomechanics and applying technology can help to improve rowing performance. The presenters will introduce some concepts, controversial issues, demonstrations and methodologies which will be the focus for discussion among the participants.

The Basics
The rower works against hydrodynamic and aerodynamic drag. Drag is roughly proportional to the square of the boat velocity, and the greater the net force the rower applies to the boat the greater will be the boat velocity. The mechanism for applying theses forces is made all the more intriguing due to the oscillatory motion of the rower in producing the drive and recovery phases of the stroke. The rower applies force to the pins via the oar and the stretcher. During the drive phase, the net force applied to the boat is the difference between the pin force and the stretcher force while during the recovery phase it is mainly the stretcher force. Blade and boat hydrodynamics constantly change due to the changing position of and application of forces by the rower. The aim of our biomechanical studies is to understand how all these variables affect performance and channel the power output of the rower optimally into boat propulsion.

Firstly, in this Applied Session we will assess how to get the best value from various technologies that can supply the quantitative information needed for informed decision making about technique. We do this first as the following sections depend on collection of data using the equipment. The next two sections will examine various combinations of these variables as they apply to large and small boats and simulation of rowing.

Section One – Technology (Richard Smith and Valery Kleshnev)
What is the best value for money in rowing instrumentation?
We will look at the value of simple instrumentation such as accelerometry, oar angle measurement, handle and pin force through to complex three dimensional measurement of pin and stretcher forces, rower segment and oar position.

Section Two – On-water Rowing Performance (Valery Kleshnev and Richard Smith)
Large boats
What are the determinants of performance peculiar to the large boats?
Are there seat-specific differences in performance requirements in fours and eights?
Small boats
The pair - Should a pair rower learn to row in a given seat or should they be selected for their 'natural' force production technique?
Sculling - Even sculling is asymmetric. Is this detrimental ? How do we deal with asymmetry in rowing?

Section Three – Rowing simulation: (Peter Sinclair)
How specific are popular rowing ergometers to on-water rowing?
How do handle force, stretcher force and handle velocity differ between ergometer and on-water rowing?
Sliding vs fixed ergometry and discussion on the indoor rowing tank.

Rowing Applied Session @ ISBS 2009

Presenter Biographies

Richard is currently working with the Australian Institute of Sport and other partners to advance the quality of rowing performance measurement systems. He is an Associate Professor at The University of Sydney, Australia and also works with the New South Wales Institute of Sport on projects in the biomechanics of water sports. He is particularly interested in the mechanisms of rower power generation and its efficient transfer to boat propulsion. He has been studying the biomechanics of rowing for twenty years, published in international journals and has been successful in attracting competitive grant funding for research projects including rowing.

Peter Sinclair has a PhD degree from The University of Sydney and completed earlier studies at The University of Western Australia. He has lectured in Biomechanics at The University of Sydney since 1990, and is currently sub-dean of Graduate Coursework and Students. Peter was the founding chair of the Australian and New Zealand Society of Biomechanics. His research crosses a range of topics in Sports Biomechanics and has been conducted in collaboration with both the Australian and New South Wales Institutes of Sport. Peter is also interested in computer simulations of muscle performance, particularly where applied to cycling exercise for people with spinal cord injury.

Valery Kleshnev spent 10 years as a member of USSR National Rowing team, achieving Junior World Champion, Olympic Silver, World Bronze, and four National Championship titles. Valery has a Masters Degree from Leningrad Academy of Physical Education and a PhD in biomechanics of rowing and coaching science from Leningrad Research Institute of Sport. He has worked in Saint-Petersburg Institute of Sport, the Australian Institute of Sport, and the English Institute of Sport. Since April 2009 he continues his support to British Rowing as a private consultant. The main area of his research interest are rowing biomechanics and its application in coaching. He has published more than 120 papers, publishes the Rowing Biomechanics Newsletter, and publishes www.biorow.com. Valery has invented and developed more than 100 devices and programs for measurement, data analysis and feedback. The current research directions are rowing efficiency and ergonomics.

ISBS 2009

Applied Session AS3

Strength & Conditioning

APPLICATION OF COMPLEX TRAINING WITHIN STRENGTH AND CONDITIONING PROGRAMMES

Thomas M. Comyns

Munster Rugby, Irish Rugby Football Union, Dublin 4, Ireland

Alternating a resistance exercise with a plyometric exercise is referred to as "complex training". Post-activation potentiation is the physiological rationale for complex training. This form of training is widely used to improve strength, power and stretch-shortening cycle (SSC) function. Enough is not known, however, on how this form of training can be optimised. Many factors such as the rest interval and the magnitude of the resistance exercise can have an influence on the efficiency of this training modality. More research is needed so that strength and conditioning coaches can best apply the principles of complex training to athletes' strength and power programmes.

KEYWORDS: jumping, stretch-shortening cycle, post-activation potentiation

INTRODUCTION & OVERVIEW: Complex training involves the completion of a resistance exercise prior to a plyometric exercise. A classic example is to perform vertical jumps or depth jumps after the completion of a back squat exercise. The term 'complex training' is credited to Verkhoshansky et al. (1973). It is postulated that the resistance exercise will have a performance enhancing effect on the plyometric activity (Ebben and Watts, 1998), resulting in increased power output, increased performance outcome and enhanced efficiency of the SSC behaviour.

Complex training is widely used in the practical setting and is a popular training modality. A number of reviews have been written on complex training (Docherty et al., 2004; Ebben, 2002; Ebben & Watts, 1998; Jeffreys, 2008). The conclusions regarding the effectiveness of complex training as an appropriate training modality are somewhat guarded and equivocal. Research has found the complex training can be beneficial to athletic performance (Comyns et al., 2007; Evans et al., 2000; Güllich and Schmidtbleicher, 1996; Young et al., 1998), while the opposite has also been reported (Jones & Less, 2003; Scott & Docherty, 2004). A possible explanation for this contradiction in findings is that many variables have an influence on the research outcomes and thus the efficiency of complex training, such as the magnitude and mode of the preload exercise as well as the rest interval between the preload and the plyometric components of complex training. In addition, the gender, training status, training age, strength levels of the participant may influence the potentiation benefits of complex training (Docherty et al., 2004; Robbins 2005).

Postactivation potentiation (PAP) is the physiological rationale for complex training (Docherty et al., 2004). PAP results in an enhancement in the explosive capability of the muscle due to prior contractile activity (Docherty et al., 2004; Robbins, 2005). Examples of contractile activity that have been used in PAP research include maximum voluntary contractions, (Güllich and Schmidtbleicher, 1996) and the execution of resistance exercises (Young et al., 1998; Evans et al., 2000). Two mechanisms have been put forward to explain the workings of PAP. Firstly the enhancement in plyometric performance after performing the contractile activity may be due to an increase in neural excitability (Güllich and Schmidtbleicher, 1996). Alternatively the phosphorylation of the myosin light chain has been proposed as a mechanism attributed to the PAP (Sale, 2002). Docherty et al. (2004) noted, however, that it is possible that PAP is the result of interactions between both the neural and muscular mechanisms.

The fitness-fatigue paradigm (Plisk & Stone 2003) provides a framework for explaining the potentiation effect associated with some complex training studies. Fitness is the term used to explain the positive response and adaptations that occur as a result of exposure to a training stimulus. Fatigue is a general term describing a loss of capability to generate force or an inability to sustain further exercise at the required level (Strojnik & Komi, 1998). Exposure to a training stimulus will result in both a fitness and fatigue response (Plisk & Stone, 2003). The fitness response is the desired response from the training stimulus, while the fatigue is the side effect. While the fitness and fatigue responses have similar traits in that they stem from the same source, co-exist (Rassier & MacIntosh, 2000), rise sharply after the training stimulus and dissipate thereafter, they exert opposite effects on the potential for potentiation and performance preparedness. The fitness-fatigue model, which is illustrated in figure 1, details the interplay between fitness and fatigue. This model can be used to explain the PAP effect related to complex training.

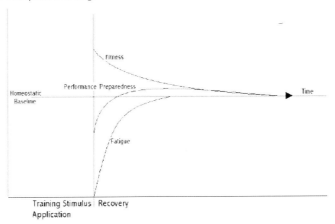

Figure 1. Fitness-fatigue model adapted from Plisk & Stone (2003:21)

An athlete's preparedness for performance and potentiation is defined as the summation of the two after-effects of training, i.e. fatigue and fitness. Preparedness, according to Plisk & Stone (2003), can be optimised with strategies that maximise the fitness response to training stimuli while minimising fatigue. Complex training is possibly one of these strategies that can result in realising this performance preparedness. Examples of research into how both the fitness and fatigue responses interplay as a result of complex training are outlined in this paper. Further research is needed, however, on the optimisation of potentiation and performance preparedness as a result of exposure to complex training protocols.

COMPLEX TRAINING REST INTERVAL: The rest interval between the resistance and plyometric components of complex training has been referred to as the intra-complex rest interval. In complex training research this rest interval has ranged from 10 seconds (Jensen & Ebben, 2003) up to 20 minutes (Jones & Lees, 2003). Research has indicated that 3 to 4 minutes may be optimum (Evans et al 2000; Güllich and Schmidtbleicher, 1996; Young et al., 1998). Jensen & Ebben (2003) investigated the effect of 5 repetition maximum (5RM) squat on countermovement jumps (CMJ) that were performed 10 seconds, and 1, 2, 3 and 4 minutes after the squat. Results revealed no significant difference in jump height from pre-squat to post-squat for any of the rest intervals. The jump performance, however, at the 10 seconds interval was reduced but not significantly. A nonstatistically significant trend of improvement in jump height occurred from 10 seconds up to 4 minutes. Jones and Lees (2003) adopted a similar approach to Jensen & Ebben (2003) by manipulating the length of the rest interval. They investigated the effect of 5RM back squatting on CMJs and drop jumps (DJs) that were performed immediately, 3, 10 and 20 minutes post-lifting. While no statistical significance was found, suggesting that complex training did not enhance plyometric performance, it was also noted that no adverse effects occurred. One possible reason for the above findings could be that the potentiation window may differ for individuals, as suggested by Docherty et al. (2004), and thus mask any ergogenic response at the different intervals. Comyns et al. (2006) addressed this issue in their optimum intracomplex rest interval study by analyzing the greatest improvement and reduction in the dependent variables' scores compared to the baseline scores regardless of when they occurred to assess if complex training had an ergogenic effect.

Comyns et al. (2006) investigated the effect of 5RM back squatting on CMJs that were performed at rest intervals of 30 seconds, and 2, 4 and 6 minutes post-lifting. Flight time and peak ground reaction force (GRF) were the dependent variables. Repeated measures ANOVA found a significant reduction in flight time at the 30 second and 6 minute interval ($p < 0.05$). No significant difference was found between men and women. Only the men showed an enhancement in jump performance after the 4 minute interval. The improvement window was different for each subject and an analysis of the greatest increase and decrease in flight time and peak ground reaction force was conducted, showing a significant decrease for men and women and a significant increase in flight time for men and peak ground reaction force for women. The results suggest that complex training can benefit and/ or inhibit countermovement jump performance depending on the rest interval. The individual determination of the intracomplex rest interval may be necessary in the practical setting.

Clearly the efficiency of complex training relies heavily on the rest interval. The current findings on the optimal complex training rest interval are ambiguous. Jeffreys (2008) highlighted this finding and noted that it is likely that optimal PAP will only be evident at a given window of opportunity and outside this performance may be impaired, have no impact on performance or show limited benefit. The results from Comyns et al. (2006) support this viewpoint. In addition, the optimal timeframe for this window of potentiation opportunity is individual so the identification of an optimal intracomplex rest interval for group situations is not appropriate.

COMPLEX TRAINING OPTIMAL LOAD: When squats are used as the preload component of complex training a key variable in determining the efficiency of the complex training protocol is the magnitude of the load. Ambiguity exists in complex training research about the optimal load that needs to be lifted in the resistance exercise to maximise the benefits of PAP. Jeffreys (2008) suggests that as PAP is associated with Type II muscle fibres, then the preload component has to stimulate an appropriate number of Type II fibres, whether through high resistance or high velocity. Consequently, a 5RM squatting protocol has been used in the majority of complex training studies. The results using this approach have been mixed. Some researchers have found that this resistive load had a significant effect on the performance of the plyometric exercise (Young et al., 1998; Evans et al., 2000). Others found the 5RM did not produce statistically significant results for the dependent variables associated with the plyometric exercise (Jensen & Ebben, 2003; Scott & Docherty, 2004).

Research into loads outside of this 5RM range is sparse. Baker (2003) investigated the effect of lifting six repetitions at 65% of 1RM for the bench press on an explosive bench press-style throw (plyometric exercise) and showed a significant increase of 4.5% in power output from pre- to post-test for the experimental group. This finding is important as it suggested that a relatively light load of 65% could produce an enhancement in performance in a subsequent plyometric exercise. However, Hanson et al. (2007) investigating the effects of squats at 40% 1RM and 80% 1RM on repeated CMJ performance, found no change suggesting that these loads were too light to provided a PAP effect.

Comyns et al. (2007) addressed the optimal load issue by examining the effect of a 65% 1RM, 80% 1RM and 93% 1RM back squat on the DJ performance. Flight time, ground contact time, peak ground reaction force, reactive strength index and leg stiffness were the dependent variables. Repeated measures ANOVA found that all resistive loads significantly reduced ($p < 0.01$) flight time, but lifting at the 93% load caused a significant improvement ($p < 0.05$) in ground contact time and leg stiffness. From a training perspective, the results indicate that the heavy lifting will encourage the DJs to be performed with a stiffer leg spring action, which in turn may benefit performance. Although the findings from complex training optimal load studies are somewhat contradictory, it would appear that PAP is greatest when heavy loads are utilised.

CONCLUSION: Many factors contribute to the effectiveness of complex training as a practical training modality. While PAP is a very real physiological phenomenon, attaining it through complex training protocols is very much dependent on the rest interval, magnitude and mode of the preload activity, and the training status, training age, gender and strength levels of the participant. While heavy loads appear to elicit a PAP effect compared to lighter loads the optimal rest interval would seem to be individual. The optimal rest interval depends on the interaction between potentiation and fatigue and the latter is related to the magnitude and mode of the preload activity as well as the individual's tolerance to fatigue, which may change over time. All of this needs to be taken into consideration when designing strength, power and plyometric programmes. Further research is still needed, especially on long term benefits of complex training programmes, to provide greater insight into the value of applying PAP training modalities to athletes' training programmes.

REFERENCES:

Baker, D. (2003). Acute effect of alternating heavy and light resistances on power output during upper-body complex power training. *J Strength Cond Res.* 17, 493-497.

Comyns, TM, Harrison, AJ, Hennessy, LK, and Jensen, RL. (2006) The optimal complex training rest interval for athletes from anaerobic sports. *J Strength Cond Res.* 20, 471–476.

Comyns, TM, Harrison, AJ, Hennessy, LK, and Jensen, RL. (2007). Identifying the optimal resistive load for complex training in male rugby players. *Sports Biomech*, 6, 59-70.

Docherty, D, Robbins, D, and Hodgson, M. (2004). Complex training revisited: A review of its current status as a viable training approach. *Strength Cond J*, 26, 52 –57.

Ebben, WP. 2002. Complex training: a brief review. *Journal of Sports Science and Medicine*, 2, 42-46.

Ebben, WP and Watts, PB. (1998). A review of combined weight training and plyometric training modes: Complex training. *Strength Cond J*, 20, 18-27.

Evans, AK, Hodgkins, TD, Durham, MP, Berning, JM, and Adams, KJ. (2000). The acute effects of a 5RM bench press on power output. *Medicine and Science in Sport and Exercise.* 32(5), S311.

Güllich, A, and Schmidtbleicher, D. (1996). MVC-induced short-term potentiation of explosive force. *New Studies in Athletics.* 11(4), 67-81.

Hanson, ED, Leigh, S. and Mynark, RG. (2007). Acute effects of heavy and light load squat exercise on the kinetic measures of vertical jumping. *J Strength Cond Res*, 21, 1012-1017.

Jeffreys, I. (2008). A review of post activation potentiation and its application in strength and conditioning. *Professional Strength and Conditioning.* 12, 17-25.

Jensen, RL and Ebben, WP. (2003). Effect of complex training rest interval on vertical jump performance. *J Strength Cond Res.* 17, 345 - 349.

Jones, P., and Lees, A. (2003). A biomechanical analysis of the acute effects of complex training using lower limb exercises. *J Strength Cond Res.* 17, 694 - 700.

Plisk, SS and Stone, MH. (2003). Periodization Strategies. *Strength Cond J*, 25, 19-37.

Rassier, D.E. and MacIntosh, BR. (2000). Coexistence of potentiation and fatigue in skeletal muscle. *Brazilian Journal of Medical and Biological Research*, 33, 499-508.

Robbins, DW. (2005). Postactivation potentiation and its practical applicability: a brief review. *J Strength Cond Res*, 19, 453-458.

Sale, D. (2002). Postactivation potentiation: Role in human performance. *Exerc Sport Sci Rev.* 30, 138-143.

Scott, SL and Docherty, D. (2004). Acute effects of heavy preloading on vertical and horizontal jump performance. *J Strength Cond Res*, 18, 201–205.

Strojnik, V. and Komi, PV. (1998). Neuromuscular fatigue after maximal stretch-shortening cycle exercise. *J Appl Physiol*, 84, 344-350.

Verkhoshansky, Y and V. Tatyan, (1973). Speed-strength preparation of future champions. *Logkaya Atletika*, 2, 2-13.

Young, WB, Jenner, A, and Griffiths, K. (1998). Acute enhancement of power performance from heavy load squats. *J Strength Cond Res*, 12, 82–84.

UNDERSTANDING AND OPTIMISING PLYOMETRIC TRAINING

Eamonn P. Flanagan

Biomechanics Research Unit, University of Limerick, Limerick, Ireland

Plyometrics are exercises involving rapid, powerful movements preceded by a preloading countermovement. The stretch shortening cycle (SSC) is the basis of plyometric exercises. The precise mechanisms that underpin any SSC activity may be determined by the demands of the criterion task. As a result the SSCs can be classified as slow or fast. Accordingly, different plyometric exercises or the manner in which exercises are performed may elicit different mechanisms of SSC action. Slow SSC plyometrics are useful to teach athletes appropriate jumping and landing techniques and to develop maximal jumping ability. Fast SSC plyometrics develop athlete's ability to generate high power outputs and fast rates of force development.

KEY WORDS: force, jumping, reactive strength index.

INTRODUCTION & OVERVIEW: Plyometrics are exercises involving rapid, powerful movements preceded by a preloading countermovement. Examples of plyometrics include depth jumps, hurdle jumps and bounding. Plyometric training has been shown in the literature to have a number of beneficial effects for athletes. These range across injury prevention, power development, sprint performance, agility development, and running economy. Little to no morphological changes occur in response to plyometric training. The adaptation which takes place is primarily on a neural level (Markovic et al., 2005).

The stretch shortening cycle (SSC) is the basis of plyometric exercises. The SSC is a natural type of muscle function in which muscle is stretched immediately before contraction. This eccentric/concentric coupling of muscular contraction produces a more powerful contraction than that which would result from a purely concentric action alone (Komi, 1992). In real-life situations, exercise seldom involves a pure form of isometric, concentric, or eccentric actions (Komi, 2000). The SSC appears to be the natural form of muscle function, and it is evident in everyday activities, such as running, throwing and jumping.

A number of biomechanical mechanisms are thought to contribute to the SSC. It has traditionally been thought that the SSC causes an enhancement during the concentric phase due to the storage and reutilization of **elastic energy** (Cavagna et al., 1968). During the eccentric phase, the active muscles are pre-stretched and absorb energy. Part of this energy is temporarily stored and then reused during the concentric contraction phase of the SSC (Bobbert et al., 1996). A short transition between the eccentric and concentric phase is necessary for this elastic energy to be used optimally.

Additional mechanisms of action also have been proposed. It has been proposed that the prestretch in the SSC may enhance the concentric contraction through **neural potentiation** of the muscle contractile machinery during the eccentric phase. This would allow for a greater number of motor units to be recruited during the concentric contraction (Van Ingen Schenau et al., 1997). This potentiation effect is thought to increase with the speed of the eccentric action and is maximised when a short transition time between the eccentric and concentric phases occurs (Bobbert et al., 1987a).

Bobbert et al. (1996) determined that in tasks such as maximal effort vertical jumps, in which eccentric-concentric coupling is used compared with purely concentric squat jumps, the performance enhancement in the SSC is likely caused by the eccentric phase, allowing **an increased time to develop force**. A slow eccentric phase allows muscles to develop a high level of active state (more attached crossbridges) before the start of concentric action. As a result, developed force and joint moments are greater at the beginning of the concentric phase and more work is produced through the first part of the concentric motion compared to concentric only squat jumps.

Reflex contributions of the muscle spindle mechanoreceptor can also contribute to the enhanced work output observed in the SSC. **The muscle spindle reflex** reacts to rapid changes in a muscle's length to protect the muscle–tendon complex. As eccentric stretching approaches a rate that could potentially damage the muscle–tendon complex, the muscle spindle activates and reflexively stimulates an opposite contraction of the agonist. Contributions from the muscle spindle are one mechanism that accounts for the performance enhancement observed in SSC activities which involve very rapid eccentric phases (Bobbert et al., 1987b).

The precise mechanisms that underpin any given SSC activity may be determined by the demands of the SSC criterion task (Flanagan et al., 2007). For example:

- For the muscle spindle reflex to be initiated, a fast rate of eccentric stretching must occur.
- For elastic energy to contribute, there must be a short transition period between the eccentric and concentric phase
- For neural potentiation to contribute there must be a fast eccentric phase and a short transition period between the eccentric and concentric phase.
- In order to allow an increased time to develop force, the eccentric phase must be slow.

Considering this, Schmidtbleicher (1992) has suggested that the SSC can be classified as either slow or fast. The fast SSC is characterized by short contraction or ground contact times (<0.25 seconds) and small angular displacements of the hips, knees, and ankles. A typical example would be depth jumps. The slow SSC involves longer contraction times (>0.25 seconds), larger angular displacements and is observed in maximal effort vertical jumps. Data from Flanagan and co-workers (2007) have shown that slow SSC actions such as maximal effort countermovement jumps generally produce ground contact/contraction times far greater than Schmidtbleicher's (1992) threshold of 0.25 seconds and that the difference between slow and fast SSC activities is easily discernable.

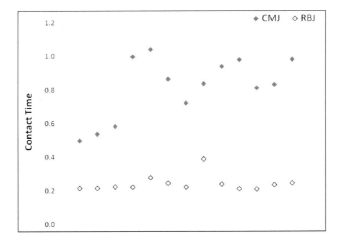

Figure 1: Data from Flanagan et al. (2007). Each subject produces significantly shorter ground contact/contraction times (s) in the fast SSC action of the rebound jump (RBJ) compared with the slow SSC action of the countermovement jump (CMJ).

For example, the muscle spindle reflex is dependent on a fast rate of eccentric stretching and elastic energy contribution may rely on a short transition period between eccentric and concentric phases (Bobbert et al., 1987a). Decay in the magnitude of potentiation has been observed as the transition time between eccentric and concentric contraction increases (Wilson et al., 1991). These mechanisms then are more likely to contribute to the fast SSC which has a faster eccentric velocity and a shorter transition period than the slow SSC (Bobbert et al., 1987a).

Performance enhancement in slow SSC activities may be primarily due to the slow eccentric phase allowing an increased time to develop force (Bobbert et al., 1996). The slower, longer eccentric phases and the greater transition times between eccentric–concentric coupling observed in slow SSC activities cast doubt as to whether mechanisms such as the muscle spindle reflex, elastic energy contributions, and potentiation could be as active in slow SSC tasks compared with fast SSC activities (Flanagan et al., 2007). As a result, it has been hypothesized that the slow and fast SSC may represent different muscle action patterns that rely on differing biomechanical mechanisms, which can affect performance in different ways (Flanagan et al., 2007).

The primary differences between the fast and the slow SSC are that the fast SSC involves a fast, short eccentric phase and a rapid transition between the eccentric and concentric phase. The slow SSC is identifiable by a longer eccentric phase and a slower transition. The magnitude of potentiation increases dependent on the speed of the eccentric phase and decreases with transition time (Bobbert et al., 1987a). As a result much greater joint moments, power outputs and rates of force development are observed in fast SSC plyometrics. Alternatively, greater jump heights can be observed in slow SSC plyometrics due to the increased time allowed to develop force. Slow SSC plyometrics are also very useful coaching tools to teach athletes appropriate jumping and landing techniques before progressing to the more challenging fast SSC plyometrics.

This has implications for strength and conditioning practitioners. Different exercises or the manner in which exercises are performed may elicit different mechanisms of SSC action. Training slow SSC activity may not by as beneficial for athletes who primarily rely on the fast SSC in their chosen sports (and vice versa). To adhere to the principle of specificity, coaches must be able to distinguish which plyometric exercises utilise the fast or slow SSC. Careful consideration must be made to select modes of training which incorporate the appropriate SSC action for the athlete's specific needs.

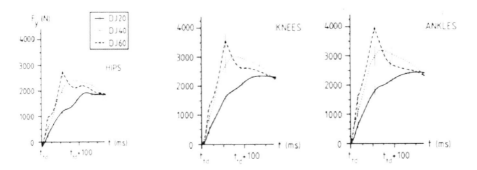

Figure 2: Data from a single subject in Bobbert et al. (1987a). Peak joint reaction force is recorded at the hips knees and ankles during depth jumps from heights of 20, 40 and 60cm.

Due to the greater forces, joint moments and power outputs which occur in fast SSC plyometrics there is a degree of injury potential if training is not carried out sensibly. Bobbert et al. (1987b) provided some data describing one process through which the inappropriate use of fast SSC plyometrics could potentially cause injury.

In this research, subjects performed depth jumps from heights of 20, 40 and 60cm. Joint reaction forces were recorded during the jumps at the hips, knee and ankle. In depth jumps subjects are instructed to minimize ground contact times, to utilize the fast SSC. To do this they use a stiff leg action and stay on the balls of their feet. Bobbert et al. (1987a) observed that certain depth jump heights (60cm in this case) can be too great. At an athlete's critical threshold, the downward velocity becomes too great and the athlete can lack the requisite strength to overcome this eccentric loading and transition effectively to a powerful concentric phase. As a result, Bobbert et al. (1987a) report that subjects were unable to stay on the balls of their feet and their heels came down on the ground. This caused sharp peak forces to be generated resulting in high joint reaction forces at the ankle, knee and hip. These sharp, high joint reaction forces can potentially cause damage to passive structures such as the joint surfaces.

This is not optimal from a safety perspective but also, from a training effect standpoint, can violate the principle of specificity. With subjects unable to react against the high eccentric velocities, their heels come down to the ground and they spend longer on the ground in order to generate enough force to perform the concentric phase. This elongation of the ground contact phase may mean the athlete is now performing a slow SSC movement when the objective of training may be to improve fast SSC performance.

PRACTICAL APPLICATION: Coaches must be able to identify poor technique and choose appropriate training intensities in fast SSC plyometrics to optimize the training from both a safety and performance perspective. The practical portion of this presentation will demonstrate the following:

- the differences between slow and fast SSC plyometric exercises.
- how slow plyometrics can be used to train proper jumping and landing techniques.
- the appropriate execution of fast SSC plyometric exercises.
- how fast SSC plyometrics can be progressed and increased in intensity.
- how slow and fast plyometrics can be used in tandem with training sessions.

REFERENCES:

Bobbert, M.F., Huijing P.A., Van Ingen Schanau G.J. (1987a) Drop Jumping I. The influence of jumping technique on the biomechanics of jumping. *Medicine and Science in Sports and Exercise.* 19: 332 – 338.

Bobbert M.F., Huijing P.A., Van Ingen Schanau G.J. (1987b) Drop jumping II: The influence of dropping height on the biomechanics of drop jumping. *Medicine and Science in Sports and Exercise* 19: 339–346.

Bobbert, M.F., Gerritsen, K.G.M., Litjens, M.C.A., Van Soest, A.J.V. (1996) Why is countermovement jump height greater than squat jump height? *Medicine and Science in Sport and Exercise*, **28**, 1402-1412.

Cavagna G.A., Dusman B., Margaria, R. (1968) Positive work done by a previously stretched muscle. Journal of Applied Physiology 24, 21–32.

Flanagan, E.P., Ebben, W.P., Jensen, R.L. (2007) Reliability of the reactive strength index and time to stabilization during depth jumps. *Proceedings of the XXV International Symposium of Biomechanics in Sports.* 509-512.

Komi, P.V. (1992) Stretch-shortening cycle. In P.V. Komi (Ed.), *The Encyclopeadia of Sports Medicine. Vol 3: Strength and Power in Sport* (pp. 169-179). Oxford, UK: Blackwell. pp. 169–179.

Komi P.V. (2000) Stretch-shortening cycle: a powerful model to study normal and fatigued muscle. *Journal of Biomechanics,* **33**, 1197–1206.

Markovich, G., Jukic I., Milanovic D., Metikos D. (2005) Effects of sprint and plyometric training on morphological characteristics in physically active men. *Kinesiology.* 37: 32-38.

Schmidtbleicher, D. (1992) Training for power events. In P.V Komi (Ed.) *The Encyclopeadia of Sports Medicine. Vol 3: Strength and Power in Sport* (pp. 169-179). Oxford, UK: Blackwell.

Van Ingen Schenau G.J., Bobbert M.F., De Hann A. (1997) Does elastic energy enhance work and efficiency in the stretch shortening cycle? *Journal of Applied Biomechanics* 13: 389–415.

Walshe, A.D., Wilson, G.J. Ettema, G.J.C. (1998) Stretch-shorten cycle compared with isometric preload: contributions to enhanced muscular performance. *Journal of Applied Physiology.* 84, 97-106.

Wilson, G.J., Elliott, B.C., Wood, G.A. (1991) The effect on performance of imposing a delay during a stretch-shorten cycle movement. *Medicine and Science in Sport and Exercise*, 23: 364 – 370.

Young, W. (1995) Laboratory strength assessment of athletes. *New Studies in Athletics.* 10, 88 - 96.

Zatsiorsky, V.M., Kraemer, W.J. (2007) *Science and Practice of Strength Training.* Champaign, IL: Human Kinetics.

MEASUREMENT TECHNIQUES IN ASSESSING ATHLETIC POWER TRAINING

Randall L. Jensen
Dept. HPER, Northern Michigan University
Marquette, MI, USA

Athletic performance can be altered via various training regimens and an important component to increase performance is the intensity of the training stimulus. Plyometric and complex training have been suggested to be useful training methods to improve athletic power, but standardization of techniques to assess training changes can be difficult. Assessments of activity have been made using electromyographic, kinematic, and kinetic measures. Practitioners must be able quantify the effect of their training programs; ideally with measures made in the field. This presentation will review methods to measure outcomes of athletic power training and make suggestions for implementing these measures in the training venue.

KEYWORDS: jumping, reactive strength index, ground reaction force, muscle stiffness index

INTRODUCTION & OVERVIEW: The improvement in performance due to resistance training can be easily quantified because weight training devices usually have clearly labeled masses and thus an athlete's repetition maximum (RM) or 1RM can be established. While resistance training has been shown to improve strength, plyometric training appears to have a greater impact on athletic power (Potach & Chu, 2000). The intensity of plyometric exercises is more difficult to verify, but has been evaluated using many variables including muscle activity, ground and joint reaction forces, number of points of contact during landing, the speed of the drill, the height of the jump, and the athlete's weight as well as other measures (Ebben et al., 2008; Jensen and Ebben, 2007; Potach & Chu, 2000). Some of the measures used in assessing plyometric exercises can also be used to determine gains made from training. Examples of these include vertical jump height, flight time, contact time, reactive strength index, speed strength index, rate of force development, leg spring stiffness, and starting strength (Comyns et al. 2006; Ferris & Farley, 1997; Harrison & Bourke, 2009; McClymont, 2003; Young, 1995).

VERTICAL JUMP HEIGHT: Vertical jump height is an easily obtained measure of performance and while it can be a performance itself (e.g. in high jumping), it has been moderately correlated to sprint and agility performance. Thus additional measures of jumping performance are often used to quantify changes in training.

FORCE MEASURES: Peak force is the highest force obtained during an activity and in jumping activities is usually assessed via ground reaction forces (GRF) on a force platform, but could include forces that take place within the body (e.g. between joints). It can be expressed in absolute terms the raw force obtained in Newtons or scaled relative to body mass (N/Kg). Scaling allows for comparison of different sized individuals as well as to compare force generation across active muscle mass, which can facilitate comparisons across body parts or athletes of widely varying body fat. Peak force is useful in establishing the force the body can exert or the amount of force the body must withstand during landing from jumps (Jensen and Ebben, 2007). The latter while not important in performance can be helpful in determining the risk of injury.

RATE OF FORCE DEVELOPMENT: Through use of a force platform the start of the concentric contraction has been defined as the point where force readings become 10N greater than the average of the force readings when the subject is static in the squat jump starting position (see Figure 1) (Harrison & Bourke, 2009). Maximum Rate of Force Development (max RFD) is then calculated as the greatest rise in force during 5-millisecond periods from the start of the concentric contraction (Wilson et al., 1995; Harrison & Bourke, 2009). Time to peak GRF is calculated by finding the difference in time between the point of the first concentric contraction and peak GRF on the force-time curve (Harrison & Bourke, 2009) and has also been used to estimate the average RFD (Jensen and Ebben, 2007). As most sporting activities require not only a large degree of force, but for it to be generated rapidly, RFD is an important component of sport (Wilson, 1995). Although this time has been used by a number of researchers, it has recently been questioned by Jensen and colleagues (2009) who have noted that simultaneous video analysis indicates the concentric contraction may take place later than this point on the force time tracing. Regardless of when the RFD takes place it has been used as an indication of power production in a variety of movements.

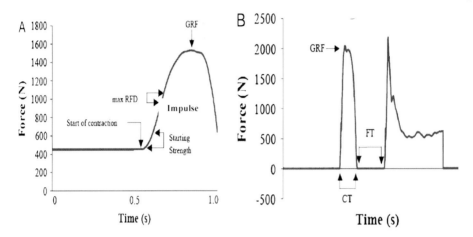

Figure 1. A) Force-time trace for a squat jump illustrating: start of contraction, maximum rate of force development (max RFD), starting strength, impulse, and peak ground reaction force (GRF); B) Force-time trace for a drop jump illustrating ground contact time (CT), flight time (FT), and ground reaction force (GRF) (adapted from Harrison & Bourke, 2009).

STARTING STRENGTH: The force produced 30 ms after the start of the concentric contraction (see Figure 1) has been defined as the starting strength and is describes as the ability to produce force very rapidly (Young, 1995). Starting strength has been correlated to initial acceleration and is therefore important for athletes needing to accelerate quickly from a stationary position (Harrison & Bourke, 2009; Young, 1995).

LEG SPRING STIFFNESS: Stiffness of the leg spring in the vertical direction (K_{vert}) has been suggested by Ferris and Farley (1997) to represent the stiffness of the integrated musculoskeletal system during locomotion. As K_{vert} regulates the interaction of the musculoskeletal system and the external environment during the ground-contact phase of locomotion it has been related to running (Ferris & Farley, 1997; Harrison et al., 2004) and hopping/jumping performance (Comyns et al., 2006). K_{vert} is calculated during the ground-contact phase by taking the ratio of the peak vertical ground reaction force (F_{peak}) to the maximum vertical displacement of the center of mass of the body at the instant that the leg spring was maximally compressed (Ferris & Farley, 1997). The measure requires a force platform and double integration of the force or video analysis along with force platform measures. Harrison and colleagues (2004) found that K_{vert} during drop jumps was significantly higher (42%) for national level sprinters compared to endurance runners. Therefore leg stiffness may be an important predictor of sprint ability.

STRETCH SHORTENING CYCLE & CONTACT TIME: In addition to RFD, aspects of the stretch shortening cycle (SSC), the coupling of a rapid eccentric/concentric muscular contraction, can be useful to assess training intensity as well as how the athlete responds to training (Young, 1995). The classification of the SSC as fast (<250 ms) or slow (>250 ms) has also been suggested to be useful in identifying performance in athletes (Schmidtbleicher, 1992). A fast SSC is characterized by a drop jump (DJ), while a countermovement jump (CMJ) illustrates a slow SSC. Harrison and co-workers (2004) obtained an SSC performance index by dividing the subject's respective CMJ or DJ flight time by their average squat jump flight time. Contact time (CT) is defined as the amount of time spent in contact with the ground prior to a jump (Ferris & Farley, 1997). The time of the SSC is determined using the CT following a drop jump. Because the individual is in contact with the ground during a CMJ, contact time cannot be determined for a CMJ. McClymont (2003) notes that determining if an athlete can attain a fast CT may be useful in making suggestions for their training program. Those athletes that need to incorporate more strength into their exercise, such as front line players in rugby, typically display a longer CT and slow SSC; while sprinters and backs in rugby or American football usually exhibit a fast CT and SSC.

REACTIVE STRENGTH INDEX: The reactive strength index (RSI) is calculated by dividing the height jumped by the ground contact time (Young, 1995) and similar to the CT has been used as an indicator of performance. The RSI describes an individual's ability to change quickly from an eccentric to concentric muscular contraction and expresses the athlete's explosive capabilities in dynamic jumping activity (Flanagan et al., 2008). Thus it can provide useful information for coaches of athletes that need to make rapid movements and change of direction. Because CT, RSI, and jump height can be easily obtained using not only force platforms, but also with contact mats or accelerometers and are highly reliable (Flanagan et al., 2008); they can be a valuable tool in assessing athletes' performance. Indeed, McClymont (2003) has noted that RSI testing provides an effective and useful tool in the preparation of elite athletes. Harrison and Bourke (2009) have stated that the relationship of RSI to sprinting performance is similar to that of leg-spring stiffness and sprint performance.

FIELD VERSUS LAB TESTING: As noted above the measures used to assess athletic power performance can be obtained via a force platform. Although this technology will usually be used within a laboratory to insure high reliability, the recent advent of portable force platform devices has allowed these assessments to be brought to training venues. In addition, because they can be used to provide virtually all the measures mentioned in this review, they may be very useful. Never-the-less because these devices typically cost more than $10000, they may be beyond the reach of coaches and strength training practitioners. However, a number of the assessments can be made in the field with less costly alternatives including contact mats, linear position transducers, and accelerometers. In particular contact time, reactive strength index, rate of force development and height jumped (calculated via jump flight time) can be obtained with these tools by using the athlete's mass and acceleration through inverse dynamics. Previous research has shown that measurements using these methods can be reliable within controlled situations (Jennings et al., 2005; McClymont, 2003; Wilson et al., 1997; Young, 1995).

PRACTICAL APPLICATIONS: A variety of measures can be useful in assessing the athlete in the training of athletic power. Because the coach or practitioner in a training venue may be limited by not having extensive equipment, such as a force platform and video analysis that are available in the laboratory setting, alternatives are needed. While vertical jump height can be easily assessed using a jump and reach test with virtually no equipment, this methodology does not allow for more detailed measures. Measurements obtained using contact mats, linear position transducers, accelerometers and force platforms have been used to provide more extensive information on contact time, jump flight time, reactive strength index, rate of force development, and leg spring stiffness, as well as jump height.

REFERENCES:

Comyns, TM, Harrison, AJ, Hennessy, LK, and Jensen, RL. (2006) The optimal complex training rest interval for athletes from anaerobic sports. *J Strength Cond Res 20*, 471–476.

Ebben, W.P., Simenz, C. and Jensen, R.L. (2008) Evaluation of plyometric intensity using electromyography. *J Strength Cond Res 22*, 861-868.

Ferris, DP, and Farley, CT. (1997) Interaction of leg stiffness and surface stiffness during human hopping. *J Appl Physiol 82*, 15–22.

Flanagan, EP, Ebben, WP, and Jensen, RL. (2008) Reliability of the reactive strength index and time to stabilization during plyometric depth jumps. *J Strength Cond Res 22*, 1677-1682.

Harrison, AJ, and Bourke, G. (2009) The effect of resisted sprint training on speed and strength performance in male rugby players. *J Strength Cond Res 23*, 275–283.

Harrison, AJ, Keane, SP, and Coglan, J. (2004) Force-velocity relationship and stretch-shortening cycle function in sprint and endurance athletes. *J Strength Cond Res 18*, 473–479.

Jennings, CL, Viljoen, W, Durandt, J, and Lambert, MI. (2005) The reliability of the FitroDyne as a measure of muscle power. *J. Strength Cond. Res. 19*, 859–863.

Jensen, RL and Ebben, WP. (2007) Quantifying plyometric intensity via rate of force development, knee joint and ground reaction forces. *J Strength Cond Res 21* 763-767.

Jensen, RL, Leissring, SK, Garceau, LR, Petushek, EJ and Ebben, WP (2009) Quantifying the onset of the concentric phase of the force-time record during jumping. *In Proceedings of the XXVII International Symposium of Biomechanics in Sports;* (AJ Harrison, I Kenny, and R Anderson, editors) In Press.

McClymont, D. (2003) Use of the reactive strength index (RSI) as a plyometric monitoring tool. Available at: http://coachesinfo.com/index.php?option=com_content&view=article&id=146:rugby-rsi&catid=47:rugby-general&Itemid=77.

Potach, DH and Chu, DA. (2000) Plyometric training. In: *Essentials of Strength Training and Conditioning*. R.W. Earle and T.R. Baechle, eds. Champaign, IL: Human Kinetics, pp. 427–470.

Schmidtbleicher, D. (1992) Training for power events. In P.V Komi (Ed.) *The Encyclopeadia of Sports Medicine. Vol 3: Strength and Power in Sport* (pp. 169-179). Oxford, UK: Blackwell.

Wilson, GJ, Lyttle, AD, Ostrowski, KJ, and Murphy, AJ. (1995) Assessing dynamic performance: A comparison of rate of force development tests. *J Strength Cond Res 9* 176-181.

Young, W. (1995) Laboratory strength assessment of athletes. *New Stud Athletics 10* 88–96.

Acknowledgements: Thanks to Danny Rutar of Redback Bioteck for providing the Myotest system for demonstration. Thanks also to Tom Comyns of Munster Rugby and the Irish Rugby Football Union for providing the FitroDyne and ??? contact mat.

BASIC PERFORMANCE CUES FOR TEACHING THE POWER SNATCH AND POWER CLEANS TO NON-OLYMPIC WEIGHTLIFTING ATHLETES

Andrew Tysz
Head Weightlifting Coach
United States Olympic Education Center at Northern Michigan University
Marquette, MI, USA

OVERVIEW: The intent of this article is to introduce the basic tenets of performing the movements of the derivatives of Olympic Weightlifting. More specifically, the two movements derived directly from the competition exercises, the Power Snatch (PS) and Power Clean (PC). Precise performance of these exercises requires a multitude of physical qualities coordinated at very high rates of speed. Through my experience, many strength and conditioning coaches throughout the world utilize these lifts for enhancing the overall productivity of athletes under their charge. These exercises are traditionally based in the strength and power sports, but I have used these exercises with athletes who participate in endurance activities, as well with very good results. Most competitive activities will have some aspect of power which can be enhanced by applying the PS and/or PC in the training regimes.

The movements are very similar in their performance; the main differences being final resultant bar accelerations and velocities, hand spacing and where the bar is fixed in relation to the body at the completion of the movement. The following is a brief synopsis of the main coaching points to teach athletes at the onset of applying these movements into the training sessions.

Coaching Points: The Starting Position and pull are the same for each exercise with the exception of hand spacing on the bar and the height at which the bar will "brush" against the upper thigh or lower hip region as it accelerates upward to the receiving position. In the PS, the hands will be spaced approximately the width between elbows when extending the humerus laterally to a 90 degree angle in relation to the torso. The PC will have hand spacing equal to the width of holding the hands just lateral of the hips, similar to the traditional anatomical position.

- All Body Levers Are "Tight"
- Feet Slightly Turned Out and in the "Vertical Jump" Position
- The Back Is "Flat" and Even Concave
- Arms Are Straight and the Elbows Are Out
- The Head Is Up and the Eyes Are Focused Straight Ahead
- The Hips Are Higher Than the Knees
- The Shoulders Are In Advance of the Barbell

The Pull
- The Barbell Moves Back Toward the Athlete
- The Hips and Shoulders Rise at the Same Time
- The Head Stays in a Level Position
- The 2nd Pull Must Be Faster Than the 1st Pull
- The Athlete Should Try To Stay "Flat-footed" as Long as Possible
- The Arms Bend Only To Pull the Athlete Under the Bar
- The Feet Move From a Pulling Position To a Receiving Position

Power Snatch
- The lifter approaches the barbell and sets the feet
- Then adopts the starting position
- Inflate chest; set back
- Shoulders are in advance of the bar

- Arms are straight
- Eyes are focused straight ahead
- Weight is distributed evenly
- It is imperative to push with the feet initially and as the barbell passes the knees acceleration should constantly increase
- The lifter then extends the body upward in a violent motion
- The shoulders shrug, the arms are straight and the weight shifts from the heels to the ball of the feet
- The lifter will exert so much force that it will continue to rise while he pulls herself underneath the barbell
- After jumping under the bar, the lifter will receive the barbell at arms' length

Power Clean
- The lifter approaches the barbell and sets the feet
- Then adopts the starting position
- Inflate chest; set back
- Shoulders are in advance of the bar
- Arms are straight
- Eyes are focused straight ahead
- Weight is distributed evenly
- It is imperative to push with the feet initially and as the barbell passes the knees acceleration should constantly increase
- The lifter then extends the body upward in a violent motion
- The shoulders shrug, the arms are straight and the weight shifts from the heels to the ball of the feet
- After the lifter finishes the pull, she pulls herself under the bar and catches it in the receiving position
- The bar should rest across the shoulders and clavicles while keeping the chest and elbows elevated

PRACTICAL APPLICATION: These are the basic principles involving performance of the two movements. If applied correctly, the exercises can be included in training regiments of all sports to enhance the various physical qualities needed for a higher level of performance production in the sport. The practical portion of this presentation will demonstrate the principles, techniques and performance of the Olympic lifts.

ISBS 2009

Applied Session AS4

Data Analysis Techniques

SYMPOSIUM ON RECENT DEVELOPMENTS IN DATA ANALYSIS

J. Hamill[1], R. Van Emmerik[1], R. Miller[1], K. O'Connor[2], N. Coffey[3], D. Harrison[4]

Biomechanics Laboratory, University of Massachusetts, Amherst, MA U.S.A.[1]
Human Performance Laboratory, University of Wisconsin, Milwaukee, WI, USA[2]
Department of Statistics, University of Limerick, Limerick, Ireland[3]
Department of Physical Education and Sports Science, University of Limerick, Limerick, Ireland[4]

The purpose of this symposium is to present recent developments in biomechanical data analyses in two areas. First, current methods used in a dynamical systems approach will be described. Second, two statistical approaches, Principal Components Analysis and Functional Data Analysis, will be presented. The emphasis in this symposium will be on how to use each of these recent analysis techniques.

KEY WORDS: dynamical systems, principle components analysis, functional data analysis

INTRODUCTION:

Biomechanics has progressed significantly from both technological and analytic viewpoints. The development of new technologies has allowed biomechanists to become ever more sophisticated and ask much more complicated questions. It has also allowed biomechanists to generate large quantities of higher dimensional data. In this symposium, we will present approaches to the analysis of biomechanical data from two perspectives that are not in common use in biomechanics research. First, we will present methods of analysis that are used in a dynamical systems approach. Second, we will present two statistical methods that can be used to analyze large sets of continuous data.

DYNAMICAL SYSTEMS (Van Emmerik, Miller and Hamill):

The goal of this presentation is to present a review of current methods in the assessment of movement coordination from a dynamical systems perspective (Glass, 2001). The data analysis techniques that will be presented are essential in the assessment of stability and adaptability of movement patterns. These techniques are also important in assessing the role of movement variability in expert performance, learning/development and disease. Although traditional perspectives in biomechanics and motor control have highlighted the negative role of movement variability, dynamical and complex systems approaches have emphasized the functional role of variability in creating adaptive and stable movement patterns.

Measures of movement coordination that will be presented include relative phase and vector coding techniques (Hamill et al., 2000; Chang et al., 2008). Vector coding analysis of coordination has been primarily applied to angle-angle diagrams of lower extremity segmental or joint motions during locomotion. Relative phase techniques have traditionally been applied to bimanual coordination and lower and upper body movements during locomotion. Both continuous and discrete relative phase techniques will be discussed. Discrete relative phase (DRP) is a valuable tool to assess more complex coordination patterns containing multiple frequencies, as for example in the coordination between the respiratory and locomotor systems. Continuous relative phase (CRP) techniques are based on higher dimensional state space reconstructions (position/velocity phase plane). Issues that will be discussed in CRP analysis will include normalization of the phase plane and the use of circular statistics in the assessment of coordination patterns. Also, a comparison of the different coordination measures and their limitations will be provided.

Movement coordination is also strongly associated with the perceptual variables that may play a role in the detection of stability boundaries for upright stance or during locomotion. It is argued that a stronger emphasis on these perceptual variables is needed to assess conditions prone to postural instability. This presentation will discuss research that highlights the importance of a systematic investigation of the role of perceptual control variables such as time-to-contact (Haddad et al., 2006) as a necessary prerequisite to understand postural and gait control.

The final part will include a review of new 'complexity' methods to assess the structure and variability of human movement. These techniques include different measures of entropy, fractal structure and recurrence quantification analysis (RQA). Complexity analysis has been used to demonstrate changes in movement coordination and use of degrees of freedom as a function of development, disease and movement expertise.

STATISTICS:

The development of highly sophisticated data collection tools in many real-world applications (e.g. biomechanics, imaging, etc.) has resulted in the production of high dimensional data. In order to analyze these data sets, two statistical approaches have been proposed: 1) principal components analysis (Jolliffe, 2005; O'Connor and Bottum, 2009; Wrigley et al., 2006); and 2) functional data analysis (Ramsay and Dalzell, 1991; Ramsay and Silverman, 2002). In the symposium, the same data set will be analyzed to contrast the two approaches.

Principal Component Analysis (O'Connor and Hamill):

In a Principal Components Analysis (PCA), the time series of each trial serves as input. PCA can be utilized to identify the dominant modes of variation within waveforms. This approach allows examinations of time series data without making a priori assumptions regarding critical dependent variables, such as maximum or minimum angles. This technique can also illuminate patterns that may not be readily obvious in examining the original waveforms. In order to facilitate data processing, each trial is time scaled to 101 data points. An $n \times p$ matrix is created with n = the number of time series and p = 101. The analysis is performed through an eigenvalue analysis of the covariance matrix, $S_{101 \times 101}$ which yields eigenvectors ($U_{101 \times 101}$) and eigenvalues ($L_{1 \times 101}$). The eigenvector matrix, $U_{101 \times 101}$, contains the coefficients for each of the 101 principal components (PC) that are extracted. The eigenvalue matrix, $L_{1 \times 101}$, contains the relative contribution of each PC to the total variation. An analysis is then performed to retain only those principal components that contribute modes of variation greater than an equivalently sized input matrix of randomly generated numbers. The PC scores (Z) for each of the n individual times series are calculated by multiplying each individual trial's variation about the overall mean with the transpose of the eigenvector matrix:

$$Z_{n \times 101} = (X - (1 - x_{1 \times 101})) \times U'$$

where $x_{1 \times 101}$ is the mean waveform of all trials. The Z scores for each retained PC can then be compared using traditional statistical techniques (e.g., gender differences).

In order to assess how well the retained PCs represent the original input data, the Q-statistic is calculated. The Q-statistic is the sum of squares of the residuals between the individual trial and the reconstructed profile based on the retained PCs. A critical Q alpha value of 0.05 is used in this analysis. Reconstructed trials with a Q statistic less than 0.05 indicate that the original data were adequately represented.

PCA also provides a unique method of investigating between-subject variability. To begin, correlations (r_{ij}) are calculated between the ith principal component and the jth time sample:

$$r_{ij} = \frac{U_{ji}\sqrt{L_i}}{s_j}$$

where s_j is the standard deviation at a given time in the input series. The value r_{ij}^2 is the percent explained variance across time for a given PC. Summing the variance across time provides the ability to separate the overall variation in the data into random and deterministic components. This may provide a powerful new approach to examining the nature of movement variability.

Functional Data Analysis (Coffey and Harrison):
Functional data analysis (FDA) is a statistical methodology used to analyze such data. Functional data is usually measured at a discrete number of time points but it is assumed that some underlying function $x_i(t)$ generates the observed data ($y_{i1}, \ldots y_{in}$) for individual i. A key idea in FDA is that the underlying function is *smooth*, i.e. pairs of adjacent values are "linked" and do not differ in value from each other by a large amount. Thus the ordering of the observed values is important. As a result, the basic idea behind FDA is to treat the entire sequence of measurements for a particular experimental unit as a single functional entity. In addition, FDA often makes use of the derivatives of the curves which greatly extends the power of FDA methods and leads to functional models as defined by differential equations (dynamical systems). It also facilitates the creation of phase plane plots which may prove highly informative about the processes generating the functions.

As stated above, functional data is typically measured at a finite number of time points, possibly with some measurement error. Therefore, the observed values ($y_{i1}, \ldots y_{in}$) can be represented as $y_{ij} = x_i(t_j) + \varepsilon_{ij}$, where $x_i(t)$ is a smooth function and ε_{ij} is measurement error. As a result, the raw data need to be converted to the smooth function $x_i(t)$. This is achieved using basis function expansions and one of several possible smoothing techniques such as regression splines, smoothing splines, etc. Regression splines estimate $x_i(t)$ as a linear combination of K basis functions ($\phi_1(t),\ldots,\phi_K(t)$), e.g. B-splines, Fourier splines, wavelets such that $x_i(t) = \sum_{k=1}^{K} c_{ik}\phi_k(t)$. The coefficients c_{ik} are estimated by minimizing

$$\sum_{i=1}^{N}\sum_{j=1}^{n}[y_{ij} - x_i(t_j)]^2 = \sum_{i=1}^{N}\sum_{j=1}^{n}[y_{ij} - \sum_{k=1}^{K} c_{ik}\phi_k(t)]^2$$

Choosing the number of basis functions is important since if $K < n$ the data are smoothed, while if $K = n$ the data are interpolated. This is a difficult problem and an alternative approach is to use smoothing splines which set $K = n$ (interpolating the data) and control any over-fitting using an additional penalty term. The penalty term penalizes the curvature of the fitted function and thus curves that are more variable will be penalized more than functions that are smoother. The coefficients are now determined by minimizing

$$\sum_{i=1}^{N}\sum_{j=1}^{n}[y_{ij} - \sum_{k=1}^{K} c_{ik}\phi_k(t)]^2 + \lambda \int D^2 x_i(t) dt$$

where $D^2 x_i(t)$ is the second derivative (curvature) of $x_i(t)$ and λ is a smoothing parameter controlling the trade-off between fidelity to the data and the smoothness of the resulting fitted curve. If λ is large the fit will be smoother while if λ is small the fit will be less smooth. The figures below display the result of changing the value of λ. Smoothing splines allow the user to have more control over the amount of smoothing that is achieved and λ can be chosen using several techniques, e.g. cross-validation.

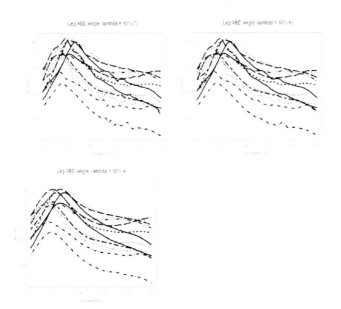

Once the raw data have been smoothed, it is possible to carry out further analyses, e.g. functional principal components analysis (FPCA), functional canonical correlation analysis, functional discriminant analysis, principal differential analysis, functional regression, etc. FPCA is an extension of multivariate principal components analysis to functional data which determines the main modes of variation in a set of *curves*. The extracted components are now functions rather than vectors, and are used to identify the characteristic features of a set of curves throughout an entire time interval. As in the multivariate case, the first few functional principal components usually account for the majority of the variation in the set of curves providing a way of looking at the variance structure which can often be more informative than a direct examination of the variance-covariance function. Functional principal component scores can also be determined for each individual and these provide a means of determining the characteristic behaviour of specific cases. They are also useful for identifying outlying observations, i.e. individuals who score very differently from the remaining individuals in a sample of curves.

REFERENCES:
Chang, R., Van Emmerik, R.E.A., & Hamill, J. (2008). Quantifying rearfoot-forefoot coordination in human walking. *Journal of Biomechanics*, 41:3101-3105
Glass, L. (2001). Synchronization and rhythmic processes in physiology. *Nature*, 410:277-284.
Haddad, J. M., Gagnon, J., Hasson, C. J., Van Emmerik, R. E. A., & Hamill, J. (2006). The use of time-to-contact measures in assessing postural stability. *Journal of Applied Biomechanics*, 22:155-161.
Hamill, J., Haddad, J.M., & McDermott, W.M. (2000). Issues in quantifying variability from a dynamical systems perspective. *Journal of Applied Biomechanics*,16:407-419.
Jolliffe, I. (2005). Principal Component Analysis. New York: John Wiley & Sons.

O'Connor, K.M., Bottum, M.C. (2009). Differences in cutting knee mechanics based on principal components analysis. *Medicine and Science in Sports and Exercise* 41:867-878.
Ramsay, J.O., Dalzell, C.J. (1991). Some tools for functional data analysis. Journal of the Royal Statistical Society 53:539-572.
Ramsay, J.O., Silverman, B.W. (2002). Applied Functional Data Analysis. New York: Springer.
Ramsay, J.O., Silverman, B.W. (2005). Functional Data Analysis. New York: Springer.
Wrigley, A.T., Albert, W.J., Deluzio, K.J., Stevenson, J.M. (2006). Principal component analysis of lifting waveforms. *Clinical Biomechanics* 21:567-578.

ISBS 2009

Applied Session AS5

Swimming

PREVENTING INJURIES IN SWIMMING

Kevin T. Boyd MBBS FRCS(Tr&Orth) FFSEM(UK) DipSportsMed

Consultant Orthopaedic & Sports Surgeon, University Hospitals of Leicester NHS Trust, United Kingdom
Chairman, British Swimming Medical Committee

Statistics tell us that overall injury rates in swimming are very low in comparison with other sports. It is the elite swimmer that suffers the majority of injuries. This is largely due the substantial training loads undertaken and the large reliance on the upper limbs for propulsion. The four strokes differ in subtle ways in their injury patterns.

Acute injuries are relatively rare due to the lack of bodily contact and the relatively slow speeds. Discipline in and around the pool and caution when diving are important. Blunt injuries for the majority are minor and self-limiting. Indirect muscle strains can occur following failure to perform a suitable warm-up.

Overuse injuries are often multi-factorial and present the greatest challenge. They result when biomechanical demands are not matched by appropriate adaptation and recovery. Training errors may be factors but these can be minimised by individualised, responsive programmes with in-built recovery periods for each physiological system. Any external demands on the athlete must not be underestimated.

Shoulder problems are most common. The joint relies heavily on dynamic control of the rotator cuff and the scapular stabilising muscles. Instability is exacerbated as muscles fatigue with activity. Specific strengthening exercises should be part of an overall training programme. The knees of Breaststrokers are vulnerable to combination stresses. Thigh strengthening should focus on closed-chain, terminal-range exercises. The streamline position encourages repeated hyperextension of the lumbar spine, particularly in Butterfly and Breaststroke. Such actions focus stress on the posterior structures of the spine. Core stability programmes concentrating on the endurance and tone of the major muscle masses are key. Inherent or acquired anatomical variations may make some athletes more susceptible to injury than others.

Understanding the causes of injury allows doctors and coaches to minimise risks and allow prompt intervention to prevent chronicity and underperformance.

BRITISH SWIMMING - CREATING A PLATFORM FOR ELITE SWIM PERFORMANCES

Andrew Logan

British Swimming Manager Sport Science Medicine

In 2008, British Swimming (BS) restructured the World Class Programme (WCP) for swimming with the establishment of five Intensive Training Centres (ITC) throughout the United Kingdom. The aim being to provide an optimal environment for elite athletes to best prepare for winning performances at international competitions. Integral to each ITC is a sport science and medical service delivery platform whereby coaches and athletes are provided access to a comprehensive array of science and medical support within their daily training and competition environment.

The purpose of this talk is to provide to you an overview of the integration and applied application of swimming sciences, by coaches, within the ITC daily training environment. Particularly, the focus will be on the processes established within the ITC to identify the best use of swimming science. As well, the talk will provide an insight into the current tools and practices that the ITC coaches have embraced to create an environment that will impact and

RECOGNISING AND AVOIDING OVERTRAINING

Francisco Alves

Faculty of Human Kinetics, Technical University of Lisbon, Portugal

Training induces both physiological adaptations that improve performance and fatigue that decreases performance. The purpose of recovery or peaking (taper) periods is to eliminate the fatigue and allow the full benefits of the adaptations that should have occurred during overload phases to emerge at the right time.

Stress encompasses all aspects of training, competition and non-training factors. Stress can have both positive and negative effects depending on the state of the athlete and recovery process. Increased exercise stress is manifested in physiological and biochemical changes and is often in conjunction with psychological alterations, all of which result from an imbalance in homeostasis. However, the quantity of training stimuli that result in either performance enhancement or a chronic fatigue state is presently unknown.

It is usual in training regimes of endurance athletes the integration of high training volumes combined with limited recovery periods. This may disrupt the fragile balance in the interaction between fatigue and adaptation, and the accumulation of exercise stress may exceed an athlete's finite capacity of coping with the demands put on him.

Because it is difficult to ascertain the volume of training that will result in overreaching or overtraining, it would be important to identify markers that distinguish between acute training-related fatigue and overreaching. However, a "golden standard" to diagnose overreaching or early detection of overtraining does not exist. The combination of several criteria as maximal lactate concentration, OBLA or other submaximal markers, and RPE require intraindividual comparative data to be meaningful. On the other hand, fatigue and state moods inquiries seem to be very helpful in checking how the swimmers are coping with training-induced stress.

Keeping a training log is an easy way to track your progress and watch for symptoms of overtraining. Indicators that are easy and inexpensive to obtain, are exactly the ones that may prove most suitable for inclusion in a training diary based system for monitoring adaptation to training.

GREEN SWIMMING: GETTING INTO THE ENERGY SAVING RHYTHM

Ross Sanders

Centre for Aquatics Research and Education, PESLS, The University of Edinburgh, Edinburgh, UK

Why is butterfly almost as fast as freestyle? By developing good rhythm and timing of body actions skilled butterfly swimmers are able to swim more economically than less skilled swimmers – almost as economically as freestyle! Implications for coaching and identification of swimmers suited to competitive butterfly swimming are discussed.

BODY ROLL: WHAT WE NOW KNOW

Carl J. Payton

Department of Exercise & Sport Science, Manchester Metropolitan University, Cheshire, England

This presentation will discuss the importance of body roll in competitive swimming. It will critically evaluate the scientific evidence supporting the potential benefits associated with body roll and provide some practical recommendations for coaches.

When swimmers rotate about their longitudinal axis in the front crawl and backstroke, this is commonly referred to as body roll. This rolling movement is considered an essential component of these two strokes. Body roll ocurs as a consequence of the asymmetrical movements of the lower and upper limbs, and gravitational effects. Studies have shown that the shoulders and hips do not roll as one unit and that the timing and the magnitude of hip and shoulder roll depends on a number of factors including the swimmer's speed, stroke rate, kick technique, breathing action and skill level. The key findings from these studies will be presented and their relevance to coaching discussed.

There is some speculation that body roll can enhance the amount of propulsion created during front crawl swimming. Several studies have attempted to quantify the relationship between body roll and the underwater actions of the swimmer's arm, shedding some light on the possible links between body roll and propulsion. The implications of this research for coaching will be discussed.

It seems likely that body roll would have a considerable influence on the amount of hydrodynamic drag experienced by swimmers, although the precise nature of this influence is unclear. The potential links between body roll and drag will be addressed.

Swimmers can reduce the risk of shoulder impingement injury by altering their stroke mechanics. One of the most common recommendations given to front crawl swimmers that suffer from impingement syndrome is to increase the amount of body roll they use. The scientific evidence supporting the proposed link between body roll and shoulder impingement will be discussed.

HOW TO START IN BACKSTROKE CONSIDERING THE NEW RULES?

J. Paulo Vilas-Boas, Karla de Jesus, Kelly de Jesus, Pedro Figueiredo, Suzana Pereira, Pedro Gonçalves, Leandro Machado, Ricardo Fernandes

University of Porto, Faculty of Sport, CIFI^{2}D, Porto, Portugal

FINA recently changed the rule that governs the starting position for the backstroke starting technique. With this change, swimmers may now decide to start with the feet emerged, which was previously strictly forbidden. This new liberalization naturally determines the rise of a new technical question for swimmers and coaches: do a starting position with the feet emerged allow a better performance, or better performance conditions for the following actions?

To our knowledge no previous scientific results are available concerning this question, allowing to support any technical decision. During this presentation we will try to deliver arguments to support a technical option in this particular topic, based on the research developed by our group, comparing the traditional starting technique with the feet immerged (BSFI), with the one allowed nowadays, with both feet totally, or partially, emerged (BSFE).

To fulfil this purpose, we studied six experienced male swimmers that maximally performed 4 repetitions of each technique over a distance of 15 meters. All performances were dual-media videotaped (50 Hz) in the sagittal plane, synchronized with kinetic and EMG data simultaneously registered. Kinetic data were assessed using an underwater force plate mounted on a special support on the wall of the pool, allowing the registration of the horizontal component of the forces exerted by the swimmers' feet. The handgrip system was adapted to reproduce its legal position and configuration, but instrumented with a load cell (Globus, Italy) to allow the assessment of the horizontal component of the forces exerted by swimmers' upper limbs. Findings pointed out that BSFI was significantly faster till the 5m reference, with less muscular activity, and with a tendency to produce higher forces against the starting wall. No argument was obtained to support the use of the BSFE in swimming competitions.

A GUIDE FOR THE CO-ORDINATION IN THE FRONT CRAWL VARIANTS

Ulrik Persyn[1] & Filip Roelandt[2]

1 - Faculty of Kinesiology, K.U.Leuven, Belgium
2 - Universiteit Gent, Belgium

In this presentation an interactive cd-rom, describing and discussing the front crawl of high level swimmers, is introduced; more specifically several variants (e.g. 6-beat, 2-beat, 2 beat crossover) and different technique aspects. The intention is that the coach corrects the own competition swimmers and triathletes, based on a careful video observation. Therefore, only practical knowledge is collected, confirmed by expertise and applied research at the K.U.Leuven Evaluation Centre. Different variants and technique aspects are clarified by video images of recent swimmers at national level in different speeds, obtained with 5 rotating cameras from 5 points of view.

To be able to start with an Evaluation Centre, in the movement analysis of high level swimmers special attention was given to the co-ordination of the arm, leg and head movements (relative to the body) and of the body movements (relative to the water surface). Since the Olympic Games in Munich (1972), for each segment clearly delimited phases were studied in the stroke cycle of each variant. An interesting criterion for propulsion is the speed variation of the body from phase to phase. Each swimmer can feel propulsion per phase, estimated by the coach on video, from bubbles displaced backward; each swimmer can also feel the coordination, observable by the coach on video.

Although the optimal variant per distance of a swimmer could be determined from his physical profile (e.g.; body structure, buoyancy, flexibility, strength,,,), each individual is advised to experiment with the different variants and technique aspects. Speed and/or economy remain the essential criterions. The body control of swimmers must thus be very adaptable.

PHYSIOLOGICAL MONITORING OF SWIMMING

John Bradley

Department of Education, University College Cork

This talk will discuss the Why? What? and How? of the physiological monitoring of swimming. The goal of any support programme is to enhance swim performance. The role of an exercise physiologist and the impact of exercise physiology in that support programme can be very valuable. The talk will be very practically orientated, illustrated by examples taken from a variety of physiological support programmes including from a successful support team for swimmers preparing for the Paralympic Games in Beijing.

The presentation will start by outlining why a coach may want to include physiological monitoring in his/her swimming programme. What does a coach/swimmer do with the results of testing? The rationale behind physiological testing and why certain tests may be used in particular situations will be discussed. Examples of how the results of physiological tests can be interpreted and how they can then be incorporated into the coaching programme to enhance the preparation of swimmers will be shown.

With particular reference to the Paralympic preparation programme the talk will present how a physiologist can successfully interact with a swimmer-coach unit and how physiological results can be related to coaching objectives.

UNDERWATER DOLPHIN KICKING IN STARTS AND TURNS

Raúl Arellano

Faculty of Physical Activity and Sports Science, University of Granada, Granada, Spain

Our studies related with underwater dolphin kicking helped us to understand the complexity of this propulsive action (applying kinematic analysis and flow visualization) to develop more appropriate prescriptive information to be applied in application of this technique in the performances of national and international swimmers.

Under the rules limitations (15m), the swimmers can apply properly this technique and reduce the starting time until times close to 5 s. (considered impossible in the past). New swimsuit technology improves particularly the performance in this technique.

Some aspects need to be considered based on our research in this technique: a) when start the dolphin kick after the start or turn [how long should glide]; b) influence of the previous gliding velocity; c) undulating or oscillating; d) influence of the morphological factors; e) different models and variables to measure its efficiency; f) how long should keep the dolphin kick and; g) when to finish the dolphin kicking.

In our talk we will try to advice on the aspects mentioned and to show the kind of methodology we are applying to improve the use of this technique to the Spanish swimmers.

CFD AND SWIMMING: PRACTICAL APPLICATIONS

António J. Silva[1,2], Daniel A. Marinho[2,3]

Department of Sport, Health and Exercise, University of Trás-os-Montes and Alto Douro, Vila Real, Portugal[1]

Research Centre in Sports Sciences, Health Sciences and Human Development, Vila Real, Portugal[2]

Department of Sport Sciences, University of Beira Interior, Covilhã, Portugal[3]

In this presentation topics in swimming simulation from a computational fluid dynamics perspective are discussed. This perspective means emphasis on the fluid mechanics and computational fluid dynamics methodology applied in swimming research.

This talk presents new information based on recent scientific research conducted at the Research Centre in Sports Sciences, Health Sciences and Human Development (CIDESD, Vila Real, Portugal). We concentrated on numerical simulation results, considering the scientific simulation point-of-view and especially the practical implications with swimmers and coaches.

Computational Fluid Dynamics has been applied to swimming in order to understand its relationships with performance. The numerical techniques have been applied to the analysis of the propulsive forces generated by the propelling segments and to the analysis of the hydrodynamic drag forces resisting forward motion.

"I ALWAYS SWIM BADLY IN THE FINAL"

Brian Daniel Marshall

Icelandic Swimming Association and Reykjavík University

As coaches we have all experienced the situation where a swimmer starts making destructive or negative comments that impact their ability to perform to their maximum. Such destructive thinking is well known in psychology, for example, amongst clients who experience anxiety or are depressed. In swimmers, we hear them blaming others, feeling guilty, predicting (negative) results before they happen and exaggerating previous negative results. Furthermore, swimmers can experience a mental block in terms of how to overcome an obstacle which, in their minds, seem impossible to overcome.

In this talk I will introduce several practical tools and techniques that can be used by the coach to aid the swimmer in thinking in a constructive way. These tools and techniques have been developed in the fields of *Cognitive Behavioural Therapy* and within *Life Coaching* (sometimes referred to as Executive Coaching) and can be applied over a period of weeks or months, or at the competition site itself. By developing the practical tools that I introduce here, coaches should feel more empowered in dealing with these problems rather than feeling exasperated or annoyed at the negative thoughts of their swimmers.

CORRECTING PHYSICAL AND TECHNICAL ASYMMETRIES OF SWIMMERS

Alison Fantom

Inverurie Chartered Physiotherapy Clinic, Aberdeenshire, Scotland

This talk will look at the importance of maintaining a sound postural base and its links to efficiency when in the water. The common postural habits of swimmers including typical patterns of asymmetry will be discussed and their possible impact on performance.

A case study of an Olympic swimmer will be presented with video footage of the programme she followed as she prepared for the 2008 Olympic Games in Beijing. The work with this athlete from initial screening and basic postural control through to high level challenges will be discussed. The importance of strong communication links between swimmer, coach and physiotherapist will be highlighted.

The possible implications on technique for both elite and non elite swimmers will be discussed.

BREATHING AND STROKE FREQUENCY STRATEGIES FOR TOP PERFORMANCE

Per-Ludvik Kjendlie

Department of Physical Performance, The Norwegian School of Sport Sciences, Oslo, Norway

How do you optimise the stroke frequency in swimming? The answer to this question is important in the pursuit of higher performances in the pool. Selecting the right stroke frequency is detrimental for an optimal performance. It should be adjusted to each individual athlete's characteristics.

For running and cycling, we know much on energy consumption and stride frequency. In swimming though, which is a relatively new activity for humans, the optimum stroke frequency is less researched. In this lecture we will look briefly to running and cycling, and the models that explain the optimal human cadence. How these models can be transferred to swimming will be discussed. What we know from swimming studies, optimising stroke frequency for different age- and performance levels swimmers in different strokes will be reviewed. Finally, our research shows that certain stroke rate strategies seem to be ideal during a race. This lecture will examine what stroke rates strategies to choose during a race in order to win.

SPRINT AND DISTANCE SWIMMERS: THE SAME ANIMAL?

Carla B. McCabe

Centre of Aquatics Research and Education, University of Edinburgh, Edinburgh, Scotland

In this talk the techniques of sprint and distance swimmers are discussed with a view to informing coaches of the similarities and differences between these groups. In the past it has been reported that sprint and distance swimmers are different in several aspects of technique. However, previous comparisons were at the respective race pace and sprint and distance specialists have not been compared when swimming at the same pace. Therefore it is difficult for coaches to know whether to teach the swimmers the same way when developing good technique.

This talk presents new information based on recent scientific research conducted at the Centre for Aquatics Research and Education (CARE). The variables of interest were: average swim speed, stroke length, stroke frequency, stroke index, hand stroke pattern, foot range of motion, elbow angle, shoulder and hip roll angle and stroke phase durations.

Interesting and unexpected findings emerged that have implications for the way specialist sprint and distance swimmers should be coached.

SUPPORTING THE COACH WITH SCIENCE: THE IRISH EXPERIENCE

Conor Osborough

School of Science and Technology, Nottingham Trent University, Nottingham, UK

Recognising the need for a strong sport science support system, a programme of biomechanical support was set up by Swim Ireland (the Irish Swimming Federation) in 2006. This programme, which forms part of a larger overall inter-disciplinary sport science and medicine service, aims to provide a comprehensive level of biomechanical support for Irish high-performing swimmers and their coaches to enhance performance at elite level. At the centre of the biomechanics programme is the swimmer-coach unit; their performance needs are critical. The scientist's role is to provide the coach with useful, user-friendly and objective evidence upon which the coach can make effective informed training decisions. Both qualitative and quantitative biomechanical assessment techniques are used to analyse swimmers in training and competition, which typically require the use of above and below water digital video recordings. Suitable intervention strategies for modifying swimming techniques, where necessary, can then be implemented and regularly monitored for effectiveness. The aim of this presentation is to provide an overview of the biomechanics support programme being provided to elite Irish swimmers and their coaches, with particular reference to: 1) competition analysis; and 2) two-dimensional video analysis. Such examples encourage the use of evidence-based coaching.

LAND TRAINING FOR SWIMMING

Neil Donald

Lead Strength & Conditioning Coach for Swimming
SportScotland Institute of Sport and Stirling Intensive Training Centre (ITC)

This presentation looks at the role land training plays in the physical preparation of swimmers for performance, covering the following areas: identification of postural dysfunction and compensation patterns, effects it has on the swimmer on land and in the water, strategies to correct these patterns and in turn reduce injury potential and improve performance.

The aim of the presentation is to give coaches practical information they can take away and apply immediately within the various aspects of their land programme to enhance performance and reduce injuries through exercise modifications and training strategies.

USING CRITICAL VELOCITIES TO SET TRAINING INTENSITIES

Jeanne Dekerle

Chelsea School, University of Brighton, Eastbourne, England

Determining training intensities is a real challenge for a swimming coach because of the few physiological variables measurable on pool-side. The use of blood markers such as lactate can help in the assessment of a swimmer's aerobic endurance through the identification of a lactate threshold although a) blood sampling is not necessary an option and b) lactate threshold is such a low intensity (maintainable for hours) that it is not necessary very pertinent for setting training intensities. Similarly, performances over long distances (2-km or 3-km time trials) have been suggested to help defining training intensities but have their own limitations (pacing issues; physiological meaning).

This talk will focus on the critical velocity concept, which in swimming research, and since the early nineties, has been suggested to be a valuable tool to assess aerobic endurance. A stop watch is the only equipment required to determine a swimmer's critical velocity. The method relies solely on the measure of two or more performances (from 3 to 15-20 min) from which a distance vs time relationship is plotted and modelled using a 2-parameter model (y=ax+b). The slope of this relationship (a) is recognised as critical velocity, an intensity a swimmer would maintain, in theory, indefinitely. In reality, critical velocity can be sustained for around 30 minutes.

This presentation will focus on the latest findings on critical velocity and the reasons why it can be seen as an attractive tool to set training intensities. Critical velocity will be compared with more classical "thresholds" and the findings will lead the audience to consider their own ways of setting their aerobic training zones. Some concepts such as aerobic power and capacity will be challenged in an attempt to gain an appreciation of the physiological mechanisms behind swimming endurance.

COORDINATION: HOW ELITE SWIMMERS DIFFER FROM SUB-ELITE

Didier Chollet & Ludovic Seifert

Centre d'Etude des Transformations des Activités Physiques et Sportives (CETAPS), University of Rouen, Faculty of Sports Sciences, France

In this talk the inter-limb coordination of elite and sub-elite swimmers are discussed with a view to informing coaches of the similarities and differences between these groups. In the past it has been reported that the inter-limb coordination should show an opposition mode, i.e. a propulsive continuity between the propulsion of one limb and those of the other limb, in order to minimize the intra-cyclic velocity variations. However, the research of our centre of research highlighted the fact that the inter-limb coordination mode adopted by the swimmers corresponds to three types of constraint defined by Newell (1986): organismic, task and environmental constraint. The skill level of the swimmers, the specialty, the gender, the handedness and the breathing laterality act as organismic constraints; the imposed race pace, the stroke frequency, the number of strokes, the breathing frequency and pattern could be consider as task constraints while the active drag and his correspondent velocity relate to the environmental constraints. Inter-limb coordination was found to vary from catch or glide coordination mode to superposition mode, showing that the opposition mode is only the best "theoretical" mode and the glide mode is not a technical mistake. Therefore it is advised for coaches to don't consider an ideal coordination mode in the absolute but to teach the swimmers in different ways when developing coordination.

This talk presents new information based on recent scientific research conducted at the CETAPS. The variables of interest were: average swim speed, stroke length, stroke frequency, intra-cyclic velocity variations, breathing laterality, relative duration of arm and leg stroke phases, time gap between propulsive actions assessed by total time gap (TTG) in the simultaneous strokes and by index of coordination (IdC) in the alternate strokes.

Interesting findings emerged that have implications for the both elite and sub-elite swimmers should be coached.

SUCCESSFUL INTERVENTIONS FOLLOWING ANALYSIS

Bruce R. Mason

Aquatics Testing, Training and Research Unit, Australian Institute of Sport, Canberra, Australia

Of all sports, swimming is probably the most challenging to provide biomechanical analysis that will result in improved performance. The three main reasons for this are: biomechanical equipment will not perform well or possibly even survive for any length of time in an aquatics environment; in free swimming, propulsion occurs as a consequence of the swimmer's body reacting with water, which makes the measurement of force almost impossible; because swimming occurs at the interface between water and air it is difficult to capture data from measuring apparatus and provide immediate feedback to the swimmer. Consequently, much of the technique analysis in the sport generally involves video cameras to obtain footage of the swimmer's performance. As a consequence of such video usage, both biomechanists and coaches often rely upon subjective analysis based on just moving images to provide advice to the swimmer. If it looks good and fits the mould it must be good.

At the Australian Institute of Sport, the Australian government has invested in an aquatics testing, training and research centre. The centre was opened in 2006 at a cost of AUD17million. Purpose built analysis systems have been developed to provide immediate feedback to swimmers and coaches concerning performance. Here the biomechanists operate the analysis systems and the coaches are encouraged to provide the feedback to the swimmers on pool deck. As well as provide a high quality image of the performance, the analysis systems also provide quantitative parameters to objectively assess the performance. Information such as the magnitude and the direction of forces generated by the swimmer, the velocity of the swimmer at various stages of performance, the angles of movement of the swimmer's body and the times taken to reach various points in the activity are provided so that objective assessment may be made to evaluate the performance. Here the coach can objectively identify not only if the performance has improved but also the degree of such improvement. This talk will focus on technique inefficiencies that have been disclosed in Australia's top swimmers and how they were dealt with using the analysis systems available at the institute.